D1480734

THE PICTURE OF THE TAOIST GENII PRINTED ON THE COVER
of this book is part of a painted temple scroll, recent but traditional, given to
Mr Brian Harland in Szechuan province (1946). Concerning these four divinities,
of respectable rank in the Taoist bureaucracy, the following particulars have been
handed down. The title of the first of the four signifies 'Heavenly Prince', that
of the other three 'Mysterious Commander'.

At the top, on the left, is Liu *Thien Chün*, Comptroller-General of Crops and
Weather. Before his deification (so it was said) he was a rain-making magician
and weather forecaster named Liu Chün, born in the Chin dynasty about +340.
Among his attributes may be seen the sun and moon, and a measuring-rod or
carpenter's square. The two great luminaries imply the making of the calendar, so
important for a primarily agricultural society, the efforts, ever renewed, to reconcile
celestial periodicities. The carpenter's square is no ordinary tool, but the gnomon
for measuring the lengths of the sun's solstitial shadows. The Comptroller-General
also carries a bell because in ancient and medieval times there was thought to be
a close connection between calendrical calculations and the arithmetical acoustics
of bells and pitch-pipes.

At the top, on the right, is Wên *Yuan Shuai*, Intendant of the Spiritual Officials
of the Sacred Mountain, Thai Shan. He was taken to be an incarnation of one of
the Hour-Presidents (*Chia Shen*), i.e. tutelary deities of the twelve cyclical characters
(see Vol. 4, pt. 2, p. 440). During his earthly pilgrimage his name was Huan Tzu-Yü
and he was a scholar and astronomer in the Later Han (b. +142). He is seen
holding an armillary ring.

Below, on the left, is Kou *Yuan Shuai*, Assistant Secretary of State in the Ministry
of Thunder. He is therefore a late emanation of a very ancient god, Lei Kung.
Before he became deified he was Hsin Hsing, a poor woodcutter, but no doubt an
incarnation of the spirit of the constellation Kou-Chhen (the Angular Arranger),
part of the group of stars which we know as Ursa Minor. He is equipped with
hammer and chisel.

Below, on the right, is Pi *Yuan Shuai*, Commander of the Lightning, with his
flashing sword, a deity with distinct alchemical and cosmological interests. According
to tradition, in his early life he was a countryman whose name was Thien Hua.
Together with the colleague on his right, he controlled the Spirits of the Five
Directions.

Such is the legendary folklore of common men canonised by popular acclamation.
An interesting scroll, of no great artistic merit, destined to decorate a temple wall,
to be looked upon by humble people, it symbolises something which this book has
to say. Chinese art and literature have been so profuse, Chinese mythological
imagery so fertile, that the West has often missed other aspects, perhaps more
important, of Chinese civilisation. Here the graduated scale of Liu Chün, at first
sight unexpected in this setting, reminds us of the ever-present theme of quantitative measurement in Chinese culture; there were rain-gauges already in the Sung
(+12th century) and sliding calipers in the Han (+1st). The armillary ring of
Huan Tzu-Yü bears witness that Naburiannu and Hipparchus, al-Naqqāsh and
Tycho, had worthy counterparts in China. The tools of Hsin Hsing symbolise that
great empirical tradition which informed the work of Chinese artisans and technicians all through the ages.

SCIENCE AND CIVILISATION IN CHINA

Among the Chinese frequent examples are to be found of discoveries, especially in the arts, which other nations made independently whereas the Chinese had come upon them long before.
WILLEM TEN RHIJNE
De Arthritide (+1683)

And if we look so far as the Sun-rising, and hear *Paulus Venetus* what he reporteth of the uttermost Angle and *Island* thereof, wee shall finde that those Nations have sent out, and not received, lent knowledge, and not borrowed it from the West. For the farther East (to this day) the more civill, the farther West the more salvage.
SIR WALTER RALEIGH
'History of the World', +1614 (+1652)
Pt. I, Bk. 1, ch. 7, §10, sect. 4, p. 98

I take my own intelligence as my teacher.
A-NI-KO, master-artisan of Nepal,
addressing the emperor Shih Tsu, +1263
(*Yuan Shih*, ch. 203, p. 12*a*)

I hear, and I forget.
I see, and I remember.
I do, and I understand.

Ten thousand words are not worth one seeing.
Chinese proverbs.

I am not yet so lost in lexicography, as to forget that words are the daughters of earth, and that things are the sons of heaven.
SAMUEL JOHNSON
Preface to his 'Dictionary of the
English Language' (+1755)

By studying the organic patterns of heaven and earth a fool can become a sage.
So by watching the times and seasons of natural phenomena we can become true philosophers.
LI CHHÜAN
Yin Fu Ching (c. +735)

中國科學技術史

李約瑟 著

冀朝鼎

SCIENCE AND CIVILISATION IN CHINA

BY

JOSEPH NEEDHAM, F.R.S., F.B.A.

SOMETIME MASTER OF GONVILLE AND CAIUS COLLEGE, CAMBRIDGE
FOREIGN MEMBER OF ACADEMIA SINICA

With the collaboration of

HO PING-YÜ, PH.D.

PROFESSOR OF CHINESE AT GRIFFITH
UNIVERSITY, BRISBANE

and

LU GWEI-DJEN, PH.D.

FELLOW OF ROBINSON COLLEGE, CAMBRIDGE

and a contribution by

NATHAN SIVIN, PH.D.

PROFESSOR OF CHINESE IN THE UNIVERSITY OF
PENNSYLVANIA, PHILADELPHIA

VOLUME 5

CHEMISTRY AND CHEMICAL TECHNOLOGY

PART IV: SPAGYRICAL DISCOVERY AND
INVENTION: APPARATUS, THEORIES AND GIFTS

CAMBRIDGE UNIVERSITY PRESS

CAMBRIDGE
LONDON   NEW YORK   NEW ROCHELLE
MELBOURNE   SYDNEY

Published by the Press Syndicate of the University of Cambridge
The Pitt Building, Trumpington Street, Cambridge, CB2 IRP
32 East 57th Street, New York, NY 10022, USA
296 Beaconsfield Parade, Middle Park, Melbourne 3206, Australia

© Cambridge University Press 1980

Library of Congress catalogue card number: 54-4723

ISBN: 0 521 08573 X

First published 1980

Printed in Great Britain at the
University Press, Cambridge

British Library Cataloguing in Publication Data
Needham, Joseph
Science and civilisation in China.
Vol. 5: Chemistry and chemical technology.
Part 4: Spagyrical discovery and invention:
apparatus, theories and gifts
1. Science – China – History
2. Technology – China – History
I. Title   II. Wang Ling   III. Lu Gwei-djen
IV. Ho Ping-yii
509'.51   Q127.C5   54-4723
ISBN 0 521 08573 X

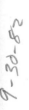

To
WU HSÜEH-CHOU
*sometime Director of the Chemical Institute of
Academia Sinica*

and

CHANG TZU-KUNG
*sometime Chemical Adviser to the National Resources
Commission*

in warmest recollection of long
and enlightening discussions at
Kunming and Nan-wên-chhüan
1942 to 1946

as also in memory of an earlier friend

LOUIS RAPKINE
*sometime Professor of Biochemistry at the Institut de
Biologie Physico-Chimique, Paris*

in our youth
colleague at the Marine Station at Roscoff
compass-needle for truth and social justice
Spinoza redivivus

this volume is dedicated

CONTENTS

LIST OF ILLUSTRATIONS

LIST OF TABLES

LIST OF ABBREVIATIONS

The following abbreviations are used in the text and footnotes. For abbreviations used for journals and similar publications in the bibliographies, see pp. 511 ff.

B	Bretschneider, E. (*1*), *Botanicon Sinicum*.
CC	Chia Tsu-Chang & Chia Tsu-Shan (*1*), *Chung-Kuo Chih Wu Thu Chien* (Illustrated Dictionary of Chinese Flora), 1958.
CCIF	Sun Ssu-Mo, *Chhien Chin I Fang* (Supplement to the Thousand Golden Remedies), between +660 and +680.
CCYF	Sun Ssu-Mo, *Chhien Chin Yao Fang* (Thousand Golden Remedies), between +650 and +659.
CHS	Pan Ku (and Pan Chao), *Chhien Han Shu* (History of the Former Han Dynasty), *c.* +100.
CJC	Juan Yuan, *Chhou Jen Chuan* (Biographies of Mathematicians and Astronomers), +1799. With continuations by Lo Shih-Lin, Chu Kho-Pao and Huang Chung-Chün. In *HCCC*, chs. 159 ff.
CLPT	Thang Shen-Wei *et al.* (ed.), *Chêng Lei Pên Tshao* (Reorganised Pharmacopoeia), ed. of +1249.
CSHK	Yen Kho-Chün (ed.), *Chhüan Shang-Ku San-Tai Chhin Han San-Kuo Liu Chhao Wên* (Complete Collection of prose literature (including fragments) from remote antiquity through the Chhin and Han Dynasties, the Three Kingdoms, and the Six Dynasties), 1836.
CTPS	Fu Chin-Chhüan (ed.), *Chêng Tao Pi Shu Shih Chung* (Ten Types of Secret Books on the Verification of the Tao), early 19th cent.
EB	*Encyclopaedia Britannica*.
HCCC	Yen Chieh (ed.), *Huang Chhing Ching Chieh* (monographs by Chhing scholars on classical subjects), 1829, contd. 1860.
HCSS	*Hsiu Chen Shih Shu* (Ten Books on the Regeneration of the Primary Vitalities, physiological alchemy), *c.* +1250.
HFT	Han Fei, *Han Fei Tzu* (Book of Master Han Fei), early −3rd cent.
HHPT	Su Ching *et al.* (ed.), *Hsin Hsiu Pên Tshao* (Newly Improved Pharmacopoeia), +659.
HHS	Fan Yeh & Ssuma Piao, *Hou Han Shu* (History of the Later Han Dynasty), +450.
HNT	Liu An *et al.*, *Huai Nan Tzu* (Book of the Prince of Huai-Nan), −120.
ICK	Taki Mototane, *I Chi Khao* (*Iseki-kō*) (Comprehensive Annotated Bibliography of Chinese Medical Literature [Lost or Still Existing]), finished *c.* 1825, pr. 1831; repr. Tokyo 1933, Shanghai 1936.

ITCM	Wang Khên-Thang & Chu Wên-Chen (ed.), *I Thung Chêng Mo Chhüan* (Complete Collection of Works on Medicine and Sphygmology), +1601.
K	Karlgren, B. (1), *Grammata Serica* (dictionary giving the ancient forms and phonetic values of Chinese characters).
KCCY	Chhen Yuan-Lung, *Ko Chih Ching Yuan* (Mirror of Scientific and Technological Origins), an encyclopaedia of +1735.
KHTT	Chang Yü-Shu (ed.), *Khang-Hsi Tzu Tien* (Imperial Dictionary of the Khang-Hsi reign-period), +1716.
Kr	Kraus, P., *Le Corpus des Écrits Jābiriens* (*Mémoires de l'Institut d'Égypte*, 1943, vol. 44, pp. 1–214).
LPC	Lung Po-Chien (*1*), *Hsien Tshun Pên Tshao Shu Lu* (Bibliographical Study of Extant Pharmacopoeias and Treatises on Natural History from all Periods).
LS	Tsêng Tshao (ed.), *Lei Shuo* (Classified Commonplace-Book), +1136.
MCPT	Shen Kua, *Mêng Chhi Pi Than* (Dream Pool Essays), +1089.
N	Nanjio, B., *A Catalogue of the Chinese Translations of the Buddhist Tripiṭaka*, with index by Ross (3).
NCCS	Hsü Kuang-Chhi, *Nung Chêng Chhüan Shu* (Complete Treatise on Agriculture), +1639.
NCNA	New China News Agency.
PPT/NP	Ko Hung, *Pao Phu Tzu* (*Nei Phien*) (Book of the Preservation-of-Solidarity Master; Inner Chapters), *c.* +320.
PPT/WP	*Idem* (*Wai Phien*), the Outer Chapters.
PTKM	Li Shih-Chen, *Pên Tshao Kang Mu* (The Great Pharmacopoeia), +1596.
PWYF	Chang Yü-Shu (ed.), *Phei Wên Yün Fu* (encyclopaedia), +1711.
R	Read, Bernard E. *et al.*, Indexes, translations and précis of certain chapters of the *Pên Tshao Kang Mu* of Li Shih-Chen. If the reference is to a plant see Read (1); if to a mammal see Read (2); if to a bird see Read (3); if to a reptile see Read (4 or 5); if to a mollusc see Read (5); if to a fish see Read (6); if to an insect see Read (7).
RBS	*Revue Bibliographique de Sinologie.*
RP	Read & Pak (1), Index, translation and précis of the mineralogical chapters in the *Pên Tshao Kang Mu.*
S/	Stein Collection of Tunhuang MSS, British Museum, London, catalogue number.
SC	Ssuma Chhien, *Shih Chi* (Historical Records), *c.* −90.
SF	Thao Tsung-I (ed.), *Shuo Fu* (Florilegium of (Unofficial) Literature), *c.* +1368.
SHC	*Shan Hai Ching* (Classic of the Mountains and Rivers), Chou and C/Han.

SIC   Okanishi Tameto, *Sung I-Chhien I Chi Khao* (Comprehensive Annotated Bibliography of Chinese Medical Literature in and before the Sung Period). Jen-min Wei-shêng, Peking, 1958.

SKCS   *Ssu Khu Chhüan Shu* (Complete Library of the Four Categories), +1782; here the reference is to the *tshung-shu* collection printed as a selection from one of the seven imperially commissioned MSS.

SKCS/TMTY   Chi Yün (ed.), *Ssu Khu Chhüan Shu Tsung Mu Thi Yao* (Analytical Catalogue of the *Complete Library of the Four Categories*), +1782; the great bibliographical catalogue of the imperial MS. collection ordered by the Chhien-Lung emperor in +1772.

SNPTC   *Shen Nung Pên Tshao Ching* (Classical Pharmacopoeia of the Heavenly Husbandman), C/Han.

SSIW   Toktaga (Tho-Tho) *et al.*; Huang Yü-Chi *et al.* & Hsü Sung *et al. Sung Shih I Wên Chih, Pu, Fu Phien* (A Conflation of the Bibliography and Appended Supplementary Bibliographies of the History of the Sung Dynasty). Com. Press, Shanghai, 1957.

STTH   Wang Chhi, *San Tshai Thu Hui* (Universal Encyclopaedia), +1609.

SYEY   Mei Piao, *Shih Yao Erh Ya* (The Literary Expositor of Chemical Physic; or, Synonymic Dictionary of Minerals and Drugs), +806.

TCTC   Ssuma Kuang, *Tzu Chih Thung Chien* (Comprehensive Mirror (of History) for Aid in Government), +1084.

TFYK   Wang Chhin-Jo & Yang I (eds.), *Tshê Fu Yuan Kuei* (Lessons of the Archives, encyclopaedia), +1013.

TKKW   Sung Ying-Hsing, *Thien Kung Khai Wu* (The Exploitation of the Works of Nature), +1637.

TMITC   Li Hsien (ed.), *Ta Ming I Thung Chih* (Comprehensive Geography of the Ming Empire), +1461.

TPHMF   *Thai-Phing Hui Min Ho Chi Chü Fang* (Standard Formularies of the (Government) Great Peace People's Welfare Pharmacies), +1151.

TPKC   Li Fang (ed.), *Thai-Phing Kuang Chi* (Copious Records collected in the Thai-Phing reign-period), +978.

TPYL   Li Fang (ed.), *Thai-Phing Yü Lan* (the Thai-Phing reign-period (Sung) Imperial Encyclopaedia), +983.

TSCC   Chhen Mêng-Lei *et al.* (ed.), *Thu Shu Chi Chhêng* (the Imperial Encyclopaedia of +1726). Index by Giles, L. (2).
References to 1884 ed. given by chapter (*chüan*) and page.
References to 1934 photolitho reproduction given by *tshê* (vol.) and page.

TSCCIW   Liu Hsü *et al.* & Ouyang Hsiu *et al.*; *Thang Shu Ching Chi I Wên Ho Chih*. A conflation of the Bibliographies of the *Chiu Thang Shu* by Liu Hsü (H/Chin, +945) and the *Hsin Thang Shu* by Ouyang Hsiu & Sung Chhi (Sung, +1061). Com. Press, Shanghai, 1956.

TSFY   Ku Tsu-Yu, *Tu Shih Fang Yü Chi Yao* (The Historian's Geographical Companion), begun before +1666, finished before +1692, but not printed till the end of the eighteenth century (1796 to 1821).

TT   Wieger, L. (6), *Taoïsme*, vol. 1, Bibliographie Générale (catalogue of the works contained in the Taoist Patrology, *Tao Tsang*).

TTC   *Tao Tê Ching* (Canon of the Tao and its Virtue).

TTCY   Ho Lung-Hsiang & Phêng Han-Jan (ed.). *Tao Tsang Chi Yao* (Essentials of the Taoist Patrology), pr. 1906.

TW   Takakusu, J. & Watanabe, K., *Tables du Taishō Issaikyō* (nouvelle édition (Japonaise) du Canon bouddhique chinoise), Index-catalogue of the Tripiṭaka.

V   Verhaeren, H. (2) (ed.), Catalogue de la Bibliothèque du Pé-T'ang (the Pei Thang Jesuit Library in Peking).

WCTY/CC   Tsêng Kung-Liang (ed.), *Wu Ching Tsung Yao* (*Chhien Chi*), military encyclopaedia, first section, +1044.

YCCC   Chang Chün-Fang (ed.), *Yün Chi Chhi Chhien* (Seven Bamboo Tablets of the Cloudy Satchel), Taoist collection, +1022.

YHL   Thao Hung-Ching (attrib.), *Yao Hsing Lun* (Discourse on the Natures and Properties of Drugs).

YHSF   Ma Kuo-Han (ed.), *Yü Han Shan Fang Chi I Shu* (Jade-Box Mountain Studio collection of (reconstituted and sometimes fragmentary) Lost Books), 1853.

ACKNOWLEDGEMENTS

LIST OF THOSE WHO HAVE KINDLY READ THROUGH SECTIONS IN DRAFT

The following list, which applies only to Vol. 5, pts 2–5, brings up to date those printed in Vol. 1, pp. 15 ff., Vol. 2, p. xxiii, Vol. 3, pp. xxxix ff., Vol. 4, pt. 1, p. xxi, Vol. 4, pt. 2, p. xli and Vol. 4, pt. 3, pp. xliii ff.

Dr F. R. Allchin (Cambridge)	Apparatus (alcohol).
Dr M. R. Bloch (Beersheba)	Nitre.
Prof. Derk Bodde (Philadelphia)	Introductions.
Dr C. S. F. Burnett (Cambridge)	Comparative (Latin).
Dr Anthony Butler (St Andrews)	Solutions.
Mr J. Charles (Cambridge)	Metallurgical chemistry.
Mr W. T. Chase (Washington, D.C.)	Apparatus.
Prof. A. G. Debus (Chicago)	Modern chemistry (Mao Hua).
The late Prof. A. F. P. Hulsewé (Leiden)	Theories.
Dr Edith Jachimowicz (London)	Comparative (Arabic).
Dr Felix Klein-Franke (London)	Comparative (Arabic).
Mr S. W. K. Morgan (Bristol)	Metallurgy (zinc and brass).
The late Prof. Ladislao Reti (Milan)	Apparatus (alcohol).
Dr Kristofer M. Schipper (Paris)	Theories.
Prof. R. B. Serjeant (Cambridge)	Apparatus (Arabic).
Mr H. J. Sheppard (Warwick)	Introductions, and Comparative (Hellenistic).
Prof. Cyril Stanley Smith (Cambridge, Mass.)	Metallurgy, and Theories.
Mr Robert Somers (New Haven, Conn.)	Theories.
Dr Michel Strickmann (Kyoto)	Theories.
Dr Mikuláš Teich (Cambridge)	Introductions.
Mr H. G. Thurm (Rüdesheim-am-Rhein)	Ardent Water.
Mr R. G. Wasson (Danbury, Conn.)	Introduction (ethno-mycology).
Prof. R. McLachlan Wilson (St Andrews)	Comparative (Gnostic), and Theories.
Dr John Winter (Washington, D.C.)	Apparatus and Lacquer.
Mr James Zimmerman (New Haven, Conn.)	Theories.

AUTHOR'S NOTE

IT is now some sixteen years since the preface for Vol. 4 of this series (Physics and Physical Technology) was written; since then much has been done towards the later volumes. We are now happy to present a further substantial part of Vol. 5 (Spagyrical Discovery and Invention), i.e. alchemy and early chemistry, which go together with the arts of peace and war, including military and textile technology, mining, metallurgy and ceramics. The point of this arrangement was explained in the preface of Vol. 4 (e.g. pt. 3, p. l). Exigences not of logic but of collaboration are making it obligatory that these other topics should follow rather than precede the central theme of chemistry, which here is printed as Vol. 5, parts 2, 3, 4 and 5, leaving parts 1 and 6 to appear at a later date.

The number of physical volumes (parts) which we are now producing may give the impression that our work is enlarging according to some form of geometrical progression or along some exponential curve, but this would be largely an illusion, because in response to the reactions of many friends we are now making a real effort to publish in books of less thickness, more convenient for reading. At the same time it is true that over the years the space required for handling the history of the diverse sciences in Chinese culture has proved singularly unpredictable. One could (and did) at the outset arrange the sciences in a logical spectrum (mathematics—astronomy—geology and mineralogy—physics—chemistry—biology) leaving estimated room also for all the technologies associated with them; but to foresee exactly how much space each one would claim, that, in the words of the Jacobite blessing, was 'quite another thing'. We ourselves are aware that the disproportionate size of some of our Sections may give a mis-shapen impression to minds enamoured of classical uniformity, but our material is not easy to 'shape', perhaps not capable of it, and appropriately enough we are constrained to follow the Taoist natural irregularity and surprise of a romantic garden rather than to attempt any compression of our lush growths within the geometrical confines of a Cartesian parterre. The Taoists would have agreed with Richard Baxter that ''tis better to go to heaven disorderly than to be damned in due order'. By some strange chance our spectrum meant (though I thought at the time that the mathematics was particularly difficult) that the 'easier' sciences were going to come first, those where both the basic ideas and the available source-materials were relatively clear and precise. As we proceeded, two phenomena manifested themselves, first the technological achievements and amplifications proved far more formidable than expected (as was the case in Vol. 4, pts. 2 and 3), and secondly we found ourselves getting into ever deeper water, as the saying is, intellectually (as will fully appear in the Sections on medicine in Vol. 6).

Alchemy and early chemistry, the central subjects of the present volume, exemplified the second of these difficulties well enough, but they have had others of their own.

At one time I almost despaired of ever finding our way successfully through the inchoate mass of ideas, and the facts so hard to establish, relating to alchemy, chemistry, metallurgy and chemical industry in ancient, medieval and traditional China. The facts indeed were much more difficult to ascertain, and also more perplexing to interpret, than anything encountered in subjects such as astronomy or civil engineering. And in the end, one must say, we did not get through without cutting great swathes of briars and bracken, as it were, through the muddled thinking and confused terminology of the traditional history of alchemy and early chemistry in the West. Here it was indispensable to distinguish alchemy from proto-chemistry and to introduce words of art such as aurifiction, aurifaction and macrobiotics. It is also fair to say that the present subject has been far less well studied and understood either by Westerners or Chinese scholars themselves than fields like astronomy and mathematics, where already in the eighteenth century a Gaubil could do outstanding work, and nearer our own time a Chhen Tsun-Kuei, a de Saussure, and a Mikami Yoshio could set them largely in order. If the study of alchemy and early chemistry had advanced anything like so far, it would be much easier today than it actually is to differentiate with clarity between the many divergent schools of alchemists at the many periods, from the -3rd century to the $+17$th, with which we have to deal. More adequate understanding would also have been achieved with regard to that crucial Chinese distinction between inorganic laboratory alchemy and physiological alchemy, the former concerned with elixir preparations of mineral origin, the latter rather with operations within the adept's own body; a distinction hardly realised to the full in the West before the just passed decade. As we shall show in these volumes, there was a synthesis of these two age-old trends when in iatro-chemistry from the Sung onwards laboratory methods were applied to physiological substances, producing what we can only call a proto-biochemistry. But this will be read in its place.

Now a few words on our group of collaborators. Dr Ho Ping-Yü,[1] since 1972 Professor of Chinese and Dean of the Faculty of Asian Studies at Griffith University, Brisbane, in Queensland, was introduced to readers in Vol. 4 pt. 3, p. lv; here he has been responsible for drafting the major part of the sub-section (e) on the history of alchemy in China. Dr Lu Gwei-Djen,[2] my oldest collaborator, dating (in historian's terms) from 1937, has been involved at all stages of the present volumes, especially in that seemingly endless mental toil of ours which resulted in the introductory subsections on concepts, definitions and terminology (b), with all that that implies for theories of alchemy, ideas of immortality, and the physiological pathology of the elixir complex. But her particular domain has been that of physiological alchemy, and it was her discoveries, just at the right moment, of what was meant by the three primary vitalities, mutationist inversion, counter-current flow, and such abstruse matters, which alone permitted the unravelling, at least in the provisional form here presented (in the relevant sub-section j) of that strange and unfamiliar system, quasi-Yogistic perhaps, but full of interest for the pre-history of biochemical thought.[a]

[a] Some of her findings have appeared separately (Lu Gwei-Djen, 2).

[1] 何丙郁   [2] 魯桂珍

A third collaborator is now to be welcomed for the first time, Dr Nathan Sivin, Professor of Chinese in the University of Pennsylvania at Philadelphia, who has contributed the sub-section on the general theory of elixir alchemy (*h*).

Although Prof. Sivin has helped the whole group much by reading over and suggesting emendations for all the rest, it is needful to make at this point a proviso which has not been required in previous volumes. This is that my collaborators cannot take a collective responsibility for statements, translations or even general nuances, occurring in parts of the book other than that or those in which they each themselves directly collaborated. All incoherences and contradictions which remain after our long discussions must be laid at my door, in answer to which I can only say that the state of the art is as yet very imperfect, that it will certainly be improved by later scholars, and that in the meantime we have done the best we can. If fate had granted to the four of us the possibility of all working together in one place for half-a-dozen years, things could have been rather different, but in fact Prof. Ho and Prof. Sivin were never even in Cambridge at one and the same time. Thus these volumes have come into existence the hard way, drafted by different hands at fairly long intervals of time, and still no doubt containing traces of various levels of sophistication and under-standing.[a] Indeed it would have been reasonable to mark the elixir theory sub-section 'by Nathan Sivin', rather than 'with Nathan Sivin', if it had not been for the fact that some minor embroideries were offered by me, and that a certain part of it, not perhaps the least interesting, is a revised version of a memoir by Ho Ping-Yü and myself first published in 1959. Lacking the unities of time and place, complete credal unity, as it were, has been unattainable, but that does not mean that we are not broadly at one over the main facts and problems of the field as a whole; so that rightly we may be called co-workers.

Besides this I am eager to make certain further acknowledgements. During the second world war I was instrumental in securing for Cambridge copies of the *Tao Tsang* and the *Tao Tsang Chi Yao*. At a somewhat later time (1951–5) Dr Tshao Thien-Chhin,[1] then a Fellow of Caius, made a valuable pioneer study of the alchemical books in the Taoist Patrology, using a microfilm set in our working collection (now the East Asian History of Science Library, an educational charity). After his return to the Biochemical Institute of Academia Sinica, Shanghai, of which he has been in recent years Vice-Director, these notes were of great help to Dr Ho and myself, forming the ultimate basis for another sub-section (*g*), on aqueous reactions. Secondly, before he left Cambridge in 1958, Dr Wang Ling[2] accomplished a good work by making an analytical index of the names and synonyms of substances mentioned in the *Shih Yao Erh Ya*. Third, when we were faced with the fascinating but difficult study of the evolution of chemical apparatus in East and West, Dr Dorothy Needham put in a

---

[a] No less than eight years have now elapsed since Prof. Sivin first drafted the theoretical sub-section in this volume, and it could hardly be expected that during such a period insights and understanding would not mature and grow. Consequently this material should be supplemented by reference to Sivin (14), which gives a concise summary of his present views.

[1] 曹天欽   [2] 王鈴

considerable amount of work, including some drafting, in what happened to be a convenient interval in work on her own book on the history of muscle biochemistry, *Machina Carnis*. She has also read all our pages—perhaps the only person in the world who ever does so!

While readers of sub-sections in typescript and proof have not been as numerous, perhaps, as for previous volumes, a special debt of gratitude is due to Mr J. A. Charles of St John's College, chemist, metallurgist and archaeologist, whose advice to Prof. Ho and myself from the earliest days was extremely precious. Valuable consultations also took place with Mr H. J. Sheppard of Warwick, especially during his time in Cambridge as a Schoolmaster-Fellow of Churchill College. The late Dr Ladislao Reti was prodigal in his helpful advice on all aspects of the history of distillation, based on a lifetime's experience in chemical industry. Dr F. R. Allchin later communicated to us valuable unpublished information on the Gandhāran stills of Taxila and Push-kalāvatī. Subsequently we were able to benefit by an extensive correspondence on spirit production with Mr H. G. Thurm of the Asbach Distillery at Rüdesheim-am-Rhein; while Prof. E. J. Wiesenberg of London guided us with great expertise through the labyrinths of possible Semitic connections with the origin of the root 'chem-'. Few chemists in Cambridge, by some chance, happen to be interested at the present time in the history of their subject, but if Dr A. J. Berry and Prof. J. R. Partington had lived we could have profited greatly from their help. With the latter, indeed, we did have fruitful and most friendly contact, but it was in connection mainly with the gun-powder epic, Prof. Wang Ling and I endeavouring, not unsuccessfully, to convince him of the real and major contribution of China in that field; those were days however before any word of the present volumes had been written. In 1968, well after it had been started, there was convened the First Conference of Taoist Studies at the Villa Serbelloni at Bellagio on Lake Como; Ho Ping-Yü, Nathan Sivin and myself were all of the party, and here much stimulus was obtained from that remarkable *Tao shih* Kristofer Schipper—hence the unexpected sub-section on liturgiology and alchemical origins in our introductory material (*b*). In addition to the invaluable advice of many other colleagues in particular areas, we record especially the kindness of Professor Cyril Stanley Smith in commenting upon the sub-sections on metallurgy (*c*) and on the theory of elixir alchemy (*h*). Dr N. Sivin also expresses gratitude to Prof. A. F. P. Hulsewé and his staff for the open-hearted hospitality which they gave him during the gestation of the latter study, carried out almost entirely at the Sinologisch Instituut, Leiden.

It is right to record that certain parts of these volumes have been given as lectures to bodies honouring us by such invitations. Thus various excerpts from the intro-ductory sub-sections on concepts, terminology and definitions, were given for the Rapkine Lecture at the Pasteur Institute in Paris (1970) and the Bernal Lecture at Birkbeck College in London in the following year. Portions of the historical sub-sections, especially that on the coming of modern chemistry, were used for the Ballard Matthews Lectures of the University of Wales at Bangor. A considerable part of the physiological alchemy material formed the basis of the Fremantle Lectures at Balliol

College, Oxford,[a] and had been given more briefly as the Harvey Lecture to the Harveian Society of London the year before. Four lectures covering the four present parts of this volume were given at the Collège de France in Paris at Easter, 1973, in fulfilment of my duties as Professeur Étranger of that noble institution. Lastly, the contrasts between Hellenistic proto-chemistry and Chinese alchemy, with the spread of the elixir concept from China throughout the Old World, were expounded at the Universities of Hongkong and at the University of British Columbia at Vancouver, in 1975; while further aspects of the chemical relationships of the civilisations, east and west, formed the subject of the Bowra Lecture at Wadham College, Oxford, in 1976.

If there is one question more than any other raised by this present Section 33 on alchemy and early chemistry, now offered to the republic of learning in these volumes, it is that of human unity and continuity. In the light of what is here set forth, can we allow ourselves to visualise that some day before long we shall be able to write the history of man's enquiry into chemical phenomena as one single development throughout the Old World cultures? Granted that there were several different foci of ancient metallurgy and primitive chemical industry, how far was the gradual flowering of alchemy and chemistry a single endeavour, running contagiously from one civilisation to another?

It is a commonplace of thought that some forms of human experience seem to have progressed in a more obvious and palpable way than others. It might be difficult to say how Michael Angelo could be considered an improvement on Pheidias, or Dante on Homer, but it can hardly be questioned that Newton and Pasteur and Einstein did really know a great deal more about the natural univers e than Aristotle or Chang Hêng. This must tell us something about the differences between art and religion on one side and science on the other, though no one seems able to explain quite what, but in any case within the field of natural knowledge we cannot but recognise an evolutionary development, a real progress, over the ages. The cultures might be many, the languages diverse, but they all partook of the same quest.

Throughout this series of volumes it has been assumed all along that there is only one unitary science of Nature, approached more or less closely, built up more or less successfully and continuously, by various groups of mankind from time to time. This means that one can expect to trace an absolute continuity between the first beginnings of astronomy and medicine in Ancient Babylonia, through the advancing natural knowledge of medieval China, India, Islam and the classical Western world, to the break-through of late Renaissance Europe when, as has been said, the most effective method of discovery was itself discovered. Many people probably share this point of view, but there is another one which I may associate with the name of Oswald Spengler, the German world-historian of the thirties whose works, especially *The Decline of the West*, achieved much popularity for a time. According to him, the sciences produced by different civilisations were like separate and irreconcilable works

---

[a] The relevant volume is therefore offered to the Trustees of the late Sir Francis Fremantle's benefaction in discharge of the duty of publication of his Lectures.

of art, valid only within their own frames of reference, and not subsumable into a single history and a single ever-growing structure.

Anyone who has felt the influence of Spengler retains, I think, some respect for the picture he drew of the rise and fall of particular civilisations and cultures, resembling the birth, flourishing and decay of individual biological organisms, in human or animal life-cycles. Certainly I could not refuse all sympathy for a point of view so like that of the Taoist philosophers, who always emphasised the cycles of life and death in Nature, a point of view that Chuang Chou himself might well have shared. Yet while one can easily see that artistic styles and expressions, religious ceremonies and doctrines, or different kinds of music, have tended to be incommensurable; for mathematics, science and technology the case is altered—man has always lived in an environment essentially constant in its properties, and his knowledge of it, if true, must therefore tend towards a constant structure.

This point would not perhaps need emphasis if certain scholars, in their anxiety to do justice to the differences between the ancient Egyptian or the medieval Chinese, Arabic or Indian world-views and our own, were not sometimes tempted to follow lines of thought which might lead to Spenglerian pessimism.[a] Pessimism I say, because of course he did prophesy the decline and fall of modern scientific civilisation. For example, our own collaborator, Nathan Sivin, has often pointed out, quite rightly, that for medieval and traditional China 'biology' was not a separated and defined science. One gets its ideas and facts from philosophical writings, books on pharma- ceutical natural history, treatises on agriculture and horticulture, monographs on groups of natural objects, miscellaneous memoranda and so on. He urged that to speak without reservations of 'Chinese biology' would be to imply a structure which historically did not exist, disregarding mental patterns which did exist. Taking such artificial rubrics too seriously would also imply the natural but perhaps erroneous assumption that medieval Chinese scientists were asking the same questions about the living world as their modern counterparts in the West, and merely chanced, through some quirk of national character, language, economics, scientific method or social structure, to find different answers. On this approach it would not occur to one to investigate what questions the ancient and medieval Chinese scientists themselves were under the impression that they were asking. A fruitful comparative history of science would have to be founded not on the counting up of isolated discoveries,

---

[a] Just recently a relevant polemical discussion has been going on among geologists. Harrington (1, 2), who had traced interesting geological insights in Herodotus and Isaiah, was taken to task by Gould (1), maintaining that 'science is no march to truth, but a series of conceptual schemes each adapted to a prevailing culture', and that progress consists in the mutation of these schemes, new concepts of creative thinkers resolving anomalies of old theories into new systems of belief. This was evidently a Kuhnian approach, but no such formulation will adequately account for the gradual percolation of true knowledge through the successive civilisations, and its general accumulation. Harrington himself, in his reply (3), maintained that 'there is a singular state of Nature towards which all estimates of reality converge', and therefore that we can and should judge the insights of the ancients on the basis of our own knowledge of Nature, while at the same time making every effort to understand their intellectual framework. In illustration he took the medieval Chinese appreciation of the meaning of fossil remains (cf. Vol. 3, pp. 611ff.). We are indebted to Prof. Claude Albritton of Texas for bringing this discussion to our notice.

insights or skills meaningful for us now, but upon 'the confrontation of integral complexes of ideas with their interrelations and articulations intact'. These complexes could be kept in one piece only if the problems which they were meant to solve were understood. Chinese science must, in other words, be seen as developing out of one state of theoretical understanding into another, rather than as any kind of abortive development towards modern science.

All this was well put; of course one must not see in traditional Chinese science simply a 'failed prototype' of modern science, but the formulation here has surely to be extremely careful. There is a danger to be guarded against, the danger of falling into the other extreme, and of denying the fundamental continuity and universality of all science. This could be to resurrect the Spenglerian conception of the natural sciences of the various dead (or even worse, the living) non-European civilisations as totally separate, immiscible thought-patterns, more like distinct works of art than anything else, a series of different views of the natural world irreconcilable and unconnected. Such a view might be used as the cloak of some historical racialist doctrine, the sciences of pre-modern times and the non-European cultures being thought of as wholly conditioned ethnically, and rigidly confined to their own spheres, not part of humanity's broad onward march. However, it would leave little room for those actions and reactions that we are constantly encountering, those subtle communicated influences which every civilisation accepted from time to time.

In another place Nathan Sivin has written: 'The question of why China never spontaneously experienced the equivalent of our scientific revolution lies of course very close to the core of a comparative history of science. My point is that it is an utter waste of time, and distracting as well, to expect any answer until the Chinese tradition has been adequately comprehended from the inside.' The matter could not be better put; we must of course learn to see instinctively through the eyes of those who thought in terms of the Yin and Yang, the Five Elements, the symbolic correlations, and the trigrams and hexagrams of the *Book of Changes*. But here again this formulation might suggest a purely internalist or ideological explanation for the failure of modern natural science to arise in Chinese culture. I don't think that in the last resort we shall be able to appeal primarily to inhibiting factors inherent in the Chinese thought-world considered as an isolated Spenglerian cell. One must always expect that some of these intellectual limiting factors will be identifiable, but for my part I remain sceptical that there are many factors of this kind which could not have been overcome if the social and economic conditions had been favourable for the development of modern science in China. It may indeed be true that the modern forms of science which would then have developed would have been rather different from those which actually did develop in the West, or in a different order, that one cannot know. There was, for example, the lack of Euclidean geometry and Ptolemaic planetary astronomy in China, but China had done all the ground-work in the study of magnetic phenomena, an essential precursor of later electrical science;[a] and Chinese culture was permeated by conceptions much more organic, less mechanistic, than that of the

---

[a] See our discussions in Vols. 3 and 4, pt. 1.

West.[a] Moreover Chinese culture alone, as we shall see, perhaps, provided that materialist conception of the elixir of life which, passing to Europe through the Arabs, led to the macrobiotic optimism of Roger Bacon and the iatro-chemical revolution of Paracelsus, hardly less important in the origins of modern science than the work of Galileo and Newton. Whatever the ideological inhibiting factors in the Chinese thought-world may turn out to have been, the certainty always remains that the specific social and economic features of traditional China were connected with them. They were clearly part of that particular pattern, and in these matters one always has to think in terms of a 'package-deal'. In just the same way, of course, it is impossible to separate the scientific achievements of the ancient Greeks from the fact that they developed in mercantile, maritime, city-state democracies.

To sum it up, the failure of China to give rise to distinctively modern science while having been in many ways ahead of Europe for some fourteen previous centuries is going to take some explaining.[b] Internalist historiography is likely to encounter grave difficulties here, in my opinion, because the intellectual, philosophical, theological and cultural systems of ideas of the Asian civilisations are not going to be able to take the causal stress and strain required. Some of these idea-systems, in fact, such as Taoism and Neo-Confucianism, would seem to have been much more congruent with modern science than any of the European ones were, including Christian theology. Very likely the ultimate explanations will turn out to be highly paradoxical—aristocratic military feudalism seeming to be much stronger than bureaucratic feudalism but actually weaker because less rational—the monotheism of a personal creator God being able to generate modern scientific thought (as the San Chiao could never do) but not to give it an inspiration enduring into modern times—and so on. We do not yet know.

A similar problem has of late been worrying Said Husain Nasr, the Persian scholar who is making valuable contributions to the history of science in Islam. He, for his part, faces the failure of Arabic civilisation to produce modern science. But far from regretting this he makes a positive virtue of it, rejecting belief in any integral, social-evolutionary development of science. Opening one of his recent books we read as follows:[c]

The history of science is often regarded today as the progressive accumulation of techniques and the refinement of quantitative methods in the study of Nature. Such a point of view considers the present conception of science to be the only valid one; it therefore judges the sciences of other civilisations in the light of modern science, and evaluates them primarily with respect to their 'development' with the passage of time. Our aim in this work however, is not to examine the Islamic sciences from the point of view of modern science and of this 'evolutionist' conception of history; it is on the contrary to present certain aspects of the Islamic sciences as seen from the Islamic point of view.

[a] This was emphasised in Vol. 2, *passim*.
[b] We set forth in a preliminary way what is at issue here in Vol. 3, pp. 150ff. Some 'thinking aloud' done at various times has also been assembled in Needham (65).
[c] (1), p. 21.

Now Nasr considers that the Sufis and the universal philosophers of medieval Islam sought and found a kind of mystical *gnosis*, or cosmic *sapientia*, in which all the sciences 'knew their place', as it were (like servitors in some great house of old), and ministered to mystical theology as the highest form of human experience. In Islam, then, the philosophy of divinity was indeed the *regina scientiarum*. Anyone with some appreciation of theology as well as science cannot help sympathising to some extent with this point of view, but it does have two fatal drawbacks: it denies the equality of the forms of human experience, and it divorces Islamic natural science from the grand onward-going movement of the natural science of all humanity. Nasr objects to judging medieval science by its outward 'usefulness' alone. He writes:[a] 'However important its uses may have been in calendrical computation, in irrigation or in architecture, its ultimate aim always was to relate the corporeal world to its basic spiritual principle through the knowledge of those symbols which unite the various orders of reality. It can only be understood, and should only be judged, in terms of its own aims and its own perspectives.' I would demur. It was part, I should want to maintain, of all human scientific enterprise, in which there is neither Greek nor Jew, neither Hindu nor Han. 'Parthians, Medes and Elamites, and the dwellers in Mesopotamia, and in Judaea and Cappadocia, in Pontus and Asia...and the parts of Libya about Cyrene... we do hear them speak in our tongues the marvellous works of God.'[b]

The denial of the equality of the forms of human experience comes out clearly in another work of Said Husain Nasr (2). Perhaps rather under-estimating the traditional high valuation placed within Christendom upon Nature—'that universal and publick manuscript', as Sir Thomas Browne said,[c] 'which lies expans'd unto the eyes of all'— he sees in the scientific revolution at the Renaissance a fundamental desacralisation of Nature, and urges that only by re-consecrating it, as it were, in the interests of an essentially religious world-view, will mankind be enabled to save itself from otherwise inevitable doom. If the rise of modern science within the bosom of Christendom alone had any causal connections with Christian thought that would give it a bad mark in his view. 'The main reason why modern science never arose in China or Islam', he says,[d]

is precisely because of the presence of a metaphysical doctrine and a traditional religious structure which refused to make a profane thing of Nature....Neither in Islam, nor India nor the Far East, was the substance and the stuff of Nature so depleted of a sacramental and spiritual character, nor was the intellectual dimension of these traditions so enfeebled, as to enable a purely secular science of Nature and a secular philosophy to develop outside the matrix of the traditional intellectual orthodoxy....The fact that modern science did not develop in Islam is not a sign of decadence [or incapacity] as some have claimed, but of the refusal of Islam to consider any form of knowledge as purely secular, and divorced from what it conceived to be the ultimate goal of human existence.

[a] (1), pp. 39–40.
[b] Acts, 2. 1.
[c] *Religio Medici* I, xvi. 'Thus there are two Books from whence I collect my Divinity; besides that written one of God, another of his servant Nature....'
[d] (2), p. 97.

These are striking words,[a] but are they not tantamount to saying that only in Europe did the clear differentiation of the forms of experience arise? In other terms, Nasr looks for the synthesis of the forms of experience in the re-creation of a medieval world-view, dominated by religion,[b] not in the existential activity of individual human beings dominated by ethics. That would be going back, and there is no going back. The scientist must work *as if* Nature was 'profane'. As Giorgio di Santillana has said:[c]

> Copernicus and Kepler believed in cosmic vision as much as any Muslim ever did, but when they had to face the 'moment of truth' they chose a road which was apparently not that of *sapientia*; they felt they had to state what appeared to be the case, and that on the whole it would be more respectful of divine wisdom to act thus.

And perhaps it is a sign of the weakness of what can only be called so conservative a conception that Nasr is driven to reject the whole of evolutionary fact and theory, both cosmic, biological and sociological.

In contemplating the estimate of modern physical science as a 'desacralisation of Nature' many ideas and possibilities come to mind,[d] but one very obvious cause for surprise is that it occurred in Christendom, the home of a religion in which an incarnation had sanctified the material world, while it did not occur in Islam, a culture which had never developed a soteriological doctrine.[e] This circumstance might offer an argument in favour of the primacy of social and economic factors in the breakthrough of the scientific revolution. It may be that while ideological, philosophical and theological differences are never to be undervalued, what mattered most of all were the facilitating pressures of the transition from feudalism to mercantile and then industrial capitalism, pressures which did not effectively operate in any culture other than that of Western, Frankish, Europe.

In another place Nasr wonders what Ibn al-Haitham or al-Bīrūnī or al-Khāzinī would have thought about modern science. He concludes that they would be amazed at the position which exact quantitative knowledge has come to occupy today. They would not understand it because for them all *scientia* was subordinated to *sapientia*. Their quantitative science was only one interpretation of a segment of Nature, not the means of understanding all of it. '"Progressive" science', he says,[f] 'which in the Islamic world always remained secondary, has now in the West become nearly everything, while the immutable and "non-progressive" science or wisdom which was then primary, has now been reduced to almost nothing.' It happened that I read these words

---

[a] Views such as this are by no means restricted to eastern Muslim scholars. From within the bosom of the West a very similar attitude is to be found in the book on alchemy by Titus Burckhardt (1), cf. esp. pp. 66, 203.

[b] It seems very strange to us that he should regard Chinese culture as having been dominated by religion at any time.

[c] In his preface to Said Husain Nasr (1), p. xii.

[d] This, for example, is one of the outstanding questions in the attack on the uncontrolled manifestations of modern science and technology led by today's 'counter-culture'. There is much food for thought in the books of Roszak (1, 2) and Leiss (1); while Pirsig's famous meditation (1) inspired the defence of science (though not of misapplied technology) in Waddington's Bernal Lecture (5).

[e] This point was made by the Rev. D. Cupitt in discussion following a lecture for the Cambridge Divinity Faculty (1970) in which some of these paragraphs were used. It was afterwards published in part (Needham, 68). The contrast may be to some extent a matter of degree, since Islamic philosophy tended to recognise the material world as an emanation of the divine. 　　　[f] (1), p. 145.

at a terrible moment in history. If there were any weight in the criticism of the modern scientific world-view from the standpoint of Nasr's perennial Muslim *sapientia* it would surely be that modern science and the technology which it has generated have far outstripped morality in the Western and modern world, and we shudder to think that man may not be able to control it. Probably none of the human societies of the past ever were able to control technology, but they were not faced by the devastating possibilities of today, and the moment I read Nasr's words was just after the Jordanian civil war of September, 1970, that dreadful fratricidal catastrophe within the bosom of Islam itself. Since then we have had the further shocking example of Bengali Muslims being massacred by their brothers in religion from the Indus Valley. *Sapientia* did not prevent these things, nor would it seem, from the historical point of view, that wars and cruelties of all kinds have been much less within the realms of Islam or of East Asia than that of Christendom. Modern science, at all events, is not guilty as such of worsening men's lot, on the contrary it has immensely ameliorated it, and everything depends on what use humanity will make of these unimaginable powers for good or evil. Something new is needed to make the world safe for mankind; and I believe that it can and will be found.

In later discussions Nathan Sivin has made it clear that he is just as committed to a universal comparative history of science as any of the rest of us. That would be the ultimate justification of all our work. His point is not that the Chinese (or Indian, or Arabic) tradition should be evaluated only in the light of its own world-view, then being left as a kind of museum set-piece, but that it must be understood as fully as possible in the light of this as a prelude to the making of wide-ranging comparisons. The really informative contrasts, he suggests, are not those between isolated discoveries, but between those whole systems of thought which have served as the matrices of discovery.[a] One might therefore agree that not only particular individual anticipations of modern scientific discoveries are of interest as showing the slow development of human natural knowledge, but also that we need to work out exactly how the world-views and scientific philosophies of medieval China, Islam or India, differed from those of modern science, and from each other. Each traditional system is clearly of great interest not only in itself but in relation to our present-day patterns of ideas. In this way we would not only salute the Chinese recording of sun-spots from the − 1st century,[b] or the earliest mention of the flame test for potassium salts by Thao Hung-Ching in the + 5th century,[c] or the first correct explanation of the optics of the rainbow by Quṭb al-Dīn al-Shīrāzī in + 1300,[d] as distinct steps on the way to modern science, but also take care to examine the integral systems of thought and practice which generated these innovations. Modern science was their common end, but their evolution can only be explained (that is to say, causally accounted for) in the context of the various possibilities opened and closed by the totality of ideas, values and social attitudes of their time.

Section 33(*h*), on the theoretical background of proto-chemical alchemy, may be

[a] Cf. Sivin (10). [b] Cf. Vol. 3, p. 435. [c] Cf. Vol. 5, pt. 3, p. 139.
[d] Cf. Vol. 3, p. 474.

taken as an exemplification and a test of this way of looking at early science.[a] Nathan Sivin's contribution deals with an abstract approach to Nature which has little to do with post-Galilean physical thought. Looking at the aims of the theoretically-minded alchemists as expressed in their own words, they turn out to be concerned with the design and construction of elaborate chemical models of the cyclic Tao of the cosmos which governs all natural change. A multitude of correspondences and resonances inspire the design of these models. One can distinguish as elements in their rationale the archaic belief in the maturation of minerals within the earth, the complex role of time, and the subtle interplay of quantity and numerology in ensuring that the elaboratory would be a microcosmos. Once we have reached at least a rough comprehension of the system which unites these elements, we can apprehend the remarkable culmination envisaged by the Chinese alchemists: to telescope time by reducing the grand overriding cycles of the universe to a compass which would allow of their contemplation by the adept—leading, as we have phrased it, to perfect freedom in perfect fusion with the cosmic order. But in the course of our reconnaissance we gather a rich harvest of ideas worth exploring and comparing with those of other cultures, including those of the modern world—for instance, the notion of alchemy as a quintessentially temporal science, springing from a unique concept of material immortality, a sublime conviction of the possibility of the control of change and decay. And we make a beginning towards understanding how the alchemist's concepts determined the details—the symmetries and innovations of materials, apparatus, and exquisitely phased combustions—of his Work, and how new results were reflected in new theoretical refinements as the centuries passed.

It is no less important to be aware that every anticipatory feature of a pre-modern system of science had its Yin as well as its Yang, side, disadvantages as well as advantages. Thus the polar-equatorial system of Chinese astronomy delayed Yü Hsi's recognition of the precession of the equinoxes by six centuries after Hipparchus, but on the other hand it gave to Su Sung an equal priority of time over Robert Hooke in the first application of a clock-drive to an observational instrument; and the mechanisation of a demonstrational one by I-Hsing and Liang Ling-Tsan was no less than a thousand years ahead of George Graham and Thomas Tompion with their orrery of 1706.[b] In a similar way, perhaps, the conviction of the existence of material life-elixirs cost the lives of untold numbers of royal personages and high officials no less than of Taoist adepts, but it did lead to the accumulation of a great fund of knowledge about metals and their salts, in the pursuit of which such earth-shaking discoveries as that of gunpowder were incidentally made. So also the ancient idea of urine and other secretions as drugs might easily be written off as 'primitive superstition' if we did not know that it led, by rational if quasi-empirical trains of thought, combined with the use of chemical techniques originally developed for quite different purposes, to the preparation of steroid and protein hormones many centuries before the time of experimental endocrinology and biochemistry.

---

[a] Another attempt at this approach, applied to mathematical astronomy, will be found in Sivin (9).
[b] On all these subjects see Vol. 3 and Vol. 4, pt. 2.

The only danger in the conception of human continuity and solidarity, as I have outlined it, is that it is very easy to take modern science as the last word, and to judge everything in the past solely in the light of it. This has been justly castigated by Joseph Agassi, who in his lively monograph on the historiography of science (1) satirises the mere 're-arranging of up-to-date science textbooks in chronological order', and the awarding of black and white marks to the scientific men of the past in accordance with the extent to which their discoveries still form part of the corpus of modern knowledge. Of course this Baconian or inductivist way of writing the history of science never did justice to the 'dark side' of Harvey and Newton, let alone Paracelsus, that realm of Hermetic inspirations and idea-sources which can only be regained by us with great difficulty, yet is so important for the history of thought, as the life-work of Walter Pagel has triumphantly shown. One can see immediately that this difficulty is even greater in the case of non-European civilisations, since their thought-world has been even more unfamiliar. Not only so, but the corpus of modern knowledge is changing and increasing every day, and we cannot foresee at all what its aspect will be a century from now. Fellows of the Royal Society like to speak of the 'true knowledge of natural phenomena', but no one knows better than they do how provisional this knowledge is. It is neither independent of the accidents of Western European history, nor is it a final court of appeal for the eschatological judgment of the value of past scientific discoveries, either in West or East. It is a reliable measuring-stick so long as we never forget its transitory nature.

My collaborators and I have long been accustomed to use the image of the ancient and medieval sciences of all the peoples and cultures as rivers flowing into the ocean of modern science. In the words of the old Chinese saying: 'the Rivers pay court to the Sea'.[a] In the main this is indubitably right. But there is room for a great deal of difference of opinion on how the process has happened and how it will proceed. One might think of the Chinese and Western traditions travelling substantially the same path towards the science of today, that science against which, on the inductivist view, all ancient systems can be measured. But on the other hand, as Nathan Sivin maintains, they might have followed, and be following, rather separate paths, the true merging of which lies well in the future. Undoubtedly among the sciences the point of fusion varies, the bar where the river unites at last with the sea. In astronomy and mathematics it took but a short time, in the seventeenth century; in botany and chemistry the process was much slower, not being complete until now, and in medicine it has not happened yet.[b] Modern science is not standing still, and who can say how far the molecular biology, the chemistry or the physics of the future will have to adopt conceptions much more organicist than the atomic and the mechanistic which have so far prevailed? Who knows what further developments of the psycho-somatic conception in medicine future advances may necessitate? In all such ways the thought-complex of traditional Chinese science may yet have a much greater part to play in

[a] *Chhao tsung yü hai.*[1] Cf. Vol. 3, p. 484.
[b] This picture has been elaborated elsewhere; Needham (59), reprinted in (64), pp. 396ff.

[1] 朝宗于海

the final state of all science than might be admitted if science today was all that science will ever be. Always we must remember that things are more complex than they seem, and that wisdom was not born with us. To write the history of science we have to take modern science as our yardstick—that is the only thing we can do—but modern science will change, and the end is not yet. Here as it turns out is yet another reason for viewing the whole march of humanity in the study of Nature as one single enterprise. But we must return to the volume now being introduced.

Although the other parts of Vol. 5 are not yet ready for press we should like to make mention of those who are collaborating with us in them. Much of the Section on martial technology for Vol. 5, pt. 1 has been in draft for many years now,[a] but it has been held up by delays in the preparation of the extremely important sub-section on the invention of the first chemical explosive known to man, gunpowder, even though all the notes and books and papers necessary for this have long been collected.[b] At last we can salute the advent of a relevant draft of substantial size from Dr Ho Ping-Yü at Brisbane, recently Visiting Professor at Keio University in Tokyo, aided by Dr Wang Ling (Wang Ching-Ning[1]) of the Institute of Advanced Studies at Canberra. Meanwhile Prof. Lo Jung-Pang,[2] of the University of California at Davis, spent the winter of 1969–70 in Cambridge, accomplishing not only the sub-section on the history of armour and caparison in China, but also the draft of the whole of Section 37 on the salt industry, including the epic development of deep borehole drilling (Vol. 5, pt. 6). Other military sub-sections, such as those on poliorcetics, cavalry practice and signalling, we have been able to place in the capable hands of Dr Korinna Hana of München. About the same time we persuaded Dr Tsien Tsuen-Hsuin (Chhien Tshun-Hsün[3]), the Regenstein Librarian at the University of Chicago, to undertake the writing of Section 32 on the great inventions of paper and printing and their development in China; this is now more than half done. For ceramic technology (Section 35) we have obtained the collaboration of Mr James Watt (Chhü Chih-Jen[4]), Curator of the Art Gallery at the Institute of Chinese Studies in the Chinese University of Hongkong. The story of these marvellous applications of science will be anticipated by many with great interest. Finally non-ferrous metallurgy and textile technology, for which abundant notes and documentation have been collected, found their organising genii in two other widely separated places. For the former we have Prof. Ursula Martius Franklin and Dr Hsü Chin-Hsiung[5] at Toronto; for the latter Dr Ohta Eizō[6] of Kyoto and Dr Dieter Kühn. When their work becomes available, Volume 5 will be substantially complete. This by no means exhausts the list of our

[a] Including an introduction on the literature, a study of close-combat weapons, the sub-sections on archery and ballistic machines, and a full account of iron and steel technology as the background of armament. The first draft of this last has been published as a Newcomen Society monograph; Needham (32), (60).

[b] A preliminary treatment of the subject, still, we think, correct in outline, was given in an article in the *Legacy of China* thirteen years ago; Needham (47). This has recently been re-issued in paper-back form.

[1] 王靜寧　　[2] 羅榮邦　　[3] 錢存訓　　[4] 屈志仁　　[5] 許進雄
[6] 太田英藏

invaluable collaborators, for several others are concerned with Volumes 6 and 7; but they will be introduced to readers in due time.

As has so long been customary, we offer our grateful thanks to those who try to keep us 'on the rails' in territory which is not our own: Prof. D. M. Dunlop for Arabic, Dr Sebastian Brock for Syriac, Dr Charles Sheldon for Japanese, Prof. G. Ledyard for Korean, and Prof. Shackleton Bailey for Sanskrit and Tibetan.

A couple of years ago it became clear that our working library and its operations had grown so much in size and complexity that a full-time Amanuensis (*chêng chen shu tshao*[1]) or Librarian was needed. For this we first recruited a physical chemist, Dr Christine King (Ting Pai-Fu[2]), who gave us much assistance; being succeeded after some time by a valued former associate, the Japanologist Miss Philippa Hawking. Her organising abilities stood us in good stead during the moves of the library mentioned below. The best librarians are born, not made, and she is of that company.

Next comes our high secretariat—Miss Muriel Moyle, who continues to give us impeccable indexes; and Mrs Liang Chung Lien-Chu[3] (wife of another Fellow of Caius, the physicist Dr Liang Wei-Yao[4]), who has inserted many a page of well-written characters and made out many a biographical reference-card, as well as editing the typescripts of collaborators to conform with project conventions. We also offer appreciative thanks for skilled and accurate typewriting by Mrs Diana Brodie and Mrs Evelyn Beebe; and for editorial work by Mrs Janin Hua Chhang-Ming[5], Mrs Margaret Whetham Anderson and Major Frank Townson.

All that has been said in previous volumes (e.g. Vol. 4, pt. 3, p. lvi) about the University Press, our treasured medium of communication with the world, and Gonville and Caius College, that milieu in which we used to live and move and have our being, has become only truer as the years go by—their service and their encouragement continues unabated and so does our heartfelt gratitude. If it were not for the devotion of the typographical—and typocritical—masters, and if one could not count on the understanding, kindness and appreciation of one's academic colleagues, nothing of what these volumes represent could ever have come into existence. We have taken pleasure on previous occasions of paying a tribute to our friend Mr Peter Burbidge of the University Press, and as we do so again we would like to associate with his name all those in that unique organisation who deal so faithfully, accurately and elegantly with our very difficult work.

Down to the summer of 1976 the library which constitutes the engine-room of the project was housed in Caius, but upon my retirement from the Mastership it was moved to a temporary building in Shaftesbury Road just outside the 'compound' (as one would say in Asia) of the University Printing House. Later we were installed in a spacious house in Brooklands Avenue. This building belongs to the Press, and is lent by the Syndics to the Trustees of the East Asian History of Science Library, *pro tem*. We acknowledge with warmest thanks a generous installation grant from the British Museum Library Ancillary Libraries Fund, and a special grant from the Sloane Foundation in America. Particular continuing gratitude is due to the Wellcome Trust of London, whose generous support has upheld us throughout the

[1] 正眞書曹 [2] 丁百馥 [3] 梁鍾連杼 [4] 梁維耀 [5] 華昌明

period of preparation of these chemical volumes. Since the history of medicine is touched upon at so many points in them we feel some sense of justification in accepting their unfailing aid. It can hardly be too much emphasised that in China proto-chemistry was elixir alchemy from the very beginning (as it was not in other civilisa-tions of comparable antiquity), and by the same token alchemists there were very often physicians too (much more so than they tended to be in other cultures). For the basic elixir notion was a pharmaceutical and therapeutic one, even though its optimism regarding the conquest of death reached a height which modern medical science dare not as yet contemplate. All this will be clarified in what follows.

More recently our project received a notable benefaction from the Coca-Cola Company of Atlanta, Georgia, through the kind intermediation of Dr C. A. Shilling-law, and for this also our grateful thanks are due. The support of their benevolent fund is being continued for the expenses of Dr Li Li-Shêng[1], who has spent some months in Cambridge completing Section 34 on the chemical industries, a first draft for which was made some years ago by Prof. Ho Ping-Yü. To Thames Television we acknowledge a useful grant for the support of our amanuensis, and to the Lee Founda-tion of Singapore (founded in memory of the late Dato Lee Kong-Chian[2]) several most welcome grants for general project expenses. Help on a lesser scale has also been forth-coming from the American Philosophical Society. Certain private persons, too, have sent us truly notable donations from time to time; and here we cannot forbear from offering our warmest thanks to Mrs Carol Bernstein Ferry and Mr W. H. Ferry of Scarsdale, N.Y., as also to Mr and Mrs P. L. Lamb of Hongkong. Lastly Dr N. Sivin wishes to acknow-ledge financial assistance from the National Science Foundation in Washington, D.C., and the Department of Humanities at the Massachusetts Institute of Technology.

Let us end with a few words of help to the prospective reader, as on previous occasions, offering some kind of waywiser to guide him through those pages of type not always possible to lighten by some memorable illustration. This is not intended as a substitute for the contents-table, the *mu lu*, or as any enlargement of it; but rather as some useful tips of 'inside information' to tell where the really important paragraphs are, and to distinguish them from the supporting detail secondary in significance though often fascinating in itself.

First, then, we would recommend a reader to study very carefully our introduction (Sect. 33*b*, in Vol. 5, pt. 2) on concepts, terminology and definitions, especially pp. 9–12; because once one has obtained a clear idea of the distinctions between aurific-tion, aurifaction and macrobiotics (already referred to, p. xxxii above), everything that one encounters in the proto-chemistry and alchemy of all the Old World civilisations falls into place. There is a parallel here with the history of time-keeping, for the radical gap between the clepsydra and the mechanical clock was only filled by half-a-dozen centuries of Chinese hydro-mechanical clockwork. So in the same way the radical gap between Hellenistic aurifictive and aurifactive proto-chemistry at one end, and late Latin alchemy and iatro-chemistry at the other, could only be explained by a know-ledge of Chinese chemical macrobiotics.

[1] 李勳生   [2] 李公健

After that the argument develops in several directions, among which the reader can take his choice. How could belief in aurifaction ever have arisen when the cupellation test had been known almost since the dawn of the ancient empires? Look at 33*b*, 1–2, and especially p. 44 of pt. 2. What was the position of China in this respect, and what were the ancient Chinese alchemists probably doing experimentally? Read 33*b*, 3–5; and *c*, 1–8. Why were they so much more occupied with the perpetuation of life on earth, even in ethereal forms, than with the faking or making of gold? We try to explain it in 33*b*, 6. Such an induction of material immortality was indeed the specific characteristic of Chinese alchemy, and our conclusion is that the world-view of ancient China was the only milieu capable of crystallising belief in an elixir (*tan*[1]), good against death, as the supreme achievement of the chemist (see esp. pt. 2, pp. 71, 82, 114–15).

This is the nub of the argument, and in the present part (pt. 4, Sect. 33*i*, 2–3) we follow the progress of that great creative dream through Arabic culture and Byzantium into the Latin Baconian and Paracelsian West. Differences of religion, theology and cosmology did not stop its course, but there can be no doubt that it was born within the bosom of the Taoist religion, and hence the reader is invited to participate in a speculation that the alchemist's furnace derived from the liturgical incense-burner no less than from the metallurgical hearth (pt. 2, Sect. 33*b*, 7, see esp. pp. 127, 154). Finally something is said on the physiological background of the ingestion of elixirs (33*d*, 1, see esp. p. 291); why were they so attractive to the consumer initially and why so lethal later? Here belongs also the conservation of the body of the adept after death, important in the Taoist mind in connection with material immortality (pt. 2, Sect. 33*d*, 2, see esp. pp. 106, 207–8).

In the sub-section giving the straight historical account of Chinese alchemy from beginning to end, *chi shih pên mo*,[2] as the phrase was (pt. 3, Sect. 33*e*, 1–8), no part is really more significant than any other. Yet special interest does attach to the oldest firm records of aurifiction and macrobiotics expounded in (1), and to the study of the oldest alchemical books in (2) and (6, i). Now and then the narrative is interrupted by passages of detail, especially in (1), (2), (3, iii) and (6, vii) which readers not avid for minutiae may like to pass over (esp. pt. 3, pp. 42–4, 52–6, 76–8, 111–13, 201–5); such is the wealth of information not previously available in the West. The following sub-sections in the present part on laboratory apparatus, distillation, aqueous reactions, and alchemical theory (pt. 4, sects. 33*f*, *g*, *h*) explain themselves from the contents table, and again no passage stands out as particularly crucial; unless it were the relation of the Chinese alchemist to time (33*h*, 3–4). His was indeed the science (or proto-science) of the Change and Decay Control Department, as one might say, for he could (as he believed) accelerate enormously the natural change whereby gold was formed from other substances in the earth, and conversely he could decelerate asymptotically the rate of decay and dissolution that human bodies, each with their ten 'souls' (*hun*[3] and *pho*[4]), were normally subject to (cf. Fig. 1306). Thus in the words of the ancient Chinese slogan (33*e*, 1) 'gold *can* be made, and salvation *can* be

---

[1] 丹   [2] 紀事本末   [3] 魂   [4] 魄

attained'. And the macrobiogens were thus essentially time- and rate-controlling substances—a nobly optimistic concept for a nascent science of two thousand years ago.

Lastly, in part 5, we pass from the 'outer elixir' (*wai tan*[1]) to the 'inner elixir' (*nei tan*[2]), from proto-chemistry to proto-biochemistry, from reliance on mineral and inorganic remedies to a faith in the possibility of making a macrobiogen from the juices and substances of the living body. For this new concept we coin a fourth new word, the enchymoma; its synthesis was in practice the training of mortality itself to put on immortality. This 'physiological alchemy' will be explained in the next part of Volume 5 (Sect. 33*j*, 1–8), and the basic ideas will be found in two places, (2) especially (i, ii), and (4). It was not primarily psychological, like the 'mystical alchemy' of the West, though it made much use of meditational techniques, as did the Indian *yogacāra* with which it certainly had connections. Our conclusion is, at the end of (4) and in (8), that most of its procedures were highly conducive to health, both mental and physical, even though its theories embodied much pseudo-science as well as proto-science.

In the end, the iatro-chemistry of the late Middle Ages in China began to apply *wai tan* laboratory techniques to *nei tan* materials, bodily secretions, excretions and tissues. Hence arose some extraordinary successes and anticipations (33*k*, 1–7), but we must not enlarge on them now. And this may suffice for a reader's guide, hoping only that he may fully share with us the excitement and satisfaction of many new insights and discoveries.

[1] 外丹   [2] 內丹

33. ALCHEMY AND CHEMISTRY

(f) LABORATORY APPARATUS AND EQUIPMENT

It will readily be allowed that the history of chemical apparatus and equipment must constitute a sector of cardinal importance in any history of alchemy and early chemistry. Was it not after all the foundation of the techniques of modern chemical science? We have already quoted the words of Francis Bacon (pt. 2, p. 32) on the husbandman and his buried gold in Aesop, and later on (in pt. 5) we shall give those of Hermann Boerhaave and Albrecht von Haller, all recognising the immeasurable debt which true chemists owe to their alchemical, iatro-chemical and artisanal ancestors. How far we shall have to accept a special indebtedness to those of them who were Chinese will appear as the following pages pass. The full appreciation of the facts will be assisted by a reference to the contents of Sect. 26g, 5 on the history of glass in China (Vol. 4, pt. 1, pp. 101 ff.), and when the Sections on ceramics (35) and metallurgy (36) become available, further light will be thrown on what possibilities were open; for all these practical arts were necessarily laid under contribution by the Taoist alchemists of the Middle Ages in fitting out their elaboratories.

Tshao Yuan-Yü (1) was the first to make a study (1933) of Chinese alchemical apparatus. His remarkable paper on the 'Apparatus and Methods of the Ancient Chinese Alchemists' aroused such interest that English abridgements were made by Barnes (1) and Wilson (2b, c) in the following years. Li Chhiao-Phing (1, 1) also devoted a few pages to alchemical apparatus in his book. Further short descriptions in German and Chinese were later given by Huang Tzu-Chhing (1) and Yuan Han-Chhing (1) respectively, and then in 1959 the subject of laboratory equipment was extensively reviewed by Ho Ping-Yü & Needham (3). Since then there has been little save the book of Sivin (1) which has touched illuminatingly on certain special aspects of the subject.[a] In the present sub-section we have drawn materials more exhaustively both from the Taoist patrology and from ethnographical data, designing to treat the matter more thoroughly, and more comparatively, than anyone has so far done. Some of our interpretations, such as that of the East Asian types of still, were already essentially different from previous suggestions, and some techniques, like the method of *destillatio per descensum* using a bamboo tube, had not been mentioned before our first review.

Our main sources here include more than twenty different alchemical texts, all from the *Tao Tsang*, and many of them illustrated. Since they must be referred to very often it will be convenient to tabulate them in a list with the names of their writers (if possible), their approximate dates of composition, and their numbers in the standard catalogues (Table 114).

---

[a] There are also interesting discussions in Yoshida Mitsukuni (7), pp. 223 ff., 249 ff., 252 ff. His interpretations differ but little from ours, though we cannot quite follow him in his generalisations about the development of early chemical equipment in East and West.

Table 114. *Names and details of 'Tao Tsang' texts useful in the study of chemical apparatus*

Catalogue numbers			Approx. date	Author
Wieger (6) TT]	Ong Tu-Chien (1)			
229	232	*Huan Tan Pi Chieh Yang Chhih-Tzu Shen Fang* 還丹祕訣養赤子神方 (The Wondrous Art of Nourishing the (Divine) Embryo (lit. the Naked Babe) by the use of the secret Formula of the Regenerative Enchymoma)	Sung, late +12th	Hsü Ming-Tao 許明道
874	880	*Thai-Chhing Shih Pi Chi* 太清石壁記 (The Records in the Rock Chamber (lit. Wall); a Thai-Chhing Scripture)[a]	Liang, early +6th, but including material as old as the late +3rd (Chin)	ed. Chhu Tsê 楚澤 orig. writer: Su Yuan-Ming 蘇元明 (Chhing Hsia Tzu) 青霞子
878	884	*Huang Ti Chiu Ting Shen Tan Ching Chüeh* 黃帝九鼎神丹經訣 (The Yellow Emperor's Canon of the Nine-Vessel Spiritual Elixir, with Explanations)	Thang or Sung, but incorporating some material as old as the +2nd (H/Han)	unknown
884	890	*Ta-Tung Lien Chen Pao Ching, Chiu Huan Chin Tan Miao Chüeh* 大洞鍊真寶經九還金丹妙訣 (Mysterious Teachings on the Ninefold Cyclically Transformed Gold Elixir, supplementary to the Manual of the Making of the Perfected Treasure; a Ta-Tung Scripture)[b]	Thang, perhaps c. +712	Chhen Shao-Wei 陳少微
885	891	*Thai-Shang Wei Ling Shen Hua Chiu Chuan Huan Tan Sha Fa* 太上衛靈神化九轉還丹砂法 (Methods of the Guardian of the Mysteries for the Marvellous Thaumaturgical Transmutation of Ninefold Cyclically Transformed Cinnabar; a Thai-Shang Scripture)[c]	uncertain, probably Sung	unknown
886	892	*Chiu Chuan Ling Sha Ta Tan* 九轉靈砂大丹 (The Great Ninefold Cyclically Transformed Numinous Cinnabar Elixir)	unknown	unknown
889	893	*Yü Tung Ta Shen Tan Sha Chen Yao Chüeh* 玉洞大神丹砂真要訣 (True and Essential Teachings about the Great Magical Cinnabar of the Jade Heaven)	Thang, early +8th	Chang Kuo 張果
893	899	*Tan Fang Hsü Chih* 丹房須知 (Indispensable Knowledge for the Chymical Elaboratory)	Sung, +1163	Wu Wu 吳悞
894	900	*Shih Yao Erh Ya* 石藥爾雅 (The Literary Expositor of Chemical Physic; or, Synonymic Dictionary of Minerals and Drugs)	Thang, +806	Mei Piao 梅彪

		Title	Date / Ascription	Attribution
895	901	*Chih-Chhuan Chen-Jen Chiao Chêng Shu* 稚川真人校證術 (Technical Methods of the Adept (Ko) Chih-Chhuan (i.e. Ko Hung), with Critical Annotations)	Ascr. Chin, c. +320, but most of it probably a good deal later	Attrib. Ko Hung 葛洪
902	908	*Lung Hu Huan Tan Chüeh* 龍虎還丹訣 (Explanation of the Dragon-and-Tiger Cyclically Transformed Elixir)	probably Sung	Chin Ling Tzu 金陵子 (ps.)
904	910	*Kan Chhi Shih-liu Chuan Chin Tan* 感氣十六轉金丹 (The Sixteen-fold Cyclically Transformed Gold Elixir preapred by the 'Responding to the Chhi' Method)	Sung	unknown
905	911	*Hsiu Lien Ta Tan Yao Chih* 修鍊大丹要旨 (Essential Instructions for the Preparation of the Great Elixir)	Sung	unknown
907	913	*Chin Hua Chhung Pi Tan Ching Pi Chih* 金華沖碧丹經祕旨 (Confidential Instructions on the Manual of the Heaven-piercing Golden Flower Elixir)	Sung, +1225	Phêng Ssu 彭耜 & Mêng Hsü 孟煦
908	914	*Huan Tan Chou Hou Chüeh* 還丹肘後訣 (Oral Instructions on Handy Formulae for Cyclically Transformed Elixirs)	Ascr. Chin, c. +320, but actually by a Thang writer between +874 and +879	Attrib. Ko Hung 葛洪
911	917	*Chu Chia Shen Phin Tan Fa* 諸家神品丹法 (Methods of the Various Schools for Magical Elixir Preparations)	Sung	Mêng Yao-Fu 孟要甫 (Hsüan Chen Tzu) et al. 玄真子
912	918	*Chhien Hung Chia Kêng Caih Pao Chi Chhêng* 鉛汞甲庚至寶集成 (Complete Compendium on the Perfected Treasure of Lead, Mercury, Wood and Metal)	Thang, +808	Chao Nai-An 趙耐菴
935	941	*Thung Hsüan Pi Shu* 通玄祕術 (The Secret Art of Penetrating the Mystery)	Thang, +864	Shen Chih-Yen 沈知言
939	945	*Thai-Chi Chen-Jen Tsa Tan Yao Fang* 太極真人雜丹藥方 (Tractate of the Supreme-Pole Adept on Miscellaneous Elixir Recipes)	unknown, but probably Sung on account of the philosophical pseudonym in the title	unknown
946	952	*Kêng Tao Chi* 庚道集 (Collection of Procedures on the Golden Art)	Sung or Yuan, date unknown but after +1144	unknown
990	996	*Chou I Tshan Thung Chhi Chu* 周易參同契註 (The Kinship of the Three and the Book of Changes, with Commentary)	tradit. date of orig. text, H/Han, +142; this comm. ascr. H/Han, c. +160, but more probably Sung.	Attrib. ed. & comm. Yin Chhang-Shêng 陰長生
1020	1026	*Yün Chi Chhi Chhien* 雲笈七籤 (itself a collection) (The Seven Bamboo Tablets of the Cloudy Satchel)	Sung, c. +1022	ed. Chang Chün-Fang 張君房
1054	1060	*Chin Tan Ta Yao Thu* 金丹大要圖 (Illustrations for the Main Essentials of the Metallous Enchymoma: the true Gold Elixir)[d]	Yuan, +1333; but based on drawings and tables of the Sung, +10th century, onwards by	Chhen Chih-Hsü 陳致虛 (Shang Yang Tzu) 上陽子 Chang Po-Tuan 張伯端 Lin Shen-Fêng et al. 林神鳳

[a] Tr. Ho Ping-Yü (8).   [b] Tr. Sivin (4).   [c] Tr. Spooner & Wang (1); Sivin (3).   [d] Tr. Ho Ping-Yü & Needham (2).

To describe their experiments the medieval Chinese alchemists and proto-chemists employed a host of technical terms.[a] Unfortunately, in contrast with those used in astronomy, definitions of such terms, so far as we know, have not been found in the literature.[b] A study of some of them has been made in recent times by Yuan Han-Chhing (*1*)[c] and Ho Ping-Yü (15, 18). Their results are further elaborated in the following list:

Table 115. *Technical terms of operations*

an[1] (lit. to place)	to set up, to place in position.
chêng[2] (lit. steaming)	to steam grain, food, ferment or any other material in a steamer.
chêng[3] (lit. steaming)	distillation.[d]
chiao[4] (lit. to water)	to pour out a hot liquid and allow it to cool down or solidify slowly.
chieh[5] (lit. to tie up, to form an alliance)	to congeal, or solidify, generally by evaporation. Also used to refer to the formation of crystals.
chien[6] (lit. to fry)	to heat while stirring, either dry or with oil.
chien lien[7] (lit. to fry and refine)	to recrystallise.
chih[8] (lit. to control)	to prevent or delay the process of volatilisation, sublimation or distillation (i.e. fixation); to produce a change (cf. *fu* and *sha*).[e]
chih[9] (lit. to broil, stew, or toast; also to cauterise)[f]	to apply heat locally; to make an aqueous extract by heating; to dry by heating.
ching,[10] *ching hua*[11]	crystal; to crystallise or make to crystallise.
chu[12] (lit. to boil)	to heat a substance in water, to simmer.
chuan[13] (lit. turn)	a cycle of changes, usually several times repeated. Cf. *huan*.
chhou[14] (lit. to draw out or pull up)	to distil, especially of mercury.[g]
fei[15] (lit. to fly)	sublimation; distillation (especially in the case of mercury);[h] vaporisation in general.

[a] Hopkins gave a striking example of the incomprehensibility of technical terms to laymen, (1), p. 91. He took as his text a sample of instructions to a seamstress in 1934: 'Cross-cut bands are the medium turned in even to face and tacked at the edges; holes are used instead of eyes, made with a stiletto, and fan-stitch is used to fix the bones.' He added that not everyone, even at that time, would know the meaning of other terms such as herring-boning, fagotting, shirring, easing, piping, basting, overcasting, coarse-running and tacking out. Furthermore, he said, a foreigner would find the words hard to translate, especially if contemporary literature had disappeared, a thousand years hence. The same applies, of course, to all the arts and trades, so one cannot be surprised that there are still problems in the technical terms of alchemy and chemistry in the different cultures.

[b] There is similar need for a glossary of technical terms in pharmacology and this we propose to provide in Sect. 45 (Vol. 6).

[c] Pp. 207ff.   [d] Mod. *chêng liu*.[16]

[e] For example, heating sal ammoniac with tin so as to produce stannous chloride.

[f] By confusion with *chiu*,[17] the correct medical term for moxibustion and other forms of cautery.

[g] Cf. e.g. TT893, p. 7*a*, *b*.   [h] Cf. again TT893, p. 7*a*, *b*.

1 安	2 烝	3 蒸	4 澆	5 結	6 煎	7 煎錬
8 制	9 炙	10 晶	11 晶化	12 煮	13 轉	14 抽
15 飛	16 蒸餾	17 灸				

Table 115 (continued)

fu[1] (lit. rotten, corrupt) — putrefaction; but also certain special fermentations.

fu[2] (lit. to subdue, to make to lie prostrate) — to extract; to separate out from; to purify; esp. to prevent or delay the process of volatilisation, sublimation or distillation (i.e. fixation); to inhibit the potency of some other substance; cf. *chih* and *sha*.

fu chi[3] (lit. cover and bed) — a layer of mineral substance placed below and above the reactants in the vessel.

fu huo[4] (lit. to subdue in the fire) — to heat until the substance is subdued (i.e. fixed).

hua[5] (lit. to change) — to undergo, or make to undergo, chemical change; to melt or to solidify.

hua chhih[6] (lit. radiant or flowery pool) — a bath of strong acetic acid (vinegar, with additions), in which is immersed a substance or substances, sometimes contained within a pared and sealed bamboo tube.[a]

hua khai[7] (lit. to change so as to separate) — fusion; melting; thawing; digestion.

huan[8] (lit. return) — a cyclical operation several times repeated. Cf. *chuan*.

hui chhih[9] — ash-bath (sand-bath)

hsia[10] (lit. down) — to put an ingredient in a vessel; to drive down; precipitation; descensory distillation.

hsiao,[11] *hsiao hua*[12] (lit. to disperse, dissipate) — to dissolve; to digest.

jou[13] (lit. weak, to weaken) — to soften; to macerate; ceration.

jung[14, 15, 16] (lit. to melt) — to smelt; to melt; to fuse; to blend; to dissolve.

kang[17] (lit. hard) — to harden.

Khan kua[18] (*Khan* trigram)[b] — to boil in water; to heat over a water-bath.

kou[19] (lit. to hook) — to extract (e.g. a metallic *chhi* from its ore).

ku chi[20] (lit. firmly enclosed) — sealing the parts of a vessel together, with the aid of a lute, to make it as gas-tight (or water-tight) as possible, so that processes of change, especially those involving ascent and descent (as in sublimation and distillation), can go on in the interior, isolated thus as far as possible from its surroundings.[c]

[a] One must be on the watch for very different meanings of this term, partly in pharmacy and medicine where various drugs might be combined with salts and vinegar, but especially in physiological alchemy, where (as with other *wai tan* terms) the significance is entirely different (cf. pt. 5 below).

[b] Cf. Vol. 5, pt. 5 below, in our discussion of physiological alchemy.

[c] This expression is an obscure one, and has given rise to some misunderstandings. Its second character evokes the hexagrams *Chi Chi* and *Wei Chi* (cf. pp. 68, 70–1); and here implies a perfecting of Yin–Yang relationships in compensation and equilibrium. That this could be done by moving things up and down was mirrored in the origin of these two *kua* themselves from the trigrams *Khan* and *Li* by

[1] 腐	[2] 伏	[3] 覆藉	[4] 伏火	[5] 化	[6] 華池	[7] 化開
[8] 還	[9] 灰池	[10] 下	[11] 消	[12] 消化	[13] 揉	[14] 熔
[15] 鎔	[16] 融	[17] 剛	[18] 坎卦	[19] 勾	[20] 固濟	

Table 115 (*continued*)

kuan[1] (lit. portal) — to bury in a container under the ground and allow slow chemical change to proceed without heating.

Li kua[2] (*Li* trigram)[a] — to heat directly in the fire of a stove.

lien[3] (lit. to refine) — to heat a substance (especially a metal) without water; more broadly, to effect any chemical transformation.

lin[4] (lit. to soak) — to dissolve part of a substance (e.g. a mixture of salts) in water; to separate a solution from a precipitate or residue by filtration or decantation.

liu[5] (lit. steamed food) — some preparation submitted to the action of steam.

lo[6] (lit. gauze) — to sift through a sieve of cloth.

lu[7,8] or *lü*[7] (to strain) — filtration; to filter.

mu yü[9] (lit. to bathe) — to grind in the presence of water or some other liquid.

niang[10] (ferment) — to ferment; fermentation.

ning[11] (lit. to congeal) — to solidify; to harden; coagulation.

o[12] — see *wu*.

phu[13] (to spread) — to spread out a bed of mineral material.

san[14] (lit. to scatter) — to separate; to disperse; to comminute; a medicinal powder.

sha[15] (lit. to kill) — to change a substance so that it is no longer volatile (cf. *chih* and *fu*).

shai[16] (lit. to sift) — to sift through a sieve of hair or rattan.

shang[17] (lit. up, above) — to drive up; sublimation; distillation.

shêng[18,19] (lit. to rise or raise) — to sublime; to distil; to evaporate and vaporise in general.

shêng hua[20] (lit. rising flower, ascending floreate essence) — sublimate; distillate; condensate.

shih[21] — see *wei chhi shih*.

changes in the position of their central lines (cf. p. 271). In the simplest *wai tan* usage, therefore, *ku chi* gave the instruction 'seal and sublime' (cf. pp. 47, 79); though in some contexts the first half might perhaps predominate over the second (cf. Tshao Yuan-Yü (1), pp. 43, 52 (78, 85); Yuan Han-Chhing (1), p. 209; Sivin (1), p. 185).

But the phrase was also adopted in *nei tan* terminology, referring then to the sealing in of secretions normally lost from the body (cf. pt. 5), and to the ascent and descent of *chhi* and secretions within it; hence further to the ultimate retention of the enchymoma when formed. As we read in *Chin Tan Ta Chhêng* (*HCSS*, ch. 10, p. 9b): 'Thai-I Chen Jen says: "Seal the container (lit. womb, *thai*[22]) firmly; then the chemical transformations (of the various materials inside) will take place with celerity". He is speaking of "water" and "fire" combining to form the *kua Chi Chi*. Close the doors of the mysterious chamber and let nothing escape.' Later on, the 'hermetic' sealing idea was applied to the sealing out of sense impressions and wandering thoughts (cf. pt. 5). 'Forgetting forms and abandoning desires and memories, that is called *ku chi*' (*HCSS*, ch. 1, p. 3b).

Finally, *ku chi* was also used in medical language, again with the nuance of ascent and descent within. Fang I-Chih explains the method that went by that name (*Wu Li Hsiao Shih*, ch. 4, p. 17b) as applying drugs which would drive up or down the malign *Yang chhi* according to the illness concerned.

---

[a] Cf. Vol. 5, pt. 5 below, in our discussion of physiological alchemy.

[1] 關   [2] 離卦   [3] 煉   [4] 淋   [5] 餾   [6] 羅   [7] 瀘
[8] 漉   [9] 沐浴   [10] 釀   [11] 凝   [12] 惡   [13] 鋪   [14] 散
[15] 殺   [16] 篩   [17] 上   [18] 升   [19] 昇   [20] 昇華   [21] 使   [22] 胎

Table 115 (*continued*)

shui fa[1] (lit. water method)   solubilisation; bringing substances into aqueous solution.

shui fei[2] (lit. flying on water)   purification of a powdered mineral by flotation on water (cf. *fei*).

shui hai[3] (lit. water sea)   a cooling-water reservoir or condenser vessel.[a]

shui kuan[4] (water pipe)   a cooling-water tube or coil.[a]

ssu[5] (lit. death, to die)   change of a substance so that it loses its original form or properties; to detoxicate; to decompose.

tao[6] (to beat)   to pound (as in a mortar).

tê[7] (lit. to obtain)   'going well with', the synergistic action of substances chemically or pharmacologically; one thing enhancing the action of another (an expression which could have covered cases of what we should now call catalysis). Cf. *wei chhi shih*.

thi ching[8] (to cleanse)   to purify; to separate a metal from an alloy.

thi lien[9]   to refine.

tien[10, 11] (lit. a spot)   a pinch, a speck, a knife-point, 'a spot of'; and to put such a small amount into a larger body of something else; projection.

tien hua[12]   projection; a small quantity of one substance producing change in a much larger quantity of another substance.

tuan[13] (lit. to forge)   to heat at a high temperature.

wei[14]   'to have a fear of', i.e. to be capable of dissolving in, some solvent.[b]

wei chhi shih[15]   'acting as its envoy (or adjutant)',[c] said when one substance enhances or activates the effect of another, chemically or pharmacologically. Cf. *tê*.

wu[16] (lit. to hate)   to inhibit the potency of some other substance.

yang[17] (lit. to nourish)   to apply heat gently over a long period, as by dung fire, charcoal embers, the water-bath, bed of ashes, or sand-bath (athanor).

yen[18] (lit. to grind)   to comminute, to powder.

yü yen[19] (lit. fish eyes)   bubbles appearing on the surface of a heated liquid, like fish eyes.

yü yen fei[20] (fish-eye boiling)   a particular stage in the boiling process (cf. Vol. 4, pt. 1, p. 69).

yung[21, 22, 23]   see *jung*.

[a] On these two expressions see particularly *TT*907, discussed on pp. 35 ff.
[b] Said, for example, of gold with respect to mercury, because of the formation of amalgams.
[c] Cf. our account of the most ancient Chinese pharmacological classification system in Sect. 38 (Volume 6).

[1] 水法	[2] 水飛	[3] 水海	[4] 水管 (筦)	[5] 死	[6] 擣	[7] 得
[8] 提淨	[9] 提煉	[10] 點	[11] 点	[12] 點化	[13] 煅	[14] 畏
[15] 爲其使	[16] 惡	[17] 養	[18] 研	[19] 魚眼	[20] 魚眼沸	[21] 熔
[22] 鎔	[23] 融					

From this it can be seen that the armamentarium of technical terms available to the ancient and medieval Chinese alchemists, proto-chemists and pharmacists was quite parallel with those used by the Greeks[a] and Latins[b] in the West. Lists of standard operations are often found in the occidental texts,[c] and it may be worth looking at them for a moment by way of comparison. One can tabulate them as follows, in accordance with the changes of state which they implied:[d]

solid——→solid
    Calcination   G/4
    Fixation   G/7
    Ceration   G/8
solid——→liquid
    Fusion   G/5
    Solution   G/2
    Descension
liquid——→solid
    (Crystallisation)
    Coagulation   G/6
    (Precipitation)
    (Filtration)
solid——→gas
    Fermentation or Putrefaction
solid——→gas——→solid
    Sublimation   G/1
liquid——→gas
    (Evaporation)
gas——→liquid
    (Condensation)
liquid——→gas——→liquid
    Distillation   G/3

In the old lists of definitions the terms occur in a variety of different orders, sometimes with omissions, sometimes with additions, and they do not include all of the modern operational ideas which one would expect. A few words of further explanation will suffice to assist comparisons with the Chinese terms.

By Calcination[e] was meant the reduction of any solid to a powder by chemical means (e.g. a metal to its oxide)—'the pulverisation of a thing by fire'.[f] Fixation,[g] reminiscent of *chih*,[1] was 'the convenient disposing of a fugitive thing to abide and

---

  [a] See Berthelot (2), i.e. Berthelot & Ruelle (1), pp. 263–4.
  [b] See e.g. Holmyard (1), pp. 43 ff.
  [c] As also in Syriac and Arabic MSS. For such lists of operations see Berthelot & Duval (1), pp. 165 ff.; Stapleton, Azo & Husain (1), pp. 326 ff., 356 ff., 366 ff., 385 ff.
  [d] Terms primarily modern, though occasionally used in medieval times, are placed in brackets. The numbers marked G show the order of description in the Geberian *Summa Perfectionis*, c. +1290.
  [e] *Summ. Perf.*, ch. 51 (like all the other chapters here quoted, in bk. 4).
  [f] *Ibid.*, Russell tr., p. 101.
  [g] *Summ. Perf.*, ch. 54.

  [1] 制

sustain the fire'.[a] Ceration,[b] softening or 'waxifying', was 'the mollification of a hard thing, not fusible unto liquefaction',[c] e.g. the formation of amalgams and sulphides. Fusion applied generally to all smelting and melting; it might have been regarded by the Geberian writer as a form of Solution,[d] 'the reduction of a dry thing into water',[e] as happens when a salt is dissolved. Descension[f] was simply what we shall shortly discuss as *destillatio per descensum*,[g] the liquefying of mercury or an oil by heat and its descent into a receiver below. Coagulation[h] was defined in Geber as 'the reduction of a thing liquid to a solid substance by privation of the humidity',[i] as when mercury is combined with sulphur to form vermilion.[j] Crystallisation, Precipitation and Filtration were processes known, of course, to all the proto-chemists from Hellenistic times onwards, and China also; though not often listed in the Western medieval categories of operations. These do generally include, however, Fermentation or Putrefaction, names referring to the natural changes occurring in dead organic materials under the action of bacteria, yeasts and moulds, often with the evolution of gases; as also the formation of gases from inorganic substances in certain reactions— but the terms were commonly applied as well to any chemical change brought about by long subjection to mild heat. Sublimation,[k] on the other hand, was a term always used in much the same way as we ourselves use it, vaporisation with condensation above in solid form; and this 'elevation of a dry thing by fire, with adherency to its vessel'[l] was the process which occasioned the lengthiest descriptions in the *Summa Perfectionis*. Evaporation and Condensation are terms rather more modern, but Distillation[m] necessarily gave rise also to a long discussion.'The cause why distillation was invented' said the Geberian writer, 'and the general cause of the invention of every distillation, is the purification of liquid matter from its turbulent faeces and the conservation of it from putrefaction'.[n] The term included also a medieval process which has few remains in modern technique, *destillatio per filtrum*, where a siphon is made of a piece of cloth hanging across the edge of a pan to take the solvent over by capillary attraction into a separate receiver.[o] Finally, besides all these we must remember the characteristically alchemical processes of Projection, clearly recognisable in *tien hua*[1] (pt. 3, pp. 38, 88, etc.); as also Separation, Mortification, Ablution, *Nigredo*, *Albedo*, *Citrinitas* and *Rubedo* (pt. 2, p. 23), about which no more need be said here. Thus, all in all, an inspection of the two lists of technical terms will show considerable parallelism in the development of chemical technique in the Far East and the Far West.

We are now in a position to make a tour of the Chinese medieval alchemical and

[a] *Ibid.*, Russell tr., p. 116.
[b] *Summ. Perf.*, ch. 55.
[c] *Ibid.*, Russell tr., p. 119.
[d] *Summ. Perf.*, ch. 52.
[e] *Ibid.*, Russell tr., p. 107.
[f] *Summ. Perf.*, ch. 49.
[g] Pp. 55 ff. below.
[h] *Summ. Perf.*, ch. 53.
[i] *Ibid.*, Russell tr., p. 110.
[j] Cf. pp. 262–3 below, and pt. 3, pp. 126, 198.
[k] *Summ. Perf.*, chs. 39 to 48 incl.
[l] *Ibid.*, Russell tr., p. 74.
[m] *Summ. Perf.*, ch. 50.
[n] *Ibid.*, Russell tr., p. 96.
[o] Whether or not this practice was also current in medieval Chinese alchemy we are at present unable to say.

[1] 點化

iatro-chemical laboratory,[a] and to examine systematically the pieces of apparatus that were used there.[b]

(1) THE LABORATORY BENCH

The Chinese alchemist's version of the modern laboratory bench was the *than*[1] (lit. platform or altar). No specific rules were laid down with regard to its dimensions and constructions. *TT*904, a Sung book, gives an illustration of it (Fig. 1374)[c] together with the following explanatory notes:

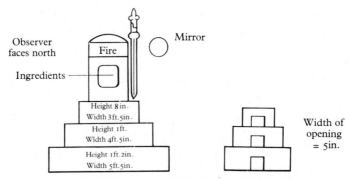

Fig. 1374. Stove platform from *Kan Chhi Shih-liu Chuan Chin Tan* (a Sung text).

The *tsao*[2] (furnace) is the *yao lu*[3] (chemical stove). The *ting*[4] (vessel) is called the *sha ho*[5] (cinnabar enclosure). The *shen shih*[6] (magical reaction-chamber) is the *hun tun*[7] (world of chaos).

The same text describes the diagram, saying:[d]

Build a *than* (platform) of three stages, with a (total) height of 3 ft. 6 in. The platform is square with a perimeter of 10 ft.

One notes that this description does not coincide with the dimensions given in the diagram itself.

---

[a] Here we are concerned primarily with the 'hardware', but later more will be said about the liturgical and magical aspects of the matter (cf. pp. 289ff. below, and Fig. 1521). The sword, the mirror, the jars of pure water, the peach-wood talismans, all have to be borne in mind along with the aludels, matrasses and stills; cf. Yoshida Mitsukuni (7), pp. 250, 257.
[b] We may be excused from offering any complete guide to the lists of apparatus and instruments used in other culture-areas. For the Hellenistic apparatus Berthelot (1, 2) and Sherwood Taylor (2, 5) are of course indispensable. Arabic apparatus (*tadābīr*) is listed and described, *inter alia*, in Berthelot & Duval (1), pp. 150ff. (Syriac MSS); and in Stapleton, Azo & Husain (1), pp. 324ff., 353ff., 362ff., 378ff.; Stapleton & Azo (1), pp. 60ff. Wiedemann (22) consecrated a special paper to the apparatus of the Arabic chemists such as al-Rāzī (*c.* +900); and names of the parts of apparatus can be found in the dictionary of Siggel (2). Holmyard (18) covers clearly and succinctly the whole range from the Hellenistic proto-chemists through the Arabic writers and the Latin West to the 17th century in Europe.
[c] P. 8*a*.   [d] P. 7*b*.

¹ 壇   ² 竈   ³ 藥鑪   ⁴ 鼎   ⁵ 砂合   ⁶ 神室   ⁷ 混沌

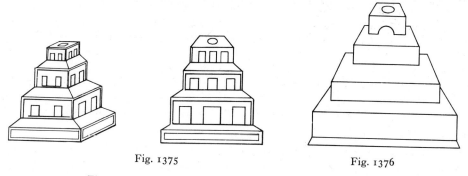

Fig. 1375   Fig. 1376

Fig. 1375. Stove platforms from *Tan Fang Hsü Chih* (+1163).
Fig. 1376. Stove platform from *Yün Chi Chhi Chhien* (+1022).

In *TT*893, also a Sung book, we find two illustrations of the *lung hu tan thai*[1] (Dragon-and-Tiger platform), (Fig. 1375) and also the following description:[a]

The *Tshan Thung Lu*[2] (Records of the Kinship of the Three)[b] states: 'below the *lu*[3] (stove) is the *than* (platform), which consists of three stages put one over the other. Each stage faces the eight directions and has eight openings'.

The construction of the *than* was by no means standardised in the Sung as can be seen from another example in Fig. 1376 taken from *TT*1020. The text says, 'The *than* can be so constructed as to suit one's convenience'.[c] This implies that there were no fixed rules in the construction of the *than*, but that it could be made to fit the circumstances, such as the size of the stove and the size of the laboratory.

It is curious to see a stepped stove platform looking very like these in the Syriac alchemical texts of the +10th or +11th century (though the MSS we have were not written till the +16th).[d] It would not be at all unreasonable to take this as suggestive of Chinese influence.

(2) THE STOVES *LU*[3] AND *TSAO*[4]

Although the word *tsao* generally refers to the kitchen stove,[e] the two words *lu* and *tsao* both mean the heating apparatus of the alchemists which took many different forms. As the texts do not employ consistent terminology, in certain cases the heating apparatus may be taken to mean a stove or furnace, while in other cases it must mean an oven or combustion-chamber.

---

[a] P. 5*b*.
[b] Presumably a reference to the *Tshan Thung Chhi* or some commentary on it.
[c] Ch. 72, p. 12*b*. In these drawings note one on the usual axonometric projection (left) and two in optical perspective (right). Cf. Vol. 4, pt. 3, pp. 113 ff. and Figs. 758, 776, 778.
[d] Berthelot & Duval (1), p. 113.
[e] Pottery tomb-models of stoves with as many as nine or ten openings carrying vessels of various kinds are extremely common in Chinese museums. Some particularly good ones are in the Archaeological and Historical Museum at Canton.

[1] 龍虎丹臺   [2] 參同錄   [3] 爐   [4] 竈

*TT*1020, a Sung text, describes one form of stove, saying:[a]

The *lu* (stove) forms the walls of defence for the *ting*[1] (reaction-vessel). Without the walls there would be evil influences (coming from outside). From top to bottom it resembles the *phêng hu*[2] pot and from side to side it symbolizes the Five Sacred Mountains (*wu yo*[3]). The platform consists of three stages, while the combustion-chamber has eight openings. The twelve cyclical signs and the months follow the (*Pei-*) *Tou*[4] (the Great Bear)....The *hua chhih lu*[5] (stove for digestion in vinegar) is 4 ft. high, 6 in. thick and has an internal circumference of 3 ft. 5 in. The openings measure 2 in. and they are eight in number.

The same text also tells us about the *thai i lu*[6] stove. It says:[b]

The *thai i lu* is placed over the platform. It is 2 ft. high, 6 in. thick and has an internal circumference of 3 ft. 5 in. Each opening is 2 in. high and half an inch wide. The 12 projections (*chih*[7]) are one inch wide all round. The platform can be made to suit one's convenience. Again, the *hua chhih lu* is 4 ft. high, 6 in. thick and has eight openings. It also has a two-inch rim....

Here is another description from *TT*904:[c]

On the platform is the *tsao*[8] (stove), on which is placed the *ting* (reaction-vessel). Within the *ting* is placed the *shen-shih*[9] (magical reaction-chamber).

A text of the Liang period, *TT*874, gives an account of the construction of the *tan lu*[10] (elixir stove) saying:[d]

Iron rods are fixed at the bottom of the stove. There should be, say, twelve or thirteen of them, each being 1 ft. in length and with a cross-section of 0·4 in. square. They are put in position (so as to form a grate) over the hollow space (*chhien*[11]) at a distance of 0·2 in. from one another. There is an empty space beneath the rods, which are placed two inches above ground (or rather above the base of the stove). The *tan lu* has an opening four and a half inches wide at the centre. The openings in front of and behind (the stove) enable air to pass in and out. The fire is lit above the rods and is fanned by the air current....

Figure 1377 shows the *yen yüeh lu*[12] (inverted-moon stove), taken from *TT*1054, a Yuan work.[e] The stove has a flat top, at the centre of which is an opening for the container or crucible and for the emission of flame. The text says:

The (upper) surface of the *lu* (stove) has a circumference (perhaps it should mean diameter) of approximately 1 ft. 2 in. It has a central opening measuring 1 ft. across. The rim all round is 2 in. wide and 2 in. thick. The opening faces upwards (to hold) the *kuo fu*[13] (pot and crucible) resembling an upturned moon. Hence the name *yen yüeh lu* (inverted-moon stove). In Chang Sui's[14] annotations it is also known by the name *wei kuang ting*[15] (reaction-vessel of intense brightness).[f]

It seems likely that the *yen yüeh lu* was the stove referred to by Wei Po-Yang[16] in the mid +2nd century in *TT*990.[g]

---

[a] Ch. 72, p. 11*a*.   [b] P. 11*b*.   [c] P. 7*b*.   [d] P. 14*a*.

[e] P. 9*b*. This term had also a special significance in physiological alchemy quite different from the plain meaning here (cf. Vol. 5, pt. 5 below).

[f] We have no further information about this adept. The same description first occurs in the preliminary material of the *Wu Chen Phien* of +1075, but only in the version of this contained in *Hsiu Chen Shih Shu* (*TT*260), ch. 26, p. 7*a*.

[g] Ch. 1, p. 32*b*.

[1] 鼎	[2] 蓬壺	[3] 五岳	[4] 北斗	[5] 華池鑪	[6] 太一鑪
[7] 支	[8] 竈	[9] 神室	[10] 丹鑪	[11] 塹	[12] 偃月鑪
[13] 鍋釜	[14] 張隨	[15] 威光鼎		[16] 魏伯陽	

*TT*878 explains how the crucible was placed over a tripod inside the *tsao* (combustion-chamber; oven). It says:[a]

...Within the *tsao* (combustion-chamber) is placed an iron tripod, which is best made of cast iron (*shêng thieh*[1]). The *yao fu*[2] (closed vessel) is placed over the tripod and adjusted until it is in the centre of the chamber. Care should be taken so that it does not incline to one side. The four sides should be about three and a half inches away from the wall of the chamber. The chamber should be two inches higher than the vessel. Rice-husk (fuel) should be regularly placed around the four sides of the crucible and more must be added as heating progresses. This is necessary for fear of uneven heating due to the varying intensity of the fire.

For ordinary heating the containing vessel was simply placed over the stove and fire was applied below. This is shown in Fig. 1378, taken from the Sung text *TT*1020.[b]

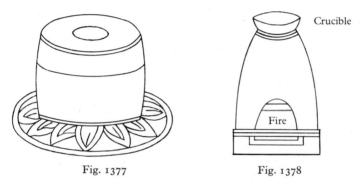

Fig. 1377   Fig. 1378

Fig. 1377. The 'inverted-moon stove' (*yen yüeh lu*) from *Chin Tan Ta Yao Thu* (+ 1333).
Fig. 1378. Stove depicted in *Yün Chi Chhi Chhien* (+ 1022).

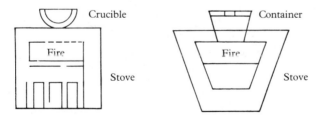

Fig. 1379. Stoves drawn in *Thai-Chi Chen-Jen Tsa Tan Yao Fang* (a text probably of the Sung).

Another example is given in *TT*939.[c] It shows *yang lu*[3] stoves (Fig. 1379), which employed a strong fire (*wu huo*[4]).

Curious stoves which may well have had alchemical use are to be seen in Chinese museums. For example the Archaeological Institute of Academia Sinica in Peking has a flat conical pottery object shaped rather like a hollow mountain having holes for escape of fumes from burning charcoal or other fuel, and four cupped holders to take

---

[a] Ch. 7, p. 4*a*.   [b] Ch. 72, p. 20*b*.   [c] P. 5*a*.

[1] 生鐵   [2] 藥釜   [3] 陽爐   [4] 武火

Fig. 1380*a*. Pottery stove designed to heat four containers at one time, in an excavated tomb of the Thang period, *c*. +760 (Chhen Kung-Jou, *3*).

Fig. 1380*b*. As it was found on the floor of the tomb chamber.

lidded pots, surrounding a chimney at the top of the stove. This apparatus, which was exacavated from a Thang tomb[a] near Yü-hsien[1] in Central Honan,[b] could have served well for continuous slow heating (Figs. 1380*a*, *b*). A very different stove for gentle charcoal heating is the − 11th-century 'hot-plate' comprised in the Tuan-Fang altar set (Fig. 1380*c*).

Over and over again, not only in alchemical texts, but in descriptions of industrial fermentations, tea-making, etc., emphasis is laid on the indispensability of careful temperature regulation—without which everything will fail.[c] Sivin (2) has drawn

---

a Dated by coins in the neighbourhood of +760, not earlier.
b It has been figured and described by Chhen Kung-Jou (*1, 3*).
c Cf., for instance, the explanation of the term *wên huo*[2] in *Wei Lüeh*, ch. 11, p. 5*b*.

[1] 禹縣   [2] 文火

Fig. 1380c. A bronze 'hot-plate' of the −11th century, for warming sacrificial wine in liturgical vessels (photo. Metropolitan Museum of Art). This is the Tuan-Fang altar set, so named after the enlightened Governor of Shensi in whose time (1899 to 1901) it came to light at Tou-chi-thai in that province. Besides its own vents, the stove platform has a cubical chimney at one side also fitted with slits. The vessels are probably not all of the same date, but were assembled by some early Chou ruler; cf. Li Chi (5).

attention to the concern shown in many of the alchemical writings for the precise control of the intensity and duration of the heat. A quotation from the *Chu Chia Shen Phin Tan Fa* runs as follows:[a]

The amounts of fuel to be weighed out are increased and decreased in cyclical progression according to the phases of *Yin* and *Yang*. They must conform to the (order of the) symbols of the *Book of Changes*,[b] to the threefold concordance,[c] to the (correspondences of the) four, eight, twenty-four and seventy-two seasonal divisions of the year, and to the implicit configuration and proper activity of the year, month, day and hour, without one jot or tittle of divergence.

This quantitative aspect is also seen, as Sivin points out, in the monograph of Chhen Shao-Wei on the careful assessment of the amount of cinnabar to be obtained from cinnabars of different quality. The yield recorded from 'lustrous cinnabar' (14 ozs. from one 1 lb.) comes very close to the theoretical yield from pure cinnabar—13·8 ozs. We shall return to this subject on p. 300.[d]

About the means used to ensure a good draught for the furnace not much is known. The expression *fêng lu*[1] occurs frequently enough, as in the writings of Sun Ssu-Mo,[e] and this is explained in the *Thang Yü Lin* as meaning a stove or brazier pierced with

[a] *TT*911, ch. 4, p. 1*b*; a Sung work by Mêng Yao-Fu and others. Tr. Sivin (2), p. 14*a*.
[b] I.e. the trigrams and the hexagrams. See Sect. 13 in Vol. 2.
[c] Heaven, earth and man? More probably, whatever was implied in the title of the *Tshan Thung Chhi*.
[d] The great importance attached to weights in chemical operations by the Arabs is well known (cf. pp. 393–4). Stapleton & Azo (1), in their study of the treatise of Ibn 'Abdal-Malik al-Kāṭī (+1034) emphasise this point.
[e] E.g. *Tan Ching Yao Chüeh*, pp. 18*a*, *b*. See Sivin (1), pp. 206–7, and also pt. 3, pp. 132 ff. above.

[1] 風 鑪

numerous holes like an iron beacon basket so as to catch the wind from whatever quarter it might be blowing.[a] Some processes needing strong and continuous heat called, no doubt, for the use of the cylindrical box-bellows (*fêng hsiang*[1]),[b] and this would have been available throughout the medieval period. *Fêng lu* could also be of this kind, for Sun was able to melt cast-iron in such 'blast'-furnaces.

(3) THE REACTION-VESSELS *TING* (TRIPOD, CONTAINER, CAULDRON) AND *KUEI* (BOX, CASING, CONTAINER, ALUDEL)

The word *ting*[2] normally refers to the tripod cauldron, so familiar among the bronzes in Chinese archaeology, but the alchemists' apparatus known by this name included not only pots of this kind but also various other forms of reaction-vessel to which fire was applied externally. Perhaps the best distinction between the *lu*[3] (stove; combustion-chamber) and the *ting* is that the former had fire within it whereas the latter was surrounded by fire. The *ting* itself might contain an inner reaction-chamber in which the ingredients were placed.

The earliest account[c] of the reaction-vessel is found in the +2nd century *Tshan Thung Chhi* of Wei Po-Yang.[4] It says:[d]

The Song of the *Ting* (reaction-vessel): Its circumference is three-five (i.e. 1 ft. 5 in.) (and its thickness is) one inch and a tenth (i.e. 1·1 in.). The circumference of the mouth is four and eight (i.e. 12 in.). Its lips are 2 in. thick. The total body height is 12 in. (i.e. 1 ft 2 in.), with an even thickness throughout. With its belly set erect it is to be warmed gently (over the stove). (The *chhi* of) Yin (i.e. the reaction-vessel) stays above, while (that of) Yang (i.e. the fire) runs below. Use a strong fire during the end and the beginning of each (lunar) month, but a gentle fire during the middle of it. Begin heating for 70 days, and after the end of another 30 days the contents are to be properly mixed and heated for a further 260 days (making a total of 360 days, i.e. twelve lunar months)....[e]

*TT*889 mentions five types of *ting*:[f]

One is called *chin ting*[5] (gold vessel), the second is called *yin ting*[6] (silver vessel), the third is called *thung ting*[7] (copper vessel), the fourth is called *thieh ting*[8] (iron vessel) and the fifth is called *thu ting*[9] (pottery vessel).

Figure 1381, taken from a Thang or Sung text in *TT*1020, shows a *chin ting* ("gold" vessel).[g] The text gives the following description:

According to rule the *ting* measures 1 ft. 2 in. in height and weights 72 oz. The number is nine (alternative translation: 'there are nine of them'). The inner circumference is 1 ft. 5 in. The *ting* is supported by legs so that it stands two and a half inches above ground. The base has a thickness of 2 in., while the body is one and a half inches thick. It has a capacity of

---

[a] Ch. 8, p. 22*a*.   [b] Cf. Vol. 4, pt. 2, pp. 135 ff.   [c] *TT*990, ch. 3, p. 11*b* ff.

[d] We depart here from the translation made formerly by Wu Lu-Chhiang & Davis (1), p. 260.

[e] In interpreting the above measurements it must be remembered that the Chinese foot was one of 10 inches, not 12.   [f] P. 2*b*.

[g] YCCC, ch. 72, p. 10*a*, *b*. Sep. ed. p. 11*a*, *b*.

¹ 風箱   ² 鼎   ³ 爐   ⁴ 魏伯陽   ⁵ 金鼎   ⁶ 銀鼎
⁷ 銅鼎   ⁸ 鐵鼎   ⁹ 土鼎

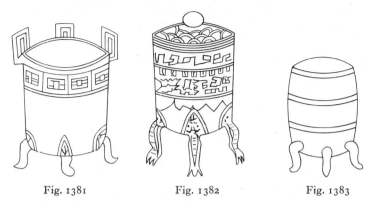

Fig. 1381   Fig. 1382   Fig. 1383

Fig. 1381. Gold reaction-vessel, a drawing from *Yün Chi Chhi Chhien* (+ 1022).
Fig. 1382. Covered reaction-vessel, a drawing from the same work.
Fig. 1383. 'Suspended-womb' reaction-vessel, also with three legs (*hsüan thai ting*), from *Chin Tan Ta Yao Thu* (+ 1333).

three and a half *shêng*[1] (approx. 100 cu. in.) when filled to a depth of 6 in. The cover is 1 in. thick and the ears are one and a half inches high.

In the same chapter of this text we find another picture of the *ting* (Fig. 1382).[a] It has a lid and the decorations are more elaborate.[b]

The Yuan text *TT1054* gives an illustration (Fig. 1383) of the *hsüan thai ting*[2] (suspended-womb vessel) with the following description:[c]

The *ting* has a circumference of 1 ft. 5 in. and is hollow inside for 5 in. It is 1 ft. 2 in. high like a *phêng hu*[3] pot.... It also symbolizes the human body. It consists of three layers corresponding to the Three Powers (*san tshai*[4]—Heaven, Earth and Man). The upper and the middle section of the *ting* are connected by the same vertical passage. The upper, middle and lower sections must be evenly set. (The *ting*) is put into a *lu*[5] (stove) to a depth of 8 in. or is suspended inside a *tsao*[6] (combustion-chamber) so that it does not touch the base. Hence the name *hsüan thai* (*ting*) (suspended-womb vessel). It is also called the cinnabar vessel (*chu sha ting*).[7] In Chang Sui's[8] annotations it also receives the name *Thai I shen lu*[9] (magical vessel of the Great Unity).[d]

From the above quotations we can see that sometimes it is difficult to distinguish between a *ting* (reaction-vessel) and a *lu* (stove) as the two terms may refer to the same apparatus.

[a] P. 24*b*. Sep. ed. p. 25*b*.
[b] The title of the tractate is *Ta Huan Tan Chhi Pi Thu*;[10] we discuss it in other contexts elsewhere (pt. 5). One should be warned here that some of these works may really be talking about physiological alchemy (*nei tan*,[11] cf. pt. 5 below); nevertheless their illustrations draw on the equipment of the laboratory alchemists and proto-chemists.
[c] P. 8*b*. The same text first occurs in the preliminary material of the *Wu Chen Phien* of +1075, but only in the version of this contained in *Hsiu Chen Shih Shu* (*TT*260), ch. 26, p. 6*b*.
[d] Tshao Yuan-Yü (*1*) confused the *hsüan thai ting*[2] with the *yen yüeh lu*[12] by calling the former *wei kuang lu*[13] and the latter *Thai I shen lu*.[9]

[1] 升   [2] 懸胎鼎   [3] 蓬壺   [4] 三才   [5] 爐   [6] 竈
[7] 朱砂鼎   [8] 張隨   [9] 太一神鑪   [10] 大還丹契秘圖   [11] 內丹
[12] 偃月鑪   [13] 威光鑪

In all the examples so far mentioned the *ting* has been represented by cauldron-like forms with three legs, i.e. tripods. However, the legs were often omitted, for one may see in Chinese museums (e.g. at Sian and at Chêngchow) large cast-iron cauldrons about 1 ft diameter at the mouth dating from the Han period and almost certainly used for alchemical or technological preparations. Moreover, as we have pointed out, the word *ting* has a wider meaning. For example, no legs are attached to the *hun tun ting*[1] (chaos vessel), taken from the Sung book *TT*904 and shown in Fig. 1384.[a] In fact, it has now become plainly a reaction-chamber.

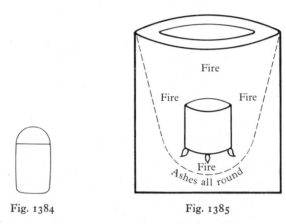

Fig. 1384   Fig. 1385

Fig. 1384. Aludel from the *Kan Chhi Shih-liu Chuan Chin Tan* (a Sung text).
Fig. 1385. Furnace and reaction-vessel, from *Chhien Hung Chia Kêng*...(+ 808 or later).

In one method of *yang huo*[2] (conserving the glowing fire) we shall see (p. 57) that fire was applied above and around the reaction-chamber, below which was placed a receiver containing water, and that the space inside the combustion-chamber was packed with ashes. In another method described in *TT*912, a Thang text,[b] we find that fire was applied all round the reaction-vessel. Fig. 1385 shows the *lu* (stove) with a *ting* (vessel) and how the glowing fire was conserved with a lagging of ashes.[c] The fuel used in this case was charcoal.

Next we have what was known as the *kuei*[3] (box, casing, container) the function of which was rather similar to that of the *ting* (reaction-vessel), because within the *kuei* was placed the reaction-chamber. Sometimes the *kuei* itself formed the reaction-chamber. Broadly speaking, *kuei* had lids while *ting* were open at the top, though the terms were not consistently used. Several types of *kuei* are illustrated in *TT*939 (Fig. 1386).[d] Some of them appear to have been containers pure and simple. The same text also mentions the following types:[e]

[a] See above, p. 17 note d.
[b] This is the book supposedly of + 808 by Chao Nai-An (see pt. 3, pp. 158–9), but most of which may date rather from Wu Tai or early Sung.
[c] From ch. 1, p. 9b of this text.
[d] In Fig. 1386(*a*) is taken from p. 4b, (*b*) from p. 5b, (*c*) from p. 8b and (*d*) from p. 10b.
[e] Pp. 14b and 15a.

[1] 混沌鼎   [2] 養火   [3] 匱

huang ya kuei[1] (yellow sprout casing)
pai hu kuei[2] (white tiger casing)
hei hu kuei[3] (black tiger casing)
huang kuei[4] (yellow casing)
hsüan chen kuei[5] (suspended needle casing)[a]
li chih kuei[6] (immediate fixing casing)
san chih kuei[7] (rice cake casing)
yung chhüan kuei[8] (bubbling spring casing)
thien-shêng huang ya kuei[9] (natural yellow sprout casing).

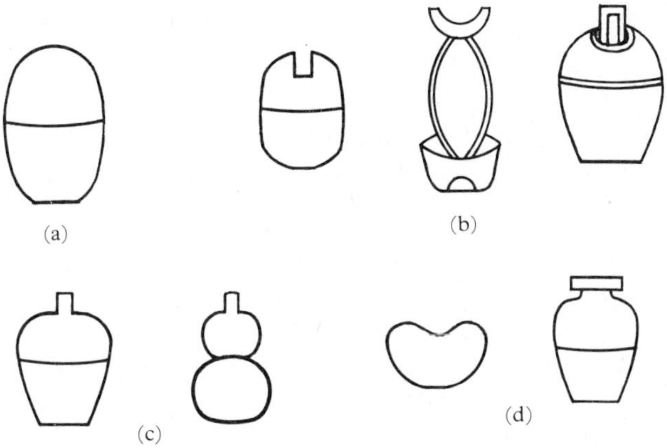

Fig. 1386

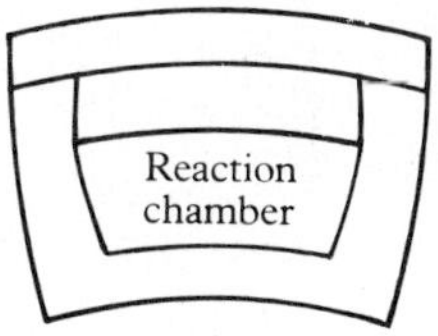

Fig. 1387

Fig. 1386. Types of aludels and reaction-vessels from *Thai-Chi Chen-Jen*...(probably Sung). (*a*) P. 4*b*, (*b*) p. 5*b*, (*c*) p. 8*b*, (*d*) p. 10*b*.
Fig. 1387. Reaction-chamber from the *Kêng Tao Chi* (+1144 or later).

Further accounts of other types of *kuei* are given in *TT*946. Fig. 1387 is an example taken from this text.[b] In the same work we also find a description of a *phi kuei*[10] (arsenical lead casing):[c]

Make *tzu ho chhê*[11] (lead) into powder. For every ounce of *phi mo*[12] (arsenic powder) use one and a half oz. of *ho chhê*[13] (lead) powder. After stirring and mixing they are put into a *kan kuo*[14] (crucible).[d] Begin with a gentle fire and gradually increase its intensity. When calcination is over use the residue to make a *kuei* (casing). This is most useful for subliming (*yang*)[15] calomel (*fên shuang*[16]).[e]

[a] This 'suspended needle aludel' we have encountered before, in Vol. 4, pt. 1, p. 275, where its name provided evidence of value from the mid +11th century for the history of the magnetic compass. Presumably the significance of its figurative appellation here was that it was intended to stand bolt upright in the furnace.
[b] Ch. 1, p. 3*a*.
[c] Ch. 2, p. 15*a*.
[d] Also called *khan kuo*.[17]
[e] Lead and arsenic alloy readily, hardening the metal yet increasing its fluidity when molten. Hence the use of arsenic in spherical lead shot. Cf. Gowland (9), p. 133.

1 黃芽匱 2 白虎匱 3 黑虎匱 4 黃匱 5 懸針匱 6 立制匱
7 糝制匱 8 湧泉匱 9 天生黃芽匱 10 砒匱 11 紫河車
12 砒末 13 河車 14 甘堝 15 養 16 粉霜 17 坩堝

In another example mentioned in the same treatise[a] a *kuei* was made from *chhing yen*[1] (blue salt; rock salt), *pai yen*[2] (white salt) and the juice extracted from arrowroot. This recipe must originally have included a refractory clay. It is interesting in this connection that accounts of blast-furnaces in late medieval China generally mention 'salt' as well as clay, lime and sand. The suggestion is elsewhere made that gypsum (calcium sulphate) was really meant in these cases, for among its traditional names are

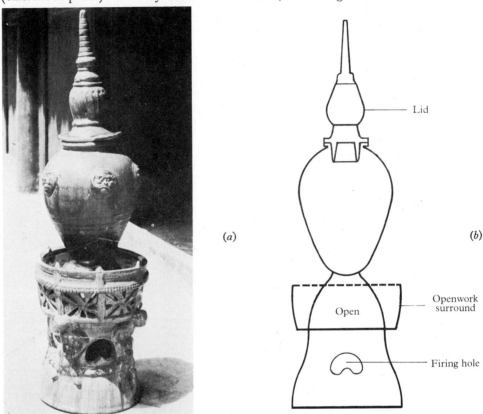

(a)   (b)

Lid

Openwork surround

Open

Firing hole

Fig. 1388. 'Precious vase' (*pao phing*), probably of Thang date, in the Provincial Historical Museum, Thaiyuan, Shansi (orig. photo. 1964). Recovered from a tomb excavation at Lin-fei-chhang to the west of the city. As the cross-section line drawing (b) shows, it was probably used as a digester with gentle heat.

yen chen[3] ('salt's pillow') and *yen kên*[4] ('salt's root'), both probably derived from stratigraphic relationships.[b] Gypsum is still used in the making of mortar and cement. Unless certain characters have dropped out of the text, the bald statement we have here may have been meant to deceive the uninitiated while being perfectly comprehensible to those with alchemical training. Other examples of *kuei* are given in *TT*912, an important Thang text.

A very different type of reaction-vessel, if that is what it was, is constituted by the *pao phing*[5] or 'precious vases', one example of which, conserved in the Provincial

Fig. 1389. Bronze reaction-vessel (*ting*) with clamp handle mechanism to permit tight sealing (Anon. (*106*), pl. 7). From the tomb of Liu Shêng (d. −113) at Man-chhêng. Ht. 17·8 cms., diameter at opening 17·2 cms. Two of these vessels were found in the tomb.

Historical Museum at Thaiyuan in Shansi, is illustrated in Fig. 1388. Of green and brown glazed pottery, it is considered Thang in date. As can be understood from the cross-section appended, the bottom part is a charcoal stove with firing openings and a solid ceiling, then above that a separable inverted hemisphere bears an egg-shaped reaction-vessel stoppered at the top by a bung in the form of a miniature pagoda. It seems unlikely that Taoist apparatus of this kind could have been employed for operations requiring any great degree of heat, but one could imagine it in use for the slow oxidation of mercury, or still better for the *hua chhih*[1] bath of strong vinegar (perhaps distilled or otherwise concentrated acetic acid) and saltpetre, in which various rather insoluble minerals could be dissolved by the dilute nitric acid (and perhaps also hydrochloric, if salt were present) formed.[a]

Yet another type of bronze vessel was like a *ting* with legs, but provided with a flat cover which could be held down tight with clamp handles. These pieces are rare, but two were found in the tomb of Liu Shêng,[2] Prince Ching of Chung-shan[3] (d. −113), the same that provided the two complete jade-plate body-cases shown in Fig. 1332 (Vol. 5, pt. 2, Pl. CDLII). We reproduce a photograph of one of these pressure vessels here (Fig. 1389).[b]

[a] Cf. pp. 167 ff. below.
[b] Anon. (*106*), pl. 7; Hsia Nai, Ku Yen-Wên *et al.* (1), pp. 8 ff., 13 ff.

[1] 華池 [2] 劉勝 [3] 中山王靖

(4) THE SEALED REACTION-VESSELS *SHEN SHIH* (ALUDEL, LIT. MAGICAL REACTION-CHAMBER) AND *YAO FU* (CHEMICAL PYX)

Besides the more open crucible or bowl-like forms of reaction-vessel, whether with lids or not, many kinds of sealed containers (*shen shih*[1]) were employed. These corresponded also in some degree, no doubt, to the aludels of Arabic-Western alchemy. In some forms, especially when they were made of metal, pressures considerably higher than atmospheric could be generated in them;[a] in other forms they were used for sublimation. Fig. 1390 taken from *TT*907 illustrates one of these reaction-chambers, as used in the Sung.[b]

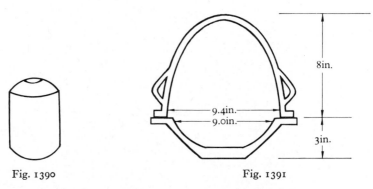

Fig. 1390.   Fig. 1391

Fig. 1390. Reaction-chamber from the *Chin Hua Chhung Pi Tan Ching Pi Chih* (+ 1225).
Fig. 1391. Reconstruction of the *yao-fu* 'bomb' from the *Thai-Chhing Shih Pi Chi* (Liang, + 6th century, or earlier).

Another important closed vessel was the pyx or 'bomb', *yao fu*[2] (a vessel composed of two more or less hemispherical crucible-like bowls with flanges placed mouth to mouth). The following is an account of its construction taken from *TT*874, a text written probably in the Liang period (+ 6th century):[c]

Method of making a *yao fu*:
The lower iron bowl (*thieh fu*[3]) has a capacity of one peck (*tou*[4]), a diameter of 9 in. and a height of 3 in. At the base, which comes in contact with the fire, the thickness is 0·8 in., but around the four sides the thickness is 0·3 in. The upper and lower bowls are of equal thicknesses. The base is flattened. The flange all round is one and a half inches wide and 0·3 in. thick; it is also flattened. The two handles at the side are three inches long and three and a half inches wide; they are situated above the flange. The upper bowl (i.e. the cover) is made of pottery (*shao wa*[5]). It has a diameter of 0·4 in., a height of 8 in. and a thickness of 0·3 in. The cover thus has a greater curvature (than the lower bowl). Its flange is also made flat. The *yao fu* is used for the preliminary treatment of the ingredients and hence its size. After the ingredients have become refined they should be transferred to a *hsiao fu*[6] (small vessel), which measures $2\frac{1}{2}$ inches across at the mouth and 6 inches in height. Apart from

---

[a] This is certain because of the directions which often occur in the texts to bind the parts of the apparatus together with iron wire. Examples of such directions are given on p. 40.
[b] Ch. 2, p. 2*a*.   [c] P. 14*a, b*.

[1] 神室   [2] 藥釜   [3] 鐵釜   [4] 斗   [5] 燒瓦   [6] 小釜

this the shape and other dimensions (of the lower bowl) are the same as (those for the *yao fu*). For the cover the diameter is 6·2 inches and the height 6 inches. Apart from this the shape and other dimensions do not differ from those of the bigger vessel.

A conjectural diagram of the *yao fu* is given in Fig. 1391.

The dimensions of the *yao fu* were by no means standardised. For example different values are given in descriptions of the *yao fu* in the *Yün Chi Chhi Chhien*.[a] Another specification is to be found in Sun Ssu-Mo's *Tan Ching Yao Chüeh*[b] of c. +640, studied and reconstructed by Sivin.[c] In general the design is the same, but Sun's has thicker walls, a narrower lip, and a taller upper compartment. Its material is cast iron, as was probably most usual.

Fig. 1392. A bronze *tui*, possibly used as a reaction-vessel, Chou period, c. −6th century, from Chia-ko-chuang, near Thangshan in Hopei (National Institute of Archaeology, Peking, cf. Watson & Willetts (1), p. 8). Ht. 21·9 cms.

The description of the *yao fu* suggests the use of close-fitting surfaces by the Chinese alchemists. Iron bowls with smooth lapped edges were produced early in China. Li Kao,[1] a Thang prince (+752 to +820), experimented with bowls and

---

[a] E.g. ch. 68, p. 27a. *TT*1020.   [b] P. 8a, b (only in the *Tao Tsang* version).
[c] (1), pp. 166–7.

[1] 李皋

Fig. 1393. Bronze *tui*, usable as a reaction-vessel, Chou period, between −493 and −447, from the tomb of a Marquis of Tshai, near Shou-hsien in Anhui (National Institute of Archaeology, Peking, cf. Watson & Willetts (1), p. 8). Ht. 33 cms.

plates fitting so well that no air could enter and displace liquids contained within them. Li Yuan[1] used iron bowls ground very smooth at the edges. Similar experiments were carried out by Jen Shih-Chün[2] about +780. The context of this was the search for accuracy in tuning musical instruments by filling sets of precisely fashioned vessels with different amounts of water.[a] But in due course a much more sinister development occurred when the first gunpowder bombs were made from these opposed hemispheres or 'coquilles' in the +12th and +13th centuries. The story is told elsewhere in this work.[b]

These close-fitting bipartite vessels go back very much further than the *yao fu* of the medieval alchemists. Spherical or nearly spherical tripod pyxes of bronze from the

[a] Vol. 4, pt. 1, pp. 38, 192, 194.   [b] Sect. 30 in Vol. 5, pt. 1.

[1] 李琬   [2] 任使君

Fig. 1394. Stoppered aludel of silver, from the hoard of the son of Li Shou-Li, buried at Sian in +756 (Anon. (*106*), pl. 63 c).

Chou and Han periods with removable enantiomorphic lids are quite common in museums (Figs. 1392, 1393).[a] Apart from other projecting ornaments or feet, they generally have ring handles on both parts, perhaps for lifting off the lid with chains, perhaps for binding the two halves tightly together. The ancient name for these vessels is *tui*[1]. They have mostly been dubbed 'food vessels' by the archaeologists, but they seem curiously armoured for such a purpose. Imitations in pottery for tomb-goods are also common.

Still another type of vessel which may belong to the category of *yao fu* are the little ovoid silver bottles with well-fitting stoppers standing about 4 or 5 ins. high (Fig. 1394).[b] At least three of these were found in the hoard excavated at Hsing-hua Fang[2] in Sian which gave us the named and labelled specimens of chemicals already described in Vol. 5, pt. 2, p. 161, so they must date from the near neighbourhood of +750. These metal bottles would withstand considerable pressure, especially if the stoppers were wired down, and Chinese archaeologists regard them as alchemical in purpose.

[a] See, e.g. Anon. (*11*), pp. 27, 31, 39, 40, pls. 9, 12, 63; Anon. (*17*), p. 7, pl. 6; O. Fischer (*1*), p. 303; Willetts (*3*), pp. 86, 89 and pl. 18.
[b] Anon. (*106*), pl. 63c; Hsia Nai, Ku Yen-Wên *et al.* (*1*), pp. 3ff.

[1] 敦   [2] 興化坊

(5) STEAMING APPARATUS, WATER-BATHS, COOLING JACKETS, CONDENSER TUBES AND TEMPERATURE STABILISERS

At this point our exposition must commence in prehistoric times when alchemy and chemistry had not yet developed from the techniques of cooking. During neolithic times (before -1500) the Chinese people invented a peculiar type of vessel, the li^1,[a] in shape like a substantially built pottery jug, with or without handles but always having the bottom running smoothly into three hollow, bulbous legs, often strikingly resembling breasts. The purpose of this cooking-pot was presumably to bring the food into closer contact with the heat of the fire rather than to be able to cook three

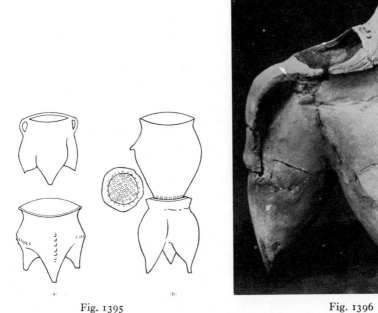

Fig. 1395 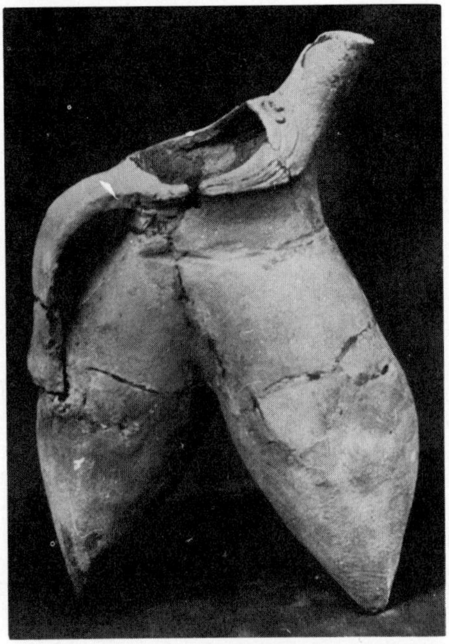   Fig. 1396

Fig. 1395. Neolithic pottery vessels connected with the origins of chemical apparatus, all from the -3rd millennium.

   a   On the left, two types of *li* (from de Tizac, 1). The form of the *li* gave it a large effective heating surface, and of course permitted the cooking of more than one foodstuff at a time, but it must have required great skill on the part of the potter. Moulds for the bulbous legs have survived. The shape was perpetuated in the later bronze tripod cauldrons or reaction-vessels (*ting*), but their legs are usually solid, and as transitional types show, these arose by the collapse of the hollow ones.

   b   On the right, a *li* surmounted by a pot with perforations in the bottom, as shown in the middle above (from Andersson, 6). The combination of the two vessels was called a *tsêng*, though the term could also be applied to the upper one alone, if separate. Later on, the two vessels were combined into one, having a grating between the two parts, generally removable. Vessels of this type were called *hsien*, and became much more preponderant when bronze replaced pottery. As we shall see later on (p. 97), the *tsêng*, or perhaps rather the *hsien*, generated the characteristically Chinese types of distillation apparatus.

Fig. 1396. Three-lobed spouted jug (*li*) of red pottery, from a neolithic site at Shih-chia-ho in Thien-mên Hsien, Hupei, *c*. -2000 (National Institute of Archaeology, Peking, cf. Watson & Willetts (1), p. 3). Ht. 18·4 cms.

   [a] K855. The ancient pictograph has already been given in Vol. 1, p. 81.

   [1] 鬲

Fig. 1397. A *li* vessel of grey pottery, in the style of the Hsiao-thun culture, excavated from neolithic levels at Erh-li-kang, near Chêngchow in Honan, dating from the −23rd to the −20th century (National Institute of Archaeology, Peking, cf. Watson & Willetts (1), p. 5). An early stage in the solidification of the legs to give the *ting* form. Ht. 23 cms.

different foods separately at the same time. Fig. 1395*a* shows two characteristic *li* of pottery (after de Tizac, 1); abundant specimens are preserved in museums all over the world. The shape continued into the Shang and Chou periods made in bronze instead of pottery. Sometimes the vessel may have a fixed cover with a hole for filling and a spout for pouring out (Fig. 1396). There is reason for believing that the tripod cauldron (*ting*) originated by the collapsing of the bulbous legs of the *li* (Figs. 1397, 1398).

As is well known, the typically Chinese method of making bread from cereals throughout the ages was steaming, not baking. And so the *li* generated a form of double vessel in which it was surmounted by a simple pot having holes in the bottom through which the steam could mount to cook the dough. While we must not here go too far into the typology of these vessels, it should be said that they are distinguished by whether the top vessel is separable from the bottom one or not.[a] Steamers of the former type are known as *tsêng*,[1] those of the latter type as *hsien*[2] or *yen*[3].[b] Again the early crude pottery forms were perpetuated in much more elegant bronze, and with

[a] See e.g. Tzu Chhi (*1*) and Willetts (1), vol. 1, pp. 125ff., (3), pp. 85–6, 87–8.

[b] K252. The ancient pictograph has already been given in Vol. 1, p. 81. Hopkins (25), p. 475, gives other forms, with an ingenious explanation of the ancient scribal mistake by which the 'tiger' radical came to be incorporated in the left-hand component of the character.

[1] 甑   [2,3] 甗

free ring handles, during the Shang and Chou periods. In the pottery stage the *tsêng*
are much more common than the *hsien*; Andersson (6) in 1947 found upper vessels
with perforated bottoms obviously intended for placing over *li* with flanged mouths
(see Fig. 1395 *b*). Examples of these are to be seen today in Chinese museums; one for
instance is in the Archaeological Laboratory of Academia Sinica at Sian.

A Shang example of a bronze *hsien* from about −1300 is seen in Fig. 1399, and an
early Chou type in Fig. 1400.[a] By the Warring States period the ornamentation

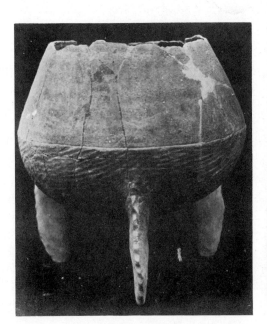

Fig. 1398   Fig. 1399

Fig. 1398. Tripod vessel (*ting*) of coarse dark grey pottery, from a neolithic site at Shih-chia-ho in
Thien-mên Hsien, Hupei (National Institute of Archaeology, Peking, cf. Watson & Willetts (1), p. 3).
Its date may be as early as −2000. Ht. 28·5 cms.
Fig. 1399. Bronze *hsien* of the Shang period, *c.* −1300 (Bushell (2), vol. 1, p. 67; cf. Chêng Tê-
Khun (9), vol. 2, pl. 44a).

becomes less florid and the vessel looks more and more like a piece of apparatus
(cf. Fig. 1401). Whether the bottom of the upper vessel is perforated (like a colander)
or whether there is a removable plate (when the whole vessel is in one piece) the
grating takes a wide variety of forms, with round holes, slits, crosses, etc. in multifarious patterns.[b]

[a] The former from Bushell (2), vol. 1, fig. 45; the latter from White (3), pl. 65, following Mizuno
Seiichi (3), pl. 11, fig. 5.
[b] A very good Warring States bronze example with radiating slots was on view in the Chhêngtu
Archaeological and Historical Museum in the summer of 1972.

Fig. 1400 Fig. 1401

Fig. 1400. Early Chou bronze *hsien, c.* −1100, photographed to show the grating. Now in the Royal Ontario Museum, Toronto. Mizuno Seiichi (3), pl. 11, no. 5; White (3), p. 131, pl. 65.

Fig. 1401. Bronze *tsêng*, i.e. an apparatus of two separate vessels with a grating between them, and fitting well together. From a tomb of the −4th or −3rd century at Chao-ku near Hui-hsien in Honan (National Institute of Archaeology, Peking, cf. Watson & Willetts (1), p. 8). Ht. 60 cms.

One example of a bronze *ting* which might have been the bottom of a *tsêng*, from about − 500 (in the Shantung Provincial Museum at Chinan), has a very well moulded water-seal rim round the top of the lower bulbous-legged vessel (Fig. 1402). This was probably an ancestral form of one that is still common in Chinese kitchens today, the *chi tshai kuan*[1] (colloquially *phao tshai kuan*[2]). A simple round vessel of red pottery is provided with an annular rim round its mouth which acts as a water-seal, supporting an outer domical cover under which there is an inner flat stopper which just covers the hole (Fig. 1403). This is used for making pickled vegetables. Chinese cabbage, carrots, celery, cucumber, turnips, peppers, etc. are first sun-dried, then cut up and allowed to turn pleasantly sour by natural fermentation in half-saturated salt solution. The quasi-anerobic conditions prevent the growth of moulds which otherwise give the mass a white frothy scum and a disagreeable taste and smell.[a] These annular troughs

[a] We shall discuss traditional fermentation techniques at length in Sect. 40 (Vol. 6).

[1] 漬菜罐   [2] 泡菜罐

are of much cultural and evolutionary interest in connection with what we shall have to say presently about the ancient history of distillation, though that technique in China developed quite different apparatus. Indeed we shall show (p. 97) that the characteristic Chinese still form was derived from the *tsêng* and the *hsien* surmounted by a bowl of cooling water and with a smaller receiver bowl placed on the grating. Though the annular trough went no further in China it generated on the other hand the characteristic Hellenistic still form at the Western end of the Old World. As for the

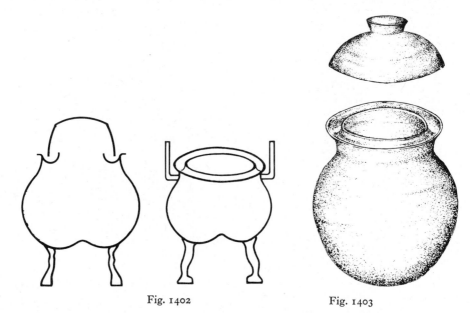

Fig. 1402   Fig. 1403

Fig. 1402. Bronze *ting* with well moulded water-seal rim, c. −500, from Huang-hsien (in the Chinan Provincial Museum, Shantung; orig. drawing).

Fig. 1403. Characteristic pickling pot in common use in China, usually in red pottery, with annular water-seal for maintaining almost anaerobic conditions (*phao tshai kuan* or *chi tshai kuan*; orig. drawing).

developed *phao tshai* pot, it cannot be later than the +3rd century, for the Nanking Museum has one taken from a tomb of the Western Chin dynasty (+265 to +316).

By the Chhin and Han there are plenty of literary references to the steamer vessels. For example, in the *Chou Li*[1] (Record of Institutions of the Chou Dynasty), a text of perhaps the −4th, certainly of the −2nd century, the *tsêng* is mentioned together with its close relation the *hsien* steamer.[a] Many of these bronze vessels dating back to the Shang, Chou and Han periods still exist today in museums and private collections.

Han funerary stoves of pottery or bronze frequently have a special hole in the top designed to take the lower vessel of a *hsien* or *tsêng* resting upon a flange around its middle. One of these in the Archaeological Institute of Academia Sinica at Sian has a set of three top vessels, one with coarse holes for steaming, another with fine holes, and

---

[a] Ch. 12, p. 7*a*; Biot's translation vol. 2, p. 537.

[1] 周禮

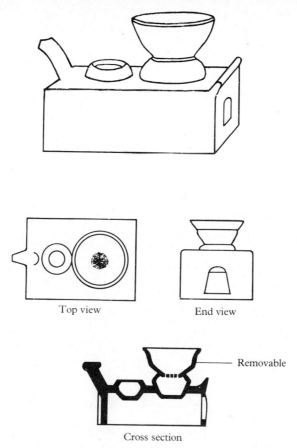

Top view   End view

Removable

Cross section

Fig. 1404. Tomb-model of stove and steamer from Chhangsha; Earlier Han period, −2nd or −1st century. These are very common in Han tombs, and usually of rough pottery or terra cotta.

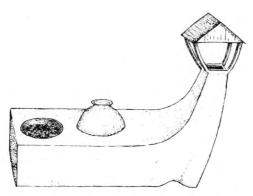

Fig. 1405. Tomb-model of stove, and chimney with roof, from a Later Han grave at Chhangchow. The pot, which could be the lower part of a *tsêng*, is a miniature in bronze (Chiangsu Provincial Museum, Nanking; orig. drawing).

a third with no holes at all for water-bath heating.[a] Since these date from between the −2nd and the +2nd century they parallel the literary references of early times to the water-bath (the 'bain-marie') in the Mediterranean region.[b] Another example, in the Kansu Provincial Museum at Lanchow, has a slit-perforated steaming basin on three legs to fit into a larger basin on a stove, and above that again a smaller basin on three legs with a solid bottom for pure steam heating. The close connection between all these cooking utensils and the equipment of the alchemists' laboratory is obvious.

Fig. 1406. Elaborately ornamented water-bath, with grate underneath, in bronze, of middle or late Chou date (c. −6th century). No inscription. Supplement to the Catalogue of the Sumitomo Collection at Kyoto; Umehara Sueji (3), no. 245, pl. 6 and fig. 5, descr. pp. 5–6.

[a] Other examples from the −1st century are illustrated in Anon. (11), pp. 100, 101, pl. 53. We give one as Fig. 1404. Cf. also the photograph in Graham (4). In others again (Fig. 1405), a roofed chimney is prominent, as in a model of Han date exhibited in the Chiangsu Provincial Museum at Nanking.

[b] This topic was the subject of a special paper by von Lippmann (19). In spite of the common name of the water-bath there are unfortunately no grounds for attributing the invention to the Alexandrian proto-chemist Mary the Jewess, who belongs to the +1st century. As von Lippmann pointed out, the earliest Western mentions occur in the Hippocratic Corpus (De Morbis, III), c. −350, and in the fragmentary book of Theophrastus on perfumes (v, 22), c. −300; then in Cato (De Re Rustica, ch. 81), c. −170, and many other writers. Oil-baths are mentioned by Galen (De Sanitate Tuenda, IV, 8), c. +170. Ash-baths and sand-baths as well as water-baths were of course used by the Alexandrians. But drawings in MS. Marc. 299 which Berthelot (2), pp. 146–7 labelled as forms of the 'bain-marie' seem much more like perforated stoves under the kerotakis. On the other hand, an enigmatic diagram (op. cit., p. 141) also from this MS., with no accompanying explanation in the text, has a cup on the right labelled pontos (πόντος), 'the sea', but it seems rather to be the receiver of a still, possibly a basin supported in a bath of cold water. Nevertheless this gave von Lippmann the clue for his alternative explanation that the name 'bain-marie' derives from mare, the sea, and its Graeco-Egyptian goddess Isis-Cypris or Pelagia-Marina, the Aphrodite of the sailors. Subsequent confusions between this Marina, Miriam the sister of Moses, and the B.V.M., need not detain us.

A rather unusual but striking water-bath in bronze from the Chou period, some five hundred years before the Alexandrians, is illustrated in Fig. 1406. In the Sumitomo Collection at Kyoto,[a] it is rectangular in shape, with an elaborate fire-grate underneath, and large enough to take quite a number of pots, flasks or other containers.

There can be no doubt that the *tsêng* and the *hsien* were of cardinal importance for the invention of distillation in its East Asian form. But the *tsêng* found a fiercer use in metallurgy (especially non-ferrous) when it took the form of two superimposed crucibles, the upper one having a perforated bottom.[b] From the ingredients placed above, the molten metals descended below when the heating was sufficient, leaving the less fusible oxide scoriae and slag in the upper vessel.[c] The whole was known in Syriac as the *bot-bar-bot* ('the crucible and the son of the crucible') a term which got into Latin as the unintelligible *botus barbatus*.[d] It was used for preparing alloys of copper, lead, iron, tin, arsenic, etc., some of which closely resembled silver.[e] More than a hint as to the origin of this apparatus is contained in the untitled +9th to +11th-century Arabic–Syriac alchemical treatise translated by Berthelot & Duval, which says, 'it will come down to form an ingot like Chinese iron'. This *khār-ṣīnī* was probably not metallic zinc but rather the famous alloy paktong (*pai thung*[1]), a mixture of copper, zinc and nickel which had been made in China for many centuries before it became known in the West.[f] What is probably one of the earliest references to it is in the *Kuang Ya*[2] dictionary by Chang I[3] c. +230. Its appearance in the Arabic Jābirian corpus of the +9th and +10th centuries is one of the pieces of evidence pointing to Chinese influence at that time. But this may have been exerted still earlier, as another, purely Syriac, treatise, also translated by Berthelot & Duval, dating from the +7th to +9th centuries and closely related to the earlier Greek texts, speaks of taking 'two amphorae, one being pierced with holes'.

The texts classify *ting* into two types—*huo ting*[4] (heating reaction-vessel) and *shui ting*[5] (cooling vessel). For *huo ting* heat was applied externally, sometimes all round, sometimes only underneath. This is in effect the *ting* already described. But the *shui ting* vessel was a condenser full of cold water for use in conjunction with a *huo ting*. According to its position, it brought about local cooling within or above the *huo ting* to facilitate sublimation and condensation; or it prevented the rise of the temperature of the reactants above boiling water and steam level. Fig. 1407, taken from *TT*908, a work perhaps of the Chin period (+256 to +420) but more probably of the late +9th century,[g] indicates diagrammatically the position of the *shui ting* as a condenser

[a] Supplementary Catalogue, Umehara (*3*), no. 245, pl. VI.
[b] Thus the ancestor of the Gooch crucible.
[c] The process was obviously a special case of *destillatio per descensum*, on which we shall have more to say presently (p. 55).
[d] Berthelot & Duval (1), pp. 58, 149–50; Wiedemann (15).
[e] On aurifictive and argentifictive metallurgy in ancient and medieval China cf. our extended discussion, pt. 2, pp. 188 ff.
[f] On the history of cupro-nickel see pt. 2, pp. 225 ff. and on *khārṣīnī* pp. 428 ff. below.
[g] Ch. 1, p. 25a.

[1] 白銅　　　[2] 廣雅　　　[3] 張揖　　　[4] 火鼎　　　[5] 水鼎

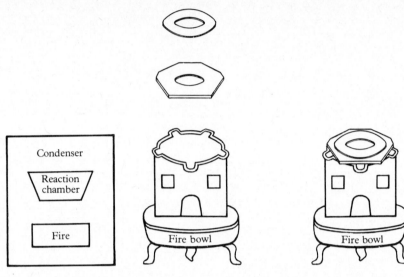

Fig. 1407   Fig. 1408

Fig. 1407. Diagram of furnace with condenser, from the *Huan Tan Chou Hou Chüeh*, a Thang text of the +9th century. The *shui ting* above, the fire (*huo mên*) below, and the reaction-chamber in the middle, marked *chin hung* (gold and mercury).

Fig. 1408. Stove and steamer-support, from *Chin Hua Chhung Pi Tan Ching*..., a Sung text, + 1225. The picture shows only the bowl for the fuel, and the chimney, on the top of which there are rings giving openings of different sizes to accommodate vessels, steamers or water-baths large or small.

in one of the alchemical processes. The adjacent text gives no explanation. This arrangement brings us very near to the beginnings of distillation in its Chinese form.[a]

*TT*907 describes the water-bath as used by Sung alchemists under the name *huo phên*[1] (fire bowl), though this was simply a basin with a three-legged support containing charcoal or other fuel. Fig. 1408, copied from this text,[b] purports to show a fire-bowl together with what is called a *tsêng*, though actually the stove and chimney only are shown, and the water-bath or steamer is left to the imagination. The text describes the apparatus as follows:

Below (the steamer-chimney) is placed a *huo phên* (fire-bowl). Bricks are laid until level as a support. On top (of the fire-bowl) is built a *tsêng* (steamer-chimney) 1 ft. 5 in. high and 1 ft. 2 in. in diameter. Four openings are made in the middle facing north, south, east and west (*tzu*,[2] *wu*,[3] *mao*,[4] *yu*[5]) and in communication with the mouth of the steamer-chimney at the top. These five openings enable fire to be emitted. (The walls) should be quite thick. The mouth for the steamer is circular, 5 inches in diameter. A piece of tile is cut into two and

[a] In fact if it was a bowl of water continually renewed with cold supplies, the apparatus was indeed engaged in reflux distillation. It might even suggest the presence of a central catch-bowl or a catch-bowl with side-tube already at the early Chin date. This piece of evidence should be remembered when considering the antiquity of the Chinese still type.

[b] Ch. 1, p. 4a. The picture shows at the top of the chimney a couple of rings giving openings of different sizes so as to accommodate water-baths large or small—just the same device as we find in all chemical laboratories today.

¹ 火盆   ² 子   ³ 午   ⁴ 卯   ⁵ 酉

placed squarely on the heater, while a second piece with side measuring 1 ft. 2 in. is put over the top. (The apparatus) uses both water and fire. The *tan ting*[1] (reaction-vessel) is suspended at the centre (presumably in the steam).

In the alchemical books of the Sung period very peculiar and complicated combinations of condensers and water-jackets are described. How the different parts of the apparatus were assembled for these more complicated experiments can be seen most conveniently from *TT*907. Here Phêng Ssu wrote:[a]

Method of constructing the *shen shih*[2] (magical reaction-chamber):

Eight ounces of pure gold (lit. full-coloured gold) are cast into a *hun-tun thai-yuan ho-tzu*[3] (chaos womb-shaped closed vessel; reaction-chamber), which is shaped like an egg or a round ball. Then take another ounce of pure gold to make a tube (*chhi kuan*[4]) with a bore the size of the hole in a coin. The length of the tube should project beyond the reaction-chamber by about half an inch. (The capacity of) the reaction-chamber should be just sufficient for holding the *tan phi*[5] (elixir embryo, i.e. the ingredients),[b] and should be neither too big nor too small. After the ingredients have been introduced the joints are sealed by means of *chhih shih chih*[6] (red bole clay) mixed with *chin thu*[7] (earth) and vinegar and left to dry.

Eight ounces of silver (*pai chin*[8]) are used to make a *shui hai*[9] (funnel-shaped reservoir). The lower end is set on the opening of the *ting* (reaction-vessel) and goes down to a depth of about 2 in. When the two are found to fit, the lower end of the *shui hai* reservoir is connected to the gold tube (*chhi kuan*) so that water can flow (down into the latter). The joint is firmly sealed by means of *chhih ni*[10] (a lute or sealing mixture, possibly of red bole clay and mud) and left to dry before water is poured in.

The *wai ting*[11] (outer vessel):

This is made of pottery. Its capacity should be just sufficient for inserting the reaction-chamber and 2 lb. of silver (*pai chin*[12]) and must be neither too spacious nor too narrow. If it is too large the space can be filled in, using yellow earth mixed with vinegar and left to dry. Then half a handful of silver is put in and adjusted until it is packed and level before the gold reaction-chamber is inserted. Silver is again put over the top until (the reaction-chamber) is covered up. A paper ring is put over the reaction-chamber to mark the position for the *shui hai* reservoir. After filling up (the *ting*) with silver, the silver *shui hai* reservoir is put in position through the hole as marked by the ring. (The *ting*) is then shaken up until everything is properly packed. The outside has to be tightly sealed before (the vessel) is suspended inside the combustion chamber.

Fig. 1409, taken from this text,[c] illustrates the reaction-chamber (*b*) and the *shui hai* reservoir (*a*), and also how they were assembled inside the *huo ting*[12] (heating vessel, *c*). The description of the diagrams is rather similar to what we have just read. It says:

Eight ounces of pure gold are cast into a *hun-tun chi-tzu shen shih*[14] (magical chaos egg-shaped reaction-chamber). Again one ounce of gold is used to make a water-tube (*shui kuan tzu*[15]), the lower end of which is closed and does not allow water to pass through.[d] It is about

---

[a] Ch. 1, p. 1*a*.

[b] On this very ancient appellation, which goes back, perhaps, to Babylonian antecedents, cf. pp. 293, 296.

[c] Ch. 2, p. 2*a*.

[d] This is curiously evocative of the 'cold-finger' still types in modern organic chemistry which we shall mention in connection with the principle of the Chinese still, pp. 100 ff. below.

[1] 丹鼎	[2] 神室	[3] 混沌胎元合子	[4] 氣管	[5] 丹胚
[6] 赤石脂	[7] 金土	[8] 白金 [9] 水海	[10] 赤泥	[11] 外鼎
[12] 白金	[13] 火鼎	[14] 混沌雞子神室	[15] 水管子	

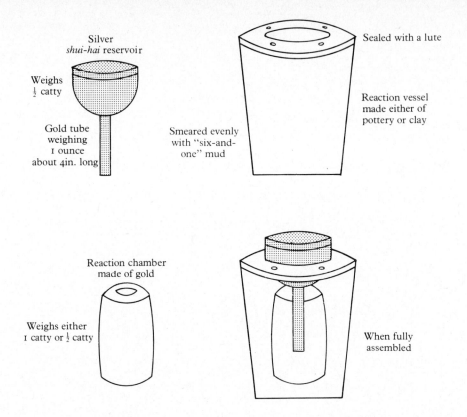

Fig. 1409. Water-cooled reaction-vessel from the *Chin Hua Chhung Pi Tan Ching...*, +1225. The upper reservoir is prolonged into a blind finger below, so that the cold water can moderate the temperature of the reaction proceeding in the chamber.

4 inches long and is inserted down to the base of the *hun-tun ho*[1] (closed chaos-vessel; reaction-chamber). The top end is attached to the *shui hai* reservoir, which is made of eight ounces of silver. The joints are all tightly sealed with (*chhih shih*) *chih*[2] (red bole clay) and *fan* (*-shih*)[3] (alum) and allowed to dry. The portion (of the tube) inside the reaction-chamber is filled with water.[a]

The text then describes the process that followed after the ingredients were introduced:

The joints (of the reaction-chamber) are tightly sealed. It is then put into a pottery vessel (*thu ting*[4]) and the space inside filled with silver beads (*yin chu*[5]) so that no gap is left. After the silver *shui hai* reservoir is set on top the outside is smeared (with a lute) to a thickness of a finger's width. After drying for half a day (the vessel) is suspended inside the combustion chamber. Fire is applied below and round all sides (of the vessel). At first 5 lbs of charcoal are used, and when more than half that amount is burnt another 5 lbs of charcoal are needed. These are to be added twice or thrice within the period of one day and night...

a P. 2b. Some of these statements can be seen in the original captions of Fig. 1409.

1 混沌合   2 赤石脂   3 礬石   4 土鼎   5 銀珠

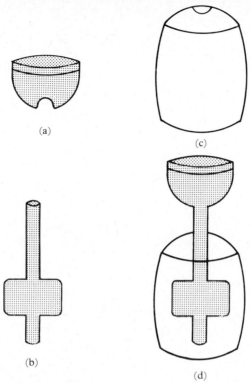

Fig. 1410. Water-cooled reaction-vessel, from the same work (+ 1225). The central tube descending from the cooling reservoir is enlarged into a single bulb. (a) Cooling reservoir, (b) blind finger with bulb, (c) reaction-chamber, (d) full assembly.

Another device of closely similar character is illustrated in the same text,[a] but while the former is called a *chi-chi*[1] apparatus (cf. p. 68), the latter is not. What is particularly noteworthy about these pieces of apparatus is the large vessel of water at the top of the system, for it corresponds closely with the position of the main condenser basin in the characteristic East Asian type of still.[b]

The purpose of the reservoir (*shui hai*[2]) and the attached tubes (*shui kuan*[3, 4]) must have been to exert a thermostatic control by ensuring the presence of water at boiling temperature isolated from the reactants. The insertion of a blind tube into the centre of the reaction-chamber meant that the reactants would be cooled centrally. We find many other approximations to constant-temperature technique in the same text. The first elaboration was to introduce a bulb in the central cooling columns, as in Fig. 1410.[c] The next thing was to introduce two bulbs as shown in Fig. 1411.[d] Still more complicated is the system given in the same chapter where the central tube connects the upper reservoir or condenser with a water space below built in to the side-walls of the reaction-chamber (Fig. 1412).[e]

[a] Ch. 2, p. 18*a*.　　　[b] Cf. p. 63 below.　　　[c] Ch. 2, p. 16*b*.　　　[d] Ch. 2, p. 10*a*.
[e] Pp. 5*b* and 6*a*. It is not quite clear from this and some of the following illustrations whether cross-sectional elevations are intended, so one cannot always be sure whether the coolers were tubes in the plane of the paper or all-round water-jackets with concentric walls.

[1] 既濟　　　[2] 水海　　　[3] 水管　　　[4] 水甕

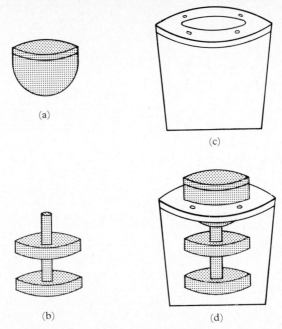

Fig. 1411. Water-cooled reaction-vessel, from the same work (+ 1225). The central tube has two bulbs flattened like radiator fins. (*a*) Cooling reservoir, (*b*) blind finger with two flattened bulbs, (*c*) reaction-vessel, (*d*) full assembly.

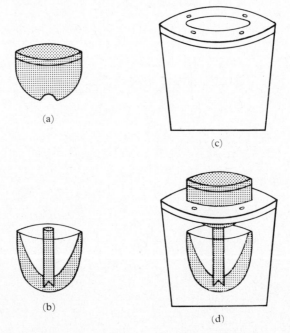

Fig. 1412. Water-cooled reaction-vessel, from the same work (+ 1225). The central tube connects the cooling reservoir with a peripheral water-jacket in the wall of the reaction-chamber. (*a*) Cooling reservoir, (*b*) tube connecting with jacket, (*c*) reaction vessel, (*d*) full assembly.

Fig. 1413. Water-cooled reaction-vessel, from the same work (+ 1225). Here the peripheral water-jacket is provided with a filling tube at one side as well as the connection with the cooling reservoir through the central tube. The assembly can be understood in the light of the preceding drawings.

Several other types of cooling system related to those already described and taking the form of veritable water-jackets are given in the same text and are shown in Figs. 1413 to 1417. One sees how the aim of Mêng Hsü and his friends was to increase the cooling surface. It was evidently a prime concern of the + 13th-century alchemists to control the temperature of their reactants and prevent it from rising too high. By varying the extent of the water-jacketing they could choose a wide variety of temperatures for their reactions. Fig. 1413 shows a water-jacket with an upper filling tube at one side.[a] Fig. 1414 shows another version of the external water-jacket.[b] A double water-jacketed hood is seen in Fig. 1415 forming a more complicated type of condenser.[c] Fig. 1416 shows a complete water-jacket.[d] Finally, in the *san shui-kuan*[1] (triple water-tube) reaction vessel, water-cooling was applied externally by means of three meridional tubes.[e] This is shown in Fig. 1417.

In this account we have followed an ascending order of apparent complexity natural

[a] Ch. 2, pp. 8b, 9a.   [b] Ch. 2, p. 7a.
[c] Ch. 2, pp. 11b, 12a.
[d] Ch. 2, p. 13b. Or possibly two tubes connected at the bottom as well as the top.
[e] Ch. 2, pp. 3b, 4a.

[1] 三水筦

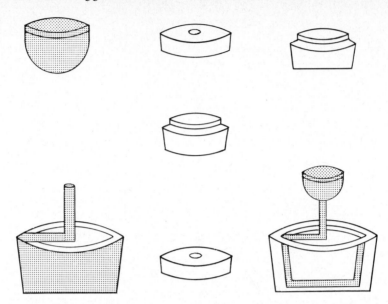

Fig. 1414. Water-cooled reaction-vessel, from the same work (+ 1225). There is no central tube, and the cooling reservoir connects directly with the peripheral water-jacket at one point on its circumference above.

to us with our hindsight knowledge of modern chemical apparatus. But in Mêng Hsü's time (+ 1220) there was no thought of any such order. The pieces of apparatus just described were all used at appropriate stages in a long gold-elixir preparation consisting of nine 'turns' (chuan[1]) or chemical transformations, plus two preliminary processes, most of which employed each a different system of reservoir cooling with its tubes or jackets. The order of use of the apparatus is shown in the accompanying Table 116.[a] Exactly what was happening at each stage and what the end-product was will only be elucidated in the course of further research, which could be quite rewarding; all we can say at present is that gold and silver, lead, cinnabar, sulphur, mercury, vinegar, alums and the arsenic sulphides were all involved at one phase or another. It may also be relevant to add that there are numerous mentions of binding the parts of the apparatus together with iron wire (thieh hsien[2]), just as one used to do with modern pressure-tubing connections;[b] this must mean that considerable pressures were liable to be generated during the reactions.[c]

The exact place of these extraordinary developments in the history of chemical apparatus technology as a whole, whether in relation to water-tube and fire-tube boilers,[d] or to heating and refrigerating and condensing coils in chemical industry,

---

[a] A similar analysis of the text has been made by Yoshida Mitsukuni (7), pp. 252ff.

[b] E.g. ch. 2, pp. 6b, 15b, 17a. In other books the same is found, as, for example, in TT912, ch. 1, pp. 5b, 7b.

[c] This is not without significance in connection with the earliest proto-gunpowder mixtures, on which see pt. 3, p. 159 above.   [d] Cf. Needham (48), and (64), p. 153.

[1] 轉   [2] 鐵線

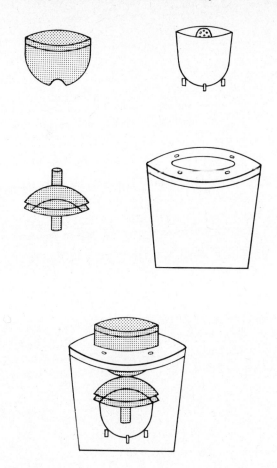

Fig. 1415. Water-cooled reaction-vessel, from the same work (+ 1225). Cooling reservoir, central tube, and two water-jackets in the form of hoods, but no peripheral cold-water walls.

remains to be determined in future historical studies, but it would seem highly unlikely that they constituted a unique phenomenon without antecedents or later repercussions in the history of techniques. There is much room for thought in considering what may have derived from these strange circulatory systems.

One fairly clear descendant could be descried in the Japanese *daki*[1] temperature stabiliser used in the fermentation industry. As Shinoda Osamu (*1*) has shown, this takes the form of containers, now of pottery, formerly of wood, which are suspended in the saccharification vats and filled with water of a temperature appropriate to the season, cold in summer and hot in winter. Such a temperature stabilisation helps the moulds to break down the polysaccharides, and discourages the yeasts at this stage.[a]

Again, later on (pp. 124, 127), when studying the limiting factors for the discovery of

[a] At a visit to the Ōkura Sake Brewery in Kyoto in 1964 this technique was seen by us in operation.
[1] 暖氣

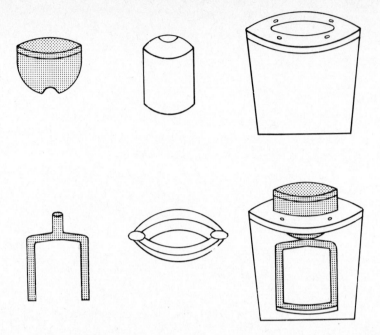

Fig. 1416. Water-cooled reaction-vessel, from the same work (+ 1225). Cooling reservoir with complete water-jacket horseshoe-shaped in cross-section but continuing under the floor of the chamber.

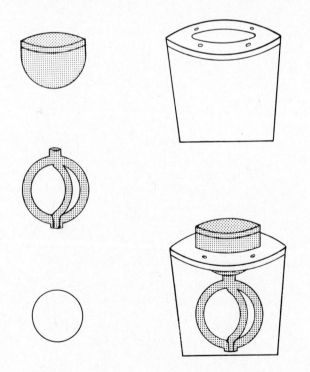

Fig. 1417. Water-cooled reaction-vessel, from the same work (+ 1225). Cooling reservoir prolonged below into three meridional tubes passing around the chamber and meeting at the bottom.

Table 116. *Order of operations in the elixir preparation of the*
'Chin Hua Chhung Pi Tan Ching Pi Chih', ch. 2 (TT907)

Turn (*chuan*)	Ch./page	Apparatus used	Fig.	Ho Ping-Yü & Needham(3), fig.
—	1/4a	stove, steamer-chimney and rings for giving openings of different sizes.	1408	19
prelim. process (a)	2/1b, 2a	chi-chi[1] app. (cf. p. 68); cooling reservoir and central tube.	1409	20
prelim. process (b)	2/3b, 4a	ting chhi[2] app. (a term also applied to all the rest); cooling reservoir and three meridional tubes below meeting at the bottom.	1417	29
1	2/5b, 6a	cooling reservoir, central tube and peripheral jacket below.	1412	24
2	2/7a	cooling reservoir and peripheral jacket below.	1414	26
3	2/8b, 9a	cooling reservoir, central tube and jacket with additional filling tube at one side.	1413	25
4	2/10a	cooling reservoir and central tube with two bulbs shaped like radiator fins.	1411	23
5	2/11b, 12a	cooling reservoir and double water-jacketed hood.	1415	27
6	2/13b	cooling reservoir and complete water-jacket horseshoe-shaped in cross-section.	1416	28
7	2/14b, 15a	cooling reservoir and central tube in connection with some kind of sublimation or distillation apparatus (cf. p. 72).	1448	49
8	2/16b	cooling reservoir and central tube with one bulb.	1410	22
9	2/18a	cooling reservoir and central tube; similar to prelim. process (a).	1409	21
ancillary process	2/4a, b	ambix and *lopas* for the descensory distillation of mercury	1433	37

alcohol in the West, we shall find that adequate cooling by water was probably the essential thing, and since there seem no antecedents for this among the chemists of Islam the Chinese still with its water-cooled head may have been influential. The oldest cooler in the West, however, seems to have been a serpentine side-tube rising through a barrel of cold water, and the question may therefore be posed whether this could have been stimulated (even as a matter of hearsay) by the tubes, bulbs, and

[1] 既濟   [2] 鼎器

layers which had developed in China. The time when Mêng Hsü was working was just half a century before that of Taddeo Alderotti (cf. pp. 92, 122) and the first indubitable condenser coil in the West. Figs. 1414, 1416, and especially 1417, show apparatus which could conceivably have exerted some influence. Mêng, to be sure, was active half a century later than the Salernitan Masters and the writer of the recipe in the *Mappae Clavicula* (cf. p. 123), but what apparatus they used is not known, and the complexity of the devices described by Mêng points to a longish previous development, going back to the Northern Sung (+11th century) if not indeed to the Thang. Moreover, the +13th century was a time of intensified intercourse between East and West, marked not only by Marco Polo and other merchants, but also by the Franciscan friars in Mongolia and China;[a] nor do we lack outstanding examples of other westward transmissions in the +12th century (cf. p. 403).

These speculations are strengthened perhaps by the fact that the *daki* temperature stabiliser is described already in the text of the *Pei Shan Chiu Ching*[1] (Northern Mountain Wine Manual), written by Chu I-Chung[2] in +1117, where it has the name of *chui hun*,[3] or 'recovering the soul' (which otherwise was chilled and lost, or driven away by over-heating).[b] This was a whole century before Mêng Hsü's descriptions, and half a century before the Masters of Salerno.

Lastly, it is obvious that the principle of increase of surface which these devices embodied is constantly used in modern scientific apparatus. Fractionation-columns and scrubbing-towers come at once to mind.[c] As another instance Ratledge (1) recently described flasks with deep finger-like indentations designed to cool bacterial suspensions more effectively during the ultrasonic disruption of the cells, when immersed in brine at −20 °C.

(6) SUBLIMATION APPARATUS

The simplest form of vessel for this purpose was nothing but a pot inverted and suspended over a glowing fire. Substances were cast in small quantities on to the red charcoal or hot ashes, and the sublimate caught in the receiver above. This procedure is little mentioned in histories of chemical technology, but it was used a great deal, even in the +17th century by J. R. Glauber, though by then there was much refinement of superimposed receivers.[d] With this system he could make many things commercially including not only flowers of antimony but hydrochloric acid. If simplicity is an adequate criterion of age, the inverted pot must be considered very old, and it would be surprising if the alchemists in ancient China made no use of it. The next simplest thing was a pot with a removable lid placed on the first mouth upward, heating being applied locally at the bottom so as to allow for the condensation of the volatile substance on the under surface of the lid, whence it could be easily removed.

The first Western description of the process of obtaining metallic mercury from

---

a Cf. Vol. 1, pp. 188 ff. and Needham (64), pp. 61, 201, 300.   b Ch. 2, p. 11b.
c Cf. Morton (1), pp. 75 ff., 91 ff.   d See Greenaway (5).

1 北山酒經   2 朱翼中   3 追魂

cinnabar by the method of sublimation is generally attributed to Dioscorides (*c.* +50), who said that cinnabar was heated on an iron saucer contained in a pot and covered by another pot.[a] Throughout history mercuric sulphide in the form of natural cinnabar was perhaps the single most important raw material used by the Chinese alchemists, its name (*tan*[1]) being identical indeed with the very word for elixir itself. Although we do not know exactly when they first began converting cinnabar to mercury,[b] it must have been at least as early as the Warring States period, and the first textual mentions come from the beginning of the Former Han (−2nd century). The *Shen Nung Pên*

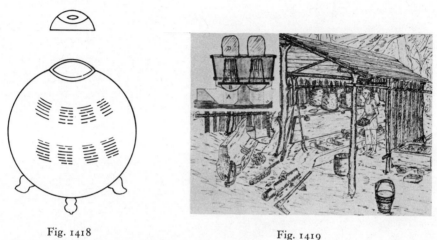

Fig. 1418   Fig. 1419

Fig. 1418. A sublimatory vessel in the shape of a narrow-mouthed *ting*, from the *Hsiu Lien Ta Tan Yao Chih* (Sung).

Fig. 1419. Traditional Thaiwanese sublimatories for camphor (Davidson). A, fireboxes; B, pans of water for the steam; C, containers for the wood chips piled on a grating; D, inverted jars to collect the sublimed crystals.

Tshao Ching[2] (Pharmacopoeia of the Heavenly Husbandman) states clearly that cinnabar can be converted into mercury, taking it indeed as a matter of course.[c] Then the same statement is found in the *Huai Nan Wan Pi Shu*[3] (Ten Thousand Infallible Arts of the Prince of Huai-Nan), first compiled about −120 or very little later.[d] A variety of vessels, made of pottery or metal, and having removable lids, are illustrated in Chinese alchemical books. Fig. 1418 shows what is called a 'mercury vessel'—*hung ting*[4]— in *TT*905, a Sung book.[e] It was certainly used for sublimation.[f] The sublimate adhering

---

[a] *Mat. Med.* v, 110; Gunther tr. p. 638. See Sherwood Taylor (4), p. 52; Forbes (9), p. 17. The method was mentioned by Agricola in +1556 (Hoover & Hoover tr. p. 427).

[b] We give a discussion of the history of mercury in China elsewhere (pt. 3, pp. 4 ff.).

[c] Mori Tateyuki's ed. (1845), p. 22; Miu Hsi-Yung's (+1625), ch. 3, p. 1*b*. The text is also quoted by Liu Wên-Thai in *Pên Tshao Phin Hui Ching Yao*[5] (Essentials of the Pharmacopoeia classified according to Nature and Efficacity), +1505, ch. 1, p. 1*a*, and by nearly all the other writers of pharmaceutical natural histories.

[d] *TPYL*, ch. 988, p. 6*a*; Yeh Tê-Hui's reconstruction, no. 83.   [e] Ch. 2, p. 3*b*.

[f] Vermilion from mercury and sulphur (pt. 3, p. 74) is almost as old; cf. Gettens, Feller & Chase (1).

[1] 丹   [2] 神農本草經   [3] 淮南萬畢術   [4] 汞鼎   [5] 本草品彙精要

Fig. 1420. Camphor sublimation with steam as carried out in Japan, from the *Nihon Sankai Meibutsu Zue* (+ 1754).

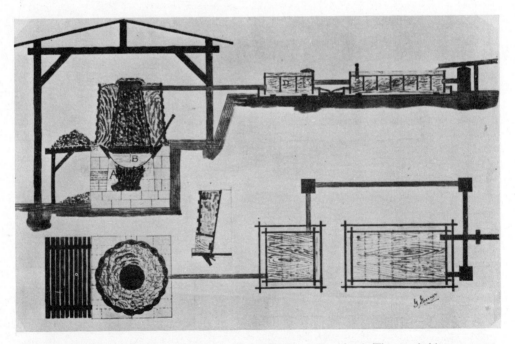

Fig. 1421. Camphor still of Japanese type as used in Thaiwan (Davidson). The wood chips are steam-distilled, and the vapours pass over into long shallow boxes fitted with partitions and covered by a long trough-like lid within which the cooling water is made to flow. The oil and the camphor both condense in the box receiver, but the oil floats below on the water from the steam while the camphor crystals gather on the sides and under-surface of the cooling trough. A, firebox; B, pan of water; C, the chip 'retort'; D, cooling box; E, crystallisation box.

to the cover—the *yin kai*[1]—was collected at the end of the process by scraping with a feather. It would seem that the *yao fu*[2] types of vessel (see p. 22 above) were commonly used for this operation. But in descriptions one may encounter almost any term applicable to an aludel-like vessel capable of being hermetically sealed, e.g. *kuan*[3] (cf. p. 58), and the phrase *ku chi*[4] often occurs in connection with it (cf. pp. 5, 6, 79).

Another sublimation widely practised in China for centuries, indeed on a commercial scale, at least since the times of Thao Hung-Ching and Sun Ssu-Mo, was that of the chlorides of mercury.[a] We were able to illustrate this in Fig. 1357, taken from a rare MS. of the *Pên Tshao Phin Hui Ching Yao* (+1505) with pictures in colour,[b] which shows the age-old pharmaceutical preparation of calomel.

One substance early obtained by sublimation was camphor, an indigenous aromatic of great antiquity in Chinese culture.[c] We have not found any drawings in traditional style of the apparatus used, but in 1903 Davidson (1) gave a sketch of the traditional Chinese sublimatory stoves then working in Thaiwan (Formosa). This is reproduced in Fig. 1419. The chips of camphor-tree wood were renewed twice daily, and the camphor was carried upwards in the steam to condense like snow in the inverted earthenware jars above, whence it was collected by hand every ten days. This was very like the procedure (Fig. 1420) which the Jesuit F. X. d'Entrecolles had described[d] in +1736. Davidson also studied a composite process of sublimation and distillation which produced not only solid camphor crystals but also camphor oil floating on the condensed water. He gave an interesting drawing of a 'Japanese' still, in use on the island in his time, for the preparation of both the solid and the liquid products (Fig. 1421).[e] Steam generated below a perforated plate on which the chips are piled extracts the aromatic substances from them and fills the space above, then passing out through a long bamboo tube from the top of the 'retort' reaches a long shallow box fitted with partitions and covered by a trough-like lid within which cooling water is made to flow. The oil and the camphor crystals condense in the box, the oil floating on the aqueous phase, the camphor crystals gathering mainly on the sides and under the top of the box.

Bryant (1), writing in 1925, described a similar sublimation-distillation apparatus used in Chiangsi, Kuangtung and Hainan (Fig. 1422).[f] This arrangement differs from

[a] On this important chemical discovery see Vol. 5, pt. 3, pp. 123ff. above.

[b] This was on sale in the market at Hongkong in 1959, and we owe our knowledge of it to Dr S. D. Sturton, who kindly supplied some photographs. The same illustration, from another MS. copy, has been reproduced by Bertuccioli (2).

[c] This we discuss elsewhere (pt. 2, pp. 135ff.) in relation to liturgy and alchemy. The typical Chinese camphor was 'chang camphor', from *Cinnamomum Camphora*. Later, other camphors were imported, and some other camphor-producing species acclimatised. A glimpse of the camphor industry in late Ming times is found in Fernaõ Mendes Pinto (1), p. 118, (ch. 30, sect. 3). He was in China about +1543 and his book was first printed in +1614.

[d] (2), pp. 232ff. The only difference was that the chips were then first submitted to extraction with boiling water, and the cooled aqueous mass piled in layers with earth before subliming. This purification process was still in use in 1867 (Julien & Champion (1), p. 229).

[e] His elaborate chapter on the Thaiwan camphor industry is well worth reading. Cf. Schelenz (2), fig. 119. An earlier description (1895) was given by Grassmann (1).

[f] A brief recent account of camphor sublimation-distillation among the Thai people of the Hsi-shuang-bana autonomous region in Yunnan has been given by Alley (9).

[1] 銀蓋   [2] 藥釜   [3] 罐   [4] 固濟

Fig. 1422. Traditional sublimation-distillation apparatus for camphor used in South China (Bryant; photo. Brisker).

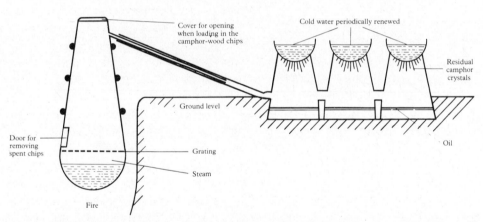

Fig. 1423. Diagram to illustrate the operation of this apparatus. On the left, the 'retort' with grating separating the wood chips from the source of steam; on the right, a succession of chambers each like a Mongolian still (cf. p. 62) with a cooling reservoir above, acting as a receiver into which the side-tube delivers. The camphor oil floats on the condensed water below, the camphor itself sublimes as crystals round the bottoms of the coolers and on the sides of the chambers.

that shown in the previous illustration because the condensing box and trough is replaced by a series of three tubs forming a continuation of the bamboo side-tube and each capped by a 'Mongolian'[a] water-basin still-head cooler.[b] This description

a The terms 'Mongol' and 'Chinese' as applied to stills are explained in the course of the following pages.
b Sublimation technique just like this is current in modern organic chemistry, e.g. for anthracene, when a flask with circulating cold water is suspended above a perforated plate over a beaker (Robinson & Deakers (1); Morton (1), fig. 116, p. 215). See inset diagram on p. 49. The crystals accumulate on the lower surface of the flask.

will be understood from Fig. 1423 but better appreciated from p. 62 below, whence it will be seen that this camphor apparatus is likely to be much older than the trough-and-box system. The ingenuity of both lies in the fact that they solved the problem of handling vapour liable to deposit a solid which could clog narrow condenser tubes and develop a dangerous pressure in the 'still'. The train of tubs attached here to each side-tube is reminiscent of the trains of three to six mercury-condensing aludels at the famous works of Almadén in Spain. Since the Arabic world once stretched from Cadiz to Canton there might even be a genetic connection. Bryant also gave details of a complicated process for purifying the commercially valuable camphor oil (containing more than twenty important compounds such as pinene, limonene, cineol and terpineol) nd obtaining considerable amounts of solid camphor (up to 40%) from what had been regarded as the waste material

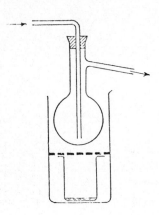

of the method. In this further operation, said to be of Japanese origin, the oil is placed in simple pot stills connected to long coiled brass condenser tubes cooled by flowing water. The various fractions coming out of this are carefully examined, filtered, re-distilled, etc. Such an apparatus seems to be of essentially modern type with a Western background (cf. p. 93 below).

Sublimation of camphor was also carried out in India, Southeast Asia, and by the Arabs. The +9th-century text of al-Kindī, translated and discussed by Garbers (1), though mainly devoted to distillation, contains several recipes for the purification of camphor by sublimation. Garbers gives redrawings of some of the original illustrations.[a]

If we are right in our interpretation, one of the most interesting of all ancient pieces of Chinese chemical apparatus was used for sublimation. During the second world war (in 1943) the Institute of Cultural Studies of the University of Nanking (then located at Chhêngtu in Szechuan) exhibited a remarkable bronze self-named from its inscription as a 'rainbow vessel' (kung têng[1]).[b] We reproduce it in Fig. 1424.[c] Above a round tripod vessel in the form of a flattened sphere there is a cylindrical compartment with sliding walls capped by a hemispherical dome. Two tubes of ample lumen arise one on each side from the lower vessel and join at the top at the crown of the dome. The upper halves of the tubes and the dome are removable, after which the central cylinder, which has no direct communication with the bottom, can be lifted off by the side handle. The inscription says Yen Ong Chu thung kung têng i chü,[2] i.e. 'One rainbow

[a] (1), pp. 19–21, 95.

[b] Normally the word for rainbow (hung[3]) has the 'insect' radical (K 1172j), and the lexicographers' meanings for kung[4] are the iron bearing of a chariot-hub or an ornament worn by women at their belts. But the present context is unique. Similarly têng[5] is the word later used for a stirrup, not a lamp or vessel.

[c] By the kindness of Dr Li Hsiao-Yuan. It was President Hsiang Hsien-Chiao who identified it as a piece of chemical apparatus, thinking it might have been one of the instruments obscurely referred to in the Huai Nan Tzu book.

¹ 釭鐙   ² 閼翁主銅釭鐙一具   ³ 虹   ⁴ 釭   ⁵ 鐙

Fig. 1424. Bronze 'rainbow *têng*' of Later Han date (+ 1st or + 2nd century) (photo. Nanking University Museum, 1943). A tripod *ting* below is connected by two side-tubes of ample lumen to the top of a cylindrical space with sliding walls situated on top of the 'boiler' below but having no connection with it. Probably apparatus of this type was used for subliming volatile substances such as camphor.

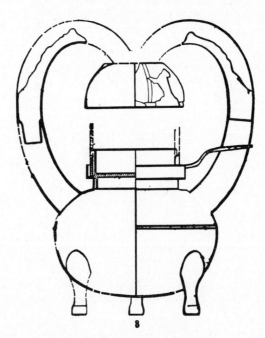

Fig. 1425. Cross-section of another example of this apparatus, found at Chhangsha and dating from the Early Han period, *c.* − 1st century (Anon. (*11*), p. 115, fig. 94).

vessel belonging to Old Master Yen'. The object, which was found in a tomb at Chhangsha in Hunan, is considered to be of Hou Han date (+1st or +2nd century).

More recently, further excavations at Chhangsha have brought to light another example (see Fig. 1425), in dimensions closely similar to the former (e.g. height about 34 cm.), but in date somewhat earlier (−1st century).[a] It has no inscription. The authors of the report regarded the object as some kind of lamp, but this is hard to believe since no space exists from which light could be irradiated. We much prefer the view that substances to be volatilised were placed in the lower part of the vessel, so that on heating the vapour would ascend through the tubes and condense in the upper compartment, perhaps with the assistance of a sponge or cold wet rags outside. The fact that according to the writers of the report some waxy material was found in the upper compartment when the vessel was excavated must surely suggest that volatile organic substances had been treated in it.

Other examples were illustrated long ago in archaeological works. The Sung imperial collection possessed one, with an inscription saying: *Wang shih thung hung chu ting*[1] (Mr Wang's bronze rainbow lamp heater), but we do not know the date of it.[b] Two others were in the Chhing imperial collection, as we see in the *Hsi Chhing Ku Chien*[2] (Catalogue of Ancient Mirrors (and Bronzes of the Imperial Collection in the Library of) Western Serenity), compiled by Liang Shih-Chêng.[3] One occurs in the original work (+1751),[c] the other in the second supplement (+1793).[d] The latter had lost its central cylindrical walls. The editors called the first a *têng*,[4] but this one a *ting*,[5] i.e. a hot-plate.[e]

There has been some argument among philologists about *ting* and *têng*; the question at issue being the identification of one of the pictographs on the Shang oracle-bones—a semi-circular amphisbaema or two-headed dragon. Takada Tadasuke believed that it meant *hung ting*, the 'rainbow heater' or sublimatory vessel itself (afterwards shortened to *ting* alone), but Hopkins (26) preferred to see in it the word for rainbow as such,[f] and on this view he could make sense of certain bone inscriptions otherwise difficult to translate. While it would certainly be interesting if our sublimatory design were as old as the −2nd millennium there is as yet no archaeological evidence in support of that, and it is more likely that the rainbow vessel with its over-arching side-tubes got its quite appropriate name in Warring States or Early Han times.[g]

---

[a] Anon. (*11*), p. 115.

[b] *Hsüan-Ho Po Ku Thu Lu*[6] (+1111 to +1125), ch. 18, p. 41 *a*. On the stormy career of the curator who made this catalogue, Wang Fu,[7] cf. Vol. 4, pt. 2, p. 500.

[c] Ch. 30, p. 27 *a*.   [d] Ch. 13, pp. 33 *b*, 34 *a*.

[e] This was one of the most ancient meanings of the word, which later on came to signify an ingot (cf. pt. 2, pp. 67–8, pt. 3, p. 102) and eventually an anchor (cf. Vol. 4, pt. 3, p. 657).

[f] Cf. Vol. 3, p. 473, where we gave a cut of the ancient graph.

[g] Figurines of serpent-like creatures arching their backs with a human head at each end are not uncommon in archaeological collections; the Royal Ontario Museum at Toronto has one dating from the Wei dynasty made of grey pottery and painted white, light green and red (Fig. 1426). Doubtless they personified the rainbow.

[1] 王氏銅虹燭錠   [2] 西清古鑑   [3] 梁詩正   [4] 鐙   [5] 錠
[6] 宣和博古圖錄   [7] 王黼

Fig. 1426. Figurine of an amphisbaena or two-headed serpent, probably personifying the rainbow and representing a visible rain-bringing dragon, generally beneficent (photo. Royal Ontario Museum, Toronto). Of grey pottery, painted white, light green and red, in stripes; date Wei. Such representations, not uncommon, throw light on the naming of the apparatus with its double over-arching side-tubes.

Fig. 1427. Rainbow *têng* with only one side-tube, Sung in date, but the bronze inlaid in Chhin or Han style (photo. British Museum, 1960).

Fig. 1428. Modern still for the purification of mercury in high vacuum, lined with glass and porcelain to prevent contact with the metal (photo. Multhauf, 1961, Smithsonian Museum, Washington). Although the heating chamber is above and the receiver below, the design with double side-tubes perpetuates the ancient Chinese sublimatory system.

Two further variations of vessels of this type are now known. In one, from another tomb at Chhangsha, described by Kao Chih-Hsi (1), the boiler below is in the form of an ox; the walls of the upper chamber or receiver have been lost, but there are two tubes rising up as continuations of the backward-curving horns.[a] An inscription says: 'Made for the Imperial Temple (*chhih miao*[1]) by the Chief Intendant of the Four Ceremonials', and accompanying objects were dated as from Chhangsha 'in the first year'. This cannot be further identified but is considered to have been some time during the Former Han period (−2nd or −1st century). The other vessel, different in having only one rising tube, is in the British Museum; it is the latest of all, dating from the Sung period, but its bronze is inlaid in Chhin and Han style, and it has gilded feet in the shape of bears after the Han fashion (Fig. 1427). The general ornamentation is

[a] Kao's own interpretation does not commend itself to us; he suggested that the lost enclosing walls were to protect a lamp flame from draught, and that the tubes were to convey its smoke down into the interior, thus keeping the room clean.

[1] 敕 廟

Fig. 1429. Single-tube rainbow *têng* in the form of a serving-maid holding a lamp, from the tomb of Liu Shêng, Prince Ching of Chung-shan, d. −113 (photo. Hsinhua Thung-hsün Shih, 1971).

much more elaborate than that of the ancient examples, but like some of them it has lost the walls of the central chamber.[a]

This last object was catalogued by the curators as a 'perfume still',[b] and that raises the interesting problem of what they were all used for. In personal correspondence Dr R. P. Multhauf has discussed with us the possible uses of the apparatus. First he brought to our attention a modern still for purifying mercury (made about 1925) which has a strange structural resemblance to the ancient 'rainbow *têng*' (Fig. 1428), only in reverse, the electrically heated chamber being above and the receiver for the distillate below.[c] It is a vacuum still, lined with glass and porcelain, so mercury and metal do not come in contact. Cinnabar mixed with fuel (and possibly with metal filings or lime to release the mercury better) could of course have been heated in the lower chamber of the rainbow vessels, but the likelihood of attack on the bronze walls by the mercuric vapour with the formation of amalgam,[d] seems to all of us to suggest that this was not their use. Multhauf inclined to the view that they were used in some distillation process, perhaps of essential oils in connection with drug and perfume preparations. But even more attractive is the possibility that they were sublimatories for some substance which would condense in crystals on the inside surface of the dome

[a] We know now that the design with one rising tube only was also current in the Han, for an example of this, with its sliding side-walls complete was found in the tomb of Tou Wan alongside that of Liu Shêng at Manchhêng (−113). A photograph is reproduced in Watson (5), no. 165, p. 107; Capon & McQuitty (1), p. 17.   [b] Anon. (94), no. 237.

[c] This morphological identity with physiological or functional inversion reminds one of the parallel case of the 'water-powered reciprocator' and the steam-engine (Vol. 4, pt. 2, p. 387, also Needham (64), p. 200).   [d] Apart from the obvious danger of clogging the tubes in any sublimation process.

and upper parts of the walls kept cool by a wet sponge. It may therefore be suggested that the purification of camphor by sublimation is the clue to the problem.[a]

Before leaving the subject of these interesting pieces of apparatus a word should be said about a remarkable 'lamp' of gilt bronze excavated in China in 1968 and since then depicted in several exhibitions.[b] It shows a serving-maid holding up an object which closely resembles the 'rainbow *têng*' in several ways, the cylindrical central compartment with place for a sliding door, the hemispherical dome, and one side-tube of ample lumen, disguised as the maid's right arm and sleeve, entering into its crown (Fig. 1429).[c] This object is precisely datable, for it was found in the tomb of Liu Shêng,[1] Prince Ching of Chung-shan,[2] who died in -113.[d] The words *chhang hsin*,[3] inscribed on the maid's left sleeve, indicate that the apparatus was part of the dowry given by the Empress Dowager Tou (imperial concubine of Han Wên Ti) to her grand-niece Tou Wan[4] (wife of Liu Shêng), and came from the Chhang-Hsin Palace, royal residence of empresses. Could the designation 'lamp' perhaps be wrong? Might it not be an ornamental one-tube rainbow *têng*?

(7) DISTILLATION AND EXTRACTION APPARATUS

(i) *Destillatio per descensum*

We have spoken above of the sublimation of mercury formed in a closed space from its sulphide, cinnabar. The other ancient and early medieval method of preparing it was known as *destillatio per descensum*. A number of flask-shaped pots were filled with cinnabar ore, plugged loosely with moss, and inverted over a second series of pots buried in the earth. Under the influence of strong heat from above, the mercury liberated by oxidation dropped down into the lower receivers while the sulphur dioxide escaped through the porous plugs and walls of the 'retorts'. We do not know when this process was first used in Europe.[e] However, it clearly appears in *TT*884, a text of the Thang period. Writing about $+690$, Chhen Shao-Wei says:[f]

[a] The inscription just given points rather clearly to a connection with temple incense. On this see pl. 2, pp. 134 ff.

[b] We saw it first in Hongkong in Sept. 1971, and are much indebted to Mr Li Tsung-Ying for the photograph here reproduced. It was soon afterwards published in Hsia Nai (6). The tomb of Liu Shêng and Tou Wan, at Manchhêng in Hopei, contained rich finds including certain objects previously unknown in Chinese archaeology. Cf. Anon. (*106*), pl. 1.

[c] The 'lantern' and the head of the maid of honour holding it can be taken apart.

[d] He had thus been a contemporary of Liu An, Prince of Huai-nan.

[e] It must surely have some connection with the age-old winning of tar and pyroligneous acid from the kilns of charcoal-burners. In the form of the *bot-bar-bot* (p. 33 above) it is in al-Rāzī ($+9$th century). It is seen again in the apparatus called *koṣṭhi-yantra* and *adhaspātana-yantra* described in the *Rasaratna-samucchaya*, and it figures also in Geber, both works being of $+1300$ or a little before. For the former see Ray (1), 2nd ed., fig. 30*a* ii and p. 172; fig. 30*b* ii and p. 189. For the latter see Darmstädter (1), pp. 50–1, 115 and pl. IX, from *Summa Perfectionis*, ch. 48 and *Liber Fornacum*, ch. 5. Conrad of Megenburg in the $+14$th century describes the use of descensory distillation for the preparation of juniper oil. On the whole subject see Schelenz (2), pp. 14, 29, and fig. 5. Could it not be that the perfora-tion of the middle septum of Chhen Shao-Wei's bamboo was done in direct analogy from the ancient steamer vessels (p. 27 above)?

[f] P. 1*b*, tr. auct. The passage is also quoted in *YCCC*, ch. 68, p. 9*a*, from a tractate with the title (probably abbreviated) *Chiu Huan Chin Tan*.[5]

[1] 劉勝   [2] 中山靖王   [3] 長信   [4] 竇綰   [5] 九還金丹

From 1 catty (16 ozs.) of *kuang ming sha*[1] (a high-grade cinnabar) 14 ozs. of mercury can be extracted.[a] The method is to make a tube from a stem of young bamboo so that it has three septa in all. Pellet-sized perforations are made (in the uppermost septum), and small holes about the size of the thick end of a chopstick in the middle one, to enable the mercury to flow downwards. First two layers of waxed paper are placed over the middle septum. Then the finely ground cinnabar is introduced into the (upper section of the) tube.[b] The whole is next wrapped round with hempen cloth and steamed for one day before being plastered over with yellow clay to a thickness of about 3 ins. It is buried underground so that its upper end comes level with the surface. The tube must be tightly sealed all round to prevent leakage. Firewood is then piled on top and burnt for one day and one night until the heat has thoroughly penetrated the upper section (of the tube). Mercury will flow into the lower section without any loss.

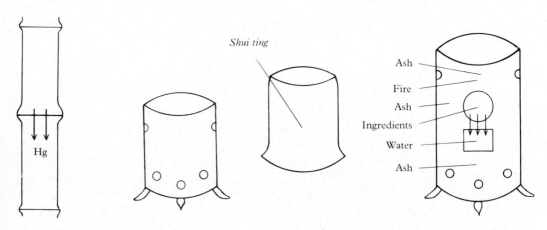

Fig. 1430   Fig. 1431

Fig. 1430. Bamboo tube for descensory distillation of mercury, from *Ta-Tung Lien Chen Pao Ching...* (Thang).

Fig. 1431. Stove arranged for descensory distillation; drawings from the *Kan Chhi Shih-liu Chuan Chin Tan*, a Sung work.

A very similar method appears in *TT*902, most probably a Sung text.[c] A conjectural diagram of the apparatus is seen in Fig. 1430.

This bamboo method is so evidently ancient and rustic that it invites a reconsideration of the antiquity of the knowledge and applications of mercury in China,[d]

---

[a] Note the quantitative figures, not at all uncommon in medieval Chinese texts on alchemy and chemical technology; cf. pp. 300ff. and the discussion in Sivin (2). The theoretical value is 13·8.

[b] Presumably the uppermost septum was replaced as a lid, its holes permitting the entry of air and the escape of SO_2.

[c] We should like to draw particular attention to this ingenious use of the ever-helpful bamboo in technology (cf. Vol. 4, pt. 2, pp. 61ff.). We shall refer again shortly (p. 164) to the value of bamboo tubing for the Chinese alchemists and chemical technologists. As a natural tube its availability doubtless led to the widespread and characteristic employment of the sighting-tube in Chinese astronomy (Vol. 3, p. 352), and it was also responsible for the first 'barrel-guns' in the shape of 'fire-lances' which developed in the Wu Tai and Sung periods (see Sect. 30 in Vol. 5, pt. 1).

[d] See Vol. 5, pt. 3, pp. 4ff.

[1] 光明砂

especially in amalgamation gilding and silvering.[a] All the evidence we have presented in previous volumes points to the beginning of this art about the −4th century, the late Chou period and the time of Tsou Yen; concrete chemical-archaeological results now available support this. Lins & Oddy (1), applying emission spectrography to gilded metal objects of known date in the British Museum, have found mercury traces on Chinese belt-hooks and ornaments from the −3rd century onwards, but none on Greek and Roman articles before the +2nd.[b] Andersson has reported amalgamation gilding on a bronze as early as the Shang period in date,[c] but this has not yet been chemically confirmed. Amalgamation silvering on mirrors may be accepted fairly safely for the Chhin and early Han, however.[d] It seems that at present other culture-areas do not compete; Sassanian Persia starts only in the +3rd. Thus the simple bamboo descensory still, so pure and primitive, may be a kind of relict witness of the capabilities of China's early chemical craftsmen.[e]

In TT904, another Sung text,[f] we find an illustration of a lu[1] (stove) with a three-legged support, and a second diagram to show how fire was conserved in the apparatus, presumably to produce a low but uniform and sustained heat. The same set of drawings, reproduced in Fig. 1431, also includes the diagram of a shui ting,[2] which in this case was probably a receiver partly filled with water. The text explains:[g]

The lu[1] (stove) is made of earth or baked clay, hollow inside, and measures 2 ft. 2 in. high and 1 ft. 2 in. in diameter. There are two openings near the top and three openings near the base. The shui ting[2] (water-receiver) is made of porcelain with a capacity of about 3 pints, and the mouth fits exactly that of the reaction-vessel. Whenever the shui ting (water-receiver) is used it should be filled to 6/10th or 7/10th of its capacity with boiling water. The ash inside the combustion chamber is obtained from burnt paper.

All this is clearly another method of destillatio per descensum. Incidentally it may also give further evidence of the close-fitting surfaces or flanges which we referred to above.[h]

In TT912, a text which may be of +808, we find another illustration[i] showing the method of yang huo[3] (conserving the fire); cf. Fig. 1432. Here the reaction-chamber is

[a] See Vol. 4, pt. 1, p. 91; Vol. 5, pt. 2, pp. 62, 67, 206, 232–3, 243–4, 246–9, 274ff.; Vol. 5, pt. 3, pp. 123, 207.

[b] They therefore throw a little doubt on the passages from Pliny and Vitruvius which we accepted in Vol. 5, pt. 2, p. 248.

[c] (8), p. 37.

[d] Here Lins & Oddy were right, but for the wrong reason. They cited the Khao Kung Chi chapter of the Chou Li (ch. 11, p. 20b, probably of the −2nd century), which they knew from a quotation in the text of the Thien Kung Khai Wu (ch. 8, p. 4b, +1637), translated by Sun & Sun, p. 165, but they did not notice where the quotation ended. In fact there is nothing about mercury gilding or silvering in the Chou Li here (cf. Biot's tr. (1), vol. 2, pp. 491 ff.). But another work of similar date, the Huai Nan Tzu (c. −120), has a passage (ch. 19, p. 7b) which has always been interpreted as referring to amalgamation silvering (cf. Morgan's tr. (1), p. 231). Tinning, however, cannot be excluded; cf. Vol. 5, pt. 2, pp. 232–3. It would be extremely interesting to apply the technique of Lins & Oddy to the numerous Chou and Han mirrors available in museums.

[e] On the history of the winning of gold and silver by amalgamation see Teich (3). The question of evidence for the use of the process in the Roman Empire, or in the Hellenistic proto-chemical Corpus, or in ancient China, needs a good deal more study.

[f] P. 2a, b.   [g] Tr. auct.   [h] P. 24.   [i] Ch. 1, p. 10b.

[1] 爐   [2] 水鼎   [3] 養火

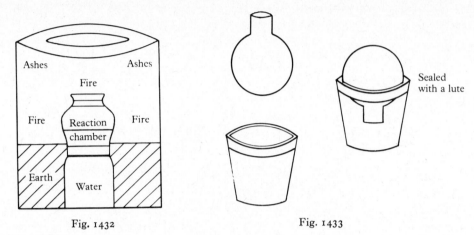

Fig. 1432   Fig. 1433

Fig. 1432. Furnace arranged for descensory distillation, from the *Chhien Hung Chia Kêng*... of +808 (Thang).

Fig. 1433. 'Pomegranate' flask (an ambix) used in descensory distillation, from *Chin Hua Chhung Pi Tan Ching Pi Chih* of +1225 (Sung).

surrounded on all sides, except the base, by fire and ashes, while the bottom is in contact with a buried receiver containing water.

In *TT*907, the Sung text so often quoted here, we can see a further example[a] of *destillatio per descensum* closely similar to the classical description in Agricola.[b] It shows a diagram of a *tzhu shih-liu kuan*[1] (pomegranate-shaped porcelain vessel); see Fig. 1433. This was simply a porcelain flask (in fact an *ambix*, ἄμβιξ) with a porous plug, used for converting cinnabar to mercury by downward distillation. The flask containing cinnabar was inverted over a *kan kuo tzu*[2] ('crucible'), analogous to the *lopas* (λοπάς), which contained some vinegar. The joint was made air-tight using a lute (equivalent to the *pēlos*, πηλός).[c] Fire was then applied above. The mercury formed passed through the porous plug and fell into the crucible.[d]

To end our discussion of descensory distillation we may give two further examples from the Sung period. First, in the *Pên Tshao Thu Ching*[3] of +1062 the following passed the keen scrutiny of Su Sung's editorial eye:[e]

Take a jar made at Yang-chhêng[4] and fill it with cinnabar mixed with small pieces of hard charcoal. Cover the mouth of the jar with a piece of iron sheet that has been perforated with small holes. Hold the iron sheet in position by fixing a length of iron wire around the jar.

---

[a] Ch. 2, p. 4*a, b.*   [b] Hoover & Hoover tr., pp. 426 ff.

[c] For the background of the Greek terms see Sherwood Taylor (5), p. 188. They were equally applicable to the Dioscoridean sublimation process.

[d] The whole apparatus was called a *wei-chi*[5] (cf. p. 68).

[e] Tr. auct., from the quotation given in *Pên Tshao Phin Hui Ching Yao*, ch. 3, (p. 155). This seems a better text than that in *PTKM*, ch. 9, (p. 56) or in *CLPT*, ch. 4, (p. 107. 1), though the latter adds an illustration (Fig. 1434).

[1] 磁石榴罐   [2] 甘堝子   [3] 本草圖經   [4] 陽城   [5] 未濟

Then invert the jar and place it over another similar jar containing water in such a way that the two come into contact mouth to mouth. Apply a lute composed of salt, clay and pig's hair all over the upper jar, and especially round the rim where the two jars meet. After the lute has dried bury the lower jar in the ground so that the rim appears about an inch above the earth. Then build a stove surrounding the upper jar so that fire can be applied all about it to heat the contents. Let four openings be made, one on each side of the stove, to supply air for the burning. After heating for two hours the mercury will trickle down into the lower jar.

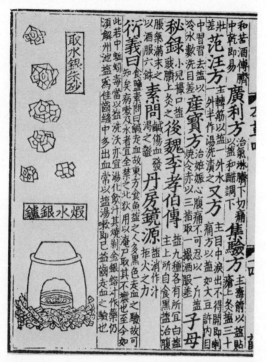

Fig. 1434. Mercury distillation *per descensum* depicted in the *Chêng Lei Pên Tshao* of +1249 (Sung), ch. 4, p. 14b. The text on the right describes various uses of edible salt.

Secondly, we have an account in *Ling Wai Tai Ta*[1] (Information on What is Beyond the Passes), a book on the products and practices of the southern provinces and the South Seas by Chou Chhü-Fei[2] in +1178. He says:[a]

The people of Yung(-chou)[3] (in Kuangsi) turn cinnabar into mercury as follows. Iron is used to make an upper and a lower bowl-like vessel (*fu*[4]). The upper vessel holds the cinnabar, which is separated (from the lower one) by an iron plate with small perforations. The lower vessel contains water and is buried in the ground. The two are joined mouth to mouth and sealed together just at ground level. A strong fire is then applied. On being heated the cinnabar changes into vapour, and on coming into contact with water it condenses, descending thus in the form of (liquid) mercury.

[a] Ch. 7, p. 11a, b, tr. auct. Very likely the apposed iron bowls were flanged (cf. p. 24 above).

[1] 嶺外代答   [2] 周去非   [3] 邕州   [4] 釜

Thus by this time it had become customary to make both vessels of iron. A still later description seems to imply the same thing, i.e. that of Hu Yen[1] in his *Tan Yao Pi Chüeh*[2] (Confidential Oral Instructions on Elixirs and Drugs), a pharmaceutical and chemical work ascribable to the Yuan or the early Ming.[a]

(ii) *The distillation of sea-water*

Perhaps the oldest thoughts of ascending distillation arose in connection with the question of obtaining fresh water from sea-water. This was a subject much discussed through many centuries.[b] Thales of Miletus in the −6th century recorded his belief that fresh water resulted from the filtration of sea-water through the earth, and the same view was among those mooted by Aristotle. In the *Meteorologica* he notes that when the vapour from salt water condenses it does not give salt water again.[c] The idea, expressed for example by Hippocrates, that liquids including sea-water could be made sweet by boiling,[d] may have arisen from some misunderstanding of a primitive distillation. Aristotle also believed that a closed wax bottle left in the sea would be found to contain fresh water,[e] and this idea of the efficiency of filtration through wax persisted down even to the late +18th century, though many times experimentally disproved.

We are brought nearer to ideas of true distillation by the observation of Pliny in the +1st century that fleeces spread out round a ship became moist with evaporated water, and that fresh water might be wrung out of them.[f] St Basil on his travels in the +4th century reported that sailors boiled sea-water in a vessel over a fire, suspending sponges above to catch the condensing vapour.[g] The method long endured, discussed by Abū al-Qāsim al-Zahrāwī (Abulcasis) in the +10th century, and illustrated by Conrad Gesner in +1555 (Fig. 1435).[h]

The first description of sea-water distillation seems to be due to Alexander of Aphrodisias in the +3rd century, in his commentary on Aristotle's *Meteorologica*.[i] He writes of 'condensing and collecting the vapour in appropriate covers'. The +10th-century Persian physician Abū Manṣūr al-Harawī indicated distillation as a method for water desalinisation,[j] and in the +14th century John of Gaddesden wrote in his *Rosa Anglica* of four available methods: filtration of sea-water through earth, boiling and condensation of the vapour on linen, distillation with alembics, and filtration

---

[a] Quoted in *PTKM*, ch. 9, (p. 56). One would like to know more about this work and its author. An account of the industrial production of mercury by descensory distillation as traditionally practised at Ise in Japan will be found in Yoshida Mitsukuni (7), p. 251.

[b] A good account is given by Nebbia & Nebbia-Menozzi (1).

[c] 358*b* 15, Lee tr., p. 157.   [d] *Airs, Waters and Places*, 8.

[e] 358*b* 34, 359*a*, Lee tr., p. 159.   [f] *Hist. Nat.*, XXXI, 70.

[g] *Homilies*, IV. This is repeated elsewhere, as by Olympiodorus, the commentator of Aristotle, in the +6th century.

[h] In the *Thesaurus Euonymus Philiatri*; Forbes (9), fig. 48; Underwood (1), fig. 29; Schelenz (2), fig. 6.

[i] 384*a* 3; Venice ed. of +1527, p. 97. Cf. Düring (1), p. 45.

[j] See Sarton (1), vol. 1, p. 678. Further references will be found in Nebbia & Nebbia-Menozzi (1), p. 135.

[1] 胡演   [2] 丹藥祕訣

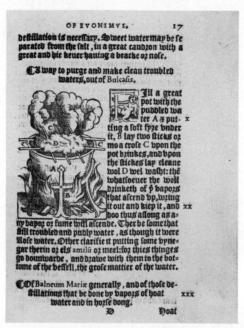

Fig. 1435. Condensing the distillate from sea-water in a suspended fleece, an ancient method illustrated by Conrad Gesner in his *Thesaurus Euonymus Philiatri* (+1555); from the English edition of +1559. He had it, he says, from Abū al-Qāsim al-Zahrawī (d. *c.* +1013), the great Moorish medical encyclopaedist.

through a wax vessel.[a] Two hundred years later the Spanish physician Andrés Laguna listed the same four methods.[b] During the +16th, +17th and +18th centuries many descriptions of stills for the production of fresh from sea-water on board ship or in states of siege were written. One difficulty often remarked upon was the unpleasant bitter taste of the distillate. Attempts to overcome this, sometimes successful, were made by adding to the sea-water ingredients (perhaps of an alkaline nature) the composition of which was often kept secret. Stephen Hales in +1739 described a fractional distillation, finding that only the first fractions coming over really tasted good. He therefore advised drinking only a part of the water.

So far we have not encountered any descriptions of the desalinisation of sea-water by distillation in the Chinese literature, but very possibly it was done during the great voyages of the fleets of Chêng Ho during the +15th century[c] if not at other times. Paradoxically nevertheless the old technique of the fleeces and sponges may have been relevant to the development of the characteristically Chinese and Mongol types of still, as will shortly appear.

While on the subject of distilled water, it is worth mentioning, perhaps, that this universal desideratum of every modern laboratory bench has appeared in most of the Chinese pharmaceutical natural histories since the beginning of the +8th century.

---

[a] Sarton (1), vol. 3, pp. 880ff. [b] Dubler (1), vol. 3, p. 514.
[c] Vol. 4, pt. 3, pp. 487ff.

First introduced in the *Pên Tshao Shih I* of +725, it occupies a modest place in the *Pên Tshao Kang Mu*,[a] where it is recommended for use in various children's diseases. It hides under the name of *tsêng chhi shui*,[1] water condensed in the upper parts of culinary steamers, but it was recognised long ago by Geerts.[b] Of course the chemical significance of using pure water in laboratory reactions and experiments was not understood until the period of modern chemistry.

(iii) *East Asian types of still*

The process of distillation today is unambiguously called *chêng liu*.[2] But in traditional usaget here was no one word for it, since *chêng*[3, 4] alone meant also 'to steam', as in cooking, and the name of the apparatus, *tsêng*[5],[c] so often used for a still, could also mean any kind of 'steamer' (cf. p. 26 above). The difficulties which this vagueness can create for the study of the history of the chemical arts will make themselves sufficiently apparent later on (pp. 179 ff.), but fortunately there are ways of getting round them.

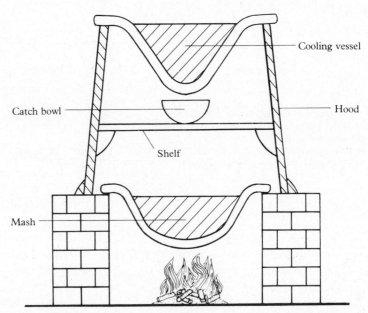

Fig. 1436. The 'Mongolian' still (after Hommel). The condensing surface is convex and the distillate drops centrally into a receiver within the still body; there is no side-tube. The shelf is of course perforated. The appellation originated ethnologically, but the assumption is natural that this was the most primitive and ancient of the East Asian types.

We must now describe the two most characteristic types of East Asian stills as found in traditional use down to the present day. The simplest form, known as the

---

[a] Ch. 7 (p. 54).   [b] (1), no. 33, p. 155.

[c] The lexicographers give *ching* or *chêng* as a special alternative pronunciation of this word when it is used for a distillation apparatus, but we have not felt sufficiently sure of the general validity of this usage to adopt it here.

[1] 甑氣水   [2] 蒸餾   [3] 蒸   [4] 烝   [5] 甑

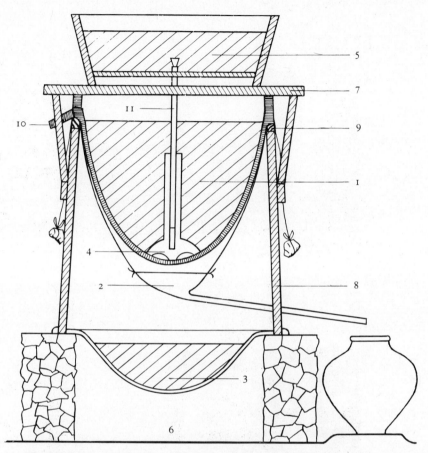

Fig. 1437. The 'Chinese' still (after Hommel). The principle is the same as that in the previous diagram, but the receiver is provided with a side-tube which draws off the distillate into a cool, or cooled, vessel outside the still body.

1. Pewter cooling reservoir.
2. Pewter catch-bowl and side-tube, suspended by cords.
3. Cast-iron bowl forming the bottom of the still and containing the mash to be distilled.
4. Inverted funnel of pewter to guide the cold water to the bottom of the cooling reservoir.
5. Wooden reservoir of cold water, a shallow tub.
6. Fire and grate.
7. Wooden frame to support the tub.
8. Wooden barrel-work forming the side of the still.
9. Tube-like ring of sewn cloth filled with sand and serving as a gasket.
10. Overflow pipe for heated water.
11. Wooden pipe with wooden stopper for letting cold water flow down into the cooling reservoir or condenser at the still-head.

Again the appellation originated ethnologically, but this type must surely be regarded as the more developed, and hence the later, of the two.

Mongol still, is seen in Fig. 1436, taken from Hommel (1). The vapours from a boiling liquid in a pan (*fu*[1]) below are condensed on the under surface of a similar pan of cold water placed above, and caught in a bowl resting on a shelf in the middle of the space formed by a wooden cylindrical barrel-like wall (*thung*[2]). Such stills are used for preparing the spirit distilled from fermented mare's milk. The more developed form

Fig. 1438. Alcohol still at Tung-chêng in Anhui (photo. Hommel), showing the external appearance of a traditional Chinese distillation apparatus. The overflow pipe for the warmed water is on the far side of the still, but the bamboo gutter into which it discharges can be seen on the right. On the left is the pewter side-tube for the distillate, prolonged by a wooden pipe which is suspended in position by a cord.

Fig. 1439. Pewter cooling reservoir or condenser vessel of a traditional liquor still, photographed upside down, at Lin-chiang in Chiangsi (Hommel). Two handles project, one on each side; that on the left is hollow and serves as an overflow pipe for conveying away the warmed water.

(Fig. 1437), known throughout China and used for making vodka-like spirits from fermented glutinous rice, kao-liang, millet or other cereals, is essentially the same except that the pewter catch-bowl is provided with a side-tube (*lou tou*[3]) forming something like an old-fashioned 'churchwarden' clay tobacco pipe, and conveying away the distillate through the wooden wall into a receiver. A wooden pipe (*mu thung*[4]) may guide the distillate into a pottery vessel standing in a small tub of cold

¹ 釜   ² 桶   ³ 漏斗   ⁴ 木筒

Fig. 1440. Pewter catch-bowl and side-tube of a traditional liquor still, photographed upside down, at Lin-chiang in Chiangsi (Hommel).

Fig. 1441. Traditional Chinese liquor still at Chia-chia-chuang Commune, Shansi (orig. photo. 1964). The side-tube delivering *pai-kan-erh* spirits from kao-liang grain is pointing towards the camera, and the distillery worker is holding another catch-bowl and side-tube with one hand while stirring the cooling water reservoir with the other.

water. This we shall refer to as the Chinese still. Both types are of course seated on a stove (*tsao*[1]). Fig. 1438 shows a Chinese still photographed near Tung-chêng in Anhui province; the delivery tube is seen on the left, and an overflow pipe from the water-cooler, assuring a constant level of cooling water, is seen on the right.[a] Fig. 1439 from Lin-chiang in Chiangsi province shows the condenser vessel upside down, with its two handles, one of which is hollow and serves as the overflow pipe. Fig. 1440 from the same place shows the catch-bowl and side-tube upside down. A more recent

---

[a] On the matter of overflow pipes to secure constant-level conditions in reservoirs, see Vol. 3, pp. 316 ff., 324, for the great importance of this development in China with regard to water-clock (clepsydra) technology.

[1] 竈

photograph of a still and a spare catch-bowl held by one of the workers (Fig. 1441) was taken at the Chia-chia-chuang Commune near Thaiyuan in Shansi in 1964. It was producing *pai-kan-erh* spirits (cf. p. 142) of 55 % alcohol content from kao-liang grain.

Representations of these stills in literature appear to be rare. The only Chinese-style illustration of a Chinese still which we have found is contained in the *Nung Hsüeh Tsuan Yao*[1] (Essentials of Agricultural Technology) written as late as 1900 by Fu Tsêng-Hsiang[2] but rather traditional in character.[a] We reproduce this in Fig. 1442. Here it is shown in connection with the making of the essential oil of peppermint (cf. p. 117). It is unfortunate that ch. 17 of the famous *Thien Kung Khai Wu*[3] (Exploitation of the Works of Nature) written by Sung Ying-Hsing[4] in +1637, which deals with wine-making, does not include a picture of a Chinese-type still.

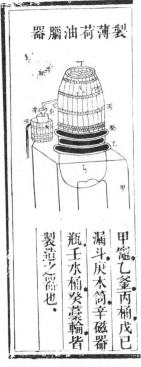

Fig. 1442   Fig. 1443

Fig. 1442. Traditional still from the *Nung Hsüeh Tsuan Yao* (1901) by Fu Tsêng-Hsiang; one of the rare Chinese-style illustrations of the apparatus. The dotted lines which show the side-tube within the still body are in the original. The drawing occurs in connection with the distillation of the essential oils of peppermint, *Mentha arvensis* and other species. The text below describes the parts; the catch-bowl and side-tube being called *lou-tou*.

Fig. 1443. A wine-distilling scene from the frescoes at the cave-temples of Wan-fo-hsia (Yü-lin-khu) in Kansu, dating from the Hsi-Hsia period (+1032 to +1227). Copy-painting by Tuan Wên-Chieh. If there is no side-tube in the original, the still must be of the Mongolian type.

a Ch. 2, p. 31*b*.

[1] 農學纂要   [2] 傅增湘   [3] 天工開物   [4] 宋應星

Apart from the drawings in the alchemical books in the *Tao Tsang*,[1] which we shall study presently, we know of no other medieval Chinese literary representations of stills. But it is to be hoped that further studies of the frescoes depicting daily life on the walls of temples and cave-temples in various parts of China may bring to light some paintings of distilling apparatus. We can here offer but one (Fig. 1443), a scene of wine-making from the frescoes of the cave-temples at Wan-fo-hsia[2] (Yü-lin-khu[3]), in Kansu province.[a] This dates from the Hsi-Hsia[4] period (+1032 to +1227). We suspect that a closer study of the painting would show a side-tube leading to the cooling bucket on the right; if not, it is a painting of a Mongol rather than a Chinese still and has a catch-bowl set centrally.

Fig. 1444. A late + 18th-century coloured MS. drawing of a Chinese wine still (Victoria and Albert Museum). Side-tube on the left, filling-tube with bung on the right.

Pictures of what are presumably Chinese stills drawn by Europeans, or in European style by Chinese, during the past two hundred years, are not uncommon. Usually one sees only the side-tube and the large bowl of cooling water at the top of the apparatus. Fig. 1444 shows a late +18th-century coloured MS. drawing in the Victoria and Albert Museum[b] which was undoubtedly the original for the coloured print embodied in the book of G. H. Mason (1800).[c] The vessels on the left bear the characters *chhang li*[5][d] (ordinary spirits for formal occasions).[e]

---

[a] From the booklet of reproductions issued by the Tunhuang Research Institute in 1957, ed. Tuan Wên-Chieh (1).

[b] Print Room II-20; no. 31, D 83–1898.　　　　　　[c] (1), pl. xxiv.

[d] This *li* properly means a sacrificial vessel.

[e] A small picture in exactly the same iconographic tradition, but cruder, will be found (in the English edition but not in either of the Chinese editions) of Li Chhiao-Phing (1), p. 199. Such representations, with varying degrees of artistry, are quite common; e.g. Gray (1), vol. 2, p. 140 (Fig. 1445). No sources

[1] 道藏　　　[2] 萬佛峽　　　[3] 楡林窟　　　[4] 西夏　　　[5] 常豐

Fig. 1445. Traditional-style drawing of a Chinese liquor distillery, from Gray (1878).

We have not made any great search to establish the customary alcohol-content of distilled Chinese spirits, but Deniel in Vietnam found as much as 54%,[a] while Li Chhiao-Phing gives 65% as a usual figure for North China kao-liang spirit.[b] About alcohol and its history we shall have a good deal more to say later on (pp. 121 ff.).

(iv) *The stills of the Chinese alchemists*

We are now in a position to examine the distillation apparatus described and figured in the medieval Chinese alchemical texts. In general we have two widely-mentioned forms of equipment, the *wei-chi lu*[1] (imperfect accomplishment stove) and the *chi-chi lu*[2] (perfect accomplishment stove).[c] Fig. 1446 (*a*) and (*b*) are taken from *TT*893, a work of the +12th century.[d] Unfortunately, the text, as it has come down to us, does not describe the function of these two complicated pieces of apparatus. According to Tshao Yuan-Yü (*1*), Li Chhiao-Phing (*1*) and Huang Tzu-Chhing (*1*), however, in the *wei-chi lu* (*a*) the ingredients were contained in A, cold water was contained in B, and the left-hand object C served as an inlet for water and also as an outlet for steam. The upper section where they think fire was applied has indeed a perforated top. Heating would thus have occurred around A while the space surrounding B was filled with ashes. These interpretations were those of the three modern authors and do not arise directly from anything in the accompanying text, while no explanation for the right-hand object was forthcoming from either. The nature of an operation carried on in this way would thus remain extremely puzzling. On the other hand, the second apparatus, the *chi-chi lu* (*b*), occasions little uncertainty. In the *chi-chi lu* the ingredients were contained in B and cold water in A, while fire, they think (we believe rightly), was

are ever stated. The fact that the side-tube generally comes off at what may seem a very low position is not inconsistent with the Chinese still pattern, as witness Fig. 1437.

---

[a] (*1*), p. 86. This figure is about that for ordinary brandy, whisky and rum. The wines which he analysed gave very variable results, and it is not clear whether all of them were in fact distilled. Some had only about 20% alcohol, equivalent to the 'fortified' wines of Europe.

[b] (*1*), p. 208. This would approach vodka in alcohol-content. Useful information on the strengths of wines and spirits in general can be found in Ure (*1*), vol. 1, p. 59, and in the book of Ribereau-Gayon & Peynard (*1*).

[c] We shall return in a moment to the significance of these strange names.

[d] P. 9*a*.

[1] 未濟爐   [2] 既濟爐

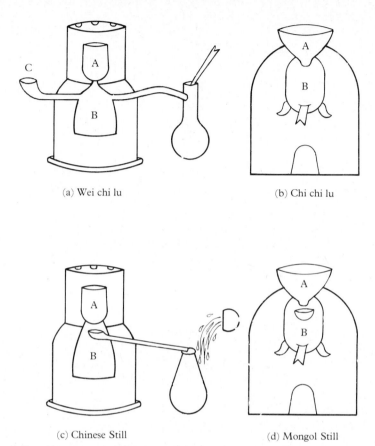

(a) Wei chi lu

(b) Chi chi lu

(c) Chinese Still

(d) Mongol Still

Fig. 1446. East Asian stills (*a, b*) from the *Tan Fang Hsü Chih* of +1163 (Sung). Below (*c, d*) re-drawn in interpreted form (see text).

applied underneath B. All these forms of apparatus were sealed by using a lute or sealing-compound.

Although it might seem strange to do a great deal of heating from above, there can be no doubt from a general survey of these texts that the Thang and Sung alchemists did practise alternate heating from above and below, thus inverting (as they believed they had to do) the positions of Yin and Yang in their apparatus. They might even be told to alternate the procedure each day for two months.[a] The two arrangements certainly derived from the ancient techniques of descensory distillation on the one hand, and of sublimation to a cold surface (later, distillation) on the other. It may at the outset be admitted that not every procedure carried out by the medieval Chinese alchemists will necessarily make sense in modern terms when all the results are in, because many of their ideas and presuppositions were not the same as those legitimated by modern science; but nevertheless it is our duty in the meantime to interpret them as reasonably as possible. By a kind of extension from William of Ockham one

[a] Cf. *TT*907, p. 15*b* and pp. 74, 283 below.

should always assume science until proto-science is inescapably proven. We have to allow also for some distortion and misunderstanding in the transmission of the illustrations, and some corruptions in the texts. We may also be able to descry, through the apparatus depicted by the alchemists, techniques perhaps more practical used by the brewers, food industry workers, pharmacists and physicians contemporary with them. Their designs, in a word, did not come out of the blue.

From the explanation of Tshao, Li and Huang it is difficult to see what purpose was served by the vessel shown so clearly on the right-hand side of Fig. 1446 (a). It must surely have been the receiver of a still. Moreover, the unexplained slanting lines on the right which at first light look like a tapering tube are plausibly to be interpreted as a stream of water poured on to the receiver to cool it.[a] One of the Indian pictures referred to below (p. 104) has exactly this.[b] Our Chinese diagram may well have become distorted by draughtsmen who did not quite understand what they were drawing or copying. On the other hand the 'inlet' C may have been drawn projecting outwards and the horizontal tube shown as continuous with the body of the still not so much for the purpose of 'deceiving the experts' (as practised by the Greek metallurgical artisans), but to 'discourage the layman'. To quote from the preface of TT894 (a text of +806):[c]

> Whereupon I realised that the sages do not desire their subtle and efficacious methods to be understood by the common people who happen to come across them by chance. They have intentionally made their processes involved so that the wise would diligently pursue them, while the average person would leave them alone and indeed scorn them.

Alternatively, again, the catch-bowl and side-tube was hung up outside and in front of the still when the draughtsman uncomprehendingly drew it. If our guess is correct then, we obtain immediately the two types of East Asian still, as shown in the two conjectural diagrams in Fig. 1446 (c) and (d). A would then be the water-condenser vessel in both cases and B would be the body of the still. The *chi-chi lu* would then be in effect the Mongol still, with a simple catch-bowl at the centre, while the *wei-chi lu* would be the more developed type or Chinese still in which the distillate is conducted away from the catch-bowl by a side-tube.[d]

To understand these names better it is necessary to turn to our discussion of the *I Ching* (Book of Changes) in Vol. 2. There it will be seen that Chi-Chi and Wei-Chi are the last two of all the *kua* (the hexagrams),[e] bringing up the rear like St Sylvester in the calendar of the Western Church, and not without some of the associations that tend to gather round his name. Chi-Chi signifies consummation or perfect order,

---

[a] Another possibility which has been suggested is that they represent a feather such as was used by the alchemists for collecting sublimates from the surfaces on which they had condensed. This seems much less convincing.

[b] The *tiryakpātana yantra* (Ray (1), 2nd ed., fig. 30e i, and p. 190).

[c] Tr. auct.

[d] The only other conceivable possibility which has occurred to us for the interpretation of Fig. 1446(a) is that the central part could be a two-armed dephlegmator vessel like that in Fig. 1473(a), with the still itself invisibly on the left and the final receiver shown on the right. But this is negatived by many things, the absence of any other reference to such a device in China, and the stated presence of fire within the stove-like container.

[e] See Table 14 at Vol. 2, p. 320, as also, for the relations with alchemy, p. 331.

completion, equalisation and successful accomplishment, but in a figure full of symbolism it cedes the final place to Wei-Chi, the *kua* which signifies disorder capable of consummation, order, equalisation and perfection, the position, in fact, when all has not yet quite been successfully accomplished. Why then should the 'Mongol' still have been called a *chi-chi lu*? Perhaps because it formed a unitary and perfect whole, like the philosophers' egg,[a] the distillate condensing at the very heart of the system. When the catch-bowl was provided with a side-tube (as in the 'Chinese' still), chemical efficiency was assuredly multiplied, but the pattern lost its symmetrical perfection, the distillate being conveyed away to a receiver outside the microcosmos. Perhaps the name *wei-chi lu* thus betrays a certain disapproval on the part of the Taoist mutationists and symbolists, even though their alchemical colleagues were not thereby deterred from accepting a radical technical improvement. But it would be premature to accept this interpretation, and one must leave open the possibility of some alternative connection, not so far clear, with the relative positions of the heat source and the cooling system. Certainly in *TT*907 one of the systems with upper cooling, where the fire must clearly have been applied below, was called a *chi-chi* (see pp. 37, 43 and Fig. 1409), while the simple apparatus for descensory distillation (see p. 58 and Fig. 1433) was called a *wei-chi*.[b] It remains to be seen whether these terms were invariably used in this way or not; they certainly do not always appear where (on this view) they might be expected.[c] In any case, whatever the explanation of the names may be, the drawings with the long side-tubes do, we believe, betray the existence of the Chinese still at the time when the texts were written.

In another book (*TT*895), we find a further diagram of what seems to be a *wei-chi lu*, reproduced in Fig. 1447(*a*).[d] This work is ascribed to the +4th century (Chin), the time of Ko Hung and Zosimus, though presumably it must be somewhat later.[e] The name *wei-chi lu* does not actually appear in the text, and we use it only by analogy. Now if we reconstruct this drawing on the assumption that the Chinese still was what was being depicted, we get Fig. 1447(*b*) showing the side-tube in place. Thus one might suppose (as in the case of Fig. 1446) that the churchwarden-pipe-like bowl and side-tube had been hung up in front of the apparatus before the artist drew it, no doubt in the elaboratory of some Taoist temple. Unfortunately no help whatever is

[a] Cf. Cheppard (2, 7), and pp. 17 ff. above. Actual egg-shells were sometimes used as containers or reaction-vessels by the Chinese alchemists (cf. pp. 292 ff.).

[b] Perhaps it was considered more high and philosophical to make things go up rather than persuading them to flow down.

[c] The Chi-Chi and Wei-Chi hexagrams will be encountered again in our presentation of the theory of physiological alchemy (Vol. 5, pt. 5 below) where they play an important part. One feels also that there must have been some connection between the alternation of the positions of water and fire (Yin and Yang), above and below, in these physical apparatuses, and the principle of inversion (*tien tao*) so fundamental in physiological alchemy (see pt. 5).

[d] P. 1*b*. On the fire-water theory the name would imply that the heating should be from above, but the apparatus is then quite incomprehensible.

[e] There might be philological justification for considering it of the Liu Chhao period or at least Thang, rather than Sung or Yuan. Further study of the text could perhaps date it more closely by internal evidence. The question is of much importance because of the obscurity which veils the historical origins of the Chinese still (cf. pp. 78–9, 155), and the strong probability arising from textual evidence that alcohol was being regularly distilled from the +7th century onwards (cf. p. 162).

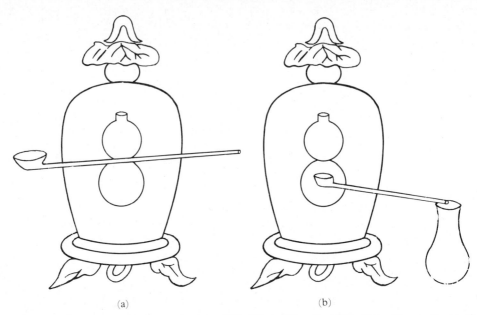

(a)   (b)

Fig. 1447. Another Chinese apparatus (a) from *Chih-Chhuan Chen-Jen Chiao Chêng Shu*, ascribed to the Chin period (+3rd or +4th century), but more probably Thang (+8th or +9th). Redrawn in (b) interpreted as a still of 'Chinese' type.

forthcoming from the accompanying text, which is concerned only with generalities and poetical cover-names for certain substances and processes.

Yet another example occurs in *TT*907, a text definitely of the Sung (+13th cent.), the very same from which we drew above so many examples of cooling-'coils' and water-jackets. Here we find two diagrams, reproduced in Fig. 1448(a) and (b).[a] They look like *wei-chi lu* and *chi-chi lu* respectively, though these names are not used in text or captions; and they represent the apparatus to be used in Turn 7 of the elixir preparation (cf. Table 116 above). Together they constitute perhaps the hardest nut to crack in the study of medieval Chinese alchemical apparatus. One could of course argue, in the case of (a), that it was a variant of the Chinese still in which the central collecting-bowl had two side-tubes (as in the reconstruction of Fig. 1448c); or alternatively that two separate catch-bowls and side-tubes had been hung up in front of the apparatus before the artist drew it. If the former interpretation were to be admitted there would be a remarkable parallel with the apparatus of the Hellenistic proto-chemists, for among them one repeatedly encounters a still having two side-tubes, the *dibikos* (δίβίκος).[b] This occurs in fact in one of the oldest texts, the *Chrysopoia* of Cleopatra, now consisting of only one page of diagrams.[c] The arrangement was presumably adopted with the object of removing distillate as quickly as possible from the hot atmosphere of the interior of the still.[d] But since the *dibikos* lacked all future

[a] Pp. 14b, 15a.
[b] Berthelot (2), pp. 132, 138; Sherwood Taylor (2), pp. 117, 136, 137.
[c] Berthelot (1), pl 1, (2), fig. 11.
[d] Sometimes indeed three side-tubes are shown (Berthelot (2), pp. 141, 161, 163).

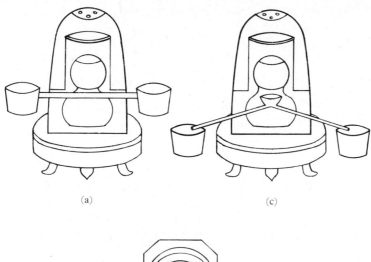

(a)   (c)

(b)

Fig. 1448. East Asian stills (*a, b*) from *Chin Hua Chhung Pi Tan Ching Pi Chih* of + 1225 (Sung). The latter presumably depicts the 'Mongolian' type; the former is redrawn in (*c*) interpreted as a *dibikos* with two side-tubes and two receivers.

in the West,[a] and since no other example of a *dibikos* in the Chinese style of still is known, either alchemical or technological, and since furthermore a dozen centuries separated the two devices, such an interpretation would imply a possible rapprochement between Hellenistic and Chinese chemical technique as fascinating as it would seem unlikely.

No, there is much that we do not yet understand about this apparatus. First, the caption for the '*wei-chi lu*' with the assumed double catch-bowls, Fig. 1448(*a*), distinctly says that the fire is to be applied above and that there is to be water below.[b] Secondly, it is clear from the accompanying diagrams that in the '*chi-chi lu*' the furnace contains an arrangement of upper water-reservoir (*shui hai*[1]) and central

[a] A badly drawn one has been illustrated from late (+ 16th), but archaic, Armenian alchemical MSS (see Kazanchian, 1). Curiously enough, as we shall see (pp. 103, 105–6), there are echoes of it in Mongolia, and several in India.

[b] And one can see furnace holes for the escape of fuel vapours at the top.

[1] 水海

cooling-tube closely analogous (with fire below) to those already shown in Figs. 1409 and 1413, and used for Preliminary Process (*a*) and Turn 9 of the elixir preparation (cf. Table 116). It therefore cannot be a Mongol still in the usual sense, though it could of course have permitted the condensation of a sublimate on the water-reservoir and its tube. This is in fact mentioned in the accompanying text, which prescribes that it should be collected and used for subsequent stages in the whole preparation. Once again, unfortunately, it gives us very little other help beyond prescribing alternate heating in the two types of apparatus each day for two months. It also makes clear

Fig. 1449. One of the forms of the *kērotakis* (reflux distillation) apparatus of the Greek proto-chemists (Marc. 299, fol. 195*v*). See discussion in text.

that the starting material is the product of Turn 6, and that the end-result of the present stage (7) is an elixir one ounce of which will convert by projection (*tien*[1]) five ounces of ordinary silver into yellow gold. One has to conclude that what exactly was going on here will not be elucidated until the book of Phêng Ssu & Mêng Hsü has been studied at leisure and translated in full.

Our assumption that the drawings of Figs. 1446(*a*), 1447(*a*) and perhaps 1448(*a*) represent (or were derived from) Chinese stills not assembled for operation seems in the context not unjustifiable. But there remains a disturbing resemblance between them and one of the Greek drawings of the *kērotakis* (κηροτακίς) apparatus (Fig. 1449).[a] As is well known, the primary form of this was a long cylindrical vessel having boiling mercury below and a palette (*kērotakis*) at the top on which was placed small pieces of copper or copper alloys; then acting as a reflux extractor the apparatus served for the preparation of the golden-coloured alloy or amalgam of copper with 13 per cent of mercury (Fig. 1450).[b] Globular forms of this vessel occur in various places in the

---

[a] Marc. 299; Berthelot (2), p. 146; Sherwood Taylor (2), p. 134.
[b] Berthelot (2), p. 143; Sherwood Taylor (2), p. 132. The above interpretation was one of Sherwood Taylor's; in later writings (e.g. (3), pp. 46ff.) he joined with others, notably E. J. Holmyard (esp. (1),

[1] 點

CONDENSING COVER

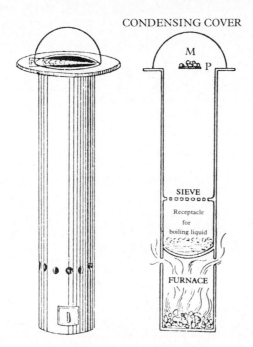

Fig. 1450. Sherwood Taylor's conjectural reconstruction of the long type of *kērotakis* apparatus. P, the 'palette'; M, the metal or other material to be acted upon by the vapours.

MSS[a] but only in one drawing does the 'palette' appear as the very elongated object bisecting the two round ones which we see in Fig. 1449. The legends in this drawing are also a little different from those on the others. The uppermost circle is labelled *phialē* (φιάλη), the standard term for a still-head, and the lowest *palaistiaion kaminion* (παλαιστιαῖον καμίνιον), heater 'a hand's length in diameter'. Below this space a fire-holder in the lowest part of the vessel, pierced with holes for ash, may be assumed. Then the 'palette' bears the words *pharmakon kērotakēs* (φαρμακον κηροτακης), 'the *kērotakis* of the drug', and immediately above it the small inner circle is labelled *kumbanē* (κυμβάνη), the cup. The thought arises therefore that this 'cup' might be the same thing as the rounded left-hand end of the long horizontal component itself, in which case the drawing of the apparatus could be considered very similar to those of the Chinese stills which we have been examining. That the word *kērotakis* could be very loosely used by the Greek proto-chemists is suggested by other places, e.g. the dialogue of Synesius (+4th century) where he directs that the still shall be placed on a hot ash bath 'which is a *kērotakis*';[b] there the word can only mean a platform like a modern

p. 47), in supposing that boiling sulphur rather than mercury was the substance at the bottom of the reflux apparatus. It is noteworthy that the analogous Indian *dhūpa yantra* is distinctly stated to have sulphur and the sulphides of arsenic as the 'solvent' (Ray (1), 2nd ed., fig. 30 *a* i, and p. 191).

---

[a] Berthelot (2), pp. 148, 149; Sherwood Taylor (2), p. 134
[b] Berthelot (2), p. 164; Sherwood Taylor (5), p. 197.

sand-bath.[a] Thus we have the possibility at any rate that the horizontal component was really a catch-bowl and side-tube (or even for part of its length a side-trough), and that the Chinese form of the still was known to the Greek proto-chemists. If so, it did not develop in later Greek or Latin tradition,[b] and it would certainly be simpler to assume two quite separate lines of evolution as suggested in Fig. 1454 below. Perhaps those who are more familiar with the Greek proto-chemical texts than we are may be able to settle this point.

Alternatively, there is the remote possibility that the Chinese pictures are representations of reflux *kērotakis* extractors intended for preparing golden-coloured Cu–Hg alloys. There is no doubt that this amalgam was known to the Chinese alchemists from an early time (cf. pt. 2, p. 243). The drawing of the Chinese apparatus, however, in all cases so clearly tubular, and also in some, e.g. Fig. 1447(a), showing the catch-bowl so clearly, seems to us to make it even less likely that the stills in the Chinese drawings were reflux extractors than that the special form of the Greek *kērtoakis* was a Chinese still.

Here one perceives a certain morphological similarity, however, between the *kērotakis* and the East Asian still. Both have a boiling liquid below (the 'Hades'), and an object held in the upper central part of the vapour zone. In the *kērotakis* the purpose of the arrangement is the effecting of chemical alteration in the object's substance or contents by the action of the vapour, as in the formation of a coloured surface-film, with the possibility also of gradual extraction of the product formed and its accumulation in the solvent at the bottom, or simply indeed the extraction of any extractable component. The *kērotakis* might therefore be considered the ancestor of the Soxhlet siphon and all other modern reflux extractors (see inset). The East Asian still on the other hand uses the central object to collect the drops that fall from the uppermost ('heavenly') condenser like rain in the meteorological water-cycle, and even to remove them from the system. It is thus the ancestor of many molecular stills of today.[c] But whether this morphological similarity of arrangement could mean anything in terms of intercultural borrowing (cf. Fig.1454h) it would be very hard to say. As we shall see (p. 331) there is a certain Chinese priority in the first beginnings of alchemy and chemical technology as compared with the Hellenistic world, and mutual influences remain possible. As a focal date we would only recall here the opening of the Old Silk Road in −110 and the facilitation of cultural and intellectual interchange so permitted. But as yet there is no positive evidence that the Mongol or Chinese still was in use as early as the Chhin and Han, nor is it obvious how the *kērotakis* apparatus, with its total absence of water-cooling, could have been stimulated by it.

Among later drawings of chemical apparatus from the Mediterranean region we have come across one which could conceivably be analogous to the drawings of

----

[a] It is true that Berthelot & Ruelle (vol. 3, p. 65), translated differently, 'such as is used for a *kērotakis*', *Corp. Alchem. Gr.* II, iii, 6.

[b] But cf. the surmise on p. 77.　　　　　　　　　　[c] Cf. p. 101 below.

Chinese stills not assembled for operation which we have been discussing. It occurs in the *Liber Florum Geberti*, that curious Arabic–Byzantine MS. in Latin, the text of which seems to be earlier than the +14th century (cf. p. 94). As Fig. 1451 shows,[a] the object seems to be a still on a tripod base, and the S-shaped tube looking like the arms of a flag-signaller might possibly be the catch-bowl and tube of a Chinese still hung up in front of it. However, it would seem very unlikely that draughtsmen would have been confronted with just such an arrangement in Europe as well as in China, and besides, there is no other evidence for the appearance of the Chinese still in the Mediterranean culture-zone.[b]

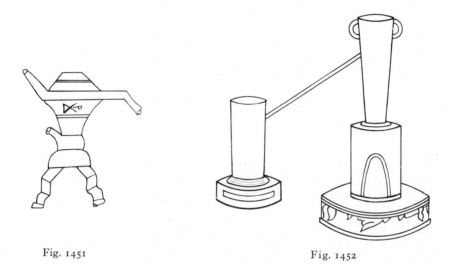

Fig. 1451   Fig. 1452

Fig. 1451. A drawing from the *Liber Florum Geberti*, a pre-Geberian Arabic-Byzantine MS. in Latin, probably of the + 13th century (München Staatsbib. Cod. Lat. 25, 110, from Ganzenmüller, 3). The apparatus could possibly be a still of Chinese type, on a tripod or four-footed base, with the catch-bowl and side-tube strung up in front of the still body, as suggested for the Chinese drawings in the preceding Figures. The cooling reservoir with its convex bottom would then be the uppermost component, and the short tube on the left would have been for replenishing the liquid to be distilled. But Ganzenmüller himself could not decide the true purpose of the apparatus, and the significance of the alchemical symbol marked on it is obscure as this does not occur elsewhere.

Fig. 1452. Mercury still from the *Tan Fang Hsü Chih* of + 1163 (Sung). The catch-bowl at the top of the side-tube is not seen, but the shape of the still-head indicates fairly clearly a cooling water reservoir or basin. Indeed, this is actually referred to in the accompanying text, which purports to be of the + 3rd century (San Kuo period), but could well be of Thang date (+ 7th to + 9th century).

The only distillation apparatus depicted in the *Tao Tsang* as set up ready for use occurs in *TT893*, a +12th-century text (see Fig. 1452). The following explanation is given:[c]

Ko the Immortal Elder (*hsien ong*[1]), (i.e. Ko Hsüan,[2] *fl.* +238 to +250, the great-uncle of Ko Hung) says: 'For the distillation of mercury the stove has a wooden frame (*chhuang*[3])

---

[a] Ganzenmüller (1), p. 293, fig. 19, no. 9, repr. from (3).
[b] Except the haunting conjecture referred to on p. 74.
[c] P. 7*a*, *b*, tr. auct.

[1] 仙翁   [2] 葛玄   [3] 床

measuring 4 ft. (in circumference, i.e. a base). The wooden legs supporting the stove are more than 1 ft. high so as to avoid the dampness of the ground. A hollow space is cut out (at the top, for the still to sit on). The closed vessel (or still, *fu*[1]) has a capacity of two pecks. Fire must be kept up at a distance of not less than 8 ins. from the vessel. The stove on the frame base should be made in accordance with the size of the closed vessel.' The commentary says: 'The uppermost part of the vessel is well covered with clay (as a lute) and rendered leak-proof. A tube for the vapours (*chhi kuan*[2]) is attached to the cover as usual. Water is filled in to the water reservoir at the top. This prevents the escape and loss of mercury'.

Fig. 1453. A retort still for mercury, from *Thien Kung Khai Wu* (+ 1637) by Sung Ying-Hsing.

These words thus imply that Ko Hsüan was familiar with the distillation of mercury in the + 3rd century. They could thus be extremely important for the history of the still in China, and their veracity is not necessarily impugned by the relatively late date of the text in which they occur because of the marked tendency of alchemical writers to copy from one another century after century. Even a cautious estimate, however, could place the text in the Thang, and that alone would be highly significant (along with Figs. 1407 and 1447) in the context of the literary evidence for the distillation of alcohol during the + 7th, + 8th and + 9th centuries (cf. pp. 141 ff. below). All indications justify the Liu Chhao period as the time when we can first be reasonably sure of distillation in China; doubt only remains concerning the Chhin and Han. Although the still-head end of the side-tube is not shown in the drawing (Fig. 1452). the fact that a water-container is clearly mentioned indicates that there must have been a central catch-bowl underneath it.[a]

We come lastly to types of still depicted in Chinese books which have no catch-bowl and depend upon the withdrawal of the vapour for condensation in a separate recipient. The mercury 'retort' in the *Thien Kung Khai Wu*[1] of +1637 (Exploitation of the Works of Nature) is perhaps the best example of this.[b] As Fig. 1453 shows, no water-cooling is provided at the still-head, and the shape of the side-tube precludes the presence of any catch-bowl, but cooling water in, and perhaps around, the receiver is vouched for by the caption adjacent in the picture. The caption on the still itself, *ku chi*,[2] means 'solidly fitted together'. It is a term often found in the alchemical and chemical literature,[c] and implies that optimal fit of adjoined edges, often but not necessarily secured with the aid of lute (*liu i ni*[3]), which we have spoken of already when discussing the flanges of the apposed halves of reaction-vessels (pp. 22, 35–6). The word *chi* clearly has a connection with the *kua* which gave their names to the two types of East Asian still (pp. 5, 6, 68, 70–1 above), signifying that the Yin and the Yang were 'completely' compensated or equalised, fitting together like a tally without a hairs-breadth of inaccuracy. As for the caption on the side-tube, it simply says 'empty bow-shaped iron tube'.

A perfect description of the apparatus is given in a much older text, from the Sung, the *Ling Wai Tai Ta*[4] of +1178. This demonstrates that the retort type of still was then employed for the purification of mercury, and therefore presumably also for its preparation from the sulphide ore. It runs as follows:[d]

The people of Kuei(-chou), (modern Kuangsi province) heat mercury to make vermilion (*yin chu*[5]). They use an upper and a lower vessel made of iron. The lower vessel is like a bowl and holds the mercury. The upper vessel acts as the cover and has a hole at the top through which a tube passes. The tube bends over and curves downwards away from the vessels. The two vessels are closely and tightly fitted together (*ku chi*[6]). The open end of the tube is made to dip into water (in a receiver). The fire is applied below the lower vessel. Under the influence of the heat, the mercury distils (lit. flies up, *fei*[7])[e] but on coming into contact with the water (the process) is arrested (i.e. it condenses)...

Chou Chhü-Fei then goes on to speak of the two grades of vermilion on the market, omitting the later sublimation of the sulphide from the purified mercury.[f] Perhaps

---

[a] Since mercury boils at 357 °C. this water-cooled head was quite unnecessary and the water in it would have been boiling fiercely all the time. Its retention here, if by inertia, would show the fixity of pattern of the Chinese still, in which cooling by water had been an essential element from the beginning. But is it not more probable that the still in question was in fact used for quite different things, such as wine, vinegar or essential oils, and that the drawing became associated with the text by mistake?

[b] Ch. 16, p. 5*b*, text p. 2*a*, tr. Sun & Sun (1), p. 280.

[c] Cf. Tshao Yuan-Yü (1), pp. 43, 52 (pp. 78, 85); Sivin (1), p. 185.

[d] Ch. 7, p. 12*a*, tr. auct. Cf. p. 59 above.

[e] This term was almost always reserved for sublimation processes (cf. pp. 4, 134–5), but perhaps mercury was an exception. If we allow this, then the distillation of mercury was certainly known to Ko Hung, *c*. +300, as we shall see. Cf. pt. 2, p. 65, pt. 3, p. 103.

[f] This is clearly described in *TKKW*, ch. 16, pp. 2*b*, 6*a* (Sun & Sun tr., pp. 280, 283, 285).

[1] 天工開物　　[2] 固濟　　[3] 六一泥　　[4] 嶺外代答　　[5] 銀朱　　[6] 固濟　　[7] 飛

a sentence or two fell out. All this may not necessarily mean that Western still-types had reached China by the +12th century, for there is no trace of the characteristic peripheral rim (cf. p. 84), but it does recall the much older Indian, Gandhāran, still-type (cf. p. 86 and Fig. 1460), which had probably had water-cooling at the receiver only. It looks, therefore, as if some time between the +7th and the +12th centuries this was recognised in China as more practical for the purpose than the stills of Mongol-Chinese type, and adopted accordingly.

As a pendant to the foregoing vistas of the industrial preparation and purification of mercury in the middle ages,[a] we may quote the words of one of the alchemical works of the Thang. In *TT878* we read that according to the *Hu Kang Tzu*[1] book:[b]

> People who make mercury by roasting (and distilling) eat much pork and drink much wine. If they did not eat this the *chhi* of the mercury would enter their stomachs and their five viscera would become stopped up. They would become unable to take food and drink, and after a long period they would suffer serious injury. Great care should be taken in these matters.

It is interesting to read this ancient warning of an industrial hazard, an occupational disease dangerous for artisans and alchemists alike.[c] The pork was doubtless prescribed on account of the fat, for mercury tends to form compounds with fatty acids, and this is the probable method of its normal absorption.[d]

(v) *The evolution of the still*

How do all these facts fit in with what is known about the general history of distillation apparatus? It is curious that the two East Asian types have never been taken into consideration in the classical theories of the evolution of the still, such as those of Berthelot[e] and Sherwood Taylor.[f] As will be remembered, they took as their starting-point the process of sublimation of mercury described by Dioscorides, where a flask-shaped vessel, the *ambix* ($\breve{\alpha}\mu\beta\iota\xi$), was inverted over an iron saucer of cinnabar resting inside an earthenware pot, the *lopas* ($\lambda o\pi\acute{a}s$); cf. Fig. 1454(*b*). This stage itself would have derived from the simplest possible combination of heated pot and lid; cf. Fig. 1454(*a*). The next development was the better fitting together of the mouths of the two vessels, and the turning-in of the rim of the upper one so as to form an annular channel for the reception of the condensate; this might be considered the

[a] On the traditional technology cf. Geerts (5).

[b] *TT878*, ch. 11, p. 4*a*, tr. auct. We have already met with the obscure personality of Hu Kang Tzu, whoever he was; cf. pp. 188, 302 and also Vol. 4, pt. 1, p. 308. The oldest work bearing his name is in the Sui bibliography, but none of them have survived.

[c] We shall have more to say of industrial diseases and their recognition in medieval China in Sect. 45 in Vol. 6.

[d] Cf. Clark (1), p. 611. Milk has always been the classical antidote for oral poisoning by mercury, which causes, too, fatty degeneration in many viscera; cf. Sollmann (1), 1st ed., p. 634.

[e] (2), p. 165. See also Berthelot (10); Berthelot & Houdas (1), *passim*.

[f] (5). It is indeed the case, broadly speaking, that nothing resembling either of the East Asian stills has made its appearance so far in Greek, Syriac or Latin alchemical or proto-chemical texts and diagrams, nor (so far as we can see) in Arabic sources either.

[1] 狐剛子

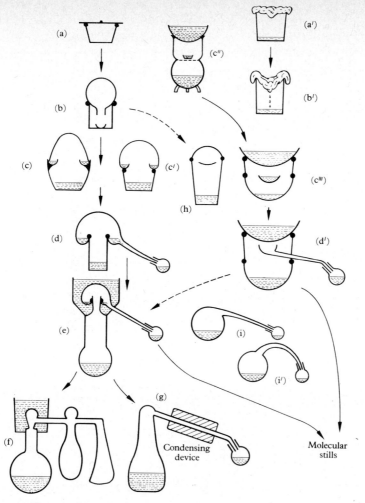

Fig. 1454. Chart to illustrate the evolution of the still (see text).

(a) Simplest combination of heated pot and lid, for sublimation.

(b) Flask-shaped vessel (*ambix*) inverted over an earthenware pot (*lopas*) containing a saucer with the substance to be heated for sublimation.

(c) Ancient Mesopotamian pot with annular rim surmounted by an inverted pot. The distillate running down is collected in the channel.

(c') A more developed form in which the edges of the still-head are turned inward to form an annular gutter.

(d) Typical Hellenistic still in which the gutter is provided at one point with a side-tube leading off to a receiver.

(a') Collection of distillate in a fleece or ball of floss above the liquid to be distilled.

(b') Conjectural central drip of distillate from such a fleece.

(c") Reconstruction of the most ancient Chinese (Mongolian) still-type; a small bowl placed on the grating of a *tsêng* or *hsien* (cf. pp. 27, 97) receives the distillate condensed on the convex bottom of a basin of cooling water placed over the mouth.

(c''') The Mongol still, with the catch-bowl held centrally within the still body in a variety of ways.

(d') The Chinese still, with catch-bowl, side-tube and receiver.

(e) The 'Moor's Head' helm or still-top, in which cooling water surrounds the Hellenistic annular rim and side-tube.

(f) Dephlegmator of medieval Europe; a second vessel intervenes between the cooled still-head and the receiver so as to condense the less volatile fractions and separate components of the distillate.

(g) Cooling condenser applied to the side-tube of the still, with no cooling at the head.

(h) The Hellenistic *kērotakis*, a reflux distillation apparatus with concave head and no cooling.

(i) Retort with cooled receiver deriving from the Gandhāran tradition.

(i') Retort with cooled receiver used in China in the Ming period.

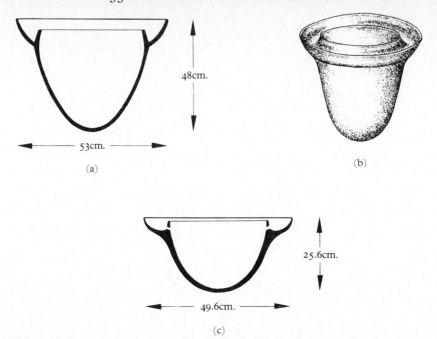

48cm.

53cm.

(a)

(b)

25.6cm.

49.6cm.

(c)

Fig. 1455. Distillation and extraction apparatus from pre-Akkadian times in Northern Mesopotamia, *c.* −4th and −3rd millennia (Levey).

(*a*) Still body of brown earthenware with annular rim for the collection of the distillate from the inside surface of another pot inverted above it. The gutter shown in the cross-section contains about 2 litres in comparison with the body capacity of 37 litres.

(*b*) Drawing of such a still body and its rim.

(*c*) Cross-section of a similar pot having holes which connect the rim with the still body. This would have been used as a continuous extractor, vegetable or animal substances being placed in the rim, and the solvent returning to the pot for re-circulation until the extraction was completed.

ancestor of Western still-heads; Fig. 1454(*c′*). Although we have shown it thus arranged, the same objective was in fact probably first achieved by giving the lower vessel an annular rim into which a domical cover would direct the condensate on all sides (*c*). Mesopotamian 'stills' of this type (Fig. 1455*a*, *b*), dating as far back as −3500, have indeed been recovered by Speiser (1) and Tobler (1) from Tepe Gawra,[a] and studied by Levey (1–4). It may possibly be that the Chou bronze tripod container with an annular rim (from about −500) already mentioned[b] was an apparatus of this kind, though, like the Tepe Gawra pots, it has lost its original cover. These Assyriologists also describe companion pieces having holes connecting the annular gutter with the body of the pot (Fig. 1455*c*); such vessels were doubtless used as extractors, the plant material to be extracted being placed in the rim.

Levey has discussed[c] the interpretation of Ebeling's translation of a group of Akkadian cuneiform tablets dating from the later half of the −13th century and around −1100, and dealing with the preparation of perfumes. It seems that myrrh,

[a] A mound site, some 15 miles NE of Mosul in Iraq.   [b] P. 30 above.
[c] (2), pp. 36 ff., 132 ff. (4).

sweet grasses, incense gums and balsams were among the materials treated with steam and hot oils. Emphasis is laid on the repeated wiping out of the inside of the pot with a handcloth and the replacement of the cover. It is suggested that this refers not to the main walls of the pot but to the upper built-in annular receiving channel.[a] If the channel had holes, soluble constituents would of course gradually accumulate in the solvent. Doubtless the type of vessel without holes was the ancestor of all those aludels with 'shelves (*itrīz*)', used as sublimatories or 'stills' by the Arabic alchemists.

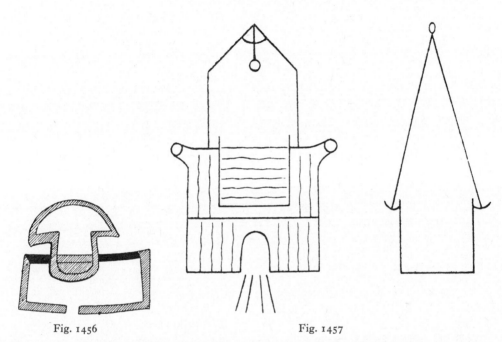

Fig. 1456   Fig. 1457

Fig. 1456. Aludel with annular shelf used by the Arabic alchemists as a sublimatory or 'still', from a text of al-Kāṭī (+1034) reproduced by Stapleton & Azo (1).

Fig. 1457. Aludels with annular shelves or gutters, from a late +13th-century MS. of Geber, *Summa Perfectionis Magisterii* (Bib. Nat. Paris, Cod. Lat. 6514; reproduced by Stapleton & Azo).

There is a drawing of one of these in the treatise of Ibn 'Abd al-Malik al-Kāṭī, written in +1034. Fig. 1456 is taken from this work.[b] Much is made of the good fit of the cover achieved by careful polishing as well as by luting the edges with clay.[c] The 'shelf' is mentioned also about +900 by al-Rāzī,[d] and diagrams of aludels with a similar gutter occur in late +13th-century MSS of Geber (Fig. 1457).[e]

[a] In view of the prominence of women in Hellenistic proto-chemistry (Mary the Jewess, Cleopatra, Paphnutia, Theosebeia), and in Chinese alchemy (cf. pt. 3, pp. 38, 42, 169, 191), it is interesting that these texts sometimes give women perfume-craft mistresses as their authorities; for example Tapputi-Bēlatēkallim of the −13th century.

[b] A translation of the accompanying text is given by Stapleton & Azo (1).

[c] Cf. p. 23.

[d] Stapleton, Azo & Husain (1), p. 386.

[e] Stapleton & Azo (1), p. 49; Berthelot (10), pp. 149, 150. Cf. Ahmad & Datta (1). Another is depicted in the late Armenian MSS discussed by Kazanchian (1).

Next came the addition of one or more side-tubes to convey away the distillate from the annular channel into cooler surroundings (Fig. 1454d); or from a cup in which it was caught (Fig. 1454d'). Perhaps the art of distillation may be said to have really begun at this point, when it was realised that by withdrawing the condensate as quickly as possible from the heated vapour the process could be made to run con-

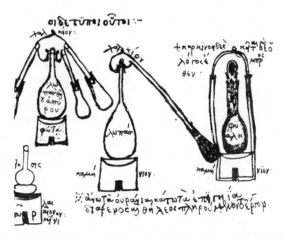

Fig. 1458. Illustration of Hellenistic apparatus from MS. Paris 2327, a copy made in + 1478 of the proto-chemical Corpus in Greek compiled by Michael Psellus in the + 11th century, and containing material from the ± 1st century onwards (cf. Vol. 5, pt. 2, pp. 16–17 and pp. 324 ff., 501 below). In the centre, a characteristic still, with its flask or still-body (*lopas*) and still-head (*chalkion*), heated on a stove (*kamēnion*). On the left, a similar still, but provided with three side-tubes leading off from the annular gutter, hence called a *tribikos*. Under this alembic, fire (*phōta*) is marked. On the right a flask (*phialē*) for digestion or reflux distillation, derivative from the *kērotakis*. For further explanation of the inscriptions see Berthelot (2), pp. 160 ff. Fol. 81 v.

tinuously until completion. By the time of the first group of Hellenistic proto-chemists such as Mary the Jewess, Cleopatra, Pammenes and pseudo-Democritus (+ 1st century), the technique had already developed as far as this (Fig. 1458).[a] Similar designs continued in use essentially unchanged until the + 18th century,[b]

[a] See Sherwood Taylor (2, 5); Berthelot (2), pp. 127ff., esp. pp. 132, 136, 161, 163.

[b] This is well seen in a page from a + 14th-century MS. in the Library of Caius College, Cambridge (181/214), p. 441, where the drawings of still-heads with annular gutters are strongly reminiscent of the Greek proto-chemical tradition (Fig. 1459). It may be desirable to recall here that we have no illustrations of these pieces of apparatus earlier than the + 11th century, the date of the most important Greek proto-chemical MS. (Marc. 299). Next in importance are the two Paris MSS (2325 and 2327) of the + 13th and + 15th centuries respectively. It is of course recognised that the apparatus they depict corresponds well with the texts of the Hellenistic proto-chemists themselves. By contrast in China we have printed illustrations going back to a *printed text* of the early + 12th century, though none of the early editions has survived. These too correspond well with texts which derive in some cases from as far back as the Later Han (+ 2nd century). Of chemists contemporary with Mary and pseudo-Democritus, and even rather earlier, we have no lack in China, as we point out in more detail elsewhere (p. 330), e.g. Li Shao-Chün, Liu An, Liu Hsiang, Mao Ying, etc., but before Wei Po-Yang in the + 2nd century and Ko Hung in the + 4th, they did not leave proto-chemical or alchemical writings which have come down to us. There is however a certain parallelism between the metallurgical material in the *Khao Kung Chi* (Artificers' Record), originally of the − 4th century, though based on traditions of much earlier times, and incorporated into the *Chou Li* in the − 2nd or − 1st; and the material on alloys in the + 3rd-century Leiden and Stockholm papyri (cf. pt. 2, pp. 15 ff.).

Fig. 1459. Page from a + 14th-century MS. of the *Turba* and of Geberian writings, in the library of Gonville & Caius College, Cambridge (181/214), p. 441. The typical Hellenistic still-head with its annular gutter and side-tube surmounting the long-necked flask and receiver can be seen in the side-drawings. The lower flask has re-entrant tubes which were probably intended to have a dephlegmator effect, returning heavier fractions to the liquid being distilled.

and the basic pattern, with a thousand modifications, permeates all chemistry and chemical industry at the present day.[a]

The only developed form of still (i.e. with a side-tube) which has come down to us as an object anything like contemporary with the Hellenistic proto-chemists is that deduced from the pottery pieces found in the excavations at Taxila in Northern India, about 1930, and described by Marshall as a 'water-condenser'.[b] The constituent

[a] Actual fragments of alembics or still-heads in glass and pottery from a number of sites in England and datable in the +15th century have been described by Moorhouse, Greenaway *et al.* (1).

[b] (1), vol. 1, pp. 149, 180, 193, 402–1 and pl. 125, nos. 127, 128, 129 and 129*a*. We are indebted to Mr John Dearlove for our first knowledge of this still, and to Dr M. Sharif, Custodian of the Taxila Museum (Pakistan Government Department of Archaeology) for correspondence on the subject.

still-heads and receivers, made of baked clay, were unearthed at Sirkap, the site on the north-west frontier of the Punjab to which the city of Taxila was transferred in the −2nd century, and which remained in occupation for three hundred years. The still belongs to levels of Śaka times, c. −90 to +25. Fig. 1460 shows the characteristic receiver-bottle and the helm or alembic, intended for fitting over the mouth of a *haṇḍī* pot, with the suggested assembly.[a] If Marshall is right and water-cooling of the receiver was actually practised, this was a set-up much in advance of its time, as will emerge shortly. The tapered character of the side-tube has also a strangely modern air.

Some twenty years later Ghosh reported a further example from Taxila,[b] and since that time several others have come to light. Then, in his recent excavations at Shaikhān Dherī, Allchin (1) has found one more, together with no less than 130 receiver-bottles, and many pots with soot on them which could have served as still-bodies. The receivers are quite capacious, holding just under 8 litres each. There were also many basins of a size which would have been suitable for cooling the receiver-bottles, and one pottery tube, ribbed as if to imitate bamboo, which would have connected a receiver with an alembic.[c] This site is at Charsada in the vale of Peshawar, and corresponds with the ancient city of Pushkalāvatī, one of the two capitals of Gandhāra, Taxila being the other. Stratigraphic and other evidence points to a date between −150 and +350. It can be seen at a glance that these stills, which may be called Gandhāran, are closely related to the Western or Hellenistic stills in that they have a concave roof at the still-head, but they lack the built-in annular peripheral rim or gutter. True, the alembic has a lower bevelled edge, and some distillate might collect between that and the mouth of the pot, but the position of the side-tube shows that there was no intention of drawing it off. The space thus served the purpose of a dephlegmator (cf. pp. 81, 93), conserving the heavier fractions. Condensation in the still-head itself was not envisaged in this design; only the vapour passed through the side-tube to condense in the receiver. The Gandhāran stills were thus essentially 'retorts', and they may well be the origin of all such forms of still.[d]

Greater cooling and the collection of more distillate would come about of itself if the side-tube were made sufficiently long. Thus we find that the still-head was dropped as a separate entity in certain cases very early among the Greek proto-chemists, for an apparatus appears in which the neck of the still is greatly elongated, and turning at a

---

[a] The still-head component looks remarkably like an inverted bed-pan, a fact which might invite the curious to embark upon a comparative history of bed-pans, were it not for their radical incompatibility with all Indian ideas of hygiene and nursing. Another possibility is that what we take to be a side-tube could have been intended for the wedging-in of a wooden handle. But the 'alembic' cannot be considered in isolation from the receiver-bottles, which would be incomprehensible if not still-parts.

[b] (1), p. 63.

[c] Occasionally the side-tubes of the alembics are fitted with studs, which would have been convenient for binding on the longer tube and the receiver-bottle; Dani (1), fig. 34.3.

[d] Wheeler (8) has described (pp. 224, 226, 228–9) similar pottery still-heads from a Mysore culture; the Brahmagiri neolithic. This means a date in the −2nd millennium, though coming down to as late as the −2nd century. At first sight it would violate one's sense of historical perspective to imagine distillation in India as far back as that, yet we must remember the Babylonian pots with annular rims (p. 82), ancestors of the Hellenistic still, as also the grated steamers of the Chinese neolithic (p. 27), in which we descry the origins of the Mongol and Chinese stills.

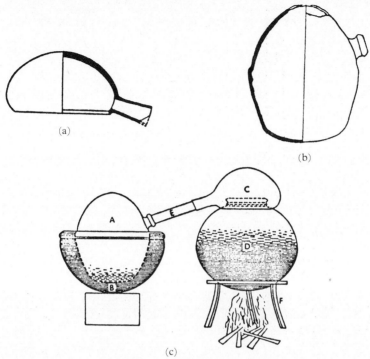

Fig. 1460. Distillation equipment found at Taxila in the Punjab, India, and dating from the ± 1st century (Marshall (1), vol. 1, pl. 125). All in pottery, the pieces consist of a helm or alembic (a, C) fitting over the mouth of a *haṇḍī* pot (D), and delivering into a receiver-bottle with only one opening (b, A). The connecting tube may well have been of bamboo. The water-bath cooling (B) of the receiver was a conjectural addition of Marshall's, but stills of the same type used traditionally in India today have it.

right angle is prolonged into a tube at the end of which is a receiver (Fig. 1461).[a] Such an arrangement, however, would have favoured the deposition of distillation-products in the side-tube, whence it would be difficult to remove them. Nevertheless it must have aided the development of the type of distillation apparatus classically known in chemistry as the retort, also called 'pelican' or 'cucurbit' because of its bird- or gourd-like shape (Fig. 1454 *i, i'*). Many figures in the herbal of Lonicerus and the books of Brunschwyk show retorts of this kind.[b] At what date it reached its most typical one-piece form in the West is not clear, but that probably started among the Arabs.[c] There is a good drawing of it in one of the Syriac MSS studied by Berthelot &

[a] See Berthelot (2), pp. 140, 163; Sherwood Taylor (5), p. 192.
[b] Cf. Forbes (9), figs. 31, 67.
[c] Cf. Schelenz (2), figs. 14, 15; Forbes (9), figs. 13, 16, 17, 18. But some of these were certainly in two parts (like Indian, Gandhāran, stills), the body (*qar'a*) and the helm (*anbīq*), as well as the receiver (*qawābil*). Retorts occur among the very Moorish illustrations in the late + 15th-century MS. *Liber Florum Geberti* studied by Ganzenmüller (1), figs. 21, 24, nos. 29, 60. This text itself seems to be Byzantine in origin with much Arabic influence, and internal evidence suggests that it dates from some time before the + 14th century (+ 1000 to + 1300). There is much on retorts also in the *Kitāb Nukhbat al-Dahr* of al-Dimashqī, written about + 1320 (cf. the translation of Mehren). Drawings of them occur, too, in late Armenian MSS (Kazanchian, 1).

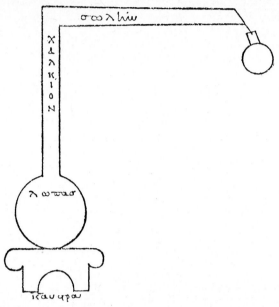

Fig. 1461. Hellenistic still with no guttered still-head but a single side-tube of large diameter leading to a receiver—the first of all Western 'retorts'. The ascending neck is still called *chalkion*, however, and the side-tube *solēn* as usual. From a description by Mary the Jewess, in Cod. Marcianus 299, fol. 194v; cf. Berthelot (2), p. 140; Sherwood Taylor (5), p. 192.

Fig. 1462. Drawing of a retort in a Syriac MS. copied in the + 16th century but containing material going back to the + 1st century, and much from the + 2nd to the + 6th (BM Egerton 709; from Berthelot & Duval (1), p. 120). The inscription says 'curved like a bow', and on the receiver (not shown here) is written 'place where it stops'.

Duval.[a] Unfortunately, although these codices date from about the + 10th century and contain some old Hellenistic material, the MS. concerned is not older than the + 16th, so that the picture in question (Fig. 1462) may be as late as that time, and indeed seems to be drawn in a rather different style from the majority of the illustrations. We know of no Chinese drawing of the retort in its typical Western shape, but the mercury still of Fig. 1453 is essentially the same thing.[b]

[a] (1), p. 120. By some remarkable inadvertence this was illustrated as Chinese by Huard & Huang Kuang-Ming (2), pl. opp. p. 24, but of course the slip was evident. It may be of interest to record that retorts of the classical form were always held among the glassware stores of biochemical and chemical laboratories in my young days, but no-one ever used them.

[b] Further research will be required to determine whether the retort in China (Fig. 1454 *i'*) was an introduction from the Hellenistic, the Arabic, or the Indian chemical traditions, or whether it arose independently from the East Asian still type as an abandonment of water-cooling for specific purposes.

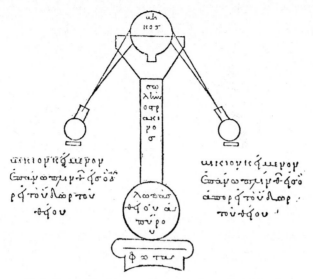

Fig. 1463. A *dibikos* or Hellenistic still with two side-tubes, from Cod. Marcianus 299, fol. 193 v. For the translation of the inscriptions see Berthelot (2), pp. 137, 139.

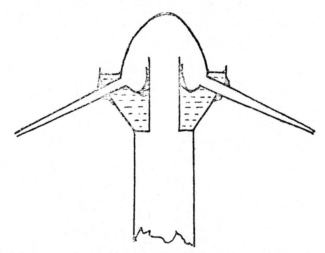

Fig. 1464. The same as reconstructed by Sherwood Taylor (5), p. 196, proposing that the uppermost funnel-shaped structure was a trough in which cold water could be placed and renewed. The Greek text, however, makes no mention of water-cooling at the still-head, and Berthelot did not assume it.

The greatest secret of the art of the still may thus be said to lie in the withdrawal of the cooled distillate as rapidly as possible from the hot vapours in its body. We have seen something of how this was achieved by the Chinese, but for the moment must continue the consideration of the course of development in the West. There are three points at which arrangements for the application of cooling water may be incorporated in the apparatus—the receiver, the side-tube or the still-head itself. The presence of

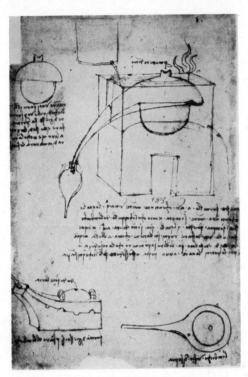

fcat capitellum
D. Capitellum.
E. Vas circundãs
capitellum , in
quod continuē
aqua frigida il-
labitur.
F. Recipiens ob-
longum.
G. Epiſtomium,
aquam calefa-
ctam educens.
Hanc fornacis
figuram , accepi
nuper à D. Felici
Platero Medico
Baſilienſi.

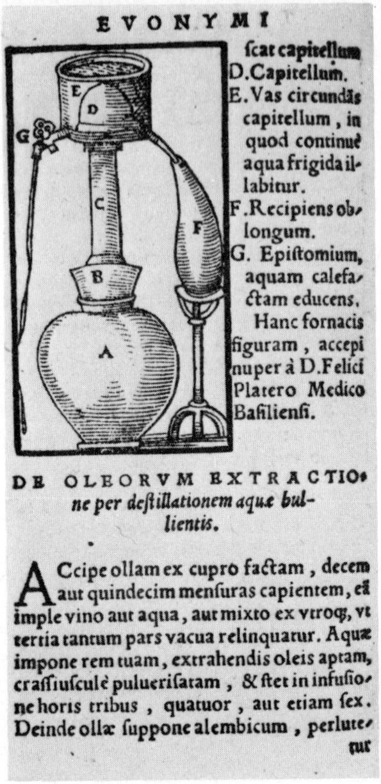

DE OLEORVM EXTRACTIO-
ne per deſtillationem aquæ bul-
lientis.

Ccipe ollam ex cupro factam , decem
aut quindecim menſuras capientem, eã
imple vino aut aqua, aut mixto ex vtroq, vt
tertia tantum pars vacua relinquatur. Aquæ
impone rem tuam, extrahendis oleis aptam,
craſſiuſculè puluerifatam , & ſtet in infuſio-
ne horis tribus , quatuor , aut etiam ſex.
Deinde ollæ ſuppone alembicum , perlute-
tur

Fig. 1465   Fig. 1466

Fig. 1465. The earliest representation of a Moor's head cooling bath, a drawing by Leonardo da Vinci, *c.* +1485 (Codex Atlanticus, fol. 400 v).

Fig. 1466. The Moor's head condenser as depicted by Conrad Gesner (*De Remediis Secretis,* +1569). A tap at the base allows for replenishment with cold water.

a water-cooled receiver in the Taxila still was conjectural.[a] There has also been much uncertainty about the structure of some of the still-heads of the Alexandrian proto-chemists. Apparatus of the type of the *dibikos* in Fig. 1463[b] was first interpreted by Sherwood Taylor[c] in 1930 as meaning that the globular still-head was luted into a funnel-shaped enlargment of the still flask itself, but fifteen years later he proposed[d] that it might represent a built-in cooling bath at the head of the still (Fig. 1464; cf. Fig. 1454 e). The only passage in the Corpus which he could adduce to justify this is in one of the Zosimus texts[e] which says that 'one should have a cup full of water at the top (or, in general), and wipe the vessel all round with a sponge'. Obviously this does

[a] But stills of just this type, with the receiver cooled in water, have been common in modern India; Mahdihassan (56), fig. 42. And those in the Tantric alchemical books are the same.

[b] Berthelot (2), p. 138. It is interesting that this feature only appears in one of the two iconographic traditions of the Greek chemical MSS, and that the annular rim only occurs in the other, the still-head in the first being always represented as a simple globe.

[c] (2), p. 137.   [d] (5), pp. 195–6.

[e] *Corp. Alchem. Gr.* III, xlvii, 2; tr. Berthelot & Ruelle (1), vol. 3, pp. 216–17

not imply a proper water-cooled jacket, but there may possibly have been some enclosure there to prevent the sponge water from running down the hot still into the furnace.

It is not until we come to the time of Geber that effective devices for cooling the parts of stills were definitively introduced in Latin Europe. In the +14th and +15th centuries two methods of cooling were used; first to condense the whole of the distillate, and secondly to divide it into fractions of higher and lower boiling-point. The

Fig. 1467. The elegant picture of the Moor's head in Mattioli's *De Ratione Distillandi*, + 1570.

so-called 'Moor's head' was a cooling-bath so fashioned as to embrace the whole of the upper part of the still, allowing all fractions to run into the internal annular channel on condensation, and thence out into a receiver (Fig. 1454 *e*). The 'dephlegmator' was a variable arrangement in which the distilled vapours were carried through air- or water-cooled tubes or vessels which detained the heavier fractions or returned them to the still, only the lighter ones passing on into the receiver (Fig. 1454 *f*). The earliest

THESAVRVS.   15

in ea funt fententia, vt exiftiment inftrumen-
ta fornifecus nul-
lo pacto refrige-
randa, neq; capi-
tellum neque ro-
ftrum, quia olea
rurfum reprimit
& in cucurbitam
decidunt , vnde
nunquam poftea
eleuari poffunt,
     Roftrum nõ fit
longius fpitamo vno, aut vno cum dimidio,
priusquam aquam contingat, alioquin cana-
li longiore exiftente, tum olea, tum aquæ nõ
nihil confumuntur.

Fig. 1468. Bladder still-head cooler, from Gesner's *De Remediis Secretis*, + 1569.

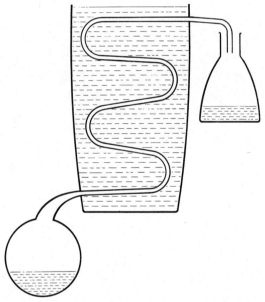

Fig. 1469. Reconstruction by Ladislao Reti of the dephlegmator described by Taddeo Alderotti in his *Consilia Medicinalia* (late + 13th century). It is simply a rising water-cooled serpentine coil connecting the still-body with the receiver.

extant representation of the Moor's head can be seen in Leonardo da Vinci's drawing, *c.* + 1485 (Fig. 1465);[a] and from the + 16th century come those of Gesner (Fig. 1466), Mattioli (Fig. 1467) and others.[b] Intermediate between Zosimus' sponge and the

---

   [a] *Codex Atlanticus*, 400 v, *c*. See Reti (7), figs. 4, 5, 6, (10), fig. 5.
   [b] Forbes (9), figs. 52, 60; Geisler (1), fig. 2. For later examples see Ferchl & Süssenguth (1), figs. 86*a*, *b*, 96, 140 (57), 143, 145.

Moor's head comes the wrapping of linen sheets periodically moistened with cold water round the top of the still, as described by Michele Savonarola, *c.* +1440.[a] Theoretically intermediate too in its simplicity (though not exactly so in time) was the system of surrounding the still-head with a large tightly-fitting bladder in which cold water was renewed as distillation proceeded. This was mentioned by Lonicerus in +1555 and illustrated by Gesner in +1583 (Fig. 1468).[b]

The earliest account of a dephlegmator (without any drawing) appears to be that of Taddeo Alderotti[c] in the late +13th century.[d] The description in his *Consilia Medicinalia* is interpreted by Reti as shown in Fig. 1469, a water-cooled serpentine side-tube rising above the still, so that all the heavy fractions would be returned into the still body.[e] A dephlegmator of this kind[f] must not be confused with the simple cooling of a descending side-tube, where all fractions will pass into the collecting vessel (Fig. 1454 *g*). The earliest extant European illustration of such a side-tube cooled by a water-jacket seems to be that of Johannes Wenod de Veteri Castro in a MS. of +1420, showing the distillation of alcohol from beer (Fig. 1470).[g] The arrangement of Brunschwyk (soon after +1500, Fig. 1471),[h] however, has the same effect as that of Alderotti; the water-cooling is applied at the upper part of the still in Moor's head style, but since there is no annular channel, the heavy fractions fall back, only the lighter going on.

The dephlegmation method was soon applied to fractionation. An early drawing of a set-up with an intermediate vessel for this purpose occurs in a Bavarian MS. of +1519 in the Jagellonian Library at Cracow, recently studied by Ameisenova.[i] As

[a] MS. entitled *Ad Divum Leonellum Marchionem Estensem Libellus de Aqua Ardenti*, not printed till +1484.

[b] *Ander Theil des Schatzes Euonymi*, pp. 14–15. The turban-like shape of this bladder and the hat-like forms of the Moor's heads suggested to the late Dr Reti an origin for the latter expression—'oriental headgear'. But to admit an 'oriental' influence is to open the door to the possibility of a deeper form of it (cf. p. 120). His impression is that the term arose later in Europe than the invention, as it seems not to occur in Gesner, della Porta, Besson, Mattioli or Lonicerus. On the other hand it may have been considered 'lab. slang' and so excluded from books.

[c] +1223 to +1303, one of the Papal physicians.

[d] Text in von Lippmann & Sudhoff (1); partial translation in von Lippmann (13), p. 1359. Cf. Mieli (3), p. 132.

[e] Personal communication in correspondence. Dr Ladislao Reti told us that he could obtain experimentally 90% alcohol in a single run with an apparatus of this kind. Yet Alderotti, as we shall see (p. 123), redistilled, taking off $\frac{5}{7}$ at each rectification. Cf. Reti (8); Forbes (9), pp. 60–1. It is puzzling to speculate on what inspired Alderotti to use a rising serpentine. He may have observed, as Dr Reti wrote to us, the partial reflux of the condensed vapours in a glass still, and felt that a longer path would offer opportunity to the 'lighter spirits' to escape from the slow and heavy 'phlegm'. But we suggest elsewhere (p. 44) influences of a more concrete kind as well.

[f] Subsequent use of rising serpentine coils is seen in the Oldanis MS. (+15th century but probably recording practices of the +14th) described by Carbonelli (1), p. 136; in Ulstadt (+1526) and the later editions of Brunschwyk (Forbes (9), fig. 56), an apparatus often misunderstood but correctly interpreted by Sudhoff (3) and Egloff & Lowry (1). Cf. also Lonicerus' set-up (+1578) in Forbes (9), fig. 67), and the account in Biringuccio (Smith & Gnudi tr., p. 348, fig. 65).

[g] Sudhoff (3), often afterwards reproduced, as in Ferchl & Süssenguth (1), fig. 30; Forbes (9), fig. 31.

[h] Ferchl & Süssenguth (1), fig. 32; Forbes (9), fig. 30; Underwood (1), fig. 13; Schelenz (2), fig. 27. But Fester (1), p. 101, and Reti (8) have alone correctly explained its purpose.

[i] Personal communication from the late Dr Zofia Ameisenova, recorded with gratitude here. The MS. (35/64) is a universal encyclopaedia, something like the *Liber Floridus*, containing coloured illustrations of many scientific and technical subjects (our picture is on fol. 52v). Fol. 82v has a remarkable mining painting with diagrammatic plans of underground workings and of a geodetic compass used

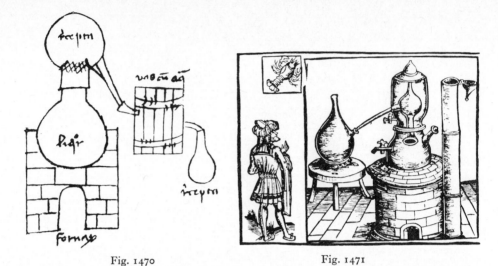

Fig. 1470   Fig. 1471

Fig. 1470. The oldest representation of a water-cooling condenser surrounding the side-tube between the still and the receiver, in a MS. of + 1420 by Johannes Wenod de Veteri Castro.

Fig. 1471. A form of Moor's head described by Hieronymus Brunschwyk soon after + 1500 in *Liber de Arte Distillandi de Compositis* (1st ed., + 1512). Here the water-cooling is applied at the still-head rather than the side-tube, but as there is no peripheral channel, the heavy fractions fall back and only the more volatile ones go on. Note the cylindrical automatic stoker on the right ('Slow Harry').

will be seen from Fig. 1472, it incorporated a rimless Moor's head cooler, which presumably restored the heaviest fractions to the still, allowing intermediate ones to be caught in the intermediate vessel. Perhaps the earliest attempt was the set of flasks along an extended side-tube drawn in a + 14th-century MS. on potable gold attributable to the court physician Albini di Moncalieri.[a] Later examples are those described by Gesner in the middle of the century (+ 1552, Fig. 1473 *a*),[b] by Lonicerus in + 1578 (Fig. 1473 *b*),[c] and by della Porta (+ 1609).[d] Vessels which might perhaps be intermediate dephlegmators of this kind occur also in the rather enigmatic *Liber Florum Geberti*, apparently an Arab–Byzantine work of the + 13th century, the time of Alderotti himself.[e]

Most of this sequence of developmental stages (quite acceptable as far as it goes), which we have diagrammatically depicted in Fig. 1454, presupposes that the cover which gave rise to the still-head was originally concave to the distillation space. The condensing distillate would then necessarily run down in all directions to the periphery, stimulating the invention of the annular gutter—something very ancient indeed, if we are to judge from the Babylonian rim-pots (p. 82). But if the original cover had been convex to the distillation space the condensing distillate would have

in them. The publication of this MS. in facsimile was planned by the late Dr Ameisenova, and will still be eagerly awaited.

---

[a] Sudhoff (4).   [b] Cf. Forbes (9), fig. 54.   [c] Cf. Forbes (9), fig. 68; Schelenz (2), fig. 51.
[d] Cf. Forbes (9), fig. 41, and from Libavius, fig. 83. For other and later examples see Ferchl & Süssenguth (1), figs. 55 (8*a*, *b*), 65, 82 (4), 140 (61), as also Geisler (1), fig. 3.
[e] See Ganzenmüller (1), figs. 18, 25, nos. 7, 69.

Fig. 1472. An early drawing of a dephlegmator arranged for fractionation, from a MS. (35/64) in the Jagellonian Library at Cracow. Bavarian, dated + 1519, it is a universal encyclopaedia of the sciences. In this picture (fol. 52 v) there can be seen at the top a rimless Moor's head cooler, the side-tube from which enters the top of an intermediate vessel before passing on to the receiver (cf. Fig. 1454 f). In such an apparatus the heaviest fractions would return to the still-body, intermediate ones would condense in the second vessel, and only the lightest would collect in the third. Apart from retorts and aludels, the lower part of the picture shows a curious reticulate arrangement of dephlegmator type but the text has no explanation.

been seen running down to the central and lowest point before dropping back into the still body, so that there would have been a stimulus to provide first a catch-bowl (Fig. 1454 c'''), i.e. the Mongol still; and then a catch-bowl with a side-tube (Fig. 1454 d'), i.e. the Chinese still. That the convex roof was traditionally a bowl (kuo^1) of cold water we know, but what could have been the origin of such a device?

Perhaps the answer was that already suggested by Sherwood Taylor (5) when he referred to the technique of obtaining *pisselaion* ($\pi\iota\sigma\sigma\acute{\epsilon}\lambda\alpha\iota\sigma\nu$), oil of pitch, mentioned by Dioscorides.[a] This consisted of stretching a clean fleece over the heated pitch and recovering the distilled oil by squeezing (Fig. 1454 a'). We have also noted already the collection of the condensate of sea-water in fleeces and sponges mentioned by Pliny and St Basil.[b] Such a method could have been known in ancient times throughout the

[a] *Mat. Med.* I, 95; Gunther tr., p. 51. The same method was used for cedarwood oil, see I, 105; Gunther tr. p. 57. Pliny discusses oil of pitch and its collection by fleeces in *Hist. Nat.* XV, vii, 28, 31; cf. his *picea resina stillaticia* in XVI, xxii, 54. Cf. Schelenz (2), p. 15.

[b] P. 60 and Fig. 1435 above. Alexander of Aphrodisias also mentions this.

1 鍋

FIGVRA RARA QVAEDAM
Alchymiſtarum, deſumpta ex libro veteri
Alchymico manuſcripto

VAſa ſeparationis; vbi quinta eſſentia; ſeu ſpiritus occultus, vna deſtillatione capitur, & eſt formaliſsimus modus, ſimulq; abbreuiatio operis maxima: conducit autem tam ad aurum potabile, quàm pro lapide philoſophorum.

In iſto furnello ad dextrã tres habẽte flammas, debet eſſe arena minuta & cribrata, & ignis ſuccenſus ſit trium candelarum.

C 2

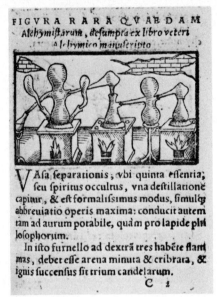

Fig. 1473*a*. A complex dephlegmator of five vessels depicted by Conrad Gesner in + 1552, from *De Remediis Secretis* (+ 1569). Three of the 'vases of separation' being heated on sand-baths, opportunity is afforded for the heavier fractions to condense in the successive vessels, and only the lightest fractions to accumulate in the receiver, though no still-head cooling is shown at any point. The bulbous objects on the left represent smoke from the stove underneath.

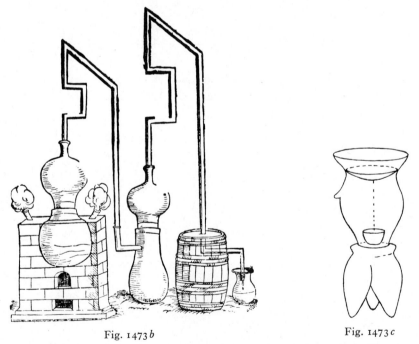

Fig. 1473*b*

Fig. 1473*c*

Fig. 1473*b*. Dephlegmator from the *Kräuterbuch* of Adam Lonicerus (+ 1578).

Fig. 1473*c*. Conjectural design of the most ancient Mongolian-Chinese still type. A bowl of cooling-water caps the upper vessel (*tsêng*) and a collecting-bowl stands upon its perforated bottom (or of course on the grating of a *hsien* when the two vessels were combined into one); cf. Fig. 1395. The *li* below provides the vapours, and the distillate collects in the bowl, equipped subsequently with a side-tube. An earlier sketch of this by us has been reproduced in Thurm (1), p. 18.

length and breadth of the Old World. An arrangement of this sort could easily have led to the convex surface for which we are looking, and it is not difficult to imagine the observation of the dropping back of the distillate from the central part of such a large plug of fleece, felt or floss silk (Fig. 1454 *b'*). From this the characteristic East Asian still forms (Fig. 1454 *c'''*, *d'*) could have derived. To support this argument one might adduce a piece of concrete evidence from that culture-area, namely the use of silk floss for plugging bamboo tubes used in solubilisation reactions. This is described in the *San-shih-liu Shui Fa*[1] (Thirty-six Methods for bringing Solids into Aqueous Solution), a text of approximately Liang date (early +6th century) which we shall discuss more fully in the next sub-section (p. 167). Silk floss, therefore, could have been the Chinese equivalent of fleeces and sponges.

But this idea does not throw any light on the nature of the oldest East Asian still bodies (or barrels), and it is not necessary to invoke fleeces or flosses as the ancestors of all Mongol and Chinese still-head condensers, for they could have been basins filled with cold water from the very beginning. At an earlier stage (pp. 26 ff.) we described the 'steamers' (*tsêng*[2] and *hsien*[3]), with their perforated gratings,[a] which were so characteristic of the cuisine of the Shang and Chou periods,[b] and we hinted that they may have had much importance in the beginnings of Chinese chemical technique. Now Mahdihassan (56) drew attention to the colanders (pottery basins with holes pierced in the bottom)[c] which are common in India, and which form the middle vessel of the Mongol still as used by the forest-dwelling tribal people in Bihar. These perforated bowls must be quite ancient, since examples have been found in the excavations at Hastināpura near Meerut.[d] As we knew already in Vol. 1,[e] forms of this kind were at least as ancient in China, going back to Neolithic times and the Shang period (−2nd millennium).[f] Consequently it becomes evident that if a small catch-bowl was set on the grating of a *tsêng* or *hsien*, and a bowl of cold water then placed over the mouth, the pattern of the Mongol still would immediately appear (Fig. 1454, *c''*). What more natural and convenient support for the catch-bowl could there be than the grating at the waist of the *hsien* or the perforated bottom of the colander half of the *tsêng*? Vapours could rise readily past the receiver. That this must be the origin of the Mongol and Chinese stills becomes almost certain when one remembers[g] that the perennial term for distillation is *chêng liu*,[4] i.e. 'steaming', and nothing more. But at what date this simple and elegant invention was made (cf. Fig. 1473 *c*) remains at present beyond conjecture.

The final stage of development of the Moor's head (Fig. 1454 *g*) where the con-

[a] Detachable in the former, built in as part of the body in the latter. See Tzu Chhi (*1*). For the bone and bronze forms of the characters *hsien* and *li* see K 252 and 855.

[b] Cf. Figs. 1398 to 1401 above.

[c] From Fr. *couler*, to flow, and Prov. *couladour*, a vessel for straining.

[d] Lal (*1*) pp. 58–9. A −6th-century date would be reasonable, according to Bose, Sen & Subbarayappa (*1*). [e] Fig. 9, p. 82.

[f] Their antiquity is indicated by the fact that the characters always kept the pottery radical (Rad. no. 98) even when the object was made of bronze or other metal. There are characters with these phonetics and the metal radical (Rad. no. 167), but they were used for quite different words and meanings—the clashing sound of metal, or parts of horse-bits. [g] Cf. pp. 132 ff. below.

[1] 三十六水法　　　[2] 甑　　　[3] 甗　　　[4] 蒸餾

denser device is applied wholly to the side-tube, would have arisen naturally enough. This is the principle seen in the still of Joh. Wenod (Fig. 1470), and the predecessor of the familiar Liebig condenser. It would not be likely to have sprung from the Chinese still, because there the water-cooling of the head above the catch-bowl or cup was a *sine qua non* from the very beginning. The coiled or serpentine side-tube descending within a barrel of cold water would also seem to have been a development of the +15th century, though not commonly illustrated until the following one, as in the works of Biringuccio (+1540),[a] Gesner (Fig. 1474),[b] Hermann (+1552)[c] and Lonicerus (+1578).[d] The modern counter-current condenser came towards the end of the +18th century,[e] due among others to Poissonnier for shipboard fresh water supplies (+1779)[f] and to C. E. von Weigel for laboratory use (+1773).[g]

Thus it would seem that there were three entirely different lines of development of the still, each starting from a different kind of primitive apparatus (Fig. 1454). One cannot but agree with Hommel's conclusion that the Mongolian and the Chinese types[h] 'are distinctly different from the Mediterranean types and cannot by any stretch of the imagination be explained as related to or derived from them'.[i] He went on to say that the former were an indigenous 'development of inner Asia, and may be closely linked with the discovery of alcohol'. To what extent this could be true will depend on data to be examined in the sequel.

The filiation of modern chemical apparatus with its medieval antecedents back to the equipment of the Alexandrian proto-chemists has been traced by many writers. But it would be quite a misapprehension to think that the East Asian still types play no part in modern chemical technique, and that all stills are descendants of those of the Greeks. An interesting application of the principle of the Mongol still is seen in the method introduced by Jackson & van Bavel (1) for the collection of water from soil and plant materials as a means of survival under semi-desert conditions. The convex cover is here represented (Fig. 1475) by a sheet of plastic and the heat is supplied by the sun-warmed earth. The apparatus devised by Brailsford Robertson & Ray (1) in 1924 for the continuous extraction of solids at the boiling temperature of the solvent (Fig. 1476) was identical with the Mongol still except that the distillate was allowed to drip back into the still body after extracting the solid held centrally in a Buchner funnel.[j] And the same idea has been applied on a minute scale to micro-Soxhlet siphon extractors by Wasitzky (1).[k]

a Schelenz (2), fig. 37; Forbes (9), fig. 57.   b Forbes (9), fig. 49.   c Forbes (9), fig. 87.
d Underwood (1), figs. 24, 25; Forbes (9), figs. 65, 66.
e Though Leonardo da Vinci had in his youth proposed something very similar investing both the side-tube and the head of a still (*Cod. Atl.* 400v, *c*; Reti (7), p. 656).
f Underwood (1), fig. 50; Forbes (9), fig. 133.
g Forbes (9), fig. 134. Was not the beautiful shrub *Weigelia* named after him?
h It is interesting that both of their most characteristic forms gave perfect equilibrium vapour-pressure between the two phases, no 'bottle-neck' creating slight pressure in the still-body. But the disadvantage was that with irregular boiling some splashings would come over; this is better avoided in the traditional flask and Liebig condenser of modern times.
i (1), pp. 146–7. Dr Ladislao Reti also expressed to us (in private correspondence) his conviction of the originality of the Mongol and Chinese stills.
j These workers might have been surprised at the similarity of their device to some of the traditional essential-oil stills of East Asia (cf. pp. 116ff. below).   k Fig. 1477a; cf. Morton (1), p. 202.

Fig. 1474. Coiled serpentine side-tube descending through a reservoir of cold water, from Gesner's *De Remediis Secretis* (+ 1569). It was probably used, though not illustrated, in the previous century.

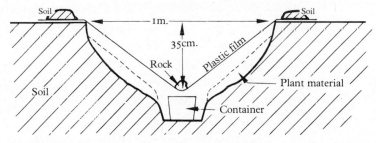

Fig. 1475. Solar still for purifying water described by Jackson & van Bavel (1). A direct derivate from the Mongolian still.

The principle of the Chinese still, moreover, is the basis of many interesting modern high-vacuum molecular stills. A molecular still may be defined as a still in which the distance between evaporating surface and condenser surface is less than the mean free path of the molecule.[a] Hence no re-condensation can occur on the evaporating

[a] Morton (1), p. 118.

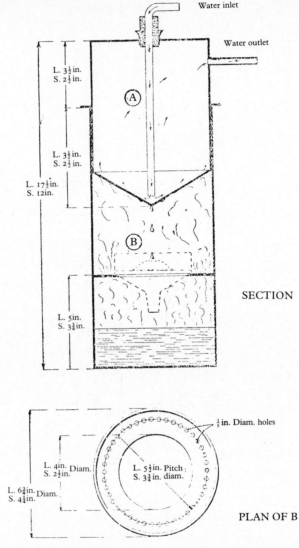

Fig. 1476. Apparatus for the continuous extraction of plant or other material at the boiling temperature of the solvent described by Brailsford Robertson & Ray (1). This is again a direct derivate from the principle of the Mongol still.

surface. Under such conditions the fractional distillation of organic compounds in non-aqueous or nearly non-aqueous medium can be accomplished because the molecules of each substance have their own characteristic mean free path. By adjusting the distance between the donator and receptor surfaces very delicate separations can be carried out. Such stills are often called 'cold-finger stills', and may employ liquid nitrogen as the coolant. In one of the simplest, the Washburn still (Fig. 1477 b), the substance sublimes on to the convex bottom of a cooling tube; this can distil paraffin

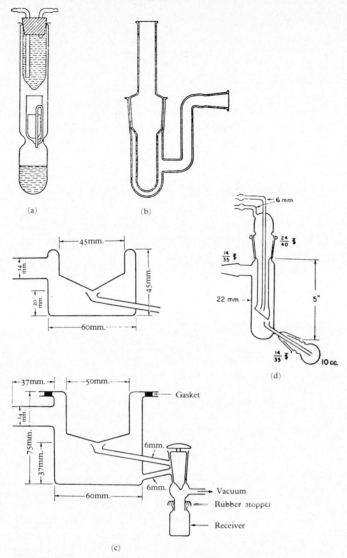

Fig. 1477. The Mongolian and Chinese stills in modern chemical practice.

(a) Wasitzky's apparatus for the continuous extraction of micro-quantities of material. Here the Soxhlet system is combined with central drip from a cooling finger. Morton (1), p. 202.

(b) Molecular still for use with high vacua; the Washburn type. The disti-sublimate collects on the convex surface of the cooling tube. Morton (1), p. 119.

(c) Molecular stills; the Hickman all-glass pot type, with exchangeable receivers. Perry & Hecker (1), p. 528. Dimensions in millimetres.

(d) Molecular stills; a Riegel pot still embodying, like the preceding one, a Chinese catch-bowl and side-tube. Perry & Hecker (1), p. 530.

wax at 55 °C. and sucrose at 120°. Tetrasaccharides have been separated and purified by distillation at about 280°.[a] On the other hand, at these very low pressures (e.g. 2 to 6 × 10⁻⁵ mm. Hg), a substance like α-bromo-naphthalene, normally boiling at 281°, will distil at 19°.[b] In the Hickman still (Fig. 1477 c), called by him a 'vacuum alembic with collecting arm', a catch-bowl and pipe exactly in the Chinese style is used to remove the 'disti-sublimate' condensing on the upper cooler and running down to its central point.[c] The Riegel pot still, on an even smaller scale, is similarly conceived

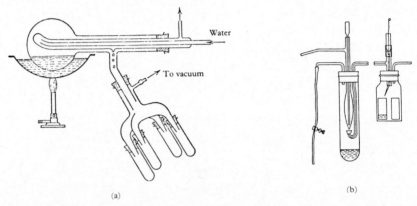

Fig. 1478. Mongolian and Chinese principles in molecular stills.
(a) Horizontal-flask molecular still with bent-finger condenser, drip point and four-fold rotating receiver (Morton (1), p. 120).
(b) Still of Ellis, in which the catch-bowl transmits to a capillary tube connected to a simple apparatus delivering as rotated to a series of receivers (Morton (1), p. 111).

(Fig. 1477 d).[d] Or the distillate may drip off a nipple on a horizontal cold-finger[e] (Fig. 1478 a); or be made to pass out of a micro-still by a capillary tube for collection in a series of receivers (Fig. 1478 b).[f] General descriptions and further diagrams of these stills are easy to find,[g] and we have noticed a good photograph of a high-vacuum still of 'Chinese' type in Morrison & Morrison (1).[h]

[a] Freudenberg, Friedich & Burmann (1).   [b] Watermann & Elsbach (1).
[c] See Hickman (1, 3); Hickman & Sanford (1, 2). In some forms, the collecting tube passes vertically down through the still centrally and ends in a set of several fraction-collecting flasks. In others a central glass rod built in acts as a strut strengthening against implosion, and as a guide to carry the distillate down. It is also interesting (as Reti (8) has pointed out) that yet other types of molecular stills revive the ancient peripheral rim gutter of the West; cf. Hickman (1, 2, 3); Hickman & Trevoy (1); Hickman & Weyerts (1). Some, moreover, have a 'hot finger' instead of a cold one. It would be interesting to know whether any of the pioneers of the modern technique of molecular distillation were aware of the age-old Chinese precedents.   [d] Riegel, Beiswanger & Lanzl (1).   [e] Morton (1), p. 120.   [f] Ellis (1).
[g] E.g. Perry & Hecker (1), pp. 527 ff.; Morton (1), pp. 118 ff.; Fieser & Fieser (1), p. 34.
[h] We are indebted to Dr W. E. van Heyningen for first pointing out the extent to which the principle of the Chinese still has been incorporated in the modern practice of organic chemistry.

(vi) *The geographical distribution of still types*

A special investigation should be devoted to the history and geographical distribution of all the possible forms of distillation and extraction apparatus. Of these the East Asian types constitute one branch of considerable interest, not hitherto adequately taken into account by historians of chemistry. We have now given some description of their traditional use in modern times (pp. 62 ff. above), and we have traced them back into the past so far as is at present feasible (pp. 68 ff.).[a] Here we intend to sketch very briefly the extent of their geographical spread.[b]

On distillation among the Mongol peoples perhaps the best paper is that of Montell (2), who describes three types of stills (*burchur*), the Mongol one proper with the central catch-bowl, the Chinese one with the side-tube originating from a flat cup or shallow grooved rectangular wooden plate,[c] and lastly one of retort type in which the heated pot is simply connected by an arched wooden leather-covered tube with an iron receiver jug standing in a basin of cold water.[d] Notable is the use of clay-daubed felt for making joints steam-tight, since felt was one of the most characteristic Mongol inventions.[e] All these types were and are used for obtaining spirits (*arihai, airiki*) or twice or thrice distilled spirit (*arsa, chorsa*) from fermented mare's milk (*kumys, airak, arik*) containing only some two per cent of alcohol.[f]

For the Mongol still proper we need not adduce further sources, except the milk-can-like example (filled at the top with snow and ice), reported by Krünitz in +1781 (Fig. 1479).[g] But it is interesting that Pallas, in his famous travel book of +1776, figured two sorts of stills for making spirits from fermented milk.[h] The Kalmuk one, lacking still-head and catch-bowl, simply had two 'kettles' connected by a tube

<hr>

[a] On various occasions in the past we have referred to what we call the Department of Face-Saving Re-definitions (Vol. 4, pt. 2, p. 545, pt. 3, pp. 564, 651). Here is another: Arntz (1), p. 203, describing briefly the Mongol still, says that it should not make us undervalue 'Destilliergeräten im echten Sinn'. But why should Western-type stills be regarded as any more 'genuine' than Mongol and Chinese stills?

[b] If we are right in our explanation of the origin of the Mongol and Chinese stills (p. 97), all their forms everywhere must have derived from ancient China, since it was only in that culture that the steamers (*tsêng* and *hsien*) existed.

[c] Rather like a butter-pat board, but scooped out and with grooves, the handle constituting the runnel which takes the distillate away.

[d] A similar apparatus is shown in the photograph from Hoernes (1), vol. 2, p. 24, fig. 9, reproduced by Schelenz (2), fig. 114, of an arrack still with two side-tubes used by the Sagai Turks in Southern Siberia. It is apparently non-cooled save that the two receivers are set in a trough which was presumably filled with cold water. Schelenz dubs it 'äusserst urwüchsig', but sees a connection with the Hellenistic style; presumably he had in mind the *dibikos* with its two side-tubes (cr. Berthelot (2), pp. 132, 138, 141; Sherwood Taylor (2), p. 137). Apart from the fact that the tubes come out near the top of the still, showing that it cannot have any internal rim or gutter, this strange survival really does seem a descendant of the *dibikos*, transmitted presumably through the Asian descendants of the Bactrian Greeks. A closely similar drawing given in Wiberg (1), fig. 4, from Maurizio (1), is ascribed to 'K. Stalywho, 1913' and its exact provenance not stated, but it must be also from Turkic Siberia. Presently we shall see other eastward penetrations of Hellenistic and later European still designs (p. 113).

[e] Cf. Olschki (7).

[f] References to this are legion. We may simply add: Buckland (1), p. 263; Maurizio (1), pp. 217 ff.; Gmelin (1), vol. 2, pp. 126 ff.; Wiberg (1), pp. 74 ff. and figs. 4, 5, 6. These refer to Kalmuks, Kirghiz, Baschkirs and Astrakhan Tartars.   [g] (1), figs. 470, 471. From Schelenz (2), fig. 113.

[h] It is unfortunate that although the great Alexander von Humboldt personally saw distillation going on among the Kalmuks in 1829, it did not occur to him to give in his long discussion on the history of distillation (3), a precise account of the apparatus used.

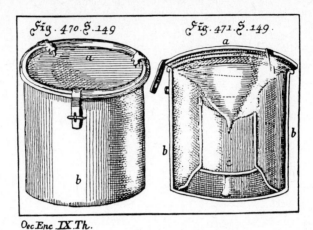

Fig. 1479. The Mongol still in folk use; the can of the Crimean Cossacks for distilling *arak* from *kumiss* (Krünitz, +1781). (*a*) Cavity for coolant, (*b*) still body.

(Fig. 1480 *a*),[a] but that shown as in use among the Mongols, Buriats and Tungus was clearly provided with the upper bowl of cooling water and the side-tube with catch-bowl (Fig. 1480 *b*); it was in fact a 'Chinese' still.[b] A larger one of the same kind appears in a more recent book resulting from the travels of Maenchen-Helfen in Tannu-Tuva (Fig. 1480 *c*).[c] Hermanns has extended its range in his account of the Tibetan nomads, whose economy is quite similar to that of the Mongols.[d] We have not been able to find much information on other marginal zones of the Chinese culture-area such as Korea or the lands of the Thais.

For India the information is complicated. As regards the Tantric alchemical litera-ture, Ray gives excerpts from the *Rasaratna-samucchaya*, probably compiled soon after +1300, which apparently copied its section on apparatus largely from the +12th- or +13th-century *Rasendra-chūḍāmaṇi* of Somadeva. The descriptions and illustra-tions[e] show only two regular stills (the *dheki-yantra* and the *tiryakpātana-yantra*) with side-tubes and receivers but no peripheral gutters and no cooling at the still-head, certainly no evidence of catch-bowls.[f] They are in fact retorts, like the Gandhāran

---

[a] (1), p. 3 and pp. 205 ff. A Gandhāran type.

[b] (1), pl. 7 and p. 272. The description was carefully considered by Huber (1), who drew attention to the presence of a perforated board above the boiling water at the bottom of the still, and the suspension of the fermented curds on a thick hempen cloth above this and below the position of the catch-bowl and side-tube. He also noted how in travellers' descriptions the Greek type of still tended to predominate west of the longitude of Lake Baikal (*c.* 105°) and the Chinese type east of it. Some of the peoples, according to Pallas, measured the strength of their 'Milchschnaps' in terms of the number of times the condenser bowl was renewed with cold water during the distillation.

[c] (4), opp. p. 53; cf. also p. 57.

[d] (1), pp. 66 ff.

[e] (1), 2nd ed., pp. 151, 158 ff., 189 ff. and fig. 30*d*, *e*. The drawings are often copied, as in Schelenz (2), figs. 19, 20, 21; Ferchl & Süssenguth (1), figs. 8, 9. In Ray's 1st ed. the titles of the *tiryakpātana-yantra* and the *vidyādhara-yantra* are inadvertently interchanged.

[f] The second of these is the one that has the cooling of the receiver by a stream of water poured from a jug (pp. 69–70 above). A photograph of just this being done during the distillation of palm wine or toddy in South India will be found in Hemneter (1).

stills (p. 87), and therefore allied with the early Greek still-types (p. 84). Hellenistic influence, as might be expected,[a] appears strikingly in the *dhūpa-yantra*, a remarkably precise echo of Mary's *kērotakis*.[b] And, as we shall see in a moment, the *dibikos* appears in India. A link of some interest with the Mediterranean area is the fact that stills of closely similar proportions and dimensions to these were used in the early decades of the present century for the distillation of *araq* among the Arabs of Palestine, where they were studied by Dalman.[c] More in accordance with expectation are the stills of Gandhāran and Tantric type used to this day at Pabna and other places in Bangladesh for illicit spirit distillation.[d] Moreover, in Grierson's description of arts

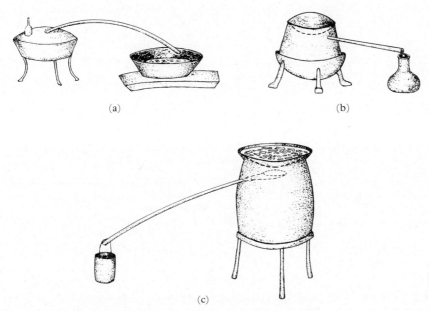

(a)   (b)

(c)

Fig. 1480. Forms of still in use among Siberian peoples for *arak*.
  (*a*) Two vessels connected by an arching side-tube.
  (*b*) Chinese still with catch-bowl and side-tube used by Mongols, Buriats and Tungus.
  (*c*) Larger Chinese still employed by the people of Tannu-Tuva.

and crafts in Bihar, which gives the technical terms for all the parts of the traditional spirit still,[e] the apparatus is clearly of Western or Gandhāran type. A similar single side-tube alcohol still is customary among the Chenchu jungle people of Andhra Pradesh in Eastern India;[f] but much more extraordinary we find a veritable *dibikos* (with two side-tubes though no annular gutters) in use among the Baiga in Central

---

  [a] From the parallel dissemination of Hellenistic mathematics and astronomy, so well known (cf. Vol. 3, pp. 146 and 176).
  [b] There are also four sand-baths, two sublimatories, two extractors, and (as we have seen, p. 55) two arrangements for descensory distillation.
  [c] (1), vol. 4, p. 368 and fig. 112. The side-tube, however, was usually passed through an amphora pierced with two holes and filled with cold water.   [d] Mahdihassan (56), fig. 42.
  [e] (1), pp. 77 ff., but unfortunately no drawing.   [f] Von Fürer-Haimendorf (1), fig. 61.

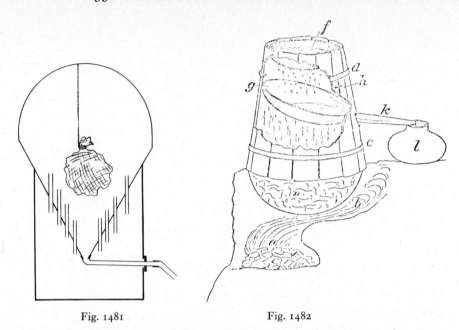

Fig. 1481   Fig. 1482

Fig. 1481. A combination of Chinese and Western designs: the extraction apparatus for materia medica used among the Algerians (Hilton-Simpson (1), p. 21). The head is convex like a Hellenistic helm, but the extract is collected in a Chinese catch-bowl with side-tube, penetrated by tubes to conduct the ascending vaporised solvent.

Fig. 1482. A Tarasco still from the Lake Patzcuaro region in Mexico (Bourke (1), p. 67). Of pure Chinese type, the catch-bowl is particularly large. (a) fire, (b) chimney, (c, d) hoops confining the barrel forming the still body, (e) maguey mash in a large earthenware bowl, (f) cooling reservoir replenished with cold water, (g) catch-bowl of metal, (h) barrel walls, (k) side-tube, (l), receiver.

India.[a] Thus so far nothing of East Asian style is to be seen.[b] On the other hand, among certain primitive peoples, such as the Bhīls, who distil from fermented *mahua* flowers[c] a spirit greatly used in all their ceremonies, the only ritually correct type of still is the Mongol one with the catch-bowl.[d] How did this technique reach Rajputana, Gujarat and Malwa? And why do the Bhīls honour it above the *dibikos*, which they also possess and use?[e] Finally, the Mongol still is commonly employed by the Newars of the Nepal valley, though that is less surprising.[f]

For Mogul India we have an important source in the *Ā'īn-ī Akbarī* (The Administration of the Emperor Akbar) written about +1590 by the great historian Abū'l-Faẓl 'Allāmī. In his account of the imperial still-room, he describes three types of stills, the Mongol suspended catch-bowl type, the Chinese type having a 'large spoon

[a] Verrier Elwin (1), pp. 44–5.
[b] With the partial exception of the *vidyādhara-yantra*, a sublimatory for mercury which has a basin of cold water set above the space in which the metal sublimes; see Ray (1), 2nd. ed., pp. 190–1 and fig. 30*e*; Roy & Subbarayappa (1), pp. 7, 68.
[c] These are the flowers of *Madhuca* (or *Bassia*) *latifolia* (Burkill (1), vol. 2, pp. 1387ff.). The collecting cup or catch-bowl (*dōī*) hangs inside a *haṇḍī* pot surmounted by the bowl of cold water (*vāṭkā*).
[d] See Pertold (1); and Mahdihassan (56), fig. 41, for the forest tribes of Bihar.
[e] Doshi (1), p. 107.   [f] Regmi (1), pp. 787–8.

with a hollow handle leading into a jar', and the Greek still-head with two pipes and two receivers, in fact the *dibikos*.[a] At such a late date, of course, it is difficult to trace anything of technical inter-change in earlier times.[b]

From the close commercial contacts of Arabic and Chinese merchants all through the Middle Ages one would expect to find the East Asian still types in the Islamic culture-area. At present, however (apart from the instance just given), we cannot prove this. But we should like to call attention to the Algerian pharmaceutical distil-

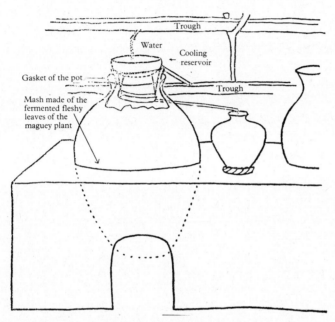

Fig. 1483. Zapotec still, also from Mexico, one of a series set along a bench-like stove (de la Fuente (1), p. 97). Again of pure Chinese type, but with an arrangement for a constant current of cold water through the upper reservoir. For alcohol from maguey mash.

lation and extraction apparatus described by Hilton-Simpson (Fig. 1481).[c] This is a curious combination of Western and Eastern designs in that it has a concave roof but a central collecting point for the condensed extract, surrounded by tubes permitting the rise of the vapour into the upper space. Such an arrangement is reminiscent of the pipes that sometimes rise through the Chinese catch-bowl in curious intermediate forms of apparatus which we shall examine in a moment (p. 119), but it lacks all water-cooling at the top. Hilton-Simpson did not actually see one of these stills, but his Shawiya medical friend approved the drawing of it.

[a] Blochmann (1), vol. 1, p. 69. It is rather remarkable that the only reference to the Chinese still in Forbes' substantial treatise on the history of distillation should be to Abū'l-Faẓl, (9), p. 54. And then he did not recognise it for what it was.
[b] Certain authors, e.g. Wiberg (1), have been inclined to look upon India as the original home of all stills, but without much positive evidence. It now looks rather as if there were three foci for the invention of distillation—the Babylonian-Hellenistic, the Gandhāran, and the Mongol-Chinese.
[c] (1), p. 21.

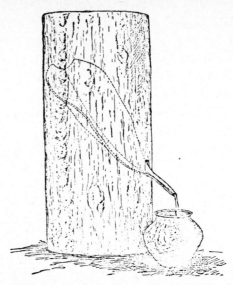

Fig. 1484. Part of a Chinese still used by the Cora and Tarasco Indians in Mexico. A roll of mountain-cedar bark, over a yard tall, forms the sides of the still, its edges being jointed with glue. The catch-bowl is a maguey leaf cut into the shape of a spoon (hence its name, *cuchara*), and its stem, passing outside through a hole, forms the side-tube. From Lumholtz (1), vol. 2, p. 186.

The typical Chinese still turns up in a rather unexpected place, namely Mexico, where the characteristic apparatus for distilling *mezcal* spirit from fermented maguey juice is provided with an upper cooling water-basin and a catch-bowl and side-tube underneath it. This was seen by Bourke (1) in 1893 on the island of Tzintzontzin in Lake Patzcuaro (Fig. 1482).[a] Further descriptions have been given by de la Fuente (1) for the Zapotecs;[b] and Lumholtz (1)[c] for the Cora Indians, who use a roll of mountain-cedar bark as the still sides, with a maguey-leaf cut in the shape of a spoon for the catch-bowl, its stem forming the side-tube (Fig. 1484). Besides this, Lumholtz added evidence that the true Mongol still also is found in Mexico, among the Huichol Indians (Fig. 1485 *a*).[d] It would seem at first sight that the only route by which these could have arrived there was through Muslim influence in Spain. One would hesitate

[a] He called the spirits *mescal*, but it is better to spell it *mezcal*, as the Mexicans do, distinguishing it thus from the 'mescal buttons' derived from the famous *peyotl* cactus *Lophophora Williamsii*. These contain psychotropic active principles (cf. la Barre, 1), especially the alkaloid mescaline, long used in certain Amerindian religious cults, and perhaps the first hallucinogen which received modern scientific study. The similarity of terms no doubt arose because of a confusion between different forms of intoxication. As de Barrios (1) points out, it is not quite true to say that *mezcal* (and the well known *tequila*, a particular regional form of it) is distilled from *pulque*. *Pulque* is the beer fermented from the sap or *aguamiel* of the maguey plant, *Agave atrovirens* (and several other species), in a natural process (i.e. without added sugar) wherein the yeasts are supplemented by an unusual alcohol-producer, *Thermobacterium mobile*. *Mezcal*, on the other hand, is distilled from the press-juice of the hearts of *Agave tequilana* (and several other species), with the addition of sugar, and primarily by yeasts; it is always rectified by double distillation. The agaves belong to the Amaryllidaceae, and are also grown for their valuable sisal fibre.

I cannot refrain from recalling here my first introduction to the excellent *tequila* of Mexico by my late friend Miguel Covarrubias.

[b] His fig. 11, p. 97, is here reproduced (Fig. 1483).   [c] In his vol. 2, p. 186.

[d] (1), vol. 2, p. 184. His diagram seems to have been redrawn erroneously in Wiberg (1), fig. 7.

perhaps in a case like this to assume a direct trans-Pacific pre-Columbian passage from Asia.[a] Yet Lumholtz was much inclined to regard the Huichol method as pre-Columbian because of its simplicity,[b] and Bourke (2) was able to add one positive argument in favour of such a belief. He drew attention to an edict of +1529 by Charles V against the use of distilled *pulque* by the Indians of New Spain.[c] The key word was specifically used—*que destilan los magueyes*—but the complaint was largely

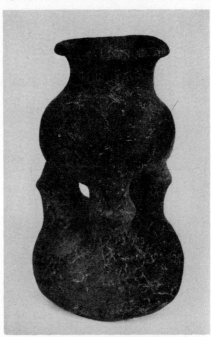

Fig. 1485*a*   Fig. 1485*b*

Fig. 1485*a*. A Mongol still used by the Huichol Indians in Mexico. A mound of stone and earth is built as an oven around a large pottery jar or boiler, thick rings of grass making tight the space above it. The funnel formed by the top of the mound supports a copper cooling-basin, and a central pottery catch-bowl is suspended underneath by two cords of yucca fibre. From Lumholtz (1), vol. 2, p. 184.

Fig. 1485*b*. 'Trifid' pottery vessel from the Colima culture of north-western Mexico, c. −1450 (after Isabel Kelly, 1, 2). With a cooling-water basin above and a catch-cup inside, such pots could have been used for alcohol distillation in pre-Columbian times.

against substances added to the liquor from roots and berries with stupefying, excitatory or hallucinogenic effects, customary in 'heathen' ceremonies. The strongest measures were to be taken against such practices.[d] Now this was less than ten years after the conquest, a remarkably short time if knowledge of distillation had been brought only by the Spaniards. An imported drink-preparation would hardly have been adopted so quickly into the service of gods and ancestors. Supposing therefore

[a] This difficult subject has been looked into already in Vol. 4, pt. 3, pp. 540ff. Cf. Needham & Lu (12).

[b] (1), vol. 2, pp. 183, 185. Diguet (1), p. 610, supported him in this, but Seler (1) preferred to think of a borrowing from Spanish-American mestizo sources.

[c] Statute xxxvii in de Paredes (1). Cf. Wiberg (1), pp. 109ff.

[d] But in fact they still continue to this day.

that the words of the edict mean just what they say, the question of a possible pre-Columbian origin of distillation in Central America remains quite open.[a] The fact that the two methods used were distinctively Asian and not European must give some pause for thought.[b]

An intermediate alternative would be that the stills were indeed Asian but not pre-conquest, and this is the view put forward by Bruman (1, 3) who also visited the Huichol people himself, and found that their name for the fermented must before distillation was *tuba*. Since this same Tagalog word is used for palm toddy in the Philippines, Bruman proposed that the Asian stills were brought by the Filipino sailors who worked the Manila galleons, the first voyage of which took place in +1565. As would be expected, both the Mongol and the Chinese still-types have long been common in the Philippines.[c] Coconut culture later became an important industry on the littoral of Western Mexico, and Filipino influence certainly had much to do with its development.[d] But the Huichol distilled the fermented mash of the *sotol* plant[e] to make their *tuchi*, not palm toddy; and landfalls of Chinese sailors on rafts or dismasted junks during previous centuries cannot be ruled out. So judgment should perhaps be reserved for a while yet.

The Mongol still has also appeared, believe it or not, among the Irish peasantry. Because of economic depression and heavy taxes, illicit distillation has been a prominent feature of that countryside ever since the end of the +17th century,[f] as we can learn in the interesting survey of Connell[g] and the colourful descriptions of Hanna Bell.[h] Though generally a cooled coil or worm is used for the condenser, all kinds of tar-barrels, milk-churns, oil-drums and potato-pots being pressed into service, the poteen or *usquebeatha* ('water of life', the original Erse form of the word whisky) has

---

[a] Maize beer could also have been distilled. Was this perhaps the *yolatl* or 'heart-water' with which the Aztec captains of Axayácatl consoled themselves after their defeat by the Tarascans in +1478? See Davies (1), pp. 147, 331; Durán (1), vol. 2, p. 283 (xxxvii, 13), Heyden & Horcasitas tr., p. 167. Durán called it a *caldo esforzado*, which could mean 'strength-giving wine'. Today spirits (*chicha*) made from maize are common in many parts of Mexico.

[b] In her excavations of the Capacha phase of the Colima culture of the north-western coast of Mexico, Isabel Kelly (1, 2) found many double gourd-shaped pottery vessels like 'steamers', the two parts, upper and lower, being connected not by a grating but by two or three tubes (Fig. 1485*b* illustrates a typical 'trifid'). The date of these would be about −1450. They are mostly small, with a diameter of some 7 cms. at the mouth, but if surmounted by a cooling bowl and provided with a little catch-cup inside, alcohol could certainly have been distilled in them. We are much indebted to Dr Kelly for knowledge of these pots and discussions concerning them.

[c] See Feliciano (1).   [d] Bruman (2).   [e] *Dasylirion* spp. (Liliaceae).

[f] There has long been a persistent claim that the people of Ireland were able to distil spirits at a very early date. The most reasonable statement of the case is that when the army of Henry II invaded Ireland in +1172, they found the Irish using some kind of distilled wine or beer. Older works, such as that of Scarisbrick (1), p. 44, tend to accept the story; but more recent studies, e.g. McGuire (1), p. 91, are sceptical, though not completely dismissive. As we shall shortly see (p. 123), the earliest certain alcohol distillation in Europe was carried out by the Salernitan Masters in Italy in the neighbourhood of +1160, so that if the process was known in Ireland only some twelve years later it must have travelled thither with bizarre speed. Wandering Irish monks could have made that just possible, but it remains very improbable. Moreover, no mention of spirits has been found in any primary source or basic authority for Henry II's invasion of Ireland (priv. comm. Dr Roger Lovatt). Of course, if the distillation of alcohol was really known in late +12th-century Ireland it would presumably have been done with stills of Greek, not Mongolian, type. But judgment is best suspended until further evidence appears.

[g] (1), pp. 1–50.   [h] (1), pp. 50ff.

also been distilled in simple kettles.[a] A kettle was partly filled with the fermented liquor and a pint mug placed inside it (presumably on some support to lift it from the bottom); the spout of the kettle was then sealed or corked and its lid turned upside down so that the knob hung above the mug and cold water could be added to the hollow of the lid. Thus the vapours condensed on the under surface of the lid and the concentrated alcohol dripped into the mug.[b] Did some Irishman think this up on his own, or had he been travelling among the Mongolians?

Actually, he need not have wandered so far, since the Mongol still design exists to this day all over Russia, Poland, Hungary, Rumania, Czechoslovakia and other parts of Eastern Europe. The fact that distillation by the peasants has generally been illegal has rendered this fact less well known than it might be.[c] *Pálinka* (the 'mountain dew' again) is made by setting a bowl of cold water atop a large tin pail, and supporting a catch-bowl below it by means of pieces of stone or a wooden frame or iron tripod. The must (*cefre*) is usually derived from cheap pure sugar and yeast with flavourings subsequently added. The ubiquity of the arrangement throughout the Eastern European countries may well have had something to do with that persistence of Mongol and Chinese still designs into the sophisticated apparatus of modern chemistry which we discussed above (pp. 99 ff.).

Conversely, clear derivatives from the Hellenistic or Gandhāran still, mostly rimless, and if possessed of water-cooling only round the receiver, have penetrated widely over the world, including parts of Asia. In traditional Ethiopia a still of this kind was used for beer and mead,[d] while that for palm toddy in South-east Asia often had a long bamboo side-tube,[e] and as for Africa, the Anyanja of Nyasaland used a simple 'pot and gun-barrel'.[f] Typical Hellenistic stills with annular gutters are seen in Armenian documents of the +16th century.[g] More advanced types with side-tubes passing through a condenser trough or barrel analogous to Liebig's, are reported for the Ostjaks,[h] the Wotjaks (cf. Fig. 1486),[i] the Palestinian Arabs[j] and the Madagascans, who distilled from fermented sugar-cane, mead, the berries of *Buddleia madagascarensis* and other musts.[k] Some of these forms are irresistibly reminiscent of the picture of Johann Wenod de Veteri Castro (p. 94), and presumably there was plenty of time for them to spread outwards from Europe since the early +15th century.

That the Hellenistic or Indian still-type penetrated far to the east rather earlier

---

[a] E.g. in Co. Longford.

[b] Irish Folklore Commission MS. 1458, pp. 457–8. At least one such apparatus can be seen in the collection of H. M. Customs and Excise at King's Beam House in the City of London, and a label says that it was seized at Edinburgh about 1950. We are indebted to Mr Arthur Slater for a knowledge of this museum, and to Mr T. Graham Smith the Librarian, with his colleague Mr Trevor Machin, for their kindness in showing it to us.

[c] We ourselves gained this valuable information from our friend Prof. Horváth Árpád of Budapest (Aug. 1970).

[d] See Huber (2); Maurizio (1), reprod. in Wiberg (1), fig. 2. Cf. Ratzel (1), vol. 3, p. 228.

[e] See Weule (1); Maurizio (1), reprod. in Wiberg (1), fig. 3.

[f] Stannus (1).   [g] Figured in Kazanchian (1).

[h] Maurizio (1), reprod. in Wiberg (1), fig. 5.

[i] From Buch (1), p. 505, an account of 1883. There is a bad redrawing in Wiberg (1), fig. 6.

[j] Dalman (1), vol. 4, fig. 112 and p. 368.

[k] See Ellis (1), vol. 1, p. 211; cit. Crawley (1), pp. 185 ff.

than that is indicated by an interesting passage in a Chinese encyclopaedia compiled about +1301. The *Chü Chia Pi Yung Shih Lei Chhüan Chi*[1] (Collection of Certain Sorts of Techniques necessary for Households),[a] probably put together by Hsiung Tsung-Li,[2] contains the following passage on 'The Burnt-wine Method of the Southern Tribesfolk' (*Nan fan shao chiu fa*)[3] which they used to make *a-li-chhi*,[4] i.e. *araqi*. The rather detailed description can be followed more easily by the aid of the drawing (Fig. 1487 *a*). Hsiung Tsung-Li says:[b]

For this item you can use all sorts of wine, sour or sweet, weak or insipid, or wine that doesn't have a proper taste.[c] (Pour the wine into) a pot so that it is eight-tenths full, and place another pot above it so that the two mouths correspond but the upper one slants to one

Fig. 1486. Western-type still with cooling-barrel condenser on the side-tube used for the distillation of *arak* by the Wotjaks in Siberia (Buch (1), p. 505).

side. Have a hole at the side of the upper empty vessel, with a bamboo tube coming out of it, and under the end (lit. the beak) of this tube place another empty pot (as receiver). Opposite the side where the rims of the two pots come close and the bamboo tube projects, fill the gap with one or more pieces of white porcelain from a broken basin, or else of pottery or tile, so that (the still) will be air-tight when cemented with the lute. Make the lute by pounding paper to a pulp and mixing it with lime, then seal all chinks carefully as thick as four fingers. Then put (the still) into a large new earthenware vat, embedding its base firmly in a mass of the same paper-lime lute, and heat it all around with two or three catties of hardwood charcoal, but not above (the level of the wine in) the still. When the wine boils the vapours mount up into the empty pot, and from this (still-head) they flow down through the bamboo tube into the pot below (the receiver). The colour (of the spirit) is quite white, no different from that of pure water. Sour wine will give an acrid (distillate), but that from sweet and insipid wine will be agreeable; in any case, one part of good spirits is obtained from three parts of wine. This technique can be used with winter-sacrifice wine, or heated wine;[d] in fact all wines can be 'burnt' (i.e. distilled) in this way.

So here was a 'pot and gun-barrel' still of simple construction, and its southern environment indicates that it had come from the Indian culture-area.

[a] *SKCS/TMTY*, ch. 130, p. 75a.
[b] Ch. 12, pp. 42b, 43a, reproduced in Shinoda & Tanaka (1), p. 345.
[c] Cf. p. 135, note e.   [d] Cf. p. 67.

[1] 居家必用事類全集   [2] 熊宗立   [3] 南番燒酒法   [4] 阿里乞

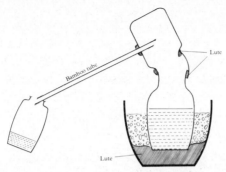

Fig. 1487 a. Reconstruction of the 'pot and gun-barrel' still described in the *Chü Chia Pi Yung Shih Lei Chhüan Chi* of +1301. An Indian or Western design with no cooling for either the helm or the side-tube.

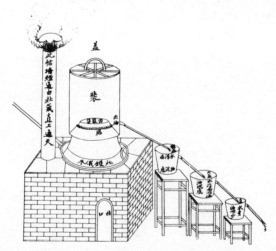

Fig. 1487 b. Still of Hellenistic type used in China for the industrial preparation of essential oil of cassia, from the leaves and flowers of *Cinnamomum Cassia* (Schelenz). As the inscriptions say, the oil sinks to the bottom in the successive decantation receivers, an emulsified distillate having been drawn off from the peripheral gutter. The chimney is on the left, not to be confused with the automatic fuel stoker shown in Fig. 1471.

Furthermore, stills of Western or Hellenistic type, especially for preparing essential oils, are found in modern China. Or so we must assume from the illustration of a cassia oil still with annular rim, no water-cooling at the head or for the side-tube, and three successive automatic decantation buckets as receivers beyond,[a] presented by Li Chhiao-Phing.[b] We give the better version of Schelenz in Fig. 1487 b.[c] It is clear that

[a] The heavy oil sinks in the successive decantations, as the Chinese captions say.

[b] (1), fig. 73, opp p. 148; again this is in the English edition only, and again without statement of source.

[c] (2), fig. 116. The drawing appears again in Forbes (9), p. 7, where it is credited to the Indian Institute at Amsterdam. The probable source is a commercial report of 1893 (Anon. 95), whence it also got into Gildemeister & Hoffmann (1).

this would not work well for a low boiling-point liquid such as alcohol, where some form of water-cooling is desirable. Yet Guppy's description (1) of the distillation of 'samshu' (*san shao*,[1] cf. p. 149 below), thrice-distilled spirits, in North China in the eighties of the last century might be thought to point to the same thing, as he speaks of a gutter. 'The fermented millet', he wrote, 'is placed in a large wooden vat or tub, the bottom of which is made of a kind of grating, and beneath the vat there is a large boiler of water heated by an adjacent furnace. The steam ascending through the grating and passing through the fermented millet finally comes in contact with a cylinder

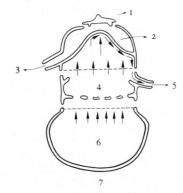

Fig. 1488a   Fig. 1488b

Fig. 1488a. Japanese *rangaku* or + 18th-century pharmaceutical extractor still (Elm), a very compact device.

Fig. 1488b. Cross-section of the same (Elm). 1, lid; 2, cooling water reservoir; 3, outlet for renewing the water; 4, still body, with the plant or other material to be extracted supported on a grating; 5, side-tube from annular gutter delivering the extract in the condensed solvent; 6, solvent to be distilled; 7, source of heat, a charcoal fire.

of cold water; it is there condensed and trickling off into a little gutter, finds its way out through a long spout in a clear stream of veritable *samshu*.' But it is perhaps more likely that by 'cylinder' he meant the cooling-bowl top of a Chinese still rather than any kind of Western Moor's head, and by 'gutter' he referred to the catch-bowl, very possibly elongated, at the end of its side-tube.[a]

A more certain example of the penetration of the annular rim Western still-type into the Chinese culture-area can be seen in the pottery extractor-still used until recently by Japanese pharmacists and physicians. Fig. 1488a shows one of these, and from the sectional diagram (Fig. 1488b) it is clear that the cooled head has a concave roof and that the distillate is carried away by a true peripheral gutter.[b] Similar stills

[a] According to Guppy, the *samshu* produced had an alcohol-content of 48–54%.

[b] The object here shown belonged to a family which had practised medicine since the + 15th century in Yamato province near Nara. Starting as physicians of the Chinese school, they took up Rangaku learning towards the end of the Tokugawa period, hence the Western influence. A complete collection of their instruments and appliances was described by Messrs. Elm & Co. of Osaka in their Catalogue MB 1967. We are much indebted to this firm for kindly providing us with the cross-section diagram. In 1964 we had been able to examine personally a similar still in the Museum of the Takeda Chemical Company's factory at Osaka, for which we also record thanks to our kind hosts, Dr K. Watanabe and Dr Miyashita Saburo. These compact pottery stills are hard to date, but the famous firm of Wedgwood

[1] 三燒

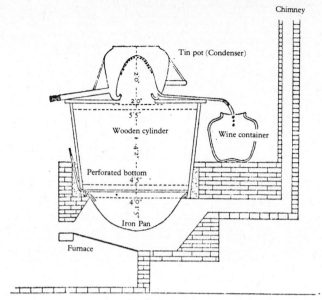

Fig. 1489. Chinese industrial still of Western type for the distillation of kao-liang wine.
From Li Chhiao-Phing (1), p. 209.

on a much larger scale came into use in modern times for distilling kao-liang spirits in North China (Fig. 1489).[a] It is not necessary for us here to expatiate on the spread of the Western still-type to other cultures, but it may be worth recording that stills not unlike those of the Japanese and Moroccan pharmacists (cf. p. 130) occur in Mexico made of earthenware, doubtless introduced by Jesuit rather than by Dutch or Arabic influence, while in Brazil such Moor's head stills are carved entire from the local soapstone.[b]

Another possible penetration of the Western annular rim or gutter concept into the Chinese culture-area is to be found in the realm of the kitchen (that foster-home of so much chemical technology),[c] and in the province of Yunnan—though somewhat distorted topologically. Those who have had the good fortune to tread the streets of Kunming, Tali or Kochiu may have enjoyed in restaurants a special dish in which

was making similar types as late as 1802, when one is depicted in the firm's 'Shape Book' or 'Drawing Book', no. 814. The Dutch virtuoso and physicist Martinus van Marum bought a couple of them from Wedgwood in 1790 and these are still in the Museum founded by Pieter Teyler van der Hulst at Haarlem (see Turner & Levere (1), vol. 4, p. 350, no. 346, fig. 309). For this information we are indebted to Mr John Chaldecott of the Science Museum Library and the Curators of the Wedgwood Museum at Barlaston. At what time these distillation devices in pottery or porcelain reached Japan remains obscure, but doubtless it was through Dutch intermediation.

[a] Li Chhiao-Phing (1), fig. 83, p. 209. Not in the 2nd (illustrated) Chinese ed. (1). The description of 1867 for a Hankow distillery, in Julien & Champion (1), pp. 201–2, corresponds in all particulars with this diagram. So does that of Yang Tzu-Chiu (1) for Canton establishments which in 1919 were producing thrice-distilled rice spirits of c. 45 % alcohol content. In Kuangtung at this time the still-head was sometimes made of pottery rather than of metal.

[b] Dr L. Reti (private comm.).

[c] Cf. here what has been said above (p. 30) on annular rim-trough water-seals.

chicken, ham, meat balls and the like have been cooked in water just condensed from steam.[a] This is done by means of an apparatus called *chhi kuo*[1] (or formerly *yang li kuo*[2]),[b] made especially at Chien-shui[3] near Kochiu. It consists simply of a red earthenware pot with a domical cover, the bottom of the pot being pierced by a tapering chimney so formed as to leave on all sides an annular trough (Fig. 1490). The *chhi kuo* once placed on a saucepan of boiling water, steam enters from below and

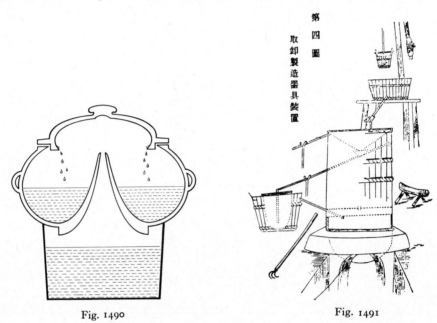

Fig. 1490   Fig. 1491

Fig. 1490. Cross-section of a *chhi kuo*, for cooking foods in water just condensed from steam, without loss of volatile flavouring substances. The edges of the domical lid direct the distillate inwards, and the food lies in a large annular space surrounding a central chimney with a narrow mouth. A topological relationship with the rim-pot stills of ancient Mesopotamia is clear (cf. Fig. 1455). Vessels of similar shape, but with a spout instead of a lid, are known from classical antiquity (Kenny (1), p. 252); their purpose is uncertain.

Fig. 1491. Japanese industrial still for peppermint oil (from Schelenz (2), p. 129, after Tanaka Setsu-saburo and Schimmel). Fully Chinese in type, it has an arrangement for renewing the cooling water at the top, and for returning (by means of a siphon, not seen) the imperfectly separated emulsion of oil and water to the still body for re-distillation.

is condensed so as to fall upon and cook the viands in the trough, resulting thus after due process in something much better than either a soup or a stew in the ordinary sense. Since the chimney tapers to a small hole at its tip no natural volatile substances are lost from the food, hence the name of the object and the purpose of the exercise. The *chhi kuo* must claim to be regarded as a distant descendant of the Babylonian rim-pots (for it has and needs no Hellenistic side-tube) with the ancient rim expanded

a In recent times this technique has become known all over China. It is fully described, with photographs, in Anon. (*101*), vol. 11, p. 6.
b From the dripping down of the steam condensate infused with fiery Yang.

¹ 汽鍋   ² 陽瀝鍋   ³ 建水

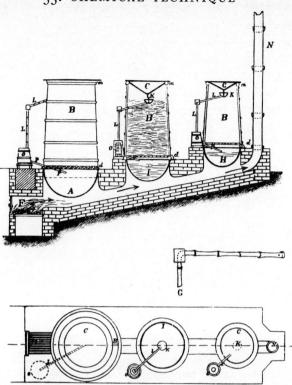

Fig. 1492. Triple still for peppermint oil, also from a Japanese source (Schelenz (2), p. 128). The three stills are set up in a row above a rising flue like a 'hillside' pottery kiln, and all have the return pipes from the receivers, in this case low so as not to disturb the supernatant oil. A, H, I, water for the steam distillation; d, d, d, gratings for the vegetable material; B, B, B, still bodies; C, C, cooling reservoirs; F, fire; G, manner of jointing bamboo pipes so as to negotiate corners; K, catch-bowls; L, side-tubes; N, chimney; O, receivers; P, return-pipes.

to form a trough, compressing the 'still'-body to a narrow chimney. But how the idea found its way down through the ages, and from Mesopotamia to Yunnan, might admit of a wide conjecture.

The further pursuit of the Chinese still and its variations by means of eye-witness accounts of traditional technology leads to some very remarkable transitional forms.[a] Two Japanese illustrations of stills for peppermint oil are given by Schelenz (2); see Figs. 1491,[b] 1492.[c] In both of these the wooden tub- or barrel-like towers are

[a] Certain of those reported seem to need further investigation. Deniel (1) in 1954 gave drawings of stills of Chinese type used in Vietnam for the preparation of alcohol from fermented rice, sorghum or kao-liang. But they raise some doubts about their accuracy. In one case (Fig. 1494 a), the catch-bowl is very large, almost as large as the still-body, while the side-tube extends uselessly beyond its apex to the opposite side of the still. There are other examples of very large catch-bowls (cf. Figs. 1482, 1484), but the point of the latter arrangement is not obvious unless it was a way of fixing the side-tube more firmly. It occurs also in Fig. 1491. Then in another case (Fig. 1494 b), the side-tube descends as a spiral tube through the hot vapours before leaving. This seems incomprehensible, unless the coil was really a cooling-coil outside the still, misplaced by a misunderstanding.

[b] His fig. 112. This also appears in the book of Gildemeister & Hoffmann (1), vol. 3, p. 533, where it is attributed to E. Marx (2).

[c] His fig. 111. Schelenz appears to attribute this to an 'excellent report' made by Tanaka Setsusaburo to Messrs. Schimmel & Co. of Miltitz in 1908.

strengthened with hoops and have at the bottom gratings above the steam boiling pan on which the peppermint plants are piled up, filling most of the still body. This is reminiscent of the set of sieves piled one upon another in which the familiar *pao-tzu*[1] (filled dumplings) are steamed; and of course of the 'steamers' of the Chou and Han (p. 27 above). In both of the stills depicted in Figs. 1491 and 1492 a bamboo tube leading from the receiver back into the still under the grating returns the aqueous phase for re-distillation. The apparatus in Fig. 1492, however, has the further interest

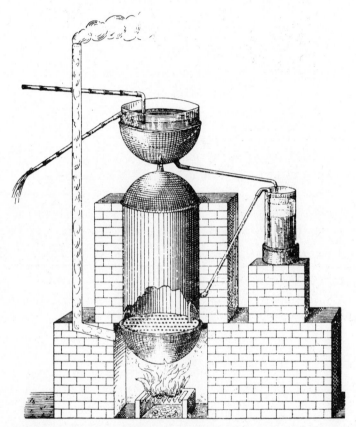

Fig. 1493. Vietnamese industrial still for the essential oils of star anise, *Illicium verum* (from Schelenz (2), p. 134, after Gildemeister & Hoffmann). This is a form transitional between the Chinese and the Western systems; for the steam and the distilled oil rise from a concave-headed still through a short pipe penetrating what is essentially a Chinese catch-bowl, so that the distillate is taken off through the side-tube from an annular channel. The cooling reservoir at the top is typically Chinese. There is also the siphon to return the oil–water mixture for re-distillation.

that a succession of stills is arranged one behind the other upon a rising flue closely reminiscent of the 'hillside' kilns of the pottery and porcelain industry.[a] So far, these forms are of the standard Chinese pattern.

[a] Cf. Sect. 35.
[1] 包子

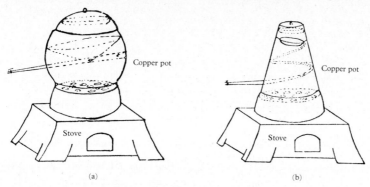

Fig. 1494a. Vietnamese alcohol still (from Deniel (1), p. 43). A wholly Chinese design, with a particularly large catch-bowl (*mai rua*); cf. Figs. 1482, 1484.

Fig. 1494b. Another Vietnamese alcohol still (from Deniel (1), p. 43). Here the side-tube is depicted as circulating in a coil through the still body (see text, p. 117).

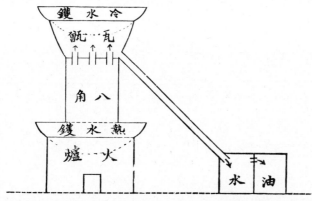

Fig. 1495. Chinese industrial still for star anise oil (*pai chio yu*), from Schelenz (2), p. 135. It is similar to that shown in Fig. 1493, but there are three rising pipes through the 'catch-bowl', and instead of a return tube from the receiver there is a decantation arrangement which separates off the supernatant oil.

But Schelenz (2) also gives pictures of two very strange developments of this type of still. Fig. 1493 shows a Vietnamese one used in Tongking for star anise oil.[a] Here the plant material is placed upon a grating as usual, but the distillation vapour rises through a short tube which passes into a bowl-shaped space surmounted by the characteristically Chinese convex still roof, above which cooling water circulates in a higher bowl. This then is as if a rising vapour tube were to pass through the bottom of a Chinese catch-cup. Thus there is here a remarkable mixture of the Chinese and Western still types, for the cooling is done by a bowl at the still-head, yet the distillate is removed by a side-tube leading off from an annular channel. But since this trough

[a] His fig. 117, from Gildemeister & Hoffmann (1), vol. 2, pp. 379ff., based on Anon. (97), p. 659 and (96), p. 85. On star anise, *Illicium verum*, see pt. 2, pp. 136–7. The 'badiana oil' got from it is an important raw material providing most of the anethol used extensively in the liqueur and cosmetic industry.

space is much larger than the entry pipe it is almost as if the Yunnanese *chhi kuo* had been provided with a side-tube as well as the usual convex water-cooled head. Fig. 1495 gives a further variation,[a] the bottom of the Chinese 'spoon' or catch-cup and side-tube, under the bowl cooler, being pierced by tubes in three places to allow the entry of the vapour.[b] This type was apparently in use in Kuangtung and Kuangsi during the last century. We are here very close to the Moorish (Algerian) extractor still of Fig. 1481, with the difference that the take-off is peripheral rather than central, and the head is a convex water-cooled one rather than a concave one with no special cooling. The receiver of the Chinese apparatus is compartmented to allow the automatic decantation of the oil which floats on the water.[c] In these arrangements, as in other extractor systems which we have already mentioned (pp. 113 ff.), it is worth noting that they all depend on steam distillation; it must have been found out very early that attempting to distil essential oils at temperatures near their boiling-points would only decompose them and lose their fragrance as well as giving a poor yield.

The two pieces of apparatus just described constitute remarkable transitional forms between the Chinese and Western types of still. After all, if a pipe is led up through the bowl of the Chinese catch-cup or spoon collector on its long arm, the structure becomes topologically similar to the ancient Western annular rim channel. Is it conceivable, perhaps, that these types of still were invented by people who had been brought up in the Chinese tradition and then came to know of the Western peripheral gutter? Conversely, could the Moor's head still top of the West have arisen because a knowledge (or even a rumour) of the Chinese cooling-bowl reached the occidental world some time between the +12th and the +15th centuries? This surmise will acquire particular significance in the light of facts recorded in the ensuing sub-section.

It is thus evident that much further research will be needed before we can hope to translate the theoretical diagram of Fig. 1454 into terms of the concrete appearance and development of the different forms among the various peoples in successive ages. But the task will be a truly fascinating one. Here we have only touched the surface layers.

Nevertheless, it does begin to look as if the two polar types of still, the Western and the Eastern, started out from two entirely different instruments of the simplest description, the pot-lid with turned-up edges and the steamer with its grating. The former led to the concave still roof and annular rim collection, the latter to the convex still roof and central collection. Neither of these types could easily affect the other because of their complementary logic, yet (as we have just seen) inter-specific 'crossing' did eventually occur. The most important transmission seems to have come (so far as we can see at present) from the 'Moors' of the East as an idea only, skipping the Arabic culture-area and generating a necessarily different form of still-head cooling in the

---

[a] Schelenz (2), fig. 118, from Gildemeister & Hoffmann (1), vol. 2, pp. 379 ff., based on Anon. (97), p. 659 and (96), p. 85. This still also was used for star anise oil.

[b] Both these still types seem open to the criticism that some of the condensate would drop directly into the tube or tubes and so back into the still, but this may have been obviated in practice by a short right-angle turn at their tops, like the cowled ventilators of ships.

[c] A 'verbluffend einfache und zweckmässige Vorlage', remarked Schelenz.

West. But there was another source from which the principle of water-cooling for condensation could have come, namely the third tradition, geographically intermediate, Indian or Gandhāran. This still-type, the probable ancestor of all retorts, was allied with the Western still because of its concave still-head, but also with the Mongol-Chinese because of the water-cooling which its receiver probably had from an early date. Eventually of course all three traditions have been incorporated among the myriad devices of modern chemistry.

(8) THE COMING OF ARDENT WATER

Of all the evocative names in the history of alchemy and chemistry none can be considered more striking than *aqua ardens*—the water that burns—for burning is an attribute of fire, not water. No one can fully appreciate the strangeness of what this meant to medieval people unless he is aware that the work of the alchemist and proto-chemist, almost from Hellenistic times onwards, and certainly also in China, had concerned the *conjunctio oppositorum*,[a] the hierogamic union of sun and moon, the Tao of the synthesis of Yang and Yin, indeed the 'marriage of fire and water'.[b] True, the light fractions of petroleum had been distilled into 'Greek Fire' by Callinicus at Byzantium about +670, but the petrol- or gasoline-like liquids which he and his successors produced never got the name of 'waters', either in West or East, presumably because although whitish and fairly transparent they smelt so differently and were immiscible with water. They were in fact thought of more as oils, quite rightly; and actually received in China that designation from the time of the first knowledge of them.[c] It is therefore a matter of great interest to discover, if we can, who first became familiar with the taste and smell of strong distilled alcohol,[d] and where this was.[e]

With the praises of the drug indited by European writers we can well dispense, yet the words on *aqua vitae* written by Conrad Gesner of Zürich, the great naturalist and chemist, and a personal friend of John Caius, are too seductive to omit.

The taste of it [saith he] excedeth all other tastes and the smel all other smelles. It comforteth the natural heat more than any other remedye; it is most holsom for the stomake, the

---

[a] Cf. pt. 3, pp. 69, 70, 149.

[b] A fascinating account of the 'submarine mare' in the mythology of Śiva has been given by O'Flaherty (1). The fire-breathing mare, *Aurva vaḍavā*, at the bottom of the sea, is like the Yang within the Yin (cf. Vol. 5, pt. 5), and symbolises Śiva and his *shakti Pārvatī*. He is essentially fire and she is water, but they are eternally one. 'He is the ascetic fire which rages against the erotic power, but also the fire of passion that cannot be controlled by asceticism.' She is the giver of the energy of both, and together they image the balance of forces in the universe, flame that can never be quenched, oceans that can never dry. Psychologically it would be hard to go deeper, with the dynamism of the id expressing itself in inwardly-directed mortido and outwardly-directed libido. On the symbolism of the mare and the dragon cf. Rousselle (8).   [c] Cf. pp. 158 ff.

[d] The term originated, as is well known, from the *al-kuḥl* of Arabic authors, antimony sulphide in very finely comminuted form, an impalpable powder used as a kind of mascara by girls and women in the Middle East (cf. Vol. 5, pt. 2, pp. 267–8). Paracelsus applied it about +1535 to the distillate of wine because alcohol was such an extremely subtle thing, so there was *alcool vini* as well as *alcool antimonii* (cf. Partington (7), vol. 2, p. 149).

[e] The best history of alcohol-distillation, perhaps, is that of Arntz (1), but it is lamentably deficient on the Chinese (and indeed the Indian) evidence. There is a short general account of alcohol and its history in a lecture by a distinguished chemist; Armstrong (1).

harte and the liver, it noryseth blud, it agreeth mervelously and most with mans nature, it openeth and purgeth ye mouthes and entrances of the membres, vaines and poores of the body every one, it avoydeth all obstruction and comforteth them—yea, it changeth the affections of the minde, it taketh away sadnes and pensivenes, it maketh men meri, witty, and encreaseth audacitie...It avoydeth and kepeth a man from gray heares, ...it encreaseth the ability of accompanying with women, ...it maketh women apt to conceive but anoyeth them that be greate with childe...[a]

And so on. Who were really the first to prepare this stuff?

(i) *The Salernitan quintessence*

Any definitive picture of the development of the means of distillation of volatile substances in the West in the period between the +2nd century and the late +13th century, the time of Taddeo Alderotti, will have to include evidence which may be gleaned about the idea of a 'Quintessence' derived from wine.[b] This appellation, designating ethyl alcohol as of a quasi-spiritual nature,[c] and adding a fifth (*quinta essentia*) to Aristotle's four elements, is fully present in the writings of the followers of Raymond Lull[d] in the mid or late +14th century.[e] Already before this time alcohol (*aqua ardens* or *aqua vitae*, the 'burning water' or 'water of life') had become important in medicine,[f] and in techniques for the preservation of organic substances; for example John of Rupescissa welcomed it as a quintessential stabiliser which would defend the body from corruption until the last day.[g] Alderotti, in his *Consilia*

---

[a] *Treasure of Euonymus* (+1559), p. 85.

[b] Here the article of Sherwood Taylor (6) is an indispensable aid. In the Jābirian Corpus there is a *Kitāb al-Ṭabīʿa al-Khāmisa* (Book of the Fifth Nature) Kr/396, and while this is mostly explained as a basic materia prima, the word *rūḥ* (pneuma) is occasionally applied to it. There may therefore have been Arabic influence on the Lullian Corpus.

[c] The whole thought-complex here is extremely suspicious of East Asian ideology. Not only were there always five elements in Chinese natural philosophy, unlike the four of the Greeks (cf. Vol. 2, *passim*), but as we have frequently pointed out, Chinese thinkers were averse to any such sharp distinctions between matter and spirit as were customary in European thought from the beginning (cf. here, pt. 2, pp. 86, 92–3, pt. 3, p. 149). If the idea of the quintessence was really in earlier Arabic texts, we should have yet another reason for suspecting Chinese influence on the Jābirian Corpus.

[d] Himself +1235 to +1315, but probably never a practising alchemist.

[e] Especially the treatise entitled *De Secretis Naturae seu de Quinta Essentia*, which describes the distillation of alcohol as well as the theory, and gives pictures of stills. The receivers are cooled.

[f] In the late +13th century Vitalis du Four (cf. p. 196) and Taddeo Alderotti himself used it therapeutically. In +1288 stills were banned from Dominican friaries, pharmacy being considered a secular calling. About +1250 the physician Gilbertus Anglicus recommended *aqua vitae* for travellers (Sarton (1), vol. 2, p. 658)—very reasonably, if one considers the rigours of winter journeys in those days.

[g] *De Consideratione Quintae Essentiae*, written in the first half of the +14th century. This idea was quite a legitimate one; it may well have derived from actual observations of the preservation of perishable plant and animal matter in alcohol. Thence the road lay straight to the use of alcohol and other liquid organic compounds as fixatives, first noted by Robert Boyle, who in +1666 described 'a way of preserving birds taken out of the egge, and other small faetuses' (cf. Needham (2), p. 137). John of Rupescissa quenched gold leaf in his alcohol 'to fix Sol in it', and the numerous complex liqueurs which the religious orders developed in the +16th century were really examples of microcosmic 'philosophical heavens' in which the stellar influences generating the virtues of many plants had been, as it were, captured on earth and 'fixed' for human benefit. On John of Rupescissa himself see Multhauf (1). On ideas of the permanent preservation of the body in relation to material immortality in China, see pt. 2, pp. 294 ff.

Medicinalia[a] about +1280 described how he used the four times re-distilled wine for medical purposes. The ten times distilled, obtained only in small quantities, he called *perfectissima*;[b] it would burn completely away when ignited, and moreover any cloth soaked in it would burn also. Von Lippmann quotes Alderotti's detailed description of his procedure and calculated that at least 90% alcohol could be obtained by his method.[c]

Going back, we find accounts of the distillation of wine in the +13th-century text of Marcus Graecus,[d] as also in Salernitan texts[e] and the +12th-century version of the *Mappae Clavicula*.[f] The latter description is in the form of a cryptogram which was solved without difficulty by Berthelot (10).[g] Little is said about the distillation apparatus, and there is no clear mention of cooling in any text. The distillate, 'burning water', caught fire when a flame was applied, but combustible things dipped in it would not themselves burn.[h] The same observation, using a linen cloth, is made in Marcus Graecus.[i] Berthelot[j] and Diels (3) both emphasised the great similarity in style and content of these two treatises to that of the Alexandrian and Byzantine

---

[a] Now published in full by Nardi (1). Cf. Siraisi (1).

[b] It is not clear why Alderotti redistilled, since his cooled rising coil will give 80% or 90% alcohol in one operation (Reti). It would seem that his text may be corrupt, perhaps containing insertions made by his students who were used to having to re-distil. Without dephlegmatory or fractionation devices multiple distillation is mandatory, with or without water-cooling. The Chinese certainly did it, as is evident from traditional phrases like *san shao*,[1] 'thrice burnt'. About five distillations would be required for 80% alcohol; it is unlikely that in traditional China anyone ever went beyond this level towards absolute, and probably not usually beyond 70%.

[c] (13), p. 1359. Cf. p. 93 above. Reti's reconstruction of the apparatus Alderotti used has been given in Fig. 1469. Cf. Arntz (1), pp. 227ff.

[d] The famous *Liber Ignium ad Comburendos Hostes*, so important for Greek fire, and the beginnings of gunpowder in the West (cf. Leicester (1), pp. 78–9; Berthelot (10), vol. 1, pp. 89ff., tr. pp. 100ff., alcohol recipes, pp. 117, 122, 134, 136ff., 142; Multhauf (5), p. 205; Arntz (1), pp. 218ff.). The present view is that the text as we have it now is of the late +13th century, with much Arabic influence. No Greek version has been found, in spite of the writer's ethnikon; and the background of the text, from internal evidence, is Spanish rather than Byzantine. The earliest entries may go back to the +8th century (cf. Partington (5), p. 60), but both alcohol and gunpowder belong to the last versions.

[e] The writings of the physicians of the School of Salerno (see Multhauf (5), p. 205–6; Leicester (1), p. 76; Forbes (9), pp. 57–8; von Lippmann (15) and (1), vol. 3, p. 32). They fix the discovery not long before +1170, possibly as early as +1155. The Salernitan Corpus from the first half of the century (ed. Sudhoff) has much to say on the distillation of essential oils, but does not mention alcohol. A MS. copy of *c.* +1200, *Aqua ardens sic fit*, in the Library of Gonville & Caius College (451 (392), 15 c) has been reproduced and translated by Arntz (1), pp. 223, 225.

[f] 'Little Key to Painting', one of the Latin practical compendia of metallurgy, dyeing and other chemical arts, probably first begun about +820, but now available only in MSS. of the +10th and +12th centuries (cf. Leicester (1), p. 76; Berthelot (10), vol. 1, pp. 23ff., tr. pp. 31ff., alcohol recipe, p. 61; Multhauf (5), p. 205; Smith & Hawthorne (1), p. 59; Arntz (1), pp. 214ff.). The alcohol recipe (a cryptogram) is not in the +10th-century version.

[g] This is not, of course, the only instance of that kind of discreet self-protection by the 'subtile clerkes' and craftsmen of those centuries. Cf. Vol. 5, pt. 1.

[h] This was clearly on account of the relatively large percentage of water present. Beckmann did an experiment for Degering (1) in which he showed that alcohol–water mixtures containing less than 35% alcohol would quench burning sulphur. Such tests as these, made in medieval times no doubt, would have been the origin of the 'proof-spirit' system, a little gunpowder being placed in a spoon with the alcohol, and set off, or not, as the case might be, when the alcohol burnt down. The technique remained in use until comparatively recent times, and the name lingers still in the excise and commercial worlds, though nowadays differently defined. Cf. Sherwood Taylor (4), p. 156.

[i] Berthelot (10), p. 142.   [j] *Op. cit.*, pp. 89ff.

[1] 三燒

manuscripts,[a] some passages even reproducing the Greek text in translation. This is perhaps truer of the *Mappae* than of the *Liber Ignium*, but it is not untrue of the latter, which also draws, however, on military writers such as Julius Africanus of the +2nd century.[b] Berthelot and Diels were thus inclined to see a great influence of the earlier Greek proto-chemical writings on the *Mappae Clavicula* authors,[c] as indeed on other early medieval Latin recipe-books also.

Diels alone, however, went on to maintain (3) that alcohol had been known to the Hellenistic or Byzantine proto-chemists. As Diels remarked, Aristotle knew that the vapours of wine would ignite, though according to his theory wine, like all other aqueous substances, would only give off water.[d] Diels attached much importance to the recipe mentioned by Hippolytus (d. +235) in which sweet wine was heated with 'sea-foam' and sulphur, producing in some way a lambent nimbus of flame when sprinkled on the head in temple rites.[e] Hippolytus was concerned to unmask the arts of the magicians and thaumaturgic priests, as in Egypt; and Diels suggested that a distillate of weak alcohol, facilitated by the additions to the wine, would have been very suitable for their purpose. He found it significant that two points of the Hippolytus recipe—the additions to the wine, and the failure of the alcohol to burn away completely, igniting material in contact with it—appear in the *Liber Ignium* of Marcus Graecus.[f] Von Lippmann, on the other hand, in a series of papers (13, 14, 15, 16, 17, 18), argued strongly that the discovery of alcohol occurred in Italy and did not antedate the +12th century, i.e. the time just before the last version of the *Mappae Clavicula*, emphasising especially the dependence of the discovery on improvement in the methods of cooling.[g] He pointed out that Hippolytus made no mention of alcohol under any name or of distillation in any form, and that the nimbus effects could have been obtained by sprinkling the hot wine itself, its readiness to evolve alcohol being increased by its salt content. In a letter to Diels, von Lippmann described his quite negative results on attempting to obtain alcohol by means of the Alexandrian still (*ambix* or *dibikos*).[h] There was no trace of alcohol in the distillate.

Since alcohol boils at a temperature more than 20° lower than water it is indeed

a Both the papyri and the Corpus.

b Whose *Kestoi* contains much on incendiary compositions (cf. Partington, 5). The phosphorescence recipes in Marcus are clearly Hellenistic.

c Whom Diels from philological evidence was inclined to locate in Carolingian France. The more usual view is that they were Italians.

d *Meteorologica*, 387*b* 9; Didot ed., vol. 3, p. 622. So also Theophrastus *De Igne*, 67, Wimmer ed., vol. 2, p. 70.

e *Refutatio Omnium Haeresium*, IV, 31; McMahon & Salmond tr., vol. 1, p. 98; Legge tr., vol. 1, p. 96. The sea-foam was probably crude salt, which raised the boiling-point of the wine and reduced the proportion of water in the evaporate. Arntz (1), p. 207, describes experiments which demonstrate the effect. De Rochas d'Aiglun (1) was one of the first to call attention to the passage.

f In Berthelot (10), pp. 117, 142. Diels' belief in the distillation of alcohol by the later Alexandrian proto-chemists was supported to some extent by Degering (1) who urged (on philological grounds) that an alcohol recipe in one of the *Mappae Clavicula* MSS should be placed in the +8th century. This was strongly contested by von Lippmann (9), vol. 1, and has not won general assent. Cf. Forbes (9), pp. 88–9.

g Especially the introduction of water-cooling. Ruska (17); Sudhoff (2) and Sherwood Taylor (4), p. 156, concurred in these views of von Lippmann. So at one time did Reti (7), p. 656. With some reservations (p. 127) we still do, and so does Arntz (1), pp. 205, 211.

h In Diels (3), pp. 30–1. In spite of this, Diels never gave up his point of view; cf. (1), p. 153.

easily lost. Nevertheless the dependence of alcohol-production on still-head, condenser or receiver cooling must not be stated in too extreme a way. Much depends on the exact technique used. Suppose wine is distilled in an Alexandrian still with peripheral catch-rim, cooled by air alone. As long as the rising vapours are rich in alcohol the still-head warms up only slowly, and the condensate is collected reasonably well, but as the alcohol-content of the wine in the still falls the temperature at the roof starts to climb, and much of the vapour escapes uncondensed. This results from the specific heat of alcohol, which is only 58% of that of water, and from its heat of vaporisation,

Fig. 1496. 'Rosenhut' still for alcohol distillation, from Puff von Schrick's book, first published in +1478. When there is no still-head cooling, as here, the design and dimensions of the still have to be rather special, and particular conditions have to be fulfilled.

only 37% of that of water. Hence it will come off while the solution is still relatively cool, and it will not heat the still-head too much when it gets there. The still top also has to be large relatively to the size of the bottom containing the wine (cf. p. 126), the room temperature has to be cool, and moreover the operation has to be carried out gently and slowly, never letting the still-contents boil—as indeed many of the old distillation manuals recommend.[a] All this probably explains why the metal 'Rosenhut' still with a high conical un-cooled head is so often depicted during the +15th and +16th centuries in use for the distillation of strong spirits.[b]

This last still seems rather similar to one of those described in the Greek proto-chemical Corpus.[c] In the +4th century Zosimus (Ko Hung's contemporary) added to the other forms a still with a body of large diameter so as to give a greater evaporating surface suitable for distillation at relatively low temperatures, as on a water-

[a] Dr L. Reti told us that he had confirmed in personal experiments the feasibility of the method if carefully handled.

[b] E.g. by Puff von Schrick (Fig. 1496), Brunschwyk, Biringuccio, etc., cf. Forbes (9), figs. 29, 40, 57.

[c] Here Dr Reti dissented, as he felt that the height and conical shape of the Rosenhut were particularly important, though with sufficient care alcohol can be got off from any still.

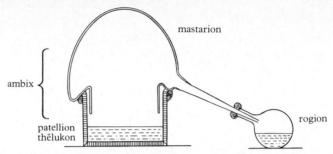

Fig. 1497. The *mastarion* or 'cold-still' of Zosimus (reconstructed by Sherwood Taylor (5), pp. 199 ff.). The *mastarion* is the breast-shaped still-head; and the pan which formed the still body was called the *patellion thēlukon* or female dish, presumably because the still-head was inserted into it. The two together formed an *ambix*. This apparatus was used at low temperatures on a water-bath or dung-bed, and its form permitted a rapid diffusion of vapour to the still-head, as is necessary for distillation in such conditions.

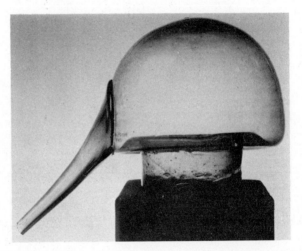

Fig. 1498. Glass still-head of *mastarion* type from Egypt, datable from the + 5th to the + 8th century (photo. Royal Ontario Museum, Toronto).

bath (Fig. 1497).[a] The conical or breast-shaped cowl was called *mastarion* (μαστάριον), and the whole still (*ambix*, ἄμβιξ) with its two parts 'hermaphrodite' (*arsenothēlu*, ἀρσενόθηλυ).[b] The process described was the destructive distillation of eggs to get ammonium sulphide and calcium polysulphides;[c] and the passage seems to be the oldest in the West in which the 'bain-Marie' or water-bath is mentioned. Actual examples of glass still-heads apparently of this type, dating from the + 5th to the + 8th centuries, have been recovered in Egypt and Syria (Fig. 1498).[d] In later times, such pieces of apparatus were known as 'cold-stills'.[e]

[a] *Corp. Alchem. Gr.* III, viii, 1, tr. Berthelot & Ruelle (1), vol. 3. p. 143.
[b] Sherwood Taylor (5), pp. 198 ff.   [c] Cf. Vol. 5, pt. 2, pp. 252, 271, pt. 3, p. 103.
[d] Cf. Erman & Ranke (1); Sherwood Taylor (4), p. 155 and pl. 10*a* opp. p. 161; Davies (3). Another example is in the Victoria and Albert Museum, London; see Moorhouse, Greenaway *et al.* (1), p. 101.
[e] As, e.g. in French (1), p. 17, reproduced by Sherwood Taylor (5), fig. 12 (cf. Fig. 1496).

These important considerations, however, do not necessarily invalidate von Lippmann's view that still-head water-cooling was historically the basic limiting factor for
the discovery of alcohol. The Salernitan Masters and the writer of the recipe in the
last version of the *Mappae* may conceivably have used no water-cooling—though after
all a wet sponge had been mentioned by Zosimus—but the working hypothesis
remains open that efficient still-head cooling was the trigger for the discovery of
alcohol in the West. Moreover there was a part of the Eastern world where the
practice was traditional and where the idea could have come from.[a] Otherwise one
has to explain why eleven centuries elapsed between the first stills of the Alexandrians
and the first successes in wine distillation by the Salernitan Masters.[b]

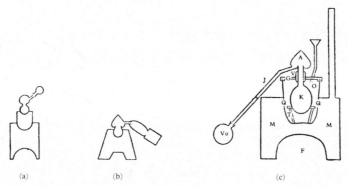

Fig. 1499. Drawings of stills in al-Kindī's *Kitāb Kīmiyā' al-'Iṭr wa'l-Taṣ'īdāt* (Book of Perfume
Chemistry and Distillations).
   (*a*) Still of retort type, with no annular rim but set upon a water-bath above the stove (from Garbers
(1), p. 94).
   (*b*) Still of Hellenistic type with annular rim, set in a stove gently heated with charcoal or coal (from
Garbers (1), p. 95).
   (*c*) Reconstruction of the latter drawing by Garbers (1), p. 19. F, the fire-place (*mauḍi' al-wuqūd*);
M, the stove (*mustauqad*); Q, tall basin of pottery or stoneware (*qidr birām au fakhkhār*); T, ring support
of wood (*ṭauq*); K, still body. V, lute and bands connecting it with A, the still-head, of Hellenistic
type, delivering to J, the side-tube (*iḥlīl*), and Vo, the receiver (*qabila*); G, a wooden cover to keep in
the heat (*ghiṭā'*); O, funnel for adding hot water. The chimney is on the right.

What exactly the Arabs got when they put wine in their stills remains a moot point.
That they did do this we know from several statements, but the usual view is that
although they distilled wine they did not find the product very interesting. Often
they do not mention wine at all.[c] For example al-Anṣārī al-Dimashqī (d. +1327)
devotes much space in his cosmography (*Nukhbat al-Dahr*) to the distillation of

---

   [a] Not only so, but as we shall shortly try to demonstrate (pp. 141 ff.), precisely that part of the world
had been producing strong alcohol four or five centuries before people in Europe did it. The possibility
therefore presents itself that the use of a Rosenhut or cold-still was a secondary development, trickier
to carry out but avoiding the greater complexity of water-cooling devices. In pondering these problems
we have been much helped by private correspondence with our late friend Dr Ladislao Reti, for which
we record our warmest thanks.
   [b] And even since Zosimus' sponge and his *mastarion* cold-still eight centuries had passed. Surely
there must have been some new impetus.
   [c] There was of course the Koranic prohibition of wine, which Bernal (1), p. 203, saw as the main
limiting factor, but wine was very well known in Islamic culture, especially after the conquest of Persia.

essential oils and naphtha, but none to wine.[a] Abū'l-Qāsim al-Zahrāwī (Abulcasis, d. c. +1013) described the distillation of vinegar 'for whitening'[b] in an apparatus similar to that in which rose-water was distilled, adding that wine can be distilled in the same way.[c] Much earlier, Ya'kub ibn Isḥāq al-Kindī (fl. c. +803 to c. +870) had alluded to the same thing in his treatise on essential oils, *Kitāb Kīmiyā' al-'Iṭr wa'l-Taṣ'īdāt* (Book of Perfume Chemistry and Distillations).[d] Just beside a picture of an alembic on a stove (Fig. 1499a) he says: 'In the same way one can drive up date-wine (*nabīdh*) using a water-bath (*fī'l-ruṭūba*), and it comes out the same colour as rose-water'.[e] In the same century the Jābirian Corpus contains statements that the vapour of boiling wine would catch fire,[f] but that was not going beyond Aristotle.

Perhaps the reason why the great Arabs found the distillate of wine uninteresting was because its alcohol-content was so low. None of them make any reference to cooling, either of still-head, receiver or side-tube, and they were probably quite unaware that there could be any 'spirits' with (as we should say) a boiling-point lower than water,[g] and therefore likely to be lost in the surrounding air if not expressly cooled. Their alembics were evidently quite capable of distilling in quantity not only the essential oils of plants and flowers for civilised life,[h] but also the 'naphtha' or

<hr/>

[a] Tr. Mehren (1), pp. 58, 264; cf. Forbes (9), pp. 48ff.; Wiedemann (22), pp. 246ff. Ruska (22) was convinced that al-Rāzī never knew alcohol, nor Ibn al-Baiṭār either. And if al-Khāzinī (c. +1120) had done so, he would hardly have listed olive oil (sp. g. 0·915) as the lightest of known liquids (cf. von Lippmann, 7). Ruska (23), continuing his search, examined the MSS of the most important Arabic treatises on agriculture, and found that although all aspects of wine-growing were dealt with at length there was never any mention of distilled spirits.

[b] This had also been done in a previous generation by al-Rāzī (d. +925) according to Haschmi (priv. comm.). Since the boiling-point of acetic acid is 118° it was theoretically feasible, but in fact the operation is very difficult. That the Chinese may have tried to do this in the +6th century we suggest elsewhere (p. 178). On the 'sharp waters' of the Arabs, which included caustic alkalies, see Ruska & Garbers (1); Leicester (1), pp. 69, 72; Multhauf (5), p. 140. In China *shih hui*[1] (lime) had been slaked to calcium hydroxide (*shu shih hui*[2]) since time immemorial (RP71); while potassium carbonate (*tung hui*,[3] *hui chien*[4]) was got from wood ash, and sodium carbonate (*chien*[5]) from natural deposits (cf. p. 180, and *PTKM*, ch. 7, (pp. 90–1), etc.). Caustic alkalies made by the action of slaked lime on these carbonates were therefore available to the Chinese alchemists also, though we do not hear much about them. Cf. pp. 395, 398.

[c] *Liber Servitoris*, a translation of ch. 28 of his system of medicine, the *Kitāb al-Taṣrīf*. On this see Hamarneh & Sonnedecker (1). For the reference to the distillation of wine see also Sherwood Taylor (6), p. 254; Forbes (9), p. 41.

[d] Dunlop (6), pp. 229ff., voices doubts about the authorship of this work, mainly because of some recipes for the falsification of essential oils; but concedes that parts of it are probably rightly attributed, the others being by a pupil or pupils.

[e] Garbers tr. (1), p. 95; discussed by Haschmi (2, 3). Presumably this means colourless or almost so, not pink. Al-Zahrāwī also made much use of water-baths, including one, the *berchile*, which did not stand on the furnace itself, but was kept at just under boiling-point by a supply of boiling water from a neighbouring vessel directly heated; on this see the discussion of Speter (1) and Ruska (18).

[f] Haschmi (priv. comm.), from the work of Kraus and Steele.

[g] The boiling-point of ethyl alcohol is just under 78° C.

[h] The aromatic constituents of these oils, terpenes and polyterpenes, have boiling-points ranging between 150° and 275° (e.g. pinene at 155°, camphene at 160°, limonene at 175°, caryophyllene at 255° and cadinine at 274°). Similarly geraniol boils at 229°, linalool at 198° and citral at 225°. But of course they will come over well below their boiling-points, if their vapour-pressures are suitable, especially when accompanied by steam. It is fairly clear from the book of al-Kindī that all the perfume-oil processes were steam-distillations (cf. p. 120) even if no water was added to the fresh plant material; and generally done on the water-bath, i.e. much below the boiling temperatures of the oils themselves, thus

<hr/>

¹ 石灰　　² 熟石灰　　³ 多灰　　⁴ 灰鹼　　⁵ 鹼

lighter fractions of petroleum for military use in Greek fire flamethrowers.[a] Yet the apparatus used seems no advance at all on that of Mary the Jewess. Fig. 1499 *a, b* shows al-Kindī's own drawings,[b] and Fig. 1499 *c* Garbers' reconstruction of the still he used in the +9th century for the preparation of all kinds of perfumes and essential oils.[c] Von Lippmann's final opinion was that lack of special cooling arrangements among the Arabs was beyond doubt, so that separation of substances of low boiling-point would have been impossible.[d] This would be true also of the later Arab chemists such as Abū Bakr ibn Zakarīyā al-Rāzī (+865 to +925)[e] or Abū'l-Qāsim al-Sīmawī al-'Irāqī about +1270.[f] But then we are faced with a curious paradox—the most effective still-head cooling device in Renaissance Europe was always called, as we have seen, the 'Moor's head' (Fig. 1466).[g] No explanation of the origin of this term seems to have been

preventing any decomposition. But since their boiling-points are so much higher than that of water they condense very easily without any cooling devices.

---

[a] Here just the same thing applies. When crude petroleum is distilled, the largest fractions obtained have boiling-points higher than that of water—'ligroin' from 100 to 120°, 'cleaning oil' from 120 to 150°, and kerosene or 'burning oil' from 150 to 300°. Hydrocarbons like dodecane (214°) and cetene (274°) would be in this last fraction, and the octane (126°) would also have been obtained; but the Byzantine and Arabic chemists would have lost all their 'petroleum ether', i.e. pentane (37°) and hexane (69°); probably a good deal of their heptane (98°) as well. The higher b.p. lubricating oils, vaselines and waxy paraffins would have remained behind in their stills. Multhauf (2) makes the interesting point that through many centuries of early distillation there was a prejudice against residues, not overcome (with a better appreciation of the meaning of chemical separation) until Libavius' time at the end of the +16th century. All in all, the preparation of 'Greek fire' or *naft* must have been a hazardous exercise, and one would like to know just how the dangers of fire were overcome (cf. Forbes (9), pp. 52–3). Presumably water-baths were not used, since the Byzantines and Arabs did not much mind what particular mixture of light fractions they got, so long as it would burn fiercely—like our petrol or gasoline —even on water. A certain amount of decomposition did not matter, but even so the furnaces must have been regulated very carefully, especially if the side-tubes were relatively small in diameter.

[b] Garbers (1), pp. 94, 95.   [c] (1), p. 19.

[d] (1), vol. 3, pp. 32, 54, 113. See also Wiedemann (15, 22, 29), outstanding for his knowledge of Arabic chemical and technical texts, who could adduce no cooling systems. We say 'special cooling arrangements' having in mind primarily water-cooling, though presumably the Arabs could have achieved something by the use of the cold-still or Rosenhut (cf. p. 125). Although this seems to go back to Zosimus, the fact that neither the Alexandrians nor the Arabs successfully prepared alcohol strengthens the suggestion already made that distilling it without water-cooling was a rather tricky procedure requiring several special conditions, and may therefore have come a good deal later than the first discovery. Dr Reti was loth to accept this, feeling that air-cooling must have preceded water-cooling, but the logical order may not always have been the historical one, as we have found before in the sequence of power-sources (Vol. 4, pt. 2, pp. 192–3) where animal power may not necessarily always precede water-power.

[e] Most of what al-Rāzī says about apparatus is in his *Kitāb Sirr al-Asrār* (Book of the Secret of Secrets), tr. Ruska (14), comm. Ruska (15, 16) with notes on his full bibliography.

[f] See Mieli (1), 2nd. ed., p. 156. Most of the apparatus used by him is described in the *Kitāb al-Kanz al-Afkhar wa'l-Sirr al-A'ẓam fī Taṣrīf al-Ḥajar al-Mukarram* (Book of the Most Glorious Treasure and Greatest Secret on the Transmutation of the Philosopher's Stone), on which see Ruska & Wiedemann (1). His more famous book, *Kitāb al-'Ilm al-Muktasab fī Zirā'at al-Dhahab* (Book of the Knowledge of the Cultivation of Gold) is available in translation by Holmyard (5).

[g] It is true that the oldest device, as used by Alderotti, and presumably therefore also by the +12th-century Salernitan Masters and the shadowy distillers of the *Mappae* and the *Liber Ignium* (though we have no positive knowledge of what they used), seems to have been the rising water-cooled serpentine side-tube (Fig. 1469), but one cannot find Arabic antecedents for that either. The possibility remains open that the men of the +12th century used 'Chinese' stills; both serpent and 'Moor's head' being later developments, neither exactly like the Chinese still-head cooling-bowl with its central collection system.

This might be a rather hazardous hypothesis if no relict traces of the Chinese still were to be found in the Western world. But the fact is, as we have seen (pp. 106, 111 above), that there are such traces both in Europe and the Americas, how and when mediated, and by whom, remains a mystery.

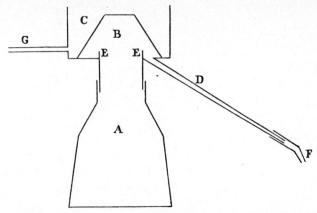

Fig. 1500. A 'Moor's head' still from a Moorish land, cross-section of the Algerian steam distillation apparatus described by Hilton-Simpson (1), p. 20. A, the still body; B, still-head; C, cooling reservoir; D, side-tube; E, beading forming the annular gutter; F, receiver; G, tube for withdrawal of heated water. All parts are made of tinned copper.

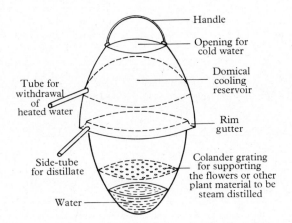

Fig. 1501. Steam distillation apparatus of compact form in tinned copper from Fez in Morocco (orig. drawing, from the specimen in the collection of Dr Stephen Toulmin and Dr June Toulmin, 1963). Here the Moor's head condenser approaches closely to the Chinese type, yet its base which forms the roof of the still is concave in true Western style, not convex, and thus the Hellenistic annular gutter is retained.

attempted by historians of chemistry, but it is hard to believe that it was purely pejorative and does not betray to us some influence from the world of Islam. It is at least a fair comment that Moor's heads do occur in Moorish lands now, as witness the traditional *qaṭṭāra* stills of Biskra and Timgad in Algeria,[a] of Fez in Morocco,[b] and of

[a] Hilton-Simpson (1), p. 20 and pl. III (see Fig. 1500). Thomas Shaw, travelling in the Levant in +1720, saw stills of this kind.

[b] See Goodfield & Toulmin (1), with photographs. I had the pleasure of examining one of these, all of tinned copper, brought back to London from Fez in 1963 by Dr June Goodfield Toulmin, to whom our thanks are due (see Fig. 1501). Its cross-section is quite similar to those of Baumé in +1777 (Schelenz (2), fig. 40). In 1909 Wiedemann (22) published a cross-section of a still then used in Damascus for the preparation of rose-water; it does not make sense as it stands, but the most likely interpretation would approximate it to those just mentioned.

Karachi in Pakistan.[a] If Islamic influence can nowise be substantiated in new researches we may have to look further east, and suppose that 'Moor' meant anyone from the 'Farther Indies', remembering of course that the Chinese still had a water-cooled head built in to its pattern of necessity from the beginning.[b] Here we almost reach a parallel with the case of the magnetic compass related in Section 26, long known in China before its first appearance in Europe c. +1180, yet showing no trace in either Arabic or Indian literature of its passage through those intermediate culture-areas.

And what indeed can be said of India? A century ago Rajendralala Mitra (1) claimed that the *surā* of the Saṃhitās, Brāhmaṇas and epics (−8th to +2nd centuries) was wine distilled rather than only fermented, but he produced no cogent proof. For the Tantric period, of course (+6th century onwards), he may well have been right; but for antiquity the last word long remained with von Lippmann, who argued convincingly that *surā* was wine prepared by fermentation, not spirits.[c] It has also been thought that the *Arthaśāstra* (which reached its present form in the +3rd century) mentions spirits among the many kinds of alcoholic drink under the control of the Fermented Beverages Superintendent;[d] but here again an inexcusably loose use of the word 'liquor' was liable to give a false impression, and von Lippmann concluded that we have really no ground for assuming distillation in this source.[e]

The question has now been reopened by Allchin (1), who assembles an argument indicating that distilled strong alcohol was known in Indian antiquity, and that the Gandhāran stills were in fact used for this purpose.[f] The first part is philological. Allchin noticed that certain words have acquired a curious and persistent set of double meanings. Thus *śuṇḍa* means both 'elephant trunk', 'alcoholic drink' and 'tavern'; *śuṇḍin* 'possessing a trunk' and 'maker of alcoholic beverage'; *śuṇḍika* 'seller of alcoholic wares' and 'tavern'. These senses occur as early as the grammarian Patānjalī in the −2nd century, as well as in the *Ramayana* and the *Mahābharata* (−2nd to +3rd centuries). They are never explained, but the association with the side-tube of the still is, Allchin urges, too obvious to be disregarded.[g] At the other end of history Molesworth (1) in his Marathi dictionary of 1857 gives *śuṇḍa-yantra* as meaning baldly an alembic or retort. Furthermore, the oldest certain reference to distillation in an Indian text, though very late (+16th or +17th century), brings in elephants again, for Govinda Dāsa calls the still which he describes[h] *gaja-kumbhavat*, 'resembling an

---

[a] Jamshed Bakht & Madhihassan (2). They are used mainly for preparing essential oils (*araqiath*), and the *ḥammām-i Mariya* is often used.

[b] In this we are emboldened by the intuition of Schelenz (2), p. 39, favouring westward transmissions from China and India in the history of distillation. The whole question was raised for us in a stimulating correspondence with Dr Ladislao Reti (1965–6).

[c] (9), vol. 2, pp. 204 ff.   [d] Ch. 25, Shamasastry tr., pp. 131 ff.

[e] *Op. cit.*, pp. 29 ff.

[f] Such a view had already been maintained by Mahdihassan (56), p. 164.

[g] Cf. terms derived from shape, such as 'cucurbit' and 'pelican' in the West. The morphological resemblance was first noted by an Indian scholar, V. S. Agrarvala. There are other words with some relevance, such as *pariśrutā*, 'trickling down', in the *Śatapatha-brāhmaṇa* for example, but the significance of this is hard to distinguish from filtration.

[h] Strangely enough, a *dibikos*.

elephant-pot', this referring not so much to the trunk as to the two protuberances on the elephant's head which exude a secretion during rutting. This is the *Bhaiṣajya-ratnāvalī* (Jewel Necklace of Materia Medica), used by Mitra though not clearly mentioned by him.[a]

Secondly, the archaeological. At Pushkalāvatī there were found a large number of receiver-bottles though only one or two alembics or their fragments, and many of these receivers were stamped with *tanga* monograms, sometimes royal seals (*raja-mudrā*), suggesting that the receivers were corked[b] and marketed as such, almost as if 'appellation controlée'. Such seals are mentioned in the *Arthaśāstra*.[c] Large numbers of drinking-cups were also found piled up at the site. As for the chemical technology, although some *ad hoc* experiments would be desirable, it seems fairly sure that stills of Gandhāran type could be used for the preparation of alcohol if small amounts of wine were used, with carefully controlled heating, and the receiver cooled in water. If so, we should have another Asian source for the stimulus of the Moor's head. On the other hand the possibility has to be kept open that the main use for these stills was for mercury; that would be much more in line with the *rasāyana* of the later Tantric alchemical books, and perhaps with the marked opposition of the mainstream Indian religious traditions to wine, *a fortiori* to distilled spirits. The royal seals would be equally appropriate for the purified 'precious' metal, but the disproportion of receivers and alembics would be more bizarre, as also the presence of so many vessels that could have been used as drinking-cups.[d] So perhaps judgment should be suspended for a while, though there is certainly a strong case for Gandhāran alcohol-distillation. But now the moment has come to look more closely at the evidence for alcohol-distillation in early medieval China itself.

(ii) *Ming naturalists*

This has been an intricate story, beset by lack of explicit early descriptions, by the absence of a distinctive technical term for distillation in classical Chinese, and by the misunderstandings not only of Western scholars but of the Chinese naturalists themselves. Nevertheless with patience a coherent picture can be made to come into focus. Li Shih-Chen, in his *Pên Tshao Kang Mu* of +1596, gave two rather confused accounts, the wording of which has given rise to many misapprehensions that need correcting with care, but a close study of his words reveals much. In modern scientific language distillation is called *chêng liu*,[1] an unambiguous binome, but in earlier times one may have to recognise the process by the first word alone, which also means 'to steam' in general, and the apparatus by the word *tsêng*,[2] which can also mean any kind of 'steamer', such as those ancient ones already described (p. 27 above). Other words

---

[a] Cf. Majumdar (2), p. 251.   [b] With what, one wonders? Cf. p. 143.
[c] Ch. 21, Shamasastry tr., pp. 121, 123.
[d] It should be possible to settle the matter by using modern methods for detecting traces of metallic mercury in the pottery.

[1] 蒸餾   [2] 甑

therefore acquire particular importance as indicators of what was going on, e.g. *shao chiu*[1] or 'burnt-wine' (entirely analogous to *Branntwein* and brandy),[a] for which there are a number of references in Thang literature none of which were taken into account by Li Shih-Chen. The question of such a distillation of *aqua ardens* between the +6th and the +12th centuries has been much debated by Chinese historians of science. Yuan Han-Chhing (*1*) was the first to collect the literary references, Tshao Yuan-Yü (*2*) was convinced to begin with but later changed his mind (*3*), while since then Wu Tê-To (*1*) has provided more pieces of evidence, and we shall here adduce a greater number still.[b] One must of course in assessing probabilities always remember the water-cooling embodied in the basic design of the Mongol and Chinese apparatus. Furthermore, the subject of strong alcohol in Chinese culture is intimately bound up with an entirely different method of getting it, not by distillation but by freezing the water with which it was mixed. We shall leave the discussion of this simple and primitive technique, assuredly originating in Northern or Central Asia, till the end of our survey, not as a half-relevant appendix, but because it may perhaps be the foundation-stone of the story of all alcohol production everywhere, the starting-point from which it may be possible to trace step by step its gradual spread through all the Old World cultures. But before going further, however, it is necessary to purge the sinological literature of a number of mistranslations which have purported to indicate the existence of distillation in China before the +6th century. We have no wish at all to deny the possibility of its existence as far back as Li Shao-Chün or Liu Hsiang; we can only say that there is as yet no positive evidence for it in the Chhin, Han and Chin periods, such evidence not being provided by the words which the eminent sinologists were translating.

Thus, as the first example, let us look at passages in the book of Wang Chhung[2] written about +83, *Lun Hêng*[3] (Discourses Weighed in the Balance), as translated by Forke (*4*). 'From cooked grain wine is distilled', or 'Distilled wine has different flavours', or 'The cook and the distiller';[c] but in all cases the term in the text is *niang*,[4] which invariably means fermented, never under any circumstances distilled.[d] Elsewhere the characters *chih niang*[5] come together,[e] but they belong to two different grammatical phrases, so that instead of 'Fragrant grass can be used for the distillation of spirits, its perfume being very intense' one should read:'...can be burnt (as incense); and fermented (wine) can be perfumed with it.'[f] In the following century

---

[a] On the origin of these terms from Lat. *coctum*, and Ger. *gebrannt*, cf. Schelenz (*2*), p. 65.

[b] The positive view has been warmly supported by Schafer (*16*), p. 190, and earlier by W. Eberhard in his review of Hermanns (*1*). Shang Ping-Ho (*1*) concurs. Shinoda Osamu (*3*) and Ōtani Shō (*1*) are more uncertain, but incline to the positive view.   [c] Ch. 5, tr. Forke (*4*), vol. 1, p. 154.

[d] As we already pointed out, following Dudgeon (*2*), in Vol. 1, p. 7. Could it be that a similar misunderstanding led Frodsham (*1*), p. 109, to write (commenting on his translation of the poems of Li Ho, *c.* +810), 'Hsin-fêng was a suburb of Chhang-an where wines were distilled'? In any case, evidence which will be presented below goes to demonstrate that for this period (though not very long before it), his expression could be well justified.   [e] Ch. 18, tr. Forke (*4*), vol. 2, p. 167.

[f] Or just 'to give a perfume', since *niang* can sometimes mean 'to cause'. Legge (*8*), pp 152–3, had an *idée fixe* that *chiu* (wine) in the *Shih Ching* always meant distilled spirits; this was supported by a mistranslation due to de la Charme (*1*) in +1733, but it has no basis in fact.

[1] 燒酒   [2] 王充   [3] 論衡   [4] 釀   [5] 爇釀

came the life of Tshui Shih,[1] studied by Balazs (1), who in a moment of aberration called him at one point a 'destillateur d'eau-de-vie'.[a] The text says that after being Governor of Liaotung he fell upon bad times and gained his living *i ku niang fan chou wei yeh*,[2] i.e. by running a wine-making business and selling congee or rice porridge.[b] Then, as might be expected, the *Pao Phu Tzu* book has caused many to stumble. Elsewhere we refer to the double impossibility of Feifel (1), translating Ko Hung's remark about crabs and hemp: '...if hemp sets distilled spirits in fermentation',[c] which ought to read 'But as for crabs affecting (the setting of) lacquer,[d] or hempseed (oil) spoiling wine (*ma chih huai chiu*[3]), these are matters for which one cannot deduce the (causative) pattern-principle.'[e] In another place Feifel rendered *chiu yün chih shun*[4] as 'wine distilled nine times',[f] but this was a confusion with an ancient process whereby further supplies of carbohydrate in the form of freshly steamed grain were added successively to a fermentation process which would otherwise come to a stop. The context is as follows:[g]

> There are many lesser recipes for consuming elixirs, but since the skill with which they are prepared varies their efficacy is also different. And whatever such factors there may be, an elixir which has not undergone an adequate number of cyclical transformations is like wine that has been just fermented once—it cannot be compared with good strong wine which has been fermented nine times (i.e. by eight additions of steamed grain), (*i tou chih chiu pu kho i fang chiu yün chih shun*[5]).[h]

Finally, even Ware (5), generally so trustworthy, went beyond what we at present dare, in translating the word *fei*,[6] to fly up, as distillation, in the context of mercury. For example, Ko Hung seems to say:[i]

> I know for a fact that at the present day one can become an immortal. I myself can renounce cereal foods, and other things that people normally eat. I aver that mercury can be distilled (*wo pao liu-chu kho fei yeh*[7]), and that gold and silver can be sought out (i.e. made artificially), (*huang pai chih kho chhiu yeh*[8])...

Now although we know that mercury was being distilled in plenty in the ordinary way later on, it was probably being prepared in Han and Chin times mainly by *destillatio per descensum* (cf. p. 55), and therefore it would be safer to translate Ko Hung here with the meaning of sublimation. Moreover it seems significant that descriptions of

---

[a] P. 107. Cf. Vol. 5, pt. 2, p. 140 and Fig. 1341 *b*.

[b] It is true that late Western dictionaries associate the word *ku* with spirits, but this was a development of recent times, and anciently— as can be seen from the Khang-Hsi Dictionary—the meaning was always wine-shops and the wine-trade.　　　[c] P. 197, from ch. 3, p. 5*b*.

[d] On this see p. 207 below.　　　[e] Tr. auct. adjuv. Ware (5), p. 61.

[f] (2), p. 6.

[g] Ch. 4, p. 3*a*, tr. auct., adjuv. Ware (5), p. 72. Wu Lu-Chhiang & Davis (2), p. 237, sensed the pitfall, and contented themselves with 'processed'.

[h] *Tou* is the technical term for submitting wine to successive fermentations. Couvreur in his dictionary got this right, but Giles put carelessly 'to redistil'.

[i] Ch. 3, p. 5*b*, tr. Ware (5), p. 60, mod. auct.

[1] 崔寔　　　　[2] 以酤釀販鬻爲業　　　[3] 麻之壞酒　　　[4] 九醞之醇
[5] 一酘之酒不可以方九醞之醇　　　[6] 飛　　　[7] 吾保流珠可飛也
[8] 黃白之可求也

distillation in subsequent centuries rarely use the word *fei*, even for mercury,[a] though we have seen an example of it in the +12th-century text quoted on p. 79 above. Therefore on the whole we should prefer to read: 'I aver that mercury can be volatilised, and that yellow and white (metal, i.e. gold and silver) can be produced (successfully by chemical means).'[b] And the same applies to what Huang Ti did;[c] we should be happier to say for *tao Ting-hu erh fei liu-chu*[1] 'When he came to Tripod Lake he volatilised (or sublimed) mercury'.

We may now consider the best known passage in the *Pên Tshao Kang Mǔ*.[d] Giving as synonyms for *shao chiu* (burnt-wine) *huo chiu*,[2] 'fire-wine', and *a-la-chi chiu*,[3] Mongol *araki* (which he thought had first been mentioned in the *Yin Shan Chêng Yao*, c. +1330), Li Shih-Chen goes on to say:

The making of burnt-wine was not an ancient art. The technique was first developed in Yuan times (+1280 to +1367). Strong wine (*nung chiu*[4]) is mixed with the fermentation residues (*tsao*[5]) and put inside a still (*tsêng*[6]). On heating (*chêng*[7]) the vapour is made to rise, and a vessel is used to collect the condensing drops (*ti lu*[8]). All sorts of wine that have turned sour can be used for distilling (*chêng shao*[9]).[e] Nowadays in general glutinous rice (*no-mi*[10]) or ordinary rice (*kêng mi*[11]) or glutinous millet (*shu*[12]) or the other variety of glutinous millet (*shu*[13]) or barley (*ta mai*[14]), are first cooked by steaming (*chêng shu*[15]), then mixed with ferment (*chhü*[16]) and allowed to brew (*niang*[17]) in vats (*yung*[18]) for seven days before being distilled. (The product) is as clear as water and its taste is extremely strong. This is distilled spirits (*chiu lu*[19]).[f]

Not much commentary is necessary here except to draw attention to the ferment *chhü*,[16] highly characteristic of Chinese brewing; it is a mixture of *Aspergillus* moulds together with yeasts grown under special conditions on a prepared cereal basis, and the moulds have the function of breaking down the polysaccharides (just as the intrinsic enzymes of malted grain do in the West) so that the yeasts can act upon them to produce the alcohol. An interesting account of the preparation of *chhü* has been translated from the *Chhi Min Yao Shu* of c. +540 by Huang Tzu-Chhing & Chao Yün-Tshung (1); we shall quote from this in due course when dealing with the fermentation industries in Sect. 40 (Vol. 6). A little later, among his remarks on the medical uses of strong alcohol, Li Shih-Chen says that it is of the same nature as fire, and that it can easily be set alight, disappearing as it burns away. He concludes the

---

[a] *TKKW*, ch. 16, p. 2*a*, has *shêng*,[20] to rise up (cf. Sun & Sun tr. (1), p. 280).

[b] Feifel (1), p. 197, was much further off the rails than Ware—'I warrant that one can make people fly (with the aid of) the "flowing pearls" (i.e. mercury)'.

[c] Ch. 13, p. 3*a*, cf. Ware (5), p. 215.

[d] Ch. 25, p. 41*b* (p. 34), tr. auct.

[e] This remark recalls the tradition recorded by Cibot (15) that spirits from cereal wine were first found out by a Shantung peasant-farmer who used mash from a mould fermentation which had gone wrong and was not going to yield a drinkable result.

[f] The use of the word dew (*lu*) here is strangely reminiscent of the 'mountain-dew' beloved in Scots and Irish use for designating whisky, especially when illicit.

[1] 到鼎湖而飛流珠	[2] 火酒	[3] 阿剌吉酒	[4] 濃酒	[5] 糟
[6] 甑	[7] 蒸	[8] 滴露	[9] 蒸燒	[10] 糯米	[11] 粳米
[12] 黍	[13] 秫	[14] 大麥	[15] 蒸熟	[16] 麴	[17] 釀
[18] 甕	[19] 酒露	[20] 升		

above passage by quoting from Wang Ying's[1] *Shih Wu Pên Tshao*[2] (Nutritional Natural History) of *c.* +1520 about a particularly strong variety of spirits imported from Hsien-Lo[3] (Siam). There, it seems, they rectified, distilling the wine twice and adding to it aromatic substances. Stored in casks made lacquer-shiny inside by the smoke of burning sandalwood, sealed with wax and matured for two or three years, such 'whisky' was imported by sea; it was expensive but very potent, not only as a drink but as a medicine.[a] This description was in fact several centuries earlier than Wang Ying himself (who had only copied it), or even the +15th-century navigations, so that it will figure presently in our account of pre-European alcohol distillation (p. 145 below).

Many Western scholars, such as Laufer[b] and von Lippmann,[c] have known about this first passage, and were only too happy to accept the late dating of so high an authority, but things are not so simple. There is a second passage in the *Pên Tshao Kang Mu*, even more important, which has generally been overlooked by them. It runs as follows:[d]

Mêng Shen[4] in his *Shih Liao Pên Tshao*[5] (Nutritional Therapy and Natural History, *c.* +670) says: 'Grapes can be used for making wine (*kho niang chiu*[6]). The vine juice (*thêng chih*[7]) is also suitable (for the same purpose).'

There are, in fact, two sorts of grape wine, that obtained by fermentation, which has an elegant taste, and that made like *shao chiu*[8] (by distilling), which has a powerful action (*ta tu*[9]). The makers mix the juice with *chhü*[10] just as in the ordinary way for the fermentation of glutinous rice. Dried raisins ground up can also be used in place of the juice. This is what the emperor Wên of the Wei Dynasty meant when he said that wine made from grapes was better than that from *chhü*[10] and rice, because the intoxicating effect of the former fades away more quickly. In the distillation method many dozens of catties of grapes are first treated with the 'great ferment' (*ta chhü*[11]) such as is used to make vinegar, and then put into the still and heated. A receiver is used to collect the distillate (*ti lu*[12]). This is a beautiful pink colour. Anciently, such brandy was made in the Western countries (*hsi yü*[13]). It was only when Kao-chhang[14] (i.e. Turfan, in mod. Sinkiang) was captured during the Thang period that this technique was obtained.

According to the *Liang Ssu Kung Chi*[15] (Tales of the Four Lords of Liang), Kao-chhang presented 'frozen-out wine' (*tung chiu*[16]), made from dried grapes, to the imperial court. According to (Wan) Chieh Kung[17] the grapes with thin skins tasted best, while those with thick skins tasted tart. He also said that the 'frozen-out wine' made in the Pa-fêng Ku[18] (valley) would keep for years.[e]

---

[a] There are further references to the distilled spirits of South Asia, in Chêng Ho's time, in Ma Huan's[19] *Ying Yai Shêng Lan*[20] of +1433. Of the Siamese (Hsien-Lo[3]) people he says that they have burnt-wine (*shao chiu*[8]) distilled from both rice and coconuts, presumably palm toddy, sold very cheap (p. 32; Mills (11), p. 107). In Bengal (Pang-Ko-La[21]) there are four kinds of burnt-wine, from rice, from some kind of tree, and from two sorts of palm, the coconut and the *chiao-chang*[22] (*Nipa fruticans*), all abundantly available (p. 76; Mills (11), p. 161).

[b] (1), p. 238.   [c] (9), vol. 2, pp. 65–6. Cf. also Goodrich (1), p. 176.
[d] Ch. 25, p. 43*a* (p. 35), tr. auct.   [e] The full text is in *TPYL*, ch. 845, pp. 5*b*, 6*a*.

[1] 汪穎   [2] 食物本草   [3] 暹羅   [4] 孟詵   [5] 食療本草
[6] 可釀酒   [7] 藤汁   [8] 燒酒   [9] 大毒   [10] 麴
[11] 大麴   [12] 滴露   [13] 西域   [14] 高昌   [15] 梁四公記
[16] 凍酒   [17] 艷杰公   [18] 八風谷   [19] 馬歡   [20] 瀛涯勝覽
[21] 枸葛剌   [22] 菱萆

Then Yeh Tzu-Chhi,[1] in his *Tshao Mu Tzu*,[2] says that 'under the Yuan Dynasty grape wine was made in the districts of Chi[3] and Ning.[4] During the eighth month people used to go into the Thai-hang Shan[5] (mountains) to test whether their (distilled spirits) were genuine. Genuine (*araqi* spirit) will not freeze, and if you tilt (the vessel) it will flow; while adulterated kinds, which have been mixed with water, freeze solid at the middle. But if they are kept exposed for a long time there is a quantity which will never freeze, however cold it is, even when everything else has frozen solid.[a] This is the essence (*ching i*[6]) of the wine; when taken it can cause death after intense perspiration (lit., after penetrating through the armpits, *i*[7]). Such wine remains very powerful even when two or three years old.'

The *Yin Shan Chêng Yao*[8] (Principles of Correct Diet) says: 'There are many different sorts of (distilled grape) wine. The strongest comes from the Ha-la-huo[9] country (Qara-Khoja), the next from the Western tribes (Hsi fan,[10] i.e. Mongols or Tibetans), and the next from Phingyang[11] and Thaiyuan[12] (in Shansi).'

Some people hold that when grapes are kept for a long time they turn of their own accord into wine. This wine is fragrant, sweet and strong, and it is really the true grape wine.

This passage needs a considerable amount of exegesis, and what has to be said about it is too important to be relegated to footnotes. The thing to do is to take its statements one by one as they come. First, as to grapes in general, there is no reason to doubt that *Vitis vinifera* was introduced from Bactria to China about −126 by Chang Chhien,[13] whose story we told at the outset of the present work.[b] But they may have been in Lung-hsi (mod. Kansu) somewhat earlier, since the *Shen Nung Pên Tshao Ching*, first of the pharmaceutical natural histories, and probably a work of the Early Han, mentions them and says that wine can be made from them.[c] Mêng Shen's remark about the juice or sap is obscure,[d] but it may point to the utilisation of other vines native to China for wine from early times. Li Shih-Chen quotes older treatises[e] as saying that wine can be fermented from 'mountain grapes' (*shan phu thao*[14]) or wild vine species called *ying yü*[15] and *yen yü*,[16] and these have been identified as *Vitis Thunbergii*[f] or *V. filifolia*.[g] Be this as it may, grape wine as we know it was certainly made and drunk, if intermittently, in China from the Han period onwards.[h]

---

[a] It has been necessary to correct the *PTKM* text here by that of the original book (ch. 3B, p. 80*a*, *b*) for it is not exactly as Li Shih-Chen abridged and modified it. One thing he left out was that Yeh Tzu-Chhi began by speaking of *araqi* (*ha-la-chi*[17]), or brandy, distilled from grape-wine, not about grape-wine alone. The spirits were, Yeh said, as clear as water, a distillate of wine (*chiu lu*[18]). The passage as amended makes excellent sense. A page later (p. 81*b*) Yeh gives his opinion that distilled spirits from grape-wine first became known in Yuan times.

[b] See Vol. 1, pp. 174 ff. His introduction of the cultivated grape-vine was universally recognised in Chinese literature, cf. e.g. *Pên Tshao Phin Hui Ching Yao* (+1505), ch. 32, (p. 771).

[c] Mori ed. (p. 48). Li Shih-Chen noticed this and drew the same conclusion, *PTKM*, ch. 33, p. 9*b*, (pp. 54–5).

[d] The text is correct none the less, as we see from the Tunhuang MS. published by Nakao (*1*), no. 144, pp. 153–4, no. 174, pp. 178–9. Of course he may have meant just the fruit pulp without the skins.   [e] *PTKM*, ch. 33, p. 11*b*, (p. 56).   [f] CC 768.

[g] Laufer (1), p. 243.

[h] For example the *Hsü Han Shu*[19] records that between +168 and +184 Mêng Tho[20] gave a generous present of grape wine to Chang Chih,[21] and got himself appointed Prefect of Liangchow in consequence.

[1] 葉子奇	[2] 草木子	[3] 冀	[4] 寧	[5] 太行山	[6] 精液
[7] 腋	[8] 飲膳正要	[9] 哈喇火	[10] 西番	[11] 平陽	[12] 太原
[13] 張騫	[14] 山葡萄	[15] 蘡薁	[16] 燕薁	[17] 哈剌基	[18] 酒露
[19] 續漢書	[20] 孟佗	[21] 張訨			

Fig. 1502. Wên Ti, ruler of the State of Wei in the Three Kingdoms period, pictured with his colleagues of the States of Shu and Wu in a MS. of + 1314, the *Jāmi'al-Tawārīkh* (Collection of Histories) by Rashīd al-Dīn al-Hamdānī (cf. Vol. 1, p. 218), the Chinese portion of which was completed in + 1304. Roy. Asiat. Soc. A 27, reproduced in Jahn & Franke (1), pl. 46, cf. tr. p. 52.

What Li says about the technique of distillation calls for no special remark,[a] but Wei Wên Ti's words do. This really was Tshao Phei,[1] of the Three Kingdoms period (r. +220 to +226),[b] and his statement has fortunately been preserved elsewhere.[c] Writing to a friend, Wu Chien,[2] about the fruits of China, he alludes to the grape, saying:

Grapes...can be fermented (*niang*[3]) to make wine. It is sweeter and more pleasant than (the wine from cereals) made using *chhü* (the moulds and yeasts mixture) or *nieh* (sprouted malt). One recovers from it more easily when one has taken too much. (*Kan yü chhü nieh, shan tsui erh i hsing*[4]).

The meaning of this is not entirely sure. It is evident from Li Shih-Chen's own words, and from much other evidence, that a great deal of grape wine was made in China through the centuries with moulds-and-yeasts mixtures similar to those used im-

[a] Except on the pink colour of the distillate, a subject to which we shall shortly return at a more convenient place.
[b] We take this opportunity of reproducing his picture (Fig. 1502) from the history of China in Rashīd al-Dīn's *Jāmī'al-Tawārīkh* (+1304). It comes from an almost contemporary manuscript (Roy. Asiat. Soc. A 27); see Jahn & Franke (1); Meredith-Owens (1).
[c] *CSHK* (San Kuo sect.), ch. 6, p. 4a, b; from *I Wên Lei Chü*, ch. 87, *Ta-Kuan Pên Tshao*, ch. 23, and *TPYL*, ch. 972. Also in *PTKM*, ch. 33, (p. 55). Tr. auct.

[1] 曹丕   [2] 吳監   [3] 釀   [4] 甘于麴蘗善醉而易醒

memorially for cereal wines, but the spontaneous fermentation by naturally occurring yeasts from the vines was also known. This is proved by statements such as that in the *Hsin Hsiu Pên Tshao* of +659: 'Unlike other wines, grape wine and mead do not require *chhü*'.[a] As we have phrased the emperor's words above, then, he was saying that 'naturally' fermented grape wine was better than the cereal wines made with moulds as the saccharifying agents; but Yuan Han-Chhing may well be right in interpreting his preference as one for grape wine made with *chhü* as against cereal wine made with *chhü*.[b] This view rests on Yuan's interpretation of what happened in +640, the year of the conquest of Kao-chhang (Turfan),[c] and to that we must now turn.

The most significant passage concerning it has been preserved in the *Thai-Phing Yü Lan*[d] and runs as follows:

Grape wine was always a great thing in the Western countries.[e] Formerly they sometimes presented it (as tribute) but it was not until Kao-chhang was captured that the seeds of the 'horse-nipple grapes (*ma ju phu-thao*[1])'[f] were obtained, and planted in the imperial gardens. The method of wine-making was also obtained, and the emperor himself took a hand in preparing it. When finished it came in eight colours, and with strong perfumes like those of springtime itself; some sorts tasted like a kind of whey (*thi ang*[2]).[g] (Bottles of) it were given as presents to many of the officials, so people at the capital got to appreciate the taste of it.

This passage, which Yuan Han-Chhing was the first to notice, contributes a good deal to the elucidation of Li Shih-Chen's vague way of talking. When he said that 'this technique was obtained' in +640 or soon afterwards, what did he mean? Grape wine as such, or grape wine made 'naturally' without *chhü*, or the distillation of grape wine? And when he said 'obtained', did he mean that the technique was brought eastwards from Turfan, or that it just grew up in China as the result of some stimulus

---

[a] Ch. 19, p. 8*a*, (p. 301).

[b] (*1*), p. 97. On wine in general see pp. 73ff., on distilled wine, pp. 94ff.

[c] For the attendant circumstances see Cordier (*1*), vol. 1, p. 419.

[d] Ch. 844, p. 8*a*, tr. auct. It purports to come from the *Thang Shu*,[3] but we have not been able to find it in either of the two great current versions. It is in *Thang Hui Yao*, ch. 100, (p. 1796).

[e] This of course does not mean Europe, but as the next sentence shows, the small States of Central Asia. It will be remembered that Ssuma Chhien had a good deal to say in the *Shih Chi* (ch. 123) on the grape wine of Ta-Yuan (Ferghana) and An-Hsi (Parthia). There are three separate mentions in this chapter (tr. Watson (*1*), vol. 2, pp. 266, 268, 279ff.). Chang Hsing-Yün (*1*), who wrote a book on dietetics at the beginning of the nineteenth century, was most apprehensive of the effects of alcoholism, and believed that all Chinese wines and spirits were poisonous. He ascribed the good keeping properties of the Ferghanese wine to the climate, but it may have been in fact strong alcohol obtained by the freezing-out process (cf. p. 151), or alternatively these people possessed adequate means of corking the containers (cf. p. 143).

[f] This is a stumbling-block in the literature, since *ma ju* can also, and often does, mean mare's milk, from which the Mongols made their *kumiss* and distilled their *araki*. Hence translators have to be watchful.

[g] This translation, the most obvious one, is probably wrong, for it seems that *thi* stands for *thi chiu*[4] here, explained as a kind of red wine, and *ang* for *ang chiu*,[5] a kind of greenish-white wine (Morohashi dict. vol. 11, p. 384). So they might have been like our red and white Burgundies, but then that would duplicate the remark just previous about the eight colours. Normally *thi* is associated with milk products (on which see Pulleyblank (*11*) and the paper of Suzuki Shigeaki (*1*) for Middle Eastern parallels). *Thi-hu*[6] in Buddhist texts means *ghee* (butter-oil), but where Mongol affairs are concerned more likely whey. The question remains open.

[1] 馬乳萄桃   [2] 醍盎   [3] 唐書   [4] 醍酒   [5] 盎酒   [6] 醍醐

from Turfan? Yuan Han-Chhing took the view that his words here did not concern distillation at all, but grape wine made without the moulds-and-yeasts mixture, so that it pleased by a floral rather than by a mycological flavour.[a] If that was the technique that came from Turfan, the statement in the *Hsin Hsiu Pên Tshao* just mentioned would not refer to centuries before the +7th. Other scholars, such as Shih Shêng-Han,[b] have preferred to join with old Edkins (18) in thinking that what Li Shih-Chen had chiefly in mind here was the distillation of brandy. But what those who have mostly emphasised the Western origin of this, such as Laufer,[c] have failed to see, is that nowhere in the West at that time was it possible to produce alcohol by distillation, because the Hellenistic–Byzantine stills lacked all head- or condenser-cooling.[d] Therefore that technique cannot have been transmitted to China in the mid +7th century, though it may well have originated there at that time as the result of some stimulus from Turkestan.

What this was may be lying under our very nose in the quotations which Li Shih-Chen gives immediately following about the 'frozen-out wine', i.e. alcohol which has been concentrated by the freezing of its accompanying water. It will be more convenient, however, to postpone the discussion of this, assembling all the relevant information, until the end of the argument. Here it need only be said that while the *Tshao Mu Tzu* (Book of the Fading-like-Grass Master) is a work of the Ming, finished in +1378, the other reference, *Liang Ssu Kung Chi* (Tales of the Four Lords of Liang), is a Thang one, written by Chang Yüeh[1] in +695, but dealing with events of the period +500 to +520.[e] This shows at any rate one thing, over how long a period people in China were familiar with the congelation method of making strong alcoholic solutions, first practised though no doubt it was by the neighbouring peoples of the black North, living on the Thien Shan or beyond the Gobi Desert.

With this we come to Hu Ssu-Hui,[2] the great nutritionist and first discoverer of deficiency diseases,[f] who speaks in his *Yin Shan Chêng Yao* about the brandies of

---

[a] Yuan adduces supporting opinions on this from later scholars, such as Kao Lien in his *Tsun Shêng Pa Chien* (+1591).

[b] Priv. comm., July, 1958.   [c] (1), pp. 237–8.

[d] This statement needs qualification by what has been said above (pp. 125 ff.), but it remains essentially true. We know of no Hellenistic, Byzantine or Arab alcohol.

[e] This is a queer and intriguing book, perhaps the best critique of which was written by Pelliot (47), vol. 2, pp. 677 ff. It cannot be earlier than the +7th century, and Chang Yüeh's authorship is doubtful, but if Lu Shen[3] or Liang Tsai-Yen[4] or Thien Thung[5] wrote it (and all are suspected), the date might well be nearer +650 than +700. In any case, it records traditions in a romancé manner, and none of the Four Lords is a historical character attested from other sources. Nevertheless (to use an adage perversely inappropriate here) there is no smoke without fire, and the freezing-out process could not have been imagined without an empirical basis. Conceivably the 'frozen-out wine' came first to China in +640 or thereabouts rather than +540—but that would not affect our general line of argument.

The original text of the *Liang Ssu Kung Chi* has been preserved only in lengthy quotations. We may find the passage about the frozen-out wine from Kao-chhang (Turfan) also in *Thai-Phing Kuang Chi*, ch. 81, p. 5 a (vol. 1, p. 336.2), in *TSCC, Shen i tien*, ch. 311, p. 4 a, and in some editions of *Shuo Fu*, ch. 113. The full form contains various arguments about the genuineness of what the ambassadors presented, but states clearly that the freezing-out process was performed expecially in the Pa Fêng Ku[6] (Eight Winds Valley).   [f] Cf. Lu Gwei-Djen & Needham (1), as also Sect. 40 in Vol. 6.

[1] 張說   [2] 忽思慧   [3] 盧詵   [4] 梁載言   [5] 田通
[6] 八風谷

Qara-Khoja[a] and of Shansi.[b] This was in +1330. Half a century earlier, Marco Polo had been in those parts, and mentioned the vineyards of Thaiyuan. 'There grow here (at Taianfu), [he said], many excellent vines, supplying great plenty of wine; and in all Cathay this is the only place where (grape-) wine is produced, being carried thence all over the country.'[c] That he did not mention distilled spirits has been fastened upon by some,[d] disinclined on other grounds to believe in their existence in China at that time, but everyone knows that there were many interesting things which Marco Polo did not mention,[e] and by the end of the +13th century *aqua ardens* may have been so commonplace for him that he was not surprised to find it in Cathay.

(iii) *Thang 'burnt-wine'*

We can now take a look at those writings of the Thang and early Sung which seem to indicate the existence of alcohol distillation from the +7th to the +12th centuries, in all cases prior to the first appearance of *aqua ardens* in the West. Though none may be considered to offer decisive proof, they give rise to a profound suspicion. Let us take them approximately in order of date, as we have done on previous occasions, when for instance we were elucidating the origin of the stern-post rudder.[f] First then we may quote from someone recently mentioned, Mêng Shen in his *Shih Liao Pên Tshao* of +670. There are words in the Tunhuang MS. which could mean:

Take *chhing liang mi* (some kind of grain) and add one *tou* (10 pints) of pure bitter wine, soak it for three days (to undergo some further fermentation?), then distil (?) over a strong fire many times (*chhing liang mi i shun khu chiu i tou, tzu chih san jih, chhu pai chêng pai pao*[1]). Keep this good stuff, and if you go on a long journey take some; it will ward off hunger for ten days.

Again, in a neighbouring place he says:

According to the (*Tung Hsüan*) *Ling Pao Wu Fu Ching*[2],[g] one should take *pai hsien mi* (another kind of grain) and (after fermentation?) distil (?) over a strong fire nine times. One can use this as a help for avoiding the eating of cereal foods. (*Pai hsien mi chiu chêng chiu pao, tso pi ku liang*[3]).[h]

But the language is obscure and peculiar, and one cannot be sure that he was not

---

[a] Pelliot (47), pp. 161 ff., has discoursed on this; it was none other than the ancient Uighur capital some 17 miles east of Turfan. Khoja and Kao-chhang are related forms.

[b] Under the head of the former, he says: 'Take good wine, distil it as it boils, and collect the distillate (*yung hao chiu, chêng ao, chhü lu*[4]).'

[c] See Cordier, in Yule (1), vol. 3, p. 75, from Bk. 2, ch. 37, p. 13, or Moule & Pelliot (1), vol. 2, p. xxvi.   [d] E.g. von Lippmann (5), (9), vol. 2, p. 66.

[e] For instance, scientific astronomy (cf. Vol. 3, p. 378), the magnetic compass (cf. Vol. 4, pt. 1, p. 245), advanced textile machinery (cf. Vol. 1, p. 189) and printed books (Carter, 1).

[f] Vol. 4, pt. 3, pp. 638–9.

[g] 'Manual of the Five Talismans, a Tung-Hsüan Ling-Pao Scripture'. A book with this title was in the *Tao Tsang* once, but it is now lost. The MS. writes Chih[5] for Pao.

[h] Both these passages will be found in Nakao (1), no. 178, pp. 181–2. Tr. auct.

[1] 青粱米以純苦酒一斗漬之三日出百蒸百暴   [2] 洞玄靈寶五符經
[3] 白鮮米九蒸九暴作辟穀粮   [4] 用好酒蒸熬取露   [5] 窆

talking about some preparation of fermented and dried grain—though it would be odd if that were to assist one to abstain from cereals.[a]

In the following century there is another reference to white wine, in the poems of the famous Li Pai[1] (+701 to +762).[b]

> Returning from the mountains one finds that the *pai chiu*[2] has just matured.
> And the yellow chickens, fattened by millet, are ready for the pot,
> Now that autumn has come....

The significance of this and other mentions of *pai chiu* in Thang poetry is not obvious unless one knows that through many recent centuries one of the commonest names for distilled spirits is 'white-and-dry' (*pai kan (erh)*[3]),[c] especially in North China. But there is no certainty that this was what Li Pai was talking about—only the suspicion.[d]

Suspicion grows stronger, however, from the +9th-century references, of which there are a good number.[e] Li Chao[4] (*fl.* +810), talking of wine in his *Kuo Shih Pu*[5] (Emendations to the National History), mentions a 'burnt spring wine' (*shao chhun chiu*[6]) that was made at Chien-nan.[f] Pai Chü-I,[7] the celebrated poet (+772 to +846)

---

[a] We have not come across the word *pao*, heating, scorching, sun-drying, elsewhere in connection with distillation. Wu Tê-To (*1*), however, felt that these statements should be carefully considered.

[b] Cit. with refs. by Tshao Yuan-Yü (*3*), p. 24. Tr. auct.

[c] Alcohol content *c.* 55 %. It is conjectured from the other way of writing *pai kan erh*[8] that the expression may have derived from the *han chiu*[9] of Yuan times. This most probably meant 'sweated wine' (i.e. distilled), and was therefore a direct translation of the Ar. *al-araq*; but it could also perhaps have signified 'wine of the Khan', since *han* was used to transliterate that title. This puts one in mind of the silver fountain with four spouts that William Boucher of Paris built for Mangu Khan at Karakoron in +1254 (cf. Vol. 4, pt. 2, p. 132, and Olschki (*4*), pp. 57, 63). They are said to have served for the guests four different alcoholic drinks, but kumiss distilled was not one of them. The drinks were grape-wine, fermented mare's milk (*caracosmos, kumiss*), mead (*bal*) and rice-wine (*terracina, cervisia*); cf. Rockhill (*5*), pp. 207–8, quoted also in Lattimore & Lattimore (*1*), p. 77; and elsewhere. The distilled *kumiss* could probably not have been made in sufficient quantity. Pien Ssu-I[10] had a poem about *han chiu* in his *Thieh Ti Shih*,[11] *c.* +1338; and in +1751 Tsê Hao[12] stated in his book on popular terms, ideas and customs, *Thung Su Pien*,[13] that the *han chiu* of the Mongols was the same as *shao chiu*.[14] We are grateful to Dr Chêng Tê-Khun for drawing our attention to this point.

[d] Tshao Yuan-Yü (*3*) did not dare to accept the identity, and his caution was echoed by Shinoda Osamu, in Yabuuchi (*11*), pp. 79, 90. Eberhard, reviewing Hermanns (*1*), did, however.

[e] Though some are unsubstantial. Poets, when translating, tend to use technical terms in a way which might be deceptive if the reader were not circumspect. For example, Chang Hsin-Tshang (*4*), p. 112, in his rendering of a poem of about +808 by Liu Tsung-Yuan, writes: 'We drink the river-water, purer than the best distillation...' Yet the text has only 'pure goblets', filled no doubt with clear wine, but not significant in the present context (see *Chhüan Thang Shih*, ch. 352, in vol. 6, (p. 3941) for the text).

[f] Ch. 3, p. 11*a*, *b*. There are other references to this wine, for example in the poetry of Wei Chuang,[15] who exclaimed:

> 'How beautiful the Chin-chiang (river) in the spring,
> And lovely are the girls of Szechuan too,
> Pouring out elegant "burnt spring wine" beside it.'   (*Huan Hua Tzhu*[16])

This would have been between +880 and +920. A century later Su Tung-Pho explained, saying: 'In the Thang period there was a wine called "burnt spring", but it was the same as what is now called *shao chiu*,[17] burnt-wine' (Morohashi dict., vol. 7, p. 525). And there is mention of it again in Fan Chhêng-Ta's ode on lichis (*c.* +1175); cf. Shinoda (*3*), p. 305, Ōtani (*1*), p. 74. Schenk (*1*), p. 122, was perhaps the first sinologist to recognise that 'burnt spring wine' must have been distilled, but misled by Li Shih-Chen he supposed that it must have been imported from Mongolia.

[1] 李白	[2] 白酒	[3] 白乾 (兒)	[4] 李肇	[5] 國史補
[6] 燒春酒	[7] 白居易	[8] 白干兒	[9] 汗酒	[10] 卞思義
[11] 鐵笛詩	[12] 翟灝	[13] 通俗編	[14] 燒酒	[15] 韋莊
[16] 浣花詞	[17] 燒酒			

wrote some verses entitled Li-chih Lou Tui Chiu[1] which include the following lines:[a]

> The lichis are newly ripe, the colour of a cock's crown,
> One catches the first whiff of a perfume like amber from the burnt-wine (*shao chiu*[2]),[b]
> How one would like to pluck a branch, and to drink a cup!
> But there is no one here in the West with whom to share this beauty.

This could be dated about +820 and starts the run of references to 'burnt-wine'. Less than a decade later Fang Chhien-Li[3] was writing his *Thou Huang Tsa Lu*[4] (Miscellaneous Jottings far from Home), but he was stationed in the far south rather than the far west. Discussing drinks, he spoke of an 'after-burning wine' (*chi shao chiu*[5]), which was kept in sealed pots and pipetted out as needed. What he said was this:[c]

> In the south they drink 'after-burning wine' (*chi shao*) taken from pots filled up and then sealed with clay; it is prepared by 'burning' with fire, and matured; if not, it is not good to drink. When the pots are opened there is a tendency for them to be somewhat empty, although the clay seal is still there. When merchants want to know the quality of the wine they bore a very small hole through the seal and insert a reed, withdrawing some of the wine by means of this pipette so that they can see what the taste is like.[d]

They could also see something else, namely the extent to which evaporation had occurred, and what sum, therefore, they ought to offer for the jar. This is redolent indeed of the behaviour one would expect of strong alcohol in a hot climate when people had not found out an adequate form of corking.[e] Again, about a decade later, *c.* +840, Yung Thao[6] wrote a poem in Szechuan in which he said:[f]

> Since I reached Chhêngtu the burnt-wine (*shao chiu*) has matured.[g]
> I doubt if I shall ever go back to Chhang-an again...

And then there is a highly suspicious phrase in one of the poems of Li Ho[7] (+791 to +817) incorporated in his *Chhang Ku Chi*.[8]

[a] *Chhüan Thang Shih*, han 7, tshê 4, ch. 8, p. 11a, tr. auct. adjuv. Schafer (16), p. 190.

[b] Some versions read *kuang*,[9] 'the amber sparkle', instead of *hsiang*;[10] which makes better sense So we could read: 'One catches the first glint of amber sparkle in the burnt-wine'.

[c] Cit. *Thai-Phing Kuang Chi*, ch. 233, p. 34a (ed. Li Fang in +978). Cf. Wu Tê-To (1), p. 54.

[d] Tr. auct. The passage could be read as if the burning was done in the same vessels as the keeping, but either this was a misapprehension or there has been some corruption in the text.

[e] The history of bungs and corks, perhaps not yet adequately written, is one of those subjects which may seem trivial yet have far-reaching cultural repercussions. It was the main burden of Warner Allen's book (1) on the history of wine in Europe that the exquisite tastes of vintage qualities was unknown in the Middle Ages because casks only were used for storage, the well-stoppered amphora having gone out, and the corked bottle not having come in. The Chinese nearly always used ceramic jars, how well bunged remains to be seen, probably with lacquer as well as wax. In Allen (1) see pp. 88–9, 162, 169, 190–1, 206.

On amphoras see for an introduction Grace (1), though nothing is said of the stoppering. They seem to have been a Canaanite (Phoenician) invention of the −14th century.

[f] *Chhüan Thang Shih*, han 8, tshê 6, p. 4b, tr. auct., adjuv. Schafer (16), p. 190. Elsewhere Yung Thao spoke of *shao shen chhou*,[11] 'sacrificial burnt-wine for the spirits' (cf. Ōtani Shō (1), p. 73).

[g] The word here used, *shu*,[12] as in the case of Li Pai, has a distinct undertone of cooking over fire.

[1] 荔枝樓對酒	[2] 燒酒	[3] 房千里	[4] 投荒雜錄	[5] 既燒酒
[6] 雍陶	[7] 李賀	[8] 昌谷集	[9] 光	[10] 香
[11] 燒神酎	[12] 熟			

>   In the crystal cup the amber (-sparkling) juice is thick,
>   And the lovely wine runs out like a rivulet of pearly red.[a]

But the Chinese words are more suggestive than this rendering would convey, for they include the expression *hsiao tshao*, meaning a little trough (*Liu-li chung, hu-pho nung, hsiao tshao chiu ti chen chu hung*[1]). Could this not have been a reference to the side-tube of the still? Perhaps it would be better translated:

>   And the wine-drops from the little channel are a pearly red.

Tshao has overwhelmingly often the meaning of gutter, conduit or flume;[b] and the case is still more strengthened when one knows that in relatively modern times at least *tshao fang*[2] has been the common name for a distillery. So a trickle of wine on a not too clean surface may be more far-fetched than to think that Li Ho knew the side-tubes of stills, and had them in mind when he wrote about his wine-drops.[c]

The last +9th-century witness to be called is Liu Hsün,[3] whose *Ling Piao Lu I*[4] (Strange Things Noted in the South) would have been in the writing about +880. There he said:[d]

In the South it is warmer, and in the spring and winter fermentation takes (only) seven days, in summer and autumn (only) five. When it is ready it is put into an earthenware container, placed over a fire of dung, and 'burnt' (*shao chih*,[5] i.e. heated). [Author's comm.][e]

>   There is also a kind which is not 'burnt' (or, heated) and this is called *chhing chiu*[6] ('pure' or 'plain' wine).

Chhing chiu is often mentioned in other texts, such as the *Yu-Yang Tsa Tsu* (+683),[f] and Liu Hsün's statement seems to demonstrate that *shao chiu* was made from it. Furthermore it closes one escape route very neatly. For most of these early mentions of *shao chiu* it would be possible, if not very plausible, to maintain that mulled wine rather than distilled wine was in question, but here Liu Hsün seems clearly to be talking about two kinds of wine in the same sense as Li Shih-Chen, one fermented and stored as such, the other distilled.[g]

Towards the end of the following century, *c.* +990, soon after the beginning of the Sung dynasty, an eminent scholar, Thien Hsi,[7] wrote an interesting little work, the *Chhü Pên Tshao*[8] (Natural History of Yeasts and Fermentations). In this he gave the

---

   a Quoted by Lu Yu in *Lao Hsüeh An Pi Chi*, ch. 5, p. 15b, and by Fêng Shih-Hua in *Chiu Shih*, ch. 2, (p. 35). Tr. auct.
   b Many examples have been encountered in Sect. 28 on civil engineering in Vol. 4, pt. 3.
   c Frodsham (1), p. 239, translated: 'From a little vat the wine drips down.' But vats do not drip, unless leaky and useless. The image of dripping is continued in the lines immediately following, which speak of the fat of 'boiling dragons and roasting phoenix'. Graham (8), p. 102, brings in a wine-cask, which is open to the same objection, and not in the text either.
   d Cit. *TPYL*, ch. 845, p. 7a, tr. auct. Liu is almost certainly talking about rice wine, not grape wine.
   e This is found only in the *TPYL* version. Even if it were an insertion of the editors, that would make it date to +983, which is also a long time before alcohol distillation in the West.
   f Cf. Wu Tê-To (1), p. 53.
   g Earthenware as a practical material for low boiling-point distilling has already been met with in the Japanese pharmaceutical stills (p. 114 above).

¹ 琉璃鍾琥珀濃小槽酒滴眞珠紅   ² 槽坊   ³ 劉恂   ⁴ 嶺表錄異
⁵ 燒之   ⁶ 清酒   ⁷ 田錫   ⁸ 麴本草

description of the distilled Siamese toddy just as we read it on p. 136 above.[a] Although that was an imported article of commerce,[b] it is obvious from the words of Thien Hsi that in Wu Tai and early Sung times Chinese people understood perfectly clearly what it was that they were drinking, and how it was made. One significant point is that the wine was said to have been 'burnt' twice (*fu shao erh tzhu*[1]), which would not make sense if it was merely warming or mulling.

We still have plenty of time before us ere we reach the mid + 12th century, time of the first alcohol distillations in the West, but from the intervening period we shall quote only two more pieces of evidence. About + 1080 Su Tung-Pho,[2] the great poet-scholar (+ 1036 to + 1101) wrote a rhapsodic ode entitled *Tung-thing Chhun Sê Fu*[3] (Spring Colours by the Tung-thing Lake), in which he thanked and praised the Prince of An-ting, who had made some wine from oranges and had presented him with several bottles of it. Although the language is rather obscure and allusive, the meaning appears to be unmistakable. Su wrote:[c]

> To blend the mixture they use the double-kernelled millet,[d]
> And call on the help of a tube of the triple-ridged reed,[e]
> Suddenly the cloudy vapour condenses like melting ice,
> Whereupon tears come forth dripping down like liquid pearls.

As Alice or somebody said, this is curiouser and curiouser. The side-tube of a still might well be the size of a thick graminaceous stem, the distillate would indeed run from the end of it like a trickle from melting ice, and 'tears' would well describe what dropped into the receiver. One cannot help concluding that what Su Tung-Pho enjoyed was a liqueur something like the *cédratine* of Corsica,[f] though made from fermented millet rather than the trodden grape.

Finally, in + 1117 comes the book of Chu Kung[4] on wine, the *Pei Shan Chiu Ching*[5] (Northern Mountain Wine Manual). None of the tractates on the different sorts of wine which have come down to us have much to say about distilled wine (perhaps for reasons to be mentioned presently), but there is one passage in Chu Kung's text which

[a] In *Shuo Fu* (Ming ed.), *han* 20, ch. 94, tractate 19, p. 2a. Reproduced in Japanese by Shinoda Osamu (3), p. 305; Ōtani Shō (1), p. 73. So far as we know, Thien Hsi's was the earliest description of the Siamese 'whisky'.

[b] Excavations at the site of a port town north of Songkhla in Thailand on the eastern coast of the Malayan peninsula have unearthed large quantities of empty Chinese pottery bottles which might well have been used in this spirits trade. They date from the +7th to the +12th centuries. We owe this information to Prof. Janice Stargardt and Prof. Wolfgang Stargardt. The former goes on to say that Chinese trade with this Satingpra region began early in the Thang and continued for centuries. The toddy was made from the palms *Nipa fruticans* and especially *Borassus flabellifera*, which have juices extremely rich in sugar. Other items of this trade were camphor crystals, oils and perfumes.

[c] *TSCC, Tshao mu tien*, ch. 226, *i wên* 1, p. 3a; also cit. *Chiu Shih*, ch. 1, (p. 8); tr. auct., adjuv. M. J. Hagerty MSS unpub.

[d] *Erh mi chih ho*[6]; an auspicious sign.

[e] *San chi chien*;[7] the numbers complement one another. The plant referred to is probably the white grass', *pai mao*,[8] i.e. *Imperata arundinacea*. *PTKM*, ch. 13, (p. 64).

[f] This is flavoured with the citron, *Citrus medica*, a citrous fruit little seen in Northern Europe, but in fact the first to arrive in the West. It was known to Theophrastus, and the 'Buddha fingers' (*fo shou kan*[9]) of China is a variety of it. Other citrus perfume-oils have also, of course, been used for flavouring liqueurs. See Sect. 38 in Vol. 6 for much further information on the *Citrus* genus in China.

[1] 復燒二次	[2] 蘇東坡	[3] 洞庭春色賦	[4] 朱肱	[5] 北山酒經
[6] 二米之禾	[7] 三脊菁	[8] 白茅	[9] 佛手柑	

arouses suspicion. This describes 'fire-pressured wine' (*huo pho chiu*[1]), but very obscurely.[a] Take good *chhing chiu*,[2] he says, and let it settle for three days, then build up a stove of five layers of bricks in a windless room, and put the earthenware vessel on it. Use three *chhêng*[3] (steelyard-weighed lots) of charcoal, and 'put the *lung*[4] right in the centre, with half of the already glowing charcoal underneath.' *Lung*[4] normally, of course, means basket, but it is tempting here to interpret it as a technical term of those days for the catch-bowl of the Chinese still, more especially as *chêng lung*[5] in later times could mean various kinds of steamers for cooking.[b] Anciently, a kind of *lung* called *ling*[6] was a wickerwork net protecting an earthenware vessel like the glass wine-bottles of present-day Italy, so it is not too difficult to see how the catch-bowl could have acquired the name of *lung*. The only other instruction given by Chu is to let it go on heating (i.e. as we suspect, distilling) in a quiet place for seven days. He also speaks of a pipette for sucking up the clear supernatant layer of wine, and says of the 'fire-pressured wine', significantly enough, that it is much better than mulled wine, *chu chiu*[7].[c]

It was about this time (or rather earlier) that Chu Fu[8] wrote his *Chhi Man Tshung Hsiao*[9] (Amusing Anecdotes of the Chhi Man Tribesfolk, in Southern Hunan). In this he has a passage about *tiao thêng chiu*,[10] 'hooked vine wine'.[d]

This wine [he says] is perfected by fire; it is not any kind of vinegar nor rough-tasting stuff. There are two vessels east and west, and the wine is collected and sucked through a hollow stem.

The difficulty here is the interpretation. The fire might mean only warming or mulling and the last phrase is most probably a reference to the well-known practice of the southern Chinese tribal peoples and the Indo-Chinese cultures in general of sitting round a pot of wine and each person sucking it up through a straw, as we should say. On the other hand the purpose of the two vessels is not clear, and conceivably it might be a reference to the still and the receiver standing side by side, while the 'collecting' hollow stem could refer to the side-tube. But the probability is that distillation is absent here.

To sum it up, from first to last, none of the pieces of evidence for alcohol distillation in the Thang and the early Sung is quite decisive; some are relatively convincing, others less so. But there are times when probabilities accumulate to such an extent as to change quantity into quality and justify a circumstantial conclusion. Before adopting this we ought to look at some of the arguments pro and con which have been advanced in this problem of Thang distillation. For example it has been felt (as by Tshao Yuan-Yü) that the instances are insufficiently numerous. But apart from the likelihood of many more being found as research continues, there may have been rather

---

[a] Ch. 3, pp. 16*b*ff., Shinoda & Tanaka repr. pp. 135–6. Cf. Wu Tê-To (*1*), p. 54.
[b] By this time the wickerwork must have disappeared, or it would have been burnt by the fire.
[c] If it were not for this last statement, one would be inclined to regard the whole process as some kind of pasteurisation to improve the keeping properties of the wine.
[d] Hsü Pai Chhuan Hsüeh Hai ed., vol. 3, p. 1601.

| [1] 火廹酒 | [2] 清酒 | [3] 秤 | [4] 籠 | [5] 蒸籠 | [6] 笭 |
| [7] 煮酒 | [8] 朱輔 | [9] 溪蠻叢笑 | | [10] 釣藤酒 | |

good reasons why the Thang scholars did not expatiate too much on the mountain dew, any more than the peasant-farmers of Co. Longford. In a word, the Excise was at hand—as could only be expected in so bureaucratic a civilisation as that of China. As the study of Wang Chin (7) has conveniently shown, wine taxation, government monopolies, and even prohibition, were features of all the early Chinese dynastic periods. Wang Mang between Early and Later Han established a brewing monopoly or 'nationalisation' of the industry, as in the better known cases of salt and iron; in the San Kuo time there was a prohibition of drinking under the Wei State (*chin chiu*[1]), and later on the Northern Wei exacted the death penalty for infraction of the government brewing monopoly. In the Thang Dynasty, which particularly concerns us, there was a strict prohibition on private wine-making (and no doubt *a fortiori* distilling), while in +847 heavy taxes were imposed on those who were licensed to carry it on. Naturally, therefore, popular cover-names have come down to us. The *Chiu Shih*[2] by Fêng Shih-Hua[3] (+16th century) records that *chhing chiu*[4] was known as 'the sage' (*shêng jen*[5]), and cloudy wine, *cho chiu*,[6] was called 'the worthy' (*hsien jen*[7]);[a] while Buddhist monks (says Tou Phing[8] in his *Chiu Phu*[9] of +1020), to whom it was doubly forbidden, would invite favoured guests to 'take a drop of "wisdom soup"' (*pan-jo thang*[10]).[b] All this, it may be admitted, tends to impede the search for decisive Thang evidence.

Secondly we have seen a number of cases where the texts speak of the colour of the distillate as pink, brown or red, though in some cases (e.g. Li Pai) the white or transparent colour is emphasised. About this there are two things to be said. First one has to reckon with splash in these relatively primitive stills, and if the catch-bowl was large and the still-contents 'bumping' considerably, it would be only too likely that doses of anthocyanin would find themselves in the side-tube. This would easily account for the colours mentioned. But besides this there is the certain fact that from early times down to the present day distillers have added colouring matters artifically to their liqueurs to improve their appearance. Such is the case with the pink *mei kuei chiu*[11] and the delicious green *chu yeh chhing*[12] which we can enjoy in China today;[c] nor is it difficult to find Arabic parallels, not indeed for alcohol, but for the perfumed essential oils, which according to the sure witness of al-Kindī in the +9th century, were coloured artificially with various fat-soluble dyes before appearing in the market.[d]

Certainly the crux lies in the meaning of *shao*.[13] While it could imply no more than heated, mulled or boiled,[e] its universal use in recent centuries to designate distilled must carry a certain authority with it as we trace it back in time to earlier ages.[f] Of

---

[a] Ch. 2, (p. 45). Based on *Wei Lüeh*, quoted in *TPYL*, ch. 844, p. 1a.

[b] P. 2a; also in *Chiu Shih*, ch. 2, (p. 47). Lewin (1), pp. 122–3, has collected much information on the Fermented Beverages Authority in the +9th and +10th centuries; it controlled distilled spirits as well as wine.

[c] Alcohol content c. 45%. This usage still exists in Japan, as we found in Kyoto in 1971.

[d] See Garbers (1), p. 16.   [e] This was sometimes done to stop the fermentation.

[f] There is a complete continuity. Burnt-wine (*shao chiu*[14]) is listed among the wine-shop wares in *Mêng Liang Lu*,[15] Wu Tzu-Mu's[16] description of Hangchow in +1275, (ch. 16, p. 5b). And at the dawn

[1] 禁酒   [2] 酒史   [3] 馮時化   [4] 清酒   [5] 聖人
[6] 濁酒   [7] 賢人   [8] 竇苹   [9] 酒譜   [10] 般若湯
[11] 玫瑰酒   [12] 竹葉清   [13] 燒   [14] 燒酒   [15] 夢梁錄   [16] 吳自牧

course many East Asian wines, such as *huang chiu*[1] or Japanese *saké* are normally drunk warm today, but no one could examine the passages adduced above and come away with the feeling that nothing more was involved than 'chambrage' for the dinner-table. Besides, the word *shao* tends to occur in couplet form, as in Pai Chü-I's poem about the lichis, complementary to something else, suggesting the name of a special wine rather than the adjective for any kind of wine heated. To sum it up, we have a similar position here to that encountered at an earlier stage with regard to the stern-post rudder—a cumulative case which carried considerable conviction short of full proof. And perhaps there is a lesson in the fact that years after Wang Ling and I had built up the textual rudder case, Lu Gwei-Djen and I found in Canton the Han tomb model which settled the matter. So we dare to entertain the hope that somebody will one day find a tin or pewter catch-bowl and side-tube in a Thang tomb.

So far we have said almost nothing about the *kumiss* of the Mongols and the *araki* which they distilled from it, except in connection with the history of the still (pp. 103, 105). We have not sought for early Chinese descriptions of *araki* distillation, though they could doubtless be found, because we do not think that it had much influence on Chinese wine distillation. Indeed it is more likely that any transmission went the other way. Here, however, we may quote what Hsiao Ta-Hêng[2] said in his *I Su Chi*[3] of +1594.

Mare's milk at the beginning is too sweet to drink, and in two or three days it has gone sour and cannot be taken then either. You can use it only for wine, and this is no different from 'brandy' (*shao chiu*[4]). First the (fermented) milk is distilled, then the wine is again distilled, and when this has been done three or four times the taste of it is exceedingly good.[a]

Another point arises now. Earlier in this survey we examined some of the illustra¯tions of stills which can be found in the medieval alchemical books (pp. 68 ff.). The work of Phêng Ssu (*TT*907) is too late (+1225) to be of importance for the present argument, but that of Wu Wu (*TT*893), illustrated in Fig. 1446, is significant, for its date, +1163, is just about that of the first still-cooling in the West which permitted the successful distillation of alcohol. We have also illustrated a highly developed still, however (Fig. 1447), from a treatise (*TT*895) which claims to date from the Chin period, and while that one might hesitate to allow, one could suppose reasonably enough that it may be of Thang date, in which case it could be contemporary with the texts which we have been studying. The same applies to the diagram in Fig. 1407 from *TT*908, a text which has been considered Chin in date, but which is now thought to belong rather to the late +9th century.[b] All this, of course, is a separate matter from

market one could buy *shui ching hung pai shao chiu*[5] 'crystal burnt-wine, red and white', which, he said, had a gentle fragrant taste, and evaporated as soon as it entered the mouth (ch. 13, p. 7b).

[a] Tr. auct. from *KCCY*, ch. 22, p. 11a, adjuv. Serruys (1). On Hsiao Ta-Hêng's book about Mongolian customs see W. Franke (4), p. 213.

[b] It is difficult to impose unitary datings on Chinese alchemical texts, except in certain cases, for so many of them consist of a core to which accretions were added in successive periods. The diagrammatic simplicity of the picture in *TT*908 might well plead for an earlier dating (Liang, if not Chin) than the elegant drawing in *TT*895.

[1] 黃酒   [2] 蕭大亨   [3] 夷俗記   [4] 燒酒   [5] 水晶紅白燒酒

those workaday stills of the countryside which we can imagine by projecting backward the designs familiar in our own and recent times (cf. pp. 63 ff.).

Of these the Jesuit Cibot (5), writing in +1780, said that 'the Chinese alembics, heated by millet straw, are so simple, or rather so rustic, that we would not dare to give a description of them.' Historians of science and technology could wish that he had not been so coy. Nevertheless, in his special notice on *eau-de-vie*, he remarked that 'we find the brandy of grape(-wine) celebrated in poems of the seventh century, and also indicated in medical books of the eleventh, and perhaps earlier, as an excellent remedy against wounds, bruises and several internal diseases...But if we are to believe the author of the *Pên Tshao Kang Mu*, the invention of brandy from cereal (wine) is not ancient in China, going back only to the Yuan dynasty, that is to say, the end of the thirteenth century...' He then goes off into a rambling soliloquy on the difficulty there is in one invention leading to another—evidently puzzled how to reconcile these two opinions. He also adds the interesting incidental intelligence that 'the Chinese *eau-de-vie* has a very disagreeable taste,[a] but in spite of that the people are used to it and like to drink it warm, more of it indeed than one would dare to mention. Moreover there are those who will only drink that which has been re-distilled in the alembic, and which is so strong that it burns almost like "spirits of wine"...Our European pharmacists are all agreed that it is as good or even better than that from grape-wine for all external uses.' But what is valuable about Cibot's notice is that it raises the question of tradition; evidently in his time scholars took the affirmative view about the passages in the Thang poets.[b]

Going back further than Cibot and his contemporaries, there arises the question of Chinese exports during the Sung, for in no less than eight cases Chao Ju-Kua in his *Chu Fan Chih* (+1242 to +1258, if not earlier, c. +1225) mentions wine as carried by the merchants from China to as many countries in the South Seas and Indian Ocean.[c] The word used is *chiu*[1] every time, but Hirth & Rockhill (1) may not have been meaningfully wrong when they translated persistently[d] by the Old China Hand word *samshu* (i.e. *san shao*,[2] thrice distilled),[e] for transport economics would obviously have dictated the carriage of the stronger rather than the weaker liquid.[f] As for distilleries in China, Fang Hsin-Fang[3] has made a special study of the very strong *fên chiu*[4] spirits[g] for which the village of Hsing-hua Tshun[5] south of Thaiyuan in Shansi is famous, and he recounts that all local tranditions say that the distillation started in the Thang time.[h] We ourselves can bear this out because of a personal visit

---

[a] Not our own experience, it must be said.

[b] Deniel (1), p. 15, saw the point of this, though he himself did not know Cibot's original paper. Huber (1), p. 147, who maintained the same, probably did, though not quoting it. Jesuit relations of this kind must assuredly account for the statements of certain European chemists, such as Demachy (1) in +1773 and later, that the origin of all alcohol-distillation was to be sought in China.

[c] Ch. 1, pp. 3*a*, 4*b*, 7*a*, 8*a*, *b*, 36*b*, ch. 2, p. 16*b*.

[d] Pp. 49, 53, 61, 67, 68, 69, 158, 177.   [e] See Giles (14), p. 245.

[f] Schelenz (2) says that Hirth (7) mentions the export of *samshu* in the Sung; we can only find that he says 'wine' (p. 58).

[g] Alcohol content 62%. Named after the river of the province flowing near by.

[h] Yuan Han-Chhing (1), p 96.

[1] 酒   [2] 三燒   [3] 方心芳   [4] 汾酒   [5] 杏花村

Fig. 1503. Modern alcohol vat still for *fên chiu* at Hsing-hua Tshun, Shansi (orig. photo. 1964).

Fig. 1504. Modern vat stills at Shao-hsing, Chekiang (orig. photo. 1964).

made in 1964,[a] on which occasion we were able to read an inscribed stele dated +555 to +557 and connected with a Northern Chou wine official (Chiu Kuan[1]) named Wu Chhêng,[2] who appears to have been the founder of the local industry.[b] It used millet then, and uses kao-liang now, but it is very near the famous vineyards which Marco Polo described and which we also visited, so grape-wine may well have been distilled there in earlier times. The +6th century would seem just not too early, in the light of all the other evidence for the distillation of alcohol in China. Now of course the Chinese

---

[a] With Dr Dorothy Needham.
[b] This stands outside the Shen-Ming Thing[3] pavilion, inside which is the celebrated well (Ku-Ching Thing[4]) which provided the water.

[1] 酒官   [2] 武成   [3] 申明亭   [4] 古井亭

factories mostly use vat stills of modern type with wide-bore retort tubes and spiral condensers sunk in water-tanks (cf. Fig. 1503 from Hsing-hua Tshun and Fig. 1504 from Shao-hsing in Chekiang, another very famous centre).[a]

(iv) *Liang 'frozen-out wine'*

We must now go back to take a closer look at the stone which may turn out to be the keystone of our arch, namely the story about the preparation of a strong alcohol solution by freezing. The *Liang Ssu Kung Chi* (Tales of the Four Lords of Liang), though written towards the end of the +7th century, deals with events then comparatively recent since they occurred in the early part of the +6th. It gives us the previous information that Kao-chhang (Turfan) presented 'frozen-out wine' (*tung chiu*[1]) to the imperial court about +520, most probably on a number of successive tribute visits. Then, much later, the *Tshao Mu Tzu* book describes, towards the end of the +14th century, how people used to test their spirits by going to huts in the high mountains and leaving it out to freeze; in such conditions genuine strong alcohol solutions would not do so, but imitations, diluted or faked perhaps with piquant herbs, would. Laufer annotated the passage with the words: 'This is probably a fantasy. We can make nothing of it, as it is not stated how the adulterated wine was made.'[b] Yet it is perfectly comprehensible when one knows that freezing-out methods live on in common practice and in scientific work at the present day;[c] if the process is carefully done, the ice formed will consist of pure water and all the solutes will be concentrated in a central liquid phase which does not freeze.

This chain of references is enlarged by other important links. Chang Hua[2] in his *Po Wu Chih*,[3] written about +290, remarks that 'the Western regions have a wine made from grapes which will keep good for years, as much as ten years, it is commonly said; and if one drinks of it, one will not get over one's drunkenness for days'.[d] Clearly this was a description of spirits, not ordinary wine, but in view of the date we believe that it was frozen-out wine, not distilled wine. A very similar thing is said in the biography of Lü Kuang[4] (d. +399), the conqueror of Kucha in Sinkiang in +384, which describes, quoting from his report,[e] the wealth of the citizens' families, many of which 'had as much as a thousand *hu*[5] of grape-wine in their houses. Even after ten years it did not go bad.'[f] But this kind of information about the Western regions was

---

[a] At times in Chinese history the distilleries have played an unusually important sociological and economic role; cf. the study of Kawakubo Teiro (1).

[b] (1), p. 237. He probably read only the *PTKM* version.

[c] We have to thank Prof. Stephen Mason and Dr J. H. Lindsay for reminding us here of the place of 'applejack' in Canadian folklore. That was the country where grandad always left half-a-dozen casks of cider out in the snow and ice during winter-time; then at Christmas a tube would be inserted and the liquor drawn off. The excisemen supposedly turned a blind eye on this. Professor Lynn White has told us equally of 'New Jersey lightning'. Concentrates can also be further concentrated, as in the case of the bottles of Grand Marnier left out in stores through a few Antarctic winters, according to an experience related to us by Dr Launcelot Fleming, formerly Bishop of Norwich.

[d] Ch. 5, p. 4*b*.   [e] *Chin Shu*, ch. 122, p. 2*b*, tr. auct., adjuv. Liu Mao-Tsai (1).

[f] About 3000 gallons in our reckoning. This would mean some 18,000 bottles of the size commonly used for wine today.

[1] 凍酒   [2] 張華   [3] 博物志   [4] 呂光   [5] 斛

quite traditional, for as early as −90 Ssuma Chhien had told how in Ferghana (Ta-Yuan) and its neighbourhood 'wine is made from grapes, the wealthier inhabitants keeping as many as ten thousand or more *tan*[1] of it stored away. It can be kept for as long as twenty or thirty years without spoiling. The people love their wine, just as their horses love their alfalfa (fodder).'[a] If these two statements really imply strong alcohol, then the freezing-out method may go back to the −2nd century at least, which is not at all impossible; but the difficulty is to distinguish it from the effects of adequate corking on the preservation of unconcentrated wine (cf. p. 143). We cannot be quite sure. The rather large amounts would plead for the stoppering interpretation, but it is doubtful how effective this was among any ancient people; so 'frozen-out wine' seems on the whole the more probable.[b] Again, Li Chao (fl. +810), whom we encountered on p. 142 above, added to his list of famous wines a *shih tung chhun chiu*,[2] 'spring wine frozen-out on the crags', from Fu-phing in Shensi.[c] Thus between the −2nd and the +14th centuries we have at least six references.[d]

The two passages quoted by Li Shih-Chen, and the others, are probably the oldest on this phenomenon in any world literature.[e] The earliest description in Europe, so far as we can see, occurs in the sixth book of the *Archidoxis* of Paracelsus, written about +1527 but not printed till +1570.[f] This had great repercussions in Europe as an extraordinary fact of Nature. About +1620 Francis Bacon wrote: 'Paracelsus reporteth, that if a glass of wine be set upon a terras in a bitter frost, it will leave some liquor unfrozen in the centre of the glass, which excelleth *spiritus vini* drawn by fire.'[g] And in +1646 Sir Thomas Browne noted that 'Paracelsus in his *Archidoxis*, extracteth the magistery of wine; after four moneths digestion in horse-dung, exposing it unto the extremity of cold; whereby the aqueous parts will freeze, but the Spirit retire and be found congealed in the centre.'[h] There were several other mentions before the end

---

a *Shih Chi*, ch. 123, p. 15*a*, tr. auct., adjuv. Hirth (2), Watson (1). The estimate is about twenty times the preceding one, but there is no need to take either of them *au pied de la lettre*.

b Unless of course the people of Sinkiang were distilling in Gandhāran retorts (pp. 86–7, 121).

c Ch. 3, p. 9*b* in the Thang Tshung Shu ed.

d Doubtless more will come to light. Wang Chia, in his *Shih I Chi* (Memoirs on Neglected Matters), written about +370, has a curious passage about 'gut-rotting wine'. He starts off by telling how Chang Hua, the naturalist (+232 to +300) and author of the *Po Wu Chih* just quoted, used to make a special kind of wine with ferments which he got from the Western Chhiang and Northern Hu tribal peoples. He then goes on to say that the Hu foreigners have a *chih hsing mai*[3] (variety of cereal) which can be malted, and makes a wine that causes chattering of the teeth and apparent drunkenness without shouting or laughter, injuring the liver and intestines. Hence the ordinary people called it 'gut-rotting wine' (*hsiao chhang chiu*[4]). Nevertheless some people experienced pleasure from it, not caring to preserve their lives. We should hesitate to follow Eberhard (in his review of Hermanns, 1) in his interpretation of this as distilled wine. The story suggests to us rather a wine containing some kind of toxic substance. But it could possibly be an early reference to the outlandish and surprising effects brought about by the 'frozen-out wine', especially as it came from some tribal or city-state people in the north or north-west.

e Seneca, in his *Quaest. Nat.*, written in +64, has a curious passage about 'wine frozen by lightning', which, when re-liquefied, 'kills or drives mad those who drink of it' (II, lii, liii, Clarke tr. p. 97). We can probably neglect this as fabulous.

f Cf. Pagel (10), p. 274; Debus (15), p. 33. The text is to be found in the Sudhoff ed. vol. 3, pp. 165–6; Strebel ed., vol. 8, pp. 358–9.

g *Inquisitio Legitima de Calore et Frigore*, in *Works*, Montagu ed., vol. 1, p. 333.

h *Pseudodoxia Epidemica*, Sayle ed., vol. 1, pp. 204–5. He meant, of course, concentrated, the opposite of congealed in this case. Cf. Pagel (10), p. 274; Debus (15), p. 33, (16), p. 71.

¹ 石   ² 石凍春酒   ³ 指星麥   ⁴ 消腸酒

of the $+$16th century, notably one by Conrad Khunrath[a] in $+$1594, who mixed his frozen-out alcohol with *aqua vitae* distilled in the usual way.[b] Glauber again, in $+$1657, found that he could concentrate acetic acid by freezing out the water.[c]

The question then acquired considerable theoretical importance in the 'Sceptical Chymist' of Robert Boyle ($+$1661), who after giving Paracelsus' Latin text embarked upon one of his long-winded but charming discourses, making much of the experiences of 'the Dutch men that Winter'd in Nova Zembla'. In their own words, which he quoted: 'There was scarce any unfrozen Beer in the barrel; but in that thick Yiest that was unfrozen lay the Strength of the Beer, so that it was too Strong to drink alone, and that which was frozen tasted like Water...'[d] And Boyle went on to say that he 'might confirm the Dutchmen's Relation, by what happen'd a while since to a neere Friend of mine, who complained to me, that having Brew'd some Beer or Ale for his own drinking in Holland (where he then dwelt), the Keenness of the late bitter Winter froze the Drink so as to reduce it into Ice and a small proportion of a very Strong and Spirituous Liquor...'[e] What all this was in aid of was the criticism of the spagyrical tradition that fire alone would analyse mixed bodies.[f] Boyle (in the person of his character Carneades) was setting out to make 'the common Assumption of our Chymists and Aristotelians appear Questionable'.[g] Cold had been considered *tam Homogenea quam Heterogenea congregare*, but now it seemed, like heat, also to be able *congregare Homogenea, et Heterogenea segregare*. Thus did the observation known to Wan Chieh and the other Lords of Liang find its place in the theoretical cogitations of nascent modern chemistry.

Naturally the process could be viewed in reverse, not for the concentration of spirituous liquor, but for the winning of pure water from the sea[h] or from any un-drinkable aqueous solution. This seems to have occurred first to the Danish physician,

[a] Brother of the more famous Heinrich Khunrath, whom we shall meet in Vol. 5, pt. 5. Cf. Partington (7), vol. 2, p. 88.

[b] *Medulla Destillatoria et Medica*, edition of $+$1680, vol. 2, p. 304. Cf. Arntz (1), p. 202, who quotes the passage in full.

[c] *Miraculum Mundi Continuatio*, p. 215 in the $+$1658 edition of *Opera Chymica* (Frankfurt). His words are worth recording: 'Wann man diesen Holz-Essig in Fassern im kalten Winter gefrieren lasst, so gefrieret nur das phlegma, und wird zu Eis, der scharffe Spiritus mit dem Oel geht hineinwarts, und frieret nichts, wird so starck, dass er die Metallen mit Gewalt angriefft, wie ein Aqua Fortis.' The question is an important one, because there is reason to think, as we shall later find, that rather strong acetic acid was known and used in medieval China, yet it is very difficult to produce by ordinary distillation (cf. pp. 178ff. below).

[d] This was the party of Gerard de Veer, who had wintered with Barents. His account was included in 'Purchas his Pilgrimes' (1625 ed., vol. 3, pt. 2, bk. iii, p. 493; McLehose ed. vol. 13, p. 91). It is interesting that they were trying to find a north-east passage to Cathay. [e] Pp. 95–102.

[f] In his *New Experiments and Observations touching Cold* ($+$1683 ed.) Boyle (5) described his further studies on freezing. He concentrated leaf extracts, getting all the coloured and flavoured constituents in the unfrozen part (p. 256, and 2nd app., p. 14), and he could do the like with pigments such as gentian and cochineal (1st app., pp. 11, 19). Inorganic colours like green or blue vitriol behaved in the same way (p. 54), and the ice of their solutions was the same as ordinary ice. Also he could ignite the unfrozen alcohol from sack and sherry (pp. 56–7), showing that whatever would not burn off would freeze.

[g] On this topic, see the interesting survey of Debus (14). It was the late Sir Ronald Fisher who often used to remind us of the part played in Boyle's thought by the freezing-out technique.

[h] See the review of Nebbia & Nebbia-Menozzi (2). At the time of writing (Feb. 1971), the Atomic Energy Authority in this country was planning to construct a large-scale plant for the freeze desalinisation of sea-water on the coast of East Anglia; this could be of great help to the U.K.'s water resources always under strain. Cf. Snyder (1).

Thomas Bartholinus,[a] in the same year that Robert Boyle published his book, and Boyle himself recommended the same method to sailors in cold latitudes a few years later.[b] Before long it was widely used; Captain Cook, for instance, supplied fresh water to his crew by melting sea ice in +1773. A scientific study of desalinisation by this means was made in +1786 by Lorgna (1), who carried out a series of successive freezings of the liquid obtained by melting the ice at each stage. After four freezings the ice contained only a trace of salt, but the yield was low. He also tested the process successfully on urine, obtaining a concentrate 'of a very deep red colour' because of the urobilin and other pigments, as well as almost pure water to drink. Today this is proposed—perhaps utilised—in the technique of space travel, a development which would have greatly surprised the Four Lords of Liang. Finally it should be repeated that the 'frozen-out wine' technique plays a growing part in modern chemistry as a safe and delicate procedure for the concentration of dilute solutions.[c] Describing this 'exceedingly useful but hitherto neglected method', Shapiro (1) says that mechanical stirring or shaking is essential to prevent supercooling and the sudden freezing of boundary layers rich in solute. It is not confined to aqueous solutions but will work with any which have a suitable freezing-point,[d] and it is 'almost above reproach with regard to the chemical or physical alteration of the substances being concentrated'.

Finally, the freezing-out phenomenon supplied one of our favourite poets with a useful metaphor. Byron was trying to describe girls who hide a passionate nature under a rather cold exterior manner—but he found the celebrated volcano motif hackneyed and absurd, so he thought of something else.

> I'll have another figure in a trice:
>   What say you to a bottle of champagne?
> Frozen into a very vinous ice,
>   Which leaves few drops of that immortal rain,
> Yet in the very centre, past all price,
>   About a liquid glassful will remain;
> And this is stronger than the strongest grape
> Could e'er express in its expanded shape.
>
> 'Tis the whole spirit brought to a quintessence.
>   And thus the chilliest aspects may concentre
> A hidden nectar under a cold presence.
>   And such are many, though I only meant her,
> From whom I now deduce these moral lessons,
>   On which the Muse has always sought to enter.
> And your cold people are beyond all price,
>   When once you've broken their confounded ice.[e]

---

[a] (1), ch. 4, p. 42.   [b] (5), p. 59.

[c] Thus Mellanby (1) used it in 1908 for concentrating a protein, diphtheria antitoxin, Palmer (1) later on for milk globulin, and Bawden & Pirie (1, 2) for tobacco-plant viruses.

[d] It may be worth remembering that one of the foundations of the aromatic chemical industry was Mansfield's separation of toluene from benzene by freezing (Campbell (1), p. 84), in 1847.

[e] *Don Juan*, canto xiii, 37, 38; Steffan, Steffan & Pratt ed., p. 452. The heroine here was Adeline, the girl of English aristocratic stock.

(v) *From icy mountain to torrid still*

The moment has now come when we can survey the whole problem and plan perhaps a provisional working hypothesis about the history of alcohol East as well as West. Cibot and Laufer both made the point that it was very odd if the distillation of grape-wine should have come to China from Turfan in +640 and that the Chinese should then have waited seven centuries before applying it to their own indigenous cereal wines. The studies of von Lippmann make it clear that no sort of wine distillation could have come from Turfan at that period because no Western stills had a cooling system until five centuries later.[a] Though the Gandhāran ones probably did, it is not certain that they were used for alcohol.[b] But we know that 'frozen-out wine' did come from Turfan, and had already done so for at least a century before its annexation. Surely then the most likely thing is that the 'frozen-out wine' triggered the first distillation of wine in the East, and that this process was continued in China both for grape and cereal wines thenceforward. The sending of the first alcohol concentrates is the crucial point, for the Chinese would then have acquired (like the Uighurs) the taste for relatively strong alcohol—it may not have tasted very good but at least it made them merry quickly and this was the important thing. That the catch-bowl still, whether 'Mongol' or 'Chinese' in type, already existed in China we cannot yet fully prove, but it may be assumed without undue risk from the time of Thao Hung-Ching onwards, and indeed we suggest elsewhere (p. 178) that the concentration of acetic acid by distillation may have played a part in the solubilisation methods of the *San-shih-liu Shui Fa*, which are otherwise rather difficult to explain. The stills of China could also reasonably have been used in the +6th century for essential oils (a subject on which we have yet to say a few words). One source of obscurity is that we do not know enough about the first origin of the catch-bowl still without the side-tube; after all 'Mongol' and 'Chinese' are only ethnographical expressions, and between +500 and +800 the still used in China may well have been the simple 'Mongol' type. But the side-tube must surely have been present by +900, and probably a good deal earlier, because of several allusions which we have noted in the literary references (pp. 144–5), quite apart from the drawing in the *Chih-Chhuan Chen-Jen Chiao Chêng Shu* (*TT* 895), Fig. 1447.

'Frozen-out wine' then, in all its primitive simplicity, was, we would suggest, an important step on the road from beer or wine as such to distilled 'strong liquor'. Its origin in empirical experience among the snows of the Thien Shan or the bitter winds of the Gobi is easy enough to picture, but how on earth could any Taoist in +6th- or

[a] Or, more precisely, no one in the West had discovered how to get alcohol from wine either by water-cooled stills or the subtle use of air-cooled ones, and would not do so for another five centuries. We should perhaps leave open the possibility that the Uighurs of Turfan were using the Mongol or Chinese catch-bowl still, but that would place them firmly within the Chinese culture-area. It is also in a way a superfluous assumption if their 'frozen-out wine' was as potent as we think it was; and we know that that went east.

[b] If that could be proved, then a diffusion stimulus for Chinese alcohol-distillation could have come from northern India, yet it would have been only a stimulus since the Chinese proceeded to use their own stills of quite a different type.

+7th-century Chhang-an have imagined that by submitting wine to intense heat one could get the same result as that brought about by its exposure to intense cold?

One ideological possibility presents itself in the 'similarity of extremes', almost, one might say, the 'identity of opposites'. The 'frozen-out wine' was connected with a *shêng Yin*[1] condition, so perhaps something equally interesting would happen if it were submitted to a *shêng Yang*[2] process. As Li Shih-Chen remarked: 'Burnt-wine is a powerful drug, for it partakes of the nature of pure Yang; its character is similar to that of fire (*Shao chiu shun Yang tu wu yeh, yü huo thung hsing*[3]).'[a] Just so might it have been argued that 'frozen-out wine' partook of the nature of pure Yin, also very dangerous in its way, like frostbite, snow-blindness, and the caustic feel of intense cold. This would have been another aspect of the 'marriage of fire and water'; and it is easy to show that the Chinese elements Fire and Water were indeed related to each other as Yang-Yin opposites.

About +640, the great scholar Khung Ying-Ta,[4] commenting on the Old Text version[b] of the *Shu Ching*'s Hung Fan chapter,[c] wrote as follows:[d]

'Wood can' to 'usefully transformed (*kai pien*[5])'.

[Comm.] These words express its nature; it can be softened and made curved or straight, as is necessary for the making of implements (and objects). Transformation for human convenience is like the fusion and melting (of metal) for implements (and vessels). Just as Wood can be softened and made curved or straight, so Metal can also be remoulded to the heart's desire, in accordance with its use for mankind—that's the meaning of it. From this one can see that the usefulness of Water for irrigation lies in its tendency to seep down and enrich the earth; so also the usefulness of Fire is that it goes upwards, giving combustion and heat. This is quite understandable. Since Water is pure Yin (*shun Yin*[6]), it (naturally) soaks, moistens, enriches) and descends, tending towards the Yin (of the earth). But Fire is pure Yang (*shun Yang*[7]), so (naturally) when it burns and blazes it rises upwards, tending towards the Yang (of the heavens). As for Wood and Metal, they are composed of Yin and Yang mixed together. Therefore their form can be made crooked or straight, that is, altered and changed, transformed for human convenience.

This is a classical statement of the proportions of Yin and Yang in the Five Elements. Earth alone is not specifically mentioned, but from its central position among the four

---

[a] *PTKM*, ch. 25, (p. 35).

[b] Something has already been said of this famous philological controversy in Vol. 2, p. 248, and later in Needham (56), p. 30. Briefly, texts of the Confucian classics written in ancient characters were supposedly discovered in −148 or −135, and since they differed somewhat from those currently accepted they were commented on about −100 by Khung An-Kuo[8] and later other famous scholars—then they were again lost. Subsequently, between +317 and +322, Mei Tsê[9] claimed to have found both text and commentary, and these were the versions on which Khung Ying-Ta wrote his commentary early in the Thang. Sung scholars were sceptical of the authenticity of these texts, Mei Tsu[10] demolished it in +1513, and the *coup de grâce* was delivered by Yen Jo-Chü[11] in +1745. Nevertheless, this does not mean that many ancient fragments were not incorporated into his pastiche by Mei Tsê in the +4th century, nor that it is without value for statements on points in the perennial natural philosophy of China which may not happen to occur elsewhere. For a succinct statement of the case cf. Hummel (2), p. 909.

[c] See Vol. 2, pp. 242-3.   [d] *Shang Shu Chêng I*,[12] ch. 11, p. 7b.

[1] 盛陰	[2] 盛陽	[3] 燒酒純陽毒物也與火同性	
[4] 孔穎達	[5] 改便	[6] 純陰 [7] 純陽	[8] 孔安國
[9] 梅賾	[10] 梅鷟	[11] 閻若璩 [12] 尚書正義	

directions, it must evidently be equally balanced between Yin and Yang. The former predominates in Metal, which can melt to the liquid state; the latter in Wood, which cannot. A table may elucidate.[a]

M	mixed	Yin > Yang (Yang in Yin)	shao Yin
W	mixed	Yang > Yin (Yin in Yang)	shao Yang
w	pure Yin		thai Yin
F	pure Yang		thai Yang
E	mixed equal proportions		equal balance

Once the product of distillation had been tested, the similarity with alcohol made the other way would have been obvious, and since fire was much more easily and widely obtainable than snow and ice it naturally become the dominant process.[b] Then, much later in time, and by ways which as yet we cannot discern (though Central Asian intermediation may be more likely than Islam and India), the still-cooling of China could have made its way westwards to influence the +12th-century Masters of Salerno and their Italian friends.[c] In this manner for the first time a coherent scheme of the development of means for making alcohol solutions above 40% in strength, throughout the Old World cultures, presents itself, at least as a working hypothesis.[d]

We can now look back once more at Li Shih-Chen's passages and see what misunderstandings arose from them. Distillation of cereal wines did *not* begin first under the Yuan dynasty; Li took Yeh Tzu-Chhi and Hu Ssu-Hui (talking about grape brandy) as his starting-point and ignored most of the Thang and Sung literature. Distillation of grape-wine (or wine of any sort) did *not* come from Western sources through Turfan in the early Thang because it had not been accomplished anywhere at that time without the Asian types of water-cooled still. Distillation of any sort of wine was *not* ancient in the Far West, as might too readily be assumed from Li's words.[e] The Uighurs may well have provided China in +640 with mare's nipple grapes and the recipe for making grape-wine without *chhü* ferment, but what really mattered was the *tung chiu*,[1] the 'frozen-out wine', which they had already been sending for some time previously. Surely this was the father and mother of all 'strong liquor', and the ancestral inspira-

---

[a] From Table 12 in Vol. 2, p. 263.

[b] There might be a parallel here with certain other deductions from medieval natural philosophy where thinkers were faced by a problem of opposites. In Sect. 45 we shall show how the physicians, at least as early as the +7th century, treated goitre with thyroid glands from domestic animals. To heal an enlargement of an organ by giving more of the same thing might seem a strange procedure, but the physicians had the insight to realise that thyroid hyperplasia was in many cases the sign of a basic deficiency (*hsü*[2]), indicating what we should now call hypothyroidism. In the meantime see further on this Needham & Lu Gwei-Djen (3).

[c] It looks therefore as if we must place alcohol distillation in what has been called the +12th-century cluster of transmissions (Needham (64), p. 61). These include the magnetic compass, the stern-post rudder and the windmill. And for the first two of these no evidence of Islamic and Indian way-stations is perceptible.

[d] Great as the contributions of von Lippmann were, he placed all his money on a loser in declining persistently to admit any creative influence of Chinese culture in the history of chemistry. It would have been better to admit ignorance—which nearly everyone in the West then shared.

[e] For example by Laufer (1), pp. 220ff., esp. pp. 235ff., followed by Shinoda Osamu (2).

[1] 凍酒   [2] 虛

tion of all Chinese distillers. So Li Shih-Chen was probably quite right in saying (if that was what he meant to say) that grape-wine brandy started early in the Thang; and lacking as he did so much of what we know now, he was only wrong in implying (if indeed he did) that it had come from somewhere else.

(vi) *Oils in stills; the rose and the flame-thrower*

By way of an appendix to all the foregoing something remains to be said about the distillation of essential oils.[a] What we should like to have would be a few texts about this practice dating from the +5th, +6th or +7th centuries, i.e. before the appearance of strong alcohol solutions in China; but we have not found anything of this antiquity, and the relevant certain mentions begin in the +10th, i.e. well after the period during which, as we think, the distillation of wine was being successfully carried out. Perhaps the parallel with the Arabic culture-area is deceptive, and one should not assume that because the Arabs and Byzantines did so much distilling of essential vegetable oils and petroleum oils before alcohol distillation started in the West, the same sequence took place in China—it may have been that Chinese stills were first used for vinegar (cf. pp. 128, 178) and other substances in which the alchemists were particularly interested, e.g. mercury. The dry distillation of eggs (as among the earlier Hellenistic proto-chemists), or hair (as among the Arabs subsequently), may have played a part here, though it might be hard to point to any overt evidence of it. But even if we cannot at present carry back the story of the volatile oils beyond that of alcohol, it will be worth while to give a few later quotations concerning it.

Rather strikingly, when we first come upon essential oils as a notable import, they are in close juxtaposition with light petroleum fractions of the 'Greek fire' type. In +958 the King of Champa (Chan-Chhêng,[1] mod. Annam and Tongking), Śri Indravarman, sent as ambassador to China an Arab, or at least an envoy with an Arabic name, Abū'l-Ḥasan (Phu-Ko-San[2]), who presented fifteen bottles of rose-water (*chhiang-wei shui*[3]) and eighty-four glass bottles of Greek fire (*mêng huo yu*[4]).[b] The former came from the Western countries (Hsi Yü[5]) and was intended for sprinkling on clothes, the latter was for pyrotechnics or war, and burnt even better when spread upon water.[c] This was in the Later Chou dynasty (+951 to +960), but rose-water, like the 'petrol', had been prominent in China some twenty or thirty years earlier, under another of those ephemeral dynasties of the Wu Tai period, the Later Thang (+923

---

[a] On the principles involved something has been said already, pp. 128–9.

[b] *Thai-Phing Huan Yü Chi* (c. +980), ch. 179, p. 16b, cit. in *Tshê Fu Yuan Kuei* (+1013), ch. 972, p. 22a, b. Another almost contemporary account is in the *Chuang Lou Chi*[6] of Chang Mi.[7] Cf. Schafer (13), p. 173. See also *Wu Tai Shih Chi*, ch. 74, p. 17a, discussed by Fêng Chia-Shêng (2), p. 17.

[c] Greek fire 'petrol' had been available in China since at least +917, and probably rather earlier; the first appearance of gunpowder is in the form of slow match for a flame-thrower using it (cf. Sect. 30), in +919. The *Wu Tai Shih Chi* adds a record that the 'fierce fire oil' was useful for removing stains from clothes. One need make no great claim for Chinese originality in this, but it is rather startling to find 'dry cleaning' practised along the Arabs, Annamese and Chinese in the +10th century. For a brief account of the industry at the present day, cf. Popham (1). On its history in the West, Edelstein (1, 2).

[1] 占城   [2] 蒲謌散   [3] 薔薇水   [4] 猛火油   [5] 西域
[6] 妝樓記   [7] 張泌

to +936). Thao Ku[1] tells us in his *Chhing I Lu*[2] (Records of the Unworldly and the Strange) that about +930 the emperor had a miniature city and gardens made of unusual materials laid out in one of the palace halls.[a] This Ling Fang Kuo,[3] as it was called (Country of Numinous Fragrances), had hills and mountains made of lignaloes wood, lakes and rivers of storax and rose-water, trees of cloves and other aromatics, walls and ramparts of frankincense, buildings of rosewood and sanderswood, and human figures carved in sandalwood.[b] Thao himself had probably seen this master-piece, the constituents of which were said to have come in part from the conquered State of Shu in Szechuan. This reminds us that during the first thirty years of the century Shu had been the home of two outstanding experts on perfumes and aromatic drugs, Li Hsün,[4] the writer of the *Hai Yao Pên Tshao*[5] (Natural History of the Southern Countries beyond the Seas), and his younger brother Li Hsien,[6] alchemist, naturalist, chess master and like Li Hsün a poet.[c] The family was of Persian origin, and it is hard to believe that they were ignorant of the distillation of essential oils. Peppermint oil (*po ho yu*[7]) is said to be mentioned in the *I Hsin Fang* (*Ishinhō*[8]) of +982, which would imply steam distillation.[d]

Pushing further back, there are references to rose-water, in the form of 'rose dew' (*chhiang-wei lu*[9]), as early as about +800. Fêng Chih[10] in his *Yün Hsien Tsa Chi*[11] of *c.* +904 says that whenever Liu Tsung-Yuan[12] (+773 to +819) received a poem from the great scholar Han Yü[13] (+768 to +824) his admiration was such that he insisted on washing his hands in rose dew before reading it.[e] The difficulty about interpreting these early references is that three things have commonly been confused under the name 'rose-water'. There is a way of making a kind of press-juice of the rose petals which may be quite old in China, and is still used at the present time for a flavouring perfume and a cooling drink. The *chhiang-wei* petals[f] are ground to a paste with water, excess water filtered off, and sugar added to form a kind of jam which can be durably stored in porcelain pots; dilution of this then gives a fragrant solution.[g] Secondly, there is the distillation of the essential oil such as was practised in the Arabic culture-area throughout the +9th century (cf. p. 128 above).[h] Third (though of less concern to us here) is the attar of roses, an oil which separates spontaneously on the surface when an aqueous extract of rose-petals is left to stand.[i] At present it is not possible to say what kind of rose-water it was in which Liu Tsung-Yuan washed his hands,[j] and

---

[a] Ch. 2, p. 58b. Cf. Schafer (13), p. 173.

[b] For further information on these perfume sources see pt. 2, pp. 136ff., and on the coloured woods Schafer (8). [c] See further in Sect. 38 in Vol. 6, and p. 421 below.

[d] Schelenz (2), p. 129, but his reference is garbled, and we have not so far been able to locate it in the great work of Tamba no Yasuyori. The plant, presumably *Mentha arvensis* (R129; CC337), and probably var. *piperascens*, was first mentioned in the *Hsin Hsiu Pên Tshao* of +659.

[e] Ch. 6, p. 46a, b. Cf. Schafer (13), p. 174.

[f] This is *Rosa multiflora* (CC1144), the ancestor of all rambler rose varieties (cf. Li Hui-Lin (8), pp. 92ff. [g] Li Hui-Lin (8), pp. 95–6. [h] Cf. Hanbury (9).

[i] This has been mainly a product of India, and traditionally ascribed to Nur Jehan, the queen of the Mogul Emperor Jehangir, as an invention of about +1612. See Burkill (1), vol. 2, p. 1915.

[j] But 'dew' does, after all, imply drops, and hence a distillate.

[1] 陶穀	[2] 清異錄	[3] 靈芳國	[4] 李珣
[5] 海藥本草	[6] 李玹	[7] 薄荷油	[8] 醫心方
[9] 薔薇露	[10] 馮贄	[11] 雲仙雜記	[12] 柳宗元	[13] 韓愈

we must hope that further Thang allusions will be found whereby we can decide whether the distillation of essential oils was then going on. It certainly went on later, though perhaps rather for other plants, since one gets the impression that rose oil remained a valued import from the Arabic culture-area. Here, for example, is Tshai Thao[1] talking just before +1115 in his *Thieh Wei Shan Tshung Than*[2] (Collected Conversations at Iron-Fence Mountain).[a]

According to an old idea, rose-water (*chhiang-wei shui*) was obtained by collecting the dew from the *chhiang-wei* flowers (roses) in some foreign country. This is in fact not true, for a still (*tsêng*[3]) made of some white metal (*pai chin*[4])[b] is used. The roses are collected and heated, producing vapours which condense and form a water (*chêng chhi chhêng shui*[5]). By means of repeated collecting and repeated heating the (liquid) is strengthened and gives forth great fragrance.[c] That is why it lasts so well.

The perfume of the roses of foreign parts is particularly strong, so the rose-water of the Arabs, even if put into a glass bottle and closely sealed with wax, will still escape to a slight extent and diffuse its delightful odour. One can smell this several dozen paces away, and if sprinkled on clothes it will last for several weeks. In other places abroad where they have not got roses they make a similar liquid from the two kinds of jasmine—but that is only a slave-girl in comparison with the rose-water of the Arabs.[d]

But perhaps opinions differed, for the essential oils of citrous flowers distilled in China were greatly admired throughout the Sung, as we know from many +12th-century references.[e] In the *Yu Huan Chi Wên*[6] of Chang Shih-Nan[7] we read as follows:[f]

The people of San-shan[8] say that...the oranges of Yung-chia[9] are the best in the whole world. There is one kind called *chu luan*[10] the flowers of which have a perfume excelling that of all other citrous flowers or fruits.[g] They are placed, with shavings of a kind of lignaloes (*chien hsiang*[11])[h] and laka-wood (*chiang chen (jen) hsiang*[12]),[i] in a small steamer (*tsêng*[3] i.e. a still) made of tin so that the flowers and the bits of wood form alternate layers, only there are usually more flowers than wood. At the opening at the side of the still drops of liquid collect like sweat, and are received in a container. Then the distillation is stopped and the flowers taken out, while the distillate is put back to soak into the wood again. After being left overnight the process is repeated and a fresh distillation (*chêng*[13]) made, in all three or four changes of flowers being used. In the end the chips are dried and kept in sealed porcelain vessels. The perfume is extraordinarily elegant.

---

[a] Ch. 5, p. 20*b*ff., tr. auct.
[b] Probably tin. If in China it could well have been an alloy of zinc.
[c] This suggests re-distillation.
[d] The passage brings vividly to mind an Arabic merchant of the old tradition, proud of the pure oils which he sold, and despising all diluted forms, with whom we had much converse once at Houmt Souk in Tunisia.
[e] Cf. the passage from the *Ling Wai Tai Ta* of +1178 on enfleurage, quoted elsewhere, pt. 2, p. 146.
[f] Ch. 5, p. 7*a*, tr. auct.
[g] This is the sour, or 'Seville' orange, *Citrus Aurantium* (see Sect. 38 in Vol. 6).
[h] I.e. garroo wood, the 'sinking aromatic', *Aquilaria agallocha* or *sinensis*, from Annam or Hainan, on which see pt. 2, p. 141 above.
[i] *Kayu* or *laka*, the 'purple liana aromatic', *Dalbergia parviflora*, on which see also pt. 2, p. 141 above.

[1] 蔡絛   [2] 鐵圍山叢談   [3] 甑   [4] 白金   [5] 蒸氣成水
[6] 游宦紀聞   [7] 張世南   [8] 三山   [9] 永嘉   [10] 朱欒
[11] 箋香   [12] 降眞人香   [13] 蒸

In this process then we have both enfleurage and distillation leading to a solid preparation and an essential oil.[a] Although it is not mentioned, a water-bath must almost
certainly have been used to prevent damage by over-heating. Chang Shih-Nan's book
was finished in +1233, but what he says about the preparation of this 'flower dew'
(*hua lu*[1]) is paralleled in almost exactly the same words in Han Yen-Chih's[2] *Chü Lu*[3]
(Orange Record), written in +1178, the type-specimen, as we shall later see,[b] of the
Sung botanical and horticultural monographs.[c] This mentions also the still of tin.
A reference not much earlier is that in the *Mei-Chhi Shih Chu*[4] of Wang Shih-Phêng[5]
about +1140, who says that 'citrous flowers distilled (*chêng*[6]) make a perfume, good
for keeping insects away from clothes'.[d]

Lastly, it is interesting to read the entry specially devoted to rose-water (*chhiang-
wei shui*[7]) in the *Chu Fan Chih*[8] (Records of Foreign Peoples and their Trade), written
by Chao Ju-Kua[9] about +1225.

Chhiang-wei shui [he says][e] is the dew of flowers (*hua lu*[1])[f] in the Arab countries (Ta-Shih
Kuo[10]). In the Wu Tai period the foreign envoy Phu-Ko-San brought 15 bottles as tribute,
after which time it was not often seen. Nowadays a common substitute is made by gathering
the flowers, steeping them in water, and distilling (*chêng*[6]), the condensate (*i*[11]) being collected.
Rose-water is much counterfeited and adulterated. To test it, the liquid should be placed in
glass bottles and shaken about for a while, then if it is full of bubbles moving up and down, it
is genuine. The flower (from which it is made) is not the same as the *chhiang-wei* rose of China.

From the first sentences it is hard to say whether Chao Ju-Kua believed that the King
of Champa's bottles had contained the essential oil of roses or something still better,
but he certainly knew it in the international commerce of his own time. His words
on the test describe rather well the rapid separation of two immiscible liquids. And he
was right that the rose of the Arabic countries used for its oil was not the same as any
of those in China such as *Rosa multiflora*. It was probably then, as now, *Rosa bifera*,
derived from the wild *R. rubra* and *R. moschata* in a hybridisation long antedating
Pliny's mention of this autumn damask rose.[g]

In sum, therefore, much remains to be learnt about the distillation of essential
oils in the Chinese culture-area. Since, so far as we can see, Chinese stills had an
effective cooling device from the beginning, it may be that the essential oils followed

[a] Cf. Wu Tê-To (*1*), p. 55. On enfleurage see Hanbury (8).
[b] Sect. 38 in Vol. 6.
[c] Ch. 2, p. 3*a*, tr. Hagerty (1), p. 94.
[d] *TSCC, Tshao mu tien*, ch. 226, *tsa lu*, p. 2*a*.
[e] Ch. 2, p. 4*a*, *b*, tr. auct. adjuv. Hirth & Rockhill (1), p. 203.
[f] The difficulty of translating this is that one does not know whether it had become a technical term
for polyterpene perfume distillates by Chao Ju-Kua's time or not. He could hardly have meant dew
literally after what Tshai Thao had said a century earlier. Yet he goes on to call the distilled oil a
substitute.
[g] We shall return to the Rosaceae, and the monographs of Chinese scholars on them, in Sect. 38
(Vol. 6).

[1] 花露　　[2] 韓彦直　　[3] 橘錄　　[4] 梅溪詩注　　[5] 王十朋
[6] 蒸　　　[7] 薔薇水　　[8] 諸番志　　[9] 趙汝适　　[10] 大食國
[11] 液

alcohol instead of preceding it as they did in the West. At present we must be content to say that mercury was perhaps the first thing which the Chinese systematically distilled, then vinegar and the wine from grapes and cereals, then perhaps the vegetable and the mineral oils.[a] But this must remain for the time being a tentative conclusion.

Meanwhile a possible pattern of alcohol in the Old World seems for the first time to be emerging; no small gain brought by this sub-section. Known first from the 'frozen-out wine' of +3rd-century Central Asia, strong alcohol began to drip from the side-tubes of Chinese stills from the +7th century onwards, until eventually in the +12th effective cooling passed to the West and permitted the preparation of Ardent Water to set beside the Burnt Wine of the Chinese.

(9) LABORATORY INSTRUMENTS AND ACCESSORY EQUIPMENT

Numerous useful accessories are mentioned in *TT*874, a Liang text; and in the +7th century Sun Ssu-Mo gives a list of tools and apparatus essential for the pharmaceutical laboratory.[b] For pounding and grinding the Chinese alchemists used different types of pestle made from jade (*yü chhui*[1]) or stone (*shih chhui*[2]). *TT*886 mentions another sort carved out of willow wood (*liu-mu chhui*[3]), while *TT*885 describes the use of an antelope horn (*ling-yang chio*[4]) for grinding.[c] Fig. 1505 taken from *TT*893 shows a pestle and mortar used by the alchemists of the Sung.[d] Fig. 1506 shows a cast-iron pestle and mortar of Hou Han date, and Fig. 1507 a bronze one of Hsin or Hou Han. One should also visualise the alchemists (or rather their assistants) using the longitudinal-travel edge-runner mill (*yen nien*[5]), already discussed and illustrated (from the *Thien Kung Khai Wu*) in Vol. 4, pt. 2, pp. 195, 197. It is not quite clear how far this goes back, but one specimen of the type worked by the feet and called a *thieh tshao*[6] (iron trough mill), indubitably of the Yuan period, is preserved in the Imperial Palace Museum at Peking, having been excavated very recently. The Sung alchemists must have had these, though they may not be pre-Thang. They were always associated with pharmacy, and are still in use today.

After the process of grinding or pounding, the fine particles were separated from the coarser ones by means of various types of sieve. One made of horse-hair, called *ma-wei lo*,[7] and another of fine silk, *chhing-sha lo*,[8] are mentioned in *TT*874, written in the Liang period.

---

[a] By +1044 there is an elaborate description of a flamethrower for 'naphtha' or Greek fire which has been discussed in Vol. 4, pt. 2, pp. 145 ff. This pump of interesting design was assuredly not used solely with imported petroleum distillates, but information has not yet come to light on the time when the Chinese began preparing these themselves. It could have been after the tribute of +958, but on the other hand, so far as chemical apparatus was concerned, it could have been as far back as Callinicus himself in the +7th century.

[b] *Chhien Chin Yao Fang*, ch. 1, p. 31 *b* (p. 14.2).

[c] This is the text of which Spooner & Wang (1) made a translation, now superseded by that of Sivin (3).   [d] P. 12*a*.

[1] 玉槌   [2] 石槌   [3] 柳木槌   [4] 羚羊角   [5] 研碾
[6] 鐵槽   [7] 馬尾羅   [8] 輕紗羅

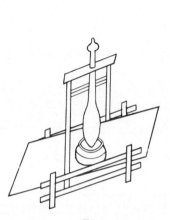

Fig. 1505   Fig. 1506

Fig. 1505. Pestle and mortar, from the *Tan Fang Hsü Chih* of +1163.

Fig. 1506. Cast-iron pestle and mortar, Later Han in date, from a tomb at Yang-tzu Shan near Chhêngtu excavated in 1957 (Chhêngtu Historical Museum, orig. photo. 1972).

For transferring or removing ingredients the same text describes the use of an iron spoon (*thieh shih*[1]) and iron chopsticks (*thieh chu*[2]). In Chinese museums one may see bronze ladles with a handle as well as three legs (e.g. at Chungking). One of the finest of these, with a collapsible handle, dates from about +750 (see Fig. 1508). It was part of the same hoard as the named specimens of chemicals described in Vol. 5, pt. 2, p. 161.[a] For collecting substances which adhered to the surface of vessels, a feather, usually from a cock, was used as a scraper.[b] *TT*885 tells us of a silver spoon (*yin pi-tzu*[3]).

Very often the alchemists had to render their reaction vessels as air-tight as they could. Several forms of luting material were used. They ranged from ordinary beeswax to the well-known 'six-and-one mud' (*liu i ni*[4]), a lute made of seven different substances. We discuss this elsewhere.[c]

For reactions in solution a bamboo tube was sometimes used. The *chu-thung*[5]

[a] Anon. (*106*), pls. 63A, B; cf. Hsia Nai, Ku Yen-Wên *et al.* (*1*), pp. 3ff.

[b] And according to *TT*878 (ch. 20, p. 16*a*) it had to be from a white male chick reared from the egg for two years 'under laboratory conditions'. In the early stages of science, correct technique and empty ritual were hardly distinguishable.

[c] Cf. pt. 3, p. 133, and p. 219 below.

[1] 鐵匙   [2] 鐵箸   [3] 銀匕子   [4] 六一泥   [5] 竹筒

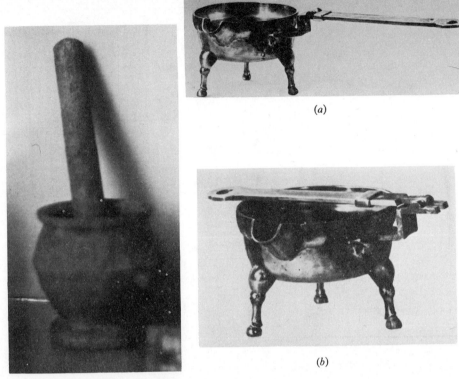

Fig. 1507   Fig. 1508

Fig. 1507. Bronze pestle and mortar, Hsin or Later Han in date, found during the construction of the Chhêngtu-Kunming Railway (Chhêngtu Historical Museum, orig. photo. 1972).
Fig. 1508. Collapsible ladle in silver, part of the hoard of the son of Li Shou-Li, probably buried in + 756, at Sian (Anon. (*106*), pl. 63 A, B).
(*a*) With handle extended.
(*b*) Handle folded in for transport.

(bamboo pipe), a section of bamboo, with walls thinned by shaving, was immersed in strong vinegar. This is widely used in the *San-shih-liu Shui Fa*[1] (*TT*923; Thirty-Six Methods for Bringing Solids into Aqueous Solutions), a book probably of the Liang period (cf. pp. 169 ff.). *TT*864, another Liang text, describes the use of a bag made of cloth (*pu tai*[2]) for the same purpose.

Bamboo was valuable also for every form of conduit to convey liquids or gases from place to place, as we have already had occasion to emphasise.[a] It also doubtless came in handy for the bubbling of gases through solutions (though the medieval alchemists and technicians would never have thought of it in that sophisticated way). Only just above we noted that the delivery tubes of Chinese mercury stills were made to dip into cold water in the receiver from the + 11th century onwards; gases would escape while the mercury condensed. But the focus of interest here, as the ancestor of all

[a] Vol. 4, pt. 2, p. 64.

[1] 三十六水法   [2] 布袋

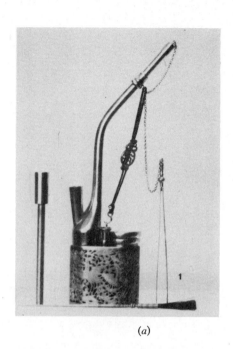

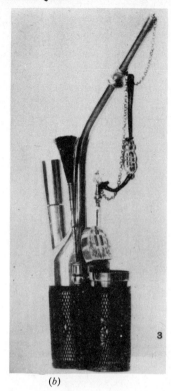

(a)   (b)

Fig. 1509. Origins of the gas bubbler or Woulfe bottle; typical Chinese tobacco water-pipes.

(a) An example from Canton, of brass or silvered copper with open-work sides. The parts are shown separately—on the left the smoke-tube with burner or bowl at the top. The scraper and brush are lying in front, and the pincers stand behind them; on the right, the tobacco box lid is closed. The design concentrated all the necessary utensils within one instrument easily transported. Laufer (42), pl. v, fig. 1; one fourth natural size.

(b) Another water-pipe from Canton, of 'tootnague' (here paktong, i.e. cupro-nickel, not zinc; cf. Vol. 5, pt. 2, pp. 212, 225 ff.). Encased in black varnished leather with cut-out patterns. From left to right, the burner or bowl, the scraper and brush, the mouthpiece and tube, the pincers in their socket, and the box for tobacco, with open lid. Laufer (42), pl. vi, fig. 3.

gas-bubblers, lies in the water-pipe (*shui yen tai*[1]) used for smoking tobacco (Fig. 1509), analogous to the *narghileh* or *hookah* of India and Islam, though characteristically much more compact in construction. Asian people liked their smoke cool, so they passed it through plain or scented water. The *narghileh* cannot antedate the first half of the +16th century, when the discovery of the Americas sent the seeds of *Nicotiana Tabacum* flying through all the civilisations of the Old World, but surely it must have been based on some previous experience of bubbling technique and on vessels of particularly suitable form.[a] One of these was no doubt the simple pot with

---

[a] This problem was seen, though not solved, by Laufer (42), esp. p. 27. For China we might remember the southern tribal, and Vietnamese, custom of drinking wine from a common pot through tubes of bamboo, cane or straw, on festival and ritual occasions (Vol. 3, p. 314, Vol. 4, pt. 2, p. 485).

[1] 水烟袋

Fig. 1510. Double-mouthed *kundika* pot or bottle, of buff clay with *ying-chhing* (shadow blue) glaze; Indian in form but Thang in date. Photo. Royal Ontario Museum, Toronto. Ht. 23·25 cms.

a necked mouth and a second necked or mammiform spout-like orifice on its flank, whence a jet of water or wine could be poured down the throat.[a] This type of drinking-pot (Skr. *kundika*, Mal. *kendi*) was also made of porcelain in China and exported all over South and South-east Asia from about +1350 onwards;[b] then in many places the convenience of it as a bubbler or base for the tobacco water-pipe was recognised, and gradually its form was adapted so as to be most suitable for that employment (Fig. 1510). This is only part of a chapter never yet written on the history of the chemical gas bubbler, the Woulfe bottle, but what more would have to go into it remains to be seen.

Taoist works, for example *TT*885, make frequent reference to weighing, but no special description of the balance or weighing-machine has yet been found in them. It is probable that the alchemists used the ordinary steel-yard, which was always the most common type of balance in China.[c] It is also interesting to note the use of the

---

[a] The practice continues notably with green glass vessels (*porrón*) in Spain, as every traveller knows.

[b] See the study of Sullivan (8).

[c] Cf. Vol. 4, pt. 1, pp. 24ff. We shall return to this subject momentarily later on (p. 266). As we shall there see, Chhen Shao-Wei in the early +8th century dealt explicitly with the problem of quantitative yield; though most of his figures were arrived at *a priori*, the foundation of his argument was an actual assay. Our collaborator Tshao Thien-Chhin called attention long ago to the proverbial expression often applied to chemical conversions (even when it could not be rigorously true)—*fên hao wu chhien*,[1] 'there is not a grain or a scruple of loss'. This recalls Maslama al-Majrītī three centuries later, who failed to note the increase of weight on calcination of mercury. Could it be that a loss of

[1] 分毫無欠

clepsydra (water-clock) and the sundial for timing the initiation and duration of alchemical experiments, as mentioned, e.g., in the Sung text *TT229*. The burning-time of incense-sticks,[a] and the interval required to cook a meal of rice, are also encountered as units of duration.

A good deal of alchemical apparatus seems to have been ordinary household and kitchen utensils pressed into service. The use of iron chopsticks, iron and silver spoons, stone pestles, etc. has already been mentioned. Other things like large jars, wooden basins, vases and copper basins, for example, are mentioned in *TT935*, a text of *c.* +864. *TT874* also tells us about such ordinary household utensils as the *chhêng* (or *tang*),[1] a vessel with feet usually used for warming wine; the *thieh chhi*,[2] a container made of iron, and the *kuo*,[3] presumably the large thin-walled cast-iron pan in such familiar use for frying or boiling food in Chinese kitchens.

With this we end our discussion of the equipment of the ancient and medieval Chinese alchemists, chemical technologists and pharmacists. It may have been rather tantalising to consider this apparatus, especially stills, in the rather abstract way to which we have been constrained here by our immediate purpose, but many pages in the rest of these volumes give an idea of what chemical reactions and processes were in fact involved in their use. As for the future, we have no doubt that the further study of the Chinese literature from the Han onwards will throw much more light on the progress of the chemical crafts and techniques in East Asia. One is left once again with the conviction that development in China went on *pari passu* with that in Europe, broadly speaking, and that Ko Hung probably knew quite as much about chemical operations as Zosimus. At any rate we hope that enough has been said about the Chinese equipment to dispel the impression given by some older authorities, who without any access to the original texts could write such words as these: 'They (the Chinese) possessed neither characteristic chemical methods of their own, nor any apparatus originating in their own culture.'[b]

(g) REACTIONS IN AQUEOUS MEDIUM

It is often supposed that the Chinese alchemists busied themselves mostly with non-aqeuous reactions of a more or less metallurgical character. But a text first studied by Tshao Thien-Chhin, Ho Ping-Yü & Needham (1) throws a considerable light on the earliest beginnings of the chemistry of inorganic reactions in aqueous medium. Weak nitric acid was employed to bring into solution a large number of inorganic substances,

mercury balanced the gain in oxygen? Goldsmiths of those times certainly had sufficiently sensitive scales, in China as well as in Andalusia.

---

[a] Cf. pt. 2, pp. 146 ff. and also Vol. 3, p. 330, Vol. 4, pt. 3, p. 570.
[b] Von Lippmann (1), vol. 1, p. 456, cf. p. 459.

[1] 鎗   [2] 鐵器   [3] 鍋

the processes being carried out either in porcelain vessels or in lengths of bamboo tubing which acted in part as a semi-permeable membrane. The text also includes mention of certain curious phenomena which were probably the effects of enzymes from organic material. It thus shows that the making of gold for the preparation of the elixir of immortality was far from being the only interest of these early medieval experimentalists. At the same time mica and certain other minerals had long been regarded in China as among the substances from which potent elixirs might be made,[a] and this no doubt explains the motive of the alchemists in their efforts to dissolve various mineral substances. The title of the text in question is *San-shih-liu Shui Fa*[1] (Thirty-Six Methods for the Bringing of Solids into Aqueous Solutions; *TT* 923). The first thing was to try to date it.

Aqueous solutions of mineral substances hard to dissolve were known to Chinese alchemists at least as early as the time of Ko Hung[2] (+283 to +343), who in his *Pao Phu Tzu*[3] gives an account of the preparation of aqueous solutions of realgar and cinnabar,[b] mentioning an earlier work called the *San-shih-liu Shui Ching*[4] (Manual of the Thirty-Six [Methods for] the Bringing of Solids into Aqueous Solution).[c] Elsewhere he knows how '...to turn the thirty-six minerals directly into aqueous solutions'.[d] Another alchemical treatise, entitled *Huang-Ti Chiu Ting Shen Tan Ching Chüeh*[5] (Explanation of the Yellow Emperor's Manual of the Nine-Vessel Magical Elixir),[e] mentions a *San-shih-liu Shui Fa*, saying that Pa Kung[6] imparted these techniques to the well-known −2nd-century alchemist Liu An,[7] Prince of Huai-nan[8].[f] It adds that 'the solubilisation of alum (*fan shih*),[9] realgar (*hsiung huang*[10]) and cinnabar (*tan sha*[11]) is based on Pa Kung's manual of the thirty-six methods for bringing solids into solution.[g] The methods all depend on the use of nitre, i.e. saltpetre (*hsiao shih*[12]). In the case of the solubilisation of cinnabar, copper sulphate (*shih tan*[13]) is (also) needed.' The *Yün Chi Chhi Chhien*[14] (Seven Tablets of the Cloudy Satchel) c. +1022, on the other hand, also attributes a book with a similar name to Thao Hung-Ching[15] (+456 to +536) the great physician and alchemist of the Liang period.[h] The title is rather revealing, namely *Fu Yün-Mu Chu Shih Yao Hsiao Hua*

---

[a] Cf. Ware (5), p. 186, Feifel (3), p. 15, translating *PPT/NP*, ch. 11, pp. 8bff.

[b] *PPT/NP*, ch. 16, pp. 7b, 8b, 9a, tr. Ware (5), pp. 272, 274.

[c] *PPT/NP*, ch. 19, p. 4a.

[d] *PPT/NP*, ch. 3, p. 1b. Feifel (1), p. 182, translates, most inadequately, '...to change the thirty-six stones suddenly into water.'

[e] *TT*878. Date Thang or Sung, but incorporating some material as old as the +2nd century.

[f] Ch. 8, pp. 1a, 2a and 4a.

[g] It is not clear whether Pa Kung was a single adept or 'the Eight Adepts'. The latter may be more probable, at any rate later literature preserved their names, which we take from Hsü Ti-Shan (1), p. 119—Chin Chhang,[16] Lei Pei,[17] Li Shang,[18] Mao Pei,[19] Su Fei,[20] Thien Yu,[21] Wu Pei[22] and Tso Wu.[23] The last of these is certainly historical. *YCCC*, ch. 109, pp. 21aff. however, makes Pa Kung a person who turned into a youth of 15 by art and gramarye at Liu An's court.

[h] *YCCC*, ch. 107, p. 9a.

---

[1] 三十六水法	[2] 葛洪	[3] 抱朴子	[4] 三十六水經
[5] 黃帝九鼎神丹經訣	[6] 八公	[7] 劉安	[8] 淮南
[9] 礬石	[10] 雄黃	[11] 丹砂	[12] 硝石	[13] 石膽
[14] 雲笈七籤	[15] 陶弘景	[16] 晉昌	[17] 雷被	[18] 李尙
[19] 毛被	[20] 蘇飛	[21] 田由	[22] 伍被	[23] 左吳

San-shih-liu Shui Fa[1] (Thirty-Six Methods for the Bringing of Substances into Aqueous Solution [by means of] Transformations caused by Nitre [with a view to the] Ingestion of Mica and all Kinds of Mineral Drugs). It was in one chapter, just as our text still is.

The *San-shih-liu Shui Fa* is in the *Tao Tsang* (Taoist Patrology)[a] but does not reveal the name of its author. It is true that it quotes a saying of an alchemist named Kao Chhi,[2] but unfortunately nothing is known about him or his date. The book is almost certainly the same as that listed in the bibliographical chapter of the *Sung Shih*[3] (History of the Sung Dynasty), +1345, and also identical with the work known to the compiler of the *Thung Chih*[4] (+1150) by the name *Lien San-shih-liu Shui Shih Fa*.[5] The book was thus well known by the Northern Sung (+11th century), but the authorship remains obscure. As we shall see, comparison with other alchemical texts strongly indicates that some at least of the methods described must have been known in the time of Ko Hung (+3rd and +4th centuries). The text now contains more than 36 recipes; and in many cases alternative methods are also given. Perhaps the best conclusion is that we have here a Corpus the beginning of which may go back to the group of Liu An in the −2nd century, but which grew as time went on. The most likely candidate for acceptance as the major contributor is in our opinion Thao Hung-Ching very early in the +6th century.[b]

Another book of interest in the *Tao Tsang* is the *Hsien-Yuan Huang-Ti Shui Ching Yao Fa*[6] (Medicinal Methods of the Aqueous [Solutions] Manual of Hsien-Yuan the Yellow Emperor).[c] So far we have not been able to trace its authorship or fix its date. It treats of all kinds of minerals subjected to more or less similar operations, e.g. triple extraction with water, hot extraction with 'bitter wine' (*khu chiu*,[7] probably vinegar containing other substances in solution), addition of copper sulphate or potassium nitrate to the extracts, mixing with numerous ingredients of vegetable origin, and burying in vessels underground. There are 32 recipes (with two missing), and besides the above methods they generally include evaporating to dryness on the water-bath. But having prepared aqueous solutions of mineral salts and the active principles of plants in a very reasonable pharmaceutical way, each recipe usually ends with the addition of the solution to metallic mercury and its conversion into silver thereby. Thus an alchemical layer seems to have been superimposed on a pharmaceutical layer. This book however, which so far has received no attention, deserves more than we are able to give it here.

In order to have an idea of the content of the *San-shih-liu Shui Fa*, the quotation of a few typical procedures is necessary. First let us look at what is said about a metal and two metallic salts.

---

[a] *TT*923. On the Patrology itself, see Vol. 5, pt. 3, pp. 113ff. above.

[b] He himself refers, in a passage from the *Pên Tshao Ching Chi Chu* (cit. *CLPT*, ch. 3, (p. 85.2), *PTKM*, ch. 11, p. 25a) to a *San-shih-liu Shui Fang*.[8]

Could this have been the name of the collection before he enlarged it?

[c] *TT*922.

[1] 服雲母諸石藥消化三十六水法　　　[2] 高起　　　[3] 宋史　　　[4] 通志
[5] 錬三十六水石法　　　[6] 軒轅黃帝水經藥法　　　[7] 苦酒　　　[8] 三十六水方

(No. 31) *Chhien-hsi shui*[1]—an aqueous solution of lead.[a]

2 lbs. of lead scrapings mixed with 4 ozs. of nitre (saltpetre), sealed (with lacquer) in a (bamboo) tube and put in vinegar will form an aqueous solution after 100 days.

(No. 41) *Chhien-kung shui*[2]—an aqueous solution of lead.[b]

5 spoonfuls of 'flying frosty snow' (*fei shuang hsüeh*[3])[c] are well mixed first with 'elegant powder of the Metal Elder' (*chin ong hua fên*[4])[d] until damp, and then with mica (powder) and brine, and (finally) nitre (saltpetre), using 2 ozs. of nitre for every lb. (of lead powder). Sealed with lacquer in a bamboo tube, placed in a jar of vinegar, and buried 3 ft. in the ground, the whole being kept warm by means of burning horse-dung; an aqueous solution will be formed after 30 days.

(No. 8) *Tzhu-shih shui*[5]—an aqueous solution of magnetite.

1 lb. of magnetite, 1 oz. of realgar and 1 oz. of copper sulphate (*shih tan*[6]) pounded together, sealed in a bamboo tube with lacquer and put in vinegar for 30 days will form an aqueous solution.[e]

Next consider formulae for the solubilisation of sulphur and several sulphides.

(No. 4*a*) *Tan-sha shui*[7]—an aqueous solution of cinnabar.

1 lb. of cinnabar with the addition of 2 ozs. of copper sulphate and 4 ozs. of nitre (saltpetre), sealed inside a freshly-cut bamboo tube with lacquer, and immersed in vinegar, will form an aqueous solution in 30 days.

(No. 9*b*) *Liu-huang shui*[8]—an aqueous solution of sulphur.

Sulphur suspended in honest vinegar (*shun tshu*[9]), with the addition of 2 ozs. of nitre (saltpetre) and enclosed in a bamboo tube as before, but buried in the ground, will turn after 15 days into an aqueous solution called *pao thien chih cho*[10] (Heaven-enveloping Potion).

(No. 2*a*) *Hsiung-huang shui*[11]—an aqueous solution of realgar.

1 lb. of realgar and 4 ozs. of nitre (saltpetre) are enclosed in a freshly-cut bamboo tube, which is sealed with lacquer and placed in vinegar for 30 days. An aqueous solution will then be formed.[f]

(No. 3*a*) *Tzhu-huang shui*[12]—an aqueous solution of orpiment.

1 lb. of orpiment and 4 ozs. of nitre (saltpetre) are sealed in a freshly-cut bamboo tube with lacquer. If this is placed in vinegar for 30 days an aqueous solution will result.[g]

(No. 3*b*) The same—another method.

(Orpiment) with the addition of 2 ozs. of alum and 2 ozs. of nitre (saltpetre) contained

---

[a] The term *chhien-hsi* can refer to either tin or lead (RP15), but the context suggests that lead was most probably meant. The acetate and nitrate of tin would have been formed just as easily, no doubt, as those of lead. A similar method for lead and tin together is given in *TT*876, p. 2*a, b*.

[b] Kung generally means the iron hub-bearing of a cart-wheel. Its exact significance here is not obvious.

[c] Undoubtedly some white sublimate, but the term does not suffice to identify it. Calomel or corrosive sublimate are perhaps more likely than anything else.

[d] Certainly lead carbonate, though the technical phrase is somewhat unusual. *Chin ong* = *chin kung*[13] = *chhien* (lead), and *chhien hua*[14] is lead carbonate.

[e] The same method is also given in *TT*878, ch. 8, p. 4*b*.

[f] This method is also given in *PPT/NP*, ch. 16, p. 8*b*, where Ko Hung specifies 'the strongest vinegar' (tr. Ware (5), p. 274); and in *TT*878, ch. 8, p. 4*a*; and in *TT*911, ch. 1, p. 9*a*—but they all use half the amount of saltpetre and give the time required as 20 days.

[g] Also given in *TT*878, ch. 10, p. 3*a*.

[1] 鉛錫水	[2] 鉛釭水	[3] 蜚霜雪	[4] 金翁華粉	[5] 磁石水
[6] 石膽	[7] 丹砂水	[8] 硫黃	[9] 淳醋	[10] 包天之汋
[11] 雄黃水	[12] 雌黃水	[13] 金公	[14] 鉛華	

(with vinegar) in a porcelain jar and buried for 20 days will form an aqueous solution which tastes sweet and has a yellow colour.[a]

Sulphates and silicates were dealt with either alone or with the addition of decaying organic materials. An example of the former would be a recipe for alum:

(No. 7*b*) *Fan-shih shui*[1]—an aqueous solution of alum.
(Alum) mixed with an aqueous suspension of greenish mica (*yün-ying shui*[2])[b] and 2 ozs. of nitre (saltpetre), placed (with vinegar) in a porcelain jar and buried for 30 days, will form an aqueous solution which tastes bitter and has a dark bluish colour.

Two examples of the latter concern calcium sulphate and potassium sodium aluminium silicate.

(No. 21) *Ning-shui-shih shui*[3]—an aqueous suspension of gypsum or calcareous spar.[c]
1 lb. of *ning-shui-shih* mixed and pounded together with the blood of green ducks (*chhing fu hsüeh*[4]), and put in a bamboo tube will form an aqueous suspension when buried 3 ft. down in damp ground for 10 days.
(No. 28*b*) *Yün-mu shui*[5]—an aqueous suspension of mica.
(Mica) thoroughly mixed with an equal quantity of aqueous extract of cinnamon (wood or bark),[d] brine, and 2 ozs. of nitre (saltpetre), sealed with lacquer in a bamboo tube and buried in the ground to a depth of 3 ft. as before, or put in a dry well and covered for 25 days, will form an aqueous suspension. This is put in a copper vessel and placed on moist ground (for keeping). It is then known as *yün-ying i*[6] (mica juice)[e] and can be used for dissolving *shih chiu ying*[7].[f]

All these are typical of the solubilisation recipes found in the *San-shih-liu Shui Fa* and scattered in many other Thang and pre-Thang alchemical texts.

Recipe no. 1 gives the technical term by which these baths of acetic acid and potassium nitrate were always known in Thang and pre-Thang writings, *hua chhih*[8] (the 'radiant pool').[g]

From the full translation of Tshao, Ho & Needham (1) it can be seen that the types of reaction described in the book are rather multifarious. They can however be divided into a number of groups which it will be convenient to discuss separately.

---

[a] Also given in *TT878*, *loc. cit.* The wide repetition of these formulae for the sulphides of arsenic (and our references must be far from exhaustive) shows how important arsenical solutions were for the elixir cult.

[b] *Yün-ying* is a kind of greenish mica (RP39), cf. de Mély (1), p. 64.

[c] *Ning-shui-shih* may be either the sulphate or the carbonate of calcium (RP119).

[d] *Cinnamomum Cassia*, the famous tree native to Kuangsi province (R495). Apart from the use of its bark, twigs, buds, peduncles and oil in pharmacy and cooking, it had an age-old reputation as an elixir constituent. It is often mentioned in this connection by Ko Hung; cf. e.g. *PPT/NP* ch. 11, p. 14*a*, *b*, the story of Chao Tho-Tzu,[9] who acquired extraordinary strength by its aid (tr. Ware (5), p. 195; Feifel (3), p. 26).

[e] There is a parallel text on this subject in *PPT/NP*, ch. 11, p. 8*b*, where Ko Hung recommends half a dozen or more substances, both inorganic and organic, for bringing mica into solution or suspension. The passage involves various difficulties of nomenclature and interpretation which it is hardly worth going into here (cf. Ware (5), pp. 186–7; Feifel (3), pp. 16–17; Davis & Chhen Kuo-Fu (1), p. 316). Cf. also *TT830*, pp. 19*b*, 20*a*.

[f] No other reference to this is known to us, but *shih ying* is quartz or rock crystal (RP37).

[g] Cf. Wang Khuei-Kho (2)

[1] 礬石水	[2] 雲英水	[3] 凝水石水	[4] 青鳧血	[5] 雲母水
[6] 雲英液	[7] 石九英	[8] 華池	[9] 趙他子	

(1) THE FORMATION AND USE OF A MINERAL ACID

As we noted just now, the *Huang-Ti Chiu Ting Shen Tan Ching Chüeh*[1] makes a very significant statement about the 36 *Shui Fa*.[2] It says that the transformation of alum, realgar and cinnabar into watery (solutions) follows the manual of Pa Kung;[3] and that all the methods depend upon the use of nitre (saltpetre) for their successful achievement. Although this makes no mention of vinegar, it is clear that in most cases explicit directions are given for its addition. Thus both substances are present in 17 recipes, and one or other of them in 31 recipes out of the total of 42. One may suspect that in the 7 where nitrate alone is mentioned and in the 7 where vinegar alone occurs, it was the original intention to direct the use of both, but one or the other dropped out in copying. At the same time there are samples where only one reagent would have been quite effective.[a] In other examples,[b] it seems certain that nitre and vinegar were originally in the text but fell out later, otherwise the procedure could not make sense. In others again neither reagent was necessary, though both are given,[c] or nitre only.[d] In any case the reactions with which we are chiefly concerned were nothing more nor less than the oxidation of the compounds or elements by dilute nitric acid, just as might be done in a modern chemical laboratory.

Most of the reactions of this group described in the text are quite plausible, involving oxidation of the insoluble sulphides of arsenic and mercury, insoluble elementary sulphur and the insoluble metals, by the action of nitrate in the presence of acetic acid. However, (i) the acidity arising from the acetic acid was extremely low, and hence the rate of reaction correspondingly sluggish, requiring weeks instead of minutes; and (ii) in most cases water and H^+ (and its anion, i.e. AcO^-) were not directly added to the solids, but reached them by diffusion through the thick (though pared-down) bamboo 'membrane', which constituted the reaction vessel, so that again the reactions were necessarily slow. No doubt some of the reaction products also diffused outwards as ions and became lost in the surrounding vinegar. But this did not worry the alchemists, because their primary concern was to see the insoluble solids disappear and change into aqueous liquids. The nature of the changes and that of the products formed was beyond their conjecture.

Where colloidal sulphur constituted part of the reaction products, this would be left inside the bamboo tube because the particle size would have been too large for free diffusion. The bamboo vessel therefore acted as a 'semi-permeable membrane', so that the alchemists were able to observe the yellow colour and the turbidity of the resulting solutions. Of course, the reactions carried out in porcelain or metal vessels involved the prior addition of vinegar and took place in essentially closed systems.

We may now deal individually with some of the aqueous solutions prepared using

---

a E.g. recipes (8) and (24).
b E.g. recipes (1b) and (17).   c E.g. recipes (36) and (40).
d E.g. recipes (37) and (38). Recipe (39) is very curious, for the acetic acid seems to have been driven off by heat before the nitrate was added, but the salt was soluble to begin with.

[1] 黃帝九鼎神丹經訣   [2] 水法   [3] 八公

nitrate and acetic acid. In order to assist the visualisation of what was probably happening, we offer a number of representations in equational form, but these are not intended to be in any way rigorous and should be taken as no more than reasonable if speculative interpretations. We shall generally omit in our equations ions that appear on both sides of the equal signs, so as to show the basic oxidation-reduction reactions. In the case of recipe no. 1 since ferric sulphate, $Fe_2(SO_4)_3$, is soluble in water anyway it would seem probable that yellow iron alum was originally the substance meant. The dissolved products would then presumably be a complex mixture of potassium and iron acetates and nitrates. Alum also occurs in recipe no. 7; presumably some would dissolve as such in the weak acid. The *Huang-Ti Chiu Ting Shen Tan Ching Chüeh* gives a method for the making of *chhing fan shih shui*[1] which would be an aqueous solution of green vitriol (copperas) or ferrous sulphate.[a] It says:

> Take some *wu fan*[2] alum and out of it select 1 lb. of the blue pieces. First keep them in good vinegar until thoroughly macerated and then put them in a container together with 2 oz. of nitre. Seal the container with lacquer and bury it 3 ft. below ground for 15 days. An aqueous solution will form spontaneously.

Here again since the salt itself is quite soluble some kind of alum was probably what was originally intended.

Another iron recipe is no. 40 which starts with ferrous acetate, not difficult to dissolve but oxidising probably under the conditions given to a blood-red solution of the basic acetate, precipitable by heating.[b] This, if dilute, might explain the colour description in the alchemical name of the product, *hsüan ling chin tzhu hsi shui*[3] (mysterious golden compassionate tide liquid).

The magnetite recipe, no. 8, gives us something different, for nitrate is absent. Here the magnetic iron oxide must have been acting as oxidising agent and the realgar as reducing agent, giving iron sulphate, arsenate or arsenite. For example it might be plausible to write an equation as follows:

$$3Fe_2O_3 + As_2S_2 + 18HOAc \longrightarrow 6Fe^{++} + 2S + 2As^{+++} + 18AcO^- + 9H_2O.$$

Since Fe_3O_4 can be regarded as $FeO \cdot Fe_2O_3$, only the oxidised part Fe_2O_3 is considered for simplicity.

Continuing our examination of the series of solutions of salts of metals we come to copper carbonate in recipes no. 5 and no. 6. Here the mercury can have performed no function. Copper carbonate is of course soluble in weak nitric acid. The recipe is repeated in the *Pên Tshao Kang Mu Shih I*[4],[c] where it is ascribed to Chu Chhüan's[5] *Shen Yin*[6].[d] It is recommended as an eye-lotion, very understandably in view of the oligodynamic action of copper.

The lead salt recipe, no. 41, is embarrassed by uncertainty as to the identification of the reactants, though lead carbonate, used in China from very ancient times as a

---

[a] *TT*878. Ch. 10, p. 2a.
[b] Cf. Vol. 5, pt. 2, pp. 292 ff. above.
[c] Ch. 1, 12a.
[d] See immediately below.

[1] 青礬石水   [2] 吳礬   [3] 玄靈金慈沙水   [4] 本草綱目拾遺
[5] 朱權   [6] 神隱

cosmetic, was certainly present. If the 'flying frosty snow' was calomel or corrosive sublimate, as seems very probable, the mercury might have acted as a catalyst in the oxidising reactions whereby basic lead acetate and a little lead nitrate were produced, besides the adjuvant action of hydrochloric acid. The process was the converse of the second stage of the classical manufacture of white lead, which appears to have been known and used in China as far back as the -4th century, i.e. at least as early as the first European description of it by Theophrastus.[a] The recipe would certainly have yielded a very poisonous drink, if ever aspirants for immortality were counselled to consume it.

The solubilisation of elementary metals by nitrate and vinegar might be represented as follows:[b]

(i) for lead and iron

$$3Pb + 2NO_3^- + 8H^+ \text{ (from vinegar)} + (2K^+) + (8AcO^-)$$
$$\longrightarrow 3Pb^{++} + 2NO + (2K^+) + (8AcO^-) + 4H_2O,$$

(ii) for silver

$$3Ag + NO_3^- + 4H^+ + (K^+) + (4AcO^-) \longrightarrow 3Ag^+ + NO + 2H_2O + (K^+) + (4AcO^-).$$

Generally, then, mixed acetates of the metals and alkali metals remained in solution, while a certain amount of nitric oxide was given off. In recipe no. 42 we find the additional presence of mercury with metallic lead. The resulting amalgam would certainly have speeded up the process. One cannot feel that the gold recipe, no. 29, gave the alchemists much satisfaction, however, even if nitrate was present. But of course if some salt had been added, as it certainly was in some of the other recipes, the resulting weak *aqua regia* might have made at least a superficial attack upon the noble metal.

Now comes the alchemically important solubilisation of mercuric sulphide, recipe no. 4a. The reaction appears to have been either

$$3HgS + 2K^+NO_3^- + 8HOAc \longrightarrow 3S + 3Hg^{++} + 2NO + 2K^+ + 4H_2O + 8AcO^-$$

or

$$3HgS + 8K^+NO_3^- + 8HOAc \longrightarrow 3SO_4^{--} + 8NO + 8K^+ + 3Hg^{++} + 4H_2O + 8AcO^-.$$

As for the function of the copper sulphate (especially mentioned in the *Huang-Ti Chiu Ting Shen Tan Ching Chüeh*) we think that it could well have been a catalyst.

The reaction in recipe no. 4b is similar to the preceding one, but the red colour demands an explanation. We suggest that it was due to the presence of selenium as an impurity. Selenium and sulphur are so similar in their properties that they belong to the same group in the Periodic Table. When acidified most selenium compounds are decomposed to form elementary selenium which is brilliantly red. Recipe no. 4b omits vinegar, perhaps unintentionally.

[a] Evidence has been given in Vol. 5, pt. 3, pp. 15 ff. Here we need only cite the complete statement in the *Chi Ni Tzu*[1] book, ch. 3, p. 1b (in *YHSF*, ch. 69, p. 24b). Partington (1) considered that the process was known and used in Greece before Theophrastus and that its age outside Greece went back still further. Perhaps it was a chemical discovery of the Fertile Crescent which spread in both directions, east and west.   [b] E.g. recipes (29), (30), (31), (42).

[1] 計倪子

So important did the alchemists consider the bringing into solution of mercuric sulphide that mention of it can be found in a number of other texts. As an example we may give the following. The *Pao Phu Tzu* about +320 gives the same recipe as no. 4 *a* but with a different proportion for nitre.[a] The same method is mentioned in the *Huang-Ti Chiu Ting Shen Tan Ching Chüeh*[b] and the *Chu Chia Shen Phin Tan Fa*.[c] Both of these are primarily Thang or Sung texts, but an earlier reference, lesser known, occurs in the *Shang-Chhing Chiu Chen Chung Ching Nei Chüeh*[1] (Confidential Explanation of the Interior Manual of the Nine Adepts; a Shang-Chhing Scripture).[d] This book bears the name Chhih Sung Tzu,[2] a pseudonym of the +4th-century alchemist Huang Chhu-Phing,[3] and there is nothing in the content to suggest a later date. Apart from descriptions of the effects of elixirs, and directions about the auspicious and inauspicious days for embarking on the Great Work, together with details of necessary sacrificial offerings, only one practical method is given, namely the bringing of cinnabar into solution by long heating with vinegar, with or without the addition of lacquer latex. In the omission of nitre this recipe resembles no. 8, but no copper sulphate is present. Perhaps however mercuric acetate was slowly formed. Conversely, the vinegar is omitted in the formula of Chao Hsüeh-Min[4] in his *Pên Tshao Kang Mu Shih I*[5] (Supplementary Amplifications of the Great Pharmacopoeia)[e] of +1769 which includes only cinnabar, copper sulphate and nitrate exactly as in no. 4 *b*. One cannot help wondering whether here again the mention of vinegar was not accidentally omitted. It is interesting that Chao quotes the recipe as coming from the *Shen Yin*[6] (Occupations for Retired Scholars) by the famous Ming prince Chu Chhüan[7] (+1390 to +1448), alchemist, botanist, geographer and musician.[f] Chao adds that the solution (mainly mercuric acetate) prolongs life, dispels evil influences, nourishes the spirits and calms the mind.

The sulphur waters in recipes no. 9 and no. 17, recall the sulphur-containing liquids beloved of the Greek proto-chemists.[g] Apparently the 'divine' or 'sulphur water' (*to hudōr tou theiou*, τὸ ὕδωρ τοῦ θείου), prepared from chalk, sulphur, vinegar, and ammoniacal urine, contained calcium polysulphides, very striking in their effects for the ancients because capable of giving coloured precipitates with metal salts, tinting the colours of metals in various ways, and even attacking the noble metals.[h] Here in the *San-shih-liu Shui Fa* we find again the use of vinegar. In no. 9 *a* the nitric acid available would only oxidise a small proportion of the sulphur present, but as the reaction-vessel was not buried, atmospheric oxidation to sulphates catalysed by oxides of nitrogen could have occurred. The change would presumably be:

$$S + 2K^+NO_3^- \longrightarrow 2K^+ + SO_4^{--} + 2NO.$$

[a] *PPT/NP*, ch. 16, p. 7*b*, 8*a*. [b] Ch. 8, pp. 4*a*, *b*.
[c] Ch. 1, p. 8*a*. [d] *TT*901, p. 1*a*.
[e] Ch. 1, p. 11*b*. This chapter would greatly repay further investigation.
[f] See Vol. 5, pt. 3, pp. 210–11.
[g] Berthelot (2), pp. 46, 47, 68, 139, etc. The recipes are first found in the Leiden papyri of the +3rd century—not much earlier than the present text.
[h] Cf. Vol. 5, pt. 2, pp. 251 ff. above.

[1] 上清九眞中經內訣   [2] 赤松子   [3] 黃初平   [4] 趙學敏
[5] 本草綱目拾遺   [6] 神隱   [7] 朱權

If the 'dew' in no. 9a contained much organic matter, bacterial reduction to sulphides might have followed. We shall return presently to the role of organic ingredients, but the effect of bacterial action may have been particularly important where sulphur was concerned. Thus in recipe no. 21, where calcium sulphate was suspended with duck blood, the putrefying bacteria would undoubtedly reduce the sulphate to calcium sulphide and then to the hydrosulphide and polysulphides, together with some carbonates.

Recipe no. 2a is a good case of getting arsenic into solution.[a] Presumably the sulphide and the nitrate formed arsenate or arsenite. It is possible to write out the oxidation-reduction reaction as follows:

$$3As_2S_2 + 22K^+NO_3^- + 4H_2O \longrightarrow 6AsO_4^{---} + 6SO_4^{--} + 22NO + 8H^+ + 22K^+.$$

Under the conditions of no. 2b the sulphur seems to have been oxidised not to sulphate, but to free sulphur. For example:

$$3As_2S_2 + 10K^+NO_3^- + 4H_2O \longrightarrow 6AsO_4^{---} + 6S + 10NO + 8H^+ + 10K^+.$$
$$\text{[colloidal sulphur]}$$

In this way the turbid yellow colour and the sweet taste might be explained. When orpiment was used, as in recipe no. 3a, the reaction was probably:

$$3As_2S_3 + 28K^+NO_3^- + 4H_2O \longrightarrow 6AsO_4^{---} + 9SO_4^{--} + 28NO + 8H^+ + 28K^+.$$

In no. 3b the yellow colour was again no doubt due to colloidal sulphur. The alum added very probably affected the pH of the medium, thereby controlling the stage of oxidation and giving an arsenite. A possible reaction would be:

$$As_2S_3 + 2K^+NO_3^- + 2H_2O \longrightarrow 2AsO_3^{---} + 3S + 2NO + 4H^+ + 2K^+.$$

It is interesting that already in 1942 Hsüeh Yü (*1*) recognised the general significance of the saltpetre and vinegar in forming a dilute solution of nitric acid, and he suspected that enough chloride was present in the cinnabar recipes (e.g. no. 4) to form weak *aqua regia*.[b] The first publication of the *San-shih-liu Shui Fa* translation by Tshao Thien-Chhin, Ho Ping-Yü & Needham (*1*) in 1959 evoked considerable interest among practical inorganic chemists, who made a number of computations suggesting that one or another of the methods could not work as they stand. Schneider believed, for example, that one could hardly expect more than 6 or 7 % of acetic acid in the vinegar used, and that only 1 % of this would be dissociated, so the extreme weakness of the nitric acid formed had to be recognised.[c] This, he felt, undermined the plausibility of the equations suggested, and indeed in reproducing them we recognise their speculative nature. Since acetic acid does dissolve some oxides and base metals Schneider was

---

[a] On this subject cf. pt. 2, pp. 282ff. above.

[b] Mercuric sulphide is readily soluble in warm dilute *aqua regia* (Durrant (1), p. 377; Partington (10), p. 401). In so far as salt was generally, or always, present in the *liu i ni* luting recipes (cf. pp. 20, 219), the chloride could often have formed HCl in the *hua chhih*.

[c] Private communication from Dr Wolfgang Schneider of Braunschweig, July 1960.

inclined to accept the possibility of the magnetite-realgar recipe (no. 8), but he found the solubilisation of metallic lead (nos. 31, 41, 42) inconceivable under the conditions stated. He mentioned only one experiment, with sulphur, to test recipe no. 9; it did not give any sulphate or nitric oxide—but it did not exactly follow the conditions prescribed either. Perhaps on the whole he under-estimated the oxidising potential of nitric acid in dilute solutions, especially in the presence of traces of nitrous acid which might be formed because of the organic matter, and indeed also the possible effects of metallic impurities in the metals used, which might set up galvanic couples.

Later on, the industrial chemist Mêng Nai-Chhang (1) made further calculations, from which he concluded that if one sticks to the actual wording of the *San-shih-liu Shui Fa* about the procedures used, metallic lead, tin and copper would perhaps be attacked, but not silver, gold or mercury; so also the sulphides of lead, tin, iron and zinc would dissolve, but not those of mercury or copper. He thought the solubilisation of the arsenical sulphides (nos. 2 and 3) might have been partly possible, but not the magnetite-realgar recipe (no. 8); and he emphasised the buffering effect of the mixture of acetic acid and potassium nitrate. Mêng recorded a single experiment in which nitrate and cinnabar were left for some weeks with 1 M acetic acid, no Hg^{++} then being demonstrable in the solution, but as with Schneider he did not try to reproduce the medieval conditions, where the presence of catalysts and impurities of importance may be surmised. Mêng, again like Schneider, found the very simplified equations of Tshao *et al.* hard to accept, but in addition put forward an alternative interpretation of the 'vinegar bath' (*hua chhih*[1]) procedure, namely that it was a way of purifying the solid inorganic substances rather than bringing them into solution.[a] Thus he noted a statement in the *Thai-Chhing Chin I Shen Tan Ching* (TT873) that strong vinegar which had been left a long time in contact with gold was a form of 'gold solution' (*chin i*[2]) and should be used for moistening the 'six-and-one lute' (*liu i ni*[3]), further-more that the gold itself became soft and crumbly to the touch. This, he suggested, was a way of purifying the gold from traces of lead; yet silver would be a more likely impurity, and as for lead, why not just complete the cupellation in the furnace? Similarly, according to Mêng Nai-Chhang, it might have been a case of purifying cinnabar from small amounts of accompanying iron sulphide—after all, weak nitric acid is still used industrially for purifying crude asbestos and talc. The whole question remains open, but it is safe to say that the best way forward will be to carry out a series of experiments imitating as closely as possible exactly what the Liang alchemists say that they did, and using reagents as impure as those which they are likely to have had. In this endeavour there is one further point which ought not to be missed.

It is curious, and perhaps significant, that the *San-shih-liu Shui Fa* and texts of similar pre-Thang date never give any hint of concentrating the vinegar. Whether distillation would have been possible at the time we discuss elsewhere (p. 155). But if this had been done and kept secret, for oral transmission only, a very different complexion would be given to the matter. Distillation of vinegar is mentioned several

---

[a] (1), p. 29.

[1] 華池   [2] 金液   [3] 六一泥

times in the Geberian books about $+1290$,[a] and may conceivably have been done already by the Hellenistic or Byzantine proto-chemists, more probably by the Arabs. Glacial acetic acid (if the Chinese alchemists could have got it) will dissolve sulphur, iodine and many other things ordinarily insoluble, and certainly it would have given with saltpetre a much stronger nitric acid solution. Could that, one wonders, have been the real explanation of most of the recipes in the *San-shih-liu Shui Fa* and its kindred texts? But if so, what are we to make of the ascription of the nitre-vinegar methods to the Eight Adepts at the court of Liu An? One may willingly accept the possibility of a distillation of vinegar in the $+6$th century while being reluctant to do so for the -2nd. Whether then the method was only fathered on them by later writers, or whether they were really the kin of Mary the Jewess, further research may reveal.

The difficulty in this whole problem is that the concentration of acetic acid by distillation is exceedingly troublesome, even with the aid of a column, because of the formation of hydrates.[b] Separation by distillation is avoided in modern chemical industry, and when concentration is necessary it is done by liquid extraction with a solvent such as ethyl acetate. In former times the weak acid was neutralised with lime, and the calcium acetate after drying was distilled with concentrated sulphuric acid, a procedure certainly not possible in $+6$th-century China. Nevertheless we may retain the hypothesis that in some way or other the alchemists of that time did manage to concentrate their acetic acid; and if so, then with higher nitric acid levels, perhaps with some free hydrochloric acid, and certainly with unconscionable durations of time, much of what they say they did may actually have been done.[c]

One possibility which presents itself is that their strong 'vinegar' might have been the pyroligneous acid obtained by the destructive distillation of green wood.[d] It would need no stretching of belief to attribute this to them, even though mention of such an operation has not so far been found in the medieval literature.[e] But pyroligneous acid usually contains no more than 10% acetic acid, together with smaller amounts of methyl alcohol and acetone,[f] so it would hardly have served their purpose better than ordinary vinegar.

[a] E.g. *Summa Perfectionis*, ch. 52 (Russell tr., p. 108, Darmstädter tr., p. 55), and *De Investigatione Perfectionis*, ch. 3 or 4 (Russell tr., p. 9, Darmstädter tr., p. 98). It was also done by Leonardo da Vinci (*c.* $+1510$) as the first stage of his remarkable preparation of acetone, in MS. K_3, 114r., noted by Reti (7), p. 665. The acetic acid was then treated with potassium and calcium carbonate (from the calcined tartar of the wine) to give the acetates, and these were then dry distilled producing acetone (*acqua risolutiva*).

[b] Some strength can be gained if the first fractions are rejected, but perhaps the main purpose of distilling vinegar in medieval times was purification rather than concentration.

[c] For advice on this subject we are most grateful to the late Dr Ladislao Reti, who was both a historian of science and a chemical industrialist of many years' experience. He agreed that only an extended series of laboratory experiments would settle the problem.

[d] This suggestion was made to us by Prof. Stephen Mason, then of Norwich. See Sudborough (1), p. 150; Perkin & Kipping (1), pp. 92, 154; Ure (1), vol. 1, p. 8, vol. 3, p. 557. The oldest descriptions in Europe seem not to be earlier than the $+17$th century (Partington (7), vol. 2, pp. 359, 419).

[e] Yet it is strange that the taste corresponding to the element Wood in the symbolic correlations (Table 12, in Vol. 2, p. 262) was *suan*,[1] sour or acidic. So much is still buried in the alchemical texts of China that one would not be at all surprised to come upon the distillation of wood.

[f] Ost (1), p. 403.

[1] 酸

However, there is one way in which acetic acid could certainly have been concentrated in those old days, namely by 'freezing out' the water, just as was done, as we have seen (p. 151 above), for wine and alcohol. Could this not have produced acetic acid at a concentration of 60 % or more? As we have seen,[a] Glauber did something of this kind successfully in + 1657. And since there is some reason to believe (p. 152) that the method was known in China as early as the − 2nd century, the Eight Adepts could have employed it after all. Just as in the case of distillation, such a proceeding would very likely have been transmitted by oral tradition and not committed to paper.[b]

Lastly one must not underrate the possible importance of the pared bamboo tube. Its walls could perhaps have constituted a semi-permeable membrane within which hydrogen ions might conceivably have tended to concentrate, sodium and potassium ions passing through in the other direction. The acetate ion would have remained outside just as most of the nitrate ions would stay inside. In this way the pH within would perhaps be much higher than could be expected if the whole reactions were taking place in a single vessel, and that would be one of the first things to test if any serious effort were made to reproduce the conditions described in the text.[c]

(2) 'NITRE' AND *HSIAO*; THE RECOGNITION AND SEPARATION OF SOLUBLE SALTS

The first question raised by the oxidation-reduction reactions described in the *San-shih-liu Shui Fa* is that of 'nitre', one of the most protean words in the history of chemistry, and paralleled in a closely similar way in China by the word *hsiao*.[1, 2] By the time of Robert Boyle and John Mayow nitre meant saltpetre, but in antiquity it was something quite different. 'A review of disputations on what salts this term comprised among the ancients would itself fill a volume' wrote the Hoovers, 'but from the properties named it was no doubt mostly soda, more rarely potash,[d] and sometimes both mixed with common salt.'[e]

In the book of the prophet Jeremiah we read:[f] 'Though thou wash thee with nitre and soap, yet thine iniquity is marked before me, saith the Lord.' And in the book of Proverbs it is said:[g]

> Confidence in an unfaithful man in time of trouble
> Is like a broken tooth, or a foot out of joint;
> As one that casteth off garments in winter, or as vinegar upon nitre,
> So is he that singeth songs to a heavy heart.

---

[a] P. 153 above. Cf. Schildknecht (1); Schildknecht & Schlegelmilch (1).

[b] Yet another way in which strong acetic acid could have been produced by very simple means was the distillation of verdigris (copper acetate). This was done by the Geberian writer (*De Inventione Veritatis*, ch. 23) who called it *oleum viridis aeris* (ct. Darmstädter tr., p. 114). No reference to this has so far appeared in a medieval Chinese text, but it would be well to look out for it. Basic copper acetate (*thung chhing*[3]) appears first in the *Chia-Yu Pên Tshao* of + 1057. Cf. RP9.

[c] We are indebted to Prof. Fridemann Freund of Cologne for emphasising this aspect of the matter in personal discussion. [d] Theophrastus, *Hist. Plant.* III, 9, mentions nitre from wood ash.

[e] (1), p. 558, in notes on Agricola; cf. p. 562.

[f] 2, 22, a text of about − 600. [g] 25, 19–20, a − 3rd- or − 2nd-century text.

[1] 滑   [2] 硝   [3] 銅青

Thus the detergent effervesced with acid, and indeed sodium carbonate it was. This salt occurs naturally, mixed with some bicarbonate, as well as the chloride (2 to 57 %) and the sulphate (1 to 70%), in parts of the Egyptian desert, notably the Wadi Natrun where there is a succession of salt lakes annually inundated, and has been gathered, purified and used for thousands of years.[a] The proper name for it was natron, a word derived from the ancient Egyptian *ntry*, hence our modern symbol for sodium; but by assimilation with Gk. *nizō* (νίζω) and *nizomai* (νίζομαι), to wash, 'nitre' resulted, via Gk. *nitron* (νίτρον) and Lat. *nitrum*. From the IVth Dynasty onwards (*c.* − 2900) natron was used, never salt, for that desiccation which was the essential process in mummification,[b] being regarded too as a great cleanser, destroying all fat and grease.[c] But it was also used in many industrial arts, such as those of incense, glass-making and the bleaching of cloth.[d] Nitre continued to have this meaning as late as Agricola, but from the beginning of the + 14th century onwards it was applied also to saltpetre— understandably enough, perhaps, since all these salts were collected from incrustations on the ground. Hence much confusion, even throughout the + 17th century, when saltpetre was often called *sal nitri*.[e]

There was natron also in China, where it went by the name of *chien*[1,2] or *shih chien*.[3] A salty incrustation found on soil surfaces,[f] it was discussed towards the end of the last century by Stuhlmann (1) and Schlegel (11), who gave analyses showing an average composition of some 12% sodium carbonate, 64% sodium sulphate and 24% sodium chloride.[g] These were all traditionally separated by differential crystallisation and filtration or decantation;[h] though sometimes the carbonate occurred naturally in purer form. The crude salt came into commerce coloured brown by organic matter, hence its name *tzu chien*[4].[i] This was much used as a desiccating agent, detergent, bleach and mordant.[j] We do not find *chien* very prominent in the pharmaceutical natural histories, probably because it belonged as 'lye' rather to the kitchen

---

[a] See Lucas (3).

[b] See Lucas (1), pp. 297ff., 307ff. and most fully in (4). Cf. also Vol. 5, pt. 2, pp. 75–6, 299 above.

[c] So it would have been, under suitable conditions, saponifying the triglycerides by alkaline hydrolysis; cf. Perkin & Kipping (1), p. 177.

[d] The very confused accounts in Pliny, *Hist. Nat.* XXXI, 106ff., XXXVI, 191, 194, have been elucidated by Bailey (1), vol. 1, pp. 49ff., 147ff., 169ff., 173, 280ff.

[e] The first critical examination of the evidence was probably that of Beckmann (1), 4th ed., vol. 2, pp. 482ff. One of the best discussions is that of Partington (5), pp. 298ff. See also Crosland (1), pp. 76, 106; Reinaud & Favé (1), pp. 14ff.

[f] Like the *tequesquite* of Chile or the *trona* of other parts of the world. Kalgan and northern Shansi were eminent sources. Cf. Torgashev (1), pp. 327ff.

[g] See also Anon. (99), reporting carbonate up to 53 % from northern Hopei and 43·6 % from Shansi. Here the chloride could be as low as 1 %, and the sulphate only a trace, though it could also rise to 31 %. But perhaps these were partially refined products. Earlier analyses were those of Abrahamsohn (1) and von Engeström (2) in + 1772.

[h] The best carbonate (*pai chien*[5]) was marketed rather pure, with only about 1 % of chloride and but a trace of sulphate. The Glauber's salt separated was also rather pure.

[i] This had up to 20 % of chloride and up to 30 % of sulphate.

[j] Also, when pure, as 'baking powder' for steamed bread. The traditional test of its purity was how much dough could be made to rise by a given quantity of the sample. Yeast of course was always used as well, but the small amount of acid formed was sufficient to produce bicarbonate and hence carbon dioxide.

---

1 鹼   2 䤅   3 石鹼   4 紫鹼   5 白鹼

and the workshop than to the pharmacy. In the *Pên Tshao Kang Mu* it is included in the entry for lake or ground salt, *lu hsien*[1];[a] as well as having a small section of its own,[b] where it means chiefly *hui chien*[2] ('ash-natron'), i.e. the potash from plants, the preparation of which by burning, solution, filtration, and crystallisation, is described.[c] The mixture of calcium and magnesium carbonates deposited as 'fur' or 'boiler scale' was also called *chien*, but given a separate account since it was recognised as something else.[d] Certainly in China there was a confusion of name as in the West, but it was a different one; potassium nitrate was never confounded with sodium carbonate, the trouble came rather with the sulphates of sodium and magnesium, as we shall now see. In interpreting the Chinese terms therefore it is wiser not to translate *hsiao*[3,4] by 'nitre', as one might be tempted to do, but rather to invent a parallel witch-word, and give it a special equivalent of its own, 'solve'. The reason for this designation will quickly become apparent; the Chinese alchemists were dealing with several soluble salts which could act as metallurgical fluxes, or pharmaceutical cathartic-diuretics, or generators of nitric acid for the solution of minerals.[e]

Sodium sulphate in its crude form was known in the pharmaceutical natural histories[f] as *phu hsiao*[5] (crude-solve), *hsiao shih phu*[6] (crude solve-stone),[g] *yen hsiao*[7] (salt-solve), and *phi hsiao*[8] (skin-solve).[h] Much confusion arose over the terms *phu hsiao* and *mang hsiao*[9] (prickle-solve), the second being regarded by pharmaceutical writers in later times as the purified form of the first, and getting its name from the 'spiky' appearance of the prismatic crystals.[i] There was also confusion among such writers between these two terms and *hsiao shih*[10,11] (solve-stone), potassium nitrate or saltpetre. A Thang specimen of *mang hsiao* (*bōshō*) preserved in the Shōsōin Treasury at Nara in Japan, where it had been deposited in +756, turned out on analysis in 1954 to be none other than Epsom salt, i.e. crystalline magnesium sulphate ($MgSO_4 . 10H_2O$);[j] and on the basis of this result Masutomi & Yamasaki very reasonably suggested that *phu hsiao* was always mirabilite, i.e. Glauber's salt, sodium sulphate ($NaSO_4 . 10H_2O$).[k] Since the two salts occur naturally mixed in Chinese deposits, and the sodium salt

[a] Ch. 11, (p. 44), cf. RP118 (correcting their *chien*[2] *to hsien*[1]). According to Li Shih-Chen, it was first brought into this context by Chu Chen-Hêng in his *Pên Tshao Yen I Pu I* of +1330, and so towards the end of the Yuan period. It could be called *lu chien*,[12] and was like *hui chien*,[2] i.e. very alkaline.

[b] Ch. 7, (p. 91).

[c] This potassium carbonate was also employed as a detergent and in baking. In medicine it was used cautiously as an antacid in hyperacidity.

[d] Ch. 11, (p. 76), cf. RP134.

[e] Schmauderer (4, 5) recalls that precisely the same sort of nomenclature arose in the West. Saltpetre, nitric acid and potassium carbonate were called by Glauber, about +1650, *menstruum universale*, because there was nothing he knew of which they would not bring into fusion or solution.

[f] Cf. RP123.

[g] This name was due to a confusion the nature of which will appear in what follows.

[h] Because of its use in tanning, as Li Shih-Chen explains.

[i] Lei Hsiao[13] (about +470) explains that crystals with the appearance of awns of wheat picked out from *phu hsiao* are *mang hsiao*; *PTKM*, ch. 11, (p. 49).

[j] There were only traces of calcium, chlorine and potassium.

[k] In Asahina (*1*), no. 35, pp. 289ff. and p. 496.

1 鹵鹹 2 灰鹼 3 消 4 硝 5 朴消 6 消石朴
7 鹽消 8 皮消 9 芒消 10 消石 11 硝石 12 鹵鹼
13 雷斆

crystallises before the magnesium one, the idea of a purification relationship could easily have arisen in the minds of those who were not actually doing the fractional crystallisation themselves.[a]

From the results of their analyses Masutomi & Yamasaki came to the following conclusions. (*a*) A Chinese origin of the specimen is well supported by the description in the Thang pharmacopoeia (*Hsin Hsiu Pên Tshao*), +659. (*b*) Since all *Pên Tshao* authorities averred that *mang hsiao* was obtained in the process of refining *phu hsiao*, it would be correct to identify *phu hsiao* with Glauber's salt (sodium sulphate) and *mang hsiao* with Epsom salt (magnesium sulphate). (*c*) *Hsiao shih*[1] ought to be considered a general term for sulphates like *phu hsiao* and *mang hsiao*. Properly speaking, the characters *hsiao shih*[2] should be used for saltpetre (potassium nitrate), and only if this is done (or assumed where textually necessary) can the statements in the pharmaceutical natural histories be brought into consistency. (*d*) The word *hsiao*[3] in the term *hsiao shih*[1] originally meant something that would dissolve readily in water, not something which would liquefy other things as a flux does. Hence *hsiao shih*[1] must have meant in the past only *phu hsiao* (sodium sulphate) and *mang hsiao* (magnesium sulphate). We accept the first two of these conclusions, but the second two will not answer. We do not think that one can find in the literature any sharp distinction between the two ways of writing *hsiao* in *hsiao shih*. Although in the majority of cases the pharmaceutical naturalists may have written the word *hsiao*[3] for *hsiao shih* with the water radical,[b] there was never any such trend among the alchemical texts that we have studied, where we find the word *hsiao*[3, 4] written freely in both forms, with the stone radical and with the water radical.[c] As for the fourth argument we fail to see any sufficient basis for it; after all, potassium nitrate too dissolves freely in water.

In late times, moreover, the term *mang hsiao* changed its meaning and came to signify sodium sulphate. Chang Hung-Chao gives a set of analyses[d] of products from Hopei and Shansi, made during the early years of this century, which showed them to contain from 36·2 to 81·1 % Na_2SO_4, the balance being mainly water of crystallisation, though magnesium sulphate was present in some specimens in the range of 2·0 to 7·3 %. Current usage continues this acceptation.[e] What brought about the change from the traditional term *phu hsiao* we do not know.

The *Shên Nung Pên Tshao Ching*[f] says that *phu hsiao* can 'transform seventy-two different minerals'.[g] Yuan Han-Chhing takes this to mean either the dehydration of

---

 [a] The two salts were not clearly distinguished in Europe till the time of Joseph Black in the mid +18th century.

 [b] But take, for example, Lei Hsiao's recipe in *CLPT*, ch. 3, (p. 85.2, 86.1); the word is written twice with the stone radical and once only with the water one.

 [c] In the *Thai-Chhing Shih Pi Chi* the stone radical is used exclusively. The *Pao Phu Tzu* book mentions saltpetre in two chapters (ch. 11 p. 10*b* and ch.16, p. 9*a*, ff.); the first uses the stone radical and the second the water one. In his index Ware (5) takes no notice of this difference in orthography, very reasonably, for in Chinese the two are in truth interchangeable.   [d] (*1*), pp. 244, 245.

 [e] As may be seen in modern pharmacopoeias, e.g. Anon. (57), vol. 4, p. 242.

 [f] Earliest of the Chinese pharmaceutical natural histories, dating from the −2nd and −1st centuries.

 [g] Mori ed., ch. 1, (p. 24). Repeated many times afterwards, as by Sun Ssu-Mo in *CCIF*, ch. 2, (p. 14.2); and cit. *CLPT* ch. 3, (p. 87.1).

 [1] 消石   [2] 硝石   [3] 消   [4] 硝

the Glauber's salt on exposure to air, leaving different mixtures of anhydrous salt and hydrates, or its action at high temperature as a flux capable of melting a number of silicates.[a] The *Thai-Chhing Shih Pi Chi* says that for *phu hsiao* one should select a specimen that has not been exposed to the wind, not dehydrated in appearance, but with a lustrous blue colour; while for *mang hsiao* one should select a specimen that looks like snow piled up in the shade, with a pleasant smooth appearance.[b] The *Ming I Pieh Lu* states that *phu hsiao* remains unchanged in the ground for a thousand years and that the refined product is white as silver.[c] Li Shih-Chen explains that the word *hsiao*[1] is included in its name because it dissolves so readily in water, brings many substances into solution or liquefaction and fuses in the furnace. After purification what is found above in the supernatant fluid (i.e. the magnesium salt) is called *mang hsiao*[2] or *phên hsiao*[3] (basin-solve) or *ma ya hsiao*[4] (horse-tooth solve),[d] while the efflorescent sodium salt[e] is called *fêng hua hsiao*[5].[f] Li Shih-Chen also cautions that *mang hsiao* should never be confused with *hsiao shih*[6] (solve-stone), saltpetre.

In its more purified form sodium sulphate was known as *hsüan ming fên*[7] (mysterious bright powder) and *pai lung fên*[8] (white dragon powder).[g] This purified form was first brought into prominence by the Taoist Liu Hsüan-Chen[9] in the time of Thang Hsüan Tsung (r. +713 to +755).[h] Liu has thus been called the Glauber of Chinese alchemy,[i] and the medicinal virtues of the purified salt were expounded in a monograph entitled *Hsüan Ming Fên Chuan*.[10] An earlier method for purifying sodium sulphate given in the *Ming I Pieh Lu* is quoted in the *Pên Tshao Phin Hui Ching Yao*.[j] It says that

some crude but clean *phu hsiao* (sodium sulphate), irrespective of quantity, should be taken (for refining) during the winter months when there is frost or snow. It is mixed with three ounces of the pods of the soap-bean tree (*tsao chia*[11]),[k] which have been heated gently for a while and pounded to a powder, then dissolved in six cupfuls of hot water. After removing the insoluble residue left at the bottom, the solution is filtered through two layers of thin paper, poured into an iron pan and evaporated until half of it is left. When cooled to a luke-warm temperature the solution is transferred to an earthenware pot and left to cool by itself in the open for a night. The next morning masses of crystals will have formed. These are dissolved in six cupfuls of boiling water and boiled together with eight ounces of large radish (*lo po*[12]) cut into pieces about two-tenths of an inch thick until the radish is cooked. The solution is again transferred to an earthenware pot, the pieces of radish (and any precipitate) having been removed (by filtration), and again left to cool by itself in the open for a night.

---

[a] (*1*), p. 242. The exact number 72 is of course not to be taken literally.
[b] *TT*874, ch. 3, p. 13*b*.
[c] *PTKM* ch. 11, (p. 49), as for the rest of this paragraph.
[d] Because of the prismatic crystal form.
[e] On standing in dry air, water of crystallisation is lost, leaving the anhydrous salt as a powder.
[f] A good detailed account of the differential crystallisation of the two sulphates can be found in the *Wai Kho Chêng Tsung*[13] (Orthodox Manual of External Medicine) of +1617 (ch. 12, p. 26*b*).
[g] *RP*124.   [h] *CLPT*, ch. 3, (p. 88.1), and *PTKM* ch. 11, (pp. 52–3).
[i] Porter Smith (2).   [j] Ch. 1, (p. 115).
[k] *Gleditschia sinensis*, R387. The pods contain much saponin; cf. Needham & Lu Gwei-Djen (1).

[1] 滑   [2] 芒滑   [3] 盆滑   [4] 馬牙滑   [5] 風化滑   [6] 滑石
[7] 玄明粉   [8] 白龍粉   [9] 劉玄眞   [10] 玄明粉傳   [11] 皂莢   [12] 蘿蔔
[13] 外科正宗

Next day further masses of crystals will have formed. After removing them from the mother liquor and drying them, these are put in a good paper bag and suspended in a place exposed to the wind. They will then turn by themselves into powder form.

Here the use of organic substances as reagents to precipitate impurities is of interest.

The *Tan Fang Ching Yuan* (+8th century) describes the chemical properties of *mang hsiao* ($MgSO_4$) saying that it can 'subdue' (*fu*[1]) orpiment, i.e. reduce the fusion point of the latter so that decomposition or sublimation will no longer take place at the usual temperature.[a] The same book also mentions among the properties of *ma ya hsiao* that it preserves (*yang*[2]) cinnabar and fixes (*chih*[3]) sal ammoniac.[b] So much for these sulphates.

Saltpetre, potassium nitrate (*hsiao shih*,[4,5] solve-stone)[c] acquired in the pharmacopoeias the synonyms *mang hsiao*[6,7] (prickle-solve),[d] *khu hsiao*[8] (bitter-solve), *yen hsiao*[9] (blaze-solve),[e] *huo hsiao*[10] (fire-solve), *shêng hsiao*[11] (natural solve) and *ti shuang*[12] (ground-frost).[f] Significantly (as we shall see), *mang hsiao*[6,7] was a name never used by the Taoist alchemists as a synonym, though they had a number of others of their own.[g] Ma Chih,[13] in the second half of the +10th century, clearly pointed out that one should not be muddled by the varying usages of the three terms; *mang hsiao*,[6,7] *phu hsiao*[14,15] and *hsiao shih*[4,5] were all definitely different things.[h] Nevertheless confusion among the pharmaceutical naturalists concerning these was not to be avoided. We have already differentiated (so far as is possible) *mang hsiao* and *phu hsiao*; now we must make clear what *hsiao shih* was.

In the *Shen Nung Pên Tshao Ching*, *phu hsiao*[16,17] and *hsiao shih*[18,19] are both listed and described separately.[i] The major emphasis is on their therapeutic and macrobiotic properties. This is evidence for the −2nd century. The name *hsiao shih*[19] itself first appears, however, in the −4th-century *Chi Ni Tzu* book (cf. pt. 3, p. 14 above) with its list of drugs and chemicals.[j] Then in the *Lieh Hsien Chuan* (Lives of Famous Hsien), the oldest parts of which date from the −1st century though the whole was not stabilised until the +3rd or +4th, we read of 'Mr Miner's Pick', the immortal Chhih

---

a Cf. pt. 3, p. 158.   b *CLPT*, ch. 3, (p. 86.2 and p. 88.2).

c RP125; Chang Hung-Chao (*1*), pp. 241ff.

d Because its crystals also are prismatic. Hence the +18th-century Western description of 'prismatic nitre' as opposed to 'cubic nitre', i.e. sodium nitrate (Crosland (*1*), p. 76; Mellor (*1*), p. 503).

e This excellent expression came in with the *Wai Tan Pên Tshao* and the *Tsao-Hua Chih Nan*, somewhere about +1040. It would reflect the first proto-gunpowder mixtures of that time.

f *PTKM*, ch. 11, p. 27*b* (p. 54).

g E.g. *pei ti hsüan chu*,[20] 'the mysterious pearl of the emperor of the North'. We need not examine further alchemical synonyms and cover-names, such as *ho tung yeh*[21] (the wilderness east of the river) and the more practical *hua chin shih*[22] (metal-changing stone).

h In the *Khai-Pao Pên Tshao*, *c.*+970, in *PTKM*, loc. cit.

i Mori ed., ch. 1, (p. 24).

j Ch. 3, p. 3*a*, in *YHSF*, ch. 69, p. 36*a*. Cf. Vol. 2, pp. 275, 554. Some scholars date the *Chi Ni Tzu* book as late as the −1st century on the ground that it cites the *Chou Pei Suan Ching*, but that argument lacks force if our dating of this mathematical and astronomical classic is acceptable (cf. Vol. 3, pp. 19ff.).

1 伏	2 養	3 制	4 硝石	5 消石
6 芒硝	7 芒消	8 苦消	9 焰消	10 火消
11 生消	12 地霜	13 馬志	14 朴硝	15 朴消
16 朴硝	17 朴消	18 硝石	19 消石	20 北帝玄珠
21 河東野	22 化金石			

Fu,[1] a temple librarian skilled in alchemy.[a] He knew how to make mercury (*shui hung*[2]) and transmute cinnabar (*lien tan*[3]), which he used to consume together with saltpetre (*hsiao shih*[4]), gaining thereby the aspect of a young man and outlasting many generations. Then in Thao Hung-Ching's *Ming I Pieh Lu* of *c*. +500 *mang hsiao*[5, 6] appears as a new item together with a statement that it is derived from *phu hsiao*[7, 8] (cf. p. 182 above). Thao Hung-Ching also pointed out certain characteristics of *hsiao shih*,[9, 4] saying:[b]

Some people formerly obtained a certain substance with a colour and nature more or less similar to that of *phu hsiao*, bright like the light of early dawn and resembling a handful of salt or snow, but not (as hard as) ice. When it is burnt or strongly heated in the fire a bluish-purple flame (*tzu chhing yen*[10]) arises, and again it turns to a limy ash, not boiling and bubbling like *phu hsiao*. This is what is said to be the genuine *hsiao shih*. There are those who say that *mang hsiao* is another name for it, but that is made nowadays by refining *phu hsiao*. Huangfu (Shih-An[11])[c] concurs with me that the matter cannot be decided by argument and examination; it is necessary to try the experiment and compare and record what happens in this method of transforming *hsiao shih*. In the *San-shih-liu Shui* (*Fa*) it is said to come from Lung-hsi (Kansu), and from Chhinchow in Szechuan; while that found at the capital, Chhang-an, comes from the Western Chhiang tribespeople. Nowadays in the mountains everywhere north of Tang-chhang there are places with a salty earth (*hsien thu*[12]) which produce it.

Since the *Ming I Pieh Lu* resumed +3rd-century knowledge, the potassium flame test probably goes back at least that far.[d] We suspect that this must be one of the oldest references to a flame test in any civilisation, for all the European mentions seem to be of the Renaissance time or at least not earlier than Latin Geber.[e] The salt must have been KNO_3, the deflagration of which on charcoal is a striking phenomenon. If heated by itself it first decrepitates, losing mechanically entangled water, melts at 338 °C. and finally gives off oxygen, leaving the nitrite. Moreover, the *San-shih-liu Shui Fa* belongs to this date, and perhaps also to this author. Since that text describes the use of acetic acid, perhaps in concentrated form, to liberate nitric acid from saltpetre and dissolve a number of inorganic substances otherwise not easily soluble (cf. pp. 172 ff.), the identity of the salt as KNO_3 may be taken as established.[f]

It looks as if confusion was sparked off by a statement of Thao Hung-Ching's earlier in the same passage that in its therapeutic effects *hsiao shih* is similar to *phu hsiao*, and that according to some, *hsiao shih* and *phu hsiao* come from the same place, though he said specifically that they were not the same thing.[g] The confusion was

---

[a] Kaltenmark (2), p. 171.

[b] *PTKM* ch. 11, p. 25 *a* (p. 54), quoting him by name, therefore primarily from the *Pên Tshao Ching Chi Chu*; also *CLPT*, ch. 3, (p. 85.2). Cf. Chang Hung-Chao (1), p. 243. Contemporary biographers of Thao Hung-Ching also give saltpetre (*hsiao shih*[12]) as an elixir constituent (Strickmann, 2).

[c] I.e. Huangfu Mi (+215 to +282).

[d] Attention has often been drawn to the importance of this text, as by Yen Tun-Chieh (20).

[e] Cf. Debus (13) and Partington (4), p. 76.

[f] It will be remembered, too, that the solubilisation formulae of the *Shui Fa* type go back quite a long time further. Even if we doubt their existence in the −2nd century among the techniques of Liu An and the Eight Masters (as tradition affirmed, cf. p. 168), they are certainly in the *Pao Phu Tzu* book, which is evidence for the +3rd (*PPT/NP* ch. 11, p. 8 *b*, ch. 16, pp. 7 *b*, 8 *b*, 9 *a*; tr. Ware (5), pp. 186, 272, 274).

[g] Cit. *CLPT*, ch. 3 (p. 85.2); *PTKM*, ch. 11, (p. 54).

| [1] 赤斧 | [2] 水䡚 | [3] 錬丹 | [4] 消石 | [5] 芒硝 | [6] 芒消 |
| [7] 朴消 | [8] 朴消 | [9] 硝石 | [10] 紫青煙 | [11] 皇甫士安 | [12] 鹹土 |

established when Su Ching[1] in the +7th century confirmed the equation of *hsiao shih* with *mang hsiao*, saying:

> Nowadays crude *phu hsiao* is refined by heating it in solution until *mang hsiao* is formed...
> This is *hsiao shih*.[a]

Identification of *hsiao shih* with *mang hsiao* seems then to have become a prevailing practice among the Chinese pharmaceutical naturalists from the Thang until the days of Li Shih-Chen towards the end of the Ming, though it was clearly pointed out by Ma Chih in the +10th-century *Khai-Pao Pên Tshao*[2] (as we have seen) that *hsiao shih* was obtained as a crystalline deposit on the ground (*ti shuang*[3]) and did not belong to the same category as *phu hsiao* and *mang hsiao*.[b] He added that Thao Hung-Ching could not have been well acquainted with the substances themselves.[c] Li Shih-Chen further quoted a statement of Su Sung to the effect that *hsiao shih* was collected by sweeping together crystals found on the earth and that all of it would burn up in the flame when tested in the fire.[d] In this connection we shall also remember from pt. 3, pp. 137, 159 above the deflagrating mixture experiment apparently attributed to Sun Ssu-Mo in the +6th century by the Sung compendium *Chu Chia Shen Phin Tan Fa* (*TT*911). The attribution is quite reasonable and potassium nitrate must have been involved.

The identification of saltpetre by observing the lilac or purple colour flame test is also mentioned in the +7th-century Thang alchemical text about the wandering monks; i.e. the *Chin Shih Pu Wu Chiu Shu Chüeh* (*TT*900), which we have given on p. 139 of pt. 3 above. It was in +664 that Chih Fa-Lin[4] recognised the presence of saltpetre in northern Shansi and referred to its 'liquidising' properties, both as a metallurgical flux and as precursor of nitric acid in the solubilisation or *Shui Fa* technique (cf. pp. 169ff.).[e] Only a few years before (+659) the *Hsin Hsiu Pên Tshao* had reproduced Thao Hung-Ching's text about the potassium flame,[f] so it was common knowledge. In connection with all this the *Huang Ti Chiu Ting Shên Tan Ching Chüeh* says:[g]

> It is difficult to procure saltpetre of good quality. Inferior specimens cannot bring realgar and cinnabar into solution. If one gets some saltpetre which looks genuine one should take a few pounds of it and try it (with vinegar) on realgar and other minerals to see if they are dissolved or not. If it does not bring them into solution it cannot be deemed to be genuine

---

[a] *Hsin Hsiu Pên Tshao*, ch. 3, pp. 9*b*, 10*a* (p. 19).

[b] *PTKM*, ch. 11, p. 27*b* (p. 54).

[c] Actually he certainly was, as other passages here quoted show.

[d] *PTKM*, ch. 11, (p. 55), from *Pên Tshao Thu Ching* (+11th century). Fused potassium nitrate is indeed a powerful oxidising agent, charcoal (as in this case), sulphur and phosphorus burn on it brilliantly, with the formation of the carbonate, the sulphate and the phosphate. The oxidation is as good as that with concentrated nitric acid, and faster, almost explosive in its effect. Decomposition gives the potassium flame, and oxides of nitrogen go off. Cf. Partington (10), p. 311; Ephraim (1), p. 590; Durrant (1), p. 334. By contrast, sodium nitrate, the 'cubic nitre' (because of its crystal form) of the +18th century, is not a powerful oxidising agent.

[e] The same book mentions the purple flame test also in another entry, probably concerned with powdered potash alum from Persia (cf. pt. 3, p. 139 above).

[f] Ch. 3, pp. 10*b*, 11*a*, *b*.

[g] *TT*878, ch. 8, p. 12*a*. Though compiled in Thang or Sung it contains some material as old as the +2nd century.

[1] 蘇敬   [2] 開寶本草   [3] 地霜   [4] 支法林

saltpetre. In appearance it looks very much like crude *phu hsiao* (sodium sulphate), and it is not rock-like but soft. One should first take a piece of it and place it upon burning charcoal. If purple smoke is emitted and the specimen turns to a kind of ash then it is of good quality, but if it fuses and bubbles for a long time then it is *phu hsiao*. It is difficult to get genuine (saltpetre), and even when you find any that appears suitable for use the test of bringing realgar or cinnabar into solution must be carried out before you can be sure (that it is genuine).

In another place the same book repeats the test to distinguish between crude sodium sulphate and saltpetre, saying that the former when heated boils and liquefies like alum, but the latter emits bluish-purple smoke and turns into a kind of ash without liquefying.[a] But the *Shen Nung Pên Tshao Ching* itself had said rightly long before that *hsiao shih* melts when heated on a fire.[b] And the *Pao Tsang Lun* of +918, remarking that saltpetre comes up like the grass of the field, says that it does not lose weight if fused some time in a crucible.[c] Yet it can turn all metals (i.e. their ores) into a soft and flowing condition.

In the *Chen Yuan Miao Tao Yao Lüeh* (TT917), that Thang work which has the oldest reference to a proto gunpowder mixture (cf. pt. 3, p. 159), there is a test for saltpetre depending on its strong oxidising property.[d] It runs as follows:

As for the 'fixation' (*fu huo*[1]) of saltpetre (*hsiao shih*[2]) it can be tested on a red-hot charcoal fire. If it fuses to an oily liquid and does not move (*tung*[3]) in the heat, then it is said to be 'fixed' or 'subdued'. Just melting it in a vessel by itself will not tell you whether it is crude (lit. raw, *shêng*[4]) or treated (lit. ripe, *shu*[5]). It always takes the shape of (lit. likes to stick to) the container which it is in. Now if you test it by putting it in the fire, that which is not 'subdued' will burst into a bright flame on meeting the charcoal.

In other words the alchemists of the +8th and +9th centuries were well able to distinguish between potassium nitrate and other salts of the alkali metals.

The *Ming I Pieh Lu* applies what the *Pên Ching* said of *phu hsiao* to *hsiao shih* (saltpetre), namely that it is capable of producing changes (*hua*[6]) in seventy-two different minerals.[e] As with Chih Fa-Lin, this probably refers both to furnace flux effects and to solution by weak nitric acid. 'Because it dissolves and transforms all ores and minerals' said Ma Chih in +970, 'it is called "solve-stone".'[f] Li Shih-Chen quotes a book by Shêng Hsüan Tzu,[7] the *Fu Hung Thu*[8] (Illustrated Manual on the Subduing of Mercury),[g] repeating this and saying that to test saltpetre one can put a little of the sample on a piece of quartz (*pai shih ying*[9]) that has just been heated over a fire, then if the sample fuses into the quartz it is genuine saltpetre. The same text

---

[a] *TT878*, ch. 16, p. 5b, 8a.

[b] *Lien chih ju kao*,[10] *CLPT*, ch. 3, (p. 85.2); Mori ed. ch. 1, (p. 24).

[c] *Thu Ching Yen I Pên Tshao*, ch. 1, p. 29b.

[d] The passage was first discovered by Fêng Chia-Shêng (4), p. 36.

[e] *PTKM* ch. 11, (p. 54); *Pên Tshao Phin Hui Ching Yao*, ch. 1, (p. 112). Again, seventy-two only means a large number. The same quotation from *CLPT* ch. 3, (p. 85.2) gives twelve instead of seventy-two.   [f] *PTKM*, ch. 11, p. 25a (p. 54).

[g] Extremely difficult to date; Wieger (6) thought Ming, but that is wrong. At least three alchemists bore this pseudonym in various forms, Tung Fêng[11] of the San Kuo period, Wang Yuan-Chih[12] of the Sui (c. +510 to +635) and Hsüeh Chih-Wei[13] of J/Chin. Perhaps the second is the most likely author.

[1] 伏火   [2] 硝石   [3] 動   [4] 生   [5] 熟   [6] 化
[7] 昇玄子   [8] 伏汞圖   [9] 白石英   [10] 鍊之如膏   [11] 董奉   [12] 王遠知
[13] 薛知微

mentions the offensive odour of nitre beds.[a] This could conceivably refer to the hydrogen sulphide of volcanic regions where nitrate might occur beside fumaroles, but it is much more likely that manure beds sodden with urine and decaying organic material, as in the stables, latrines and *salpêtrières* of Europe, were what was meant. If we could date this book more closely we might gain a better idea of the time at which 'nitre beds' were purposively worked in China, and hence of the first beginnings of the purification of the salt there.[b] *Shih phi*[1] (stone-spleen) seems to have been a mixture of saltpetre and other salts, probably from some native deposit.[c]

Whatever the exact date of Shêng Hsüan Tzu's book, it cannot have been later than about +1150, when Yao Khuan[2] wrote an account of the 'nitre' problem which is one of the most interesting we have found outside the pharmaceutical natural histories. For in his *Hsi Chhi Tshung Hua*[3] Yao started out by quoting him as follows:[d]

In his *Fu Hung Thu* (Illustrated Manual on the Subduing of Mercury) Shêng Hsüan Tzu records a method of testing *hsiao shih*[4] (saltpetre) imported from Wu-Chhang (Udyāna). He says: 'Its colour is bluish. If you heat a piece of white quartz and then put a drop of the nitre on it, it will sink in. The Taoist books say that saltpetre from Wu-Chhang can liquefy or dissolve all metals and minerals. If consumed it can prolong life. The places where it is produced have an extremely loathsome smell, so that birds cannot fly over them, but if one puts on a single garment and passes by, all the parasites in and on one's body turn to water and one will gain longevity or immortality. Pieces shaped like little goose quills are the best kind.'[e]

Then in his *Fên Thu*[5] (Illustrated Manual on Powders, i.e. Salts) Hu Kang Tzu[6] says: 'Bluish saltpetre (*chhing hsiao shih*[7]) is also called the Mysterious Pearl of the Emperor of the North (*pei ti hsüan chu*[8]).'[f] The *San-shih-liu Shui Fang*[9] (Thirty-six Methods for Making Aqueous Solutions) further describes a process for dissolving *tsêng chhing*[10] (copper carbonate) in which naturally occurring saltpetre (*chêng hsiao shih*[11]) is used.[g] From all this it may be seen that (some of) the saltpetre used nowadays is apparently not natural saltpetre.

The *Yao Ming Yin Chüeh*[12] (Secret Instructions on the Names of Drugs and Chemicals)[h] says: 'According to ancient tradition *hsiao shih*[13] (saltpetre) can liquefy and dissolve all kinds of metals and minerals; and if eaten can prolong life. But we have never found out where it is produced. We only know its name, which is as good as not having it at all. Recently Thao

---

[a] *PTKM* ch. 11, (p. 54). 'This salt comes from places which smell so horribly foul that birds can't bring themselves to fly over them....'

[b] Cf. Partington (5), pp. 314ff. The history of 'nitre-beds' in Europe has been studied by Multhauf (9), who finds that the saltpetre supply was an important limiting factor for the development of firearms in Europe between the +14th and the +18th centuries. The first reference to 'saltpetre plantations' occurs in +1406. At times there was a considerable importation from Asia, mainly Indian in origin.

[c] RP125, 135a, following *PTKM*, ch. 11, (p. 77). It was first mentioned, without pharmaceutical use, in the *Ming I Pieh Lu*.   [d] Ch. 2, pp. 36aff.

[e] Was this a reference to prismatic crystal form?

[f] We have come across Hu Kang Tzu before (Vol. 4, pt. 1, p. 308), but he is hard to date exactly. He must have been at work in the Thang or a little earlier. Cf. p. 80 above.

[g] Cf. Tshao, Ho & Needham (1), p. 126.

[h] This is perhaps another name for the *Thai-Chhing Shih Pi Chi*[14] (*TT*874), completed early in the +6th century. At any rate, the exact text of the following quotation appears in that work, ch. 3, p. 13a. This is shown by Ho Ping-Yü (8).

[1] 石脾	[2] 姚寬	[3] 西溪叢話	[4] 消石	[5] 粉圖
[6] 狐剛子	[7] 青消石	[8] 北帝玄珠	[9] 三十六水方	
[10] 曾青	[11] 正消石	[12] 藥名隱訣	[13] 消石	[14] 太清石壁記

Yin-Chü (Thao Hung-Ching) compiled a pharamaceutical natural history in which he said that *phu hsiao*[1] (plain-solve) is the *phu* (i.e. the crude unpurified form) of *hsiao shih* (solve-stone, hence the same *hsiao shih phu*[2]); and he also said that *mang hsiao*[3] (prickle-solve) and *shih phi*[4] (stone-spleen) if boiled together (and let stand in the cool) will form real *hsiao shih*[5] (saltpetre). Yet no one has been able to identify *shih phi*[4] since. Indeed, the facts have been misrepresented. We must believe that there are both naturally-occurring (*chêng*[6]) and false (*yen*[7]) forms of saltpetre.[a] The *Ching*[b] says that *hsiao shih*[5] (saltpetre) is the most magical and wonderful chemical substance in the world. Thao (Hung-Ching) says that there is none of the natural form (to be found) nowadays. He is not entirely wrong.'

Now the (*Pên Tshao*) *Thu Ching*,[c] quoting from alchemical and medical books of the Liang and Sui periods, explains that although the (saltpetre) now obtainable is not the true sort, it shows similar properties and therefore can be employed. So the substance we have today is usable.

Tshui Fang,[8] in his *Lu Huo Pên Tshao*[9] (Spagyrical natural History)[d] says: '*Hsiao shih*[10] (saltpetre) is a Yin mineral; it does not belong to the class of "rocks", and it is got by boiling (certain kinds of) lake or ground salt (*hsien lu*[11]). It is now called *yen hsiao*[12] (blaze-solve). At Shang-chhêng in Hopei and along the rivers Huai[13] and Wei[14] people scrape it up from the salty soil, and make it from (the filtered) drippings. After being boiled together with *phu hsiao*[15] (sodium sulphate) and salt (*hsiao yen*[16]) it can control and subdue (*chih fu*[17]) lead (by acting as a flux), and it can remove the "halo" (*yün*,[18] a discoloration) on copper or bronze. It is not produced at all in the South. *Phu hsiao*[15] can ripen skin or hide (in tanning), and *mang hsiao*[19] can be used as a drug (magnesium sulphate).'

Present-day commentaries on *hsiao shih*[5] say:[e] 'It lies like a kind of frost upon the ground. In (certain) mountains and marshy places this *ti shuang*[20] appears during the winter months; people sweep it up, collect it and extract and dissolve it with water, after which they boil (to evaporate) it, and so it is prepared. It is named "solve-stone" because it can dissolve and transform all kinds of ores and minerals, not because it belongs to the same class (of salts) as *mang hsiao*[19] and *phu hsiao*[15] (nitre).'[f]

The (*Pên Tshao*) *Thu Ching* further says: 'Physicians and pharmacists use only pieces that are not yet refined, slightly bluish in colour, regarding them as *phu hsiao*.[15] After refining has been carried out, the pointed (crystals) that form at the top of the pans are called *mang hsiao*,[19] while the limpid or transparent crystals (*chhêng ning*[21]) which collect at the bottom are called *hsiao shih*[5].'

Again it says: 'By refining *phu hsiao* or *ti shuang* a solid white substance like a stone is formed; this is *hsiao shih*[5] (saltpetre) and there is no other sort.'

But I maintain that there is natural *hsiao shih*[5] (saltpetre) to be got, just as is said in the

---

[a] There might be a parallel here with one of the terms for steel, *wei kang*,[22] 'false steel'. It was perfectly good steel, but made by co-fusion, not by cementation or direct decarburisation. See Needham (64), p. 110, (32), and Sect. 30.

[b] Doubtless the *Shen Nung Pên Tshao Ching*, but the words are not found in the best modern edition, Mori (*1*), ch. 1, p. 25.   [c] Edited by Su Sung in +1070.

[d] Probably another name for his *Wai Tan Pên Tshao*[23] (Iatro-chemical Natural History), *c.* +1045.

[e] Parallel passage in *CLPT*, ch. 3, (p. 85.2). It seems to be Ma Chih speaking.

[f] As already mentioned, these saline cathartic-diuretics, the sulphates of sodium and magnesium, seem to have been introduced into medicine by Liu Hsüan-Chen about +730, or at any rate popularised by him (cf. p. 183).

[1] 朴滑	[2] 滑石朴	[3] 芒滑	[4] 石脾	[5] 滑石	[6] 正	[7] 曆
[8] 崔昉	[9] 爐火本草	[10] 滑石	[11] 鹹鹵	[12] 鹻滑	[13] 懷	
[14] 葡	[15] 朴滑	[16] 小鹽	[17] 制伏	[18] 暈	[19] 芒滑	
[20] 地霜	[21] 澄凝	[22] 僞鋼	[23] 外丹本草			

Manuals of the Immortals (Hsien Ching[1]), without resorting to refining processes. What they make today by boiling and refining is of course also called *hsiao shih*, and that is what is mostly used. *Yen hsiao*[2] can indeed subdue the eight minerals, and *mang hsiao*[3] can be used as a drug. One can only use what one can get hold of. *Hsiao shih*[4] is really not some magical chemical substance (unobtainable) in the world.

Finally, the *Tan Fang Ching Yuan*[5] (Mirror of the Alchemical Elaboratory)[a] in its chapter on the solves (nitres) mentions five kinds; *ma ya hsiao*[6] (horse-tooth solve),[b] *phu hsiao*[7] (sodium sulphate), *mang hsiao*[8] (magnesium sulphate), *so sha hsiao*[9] (shrink-sand solve), and *khang hsiao*[10] (pit solve). As for *hsiao shih*[11] (saltpetre), it is included in the chapter on the various minerals, as one can see if one looks it up there.

And the (*Pên Tshao*) *Thu Ching* says that at Jen-ho[12] (in Chekiang) *yen hsiao*[13] (salt-solve) is found ten *li* east of the city, and when it is refined it gives *phu hsiao*.[7] Also in the winter months (saltpetre) effloresces on the ground in the form of transparent glittering fragments, so it is called *shuang hua*[14] (frost-flowers). Another name is *chien chi*[15] (sword-spines). When nitre is included in prescriptions it is considered to be of the same class as *hsüan ming fên*[16] (purified Glauber's salt, sodium sulphate) and *tzu hsüeh*[17] (purple snow).[c]

Reading over this one gains a vivid idea of the terminological morass in which the medieval alchemists and iatro-chemists laboured. We need not analyse the mistakes and misconceptions of each particular writer, nor yet point out how right he happened to be. But if one thing more than any other comes out crystal clear from this account, it is that methods for the collection and purification of potassium nitrate were steadily developing during the seven centuries preceding the first knowledge of the salt in Islam or the West, i.e. between +500 and +1200;[d] and probably during the last three or four of these, i.e. from the late part of the Thang period, it was being turned out on a manufacturing scale by artisans who achieved a fairly constant product but were not able to explain to the scholars exactly how they did so. Why should one then be surprised that formulae for proto-gunpowder[e] began to appear during the last half of the +9th century? Furthermore, by the beginning of the +8th at least, the sulphates of sodium and magnesium had been separated by differential crystallisation and were being used in medicine.

An interesting account of the use of saltpetre in pharmaceutical alchemy is found in the *Yu Huan Chi Wên*[18] (Things heard and seen on my Official Travels), written about +1233 by Chang Shih-Nan.[19] What he says is this:[f]

The *I Chien Chih*[20] records[g] that when Yü (Yün-Wên[21])[h] was summoned from Chhüchow, where he was Governor, to attend at the temporary headquarters of the emperor, and was

---

[a] A work of the early Thang, before +800.
[b] A name surely derived from observations of crystal form.
[c] This last salt has not yet been identified.   [d] See immediately below, p. 194.
[e] By this we mean compositions relatively low in nitrate. As is well known, the earliest formulae for true gunpowder in any civilisation appeared in the *Wu Ching Tsung Yao* of +1044 (cf. Needham (47), pp. 246–7).
[f] Ch. 1, p. 8*a*, tr. auct.
[g] This book was composed by the eminent literary critic and editor Hung Mai[22] (+1123 to +1202).
[h] The distinguished Sung general (c. +1108 to +1174); cf. Vol. 4, pt. 2, s.v.

[1] 仙經	[2] 餤消	[3] 芒消	[4] 消石	[5] 丹房鏡源	[6] 馬芽消
[7] 朴消	[8] 芒消	[9] 縮砂消	[10] 坑消	[11] 消石	[12] 仁和
[13] 躑消	[14] 霜花	[15] 劍脊	[16] 玄明粉	[17] 紫雪	[18] 游宦紀聞
[19] 張世南	[20] 夷堅志	[21] 虞允文	[22] 洪邁		

resting in the reception-hall outside the north gate, he was taken ill with a severe attack of diarrhoea (perhaps dysentery) which he had contracted on account of the great heat of the journey. This lasted for several months. On the 9th day of the ninth month he had a dream in which he found himself in some palace of the immortals. A man robed as one of their officials asked him to sit down, whereupon he noticed a rhymed inscription written on the wall which said:

> 'The poison of summer heat has gone to the spleen
> And the damp *chhi* has accumulated in the feet;
> If this be not dispelled diarrhoea will follow,
> If diarrhoea should not, there will be malaria instead.
> Only by heating realgar as the chymists do,
> Mixing with bread and taking with liquorice-root[a]
> Stirred all together in the form of a posset,
> Then indeed will blessed relief ensue.
> All other prescriptions of doctors are off the mark.'

He followed this and got well.

I, Chang Shih-Nan, when in Szechuan, visited all the Taoists at their temples in the woods searching for this 'Only-heated Method (*Tu lien fa*[1])' but I could hardly find anyone who knew it. Suddenly one day I met a Taoist from Chhing-chhêng Shan who explained it to me as follows. 'The elixir manuals say that if you can capture the dragon you can subdue the male (*cho tê lung, fu tê hsiung*[2]). This means that realgar (*hsiung huang*[3]) on meeting with the fire volatilises, giving rise to vapour and smoke, and it is most difficult to subdue (*fu*[4]). This method, therefore, grinds the realgar, any amount, to powder, and then after the crucible has become red-hot, this is put in together with powdered saltpetre. On stirring with a rod of peach-wood the mass liquefies, then it is quickly poured out on to an earthenware dish, and this is tilted a little so as to decant the clear layer. When the rest has solidified, take it out, grind it fine, and make it into a cake with steamed bread-crumbs suitable for cutting up into pills as large as peas. The dosage is from 3 to 7 pills. One should use 1/10 oz. of saltpetre for every ounce of realgar.'

This is a secret method of the iatro-chemical school (*tan tsao chia pi fa*[5]) and it is very hard to obtain. But, as the men of old said, 'It is better to give away prescriptions rather than to give away medicine'—therefore I record it here.

It is not very difficult to make sense of this passage. Both the arsenical sulphides burn when heated in air, forming SO_2 and volatile arsenious oxide (As_2O_3),[b] and this, in the form either of an arsenite or an arsenate, was what the iatro-chemists were after, for use as an intestinal disinfectant. In later centuries a mixture of saltpetre and orpiment was much used as a flux in metallurgy, the two being heated together and the melt poured out and powdered when solid.[c] Thus the arsenical sulphides and their oxidation products dissolve in potassium nitrate, which acts as a floating layer preventing arsenious oxide loss. The nitrate-based melt in Chang Shih-Nan's process could easily be decanted from the lower layer, and he would have obtained a mixture

---

[a] *Kan tshao*,[6] *Glycyrrhiza glabra*, cf. Porter Smith (1), p. 136.
[b] Partington (10), p. 629.
[c] Agricola, *De Re Metallica*, Hoover & Hoover tr., pp. 233, 236–238, cf. 245, 247.

¹ 獨煉法　　² 捉得龍伏得雄　　³ 雄黃　　⁴ 伏　　⁵ 丹竈家秘法
⁶ 甘草

of arsenates and arsenites together with some unchanged sulphide.[a] One may then reflect that potassium arsenite was nothing other than the famous Fowler's Solution (Liquor Arsenicalis) of 'modern' (or at least, recent) medicine,[b] and even more, that arsenic in organic combination, as acetarsone (stovarsol) or carbamino-phenylarsonic acid (carbasone), has been extensively used down to the present time in the treatment of amoebic dysentery.[c]

To discuss the saltpetre industry in China is to invade the province of Sect. 34, but a few words about it are indispensable here. Porter Smith (2) tells us that in late nineteenth-century China the manufacture of nitrate of potash from the efflorescent salts found on the surface of certain soils, and on walls and places charged with urine, was widely carried on, just as traditionally in Europe. Further details are available in a brief but valuable report of 1925 (Anon., 98), where analyses showed that although the industry was then still essentially a rural one, carried on by certain farmers ('saltpeterers', hsiao hu[1]) in their spare time, the traditional procedures were so good that products of up to 98·2 % purity were sold to government bureaux.[d] The account describes the collection and leaching of the soil, followed by evaporation, removal of impurities, separation from other salts, crystallisation and recrystallisation. The yield was generally only a fraction of an ounce for each catty of earth except in a few places in Heilungchiang where as much as three ounces could be obtained for each catty. About the same time Read (12) described the industry at and around Ho-chien in S.W. Hopei. The percolation was done in large brick tanks, with matting used as the filter (Fig. 1511); after the evaporation the saltpetre crystallised around maize sticks which could be lifted out, while the dark mother-liquor, lu yen shui,[2] was used for salting out soya-bean curd and as a fertilizer. Recrystallisation of the first crop, called khu yen[3] (bitter salt), gave a pure white product at a yield of 50 lbs. from 250 cu. ft. of earth. A study of the Chinese nitrate-containing soils was made in 1935 by Hou Kuang-Chao, especially the solonchaks of Honan, which are capable of producing as much as 30,000 lbs. of saltpetre per acre each year.[e]

As for traditional descriptions of the methods of preparation and purification there are a couple of pages in the *Thien Kung Khai Wu* (Exploitation of the Works of Nature), +1637, but no illustrations.[f] Here yen hsiao[4] (salt-solve) is defined as crude saltpetre from Shansi, paralleling chhuan hsiao[5] from Szechuan and thu hsiao[6] (earth-solve) from Shantung. It is explained that the recrystallised saltpetre is properly called phên hsiao[7] (basin-solve), while more elongated crystals forming at the top round the

---

a Perhaps also the polyarsenites of potassium (Partington (10), p. 627).

b Lauder Brunton (1), p. 647; Sollmann (1), p. 610; Clark (1), p. 609.

c Clark (1), p. 641.

d Out of forty analyses, from places in Hopei, Honan, Shensi and Liaoning, about one third were over 90 % and a further third over 80 %.

e A short notice of this will be found in Kovda (1), Engl. tr., pp. 121–2. During the second world war my friend Dr Wu Ching-Lieh of the 23rd Arsenal at Lu-hsien used to tell me of the substantial deposits of potassium and sodium nitrate near Khaifêng and along the Lunghai Railway. See also Torgashev (1), pp. 380 ff.

f Ch. 15, pp. 6a, b, 7a, tr. Sun & Sun (1), pp. 269–70. Cf. *PTKM*, ch. 11, (p. 49).

¹ 硝戶   ² 鹵鹽水   ³ 苦鹽   ⁴ 鹽硝   ⁵ 川硝
⁶ 土硝   ⁷ 盆硝

Fig. 1511

Fig. 1512

Fig. 1511. The saltpetre (nitre) industry at Ho-chien-fu, showing the removal of the percolated earth, with old percolating jars in the foreground. From Read (12).

Fig. 1512. A saltpetre works in Japan, from the *Nihon no Sangyō Gijutsu* (Industrial Arts and Technology in Old Japan) by Ōya Shin'ichi (1). The drawing (p. 177) is of the early nineteenth century.

edges are *ma ya hsiao*[1] (horse-tooth solve). Both these can be used for making gunpowder, but by-products also crystallising which cannot, are termed *mang hsiao*[2] and *phu hsiao*.[3] Whether these were really magnesium and sodium sulphates is a question which would require further consideration. A similar process of separation by differential crystallisation had been described by Li Shih-Chen a few decades earlier.[a] *Mang hsiao* and *ma ya hsiao* were called by him *shui hsiao*[4] (water-solve), while *hsiao shih* was called appropriately *huo hsiao*[5] (fire-solve). The only illustrated traditional account of the saltpetre industry which we have come across is that produced at Yedo by Chojiya Heibei (1) in 1863, *Shoseki Seirenho*,[6] valuable and interesting though brief.[b]

All this brings out one aspect of Chinese proto-chemistry which has so far had very little attention, namely its quasi-empirical successes in separating salts.[c] In Sect. 37 on the salt industry we shall see how even more difficult problems were solved in the use of the brine deposits of Szechuan, which necessitated dealing with borates as well as with sulphates, chlorides and nitrates. The 'nitre' complex, too, is simply one typical example of the difficulty of identifying the substances used by the medieval

[a] *PTKM*, ch. 11, (p. 56); cf. Anon. (57), p. 244.

[b] Twenty years later Kinch (1) gave some analyses of crude Japanese saltpetre, p. 115. Occasional illustrations of the equipment used for the differential crystallisations dating from the first half of the century or even the late +18th can be found; we reproduce (Fig. 1512) one given by Ōya Shin'ichi (1).

[c] On the development of solution analyses in Europe during the Renaissance, especially in connection with spa waters, from about +1200 onwards, see Debus (13).

[1] 馬芽硝   [2] 芒硝   [3] 朴硝   [4] 水滑   [5] 火硝   [6] 硝石製煉法

alchemists and pharmacists. The most helpful signs to go by are always the descriptions of the properties of the substance concerned mentioned in any given Chinese text. Thus of *hsiao shih* (which goes back as a name to the −4th century) it is often later said that it gives a bluish-purple flame when put in the fire,[a] a statement which immediately rules out salts of sodium and magnesium. The oldest description of this test comes from about +500, but it could safely be placed a couple of centuries earlier, as far back as Ko Hung. Many alchemical and pharmaceutical texts from the −2nd century onwards also say that *hsiao shih* can liquefy ores, acting as a flux, and dissolve minerals to form aqueous solutions.[b] There are also instances where *hsiao shih* is said to produce explosions or deflagrations,[c] and we have of course the gunpowder formulae with *hsiao shih* in them.[d] In such circumstances one can feel fully justified in extrapolating back the results of analyses of modern samples of *hsiao shih* which show it to be saltpetre. Rightly therefore was it called in Arabic *thalj al-Ṣīn* (Chinese snow) for it was recognised and used in China long before anywhere else.[e]

The oldest extant Arabic mention is in the *Kitāb al-Jāmiʿ fī al-Adwiya al-Mufrada* (Book of the Assembly of Medical Simples)[f] finished by Abū Muḥammad al-Mālaqī Ibn al-Baiṭār[g] about +1240. Others follow shortly after, for example Ibn abī Uṣaybiʿa, in his history of medicine mentioned on p. 226 of pt. 3 above, but as he refers back to the otherwise unknown Ibn Bakhtawayhī and his *Kitāb al-Muqaddimāt* (Book of Introductions),[h] it would be wise to place the first knowledge of saltpetre among the Arabs in the earliest decades of the +13th century.[i] On the other hand their understanding of its use in war, especially for gunpowder, belongs to the latest decades of the same century, as we know from the book of al-Ḥasan al-Rammāḥ,[j] *Kitāb al-Furūsīya waʾl-Munāṣab al-Ḥarbīya* (Treatise on Horsemanship and Stratagems of War),[k] which cannot have been composed before about +1280. The same date, as near as makes no matter, can be accepted for the completion of the *Liber Ignium ad Comburendos Hostes* of Marcus Graecus (whether or not there was ever any such individual person), and by this time both saltpetre and gunpowder, or at least proto-gunpowder, had become acclimatised in the Latin West.[l]

[a] See pp. 185, 187.   [b] Cf. p. 183 above.   [c] Cf. p. 186.

[d] First in the *Wu Ching Tsung Yao* of +1044 (*Chhien Chi*, ch. 11, pp. 27b, 28a; ch. 12, pp. 58a, b, 65a, b); and abundantly thereafter. See Sect. 30 in Vol. 5, pt. 1.

[e] In Vullers' Persian–Latin lexicon there is another phrase, *namak shūra Chīnī* (salt of the Chinese salt-marshes); Partington (5), p. 335. The other Arabic name was *bārūd*, which may or may not be connected with hailstones, recalled by the saltpetre crystals on the ground.

[f] Tr. Leclerc (1); see pp. 71, 200, 333, 420.

[g] 'The son of the Veterinary Physician' of Malaga, himself perhaps the greatest pharmaceutical naturalist of all Islam (d. +1248). See Mieli (1), p. 212; Partington (5), p. 310, following Romocki (1), vol. 1, p. 37.

[h] This was in connection with freezing mixtures, on which see pt. 3, pp. 225–6 above. Cf. von Lippmann (8); Partington (5), p. 311.

[i] Bloch (5) describes two of their probable points of production on the western shores of the Dead Sea.

[j] 'The lancer'; cf. Section 30, where he prominently appears.

[k] Cf. Partington (5), pp. 200ff. Though Al-Rammāḥ does not use the term *thalj al-Ṣīn*, his book is remarkable for the extensive use it makes of Chinese materials and the numerous pyrotechnic devices which bear the name 'of China'.

[l] Cf. p. 123 above; the gunpowder and the alcohol recipes belong to the very latest strata of the compilation. See Partington (5), pp. 42ff., 60.

(3) SALTPETRE AND COPPERAS AS LIMITING
FACTORS IN EAST AND WEST

Thus we now have a flood of light on the long-known evidence for the primary invention of gunpowder in China. It seems clear that lack of saltpetre in the West must have been the great limiting factor for this development. The oldest mentions of gunpowder in Europe are all unquestionably of the late +13th century, preceding its general introduction in the +14th. In China, on the other hand, we have the first reference to the gunpowder mixture in the +8th or +9th century, its appearance in war early in the +10th, and its widespread military use in the +11th and +12th before it reached Islam and Europe in the +13th.[a]

Secondly we have here a chapter in the earliest history of the use of the mineral acids, not as isolated and purified products, but as part of a procedure quite ancient and primitive in that distillation was probably not involved. If nitric and hydrochloric acids were only to appear as such as the products of distillation, sulphuric acid could be obtained also by the simple combustion of sulphur, and it is likely that this method preceded the distilling way.[b] It is generally accepted that mineral acids were quite unknown both to the ancients in the West[c] and to the Arabic alchemists.[d] The first

[a] The full evidence is given in Sect. 30 in Vol. 5, pt. 1. There is a résumé in Needham (47).

[b] It was not until the +17th century, however, that the 'oil of sulphur' resulting was recognised as identical with the 'oil of vitriol' produced by distilling ferrous or cupric sulphate. Neither of the 'oils' seems much to antedate +1530, the first mentions being in Brasavola, Valerius Cordus and Mattioli; cf. Partington (4), p. 47, (7), vol. 2, p. 96; Sherwood Taylor (4), pp. 95–6, 191.

[c] The question is of course not absolutely settled. The late Professor J. R. Partington drew our attention in 1959 to certain passages in the Greek proto-chemical writings, especially Zosimus (*Corp. Alchem. Gr.*, III, i, 8 and III, xlvii, 6, 7), which might conceivably be interpreted in this sense. But although the translation of these passages by Berthelot & Ruelle reads unintelligibly, Partington's versions involve excerpting the text rather forcefully and interpreting certain words very boldly (e.g. *botanē*, 'the weed', for saltpetre). One rarely encounters anything as puzzling in classical Chinese, except where textual corruption calls for massive emendation.

[d] This is true in the main, but 'not exactly' (as our Chinese friends so often have occasion to say). It is true that 'sharp waters' are frequently mentioned in the Jābirian Corpus and al-Rāzī, but in fact they all seem to be fairly caustic alkalies rather than the acids which the name might lead one to expect (cf. Ruska & Garbers (1); Ruska (14), pp. 66–7; Multhauf (5), p. 140). Nevertheless, the Arabic alchemists knew that corrosive vapours were to be obtained by distilling vitriols (sulphates) mixed with other materials. In the middle of the +10th century al-Mas'ūdī wrote: 'As for us, may God preserve us from applying ourselves to researches which weaken the brain, ruin the sight and jaundice the complexion in the midst of subliming vapours, vitriolic fumes and other mineral exhalations' (tr. de Meynard & de Courteille (1), vol. 8, p. 177, eng. auct.). Among the passages which indicate that Arabic alchemists were capable of using a mineral acid without quite knowing what it was, we may quote one from the *Kitāb Sirr al-Asrār* (Book of the Secret of Secrets) written by al-Rāzī towards +910. What he seems to be doing is making pure aluminium sulphate from alunite (the sulphate plus the hydroxide), and getting sulphuric acid in order to do it. The passage runs: 'Take white (Yemeni) alum, dissolve it and purify it by filtration. Then distil (green ?) vitriol with copper-green (the acetate), and mix (the distillate) with the filtered solution of the purified alum, afterwards let it solidify (or crystallise) in the glass beaker. You will get the best *qalqadīs* (white alum) that may be had' (tr. Ruska (14), p. 88, eng. auct.; cf. Stapleton, Azo & Husain (1), p. 373). The version given by Singer (8), p. 51, diverges much from this, and seems to be unreliable.

The late Dr E. J. Holmyard maintained that the preparation of nitric acid is to be found in the Jābirian Corpus (+9th century), but so far as we know, the text has never been identified and published. Wiedemann (27), however, found and translated a passage in al-Qazwīnī (c. +1250) which speaks of the oily nature of the fumes of heated vitriol, and perhaps of the heating of water with which they are brought into contact; the encyclopaedist also recorded that vitriolic vapours are injurious to mice and

account of the making of nitric acid (*aqua fortis, scheidewasser*) is often said to occur in the Geberian *De Inventione Veritatis*,[a] a tractate composed in the West at the end of the +13th or the beginning of the +14th century. Partington however has shown that it is also found in the *Pro Conservanda Sanitatis* of the French Franciscan Vitalis du Four, *c.* +1295.[b] It was always a matter of distilling the nitrate with alum and especially ferrous sulphate.[c] If salt was added, or sal ammoniac, a mixture of nitric and hydrochloric acids (*aqua regia*) was obtained; and this was also in Geber, by +1300 or so. Sulphuric acid (as oil of vitriol) came a good deal later, probably at the beginning of the +16th century, and hydrochloric acid not until about +1600.[d] Thus the availability of saltpetre must again have been a primary limiting factor.[e] It is interesting that the new recognition of a salt long known in East Asia,[f] and the transmission of the technique of its purification, should have permitted two developments in the West each so important as the manufacture of gunpowder and of nitric acid.[g]

The absence of nitrate from the reagents of the early medieval West restricts the significance of the references to acetic acid in the Greek proto-chemical texts. Such allusions there certainly are, e.g. in the +5th-century tractate of John the Archpriest, which speaks of the strongest white vinegar (*to leukon oxos drimutaton*, τὸ λευκὸν ὄξος δριμύτατον),[h] and in the probably +8th-century 'Practice of the Emperor Justinian'.[i] Von Lippmann was doubtless right in rejecting the speculation of Berthelot that such words concealed the use of crude mineral acids, and Greek texts which distinctly mention these, together with nitrate, are not plausibly to be dated earlier than +1300.[j]

Whether *aqua fortis* was known in China before modern times we cannot say, but the following story suggests the desirability of further researches on the borderline between Chinese and Indian chemical technology. In the former culture-area at any rate all the *dramatis personae* were ready to play their parts as early as the Thang period, from which this story dates;[k] if not considerably earlier. The alums had been

flies, driving them out from a room thus fumigated. This is found in the mineralogical section of his 'Cosmography', and the best translation of the passage is doubtless that given by Ruska (24), p. 23.

[a] Ch. 23; Russell tr., p. 223, Darmstädter tr., p. 113. On the Latin Geberian corpus in general see Multhauf (5); Sarton (1), vol. 2, p. 1044.

[b] (4), p. 40. Cf. Sarton (1), vol. 3, p. 531.

[c] The green crystals were calcined first to give the reddish anhydrous salt.

[d] First in the writings of 'Basil Valentine', then in those of Oswald Croll. But Reti (11) has provided strong evidence of an earlier preparation, from a Bologna MS. of the +15th century. Also if 'oil of bricks' was impure HCl, it might go back to the +10th century (cf. Vol. 5, pt. 3, p. 237 above). On the history of the industrial preparation of the mineral acids see Sherwood Taylor (4), pp. 90ff., 99.

[e] Later on it also entered into the manufacture of sulphuric acid, for about +1745 Joshua Ward added saltpetre to the sulphur, aiding its oxidation during the burning to make oil of sulphur, increasing yields and bringing down the price (Sherwood Taylor (4), p. 97).

[f] It will be remembered that one of the +13th–century Arabic terms for saltpetre was *thalj al-Ṣīn*, 'Chinese snow'.

[g] The central position of green vitriol (ferrous sulphate) is also worth noting. It was a raw material for all three of the strong mineral acids. Cf. p. 199 below.

[h] *Corp. Alchem. Gr.* IV, iii, 15, Berthelot & Ruelle (1), vol. 3, p. 255; cf. von Lippmann (1), p. 71.

[i] *Corp. Alchem. Gr.* V, xxiv, 2, Berthelot & Ruelle (1), vol. 3, p. 369; cf. von Lippmann (1), p. 114.

[j] *Corp. Alchem. Gr.* V, i, 18, 41, 42, Berthelot & Ruelle (1), vol. 3, pp. 312, 317, 318; von Lippmann, *loc. cit.*

[k] We gave it already in Vol. 1, p. 212, but we have retranslated it here.

known since the −4th century,[a] nitrate occurs also at least as early as the time of Ko Hung (and with certain identification in the *San-shih-liu Shui Fa*); and both ferrous sulphate (*lü fan*[1]) and sal ammoniac (*nao sha*[2]),[b] (for the making of *aqua regia*) appear in the *Hsin Hsiu Pên Tshao*[3] by +660.[c] Effective stills had by then come into existence in China.[d] There is thus some reason for anticipating that further research may uncover evidence of the preparation of nitric acid and a mixture of nitric and hydrochloric acids in China well before the +13th century and perhaps as far back as the Thang. A search in the Sung alchemical literature might well prove fruitful. Now for the story.

Wang Hsüan-Tshê[4] was an official who left China in +648 as ambassador to the court of Magadha (modern Patna) where at that time Harsha Vardhana, the friend of Hsüan-Chuang[5] the great pilgrim, was reigning. But then Harsha died, and a usurping minister (A-Lo-Na-Shun[6] in the Chinese records) thought fit to attack the Chinese party, plunder their goods, and kill most of Wang's retinue. Wang, however, was a man of resource; he escaped to the mountains, made contact with the Kings of Nepal and Tibet, who were at that time allied with China, and descending again with an army of considerable size, gave battle to the usurper and completely overthrew him. The ambassador then returned home by another route, taking with him the usurper and other Indian prisoners, whom he presented to the emperor at Chhang-an (modern Sian) with a report on his proceedings.

An account of this, written a little over two centuries later, is of great interest, as it preserves what may be one of the earliest passages on mineral acids. In the *Yu-Yang Tsa Tsu*[7] of Tuan Chhêng-Shih,[8] written in +863, we read:[e]

Wang Hsüan-Tshê[4] captured an Indian prince named A-Lo-Na-Shun. He had with him a scholar versed in arts and gramarye named[f] Na-Lo-Mi-So-Po,[9] who said he was two hundred years old. (The Emperor) Thai Tsung was very interested and invited him to live in the Chin-Yen Mên[10] (Palace), to make the drugs for prolonging life. The Emperor asked the Minister of War, Tshui Tun-Li,[11] to be in charge of it. (The Indian) said, 'In the country of the Brahmins there is a substance called *Pan-Chha-Cho Shui*[12] (Pan-Chha-Cho water)[g]

[a] *Chi Ni Tzu*, ch. 3, p. 3a (in *YHSF*, ch. 69, p. 26a).

[b] There is a +2nd-century mention of this in Wei Po-Yang (*Chou I Tshan Thung Chhi Fên Chang Chu Chieh*, ch. 2, p. 26b (ch. 30); *Tshan Thung Chhi*, Tao Tsang ed. *TT*990, ch. 2, p. 35b), where advice is given not to put ammonium chloride on ulcers.

[c] *Pên Tshao Kang Mu*, ch. 11, pp. 31a, 53b. A very important text, the *Chen Yuan Miao Tao Yao Lüeh*[13] (Classified Essentials of the Mysterious Tao of the True Origin of Things, *TT*917) mentions both nitrate and sal ammoniac. This is the text which has the first proto-gunpowder formula; cf. Vol. 5, pt. 3, p. 78. Though that itself is probably of the +8th or +9th century, the older parts of the book go back to the time of the putative author, Chêng Ssu-Yuan[14] (Chêng Yin), in the +4th century.

[d] See pp. 155ff. above.

[e] Ch. 7, p. 7a. A parallel account in the *Chiu Thang Shu*, ch. 198, p. 12b, tells how the emperor scoured the country to collect minerals and drugs for Na-Lo-Mi-So-Po's experiments, but that when the elixirs were completed they did not prove very effective, so eventually he was sent back to his own country. The *Chiu Thang Shu* was not completed till +945, but it was based on official archives and documents.

[f] The Indian form *Nārāyaṇasvāmin* has been conjectured.

[g] 'Punjab water' has been conjectured, others think *phāṇṭa* water, i.e. a liquid prepared by filtration.

[1] 綠礬	[2] 硇砂	[3] 新修本草	[4] 王玄策	[5] 玄奘
[6] 阿羅那順	[7] 酉陽雜俎	[8] 段成式	[9] 那羅邇娑婆	[10] 金颷門
[11] 崔敦禮	[12] 畔茶佉水		[13] 眞元妙道要略	[14] 鄭思遠

which is produced in the mountains in stone vessels, has seven varieties of different colours, is sometimes hot, sometimes cold, can dissolve herbs, wood, gold and iron—and if it is put into a person's hand, it will melt and destroy it. If you want to collect this water you have to use a 'camel's skull' placed in a stone vessel, and pour it out into a gourd.[a] Whenever this water is present there are also stone columns looking like men guarding it. Anyone from another mountain who shows the way to this water will die...Finally the Indian died in Chhang-an.

One of the best things about this account is its date, which is very firm. Foreshadowing perhaps the later 'alkahest' or universal solvent of Paracelsian iatro-chemistry,[b] this passage suggests at any rate that a mineral acid was known in the +7th century. It gives some colour to the hints about strong acids in Ray's history of chemistry in India. Already in the +11th century the *Rasārṇavakalpa* has much on the 'fixation' or 'killing' of metals.[c] The *Rasārṇava Tantra* (dated by Renou & Filliozat as of the +12th century)[d] speaks of the 'killing' of iron and other metals by a *viḍa* (solvent?) which is prepared from green vitriol (*kāsīsa*), pyrites, etc.[e] From the *Rasaratna-samucchaya* (compiled according to Renou & Filliozat about +1300) which reproduces material from the *Rasendra-chūḍāmaṇi* of Somadeva (+12th or +13th century), the process of 'killing' certainly seems to be the formation of salts from metals.[f] Neogi drew attention long ago[g] to the apparent presence of oil of vitriol in the later Indian alchemical treatises under the name 'essence of alum', produced by distillation. This is certainly mentioned in the *Rasaratna-samucchaya*[h] and in the *Rasaprakāśa-sudhākara*[i] of Yaśodhara (+13th century), though they do not distinctly say that the alum and the ferrous sulphate must be distilled together; but neither of these works would be older than the time of Geber in the West.

The date of Na-Lo-Mi-So-Po was not, however, so much earlier than the beginning of the 'oil of bricks' tradition there (cf. Vol. 5, pt. 3, pp. 237–8 above). And his 'camel's skull' reminds one of another curious story in Chinese literature about a special container. The word *tiao*[1] is used today for designating the whale[j] but anciently *chi tiao*[2] meant a fabulous kind of dragon (*lung*[3]), the fat of which could only be collected in eggshells. Works such as the *Pên Tshao Thu Ching* (+1062) quote a lost *Kuang Chou Chi*[4] of the Chin period by one Phei Yuan,[5] which said:[k]

The *tiao*[2] frequents Lingnan (the country south of the mountains, i.e. Kuangtung). It has the head of a serpent and the body of a tortoise, and it lives in water or in (swampy) forests. Its fat is so light that it can penetrate all vessels, whether of metal or pottery, so it has to be conserved in eggshells; then only it will not leak away.

The details are unimportant here, the content is suspicious. Might this not also be a disguised reference to corrosive liquids?

[a] Could this be a veiled reference to distillation?
[b] In an interesting study, Reti (6) has concluded that this was probably alcoholic caustic potash.
[c] Roy & Subbarayappa tr., pp. 65ff., 71ff., 87, etc.
[d] vol. 2, p. 169.   [e] Ray (1), 2nd ed., p. 138.
[f] Ray (1), 2nd ed., pp. 188ff.
[g] (1), p. 50.   [h] Ray (1), 2nd ed., pp. 173–4.
[i] Ray (1), 2nd ed., pp. 122, 153.   [j] R103.
[k] Tr. auct., from *Khang-Hsi Tzu Tien*, p. 284.

[1] 弔   [2] 吉弔   [3] 龍   [4] 廣州記   [5] 裴淵

The mention of green vitriol (ferrous sulphate, $FeSO_4.7H_2O$)[a] just above induces us to emphasise the status of this salt also as one of the great primary limiting factors of chemical advance. Although the point has not been made, so far as we know, by historians of chemistry, it came about by the necessity of things that certain substances much more than others acted like doors which would yield to the push, and let the practitioners through into a world which could generate the theoretical and experimental chemistry of today. So just as potassium nitrate led to gunpowder and to nitric acid, ferrous sulphate led to not only one, but all three, of the strong mineral acids, true foundation-stones of chemistry and chemical technology. Hence very suitably did the late Western alchemists call it by the name of *leo viridis* (the green lion).[b]

Crystalline ferrous sulphate in impure form was known to the Hellenistic proto-chemists as *chalcanthon* (χάλκανθον), *misy* or *sory*,[c] all being mixtures of the sulphates of copper, iron and aluminium in various proportions, derived from the oxidation of naturally occurring sulphide minerals (pyrites, marcasite).[d] The old Chinese name *lü fan*,[1] 'green alum',[e] paralleled our 'green vitriol', and was no further off the mark than 'flower of bronze' (*chalcanthon*),[f] not indeed so far, since copper and tin were never really involved. Yet copper did come in in a curious way, since green vitriol got the name of copperas (O.F. *couperose*), very puzzling till one realises that it probably derives from *aqua cuprosa*, i.e. the solution resulting when copper sulphate mine-waters have been passed over scrap-iron and the iron has gone into solution by exchange as sulphate, leaving the copper deposited.[g] Ferrous sulphate in its purer form was also called *atramentum* because of the black colour given with tannins and widely used for dyes and ink.[h] Here again this name was exactly mirrored in the other Chinese term, *tsao fan*.[2]

Now every one of the mineral acids needed copperas for its preparation, usually calcined from the blue-green to the anhydrous reddish form.[i] About +1300 Vitalis du Four and the Geberian writer were distilling it with saltpetre and alum to get nitric,[j] soon after +1500 Brasavola and Valerius Cordus were distilling it alone to get sulphuric,[k] and towards +1600 Thölde ('Basil Valentine') was distilling it with

---

[a] On the vague word vitriol, already in Pliny, and derived no doubt from the glassy appearance of hydrated sulphate crystals, see Crosland (1), p. 84. Later on it was sometimes synonymous with the equally vague copperas, of which there were also several sorts, distinguished by colour or place of origin.   [b] Multhauf (5), p. 195.

[c] Berthelot (2), pp. 14–15, 241–2; Crosland (1), p. 229. It was an important constituent of the cementation mixture for purifying gold, or bringing about the surface-enrichment of gold alloys (cf. pt. 2, p. 250 above).

[d] Marcasite was an old name for sulphide minerals (Partington (10), p. 853; Berthelot (2), pp. 253, 257).   [e] RP132.

[f] Doubtless so called because of a confusion with the green salts of copper.

[g] This 'wet copper' method of Cu production will be the subject of our next sub-section. There are other derivations, e.g. 'rose of Cyprus' (Mellor (1), p. 448), less convincing.

[h] Sherwood Taylor (4), pp. 82, 121. Ink is still to this day *atrament* in Polish.

[i] Reddish because of the presence of ferric oxide, Fe_2O_3, into which all the sulphate would be converted if roasted long enough.

[j] Partington (4), p. 40; Sherwood Taylor (4), p. 92, Cf. Vol. 5, pt. 3. pp. 237–8.

[k] Partington (4), p. 47; Sherwood Taylor (4), pp. 95–6, 191. It is curious that the distillation with nitre so long preceded the distillation alone. But this same simple process continued in industrial use at

[1] 綠礬   [2] 皁礬

ordinary salt and alum to get hydrochloric.[a] If this did not happen in China it may have been because the stills were not quite suitable, being mostly of the cooled head type (cf. pp. 63 ff.), true retorts appearing rather later. Moreover ferrous sulphate was not mentioned in the *Shen Nung Pên Tshao Ching*; its first special entry in a pharmaceutical natural history occurring in Ta Ming's[1] *Jih Hua (Tzu) Chu Chia Pên Tshao*[2] of +972, though some account of it had been given under another head in the *Hsin Hsiu Pên Tshao* of +659.[b] It was made in a similar way to that of Europe, sulphurous

Fig. 1513. A Japanese ferrous sulphate works, from the *Nihon Sankai Meibutsu Zue* (Illustrations of Processes and Manufactures), +1754, pp. 54–5.

coal and hepatic iron pyrites (marcasite, 'the gangue of coal, commonly called bronze-coal', *mei than wai kung shih, su ming thung than*[3]) being burnt in a heap covered over with mortar to exclude excess of air (Fig. 1513).[c] But wherever in the world doors were opened by knowledge of these salts, limiting factors it is certainly right to call them.

Nordhausen down to as late as 1900. Clow & Clow (2) have written on its central position in the industrial revolution.

 [a] Partington (4), p. 56, (7), vol. 2, p. 200; Sherwood Taylor (4), p. 99. The acid may have been produced in unrecognised form long before under the name of 'oil of bricks' (cf. pt. 3, pp. 237–8), but this is not quite certain. See also p. 197 above.
 [b] Cit. *PTKM*, ch. 11, (pp. 73–4). Su Ching and the other naturalists said that the best was got from Kuachow, near mod. Tunhuang (Kansu). All speak of its use in dyeing. Chinese ink, of course, was always something else (cf. Sect. 32).
 [c] Porter Smith (1), p. 122. The process is described in *TKKW*, ch. 11, pp. 4 a ff., tr. Sun & Sun (1), pp. 206–7, 213, but their version and notes need explanations and should be used with circumspection. In relatively late times Chinese ferrous sulphate was produced particularly pure.

 [1] 大明   [2] 日華子諸家本草   [3] 煤炭外礦石俗名銅炭

(4) THE PRECIPITATION OF METALLIC COPPER
FROM ITS SALTS BY IRON

In recipe no. 1 of the *San-shih-liu Shui Fa* we are told that 'when more vinegar is mixed with (the solution) and it is then rubbed on iron, the iron will (look) like copper'. Since the method concerns the solubilisation of iron alum, or perhaps ferric sulphate, the treatment is meaningless as it stands, but it has undoubtedly strayed in here from a discussion of copper sulphate (so often confused with alum on account of a similarity of nomenclature, *fan shih*[1] and *tan fan*[2]). Nevertheless the words are of great interest for they concern an industrial process very old in China, namely the winning of metallic copper by precipitation from solutions of its salts in the presence of metallic iron.[a]

In +1086 the great scientific scholar Shen Kua[3] (+1030 to +1094) wrote in his *Mêng Chhi Pi Than*[4] (Dream Pool Essays) the following passage:[b]

> In the Chhien-shan[5] district of Hsin-chou[6] there is a bitter spring which forms a rivulet at the bottom of a gorge. When its water is heated it becomes *tan fan*[7].[c] When this is heated it gives copper. If this 'alum' is heated for a long time in an iron pan the pan is changed to copper. Thus water can be converted into copper—an extraordinary change of substance, really unfathomable. According to the *Huang Ti Nei Ching, Su Wên*,[d] 'there are five elements in the sky, and five elements on the earth. The *chhi*[e] of Earth, when in the sky, is moisture. Earth produces metal and stone (as ores in the mountains), but water can also produce metal and stone'. These instances then are proofs (that the principles of the *Su Wên* are right). It is like water dripping in caverns and (slowly) forming stalactites, or like the formation of crystals from well and spring water at the spring and autumn equinoctial seasons, or like selenite[f] (*yin ching shih*[8]) deposits from strong brines; all show the transformations from moisture. So also the *chhi* of Wood, when in the sky, is wind, and both Wood and wind can generate Fire. Such is the nature of the Five Elements.

The passage clearly shows that Shen Kua was prevented by a too uncritical acceptance of the classical five-element theory from attaining an understanding of the true nature of solution and mixture. Yet we cannot place such an +11th-century mind in the right perspective without tracing the parallel development of thought in Europe.[g] The observation of the precipitation of metallic copper in powdery or solid form by iron, with the formation of iron sulphate, described in the opening paragraph, was an excellent one. T. T. Read (4) tells how in our own times a process for the winning of copper from mine waters by precipitation with scrap iron was developed at Butte, Montana, in ignorance of the fact that it had been well known in Moorish Spain.[h]

---

[a] On the coating of other metal surfaces by ion exchange something has already been said in the metallurgical part of the Introduction (pt. 2, p. 246). See also Haschmi (5).

[b] Ch. 25, para. 6, tr. auct. Cf. Hu Tao-Ching (1), vol. 2, pp. 792ff. Echoes of this can be found in various places, e.g. *PWYF*, Shih I, ch. 1, p. 7a (vol. 6, p. 4246.3).

[c] 'Bitter alum', lit. 'gall alum', impure copper sulphate (RP87).

[d] The Han medical classic.   [e] I.e. *pneuma*; cf. Vol. 2, p. 369.

[f] Calcium sulphate (RP120).   [g] A résumé has been given by Multhauf (7).

[h] Cf. T. T. Read (8).

[1] 礬石   [2] 膽礬   [3] 沈括   [4] 夢溪筆談
[5] 鉛山   [6] 信州   [7] 膽礬   [8] 陰精石

'Basil Valentine', in his *Currus Triumphalis Antimonii*, noted the power of iron to precipitate copper from 'an acrid ley in Hungary',[a] an effect which Paracelsus[b] and Libavius[c] still in the +16th century believed demonstrated the transmutation of metals, as also Stisser as late as +1690.[d] Van Helmont (+1624) and Nicholas Guibert (+1603)[e] surmised that the copper was in the solution beforehand, and the exchange of metals was proved by Joachim Jungius (+1630),[f] then by Robert Boyle in his 'Treatise on the Mechanical Causes of Chemical Precipitation' (+1675). It would therefore be unjust to censure Shen Kua for accepting as a transmutation of metals a process which was not properly understood until six centuries after his death.

What has not generally been appreciated is how old the technique was in China.[g] Two Han references begin the story. The *Huai Nan Wan Pi Shu*[1] says that if *pai chhing*[2] (basic copper carbonate, azurite) meets iron it turns it into copper;[h] this may not be quite as old as Liu An, the prince of Huai-Nan, himself (*d.* −122) but it will not be much later. The statement is repeated in the entry for copper sulphate (*shih tan*[3]) in the *Shen Nung Pên Tshao Ching*,[i] the first of the pharmaceutical natural histories and undoubtedly complete by the Later Han (+1st and +2nd centuries) though mainly of the Former (−2nd and −1st centuries).[j] Thus both these references are older than the remark of Pliny, *c.* +77, that iron 'if smeared with vinegar or alum, becomes coppery in appearance',[k] and also better, because both distinctly specify salts of copper while Pliny's reference could be merely to the effects of rusting.[l] Knowledge of the copper precipitation effect appears again about +300 in the *Pao Phu Tzu* book where Ko Hung says[m] that if a saturated solution of copper carbonate (*tshêng chhing*[4])[n] is placed in contact with iron the latter will take on a red colour like

---

a There has been, of course, much doubt as to the date of 'Basil Valentine'. His work is of the late +16th and early +17th centuries, not the +15th, which was the period traditionally ascribed to it. Some earlier material was no doubt contained in it. See J. Read (1), pp. 183 ff; von Lippmann (1), p. 640; Leicester (1).

b In *Chirurgia Magna* (+1536), see *Opera Omnia* (Geneva ed., +1658), vol. 3, pt. 1, p. 43; cf. Partington (7), vol. 2, p. 137.

c *Commentariorum Alchymiae* (+1606), pp. 20 ff.; *Syntagma* (+1611), pp. 280–1; cf. Partington (7), vol. 2, p. 255.

d Roscoe & Schorlemmer (1), vol. 2, p. 413. Cf. our discussion in pt. 2, pp. 24, 35, 67, 245, pt. 3, p. 207 above.

e See Duveen & Willemart (1); Partington (7), vol. 2, p. 268.

f See Kangro (1); Pagel (15), pp. 102–3; Leicester (1), p. 111.

g A brief but good account was given by Chang Hung-Chao (1), pp. 316 ff. Cf. also the remarks of Wang Chia-Yin (1), p. 60; Hung Huan-Chhun (1), p. 39.

h Cit. in *Thai-Phing Yü Lan*, the imperial encyclopaedia of +983, that great source of ancient fragments, ch. 988, p. 5a.

i Cit. in *TPYL*, ch. 987, p. 4a; also *CLPT*, ch. 3, (p. 89.2).

j The text is accepted by Mori Tateyuki in his reconstruction, ch. 1, (p. 24). So also Ku Kuan-Kuang ed., ch. 3, (p. 54); and Chang Hung Chao (1), p. 316.

k *Hist. Nat.* XXXIV, 149; see Bailey (1), vol. 2, p. 61.   l Cf. Bailey (1), vol. 2, p. 188.

m *PPT/NP*, ch. 16, p. 5a, tr. Ware (5), p. 268. We have given the whole passage in translation above, pt. 3, p. 104. Ko Hung was well aware that the copper was deposited as a layer on the iron, and did not think that the iron had all been transmuted to copper.

n In its entry for another form of copper carbonate, malachite (*khung chhing*[5]), the *Shen Nung Pên Tshao Ching* says that it will turn iron into gold (or a golden colour), cit. *CLPT*, ch. 3, (p. 90.2). Though rather more difficult to interpret, this is best taken as a further reference to the 'wet copper method', precipitation on iron.

¹ 淮南萬畢術   ² 白青   ³ 石膽   ⁴ 曾青   ⁵ 空青

copper.[a] Next in time, *c*. +500, comes the statement of Thao Hung-Ching about 'bird-droppings alum' (which must have contained copper sulphate),[b] and he too was quite clear that the copper was an external layer. Very soon afterwards would come the *San-shih-liu Shui Fa*; interesting to note is the nuance of its Liang wording which cautiously does not commit itself to an actual transmutation. Then in +659 there is a mention in the *Hsin Hsiu Pên Tshao*.[c]

In the Sung period (+10th century onward) we begin to find evidence of the industrial use of the process. We have just read the passage in Shen Kua's book of +1086. An interesting story in the *Lung Chhuan Lüeh Chih*[1] of Su Chhê[2] (+1039 to +1112; the brother of the famous poet Su Tung-Pho[3]) tells of his scepticism about it.[d] As a civil official he had to deal with a merchant who came before him and said that he had a secret method of converting iron into copper by means of copper sulphate (*tan fan*). Su Chhê said that secret methods were forbidden and if there was any value in this it ought to be disclosed to people in general, so that the public could benefit. The merchant was unwilling to do this and left, after which Su Chhê and his friends tried the effect of copper sulphate on old knives without success. This would have been about +1080. But before long the process became well known, for from about +1090 onwards, as Nakajima Satoshi (*1*) has shown in a special study, the 'wet method' (*shih shih chih lien*[4]) came into extensive use as a result of a temporary scarcity of ore for making copper cash.[e] Copper was extracted in large-scale production both from ground water containing copper salts and from solutions obtained by leaching piles of low-grade ore. The *Sung Shih* says:[f]

The method of producing 'steeped copper' (*chhin thung*[5]) is to make (lit. forge) thin plates of cast-iron and immerse them in rows in troughs of blue vitriol solution (*tan shui*[6]). After some days a layer of red powder is formed by the copper sulphate over the surface of the iron; this is collected by scraping and after three purifications in the furnace gives good copper. Broadly speaking for every pound of copper 2 lbs. 4 ozs. of iron are needed. The Hsing-li Factory at Jao-chou and the Chhien-shan Factory at Hsin-chou produced a definite amount of this 'vitriol copper' (*tan thung*[7]) each year.

This would refer to about +1100.

---

[a] Another mention of the matter, not much earlier, was that in the *Wu Shih Pên Tshao*[8] (in *TPYL*, ch. 988, p. 5*a*).

[b] *CLPT*, ch. 3, (p. 84.1), translated in full pt. 3, p. 130 above.

[c] Cit. *CLPT*, ch. 3, (p. 90.1). And in Tsan-Ning's *Ko Wu Tshu Than* of *c*. +980, (p. 28).

[d] Ch. 5, p. 3*a*.

[e] Individual installations at this time were producing outputs of some 400 tons a year purely by this method (Collins (*1*), pp. 18, 240). According to Sahlin (*1*) the process was first used in Europe, probably in Hungary, about the last decade of the +15th century, whence the distich on a famous mug:

> 'Eisen war ich, Kupffer bin ich
> Silber trage ich, Gold bedeckt mich'.

A dish bearing the same motto, and ornamented with models of minehead equipment, is illustrated by Smith (6), fig. 14. The great centre for this ware was Herrengrund in Bohemia (cf. Alexander, 1). In Sweden at the great Kopparberg mines the 'wet copper' method started only from about +1750. Cf. Lindroth (2).

[f] Ch. 180, p. 22*a*, tr. auct.

[1] 龍川略志	[2] 蘇轍	[3] 蘇東坡	[4] 濕式製鍊
[5] 浸銅	[6] 膽水	[7] 膽銅	[8] 吳氏本草

The path of the 'wet copper' men was not always easy however, for adequate supplies of the natural solution were sometimes scarce. In a book of memorabilia entitled *Chhing Po Tsa Chih*,[1] by Chou Hui[2] in +1193, we find the following passage:[a]

> In the Chhien-shan district of Hsin-chou, there used to be a stream of (blue) vitriol water (*tan shui*[3]) flowing down out of the mountains over some waterfalls. It was utilised in the 'steeping method' of making copper for the melters. The flow continued even when the weather was dry, but more abundantly in spring and summer, less so in autumn and winter. It is said that in olden times a man lost his keys in the water, and when he recovered them on the following day, they had all turned to copper. In recent years the stream almost stopped flowing, so the steeping method took longer and required more labour. Formerly there were some pits full of vitriol water, and others that were dry, but all the earth round about them contains vitriol, so it is called 'vitriol earth' (*tan thu*[4]). While it saves labour and gives more profit to use vitriol water, less satisfactory results can still be obtained by using the vitriol-containing earth, and after all, though the water can be exhausted this earth is available in plenty. So three officials of the Bureau of Forestry were appointed to search everywhere for vitriol waters and places where they had formerly been, so that profit could be obtained from both earth and water.

This place, the same as that spoken of by Shen Kua, is in northern Chiangsi quite near the Fukien border, a fact which doubtless accounts for the remark of the Jesuit Louis Lecomte in the last years of the +17th century that 'in the province of Fokien there is a spring whose water is green and changes iron into copper.'[b]

Under the Yuan dynasty in the +14th century the growing use of paper money led to the decline of the method and later sources indicate that it fell out of use, being known only from literary mentions. Yet it never died out, as we may infer from the reference in the *Thien Kung Khai Wu*[5] (Exploitation of the Works of Nature) in +1637, where we read:

> If iron objects (lit. vessels) are heated and then thrown into (lit. quenched in) copper sulphate solutions, the iron acquires the colour of copper.[c]

(5) THE ROLE OF BACTERIAL ENZYME ACTIONS

It is evident from the recipes given in the *San-shih-liu Shui Fa* that the intervention of bacteria cannot be overlooked, and was often positively encouraged. Even when organic matter was absent they may have been at work; thus in no. 5*b* denitrifying bacteria from the earth may well have reduced the nitrate, forming ammonia and giving the blue copper ammonium carbonate. The reduction of sulphates to sulphides and polysulphides has already been mentioned.[d]

Bacteria probably played a greater role however, in the putrefaction of the organic matter which had been added (often in considerable quantity) to the mixture. We note that of plant substances, expressed juice or extract occurs four times,[e] from root, wood

---

[a] Ch. 3, p. 39*b* (ch. 12, p. 3*b*), tr. auct.   [b] (1), p. 111.
[c] Ch. 11, p. 5*a*. See on this the commentary of Yoshida Mitsukuni (2). On other late encyclopaedias see de Mély (1), pp. xxiii, xxix, 114, 116, 145.
[d] See p. 176 above.   [e] Recipes (16), (28*b*), (30) and (32*b*).

[1] 清波雜志   [2] 周煇   [3] 膽水   [4] 膽土   [5] 天工開物

or fruit; while expressed oil comes once,[a] and plant sap twice.[b] Blood was a favourite ingredient on the animal side (eight mentions),[c] but we find beetle larvae once[d] and dung once.[e] It is obvious that highly colloidal solutions of partially degraded proteins would have been produced in these examples, and if the insoluble minerals such as quartz or jade[f] had been added very finely ground to an impalpable powder, then having regard to the charge on the particles, milky suspensions would probably have been produced. The alchemists could never have distinguished these from true solutions, and even today we should have recourse to the centrifuge to clear them.

One or two other features of the organic additions might be mentioned.[g] Some of the saps (as in recipe no. 25) may have been rich in tannin, which would affect the capacity to form permanent suspensions. Again, in recipe no. 30, which deals with the solution of metallic silver, some suspicion is aroused by the presence of the fruits of the *mu ching*[1] shrub. Whether or not our identification of this as *Vitex negundo*[h] is right, there is no doubt that many fruits and grasses contain large amounts of cyanogenetic glucosides. On autolysis or putrefaction it is not at all impossible that enough cyanide might be freed to effect the solution of the noble metals.[i] It may thus be significant that this recipe omits nitrate, though we have naturally assumed that this was not its intention.

(6) GEODES AND FERTILITY POTIONS

Much interest attaches to the use of conglomerate nodules or geodes (recipes no. 25 and no. 27). In the first of these, the *chiu tzu shih*,[2] 'the stone with nine little ones', appears to be a variety of nodule or geode with loose centres found in conglomerate rocks.[j] It seems to be related to the *aetites* or 'eagle-stone', an object of interest to the old European naturalists.[k] Geodes and aetites were discussed at some length by Pliny before +77 who emphasised the belief in the value of the latter as a childbirth talisman.[l] Here we reproduce an illustration of a geode from the *Chêng Lei Pên Tshao*[3] (Reorganised Pharmacopoeia) of +1249; its earthy or gravelly contents can be seen escaping from the broken-open shell (Fig. 1514). The obvious association with fertility no doubt led to the attempts of our alchemists to get its virtue into solution, almost as if it was an active principle or a biologically effective *Wirkstoff*.

The raw material of recipe no. 27 in the *San-shih-liu Shui Fa* is at first sight a difficult item. *Shih nao*[4] ('stone brain') is identified by Read & Pak (1) as paraffin,[m]

---

[a] Recipe (18). [b] Recipes (25) and (26).
[c] Recipes (6*a*), (11), (13), (21), (32*a, b, c*), (38). [d] Recipe (32*a*).
[e] Recipe (22). [f] Recipes (11), (13), (18), (21), (22), (25), (26), (28*b*), (32*a, b, c*).
[g] Recipe (34) is unusual—simply the extraction of an organic material.
[h] R148.
[i] Such a possibility has already arisen (pt. 3, pp. 88, 98–9 above) as a conceivable explanation of one of Ko Hung's potable gold elixirs. [j] According to Chang Hung-Chao (1), p. 270.
[k] See Bromehead (2); and Vol. 3, p. 652 above.
[l] *Hist. Nat.* XXVI, 140, 149 ff. Cf. Bailey (1), vol. 2, pp. 123, 127, 257, 262 ff.; Bidez & Cumont (1), vol. 2, pp. 201, 346. [m] RP67.

[1] 牡荊   [2] 九子石   [3] 證類本草   [4] 石腦

but this can only be a loose modern usage, and has no authority from the *Pên Tshao Kang Mu* itself (The Great Pharmacopoeia) of +1596, which they were abstracting. *Shih nao yu*[1] does, it is true, mean naturally occurring petroleum and its light fractions such as paraffin and naphtha[a] but with this we are not here concerned. Again, Read & Pak give *shih nao* as a synonym of *wu hsüeh yü shih*,[2] flaky arsenolite,[b] but if they had read Li Shih-Chen attentively, they would have seen that this is expressly denied in his text four times.[c] We thus isolate the proper meaning, which refers to haematitic

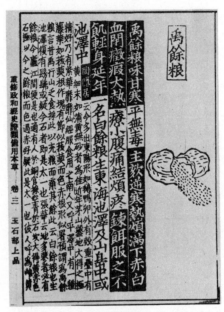

Fig. 1514. A geode (*yü yü liang*) from ferrugineous clay; a page of the *Chêng Lei Pên Tshao* (Re-organised Pharmacopoeia) of +1249, ch. 3 (p. 91).

nodules of hydrated ferrugineous geodic clay (ironstone), and so makes the term more or less synonymous with *thai i yü liang*[3] and *yü ai*[4].[d] De Mély, though giving an erroneous character for *ai*, rightly makes the connection with the rattling geodic nodules called *aetites*.[e] Now Li Shih-Chen significantly recounts the traditions concerning these stones. He says[f] that they belong to the class of stalactites, by which he means that they have to some extent an organic form, being rounded 'brain-like' *fossilia*, not amorphous mineral. He quotes Thao Hung-Ching[5] as saying that they are not to be found in the ordinary formularies but only in the 'manuals of the immortals'.[g] The general view was that they were suitable for the preparation of elixirs of longevity and immortality; Su Ching[6] is cited with reference to an adept of the Sui period (late

[a] *PTKM*, ch. 9, pp. 62*a* ff.   [b] RP90.
[c] Ch. 9, pp. 61*a*, *b*; ch. 10, pp. 26*a*, *b*.
[d] *PTKM*, ch. 10, p. 12*a*.   [e] (1), pp. 111, 225.
[f] Ch. 9, p. 61*a*.   [g] *Loc. cit.*

[1] 石腦油   [2] 握雪礜石   [3] 太一餘糧   [4] 禹哀   [5] 陶弘景
[6] 蘇敬

+6th century) Hua Kung,[1] who succeeded in this method, and a number of earlier examples are mentioned. These matters are also set forth in the *Chêng Lei Pên Tshao* of +1249,[a] and in many other works of the same character. We thus have to do again with something like *aetites*, and most probably with the preoccupation of getting a fertility-promoting and longevity-promoting virtue into solution.[b]

(7) STABILISED LACQUER LATEX AND PERPETUAL YOUTH

We now come to the curious case of lacquer. Recipe no. 33 in the *San-shih-liu Shui Fa* directs that to make lacquer solution (*chhi shui*[2]) which will still be fluid 50 days later, 18 large crabs kept overnight (*ta su hsieh*[3])[c] are to be put into each pint of the lacquer.[d] Here it is not a question of bringing something very hard and insoluble into solution but rather of preventing something liquid and creamy from rigidifying and concreting as it normally would do. Sinologists meeting in Chinese literature with statements about the harmful effects of crabs on lacquer may have thought them an old wives' tale relating to the lacquered bowls or dishes in which the shell-fish were placed, but a study of the subject shows that the references are really all to the effects on the unpolymerised latex. Thus the story is more interesting chemically and more significant philosophically.

The effect of crab tissues on lacquer is a *locus communis* in ancient Chinese literature. Probably the oldest references are in the *Huai Nan Tzu*[4] book (*c.* −120) which says that crabs spoil lacquer, so that it will not dry and cannot be used.[e] Then Chang Hua[5] in his *Po Wu Chih*[6] (Record of the Investigation of Things), written about +290, says that crabs unite with lacquer forming a 'medicine of the Holy Immortals' which can be taken orally.[f] Shortly afterwards Ko Hung refers to the effect more than once. The *Pao Phu Tzu* speaks of crabs and lacquer in the following words:[g]

Pure lacquer (-tree latex) prevented from becoming sticky (i.e. setting), if eaten, enables a man to associate with the spirits and attain longevity or immortality. The method of making it edible is to take ten large specimens of 'the Gutless Lordling' (*wu chhang kung*[7]), otherwise known as crabs,[h] and throw them into lacquer, or else use an aqueous suspension of mica or of jade. When this potion is ingested the nine parasites will quit the body and bad blood will leave by way of the nose.[i]

[a] Ch. 4, (pp. 115, 116).

[b] There might be a certain parallel here with the iatro-chemical belief in a 'natural balsamum' capable of preserving from decay and so prolonging life (cf. Mazzeo, 1). We refer to such ideas in other contexts elsewhere (Vol. 5, pt. 2, pp. 74 ff., 294 ff.), *mumia* as a drug, and material incorruptibility.

[c] Probably fresh-water *Eriocheir sinensis* (R214), the commonest species. The term *su* means 'kept overnight', as one can see from the *Hsieh Phu*,[8] ch. 2, p. 5a, so the crabs were not to be too fresh. No doubt a certain amount of autolysis enhanced the action of their tissues.

[d] Similar procedures are found in various other texts, for instance *TT*945, ch. 2, p. 8b. Sometimes the crabs are omitted; *TT*875, ch. 2, p. 6a uses only cinnabar and vinegar—or so it says.

[e] Ch. 6, p. 4a, ch. 16, p. 14b.   [f] Ch. 4, p. 6a.   [g] *PPT/NP* ch. 11, p. 10b, 11a, tr. auct.

[h] This is a very old expression, first occurring perhaps in the −2nd-century *Han Shih Wai Chuan* (cit. *SF*, ch. 80, p. 4b). We have also noticed it in the *Pei Hu Lu* (cit. *LS*, ch. 13, p. 7b, in vol. 2, p. 878).

[i] The translation of Feifel (3), p. 20 is rather unsatisfactory here, while that of Ware (5), p. 190, misses the point that the lacquer must not be allowed to set.

[1] 化公   [2] 漆水   [3] 大宿蟹   [4] 淮南子   [5] 張華   [6] 博物志
[7] 無腸公   [8] 蟹譜

Another chapter has a passing mention of the effect of crabs on lacquer.[a]

When one comes to the Sung there are many references. Fu Kung's[1] *Hsieh Phu*[2] (Discourse on Crustacea) quotes Thao Hung-Ching (+5th century) as saying:[b]

> According to the recipes of the immortals, crabs thrown into lacquer form an aqueous solution, which brings longevity when consumed.

The great naturalist-monk Tsan-Ning[3] refers twice to the phenomenon. In his *Wu Lei Hsiang Kan Chih*[4] (On the Mutual Responses of Things according to their Categories),[c] *c.* +980, he says that it is the 'fat' (*kao*[5]) of the crabs which is responsible; and Su Sung[6] in his *Pên Tshao Thu Ching*[7] (+1070) refers[d] the effect to the 'yellow' (*huang*[8]), which might be roe, but more probably means hepatopancreas, as in the words of Tsan-Ning. The latter's other reference[e] mentions the mixing of the crab material with 'damp', i.e. not set, lacquer, with the result that it remains liquid. That the *San-shih-liu Shui Fa* refers to crabs as such may be an indication of early date. Yet another Sung reference, from the +12th century, in Li Shih's[9] *Hsü Po Wu Chih*[10] (Continuation of the Record of the Investigation of Things)[f] simply says that 'after coming in contact with crabs, lacquer will not concrete.'

Since lacquer produces a well-known allergic reaction involving swelling and inflammation of the skin, etc., it is not surprising that crab tissues were appealed to as a method of therapy. In the *Pên Tshao Kang Mu* Li Shih-Chen gives crab extract or brei together with several plant drugs as an antidote.[g] Elsewhere he quotes a story from a Sung source, the *I Chien Chih*[11] of Hung Mai[12] (+1123 to +1202) to the effect that a thief having been blinded with lacquer was healed by the application of crab brei.[h]

All this is quite comprehensible if the basic facts about lacquer are recalled. After being tapped from the tree, *Rhus vernicifera*, the creamy grey latex, left to itself, gradually separates into four or five layers of different properties, but if held in complete darkness in airtight conditions can be conserved almost without change for several years. On exposure to light, warmth and a relatively damp atmosphere, however, the latex turns first to a chocolate brown colour and eventually sets to a hard brown extremely resistant substance.[i] This material, previously impregnated with various

---

[a] *Pao Phu Tzu* (ch. 3, p. 5*b*) says: 'I can guarantee that it is possible to sublime mercury and to search out (the art) of making gold and silver... As for the transformation of lacquer by crabs and the spoiling of wine by hemp (-seed oil)...'. Feifel writes (1), p. 197: 'I warrant that one can make people fly (with the help of) the *liu-chu*[13] and that one can make gold and silver... If, however, the crab heals the lacquer-sickness, if hemp sets distilled spirits in fermentation...' So this begins by missing the reference to sublimation and ends by a biochemical impossibility. Ware (5), p. 61, did see that it meant stopping the setting of the lacquer.

[b] P. 8*b*.   [c] P. 1.   [d] Quoted in *CLPT*, ch. 21, (p. 426.2).
[e] In *Ko Wu Tshu Than*,[14] ch. 1, p. 13.   [f] Ch. 9, p. 5*a*.
[g] Ch. 35, p. 20*a*.   [h] *PTKM*, ch. 45, p. 24*a*.

[i] Set lacquer is almost untouched by strong acids and alkalies, insoluble in all the usual solvents, extremely resistant to bacterial attack, heat-stable up to 400–500°, as an electrical insulator only ten times less effective than mica—in fact, as a vegetable product altogether extraordinary. It makes an excellent surface for laboratory benches in China.

[1] 傅肱   [2] 蟹譜   [3] 贊寧   [4] 物類相感志   [5] 膏
[6] 蘇頌   [7] 本草圖經   [8] 黃   [9] 李石   [10] 續博物志
[11] 夷堅志   [12] 洪邁   [13] 流珠   [14] 格物麤談

colours, black, red, gold or silver, or afterwards carved and treated in various ways, has been the basis of China's lacquer industry for more than two thousand years.[a]

Lacquer may be said to have been the most ancient industrial plastic known to man. The chemistry of the process is of considerable interest.[b] As much as 75 per cent of the latex consists of one or another catechol derivative (urushiol, laccol, moreacol),[c] having two hydroxyl groups on a single ring and one long side-chain (C_{15} or C_{17}) containing at least one or two double bonds. The oxidising and polymerising agent is the enzyme laccase, and the process needs oxygen as well as manganese as co-enzyme. The discovery of laccase by Gabriel Bertrand (1) in 1894 was one of the great milestones in the history of enzyme chemistry.[d] But the process becomes even more interesting when we set it in the broad framework of its biological significance. First the laccols are closely related to the active principles of the poison-oak and poison-ivy (e.g. lobinol), which also have a deleterious action on man. Secondly laccase as a catechol-oxidase is closely related to the polyphenol-oxidases which play such a prominent part, not only in the darkening of plant tissues, but also in the protein-tanning and melanin-blackening of the exo-skeleton or cuticle throughout the whole world of insects.[e] Indeed polyphenols and their oxidases having similar functions have a still wider distribution among invertebrates. Besides, there is the close parallel among the higher animals of the formation of melanin, the primary black and brown pigment, by the action of tyrosinase on tyrosine. Thus the lacquer process is only one special case, though of outstanding industrial importance, of a general pattern almost as widespread as plant and animal life itself.

What part, then, were the crab tissues playing? There can be no doubt that the ancient Chinese, before the -2nd century, had accidentally discovered a powerful laccase inhibitor. By preventing the action of the enzyme the darkening and polymerisation were also prevented. So great an interference with the course of nature, analogous to the arrest of a spontaneously occurring rigidification and ageing process, must have seemed highly significant to the alchemists, preoccupied as they were by the preservation of supple youth and the postponement or elimination of ankylosis and death. Moreover this action of crustacean tissues is not unique, for other researches have shown that they contain a powerful though somewhat enigmatic inhibitor for D-amino-acid oxidase.[f] Even some of the alleged therapeutic effects of crab tissues might now make sense, e.g. the case of the thief if full setting had not occurred; though the action on the dermatitis was presumably imaginary since the poisoning is due to the

---

[a] Excellently preserved and beautifully patterned lacquer bowls, boxes and other objects, not only from the Han but from the -4th-century Warring States period, are preserved in contemporary Chinese museums, e.g. at Chêngchow. [b] See the reviews of Brooks (1) and (2).

[c] The compounds are named in accordance with the species of tree which yields them, here *Rhus vernicifera* from China and Japan, *Rhus succedanea* from Annam, and *Melanorrhoea laccifera* from Cambodia.

[d] The most recent and complete study of it is due to Keilin & Mann (1, 2), who have established that it is a copper-containing protein accompanied by a blue pigment which may be its prosthetic group.

[e] See the reviews of Wigglesworth (1), H. S. Mason (1) and Dennell (1). The original discovery of the tanning was due to Pryor (1), and it was Bhagvat & Richter (1) who established the extreme richness of insects as well as plants in polyphenol oxidases.

[f] Sarlet, Faidherbe & Franck (1).

urushiol itself and not to the laccase or the polymer. As regards the belief that powdered mica or jade would also prevent the coagulation of lacquer latex, one wonders whether perhaps crab tissue was not already added. Of course if the mineral was in the form of an impalpable powder the colloidal solution of latex would be affected by the charge on the particles, and it is just possible that in such conditions the laccase enzyme protein might be prevented from gaining access to its substrate.

(h) THE THEORETICAL BACKGROUND OF ELIXIR ALCHEMY

(1) INTRODUCTION

Our focus now shifts from the Chinese alchemists' identifiable chemical and proto-chemical accomplishments to the assumptions and concepts with which they themselves sought to explain their methods and aims. This shift in point of view is perhaps more radical than might at first appear. If we wish to understand the inner coherence of alchemical theories we must, for the moment, set aside the yardstick of modern chemistry (although it will still be essential as an exploratory tool) and try to reconstruct the alchemist's abiding goals, his own standards of success and failure, as clues to how his concepts determined both what he did in his elaboratory and how he rationalised unforeseen results.

By 'theory' we mean simply the attempt to explain alchemical phenomena systematically using abstract and non-anthropomorphic concepts. In practice this means that we shall examine the application of the most fundamental and general notions of Chinese natural philosophy—the Five Elements, the Yin and Yang, the *chhi*, the trigram and hexagram systems of the 'Book of Changes', and so on—to the experience of the laboratory. We shall study how these notions were adapted to alchemical concerns either by extending their definitions, or by creating new concepts or new connections to integrate them.

It is necessary to stress that the field of alchemical theory is defined here by what alchemists did, thought, and knew about. Theoretical conceptions never exist in a vacuum; their implications and significance depend upon the matrices in which they are embedded. To pluck the 'advanced' elements out of the matrix and discard the 'retrograde' aspects is a procedure bound to lead to fundamental distortions, for the two regularly turn out to be integral and inseparable, one element defining the range of possibilities of the other. Demarcating our field of investigation so as to include any ancient Chinese activity which might fall into the area of modern chemistry would allow the casting of the net wider, but at the cost of putting many of the alchemical adept's own concerns out of bounds. Not only would we confound ideas that originally had nothing to do with each other, but we would have to reject so many central aspects of alchemy that there would be no possibility of comprehending what held it together, and no hope of ultimately making more than superficial comparisons with the traditions of other cultures.

In order to understand what the ancient Taoist adepts had in mind as they worked in their laboratories, we must examine seriously such topics as the belief in the growth of minerals within the earth, the command of time, and the role of number in establishing correspondences between the apparatus and the greater cosmos (never entirely distinct from the more familiar function of number in recording the invariant weights of reactants and products). Nor can we ignore the associated Taoist rituals, offerings, and incantations which were used in connection with every phase of the process.[a] The alchemist was applying chemical and physical procedures to the quintessentially religious end of transcending his mortality. The new observations and discoveries which today interest students of the history of science were also valued by the alchemists themselves, but not usually as the main objectives for which they were striving.

The Taoist's end in view was, one might say, perfect freedom in perfect fusion with the cosmic order. For the early Taoist philosophers this seems to have been mainly a state of heart and mind, but as we have seen, alchemists and other adherents of Taoist religion thought of perfect freedom as limited to a special state of being, that of the immortal *hsien*.[1] Immortality could be attained by a variety of means, two of which in particular mark the alchemist's Way. First there was the construction of chemical models of the cosmic process. These were apparently meant to serve as objects of ecstatic contemplation, leading to a gnosis which brought one closer to union with the Tao. Second was the production of elixirs of supramundane virtues, the action of which—upon the adept himself, upon others, or upon base metals—gave him not only personal immortality at his pleasure, but also transferable wealth and a more-than-human power to cure disease and make others immortal. The first path led the alchemist in the direction of physics, the second toward medical therapeutics, metallurgy and other technical arts.

(i) *Areas of uncertainty*

It is still too early to attempt a truly historical study of the theoretical side of Chinese alchemy, in which one could see how concepts and their relations developed and changed both through mutual influence and the pressure of wider intellectual and social currents. First, too few of the documents which have survived the attrition of successive Chinese cataclysms can yet be dated precisely with confidence, and this leaves even their logical connections obscure. Secondly, with a large part of the clearly dated literature, one cannot be sure that its vague and obscure language is in fact concerned with laboratory operations rather than with the physiological and sexual disciplines which used alchemical language.[b] We know already that most of the alchemical treatises which have been translated into Western languages actually come out of the 'dual-cultivation' régime of the Southern School of Taoism in the Sung

[a] At the same time we regret the impossibility of doing justice to the subject of alchemical ritual here. One of us (N.S.) has collected material bearing on this topic, and plans a special study. See also above, Vol. 5, pt. 2, pp. 128ff.; including our account of the *Shang-Chhing Chiu Chen Chung Ching Nei Chüeh* (*TT*901).

[b] See Vol. 5, pt. 5.

[1] 仙

and Yuan periods.[a] These practices, a blend of Internal Alchemy and sexual disciplines (*nei tan*[1]), were not in principle irreconcilable with the art of the External Elixir (*wai tan*[2]), but most devotees resembled the 'spiritual alchemists' of the European Renaissance in their explicit disdain for the actual work of the furnace.

To reduce these two fundamental areas of uncertainty will require a good deal of critical work on individual writings. In relation to the second problem, the most fruitful clues are likely to come from the study of just those sources which have the least to do with laboratory alchemy, and thus are least likely to attract students of ancient science. But the small body of sources the meanings and times of which are known does not yet provide a basis for understanding the changing character of alchemy and of its links with the other arts of Taoism. Here we can only examine the widest possible variety of evidence in order to sketch out the ideas and notions which were most general in alchemy rather than those which can be identified definitely with given periods and movements.

There is, in fact, much information in writings on 'alchemical' breath control, meditation, and sexual techniques which can be used to throw light on the intellectual background of *wai tan* alchemy, for most early adepts combined all these practices, considered them complementary, and explained them with the same concepts. However, in order to keep from losing sight of what is actually information about the Outer Elixir, it is necessary to 'presume guilt'. We consider no text chemically alchemical (i.e. *wai tan*[2]) unless it either prescribes operations so clearly that they could conceivably be carried out in the laboratory, or, if the emphasis is on theory, unless it clearly reflects knowledge of the details of laboratory procedure or the interactions of real chemical substances.

(ii) *Alchemical ideas and Taoist revelations*

Before we proceed to scrutinise the alchemists' theories, one other major limitation of our present understanding must be made explicit. One can seldom hope to reconstruct the competition of different ideas for survival and further elaboration simply on the basis of their abstract merits, without attention to their social consequences; ideas which affect the rate of social change, whether in a tiny sect or a great civilisation, are often selected or rejected for very extrinsic reasons. It is thus necessary to ask whether

---

[a] The term 'dual cultivation' was coined by Liu Tshun-Jen (1) to refer to *shuang-hsiu*,[3] 'a tendency to integrate the eugenic *fang-chung*[4] studies with the physico-mental cultivation of the Golden Pill [i.e. *chin tan*[5]], in fashion since the + 10th century', i.e. to bring together sexual practices and other *nei tan* physiological techniques. We believe—and seek to demonstrate in what follows and in pt. 5—that sexual practices were part of *nei tan* from the beginning, but also that even in the Sung many of Liu's sources still reflect first-hand knowledge of laboratory processes. For these reasons we cannot accept all his arguments, but here we retain 'dual cultivation' to designate the late movement to which Liu originally applied it; though we re-define the term to refer to an amalgamation of chemical and psycho-physiological practices in which the latter generally predominated. In referring to the history of alchemy as a whole rather than to this late movement, we use the terms 'laboratory alchemy', 'proto-chemical alchemy' or 'external alchemy' (*wai tan*[2]) on the one hand; and 'physiological alchemy' or 'internal alchemy' (*nei tan*[6]) on the other, as synonyms. For a list of treatises which have been translated, see Sivin (1), pp. 322–4.

¹ 內 丹   ² 外 丹   ³ 雙 修   ⁴ 房 中   ⁵ 金 丹   ⁶ 內 丹

alchemy was but an appendage of Taoism, neglected by all but a few specialist practitioners and non-practising patrons; or on the other hand part of a central revelation which defined the character of Taoist religion. It is clear that for early Chinese alchemy the latter is the case. Alchemy was an actual part of the founding revelation of the Mao Shan school, the group responsible for completing and putting into practice the first great intellectual synthesis of Taoism.[a] It was bound, therefore, to be affected by the application of that revelation to a particular social and historical milieu.

The chain of events which led to the establishment of Mao Shan, or Mt. Mao, as the first major permanent centre of Taoist practice began in +349 or slightly earlier with visitations by immortals to a young man named Yang Hsi[1] (traditional dates: +330 to +387) at the Eastern Chin prefectural capital, Chü-jung,[2] not far from modern Nanking. Between +364 and +370, in a series of visions, there appeared to Yang a veritable pantheon of celestial functionaries, including the Lady Wei of the Southern Peak (Nan Yo Fu-jen,[3] Wei Hua-Tshun[4]) and the brothers Mao Ying,[5] Mao Ku,[6] and Mao Chung,[7] whose names were given to the three peaks of the nearby Mt. Chü-chhü[8].[b] In the course of these interviews, aided almost certainly by cannabis,[c] Yang took down in writing a number of sacred texts which the immortals assured him were current in their own supernal realm, as well as oral elucidations and answers to Yang's queries about various aspects of the unseen world. He treasured and disseminated these scriptures as the basis of a new Taoist faith more elevated than the 'vulgar' sects of his time. He was sponsored and joined in his revelations by Hsü Mi[9] (+303 to +373), an official of the court, and his son Hsü Hui[10] (+341 to c. 370). The family connections of the Hsüs were estimable in more than the conventional sense, for Hsü Mi's uncle married the elder sister of Ko Hung,[11] the great exponent of personal access to the realm of the immortals; and they were also related to the family of Thao Hung-Ching[12] (+456 to +536), the most eminent Taoist magus of his time.[d] We have

---

[a] See Vol. 2, pp. 154 ff. and Vol. 5, pt. 2, pp. 128 ff., pt. 3, pp. 39, 41, 77, 121 above.

[b] The three Mao brothers were supposedly alchemists of the −1st century, but Wei Hua-Tshun was a contemporary. She seems to have been one of the founders of Taoist liturgiology, and a great teacher of meditation aided by psychotropic drugs. In what follows we have been greatly aided by access to an unpublished study by Strickmann (3). On dates see *Chen Kao*[13] (Declarations of Perfected (or Realised) Immortals, c. +500), ch. 20; *Chen Hsi*[14] (The Legitimate Succession of Perfected (or Realised) Immortals, +805) of Li Po,[15] in *YCCC*, ch. 5, p. 2a, and the narrative in Chhen Kuo-Fu (1), pp. 32–4. Michel Strickmann (priv. comm.) is inclined to think that Liu Phu,[16] the son of Wei Hua-Tshun, who transmitted the 'Five Amulets' (*Wu Fu*[17]) to Yang in +349 or +350 was also a real person; if so, there is nothing in the primary sources about truly visionary experiences before +364.

[c] See Vol. 5, pt. 2, pp. 150 ff. To the evidence given there about cannabis one could add a fine +6th-century example from a *Wu Tsang Ching* (Manual of the Five Viscera), attributed to Chang Chung-Ching but certainly not by him: 'If you wish to command demonic apparitions to present themselves you should constantly eat the inflorescences of the hemp plant.' Cf. Miyashita Saburō (3).

[d] See the biography by Ishii Masako (4). Strickmann has noted in the course of correspondence that in eight generations of the Hsüs, four alliances with the Kos are recorded (*Chen Kao*, ch. 20), and that Hsü Mi's own principal wife had been the daughter of Thao Hung-ching's ancestor in the seventh generation (inclusive). In much of what follows we are indebted to Dr Strickmann.

[1] 楊羲	[2] 句容	[3] 南嶽夫人	[4] 魏華存	[5] 茅盈
[6] 茅固	[7] 茅衷	[8] 句曲	[9] 許謐	[10] 許翽
[11] 葛洪	[12] 陶弘景	[13] 眞誥	[14] 眞系	[15] 李渤
[16] 劉璞	[17] 五符			

already encountered Hsü Mi's alchemist brother Mai[1].[a] In +367 Hsü Mi was informed by Mao Ying that in nine years he would be transferred from the terrestrial bureaucracy to that of the Superior Purity Heaven (*Shang-Chhing Thien*[2]). That this heaven might be available for such heady assignments had been revealed to no Taoist save Yang Hsi and his patrons.[b] Hsü apparently remained active in his post at the capital, despite repeated celestial admonitions, but his son Hsü Hui, having returned his wife to her parents, moved into the retreat his father had built at Mt. Mao, and there until his premature death he devotedly practised the operations revealed to Yang for his benefit by the immortals.[c] Yang and the Hsüs had vindicated Ko Hung's belief in the unseen world—not supernatural in Chinese terms, but concerned only with eternal things and thus more desirable than mundane society—which he had urged with such amplitude in his *Pao Phu Tzu* (*Nei Phien*).

Four generations later, when Thao Hung-Ching retired from the Chhi court in +492 to Mt. Mao, he built the Hua-yang Kuan[3] (Effulgent Yang Abbey) and proceeded to seek out the revelations and revive the spiritual experiences of Yang and the Hsüs as the basis of a religious community. The background of the Hua-Yang Abbey could hardly be better described than in the words of Michel Strickmann:[d]

What was to become the Mao Shan tradition began as the highly individual practices of three men, of whom one was a visionary and another held a full-time job. They were building upon a common base provided by the Way of the Heavenly Master (Thien Shih Tao[4]), a Taoist group specialising in the cure of disease through formalised communication with the celestial hierarchy.[e] Like most reputed founders, Yang and the Hsüs founded no order; and though between their own time and that of their eventual editor portions of their brilliant synthesis spread somewhat (first only among friends and relations), no independent organisation arose to perpetuate their names or realise the teachings of their celestial masters.[f] Thao also had the example of earlier 'abbey' (*kuan*[5]) communities, whose functions were perhaps more intimately related to their patronage than to their particular doctrines. Individual financial support involved their Taoist members with ceremonies for the well-being of their patron's family, both living and dead, and probably with the guardianship of some of his infant sons.

Thao had the wit to apprehend that analogous services, on a correspondingly grander scale, could elicit the patronage of the Liang emperor himself, thus providing the highest possible auspices for a revival of Taoism (for by Thao's time the Heavenly Master cult had fallen apart in South China). Once Thao had seen to the elaborate details of collecting, codifying, annotating, and publishing the Annunciations of the Immortals, and had thought through the problem of administrative organisation, the community was soon assembled, and

---

[a] See pt. 3, p. 76.

[b] Yang was always the intermediary, for only he was granted waking visions; anyone could dream of the immortals, of course, so dreams were given much less significance. The sequence of annunciations concerning Hsü's appointment is recorded in *Chen Kao*, chs. 1–4.

[c] See above, pt. 3, p. 121 and, for Hsü Mi's alchemical activities, pt. 2, p. 110.

[d] Priv. comm., 4 August 1970, to one of us (N.S.), edited with permission to take into account a later discussion on this point with M. Strickmann and K. M. Schipper.

[e] See above, Vol. 2, pp. 155–7. We are even less inclined now than when that was written to speak of Taoist 'monasticism'.

[f] For documentation see *Chen Kao*, chs. 19–20.

1 邁   2 上清天   3 華陽館   4 天師道   5 館

ceremonial was adopted and elaborated. Ceremonial, despite the ideological emphasis on revelation and visionary experience, must always have been the chief preoccupation of the majority at Hua-Yang Abbey. These Taoists busied themselves with ceremonies in support of the health of both Ruler and State, with the discovery of auspices, and not least with the concoction of a timely elixir. The sound fiscal basis of the enterprise enabled it to pass un-scathed through the disestablishment of Taoist organisations in +504 (this very year in fact marks the inception of Thao's alchemical operations), and in time to take hold upon the intellects (and purse-strings) of the Thang.

Thao apparently first learned of the Mao Shan writings through a few fragments in the possession of his teacher, Sun Yu-Yüeh.[1] Sun had in turn been the disciple of Lu Hsiu-Ching,[2] who had journeyed through the haunts of Taoism to be initiated into, collect, and catalogue (by +471) the major scriptures of the rival Ling-Pao[3] tradition,[a] picking up along the way some documents which emanated from Yang Hsi.[b] Since Lu was neither particularly concerned nor overly fastidious about the authenticity of the latter, most were probably poor copies or forgeries; many fakes had already been produced within the select circles which knew of the Mao Shan revelations.[c] In Thao's subsequent search, first among relatives of the Hsüs and then on a long voyage to the southeast, his acknowledged model was Ku Huan[4] (d. +485), a contemporary of Lu's. Ku had devoted much energy to seeking out (in a more limited way than Thao) the scriptural remains of Mt. Mao, and first applied a knowledge of Yang Hsi's calligraphy and that of the Hsüs to what he recognised as the essential task of separating authentic from doubtful documents.[d]

Thao Hung-Ching eventually discovered, and proceeded to edit and annotate, a remarkably intimate day-to-day record of his predecessors, including letters which had passed between them and journals of visitations by one or another immortal, often for no more exalted purpose than to offer medical advice or to negotiate some minor celestial-bureaucratic detail. In this record Thao found much of alchemical interest, which is duly preserved in his *Chen Kao*[5] (Declarations of the Perfected (or Realised)

[a] Cf. Kaltenmark (4).

[b] In +471 Lu presented his 'Tripartite Catalogue of Scriptures' to Emperor Ming of the Liu Sung dynasty. According to a rather hostile later Buddhist source, Lu claimed that there existed a total of 1228 rolls, of which 1090 were circulating in the world and 138 were still in the Celestial Palace of the immortals. This number comprised 'prescriptions' (*yao fang*[6]), which in early Taoist circles meant chiefly instructions for preparing substances which conferred immortality. See *Fa Yuan Chu Lin*[7] (+668), ch. 69, p. 5*b*; and on the subject of 'medicines', Schipper (1), p. 13. The organisation of Lu's catalogue was based on the Three Vehicles of the Buddist Tripiṭaka, and its application to Taoism goes back to a division of the celestial scriptures revealed to Yang Hsi in +364; see *Shang-Chhing Thai-Shang Pa Su Chen Ching*[8] (Realisation Canon of the Eightfold Simplicity; a Shang-Chhing Thai-Shang Scripture), TT423, pp. 4*a*–5*b*. Citations in *Chen Kao*, ch. 19–20, carry the designation *san phin mu*[9] or *san chen*[10] *phin mu*. The tripartite division was still reflected in the last great version of the Patrology, the *Chêng-Thung Tao Tsang*[11] of +1444 or +1447 (cf. pt. 3, pp. 116–17). See Holmes Welch (3), pp. 129–131; Chhen Kuo-Fu (1), vol. 1, pp. 38–46 and 106–107; and, on Lu Hsiu-Ching himself, Ōbuchi Ninji (1), pp. 259–276.

[c] See *Chen Kao*, chs. 19–20. These forgeries and their detection will be discussed in a major study of the formation of the Mao Shan corpus now under way by Michel Strickmann.

[d] Cf. *Chen Kao*, ch. 19, p. 1*a*.

[1] 孫游嶽	[2] 陸修靜	[3] 靈寶	[4] 顧歡	[5] 眞誥
[6] 藥方	[7] 法苑珠林	[8] 上清太上八素眞經		[9] 三品目
[10] 三眞	[11] 正統道藏			

Immortals), or in the now fragmentary *Têng Chen Yin Chüeh*[1] (Confidential Instructions for the Ascent to Immortality).[a]

The three progenitors of the Mao Shan cult had shared with other sects of their time a belief in an imminent apocalypse which Thao calculated would fall in +507, to be followed in +512 by the descent of the Sage to gather up the elect, the only survivors.[b] Yang Hsi had been well supplied with graphic and elegantly phrased details of the catastrophes by Wei Hua-Tshun's colleague the Lady of the Circumpolar Zone (*Tzu Wei Fu-jen*[2]), and had been assured by her that among the singular methods and supreme arts which would be practised in those latter days was alchemy:

Some will cyclically transform in their furnaces the darksome semen (*yu ching*[3]) of cinnabar, or refine by the powder method the purple ichor of gold and jade. The Lang-kan elixir will flow and flower in thick billows; the Eight Gems (*pa chhiung*[4]) will soar in cloudlike radiance.[c] The Crimson Fluid will eddy and ripple as the Dragon Foetus (*lung thai*[5]) cries out from its secret place. Tiger-Spittle and Phoenix-Brain, Cloud Lang-kan and Jade Frost, Lunar Liquor of the Supreme Pole (*Thai Chi yüeh li*[6]) and Divine Steel of the Three Rings (*san huan ling kang*[7])— if a spatulaful of (one of these) is presented to them, their spiritual feathers will spread forth like pinions. Then will they (be able to) peruse the pattern figured on the Vault of Space, and glow forth in the Chamber of Primal Commencement....[d]

Among the scriptures taken down by Yang Hsi, Thao had also found actual instructions for alchemical preparations. Two of these formulae still exist in their entirety. One, called *Thai-Shang Pa-Ching Ssu-Jui Tzu Chiang (Wu-Chu) Chiang-Shêng Shen Tan Shang Ching*[8] (Exalted Manual of the Eight-Radiances Four-Stamens Purple-Fluid Crimson Incarnation Numinous Elixir, a Thai-Shang Scripture), is preserved in the *Shang-Chhing Thai-Shang Ti Chün Chiu Chen Chung Ching*[9] (Ninefold Realised Median Canon of the Imperial Lord, a Shang-Chhing Thai-Shang Scripture);[e] a work otherwise devoted to techniques for encountering various deities

[a] There is a new critical edition of the *Chen Kao* by Ishii Masako (*1*), who has also reported favourably on its general authenticity (*2, 3*). Although no substantial portion of this has been translated, Schipper (*1*) provides a complete rendering of the *Han Wu Ti Nei Chuan*[10] (Intimate Biography of Emperor Wu of the Han, *TT*289), which, as he has demonstrated, is a product of the Mao Shan ambiance. Two thin slices of historical and legendary material about the real Martial Emperor have merely been placed outside a filling of three typical revelations originally quite unconnected with him.

Both the *Chen Kao* (*TT*1004) and the *Têng Chen Yin Chüeh* (*TT*418) were probably completed about +499, according to Strickmann (*3*), but the former was intended to arouse the interest of the Emperor, and the latter (part of which is now preserved only in *TPYL*) was meant for cultic use.

[b] *Chen Kao*, ch. 13, pp. 8*b*, 9*a*. The basic Mao Shan doctrine on the coming of this messiah is found in *Shang-Chhing Hou Shêng Tao Chün Lieh Chi*[11] (*TT*439).

[c] On the Lang-Kan gem and elixir see below, pp. 217, 268, and elsewhere, pt. 2, p. 296.

[d] *Chen Kao*, ch. 6, p. 2*b*, tr. Strickmann (*2*), mod. auct. Any such translation must still be very provisional.

[e] *TT*1357, ch. 2, pp. 8*b*–18*a*, where it is explicitly continuous with the more uncompromisingly magical portion (see p. 17*a*). It is also reprinted separately, without the ascription, in *YCCC*, ch. 68, pp. 1*a*–9*b*, under the slightly different, and probably more correct, title: *Thai-Shang Pa-Ching Ssu-Jui Tzu-Chiang Wu-Chu*[12] *Chiang-Shêng*[13] *Shen Tan Fang* (Eight-Radiances Four-Stamens Purple-Fluid Five-Peal Incarnate Numinous Elixir, a Thai-Shang Scripture).

Thao could not be sure that this treatise was part of the original Yang-Hsü corpus, since his copy was not in Yang's handwriting, and thus might have been one of the many forgeries then circulating (see

[1] 登真隱訣	[2] 紫薇夫人	[3] 幽精	[4] 八瓊	[5] 龍胎
[6] 太極月醴	[7] 三環靈剛	[8] 太上八景四蘂紫漿絳生神丹上經		
[9] 上清太上帝君九真中經		[10] 漢武帝內傳	[11] 上清後聖道君列紀	
[12] 五珠	[13] 降生			

in meditation—making them appear from within one's body, from the sun and moon, and from inside unusually coloured clouds that conceal the immortals as they travel through the sky. The elixir recipe itself, for all its twenty-four ingredients and 104 days of heating, is clearly phrased in the language of the laboratory, and could be carried out in one today. The ingredients are given elaborate cover-names, but all are defined in notes recording oral instructions (*khou chüeh*[1]) ascribed to the first Patriarch of Taoism, Chang Tao-Ling (+2nd century): e.g. Crimson Tumulus Vermilion Boy (*chiang ling chu erh*[2] = cinnabar, HgS), Elixir Mountain Solar Animus (*tan shan jih hun*[3] = realgar, As_2S_2), Arcane Belvedere Lunar Radiance (*hsüan thai yüeh hua*[4] = orpiment, As_2S_3). The formula is not dissimilar on the whole to later alchemical recipes in terminology and technique.

The second is atypical in its adaptation of vegetable processes; it falls between conventional alchemy and the art of growing the marvellous *chih* plants (*ling chih*[5]), the most famous of which is the 'magic mushroom'.[a] This is the *Tung-Chen Ling Shu Tzu-Wên Lang-Kan Hua Tan Shang Ching* (Divinely Written Exalted Manual in Purple Script on the Lang-Kan (Gem) Radiant Elixir; a Tung-Chen Scripture), originally part of a *Tung-Chen Thai-Wei Ling Shu Tzu-Wên Shang Ching*[6] (Divinely Written Exalted Canon in Purple Script; a Tung-Chen Thai-Wei Scripture).[b] A fourteen-ingredient elixir is treated in a precisely phased fire for three protracted periods,[c] after which an elixir appears inside a 'bud' of seminal essence (*ching*[7]). Planted in an irrigated field, after three years the elixir seed develops into a tree with ring-shaped fruit, one of the names of which is Supreme-Pole Arcane Chih (*thai chi yin chih*[8]). The fruit when planted yields a new plant resembling the calabash, with a peach-like fruit called the Phoenix-Brain Chih (*fêng nao chih*[9]). When this intermediate is raised to higher degrees of perfection through two further replantings, the adept harvests a fruit resembling the jujube which, when eaten, brings about assumption into the heavens. We can appreciate that this extravagantly impractical recipe is an attempt to assimilate into alchemy legends like that of the *lang-kan*[10] gems which since the Chou and Han had been said to grow on trees in the paradise of Khun-lun,[11] where also were found the peaches of immortality.[d]

Chen Kao, ch. 10, p. 5*a*). We are somewhat less reluctant to accept it, because in a number of passages parallel to the text of *Tung-Chen Ling Shu Tzu-Wên Lang-Kan Hua Tan Shang Ching* (see below), which Thao verified as in Yang's calligraphy, the later version is almost certainly derivative. Cf., for instance, *YCCC*, ch. 68, pp. 4*a*–5*b*, with *TT*252, pp. 3*b*–5*a*.

[a] See pt. 2, pp. 121 ff.

[b] But now found separately under the title *Thai-Wei Ling Shu Tzu-Wên Lang-Kan Hua Tan Shen Chen Shang Ching*[12] (Divinely Written Exalted Spiritual Realisation Manual in Purple Script on the Lang-Kan Gem Radiant Elixir; a Thai-Wei Scripture), *TT*252. Since this text is no longer incorporated in a Mao Shan scripture the history of which can be traced, its authenticity is not beyond question. The old collection, *Tung-Chen Thai-Wei Ling Shu Tzu-Wên Shang Ching*, is not in the *Tao Tsang* now, and must be lost. [c] See below, pp. 266 ff.

[d] Schafer (13), p. 246. The Lang-kan Elixir and its transformations are described in the extant Purple-Script treatise *Huang-Thien Shang-Chhing Chin Chhüeh Ti Chün Ling Shu Tzu-Wên Shang Ching*,[13] *TT*634; see also *Chen Kao*, ch. 5, p. 3*b*.

[1] 口訣　　　[2] 絳陵朱兒　　　[3] 丹山日兒　　　[4] 玄臺月華　　　[5] 靈芝
[6] 洞眞太微靈書紫文上經　　　[7] 精　　　[8] 太極隱芝　　　[9] 鳳腦芝
[10] 琅玕　　　[11] 崑崙　　　[12] 太微靈書紫文琅玕華丹神眞上經
[13] 皇天上清金闕帝君靈書紫文上經

As we shall shortly see, Thao must also have had access to other writings on alchemy, including the *Huang Ti Chiu Ting Shen Tan Ching*[1] (The Yellow Emperor's Canon of the Nine-Vessel Spiritual Elixir),[a] which Ko Hung claimed had been made public by Tso Tzhu,[2] an early denizen of Mt. Mao at the end of the Han[b]. If this is indeed the book which has been passed down in the Taoist Patrologies with a large bulk of expository material added, it is probably the oldest extant Chinese work devoted to the operational side of alchemy, paralleling the more ambiguous *Tshan Thung Chhi*.[c]

Then came a day in +504 when dreams of favourable auspices for an elixir were granted simultaneously to Emperor Wu of the new Liang dynasty and to Thao, and the question of choosing one method from among many became pressing. We do not have to depend upon hagiographic writings for the outcome of Thao's deliberations, which led to his settling upon the Ninefold Cyclically Transformed Numinous Elixir (*chiu chuan shen tan*[3]), because a surviving fragment of the *Têng Chen Yin Chüeh* records his own words. He commences with a line of transmission from the Supreme-Pole Perfected (or Realised) Immortal (*Thai chi chen jen*[4])[d] through intermediaries to Mao Ying, who he says was taught the formula in −98, and passed it on to his brothers. It was the elder of these two, Mao Ku,[e] who revealed it to Yang Hsi, and bid him show it to the Hsüs. Thao found it among the literary remains of his predecessors. He goes on to remark:

Thus all those who studied the Tao in the Han and Chin periods talked about mixing and taking Potable Gold (*chin i*[5]), and ascending to become an immortal, but they did not mention the Nine-cycle (Elixir). Thus this formula of the Realised Immortals, from the time it was first taught here below, has never been carried out.[f]

---

[a] *Huang Ti Chiu Ting Shen Tan Ching Chüeh*,[6] TT878. Ch. 1 is evidently the original canon, for the other nineteen chapters of this version explain and amplify it, and thus would be the *chüeh* (explanations meant to be orally transmitted). Although no positive evidence has been adduced for this early date, the canon corresponds to quotations in *PPT/NP*, and thus may possibly be what it claims to be, an example of the early Thai-Chhing[7] tradition into which Ko Hung had been initiated. The added chapters cannot be assigned a single date of composition, for they cite very divergent opinions (e.g ch. 11, 5b and 10a) and lump together heterogeneous material; in fact some recipes are said explicitly not to be worth using (ch. 15, p. 4b; ch. 16, p. 11a). A date of compilation in the early Sung is indicated by the statement that 'horse-tooth alum (*ma chhih fan*[8]) comes these days from Mao-chou,[9] which is within the administrative control of I-chou[10]' or modern Chhêngtu[11] (ch. 16, p. 4b). This was true through the Sung, but the name I-chou was used only during the periods *c.* +620 to +627, +977 to +988, and +994 to +1001, according to the Szechuan gazetteer *Chhung-Hsiu Ssu-Chhuan Thung Chih*[12] (revision of +1730), ch. 2, pp. 6a, 28a and ch. 5, 41a and 45b. The late Thang and most of the Sung are also ruled out by the assertion 'Now a Sage reigns, the known world (*huan yü*[13]) is united, and the Nine Provinces (or the Empire) are free of trouble' (ch. 14, p. 2a).

Chhen Kuo-Fu, after an extended comparison of this work with the two Mao Shan scriptures just cited, finds a general affinity but no evidence of mutual borrowing, (*1*), vol. 2, pp. 378–383.

[b] *PPT/NP*, ch. 4, p. 2a; tr. Ware (5), pp. 69–70.　　　　　　[c] See pt. 3, pp. 50ff.

[d] The best known holder of this title was Tso Tzhu,[14] but he comes too late.

[e] Mao Ku, who had been given the post of Certifier of Immortality Registers (*ting lu*[15, 16]) in the celestial bureaucracy, was responsible for revealing the mystical biography of his elder brother, the 'Biography of Director of Destinies Mao' (*Mao Ssu-ming Chuan*[17]) to Yang Hsi. Strickmann (2) finds indications that this scripture was the source of Thao's formula (see *Chen Kao*, ch. 5, p. 4a).

[f] *TPYL*, ch. 671, pp. 1a, b, tr. auct.

[1] 黃帝九鼎神丹經	[2] 左慈	[3] 九轉神丹	[4] 太極眞人
[5] 金液	[6] 訣	[7] 太清	[8] 馬齒礬	[9] 茂州
[10] 益州	[11] 成都	[12] 重修四川通志	[13] 寰宇
[14] 左慈	[15] 定錄	[16] 籙	[17] 茅司命傳

Lines of transmission of this sort tend to weary sinologists, and historians of science all the more, but Strickmann (2) has had the perspicacity to see Thao's point, and to link it with the statement in a biographical account by Thao's disciple Phan Yuan-Wên[1] that this was the elixir Thao decided to make. For Thao's rationale *was* this genealogy. What swayed him was that the method had descended through a series of celestial divinities to Yang Hsi, Hsü Mi, and Hsü Hui, in the very hand of one of whom Thao's copy was written. No one else had ever known of it, and the Recluse of Hua-yang would be the first mortal to prepare it.

After some notes on the ritual for the formal transmission of the canon, Thao cites a few details which clearly signify that the Medicine was indeed chemical and not physiological or mental in nature:

One who wants to mix the Nine-cycle (Elixir) first makes a Spirit Pot (*shen fu*[2]), using a clay vessel from Jung-yang[3] (Honan), Chhang-sha[4] (Hunan), or Yü-chang[5] (Chiangsi)— what is called a 'tile pot'. In antiquity the Yellow Emperor heated the Nine-Cauldron Elixir (*chiu ting*[6]) at Mt. Ching,[7] and the *Thai-Chhing Chung Ching*[8] (Thai-Chhing Median Canon) also has a Nine-Cauldron Elixir method; thus from his time onwards elixir aludels have been called 'ritual cauldrons' (*ting*[9]).[a] One uses chaff for the fire to heat them. The building for the furnace (i.e. the laboratory, *tsao wu*[10]) is constructed in an inaccessible place next to a stream on one of the Great Mountains. It must be forty feet long and twenty feet wide, with three openings towards the south, east, and west. First observe the purification rites (*chai chieh*[11]) for a hundred days, and then plaster the vessel with lute to make the Spirit Pot.... Take equal parts of these six substances: left-oriented oyster-shell from Tung-hai (Chiangsu), kaolin from Wu commandery (Chiangsu), mica powder, earth turned up by earthworms, talc, and alum.[b]

This mixture is of course the famous six-one lute (*liu i ni*[12]), which is specified in almost every elixir formula, with minor variations in ingredients, for coating reaction vessels and sealing the junctions between vessels and covers.[c]

Thao had a space cleared for a laboratory on the other side of the ridge from Hua-yang Abbey, even diverting a stream through a hole bored in the rock to provide the eastward-flowing current needed by every alchemist.[d] But there we may leave him, for his repeated failures from +505 on, and even his rather dubiously documented success in *c.* +528, are irrelevant here.

There should be no need for further proof that the history of Chinese alchemical ideas will not fall into proper perspective until much more is known of the social connections of esoteric Taoism. Thao Hung-Ching merely stands at an obvious nodal point. His predecessors had adapted and combined many of the individual medita-

[a] It was after these nine ritual cauldrons that the 'Yellow Emperor's Canon of the Nine-Vessel Spiritual Elixir' (*Huang Ti Chiu Ting Shen Tan Ching*[13]) was named. This sentence is apparently an explanatory note by Thao Hung-Ching which became incorporated in the text.

[b] *TPYL*, ch. 671, pp. 1*b*, 2*a*, tr. auct.

[c] See pp. 19ff., 35–6, 112, 163 above; and Sivin (1), pp. 160–8.

[d] Cf. *PPT/NP*, ch. 4, p. 14*a*; tr. Ware (5), p. 90.

[1] 潘淵文	[2] 神釜	[3] 滎陽	[4] 長沙	[5] 豫章
[6] 九鼎	[7] 荊山	[8] 太清中經	[9] 鼎	[10] 竈屋
[11] 齋戒	[12] 六一泥	[13] 黃帝九鼎神丹經		

tional and mediumistic practices of their time. Then on the content of their revelations, seen in the light of other traditions which he knew, and which he incorporated, Thao founded a well-patronised and enduring community dedicated to pursuing every conceivable means of co-opting individuals (especially those of the more genteel classes) into the Unseen World, and performing other conventional religious services on their behalf. Alchemy was a charter member of the Mao Shan synthesis. But medicine and astronomy too were gradually included in the Patrologies,[a] for the compilation of which the Mao Shan school was largely responsible.[b] Kristofer Schipper has called this patrician group the 'middlebrow wing of Taoism', for its concerns had not a great deal to do either with the ontological paradoxes of Lao Tzu and Chuang Tzu on the one hand, or what would later become the everyday pastoral responsibilities of the village priest on the other. Their intellectual omnivorousness was prefigured only by that of Ko Hung. Their synthesis of magic, religion, and science, doubtless too promiscuous for the taste of most modern readers when seen as a whole, was perfectly suited to that of countless Chinese enthusiasts for a millennium. The cult gradually spread to Mt. Lo-fou[1] near Canton, and other great Taoist centres. Finally a succession of Mao Shan patriarchs like the hereditary Celestial Masters (*thien shih*[2]) of the priestly Chêng I[3] tradition controlled many or most of the Taoist abbeys in China until they were taken over by the Chhüan-chen[4] sect in the thirteenth century under Mongol policy.[c]

(2) THE SPECTRUM OF ALCHEMY

Anyone who tries to sort out the relations between theory and practice has to begin by acknowledging that every possible variation in both their proportion and the quality of their connection can be found in one or another of the documents. Some alchemical writings consist only of instructions for laboratory operations, with no attempt to provide a theoretical rationale. Others are nothing but rationale, and the actual process is recapitulated only as the conceptual discussion requires. It will be convenient for heuristic reasons to consider these extremes as the ends of a spectrum, with most of the extant literature falling somewhere in between. This is not a wholly arbitrary overview, for writings near either end of the spectrum tend to have certain characteristics in common. In general the highly theoretical material reflects an attempt to construct a laboratory model of the larger cycles of change which take place in Nature, using two ingredients, or sometimes two main ingredients, which correspond to Yin and Yang. This tendency might be called scientific in the classical sense of the word, since alchemical speculation was concerned primarily with contemplating natural process rather than with manufacturing some product. At the other extreme, where the connections with both medicine and the thaumaturgical tendencies of Taoism are more obvious, we find an often purely practical concern with the manu-

---

a See pt. 3, pp. 113 ff.
b Holmes Welch (3), pp. 129–130.   c See Soymié (4); Welch (3), p. 126.

¹ 羅浮山   ² 天師   ³ 正一   ⁴ 全真

facture and employment of elixirs of immortality, agents of transmutation, and other substances—even (to reinforce the parallel with Hellenistic aurifaction and aurifiction) artificial pearls, jade, and so on. Authors of this sort were willing to countenance any possible means, any available formula, self-contradictory or impractical features notwithstanding. This latter tendency might be called technological, in the sense that the product was all-important, and we shall see that reflections of the artisan's ability to control Nature, uncommon elsewhere in Chinese thought, furnish an important part of its ideology. We shall also use the word 'pragmatic' for writings at this end of the spectrum and the approach that they imply, but it refers simply to their valuing of ends over means, and not at all necessarily to a command of laboratory practice. Nor does this term necessarily imply unconcern with the Unseen World, or for the rituals, spells, and taboos by which one paid one's respects to it.

Before going further, a caution is in order about the danger of finding in this idea of a spectrum of alchemy a real inherent structure rather than a taxonomic convenience—or, worse still, thinking of it as a 'model'. As for the genetic relations between the two extremes and the middle, at this point we can offer no more than a few scattered clues, which only a great deal of thoughtful and critical study in the future can make coherent. We do not know which tendency developed from which, and out of what necessities. The oldest extant alchemical books include both highly pragmatic and highly theoretical treatises, but they represent too tiny and accidental a remnant to encourage the conclusion that a synthesis of the two approaches came only later. There is certainly no reason to suppose that they represent different schools of alchemy. The reader interested in any aspect of esoteric thought in ancient China can hope for no better advice than that of Rolf Stein: 'I prefer to believe, not in borrowings between schools, but in a common ground, an underlying structure, which can manifest itself variously in different milieus or movements but which the majority of thinkers hold in common.'[a]

(3) THE ROLE OF TIME

In order to form a clear idea of what the theoretically oriented alchemists were doing, one must keep in mind the very special importance of time in Chinese natural philosophy, for it was all the more crucial in alchemy. In the brief review which follows we shall stress the dynamic and temporal aspects of concepts such as the Tao and the Five Elements, which are not considered in those lights by modern students of Chinese philosophy as often as they should be.[b]

Scientific thought began, in China as elsewhere, when men tried to comprehend how it is that although individual things are constantly changing, always coming to be and perishing, Nature as a whole not only endures but remains conformable to itself. In the West the earliest attempts to identify the underlying and unchanging reality tended to be concerned primarily with some basic material substrate out of which the

[a] Stein (5), p. 40, eng. auct.
[b] A more recent version of this material has appeared in Sivin (14).

things around us are formed.[a] In this way one could think of all phenomenal things, for instance, as being composed of air (or rather *pneuma*, πνεῦμα) in some state of condensation or rarefaction. Thus a tree growing out of a seed is not matter being created out of nothing, but only air, which has existed all the time, gradually taking on a new physical form. In China theories roughly of this sort, explaining material things as composed of *chhi* in one state or another, were also sketched out in the first great period of natural philosophy, though they did not play a central role in physical speculation.[b] But the earliest, and in the long run the most influential, kinds of scientific explanation, those so basic that they truly pervaded the ancient Chinese world-view, were in terms of time.[c] They made sense of the momentary event by fitting it into the cyclical rhythms of natural process, for the life-cycle of an individual organism—birth, growth, maturity, decay, and death—had essentially the same configuration as those more general cycles which went on eternally and in regular order, one fitting inside the other: the cycle of day and night which regulated the changes of light and darkness, the cycle of the year which regulated heat and cold and the farmer's growing seasons, and the greater astronomical cycles.[d]

All these cycles nested. Early Chinese cosmography, as described in the Treatises on Harmonics and Calendrical Astronomy (*lü li chih*[1]) of the dynastic histories, built up its mathematical model of the cosmos in terms of time rather than (as was more the case in the European tradition) of geometric space. The cycles of the day, the month, and the year were fitted together to form larger periods—in early astronomy, the Rule Cycle (*chang*[2]) of nineteen years, equalling 235 lunations, or the Obscuration Cycle (*pu*[3]) of seventy-six years.[e] These were defined to begin and end with the winter solstice (for the month which contained the solstice was taken by astronomers as the 'first month' for computational purposes), and the new moon (the beginning of the month) falling at midnight (the beginning of the day) of the same day. A larger cycle was needed to make them fall on a day of the same sexagenary designation (in terms of the cyclical characters, *kan chih*[4]).[f] These four cycles—day, month, sixty days, and year—were only part of a much larger system which also included eclipse and planetary cycles, in fact all cycles which were known to be periodic. The period which included them all, the Great Year (Grand Polarity Superior Epoch, *thai chi shang yuan*[5])[g] which began and ended time with a universal conjunction of sun, moon, and planets, was calculated in the Triple Concordance system (San Thung Li[6]) of Wang Mang's time (*c.* −5) to be 23,639,040 years long. A century later, in the Quarter Day system (Ssu Fên Li[7]), based on somewhat more precise values for individual periodic phenomena,

---

[a] This idea has been developed to some depth in de Santillana (2). See also Vol. 2, p. 245 and Vol. 4, pt. 1, pp. 3 ff., 13–14.

[b] See above, Vol. 2, pp. 6, 40–1, 42 ff., 371–374. The fundamental role of the *chhi* concept in early medicine is analysed in Porkert (1), on which see also the critique of Needham & Lu Gwei-Djen (9).

[c] Cf. Granet (5), pp. 86–114; Needham (55, 56); Sivin (8); van der Sprenkel (1).

[d] See Vol. 3, pp. 390 ff.

[e] See above, Vol. 3, pp. 406–407 and, for a fuller treatment, Sivin (9).

[f] See Vol. 3, pp. 396 ff.   [g] See Vol. 3, p. 408.

¹ 律曆志   ² 章   ³ 蔀   ⁴ 干支   ⁵ 太極上元
⁶ 三統曆   ⁷ 四分曆

the Great Year was of such stupendous length that it was not even calculated Practically speaking, the length of the overall cycle was so great simply because more precise fractions tend to have larger common denominators. But to work out the exact value of the Great Year cycle would in any case have been irrelevant philosophically. What mattered was the demonstration that the unending time through which the natural world remained constant (or changed gradually, according to one's theory)[a] was the sum of finite processes which were known to regulate individual cycles of growth and decay, birth and death. The life rhythms of a swarm of mayflies meshed because they occupied a certain brief phase in the round of the seasons, just as the events of a certain autumn made sense in terms of its relation to astronomical periods.

In order to make the Tao of a particular thing intelligible, its life-cycle needed to be located with respect to the greater periods. The different parts of a cycle could be analysed in terms of a number of concepts, for instance the Yin and Yang, which were the passive and active phases through which any natural cycle must pass. Another variable was the so-called Five Elements (*wu hsing*[1], for which 'Five Phases'[b] would be both a more accurate and a more literal translation). We have seen earlier that these were not material elements in the modern sense, but a finer division of the cycle into five qualitatively and functionally distinct parts.[c] The 'element' Fire, for instance, represented the phase in which activity was at its highest, and thus soon would have to begin declining; in the cycle of the year summer was the time of Fire. The trigrams and hexagrams of the 'Book of Changes' were the third set of concepts which could be applied similarly to analyse change in terms of constant cycles.[d] These concepts belong of course to the most general level of early Chinese physical theory; the various fields of Chinese science, such as medicine, geomancy or alchemy, simply applied them to different classes of phenomena.

(i) *The organic development of minerals and metals*

What was true of the mayfly was true also of the mineral, for its process of growth was time-bound too. Like thinkers in other great ancient civilisations, the Chinese alchemists believed that Nature was an organism and everything had a life-cycle; therefore minerals and metals also grew inside the earth, slowly developing along a scale of perfection over immense stretches of time.[e] This process differed from other

---

[a] Linear theories of time in China have been studied in Needham (55, 56); cf. also Vol. 7.

[b] Or 'Five Phasers', which captures the force of the original a bit better even though it does not lie as well on the tongue. Some thoughts on this difficult problem of translation have been offered by Needham & Lu Gwei-Djen (9).

[c] See Vol. 2, pp. 243 ff.   [d] See Vol. 2, pp. 304 ff.

[e] This point as well as many others essential to the argument which follows were first established as generally valid by Mircea Eliade (4, 5, 8); see also Welch (3), pp. 114–117, in which the summary of Sivin (2) is however inaccurate. Although Eliade was limited on the Chinese side by the paucity and greatly varying quality of translated documents and monographic studies available, the fundamental good sense of many of his working hypotheses will be obvious to anyone familiar with the whole literature. One exception is his opinion that the alchemists 'were "experimenters", not abstract thinkers or erudite scholastics' (8), p. 77. One can easily find examples of all three, sometimes united in a single person.

[1] 五行

kinds of growth in two respects which taken together provided the basic rationale of alchemy. First, if and only if this sequence of maturation stages continued to its end, the product, usually gold, would be invulnerable to further transformation. Since gold is not subject to decay and death, the process is not cyclical. To a man whose world-view makes cycles of change the norm, the linear perfection of gold will more or less inevitably come to signify the redemption of man. Secondly, unlike vegetable and animal growth-cycles, the mineral cycle can be not only interrupted (or, as many peoples think of it, aborted) by the miner but also speeded up by the smelter, hence, following his lead, by the alchemist. These ideas, in the specific form they took in the Chinese elixir tradition, merit close examination (cf. Fig. 1516).

The notion of the organic development of minerals and its proto-scientific explana-tion in terms of *chhi* exhalations have already been described in connection with mineralogy, and Greek parallels have been pointed out.[a] Here it will only be necessary to adduce a few relevant documents from the alchemical literature. We may begin, however, by reviewing the appearance of this idea at the beginning of systematic thought about Nature in China. The princely alchemist Liu An's[1] *Huai Nan Tzu*,[2] one of the oldest cosmological treatises (*c.* − 125), follows its primitive scheme of biological evolution with a theory of development in the mineral world; and like the speculations of the early pre-Socratics, it lies barely this side of the line which separates proto-science from myth. Here we partially retranslate, rendering rather more literally than in Sect. 25(*b*):

The *chhi* of balanced Earth[b] copulates (*yü*[3]) with Dusty Heaven. After 500 years the Dusty Heaven gives birth to (the yellow mineral) *chüeh*[4,5] which after 500 years gives birth to yellow mercury,[c] which after 500 years again gives birth to the yellow metal (gold). The yellow metal in 1000 years gives birth to the yellow dragon. The yellow dragon, entering (the earth) and going into hibernation (or pupation) engenders the Yellow Springs.[d] When the dust from the Yellow Springs ascends to become the yellow cloud, (its) Yin and (the supernal) Yang beat upon one another, produce peals of thunder, repel each other and fly out as lightning. That which was above flows downward. The running streams flow together and unite in the Yellow Sea.[e]

[a] See Vol. 3, pp. 636ff.

[b] In alchemical theory *chêng chhi*[6] means 'balanced *chhi*', i.e. balanced with respect to Yin and Yang; in opposition to '*phien chhi*',[7] 'unbalanced *chhi*', which occurs in the corresponding position in the next paragraph of this passage. The Han commentary of Kao Yu[8] refers the balanced Earth-element to the Central Land (*chung thu*[9]), which earlier in the chapter is identified with Chi-chou,[10] one of the nine archaic provinces of China, corresponding to modern Hopei, Shansi, Honan north of the Yellow River, and Liaoning west of the Liao river. Balances and imbalances of Yin and Yang in the Five Elements recall the translated passage and the tabulation on pp. 156–7 above.

[c] We follow Wang Nien-Sun[11] in deleting eight characters in this sentence which are redundant and violate parallels with the later paragraphs.

[d] This term signified, we know, something corresponding to She'ol or Hades, as well as the Plutonic regions in general, from Chou times onwards (Vol. 5, pt. 2, pp. 84–5).

[e] Ch. 4, pp. 12a ff. tr. auct. adjuv. Erkes (1), pp. 79, 80. Cf. above, Vol. 3, pp. 640–1. It is interesting that in this passage gold is not yet thought of as the terminal or immortal avatar in an age-long matura-tion, the sign and warranty of redemption or salvation. John Major of Dartmouth College has written a monograph on the ideas expressed in this chapter of the *Huai Nan Tzu* book.

[1] 劉安	[2] 淮南子	[3] 御	[4] 缺	[5] 訣
[6] 正氣	[7] 偏氣	[8] 高誘	[9] 中土	[10] 冀州	[11] 王念孫

This passage and the four which follow, all worded much like it, may be reduced to a general scheme (Table 117).

The *chhi* of X Earth $\xrightarrow{\;Y\ years\;}$ Z mineral $\xrightarrow{\;Y\ years\;}$ Z quicksilver $\xrightarrow{\;Y\ years\;}$ Z metal $\xrightarrow{\;1000\ years\;}$ Z dragon $\xrightarrow{\;Y\ years\;}$ Z springs, where X is an attribute, Y a number of years,[a] and Z a colour.

Table 117

Paragraph	X	$Y/100$	Z	Mineral	Metal	Element
1	balanced	5	yellow	realgar (or yellow jade?)	gold	Earth
2	unbalanced	8	caerulean (blue-green)	malachite	lead	Wood
3	vigorous	7	scarlet	cinnabar	copper	Fire
4	weak	9	white	arsenolite	silver	Metal
5	passive	6	black	slate (or grindstone?)	iron	Water

In this schema the deductive categories of the Five Elements have largely taken over the function of providing coherence, though the sequence of images still owes something to the looser and less logical association of mythology. The basic structure is familiar enough, for it depends on the normal number, colour, and metallic correspondences of the Five Elements, taken in a special sequence related to the Mutual Production order which characterises organic processes.[b] The mineral correlates of Earth and Water are archaic and no longer certainly identifiable, though there is no doubt that they were chosen because of their colour. By the time of Ko Hung the alchemical Five Minerals (*wu shih*[2]) had become stabilised as (in the same order) realgar, laminar malachite, cinnabar, kalinite (potassium alum) or arsenolite, and magnetite.[c]

(ii) *Planetary correspondences, the First Law of Chinese Physics, and inductive causation*

Although the planets did not play the paramount role in Chinese alchemy that they did in the West,[d] the correspondence of the Five Planets (*wu hsing*[3]) to the Five Elements naturally gave rise to schematic concordances which did not differ in spirit

---

[a] These were of course chosen with numerological considerations in mind, depending on the normal Five-Element associations, etc. Caerulean, blue-green in the Table, translates *chhing*.[1]

[b] See Vol. 2, pp. 254 ff. The order E-W-F-M-w has been called by Major (1) the 'Smelting Order'.

[c] Cf. pt. 3, pp. 86, 96. *PPT/NP*, ch. 4, p. 9*b*, gives kalinite (crude alum, *fan shih*[4]), but as the editor remarks, arsenolite (*yü shih*[5]) appears instead in ch. 988 of *TPYL*. This is a familiar confusion, since the characters *fan*[6] and *yü*[7] so much resemble each other. Nevertheless, the best available edition of *TPYL*, the Sung version reproduced in 1960, reads *fan shih*[4] at this point (ch. 988, p. 3*a*). True, that does not settle the question, for the same text from *PPT/NP* is also quoted earlier (ch. 985, p. 2*b*), and there it reads *yü shih*.[5] Both of these compounds were too important in later alchemy to allow of an *a priori* solution.

[d] See Vol. 2, pp. 351 ff.

[1] 青   [2] 五石   [3] 五星   [4] 礬石   [5] 礜石   [6] 礬   [7] 礜

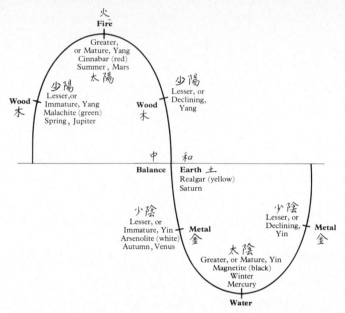

Fig. 1515. The Five Elements and the Yin and Yang as phases of a cyclical process. Cf. Fig. 277 on p. 9 of Vol. 4, pt. 1 above.

from those just discussed, since their function was the same. An important collection of elixir recipes which reached final form in the middle of the eighth century ascribes to the author of the *Huai Nan Tzu* book a method for making Five-Mineral Elixir (*wu shih tan*[1]), of which it says in a prefatory note:[a]

The Five Minerals (*wu shih*[2]) are the seminal essences of the Five Planets. Cinnabar is the essence of the mature Yang (*thai yang*[3]), Mars. Magnetite is the essence of the mature Yin, Mercury. Malachite is the essence of the young Yang (*shao yang*[4]), Jupiter. Realgar is the essence of Divine Earth (*hou thu*[5]), Saturn. Arsenolite is the essence of the young Yin, Venus. A medicine made from the essences of the Five Planets can give a man perpetual life, exempt from death for ever.

The five substances in this set of correspondences are the classical series, not those of Liu An. The 'mature' Yin or Yang is what we should call its maximum state. Having thus reached its height, its decline is about to begin, accompanied by reversion to its opposite (*wu chi pi fan*[6]). This is in accordance with what has been termed the First Law of traditional Chinese Physics (and Chemistry), namely that 'any maximum state of a variable is inherently unstable', and the process of going over to its opposite must necessarily set in.[b] Thus the winter solstice is the point when the Yin ascendancy, having reached its zenith, starts to fade, and the Yang, which will be maximal at the

a *Thai-Chhing Shih Pi Chi*[7] (*TT*/874), ch. 1, p. 13*a*, tr. auct.
b This formulation was first used by one of us (N.S.) at the Bellagio Conference on Taoist Studies, 1968. Cf. Sivin (2).

¹ 五石丹   ² 五石   ³ 太陽   ⁴ 少陽   ⁵ 后土
⁶ 物極必反   ⁷ 太清石壁記

summer solstice, begins to reassert itself. The 'young', or immature phase, represents a level intermediate between the point of balanced polarity and the maximal phase. In the cycle of the year, equal intensity of Yin and Yang is reached at the equinoxes, so the young Yang would fall betweeen spring equinox and midsummer. If we represent an ideal cyclical process by a sinusoidal curve (Fig. 1515),[a] the correspondence between the Five Elements and the five phases of Yin and Yang (mature, immature, and balance) is easily visualised. The planetary associations of the text thus turn out to be simply the usual correspondences of the planets with the Five Elements.

In the West the influence of the planets was direct; but in China it is perhaps confusing even to use the word 'influence', for the relation was one of correspondence.[b] We have just seen the association between the seminal essences of the planets and the minerals depicted not as emanation or influence, but as identity. The *chhi* of a planet could stimulate response in a metal or mineral only when they were categorically related—tuned to the same note, so to speak—within the unitary system of the physical world. The Stoic and Neoplatonic universes, which furnished the cosmic ideology of European alchemy (and to a large extent that of Islam), were organismic too, but in general influences within them proceeded in one direction, down a fundamentally linear hierarchy of value. In Chinese thought, which got along without a gradation of being based upon proximity to a Supreme Intelligence, it was possible to relate the activity of celestial bodies quite acausally to the formation of minerals in complex and interesting ways. A good example is the following excerpt from an unidentified 'Secrets of the Great Tao' (*Ta Tao Mi Chih*[1]). It is quoted in the *Huan Tan Chung Hsien Lun*[2] (Pronouncements of the Company of the Immortals on Cyclically Transformed Elixirs) dated +1052, by Yang Tsai,[3] whose graphic description of mercury-poisoning guarantees that it is concerned with the Outer Elixir:[c]

Venus, the Metal planet, is the seminal essence of Metal (*chin chih ching*[4]). It accepts the vital anima[d] of the moon, and holds within itself the *chhi* of the Earth planet Saturn.[e] Thus

[a] Cf. Vol. 4, pt. 1, p. 9. It would perhaps be more adequate to think of Yin and Yang as two sinusoidal curves out of phase by 180°, so that at any moment their sum is constant. Of course either visualisation lends Yin–Yang theory a mathematical concreteness which it usually lacked. In the few instances where interdependent measured variables were used to represent Yin and Yang forces, the actual curve was generally a so-called zig-zag function, which only approximates to a sinusoidal curve. See below, p. 275, esp. Fig. 1519a.

[b] I.e. what we called in Vol. 2 symbolic correlation.

[c] *TT*230, pp. 11a–12a, tr. auct. On mercury poisoning, see p. 80 above, as also pt. 2, pp. 282ff. and Ho Ping-Yü & Needham (4).

[d] For convenience in this translation we use anima and animus for the *pho*[5] and *hun*[6] 'souls' respectively; see pt. 2, pp. 85ff. above.

[e] In this passage, as we shall see, the planets each correspond to one of the Five Elements, and are related according to their Mutual Production order (see Vol. 2, pp. 253–261). Various verbs are used to express the relation: *han*[7] (to hold within), *chhuan*[8] (to transmit), *chiang*[9] (to descend into), *shou*,[10] *tê*[11] (to accept). The active verbs express an unambiguous sequence; A *chhuan* or *chiang* B, and B *shou* or *tê* A, mean that the element A immediately precedes B in the Mutual Production order. The static verb *han* in philosophical writing refers always to a latent aspect (e.g. the Yin concealed within the Yang during the maculine phase of a process), and does not, rigorously speaking, imply a necessary direction of evolution.

[1] 大道密旨   [2] 還丹象仙論   [3] 楊在   [4] 金之精   [5] 魄   [6] 魂
[7] 舍   [8] 傳   [9] 降   [10] 受   [11] 得

inside it, yellow in colour, is the floreate essence (or radiance) of Metal (*chin hua*[1]).[a] The stimulus of the lunar *chhi* is manifested as anima, and anima belongs to Water.[b] When subsequently (the floreate essence) has received (the *chhi* of) Metal, the Watery *chhi* will respond to Mercury (the Water planet) and give birth to lead. (E–M–w).[c]

Jupiter is the Wood planet, the vital animus of the sun and the essential *chhi* of Water.[d] This animus is scarlet, because (it corresponds to) Fire. Fire gives birth to Wood.[e] In response to the *chhi* of Mars (the Fire planet), cinnabar is born. Cinnabar holds within it the Yin *chhi* of Wood, and thus contains quicksilver. Quicksilver is called the Caerulean Dragon;[f] and the Caerulean Dragon belongs to Wood (w–W–F).

Mercury is the Water planet, and the seminal essence of Water. It transmits the *chhi* of Venus, the Metal (planet). Its flowing seminal essence responds to Earth, also receiving the vital anima of the moon, and gives birth to lead. Thus lead produces the floreate essence of Metal. The floreate essence of Metal has the Five Colours,[g] and is named 'Yellow Sprouts' (*huang ya*[2]). The *chhi* of the Water planet descends into Wood and gives birth to laminar malachite[h] (E–M–w–W).

Mars is Fire, and the seminal essence of Fire. It receives the *chhi* of the Wood planet (Jupiter) and also transmits the animus of the sun. Its flowing seminal essence enters Earth (or the earth) and gives birth to cinnabar. The animus (of cinnabar) belongs to Fire and so it is born out of Wood.[i] Since within it there is Yin, it gives birth to mercury. Fire gives birth to Earth. Earth contains the Balanced Yang,[j] and gives birth to realgar, the sapidity of which is sweet. (W–F–E).

[a] In the third paragraph this substance is identified as 'Yellow Sprouts' (*huang ya*[2]), which is not an elixir ingredient but rather an intermediary product in prototype processes. In our discussion below (pp. 256, 259, 261) we note that its chemical identity depends entirely upon the process.

[b] This sentence seems garbled, though its meaning is in any case probably much as we have rendered it. The *hun*[3] and *pho*[4] are normally the Yang and Yin personal vitalities which at death leave the body to return to the supernal and terrestrial realms respectively (see above, pt. 2, pp. 85 ff.). Here the words are used in a more abstract way (signalled by their identification with the Yang sun and the Yin moon) to refer to the cosmic Yang and Yin forces.

[c] In order to make the often indirectly expressed sequence of planet/element correspondences easier to follow, we have inserted at the end of each paragraph, in abbreviated form, the Mutual Production sequence of the elements discussed. We let W stand for Wood, and w for Water. The other capital letters are self-explanatory; cf. Vol. 2, p. 253.

[d] This appears at first sight contradictory, since Water is Yin and one would not expect a correspondence with the masculine animus (*hun*) of the sun. But the sense of the assertion that Wood (the immature Yang) is the essential *chhi* of Water is simply that the former follows the latter in the Mutual Production succession.

[e] This statement is the inverse of the usual sequence, given correctly for the same substances in the fourth paragraph. The text as a whole is corrupt enough to suggest that confusion here is not unlikely.

[f] This is merely an added Five-Elements association; the Caerulean Dragon (*tshang lung*[5]) is the eastern of the four main divisions of the sky, and thus corresponds to Wood (see Vol 3, p. 242). The Caerulean Dragon mentioned in Table 117 above has a different written character (*chhing lung*[6]), but the two are essentially of the same meaning, blue-green. Certain contexts of course dictate translation as definitely blue or green, but not this one. There is an old but good paper on this subject by von Strauss-und-Torney (1). See also Hirth (25), p. 7.

[g] The floreate essence of Metal is Earth, which mediates and reconciles the Five Elements, just as 'Yellow Sprouts' represents the unity of opposites in the alchemical process.

[h] Laminar malachite corresponds to Wood, as cinnabar to Fire, in the Five Minerals system (see above, pt. 3, p. 96).   [i] The character *tzu*[7] is obviously misplaced in this sentence.

[j] 'Balanced Yang' (*chêng Yang*,[8] whose antonym is *phien Yang*[9]), is a technical term for the pure creative Yang force which emerges from the reconciliation of opposites. See below, pp. 236, 251 ff., and above, pp. 157, 224. The phase of cosmic balance is represented by the element Earth. Sweetness corresponds to Earth in the system of the Five Sapidities (*wu wei*[10]), as does realgar among the Five Minerals.

[1] 金華   [2] 黃芽   [3] 魂   [4] 魄   [5] 蒼龍
[6] 青龍   [7] 自   [8] 正陽   [9] 偏陽   [10] 五味

Saturn is Earth. It accepts (the *chhi* of) Fire. The Earth planet holds the Balanced Yang within, and thus has realgar (F–E).

Thus the Five Planets transmit from one to another the floreate essences of sun and moon in rotation according to (the) Mutual Production (order of the elements, *hsiang shêng*[1]), each conforming to its Tao.[a]

Here, as indeed generally in alchemical writing, *chhi*[2] is not matter but a kind of configurational energy[b] which endows with structure a certain kind of matter and gives it determinate qualities. *Ching*[3] or *ching chhi*,[4] 'seminal essence' with its *chhi*, and *hua*[5] or *ching hua*,[6] 'radiance' or 'floreate essence' or 'seminal radiance', are terms for energy (in the colloquial, qualitative sense) deriving from some organised entity and applied to bring about a similar organisation in another entity.[c] In other words, these concepts come into play in order to explain change and transformation. *Hua* (lit. 'florescence') refers to the essence in its aspect of emerging from something, while *ching* (lit. 'seed, semen') refers to the essence in its function of actively forming or nurturing something else. From our point of view, it was two ways of looking at the same total phenomenon, namely the production of something with certain determinate qualities from something else, which might or might not have the same qualities.

One example of the seminal essence is the most mundane variety of *ching*,[3] namely human semen, a concentration of personal vitality which transmits characteristics from father to offspring. In other words, the configurational energy of the father imposes itself on the material basis (*chih*[7]) provided by the mother to bring about its organisation as a foetus.[d] Typical of *hua*,[8] on the other hand, is the red 'inner essence' (i.e. the oxide) which emerges as a red powder when mercury is heated in air. In the Mutual Production series of the Five Elements, analogously, the radiance or floreate essence (*hua*) of an element is the one which precedes it (i.e. its formative essence seen as emergent from its predecessor), and the seminal essence (*ching*) is the

---

[a] The order in which the Five Planets are treated in this passage is not exactly the Mutual Production enumeration order, but the so-called Modern order (see Vol. 2, pp. 253–261). The Mutual Production order, as we have seen, governs the relations between planets which correspond respectively to the outer and inner aspects of each substance—in the first paragraph, for instance, Earth and Metal. The only exception is the second paragraph, for Fire follows rather than precedes Wood in the *hsiang shêng* series. This exception is possibly due to a textual confusion, for the fourth paragraph, where the outer aspect is also cinnabar, repeats the same pair of planets, those corresponding to Wood and Fire, but in the correct Mutual Production order.

[b] The term is Manfred Porkert's. One can see how near this conception comes to Neo-Confucian *li*[14] (cf. Vol. 2, pp. 472 ff.). It may be thought to illustrate rather well the character of the Chinese mind as *anima naturaliter materialistica*, always drawing ideas away from the noumenal, the spiritual and the transcendental, to incarnate them in the immanence of the actual. Here *chhi*,[3] although itself energetic, may possess qualities which some schools attributed to *li*,[9] but even for the Neo-Confucians *li* was only manifested when incarnated in *chhi* (J.N.).

[c] The picture is clouded by ambiguities which call for constant alertness. Both *ching*[4] and *ching hua*[7] also have a wider meaning which includes the specialised functional senses of both *ching and hua*. Porkert (1) renders this sense as 'structive potential'. On all these interpretations cf. the critique of Needham & Lu Gwei-Djen (9), especially in relation to the implicit parallel with embryonic induction and determination.

[d] Note the parallel here with Aristotelian ideas of form and matter in generation. These come up for discussion elsewhere, e.g. in pt. 5; and meanwhile Needham (2).

[1] 相生   [2] 氣   [3] 精   [4] 精氣   [5] 華   [6] 精華
[7] 質   [8] 華   [9] 理

one which follows it (i.e. its forming essence as imposed upon its successor).[a] It is easy to see that this functional terminology could be applied to any stimulus–response reaction. The medieval Chinese applied it throughout the realm of scientific thought, including physics, and as we see here, chemistry; which makes apparent to us that their basic concepts of action were inclusive of the biological.[b]

The purpose of the document we are considering is to account for the dynamic relations between certain mineral substances, characterised as aspects of the Five Elements. Thus we see Yellow Sprouts (floreate essence of Metal) and mercury described as 'held within' lead and cinnabar respectively. In the eye of the alchemist's mind the inner aspect was a possible state of the outer material, and could become manifest as the result of alchemical processes. But the relations discussed in this quotation are not static, since the Five Elements are in turn functionally related to each other by the Mutual Production succession order, which governs the quasi-biological evolution of one thing or one phase of a cyclical process out of another. The genetic character of the lead/Yellow Sprouts and cinnabar/mercury relationships is established by making them correspond to the Mutual Production sequences Metal–Earth and Wood–Fire. The element sequences are not as a rule expressed directly, but more often given in terms of the planets—the Five Elements seen in their cosmological function. Only when we recognise that the fundamental level of discourse is not astronomical at all can we perceive the simple, and to the Chinese thinker familiar, sense behind the apparently very odd assertions about interactions of planets.

It would be sorely misreading the text to see in it any suggestion of physical influence exerted by planets upon terrestrial minerals. The sun and moon are no less passive in this schema than the planets. While the latter serve in the theory as aspects of the Five Elements, the former—or, to be more precise, the *hun* vitality or animus which characterises the sun and the *pho* vitality or anima of the moon—stand for the cosmological aspects of Yang and Yin.

We can thus proceed to reduce the second and most of the fourth paragraph of the text to a straightforward assertion: 'There exists the genetically related binary system mercury/cinnabar, of which mercury, corresponding to Wood, is the young (i.e. immature) Yang phase and cinnabar, corresponding to Fire, is the mature Yang phase.' The modern reader no doubt prefers a plainer formulation, for he knows how important direct statement has been in the growth of modern science. But for the ancient alchemist, the richness of association was desirable enough to be paid for in simplicity and testability. What brought the planets into alchemical theory was a motivation, in the last analysis, aesthetic.

[a] Unfortunately this distinction is not rigorous in all alchemical writing, and one can find instances of *ching* and *hua* used in senses opposite to the distinction drawn here. There is great need for a systematic philological study of the *chhi* concept in science, having regard to the remarkable beginnings made by Porkert (1) in his work on the conceptual foundations of Chinese medicine.

[b] Cf. p. 307 below.

(iii) *Time as the essential parameter of mineral growth*

The protean metalline metamorphoses of the *Huai Nan Tzu* book were avoided by later alchemists, who accepted much more straightforwardly the archaic idea of the gradual perfection of minerals within the terrestial matrix. Here the idea is expressed with pristine simplicity in one of the most influential of all alchemical writings, the supplementary instructions (*chüeh*[1]), probably of the early Sung, which now accompany the Han or pseudo-Han 'Yellow Emperor's Canon of the Nine-Vessel Spiritual Elixir':[a]

> Realgar occurs in the same mountains as orpiment, and is formed by the transformation of orpiment. (This latter) great medicine of heaven and earth (i.e., of the natural order) is called 'doe yellow' (*tzhu huang*[2]). When eight thousand years have passed, it transforms into realgar,[b] the variant name of which is 'imperial male seminal essence' (*ti nan ching*[3]). After another thousand years have passed it transforms into yellow gold, with the variant name 'victuals of the Perfected (or Realised) Immortals' (*chen jen fan*[4]).

The theory of this type most significant for the development of alchemy begins, as did Liu An's, with a hierogamy, and the time span, while still defined numerologically, is chosen more carefully for its cosmic significance. The *Tan Lun Chüeh Chih Hsin Ching*[5] (Mental Mirror Reflecting the Essentials of Oral Instruction about the Discourses on the Elixir and the Enchymoma), a theoretical treatise probably of the Thang, rationalises the preparation of the elixir of immortality by analogy with geological process:[c]

---

[a] *Huang Ti Chiu Ting Shen Tan Ching Chüeh* (TT878), ch. 14, p. 1*a*, tr. auct. The same sequence of three yellow substances is given with intervals of a thousand years between each, in *Shen Hsien Fu Erh Tan Shih Hsing Yao Fa*[6] (The Immortals' Method for Ingesting Cinnabar and (Other) Minerals and Using Them Medically), TT417. This interesting but undated treatise belongs to the quasi-alchemical tradition concerned with processing natural minerals and stones to make them edible rather than transforming them into elixirs (see pp. 168 ff. above). The first of the two transformations represents the inverse of a known geological process, for some orpiment is actually formed by the weathering of realgar; Dana (1).

[b] The common appellation of realgar, *hsiung huang*,[7] might be translated literally 'buck yellow'. *Tzhu* and *hsiung* are the general designations for genders of animals, and have no exact English equivalents; we render them by terms that at least apply to a number of species.

[c] TT928, p. 12*b*, tr. auct. The version in *YCCC*, ch. 66, p. 12*b*, is corrupt. The last character of the title is given as *chien*[8] in TT598, and *chao*[9] in the *YCCC* version, but both of these were simply means to avoid a Sung tabu. The sole basis for very provisionally assigning this work to the Thang is the unlikelihood that a Sung writer would have chosen a title which included a tabued word. This must be weighed against the statement that a disciple of Wei Po-Yang[10] who took the elixir 'is on Mt. Thai-pai[11] now, over a thousand years old', which would imply a date after the mid-twelfth century. But taking the mention of a millennium literally is ruled out by the inclusion of the treatise in *YCCC*, which was compiled early in the + 11th century.

One of us has a complete translation of this interesting source in draft (Sivin, 5). Its highly philosophical content is so divorced from the practical work of the furnace, and there is so much stress on processes within the alchemist's psyche or body, that we cannot be sure as yet how far the book is concerned with the preparation of the Outer Elixir. At the same time, its subject matter, concepts and language are so consistent with those of less ambiguous texts that relevant material need not be ignored.

---

[1] 訣 [2] 雌黃 [3] 帝男精 [4] 眞人飯 [5] 丹論訣旨心鏡
[6] 神仙服餌丹石行藥法 [7] 雄黃 [8] 鑑 [9] 照
[10] 魏伯陽 [11] 太白山

Natural cyclically-transformed elixir (*tzu-jan huan tan*[1]) is formed when flowing mercury (*liu hung*[2]), embracing Sir Metal (*chin kung*[3] = *chhien*,[4] lead), becomes pregnant. Wherever there is cinnabar there are also lead and silver. In 4320 years the elixir is finished. Realgar (*hsiung*[5]) to its left, orpiment (*tzhu*[6]) to its right, cinnabar above it, malachite (*tshêng chhing*[7]) below. It embraces the *chhi* of sun and moon, Yin and Yang, for 4320 years; thus, upon repletion of its own *chhi*, it becomes a cyclically-transformed elixir for immortals of the highest grade and celestial beings. When in the world below lead and mercury are perfected by an alchemical process (*hsiu lien*[8]) for purposes of immortality, (the elixir) is finished in one year.[a] The fire is first applied in the eleventh month, when the Single Yang (*i Yang*[9]) comes into being,[b] and the elixir is finished by the eleventh month of the next year. The natural cyclically-transformed elixir is what immortals, celestial beings, and sages of the world above gather and eat. What (the alchemist) now prepares succeeds because of its corre-spondence on a scale of thousandths (*hsiang erh chhêng chih, ta chhien chih shu*[10]).[c] Taking the product also results in eternal life, transformation into a feathered being, and power (*kung*[11]) equal to that of heaven.

We shall return shortly to the period of 4320 years in connection with the alchemist's side of the analogy between the Work of the laboratory and the Work which takes place in the womb of Mother Nature. There we shall see[d] that although the adept's period of a year is metaphysically derived from what we might call the temporal macrocosm of 4320 years,[e] historically the longer period was obviously chosen

---

[a] This phrase has a very *nei tan* (physiological alchemy) flavour; cf. pt. 5. below.

[b] The rebirth of Yang at the moment of the winter solstice is here expressed in the language of the *Tshan Thung Chhi* as the hexagram Khun ☰☰ changing into Fu ☰☰ (see Vol. 2, p. 332, Table 17, the final transition in the Diurnal Cycle).

[c] Or, to translate very literally, '(that through) correspondence it succeeds (is because of) numerical relations to the great thousands.' This understanding of the Chinese sense is only tentative, for (as Prof. A. F. P. Hulsewé has suggested in private correspondence) one or more characters are very probably missing from the middle of the text. In modern writings *hsiang*[12] often means 'symbol' or 'symbolise', but in early alchemical discourse, so far as we know, the word always refers to a relationship of correspondence. In other words, the things so related are coordinate, each partaking equally of the quality of the other, rather than the more concrete or hierarchically inferior standing for the more abstract or hierarchically superior. We do not in the least want to discourage Jungian or other psycho-logical interpretations of Chinese alchemy, but it does seem indispensable that the analytical categories of such an interpretation be allowed to grow out of a close reading and precise understanding of the sources rather than be imposed ready-made. It does not seem likely that any psychologist will succeed at this task if he does not take the trouble to learn to read medieval Chinese accurately and critically.

[d] Pp. 264–6.

[e] For an extended cosmogonic derivation in terms of four stages of 1080 years each, see *Thung Yu Chüeh*[13] (Lectures on the Understanding of the Obscurity (of Nature), *TT*906), pp. 1a to 2b. This book belongs to a group of cognate treatises the content of which overlaps: *Huan Tan Chou Hou Chüeh*[14] (Oral Instructions on Handy Formulae for Cyclically Transformed Elixirs), *TT*908, which duplicates *TT*906 to p. 20b; *Yü Chhing Nei Shu*[15] (Inner Writings of the Jade-Purity (Heaven), *TT*940), which up to p. 7a is the same as *TT*906, pp. 20b to 27a; and *Hung Chhien Ju Hei Chhien Chüeh*[16] (Oral Instructions on the Entry of the Red Lead into the Black Lead) *TT*934, which overlaps, with variations in wording, the later part of *TT*940. The content of all these tractates is extremely theoretical, and equally applicable to both chemical and physiological alchemy. Despite some conventional attempts to lend an air of antiquity, as well as the inclusion of some plausible late Thang material, they were probably all compiled in the Sung. Only *TT*940 is listed in the bibliographical treatise of the *Sung Shih*. Finally, *Tan Lun Chüeh Chih Hsin Ching*, the treatise we have just quoted, is textually cognate with part of *TT*908, ch. 2.

[1] 自然還丹   [2] 流汞   [3] 金公   [4] 鉛   [5] 雄   [6] 雌
[7] 曾青   [8] 修鍊   [9] 一陽   [10] 象而成之大千之數   [11] 功   [12] 象   [13] 通幽訣   [14] 還丹肘後訣   [15] 玉清內書
[16] 紅鉛入黑鉛訣

to correspond to the number of double-hours (*shih*[1]) in the round year of 360 days. We are quite serious in representing the two directions of correspondence as related to two distinct realities within the alchemist's universe of significance. That he did not find them contradictory testifies to the coordinate nature of correspondences as the Chinese used them. It is interesting that the writer should have expressed the relation of the two time periods in terms of order of magnitude, a concept the easy and correct use of which is far from prevalent today.

This document also alludes to two minor but not insignificant alchemical themes: the notation of the geological coupling of minerals and metals, and the idea that there exist within the earth certain substances of such quality that only immortals can have access to them. We shall postpone slightly a discussion of the second theme, since its very ample documentation makes more adequate study possible.

The regular association of certain plants with mineral deposits, and of the latter with deeper strata of metals or metallic ores, has already been considered in Section 25 in connection with geological prospecting.[a] In the text from the *Kuan Tzu* book (compiled perhaps in the late −4th century) cited there, superficial cinnabar is considered a sign of deeper gold.[b] Ko Hung (*c.* +320) agrees, again making a parallel between the evolution of gold in the mountains and in the furnace:

> When the manuals of the immortals (*hsien ching*[2]) say that the seminal essence of cinnabar gives birth to gold, this is the theory of making gold from cinnabar. That is why gold is generally found beneath cinnabar in the mountains.[c]

The coupling of cinnabar with lead ores in *Tan Lun Chüeh Chih Hsin Ching* lacks classical precedent, and we do not know from what empirical generalisation it derives. The common presence of silver in ores of lead is a commonplace in Chinese alchemy as in modern geology, and a key to one of the prototype two-element processes of the proto-scientific art.[d]

Another simple account of the subterranean evolution of metals appears in the *Chih Kuei Chi*[3] (Pointing the Way Home (to Life Eternal); a Collection) of Wu Wu,[4,5] whose manual of equipment and procedures, *Tan Fang Hsü Chih*[6] (Indispensable Knowledge for the Chymical Elaboratory), is dated +1163.[e] The former work is definitely concerned with physiological and meditational alchemy,[f] but the author was conversant with the Outer Elixir tradition and is clearly reflecting it here:[g]

---

[a] See Vol. 3, pp. 675–680. [b] See Vol. 3, p. 674.
[c] *PPT/NP*, ch. 16, p. 5a; tr. Ware (5), p. 268, mod. auct. [d] See pp. 257ff.
[e] *TT*914 and *TT*893 respectively. The latter figures prominently in our study of alchemical apparatus (pp. 11, 68ff. above).
[f] See, for instance, its sympathetic likening of Taoist concentration and breath control (*tshun shen pi hsi*[7]) to Chhan Buddhist meditation (*chhan ting*[8]); preface, p. 1b. It is more than likely that Wu's path was a blend of Inner and Outer alchemy, including other meditative techniques, and the sexual practice of 'cycling the semen'—in other words, that 'dual cultivation' which largely absorbed alchemy in Sung and later times. See below, pt. 5.
[g] *TT*914, preface, p. 2b., tr. auct. A remark on the previous page shows that Wu was also familiar with the organisation of the classical pharmaceutical natural histories.

[1] 時   [2] 仙經   [3] 指歸集   [4] 吳悟   [5] 愩   [6] 丹房須知
[7] 存神閉息   [8] 禪定

Quicksilver, under the stimulus of the *chhi* of Yin and Yang for 800 years, forms cinnabar (*sha*[1]); after 3000 years it forms silver; after 80,000 years it forms gold—the longer the firmer (*chien*[2]),[a] through a thousand metamorphoses and a myriad transformations. The sages cycle (*yün*[3]) Water and Fire,[b] following the model of the operation of the *chhi* of Yin and Yang, in order to bring to completion the virtue (of the elixir); this is what is called 'surpassing the ingenuity of the Shaping Forces of Nature (*to-tê tsao-hua chi chê yeh*[4])'.[c]

As we have seen, the archaic and ubiquitous idea of the evolution of minerals and metals along a scale of perfection was rationalised in China in terms of the Five-Element and Yin-Yang theories, provided with much concrete detail, and related to cosmic process by the choice of specific time spans.[d] It was perhaps inevitable that at least for purposes of meditation upon the creative potential of the Tao, this idea was further imaginatively extended to link it with other Chinese convictions.

One possibility was to involve the vegetable kingdom by extrapolating, so to speak, the growth of minerals backward. Philosophically this was not much of an innovation, for the idea of the fixity of species had been rejected from the start, to allow the possibility of one species metamorphosing into another, and to explain spontaneous generation.[e] Transformation was ordinarily thought of either as a binary relation, in the sense that a certain species could change spontaneously into another particular species, or as a chain relation, in which the metamorphoses form a natural series. The chain relation is represented by the *Chuang Tzu* book's renowned theory of a cycle which begins with 'germs' (*chi*[5]) in the water and evolves organically step by step to man, who in due but unspecified course reverts to the germs.[f] Where this passage is quoted in the *Lieh Tzu* book (compiled by *c.* +300), the continuity is broken, probably through late editorial inadvertence, by some typical examples of the simple binary relation:

Sheep's liver changes into the goblin sheep underground. The blood of horses and men becoming will-o'-the-wisp;[g] kites becoming sparrow-hawks, sparrow-hawks becoming cuckoos, cuckoos in due time again becoming kites; swallows becoming oysters, moles becoming quails, rotten melons becoming fish, old leeks becoming sedge, old ewes becoming monkeys, fish roe becoming insects—all these are examples of things altering and metamorphosing....[h]

---

[a] The implication is plainly that the harder the gold, the more perfect it is. That this conviction should be held by a man of Wu Wu's experience is additional evidence that the assumptions of both aurifiction and aurifaction could coexist in the mind of an alchemist without collision. See above, pt. 2, pp. 8 ff. pt. 3, p. 102.

[b] 'Water' and 'Fire' here mean both the elements, which correspond to Yin and Yang and thus to the reactants in the standard two-element processes (see below, pp. 251 ff.), and the heating fire and cooling water of the actual apparatus. These were functionally equivalent.

[c] The last sentence could also be phrased: 'robbing, or carrying off, their mechanisms (and making them work for human benefit)' (J.N.). We shall encounter this motif again in pt. 5.

[d] For a parallel in sexual alchemy, see below, pt. 5.

[e] These were not categorically distinguished from sudden transformations involving one individual, like a man changing into a woman or a were-tiger. All are intermixed under the rubric 'transformations' (*pien hua*[6]) in *TPYL*, ch. 887–888, a trove of instances. We shall discuss them systematically in Vol. 6, Sect. 39.

[f] See Vol. 2, pp. 78–79. It is interesting that the Sung scholar Chêng Ching-Wang[7] should have commented on this passage mainly in terms of binary transformations; see Vol. 2, pp. 420–2.

[g] On this see more extensively Vol. 4, pt. 1, pp. 72 ff.

[h] Ch. 1, pp. 6 ff.; tr. A. C. Graham (6), pp. 21–2, misprint corrected. The passage from the *Huai Nan Tzu* book translated earlier in this sub-section is immediately preceded by five chains which trace

---

[1] 砂   [2] 堅   [3] 運   [4] 奪得造化機者也   [5] 幾   [6] 變化   [7] 鄭景望

Another binary relation known to every physician in classical times was that between pine resin (*sung chih*[1]) and the *fu-ling*[2] fungus, a parasite upon the roots of pine trees, prized as an immortality medicine.[a] The fungus was supposed to be formed when pine resin flowed into the ground and remained there for a thousand years. When it grew especially close about the roots of the tree it was called 'pachyma spirit', or *fu shen*[3].[b] Origin from pine resin was also ascribed to amber (*hu-po*[4]) by Thao Hung-Ching,[5] who introduced amber into the pharmacopoeia; though 'an old tradition' cited by Su Ching[6] (between +650 and +659) had *fu-ling* metamorphosing into amber after a second millennium, and amber into jet (*i, hsi*[7]) after a third.[c]

Here, then, is an alchemical assimilation of these motifs into an account formally similar to the mineral sequences we have already examined:

In the great Tao of heaven and earth, what endures of the myriad phenomena is their primal and harmonious *chhi* (*yuan ho chih chhi*[8]). Of the things that exist in perpetuity, none surpass the sun, moon, and stars.[d] Yin and Yang, the Five Phases (Elements), day and night, come into being out of Earth (*i thu*[9]),[e] and in the end return to Earth. They alter in accord with the four seasons, but that there should be a limit to them is also the Tao of Nature. For instance, when pine resin imbibes the *chhi* of mature Yang for a thousand years it is transformed into pachyma fungus. After another thousand years of irradiation it becomes pachyma spirit; in another thousand years it becomes amber, and in another thousand years crystal quartz (*shui ching*[10]). These are all seminal essences formed through irradiation by the floreate *chhi* of sun and moon.[f]

This passage is not greatly innovative either in form or content; in fact it demonstrates how little originality is often needed to bring out the inherent connections of two long-established notions (in this case metamorphosis and subterranean maturation). The framework of physical explanation is perfectly typical of alchemical theory. To paraphrase as simply as possible, the cyclical processes of Nature (the Tao) can give rise to

the devolution of legendary and fabulous ancestors into men, feathered creatures, beasts, creatures with scales, and those with shells (ch. 4, pp. 11*b*, 12*a*; Erkes (1), pp. 76ff.).

---

[a] This is *Pachyma* (tuckahoe or Indian bread), the sclerotial condition of *Polyporus cocos* (R838). Cf. Burkill (1), vol. 2, p. 1618, and *PTKM*, ch. 37, pp. 3*a*ff. Besides being generated, as was thought, from pine resin, it had also a correlate, the dodder (*thu-ssu*[11]), with which it was thought to be connected as a root is with the branches and leaves of a plant. This is *Cuscuta sinensis* (R156), a Convolvulaceous phanerogam parasitic on willow branches. For further background see Vol. 4, pt. 1, p. 31.

[b] *CLPT*, ch. 12, p. 18*a* (p. 296.2), and *PTKM*, ch. 37, (p. 2). Both books cite two statements, one from a commentary to a passage which does not appear in the current text of the *Huai Nan Tzu* book (cf. ch. 17), and one from the *Tien Shu*[12] (Book of Arts) of Wang Chien-Phing[13] (+5th century). These, with others, had been collected by Chang Yü-Hsi in his *Chia-Yu Pên Tshao* (+1060).

[c] *CLPT*, ch. 12, pp. 19*b* and 20*b* (p. 297.2), and *PTKM*, ch. 37, (p. 8), entry for jet.

[d] The translation of these first two sentences is tentative, for their construction is so loose semantically that the connection of the various ideas has to depend almost entirely on interpretation.

[e] Not out of *the* earth, but as mediated by the five-elements phase Earth, in which Yin and Yang are in a perfect state of dynamic balance. The *chhi* of Earth is spoken of as primal (i.e. undifferentiated) and harmonious. Note that the Chinese natural philosopher spoke of the formation of phenomenal things in terms of both this primal *chhi* and the creative *chhi* of the Yang phase of a cycle. *Chhi* is in this context best thought of as an organising energy, and balance and creativity as two of its functional aspects.

[f] *Yü Chhing Nei Shu*[14] (Inner Writings of the Jade-Purity (Heaven), *TT*940), p. 2*a*, tr. auct. A somewhat abridged and generally inferior text appears in *Thung Yu Chüeh* (*TT*906), pp. 21*b*–22*a*.

[1] 松脂	[2] 茯苓	[3] 茯神	[4] 琥珀	[5] 陶弘景	[6] 蘇敬
[7] 瑿	[8] 元和之氣	[9] 依土	[10] 水晶	[11] 菟絲	[12] 典術
[13] 王建平	[14] 玉清內書				

things which endure, or even exist perpetually, since they have a perfectly balanced internal phasing which attunes them to the Tao's recurrent pattern.[a] The heavenly bodies embody the balance of cosmic forces (mediated by the element Earth) and are thus a paradigm of eternity.[b] At the same time the alternation of the sun and moon (and of the light and *chhi* they radiate) is identical with the cyclical domination of Yin and Yang. In the course of the cosmic cycles, exposure of pine resin underground to the configurational energy released in the recurring creative phase ('the *chhi* of mature Yang', *thai yang chih chhi*[1]) gives rise to a sequence of substances which not only endure and improve underground but are all capable of conferring immortality upon human beings.

Implicit in this sequence of ideas is a most important theme which can be glimpsed again and again in the alchemists' writings. Although the perfection of the elixir is the result of a repeated cyclical process, at each step of the treatment the intermediary product is not the same, but rather progressively exalted. Thus superimposed upon the cycle is a progressive upward tendency, which does not reverse itself.[c] The culmination of the process is irreversible—that is, no longer subject to the cyclic cosmic agencies which brought it about. In this way the adept's operations upon his materials parallel the effect of the elixir, once made, upon himself. His immortality is characterised again and again as invulnerability to the ravages of time, freedom from the cyclical attrition which governs the ageing and death—as inevitably as the birth and growth—of ordinary men. This idea is one of the crucial links between Chinese, Indian and Arabic alchemy, as well as between laboratory alchemy and other techniques of immortality in China.[d] Only our present defective comprehension of it precludes the treatment in depth which it deserves.

(iv) *The subterranean evolution of the natural elixir*

Another extension of the theory of mineral development led to positing an evolutionary branch the terminus of which was not gold, but the natural analogue to the mercuric elixir which theoreticians of alchemy valued more than any precious metal. The fact that every quality characteristic of gold varied over a certain range in native specimens of the metal encouraged early aurifactors to ignore the assayer's single standard of purity, and to envision the making of gold of still greater quintessential purity than any metal found in mines or streams. Although it is clear that the concept of the natural elixir was motivated by the desire to find a parallel for the alchemist's own Work, it was philosophically feasible because, like gold, cinnabar exists in a certain range of qualities, from very crude and irregular forms to magnificent blood-red rhombohedral crystals. The extrapolation which led to the natural elixir may be

---

[a] See pp. 404, 477 ff.

[b] The association of perfect *krasis* with perfect enduringness was an idea greatly prominent also among the Arabic alchemical theorists (cf. pp. 394, 487, 481 below). There we describe its transmission to the Latins of Western Europe from the time of Roger Bacon onwards. It certainly had one set of roots in Greek medicine, but the earlier Chinese speculations of which the Arabs received the gist could surely have been another.   [c] Cf. pp. 221 ff. above and pp. 246, 272 ff. below.

[d] See the sub-sections on physiological alchemy, pt. 5 below.

[1] 太陽之氣

followed in a remarkable extended passage from Chhen Shao-Wei's[1] *Ta-Tung Lien Chen Pao Ching Hsiu Fu Ling Sha Miao Chüeh*[2] (Mysterious Teachings on the Alchemical Preparation of Numinous Cinnabar, Supplementary to the Perfected Treasure Manual, a Ta-Tung Scripture), written perhaps *c*. +712,[a] which must be considered one of the most valuable of the surviving early treatises on account of its disquisition on the alchemy of cinnabar and its clear instructions for preparing the alchemical elixir:[b]

The highest grade of cinnabar grows in grottoes in Chhen-chou[3] and Chin-chou[4] (both in modern Hunan), and there are several types.[c] The medium grade grows in Chiao-chou[5] (centered on modern Hanoi) and Kuei-chou[6] (in Kuangsi), and is also of various sorts. The lower grade occurs in Hêng-chou[7] and Shao-chou[8] (in Hunan). That there are various grades is due to variation in purity of substance (*chhing cho thi i*[9]), diversity in perfection (*chen hsieh*[10])[d] and shadings in fineness of the *chhi* of which they are formed. Those which, stimulated by metal and mineral (influences) (*kan thung chin shih*[11]), take on a balanced *chhi*, confer, when ingested, access to the Mysteries and consecration among the Realised (or Perfected) Ones as an immortal of the highest grade (*thung hsüan chhi chen wei shang hsien*[12]). Even those composed of unbalanced *chhi* cause, when taken, perpetual life on earth.

Now the highest grade, lustrous cinnabar (*kuang ming sha*[13]), occurs in the mountains of Chhen-chou and Chin-chou upon beds of white toothy mineral (*pai ya shih chhuang*[14]). Twelve pieces of cinnabar make up one throne (*tso*[15]). Its colour is like that of an unopened red lotus blossom, and its lustre is as dazzling as the sun. There are also thrones of 9, 7, 5, or 3 pieces, or of one piece. Those of 12 or 9 pieces are the most charismatic (*ling*[16]); next are those which occur in 7 or 5 pieces. In the centre of each throne is a large pearl (of cinnabar), 10 ounces or so in weight, which is the monarch (*chu chün*[17]). Around it are smaller ones, 8 or 9 ounces (or in some cases 6 or 7 ounces or less) in weight; they are the ministers (*chhen*[18]).[e]

[a] *TT*883. In *YCCC*, ch. 69, the title appears as *Chhi Fan Ling Sha Lun*[19] (On Numinous Cinnabar Seven Times Cyclically Transformed). Sivin (1), pp. 47–48, suggests possible dates in the +6th or +8th century, but the earlier date is ruled out by a reference to O Prefecture (O-chou[20]) in the second part of the book, published under a separate title and mistakenly placed before the first part in *YCCC*; see ch. 68, p. 12b. This designation was given to the prefecture for a few years around the beginning of the +7th century, and then from the Five Dynasties period onwards. Since names once given tended to remain current among the people even after they had been officially changed, only dates prior to the end of the +6th century are ruled out. See the *Hupei Thung Chih*[21] (Historical Geography of Hupei Province), i.e. Yang Chhêng-Hsi *et al.* (1), (1921 ed.), ch. 5, p. 4b; and also above, p. 218.

[b] This translation is excerpted from an unpublished critical edition and translation of the writings of Chhen Shao-Wei by one of us (Sivin, 4). Of three available editions, the basic text was *YCCC*, ch. 69, pp. 5b–8a.   [c] Cf. Fig. 1523.

[d] The antithesis between *chen chhi*,[22] 'perfected (or realised) *chhi*', and *hsieh chhi*,[23] 'deviant *chhi*', in alchemical theories appears to be functionally equivalent to that between *chêng chhi*,[24] 'balanced *chhi*' and *phien chhi*,[25] 'unbalanced *chhi*'. For instance, earlier on, Chhen says that 'lustrous cinnabar is endowed with the clear, limpid, balanced and realised (*chen chêng*[26]) *chhi* of mature Yang' (*YCCC*, ch. 69, p. 2b). The underlying idea is that perfection in a mineral implies balance with respect to the dynamic forces of the cosmos—Five Elements, Yin and Yang, and so on. The parallel with the medical conception of health in China is obvious—and for that matter with similar conceptions in the Greek and Arabic cultures also.

[e] This is reminiscent of the drug classification in the *Shen Nung Pên Tshao Ching* (cf. Vol. 6, pt. 1).

[1] 陳少微	[2] 大洞鍊眞寶經修伏靈砂妙訣	[3] 辰州	[4] 錦州
[5] 交州	[6] 桂州	[7] 衡州	[8] 邵州	[9] 清濁體異
[10] 眞邪	[11] 感通金石	[12] 通玄契眞爲上仙	[13] 光明砂
[14] 白牙石牀	[15] 座	[16] 靈	[17] 主君	[18] 臣	[19] 七返靈砂論
[20] 鄂州	[21] 湖北通志	[22] 眞氣	[23] 邪氣	[24] 正氣
[25] 偏氣	[26] 眞正

They surround and do obeisance to the great one in the centre. About the throne are a pec[1]: (*tou*[1]) or two of various kinds of cinnabar, encircling the 'jade throne and cinnabar bed'. From among this miscellaneous cinnabar on the periphery may be picked (pieces in the shapes of) fully formed lotus buds, 'nocturnal repose', and azalea (*fu-jung thou chhêng, yeh an, hung chüan*[2]).[a] The lustrous and translucent specimens are also included in the highest class of cinnabar.

There is also a cinnabar which resembles horse teeth; that with a white lambent lustre (*pai fu kuang ming chê*[3]) is white horse-tooth cinnabar (*pai ma ya sha*[4]) of the highest grade. There is another, tabular like mica; that with a white lustre is white horse-tooth cinnabar of the middle grade. (Cinnabar) which is round and elongated like a bamboo shoot and red or purple in colour is purple numinous cinnabar (*tzu ling sha*[5]) of the highest grade. If it occurs in stony, flat prisms with a virid lustre, it is purple numinous cinnabar of the lower grade. Of (the purple numinous cinnabar) produced in Chiao-chou and Kuei-chou, only that which occurs in throne formations or which is found inside rocks when they are broken open, and is shaped like lotus buds and lustrous, is also included in the highest grade. That which is granular in form and translucent (*thung ming*[6]), three or four pieces weighing a pound, is of the middle grade. That which is laminar in form and transparent (*ming chhê*[7]) is of the lower grade. All that produced in Hêng-chou and Shao-chou is purple numinous cinnabar. Like that with a red lustre found inside rocks when they are broken open, it is lower-grade cinnabar. If creek cinnabar, granular in form and translucent, is subdued, refined, and ingested, (the alchemist) will attain perpetual life on earth, but he will not become an immortal of the highest grade. Earthy cinnabar grows in earth caves (or, mines in the earth) (*thu hsüeh*[8]), as creek cinnabar matures (*yang*[9]) in mountain rills. Because earth and mineral *chhi* are intermixed, these varieties are not suitable as ingredients of the higher kinds of medicine or for use in alchemy.

The very highest grade of cinnabar is that which occurs in throne formations. When one of the monarch pieces from the centre of the throne is obtained, subdued, refined and introduced into the viscera, the efficacy of cinnabar is particularly manifest. (This central piece) is named 'Superior Cinnabar Belvedere' (*shang tan thai*[10]).[b] It produces a permanently balanced *chhi* (i.e. bodily *pneuma*),[c] and allows one to transcend one's mundane involvements. If it is further taken in the sevenfold-recycled or ninefold-cyclically-transformed state, then without ado (*tzu-jan*[11]) the anima is transformed and the outer body destroyed, the spirit made harmonious and the constitution purified. The Yin *chhi* is dissolved, and (the persona) floats up, maintaining its shape, to spend eternity as a flying immortal of the highest grade of Realisation. Thus one knows that the realised seminal essence of the Yang[d] has

---

[a] The odd term *fu-jung thou chhêng*,[12] 'fully formed lotus buds', occurs twice more in this passage (once in the TT883 version only), and the same metaphor is also found in other writings (e.g. below, p. 242). The *fu-jung* flower is that of the lotus *Nelumbo nucifera* (R542; CC1449). The next two characters *yeh an*[13] (lit., 'nocturnal repose') must surely be the name of another flower. We have had no success at identifying it, though naturally a number of flowering plants have names beginning with *yeh*, and this one might be a textual corruption. As for *hung chüan* (interpreting it as *chüan*[14]), the name must refer to *Azalea* or *Rhododendron* spp., not now precisely identifiable (cf. R201, 203; CC524–6, 529, 530).

[b] Alternative translation: '(Taking this elixir) is called "Ascending the Elixir Belvedere".'

[c] The assumption is that taking an exceptional cinnabar, because of the internal balance of its *chhi*, will induce a corresponding balance in the adept who ingests it. As the balance in the mineral defines its perfection, the balance in the man certifies the state of immortality.

[d] See our notes earlier on *chêng Yang*[15] (p. 228) and *chen chhi*[16] (p. 237).

[1] 斗	[2] 芙蓉頭成夜安紅絹	[3] 白浮光明者	[4] 白馬牙砂
[5] 紫靈砂	[6] 通明	[7] 明徹	[8] 土穴	[9] 蕎
[10] 上丹臺	[11] 自然	[12] 芙蓉頭成	[13] 夜安	[14] 鵑
[15] 正陽	[16] 眞氣		

imbued the *chhi* (of this cinnabar) so that it exhibits a perfectly rounded nimbus, symmetrical and without imperfection. When cinnabar has been subdued and refined so that it takes the shape of a lotus bud and is translucent with a nimbus, it has become a medicine of the highest grade, which when ingested results in immortality (or, which is ingested by immortals).

The 'Canon'[a] says that cinnabar is a natural cyclically-transformed elixir, and that the vulgar are unable to gauge its fundamental principles. The uninitiated all know about 'jade throne' cinnabar. But the 'golden throne' and 'celestial throne' are cinnabars of the Purple Dragon and Dark Flower of the Most High (*thai shang tzu lung hsüan hua*[1]),[b] and not the kind which vulgar fellows can see or know about. Any devoted gentleman of the common sort, after storing up merit, can refine jade throne cinnabar alchemically and by taking it attain immortality. But as for golden throne cinnabar, a man born with immortality in his bones must first refine his spirit to a state of pure void (*chhing hsü*[2]) and live as a hermit in a cliff-bound cave. Then the immortals will gather it and feed it to him. He will forthwith be transformed into a Feathered Being (i.e., an immortal) and will bound upwards into the Lofty Purity (of the heavens). Lastly celestial throne cinnabar is collected and eaten only by the Celestial Immortals and Realised Officials in heaven. It is no medicine for lesser immortals.

When jade throne cinnabar has imbibed the pure seminal essence of Yang sentience for six thousand years it is transformed into golden throne cinnabar, the throne of which is yellow. In the centre are five pieces growing in layers, surrounded by forty or fifty small balls.[c] After 16,000 years of imbibing (essence), golden throne is transformed into celestial throne cinnabar, in which the throne is jade-green. There are nine pieces in the centre, growing in layers, pressed closely about by 72 (smaller) pieces. It floats in the midst of the Grand Void, constantly watched over by one of the spirits of the Supreme Unity (Thai I[3]). On a Superior Epoch day (*shang yuan*[4])[d] the Realised Officials descend to collect it. The mountain (on which it is found) suddenly lights up; the whole mountain is illuminated as if by fire. This celestial throne cinnabar is collected (only) by Realised Officials; people of the world can have no opportunity to gather it.

The fundamental principles of cinnabar are deep and arcane,[e] but worthy and enlightened gentlemen who have their hearts set upon floating up (to become immortals) must learn to distinguish the various qualities of the Medicine, high from low. Only then will they be ready to regulate the phases of the fire, to combine the Yin and Yang subduing methods, and then without further ado be consecrated as Perfected or Realised Immortals of high grade.

---

[a] This must be the no longer extant (and possibly non-existent) 'Canon for Making the Perfected Treasure' (*Lien Chen Pao Ching*[5]), to which Chhen Shao-Wei's works are supposed to be supplementary instructions.

[b] The Most High (Thai Shang[6]) is, like the Supreme Unity (Thai I[3]) in the next paragraph, one of the great celestial divinities of Taoism.

[c] Here we follow the reading of the *TT*883 and Chhing Chen Kuan[7] versions; the basic text has 'forty-five' (*ssu-shih-wu*[8]).

[d] The Superior Epoch day, one of three epoch days during the year, was the fifteenth day, or full moon, of the first calendar month. *TPYL*, ch. 30, p. 2a, quotes the *Shih Chi*,[9] the first of the dynastic histories (c. −90), ch. 24, p. 3a (tr. Chavannes (1) vol. 3, p. 235), to the effect that on this night offerings were made to the Supreme Unity, and comments that this is the origin of the Lantern Festival later held on the same night. On the Han cult of the Supreme Unity, see Cammann (9).

[e] The *TT*883 version reads with equal plausibility: 'The fundamental principles of cinnabar are as inaccessible as a high mountain, but superior worthies and enlightened gentlemen who have their hearts set...'

---

[1] 太上紫龍玄華　　　[2] 清虛　　　[3] 太一　　　[4] 上元　　　[5] 鍊眞寶經
[6] 太上　　　　　　　[7] 清眞館　　[8] 四十五　　[9] 史記

Of the many qualities of cinnabar enumerated above, 'creek cinnabar' and 'earthy cinnabar' were crude varieties used mainly in the commercial distillation of mercury. The kinds useful to the physician and alchemist were all exceptionally large tabular or orthorhombic crystals of substantially pure crystalline mercuric sulphide.[a] The white beds in which these minerals grew would have been drusy quartz.

Anyone who has not learned from Lynn Thorndike (1) or Frances Yates (1) to appreciate the remarkable capacity of science to coexist with magic may be troubled or even scandalised by certain tensions implicit in this text, but alchemy and even early medicine reflect them throughout. The recurring resort to scientific *chhi* and Yin–Yang explanations does not seem to sit well with the frequent reminders that the final issue of the alchemical process was expected to be an appointment to the ranks of the Spiritual Civil Service. We cannot pretend that we understand the historical dynamics of Chinese alchemy until someone has succeeded in explaining why this very real contradiction never generated sufficient dialectical voltage to be faced or resolved.[b]

It seems finally to have withered away with the ascendancy of internal or physiological alchemy in the Thang and Sung, when concern with an objective hierarchy of immortals and divinities was somewhat displaced by direct attention to the aim of what a Jungian would call psychical integration. This emphasis on personal growth is too apparent to overlook in a few lines of an alchemical poem in the 'Arcane Memorandum of the Red Pine Master' (*Chhih Sung Tzu Hsüan Chi*[1]), probably of the Thang or earlier:

> Successful means solidly building the Wall,[c]
> Indispensable to distinguish the Hard and the Soft,[d]
> Necessary that the maturing come within man,
> Due to the concentration of his heart and mind.[e]
> If his heart and mind have reached divinity, so will the Medicine;
> If his heart and mind are confused the Medicine will be unpredictable.
> The Perfect Tao is a perfect emptying of the heart and mind.
> Within the darkness—unknowable wonders.
> When the wise man has attained to the August Source,
> Then in time he will truly reach the clouds.[f]

We can only suggest for the moment that the structure of the Unseen World may have been all along in a very deep sense that of the human spirit.

[a] For a discussion of these varieties of cinnabar, see below, pp. 301ff.

[b] One must always remember the overwhelming persuasion of ancient Chinese thought that a specifically material immortality within the natural world was possible (cf. Vol. 5, pt. 2, pp. 71ff., 93ff. above). Also there is hardly any limit to the co-existence of different forms of experience within any one culture. Would Louis Pasteur, for example, have had any objection to attending mass on the Feast of St Michael and all Angels?

[c] I.e. the immurement of the spirit from the flow of chance perceptions, thoughts and images (cf. pt. 5).   [d] I.e., Yang and Yin.

[e] *Hsin*[2] is translated throughout as 'heart and mind', since neither 'heart' nor 'mind' alone would convey the sense of the original.

[f] Quoted in *Tan Lun Chüeh Chih Hsin Ching* (TT928), p. 14a; see also *YCCC*, ch. 66, p. 14a. Translation from critical edition in Sivin (5).

[1] 赤松子玄記   [2] 心

A second tension prevalent in alchemy prompts us to ask what credit Chhen Shao-Wei should be given for innovation in his account of super-cinnabar, despite his insistence that a Realised Immortal revealed the contents of his book to him one day in a mountain cave?[a] Any hope of answering this question must be greatly qualified by our inability to draw an absolute line between revelation and inspiration, but it is obviously relevant to ask how much of the information in the document was already known. We can throw light on this point to the extent that datable documents allow. Fortunately, they serve to assure us that at least the bare conception of throne formations of exceptional alchemical value was known well before the time of Chhen's epiphany.

A landmark of pharmacology, Hsü Chih-Tshai's[1] *Lei Kung Yao Tui*[2] (Answers of the Venerable Master Lei concerning Drugs), *c.* +565,[b] in the course of its enumeration of the varieties of cinnabar, makes this assertion, bland by comparison with Chhen's but an anticipation none the less:

> There is a spirit throne cinnabar (*shen tso*[3]), as well as a golden throne cinnabar and a jade throne cinnabar. If they are taken, (even) without having passed through the alchemical furnace (*ching tan tsao*[4]), they will forthwith extend one's destined span of life.[c]

As has been remarked in our study of mineralogy,[d] in the middle of the +7th century Su Ching[5] also speaks of 'lustrous cinnabar', of which

> one crystal grows separately in a 'stone shrine' (*shih khan*[6]). The largest is the size of a hen's egg, and the smallest the size of a jujube or chestnut. It is shaped like a lotus, and when broken it resembles mica, lustrous and transparent. It grows on a stone 'belvedere' inside the shrine. If he who finds it carries it on his person, it will keep him from all evil.[e]

Finally the great pharmacognostic critic Khou Tsung-Shih[7] provides an illuminating description of the mining of large cinnabar crystals at Chin-chou in his 'Dilations upon Pharmaceutical Natural History' (*Pên Tshao Yen I*,[8] preface dated +1116):

> The Old Crow Shaft (*lao ya ching*[9])...has a depth and (underground) extent of several hundred feet. First wood is piled up inside to fill the excavation and then it is set on fire. Where the dark stone cracks open there are small 'shrines'. Within each of these is a bed of white stone, which resembles (white) jade. Upon this bed grows the cinnabar, the small (crystals) like arrow-heads and the larger like lotuses. Their lustre is so great that they reflect light as well as mirrors.[f] When they are ground up their colour is a vivid red. The larger specimens of the cinnabar, together with their beds, weigh from seven or eight up to ten ounces.[g]

---

[a] *YCCC*, ch. 69, p. 1*a*. The encounter is dated in the Thien-Yuan[10] reign-period, which has not been identified with certainty. See Sivin (1), pp. 47–8.

[b] We name him as the author here, though he purports to be the commentator of an earlier Lei Kung text incorporated in his work.   [c] Cit. *CLPT*, ch. 3, p. 3*a*, tr. auct.   [d] Vol. 3, p. 649.

[e] Cit. *CLPT*, ch. 3, pp. 3*a*–3*b*, tr. auct.; cf. Okanishi (5), p. 99.

[f] 'Arrow-head' (*chien tshu*[11]) and 'mirror-face' (*ching mien*[12]) are the two most common terms used today for very high grades of cinnabar used in seal pigment. The two are mentioned together with lustrous cinnabar in the *Thien Kung Khai Wu* of +1637, ch. 16, p. 1*b*, in a passage mistranslated in Sun & Sun (1), pp. 279–80.

[g] Cit. *CLPT*, ch. 3, p. 5*b*, tr. auct.

[1] 徐之才   [2] 雷公藥對   [3] 神座   [4] 經丹竈   [5] 蘇敬
[6] 石龕   [7] 寇宗奭   [8] 本草衍義   [9] 老鴉井   [10] 天元
[11] 箭鏃   [12] 鏡面

Putting all these data together, we can reasonably posit that Chhen Shao-Wei was responsible, whether by inspiration or revelation, for adding texture to the idea of supra-normal formations of cinnabar. What interests us is that one of the conceptions which he newly applied was that of chain metamorphosis.

There is evidence that Chhen's description of super-cinnabar did not remain an utter secret after all. The *Lung Hu Huan Tan Chüeh*[1] (Explanation of the Dragon-and-Tiger Cyclically Transformed Elixir), evidently of the Wu Tai, Sung or later, follows Chhen's jade throne → golden throne → celestial throne sequence, specifying the same time-intervals between metamorphoses, and speaks of cinnabar of the highest grade as 'natural cyclically-transformed elixir'.[a] What is hardly less significant, a distant but on-pitch echo appears in the literary remains of the great Thang statesman Li Tê-Yü[2] (+787 to +849), by Taoist lights at best a 'devoted gentleman of the common sort.' His 'Essay on Smelting the Yellow', by which he means alchemy, begins:[b]

Someone asked me about the transformation involved in 'smelting the yellow'. I said: 'I have never studied these matters, so how am I to deny that there is such a thing? Still, with the aid of perfected principles one can always inquire into Nature and all its phenomena. Now lustrous cinnabar is a natural treasure of heaven and earth. It is found in rock caverns, growing on snowy beds, and resembling newly grown lotuses before the red buds have burst open. The tiny (crystals) do obeisance in a ring, while the large one occupies the centre. This corresponds to the configuration at the celestial pole, and the respective positions proper to ruler and ministers.[c] (The mineral) is lustrous and transmits light (*wai chhê*[3]). Those who gather it trace along the vein of mineral (*shih mo*[4]) (till they find it). Truly, it has been cast (*chu*[5]) by the Shaping Forces.[d]

It was not the idea of mineral evolution that interested Li, political moralist that he was. The excellence of the configuration of lustrous cinnabar lay in its resonance with the metaphysics of monarchy, which Confucius had long before illustrated with the image of the central Pole Star surrounded by genuflecting asterisms.

(4) THE ALCHEMIST AS ACCELERATOR OF COSMIC PROCESS

There is a piece of dialogue in Ben Jonson's play 'The Alchemist' (+1610) which might well serve as the text for our argument as it gradually unfolds:[e]

> Subtle: Why, what have you observ'd, Sir, in our Art,
> Seems so impossible? *Surly:* But your whole Work, no more.

a *TT*902, ch. 1, pp. 2*b*, 4*b*–5*a*.

b For a very early instance of the term *huang yeh*[6] for alchemy, see above, pt. 3, p. 36.

c An allusion to *Lun Yü*,[7] II, i, tr. Legge (2), vol. 1, p. 145: 'The Master said, "He who exercises government by means of his virtue may be compared to the north polar star, which keeps its place while all the stars turn towards it."' Cf. Vol. 3, pp. 259ff.

d *Huang Yeh Lun*,[8] in *Li Wên-Jao Wai Chi*,[9] ch. 4, pp. 6*a*–6*b*, reprinted in *Wên Yuan Ying Hua*,[10] ch. 739, p. 15*a*, tr. auct. In the preface to his 'Rhyme-prose on "Smelting the Yellow"' (*Huang Yeh Fu*[11]), which is considerably cooler toward the Outer Elixir than this essay, Li mentions the date +831 in connection with his interest in alchemy. See *Li Wên-Jao Pieh Chi*,[12] ch. 1, p. 1*a*.

e Act II, Scene iii. Duncan (1), p. 706, has pointed out that the ideas expressed are, as would be expected, Paracelsian. See also pp. 223, 231, 236 above, and p. 506 below.

[1] 龍虎還丹訣　　　[2] 李德裕　　　[3] 外徹　　　[4] 石脉　　　[5] 鑄
[6] 黃冶　　　[7] 論語　　　[8] 黃冶論　　[9] 李文饒外集
[10] 文苑英華　　　[11] 黃冶賦　　　[12] 李文饒別集

> That you should hatch Gold in a furnace, Sir,
> As they do Eggs in *Egypt*! *Subtle:* Sir, do you
> Believe that Eggs are hatch'd so? *Surly:* If I should?
> *Subtle:* Why, I think that the greater Miracle.
> No Egg but differs from a Chicken more
> Than Metals in themselves. *Surly:* That cannot be.
> The Egg's ordain'd by Nature to that end,
> And is a Chicken *in potentia*.
> *Subtle:* The same we say of Lead, and other Metals,
> Which would be Gold, if they had time. *Mammon:* And that
> Our Art doth further.

We have already seen how well Subtle's answer applies in China, and are ready to explore the transition to Sir Epicure Mammon's amplificatory remark.[a] Let us begin by summarising the next propositions which we shall endeavour to demonstrate.

Since the formation of minerals and metals is bound by time, and thus attributable to the same cosmic forces which are responsible for other life cycles, there is a very direct connection between the chemical operations of Nature and the practical techniques of the metal-working artisan. In extracting a metal from its ore, or making strong steel from brittle cast iron, he was demonstrating that man can imitate natural process, that he can stand in the place of Nature, and bring about natural changes at a rate immensely faster than in Nature's own time. The discovery that the speed of mineral growth processes, unlike those of plants and animals,[b] can be controlled by man, must certainly have been one of the main factors that led to the beginning of what we have called proto-scientific alchemy. For the alchemist went on to design processes for reproducing at a much faster rate the cyclical rhythms of Nature which controlled the maturing of minerals and metals in the earth. No man could wait 4320 years to see Nature make an elixir, but by fabricating one with his own hands in a few months or a year he would have a unique opportunity to experience and study the cyclical forces responsible for that change and thus for all natural change. No undertaking could be more quintessentially Taoist. And when the elixir acted in projection it was nothing less than a 'time-controlling substance'.[c] It accelerated the time-scale of perfection; and once the further point of perfection was reached, it cancelled time's attrition (for that is what perfection implied). Fig. 1516 has been designed to show how the deceleration of human ageing was the counterpart of the acceleration of the forming of the imperishable metal. Ko Hung says this almost in as many words: 'All the numinous fungi can bring men to longevity and material immortality—and this belongs to the same category as the making of gold'.[d] And he goes on to quote the

[a] Cf. Eliade (5), pp. 175 ff. Arabic alchemical thought, intermediate between those of China and later Europe, also had this idea. In those parts of the Jābirian Corpus and al-Rāzī's *Kitāb Sirr al-Asrār* which went to form the Latin *De Aluminibus et Salibus* a couple of hundred years later we find much talk of mercury and sulphur turning into gold and silver in hundreds of years within the earth—but also God has given power to men to accomplish the change in a few days. Fermenting like yeast, the elixirs do just this. See the translation of Ruska (21), Lat., pp. 62, 64, Germ., pp. 96, 98.

[b] The control of growth- and differentiation-rates in plants and animals had to await the development of modern biology, one of the latest of the post-Renaissance sciences.

[c] The phrase was first used by one of us (N.S.) at the Bellagio Conference on Taoist Studies (1968).

[d] *PPT/NP*, ch. 16, p. 5*a*, *b*, tr. auct., adjuv. Ware (5), pp. 268-9.

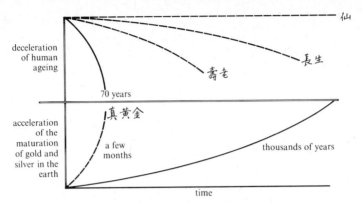

Fig. 1516. Diagram to illustrate the conception of 'time-controlling substances'. Above, the deceleration of human ageing, the attainment of prolongevity through gerontological continuance (*shou lao*) to centuries of life (*chhang shêng*) or material immortality as a *hsien*. Below, the acceleration of the normal maturation of gold and silver in the earth, 'true yellow gold' (*chen huang chin*) being produced in a few months instead of long ages.

optimistic words of Huang Shan Tzu:[1] 'Since heaven and earth contain gold, we also can make it.'[a]

What needs emphasising is that the alchemist's enterprise, as he himself defined it, was not chemistry in any usual sense of the word but physics.[b] The concern that brought his models of the cosmic process into existence was not directly with the properties and reactions of various substances. These properties and reactions were no more inherently important than the characteristics of pigments which a painter must master in order to produce the picture which exists in his mind's eye. Chemical knowledge and proto-chemical concepts were by-products, and alchemists did not lack the acumen to record and build upon them. But the aim of the process, which conditioned every step in its planning, was the model of the Tao, the cyclical energetics of the cosmos.

Looking at all the evidence impartially, one cannot escape the conclusion that the dominant goal of proto-scientific alchemy was contemplative, and indeed the language in which the Elixir is described was ecstatic. Here is one of a hundred descriptions which might be adduced to prove the point:

Open the reaction-vessel. All the contents will have taken the shapes of golden silkworms or jade bamboo shoots, or of lions, elephants, oxen, or horses, or the form of a human general of great courage. The shapes will vary, but they will all be induced by the spiritual force of the sun, planets, and stars, and the *chhi* of the heroes of sky and earth. What congeals in these amazing ways is the essence of water and fire, Yin and Yang.[c]

[a] An obscure alchemist, apparently of the Han period, but not mentioned elsewhere by Ko Hung. A Huang Shan Chün has two 'biographies' in the *Tao Tsang*.

[b] Although of course we can discern a considerable element of the former in his processes. We shall have more to say about proto-chemical ideas below, pp. 298 ff.

[c] *Chin Hua Chhung Pi Tan Ching Pi Chih*[2] (Confidential Instructions on the Manual of the Heaven-Piercing Golden Flower Elixir), +1225; *TT* 907, ch. 2, p. 16 *a*, tr. auct. This is the book by Phêng Ssu & Mêng Hsü which we draw on so much elsewhere (pp. 3, 35 ff., 43, 58, 71 ff.) in connection with chemical apparatus.

[1] 黃山子   [2] 金華冲碧丹經祕旨

In a second example we can readily identify what the alchemist was looking at:

If you wish to prepare yellow gold, take 1/24 ounce (*chu*[1]) of Cyclically Transformed Elixir and put it into a pound of lead; it will become real gold. You may also first place the lead in a vessel, heat it until it is liquefied, and then add one spatula of the Scarlet Medicine to the vessel. As you look on, you will see every colour flying and flowering, purple clouds reflecting at random, luxuriant as the colours of Nature—it will be as though you were gazing upwards at a gathering of sunlit clouds. It is called Purple Gold, and it is a marvel of the Tao.[a]

One could hardly hope for a better description of what a cupeller sees on his lead button as it oxidises and the oxide is moved by surface tension.[b] But the richness and vividness of the particulars bespeak a state of heightened awareness which one is naturally tempted to link with the alchemist's meditative practices, since we see it so widespread in the texts. We cannot rule out the possibility that drugs played a role in this tendency to perceive *multum in parvo*, many descriptions coming close to those reported by takers of hemp and other hallucinogens today, but ecstatic introspection was so common in ancient China that this is hardly a necessary hypothesis.[c]

The alchemist undertook to contemplate the cycles of cosmic process in their newly accessible form because he believed that to encompass the Tao with his mind (or, as he would have put it, his mind-and-heart) would make him one with it. That belief was precisely what made him a Taoist. As we have pointed out earlier, the idea behind Taoist ataraxy is not at all unlike one of the central convictions of early natural philosophy in the West, namely that to grasp intellectually the constant pattern which underlies the phenomenal chaos of experience is, in that measure, to be freed from the bonds of mortal finitude.[d] The idea that scientific knowledge leads to spiritual power also accounts for the extreme attention given to ritual purity—to fasting, cleanliness, invocations and spells, and the location of the laboratory in a place safe from contamination by contact with the profane.[e]

Before returning to the main thread of our exposition, it is necessary to acknowledge an obvious question. If the use of the alchemical process was contemplative, why was the adept at such pains to construct a complex object of meditation in the external world rather than in his mind? After all, a purely mental quest might well have been just as rewarding. The best answer we can offer is a reminder that laboratory alchemy was only one of many means to Taoist transcendence of the mortal condition. Each discipline had its adherents, who chose it because its style suited them (and most found it useful to choose more than one). Those who found an external object useful practised external alchemy; those who did not practised internal or physiological

---

[a] *Thai-Chhing Chin I Shen*[2] *Tan Ching* (Manual of the Potable Gold or Metallous Fluid and the Magical Elixir or Enchymoma; a Thai-Chhing Scripture), probably between +500 and +550, in *YCCC*, ch. 65, p. 15*b*, tr. auct. Another version is found in *TT*873. For other examples of heightened awareness in descriptions of alchemical products, see Sivin (1) and Maul (1).

[b] The significance of this passage was immediately apparent to the historian of metallography Cyril Stanley Smith (priv. comm. 29 October 1968).

[c] See however pt. 2, pp. 150ff. above. [d] See Vol. 2, pp. 63ff.

[e] On the further significance of such withdrawal cf. Vol. 5, pt. 3, pp. 36, 82–3 above.

[1] 銖   [2] 太清金液神丹經

alchemy; others found what they needed in sexual techniques or devotional objects;[a] those who needed no object at all sought the same end in more classical forms of meditation. External alchemy took the directions it did because some Taoists found the conjunction between spiritual perfection and the design of laboratory processes not only natural but obvious. We hope in what follows at least to begin making their reasoning accessible.

There was nothing man could do to make plants or animals grow to maturity in any but their own good time.[b] The only control the farmer exerted was to choose whether his crops or stock were to grow at all, and at the proper moment to terminate their life-cycles by harvesting or slaughtering in order to sustain the life-cycles of the human beings who were to consume them. Here one cannot but repeat Mencius' story of the man of Sung, 'who was grieved that his growing corn was not longer, so he pulled it up. Returning home, looking very stupid, he said to his people, "I am tired today. I have been helping the corn to grow long." His son ran to look at it, and found the corn all withered.'[c] This conviction that man's benefit lies in conforming to and when possible furthering the inexorably paced work of Nature lies close to the heart of Chinese quietism and Taoist ataraxy.[d] Equally, the realisation that the rate of mineral growth was controllable was one of the stoutest ideological props of the quest of the Taoist magus for a state of affairs in which, as Ko Hung[1] put it, 'my span of life is up to me, not to Heaven.'[e] That realisation was no innovation of the alchemist, although he was certainly the first to make philosophical use of it. In truth it is a distant cousin of the assertion often made about great artificers and inventors like Chang Hêng[2] (+78 to +139) that 'their ingenuity (or workmanship) rivalled (or equalled) that of the Shaping Forces.'[f] As Eliade has shown for many cultures, and Granet began to demonstrate for China, a consciousness of superhuman responsibility for interfering in the life-cycle of minerals was embodied in the rituals (often obstetrical in imagery) of the miner who delivered the ores from their womb and the smelter who converted them rapidly into metals much further along the scale of maturity.[g] Both were taking unto themselves dangerous powers, and needed all the protection that tabus and rituals could provide. This need was also urgent in alchemy.[h] That metal-workers succeeded made them magicians and heroes.[i] The alchemists, who accepted

---

[a] The most general view, however, was that sexual techniques alone could lead to lengthened life but not to immortality. See, for instance, *Huang Ti Chiu Ting Shen Tan Ching Chüeh* (*TT*878), ch. 4, pp. 1a–2b, which makes this point but also says: 'Or if he only takes the Medicine, and does not obtain the essentials of the Art of the Bedchamber, then it will be impossible for him to live for ever.' This is because if adepts 'give free rein to their emotions and desires, without knowing that they can equitably regulate the dispersion (of their *chhi*), they are hacking at the trunk of their lives.' On this whole subject see pt. 5 below.

[b] This is not to deny the existence of beliefs in the possibility, for instance, of growing plants instantly by magic. The tradition of this in Chinese culture has been traced by van Gulik (4).

[c] See Vol. 2, pp. 576–7. [d] See Vol. 2, p. 66.

[e] We discuss this slogan more fully below in pt. 5.

[f] *Chin Shu*,[3] ch. 11, p. 3b, is an example of this.

[g] Eliade (5); see also Granet (1), pp. 496–501.

[h] The most complete collection of alchemical rituals is probably that in the *Huang Ti Chiu Ting Shen Tan Ching Chüeh* (*TT*878). [i] Eliade (5), pp. 87–108.

[1] 葛洪   [2] 張衡   [3] 晉書

the reality of the magic-ritual experiential universe of the smith, were at the same time ready to apply to it an abstract proto-scientific analysis. They saw the parallel between the metallurgist's midwifery and the operation of the Shaping Forces in Nature, and adapted to cosmic concerns the arts of maturing metals.

That the apparently artificial conditions of the laboratory could be made profoundly natural and responsive to the operation of the larger Tao is an axiom of alchemy. As the 'Supplementary Instructions to the Yellow Emperor's Nine-cauldron Spiritual Elixir Canon' put it,

When earth mixes with water to form mud, and is kneaded (*hsien*[1]) (by subterranean processes) below a mountain, there will be gold, and generally cinnabar above it. When this (cinnabar) is ceaselessly metamorphosed and cycled, and once again forms gold, this is merely a reversion to the root substance, and not something to be wondered at.[a]

How could the alchemist be sure that what went on in his reaction-vessel represented a cosmic process? If it did not, like the man of Sung he would be overruled by the Tao, and his elixir would wither. Given the character of the Chinese system of thought, success was bound to be a question of establishing correspondences[b] which would ensure the identity of his process with that of Nature. The adept had at his disposal a diversity of approaches ranging from highly abstract theories to magical invocations. The complex design of most extant alchemical processes obviously depended upon so many considerations of every kind that today we can hardly begin to explain particular choices of ingredients, apparatus, and treatment. The one area in which we have at least begun to glimpse the rationale behind the concrete processes is at the same time the most abstract and very probably the most crucial, namely that which has to do with the application of correspondences by the use of qualitative or quantitative analogy.

There were three primary points of application which could be used to plan a particular process as a recapitulation of the natural evolution of metals: materials, apparatus, and the timing of combustion. We shall see that all three possibilities were actually exploited, generally in conjunction. Time was the key to all three, for the Way of Nature is cyclical. It was easily within the operator's means, through timing, to make his process a microcosm which 'succeeds because of its correspondence on a scale of thousandths.'[c] Since the cosmic cycles fall naturally into phases (generally marked by Yin–Yang states or the Five Elements), he had the option of temporally phasing some aspect of his treatment. For instance, he could vary the intensity of the fire so that it gradually increased or decreased at a measured tempo, analogous to that of the alternation of Yin and Yang in the course of the year. If this controlled variation in the temperature of the process was exerted upon two ingredients, or two main ingredients, he could expect that the phasing would set up inside his sealed vessel an

---

[a] *Huang Ti Chiu Ting Shen Tan Ching Chüeh* (*TT*878), probably compiled in the Sung (+10th cent.), ch. 13, p. 2a, tr. auct.

[b] See Vol. 2, pp. 261 ff.

[c] I.e. what we called in Vol. 2 symbolic correlations. Cf. p. 227 above and pp. 264 ff. below. We may continue to use the two terms interchangeably.

[1] 烻

alternating pattern of ascendance. First the Yin reactant would be dominant, and then the Yang one. The alternation would ensure that their qualitative correspondence to Yin and Yang became a dynamic correspondence to that rhythmic interplay of the positive and negative forces which was responsible for the maturation of metals as well as for all other growth. The alchemist could invoke further guarantees of fidelity to cosmic process by controlling the design or dimensions of the apparatus to produce a spatial microcosm as well. The furnace might be oriented with respect to earth and sky by what we would consider ritual means; then again, its measurements might be planned for numerological significance connected with the Order of Nature; or its form might be based upon that of the womb or its analogue the cosmic egg.

Let us now proceed to examine the ways in which these possibilities, and others created by their interplay, were actually applied, in order to throw more light on the ideas which evolved them. Again we can hope to do no more than demonstrate how a few basic strategies were embodied in a great variety of tactics. It is impossible to say very much about the development of these tactics when the chronological relations of so many sources can still only be guessed at. We must also remind the reader that some alchemists seem quite unconcerned with cosmic parallels, and indeed with any rationalisation of the process at all. But this almost purely pragmatic approach is the exception, and even so an acquaintance with theory is often implicit in its documents.

By the +16th century European alchemists had also come to appreciate the timing of reactions, and many a medieval Chinese adept would have agreed with the words of William Blomfield, written in +1557:

> But if thou wilt enter the *Campe of Philosophy*
> With thee take Tyme to guide thee in the way;
> For By-paths and Broade wayes, deep Vallies and hills high,
> Here shalt thou finde, with pleasant sights and gay;
> Some shalt thou meete which unto thee shall say
> *Recipe* this, and that; with a thousand things more
> To *Deceive* thy selfe, and others; as they have done before.[a]

(i) *Emphasis on process in theoretical alchemy*

Although one cannot conclude that the *Chou I Tshan Thung Chhi* (+142) was the *fons et origo* of theoretical alchemy merely because it is the oldest book of its kind which we can still examine, it was certainly considered a basic canon by later theoretically oriented alchemists, who referred to it often and adapted its idea of a chemical process based upon cosmic patterns.[b] We have seen earlier that no one can even say with confidence what the book was meant to be about. It can be—and has been—read as a poetic treatise on the inner significance of the 'Book of Changes', on cosmology,

---

[a] 'Bloomefields Blossoms', in Ashmole's *Theatrum Chemicum Brittannicum*, 1652, pp. 305–23.

[b] Studying its exegesis and adaptation in early alchemical and other Taoist treatises is indispensable for evaluating the *Tshan Thung Chhi*, since none of its early commentaries survive, and the late ones represent very tendentious readings. In particular, as we have noted (pt. 3, p. 57), they re-interpret the alchemical level entirely in terms of the enchymoma.

on breath control, on sexual techniques, on laboratory alchemy, or on any combination of these.[a] Although the uncertainty is real enough, it is not very relevant to the later development of alchemy. Those alchemists who used the book simply assumed that it really was about the Outer Elixir, and that its purpose was to describe in recondite language the metaphysics of the laboratory process. So reading it, they were no less satisfied than those who applied its concepts (and still do, for that matter) to physiological disciplines.[b]

Here we recall a comment of the bibliographer Chhao Kung-Wu[1] (d. +1171) on the 'Essay upon the Sun, Moon, and the Dark Axis' (*Jih Yüeh Hsüan Shu Lun*,[2] c. +740) of Liu Chih-Ku:[3]

In the reign of the Brilliant Emperor he was Prefect of Chhang-ming[4] in Mien-chou[5] (Szechuan). At that time there was an edict seeking out gentlemen who understood the Elixir Medicine. Chih-Ku said that of the great Medicines of the immortals, none falls outside the scope of the *Tshan Thung Chhi*. He therefore composed this essay and submitted it to the court.[c]

The *Lung Hu Huan Tan Chüeh* puts it just as unequivocally:

For the Cyclically Transformed Elixir there is no formula; the *Chin Pi Ching*[6] and the *Tshan Thung Chhi* are its formulae.[d]

The preface to the oldest extant commentary upon the latter, written c. +945 by Phêng Hsiao,[7] a priest of the Chêng-I[8] denomination of Taoism in Szechuan, sees in it the prototype of the cosmic model (though Phêng was interested in laboratory alchemy only to the extent that its ideas and imagery were incorporated in 'dual cultivation').[e] He wrote:

(Wei Po-Yang) compiled the *Tshan Thung Chhi* to show that in preparing the Elixir one's Tao is the same as that of the Shaping Forces of Nature. Therefore he drew upon the symbols of the 'Changes' to develop this point.[f]

---

[a] See Vol. 5, pt. 3, p. 74.   [b] See pt. 5.
[c] *Chao-Tê Hsien-sêng Tu Shu Hou Chih*[9] (the sequel included in *Chün-Chai Tu Shu Chih*,[10] +1151), ch. 2, p. 33b, tr. auct. This note appears in the body of the 20-chapter edition compiled by Chhao's disciple Yao Ying-Chi[11] from Chhao's 4-chapter edition with posthumous addenda, and first printed in +1249 (ch. 16, p. 10b).
[d] *TT*902, ch. 1, p. 1a, tr. auct. See also *YCCC*, ch. 70, p. 1a. The *Chin Pi Ching* remains one of the great enigmas of Chinese alchemical literature, constantly mentioned and quoted (especially from the Sung on) as a text of great authority, but without agreement as to alternative names, authorship, or provenance. For instance, the citation from it by Wu Tshêng which we have quoted earlier (pt. 3, p. 150) is to be credited to the *Chin Tan Chin Pi Chhien Thung Chüeh*[12] in the *nei tan* section of *YCCC*, ch. 73, p. 7b. As Chhen Kuo-Fu (1), pp. 287–9, has noted, a number of quotations attributed to the *Chin Pi Ching* in late works actually come from the *Tshan Thung Chhi*. Since at the moment we have nothing to add to Chhen's review of the problem, we refer the reader to his note.
[e] 'Dual cultivation' is (as we saw, p. 212) the term coined by Liu Tshun-Jen (1) to describe a late form of internal or physiological alchemy which depended greatly on a variety of sexual techniques. We do not agree with his view that laboratory alchemy played little or no part in it. See below, pt. 5.
[f] *Chou I Tshan Thung Chhi Fên Chang Thung Chen I*[13] (*TT*993), preface, p. 1b, tr. auct. Although this commentary has no date, Phêng's *Chou I Tshan Thung Chhi Ting Chhi Ko Ming Ching Thu*[14] (*TT*994) is dated +947.

[1] 晁公武   [2] 日月玄樞論   [3] 劉知古   [4] 昌明   [5] 綿州
[6] 金碧經   [7] 彭曉   [8] 正一   [9] 昭德先生讀書後志
[10] 郡齋讀書志   [11] 姚應績   [12] 金丹金碧潛通訣
[13] 周易參同契分章通真義   [14] 周易參同契鼎器歌明鏡圖

But the surest sign of the book's importance is its ubiquity. The majority of later writings (especially of the Thang and Sung) which quote any authority on theory quote it, very often citing it simply as 'the Canon'.[a]

The process of the *Tshan Thung Chhi*, when it is read on the laboratory-alchemical level, involves two ingredients which are sealed in a reaction-vessel and subjected to the cyclically regulated influence of heat. The reactants, as we have seen, are likened to Yin and Yang both directly and by the use of many Yin–Yang embodiments—dragon and tiger, fire and water, husband and wife, and so on. The equitably phased variation in the intensity of the fire is also explained in terms of the cosmic Yin–Yang cycles which condition the coming-into-being and passing away of phenomena. The sequence of steps is controlled by the use of the *I Ching* trigrams and hexagrams. The reaction-vessel is likened to the undifferentiated primordial chaos (*hun-tun*[1]) from which phenomenal things are eventually formed. Each of these themes became perennial, but there was a less obvious influence upon later generations too. In the *Tshan Thung Chhi* the emphasis is on the process, and the product is practically ignored. There are no instructions for compounding, no rituals for ingestion, and a mere couple of cursory descriptions of that immortal beatitude which to pragmatic alchemists like Ko Hung was the whole point.[b] In this sense the *Tshan Thung Chhi* was a precursor of the extreme theoretical tendency in later alchemy. Among its posterity we find occasionally such a concern with gnostic rapture, achieved by contemplating the process, that the practical steps between understanding the reaction and becoming an immortal are skipped altogether. Perhaps the clearest of many examples occurs in the *Thai Ku Thu Tui Ching*[2] (Most Ancient Canon of the Joy of the Earth),[c] an undated work, possibly Thang or earlier, on the fixing ('subduing', *fu*[3])[d] of minerals and metals:

This discussion of the Five Metals is not the great doctrine of the Perfect (or Realised) Tao. But if (the devotee) attains a clear and penetrating understanding of these Five Elements, one can proceed to a discussion of fire-subduing, and can then talk to him about the Tao of projection (*tien hua*[4]).[e] When he has comprehended every aspect of the Five Elements, he will be a man of balanced Realisation, and the Three Worms[f] will leave his body.[g]

---

[a] This point may be verified even from the few treatises which have been translated into Western languages; see e.g. Fêng & Collier (1) and Spooner & Wang (1).

[b] That the adept's life is lengthened once the elixir enters his mouth is stated in ch. 1, p. 24*a*. His journey through the void and enrolment among the immortals is described very briefly in ch. 1, p. 19*b*, and ch. 3, p. 10*b*.

[c] In this very provisional translation of the title, 'Earth' (*thu*[5]) is the median element and 'Joy' (*tui*[6]) is one of the eight trigrams of the *I Ching*. Among the metals the former is associated with gold and the latter with silver (though that is not its only meaning in alchemy; see Sivin (1), pp. 194–5 n.). On the date of this work see Chhen Kuo-fu (1), vol. 2, p. 391.

[d] Sivin (1), p. 148, notes that 'as used rigorously *fu*[7] or *fu huo*[8] means "chemical treatment of a volatile substance so that it is no longer volatile under normal conditions"'; and that the terms are also used loosely to describe certain products merely earthy in appearance, or, when mercury is fixed, merely solid. *Chih*[9] (lit. 'restrain') is used alternatively for the same kinds of process, as is *ssu*[10] (lit. 'kill') though the latter is sometimes specialised to mean ridding of toxicity (see above, pp. 4ff., 187, 191, 198 and below, p. 263).

[e] See pt. 3, *passim*.   [f] See pt. 5 below.   [g] *TT*942, ch. 1, p. 4*b*, tr. auct.

[1] 混沌   [2] 太古土兌經   [3] 伏   [4] 點化   [5] 土
[6] 兌   [7] 伏   [8] 伏火   [9] 制   [10] 死

(ii) *Prototypal two-element processes*

Although ink will continue to be spilt over the question of precisely what chemical reactions the *Tshan Thung Chhi* is describing,[a] the general outline of the process is unambiguously cosmogonic. It is recapitulated in this rhymed passage:[b]

> Cinnabar is the seminal essence of Wood;
> When it encounters Metal, they unite.
> Metal and Water conjoin,
> Wood and Fire are partners.
> These four the chaos (*hun-thun*[1]),
> Aligning as dragon and tiger.
> Dragon Yang, its number odd;
> Tiger Yin, its number even.
> Liver, caerulean, the father,
> Lungs, white, the mother,
> Reins, black, the son:[c]
> Three substances, one family,
> Reunited at the centre (*wu chi*[2]).[d]

The apparent obscurity of this text begins to dissipate as soon as we recall the correlation between the Five Elements and Yin and Yang. Fire and Water represent the maximal or mature phase of Yin and Yang respectively (Fig. 1515). Wood and Metal stand for the phase in which one of the polarities is becoming dominant but is not yet at its height—within the system of the year, naturally, the intermediate seasons of spring and autumn—and so on. Seen in another way, they are intermediate phases in the alternating dominance of the polar complements. To use a metaphor the cogency of which will shortly become clear, Wood (immature Yang) is the son of Water (mature Yin), from which it emerges, but it is also the father of the Fire phase (mature Yang) which succeeds it.[e] Chinese thinkers ordinarily referred to these emergent

[a] We do not raise again here the question of whether the *Tshan Thung Chhi* was originally meant as a work of laboratory or physiological alchemy or both (see above, pt. 3, pp. 50ff.). Our aim in what follows is simply to show how the book was understood by *wai tan* thinkers. *Nei tan* adepts had their own interpretations, which varied according to whether the treatise was being understood philosophically or in terms of respiratory or sexual practices; examples may be found throughout Vol. 5, part 3, especially pp. 150, 200.

[b] Ch. 2, p. 23*a*, tr. auct.; cf. Wu & Davis (1), p. 255. The lack of a critical edition is especially unfortunate here, for various versions differ as to the number of lines in this passage and their arrangement, as noted below. Our translation rejects all lines in which confidence is not fully warranted, but we must emphasise that it is based on a tentative reading of the text.

[c] At this point in the *Chou I Tshan Thung Chhi Fên Chang Chu Chieh*, our basic text, three additional lines occur: '(The trigram) Li,[3] scarlet, is the daughter, Spleen, yellow, is the ancestor; The motion commences on the north–south line (*tzu wu*[4]).' They appear also in the *Tshan Thung Chhi Shan Yu* of +1669 in a different order, but in Chu Hsi's *Khao I* of +1197 only the second line is found. The great commentator notes: 'The two lines "Heart, scarlet" and "Spleen, yellow" do not appear in the various editions; I do not know which version is the correct one' (pp. 20*b*–21*a*). The wording of this note indicates that versions available to Chu Hsi originally read 'Heart, scarlet' (*hsin chhih*[5]) rather than 'Li, scarlet' (*li chhih*[6]). We provisionally omit both lines partly because of their uncertainty and partly because they conflict with the idea of a triune family expressed in the next line.

[d] Literally, 'to the fifth and sixth of the ten celestial stems', which correspond to centre and Earth.

[e] We follow, as do the texts, the Mutual Production order. See also above, pp. 225, 229.

[1] 混沌   [2] 戊巳   [3] 離   [4] 子午   [5] 心赤   [6] 離赤

phases as 'the Yin within the Yang', and vice versa. Earth is the neutral phase of balance in which, as we should put it, the polarities cancel out.

Read alchemically, Wei Po-Yang's verses begin by constructing the primordial Chaos out of the four 'unbalanced' elementary phases (in which either Yin or Yang predominates). Only one of the four, Wood, is explicitly identified with a substance, though it is natural enough to speculate that Metal stands for another. But we know already that the customary association of cinnabar is not with Wood but with Fire, which follows Wood in the Mutual Production succession order. We are constrained to allow for the possibility that 'cinnabar' is meant no more concretely than the dragon's odd number a bit further on, and that the first line may be asserting nothing more than the conventional genesis of the category Fire (as 'seminal essence')[a] from the category Wood. Be this as it may, Metal and Wood (immature Yin and Yang) unite, and also merge with Water and Fire (mature Yin and Yang) through affinity of like with like, to form the Chaos. What fills the functional categories Water and Fire— whether other substances or alchemical treatment with water and fire—is left open. Indeed the point may simply be that Wood and Metal mature within the Chaos in the direction of complete differentiation as Yang Fire and Yin Water, though this explanation would seem to be based on a rather confused notion of the Chaos.

Then in the course of the treatment the 'iron law of entropy' reverses itself, and the undifferentiated contents of the vessel segregate spontaneously into Yin and Yang components (tiger and dragon), which are thought of as spatially separate. These differ from the Yin–Yang components with which the process began in that their polarities are reversed. In the cosmological tradition the status of dragon and tiger as abstractions is ambiguous. They embody Yin and Yang emergent from their opposites, but early sources differ as to whether the dragon is Yang within Yin or Yin within Yang.[b] Here we can be reasonably sure that the dragon represents Yang emergent from Yin (its odd number is merely another Yang resonance), and the tiger the opposite.

The point is reinforced by the image of a family, in which the immature Yang, or Wood (identified by its visceral and colour associations) is the father, the immature Yin, or Metal, the mother, and the mature Yin, or Water, the son. This feminine son redeems his family through a return to the Centre, that is to say through his role as an intermediary in the formation of the Yellow Sprouts from which the Elixir is grown. The line 'Spleen, yellow, is the ancestor',[c] which appears in Chu Hsi's text, affirms

---

[a] See pp. 229ff.

[b] A commentary on the *Huai Nan Tzu* book, ch. 3, p. 2*b*, which appears in *TPYL*, ch. 929, p. 7*a*, refers to the dragon as the Yin within the Yang, or immature Yin. The opposite, however, is asserted in the apocrypal *Chhun Chhiu Wei Yuan Ming Pao*[1] (*Ku Wei Shu*, ch. 7, p. 8*b*). In the *Kuan Lo Pieh Chuan*,[2] another work also mostly lost, we are told that 'the dragon is the Yang seminal essence lying concealed within the Yin'. Both these passages are cited in the encyclopaedia *Chhu Hsüeh Chi*,[3] ch. 30, p. 22*a* (pp. 740, 738) respectively. Although the dragon is conventionally associated with water (as e.g. in *Lun Hêng*, ch. 29, tr. Forke (4), vol. 1, pp. 356–7, where the relation is specified as categorical affinity), in the trigram system of the 'Book of Changes' it is associated with the fiery *kua* Chên.[4] As for Kuan Lo, he was a famous geomancer and diviner of the San Kuo period (cf. Vol. 4, pt. 1, pp. 296, 302), and his biography was written in or before the Thang.   [c] Or, 'grandparent'.

[1] 春秋緯元命包   [2] 管輅別傳   [3] 初學記   [4] 震

this point while completing the family metaphor; the ancestor, corresponding to the medial Earth phase, is the neutral organising centre to which the son returns.[a] It is easy enough to find this metamorphosis delineated explicitly in later texts, as in this example from Chang Hsüan-Tê's 'Mental Mirror':

The oral formula says: 'Use 8 oz. of lead, which is Yang, the Masculine, and the Tiger; and 9 oz. of quicksilver, which is Yin, the Feminine, and the Dragon. These two ingredients may metamorphose into a Lead which is also Yin. It corresponds to black, Water, and the number 1, and is Yin'.[b]

It is important to keep in mind that this concreteness closes many alternate avenues of opinion which the *Tshan Thung Chhi* leaves open, and which other alchemists later followed.

Thus summarised, the plot of the story incorporates the familiar separation of Yin and Yang out of the universal blend, which Wang Chhung[1] had amply expressed sixty years before Wei Po-Yang. We have already discussed the marriage of the masculine and feminine forces, which engenders the phenomenal world, as it is described in the *Huai Nan Tzu*[2] two centuries earlier still.[c] But here we see a new idea of great originality and religious depth: a double hierogamy, the first union resulting in complete undifferentiation and then complete differentiation, and the second union leading ultimately to the perfectly balanced and enduring organisation of the Elixir.[d] It would be tempting, though perhaps superficial, to point out a parallel with the basic spiritual process of Western alchemy, which unites the Stoic and Gnostic pioneers with the Christian magi of the Renaissance: the androgynous union as the Death of the Soul, and the perfect reconciliation of opposites in its resurrection.[e]

To return to our exploration of the alchemical level of meaning, in principle Yin and Yang may be brought to bear on the process in different ways and at different stages. First they can be represented by two reactants which are blended and sealed within the vessel, duality subsequently merging to constitute the Chaos. It is equally logical to apply Yin and Yang as cyclic phases which alternate in time, as the sealed vessel and its contents are subjected to the periodic variation of fire or some other

[a] One could hardly wish for a better instance of the functional character of alchemical thought. In the family image the immature elementary phases are considered the parents, which give birth to the mature Yin son, this son in turn completing them. The dragon and tiger, like the son, represent the post-Chaos stage, but they are conventionally associated with the immature polarities. The apparent contradiction disappears when we realise that the dragon–tiger image is meant to suggest not latency but emergence—in other words, to emphasise the evolution of the polarities from their opposites.

[b] *Tan Lun Chüeh Chih Hsin Ching* (*TT*928), probably Thang; p. 6a, tr. Sivin (5). It is not impossible that the oral formula ends just before the last sentence, which would thus be the author's comment.

[c] Pp. 224–5 above, and Vol. 2, pp. 371–4. Of course both these ideas may be found less distinctly expressed though fully fleshed before the Han. See, for the former, the Great Commentary of the Book of Changes (*Hsi Tzhu Ta Chuan*[3]), XI, 5, tr. Wilhelm (2), vol. 1, p. 342; and for the latter, *Lü Shih Chhun Chhiu*[4] (−239), ch. 5, sect. 2, (ch. 22); tr. Wilhelm (3), p. 56.

[d] For the *nei tan* interpretation of this same process, see below, pt. 5.

[e] Cf. pt. 2, pp. 22 ff. above, p. 361 below, and again Vol. 5, pt. 5.

[1] 王充   [2] 淮南子   [3] 繫辭大傳   [4] 呂氏春秋

treatment. The separation of the Chaos yields a 'pure' Yin and Yang, of supramundane perfection and thus no longer embodied in the ingredients. The new pair bears the same relation to the original substances as an immortal does to an ordinary mortal. Their polarities are reversed to signify the realisation of potential. Later alchemists and annotators spoke of the pair as Realised Lead (*chen chhien*[1]) and Realised Mercury (*chen hung*[2]), or Realised Metal and Realised Water, and tended to think of them as actual substances, intermediates in the preparation of the Elixir.[a] Some, less interested in theoretical rigour than in results, simply took them as cover-names for ingredients.[b] Nevertheless, once the commentaries are set aside it is hard to see this second-stage Yin and Yang represented in the text as anything but functional categories.

The implication of many commentators that the original ingredients are lead and mercury is also far from unambiguously justified, even on the assumption that the book was written to make chemical sense. In an extremely arcane argument which introduces sons and mothers, white tigers and caerulean dragons, sun and moon, purely for their categorical associations, it is perfectly possible that the metal lead is merely meant to stand for the corresponding element Water. Let us examine the only mention of metallic lead in context.[c] There it is juxtaposed with the River Chariot (*ho chhê*[3]), which in later alchemy regularly refers to mercury. It would be poor method, of course, to assume the same specific identification in the *Tshan Thung Chhi*:

> Knowing the white, cleave to the black,
> And the spirits will make their appearance.
> The white is the essence of Metal,
> The black, the fundament of Water.
> Water is the pivot of the Tao;
> Its number is called One.
> At the inception of Yin and Yang
> The Dark (*hsüan*[4]) holds Yellow Sprouts in its mouth.
> It is the Master of the Five Metals;
> The River-chariot of the North.
> Thus lead, black outside,
> Holds in its bosom the floreate essence of Metal.[d]

[a] The outstanding prominence of 'true' or 'vital' lead (*chen chhien*) and 'true' or 'vital' mercury (*chen hung*) in *nei tan* physiological alchemy we emphasise in its appropriate place below (pt. 5). Although the terms occur more rarely in *wai tan* elixir alchemy they certainly do appear in that context (cf. pp. 258ff.), and can even be found in connection with chemical industry. An example of this last usage occurs in *Ling Wai Tai Ta* (+1178), ch. 7, pp. 11b, 12a, where Chou Chhü-Fei is discussing the cinnabar mines of Yungchow and Kuei-tê. He opines that mercury produced from cinnabar is not *chen hung* (thus contrary to Mêng Yao-Fu, p. 259 below), and suggests that the term should be reserved for the native mercury which could be collected in these mines. Free mercury does in fact sometimes occur disseminated in mercuric ore beds (Gowland (9), p. 348; Mellor (1), p. 341; Partington (10), p. 393). How much Chou Chhü-Fei knew about the terminological usages of *wai tan* and *nei tan* adepts is of course a moot point. He is clearly taking '*chen hung*' in its most literal sense as 'authentic mercury'.

[b] Especially in physiological alchemy; cf. pt. 5 below.

[c] Metallic lead is also mentioned, along with white lead, in an example of transformation in ch. 12, p. 25b; see above, pt. 3, p. 68.

[d] Ch. 1, p. 16a, tr. auct. Echoes of the *Tao Tê Ching* are very obvious here; cf. Vol. 2, pp. 57, 59.

[1] 眞鉛   [2] 眞汞   [3] 河車   [4] 玄

The mention of Yellow Sprouts establishes that these verses are about the Yin and Yang which have emerged from the Chaos, not those which went into it; even in the extant commentaries, none of which is committed to a *wai tan* interpretation, we find the white and black of this passage equated with Realised Metal and Realised Water.[a] 'The Dark' and 'lead' are thus arcane ways of referring to the Black Son encountered a few pages back—the Yang Water out of which the Yellow Sprouts (and thus, in the longer view, the Elixir) is prepared. The 'floreate essence of Metal' in the last line is an already familiar way of designating the element which precedes Metal in the Mutual Production order, namely Earth, to which the balanced Yellow Sprouts corresponds.[b] The emphasis on the black, on Water as the pivot of the cosmic process, is anything but rhetoric. Of the post-Chaos Yin–Yang pair, black Metal and white Water, it is the former which becomes pregnant with the Yellow Sprouts.[c]

The early alchemist who was more concerned with finding practical instructions in this gnomic text than with plumbing its philosophical meaning could take either of two basic directions. He could interpret it as concerned with some operation involving lead and mercury, perhaps in an amalgam.[d] But he might also understand the direct reference to lead as a mere illustrative example of how a substance can be one thing on the outside (or actually) and something else inside (or potentially). On this reading the text is concerned in a very theoretical way with the metamorphoses of mercury— or cinnabar, which as we shall see amounts to the same thing—alone. Many alchemical treatises which follow in one way or another the tradition of the *Tshan Thung Chhi* merely deal abstractly with the deeper meaning of its correspondences or images, or the further development of its hexagram phasing system; these require no commitment to particular amounts of specific minerals or metals. But any Taoist who aimed to carry out a two-element elixir process had to come to a concrete chemical understanding of the *Tshan Thung Chhi*, within the limits of his access to the alchemical and technological knowledge of his time.

The intellectual history of the various choices that were made will be one of the most interesting chapters in the history of Chinese alchemy when the time comes that it can be written. In the meantime a couple of examples of actual prototypal two-element processes must serve to illustrate the interplay of theory and practice in medieval alchemy. We are here concerned primarily with practical experimentation but in considering this one should always bear in mind how deep was the impress of the two-element processes upon physiological alchemy (cf. pt. 5.). Indeed the physiological adepts were, one might say, on the whole more faithful to them than the chemical alchemists.

The oldest mercury–lead process for which we have clear directions is in the 'Yellow Emperor's Canon of the Nine-Vessel Spiritual Elixir'. One of the sources of Ko Hung's *Pao Phu Tzu* (*Nei Phien*), the 'Canon' may indeed antedate the *Tshan*

---

[a] For instance, *Tshan Thung Chhi Shan Yu* (+1729), Sect. 7.
[b] See Vol. 2, pp. 255ff.
[c] See below, p. 259, where Realised Lead is called 'the ground of the Elixir'.
[d] See Vol. 5, pt. 2, pp. 242ff.

Thung Chhi. This is the 'Canon's' recipe for 'Black-and-Yellow' (*hsüan huang*[1]),[a] an intermediary in the preparation of the Nine Elixirs:[b]

Take ten pounds of quicksilver and twenty pounds of lead. Put them into an iron vessel, and make the fire underneath intense. The lead and the quicksilver will emit a floreate essence (*ching hua*[2]). This floreate essence will be purple, or in some cases may resemble yellow gold in colour. With an iron spoon, join it together and collect it. Its name is 'Black and Yellow', and it is also named 'Yellow Essence' (*huang ching*[3]), 'Yellow Sprouts' (*huang ya*[4]), and 'Yellow Weightless' (*huang chhing*[5]).[c] The medicine is then put inside a bamboo tube and steamed a hundred times. It is mixed with realgar and cinnabar solutions and volatilised.

The point of the final instructions becomes clearer subsequently, when the alchemist is directed to dissolve the Yellow Sprouts in a weak mineral acid mixture (*hua chhih*[6]),[d] recover it by evaporation, and subdue it in the fire (*fu huo*[7])[e] by heating for 36 days in a heavily luted vessel.[f] Then it is sublimed for another 36 days over an intense fire to yield the first of the series of nine canonical elixirs, 'Elixir Flowers' (or 'Floreate Essence of Cinnabar', *tan hua*[8]).

One can only guess once again at the chemical identity of Black-and-Yellow. Its colour is not necessarily yellow. It is not necessarily a sublimate, for there is no direction that the vessel be closed. Even if sublimation was involved, the product may have included the non-volatile portion of the reactants, for the instruction to 'join' (*chieh*[9]) the product while collecting it could conceivably refer to bringing together a sublimate and a residue. All sorts of helpful details are available in the late supplementary explanations (*chüeh*[10]), which, for instance, treat the 'joining' (rather implausibly) as the formation of the amalgam, but these reflect their own time and not that of the 'Canon'.[g] Chhen Kuo-Fu has suggested that the product was a mixture of

[a] In this case black and yellow do not correspond directly to the elements Water and Earth, but to Heaven and Earth. This equation comes from the *Wên Yen*[11] commentary to the 'Book of Changes' (*sub* hexagram Khun[12]); tr. Wilhelm (2), vol 2, p. 29. Cf. also the *Thai-Chhing Chin I Shen Chhi Ching*[13] (Manual of the Numinous Chhi of Potable Gold; a Thai-Chhing Scripture, *TT*875), which makes *hsüan huang* two distinct substances, Supernal Black (*thien hsüan*[14]) and Terrestrial Yellow (*ti huang*[15]). This text, which in its flamboyant imagery and emphasis on ritual resembles the texts directly associated with Thao Hung-Ching (see above, pp. 213 ff.), also comes out of the Mao Shan milieu. Its third chapter is entirely devoted to records of visitations of Wei Hua-Tshun and her companion divinities. According to a private communication from Michel Strickmann, these records, almost all of which can also be found scattered through the *Chen Kao*[16] (*c.* +500), far antedate Thao Hung-Ching's collection. They can be identified as having been copied out by Hsü Mi's great-grandson Hsü Jung-Ti[17] (d. +435) at the end of his life, and thus make up the earliest surviving redaction of authentic Yang-Hsü manuscript materials from Mt. Mao.

[b] *TT*878, ch. 1, p. 3*b*, tr. auct., adjuv. Ware (5), pp. 78–9. This formula is also discussed above, pt. 3, pp. 83 ff.

[c] It is also called *chen sha*,[18] which could mean either 'True, or Realised granules' but more probably 'True, or Realised cinnabar', in ch. 1, p. 4*a*.

[d] See pp. 171 ff.   [e] See pp. 4 ff., 187, 191, 250.   [f] Ch. 1, pp. 3*b*–5*a*.

[g] The supplementary explication for this recipe is found in ch. 17, pp. 2*b*–3*a*, where the canonical text as repeated incorporates what must once have been a footnote. The colour of the *hsüan huang* is described in the *Chüeh* in considerable detail: 'As for the regulation of the fire, if it is too hot the colour of the flowers will be yellow; if it is too cold the colour of the flowers will be virid or purple, and they

[1] 玄黃	[2] 精華	[3] 黃精	[4] 黃芽	[5] 黃輕
[6] 華池	[7] 伏火	[8] 丹華	[9] 接	[10] 訣
[11] 文言	[12] 坤	[13] 太清金液神氣經		[14] 天玄
[15] 地黃	[16] 眞誥	[17] 許榮弟	[18] 眞砂	

yellow mercuric and lead oxides, which would be skimmed off the molten metal with the iron spoon.[a] This is possible, but oxides thus prepared would not form the good crystals which the name of the product implies. Also, the likelihood of obtaining the yellow form of HgO instead of the red under such loosely defined conditions could only be determined by experiment. An alternative possibility is that there was no sublimate, and that as the mercury was allowed to evaporate from the amalgam in an open vessel a dendritic crystalline growth of metallic lead, containing mercury in solid solution, formed on the surface. A modern chemist might not pay much heed to this phenomenon, but the alchemist, as we have seen,[b] tended to be intensely aware of subtle changes in crystalline structure and play of colour. The Yellow Sprouts might thus be a more or less oxidised form of lead. This possibility would, however, be ruled out if, as the instructions indicate, the temperature is kept above the melting-point of lead.

This procedure, or one like it, may have exerted some influence on the formation of the highly idealised type-process of the *Tshan Thung Chhi*, but cannot totally explain its operational basis. The 'Yellow Emperor's' Yellow-and-Black formula skips the crucial Yin–Yang segregation in the second stage, proceeding directly to a union of the opposites in the yellow element Earth, which represents their balance. It was much more usual in medieval alchemy to work out processes involving the conversion of two initial ingredients into Realised Lead and Realised Mercury.

By the Sung at the latest there is no difficulty about identifying the paradigmatic substances. The eclectic compendium *Chu Chia Shen Phin Tan Fa*[1] (Methods of the Various Schools for Magical Elixir Preparations), of the Sung or slightly later, includes one of the many explicit statements:

The Realised Dragon is the quicksilver within cinnabar. It is born when the solar seminal essence (*jih ching*[2]) of mature Yang pours down and its realised *chhi* enters the earth. It is named mercury. The Realised Tiger is the white silver within black lead. It is born when the lunar floreate essence of mature Yin pours down and its realised *chhi* enters the earth. It is styled lead.[c]

To put this more prosaically, mercury and silver are the essences of cinnabar and lead because the former develop from the latter within the earth under the influence of the masculine and feminine *chhi* respectively.[d]

---

will carry a *chhi* (= aroma?). If the fire is correctly adjusted, the colour of the flowers will be red or purple, or that of gold. But when it is (then) roasted until (the vessel) is the same colour as the fire (*c.* 900°), it resembles gold remarkably in appearance. When it is taken off the fire, it returns to its original substance (*pên chih*[3]).' The implication that the gold colour is merely a transient effect would not necessarily be shared by the author of the canonical text—but note the importance it gives to the alchemist's contemplation.

[a] (1), vol. 2, pp. 379, 385.   [b] See pp. 244 ff.
[c] *TT*911, ch. 1, p. 14*b*, tr. auct. Cited from a 'Gold Elixir Dragon and Tiger Manual' (*Chin Tan Lung Hu Ching*[4]). Even late writers on the enchymoma are explicit on this point. The *Nei Chin Tan*[5] (Metallous Enchymoma Within), in *Chêng Tao Pi Shu Shih Chung*,[6] *pên* 12, p. 8*a*, asserts that '"External" alchemists prepare the Gold Elixir from Realised Lead—silver—which they have extracted from lead.' See also *Ta Tan Chi*[7] (Record of the Great Enchymoma), *TT*892, p. 1*a*.
[d] It would be over-concrete to understand these *chhi* as mere sunlight and moonlight.

[1] 諸家神品丹法   [2] 日晶   [3] 本質   [4] 金丹龍虎經
[5] 內金丹   [6] 證道秘書十種   [7] 大丹記

The alchemist re-enacts this evolution when he distils or sublimes mercury from its sulphide, and extracts from crude lead the silver which often occurs in appreciable amounts as an impurity.[a] Therefore his products become the realised pair which serve as the basis of the Elixir. There is no Chaos in this practical interpretation; the Realised Lead and Realised Mercury are extracted directly and individually from the mundane ingredients. A quotation in the same compendium, from a book the title of which indicates that it was directly derived from the *Tshan Thung Chhi*,[b] connects the two perfected substances with Yellow Sprouts, after emphasising that additional ingredients, and miscellaneous processes of the kind so popular with pragmatically inclined alchemists, are to be avoided. Mêng Yao-Fu wrote:[c]

In 'using lead and mercury to make the elixir', most people erroneously take black lead for Realised Lead, or think that quicksilver[d] is Realised Mercury, or take yellow floreate essence of lead (*chhien huang hua*,[1] or massicot, PbO) for Yellow Sprout.[e] Some even boil down brine, or recrystallise salt and collect the essence (i.e. the cubic crystals), combining it with quicksilver, cinnabar, 'lead furnace' (*chhien lu*[2]),[f] and litharge, or the Two Caerulean Minerals, the Three Yellow Minerals, the Five Metals, the Eight Minerals, and that sort of thing. But the use of the Five Metals, the Eight Minerals, or any (merely) material substance (*i-chhieh yu chih*[3]) is no perfect method. As Yin Chen-chün[4] has said: 'The material is not fit to be taken as your companion; even if you succeed by force in such a preparation, when ingested it will cause damage.' It is imperative that gentlemen studying the Tao should take care.

^a Whether what he got was actually silver remains very much an open question. Although the silver content of most Chinese lead ores is low, the content of zinc in some galena and blende is very high. See above, pt. 2, p. 218, and Collins (1), especially the analyses of Hunan lead ores on pp. 100–101 and the remark about the unfeasibility of a direct process for extracting silver on p. 239. On the idea of the actual conversion of lead into silver in relation to *nei tan* theory cf. pt. 5.

^b *Chin Tan Pi Yao Tshan Thung Lu*[5] (Essentials of the Gold Elixir: Record of the Kinship of the Three) by Mêng Yao-Fu,[6] who was the main contributor to the *Chu Chia Shen Phin Tan Fa* collection.

^c Ch. 2, pp. 4*a*–5*b*, tr. auct. Despite the cautions to avoid complex recipes, there are many formulae in the collection which use numerous ingredients, e.g. ch. 3, p. 6*b*. An example of just the sort of process the author warns against occurs in one of the few alchemical writings which can be assigned with confidence a pre-Thang date, *Thai-Chhing Chin I Shen Tan Ching*[7] (Manual of the Potable Gold Magical Elixir; a Thai-Chhing Scripture), *TT*873, ch. 1, p. 15*b* (partial text also in *YCCC*, ch. 65). A lead–mercury amalgam is prepared in a heated vessel and fired with lead carbonate (*hu fên*[8]), ground with vinegar, and sublimed with cinnabar, realgar, and orpiment. Maspero (7), pp. 97–98, dated this part in the first half of the +5th century, and Rolf Stein (5) has recently shown that the latter part of the work need be no later than the middle or even beginning of the +6th.

^d I.e. commercial mercury, as distinguished from that distilled from cinnabar under proper ritual precautions and according to special procedures in the elaboratory. We use both 'mercury' and 'quicksilver' to refer to the ordinary article of commerce, corresponding to the purely verbal parallelism between *hung*[9] and *shui yin*.[10] The product of alchemical operations can similarly be referred to as 'realised (*chen*[11]) mercury' or 'realised quicksilver'.

^e Note that here 'yellow floreate essence of lead' has entered ordinary chemical parlance as the appellation of massicot. This sort of transition has happened often enough in the Chinese language to call for caution in keeping the general and the particular unentangled. Another example is the philosophical term 'mature Yang' (*thai yang*[12]), which in lay speech was specialised to refer to the most obvious manifestation of the mature Yang, the sun. But to render '*thai yang*' as 'sunlight' when it occurs in the designations of the circulatory channels of the human body, as one translator has done, makes a simple and philosophically consistent system of nomenclature incomprehensible.

^f We have not seen this term elsewhere, and the generally poor condition of the text of *Chu Chia Shen Phin Tan Fa* leads us to suspect corruption.

¹ 鉛黃華   ² 鉛爐   ³ 一切有質   ⁴ 陰眞君   ⁵ 金丹秘要參同錄
⁶ 孟要甫   ⁷ 太清金液神丹經   ⁸ 胡粉   ⁹ 汞
¹⁰ 水銀   ¹¹ 眞   ¹² 太陽

The lead and mercury of which I mean to speak are universally kept secret in the alchemical classics. If one is not told directly, there is no way to understand what they are. 'Lead' is silver; that is to say, the silver is obtained from within lead. Therefore sagely silver is Realised Lead, which is born out of the stimulus of the essential *chhi* of the moon; it is the Water essence of mature Yin.[a] If a man be able to subdue it by art to form the Elixir, and ingest it, how could he not live forever? For Realised Lead one must definitely use silver; there can be no further doubt of this.... The mercury is quicksilver which has been obtained from cinnabar, with shape but without matter (i.e. a liquid). It imbibes the *chhi* of silver and congeals to form a body.[b] Thus it is styled Realised Mercury. It is born out of the essential *chhi* of the sun; it is the Realised Fire of mature Yang.... Among the myriad phenomenal things, only from lead and mercury can the Cyclically Transformed Elixir be made; all the rest have no place in the proper method. The lead has the *chhi*,[c] and the mercury is originally without shape. The lead is Yang inside and Yin outside, so it serves as the ground of the Elixir (*tan ti*[1]). It lends its *chhi* to engender the Yellow Sprouts. We know clearly that it is through getting the Realised *chhi* that the Divine Sprout is spontaneously born, after which the Realised Lead can be discarded.[d] Chhing Hsia Tzu[2] has said:[e] 'Lead is the mother of the Sprout, and the Sprout is the son of lead.' Once this golden floreate essence has been obtained, the lead is discarded and no longer used. Mercury is originally without shape, like the state (*chuang*[3]) of *chhi*. Its inborn nature is completely Yang, and its shape completely Yin (i.e. liquid). If a hundred *hu*[4] is put into a reaction-vessel it can be boiled until the pot is dry; thus is its immateriality made manifest. If it is planted within the lead it absorbs the essential *chhi* of the lead and metamorphoses its material substance, after which it is called Yellow Sprouts. Surely this is a going over from immateriality to materiality (*tsung wu erh yu chih*[5]).[f]

So lead, irradiated by the cosmic Yin *pneuma*, becomes silver. In this realised form it serves as the passive vessel which, impregnated by the Yang mercury, bears the Yellow Sprouts. The sexual imagery could hardly be more patent, but the *Tshan Thung Chhi*'s reversal of polarity in the realised substances is obscured. Just as the mercury is Yang here because it comes from cinnabar, the silver is still spoken of as lead, for Mêng Yao-Fu is thinking of the realised substances as functionally equivalent to their sources. The idea that a substance can be Yin outside (as shown, for instance, by its liquidity) and Yang inside (determined by its function or by a product of its metamorphosis) is one more application of the old idea that there is a potential Yang within every Yin and vice versa.[g]

[a] We have already pointed out that Water is the elemental phase which corresponds to mature Yin.
[b] Or 'It can imbibe the *chhi* of silver and congeal...', since this sentence clearly refers to the formation of a silver amalgam.
[c] I.e. the energy needed to impose on the mercurial substratum ('shape') the high level of organisation of the Elixir. This is analogous to the formative energy of semen which is responsible for forming a human embryo out of the passive Yin substance of the mother.
[d] Although it is not possible to enter into the non-alchemical levels of meaning of the *Tshan Thung Chhi* type-processes here, the parallel with the adept's use of female sexual partners for his own perfection through the Art of the Bedchamber is obvious.
[e] This is Su Yuan-Lang[6] (see above, pt. 3, p. 130).
[f] Our translation of the last sentence, which we do not fully understand, is tentative.
[g] See further in pt. 5.

[1] 丹地   [2] 青霞子   [3] 狀   [4] 斛   [5] 從无而有質
[6] 蘇元朗

The identification of the two realised substances does not settle the question of the choice of process, but rather opens it in new directions. Space permits only a single example of a process for preparing Yellow Sprouts from silver and mercury. The following procedure comes also from *Chu Chia Shen Phin Tan Fa*, which does not name its source or originator.

Take realised and balanced[a] 'mountain and marsh' silver, five ounces, and with an iron pestle beat it into a cake round as the shape of the sun. Then with the iron pestle beat it a thousand and more times until it is extremely firm, in order to prevent quicksilver from contaminating the Yellow Sprouts. Then put it in an earthenware tube.[b] Put three ounces of mercury inside, and then insert the silver cake into the earthenware tube, leaving a space of two inches or so between it and the mercury. Seat it firmly and lute all around with six-one lute as in the usual method, leaving no cracks. Above it put into place a vase of water to cover the mouth of the earthenware tube completely.[c] Below it use a fire made from three ounces of charcoal to heat gently and uninterruptedly day and night for seven days. When this time has passed, open it; the quicksilver will have gone up and the silver will have grown Yellow Sprouts, shaped like needles, countless in number and all of white silver. This is called the Yellow Sprouts of the First Cycle. Again add three ounces of quicksilver and apply a nourishing heat for seven days and nights. When this time is up open the vessel and examine its contents; they will resemble the colour of young sprouts from a cut tree (*nieh*[1]). Again add three ounces of quicksilver and apply a nourishing heat for seven days. When this time has gone by, open the vessel and examine its contents. The colour will be deep brown. Do not gather anything as yet. (The crystals) will be connected (*i-li*[2]), and will have grown as if what you had planted were sprouting. On (each) seventh day open the furnace and add three ounces of mercury until seven times seven days have passed. This will have been seven cycles, and a total of 21 ounces of mercury will have been added. The product is called Purple Gold Yellow Sprouts, ... the Mother of the Cyclically Transformed Elixir. The quicksilver in the tube will still be inside the cover, and will be as red as vermilion. When it is collected there will be a couple of ounces. It is also named Son Become Mother (*tzu pien mu*[3]) or Single-bodied (*tu-thi*[4]) Vermilion.[d] This medicine, after being mixed with milk, steamed, and ground fine as flour, is made into pellets with jujube paste. Every day take three such pills with wine on an empty stomach as a tonic for the lower region of vital heat (*hsia yuan*[5]),[e] and to quiet the heart, pacify the animus, and still (*ting*[6]) the anima. It also cures cold disorders of the wind group (*fêng lêng*[7]) and other diseases. Its efficacy is so manifold that it cannot be described fully here. One can gather 12 ounces or more of the Yellow Sprouts which grow on the face (of the silver). There will be three or four ounces or more of the refractory mercury left under the silver cake.[f] That it has not been transformed is because this mercury has absorbed a sufficiency of *chhi*. It may be collected in another container, for it has its own

a This may be an over-philosophical translation, since *chen chêng*[8] in ordinary speech means simply 'genuine'.
b This must be visualised as like a test-tube or long narrow vase in shape.
c This cooling basin above is highly reminiscent of the upper condenser of the Mongol still; cf. pp. 62 ff.
d Both of these recondite names refer to the production of 'cinnabar' (actually mercuric oxide, which resembles it in colour) from mercury without the addition of sulphur.
e See pt. 5 below.
f The text is not clear as to whether the mercury is on the under-side of the silver cake or below it in the bottom of the vessel.

¹ 糵   ² 迤邐   ³ 子變母   ⁴ 獨體   ⁵ 下元   ⁶ 定
⁷ 風冷   ⁸ 眞正

utility when incorporated in medicines. The quicksilver and the Yellow Sprouts can be used as the elixir matrix (*tan mu*[1]), so they are called the Mother of the Cyclically Transformed Elixir. Cinnabar is called 'animus of the sun'; quicksilver and Yellow Sprouts are called 'anima of the moon'. There is a mnemonic verse which goes:

> 'The sage can rival the skill of the Shaping Forces;
> Raising his hand, he plucks the sun and moon from the sky
> To put in his pot....'[a]

Thus solid silver is attacked by the fumes of mercury in a sealed vessel over gentle heat for seven weeks (not, by alchemical standards, an imposingly long period). After formation of a massive beta phase, the silver gradually accumulates a needle-like crystalline growth called Yellow Sprouts. Surface oxidation of the silver 'sprouts' accounts for the gradual darkening of this colour. The formation of the red 'cinnabar' sublimate (HgO) inside the top of the vessel indicates that despite the careful closure and application of lute, the atmosphere within the vessel is oxidising, due to diffusion of air through the porous lute during the protracted firing.[b] Actually there is no vermilion inside the vessel. This alchemist was unable to distinguish red mercuric oxide from the sulphide.

The process just described makes use of a partly physical and partly chemical transformation to advance the elixir process one step, and the succeeding steps from Yellow Sprouts to Cyclically Transformed Elixir have rationales of their own (which are not germane here, but which invite investigation). An even more elegant conception is to base the whole elixir process on a single reversible chemical reaction. One might call this approach cosmological rather than cosmogonical, since it provides a model of the successive dominion of Yin and Yang in the cycles of the universe rather than of the stages in their definition out of the primal Chaos. As Chhen Ta-Shih puts it, 'That cinnabar should come out of mercury and again be killed by mercury: this is the mystery within the mystery.[c]

In the *Yin-Yang Chiu Chuan Chhêng Tzu-Chin Tien-Hua Huan Tan Chüeh*[2] (Secret of the Cyclically Transformed Elixir, Treated through Nine Yin–Yang Cycles to form Purple Gold and Projected to bring about Transformation),[d] one pound of mercury is distilled from three pounds of cinnabar in the presence of alum

---

[a] *TT*911, ch. 3, pp. 1a–2a, tr. auct. For a wet mercury–silver process, in which no strong heating is involved, see *Huan Tan Chung Hsien Lun* (*TT*230), pp. 16bff. In pt. 5 we shall illustrate this verse by a photograph of a Szechuanese temple statue.

[b] In fact the protracted firings so usual in alchemy probably developed partly because many reactions involving oxidation would not have succeeded otherwise. The possibility that long periods favoured the formation of large crystals of sublimates should also be empirically tested. See Sivin (1), p. 183.

[c] In *Pi Yü Chu Sha Han Lin Yü Shu Kuei*[3] (On the Caerulean Jade and Cinnabar Jade-Tree-in-a-Cold-Forest Casing Process), *TT*891, p. 1a. This is an annotated series of poems with esoteric commentary, followed by instructions for a fundamentally two-element process using silver-bearing lead (*yin chhien*[4]) and mercury as well as ancillary substances. It belongs to the early +11th century.

[d] *TT*888, which quotes the *Tshan Thung Chhi* often. There are parallel passages in the *Thai-Shang Wei Ling Shen Hua Chiu Chuan Tan Sha Fa*[5] (Methods of the Guardian of the Mysteries for the Marvellous Thaumaturgical Transmutation of Ninefold Cyclically Transformed Cinnabar; a Thai-Shang Scripture) *TT*885, a corrupt text based on a very similar process. This has been translated by Spooner & Wang (1) and Sivin (3).

[1] 丹母　　　[2] 陰陽九轉成紫金點化還丹訣　　　[3] 碧玉朱砂寒林玉樹匱
[4] 銀鉛　　　[5] 太上衛靈神化九轉丹砂法

and salt, and then, in the second cycle,[a] heated with four ounces of sulphur to yield cinnabar. But the cycle does not merely repeat itself. It is essential that the product reach a higher state of perfection at each step. Thus in the third cycle mercury is obtained again, but subsequently it is 'congealed' by boiling with borax, malachite, salt and alum until it loses its volatility and becomes 'subdued' (fu^1),[b] just as an immortal sheds his perishable body. The remainder of the process grows so complex chemically that one easily loses sight of the simplicity of its conception. To the alchemical theoretician the progressively more metallic products of each cycle were still in principle mercuries and cinnabars.

Just as the passage of recurrent time perfected minerals within the earth, the repetition of the simple mercury–cinnabar cycle was supposed to lead to a gradual metamorphosis, the product of which would be the Elixir of Immortality. The Chinese image of a cycle (*chuan*[2]) does not, in fact, convey the idea very adequately; since the outcome, whether geological or alchemical, is a substance both perfect and immune to decay. There is a linear component. In other words, the conception of the Tao thus implied was not a two-dimensional circle but a helix. That both cinnabar and Elixir are called *tan* does not signify their identity (the two senses were distinct in alchemy and medicine and not generally confused). Still, this sharing of a name could serve to support and preserve the idea of a genetic relationship. Some alchemists persuaded themselves that the maturation of the Elixir could be brought about by simple repetition of a cyclical treatment. This not very empirical notion appears, for instance, in the +8th-century *Ta-Tung Lien Chen Pao Ching, Chiu Huan Chin Tan Miao Chüeh*[3] (Mysterious Teachings on the Ninefold Cyclically Transformed Gold Elixir, Supplementary to the Manual of the Making of the Perfected Treasure; a Ta-Tung Scripture)[c] of Chhen Shao-Wei,[4] a sequel to the seven-chapter monograph on cinnabar quoted earlier.[d]

At one point in this treatise Chhen is discussing a basic cycle in which mercury and sulphur are first heated together in a covered and tightly luted porcelain vessel to form 'purple cinnabar' (*tzu sha*[5]), a mixture of the mercuric sulphides cinnabar and metacinnabarite. From this material, in the presence of lead and salt, mercury is recovered by sublimation (*fei*). As the process is repeated, each cinnabar develops greater powers, as indicated by the progressively exalted name. He wrote:

For instance, mercury used in the second recycling (in the chapter on Treasure Cinnabar)[e] is twice heated with sulphur to make it into cinnabar, and twice put into lead. The mercury is sublimed from (the intermediate cinnabar), added to the metal, and transformed into cinnabar. Mercury used in the third recycling (Effulgent Cinnabar) is thrice heated and sublimed before it is ready for use. Mercury used in the fourth recycling to produce Wondrous Cinnabar is sublimed and heated four times. Mercury used in the fifth recycling (Numinous Cinnabar) is sublimed and refined five times. Mercury used in the sixth recycling to produce

---

[a] In this treatise each 'cycle' (*chuan*[2]) is only half a cycle as ordinarily defined.
[b] See pp. 4ff., 187, 191, 250, 256.   [c] TT884.   [d] See pp. 237ff.
[e] I.e. in Chhen's other work, *Ta-Tung Lien Chen Pao Ching, Hsiu Fu Ling Sha Miao Chüeh*.

[1] 伏   [2] 轉   [3] 大洞鍊眞寶經九還金丹妙訣   [4] 陳少微
[5] 紫砂

Spiritual Cinnabar must correspondingly be heated and sublimed six times. Mercury used in the seventh recycling to produce Mysterious Realisation Crimson Cloud Cinnabar, just as in the previous cases, must be heated with sulphur seven times to form purple cinnabar, and lead used seven times, subliming to make it revert to Numinous Mercury. For each heating one uses three ozs. of sulphur; to reconvert it to mercury one uses one lb. of lead, heating and subliming cyclically, controlling the fire as specified earlier. In the course of these meta-morphoses brought about by heating and subliming, (the Mercury) will maintain its inner essential *chhi* of Water and Fire. Once the numerical correspondences (*ta shu*[1]) of the Seven Chapters have been satisfied, the *chhi* of the three luminaries, Water, Fire and Metal, will naturally be united in the product. When the seminal essences meet, it is transformed and becomes numinous; it attains enlightenment and becomes Realised Mercury.[a]

To the experimentally minded this airy theorising cannot have been very satis-factory. In order to bring about progressive changes in practice, the purity of the mercury–cinnabar idea had to be compromised by the use of additional reagents. Indeed, the practical instructions given by Chhen himself in his seven chapters on Numinous Cinnabar use ancillary ingredients, but he does not regard this concession as in any way a failure. The supernumerary substances are, by implication, as external to the process as the ancillary drugs used in medicine to guide the 'effective' com-ponent of a prescription to the site of the illness. Even though the products of Chhen's cycles resembled mercury and cinnabar less and less, they still corresponded functionally.

Finally it is worth while to examine briefly a technique called 'irrigation' (*chiao lin*[2] or simply *chiao*), widely used in later alchemy.[b] It involves an interesting variant of the mercury–cinnabar cycle, a sort of compromise between the methods already described. First cinnabar is made from mercury and sulphur, then treated with other minerals to fire-subdue it (*fu huo*[3]). The product is superior to mundane cinnabar because it is no longer volatile, and thus invulnerable to erosion by the fire. The novelty involves sealing it with mercury in a special mineral-lined sublimation chamber. This chamber, commonly called the Bubbling Spring Casing (*yung chhüan kuei*[4]), is then heated for days or weeks. Even though no sulphur is added, what appears to be cinnabar forms at the top of the vessel. When more mercury is added to this product more 'cinnabar' is formed, grander after each cycle, its elixir qualities more patent. From the chemical point of view it seems most likely that the initial Subdued Cinnabar is inert. It is certainly no longer cinnabar after passing through most of the fire-subduing processes. The products of the successive cycles are not cinnabar but the very similar red mercuric oxide, formed by oxidation of the added mercury within a certain range of tempera-tures.[c] If the fire is too hot (above 500°), the mercuric oxide will break down. Whether

---

[a] Pp. 2*a*–4*a*, tr. auct. in summary form.

[b] Yüan Han-chhing (*1*), p. 209, defines *chiao* as 'to pour out a liquid product and let it cool slowly'. We have never seen a text in which this meaning would fit well, and suspect that it was only inferred from the non-alchemical sense 'to pour out a libation'. For a +13th-century inventory of 'irrigation' and related procedures, see above, pp. 18–19.

[c] The preparation of 'Seven-Cycle Cinnabar' by heating mercury alone was described by Sun Ssu-Mo in the middle of the +7th century. See Sivin (*1*), p. 191.

[1] 大數　　[2] 澆淋　　[3] 伏火　　[4] 湧泉匱

some property of the red sublimate varies from cycle to cycle to support the idea of its gradual perfection could only be determined in the laboratory, because of the complexity and variety of the procedures outlined by various alchemists.[a]

The use of reagents to embody Yin and Yang and re-enact their cosmic play was not enough. Yin and Yang are in their profoundest sense temporal phases. Creating a microcosm thus involved laboratory techniques for phasing time, and these we shall now examine.

(iii) *Correspondences in duration*

The postulate that one period of time can correspond to another has already made its appearance in an excerpt from *Tan Lun Chüeh Chih Hsin Ching*. The year required by the alchemist to prepare his elixir was likened to the 4320-year term of the natural cyclically-transformed elixir which forms within the earth.[b] Let us now return to that book as it proceeds to explain the correspondence:

Query: 'How is it that one year can correspond to the constant period (*shu*[1]) required by Nature to make a cyclically-transformed elixir?'
Reply: 'One day and night in the world above is one year in the human realm. Now among men one year is twelve months, of 360 days. One month is thirty days, and one day is twelve hours,[c] so one month is 360 hours. In sum, a year is 4320 hours, which corresponds to (the time needed by) Nature to produce the natural cyclically-transformed elixir.'[d]

What would be illogical and pointless in terms of the time metric of modern science makes perfectly adequate sense once we realise that here numbers are not measures. They are being used rather to mark members of a series of things which are qualitatively related, in the mode which Granet used to call 'emblematic'. We might say 'numerological'. That a year contains 4320 double-hours *proved* that it is functionally equivalent to the natural period of maturation.

We find the same set of correlations in the *Yü Chhing Nei Shu*:[e]

A month contains 360 hours. Calculating a correspondence on the basis of hours, a year of twelve months comes out to 4320 hours. Taking one hour as equivalent to (*tang*[2]) one year, we calculate (that the year is equivalent to) 4320 years, and corresponds to the (period

---

[a] Among the more important of the treatises concerned with 'irrigation' processes are the following, all of the Wu Tai, Sung or later: *Kêng Tao Chi* (*TT*946), dating from after +1144, a large collection in which many recipes appear to have been obtained from individuals rather than books, *Lung Hu Huan Tan Chüeh* (*TT*902), and *Chhien Hung Chia Kêng Chih Pao Chi Chhêng* (*TT*912). The preface of this last collection is dated in a *ping-chhen* year, which would correspond to +836, +896, +956 or another multiple of sixty years earlier or later. The book includes an oral formula clearly dated +808, but this may not be genuine because its cyclical characters are given wrongly. At least one work included used *fên* to refer to the weight unit smaller than the *liang*, so the whole compilation may not have been assembled before the Sung. Particularly clear laboratory instructions are also found in the *Chiu Chuan Ling Sha Ta Tan* (*TT*886) and its neighbour the *Chiu Chuan Chhing Chin Ling Sha Tan* (*TT*887) which seems mostly to be paraphrased, rearranged and modified from the former, though it could equally well be a little earlier.   [b] See pp. 231 ff.

[c] I.e. Chinese double-hours (*shih*[3]). Cf. Vol. 4, pt. 2, pp. 439, 461. For further information on time units and measurements see Needham, Wang & Price (1).

[d] *TT*928, p. 13*a*, tr. Sivin (5). In the query we follow the *YCCC* edition, p. 13*a*, and emend 'five years' (*wu nien*[4]) to 'one year' (*i nien*[5]) as required by the sense and context.

[e] *TT*940, p. 19*b*, tr. auct.

[1] 數   [2] 當   [3] 時   [4] 五年   [5] 一年

of the) natural cyclically-transformed elixir. It is the conjugation of Yin and Yang, (the alternation of) winter cold and summer heat, which give rise to the correspondence.

Both of these books belong to the late tradition of dual cultivation, which made much use of time correspondences in phasing breath-control and even sexual techniques. The most elaborate scheme of time correspondences evolved in China is from an explicitly Interior Alchemy treatise, the *Huan Tan Nei Hsiang Chin Yo Shih*[1] (Golden Key to the Physiological Aspects of the Regenerative Enchymoma), written by Phêng Hsiao[2] in the middle of the +10th century. This book develops in exhaustive detail the use of the hexagrams in the *Tshan Thung Chhi* to mark periods of time, and thus to provide a terminology for phasing the breath. Each hexagram is broken into its six constituent lines to make available a system of 384 fine divisions (360 in practice). Here is part of Phêng's argument for a whole repertory of correspondences, with a year of cosmic process equated to a month, five days, 2½ days, and one day:

Thus one year of 360 days contracts (*tshu*[3]) into a month of 360 hours. Further, if within a month of thirty days, or 360 hours, we assign one hexagram to each morning and evening, we can then transfer these sixty hexagrams, with their 360 lines, collapsing them (*hsien*[4]) into five days, or sixty hours, so that this period again corresponds to a month. Two and a half days is thirty hours, which becomes thirty days, and (thus) also corresponds to a month. Having determined a hexagram for each morning and evening (in a month), again we assign 60 hexagrams, which comes to 360 lines, so that this again corresponds to a year, or 360 days. Again, if within 2½ days, or 30 hours, we separate out a period of 15 hours, this responds to (*ying*[5]) a phase (*yung shih*[6])[a] of half a month, or 15 days. Again we take this half-month, from the first to the fifteenth day (inclusive), and collapse it into the 12 hours. To the period from the (beginning of the) second half of the first hour to (the end of the) first half of the sixth hour (i.e. midnight to noon) will be assigned 30 hexagrams, which comes to 180 lines. This period therefore corresponds to that from after the winter solstice to before the summer solstice, and responds to half a year, or 180 days.... The 'Book of Changes' says that the Masculine Factor is 360.[b] When this number of days has passed the Yin will have arisen and the Yang gone down. For their cycle we use the the year of the sidereal circuit (of the sun), the great constant of the myriad phenomena. Now one year comes out to 360 days, or 4320 hours. If to the morning and evening of each day we assign two hexagrams (i.e. one to each), this will give a total of 60 hexagrams (per month). With six lines per hexagram, their entire number will amount to 360 lines.[c]

[a] *Yung shih* (lit., 'play a part in the affair') is not a noun compound, but refers to the periodic dominant activity of one agent among several within a cycle, for instance the ascendancy of Wood in the Spring of the year. Its meaning is thus close to that of 'phase'.

[b] Actually, it says nothing of the kind. This is a freely adapted reference to the Great Appendix (Hsi Tzhu[7]), I, 9, which gives the Masculine Factor as 216. See Wilhelm (2), English tr., vol. 1, p. 334, where the term *chhien chih tshê*,[8] which we translate 'Masculine Factor', is rendered 'creative total'.

[c] *YCCC*, ch. 70, pp. 3*a*–4*a*, tr. auct. Chhen Kuo-fu (*1*), vol. 2, p. 439, suggests that this is an abridgment of the original work of Phêng Hsiao, rather than a fragment. A very similar schema, using the same technical terms, appears in *Hsiu Tan Miao Yung Chih Li Lun*[9] (A Discussion of the Marvellous Functions and Perfect Principles of the Practice of the Enchymoma), *TT*228, pp. 2*a*–2*b*, which is definitely later since it refers to the Sung Taoist Master Sea-Frog (Hai Chhan Hsien-sêng[10]), Liu Tshao.[11] For another, more compressed, system of multiple time correspondences see the undated *Ta Tan Wên Ta*[12] (Questions and Answers on the Great Elixir), *TT*932, pp. 3*b*–4*a*.

[1] 還丹內象金鑰匙　　[2] 彭曉　　[3] 蹙　　[4] 陷　　[5] 應
[6] 用事　　[7] 繫辭　　[8] 乾之策　　[9] 修丹妙用至理論
[10] 海蟾先生　　[11] 劉操　　[12] 大丹問答

This is only a sample of the relevant passage, but it is enough to convey the approach and the flavour. There remains only to reproduce an annotation which appears in the text at the end of the part we have cited:

Again this appropriates (to[1])[a] a year. The 360 days, (as we see upon) calculating the number, appropriates the 4320 years that the balanced *chhi* spends within the Spirit Chamber (*shen shih*,[2] i.e. the reaction-vessel, or the *tan thien* in physiological alchemy).

This is numerology of the most extravagant kind, with its breathtaking transitions, its round number of sixty hexagrams, and its rounded-off sidereal year of 360 days.[b] In the literature of External Alchemy strictly defined we encounter nothing so elaborate, but multiple correlations are involved even if their rationale remains tacit. Phêng Hsiao's passage offers at least a hint as to why the normal quantum step in the fire-phasing cycles which we shall now examine was $2\frac{1}{2}$ days.

(iv) *Fire phasing*

Fire is the great agent that nourishes and matures the Elixir. Since the heat of the flame thus stands for the active forces, the re-creation of the cosmic process depends upon the binding of fire by time. The key to the success of the Work, the great test of laboratory skill and assiduity, particularly in the strain of alchemy concerned with ideal processes, was the technique of gradually increasing and decreasing the intensity of the fire (*huo hou*[3]) by the use of precisely weighed increments of fuel.

This is the closest thing we find in the ancient world to a quantitative conception of degrees of temperature. A constant increase in the weight of fuel does not cause a constant increase in the temperature of the thing heated, but that was beside the point before the thermometer provided a standard for testing the correlation.[c] The idea of fire control, in the sense of using an amount of fuel specified by weight, is an ancient one in the chemical arts, because to each weight of fuel, burnt in the same way, corresponds a set temperature and a predictable product. What the alchemists did was to make this concept of *huo hou* dynamic, varying the weight of fuel and thus the temperature in a regular way. They were bringing their processes under the control of one of the few exact measuring instruments at their disposal, the balance.[d]

Intensity of heat, controlled in this necessarily indirect manner, was the time-dependent variable, and the overall profile could be as precisely cyclical as the seasonal changes to which it corresponded point by point:

---

[a] This unusual use of *to*,[1] the primary meaning of which is 'to take by force or threat', very roughly parallels the widening of the meaning of the word 'abstract' in English. Cf. p. 234.

[b] The true value is of course 365·2564 days. Cf. Vol. 3, p. 181.

[c] On the beginnings of temperature measurement in East and West see Vol. 3, p. 466, Vol. 4, pt. 1, p. 63.

[d] It is not difficult to find processes in which even fractions of an ounce (*chhien*[4]) are used. See, for instance, the late *Hsiu Lien Ta Tan Yao Chih*[5] (Essential Instructions for Preparing the Great Elixir), *TT*905. Despite rather esoteric terminology, this collection of recipes using casing (*kuei*[6]) techniques (above, pp. 18ff.) reflects much practical knowledge. Take, as an example, the instructions for the lost-wax casting of a reaction-vessel (ch. 2, pp. 1*b*–2*a*).

[1] 奪   [2] 神室   [3] 火候   [4] 錢   [5] 修鍊大丹要旨   [6] 匱

The amounts of fuel to be weighed out are increased and decreased in cyclical progression according to the proper order of Yin and Yang. They must conform with the signs of the 'Book of Changes' and the 'Threefold Concordance',[a] tally with the four, eight, 24, and 72 seasonal divisions of the year, and agree with the implicit correspondences and pneumatic manifestations (*chhi hou*[1]) of the year, month, day and hour[b]—all without a jot or tittle of divergence.[c]

To see the beginnings of the notion of heat phasing we must return to the *Tshan Thung Chhi*. It devotes much space to the association of the trigrams and hexagrams of the 'Book of Changes' with temporal phases. As we have seen in a previous volume in connection with the fundamental ideas of Chinese science, six of the trigrams were used to mark off segments of the lunation cycle, and twelve of the hexagrams were assigned to the twelve double-hours of the diurnal cycle.[d] The correspondences are simple and schematic, for the diagrams were used as a kind of graphic representation of the interplay of Yin and Yang in each phase. For instance, the diurnal cycle begins with Fu,[2] Return, no. 24 in the normal order of the *I Ching*. Recalling that one reads the hexagrams from the bottom up, we see in the single solid line of Fu ☷☳ the rebirth of Yang beginning when midnight, the point of mature Yin (Khun[3] ☷☷ , Receptor, no. 2 in the textual order) has passed. The third double-hour is represented by Thai[4] ☷☰ , Upward Progress (no. 11), in which the Yang has advanced a step. Halfway around the cycle, the mature Yang phase (Chhien[5] ☰☰ , Donator, no. 1), having had its dominion, is replaced at the seventh double-hour by Kou[6] ☰☴ , Reaction (no. 44), as the Yin begins to reassert itself, and so on.[e] The thesis behind this progression is simply that the six trigrams and twelve hexagrams, chosen to represent various stages in the endlessly repeated complementary growth and decay of Yin and Yang, can be assigned to the successive phases of any temporal cycle

[a] I.e. the *Tshan Thung Chhi*.

[b] On manifestations of *chhi* phased throughout the year, see Vol. 4, pt. 1, pp. 186–92, and Bodde (17).

[c] Anonymous; cited in *Chu Chia Shen Phin Tan Fa* (*TT*911), ch. 4. p. 1*b*, tr. auct. Cf. pp. 3, 257ff.

[d] See Vol. 2, pp. 329–34, especially table 17 (p. 332). Here the hexagram Fu was wrongly incorporated with the trigrams in the lunar cycle. Probably this mistake arose because the *Tshan Thung Chhi* does say (ch. 4) that the principle of the Fu *kua* establishes all first buds of new growth (cf. Bodde (4), p. 117). The chapter numbers given on p. 331 were those of Wu & Davis (1); they should be corrected to chs. 2, 4, 18 and 19 respectively. If one writes down the trigrams one can see clearly the wave of Yang or Yin rising and falling through them. For a detailed description of the alchemical applications of the *I Ching* diagrams, see above, pt. 3, pp. 60ff.; here we recapitulate summarily for convenience. Corrections are also required for the diurnal cycle in Table 17. The order should be Fu (24), Lin (19), Thai (11), Ta Chuang (34), Kuai (43), Chhien (1); then Kou (44), Thun (33), Phi (12), Kuan (20), Po (23) and Khun (2). Cf. Vol. 5, pt. 3, p. 61 above, and Ho Ping-Yü (16).

[e] This is a binary notation, but its order is not the same as that of modern binary numbers. First Yang lines increase from the bottom toward the top, and after the hexagram has become entirely Yang, Yin lines appear at the bottom and spread upward. The twelve hexagrams used in the diurnal cycle are those the inner lines of which are all connected to top or bottom by other lines of the same sign. See Table 14 in Vol. 2, p. 315. For the same reason the trigrams Khan and Li, so important otherwise in alchemy (cf. Vol 5, pt. 5), do not come in to the six of the lunar cycle, but *Tshan Thung Chhi*, ch. 2, connects them with the change-over point between the months and indicates that they govern the whole system (cf. Fêng Yu-Lan in Bodde (4), and Wu & Davis (1), pp. 232–3).

¹ 炁候   ² 復   ³ 坤   ⁴ 泰   ⁵ 乾   ⁶ 姤

governed by the interplay of the opposites. In principle, the trigram and hexagram sequences are merely alternatives to the Five-elements phasing system, carrying rather different qualitative connotations.

There is no hint in the text of the *Tshan Thung Chhi* as to how these progressions are to be applied to laboratory operations. Looking at the text itself, all we can say for sure is that it is using the trigrams and hexagrams to divide the month and the day into qualitatively distinct phases which govern the alchemical process. The traditional view that this governance was exercised through alternate heating and cooling makes sense, but there is no evidence in the classic itself that the temperature of the furnace was graded through many steps, or controlled by weighing the fuel. Alchemists and commentators united in finding a heat phasing system in the passages on the muta- tional diagrams.[a] So, for that matter, did those who interpreted the book as respiratory or sexual alchemy, for they applied the *huo hou* concept to rhythmic cycles of breathing or sexual penetration.[b] These adepts would hardly have hesitated to read the sophisti- cated idea of heat phasing by weight into the ancient and obscure *Tshan Thung Chhi* if they had felt inclined to do so. They always found it natural to assume that the older a canonical book the deeper and more sophisticated were the ideas expressed in it. Generally, indeed, Taoists thought of the history of alchemy as a devolution rather than a progressive unfolding. They conceived the Art as something forced gradually downward by the inability of devotees to recover the austere and authentic revelations with which the tradition had begun, and by the credulity and bad faith of vulgar amateurs who contaminated the ancient doctrines.[c]

Ko Hung, early in the +4th century, maintained no more than the simple distinction between gentle and strong fires. The only securely datable early appearances of heat phasing techniques, shortly after his time, also take no notice of the *Tshan Thung Chhi*. These procedures, primitive by comparison with those popular from the Thang on, appear in the Mao Shan alchemical documents which passed through the hands of Thao Hung-Ching (*c.* +500).[d] The treatise on the 'Lang-kan Gem Floreate Essence Elixir' does not use weighings of fuel but rather varies the distance of the vessel from the chaff fire below it in the stove (*tsao*[1]). The alchemist is cautioned to keep the fire moderate, but no constant weight of fuel is specified. First the fire is maintained one foot from the vessel for 20 days, and then at distances of six and four inches for twenty days each. The flame is advanced to one inch from the vessel for ten days, and is adjusted so as just to touch it for another ten. Finally the flame is allowed to half- envelop the vessel for twenty days. A hundred days have passed and the first-stage

---

[a] See pt. 3, pp. 58 ff.

[b] As only one early example for physiological alchemy, see the *Ta Huan Tan Chhi Pi Thu*[2] (no later than *c.* +1000), in *YCCC*, ch. 72. Chhen Kuo-Fu's suggestion (*1*), vol. 2, p. 287, that this is by the Thang alchemist Chang Kuo[3] is not supported by enough evidence. Other important documents are the *Huan Tan Nei Hsiang Chin Yo Shih* (*YCCC*, ch. 70) and the *Hsiu Tan Miao Yung Chih Li Lun* cited just above (p. 265).

[c] This veneration for antiquity did not rule out occasional recognition that the ancients were fallible; see, for instance, Sivin (*1*), p. 168.

[d] See pp. 213 ff.

[1] 竈   [2] 大還丹契秘圖   [3] 張果

elixir is finished (Fig. 1517).[a] A similar scheme, with a more constant gradient applied over 120 days to a 28-ingredient elixir, is given in the 'Liquefied Gold Spiritual Chhi Canon.' Although this work is probably much later, its Mao Shan provenance is guaranteed by a chapter of revelations borrowed from Thao's *Chen Kao*.[b] A third example also comes from another of the very few texts clearly linked to the community of Mt. Mao, suggesting that heat phasing by distance was a traditional speciality of theirs.[c] The second of Thao's scriptures includes a method of the same

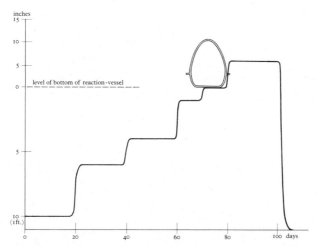

Fig. 1517. Fire-distances prescribed in the *Thai-Wei Ling Shu Tzu-Wên Lang-Kan Hua Tan Shen Chen Shang Ching* (Divinely Written Exalted Spiritual Realisation Manual in Purple Script on the Lang-Kan (Gem) Radiant Elixir; a Thai-Wei Scripture), a text of the late +4th century. The shape of the reaction-vessel derived from Needham & Ho Ping-Yü (3), p. 70.

kind, with a rather irregular gradient, but also a second technique called 'doubling the fire' (*pei huo*[1]), which ambiguously suggests quantitative regulation of fuel. Unfortunately the definition given is too opaque to allow us to judge whether this was a precursor of fuel weight regulation, or merely a sequence of timed stages (though

[a] *Thai-Wei Ling Shu Tzu-Wên Lang-Kan Hua Tan Shen Chen Shang Ching* (TT252), p. 4b. The phasing is repeated in the second stage (p. 5a).

[b] *Thai-Chhing Chin I Shen Chhi Ching* (TT875), ch. 1, pp. 4b–5a. Cf. above, pp. 245, 258.

[c] *Thai-Chi Chen-Jen Chiu Chuan Huan Tan Ching Yao Chüeh*[2] (Essential Teachings of the Manual of the Supreme-Pole Adept on the Ninefold Cyclically Transformed Elixir) TT882, p. 3a, presents a very complex scheme with irregular distance increments at nine-day intervals. The Mao Shan influence reveals itself in an account of five magic plants which grow on Mt. Mao, entitled 'Lord Mao's Formula (for Ingesting) the Five Kinds of Chih-jung' (*Mao Chün wu chung chih jung fang*[3]), pp. 6b–8a. Another schema too simple to be very significant appears in *Ling-Pao Chung Chen Tan Chüeh*[4] (TT416) p. 4a, which is evidently post-Thang because it specifies the 'large ounce' (*ta liang*[5]) measure. A vase containing reactants is heated for three days. On the first day the flame is kept two inches from the vessel; the second day, one inch; and on the third day, it must touch the vessel.

[1] 倍火   [2] 太極眞人九轉還丹經要訣   [3] 茅君五種芝茸方
[4] 靈寶衆眞丹訣   [5] 大兩

the times given in the distance-phasing schema, unlike this one, do not increase exponentially).

As for the doubling of the fire, first heat for one day, then heat to respond to (*ying*[1]) two days. After that, heat to respond to four days, and then to respond to eight days, and then to sixteen days. The constants for every period (*shih*[2]) should accord with these.[a]

The mature concept of phasing by fuel weight can be located among the handful of definitely pre-Sung works only in the writings of Chhen Shao-Wei,[3] probably not long after +713. There is no reason to believe that Chhen was its originator, since he applies the concept in a matter-of-fact way in his 'Numinous Cinnabar' treatise. But he is responsible for one of its grandest variants, found in his second work, on the 'Nine-cycle Gold Elixir'. Both are worth describing fully.

Simple *huo hou* systems involve a linear increase in fuel weight as a function of time. In order to complete the cycle, the fuel is then decreased at the same rate until the starting weight is again reached.[b] This use of two lines of constant slope (a 'zig-zag function')[c] to approximate a sinusoidal function is one of the most characteristic patterns of Chinese science. We perceive it in early figures for variation of sun shadow length with the seasons, the variation in respiration over the course of the day in breath disciplines, the rise and fall of Yin and Yang in the *Tshan Thung Chhi* series of *kua*, and so on.

Here are Chhen Shao-Wei's instructions for linear phasing as they appear in his treatise on the 'Numinous Cinnabar Seven Times Cyclically Transformed'. We are not able to comment upon the chemical reactions involved because of the large number of ingredients and, in a couple of cases, the uncertainty of their identification. They include malachite, halite, Epsom salt, *huang ying*[4] (probably a form of selenite) and *hua shih*[5] (which might be translated literally as 'fluxite', but we have been unable to determine what mineral it designates). In this particular process only the increase gradient appears, but the full cycle of increase and decrease will be reflected in a more elaborate system of Chhen's which we shall describe presently. Here he wrote:

Method of subduing by volatilisation. According to the supplementary instructions, five days is one phase (*hou*[6]); three phases is one *chhi* period.[d] In eight *chhi* periods, twenty-four phases, or 120 days, the subduing of the cinnabar is completed.[e] In the five days allotted for one phase of subduing by volatilisation, four days are governed by the *kua* Khan and one day by the *kua* Li. By 'the trigram Khan' is meant simmering in water for four days. By 'the trigram Li' is meant volatilisation over a Yang fire for one day. When the Yang fire is first laid, use seven ounces of charcoal, standing it on end below the reaction-vessel. One

---

ᵃ *Thai-Shang Pa-Ching Ssu-Jui Tzu-Chiang (Wu-Chu) Chiang-Shêng Shen Tan Fang* (*YCCC*, ch. 68), pp. 4*b*, 5*a*, 7*a*, tr. auct. Cf. p. 216 above.

ᵇ As we shall see, the return to starting weight is only approximate.

ᶜ See Neugebauer (9), pp. 110–113.

ᵈ In astronomical time-reckoning, the *chhi* or *chieh chhi*[7] are the 12 or 24 equal divisions of the tropical year. Thus one *chhi* is roughly $15\frac{7}{32}$ days. See Vol. 3, pp. 404–6.

ᵉ Cf. pp. 4ff., 187, 191, 262.

¹ 應   ² 事   ³ 陳少微   ⁴ 黃英   ⁵ 化石   ⁶ 候
⁷ 節氣

must see that there are seven ounces—no more, no less—of well-coked charcoal below the vessel at all times. After each cycle increase the amount of charcoal by one ounce and volatilise (the reactants again). Keep adding charcoal until after the fifth cycle, when suddenly a black *chhi* (= smoke) and a sublimate of mercury will come out (of the reactants). Collect the sublimate and again mix it with 1/20 ounce (*pan chhien*[1]) of previously fused halite (mineral NaCl) in a bowl. Grind lightly with a jade pestle until the mercury is completely absorbed. Then place the material in the reaction-vessel as before, and subdue by volatilisation using the Khan and Li trigrams until the twelfth cycle is completed. Add two ounces more of charcoal per cycle. Spread 1/8 ounce (*pan fên*[2]) of previously fused and powdered halite on top (before) closing (the vessel in the first place).[a] A total of two ounces or so of mercury sublimate will sublime. The void glow of the sublimate...[b] The residue in the vessel should gradually turn brown or purple. Collect the sublimate and the mercury (?), mix with 1/10 ounce of halite, and grind thoroughly in a bowl. Put the reagents into the reaction-vessel and subdue by volatilisation, phasing the heat as before until the eighteenth cycle has been completed. Increase the charcoal by three ounces (per cycle). The colour of the residue should be scarlet. Through the twentieth cycle, add four ounces of charcoal (per cycle). Only a half-ounce or less of mercury sublimate will sublime. It will be solid and hard as bronze chips (*phien*[3]), yellowish-white and lustrous. It is also to be mixed with mineral salt and ground in a mortar. Put into the reaction-vessel and subdue by volatilisation through the twenty-fourth cycle. The phasing of the cinnabar will be complete, the subduing by fire ended. (The residue) will be blazing red, lustrous and handsome; the cinnabar has been subdued.[c]

Thus one begins with a weighed amount of fuel and increases it at a rate which is kept constant for several cycles (Fig. 1518). The gradual increase in the increments is doubtless meant to accelerate the subduing of the cinnabar. The interaction of Yin and Yang is reinforced by alternately subjecting the cinnabar to wet and dry processes. The invocation of correspondences to the trigrams Khan ☵ and Li ☲ (immature Yin and Yang respectively)[d] suggests a debt to the *Tshan Thung Chhi*, which is confirmed by Chhen's habit of quoting apothegms from 'the Canon'.

The elaborate system in Chhen's second treatise (an elixir preparation for which the subdued Numinous Cinnabar was only a preliminary) not only models the ups and

---

[a] The text of *TT883* specifies 1/4 oz. (one *fên*[4]) of halite. This sentence appears defective as a whole, and our translation of it should be considered tentative.

[b] The sense is not quite complete at this point; perhaps a few characters are missing.

[c] *Ta-Tung Lien Chen Pao Ching, Hsiu Fu Ling Sha Miao Chüeh*[5] (*YCCC*, ch. 69, under the title *Chhi Fan Ling Sha Lun*[6]), pp. 9a, b, translation from critical text in Sivin (4).

[d] These are the two trigrams omitted from the lunar sequence (see Table 109, pt. 3, p. 62). In addition to their use to designate wet and dry processes, Khan and Li stand in some writings for the Yin and Yang reactants in a two-ingredient process. The two functions are combined in a story told by Shen Kua[7] late in the +11th century about a Mr Li[8] who could make a Water Elixir (*shui tan*[9]) by boiling water until it congealed to resemble caerulean jade. Asked how the process worked, he replied 'I don't use anything, but merely regulate the powers (*li*[10]) of the water and fire. If they are the least bit unequal, (the Elixir) is transformed (into water) again and escapes (as steam). This is the refined essence of Khan and Li.' And he added: 'There are set degrees (*chieh tu*[11]) for increase and decrease both daily and monthly.' Shen interpreted Mr Li's success as due to the fidelity of his temporal correspondences. See *Mêng Chhi Pi Than, Pu* sect., ch. 3, para. 13 (Hu Tao-Ching (1) no. 582). In physiological alchemy Khan and Li came to assume outstanding importance (cf. pt. 5 below).

[1] 半錢   [2] 半分   [3] 片   [4] 分   [5] 大洞鍊眞寶經修伏靈砂妙訣
[6] 七返靈砂論   [7] 沈括   [8] 李   [9] 水丹   [10] 力
[11] 節度

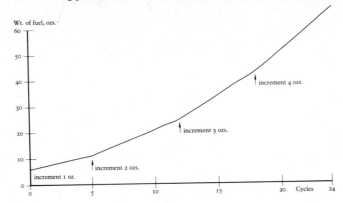

Fig. 1518. Chhen Shao-Wei's linear fire-phasing system, from his *Chhi Fan Ling Sha Lun* (On Numinous Cinnabar Seven Times Cyclically Transformed), *c.* + 713.

downs of cosmic cycles in the varying intensity of the fire, but by shifting successive cycles upward also manages to represent the gradual perfection of the Natural Cyclically Transformed Elixir in what we have no choice but to call a helical phasing scheme. Chhen begins by designing apparatus the shape and measurements of which are completely determined by cosmic correspondences. We shall study them in their place, taking note here only that the furnace of three tiers, standing for heaven, earth, and man (Fig. 1375), has in the central tier twelve doors which stand for the twelve double-hours of the day and night (*chhen*[1], to which the twelve Jupiter Stations were also functionally equivalent).[a] The fire-phasing instructions are so extended and repetitive that we shall quote only their beginning, and represent the rest schematically (Table 118). Chhen says:

Formula for fire control. The formula for the use of fire also corresponds to Yin and Yang, the twenty-four *chhi* periods, and the 72 five-day phases. Five days make up one phase, three phases make up one *chhi*, and two *chhi* make up one month. Seventy-two phases thus correspond to twenty-four *chhi*, making twelve months. Twelve months make a round year, in which the cycle of Yin and Yang reaches completion and the elixir is finished.

As for the time of firing the furnace, the fire should be applied at a midnight which is also a sexagenary hour 1, on a sexagenary day 1, in the eleventh month.[b] Begin by firing through door A for five days, using 3 *liang* of charcoal. There must always be three *liang* of well-coked charcoal (*shu than*[2]), neither more nor less, in the furnace. Then open door B and start the fire, firing for five days, using four *liang* of charcoal. Then open door C and start the fire, firing for five days, using five *liang* of charcoal. Then open door D and start the fire, firing for five days, using six *liang* of charcoal. Then open door E and start the fire, firing for five days, using seven *liang* of charcoal. Then open door F and start the fire, firing for five days, using eight *liang* of charcoal. These six doors are the Yang doors. The charcoal must be put in place vertically in order to bring the Yang *chhi* into play. Then proceed to door G and start the fire, firing for five days, using nine *liang* of charcoal. Then open door H and start the fire, firing for five days, using eight *liang* of charcoal. Then open door I and start the fire, firing

---

a Cf. the drawings of three-tiered eight- and twelve-door furnaces in Fig. 1375.
b The month containing the winter solstice, rebirth of Yang.

1 辰   2 熟炭

for five days, using seven *liang* of charcoal. Then open door J and start the fire, firing for five days, using six *liang* of charcoal. Then open door K and start the fire, firing for five days, using five *liang* of charcoal. Then open door L and start the fire, firing for five days, using four *liang* of charcoal. The charcoal must be put in place horizontally through these six doors, in order to maintain the correspondence with the phased alternation of Yin and Yang.[a] The fire has thus been started through door A and rotated through the twelve doors, using a total of seventy-two *liang* of charcoal in the furnace, corresponding to the seventy-two phases (in a year).

Thus four *chhi*, twelve phases, sixty days, and two months have passed; this is the first cycle.[b]

Table 118. *Chhen Shao-Wei's helical fire-phasing system*
(Based on *TT884*, pp. 12*a*–16*b*, and *YCCC*, ch. 68, pp. 19*b*–24*a*)
See also fig. 3 in Sivin (14).

cycle		Yang doors						Yin doors					
		A	B	C	D	E	F	G	H	I	J	K	L
1	Wt. of charcoal, ozs.	3	4	5	6	7	8	9	8	7	6	5	4
	Total wt. used Cosmic significance of wt.	72 pentadic phases (five-day periods) in a year											72
2	Wt. of charcoal, ozs.	5	6	7	8	9	10	9	8	7	6	5	4
	Increase in total wt. Cosmic significance of increase	12 nodal divisions of solar year (*chieh* 節)											12
3	Wt. of charcoal, ozs.	7	8	9	10	11	12	11	10	9	8	7	6
	Increase in total wt. Cosmic significance of increase	24 *chhi* divisions of solar year (*chhi* 氣)											24
4	Wt. of charcoal, ozs.	9	10	11	12	13	14	13	12	11	10	9	8
	Increase in total wt. Cosmic significance of increase	(as in cycle 3)											24
5	Wt. of charcoal, ozs.	11	12	13	14	15	16	15	14	13	12	11	10
	Increase in total wt. Cosmic significance of increase	(not stated)											24
6	Wt. of charcoal, ozs.	17	18	19	20	21	22	21	20	19	18	17	16
	Increase in total wt. Cosmic significance of increase	(as in cycle 1)											72

In this remarkable intellectual construction we see each cycle divided into a Yang phase of increasing intensity and a Yin phase of decrease. Even the vertical and horizontal orientations of the pieces of charcoal are meant to induce the proper action of Yin and Yang upon the reactants. Each of the sixty-day heating cycles begins at a higher level than the one before and results in a more exalted product.

The modern scientist naturally thinks of the exaltation as caused by the upward shift in the successive time–temperature curves. But resonance rather than physical

---

[a] Or: 'with Yin and Yang, the fifteen-day *chhi*, and the five-day phases (*yin-yang chhi hou*[1]).'
[b] *Ta-Tung Lien Chen Pao Ching, Chiu Huan Chin Tan Miao Chüeh*,[2] pp. 12*a*–13*a* (*YCCC*, ch. 68, under the title *Chiu Huan Chin Tan Erh Chang*,[3] pp. 19*b*–20*b*), translation from critical text in Sivin (4).

[1] 陰陽氣候   [2] 大洞鍊眞寶經九還金丹妙訣   [3] 九還金丹二章

causality is what would have been in the alchemist's mind as explaining the formation of the Elixir. Its gradual perfection was induced, he would have said, by the correspondences he had designed into the process. The rise and fall of heat, which was only one of many correspondences, paralleled the rhythmic shaping force exerted by the cosmic organism upon the Natural Elixir maturing in the bowels of the earth.

The overall symmetry of the system is not seriously compromised by minor asymmetries introduced in the interest of stronger cosmic correlations. In the first cycle only, the weight of fuel is maximal at door G, even though the transition from Yang to Yin was supposed to come after door F. This Chhen found necessary in order to correlate the total weight of fuel in the first cycle with the 72 annual pentads. The exceptionally large increment in the sixth cycle can be explained in a similar way.

In Yang Tsai's 'Pronouncements of the Immortals on Cyclically Transformed Elixirs' (*Huan Tan Chung Hsien Lun*[1]) written in +1052, such minor anomalies are unnecessary. The helical phasing system described there has attained perfect structural symmetry and regularity at the cost of a few correspondences. The alchemist is directed to choose the proper day and compass orientation, and then to build a furnace platform of pounded earth in three layers. The furnace is octagonal, with eight doors. Above the doors is a cover on which is placed the reaction-vessel, with a second vessel for cooling water resting above it. The basic cycle is an ideal month of 30 days, divided into six phases (*hou*[2]), each subdivided into two parts:[a]

First phase, first day: corresponding weight of fuel (*chih fu*[3]), 1 ounce. After $2\frac{1}{2}$ days (30 hours) increase to 2 ounces (until) 60 hours (have elapsed). Second phase, 3 ounces. After $2\frac{1}{2}$ days (30 hours) increase to 4 ounces (until) 60 hours (have elapsed)....

The total configuration over nine months is apparent from Table 119.

Table 119. *Yang Tsai's helical fire-phasing system*
(Based on *TT* 230, pp. 16b–17b)

Cycle	Fuel wt. per half-pentad, ozs.											
	1A	1B	2A	2B	3A	3B	4A	4B	5A	5B	6A	6B
1	1	2	3	4	5	6	6	5	4	3	2	1
2	2	3	4	5	6	7	7	6	5	4	3	2
3	3	4	5	6	7	8	8	7	6	5	4	3
4	4	5	6	7	8	9	9	8	7	6	5	4
5	5	6	7	8	9	10	10	9	8	7	6	5
6	6	7	8	9	10	11	11	10	9	8	7	6
7	7	8	9	10	11	12	12	11	10	9	8	7
8	8	9	10	11	12	13	13	12	11	10	9	8
9	9	10	11	12	13	14	14	13	12	11	10	9

[a] *TT*230, pp. 16b–17b, tr. auct. Cycles of this kind had considerable currency among seekers after the enchymoma, e.g. *Huan Tan Chou Hou Chüeh*[4] (*TT*908), mostly of the Sung or later, ch. 1, pp. 18b–20a; *Hung Chhien Ju Hei Chhien Chüeh*[5] (*TT*934), undated but late, pp. 3a–4a; and with wording very similar to that of the latter in *Yü Chhing Nei Shu*[6] (*TT*940), also late, pp. 11a–12b, and also pp. 14a and 18b–19b.

[1] 還丹衆仙論   [2] 候   [3] 直符   [4] 還丹肘後訣   [5] 紅鉛入黑鉛訣
[6] 玉清內書

Perhaps the last significant conceptual improvement in 'fire-times' (*huo hou*[1]) was two-variable phasing, in which the weight of both fuel and cooling water fluctuate to represent the dynamic interrelations of Yang and Yin. This we find in the 'Confidential Instructions on the Manual of the Heaven-Piercing Golden Flower Elixir' (*Chin Hua Chhung Pi Tan Ching Pi Chih*[2]), of Phêng Ssu[3] & Mêng Hsü,[4] a 'dual-cultivation' treatise of +1225 which seems to be concerned with laboratory processes, albeit in a rather abstract way.[a] Mêng outlines a symmetrical 30-day schedule to be used with sealed vessels incorporating water reservoirs and cooling tubes of various designs.[b] The cycle begins at the new moon (the maximal Yin phase) with 1 ounce of charcoal and 14 ounces of water. Each day the fuel weight is increased by 1 ounce and that of the water decreased by the same amount, until on the 14th day the proportions are reversed. The weights remain constant for 3 days, doubtless to allow leeway for the precise moment of the full moon, the point of maximal energy. Then from the 17th to the 29th the weight of fuel is decreased, and that of water increased, so that the initial weights are in force for a total of 3 days (the 29th, 30th, and 1st) around the time of the new moon. The combined weight of fuel and water is always the same, 15 ounces, and this is explained as the sum of the numbers 6 and 9, which are used for the Greater (or mature) Yin and the Greater (or mature) Yang in the 'Book of Changes' and the *Tshan Thung Chhi*. Again, if for the moment we think of Yin and Yang as two sinusoidal functions out of phase by 180° (say curves representing the sine and cosine of $\theta/2$), we see them approximated here by two interdependent zig-zag functions (see Figure 1519a).[c]

From documents of the kind just cited one can see that number and measure were being used in a way only indirectly related to their employment in modern science. The use of measurement to control the time–temperature profile of, say, an organic synthesis is familiar enough today. In ancient and medieval alchemy the specification of quantities by which the process is to be controlled derived, by way of a theory, from prior observations and measurements, just as it does in contemporary chemistry. Its elaborate cosmic phasing aside, the alchemical process did have to transform one substance into another, and so at some point certain physical and chemical conditions had to be satisfied. It is obvious enough that alchemy and chemistry differ in the number and vagueness of the links between the control specifications and the theory, and those between the theory and the original observations. But if we were to stop there we should remain unable to account for the remarkable specificity, the over-

---

[a] The second chapter of this book (*TT*907), according to the author, was revealed to him by an avatar of Pai Yü-Chhan,[5] the sixth patriarch of the Southern School of Sung Taoism, which propagated 'dual cultivation'. This book might be called an imaginative meditation upon laboratory procedures and apparatus (see above, Vol 5, pt. 3, pp. 199, 203); but there are some signs, e.g. the easy mutations in the names of the two reactants, that nothing in it was meant precisely for practical application. Nevertheless it is important for the light it throws on the actual techniques of the time. On Pai Yü-Chhan, see pt. 3, p. 202.

[b] The presence of a cold water reservoir or condenser at the upper part of Yang Tsai's apparatus is of course very relevant to the early history of sublimation and distillation, on which see pp. 44 ff., 62 ff.

[c] *TT*907, ch. 2, pp. 20a–21b.

[1] 火候　　[2] 金華沖碧丹經祕旨　　[3] 彭耜　　[4] 孟煦　　[5] 白玉蟾

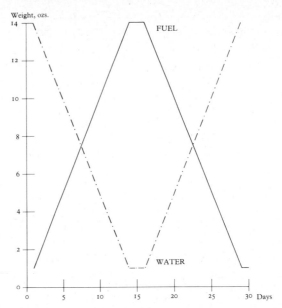

Fig. 1519a. Mêng Hsü's two-variable phasing system, from *Chin Hua Chhung Pi Tan Ching Pi Chih* (+ 1225).

determination, of the ancient formulas. The next step, therefore, is to realise that although in both alchemy and chemistry prior experience has to be shaped by a theory to evolve a new process, modern chemical theories are essentially mathematical while alchemical theories were numerological. The ancient adepts used numbers freely in a way which Granet in his classical study called 'emblematic'—ranking phenomena and things in a qualitative order which would also reflect their special qualitative values.[a] The numerals one to five could be used as strict equivalents for the Five Elements, and thus for the Five Spatial Orientations, Five Sapidities, Five Emotions, and so on.[b] Nine called up in the mind of the alchemist all the attributes and associations of the Greater or mature Yang, including such subtle notions as that of the inevitable inception of decay once the bloom of maturity is past. But Chhen Shao-Wei's use of a three-tiered furnace was not meant merely to symbolise, or allude poetically to, the classical triad of Heaven, Earth, and Man. It aimed actually to bring to bear on the process the synthetic unity which this triad embodied. The twelve firing doors (the basis of Chhen's combustion cycle) were placed in the middle tier because that tier had precisely the same significance within the system of the furnace as man had in his capacity as mediator between heaven and earth (Yang and Yin) within the cosmic Tao. The significance of a given number came from its order in a sequence, and that sequence as an organic whole derived its qualitative meaning from correspondences with other sequences. For instance, if 1 and 2 stood for primal Yin and Yang, 3, as their sum, stood not only for the synthesis which reconciled the antitheses but for the sentience of man which was capable of that synthesis. In another context,

[a] Granet (5), pp. 149–299.   [b] See Vol. 2, pp. 261 ff.

which energised another sequence, 3 might serve a different and even logically contradictory function. For instance, in the five-elements order 3, like 8, stood for the
creative phase Wood, and carried all the immature Yang associations of East, Spring,
and so on. It could be equated with, or used to mark, other things which carried the
same associations, but that was no static classification into species. The Five Elements
were functional phases of a dynamic cycle or configuration, not set qualities, and they
could only be defined individually by relation to the total system. The Aristotelian
approach, based on a rigidly structured biological taxonomy in which genera and
species were individually defined, was not present here. One can find a few static
definitions, but the only ones which had any standing were archaic.[a] Later Chinese
taxonomy might be considered a kind of 'degenerate case' of functional and dynamic
systems of correspondences,[b] just as efficient causation occurs in Chinese physics as a
degenerate case of resonance interaction. Students of ancient Chinese biological ideas
have worked so hard at digging out implicit taxonomies that they have had no time
left to explain why explicit taxonomies were so unimportant.[c] Despite the appeal of
static notions of the Five Elements and other phase-sequences to Western students of
Chinese philosophy, in so far as such conceptions are specific they are inapplicable,
and indeed their role in the historical development of the five-elements theory and its
applications was negligible.

Now a fire-phasing system is also a dynamic sequence of numbers used to induce
cyclic behaviour in a chemical process. The uniqueness of these particular systems is
due, at first glance, to the fact that they were applied quantitatively as measures. But
that can hardly be the whole story so long as Mêng Hsü feels free to define the
constant total of fuel and water weights in his two-variable phasing system as the sum
of the emblematic *I Ching* numbers 6 and 9; so we must look further. Another point
which should not be ignored is that the established qualitative associations of individual
numbers in a fire-phasing system need not come into play, though indeed latent. Some
of these correspondences may be activated, as in Chhen Shao-Wei's use of the total
weights of fuel per cycle to stand for the various divisions of the year. A third element
is that the fire-phasing systems were *ad hoc*. If the usual correlations of individual
numbers could be ignored, the alchemist was free to design his own system instead
of accepting or building on an old one. He would have no such freedom if he were
trying to explain some aspect of his process with a theory made up of five coordinate
concepts, for he would be unable to avoid the customary five-element associations.

Once all these characteristics are put together, the contrast between a classical
emblematic system such as the Five Elements and a fire-phasing system becomes much

---

[a] Even the best-known set of archaic definitions of the elements—'Water is called the soaking and
descending', etc.—are clearly concerned with modes of action rather than formal qualities. See Vol. 2,
pp. 242–6.

[b] Our 'symbolic correlations' of Vol. 2, pp. 261 ff.

[c] This does not mean, of course, that no impressive systems of botanical and zoological classification
developed as the centuries went on. The process is clearly visible in the literature of lexicography and
pharmaceutical natural history from the *Erh Ya* and the *Shen Nung Pên Tshao Ching* onwards (cf. Vol. 6).
One of us (N.S.) doubts, however, that these classifications were primarily biological in any modern
sense.

clearer if somewhat less absolute. To understand either, one has to be aware of two related kinds of emblematic significance. One is the mode of action or behaviour implied by the whole system seen as a sequence of phases. This significance Granet called 'hierarchical', for he thought of it as the construction of a hierarchy (in the case of the elements, a five-valued one) and the ranking of phenomena within it.[a] We prefer to use some such phrase as 'phase-sequential', in order to emphasise that the ranking was neither static, absolute, nor vertical, and that time was one of its most basic parameters. Spatial configurations were no less characteristic than temporal sequences, as we shall see;[b] but in place of a cycle, what they imply is a continuum.[c] The second kind of significance is determined by the qualitative associations of individual elements. These associations could be applied to phenomena on a one-to-one basis, even though they were functional and originally derived from the role of the element in the system. This is Granet's 'formal function' (*fonction protocolaire*).[d] We recognise it in the application of the number 5 to invoke the cosmic associations of Earth, which within the five-element system represents the phase of balance, neutrality, or undifferentiation.[e] In the classical systems, Yin–Yang, trigrams and hexagrams and the Five Elements, there is somewhat greater emphasis on the formal function in most theoretical applications, although both functions are generally in evidence and one tends rather clearly to imply the other. The ubiquity of qualitative associations sets a limit to the truly mensurational applicability of these systems. A fire-phasing system, on the other hand, is an *ad hoc* construction in which the formal function does not come into play unless the alchemist chooses in certain cases to assign cosmic significances. We have seen that such choices were responsible for the anomalies in Chhen Shao-Wei's schema, and that Yang Tsai achieved much greater simplicity without them. With a minimum of individual qualitative correlations, there is little in a fire-phasing system to make gravimetric applications confusing. The significance of such a system is almost completely concentrated in the sequential profile of its cycle.

Still, the metaphysical basis of a fire-phasing system was in no way different from that of, say, the Five Elements. Quantitative fire and heat control was a remarkable first step in the use of quantity to unite theory and practice, but somehow it did not carry within it the implication of further steps in the direction of mathematised science, and historically it was also the last stage of this kind in Chinese proto-science as such.[f] Nevertheless it was known and practised later in both Arabic and European alchemy, as Ben Jonson's 'Alchemist' may by itself sufficiently prove.[g]

> *Mammon:* Lungs, I will manumit thee, from the Furnace;
> I will restore thee thy complexion, Puffe,
> Lost in the Embers; and repair this Brain
> Hurt wi' the Fume, o' the Metals.

[a] Granet (5), p. 151.   [b] Pp. 286 ff.   [c] Cf. Vol. 4, pt. 1, pp. 6 ff.
[d] Granet (5), pp. 151, 566 ff. *et passim.*
[e] Cf. Vol. 2, pp. 59, 106, 112, 114 on 'the uncarved block'.
[f] For a discussion of the idea of combining weights in alchemy and its connections with numerology, see below, pp. 301 ff.   [g] Act II, Scene ii.

Face: I have blown, Sir,
Hard for your Worship, thrown by many a Coal
When 'twas not Beech, weigh'd those I put in, just
To keep your heat still even; These Bleard-eyes
Have wak'd, to read your several Colours, Sir,
Of the pale Citron, the green Lyon, the Crow,
The Peacocks Tail, the plumed Swan....

(5) COSMIC CORRESPONDENCES EMBODIED IN APPARATUS

The dearth of individual correspondences in fire-phasing did not hinder the alchemist much, for he had many other means at his disposal for bringing cosmic correspondences to bear on his process. As we have already looked at the choice of reagents in this light,[a] we can now proceed to consider cosmic correspondences in the alchemist's equipment. These were established by a variety of related means, including spatial orientations, analogical shapes, and numerologically defined dimensions of furnaces and reaction-vessels. Here we shall find many more data on the formal significance of measurements in Chinese alchemy.

Although the *Tshan Thung Chhi* is full of microcosmic correspondences, these are not developed much numerically. There is a chapter which has circulated separately under the title 'Song of the Reaction-Vessel' (Ting Chhi Ko[1]) and may indeed be later than the rest, on the canonical dimensions of the reaction-vessel;

Round three five,
Inch and a tenth,
Mouth four eight,
Two inch lips...'[b]

But the significance of these numbers, whatever it may be, remains entirely implicit. Nor is there anything of this kind in the 'Yellow Emperor's Nine-Cauldron Spiritual Elixir Canon' (late Han?), nor in the Inner Chapters of the *Pao Phu Tzu* book (*c.*+320). In the early Mao Shan documents (*c.* +500 at the latest) there are lucid instructions for centering the furnace in a space oriented by the cardinal points. A thatched elaboratory has to be built facing south. The furnace is set up precisely at its centre, and the reaction-vessel placed centrally within it.[c] But again the significance of this centering is not pointed out.

For explicit cosmic correlations we must return to the furnace and vessel within which Chhen Shao-Wei, probably in the +8th century, phased the firing of his 'Ninefold Cyclically Transformed Gold Elixir'. In *TT*884 we read:

The furnace and reaction-vessel for the Great Elixir must also be made in such a way as to incorporate (*ho*[2]) heaven, earth, and man (the Three Powers), and the Five Spirits

---

[a] See pp. 225, 251 ff.
[b] *Chou I Tshan Thung Chhi Fên Chang Chu Chieh*, ch. 33 (ch. 3), pp. 7*a*–10*b*, tr. auct. Cf. above, Vol. 5, pt. 3, p. 71.
[c] *Thai-Wei Ling Shu Tzu-Wên Lang-Kan Hua Tan Shen Chen Shang Ching* (TT252), p. 4*a*; *YCCC*, ch. 68, p. 4*b*.

[1] 鼎 器 歌   [2] 合

Fig. 1519*b*. Two of the Twelve Hour-Presidents (Spirits of the Double-hours). On the left, the patron of the hour *mao* (5 a.m. to 7 a.m.); on the right, the precentor of midnight (the *tzu* hour, 11 p.m. to 1 a.m.). A pair of Han tiles in the Royal Scottish Museum, Edinburgh.

(*wu shen*[1] = the Five Elements).[a] The vessel must be made from 24 ounces of gold from the seventh recycling, in order to respond to the 24 *chhi* periods. Sixteen ounces of it is cast into a round (or, spherical) vessel with a capacity of nine liquid ounces (*ko*[2]); and eight ounces into a cover.[b] The use of 16 ounces to make the vessel incorporates the number of (ounces in) a pound. The capacity of nine ounces embodies the Three Origins (*san yuan*[3] = the Three Powers) and the maximal Yang (number, 9).[c] The 8 ounces of the lid responds to the Eight Nodes (the beginnings and midpoints of the four seasons). The vessel and lid are thus 24, incorporating these great constants. The vessel must be emplaced according to the Eight

  [a] This meaning of *wu shen* goes back to the *Huai Nan Tzu* book, *c.* − 120 (ch. 21, p. 2*a*): 'Astrology ...conforms to the correspondences of the temporal cycles, and models itself on the constancy of the Five Spirits.'
  [b] The possible alchemical use of the ancient spherical *tui* vessels (Figs. 1392, 1393 above) has already been remarked upon.   [c] Cf. Granet (5), p. 150.

  [1] 五神   [2] 合   [3] 三元

Trigrams and the Twelve Spirits (*shih-erh shen*[1] = the 12 hours)[a] before the mixed Purple Gold Granules are placed in it. It is tightly closed and luted so that no Yang *chhi* (i.e., vapour) can escape, and is put into the furnace.

Formula for building the furnace. On a 45th sexagesimal day falling in a 41st sexagesimal decad,[b] in a place oriented toward the southwest and the ninth duodenary branch,[c] take clean earth and begin by building it up to make a platform eight inches high and two feet four inches broad. On the platform make a furnace 2 feet 4 inches high, in three levels, with free access of *chhi* from bottom to top. The upper level, 9 inches high, is Heaven. Make nine openings in it, to correspond to the nine stars (of the old Great Bear).[d] The middle level, 1 foot high, is Man. Make 12 doors, which stand for the twelve hours of the day (*chhen*[2]). A fan (*shan*[3]) must be installed in each. The lower level, 5 inches high, is Earth. Open 9 passages (*ta*[4]), which correspond to the Winds of the Eight Directions (*pa fêng*[5]). The interior of the furnace must be 1 foot 2 inches in diameter.[e]

These correlations are all paradigms of the use of number to call up qualitative associations. Their referents are rather scattered—from the balance to the Great Bear —but images patently cyclical and temporal are in the majority. All of these influences operate upon the furnace and vessel, and contribute to the formation of the Elixir. Numerological correspondences of this sort become common in later alchemy, and can even be found in physiological alchemy and magic.[f]

(i) *Arrangements for microcosmic circulation*

In one of the basic collections on 'irrigation' processes, the *Chhien Hung Chia Kêng Chih Pao Chi Chhêng*[6] (Complete Compendium on the Lead–Mercury A–G Perfected Treasure),[g] incorporating generally Sung or later materials,[h] the effect theoretically induced within the reaction-vessel by bringing the cosmic forces to bear is clearly visualised.

[a] Cf. Forke (4), vol. 1, p. 534; vol. 2, pp. 406–407. See also for the double-hour presidents, our Vol. 4, pt. 2, p. 440 and Fig. 1519 b.

[b] The practice of numbering ten-day periods (*hsün*[7]) sexagesimally was not at all common in China.

[c] According to the normal spatial associations of the duodenary series, the ninth branch is the southernmost of the three which correspond to West. See Table 34, in Vol. 3, p. 403.

[d] See Vol. 3, p. 250.

[e] TT884, pp. 11a–12a, and YCCC, ch. 68, pp. 18b–19b; tr. Sivin (4).

[f] For physiological alchemy see *Hsiu Tan Miao Yung Chih Li Lun*[8] (Sung or later, TT228), pp. 9bff. As for magic, numerologically significant numbers appear without discussion, but correspondences of furnace shape are identified, in a Thang text on the preliminaries to the casting of three magic mirrors (which themselves embody cosmic dimensions), weights, and images. This is *Shen Hsien Lien Tan Tien Chu San Yuan Pao Ching Fa*[9] of +902 (TT856). The penultimate character is sometimes given as *chao*;[10] it was altered, probably by an early copyist, to avoid a Sung taboo on the character *ching*[11] and its homonym; see Chhen Yuan (4), p. 154.

[g] *Chia*[12] and *kêng*,[13] which here we translate as A and G in this book title, are two of the ten cyclical characters (*kan*,[14] 'stems') long used for designating members of a series; and, since the +17th-century Jesuits, for translating letters of the alphabet denoting parts on scientific and engineering diagrams. Since the stems are paired two by two with the Five Elements, *chia* and *kêng* correspond to the elements Wood and Metal (immature Yang and Yin) respectively, and that is their significance here. See Vol. 3, pp. 396–8, and Vol. 5, pt. 3, pp. 158 ff., 280 above.

[h] See pp. 294 ff.

[1] 十二神　　[2] 辰　　　[3] 扇　　[4] 達　　[5] 八風　　　[6] 鉛汞甲庚至寶集成

[7] 旬　　　　[8] 修丹妙用至理論　　　[9] 神仙鍊丹點鑄三元寶鏡法

[10] 照　　　[11] 敬　　[12] 甲　　[13] 庚　　[14] 干

Upper and lower reaction-vessels (ting[1]). The body has a circumference of 12 inches to respond to the 12 months,[a] and is 8 inches long (i.e., high) to correspond with the Eight Nodes. The width of the body of the upper vessel is twice that of the lower vessel, in order to bring to bear (*an*[2]) the 24 *chhi*. The upper vessel is heaven, and the lower earth. In the upper there is ascension, so it is Yang; in the lower there is descent, so it is Yin.[b] The Yin *chhi* wants to ascend, and the Yang *chhi* wants to descend. This responds to the formative power (*thao yeh*[3]) of Yin and Yang. The length and breadth must be neither larger nor smaller than (the dimensions given). If larger, the *chhi* will disperse instead of collecting; if smaller, it will be forced to overflow. Thus (in neither case) can it be made to rise and fall equitably and harmoniously.[c]

The motion described here is unquestionably a circulation, although explained rather vaguely, apparently in terms of the automatic reversion of Yin and Yang once they have reached their limits. The author does not seem to be very concerned about the contradiction of a Yin *chhi* wanting to ascend when it is defined by its tendency to descend. Perhaps he would have explained such a movement by the urge towards creative union.

This apparent discrepancy is handled more overtly in a late text which envisions, at least for meditative purposes, a vessel with cooling water above and fire below. The explanation may not be chemical, but it is certainly physical. The writer says:

Now for the method of preparing the Elixir.[d] The reaction-vessel has three legs in order to respond to the Three Powers. The two containers, upper and lower, correspond to the Two Instrumentalities (*liang i*,[4] earth and heaven). The legs are 4 inches high to respond to the four seasons. The furnace is 8 inches deep in order to match the Eight Nodes. In the lower part 8 doors are opened to admit the Winds of the Eight Directions. The charcoal is apportioned in 24 pounds in order to arouse the 24 *chhi*.[e] Yin and Yang are inverted, with Water and Fire meeting and struggling. Above is Water, responding to the pure *chhi* of Heaven. Below is Fire, receiving the turbid *chhi* of Earth. The celestial *chhi* descends, while the terrestrial leaps upward: Heaven and Earth meeting in mutual stimulus, the primal *pneumata* (*yin yün*[5]) conjugating. They come together to form the Two *chhi* (Yin and Yang), which, once joined, blend to become one. (The product) is named the Great Elixir of the Two *chhi*, and its marvellous function depends upon these (correspondences). In a dozen hours (i.e. a day) the process asserts a power which it takes the Shaping Forces a thousand years to exert.[f]

If the task is to explain a cyclical motion within the vessel, the crux will lie in accounting for the half of the motion which is contrary to the customary sense of Yin and Yang. This tractate derives the reverse movement from the configuration of the apparatus, which from the viewpoint of Yin–Yang theory can be considered inverted, since the

---

a This refers to the lower vessel, and is given later (p. 5*a*) as diameter.
b This sentence may also be taken to mean simply 'Ascent is Yang, descent is Yin.'
c *TT*912, ch. 3, p. 4*b*, tr. auct.
d Even though the ultimate concerns of this treatise seem to be with the Enchymoma, the language of the process contemplated here is that of the Elixir, so we translate accordingly.
e I.e. the *chieh chhi* of the 24 fortnightly divisions of the year.
f *Chiu Chuan Ling Sha Ta Tan Tzu Shêng Hsüan Ching*[6] (Mysterious Sagehood-Enhancing Canon of the Great Ninefold Cyclically Transformed Cinnabar Elixir [or Enchymoma]), *TT*879, pp. 1*b*, 2*a*, tr. auct. This undated text, cast in *sūtra* form, of unknown authorship, incorporates many of the conceptions of *wai tan* (external, or chemical, alchemy), but its own rather obscure perspectives seem to be those of *nei tan* (internal, or physiological, alchemy).

¹ 鼎   ² 按   ³ 陶冶   ⁴ 兩儀   ⁵ 氤氳   ⁶ 九轉靈砂大丹資聖玄經

fire (Yang) is below and the water (Yin) is above.[a] Since the two *chhi* are out of their proper static (or rather configurational) orientations, they must move to regain them.[b] What we might for the moment visualise as the kinetic energy of their collision, responsible for the exalted level of organisation of the Elixir, is accounted for.

But if the circulation is divided into two temporally distinct halves, it becomes nothing more nor less than an oscillatory cycle. This we find a bit further on in the *Chhien Hung Chia Kêng Chih Pao Chi Chhêng* (*TT*912), as the 'four seasons' heating

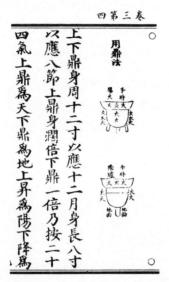

Fig. 1520. A page from Chao Nai-An's *Chhien Hung Chia Kêng Chih Pao Chi Chhêng* (Complete Compendium on the Perfected Treasure of Lead, Mercury, Wood and Metal), ch. 3, p. 4 *b*, showing the *Yang lu* and *Yin lu* furnaces. Perhaps Thang (+ 808) in date, but more probably Wu Tai or Sung.

technique (*ssu shih huo*[1]). The sealed vessel is moved back and forth each day between a 'Yang furnace' (*Yang lu*[2]), which is heated from below, and a 'Yin furnace' (*Yin lu*[3]), which has a container of water below it and is heated from above, in order to set up an oscillatory motion within the vessel (Fig. 1520). The text runs:

Use 'four seasons' heating for 7 days and nights, in order to develop the great power (*kung*[4]) (of the Elixir). Water, Fire, and the reactants: these respond to the Three Powers, heaven, earth, and man.[c] In this 'four seasons' heating the fire of spring should be mild, that of summer intense, that of autumn warm, and that of winter weak. For this method, select the first double-hour (11 p.m.–1 a.m.) of the first day of a duodenary cycle, place the vessel on its fitted three-legged support, and put it into the Yang furnace. Under the belly of the vessel pile (charcoal), extending it $\frac{3}{4}$ inch (or $\frac{3}{10}$ of the way, *san fên*[5]) up the body of the vessel. Use a gentle fire so that the contents conjugate for $1\frac{1}{2}$ hours. Next increase the fire,

[a] Note the parallel with the reversed polarity of Realised Lead and Mercury, seen earlier (p. 254).
[b] The implicit assumption—that the *chhi* of an Element will move spontaneously in order to regain the orientation proper to that element—is philosophically reminiscent of that doctrine of place on which Aristotle's physics of motion was based. See Vol. 4, pt. 1, p. 57.
[c] The reactants take the same intermediate position as does man with respect to earth and sky.

[1] 四時火   [2] 陽爐   [3] 陰爐   [4] 功   [5] 三分

piling the charcoal close to the vessel up to half its height for $1\frac{1}{2}$ hours. Then make the fire intense with a full charge of charcoal, building it $\frac{3}{5}$ of the way up the body of the vessel. After $1\frac{1}{2}$ hours gradually decrease the fire. Use a weak fire, so that the contents of the vessel will be warm and the warm *chhi* remains controlled for $1\frac{1}{2}$ hours until the sixth double-hour (11 a.m.–1 p.m.). Withdraw the vessel and start the Yin furnace, placing the belly of the vessel over the mouth of the small pot (of water which has been buried in the earth at the bottom of the furnace). Use ashes to bury the body of the vessel to $\frac{3}{5}$ of its height. Then with a cloth wipe any water off the upper vessel. After it is dry put the fitted fire-pan (*ting phan*[1]) in place, and in it put one pound of burning fuel. This is called 'inserting the spring fire', which causes the mercury inside the vessel to descend slowly for $1\frac{1}{2}$ hours. Only then may the charcoal be piled up above the ash layer and packed against the vessel as (the fire) is gradually increased for $1\frac{1}{2}$ hours. Finally the full charge of burning fuel is packed about the vessel from the fire-pan to the ash layer. This gives a strong combustion, and the flame burns intensely for $1\frac{1}{2}$ hours. Then gradually remove the fire above the vessel and heat weakly from below,[a] so that the Medicine within the vessel is kept at a controlled warm heat for $1\frac{1}{2}$ hours until midnight, the first double-hour of the day. At that time the vessel is again shifted into the Yang furnace, and as before (the Medicine) is made to go through a cycle of ascent and descent in the *chi chi*[2] and *wei chi*[3] furnaces.[b] After 7 days and nights the power of the fire will have brought the process to completion.[c]

Thus although there are four major phases, which correspond to the four seasons beginning with Spring,[d] the transitions are not at all abrupt. The phasing is moderated and made gradual by finer variations of both fire intensity and area of vessel in contact with the burning fuel, but fuel weights are not prescribed. This particular process does not happen to maintain the symmetry of water and fire throughout the cycle, but depends on fire alone during the Yang phase. Since the 'Yang furnace' (which had no water vessel and was usually heated from all sides) and the '*wei chi* furnace' (with fire above and water below) both induced the Yang mode of action within the vessel, the alchemist was constrained in choosing between them only by his taste for abstract symmetries.

In another compendium of 'irrigation' methods, the *Kêng Tao Chi*[4] (Collection of Procedures of the Golden Art),[e] written some time after +1144, there is a 'Bubbling-Spring Perpetual-Life Casing' method (*yung chhüan chhang shêng kuei*[5]) in which 'water below and fire above' is alternated with 'fire below and water above' for $2\frac{1}{2}$ days

<hr/>

a Or, punctuating differently, 'Then gradually remove the fire above the vessel and the lower fire.'

b On these pieces of apparatus see further the discussion on pp. 68 ff. and the study of the meaning of the names, p. 70. The whole conception of the Yang and Yin furnaces in this text is inextricably connected with the ancient procedures of sublimation and *destillatio per descensum* (cf. pp 44, 55), as also with the later development of true distillation in its East Asian forms (pp. 62 ff.). For Chao Nai-An and his anonymous colleagues who wrote this book the *chi-chi* and *wei-chi* apparatuses were probably not stills in any modern sense, but they may well have derived from apparatuses in artisanal or pharmaceutical use that were.

c *TT*912, ch. 3, pp. 5 *a*–6 *a*, tr. auct.

d The most common practice was to begin the process at the winter solstice, the moment of the ascendancy of mature Yin and thus of the rebirth of Yang in the cycle of the year.

e The title does not actually say 'gold' but '*kêng*',[6] the 7th of the decimal stems and thus the element Metal. But pragmatically oriented alchemists often used *kêng* as a cover name for gold, and it is possible that the compiler of *Kêng Tao Chi* meant nothing more abstract. Our rendering of the title is in any case tentative.

¹ 鼎盤   ² 既濟   ³ 未濟   ⁴ 庚道集   ⁵ 湧泉長生置
⁶ 庚

each. The fuel is increased a day at a time from 1 ounce to 16, and then decreased to 1 ounce again. We are informed that 'this is the Bubbling-Spring fire phasing system (*yung chhüan huo hou*[1]) concealed within Master Wei (Po-yang's) *Tshan Thung (Chhi)*.'[a] Here as in the earlier examples, the two-reactant system, the heat phasing, and the correspondences of the apparatus are interdependent.

By good fortune we are not limited to verbal descriptions of the two types of fire–water apparatus. In the seventh cycle of the formula for the 'Nine-Cycle Potable Gold Great Cyclically Transformed Elixir' (*Chiu Chuan Chin I Ta Huan Tan*[2]) in the *Chin Hua Chhung Pi Tan Ching Pi Chih*[3] (Confidential Instructions on the Manual of the Heaven-Piercing Golden Flower Elixir) of +1225, detailed instructions are given for heating the reactants,[b] sealed within a golden 'chaos vessel', first with 'water below and fire above' for two months, then with 'fire below and water above' for one month.[c] Although the weight of fuel remains constant within each of the two phases in this particular process, a single apparatus is converted from one function to another by substituting parts (Fig. 1448). Although, as has been pointed out above (p. 73), the 'water below and fire above' apparatus somewhat resembles a still with two side-tubes, the text indicates that the water-container with its two arms is not open to the reactants, and that they remain sealed within the egg-shaped reaction-vessel. The arms could thus serve only to lead water into the reservoir and to let steam escape. It is not unreasonable to suppose, however, that this design was prompted by know-ledge of water-cooled stills. In practice its effect would have been only to lower slightly the temperature in the vessel. On the other hand, the water reservoir in the 'fire below and water above' apparatus resembles a funnel and protrudes from the condenser head into the reaction-vessel, but its lower end is closed.[d]

With this in mind we are able to go back another 50 years in time and recognise the same alternation in two illustrations in the *Tan Fang Hsü Chih*[4] (Chymical Elaboratory Practice) of +1163.[e] The apparatus on the left in Fig. 1446, with the reaction-vessel and furnace above and the water reservoir with side-arms below, is labelled '*wei chi* furnace', while that on the right, with its water reservoir above, is marked '*chi chi* furnace', as in the Four Seasons heating technique which we have just examined.[f] In all these texts, as in others of less certain date and import,[g] we may

---

[a] *TT946*, ch. 7, pp. 16*a*–17*a*. The text is syntactically defective, and we read conjecturally *chih nei pi tshun*[5] for *pi chih nei tshun*.[6]

[b] Each cycle uses a different complex apparatus, basically a sealed gold reaction-vessel with cooling coils of various sorts; in later cycles the metal is made by transmuting mercury. For details see above, pp. 35ff. An apparatus which closely resembles the lower one in Fig. 1409 appears in ch. 2, p. 2*a*, with the superscription *chi chi*[7] (cf. Fig. 1446) as in *Tan Fang Hsü Chih* (see below).

[c] *TT907*, ch. 2, pp. 14*b*–16*a*.

[d] See above, p. 35, for an excerpt from this text which specifies that 'the lower end is closed and does not allow water to pass through'.   [e] *TT893*, p. 9*a*.   [f] See pp. 283ff.

[g] Another picture of the *wei-chi* apparatus occurs in the *Chih-Chhuan Chen Jen Chiao Chêng Shu*[8] (Technical Methods with Critical Annotations by the Perfected Immortal (Adept) Chih-Chhuan (Ko Hung), *TT895*), p. 1*b*. We reproduce it in Fig. 1447. Despite the attribution, the authorship and date of this little tractate are unknown. Nor is it critically annotated. Moreover the text is not at all oriented towards practical proto-chemical alchemy.

[1] 湧泉火候	[2] 九轉金液大還丹	[3] 金華沖碧丹經秘旨	[4] 丹房須知
[5] 之內祕存	[6] 祕之內存	[7] 既濟	[8] 稚川眞人校證術

be seeing landmarks in the history of distillation transmuted into instruments of microscosmic induction. They and their background will require much further study.

(ii) *Spatially oriented systems*

The paramount importance of time in alchemical processes has led us to give much emphasis to the significance of phased cycles. But spatial orientation was also regularly used by some alchemists in their models of the Tao. Since the Tao is the organic totality of space and time, it cannot be apprehended in its wholeness except by a mystical intuition which not every Taoist could hope to summon up at will. The adept who wanted to find his way to enlightenment through knowledge and contemplation used abstract correspondences and partial visualisations. He could concentrate on the temporal aspect of the Tao, turning his attention to the cyclical behaviour of the cosmos or one of its sub-systems. Another alternative open to him was to hold time constant, apprehending the organismic pattern as reflected in a momentary configuration. In other words, as we have suggested in passing a few pages earlier, a configuration of several organically related elements plays the same role in spatial correspondences as a cycle (a special time configuration) plays in temporal correlations. The two modes are interdependent and functionally interchangeable. As the great systematiser of five-element correspondences, Tung Chung-Shu, put it about −135, 'Thus Wood has its place in the east and has authority over the *chhi* of spring. Fire has its place in the south, and has authority over the *chhi* of summer', and so on.[a] What held spatial east and temporal spring in their equation was the sun's annual motion, of which they both represented a quarter.[b] Each of the temporal sub-systems based on the year (the 72 pentadic phases, 24 *chhi* periods, 12 months, 8 nodes) which we have seen invoked above had similar correspondences—to each *chhi* belonged $\frac{1}{24}$ of the sun's annual path,[c] and so on—which the alchemist could make explicit if he wished.

From the birth of alchemy to its demise, the Five Elements were used to characterise locations throughout space, dividing it into the four cardinal directions and a centre at which opposite modes were neutralised and harmonised. The Five Elements categorised places and at the same time organised them into a system.[d]

Examples of spatial orientation have already appeared in several of the documents above.[e] Alchemical specifications of location were so tied conceptually to temporal correspondences that they are practically never found in isolation. One of the very

---

[a] *Chhun Chhiu Fan Lu*, ch. 42, tr. Hughes (1), p. 294. See Vol. 2, p. 250. The attribution and date of this book have recently been questioned by Prof. G. Malmqvist, but the text must be an ancient one.

[b] The relation is not so simple as it may seem but we have discussed it in Vol. 3, p. 240.

[c] Cf. Yabuuchi's definition of the *chieh-chhi*[1] (the 24 *chhi* divisions) as the 'time required by the sun to move through each 15° on the ecliptic'; (9), p. 462.

[d] So far the classical account of Chinese conceptions of space lies with Granet (5), pp. 86–114, but there is urgent need for a less exclusively anthropological approach.

[e] The reader will also recall the alchemical planting of the Five Grains, each in the direction corresponding to its colour, within the precincts of Wang Mang's palace in +10. See above, Vol. 5, pt. 3, p. 37.

[1] 節氣

few exceptions is particularly interesting because it is early, its context is medical, and it is concerned implicitly but unmistakably with emplacing the reactants within the reaction-vessel in such a way as to create a microcosmic configuration. This is not an alchemical elixir but a 'Panaceal Sublimed Yellow Powder' (*kuang chi fei huang san*[1]), prescribed for sores and ulcerations in one of the great medieval compendia of medical prescriptions, Wang Thao's[2] *Wai Thai Pi Yao*[3] (Important Medical Formulae and Prescriptions revealed by a provincial governor) of +752. That its ultimate source was alchemical is more than likely. Yoshida Mitsukuni has pointed out[a] similarities to a recipe in the *Thai-Chhing Shih Pi Chi*[4] (Records of the Rock Chamber; a Thai-Chhing Scripture—before +806), a practical collection of alchemical and iatro-chemical formulae with Mao Shan associations. What Wang Thao says is as follows:

Take: Laminar malachite (*tshêng chhing*) Magnetite (*tzhu shih*)
Orpiment (*tzhu huang*) Realgar (*hsiung huang*)
Fibrous arsenolite (*pai yüshih*) Cinnabar (*tan sha*)

one ounce of each. Grind the above six ingredients to fine powders, and emplace them according to the colour correspondences of the directions: laminar malachite to the east, cinnabar to the south, white arsenolite to the west, magnetite to the north, and realgar in the central position. Two earthenware urns (*wa wêng*[5]) are coated inside with yellow clay two or three times in order to make (a lining) five- or six-tenths of an inch thick. Then place powdered orpiment in the bottom. Combine and sieve[b] the other ingredients and put them on top, afterwards laying (the other) half of the orpiment on top as a cover. Spread clay closely on the joint (between the two vessels, the mouths of which are now joined); and do not allow any of the *chhi* to leak out.[c]

Here the five-element correspondences are organised conventionally:[d]

Substance	Colour	Direction	Element
Laminar malachite	caerulean	east	Wood
Cinnabar	red	south	Fire
Arsenolite	white	west	Metal
Magnetite	black	north	Water
Realgar	yellow	centre	Earth

[a] (5), p. 220, referring to *Thai-Chhing Shih Pi Chi* (TT874), ch. 1, pp. 13a–14a. This recipe also emphasises directional correspondences, but the similarity is not quite as close as Yoshida suggests. The alchemical text has *pai fan shih*,[6] white kalinite alum, instead of arsenolite, *pai yü*[7] *shih*. Since the characters are similar, it is not impossible that one passage is corrupt, but we dare not assume it. Yoshida also adduces another similar passage. Cf. (7), pp. 227ff.

[b] The direction to combine the ingredients seems expressly to countermand the specification that they be emplaced directionally. The separation of the two instructions does not necessarily mean that they apply to two separate procedures, for the opening instruction seems to have been attached originally to the list of ingredients rather than to have been the first of several consecutive steps in the actual process. We suggest very tentatively that 'combine and sieve' (*ho shai*[8]) may be a misreading of the visually similar 'sieve separately' (*fên shai*[9]). The editor of the recent Jen-min Wei-shêng edition suspects (textual notes, p. 65b) that this recipe is incomplete, and suggests other emendations.

[c] *Wai Thai Pi Yao* (Fang), ch. 30 (p. 818.2), tr. auct.

[d] See above, p. 225, Table 117.

[1] 廣濟飛黃散 [2] 王燾 [3] 外臺秘要 [4] 太清石壁記 [5] 瓦甕
[6] 白礬石 [7] 礜 [8] 合篩 [9] 分篩

This microcosm transcends the two-dimensionality of the five-element concept by adding a higher and a lower plane. Orpiment serves for both, its yellow colour indicating the bond of the up and the down to the centre, and thus their neutrality in the scheme. If the originator of this process had wanted to bring the Yin–Yang correlations of 'up' and 'down' to bear, he would have used two different substances to represent them.[a]

Another preparation (unfortunately more confused) given by Wang Thao employs a vertical stack of five reactants in addition to the five which are horizontally emplaced:

Take cinnabar and place it to the south in an earthenware basin (*wa phên*[1]). Orpiment is placed in the centre, magnetite to the north, laminar malachite to the east, and quartz (*pai shih ying*[2]) to the west, with arsenolite above, talc (*shih kao*[3]) next in order, and stalactite (*chung ju*[4]) on the bottom. Realgar is the cover, and muscovite mica (*yün mu*[5]) is spread thinly beneath. (Use) 2 ozs. of each, first pounding and sifting into the basin. Cover it with another basin....[b]

The wording is not exact enough to determine where the horizontal and vertical axes intersected, or even the precise order of the vertical ingredients. Reasoning from the way formulae are usually worded, we might suggest the most probable order (from the top down) to be realgar, arsenolite, calcite, stalactite, and muscovite. All are white except the realgar, which the Chinese associated closely with orpiment (as does modern chemistry), and which took the central place in the other configuration.

If we trace the idea of alchemical colour associations as far back as we can, we find them being applied along with other cosmic associations not to the contents of the reaction-vessel but to the lower of the three regions of vital heat ('Fields of Cinnabar', *tan thien*[6]) in physiological alchemy.[c] The *Lao Tzu Chung Ching*[7] (The Median Canon of Lao Tzu) is a pre-Thang treatise on the physiological microcosm and its gods which maintains the pre-Mao-Shan tradition (based on a prone meditation position) with respect to the location of the *tan thien*.[d] Before noting that the divinity resident in the *tan thien* is named Confucius (*hsing Khung ming Chhiu tzu Chung-Ni*[8]), it states that

the interior of the *tan thien* is scarlet in the centre, caerulean on the left, yellow on the right, white above, and black below. It goes from square to round within (a length of) 4 inches. The reason that it is located 3 inches below the navel derives from (the triad of) Heaven, Earth, and Man. Heaven is 1, earth 2, man 3, and the seasons 4; thus it is said (that the length is) 4 inches. Based on the Five Elements, there are the 5 colours.[e]

---

a Exactly what Wang Thao ended with would be anybody's guess, but if he took the sublimate it was presumably a mixture of mercuric and arsenical compounds. Repetition in the laboratory today could alone decide what happened.

b Ch. 24 (p. 664.1), tr. auct. The Chhing xylograph which the Jen-min Wei-shêng edition reproduces is badly mispunctuated, jumbling the directional associations. The formula is attributed to one Fan Wang.[9]

c See pt. 5 below.

d Early Taoist traditions of internal cosmography and *nei tan* techniques are being studied by our colleague Prof. K. M. Schipper, to whom we owe this reference.

e *YCCC*, ch. 18, pp. 13a, b, tr. auct. It is curious that since in this case red had to be central the other colours had to be peripheral, but it is not easy to see why they were arranged as stated.

¹ 瓦盆   ² 白石英   ³ 石膏   ⁴ 鍾乳   ⁵ 雲母
⁶ 丹田   ⁷ 老子中經   ⁸ 姓孔名丘字仲尼   ⁹ 范汪

In this passage there is no attempt to assign a function to the array of colours, or to account for the fact that their directional associations are not those of the five-element convention. It may be, therefore, that the Five Colours were introduced for the simple numerological purpose which becomes plain in the final sentence of our excerpt. Still, this highly regarded treatise existed thenceforth as a precedent for deeper speculations.

Between Wang Thao's sophisticated and abstract configurations laid out within the confines of a sealed reaction-vessel and the simple centering of the vessel within a directionally oriented elaboratory in the earliest Mao Shan documents lies a great gulf.[a] In the latter we can still glimpse the ritual origin of the organisation of space. Even after the idea of configuration had been transmuted philosophically into chemical techniques, the demarcation of inviolate sacred spaces in Taoist rituals continued alongside. The following example of a rite for the protection of the alchemical furnace, included in Wu Wu's[1] *Tan Fang Hsü Chih*[2] (Chymical Elaboratory Practice) of +1163, uses at least three of the same colours and directions as Wang Thao's formula, though they are represented by different materials. It runs:

To the south, one foot from the furnace platform, bury 1 pound of crude cinnabar,[b] formed into a 5-inch long 'wire' (*hsien*[3]) after being mixed with vinegar. To the north bury 1 pound of lime; to the east, 1 pound of cast iron; and to the west, 1 pound of white silver. Above, three feet from the reaction-vessel, hang an ancient mirror, and set out lamps for the 28 lunar mansions and the Five Planets. In front set up a fine sword. Before the furnace provide a basin of water from a previously unused well, refilling (the basin) once every seven days. Use a bench of peach-wood on which to place the incense burners. Put them in every location and keep them charged day and night. By the fourth cycle the elixir will be in contact with the gods and spirits (*shen ming*[4]), and there is danger that demons (*mo*[5]) will come and encroach upon it. Guard it therefore tranquilly, saying these words of prayer:

'I respectfully call upon the Emperor of the Abscondite Origin, the Most High Ancient Lord (Hsüan-yuan Huang-ti[6]) and Thai-shang Lao-chün,[7] (an emanation of the Tao) to
> Cycle and combine the Creative and the Receptive,[c]
> Ward off invasions by diabolical beings
> Trying to touch our Perfect Medicine.
> Venerable Creative (= lead)[d] has safe refuge,
> Iron buried to the east,
> Ardent fire in the south,
> A man hidden in the west,
> A barbarian standing in the north.
> Above hangs a mirror;
> Where the Five Elements are matched
> The Ghosts and Spirits will not come.
> Let this place be tranquil,

---

[a] See pp. 216 ff.   [b] Or vermilion, artificial mercuric sulphide (*chu*[8]).
[c] I.e. the *I Ching* Chhien and Khun *kua*, which stand for Yang and Yin here.
[d] The text of this line apparently telescopes two complex associations. 'The Creative', the name of the first hexagram of the *I Ching*, is a homonym for 'lead' (*chhien*[9]), of which *chin kung*[10] is a variant. The Creative is also Yang, a symbolic correspondence of lead in two-element processes (p. 253). Note that in a story quoted earlier (Vol. 5, pt. 3, p. 26), Huan Than's pun depended upon taking *kung* in its alternate sense 'grandfather'.

[1] 吳悞   [2] 丹房須知   [3] 線   [4] 神明   [5] 魔   [6] 玄元皇帝
[7] 太上老君   [8] 硃   [9] 鉛   [10] 金公

19   NSC

Let the Realised Immortals protect me
As I hold firmly to the Perfect Tao.
Urgently, urgently, as by lawful order!'[a]

Again let us transpose these correspondences into a table:

Substance	Colour	Direction	Element
Iron	caerulean	east	Wood
Cinnabar	red	south	Fire
Silver	white	west	Metal
Lime	grey?	north	Water

Fig. 1521. Tshao Yuan-Yü's reconstruction of a mediaeval Taoist alchemical laboratory in a cave (1).

There is no fifth element, because the point of the ritual is to put the furnace and its contents in the place of Earth and thus induce the state of undifferentiation and harmonisation of opposites which Earth implies. The centering simultaneously protects the Elixir, of course, from malicious spirits.

The idea of furnishing the microcosm with its own firmament lies midway between rite and metaphysics (at a point where the distance between them is particularly

---

[a] *TT*893, pp. 5*b*–6*a*, tr. auct. This work is as important for its rites and ceremonies as for its practical instructions and illustrations of apparatus. The prayer, and especially its ending, is strongly reminiscent of the Taoist liturgies already briefly described (pt. 2, pp. 128 ff. above, with references to more detailed studies). The whole scene evoked by the passage recalls the picture sketched by Tshao Yuan-Yü forty years ago, when it appeared as the frontispiece to vol. 17 of *Kho Hsüeh*, and often afterwards reproduced, as by Barnes (1), Li Chhiao-Phing (1), Thaiwan ed. p. 27, (1), opp. p. 26. We reproduce it in Fig. 1521, leaving its evaluation to the judicious.

short). Only a post-Enlightenment man would maintain a hard distinction in principle between the ritual function of hanging up 28 lamps and the theoretical point of working the number 28 into a weight or dimension. The only practically relevant difference is that the rite, however directly derived from the Five Elements and other physical conceptions, was primarily and directly meant to produce an effect upon the Unseen World—negative upon the riff-raff of that realm, positive upon its functionaries going about their business—rather than upon the ingredients of the Elixir.

There is no reason to assume a corresponding difference in the level of abstraction of the correspondences in ritual acts and in the alchemical Work. None of the alchemist's convictions about the universe and his Art forbade even the actual depiction of earth and sky on the floor and ceiling of the elaboratory, on the furnace, or on the reaction-vessel. Given the preference of most adepts for abstraction, we can hardly expect anything so literal to be common, but one text suggests that it was not unheard of. This is the *Shang-Tung Hsin Tan Ching Chüeh*[1] (An Explanation of the Heart Elixir Canon: a Shang-Tung Scripture), undated but probably long before the middle of the +15th century when the *Chêng-Thung Tao Tsang* was printed. Despite the suggestion of the enchymoma conveyed by the title (which was meant concretely, as we shall see anon), this treatise provides clear instructions for laboratory preparations.[a] It has this to say about the design of the furnace:

If you use the Ninefold Cyclically Transformed Magical Elixir (technique just given) to treat the Three Yellow Minerals (i.e. sulphur, realgar and orpiment), the result will be equivalent to the Yellow Emperor's Nine-Vessel Sublimed Elixir. The method (is as follows). Set up the furnace platform (*than*[2]) as above,[b] carry out the purification rites, and then display the Nine Palaces and the Eight Trigrams.[c] Use a tortoise-shaped combustion-chamber (*kuei hsing lu*[3]), with the top made according to the pattern of the sky and the bottom according to the configuration of the earth (*shang an thien wên, hsia an ti li*[4]). The Three Yellow Minerals are placed on top of the cinnabar (i.e. the elixir), and covered with the 'sky-plate'.[d] After this is done sublime the ingredients with fire above and water below. Brush the Arcane Frost down, and take a dose the size of a millet grain.[e]

[a] There is in fact a section on Internal Alchemy (ch. 2, pp. 7*b*–10*a*), largely adapted from *PPT/NP*, ch. 8, but it is clearly stated that this technique is to be used just before and after taking the Outer Elixir.   [b] See pp. 10ff.

[c] This is evidently a ritual procedure meant to demarcate a sacred space. In a detailed ceremony for the previous elixir preparation (ch. 2, p. 13*a*) the adept is told somewhat more clearly, 'set out the Nine Palaces; the one who is compounding the Elixir (*ho tan chê*[5]) places himself in the central palace keeping watch over the combustion-chamber.' There is no doubt that these are the nine palaces of the Hall of Brightness (*ming thang*[6]), and thus of the old Lo Shu[7] magic square with its nine cells, three on a side. See Vol. 3, pp. 57–58, 542.

[d] It is not entirely clear whether the 'tortoise-shaped combustion-chamber' was a furnace in any ordinary sense. The combustion-chamber used for heating the casing in the preceding formula (ch. 2, p. 6*a*) was made by plastering over a wooden framework, and the casing was elevated on a short tripod over the flame. No sky-plate is mentioned. Here it is not possible to tell whether the sky-plate was a movable partition between the sealed reaction-vessel and the fire which was built above it within the combustion-chamber, or whether it was the figured lid of the chamber itself, on top of which the fire was burning. One is reminded of the 'heaven-plate' and 'earth-plate' of the Han diviner's board (*shih*[8]); cf. Vol. 4, pt. 1, pp. 262ff.   [e] *TT*943, ch. 2, p. 14*b*, tr. auct.

[1] 上洞心丹經訣   [2] 壇   [3] 龜形爐   [4] 上按天文下按地理
[5] 合丹者   [6] 明堂   [7] 洛書   [8] 弑

This could conceivably be the furnace with a round top and square base often used in similar cases,[a] but the choice of words here definitely suggests decoration, whether painted or moulded. *Thien wên*[1] in astronomy refers not to the shape of the celestial vault but to its constellations and planets,[b] and *ti li*[2] in geomancy and cartography to the lay of the land.[c] Thus images in two or three dimensions may well have been used at some point in the history of alchemy to stimulate the action of the cosmic forces upon the microcosm of the furnace.

(iii) *Chaos and the egg*

If the furnace was the Cosmos in little, the reaction-vessel was the Chaos out of which the Elixir was differentiated, the womb in which it was nourished. We have already cited in passing images like these which described the Elixir container,[d] and it is only necessary here to emphasise that for the alchemist they were not so much metaphors as identities, brought into play by the resonance of analogous configurations. In order to think of the vessel as a womb, some alchemists even took the trouble to mould it about a sphere of wax which was subsequently melted and poured away through an aperture. This sphere was called, in fact, a womb (*thai*[3]).[e]

   The Chinese conviction that correspondences implied functional identity is perhaps best illustrated by another universal organic image of Chaos, namely the egg. It was common in alchemy so to denominate the reaction-vessel. Mêng Hsü,[4] for instance, in the *Chin Hua Chhung Pi Tan Ching Pi Chih*[5] (Confidential Instructions on the Heaven-Piercing Golden Flower Elixir) of +1225, speaks of it as the 'Chaos Egg and Spirit Chamber' (*hun-tun chi-tzu shen shih*[6]),[f] noting that it was often made in the shape of an egg.[g] Phêng Ssu[7] in the same book, in a passage already quoted, did not hesitate to

---

   [a] See, for instance, *Chiu Chuan Ling Sha Ta Tan* (TT886), p. 1b. On furnaces and combustion-chambers in general see above, pp. 11 ff.
   [b] See our translation (Vol. 2, p. 326) of *thien wên* as 'the forms exhibited in the sky', where the term appears in the Great Appendix of the *I Ching*.
   [c] See Vol. 2, pp. 359 ff., Vol. 4, pt. 1, pp. 239 ff.
   [d] See pp. 16 ff.
   [e] See the alchemical miscellany *Ling-Pao Chung Chen Tan Chüeh*[8] (after +1101), TT416, p. 7a, and clearer instructions for the reaction-chamber on p. 12b. Presumably this technique had some close connection with the ancient *cire perdue* process in bronze metallurgy (see Sect. 36).
   [f] 'Spirit Chamber' was a widespread term for sealed reaction-vessels. See above, p. 22, where we translated *shen shih* as 'magical reaction-chamber'; here our rendering reflects the ritual meaning intended in the texts we cite below. But in other texts *shen shih* is a matter-of-fact technical term; e.g., in *Kêng Tao Chi* (TT946, ch. 7, p. 18b), where it is defined as 'a silver casing'. For the *nei tan* meaning of the term, see Hsü Ming-Tao's[9] *Huan Tan Pi Chüeh Yang Chhih-Tzu Shen Fang*[10] (Wondrous Art of Nourishing the Naked Babe (i.e. the primary vitalities) by using the Secret Formula for the Regenerative Enchymoma) probably of the late +12th century, TT229, pp. 1b, 2a; and a possibly pre-Sung discussion in *YCCC*, ch. 73, p. 7b.
   [g] TT907, ch. 2, pp. 2a, b and 13b. The relevant passage is translated in full above, p. 35. The shape of the Spirit Chamber is also likened to that of an egg in *Yü Chhing Nei Shu* (TT940), p. 10b; and a silver reaction-vessel is said to 'correspond to an egg, white outside and yellow inside' in *Thung Yu Chüeh* (TT906), p. 8b.

   [1] 天文   [2] 地理   [3] 胎   [4] 孟煦   [5] 金華冲碧丹經秘旨
   [6] 混沌雞子神室   [7] 彭耜   [8] 靈寶衆眞丹訣   [9] 許明道
   [10] 還丹祕訣養赤子神方

speak of the ingredients within the reaction-vessel as the 'elixir embryo' (*tan phi*[1]).[a] Similarly, a book of much earlier date, ascribed to the Chin but more probably Thang, the *Chih-Chhuan Chen Jen Chiao Chêng Shu*,[2] follows the same thought. After speaking of gold, lead, white frost of lead, potable gold, etc. it goes on to say:

Union and maternity bring completion, so that bones and flesh are formed, and the foetus or the embryo comes to the birth. In this way the potentialities of the Shaping Forces are determined, and there is the glorious manifestation of the conquest of the element Wood by the Metal element—that is the whole idea of it.[b]

Another interesting occurrence of cosmic egg images is found in that very philosophical work on the *chhi* techniques of physiological alchemy, the *Yuan Chhi Lun*[3] of the mid-Thang. Detailing the steps in its pneumatic cosmogony, it says that

before the *chhi* (of Yin and Yang in the Thai Chi[4] phase) separated, they had the configuration of a young foetus (*phi*[5]) like in shape to an egg. The original *chhi* was quite round, its shape perfect, so it is called the Grand Unity (Thai I[6]).[c]

Wu Wu,[7] writing in +1163, tells us that for success the reaction-vessel (*ting*[8]) must be as round as a hen's egg (*chi tzu*[9]).[d] A little further on he cites a shadowy predecessor as follows:

Chhing Hsia Tzu[10] says: 'The chemicals in the reaction-vessel are like the chick embryo in the egg, the child within the womb, or the fruit upon the tree;[e] when once they have received fully the requisite *chhi* they ripen and develop and come to perfection of themselves. But when the chemicals have been placed in the 'womb', it is always necessary to seal it firmly and securely for fear that any leakage of the perfected *chhi* (*chen chhi*[11]) may occur.' He also says: 'That the sealed "womb" (*ku chi thai*[12]) may not leak, and that change and transformation may proceed, it is necessary to insist that it be made spherical like heaven and earth at their beginning. If there should be any crack or seam in the vessel it must be so tightly luted that not the most minute trace can escape of the numinous cyclical evolutions (*shen yün*[13]) going on inside.'[f]

This is reminiscent of what Wang Chhung[14] had had to say about developing eggs towards the end of the +1st century. The 'formless mass' (*hung-jung*[15]) of yolk and white at the beginning was regarded by him as a harmless liquid homogeneity, organised only by Yang *chhi* during the warmth of incubation.[g]

---

[a] *TT*907, ch. 1, p. 1*a*; the full text is given in translation on p. 35 above.
[b] *TT*895, pp. 1*b*, 2*a*, tr. auct.
[c] *YCCC*, ch. 56, p. 1*a*, tr. auct., adjuv. Maspero (7), p. 207.   [d] *TT*893, p. 8*b*.
[e] Or, 'unformed as yet within the tree'.
[f] *Tan Fang Hsü Chih*, +1163 (*TT*893, p. 10*b*), tr. auct. Chhing Hsia Tzu was the Taoist appellation of the historical Su Yuan-Lang[16] (see above, pt. 3, p. 130), but one must still consider the questions of whether Su actually wrote the words attributed to him in many later sources, and whether this passage refers primarily to internal or external alchemy, entirely open.
[g] His argument was that human death and dissolution were nothing but a return to the state of formless chaos—how therefore could malevolent or dangerous ghosts and spirits exist? The relevant passage from *Lun Hêng*,[17] ch. 62, has been given in translation already (Vol. 2, p, 370).

[1] 丹胚   [2] 稚川眞人校證術   [3] 元氣論   [4] 太極   [5] 胚
[6] 太一   [7] 吳悮   [8] 鼎   [9] 雞子   [10] 靑霞子
[11] 眞氣   [12] 固濟胎   [13] 神運   [14] 王充   [15] 澒溶
[16] 蘇元朗   [17] 論衡

Not long after Wu Wu's time someone who was trying to reason out the best possible way of making a container represent an egg hit upon the unsurpassable solution—he used a hen's egg itself.[a] This may have come earlier, as early as the +9th century, but more probably it was a little later, in the Southern Sung. The 'Complete Compendium on the Lead–Mercury A–G Perfected Treasure' (*Chhien Hung Chia Kêng Chih Pao Chi Chhêng*[1]) by Chao Nai-An,[2] cites 'Secret Directions for the Yellow Sprouts Great Elixir' (*Huang Ya Ta Tan Pi Chih*[3]). One stage of the preparation goes as follows:

> Orpiment, ½ ounce
> Sal ammoniac, and
> White arsenic, ¼ ounce each

First grind the orpiment; then grind the arsenic and sal ammoniac separately, fine as flour. Take an egg and make a hole in it. Get rid of the yolk but keep the white.[b] Spread half the arsenic and sal ammoniac on the bottom inside the egg; put the orpiment in the middle, and half the arsenic and sal ammoniac to cover it. Take somewhat less than half an egg-shell to cover the hole, and seal it on with iron oxide solution (*chiang fan shui*[4]) which has been mixed (with the egg-white?).[c] Then take a pound of minium (*huang tan*[5]) and an iron reaction-vessel (*ting*[6]). Put half the minium into the vessel and place in its centre the medicines in the egg. Then cover them with the rest of the minium, applying a little pressure. Fill the vessel with lime (*shih hui*,[7] evidently raw) and lute it tightly. Using half a pound of charcoal, heat it gently in an ash bath.[d] When it is taken out it will be finished. For each ounce of *pai hsi*[8] (zinc or tin)[e] use a piece the size of a red mung bean (*hsiao tou*[9]).[f] First melt the metal, and when it is liquid project the medicine upon it. Pour it out and wait for it to cool. It will then be the colour of gold.[g]

The porosity and fragility of the egg rule out its serving both the metaphysical function of the cosmic egg and the practical function of a sturdy and impregnable container, so the two functions are separated and the latter assigned to an iron vessel. The white

---

[a] For this there were ancient precedents. Ko Hung, about +300, knew of a way of incubating mercury and its compounds inside an egg sealed with lacquer; see Table 111 above, no. 36 (*PPT/NP*, ch. 4, p. 13b; Ware (5), p. 89). He also in another formula used the blood of black crane or stork embryos; Table 111, no. 20 (*ibid.* p. 12a, p. 86). See Vol. 5, pt. 2, pp. 91–2.

[b] An alternative translation, 'Remove the yolk but leave the white inside', would imply a rather sophisticated technique, and it is more likely that the separation took place in a bowl, the albumen being reserved for later use in sealing the shell.

[c] Punctuated differently, the second half of this sentence might be rendered 'and seal it on with iron oxide which has been mixed (into a paste) with water'. Such a paste would be so deficient in adhesive qualities that this would be an obvious application for the reserved egg-white. A similar method which we cite shortly specifies the use of egg-white as a sealing agent (p. 296). We take 'red alum' to be the ferric oxide produced by the roasting of ferrous sulphate ('green alum' or melanterite).

[d] A bed of ashes, like the water-bath, was used under or about the reaction-vessel as a means of diffusing and moderating heat. For Chinese water-baths, see above, p. 32.

[e] See pt. 2, pp. 214ff.

[f] For the red mung bean (*Phaseolus mungo*) as a metrological standard in pharmacology and alchemy, cf. Sivin (1), p. 254. We have already encountered the use of black millet-grains in acoustic metrology in Vol. 4, pt. 1, pp. 200ff.

[g] *TT*912, ch. 5, pp. 12a–12b, tr. auct. We have already encountered a vaguer mention of an egg-shaped silver container in a work which may be earlier, *Chen Yuan Miao Tao Yao Lüeh*[10] (later than mid-seventh century, *TT*917); see above, pt. 3, p. 78, and Fig. 1394.

---

¹ 鉛汞甲庚至寶集成   ² 趙耐庵   ³ 黃芽大丹秘旨   ⁴ 絳礬水   ⁵ 黃丹
⁶ 鼎   ⁷ 石灰   ⁸ 白錫   ⁹ 小豆   ¹⁰ 眞元妙道要畧

arsenic and sal ammoniac and the reddish-yellow orpiment represent the albumen and yolk.

Since all three of the primary reactants were volatile, there was some danger that their vapours might explode the egg. The text is not entirely clear about whether or not the albumen was removed at some point; its presence would surely have complicated the placing of the inorganic substances and the course of the reactions. If it was there (which is unlikely), complex organo-metallic compounds might have been formed. If not, the vapours of the ingredients might have diffused gradually outwards and reacted with the lead and calcium, depending on the tightness with which the minium and lime were packed around the egg and the gentleness of the heating. The specifications given do seem designed to minimise the danger of explosion. That the procedure given by Chao Nai-An is workable we have no reason to doubt, pending a laboratory trial. At all events, it is extremely improbable that an egg would have been chosen as an inner reaction-vessel for any practical motives.

Exactly what form of aurifaction was taking place here is not immediately obvious. The reagents heated together were arsenic trisulphide, ammonium chloride (or carbonate), arsenic trioxide, ferric oxide, lead tetroxide and calcium oxide, with or without, as the case may be, a protein as source of carbon, nitrogen and hydrogen. Whether or not the tin or zinc was tinged golden only superficially is not clear from the description: if so, arsenical and other sulphides might have done just as well by themselves (cf. pt. 2, p. 252 above). If, on the other hand, copper was meant though tin or zinc actually stated, then a uniform-substrate golden alloy of arsenical copper could easily have been produced by projection as described (cf. pt. 2, p. 223 above). It will be remembered, too, that in the medieval lists of 'golds' which we studied earlier, a *pai hsi chin*[1] regularly appears (pt. 2, p. 275), which supports the practicality of what was described here, but our conclusion again has to rest upon whether after all copper was present. This is on the face of it unlikely. As for whether the metal meant to be transmuted was zinc or tin, it is difficult to see how any possible product of this formula could tint either yellow. Zinc oxide would of course be yellow when hot, but the instructions specify that the golden colour is visible after cooling. We leave this as a problem to be solved in some future laboratory devoted to the investigation of medieval alchemical procedures.

There can be no more fitting climax to this sub-section than a remarkable passage in the *Shang-Tung Hsin Tan Ching Chüeh*, even though we are unable to establish whether it is earlier or later than Chao Nai-An's collection of formulae. In it the egg as reaction bomb (*shen shih*,[2] lit., 'spirit chamber') is clearly linked with the subterranean growth of the Natural Cyclically Transformed Elixir. At the same time, the organic character of this application is emphasised by correlation with the human heart. These leitmotifs, together with an allusion to the abiding of the spirits, constitute almost a recapitulation of the chief themes of all alchemical thought.

In this process for the Heart Elixir, cinnabar which has been digested with other substances is divided among four 'spirit chambers' made by emptying eggs and

[1] 白錫金   [2] 神室

coating them with thickly ground Chinese ink.[a] The author of the supplementary instructions (*chüeh*[1]) continues:

> When I make the caps (for the holes in the eggs) I use a shoe-soling needle to pierce seven holes at equal intervals around the circumference of four eggs. These holes correspond to the seven apertures of the heart. The four egg-shell caps are also coated with ink as already specified. Eggs are white, so the spirits cannot abide there. But if they are tinted black with a black pigment, the spirits can remain secure inside. That is why the vessel is called 'the spirit chamber'.[b]

Once the appropriate rites have been carried out and the eggs have been charged,[c] they are placed in a bed of lime, arsenic and other white minerals, within a 'Five-Elements Jade Casing' (*wu-hsing yü kuei*[2]). Then to this ambiance of white material are applied the very technical terms which Su Ching[3] had used in the middle of the +7th century for the matrix out of which large cinnabar crystals grow:

> Where cinnabar grows, beneath it there are white substances; above a white bed there is a white jade 'shrine' (*khan*[4]).[d] In the preparation of this elixir both the medicines used to seal the 'spirit chamber',[e] and those within the casing which serve as the ground in which (the eggs) are planted, are white. This is in order that they may correspond to the jade bed and jade shrine of the cinnabar (growth). It is like the pericardium in human beings (because the vessel corresponds to the heart).[f]

Here one cannot forbear from a comparative glance at similar ideas in other cultures. Probably all of them have seen in the development of the cleidoic egg of fowls, so sharply bounded off from all external things, a model of the creation or evolution of Cosmos from Chaos. As is well known, the cosmic egg was a notable theme in Greek mythology,[g] but similar ideas may have been current much earlier in Babylonia.[h]

---

[a] The supplementary instructions specify that before the eggs are pierced and emptied they are to be soaked in vinegar, but how long and for what object is not made clear. The shells would certainly be softened by solution of the lime.

[b] *TT*943, ch. 1, p. 9*b*, tr. auct.

[c] Here the adept is explicitly instructed to seal the caps to the shells with egg-white (p. 10*a*).

[d] The text does not read very well at this point. If *pai pai chhuang*[5] were emended to *pai shih chhuang*,[6] it would be much improved, and the translation would read: 'Where cinnabar grows, below it there is a bed of white mineral and above it a white jade "shrine".'

[e] I.e. the egg-white.   [f] *TT*943, ch. 1, pp. 10*a*–11*a*, tr. auct.

[g] Cf. Needham (2), pp. 8ff., 50–1, with references there given.

[h] Half a century ago Zimmern (1) and Campbell Thompson (5), pp. 50, 70–1, studying cuneiform texts on the making of glass, frit and enamel, from the library of King Ashurbanipal (r. −668 to −626), reported several mentions of 'foetuses' or 'embryos' in connection with the furnaces, to which sacrifices and libations had to be made. Thompson at least was inclined to think that these were aborted foetuses which had to be propitiated as representatives of all natural processes fated not to come to term. Eisler (4), on the other hand, urged that the 'embryos' were the actual ores and other mineral ingredients themselves. The question, not yet settled, is discussed impartially by Eliade (5), pp. 43, 68ff., 75ff. On the basis of this Eisler propounded a theory of the origin of all alchemy in the Mesopotamian cultures; it caused some stir at the time but has never been generally accepted, though favoured now and again, as by Forbes (31).

What relation, if any, such practices had with the 'true and secret fertility amulets' (*chung thai chen pi fu*[7]), which Ko Hung recommended taking with one when entering the wilds of forest and mountain (*PPT/NP*, ch. 17, p. 11*b*, tr. Ware (5), p. 295), might admit of a wide solution.

[1] 訣   [2] 五行玉匱   [3] 蘇敬   [4] 龕   [5] 白白牀
[6] 白石牀   [7] 中胎眞祕符

Certainly the parallelism of the reactants-and-vessel with the foetus *in utero* was prominent in European alchemy, with its oft-pictured *vas philosophorum* or Philosopher's Egg of glass, 'Hermetically' sealed.[a] And we have already quoted the parallel between the chick's incubation and the development of gold in the earth (or in the elaboratory), as stated in Ben Jonson's play.[b] Sometimes, finally, the analogy was reversed, as in the *Secretum Secretorum* of Pseudo-Aristotle translated by Roger Bacon (on which see p. 368), where on physiognomy we read:

Thou knowest that the womb is for the embryo as the pot is for the food. Therefore whiteness, or blueness, or extreme redness (of the face) indicates imperfect coction in the matrix....Therefore beware, etc.[c]

The remarkable practice of using eggs as models for Chaos outlasted the heyday of laboratory alchemy in China, surviving like so many other alchemical methods in iatro-chemistry. We see it last in the +17th century, in Fang I-Chih's[1] collection of notes *Wu Li Hsiao Shih*[2] (Small Encyclopaedia of the Principles of Things). Fang, one of the first Chinese to pay serious attention to the whole spectrum of European knowledge then being introduced by the Jesuit missionaries, refers to the ultimately literal method of maturing inorganic medicines recorded by Ning Hsien Wang[3] (+1390 to 1448) Prince of the Ming, Chu Chhüan,[4] an amateur of every sort of arcane knowledge:[d]

Incubating medicinal eggs (*fu yao luan*[5]). For any medicine, make an egg of silver which can be opened and closed with a small cover.[e] Insert the medicine and seal it with lacquer. Put it in a nest of eggs and let the hen incubate it for exactly seven weeks [some people rotate it among several hens].[f] Its effects upon the circulation of *chhi* (in the patient's body after ingestion) are marvellously beneficial. It may also be irradiated by sunlight or nurtured over a warm fire for a hundred days. Its special virtues are due to changes stimulated when it is incubated by the female of the species. This is also the point of the procedure given by the Emaciated Immortal (Chhü Hsien[6]): 'Raise separately white cocks and hens. Take an egg (from one of the hens), extract the yolk and white, take cinnabar, grind and blend them, and put the mixture into (the shell). Seal the opening with wax. Then let one of the white hens incubate it along with its own eggs. When the chicks hatch from the others, the medicine is finished. Take it mixed with honey. Or mix realgar with the egg-contents, seal it, and heat it over a feeble fire for three days and nights.'[g]

---

[a] On this see Read (1), pp. 104, 149ff., 217–18, etc.; Ploss *et al.* (1), frontispiece and pp. 138–9, 202–3; Mahdihassan (58). Sherwood Taylor (3), pp. 44–5 has suggested that the distillation of eggs by the Hellenistic aurifactors was probably inspired by some similar symbolism. For psychological interpretations of the alchemical egg see Jung (1), Germ. ed. pp. 103, 276ff., 281, 325ff., 461; Eng. ed., pp. 192–3, 227. Finally, mem. *Corp.* I, iii.

[b] See p. 243 above.

[c] Steele (1), fasc. 5, p. 166, Ar. version, p. 219. One text runs: 'Scias ergo quod matrix est embrioni sicut olla ferculo decoquendo. Albedo ergo cum livido colore et flavus color nimis est signum diminute decoccionis embrionis in matrice.'

[d] Often encountered before, cf. Vol. 5, pt. 3, pp. 210ff. N.S. prefers *Chih* in the title.

[e] Cf. the silver egg-shaped aludel of Thang date shown in Fig. 1394 above.

[f] This is a note in the text.   [g] Ch. 4, p. 18*a*, *b*, tr. auct.

[1] 方以智   [2] 物理小識   [3] 寧獻王   [4] 朱權   [5] 伏藥卵
[6] 臞仙

(6) PROTO-CHEMICAL ANTICIPATIONS

Our understanding of alchemy places it in the mainstream of traditional Chinese scientific thought, heterodox though it was (unlike mathematics and astronomy) for conventional scholars. In its theoretical aspect it was a deductive proto-science on quite the same level as medicine, acoustics or magnetic geomancy, based on the same general laws and the same natural rhythms, its essential difference lying in the selection of phenomena which it set in order. Each of these sciences was determined by an original demarcation of a field of observation and experience, defined by imposition of the common natural philosophy, and developed partly by working deductively through the various permutations of particular facts. But if alchemy consisted wholly of this special application of an organicist philosophy of Nature on the one hand, and eclectic compendia of elixir recipes (with no indication that the reactions were understood) on the other, it would be necessary to conclude that nothing in Chinese alchemy was truly relevant to the pre-history of chemical thought.

Such a view would badly underestimate the ability of the alchemist to respond to his experience.[a] A more direct appreciation of the fact of chemical change can be documented at least as far back as Ko Hung's dictum that minium (Pb_3O_4) and white lead ($2PbCO_3 . Pb(OH)_2$) are transformations of lead.[b] If only early craftsmen had been literate, we could doubtless trace that same appreciation back to the beginnings of chemical technology in China. Even at this initial stage of research, in which our greatest accomplishment is to gauge what we do not know, it is possible to discern attempts to develop theories of substantial change. Generally these theories reflect the lack of clear distinction between physics and chemistry, inevitable so long as the language of quality and function was used for both. For example, the *Thien Kung Khai Wu*,[1] that great technical encyclopaedia of +1637, offers a physical explanation of substantial change: 'Cinnabar, mercury, and vermilion are originally the same substance. The difference in name corresponds to a difference in fineness and degree of coction (*ching-tshu lao-nun*[2]).'[c] There is every reason to believe that this idea was first worked out in an alchemical context.

Our study has given us grounds for hope that a broad and consistent theoretical picture of substantial change—though certainly very different in its definitions and assumptions from modern chemistry—can be drawn together from data scattered through the surviving literature of external alchemy. An enormous work of collation and intellectual reconstruction will be necessary, but the potential contribution to a comparative history of chemistry would more than justify it. In the meantime we can only offer a couple of clues on the approach toward chemical reasoning which we

---

[a] It would also, incidentally, ignore the great contribution of Chinese as well as Western alchemy to the development of apparatus and techniques still used daily in modern chemistry (cf. pp. 44, 101 above).

[b] This may well go back to the −4th rather than the +3rd century if the *Chi Ni Tzu* book is near the beginning of the story (cf. pt. 3, pp. 14–5 above). As for Ko Hung's understanding of chemical change, see the sub-section above on his attitude to aurifaction (pt. 2, pp. 62ff., esp. p. 70).

[c] Ch. 16, p. 1a, mistranslated euphemistically in Sun & Sun (1), p. 279.

[1] 天工開物   [2] 精粗老嫩

hope eventually to see delineated out of the Chinese sources. We shall return briefly below to the role of number in alchemical thought, first concentrating on gravimetric ideas, the use of the balance, and finally take up the development of category theories to explain the reactivity of one substance with respect to others.

Further pursuit of the sprouts of early chemical thought in China should not neglect the less obvious sources, such as for example the Neo-Confucian literature. In the thought of Chhêng I[1] (+1033 to +1107)[a] there are interesting things to be found. At one place he says:[b]

> The physicians do not sufficiently consider organic pattern-principles (li[2]); when compounding drugs in prescriptions they do not exhaustively investigate their natures (hsing[3]). They know only the therapeutic uses of each, and not what happens when the substances form combinations; how then can they understand their (real) natures? For instance, myrobalan (ho tzu[4])[c] is yellow, and alum (pai fan[5]) is white, yet when they are mixed together the mixture is black. When what is black appears, that which was yellow and that which was white have disappeared. If we put a and b together we get c, so that c manifests itself, and a and b are no longer visible. But if we get back a and b again, then c disappears. If we have c and continue to look for a and b in it, if we have black and persist in looking for yellow and white in it, then we are failing to understand the nature of things. (This is why) the ancient (sages) investigated to the utmost the organic pattern-principles of things (chhiung chin wu li[6]); they studied tastes, smelt odours, differentiated between colours, and acquired knowledge of what substances will mix or combine together (chih chhi mou wu ho mou[7]).

What he was talking about here was the production of the deep blue-black pigment formed when tannins are brought into the presence of salts of iron; the metal combines with the polyhydroxy-benzoic acid derivatives to give the colours still to this day used for inks. Chhêng I seems to have realised half-intuitively that something essentially new had been formed in the reaction. But as usual with the Neo-Confucians, he did not systematically pursue this line of enquiry. Elsewhere, however, he said:[d]

> Sound, colour, smell and taste, are all alike, in themselves empty, yet full of meaning. Every thing that has corporeal form has to have these four qualities, and out of them arise significance, appellations, images and numerical values.

These, no doubt, were numerological still, rather than quantitative in our sense. But the one could slide into the other.[e] In a third passage, directed evidently against that metaphysical idealism to which his equally eminent brother was rather addicted:[f]

> The Master said: 'To investigate exhaustively the organic pattern-principles of things is to investigate how they come to be as they are. The height of the heavens, the thickness

---

[a] Cf. Vol. 2, pp. 414, 457, 471, 479; and Forke (9), pp. 100ff.

[b] Hsing Li Ching I,[8] ch. 9, pp. 2aff., tr. auct.

[c] Myrobalan is the black fruit of Terminalia Chebula (R247), full of tannins, like those of all members of this genus of Combretaceae (Burkill (1), vol. 2, pp. 2134ff., 2139). More properly named ho li lê,[9] and by origin Indian and Burmese, it first appears in the Hsin Hsiu Pên Tshao (+659). The leaves and bark also tan.

[d] Honan Chhêng Shih I Shu, ch. 18, p. 13b, tr. auct.

[e] Cf. p. 304 below.

[f] Honan Chhêng Shih Tshui Yen, ch. 2, p. 59b, tr. auct.

[1] 程頤   [2] 理   [3] 性   [4] 訶子   [5] 白礬   [6] 窮盡物理
[7] 知其某物合某   [8] 性理精義   [9] 訶黎勒

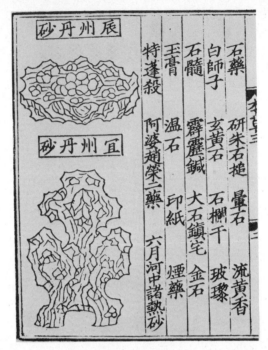

Fig. 1522. Drawings of native cinnabar from the *Chêng Lei Pên Tshao* of +1249 (ch. 3, p. 2*b*). The upper sample from Chhenchow, the lower one from Ichow. The headings on the right are the latter part of the contents table of the chapter in question (ch. 3).

of the earth, the appearance and disappearance of expansion or disaggregation, and of contraction or agglomeration, all must have some manner of coming into being. If it be said that all these things are just our way of talking about the world, and no more, then how and why did they come into existence?

In the literature of Neo-Confucian philosophy we may well find many further speculations upon distinctively chemical examples of coming-into-being and passing-away.

(i) *Numerology and gravimetry*

From the standpoint of the comparative development of chemistry, one of the most cogent themes to emerge from the study of early Chinese alchemy is its concern with quantitative factors. This is not a mere matter of the specification of amounts in formulae; Sumerian medicine had reached that point by −2500.[a] But we find evidence of a truly gravimetric application of number in the following excerpt from Chhen Shao-Wei's[1] great monograph on the alchemy of cinnabar, *Ta-Tung Lien Chen Pao Ching, Hsiu Fu Ling Sha Miao Chüeh*[2] (Mysterious Teachings on the Alchemical Preparation of Numinous Cinnabar) written, it seems, not long after +712. At this point Chhen is discussing, in descending order of purity, natural crystalline varieties

---

[a] Levey (7), pp. 61–70.

[1] 陳少微   [2] 大洞鍊眞寶經修伏靈砂妙訣

of cinnabar (Fig. 1522), and their substitution for each other in an elixir preparation. He says:

Now from 1 lb. of lustrous cinnabar (*kuang ming sha*[1]) one can distil 14 ozs. of mercury, lustrous white and free-flowing. This indicates that lustrous cinnabar of the highest quality contains only 2 ozs. of mineral *chhi*. From 1 lb. of white horse-tooth cinnabar (*pai ma-ya sha*[2]) one can distil 12 ozs. of mercury; it contains 4 ozs. of mineral *chhi*. From 1 lb. of purple numinous cinnabar (*tzu ling sha*[3]) one can distil 10 ozs. of mercury; it contains 6 ozs. of mineral *chhi*. From 1 lb. of superior translucent (commercial) cinnabar (*shang sê thung ming* [*sha*])[4] one can distil only 7 ozs. of mercury; it contains 9 ozs of mineral *chhi*.[a] Mineral *chhi* is the void *chhi* of Fire and rock (*huo shih chih khung chhi*[5]).[b] After the mercury has been extracted there will be about an ounce of Mineral Embryo (*shih thai*[6]), a greyish ash.[c]

Thus some alchemists twelve and a half centuries ago knew that 13 or 14 ozs. of mercury can be distilled from 16 oz. of the best native cinnabar. A more exact figure, according to modern calculation, would be 13·8. Chhen had learned the importance of experimentation with weighings, on principles which must at some point have come from metallurgists (unless a mercury-smelter furnished him with the figure).[d] Not only that, but he knew that this ratio must vary with the purity of the cinnabar, so that a cinnabar of lower quality will contain less mercury and more 'mineral *chhi*'— resolved by the treatment into irrecoverable *pneuma* and a residue of Mineral Embryo.[e] What is perhaps most significant, and certainly most original, is Chhen's assumption that when each kind of cinnabar is broken down into its constituents they always total 16 ozs. in weight. This was only a hypothesis, for there was no way of collecting and weighing the *chhi* which had presumably escaped, but it was just the sort of hypothesis which in much more recent times pointed the way to a chemistry based solidly on measure and number. In this way Chhen seems far ahead of his time.

But it is not quite so simple as that. Where did Chhen get his figures for the yields of mercury from different types of cinnabar? First we must ask how different in fact the four varieties were. This question can be answered, at least in a rough way. All four are, first of all, exceptionally large and rare crystalline forms, far superior to the ordinary article of commerce. It is obvious from their descriptions that they did not normally contain perceptible quantities of admixed earth and stone.

Lustrous cinnabar fits the description of translucent, nearly vitreous, rhombohedral

---

[a] These are the weights given in *YCCC*, ch. 69, pp. 18*a*, *b*. The corresponding text in *TT*883, p. 14*b*, gives the yields as 8½ and 7½ ozs. respectively. In both cases the sum is 16 ozs.

[b] Or, 'of minerals which belong to the element Fire'.

[c] *YCCC*, ch. 69, pp. 18*a*, *b*, tr. Sivin (4).

[d] A corollary of the argument which we develop below is that, if the various figures had been furnished him by an artisan, they would not have varied over so wide a range. Lustrous cinnabar and the other varieties mentioned were too precious to be used for distilling mercury; in fact transparent cinnabar crystals were often set in Chinese jewellery. On the history of gravimetry in assaying and cupellation, see above, pt. 2, pp. 36ff., 65ff.

[e] There is a parallel passage in the much later *Ling Wai Tai Ta* (+1178), ch. 7, p. 10*b*. But Chou Chhü-Fei was no chemist, and reported (unless the text is faulty) that 8 lbs. of the best cinnabar would yield 10 lbs. of mercury. He was presumably garbling information received from the mercury workers of Kuei-tê and other southern places.

[1] 光明砂   [2] 白馬牙砂   [3] 紫靈砂   [4] 上色通明砂
[5] 火石之空氣   [6] 石胎

crystals of cinnabar, such as are still used in China, whole as semi-precious stones, and pulverised as the pigment in very high grades of seal ink (*yin ni*[1]). White horse-tooth cinnabar, despite its designation, is not white in colour. Su Ching, in the mid-seventh century, points out its suitability for artist's vermilion pigment, and describes it concretely enough to allow its identification as small tubular crystals rather than granules (which are not usually translucent):

The next quality comes from within rocks or from streams, and occurs in pieces of which the largest are the size of a thumbnail and the smallest the size of apricot stones. It is lustrous and without admixed rock mineral. It is called 'horse-tooth cinnabar', and another name is 'undoubled mineral' (*wu chhung shih*[2]). It is excellent for use in drugs and also for painting, but (like lustrous cinnabar) not much of it finds its way into the possession of ordinary people.[a]

The white is explained as the colour of its lustre in the supplementary instructions to the 'Yellow Emperor's Nine-Vessel Spiritual Elixir Canon' (probably early Sung):

There are also tablets coarse as horses' teeth or like small rolls (*hsiao chüan*[3]), brilliant with radiant depths, their matter compact and their white lustre dazzling to the eye—they are styled cinnabar.[b]

This is less ambiguous than the statement of Chhen Shao-Wei earlier in his treatise that it 'shines with a radiant white light the colour of mica.'[c] Thirdly, there is nothing in the sources to deter us from considering purple numinous cinnabar as a true cinnabar of darker colour than normal.[d] Last, superior translucent cinnabar is, unlike the other varieties, a common article of commerce, although still of very high grade. The specification of translucency indicates that it is still a crystalline (and thus tolerably pure) form of mercuric sulphide. To sum up, although disparities in crystal size and transparency could have convinced the alchemist that the four varieties of cinnabar differed sensibly in their places on the hierarchic scale of maturation, their chemical purity can hardly have been very unlike. Impurities might occur from time to time in the form of mechanical admixtures, but the proportions would be random rather than constant within each type.

Since the four varieties of cinnabar do not apparently differ greatly or consistently in chemical purity, all the numbers except the first must be based, not on laboratory experience, but on the conviction that a difference in kind must be associated with a difference in number, and the further assumption that these differences must form a series of rather regularly graded steps which reflect, by implication, steps on a hypothetical mineral maturation curve.[e]

---

[a] Cited in *CLPT*, ch. 3, p. 3*b*; tr. auct.; cf. also Okanishi (5), pp. 99–100.
[b] *Huang Ti Chiu Ting Shen Tan Ching Chüeh* (*TT* 878), ch. 13, p. 3*b*, tr. auct.
[c] *YCCC*, ch. 69, p. 2*b*, tr. Sivin (4).
[d] Though we might note in passing that Brelich (1), observing Chinese mining methods seventy years ago, noted that a dark opaque red form, called 'black cinnabar' by the Kweichow miners, almost invariably contains small quantities of antimony.
[e] Another suspiciously regular gradation, based on a lower maximum yield, is attributed to the legendary Hu Kang Tzu[4] (see Sivin (1), p. 159) in the *Huang Ti Chiu Ting Shen Tan Ching Chüeh*

[1] 印泥   [2] 無重石   [3] 小捲   [4] 狐剛子

The same frame of mind shows itself in other asseverations of Chhen Shao-Wei; for instance:

One oz. of lustrous cinnabar, when taken orally, is equal in potency to 4 oz. of white horse-tooth cinnabar. One oz. of white horse-tooth cinnabar, when taken orally, is equal in potency to 8 oz. of purple numinous cinnabar. The potency of creek cinnabar (*chhi sha*[1]) or earthy cinnabar (*thu sha*[2]) is not of an order comparable with these.[a]

Yet the intensity of the physiological reaction to a dose of any of these varieties (with the probable exception of earthy cinnabar, which might contain much gangue impurity)[b] would have been very much the same. Even if it were different, it is not easy to imagine an experimental arrangement for determining the precise comparative dosage required to metamorphose experimental subjects into immortals soaring in the empyrean. There was indeed an objective verifiability of alchemical immortality, in a certain sense, but hardly the possibility of its operational quantifiability.[c]

The figures we have just seen applied to the mercury yield of cinnabar also appear metamorphosed in a discussion concerned with yields of elixir and of the intermediate 'subdued cinnabar' (*fu huo* [*tan sha*][3]):

Furthermore, when 1 lb. of lustrous cinnabar is subdued in the fire, 14 ozs. of subdued (cinnabar) are obtained, which when heated in a furnace urged with bellows yields 7 ozs. of the 'Perfect Treasure'. When 1 lb. of white horse-tooth cinnabar is subdued in the fire, 12 ozs. of subdued (cinnabar) are obtained, which when heated in a furnace urged with bellows yields 6 ozs. of the 'Perfect Treasure'. When 1 lb. of purple cinnabar is subdued in the fire, 10 ozs. of subdued (cinnabar) are obtained, which when heated in a furnace urged with bellows yields 3–5 ozs.[d] of the 'Perfect Treasure'. When 1 lb. of creek cinnabar, earthy cinnabar or other cinnabars of diverse kinds is subdued in the fire, it is possible to obtain 6 or 7 ozs. of subdued (cinnabar), which when heated in a furnace urged with bellows yields 1 or 2 ozs. of the 'Perfect Treasure'. So it is quite clear that the *chhi* with which creek cinnabar and earth cinnabar are endowed is impure—sluggish, turbid, and heterogeneous. In order to succeed in making Sevenfold-recycled (Cinnabar) or Ninefold Cyclically Transformed (Gold Elixir), a lofty and enlightened gentleman must first choose the proper cinnabar and then correctly phase the fire, regulating it to accomplish the desired end.[e]

If we look at all these quantities together (Fig. 1523), it is not hard to tell where Chhen got them. They can only be *a priori*, generated by numerological reasoning in order to construct three hierarchies—mercury yield, physiological potency, and Elixir yield—based on the fundamental hierarchy of cinnabar quality. That the first of the three was anchored to a number derived by measure out of chemical experience

(*TT*878), ch. 11, p. 4*a*: 'From 1 lb. of good vermilion one can get 12 ozs. (of mercury); from 1 lb. of medium-grade vermilion one can get 10 ozs.; from 1 lb. of inferior vermilion one can get 8 ozs.' In the same chapter (p. 2*a*) a method is given which, it is claimed, will extract a pound of mercury from a pound of cinnabar if the latter is sufficiently pure. But this is not credited to the same source.

[a] *YCCC*, ch. 69, p. 2*a*, tr. Sivin (4).
[b] According to Huang Chu-Hsün (*1*), pp. 106–7, the non-translucent type of cinnabar mined in China in the early twentieth century usually also contained some antimony.
[c] On incorruptibility, mummification, etc. see the discussion in pt. 2, pp. 249 ff. above.
[d] The text has 6 ozs., we amend to expectation.
[e] *YCCC*, ch. 69, pp. 2*b*, 3*a*, tr. Sivin (4).

[1] 溪砂   [2] 土砂   [3] 伏火丹砂

reminds us of the interpenetration of the two functions of number—mathematical and numerological—in all ancient and medieval Chinese minds. This is only one more instance of the way in which the 'advanced' aspects and the 'retrograde' or 'unscientific' aspects of early science (which once led a distinguished positivist historian to call research in alchemy, astrology, and related areas 'the study of wretched subjects') turn out to be not only balanced, but so intimately connected as to be inseparable.

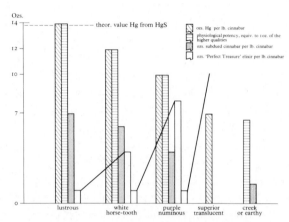

Fig. 1523. Chhen Shao-Wei's mercury yields from different varieties of cinnabar. Explanation in text.

Actually, in the case of Chhen Shao-Wei we see an important early stage in the definition of gravimetry. It is beginning to define itself out of an almost Pythagorean faith in number as a reflection of underlying reality on the one hand, and out of the metal-workers' use of the balance to control manufacturing processes on the other.[a] In the alchemical literature there are many other specifications of yield, potency, and ability to transmute base metals which must be closely examined, and if possible experimentally tested, before the history of the concept of combining weights in ancient China can fall into place.[b]

[a] Perhaps the most classical instance of the borderline between numerological and gravimetric quantification would come from the Arabic world; the *Kutub al-Mawāzīn* (Books of the Balances) in the Jābirian Corpus (+9th century) have far more to do with a theoretical assignment of elements in the composition of substances than with any real weighings (cf. p. 459). But weighings were certainly made, as in the famous mercuric oxide experiment of the *Rutbat al-Ḥakīm* (Sage's Step), attributed to Maslama ibn Aḥmad al-Majrīṭī (d. +1007); cf. Sarton (1), vol. 1, p. 669; Holmyard (1), p. 71, (11), p. 302; Leicester (1), p. 71, etc. And al-Jildakī in the +14th century maintained that substances only react according to definite weights (Holmyard, 10).

Arabic weights themselves were sometimes of remarkable accuracy; the same author gathered data showing that in the +8th century glass dinar and dirham weights had an average error of only 4 mgs. and some series agreed within ±0·3 mg. The subject is of course bound up with the history of weighing in general in the different civilisations. For Europe on this there is the book of Moody & Clagett (1); outstanding figures were Blasius of Parma (d. +1416) and Nicholas of Cusa (d. +1464), cf. Thorndike (1), vol. 4, pp. 75ff., D. Singer (1), p. 55. Unfortunately we have so far no general history of the balance and its use in Chinese culture, beyond the histories of metrology as such, and what has been said in Vol. 4, pt. 1, pp. 19ff. On the gravimetric principle in the history of chemistry there is the remarkable monograph of Walden (1), but it deals only with Europe.

[b] No less to the point is the resonance with the Daltonians' confidence in simple combining ratios, as Cyril Stanley Smith has remarked.

At this point we can only suggest that the strength and integration of numerology in Chinese thought did much to encourage truly mathematical approaches to natural phenomena, but at the same time it was difficult to see that such approaches were something different from numerology.[a] In Europe the application of a mathematical physics to earthbound experience was inevitably revolutionary, for it undercut the most fundamental metaphysical assumptions of the traditional way of looking at the universe. Since mathematics, according to Aristotle, dealt with perfect and eternal bodies and their relations, it could only be rigorously applied to astronomy, for nothing below the orb of the moon was perfect and eternal. The very success of the Aristotelian synthesis at imposing reason and coherence upon most of man's concerns over two millennia gave it a strength which could be overcome only by that total confrontation which we call the Scientific Revolution.

In China there was no such tension. The basic system of natural concepts—Yin and Yang, the five elements, and so on—were at bottom no less qualitative than Aristotle's, but number was one of the established ways of expressing quality, and no one denied that its application to terrestrial measure reflected deep realities too.[b] Thus since the +1st century a truly mathematical approach to acoustics, a human artifact, was thought of as parallel to mathematical astronomy; indeed most of the dynastic histories discuss the two together. But while the prevalent natural philosophy easily accommodated both numerology and mathematics, it did little to encourage their separation, or even to keep in view the distinction between them. In alchemy, numerologically derived quantities were treated as though they were observational; in mathematical astronomy, observational constants for periods of revolution were often metaphysically accounted for by 'deriving' them numerologically.[c] We suggest then, as a hypothesis which only many more case-studies could establish or disprove, that this very ease of accommodation meant less tension of the kind that might have led to an autonomous definition of exact science.

(ii) *Theories of categories*

About the interaction of substances and things there existed in ancient and medieval China a coherent body of doctrine springing from philosophical ideas first apparent in the writings of the famous Han scholar and thinker, Tung Chung-Shu[1] (−179 to −104).[d] In this world-order every thing and being belonged to a category (*lei*[2]), and

---

[a] That Chhen Shao-Wei and his friends did make many weighings in the late +7th and early +8th centuries can hardly be doubted. As our collaborator Tshao Thien-Chhin noticed long ago, this is suggested by the expression often used by them regarding chemical conversions—*fên hao wu chhien*,[3] 'there is not a grain or a scruple of loss'. In TT884 this is applied both to the making of mercuric sulphide (*tzu sha*[4]) from the two elements, and to the making of lead sulphide from lead metal and mercuric sulphide. This recalls Maslama al-Majrīṭī three centuries later, though the Andalusian failed to note the increase of weight on calcination. For Chinese statements about weighing in gold refining, going back to the +2nd century, cf. Vol. 5, p. 2, pp. 56–7.

[b] One thinks of the quantitative computations essential for the great works of hydraulic engineering (cf. Vol. 4, pt. 3), so closely associated since the earliest times with the fortunes of dynasties and rulers.

[c] See, for instance, Sivin (9), pp. 8–9.

[d] See Vol. 2, pp. 248–9 and *passim*.

[1] 董仲舒   [2] 類   [3] 分毫無欠   [4] 紫砂

events took place by mutual resonance between entities in the same category.[a] These classical theories took shape at the very general conceptual level of Yin and Yang and the Five Elements, so that they could be applied to the whole range of man's experience of natural transformations and natural phenomena; and their extension led to an early form of what after the Scientific Revolution were to become theories of chemical affinity. Hence though they may not have played a very large part in the detailed planning of laboratory processes they deserve close comparative study. The context of most systematic alchemical writing on categories was in fact not laboratory alchemy but rather the art of the enchymoma, which is comprehensible enough, for we have already seen by a number of indications how greatly systematic theoretical speculation flourished in physiological alchemy.[b] Perhaps the adept's experience of his own bodily and psychical states limited intellectual freedom to create order and symmetry less than his experience of chemical and physical changes in the external world. But he saw the inner and outer realms as similarly constituted, so that knowledge of one would illuminate the other.

Let us proceed by reviewing the hypotheses upon which category thinking was based. Chinese natural philosophy was given form by certain sets of concepts which could function dynamically as phases of a cyclical process, or statically as categories, that is to say as divisions within a continuum or configuration. The most universally applied of these sets of concepts were the binary Yin and Yang[c] and the Five Elements. With these five elements were aligned and associated, in symbolic correlation, everything else in the universe which could be got into a fivefold arrangement.[d] The key-word in the old Chinese thought-system was Order, but this was an order based on organic pattern, and indeed on a hierarchy of organisms. The symbolic correlations or correspondences all formed part of one colossal pattern. Things behaved in particular ways not necessarily because of prior actions or chance impulses of other things, but because their position in the ever-moving cyclical universe was such that they were endowed with intrinsic natures which made that behaviour inevitable for them. If they did not behave in those particular ways they would lose their relational positions in the whole (which made them what they were), and turn into something other than themselves. They were thus organic parts in existential dependence upon the whole world-organism. And they reacted upon one another not so much by mechanical impulsion or causation as by a kind of mysterious resonance.[e]

Nowhere are such conceptions better stated than in the fifty-seventh chapter of

[a] Later on we shall suggest that this idea was perhaps originally not unconnected with the perpetuation of likeness in fermentation and generation (pp. 364 ff. below), biological phenomena on which man must have meditated from very early times. Chinese alchemists and Alexandrian proto-chemists alike were wont to appeal to these in support of the view that similar things react with, and produce, similar things (cf. the quotation on p. 313).

[b] The theory of physiological alchemy is fully discussed in pt. 5 below.

[c] 'When the Yin and Yang unite in harmony the myriad things are begotten', *Li Chi*, ch. 11, p. 27a; Legge (7), vol. 1, p. 420. 'When Heaven and Earth combine their generative forces the changes and fermentations of the myriad things (are completed); when male and female mingle their seminal essences the transformations and births of the myriad things (are accomplished)', *I Ching* (Great Appendix), Pt. 2, ch. 5 (ch. 3, p. 33a), Baynes tr., vol. 1, p. 368; cf. Granet (5), p. 138.

[d] Cf. Vol. 2, pp. 261 ff.

[e] For a full account of this nature-philosophy see Vol. 2, pp. 279 ff., 291 ff.

Tung Chung-Shu's[1] *Chhun Chhiu Fan Lu*[2] (String of Pearls on the Spring and Autumn Annals), written about −135, which is entitled 'Thung Lei Hsiang Tung',[3] i.e. in Hughes' translation (1) 'Things of the Same Genus Energise Each Other'. We read:

If water is poured on level ground it will avoid the parts which are dry and move towards those that are wet. If (two) identical pieces of firewood are exposed to fire, the latter will avoid the damp and ignite the dry one. All things reject what is different (to themselves) and follow what is akin. Thus it is that if (two) *chhi*[4] are similar, they will coalesce; if notes correspond, they resonate. The experimental proof (*yen*[5]) of this is extraordinarily clear. Try tuning musical instruments. The *kung*[6] note or the *shang*[7] note struck upon one lute will be answered by the *kung* or the *shang* notes from other stringed instruments. They sound by themselves. This is not due to spirits (*shen*[8]) but because the Five Modes are in relation; they are what they are according to the constant relations (*shu*[9]) (whereby the world is constructed).

(Similarly) lovely things summon others among the class of lovely things; repulsive things summon others among the class of repulsive things. This arises from the complementary way in which a thing of the same class responds (*lei chih hsiang ying erh chhi yeh*[10])—as for instance if a horse whinnies another horse whinnies in answer, and if a cow lows, another cow lows in response...[a]

Similar passages occur elsewhere in the same book. In this instance one can see how each element was associated with a particular musical note and a particular animal, and in general how things of the same category (*thung lei*[11]) were conceived to act as receptors only to disturbances originating within the same category.[b]

Thus the classifiability of which Tung Chung-Shu speaks is the capacity of the various things in the universe to go into the fivefold categorisation, or others of different numerical values. Chinese thought was particularly fond of such categorisations: Mayers (1) could list 317 of them from the Two Primary Forces to the Hundred Officials. No less than eleven chapters of the *Thu Shu Chi Chhêng* encyclopaedia (+1726) are consecrated to this subject in its calendrical-mathematical section.[c] Bodde (5) has devoted a special paper to Chinese 'categorical thinking' in which he analyses the curious tabulation in the twentieth chapter of the *Chhien Han Shu* (History of the Early Han Dynasty), where nearly 2,000 historical and semi-legendary individuals were arranged in nine grades according to their virtue.

Tung Chung-Shu was elaborating a philosophy quite widespread among the scholars of the Han:[d] the idea that actions and reactions in the natural world come

---

[a] Tr. auct., cf. Vol. 2, pp. 281–2.   [b] Cf. p. 316 and p. 319.

[c] *Li fa tien*,[12] chs. 129–40.

[d] The expression *thung lei* turns up in all kinds of contexts, e.g. in a speech by Ssuma Hsiang-Ju to the emperor about bear-hunting (*CHS*, ch. 57B, p. 8b). Perhaps food-chains were at the back of his mind, for certain animals could be killed only by certain others. See also *Han Shih Wai Chuan*, ch. 1, p. 2a. It is also, as one would expect, prominent in the ancient medical texts; see, for example, *Huang Ti Nei Ching, Su Wên* (Pai-hua version), ch. 81 (p. 521)—*thung lei hsiang kan*.[13] We shall have more to say on this subject fundamental to physiology and medicine in Sect. 44 (Vol. 6).

[1] 董仲舒   [2] 春秋繁露   [3] 同類相動   [4] 氣   [5] 驗
[6] 宮   [7] 商   [8] 神   [9] 數   [10] 類之相應而起也
[11] 同類   [12] 歷法典   [13] 同類相感

about by specific stimulus and specific response of one organism upon another according to their intrinsic natures as classifiable in schemas of correspondences and categories. Of course this systematisation was based on a minimum of critical observation and no systematic experiments. And for the effects it was not necessary that the bodies should be in contact: action at a distance made no difficulty for the Chinese mind, which visualised a sort of wave-motion transmissible almost infinitely through the aetheric *chhi*.[a] Yet bodies (organisms, whether animate or inanimate) influenced one another not at random but always in accordance with their positions in the perpetually moving cyclical universe. Thus just as the astronomer had to follow the motions of his celestial bodies through the cycles of month and year, the alchemist had to pay attention to the proper alternation of the Yin and the Yang within his microcosmos. It will be clear from this that the intellectual obstacles inhibitory to a proper understanding of chemical combination in China were rather different from those in Europe. For in the West the Greek atomic theories were always waiting in the wings ready to take the centre of the stage when the time was ripe, while in China the atomic theories of the Indians, though often brought in, never seduced Chinese thinkers from their instinctive adhesion to what was essentially a prototypal wave-theory, the reciprocally dependent rise and fall of the Yin and Yang forces or the analogous succession of the five elemental phases. Thus it came about that in spite of the relatively advanced scientific character of medieval Chinese physical theory, and in spite of the numerous empirical discoveries and inventions of medieval Chinese alchemy and chemical technology, modern chemistry (like the rest of distinctively modern science) originated in Europe, and passed to China only in the later +18th and nineteenth centuries.

The deep analysis of ancient and medieval Chinese ideas on causality has hardly yet begun,[b] but the original and interesting papers of Leslie (2, 7) have been clearing the way. Using many texts from Wang Chhung, Hsün Chhing, Chuang Chou and Tung Chung-Shu, all of the Warring States and Han periods, he summarises the situation by saying that three kinds of causation were recognised: (*a*) by contact, usually of like category with like, a phenomenon in temporal succession,[c] (*b*) by action at a distance, mostly of like category with unlike, also a phenomenon involving succession in time, and (*c*) by natural harmony or 'resonance', a simultaneous or co-incidental effect resulting from the pre-established harmony or pattern of the world, and not a sequence of events in time.[d] So far as alchemy and early chemistry were concerned most of the

---

[a] Cf. Needham & Robinson (1), as also at length in Sect. 26(*b*) above (Vol. 4, pt. 1).

[b] We have oftentimes glimpsed various aspects of the matter at earlier stages, as in Vol. 2, pp. 280 ff., 288 ff., Vol. 3, pp. 415, 483 ff., Vol. 4, pt. 1, pp. 135, 233, and *passim* in all Sections. We hope to bring it all together in a compendious statement in Vol. 7 (Sect. 46).

[c] Thought of more like 'infectivity' or 'contagion', one might say, than any mechanical impulsions of particulate entities. The idea remained unanalysed. We shall consider its European parallels in Vol. 6 with a medical context.

[d] Clearly this was an ancient formulation of the synchronicity or 'acausal connecting principle' developed in our own time by C. G. Jung. Resonance could also be thought of as 'simultaneous causality', neither event of the two being distinguishable as cause or effect, but both being changes due to a higher overall dynamic pattern-principle. This is certainly not causality in any ordinary sense, and Jung thought of it more as a world-principle of 'meaningful co-incidence' working in a *mundus unus*. See Jung (2, 11);

changes would doubtless have been regarded as due to the contact of the reactants, though many other influences of a more distant, cosmic, character would be operating, or could be mobilised to operate, as well.

Though it has received almost no attention from sinologists, the ancient and medieval Chinese category theories gave rise to a literature which is of much historical interest. It forms the wider background of the book which led to the first recognition of the importance of alchemical category theory when studied and discussed by Ho Ping-Yü & Needham (2) some twenty years ago. The oldest tractate which we have now in mind derives from the early Chin period, at which time Chang Hua[1] (*fl.* +232 to +300) wrote the *Kan Ying Lei Tshung Chih*[2] (Record of the Mutual Resonances of Things). Chang Hua is better known of course for his *Po Wu Chih*[3] (Record of the Investigation of Things), a miscellany of scientific interest datable *c.* +290. Then at some point probably in the third to the seventh centuries was written the most important of these treatises from the alchemical point of view, the *Tshan Thung Chhi Wu Hsiang Lei Pi Yao* (Arcane Essentials of the Fivefold Categorisation based on the 'Kinship of the Three').[a] It was presented to the throne with a commentary by Lu Thien-Chi in +1114, or at any rate between +1111 and +1117.[b] Unlike most members of the genre, which drew their examples of transformation and interaction from the whole range of human experience, this work dealt almost exclusively with chemical changes familiar to the alchemist.[c]

Jung & Pauli (1); and an interesting recent work by Abrams (1). To what extent ancient Chinese scientific and medical thought was dominated by this conception is discussed by Needham & Lu Gwei-Djen (9) in a critique inspired by the book of Porkert (1).

[a] This was the text translated with its commentary and discussed in detail by Ho & Needham (2). All the traditional bibliographies attributed it to Wei Po-Yang of the +2nd century (e.g. *Thung Chih Lüeh, I Wên*, ch. 5; ch. 43, pp. 6*b*, 23*b*), but this was true of many other books such as the *Chhi Fan Ling Sha Ko*, and the attribution is not acceptable today without further evidence of dating. There is no intrinsic ground for denying that it is a text of the Later Han, but no positive evidence in favour either.

[b] At some time in the same period Lu also presented to the throne a tractate on psycho-physiological alchemy, the *Chih Chen Tzu Lung Hu Ta Tan Shih*[4] (Song of the Great Dragon and Tiger Enchymoma of the Perfected Master), TT266. Although the last character of Lu's name has been lost from the superscription, the detailed specification of his official titles is exactly the same as in the work which interests us more here. The reason for fixing the date of these presentations exactly at +1114 is that an imperial edict in that year appealed for the collection of all Taoist books (*Sung Shih*, ch. 21, p. 3*a*). This was brought to our attention by Dr James Zimmerman of Yale University.

[c] It may not have been the only one of this character. Certain other books with similar titles may prove relevant when they have received closer examination, though their date and authorship remain quite obscure. For example, the *Chin Pi Wu Hsiang Lei Tshan Thung Chhi*[5] (Gold and Caerulean Jade Treatise on the Similarities and Categories of the Five (Substances) and the 'Kinship of the Three'), TT897, is attributed to Yin Chhang-Shêng[6] (perhaps +120 to +210) the oldest commentator on the *Chou I Tshan Thung Chhi*. Our collaborator Tshao Thien-Chhin was inclined to think that the versified text could well be as early as the +2nd or +3rd century, while the prose commentary would be of the Liu Chhao, Sui or early Thang, thus contemporary approximately with the *Tshan Thung Chhi Wu Hsiang Lei Pi Yao*. The whole work, which has been given a preliminary study by Ho Ping-Yü (12), is distinctly *nei tan* in feeling, and though it does mention some substances and even apparatus these look like cover-names for physiological processes and techniques. Unfortunately, in its present form it contains nothing on category theory, and neither text nor commentary has an archaic air.

The other book here in mind is much more interesting for our present theme; it is called *Yin Chen Chün Chin Shih Wu Hsiang Lei*[7] (Similarities and Categories of the Five (Substances) among Metals

[1] 張華　　　[2] 感應類從志　　　[3] 博物志　　　[4] 至眞子龍虎大丹詩
[5] 金碧五相類參同契　　　[6] 陰長生　　　[7] 陰眞君金石五相類

Later, in the early Thang, the astronomer Li Shun-Fêng[1] (*fl.* +620 to +680)
followed on with his *Kan Ying Ching*[2] (On Stimuli and Responses in Nature).[a] And
we must also mention two small books deriving from the writings of the Sung monk
and lover of natural curiosities (Lu) Tsan-Ning[3] (+919 to +1001). The *Wu Lei
Hsiang Kan Chih*,[4] subject of a recent study by Yamada Keiji (*1*), has often been
attributed to the great poet and scholar-official Su Shih[5] (+1036 to +1101), since
he and Tsan-Ning shared the literary appellation Tung-Pho.[6] The present version is
an abridgement in one chapter produced (probably with some additions) by the late
Ming;[b] about the same time as another book of a like kind, the *Ko Wu Tshu Than*[7]
(Simple Discourses on the Investigation of Things), also put together from Tsan-
Ning's work and similar materials.[c] This is far from exhausting the available literature
which deals with categories and resonances in nature-philosophy, but we shall discuss
these books sufficiently as representative.

Before proceeding further, however, we must make mention of a text which might
be regarded as the *fons et origo* of the whole group. We refer to the *Huai Nan Wan Pi
Shu*[8] (Ten Thousand Infallible Arts of the Prince of Huai-nan). This strange work
dates mostly from the time of the prince of Huai-nan, Liu An[9] (*d.* −120), that great
patron of alchemists and other naturalists.[d] It is supposed to have formed a complement to the existing, and very well known, *Huai Nan Tzu*[10] book, a compendium of
natural philosophy the authenticity of which is quite unquestioned.[e] A clear tradition

and Minerals, by the Deified Adept Yin), TT899, of date and authorship so far impossible to determine.
It consists of twenty sections each headed *phei ho . . . hsiang lei*, 'the pairing and combining category of . . .'.
Each one gives synonyms somewhat in the manner of the *Shih Yao Erh Ya* but with more detailed
explanations. The first eleven sections are very chemical, dealing with (1) lead, (2) mercury, (3) laminar
copper carbonate, (4) sulphur, (5) realgar, (6) silver (the essence in lead), (7) white cinnabar, (8) gold,
(9) sal ammoniac, (10) saltpetre (*hsiao shih*,[11] also known as *chhiu shih*,[12] cf. pt. 5), and (11) nodular
copper carbonate. The remaining nine, however, are much more difficult to understand, including, for
example 'the *hun* (animus) of minium' (no. 13), or 'the male and female of the uterine palace' (no. 15),
and—a strange entry—"the Persian essence" (no. 18). This last might help to date the work, which
Tshao Thien-Chhin thought might be as old as the +3rd century but with later additions. The writer
says that all the 72 metals and minerals can be classified into Yin and Yang substances, and presumably
also into one or other of the twenty categories enumerated. This little treatise requires, and truly
deserves, further study.

   a See p. 314.
   b Probably by Chhen Chi-Ju[13] for inclusion in the third collection of his *Pao Yen Thang Pi Chi*[14]
(printed in +1615). On the history of the text and its chapter divisions see Su Ying-Hui (*1, 2*). The
question is rather complex. Sung records give Tsan-Ning as the compiler and 10 chapters as the
structure, mentioning a quadripartite organisation very different from that of the current version, which
thus cannot be taken as a simple condensation from the original work.
   c See the comments in the 18th-century imperial analytical catalogue, *Chhin-Ting Ssu Khu Chhüan
Shu Tsung Mu Thi Yao*, ch. 130, which suggests that the 18-chapter version also available at that time was
a later expansion of the 1-chapter version. Chhang Pi-Tê (*1*), p. 263, however, has noted that since
quotations in early sources correspond to the 18-chapter version, the inverse process is more likely. The
Ko Wu Tshu Than contains numerous passages which agree verbatim with the current text of Tsan-
Ning's other book, so it may be simply another abridgement of the original recension.
   d For an account of what is known of the genesis and bibliography of this book, see Kaltenmark (*2*),
p. 32. On its position in the history of practical proto-chemistry and alchemy cf. Vol. 5, pt. 3, pp. 24 ff.
above.   e Cf. Vol. 2, pp. 36 *et passim* above.

1 李淳風	2 感應經	3 錄贊寧	4 物類相感志	5 蘇軾
6 東坡	7 格物麤談	8 淮南萬畢術	9 劉安	10 淮南子
11 硝石	12 秋石	13 陳繼儒	14 寶顏堂祕笈	

among Chinese scholars going back to the + 1st century asserts that much of the *Huai Nan Wan Pi Shu* was concerned with alchemy (*shen hsien huang pai shu*[1]—the Holy Immortals' Art of the Yellow and White, i.e. elixirs and the making of gold and silver).[a] We now have only fragments of this book, but it is reasonable to look on it as the forerunner of the whole class of proto-scientific literature which we are discussing. In this connection we may recall that its date is very close to that of the *Chhun Chhiu Fan Lu*, since Tung Chung-Shu was a younger contemporary of Liu An. Broadly speaking, in this sort of literature, the later the text the less the admixture of magic and the stronger the practical technological element.

Modern scholars such as Yeh Tê-Hui[2] and Sun Fêng-I[3] have collected together the fragments of the *Huai Nan Wan Pi Shu* from numerous sources—mainly encyclopaedias and pharmacopoeias. Of 115 entries, 56 are concerned with charms and omens, but there are 32 which deal with medicine, pharmaceutics, nutrition and animal lore, while 13 involve physical phenomena and 12 alchemy or chemistry. Thus side by side with a charm to make people tell the truth by giving them insects from the bamboo plant to eat, or hanging up a piece of lodestone in a well to draw back a runaway person, one finds a clear statement of the precipitation of metallic copper from copper-containing waters[b] or a prescription for a longevity elixir using copper carbonate. Of sympathies and antipathies there are plenty—roasting crab-meat attracts rats, and horn is burnt to keep away leopards and tigers in the mountains.[c]

Ho Ping-Yü & Needham found a remarkable parallelism between this text and the *Peri Sympatheiōn kai Antipatheiōn* (περὶ συμπαθειῶν καὶ ἀντιπαθειῶν), also called the *Physica Dynamera* (φυσικὰ δυναμερά) of Bolus of Mendes, not till then pointed out. Bolus, who lived at Mendes in Egypt between −200 and −150, has generally been regarded as the initiator of that long line of Western proto-chemical literature which begins with Pseudo-Democritus and Mary the Jewess in the + 1st century and runs on continuously through Syriac and Arabic into the late Latin books.[d] Since colouring processes were so important for these practitioners of aurifaction it may be significant that the title of one of Bolus' lost books was *Baphica*, βαφικά (On Dyeing), while that of another, which we have just mentioned, was similar to the title of the truly chemical-metallurgical work of Pseudo-Democritus *Physica kai Mystika* (φυσικὰ καὶ μυστικά). Besides these there was a *Cheirokmēta* (χειρόκμητα, Prescriptions based on Sympathies and Antipathies). Bolus of Mendes, as the first Western theorist of natural phenomena in their chemical and technological aspects, applying Greek ideas perhaps

---

[a] Cf. *Chhien Han Shu*, ch. 44, p. 8*b*.   [b] Cf. p. 201 above.

[c] *PPT/NP*, ch. 3, p. 5*b*, quotes Lao Tzu to the effect that the ability of 'wildcat's head' (which is also the name of a gourd) to cure the 'rat ulcers' of lymphadenitis, and that of woodpecker's flesh to protect the teeth against cavities, can both be understood in terms of categorical resonance (*lei*[4]); cf. Ware (5), p. 61; Feifel (1), pp. 196–7. But the statement is actually a condensation of one from *Huai Nan Tzu*, ch. 16, p. 14*b*.

[d] See Berthelot (1), pp. 156ff.; Festugière (1), pp. 197ff., 219ff., 229ff.; and p. 325 below. On the social background of Hellenistic Egypt see Cumont (4). The real Democritus of Abdera, so renowned for his atomic materialism, died in −375. Bolus may have called himself 'the Democritean' (cf. pt. 2, pp. 17, 25 above).

[1] 神仙黃白術   [2] 葉德輝   [3] 孫馮翼   [4] 類

to the interpretation of Egyptian techniques, occupies a position similar to that of Tsou Yen[1] in China, though perhaps an even more shadowy figure than the great systematiser of Five-Element theory. [a]

The reconstructed book[b] of Bolus on sympathies and antipathies is much smaller than the *Huai Nan Wan Pi Shu* but the similarity is obvious. Of its 34 items, 20 concern natural antipathies, 3 tell of sympathies, and the rest have to do with marvellous properties of animals, plants or minerals. Thus with the same juxtaposition of magic and science as in the Chinese text, we find a charm to make a person tell the truth by laying the tongue taken from a live frog upon her sleeping breast, or a notice of the antipathy of serpents for the saliva of a fasting man, side by side with information about the attractive powers of amber and lodestone or about poisonous fruits. On the whole there is relatively more magic in the Greek than in the Chinese text.[c]

The significance of this whole train of thought can well be seen in an epigram which appears several times among the Greek proto-chemical writings.[d] At the beginning of the book of Pseudo-Democritus, the author relates how, being tormented by the desire to know 'how substances and natures unite and combine themselves into one substance', he invoked the shade of his master Ostanes the Mede, who had died before transmitting all his chemical learning. The answer was that the books were all hidden in a temple, but they could not be found there, and no illumination came until upon a festival evening one of the columns spontaneously split open revealing the marrow of the doctrine in an inscription: 'One nature is charmed by another nature, one nature conquers another nature, one nature dominates over another nature.'[e] Here then,[f] within the field of the metals and minerals, is just the same principle of sympathies and antipathies about which not only Bolus of Mendes but also the adepts of the Prince of Huai-nan had so much to say. However far removed this may seem from the history of chemistry, it is actually one of the most important roots of chemical thinking, for antipathies and sympathies are nothing but the prehistoric ancestors of reactivities and affinities. Tabulation of affinities according to categories was of course another and considerably later development.[g]

---

[a] On Tsou Yen (−4th century) see Vol. 2, pp. 232ff..

[b] See the learned contribution of Wellmann (2). On Bolus' possible Persian sources see Bidez & Cumont (1), vol. 1, pp. 203 ff., 244 ff.; vol. 2, pp. 311ff., 320.

[c] The whole realm of the Hellenistic magical papyri is of course relevant here. Some are bilingual, having Greek side by side with hieratic Egyptian, as in Leiden Pap. 65 and 75 discussed by Berthelot (1), pp. 82ff. There is also a marked overlap with the proto-chemical papyri (pt. 2, pp. 15ff. above), some of the magical ones containing chemical recipes.

[d] Berthelot & Ruelle (1), vol 2, pp. 42ff.; vol. 3, pp. 44ff., *Corp. Alchem. Gr.* II, 1, retranslated by Festugière (1), pp. 228ff. The aphorism recurs repeatedly in Pseudo-Democritus, cf. vol. 3, pp. 47ff., 50, 51, 52, 55; as also in Synesius, p. 61. Cf. Bidez & Cumont (1), vol. 1, pp. 203ff., 244ff.; vol. 2, pp. 311ff., 320. Similar thoughts recur in Ptolemy, *Tetrabiblos*, I, 3.

[e] It is difficult not to be reminded, in this formulation, of the Principles of Mutual Conquest and Mutual Production in Chinese Five-Element theory (see Vol. 2, pp. 253ff.), at least contemporary, indeed in some respects going back to the −4th century. To be sure, those principles were more strictly concerned with ordered succession in cycles of change. But one also senses a connection with the 'love and hate' of the pre-Socratic philosophers, especially Empedocles. This, too, has striking Chinese parallels (Vol. 2, p. 40).　　　　[f] As Festugière (1), p. 231 acutely pointed out.

[g] Without enlarging on a matter which gets separate treatment elsewhere (pp. 324ff.), we cannot refrain from pointing out here the far-reaching parallels which are being revealed in the development of

[1] 鄒 衍

One can also say that the ideas of groups, classes, categories and affinities are to be found embryonically within the writings of the Hellenistic proto-chemists, if not perhaps in the oldest texts.[a] One attributed to Zosimus, at any rate, speaks of mercury 'and its analogues', as we might say, or the substances similar to it (*kai ta homoia*, καὶ τὰ ὅμοια): and it uses the same expression for pyrites, magnesia and chrysocoll.[b] Almost in the next breath, it goes on to talk about the affinity or relationship, literally 'consanguinity' (*syngeneia*, συγγένεια), which certain substances have for others.[c] Thus magnesia and magnetite are said to have a congenital relationship, presumably an attraction, for iron, mercury for tin (obviously because of amalgamation), copper for pyrites, and lead for the Etesian stone (less obvious, but the identifications are uncertain). All this was rather unorganised thinking and it does not seem to have been followed up among the Greek writers. They also shared with the ancient and medieval Chinese (cf. pp. 317, 319) the closely related idea, derived from primitive biological observation, that like only comes from like. As is said in the tractate attributed to Isis, 'corn is engendered only by corn, man alone sows man, and only gold can give a harvest of gold', in general like generates (*genna*, γεννᾷ) like.[d]

One might wonder what happened to the lore of sympathies and antipathies apart from the stimulus which it gave to affinity theory in chemistry. In the West it petered out in the bestiaries and books on talismanic magic of the late Middle Ages and the early Renaissance, one element, at least, of considerable importance in the genesis of modern science;[e] and it may be that some of the old Chinese beliefs became incorporated, through Arabic intermediation of course, in the eventual magma. About +1056 someone in Spain put together a book on magic which bore the title *Ghāyat al-Ḥakīm* (The Aim of the Sage),[f] and this, under the name of *Picatrix*,[g] enjoyed a great success in the Latin world during the following four or five hundred years. Derivations from Arabic, Hebrew and Syriac sources, as from Greek, Sanskrit and Pehlevi, have been acknowledged, but when one finds in many of the manuscripts lists of the twenty-eight lunar mansion constellations accompanied by drawings or diagrams of them exactly in the Chinese style (the 'ball-and-link' convention)[h] one is inclined to suspect that further study would reveal sympathetic beliefs which had come down

chemistry in Mediterranean and Chinese antiquity. Not only does Bolus of Mendes parallel in time and nature the school of Liu An, but Liu Hsiang and Wei Po-Yang bracket the +1st-century developments, and then the two great synthetic writers Zosimus of Panopolis and Ko Hung appear at about the same time (*c.* +300). To say nothing of the Thang and Arabic alchemists, we end by finding the Chinese alchemical corpus first going into print in +1019, just about the date when the writings to which all our knowledge of Hellenistic proto-chemistry is due were being copied and compiled into the codices.

---

[a] As was pointed out by Hammer-Jensen (2), p. 24.

[b] *Corp. Alchem. Gr.*, III, xxviii, 9 (Berthelot & Ruelle (1), vol. 3, p. 192).

[c] *Corp.*, ibid., 10 (Berthelot & Ruelle, ibid., p. 193). Cf. p. 360.

[d] *Corp.* I, xiii, 8 (Berthelot & Ruelle, op. cit., vol. 2, pp. 30, 34, vol. 3, p. 34).

[e] Cf. Yates (2), who eloquently pleads for the thesis (which must be at any rate partly true) that the Renaissance 'magus' was the immediate ancestor of the +17th-century scientist.

[f] Cf. Ritter (1, 2); Plessner (1, 2); Yates (1), pp. 49 ff. and *passim*; translation by Ritter & Plessner (1). On account of an attribution sometimes found, the writer is often called Pseudo-al-Majrīṭī. We discuss the book further below, pp. 427 ff.   [g] From Buqrāṭis, i.e. Hippocrates, the supposed writer.

[h] Cf. Vol. 3, pp. 276 ff. Similar diagrams are found, as we learn from our friend Mr M. Destombes, on some Arabic astrolabes, MSS, and even Latin treatises on astronomy of the +12th and +13th centuries. Cf. the illustration on p. 428 below.

all the way from the *Huai Nan Wan Pi Shu*. If so, they would have had a share in the liberating Renaissance current of thought which restored to man that conviction of the possibility of dominion over Nature which the Middle Ages had minimised or deprecated,[a] and strengthened the assurance that fundamental effects could be brought about by manual operations.[b] Magic and science were still difficult to distinguish in the early days of the Royal Society, but the 'multiplication of real experiments' was already on foot to bring its ineluctable consequences.

Setting aside the *Tshan Thung Chhi Wu Hsiang Lei Pi Yao* for separate consideration presently because of its specifically chemical import, we must now briefly describe the books in the line of literary descent already adumbrated. Chang Hua's +3rd-century tractate[c] contains some matters of genuine scientific interest, set down more or less at random alongside a good deal of magical and superstitious material. Among the former we may mention his descriptions of fish poisons and of the charcoal hygrometer,[d] and his account of the optical phenomena of reflection in plane mirrors. Both this and some of the magical recipes[e] are almost identical with what is found in the *Huai Nan Wan Pi Shu*.

Chang Hua's other book also has some interesting passages. From one of them we can see that in the +3rd century 'category (*lei*[1]') was sometimes thought of as determined by the 'state' of matter, liquid, solid, gaseous, etc. He says:[f]

When lead (*chhien-hsi*[2]) is heated to make white lead (lead carbonate), (these substances) are similar in category (both being solids) (*yu lei yeh*[3]). But when cinnabar in converted into mercury there is no longer any categorical similarity (one being a solid and the other a liquid), (*tsê pu lei*[4]). (These are examples to show that) things of similar category (*thung lei*[5]) can change into categories that differ from one another.

In his +7th-century *Kan Ying Ching* Li Shun-Fêng again describes the charcoal hygrometer and has something to say on strange animals, the generation of insects from grain and of fireflies from rotting grass.[g] He says that swallows have a sense of direction, for apart from migrations their nests always face north. Unfortunately his treatise is available today only in fragmentary form.[h]

The books based upon Tsan-Ning's writings are much more advanced (though magic is not entirely absent), and each is well classified into subjects. Significantly, both mention magnetic attraction.[i] The *Ko Wu Tshu Than* opens with weather forecasting and continues with miner's lore concerning signs of ore beds, including plant indicators.[j] Tsan-Ning (or his editors) knew of poisoning by the fumes of burning coal and recommend a remedy for it. In both books there is mention of the

---

a The full argument should be read in Yates (1, 2).

b Cf. Vol. 2, pp. 34, 83, 89ff. 'However strange his operations may seem to us, it is man the operator who is glorified in the (Hermetic) *Asclepius*' (Yates (2), p. 257). The accent was on power, and that was just what the merchants of early capitalism also wanted. Did the magus go hand in hand with the entrepreneur? If so, we might have a piercing side-light on the origins of modern science.

c In *Shuo Fu*, ch. 24, pp. 18*b*ff.   d Cf. Vol. 3, p. 471.

e E.g. charms for getting a runaway person to come back.   f *Po Wu Chih*, ch. 4, p. 3*a*, tr. auct.

g The metamorphosis lore of animals and plants will be discussed in Sect. 39 below (Vol. 6).

h In *Shuo Fu*, ch. 9, pp. 1*a*ff.

i But curiously not the polarity, though it was well known in his time in China. See Vol. 4, pt. 1, Sect. 26(i).   j See further on this Vol. 3, pp. 675ff.

¹ 類   ² 鉛錫   ³ 猶類也   ⁴ 則不類   ⁵ 同類

use of lime as a dehydrator for preventing iron and steel implements from rusting. Corrosion of bronze and brass, on the other hand, can be removed by vinegar. Another interesting reference is to the so-called wet method of copper production where copper-containing mine waters are led over waste iron and the copper metal is precipitated as a powder: this technique, mentioned already, as we have just seen, in the −2nd century in the *Huai Nan Wan Pi Shu*,[a] was becoming a standard industrial process by Tsan-Ning's time.[b] One encounters all kinds of things; for example: 'For the Floating Elixir, mould camphor and vermilion into a cake; when you put it on water it will rush to and fro.'[c] Or on disinfection: 'When an epidemic comes, put the clothes of the first person falling ill in a steamer and steam them well, then the whole household will escape the infection.' On invisible ink: 'Write characters with a solution of (iron) alum, allow it to dry, then to make the writing visible wet it with an extract of gall-nuts.'

As for the *Wu Lei Hsiang Kan Chih*, more certainly Tsan-Ning's, it tells how tung oil spread on the water kills lotuses, how liquids can be clarified by filtration through sand, how spots can be bleached by the uric acid in bird droppings, how grease can be absorbed on finely powdered charcoal or talc, and how sterilised salt should be added to vinegar to prevent the formation of a white pellicle by moulds.[d]

Lastly we may refer to another Sung book, dating from the +12th century, the *Hsü Po Wu Chih*[1] (Continuation of the Record of the Investigation of Things) by Li Shih.[2] In this, Li quotes the (*Shen Nung*) *Pên Tshao Ching*[3] as follows:[e]

When the tiger roars, the wind rises.[f] When the dragon gives tongue the clouds gather.[g] The lodestone attracts needles. Amber attracts bits of straw (literally mustard seeds).[h] After coming into contact with crabs lacquer will not concrete.[i] Lacquer added to hemp (-seed oil) makes it bubble.[j] Treated with the *tshung*[4] onion, cinnamon (bark? wood?) softens. Cinnamon causes (certain) trees (or plants) to wither.[k] Crude salt preserves piles of eggs.[l] The

[a] In *Thai-Phing Yü Lan*, ch. 988, p. 5a.
[b] Cf. Vol. 2, p. 267. More detailed information has been given above, pp. 201 ff.
[c] Cf. pt. 2, p. 170 above.
[d] A complete translation of these two small books would be rewarding, especially with a commentary to bring out the widespread role of the applications of empirical science in the daily life of medieval China. [e] Ch. 9, p. 5a.
[f] A parallel statement of Han date occurs in the *Huai Nan Wan Pi Shu* (*Thai-Phing Yü Lan*, ch. 89, p. 5a).
[g] For representations of dragon–tiger resonances in ancient literature and art cf. Hawkes (1), p. 133; Riddell (2), a circular lacquer dish of +69; Chavannes (9), a rubbing of +171.
[h] The development of knowledge of magnetism and electrostatics in China is fully dealt with in Vol. 4, pt. 1, Sect. 26(i).
[i] On this *locus communis*, which involved an ancient empirical discovery of a powerful laccase inhibitor, see pp. 207ff. above.
[j] It has in fact been the practice for centuries past to add tung oil to lacquer latex as an adulterant; as Li Shih-Chen pointed out (*Pên Tshao Kang Mu*, ch. 35, p. 20a), this makes it very poisonous. The bubbling could have been disengagement of CO_2 because of an acidity difference.
[k] *Pên Tshao Kang Mu*, ch. 34, p. 15b, quotes the *Lü Shih Chhun Chhiu* (−239) as saying, 'Under the branches of the cinnamon tree no other saplings will come up.' This passage does not appear to be in the *Lü Shih Chhun Chhiu* now, but there is no reason for doubting its antiquity. The *Phao Chih Lun*[5] (c. +470) of Lei Kung[6] is also quoted in the same place as saying that if you drive a peg of cinnamon wood into the root of another tree the latter will wither.
[l] A parallel statement of Han date occurs in the *Huai Nan Wan Pi Shu* (*Thai-Phing Yü Lan*, ch. 865, p. 5a and ch. 928, p. 6b).

[1] 續博物志　　[2] 李石　　[3] 神農本草經　　[4] 葱　　[5] 炮炙論　　[6] 雷公

gall of the otter cracks (literally divides) wine-cups. (All these phenomena occur because) the *chhi* (*pneumata*) of these things are in sympathy (*chhi chhi shuang chih*[1]) and thus bring about mutual resonance (*hsiang kuan kan yeh*[2]).[a]

If this is what it purports to be it could date from the −2nd century, but the passage is not found in the version of the *Pên Ching* reconstructed by modern scholars such as Mori Tateyuki (1845).[b] Thus although the consciously chemical content of these texts is not great, they form a corpus cognate with the specifically alchemical theories worked out in the *Tshan Thung Chhi Wu Hsiang Lei Pi Yao*.

For the oldest application of the theory of categories in alchemy we have to go back to Wei Po-Yang who in his *Chou I Tshan Thung Chhi*[3] speaks as follows:[c]

Lead carbonate (*hu fên*[4]), being placed on the fire, becomes discoloured and changes back to lead. Mixed with hot liquids ice and snow melt into water (*thai hsüan*[5]). The Gold (Elixir) is mainly derived from cinnabar ([*tan*] *sha*[6]) which is naturally endowed with mercury. Transformations depend on the true nature (of the substances)—beginnings and ends are mutually related. The way to become an immortal (*hsien*[7]) through taking drugs lies in the use of substances of the same category (*thung lei*[8]). Grains are used for raising crops, hen's eggs are used for hatching chicks. With substances of (similar) categories as the assistants of natural spontaneity the formation and moulding of things is easily accomplished. Fish eyes cannot replace pearls, neither can weeds be used for timber. Things of similar category go together (*lei thung chê hsiang tshung*[9]): precious substances cannot be made from the wrong materials. This is why swallows and sparrows do not generate the *fêng*[10] (male phoenix), this is why foxes and rabbits do not suckle the horse. Flowing water does not heat what is above it, and a fire does not wet what is underneath it.

And elsewhere he says:[d]

In the activities of Nature there is never anything sinister or illusory. The *chhi* of the mountains and marshes distils into the heavens forming clouds and returns as rain. The muddiest lane becomes a dust-dry path in time, and after the fire is out all turns to dead earth-ash. Shavings of *nieh*[11] wood[e] or bark dye yellow, but if blue (indigo) be added green

---

[a] We should like to draw attention to the fact that the whole passage partakes grammatically of the nature of a sorites, each statement beginning with an agent which was the patient of the previous statement. This can be seen in the fifth and sixth, and the seventh and eighth, statements. On the sorites as a logical form see Welton (1), p. 393; Maspero (9), and Granet (5), pp. 337, 443, 487. But here the whole content is empirical, and one cannot help being reminded of the successive dominances in the Mutual Conquest Principle of the Five-Element Theory (see Vol. 2, pp. 257 ff.).

[b] As Prof. A. F. P. Hulsewé points out, the dating of this passage might hinge on the time at which cinnamon bark first became well known in China; on this see Vol. 6 (Sect. 38).

[c] *Tshan Thung Chhi Fên Chang Chi Chieh* ed., ch. 12 (ch. 1, p. 25b); *TT*990, ch. 1, p. 37a, b; cf. *Ku Wên Tshan Thung Chhi Chien Chu Chi Chieh*, ch. 6, p. 1a. We differ here from the translation by Wu & Davis (1), p. 241.

[d] Ch. 32 (ch. 3, p. 5a); *TT*990, ch. 3, p. 6b. On this the +3rd-century commentary of Yin Chhang-Shêng says that when mercury–lead amalgam is made the 'yellow sprouts' appear, and that when these are added to more mercury the cyclically-transformed elixir spontaneously develops. This is because the species are right. But trying to make gold from aqueous solutions of various minerals and drugs will effect nothing, for the categories are not congruent. Labour and skill will be vain, and success will not follow. Here also the translation of Wu & Davis (1), p. 259, misses the point.

[e] This is probably the leguminous *Pterocarpus indicus* (= *flavus*), on which see R405, CC1035 and Schafer (8). *Huang po*[12] may have been a synonym anciently, but *po* is properly the name of another plant; cf BIII 315.

[1] 其氣爽之	[2] 相關感也	[3] 周易參同契	[4] 胡粉	[5] 太玄
[6] 丹砂	[7] 仙	[8] 同類	[9] 類同者相從	
[10] 鳳	[11] 蘗	[12] 黃蘗		

girdles can be made. Cooking leather and skins gives glue; mould-leaven (*chhü*[1]), malt and yeast (*nieh* [2,3]) transform mash into wine.[a] If the categories are the same (*thung lei*[4]) it is easy to perform the work, but with discrepant species (*fei chung*[5]) difficulties will defeat the greatest skill. On this depend all the marvels of men's craft.

These words were written, we suppose, in the neighbourhood of +140. Within a few centuries thereafter, in the *Tshan Thung Chhi Wu Hsiang Lei Pi Yao*, we find a much more mature application of category theory. A large number of the common reagents are classified in categories with reference to their use in elixir processes. At one place the writer says:[b]

Now the *Tshan Thung* (*Chhi*), by means of the Five Elements, the inner and outer (aspects), and the six acoustic pitches (*liu lü*[6]), allows one to determine the proper quantities and to know (which substances are of) the same category (*thung lei*[7]).
   [Comm.] Now (in the making of) the Great Elixir, success will never be attained if the Yin and Yang principle and the Five-Element theory are not followed to classify substances into categories (*kho ting thung lei*[8]).
Now according to the Five-Element theory just explained, (one can determine) the substances which are of the same category as 'Red Marrow' (*chhih sui*[9]).
   [Comm.] The Red Marrow of Thai Yang is Thai Yang quicksilver; it is mercury extracted from cinnabar. Another name for it is *Thai Yang hung*.[10] Thai-yang is a synonym of cinnabar.[c]
The Greater, or mature, Yang, realgar, because of its male essence, contains and receives it; the Greater, or mature, Yin, orpiment, as a coagulated liquid (*ning chin*[11]) is categorically similar to it.[d] This is the great Tao of the Cyclically Transformed Elixir.
   [Comm.] Male Essence is realgar, Yin Liquid is orpiment. Ta Huan Tan (Great Cyclically Transformed Elixir) is another way of saying it.
Furthermore, realgar is of the same category as sal ammoniac (*nao sha*[12]).
   [Comm.] When realgar and sal ammoniac are heated together in the same pan a blood-coloured liquid will be formed after a short while. Heating must then be continued for a day and a night in a closed vessel to effect the subduing. From this we know that the statement is true.[e]
Without sal ammoniac, realgar will not develop the proper colour...

With the aid of Lu Thien-Chi's commentary Ho & Needham were able to recognise more or less confidently in the text mentions or descriptions of the following operations:
   (*a*) Interconversion of the metals and their sulphides (mercury and iron).
   (*b*) Amalgamation of mercury in a variety of different ways with gold, silver, copper, tin, iron, zinc and lead; in one case seemingly involving organic copper compounds, and frequently in the presence of the sulphides of arsenic.

---

[a] Cf. pp. 132 ff. above.
[b] *TT*898, p. 2*b*, tr. auct. adjuv. Ho Ping-Yü & Needham (2), p. 180.
[c] *Shih Yao Erh Ya*, ch. 1, p. 1*b*, gives Thai-yang as a synonym of mercury as well as of cinnabar.
[d] In Five-Element theory the Greater, or mature, Yang, and the Greater, or mature, Yin, correspond of course to Fire and Water respectively.
[e] Perhaps the chloride acted to prevent oxidation of the fused arsenic disulphide.

[1] 麴   [2] 糱   [3] 蘖   [4] 同類   [5] 非種   [6] 六律
[7] 同類   [8] 刻定同類   [9] 赤髓   [10] 太陽汞   [11] 凝津
[12] 硇砂

(c) Formation of the acetates of mercury and silver.

(d) Sublimation of the chlorides of mercury, and preparation of lead carbonate.

(e) Formation of sulphides and sulphates of mercury by treatment with alum.

(f) Treatment of arsenic disulphide with ammonium chloride.

(g) Association of arsenic trisulphide with alkaloid-containing liliaceous corms.

Thus some of the processes discussed were of a quasi-metallurgical nature reminiscent of the Western developments from the chemical technology of the Leiden papyri and the Hellenistic proto-chemists. But others were more characteristically Chinese.

Thus the writer of the *Wu Hsiang Lei Pi Yao* clearly believed that for two things to react there must be something similar about them. But this seems at first sight to conflict with the expectation that if all things in the universe belonged either to the Yin or the Yang, reactions might more naturally occur between things of opposite sign in this sense. And that indeed was the ground on which the late Tenney L. Davis based his favourite theory (by no means necessarily erroneous)[a] of an identity of principle in the foundations of Chinese and Western alchemy.[b]

Davis and his collaborators had no difficulty in finding occidental comparisons for the mating of contraries in the Great Work, the marriage of Sol and Luna, of sophic sulphur and sophic mercury, under a hundred synonyms; nor did they lack Western texts which emphasised the maleness and femaleness of the fundamental essences in a strangely Chinese manner.[c] Though few would now subscribe to Davis' belief that the sulphur–mercury theory occurs already in the Hellenistic proto-chemists (+1st to +4th centuries), it is certainly flourishing in the Jābirian corpus (+9th and +10th centuries), perhaps derived (as some think) from the two mineral exhalations of Aristotle,[d] perhaps rather from the antitheses of China.[e] Then it continues to flourish in the Geberian books (late +13th and +14th centuries) and so comes down to Paracelsian times. On the other hand statements of sexual type undoubtedly occur in the Greek writings. In Zosimus we find the fundamental aphorism twice—'Above what is heavenly, below what is earthly; by the male and the female the work is accomplished,'[f] and again 'Mary said: "Join the male and the female and you will find what you are seeking"'.[g] These quotations were made early in the +4th century and the latter is ascribed to the +1st. Then there is another epigram attributed by Olympiodorus (c. +420) to Mary (+1st century): If you do not make corporeal substances incor-

---

[a] Cf. pp. 454ff. below, 491, and pt. 2 above, pp. 6–7.

[b] See especially Davis & Chhen Kuo-Fu (2), and the discussions in Davis (2, 3, 4, 5). Davis (6) has been untraceable by us.

[c] Davis brought under contribution, e.g. Basil Valentine, the *Speculum Alchemiae, Le Texte d'Alchymie*, and Norton's *Catholicon*.

[d] Cf. Vol. 3, pp. 636ff.

[e] We leave on one side for the present the fascinating question of Chinese influence on the Shi'ite and Isma'ilite writers of the Jābirian books. It would have taken at least two forms, the emphasis on immortality and longevity elixirs as opposed to chrysopoiesis as such, and a powerful reinforcement of chemical dualism. See further, pp. 457ff.

[f] Berthelot & Ruelle (1), vol. 3, p. 147. This aphorism also occurs as a kind of caption below a picture of alchemical apparatus in the +11th-century Paris MS. 2327, f. 81v; see vol. 1, pp. 161, 163. *Corp. Alchem. Gr.*, III, x, end.

[g] Berthelot & Ruelle (1), vol. 3, p. 196. Berthelot considered this text as due to a Pseudo-Zosimus of the +7th century but Sherwood Taylor (2) accepted it as genuine. *Corp. Alchem. Gr.*, III, xxix, 13.

poreal and vice versa, and if you do not turn two bodies into a single one, none of the results you hope for will be produced.'a

But whatever may be the case with the Western alchemists this was always only half the story for the thought of the Chinese. Besides something like the marriage of contraries there was the firm conviction that *similia cum similibus agunt*. These two principles were combined in the thought that substances of opposite sign will react only if they belong to the same category (*lei*[1]). This is most clearly explained in the *Wu Chen Phien*[2] (Poetical Treatise on the...Primary Vitalities), written in +1075 by Chang Po-Tuan[3] (+983 to +1082), patriarch of the Southern School of Sung Taoism and one of the most authoritative exponents of internal alchemy.b It was from this work that Davis drew particular support for his theory, though the book is concerned not with laboratory operations at all but with physiological and sexual disciplines.c As Chang Po-Tuan put it, 'Yin and Yang (things), if they are of the same category, respond and interact with each other.'d On this Chhen Chih-Hsü[4] (Shang Yang Tzu[5]), two and a half centuries later,e commented:

What is meant by 'categories' is the partnership of Heaven and Earth, the complementarity of the moon and sun, and the mutuality of female and male; hence it follows that mercury must require lead as its category partner.

Later on Chang Po-Tuan says in homely analogy:f

For repairing something made of bamboo, bamboo must be used. If you want a hen to hatch chickens, they must come from eggs. No matter what you are doing, if it is not based upon the classification of things by category (*fei lei*[6]) it is a complete waste of energy.

And the commentators elaborate the point at length. Thus the Chinese alchemists had in their minds a kind of table divided one way into Yin and Yang signs and the other way into a series of categories.g

It only remains to construct such a chemical table according to the explicit statements of the *Wu Hsiang Lei Pi Yao*. No less than fifteen of these *thung lei* categories are given in the main text, and five further statements can be collected from other parts of the work.h All are shown in Table 120. When the sign of a substance is not

---

a Berthelot & Ruelle (1), vol. 3, p. 101; *Corp. Alchem. Gr.* II, iv, 40. Elsewhere, p. 124 (III, iv), the first part of the aphorism is attributed to Hermes, who is placed in the +2nd century by Sherwood Taylor (2). Cf. Festugière (1), p. 242; and further, below, pp. 360ff.

b See Vol. 5, pt. 3, pp. 200 ff. c See pt. 5 below.

d Ch. 8; cf. Davis & Chao Yün-Tshung (7), p. 104. We may remember, moreover, the pregnant statement of the +11th- or very early +12th-century commentator of our own text, the *Wu Hsiang Lei Pi Yao*, given on p. 317 above. e See pt. 3, pp. 206ff.

f Ch. 25. Cf. Davis & Chao Yün-Tshung (7), p. 106.

g An elaborate +14th-century table of this kind, constructed with the concepts of physiological alchemy by Chhen Chih-Hsü, is given in Ho & Needham (2). We shall reproduce it and discuss it in pt. 5.

h It will be remembered that the *Chin Shih Wu Hsiang Lei* lists twenty categories (p. 310 above). The entries and definitions are not by any means all the same, but the near coincidence of the total number may not be entirely fortuitous.

---

[1] 類 [2] 悟眞篇 [3] 張伯端 [4] 陳致虛 [5] 上陽子
[6] 非類

clear from the text itself, it may follow from the substance with which it is paired, or it may be obtainable from other texts. We wish to remark only on two especially interesting entries, nos. 8 and *b*, which indicate clearly there was a graduation of Yang-ness and that while mercury might be female to sulphur it would act as male to silver. Hence further vistas of complication open up, and we may expect to find in course of time due explorations of them in other medieval Chinese texts. After all, this would have been only a natural development from an idea of great antiquity, embodied to this day in the well-known Yin–Yang symbol, familiar everywhere, with a Yin heart to the Yang and a Yang heart to the Yin (cf. p. 379). It would have been related to a doctrine of the inseparability of the Yin and the Yang, the idea that there could hardly be anything in the phenomenal world so Yang as not to have some Yin within it and vice versa.[a] Bodies thus ascended to a quasi-quantitative plane, at least in theory, taking their places on a hierarchical scale in accordance with their Yin–Yang *mistio* or *krasis* (κρᾶσις), so that a given substance might act as Yin in one relation and Yang in a second, forming a succession intuitively analogous to the electrochemical series of the elements, the order in which they displace one another from their salts.[b]

Table 120. *Chemical categories from the 'Wu Hsiang Lei Pi Yao'*

Lei	Yang	Yin	Neutral
1	cinnabar	mercury	
2	realgar	orpiment	
3	realgar	sal ammoniac	
4	honey and fritillary corms[c]	orpiment	
5		arsenious acid	clay
6	sulphur	magnetite	
7	sulphur	mercury	
8	mercury	orpiment	
9	cinnabar	vinegar	
10	lead	mulberry ashes	
11	litharge	tin	five coloured clays
12	cinnabar	bronze coins	
13	Persian brass fragments[d]	mercury	
14	copper carbonate	mutton fat	
15	blue copper carbonate	red haematite	
a	lead	calomel	
b	mercury	silver	
c	red salt	alum	
d	copper carbonate	silver	
e	red salt	calomel	

[a] Cf. on this Vol. 2, p. 276 and Fig. 41; and Davis (5), p. 85.

[b] Strangely reminiscent, too, of the series of intersexes found in many invertebrates, especially molluscs, and some micro-organisms.

[c] *Pei mu*,[1] almost certainly *Fritillaria Roylei* (R678). This plant contains a number of active alkaloids and is still used in Chinese medicine; cf. Chang Chhang-Shao (1), pp. 78ff.; Lu Khuei-Shêng (1), pp. 95ff.   [d] Cf. pt. 2, p. 202, pt. 3, p. 136 above.

[1] 貝母

The question may now be asked, is there any representative in modern chemical theory of the doctrine that 'things of similar category go together'? Transmuted into terms of the deeper insights of today the doctrine in some sense still persists, for valencies in chemical combination have to be equivalent. A trivalent atom for instance will not combine wholly with a univalent one, and in so far as atoms can be octavalent there are eight categories to consider. Combining atoms also have to be similar in that they are both in a position to share an electron; some elements such as the inert gases, with their fully filled shells, are unable to co-operate in this way. The doctrine of 'similar categories' seems to have joined its opposite in the higher synthesis of modern chemistry, for while many examples of ionic bonds, some very strong, exist, there are also many examples of covalent bonds between atoms of the same kind.

At the beginning of the eighteenth century modern chemistry set out, as Metzger has so well shown, from the proposition that like attracts like.[a] This was one of the great watchwords of Stahl[b] and his disciples such as Juncker,[c] who preferred it to the Cartesian 'mechanical' system of a union of opposites. They applied their notion of affinity to the assumed indwelling 'earthy', 'aqueous', 'mercurial', or 'sulphurous' (phlogistic) principles rather than to the chemical substances themselves.[d] Newtonian gravitational attraction was then brought in to explain these affinities, and the rest of the century was occupied with the establishment and then the destructive criticism of the doctrine.[e] It thus ended with a dominance of the opposite point of view, represented, for example, in the *De Attractionibus Electivis* of Torbern Bergman (+1775);[f] not only the idea of which, but the very name, was translated back to the human sphere in the title of Goethe's famous novel *Wahlverwandtschaften*, the 'Elective Affinities'.[g]

In the nineteenth century it was only with the greatest difficulty that the prejudice against envisaging combinations of identical atoms was overcome. The tradition of Davy and Berzelius was that only atoms with opposite electrical charges could form compounds together. Thus Canizzaro had great labour in convincing chemists of the truth of Avogadro's law because it involved double molecules such as O_2, N_2, etc., i.e. combinations of like with like.[h]

But of course there is something artificial in stretching the vague conceptions of antiquity on the Procrustean bed of modern scientific theory, and it is best not to push such comparisons too far. Moreover, as we have already said, the doctrine of 'similar categories' was built up on fancied resemblances and imaginative classification, not on critical observation and experiment. Yet in its two-dimensional or 'matrix' character it did find room for both theses of the perennial contradiction—that only opposites unite, and that similar things alone react together.

---

[a] (1), esp. pp. 139 ff.

[b] Cf. Partington (7), vol. 2, p. 665 (+1660 to +1734). The most relevant passages are in his *Specimen Beccherianum* (+1738 ed.), p. 13, and *Traité des Sels* (+1783 ed.), p. 304.

[c] Partington, *op. cit.*, p. 688 (+1679 to +1759).

[d] In Stahl, mercury forms amalgams because metals contain 'mercury', metals dissolve in nitric acid because both contain phlogiston; acids unite with bases because both are 'salts', etc.

[e] Cf. Metzger (1), p. 52.; Dobbs (4), pp. 209 ff.

[f] Cf. Carlid & Nordström (1), containing the biography by Olsson.

[g] Cf. Walden (3), pp. 56, 115; and the special study of Adler (1).

[h] The history of the modern concept of chemical composition and combination has been written in

Looked at in another way, the old Chinese theory of categories is a hitherto un-recorded chapter in the prehistory of the conception of chemical affinity. According to Partington (3) the word *affinitas* was first employed with chemical meaning by the great Dominican scholar-naturalist Albertus Magnus (+1206 to +1280), a contem-porary of Phêng Ssu[1] and Mêng Hsü,[2] the writers of the famous book on the 'Golden Flower Elixir'.[a] It is evident from reading over the list of 'similar category' substances given by the *Tshan Thung Chhi Wu Hsiang Lei Pi Yao* that although the writer had no way of distinguishing between chemical reaction and physical change or mixture, he did put together pairs or groups of substances which he observed to react in one way or another. Was he not thus the ancestor of E. F. Geoffroy, whose first affinity table of +1718 served as the model for so many others,[b] preparing the way for the synthesis of Lavoisier? Men such as Guyton de Morveau (from +1772 onwards)[c] were still struggling with essentially the same problems as the writer of our text.

Again, the old Chinese theory of categories seems to take its place in the linear ancestry of the idea that things can be arranged in chemical classes the members of which are susceptible of chemically similar processes. These words are Sherlock's, in his acute study (1) of the contribution of the *Archidoxis* of Paracelsus (+1526), a book which described the preparation of a series of coloured chlorides and nitrates of the metals. Here the great advance was the first conception of a generalised method for making a number of analogous preparations.

The idea that things which belonged to the same class, or to the same phase of a cyclic process, resonated with, or energised, each other, though so characteristic of Chinese thought, was not without parallels in ancient Greece.[d] Cornford (2) has detected these in what he calls the maxims of popular belief accepted by the philo-sophers from 'common sense' without scrutiny. Take Aristotle's three kinds of change. Movement in space was explained by asserting that like attracts like; growth, by asserting that like nourishes like; and change of quality, by asserting that like affects like. To quote from Aristotle, 'Democritus held that agent and patient must be the same or alike; for if different things act upon one another, it is only accidentally by virtue of some identical property.'[e] But there was also an opposite set of maxims that like things repelled one another—'Everything desires, not its like, but its contrary' to quote from Plato.[f] All this has an evident relationship with the ideas of the pre-Socratics about 'love' and 'hatred' in natural phenomena, and it would be easy to see the origin of it in social practices, exogamy or endogamy, sympathetic magic, and so on.[g] Among the Chinese philosophers of the Warring States period closely similar

a classical monograph by Ida Freund (1). Benfey (2) has reprinted a selection of the most important papers on the subject.

[a] See Ho & Needham (3), and here, pp. 35, 44, 275, 285.
[b] Cf. Duncan (1, 2, 3). For Bergman's Tables of Affinity see Freund (1), pp. 114ff.
[c] Cf. Smeaton (1).
[d] That 'like things have an affinity for one another' is in Hippocrates, *De Morbis*, IV, 7. Cf. p. 360.
[e] *De Gen. et Corrupt.* 323b 10. His own views were complex, subtle and difficult, cf. the study by Stephanides (2).
[f] *Lysis*, 215c. But in *Tim.* 57c he puts just the opposite view, the Hippocratic, inconsistent as usual.
[g] The interesting book of Lloyd (1) on polarity and analogy in Greek thought, now available, appeared too late to be of help to us.

[1] 彭耜   [2] 孟煦

conceptions were current.[a] The point to be emphasised here is that while Greek thought moved away from these ancient ideas towards concepts of mechanical causation foreshadowing the complete break of the Scientific Revolution, Chinese thought developed their organic aspect, visualising the universe as a hierarchy of parts and wholes suffused by a harmony of internal necessities.[b] In this development the Chinese alchemists participated according to their lights, though their contributions to chemical discovery and invention, certainly not less than those of other civilisations, remained until the end of a typically pre-Renaissance character. Yet after all, the dimensional analysis of current scientific concepts[c] is showing how again and again in many different fields the two thought-patterns 'unlikes attract', and 'birds of a feather flock together', underlie the most recondite and sophisticated theories. Perhaps, as categories of thought itself, they always will.

(i) COMPARATIVE MACROBIOTICS

It seemed impossible to conclude this part of our work without some account of the general course of events throughout the Old World civilisations, for neither Chinese nor any other culture can usefully be thought of in total isolation. But do we yet know enough to demonstrate, or even to suggest, that the great intellectual adventure of proto-chemistry and alchemy, aurifiction, aurifaction and the elixir, was really one single movement, even with separate foci of origin, during the past three thousand years? Many scholars (even some among our own group of collaborators) would be inclined to say that it is too soon to attempt this, and that a good few decades must yet pass before sufficient understanding of the different traditions has been attained, and enough information is in. Nevertheless, I feel that there is already something to be said, chiefly by way of a comparative prospect of Chinese alchemy, Hellenistic proto-chemistry, and Arabic and Latin alchemy; and this is what the following pages contain.

The subject has been on the agenda for quite a long time. Theophilus Spizel, in one of the earliest European sinological books, *De Re Literaria Sinensium* (+1660), noting the ubiquity of the search for material immortality in Chinese culture, and the harm done by dangerous elixir preparations, as also the widespread belief in argentifaction,[d] agreed with H. Conring's view of +1648 (*De Hermet. Med.*, ch. 26) that they had probably got all these ideas from the Saracens.[e] As we now know, the case was just the opposite. Bernard Varenius, too, in his description of Japan and Siam, published at Cambridge in +1673, gave an account of the Taoists and their Pope of the Ciam (Chang) family, remarking how many of them worked and wrote on elixir alchemy.[f] Isaac Vossius, for his part, admired in +1685 the chemical knowledge of the Chinese, which he said had been growing for two thousand years if not the four thousand six hundred which some claimed for it; and he knew that the activities of their numerous

[a] Cf. Vol. 2, p. 39.

[b] Anthropomorphically, a harmony of wills, the spontaneous co-operation of all individual beings and things.

[c] Cf. Benfey (1).

[d] Cf. Francis Bacon on this, Vol. 5, pt. 2, p. 33 above.

[e] Pp. 259 ff.

[f] Pp. 260 ff., esp. p. 262. This was the Varenius whose *Geographia Generalis* had been edited and published by Isaac Newton in the previous year.

alchemists were directed not so much to the making of gold as to the pursuit of long life and immortality, even if the longevity of the most ancient men had been due to other causes, and the alchemists could never benefit themselves from what they so liberally promised to others.[a] Yet he joined with the new, or experimental, philosophers, in reproving the secrecy which the alchemists, both in East and West, maintained about their arcane mysteries.[b]

A century later, de Pauw devoted a whole chapter of his *Recherches Philosophiques*... (+1774) to a discussion of the state of chemistry among the Egyptians and the Chinese.[c] De Pauw did not know the Alexandrian Corpus from Pizzimenti's edition (+1572, +1717) as he might have done, so he thought of Egypt purely in terms of the technical arts such as glass and gilding, denying at the same time to the Chinese anything more than empirical industries such as gunpowder and porcelain. He believed, on the other hand, that the idea, or 'folly', of the elixir of life had come to them from the 'Tartars' their ancestors, by which he meant the Scythians and the Persians, whose *haoma* (*soma*) he knew of. As we have already seen (pt. 2, p. 121) there was a grain of truth, if only a grain, in this. Like J. C. Wiegleb, a few years later (+1777), whose *Historisch-kritische Untersuchung der Alchemie* was the book which more than any other gave the death-blow to the belief in aurifaction in Europe,[d] de Pauw knew of the attribution of alchemy to Lao Tzu.[e] But none of them had any idea of the wealth of real Chinese alchemical experimenters and writers contemporary with the Graeco-Egyptian proto-chemists, and even prior to them.

(1) CHINA AND THE HELLENISTIC WORLD

(i) *Parallelisms of dating*

It has not so far been generally appreciated by historians of chemistry that the succession of Greek-speaking proto-chemists in the Mediterranean region was closely paralleled by a line of proto-chemists, or, strictly speaking, alchemists (cf. pt. 2, p. 12) in China. This is the first point which needs to be examined. And the first problem which presents itself is that of the identity and date of the oldest writer of the Greek

---

a *De Artibus et Scientiis Sinarum*, in *Variarum Observationum Liber*, ch. 14, p. 77. The passage runs: 'Chemiam jam a bis mille annis apud Seras in usu fuisse constat. Quod si ipsos audiamus Chemicos, illi jam a sexcentis supra quater mille annis ejus arcessunt antiquitatem; nec aliunde primorum hominum longaevitatem quam hujus scientiae beneficio provenisse affirmant. Doctiores tamen Medici id genus hominum cum omnibus suis figmentis arcanis, in apertam non audentibus prodire lucem, strenue contemnunt. Nusquam plures invenias Chemicos quam apud Seras, non divitias tantum, sed et immortalitatem quoque promittentes, eaque aliis liberaliter spondentes, quae sibi ipsis praestare nequeunt.'

b Recently Rossi (2), in an interesting paper on the 'equivalence of intellects', has emphasised the importance of a democratic estimate of human capacities, and a conviction of the universal accessibility of truth, in the Scientific Revolution; as opposed to the aversion of the learned magi in all previous civilisations to revealing their knowledge to the *promiscuum hominum genus*. Hence the 'plain, naked, natural way of speaking' exacted by the early Royal Society (cf. Lyons (1), p. 54).

c In his vol. 1, pp. 376ff. Cf. pt. 3, pp. 227ff. above. The work was intended partly to explode the earlier thesis that 'the Chinese were a colony of the Ancient Egyptians' (de Guignes, +1760, J. T. Needham, +1761; cf. Vol. 1, p. 38). Partly also it was anti-Jesuit propaganda.

d Cf. Ferguson (1), vol. 2, p. 546.

e Wiegleb (1), pp. 184, 211–12, de Pauw (1), vol. 1, p. 431.

proto-chemical Corpus, Pseudo-Democritus;[a] a question which we have touched upon already at two places in the preceding sub-sections.[b]

There is no doubt that the man who stands at the head of the tradition which in due course gave rise to Hellenistic proto-chemistry is Bolus of Mendes, a city in the Nile delta. Almost certainly he called himself Bolus the Democritean,[c] a circumstance which helped to confuse him with the later proto-chemical writer. Although not one of his books has survived, he was a prolific author,[d] producing works on agronomy (*Georgica*, γεωργικά), medicine (*Technē Iatrikē*, τέχνη ἰατρική), prodigies (*Thaumasia*, θαυμάσια), entertaining magic (*Paignia*, παίγνια), military science (*Tactica*, τακτικά), morality (*Hypomnēmata Ethica*, ὑπομνήματα ἠθικά), and even a book on the history of the Jews. More important for us were his treatises on sympathies and antipathies (*Physica Dynamera*, φυσικὰ δυναμερά, or *Peri Sympatheiōn kai Antipatheiōn*, περὶ συμπαθειῶν καὶ ἀντιπαθειῶν) and on artificial sympathetic remedies (*Cheirokmēta Dynamera*, χειρόκμητα δυναμερά); these we have already compared with the *Huai Nan Wan Pi Shu*[1] in China (pp. 311 ff.). Still more important is it that Bolus wrote something on dyeing and tinting (*Baphica*, βαφικά), that subject so vital for the first proto-chemists. It may even have been in four books, like those *biblous tessaras baphicas* (βίβλους τέσσαρας βαφικάς) which Synesius in the +4th century ascribed to Ps.-Democritus,[e] dealing respectively with gold and silver colourings, the tingeing of 'gems' and glasses, and textile dyeing, especially purple. Thus some trace of incipient proto-chemistry can certainly be ascribed to Bolus of Mendes—exactly how much we may well never know.

In the −1st century Bolus was regarded as an authority of equal rank to Aristotle and Theophrastus. He must be of the −2nd century, and probably of the first half of it, for his agronomic fragments have reached us through Cassius Dionysius (−88), the botanical ones through Krateuas (*c.* −100), the zoological ones through Juba of Mauritania (*c.* −50 to +50), and the philosophical ones through Poseidonius of Apamea (−135 to *c.* −51).[f] From the +1st century a great number of writers mention and quote him. This being established, could he have been the same person as the Ps.-Democritus who heads the series of writers in the Corpus? The question has been a controversial one, and we must be content with referring to a few of the arguments on either side which have been brought forward.[g] Those who plead for

---

[a] Oldest because he quotes no one (except Ostanes and Pammenes) and is quoted by everyone. He is always called pseudonymous because no question arises of identifying him with the pre-Socratic philosopher of the early −4th century.

[b] Pt. 2, p. 17, pt. 3, p. 48.   [c] Festugière (1), p. 197.

[d] The abundant literature on the subject is discussed by Festugière (1), pp. 42 ff., 187 ff., 196 ff. See also Partington (7), vol. 1, pt. 1, pp. 211 ff.

[e] *Corp. Alchem. Gr.*, II, iii, 1. Ct. Bidez & Cumont (1), vol. 2, p. 311.

[f] The best attempt to fix his date as near as possible is that of Wellmann (3), who studied the *Georgica* minutely and collected its fragments. No one quotes Bolus earlier than Skymnos of Chios (*fl.* −185), and a recipe from the *Paignia* appears in a text of the physician Menander (*fl.* −197 to −159). The most probable conclusion is that Bolus of Mendes was a close contemporary of Aristophanes of Byzantium, a grammarian who was also interested in the sciences (d. −180), and that both of them worked in Alexandria.

[g] The state of the case has been well reviewed by Festugière (1), pp. 220 ff., 230 ff., 237 ff. Those in favour of identity have included Diels (1), pp. 127 ff.; von Lippmann (1), p. 329. Those against:

[1] 淮 南 萬 畢 術

the identity have to face the awkward fact that the name of Bolus never once appears in the writings of the Corpus; always the references are just to the name Democritus. Perhaps the true name of this first author of the group was lost because, like Bolus, he called himself 'the Democritean'. Next, from internal evidence his Corpus texts cannot be earlier than the +1st century. For example there is mention of a gold-like brassy alloy called *claudianon* (κλαυδιανόν), and this must be a reference to the emperor Claudius (r. +41 to +54).[a] So also the red dye from India called lac (*laccha*, λακχά) is referred to,[b] and this could hardly be before the first decades of the Christian era. Ps.-Democritus also complains to his colleagues (*symprophētai*, συμπροφῆται) about the young operators (*neoi*, νέοι) not wanting to follow the scriptures, but presumably to experiment on their own,[c] and these earlier writings might indeed have been those of Bolus or his time. Some of the axioms or epigrams which we shall consider a little more fully below occur in Ps-Democritus texts, e.g. 'One nature is charmed by another nature, one nature triumphs over another nature, one nature dominates over another nature';[d] and it has been argued that these crystallise exactly the doctrine of Bolus in his sympathies and antipathies of natural things,[e] but X the Democritean could have been his follower or reader, extending and applying his ideas to chemical phenomena. He certainly never mentions Bolus, ascribing his illumination rather to his master Ostanes the Mede. All in all, therefore, it seems best to place Ps.-Democritus firmly in or at the beginning of the +1st century,[f] just conceivably towards the very end of the −1st, and to conclude that his *Physica kai Mystica* (φυσικὰ

Wellmann (2) and Bidez & Cumont (1), vol. 1, p. 198, who all placed Ps.-Democritus in the +2nd century; also W. Kroll (1) and Hammer-Jensen (2), whose +5th-century date is much too late for him; Preisendanz (in Pauly–Wissowa, vol. 18, pt. 2, col. 1629); Partington (7), vol. 1, pt. 1, p. 214, and priv. comm. 1959. Festugière, *op. cit.*, gave most of the arguments (which we can only briefly summarise here) but inclined to believe that the Ps.-Democritus texts were due to Bolus, at any rate in an earlier form. Berthelot (1), p. 99, (2), p. 201, also hesitated, but he always put Bolus too late, in the −1st century.

---

a *Corp.* II, i, 7.

b *Corp.* II, i, 2. This is not absolutely decisive by itself, for as Filliozat (5) pointed out, pepper (albeit with a Persian form of the name) is mentioned in the Hippocratic Corpus ('On the Diseases of Women', Littré ed. vol. 8, p. 394). But it still has some weight. *A fortiori* the +3rd-century chemical-technological papyri show evidence of Indian connections; these have been particularly studied by Hammer-Jensen (1). In considering the numerous recipes for false gems in the Stockholm papyrus, it should be remembered that Pliny says (*Hist. Nat.* XXXVII, 79) that the art began in India, though this may well have been a confusion with the real gems of Ceylon. He also says (XXXVII, 197) that there were manuals for making such coloured glasses, bearing the specific names of authors. The *tabasios* (ταβάσιος) in the Stockholm papyrus is taken to be tabasheer, i.e. the silicic acid concretions in bamboos, which would certainly have come from India. The indigo recipes, the rice decoctions, and the use of several different kinds of milk, are all suspiciously Indian, as also the use of a basket in a steambath (like the ancient steamers so characteristic of China, cf. p. 27). All the material on pearls, Hammer-Jensen thought, is probably Indian, and Pliny certainly believed that they were the first to make artificial or imitation pearls (cf. Vol. 4, pt. 3, pp. 674 ff.). Finally Flavius Vopiscus, writing about +300, says that Aurelian, Probus and Diocletian all sent dyers to India to learn how to make the false purple, in three or four expeditions, but they could never get the secret. Since the three reigns covered the period +270 to +305, he may well be worthy of belief.

c *Corp.* II, i, 14.

d *Corp.* II, i, 3. Cf. p. 360 below. When the axiom is quoted again by Synesius in II, iii, 1, it is attributed directly to Ostanes, whose principles Ps.-Democritus adopted.

e Festugière (1), p. 231.

f Leicester (1), p. 40, concurs, though placing Bolus of Mendes also in the +1st century, which is impossible.

καὶ μυστικά) the fragments of which we still have in the Corpus, was a text quite different from anything written by Bolus of Mendes.

This being once decided, and the *floruit* of Bolus at *c.* − 175 being accepted, the rest of the Corpus falls reasonably into place.[a] The period down to + 200 is filled by a number of names, none of which can be earlier than Ps.-Democritus but none much later—Comarius, Pseudo-Cleopatra,[b] Mary the Jewess, Pelagius, Pebechius, Petasius, Petosiris,[c] Pammenes, Panseris, etc.[d] To the following, + 3rd century, belong probably a number of fragments bearing the names of Hermes, Agathodaemon,[e] Iamblichus[f] and Isis; those of Africanus certainly, for he was a perfectly historical character who died in + 232 (cf. pt. 2, p. 16). This was also the time, it will be remembered, of the chemical-technological aurifictive papyri (pt. 2, p. 20 above), the connections of which with the Corpus we have already discussed. With Zosimus of Panopolis (a city up the Nile to the south) we are again on firmer ground, for this great codifier was certainly writing between + 280 and + 320; a historical position strangely close, as we shall see in a moment, to the first great codifier of alchemy in China. The next century brings Synesius,[g] whose writings must have been completed before + 389; while a hundred years later there follows the Neo-Platonic chemist Olympiodorus, whose work must date from the close neighbourhood of + 500. Late in the + 6th century there is the Philosophus Christianus, whose personal name has been lost, and then in the + 7th there was more intense activity. For it is to this time that another great proto-chemical writer is to be ascribed, Stephanus of Alexandria (*fl.* + 620, under the emperor Heraclius),[h] as well as the Philosophus Anonymus,[i] the chemical poets Heliodorus, Theophrastes and others,[j] and in all probability whoever it was that wrote the 'Domestic Chemistry of Moses' (cf. p. 345).[k] By this time we are past + 700.[l] Exactly a century later comes the historian Georgius Syncellus, an important witness, as we shall see (pp. 339, 341); then during and after the + 9th century the tradition is in full

[a] Cf. Berthelot (1), pp. 98ff., 127ff.

[b] Berthelot (1), pp. 111, 173. In the view of Hammer-Jensen (2) the texts of these two are the earliest parts of the Corpus, but that has not been generally accepted.

[c] Berthelot (1), p. 168.

[d] 'Ostanes' (Berthelot (1), p. 163) would come in this group, but of him more later.

[e] This name fluctuates between a god or spirit, a mythical or sacred animal, and a mortal human writer; cf. pp. 344–5, 375.

[f] Not the same as the Neo-Platonic philosopher who lived under Constantine the Great (+ 306 to + 337).

[g] Probably not the same as the famous bishop of Ptolemais (Festugière (1), p. 239; Berthelot (1), p. 188).   [h] Berthelot (2), p. 287.

[i] The apocryphal fragments attributed to John the Archpriest also belong probably to this time.

[j] Possibly in fact a single writer; Festugière (1), p. 239; Hammer-Jensen (2), pp. 30ff.

[k] This by no means exhausts the list of names, some of which are curious (cf. Berthelot (1), pp. 121, 125), but these must suffice. Moses reappears in Ben Jonson's *Alchemist* (+ 1610), where Mammon says (p. 373):

> Will you believe Antiquity? Records?
> I'll shew you a Book, where Moses, and his Sister,
> And Solomon have written of the Art;
> Aye, and a Treatise penn'd by Adam. *Surly:* How?
> *Mamm.:* O' the Philosophers Stone, and in High Dutch.

[l] This is the point at which, plus or minus half a century or so, the first collections of the Corpus writings were made.

decay, preserved only by commentators and quoters like Photius or the great lexico-grapher Suidas (*c.* +98). There were one or two minor and muddle-headed writers to follow, such as Michael Psellus (*c.* +1050) and Nicephoras Blemmydes, but by this time the Corpus had been collected into forms very like those which we still have, and in the +11th century the oldest of our extant MSS were written out.[a] Such was the course of the linguistically Greek tradition of proto-chemistry. It has to be taken as representative of the Europe of those ages, for there was nothing corresponding to it in the Latin West until the period of the practical chemical-metallurgical manuals.

Of these the oldest seems to have been the *Compositiones ad Tingenda Musiva*... (Preparations for Colouring Mosaics, etc.),[b] translated into Latin by some Lombard not earlier than +750, indeed nearer +780, from an Alexandrian Greek text of *c.* +600, the time of Stephanus of Alexandria. Then came the *Mappae Clavicula* (Little Key to Painting),[c] written in Latin in the early +9th century, perhaps about +820. A work attributed to one Heraclius may again have been Greek in its first recension; this was the *De Coloribus et Artibus Romanorum* (On the Colours and Arts of the Romaioi, i.e. the Byzantines), the first Latin version of which dates from about +1050, though its final form was not reached until the end of the +12th century.[d] Fourth comes the very Latin *De Diversis Artibus* (On Various Techniques) by a monk whose name in religion was Theophilus Presbyter, almost certainly Roger of Helmarshausen;[e] this work is not as early as was once thought, but belongs to the close neighbourhood of +1130. It is important to notice how the tradition taken up by the practical Latin West direct from the Hellenistic world was the technological one of the papyri (which might involve aurifiction but was not concerned with aurifaction) —not at all the mystical one of the Corpus. This latter found its way only indirectly to the Latin West after the beginning of the great era of translations from the Arabic, and that took place, as it happened, just during the lifetime of Roger of Helmarshausen. Only thereafter did the concentration on the practical problems of gilding, dyeing, and working glass and metals, yield to the wilder, more exciting, dreams of actually succeeding in making gold from something else, and of preparing, in connection with it, an elixir medicine of at least extreme longevity.[f] Looking back on the Hellenistic proto-chemical tradition it is very striking that the collapse of pagan Mediterranean culture did not harm it at all. Even the destruction of the Library and Museum at Alexandria in +389, the Serapeum, centre of science and learning in the ancient world, failed to do it much damage. Presumably this was because the concept of aurifaction presented vistas of utility to anybody who accepted it, whether pagan or Christian, while there could be no question about the usefulness of the practical chemical-

---

[a] They first saw print in the Latin translation of D. Pizzimenti (Padua and Cologne, 1572, 1573); cf. Ferguson (1), vol. 1, p. 205 on the reprint of +1717. This was doubtless how they became known to R. Bostocke, one of the first to attempt a history of medical chemistry, in his *Difference between the Auncient Physicke...and the Latter Physicke* (London, 1585). On this see Debus (12, 19), Pagel (13). Bostocke's book was also one of the first to defend Paracelsian theories in England, so the Alexandrians came in handy.

[b] Hedfors (1); Burnam (1); Johnson (2).   [c] Phillips (1); Johnson (1).

[d] Merrifield (1); Ilg (1).   [e] Ilg (1); Dodwell (1); Hawthorne & Smith (1).

[f] Cf. pt. 2, p. 30 above, and p. 493 below.

technological arts—and the fact that they continued on their way of slow development is shown by the first successful production of 'Greek fire' (low boiling-point petroleum fractions) by Callinicus at Byzantium in the middle of the +7th century, just after the time of Stephanus.[a]

Having thus passed in review the European line of evolution of proto-chemistry we can set beside it what took place in China.[b] At the head of the list stands the natural philosopher Tsou Yen,[1] not because of any detailed alchemical writings which have come down to us, but because of the evidence that among the techniques possessed by his school was one for 'prolonging life by a method of repeated transmutation'. The chemical thaumaturgists of the −2nd century certainly traced back their filiation to him and his disciples. Thus he challenges comparison with Bolus the Democritean, though the dates differ, for Tsou Yen undoubtedly lived within the period −350 to −270.[c] If, as is widely believed,[d] Bolus of Mendes and his *symprophētai* provided a chemical 'theory', in the form of Aristotelian philosophy, for the empirical practices of the temple and palace artisanate of Egypt, Syria and Greece; so also Tsou Yen may be thought to have supplied a Taoist theory of Yin–Yang and the Five Elements for the growing Chinese conviction that material immortality was attainable, and for the empirical operations of the palace and temple artisans whose works were recorded in the Khao Kung Chi[2] chapter of the *Chou Li*.[3] Only he was about a century and a half earlier than Bolus in his activity. The relevant parts of the 'Artificers' Record', which in a way correspond to the Hellenistic chemical-technological papyri, are also relatively much older than these, for instead of the +3rd century they may be traced to the early Han period (−2nd), if not indeed to the State of Chhi in the late −4th, when Tsou Yen was a young man.[e] From him the line runs straight to Li Shao-Chün, Shao Ong, and Wei Po-Yang, but first we must pause for a moment in the first springtime of empire, the dynasty of the Chhin.

The earnest, almost desperate, searches of the first emperor, Chhin Shih Huang Ti, for the herbs or substances of immortality, and the concoctions that could be made from them, is a matter of common knowledge. Since he was reigning (as universal monarch) from −221 to −209 he and his proto-alchemical advisers (if so we might call them) again preceded the time of Bolus. We need not here recall the names of all

[a] This is discussed in Sect. 30; meanwhile Partington (5) is to be consulted. Greek fire was in China by +850 or so (cf. p. 158).

[b] In the following paragraphs cross-references will be omitted, for the persons mentioned will be readily found at their appropriate places in the historical sub-section, or by means of the general index.

[c] A more precise *floruit* might be taken as −323 to −298 (Dubs (5), pp. 75, 83).

[d] See pt. 2, pp. 21, 26 above, evidence almost over-emphasised (with racialist undertones) in Festugière (1), pp. 218–19, 222–3, 237–8. For him, Bolus added Greek *theōria* (θεωρία) to 'oriental' *praxis* (πρᾶξις). Matters, we think, were not so simple.

[e] Another link with the −4th century arises in connection with Mo Ti[4] (*fl.* −480 to −380), the philosopher of universal love, pacifism and scientific logic. Berthelot remarks in one place, (1), p. 153, how strange it was to see a man like Democritus of Abdera, a naturalist philosopher, agnostic and free-thinker *par excellence*, transformed as Pseudo-Democritus into a magician and alchemist. But that happens everywhere—the *San Kuo Chih* bibliography lists a *Mo Tzu Tan Fa*[5] (Alchemical Preparations of Master Mo). And the *Mo Tzu* book got into the *Tao Tsang*. Probably belonging to the same period is the still extant list of minerals and alchemical substances in the *Chi Ni Tzu*[6] book (cf. pt. 3, p. 14).

[1] 鄒衍   [2] 考工記   [3] 周禮   [4] 墨翟   [5] 墨子丹法   [6] 計倪子

those in charge of the searches, but there were others in his time (i.e. the Chhin) who became known as elixir-makers,[a] such as Chiang Shu-Mao,[1] Liu Thai-Pin,[2] Thang Kung-Fang[3] and Li Pa-Pai.[4] Almost nothing is known of what they did but tradition associated cinnabar and other minerals with their names.

The time of Bolus of Mendes coincided with the first decades of the Han dynasty (from −206 onwards), and here already we have firm evidence of aurifaction from the wording of the anti-coining edict, a point to which we shall return in a moment. During the first half of the −2nd century elixir-makers abounded, e.g. Huang Hua,[5] Yin Hêng,[6] Liu Jung[7] and Li Hsiu,[8] some of whom were very chemical indeed judging from expressions such as *yün shuang tan*[9] (frosty sublimate elixir) which occur in connection with them. More irrefragable evidence historically is provided by the physician Shunyü I[10] (−216 to −147), whose case histories of −167 and −154 include firm proof of the excessive taking of metallic and mineral drugs and elixirs (pt. 3, pp. 46–7). Immediately after his death there occurred events which constitute nodal points in the development of Chinese alchemy. From the official history we know that the thaumaturgical alchemist Li Shao-Chün,[11] the first in any civilisation to associate artificial gold with immortality, was at the height of his influence in −133; and when his ascendancy at the court of Han Wu Ti came to an end, he was quickly succeeded by Shao Ong.[12] This man in turn fell from grace just about the time of the judicial suicide or disappearance of Liu An[13] (−122), that Prince of Huai-nan who had gathered about him a company of scholars and adepts, the writers doubtless of the compendium of natural philosophy known as the *Huai Nan Tzu*[14] book.[b] The names of all eight of his Pa Kung[15] have in fact come down to us, eight 'venerable experts' of a strongly alchemical flavour.[c] There are besides other alchemists, such as Wang Hsing[16] and Wang Than,[17] who may have to be placed in the reign of Han Wu Ti.[d]

The −1st century was also very important in Chinese alchemy and proto-chemistry. It saw the extraordinary government-supported programme of aurifaction in the imperial workshops (−61 to −56) under the leadership of Liu Hsiang;[18] as also the first accounts of 'projection' (i.e. the use of a small amount of a chemical substance to convert a large quantity of base metal into gold or silver)[e]—one in the story of Chhêng Wei[19] and his wife, the other in traditions about Mao Ying.[20] This last, with his younger brothers Mao Ku[21] and Mao Chung,[22] exerted an incalculable influence on Chinese alchemy by their posthumous patronage of the great school of Taoism

---

[a] I.e. to later generations, for in most cases contemporary evidence about them has not been preserved. Another adept of the same name lived in +7 (pt. 2, pp. 125–6).

[b] And also of the *Huai Nan Wan Pi Shu* just now referred to. As will be remembered from pp. 311 ff., there are close parallels between this and the 'sympathies and antipathies' of Bolus of Mendes and his successors.

[c] Cf. p.168. We also have the name of Liu An's other chief chemist, Wang Chung-Kao.[23]

[d] Cf. *TT*293, ch. 9, pp. 11b ff.

[e] We shall return to this in the adjacent sub-section.

[1] 姜叔茂	[2] 劉太賓	[3] 唐公房	[4] 李八百	[5] 皇化
[6] 陰恆	[7] 柳融	[8] 李修	[9] 雲霜丹	[10] 淳于意
[11] 李少君	[12] 少翁	[13] 劉安	[14] 淮南子	[15] 八公
[16] 王興	[17] 王探	[18] 劉向	[19] 程偉	[20] 茅盈
[21] 茅固	[22] 茅衷	[23] 王仲高		

associated with the abbey on Mao Shan.[a] Among other elixir-makers were Su Lin[1] (*d.* −60) and Chou Chi-Thung,[2] while at the end of the century, during the Hsin interregnum, Su Lo[3] was prominent. These were the kind of men who corresponded in time with Anaxilaus of Larissa (*fl.* −40 to −28).[b]

This brings us to the probable date of Pseudo-Democritus himself, the +1st century, corresponding to the Later Han. Perhaps the *Kêng Hsin Ching*[4] (Book of the Realm of Kêng and Hsin, i.e. the noble metals),[c] though we do not have any of it now, should be put side by side with the *Physica kai Mystica*, and we know only the philosophical name of its author, Chiu Yuan Tzu,[5] but it would coincide closely in date.[d] From this time onwards commence those genealogical tables so characteristic of Chinese alchemy, tracing the descent of chemical secrets from master to disciple as the decades wore on. Thus Wang Wei-Hsüan[6] begat Han Chhung,[7] and Han Chhung begat Liu Khuan[8] (+121 to +186); often, as in this case, the dates of several can be estimated if those of one member of the series are known. The earliest may sometimes be of doubtful historicity, as in the sequence which leads from Ma Ming-Shêng[9] (*fl. c.*+100) to Yin Chhang-Shêng,[10] the putative master, or first commentator, of Wei Po-Yang.[11] With Wei we reach the oldest extant Chinese alchemical text, the *Chou I Tshan Thung Chhi*,[12] datable at +142, and already considered in detail by us;[e] this is entirely historical, and falls between the dates of Pseudo-Democritus and Zosimus.

Henceforward the clear chronological lead of the Chinese developments fades out, and the two traditions are fully under way, with striking coincidences of date between the greater representatives. Thus it is curious that Ko Hung,[13] author of the *Pao Phu Tzu*[14] book (+283 to +343), should have been so exact a contemporary of that other great systematiser, Zosimus of Panopolis. Similarly, Thao Hung-Ching[15] (+456 to +536) closely parallels in time Olympiodorus; while Sun Ssu-Mo[16] (+581 to +682)[f] was active just at the same time as Stephanus of Alexandria. After this the tradition continued uninterruptedly in China, declining much later than in the world of Greek culture, for while by +1000 little or nothing was left of the latter, the former was still in animated life (and not at all uncreatively so, because of the iatro-chemists) until the middle of the Chhing dynasty, about +1700. Admittedly also the Chinese tradition has preserved many more names of alchemists and records of their doings between the time of Ps.-Democritus and Stephanus than can be found in the West, and indeed the golden age of Chinese alchemy followed rather than preceded the time of

---

[a] Cf. pp. 213 ff.   [b] See Wellmann (2).

[c] Cf. pt. 3, p. 43. For *kêng* as gold and *hsin* as silver see the tabulation in *Tshan Thung Chhi*, ch. 34, p. 11 *a*.

[d] Much can often be told about a movement from its opponents, so it is relevant to be reminded that much of the book of the great sceptic Wang Chhung,[17] the *Lun Hêng*,[18] finished in +83, is directed against the belief in material immortality, hence implicitly against alchemy, which indeed from time to time is mentioned in it (cf. Vol. 2, pp. 368ff., 376).

[e] Cf. pt. 3, pp. 50ff., and pp. 248ff. above   [f] On his life-span see Sivin (1).

[1] 蘇林   [2] 周季通   [3] 蘇榮   [4] 庚辛經   [5] 九元子
[6] 王韙玄   [7] 韓崇   [8] 劉寬   [9] 馬鳴生   [10] 陰長生
[11] 魏伯陽   [12] 周易參同契   [13] 葛洪   [14] 抱樸子   [15] 陶弘景
[16] 孫思邈   [17] 王充   [18] 論衡

Stephanus. Strange is it too, that while our oldest extant MS. of the Hellenistic proto-chemical writings dates from $c. +1000$, definitive collections of the *Tao Tsang* books were made in $+990$ and $+1019$, their printing being actually accomplished in $+1115$.[a]

There remains the matter of the anti-coining edicts directed against aurification. Here again Chinese culture seems to have had a considerable lead, since one naturally tends to contrast the Diocletian proclamation of $+292$ (or $+296$)[b] with the Han edict of -144, so important, as we have seen,[c] for the problem of the origins of the aurifactive idea. But there were certainly earlier attempts in Europe to put down falsification, notably the Cornelian Law passed at Rome in -81, forbidding the making of deceitful alloys (*fingere*), the addition of superficial layers (*flare*) and the tincturing or production of superficial coloured films (*tingere*).[d] A particular study devoted to the earliest laws against metallic counterfeiting would be very useful,[e] but for the time being we seem to have a definite Chinese date older than anything in the West. This only bears out the general conclusion arising from the foregoing paragraphs, namely that in aurifiction and aurifaction both, China seems to have had a couple of hundred years' advantage over the Mediterranean region. This brings us face to face at last with the fascinating problem of possible ideological contacts and transmissions between China and Europe. What exactly these ideas could have been we must leave for the next sub-section, but first it will help to set the scene by recalling certain undoubted historical facts.

As will be remembered, the activities of the explorer Chang Chhien[1] in Greek Bactria and the neighbouring lands, with their effects, occurred broadly within the decades -140 to -110, and at the latter date approximately the first caravans of silk began their traffic over the Old Silk Road.[f] The other important Chinese traveller of those days was Kan Ying,[2] who reached the Persian Gulf in $+97$,[g] but by that time there were many visits of Gandhāran, Parthian and Roman-Syrian envoys, often more or less traders, to China (-120, -30, $+87$, $+101$, $+120$, $+134$, $+166$ and $+284$). The particularly well documented An-Tun[3] embassy[h] to the Han court, bearing its ivories and tortoise-shell, took place in $+166$. All these were possible channels of communication, and there certainly must have been a good many more the details of which have not come down to us. We even know the Chinese name of one of the leaders of these missions, Chhin Lun,[4] a Roman-Syrian merchant-envoy, who reached the Wu State (of the San Kuo period) in $+226$.[i] When one considers how intense the traffic on the Old Silk Road was during the -1st century and the first two centuries of our era, quite apart from the considerable use of the shipping

---

[a] For the full account, see pt. 3, pp. 113 ff.
[b] See further p. 340 below.   [c] Pt. 3, pp. 26 ff. above. To say nothing of the earlier edict of -175.
[d] *Corp. Jur. Civil.*, Digest, bk. 48, tit. 10, paras. 1, 8. Cf. von Lippmann (1), p. 286.
[e] There is already, for Rome, the interesting paper of Grierson (2).
[f] Vol. 1, pp. 173 ff. It is therefore strictly correct to say that Europe and its culture was discovered by China and not the reverse. Yet the Philistine view continues to dominate, as in the following example (1971): 'The Europeans roamed the world, as the Greeks had done already, and discovered India, China and the rest of the world; the inhabitants of those parts stayed at home and contemplated their navels' (Hutten, 1).   [g] Vol. 1, p. 196. Cf. Dubs (5), p. 81.   [h] Marcus Aurelius Antoninus.
[i] On this see Vol. 1, p. 198; and on the whole subject pp. 191 ff., as also Hirth (1), pp. 35 ff.

[1] 張騫   [2] 甘英   [3] 安敦   [4] 秦論

lanes to and from Indo-China and South China round India and Malaya, it would almost be surprising if no ideas connected with the chemical art travelled along them.[a] Some of its products certainly did, for example the artificial gems so prominent in the Graeco-Egyptian papyri.[b] Very soon we must look into the question of ideological parallels, but first the intermediate realm of Iranian culture demands attention.

It is a disturbing fact that Pseudo-Democritus lauds as his greatest teacher and master no Greek, no Egyptian either, but Ostanes the Mede. Uštana was a perfectly good Elamite name,[c] that is to say, characteristic of the south-western part of Persia sometimes called Susiana, region of the cities of Susa and Persepolis. Media was one of the greatest and oldest of the Persian provinces, roughly that part of northern Persia west of modern Teheran and south-east of Armenia and the Caucasus; Ecbatana (mod. Hamadhān) was its traditional capital. The original form of the name Ostanes was resumed in Arabic and later Persian, Uṣtānis, and thus he appears in the quite undatable texts which have come down to us in Arabic as his.[d] Historical geography throws much light on this Bolus–Pseudo-Democritus–Ostanes link, for as any atlas of ancient history will explain, the empire of Alexander the Great at his death in −323 constituted a vast L-shaped area, the long arm being formed by Sogdia, Bactria, the Indus Valley, Parthia, Media, Armenia and Anatolia, while the shorter one comprised Palestine and the Nile Valley. Although this was soon divided between Seleucus, Antigonus and Ptolemy, the middle of the −2nd century saw a great intensity of trade and much flow of ideas along these axes, connecting Alexandria and Rhodes in the west with Antioch, Seleucia (near Babylon), Rayy (near Teheran), Khiva, Merv, Balkh, and (after −110) all points east. Ptolemaic Egypt became a Roman protectorate after −102, and part of the empire after −45, but trade and travellers still continued as before.[e] In short, the exchange of ideas along an east–west line of communication has to be reckoned with as a very real probability from about −300 onwards, and if this is once visualised, men such as Li Shao-Chün, Liu An and Liu Hsiang on the one hand may not have been so impenetrably sundered from people like Ps.-Democritus, Comarius and Pebechius on the other as has usually been supposed, even allowing for the obvious barriers of perhaps several intervening languages.[f] At any rate, it clears the decks for an objective examination of possible contacts. As for Ostanes, he was many things, certainly a legendary character but also probably one or more living men. For the best discussion of him you turn to Bidez & Cumont.[g]

[a] At the time of elaboration of the Hermetic and Gnostic literature, wrote Filliozat (5), we should expect to find many traces of exchanges between the East and the West. On the lines of communication across Central Asia the papers of Herrmann (2, 3, 5, 6) are still authoritative. As early as 1917 Holgen (1), p. 471, drew attention to the importance of the Old Silk Road for ancient contacts of chemical thought, and later on Huang Tzu-Chhing (2) re-affirmed this.

[b] The evidence is collected in Vol. 1, p. 200.

[c] It was the name of an eminent artisan in the period −509 to −494, as Dr I. Gershevitch has informed us, in discussions on this subject for which we render our best thanks.

[d] See Berthelot & Houdas (1), pp. 116ff., cf. Festugière (1), p. 391.

[e] On daily life in Hellenistic Egypt Cumont (4) should be read; he describes the arrival of Chinese silks, and the manufacture of *ersatz* gems for export (pp. 91, 96).   [f] Cf. Vol. 1, p. 150.

[g] (1) vol. 1, pp. viff., 167ff. See also Festugière (1), pp. 42ff. and of course the encyclopaedia article of Preisendanz (1).

The 'magi' whom the Greeks knew were not the real Mazdaeans of Persia who after the Zoroastrian reform worshipped only the god of good (Ahura-Mazda) and not the god of evil (Ahriman), but *magousaioi* (μαγουσαῖοι), priests of earlier Mazdaean colonies established in Achaemenid times (−6th cent. onwards) west of Iran from Mesopotamia to the Aegean, lasting down to the end of the − 1st century and continuing still the ancient Persian tribal system of worshipping both gods.[a] Hence all kinds of theurgy and apotropaic magic, divination and astral lore, hence too connections with earlier Chaldaean science and pseudo-science which were taken over, just the sort of manual-operations medium in which mystical aurifaction could be expected to arise. The language of this 'Mazdaean Diaspora', as it has been called, became gradually Aramaic, so that the 'magi' of the Greeks could not read Avestan texts and probably had no Zend or Pehlevi sacred books, but they made up for this by acting as clearing-houses for the magical arts of all the peoples;[b] and if anyone deserved the name of *fang shih*[1] during the last three centuries of the − 1st millennium it was they.

Our Ostanes was one of them. The first of the name was supposed to have accompanied Xerxes (*r.* − 485 to − 465) to Abdera and taught Democritus when young;[c] this may be considered legendary. A second was said to have accompanied Alexander the Great in all his travels and conquests.[d] Someone of the name was referred to as the 'Prince of Magi' by Pseudo-Damigeron (himself a mage), the writer of a verse lapidary *c.* − 200. Pliny himself never saw any books attributed to Ostanes, but Bolus certainly did,[e] on magic, divination, pharmaceutical natural history and 'sympathies and antipathies'.[f] Philon of Byblos reports a work in eight volumes, *Oktateuchos* (ὀκτάτευχος), which must have been in existence by *c.* − 250 at the latest. All in all, there is no reason why Pseudo-Democritus, whatever his real name and cultural background was, could not have had a Persian teacher with the name (true or adopted) of Ostanes; and this conclusion opens gates throughout the length of Asia. Though we shall probably never know much more about him, his name does strikingly symbolise that continuity of east–west intercourse, and that general powerful Persian influence on Mediterranean culture from at least the − 4th century onwards and especially after the end of the − 2nd.[g] It certainly had some strange effects in the Graeco-Egyptian milieu, for example, as Bidez & Cumont wrote:[h] 'Thus Democritus, the pure representative of Greek philosophy, became, by a characteristic fiction, at one and the

---

[a] Cf. Benveniste (3). Some regarded infinite time (Zervan-Akarana) as the greatest god, whence the other two proceeded.

[b] Apocrypha attributed to Zoroaster circulated in Greek after − 270, for example a book on natural phenomena entitled *Peri Physeōs* (περὶ φύσεως), and there is evidence that this was used by Bolus of Mendes. See Bidez & Cumont (1), pp. 107ff., 111. It seems to have contained a lot of botany, evoked by the complex rules for liturgical rites and ceremonies.

[c] Cf. Pliny, *Hist. Nat.* xxx, ii, 8, spelling Osthanes. Cf. Bidez & Cumont (1), vol. 2, p. 267.

[d] Pliny, *loc. cit.* 11.

[e] Pliny, *Hist. Nat.*, xxiv, cii, 160, calls him *Magorum studiosissimus*. Cf. Festugière (1), p. 198.

[f] Ostanes books of various kinds are often quoted by later writers such as Tatianus, Pamphilus of Alexandria, Dioscorides, Pseudo-Apuleius, etc. For the details see Bidez & Cumont (1), vol. 2, pp. 293, 299ff.

[g] The point has been well put by Ganzenmüller (2), p. 32.

[h] (1), vol. 1, p. 204.

[1] 方士

same time the prophet of Chaldaeo-Iranian wisdom and the true chief and inspirer of the priestly colleges of Egypt.'

After all that has been said earlier in this volume we need hardly rehearse the parts played by Ostanes and Ps.-Democritus in the Corpus. The basic filiation of Ps.-Democritus centres round the vision in the temple (Ostanes, son of Ostanes, being one of the characters), when the spirit of the master is invoked[a] and a secret door in a column spontaneously opens, disclosing writings containing some of the basic proto-chemical aphorisms.[b] This scenario was the model for many later descriptions of the same kind; there is a Syriac version in a 'Letter of Pebechius to Osron'.[c] The apparatus of underground repositories of secret chemical data, with seven gates each one of a different metal etc., was much appreciated by the later Arabs, and a version in that language can be read in a 'Book of Uṣtānis' preserved in a *Kitāb al-Fuṣūl*.[d] Hence the myths of the *Tabula Smaragdina* in later times, and the *Tabula Chemica*.[e] In the Arabic story there are three such stele inscriptions, one in Egyptian, one in Persian, and one in 'Indian', thus emphasising again, though at a relatively late date, the continuity of Old World culture.

The only attributed writing of Ostanes in the Corpus is the 'Letter of Ostanes to Petasius' on the calcium polysulphides.[f] But in one of the Zosimus texts certain aphorisms or gnomic sayings are ascribed to him,[g] and in the 'Letter of Ps.-Democritus to Ps.-Leucippus' emphasis again is laid on Persian knowledge transmitted to the ancestral kings of Egypt, then confided to Phoenicians (or Mages).[h] As for Petasius, a work called 'Memoirs of Democritus' (*Dēmocriteia Hypomnēmata*, Δημοκρίτεια ὑπομνήματα) is attributed to him,[i] and in another place he is called King of Armenia,[j] a country after all eastern, and adjacent to Media. Lastly, we have already referred to that difference between Persian and Egyptian metallurgical techniques which appears to arise from the Corpus,[k] namely that the former specialised in surface films and layers while the latter worked mainly with uniform substrate alloys, and we gave reasons for thinking that there was not much reality behind it.[l] After discussing this matter, Bidez & Cumont went on to say:[m] 'Although the idea that alchemy was born in Egypt continues to spread everywhere and gain acceptance, in spite of the hesitations of specialists, what remains of our apocrypha [Ps.-Democritus in particular] serves only to show that however current this opinion, born in the shadow of the pyramids, may be, it owes its prestige to nothing more than a prejudice.'

---

[a] It is explained that the master had died by poison, either purposely or accidentally. Could we dare to understand this as the consumption of a dangerous elixir, in the Chinese style? That would indeed be an 'hypothèse hardie mais attirante', counter-indicated only by the lack of macrobiotics in Alexandrian proto-chemistry.   [b] *Corp. Alchem. Gr.*, II, i, 3.

[c] Tr. Berthelot & Duval (1), pp. 309ff. It will date any time between the +2nd and the +6th centuries.

[d] Tr. Berthelot & Houdas (1), pp. 116ff., cf. Berthelot (2), p. 216.

[e] Cf. pp. 373, 401 below.

[f] *Corp. Alchem. Gr.*, IV, ii, cf. Bidez & Cumont (1), vol. 1, p. 208.

[g] *Corp.* III, vi, 5. Cf. also II, iii, 1, 2.   [h] *Corp.* II, ii, 1.   [i] *Corp.* v, vii, 16.

[j] In the alternative title of *Corp.* II, iv.

[k] *Corp.* II, ii, 1 and more explicitly in the 'Letter of Synesius to Dioscorus', *Corp.* II, iii, 2.

[l] Pt. 2, pp. 253–4 above.   [m] (1), vol 1, p. 205.

At all events, we must be prepared to keep an open mind about possible exchanges of ideas in these early times between East and West.

In order to give a little more life to this picture of cultural continuity, and perhaps to take off any impression that the borrowings were only westwards, we may look at a few instances of things that the Chinese heard about the West. First, it is a well-known fact that Indian medical ideas began entering China towards the end of the Han or soon afterwards, accompanying of course Buddhism, and in this flow there came some constituents from further West. In view of the obvious dominance of Five-Element theory in China, the existence of 'four primes' (*ssu ta*[1])[a] in some medical writings, including those of Thao Hung-Ching in the late +5th century, struck a foreign note, as was observed already by Hsü Ta-Chhun[2] in his history of medicine, *I Hsüeh Yuan Liu Lun*,[3] written in +1757.[b] Among the translations of early Buddhist medical texts into Chinese, Sen (1) has studied and englished the *Fo Shuo Fo I Wang Ching*,[4] a *sūtra*[c] which entered the *Ta Tsang* through the hands of Chih-Chhien[5d] and an Indian collaborator in +230. This explains all the 404 diseases as caused by imbalances of the elements Earth, Water, Fire and Wind. Thus the Empedoclean and Aristotelian elements had entered Chinese thought by the San Kuo period, and if they played thereafter only a very small role relative to the indigenous theories of natural philosophy it was at any rate a striking example of cultural contact.

Still more interesting for our present theme is the information about Alexandrian aurifaction which was preserved in one of the commentaries on the *Shih Chi* (−90). There, in the chapter on Western countries,[e] Ssuma Chhien tells us about An-Hsi[6] (Parthia), saying, among other things, that the people use silver money with the face of the king stamped on it, this being changed when a new king succeeds to the throne.[f] West of the country lies Thiao-Chih[7] (Mesopotamia), and north of it Yen-Tshai[8g] and Li-Hsien[9h]. The commentator,[i] taking the last at least of these places to be more or less the same as Ta-Chhin[10] (Roman Syria, Palestine and even Egypt),[j] proceeds to quote the well-known passage in the *Hou Han Shu* on the gold and silver

---

[a] I.e. four primary or elementary constituents (*ssu ta yuan su*[11]), Skr. *catvarī-mahābhūtanī*. Further details are given in Jen Ying-Chhiu (1), pp. 42 ff.

[b] According to our colleague Nathan Sivin (priv. comm.).   [c] N 1327, TW 793.

[d] It may not be irrelevant to note that his ethnikon shows him to have been of 'Scythian', i.e. Indo-European 'Tocharian' stock, a descendant of the Yüeh-chih who overran Greek Bactria about −130 and then went on to found the Śaka kingdom in India. Cf. Vol. 1, p. 173.

[e] Ch. 123, p. 5b (p. 53), tr. Watson (1), vol. 2, p. 268.

[f] He also says that they write horizontally on strips of leather (parchment?) for their books and records.

[g] We would conjecture for this Chorasmia (Khwarizm), and its city of Khiva.

[h] Watson takes this to be Hyrcania, i.e. the parts of Persia just south of the Caspian Sea, the modern province of Mazendaran with the Elburz mountains, just north of mod. Teheran. But Hirth (1) believed that Li-Hsien (with variant orthography, Li-Kan,[12] Li-Chien[13]) was an older name for the whole of Ta-Chhin, perhaps derived from Rekem (= Petra), an *entrepôt* on the trade routes (pp. 169 ff., 180). The commentator must have thought so too, or he would not have attached so much material about Ta-Chhin to the passage. Cf. Vol. 1, p. 174, where the equation with Alexandria is discussed.

[i] Chang Shou-Chieh,[14] writing in +737.   [j] See Hirth (1), p. 180.

[1] 四大	[2] 徐大椿	[3] 醫學源流論	[4] 佛說佛醫王經	[5] 支謙
[6] 安息	[7] 條枝	[8] 奄蔡	[9] 黎軒	[10] 大秦
[11] 四大元素	[12] 黎軒	[13] 黎鞬	[14] 張守節	

deposits, asbestos, coral and amber of that region,[a] but then goes on to cite some sentences from the *Wu Shih Wai Kuo Chuan*[1] (Records of Foreign Countries),[b] a lost book by the traveller and ambassador of Wu State, Khang Thai,[2] written about +260. He again says that the people of those parts use gold and silver money, and have jewels of rock-crystal in five colours,[c] adding, however, that there are many clever craftsmen among them, who can transmute silver into gold.[d] It is quite an unexpected thing that such a rumour of Hellenistic aurifaction and aurifiction should have found its way into Chinese literature in the +3rd century.

Onc more example. In +1347 Chu Tê-Jun,[3] a distinguished literary scholar interested in scientific subjects, was sitting talking with two friends not racially Chinese (as was so common under the Yuan dynasty), the officials Yo-Hu-Nan[4] and San-Chu-Thai.[5] The former, Johanan, was probably a Christian Uighur,[e] the latter, Saljidai, a Mongolian. During their service in the imperial guard between +1314 and +1320 they had had discourse with an embassy from the West, and they retailed to Chu an alchemical fable which they had heard from the members of this embassy. Their land, they said, had a lake of mercury (*shui yin hai*[6]), from which the metal was collected in the following way. Men and horses covered with gold leaf ride along its shores, whereupon a great wave of mercury arises and pursues them, but they successfully flee away and the metal falls into pools made ready beforehand, whence the local people collect it in due course. What is more, they heat it with certain aromatic herbs so that it all turns to silver. Such was the story which Chu later recorded in his *Tshun Fu Chai Wên Chi*[7] two years later.[f] The implicit reference to Au–Hg amalgamation needs no emphasis, but the origin of the embassy is slightly puzzling. The country mentioned, Fo-Lin,[8] sounds like a variant of the standard name for Byzantium,[g] but Fuchs (7), who has gone into the matter, brings evidence that the envoys came really from Moorish Granada, and that the mercury story emanated from the famous mines of Almadén located between Toledo and Cordoba.[h]

This was only one of a series of fables with the same motif cropping up in many Old World languages. Another is found, whether concerned with tin or mercury is not quite clear, in the Syriac versions of Zosimus or Ps.-Zosimus, but in this case the glittering metal is induced to come out of its pool by a beautiful naked girl who walks past it, then runs quickly away, while young men attack it with hatchets and cut it up

[a] In ch. 118, tr. Hirth (1), p. 41.
[b] We have several times had occasion to refer to this, Vol. 3, pp. 511–12, 610, 658, Vol. 4, pt. 3, pp. 449–50, 472. For further details see Fêng Chhêng-Chün (1), pp. 11 ff.
[c] Interesting in connection with the false gems and coloured glasses of the papyri.
[d] *Jen min to chhiao nêng hua yin wei chin.*[9]
[e] Possibly one of the family recorded in *Yuan Shih*, ch. 134 (tr. Saeki (2), pp. 489 ff.).
[f] Ch. 5, pp. 11b, 12a, tr. Fuchs (7). Through various intermediaries it got into *PTKM*, ch. 9, (p. 56) and appears twice in *TSCC* (*Khun yü tien*, ch. 22, *hung pu hui khao*, p. 1a and *Pien i tien*, ch. 60, *Ta-Chhin pu chi shih*). Also in *STTH*, whence *Wakan Sanzai Zue* (de Mély (1), tr., p. 73, text, pp. 70–1) and the remarks on it in de Mély (6), p. 333.   [g] Vol. 1, p. 186.
[h] That the embassy of +1317 or thereabouts was from Muslim Spain is argued partly on the ground of the very long time they took on their journey. Also they brought with them Islamic prayer-carpets, woollen cloth, brocades, etc.

[1] 吳時外國傳   [2] 康泰   [3] 朱德潤   [4] 岳忽難   [5] 散竺台
[6] 水銀海   [7] 存復齋文集   [8] 佛菻   [9] 人民多巧能化銀爲金

Fig. 1524. A representation of the 'gold-digging ants' of Asia, from the + 1481 Augsburg edition of Mandeville's 'Travels' (Pollard ed., p. 209).

into bars.[a] A third, the most famous, is the story of the gold-digging ants (Fig. 1524), launched originally by Herodotus with reference to some Central Asian region north of India.[b] The gold particles are in the sand which they dig up to form their burrows, then when they are sheltering down below from the noontide heat, men rush to the spot with camels, fill up bags that they have with them, and make their escape even though the ants, which can run extremely fast, pursue them. Since Herodotus' time (−440) this has generated quite a literature.[c] Everybody talked about it—Strabo (−25),[d] Pliny (+75),[e] and of course Solinus (+3rd cent.);[f] and it is in the Corpus, mentioned by Olympiodorus (c. +500).[g] Thus it comes through to Vincent of Beauvais[h] and Sir John Mandeville.[i] Tibetan and Mongol versions have been published (Laufer, 41) and those of living Ladakhi folklore (Francke, 1), but so far no

---

[a] Berthelot & Duval (1), p. 245. The text would be of the +4th to +6th centuries. Amalgamation is involved again, for the mercury comes apparently from tin. The parallel was noticed long ago by de Mély (6), pp. 332ff., who added a note of a place-name in Syria, Bir al-Zeibaq, Quicksilver-Well.

[b] III, 102–105. In Vol. 1, p. 177, we mentioned Megasthenes' account, c. −300.

[c] Like the Golden Fleece of Jason and the Argonauts itself. This legend is supposedly of the −13th century (cf. J. R. Bacon, 1) but an invention of the −7th may account for it (Vol. 4, pt. 3, p. 608). Strabo already (XI, ii, 19) explained the fleece reasonably enough as referring to the placer gold of Colchian streams caught in fur or blankets. Agricola (ch. 8, Hoover & Hoover ed., p. 330) agreed. By the time of Suidas (+1000) the fleece had become a parchment (vellum) book of *chēmeia* which taught how gold could be made (*Lexicon*, vol. 1, p. 525). On both legends see Adams (1), pp. 483ff. For a Chinese reference to the placer technique see *PTKM*, ch. 8, (p. 3), quoting Chhen Tshang-Chhi (+725) on 'bran' gold washed out on felt (tr. Schafer (13), p. 251). Cf. pp. 60, 81.

[d] XV, i, 57.

[e] *Hist. Nat.*, XI, xxxvi, 111.

[f] Cf. Vol. 3, pp. 505ff.

[g] *Corp. Alchem. Gr.* II, iv, 43.

[h] *Speculum Naturale*, xx, ch. 134.

[i] 'Travels', ch. 33 Pollard ed., p. 198. On the whole legend cf. also Druce (1); Marshall (1), p. 14; Bevan (1), pp. 396, 404.

close Chinese equivalent has been found. In +1799 von Veltheim suggested as the basis of the tale the miners of the Altai or the Gobi and the burrows of the Tartary fox; Schiern (1) a century later thought the miners were Tibetan. Boni (3) scented an allegory of the levigation process for separating alluvial gold.[a] The term 'ant-gold' (*pipīlika*) occurs in the *Mahābhārata* (c. +1st century),[b] perhaps referring to the size of the alluvial grains, and some think that this was the origin of the story (Rickard, 3). Laufer, however, suggested a confusion between the name of a Mongolian clan, the Shiraighol, and the Mongol word for ant, *shirghol*. The last word has certainly not been said on the subject, but in the meantime it would hardly be possible to find a better example of the East–West cultural continuity in ancient times than these strange mining stories of the men pursued and the thing pursuing.

(ii) *The first occurrence of the term 'Chemistry'*

On this subject a great deal has been written during the last century or two, and any assured conclusion takes some digging out, but if one drives one's adit fair and true it is possible to reach the facts of the matter. The work is inescapable because of the delicate question of conceivable East Asian influence upon the Alexandrian proto-chemists. From the following pages it will emerge that the word 'chemistry', in the form of '*chymeia*', '*chēmeia*' or even '*chimeia*' (χυμεία, χημεία, χιμεία), with its enigmatic root, does not appear in the Mediterranean region before about +300. Any-one impatient of Western classical detail would be well advised to proceed directly to the next item on the agenda, but those who are willing to follow the argument closely may find some very peculiar things on the way.

The first outstanding fact is that the word is never found in the early parts of the Greek 'alchemical' Corpus, e.g. in the writings ascribable to Pseudo-Democritus, nor does it ever appear in the papyri dealing with chemical technology (cf. pt. 2, pp. 15 ff. above, on aurifaction and aurifiction). Next there are certain early references which have been claimed but must be discarded. For example, Julius Africanus wrote a book of thaumaturgical technology called *Kestoi*[c] about +230, and this was described as dealing with the 'powers of chemical preparations' (*chymeutikōn periechousan dynameis*, χυμευτικῶν περιέχουσαν δυνάμεις).[d] But that was only what was said about it in later times, by George Syncellus in his *Chronographia* (c. +800),[e] and even if he got the phrase, as he seems to have done, from the Egyptian monk Panodorus, that will not take it earlier than about +400—certainly not to the date of the original writer.

---

[a] Because Herodotus says that the male camels tire and fall victims to the ants, while the female ones, determined to regain their foals, bear the Indians and the gold swiftly to safety.

[b] II, 1860.

[c] The title was taken from the magic girdle of Aphrodite. The extant fragments are available in the edition of Thevenot (+1693). The book is important for its description of early military incendiary preparations (cf. Partington (5), pp. 7–8). The later chapters are additions made between about +550 and +800.

[d] As Hoffmann (1), p. 521, pointed out. This was an altogether remarkable study, as Hoffmann was working in the pre-Berthelot period, and went to the original MSS of the Corpus. Ruska much admired it, (11), p. 325.

[e] Goar ed., p. 359, Dindorf ed., vol. 1, p. 676.

A similar, if grosser, instance occurs in the astrological handbook of Julius Firmicus Maternus, written by +336, where the printed editions have *scientiam chimiae* as the gift of those born under the influence of the moon in the house of Saturn.[a] But it is now known that the extensive passage containing these words was inserted whole by Johannes Angelus in +1488, just in time for the first printings.[b]

Possibly admissible, however, is the traditional account of the proscription of aurifictors in +296 (or +292) by the emperor Diocletian, a measure taken, supposedly, lest the Egyptians should raise funds in this way and start a rebellion against the Roman rule.[c] The classical citation of this occurs in the *Lexicon* of Suidas (*c.* +976),[d] where we read that 'the order was given to burn all the books written in olden times on the chēmy (or chymy) of gold and silver (*ta peri chēmeias chrysou kai argyrou tois palaiois gegrammena biblia*, τὰ περὶ χημείας χρυσοῦ καὶ ἀργύρου τοῖς παλαιοῖς γεγραμμένα βιβλία).' The sources for this, of course, go further back, though not as far as we should like, for a similar text appears in the 'Acts of St Procopius' (early +8th cent.)[e] and earlier in the writings of John of Antioch (*fl.* +610),[f] who may possibly have copied from Panodorus, though this is uncertain. Nevertheless one may be inclined to accept the transmitted text of the edict as valid for the Diocletian date,[g] partly because of the parallel occurrence of words ancestral to 'chemistry' in fragments attributable to Zosimus (cf. pp. 327, 365) who was writing between +280 and +320.

These lead us into very strange country. The essential passage occurs in one of the letters of Zosimus to his sister (or *soror mystica*) Theosebeia, and runs as follows:

The holy scriptures set forth in books record, O Woman, that there was a race of daemons which coupled with the daughters of men. Hermes also says this in his *Physica*,[h] and nearly every exoteric and esoteric text reports the same. Now the ancient and divine writings say that certain angels fell in love with human women, came down (from heaven) and taught them all the operations and works of Nature, on account of which, we are told, [great offence was taken],[i] and they were excluded for ever from the celestial realms; because they had taught to mankind all things evil, and unprofitable for the soul. [From the commerce of these angels and these women, the writings also say, a race of giants was born.][j] And the first account of all these arts and techniques was that of Chēmēs (Χήμης), which is why it is called the 'Book of Chēmēs (or Chyma, Χύμα)', and why the art is called Chēmeia (χημεία, or Chymeia, χυμεία). [This book is composed of 24 sections, each having its proper name or a designatory letter. They are explained by the voices of priests. One is called Imos, another

---

a *Astron.* III, 5, ix.

b Diels (1), pp. 121–2; cf. von Lippmann (1), p. 288. Berthelot (1), p. 74, Schorlemmer (1) and Hoffmann (1), p. 522 still accepted this reference as ancient, though the usual printing was 'alchemiae'.

c As noted elsewhere, this date is to be compared with that of the Chinese edict against aurifiction, −144.

d Vol. 3, p. 669. Suidas' definition is: 'Chēmeia is the fabrication of silver and gold' (*chēmeia hē tou argyrou kai chrysou kataskeuē*, χημεία ἡ τοῦ ἀργύρου καὶ χρυσοῦ κατασκευή).

e *Acta Sanctorum* (Bollandists), Julii, II, 557 A. This uses the second form of the word in the text as given above.

f In Valesius (1), pp. 834–5, where silver precedes gold in the text, and the spelling is *chēmias*.

g Von Lippmann (1), pp. 288 ff. accepted this.

h Otherwise unidentifiable, and certainly more than semi-legendary. Festugière, it is true, suggests that our present 'Epistle of Isis to Horus' (*Corp.* I, xiii) was part of it. Cf. pp. 326–7.

i Not in the Syriac version.　　j Also not in the Syriac.

Imuth,[a] another 'Face', as we might translate it. One section is called 'Key', another 'Seal', a third 'Manual', a fourth 'Epoch'; each has its own name. One finds in this book the arts and techniques explained in thousands of words. Those who followed wrote as much by way of commentary, but nothing good. They not only spoilt the books of Chēmeia, they made a mystery of them...].[b]

Hence Zosimus' present book, addressed to Theosebeia. For it was a book, almost certainly entitled *Cheirokmēta* (χειρόκμητα) or 'Manipulations',[c] though very little of this is left now, and nothing, needless to say, of the 'Book of Chēmēs' itself. The Greek version of the passage is preserved only in the *Chronographia* of George Syncellus (c. +800),[d] and the Syriac version only in MSS of much later date,[e] but there is no compelling reason for placing either of them later than the +6th century,[f] and it is not unreasonable to believe that they come from the pen of Zosimus himself.[g]

Can we trace the 'Book of Chēmēs' any further? The obvious background to Zosimus' account of the fall of the angels is the passage in the Book of Genesis (6, 1–12) which relates how the sons of God saw that the daughters of men were fair and came down to woo them; thus the mighty men of old were born and much evil ensued, so that God repented him of his creation and sent the Flood to destroy it, only Noah with his family escaping. We are here in the presence of a corpus of legend which filtered down from Jewish sources to Essenes, Gnostics and Christians suggesting to religious minds that all the sciences and techniques were really diabolical in nature, and perhaps especially chemistry, the sources indeed of all evil, this traceable not so much to the 'sin of Adam' as to the disobedience of the 'Promethean' angels.[h] And so we are led back to the apocryphal 'Book of Enoch', one of the most interesting of the Jewish writings rejected from the canon, datable in its relevant parts at about −165, and preserved for us only in a number of Ethiopic versions. What it says is this:[i]

VI. 1 And it came to pass when the children of men had multiplied that in those days were born unto them beautiful and comely daughters.

2 And the angels, the children of heaven, saw and became enamoured of them, and said to one another: 'Come, let us choose wives from among the race of men, and beget us children.'

[a] = Imhotep, say Berthelot & Duval (1), p. xxx.

[b] The part in square brackets here is only in the Syriac version.

[c] Cf. Sherwood Taylor (8). It is said to have had 28 sections, and the passage comes in sect. VIII on tin. Why twenty-eight? Surely, Filliozat (5) suggested, because of the 28 lunar mansions (Chinese *hsiu*, Indian *nakshatra*) along the equatorial band (cf. Vol. 3, pp. 242ff., 252ff.). Here would be another remarkable instance of idea-sharing among the ancient civilisations.

[d] Goar ed., p. 13, Dindorf ed., vol. 1, p. 24. A world history from Adam onwards.

[e] Tr. Berthelot & Duval (1), p. 238. Cf. Berthelot (1), p. 9. Part of what we now have must certainly be considered Pseudo-Zosimus, for 'elixir' is mentioned on p. 258, but that does not invalidate the whole.

[f] Von Lippmann (1), p. 294 agrees.

[g] The passage was used in one of the first histories of chemistry, the *De Ortu et Progressu Chemiae* by Olaf Borrichius, printed at Copenhagen in +1668.

[h] A Chinese parallel might be found in the story of the old man in *Chuang Tzu* who would not use a swape because he felt that all ingenuity leads to evil-doing (Vol. 2, p. 124). But this anti-technology complex was never dominant in China. Of course, the problem is still with us. Cf. pp. 125–6.

[i] Tr. Charles (1), pp. 13ff., mod. auct. adjuv. Beer (1), in Kautzsch, vol. 2, pp. 238ff.; Migne (1), vol. 1, pp. 395ff. This text, written partly in Hebrew, mostly in Aramaic, but extant as a whole only in Ethiopic (Amharic), is known as 1 Enoch. The Slavonic version, which we shall mention presently, is

3   And Semjāzā, who was their leader, said unto them: 'I fear that you will not in fact perform this deed, so that I alone shall have to pay the penalty of great sin.'

4   But they answered him one and all, saying: 'We shall swear an oath, and bind ourselves by mutual imprecations not to abandon this plan but to carry it through.'

5   Then sware they all together, and bound themselves upon it.

6   And they were in all 200, who descended in the days of Jared on the summit of Mt. Hermon, and by this name it was named because they swore and bound themselves by mutual imprecations upon it.

7   And these were the names of their leaders, Semjāzā their commander, Arakiba, Aramael, Kokabael, Tamael, Araqael, Danael, Ezekael, Baraqael, Azazael, Armaros, Batarael, Ananael, Zaqael, Shamshael, Satarael, Turael, Jomjael and Sarael.

8   These were their decarchs.

VII. 1   And all the others together with them took unto themselves wives, each choosing for himself one, and they began to go in unto them and unite themselves with them, and they taught them spells and enchantments, and the lore of plants (lit. the cutting of roots),[a] and showed then the (healing properties of) herbs.

VIII. 1   And Azazael taught men to fabricate swords and knives, shields and breastplates, making known to them the metals (of the earth) and the arts of working them. He also showed how bracelets and all kinds of ornaments could be made, teaching the use of cosmetic black, and the painting of the eyes, and the knowledge of all precious stones and of all colouring tinctures.

3   Semjāzā taught enchantments and the knowledge of plant drugs (lit. the cutting of roots),[a] Armaros taught exorcism and the breaking of spells, Baraqael and Kokabael taught astronomy and astrology, Ezeqael prognostication by the clouds, Araqael prognostication by the signs of the earth, Shamshael by the sun and Sarael by the moon.[b]

2   And there arose much godlessness, and fornication, and men were led astray, and became corrupt in all their dealings.

VII. 2   And the women became pregnant, and bore great giants, whose stature was three thousand ells,

3   And who consumed all the acquisitions of men. And when men could no longer sustain them,

4   The giants turned against the men and women and devoured them.

5   So men began to sin against the birds and beasts, the reptiles and the fishes, and to feed upon one another's flesh, and to drink the blood.[c]

VIII. 4   And as men perished, they cried out, and their cry went up to heaven...

VII. 6   Then the earth laid accusation against the lawless ones.

actually a quite different text, though parallel in many ways; it is called 2 Enoch. The latest parts of the former date from about −65. In the translation we give here the order of the verses has been somewhat rearranged so as to present a more continuous story. The Aramaic and Hebrew fragments from the Qumran scrolls (c. −200 to c. +70) have been edited and translated by Milik (1). Cf. Eissfeldt (1), pp. 617ff., 622–3.

a   The 'root-cutters' (*rhizotomoi*, ῥιζοτόμοι) was the classical Greek term for the early herbalists and pharmaceutical proto-botanists.

b   In a later and different version (c. −75), reported in ch. LXIX, the fallen angels are regarded as a set of Shaitans, their names are given again, and we learn that Gadrael demonstrated weapons and Kasdeja poisons, while Penemue instructed mankind in writing with ink on paper, 'which had not been intended by the Creator'.

c   For the Jews this was a particularly horrible thing.

The general result of all this was that Michael, Uriel, Raphael and Gabriel brought the case before the Most High, significantly saying (IX. 6) that Azazael 'hath taught all unrighteousness on earth and hath revealed the eternal secrets preserved in heaven, which men were striving to learn...' Orders were accordingly issued for the arrest and eternal imprisonment of Semjāzā, Azazael and the others. Here is where the prophet Enoch comes in. He is sent to read the sentence to the fallen angels (or 'Watchers', as the text now calls them), and is asked by them to intercede for them in heaven; this he does, but unsuccessfully, and once again has to declare the irrevocable condemnation. Part of the address to the angels says (XVI. 3): 'You were in heaven, but all the mysteries had not yet been revealed to you, only worthless ones you knew, and now these in the hardness of your hearts you have made known to the women; and through these mysteries women and men work much evil upon the earth.'[a]

The whole legend is of extraordinary interest, combining, as it does, a terrifying parable of the evils which the uncontrolled use of science and technology can bring upon mankind, with the age-old fear of sex and sexual relations unauthorised by religion, i.e. by the social organisation and knowledge of the period.[b] Are not the societies for social responsibility of scientists still in the field against the giants today? But for our present purpose the value of the 'Book of Enoch' is a negative one, for the Hamlet of the piece is missing—Chymēs or Chēmēs is not one of the angels, nor is anything known in this apocryphal literature of him and his book on their teachings; there is no trace of him therefore in the −2nd and −1st centuries; we have to look later.[c] It can be said at once that outside the Greek 'alchemical' Corpus no mentions

---

[a] The Slavonic 'Book of Enoch' (2 Enoch), preserved only in Russian and Srb (see Vaillant (1); Morfill & Charles), is quite a different text from the Ethiopic, though in some places paralleling it closely. It was originally written mainly in Greek, a little in Hebrew, probably by Alexandrian Jews, −30 to +50, certainly in Egypt. The account of the fall of the angels occurs in chs. VII and XVIII, but there is little emphasis on their teaching of the arts and sciences.

[b] Some of the patristic embroideries, all in the neighbourhood of +200, are interesting. Clement of Alexandria only has a passing mention (*Stromata*, v, i, Wilson tr., vol. 2, p. 226); but Tertullian has more, attacking astrology taught by the angels, allowing metallurgy and pharmaceutical botany as useful, but emphasising the evils of gold, silver, gems, and all feminine adornments and cosmetics (*Apologeticus*, ch. 22, *De Idolatria*, IX, *De Cultu Feminarum*, I, ii, iii, II, x; Thelwall & Holmes tr., vol. 1, pp. 97, 152, 305 ff., 327). In Pseudo-Clement of Rome the angels actually transform themselves into gems, pearls, purple, gold, etc., as also beasts and reptiles, in order to tempt men and women—then resume human form to reproach them. Instead of this, however, they are themselves overcome with desire, and mate with the women, after which they leave with them the black arts of magic and metallurgy, astronomy, dyeing and plant knowledge, 'and whatever was impossible to be found out by the human mind'. The giants follow. *Homilies*, VIII, 12–18, Smith, Peterson & Donaldson tr., pp. 124 ff. In connection with an argument which will develop a page or two below, it is of interest that this pseudepigraphic work says the angels brought 'the working of gold and silver and all smelting and melting' (*chrysou kai argyrou kai tōn homoiōn chysin*, χρυσοῦ καὶ ἀργύρου καὶ τῶν ὁμοίων χύσιν, VIII, 14). Further material on the Enoch legend will be found collected in Partington (7), vol. 1, pt. 1, pp. 173 ff.

[c] In his suggestive attempt to apply some of the methods of biblical text-criticism to the Lü Hsing chapter of the *Shu Ching*, Fehl (1) has compared the rebellious angels in 1 Enoch with Chhih-Yu, Kung-Kung and others in Chinese myth (cf. Vol. 2, pp. 115, 117, the 'legendary rebels'). Strangely enough there may well have been a Chinese translation of one version of the 'Book of Enoch', namely that which was incorporated as one of the seven canonical scriptures of the Manichaeans under the title 'Book of the Giants' (*Graphē tōn Gigantōn*, γραφὴ τῶν γιγάντων); cf. Henning (3, 4). These scriptures were enumerated in an account of the Manichaean religion: *Mo-Ni Kuang Fo Chiao Fa I Lüeh*[1] (Compendium of the Doctrines and Styles of the Teaching of Mani, the Buddha of Light), prepared in the

[1] 摩尼光佛教法儀畧

have been found. While aurifaction itself is spoken of in various early texts, as in the words of Aeneas of Gaza ($+484$) already quoted (pt. 2, p. 23), none has anything to say about Chymēs or *chymeia*.

In the writings of Zosimus of Panopolis, however (or in texts plausibly attributed to him) there are four further references to the eponymous hero.[a] 'Chimēs (Χίμης) the Prophet, speaking of projection, says...';[b] 'Chimēs often proceeds by burning...';[c] 'All the writers, especially Chymēs (Χύμης) and Maria, say...';[d] or 'As Chymēs has rightly declared...' and there follows one of the proto-chemical aphorisms (cf. p. 359).[e] This evidence comes from about $+300$. Olympiodorus, writing about $+500$, has one mention: 'So also Chēmēs (Χήμης) follows Parmenides, saying...', and another form of the same aphorism follows.[f] Olympiodorus and his successors have many other related words and phrases, such as a reference to Agathodaemon, 'who wrote the book on chemistry (*biblon chēmeutikēn*, βίβλον χημευτικὴν)';[g] or, 'the chemical art (*chimmōtikē technē*, χιμμωτικὴ τέχνη)', or, 'the first of the chemists (*prōtou chimmeutou*, πρώτου χιμμευτοῦ)'.[h] Chimēs the man appears again in Stephanus of Alexandria (*c.* $+620$),[i] and his name occurs regularly in the lists of 'oecumenical philosophers' (cf. pt. 2, p. 17) given in the extant proto-chemical manuscripts,[j] as also in the *Fihrist al-'Ulūm* of Ibn al-Nadīm al-Warrāq (*c.* $+985$), where he appears twice, under the guises of Kīmās and Shīmās.[k] We need follow him no further, for it is clear that the root and its derivatives were firmly planted, even if with several variant spellings, in the $+5$th century, and probably first appeared, partly as a personal name, towards the end of the $+3$rd. It may be, as Ruska thought,[l] that some real proto-chemist wrote under this name between the time of Pseudo-Democritus and Zosimus—perhaps he was a contemporary of Mary the Jewess—but whether he took his name from the germinating word for the subject, or gave his name to it, remains entirely in the dark. In either case the etymological origin remains open. Although no book with this name has survived, either in the Corpus or outside it, the supposition would not be

College of All Sages (cf. Vol. 4, pt. 2, pp. 471–2) in $+731$, though the MS. (Stein collection, no. 3969) is more probably of about $+930$; cf. Haloun & Henning (1). The Chinese title of the book about the giants was *Ta Li Shih Ching*.[1] No Chinese text is known, but Henning (2) has transcribed and translated fragments from several Central Asian languages. None of them mentions the 'Book of Chēmēs'. Thanks are due to Dr Liu Nan-Chhiang for bringing these strange facts to our knowledge.

---

   [a] On all these loci see Hoffmann (1), p. 520; Berthelot (1), pp. 167, 193, 200, 256, 260, (2), p. 111.

   [b] *Corp. Alchem. Gr.* III, xxiv, 7.   [c] *Corp.* III, xxiv, 4.

   [d] *Corp.* III, xx, 2.   [e] *Corp.* III, xviii, 1.

   [f] *Corp.* II, iv, 27.   [g] *Corp.* II, iv, 18.

   [h] *Corp.* VI, xv, 15 (Philosophus Anonymus). Half a dozen of these are collected by Hoffmann (1), p. 525.

   [i] 9th Lect.; not in Sherwood Taylor (9), only in Ideler (1), vol. 2, p. 246. A strange rhetorical passage in the same author (*op. cit.*, vol. 2, p. 217) seems to call upon the help of chemical deities: 'Fight, copper! Fight, quicksilver! Unite the male and the female!...Fight, copper! *Chemoi* (χεμοί), help!'.

   [j] Berthelot (2), p. 111.

   [k] See Flügel ed., vol. 1, p. 353, and Fück (1), pp. 92, 118. Cf. Berthelot (1), p. 131. He also makes an appearance in an anonymous Arabic alchemical work of the previous century, the *Kitāb al-Ḥabīb*, not part of the Jābirian Corpus; where he is mentioned as 'Chymes the Sage' in a quotation purporting to be from Theosebeia, the friend or sister of Zosimus, and concerning aurifaction (Berthelot & Houdas (1), p. 114). See also Dodge (1), vol. 2, p. 849.   [l] (11), pp. 322–3.

   [1] 大力士經

unreasonable that Zosimus did actually have in his hands a 'Book of Chymēs', yet what there was to connect it with the Enoch-legends remains still in deep obscurity.

To etymology we shall shortly return, but here a word should be said about the first 'chemist' of the Western world, and the first book with the title of 'chemistry'. That the root was well implanted by the end of the + 5th century appears from the story of an aurifictor named Johannes Isthmeos, who appeared in the time of the emperor Anastasius Silentiarius (+ 504) and was consigned to prison by him.[a] Here the art is termed *cheimē* (*tēs cheimēs technōn*, τῆς χείμης τεχνῶν) and the adept appears as *cheimeutēs* (χειμευτής), the first of his line. As for the oldest book, whatever Agatho-daemon's was, that Olympiodorus mentioned, it must have been earlier than this time, yet it has often been rivalled by a book in the Corpus with that strange title usually translated 'Domestic Chemistry of Moses' (*Mōuseōs oikeia chymeutikē taxis*, Μωυσέως οἰκεία χυμευτικὴ τάξις).[b] Though unquestionably reflecting the Jewish element in Alexandrian proto-chemistry,[c] it cannot be earlier than Zosimus, as has sometimes been thought, and is probably as late as the + 7th or even + 8th century.[d] Conse-quently Agathodaemon's remains the first, though nobody knows what was in it.[e] Nor can we tell the date of its remaining fragments.[f]

The upshot of all this is that the use of the root 'chem-' for what we now broadly call chemistry seems to have started with Zosimus, or a little before him in the + 3rd century, the time of the chemical technology papyri (though words based on it never occur in them), and distinctly later than the first writings of the Corpus (Pseudo-Democritus, Cleopatra, Pebechius, etc.).[g] Clearly it was well established by the late + 5th century. The question then arises, what could have been its origin? This has been quite a controversial matter, and even today each one of the proposed solutions has grave disadvantages. Let us see what they are.

[a] The usual authority here is Cedrenus, in his *Historiōn Archomenē* of + 1059, Bekker ed., vol. 1, p. 629. The Enoch story is given on pp. 19–20, but without mention of the Book of Chēmēs. Older sources are Theophanes, *Chronographia* (c. + 800), Classen ed., vol. 1, p. 231 (the spelling is *chymeutēs*, χυμευτής, under + 499); and best of all John Malalas, *Chronographia*, Dindorf ed., p. 395, where the spelling is again *cheimeutēs*. As this historian died in | 577, he was quite close to the event. One may remind oneself that Johannes Isthmeos was a contemporary of Thao Hung-Ching.[1]

[b] As Stephanides (4) pointed out, the proper translation should be 'Suitable Classification for Chemical Substances (or Preparations), by Moses'. But the book (*Corp.* IV, xxii), though aurifactive, is very metallurgical and practical. It has been analysed by Ruska (11). The full form of the Greek title is found only in *Corp.* V, vii, 10 (an anonymous late manual on the tincture of gems): *Mōusēs ho prophētēs en tē oikeia chymeutikē taxis*. Other references (*Corp.* III, xxiv, 4, 5, xliii, 6) cite a *Hē Mōseōs Maza*, which is considered an alternative title. We shall return to this strange word later, p. 365. Cf. also Vol. 5, pt. 2, p. 74.

[c] Later Jewish alchemy can be followed in the valuable encyclopaedia articles of Gaster (1); Suler (1, 2) and Rom (1). But as these scholars were Hebraists rather than historians of science, critical reading is necessary.

[d] This was Ruska's final conclusion (pp. 425 ff.). It agrees with that of Festugière (1), p. 239.

[e] Apart from the writer who covered himself with the name, Agathodaemon was some kind of god or fabulous creature, later taken as one of their patrons by the Ṣābians of Ḥarrān (cf. p. 426, and Partington (7), vol. 1, pt. 1, pp. 330 ff.).

[f] Berthelot guessed contemporary with Zosimus or a little earlier (1), pp. 136–7, (2), p. 202. This would make the time of Ko Hung,[2] a century and a half later than the book of Wei Po-Yang[3] (pt. 3, pp. 50, 75).

[g] As Kopp saw clearly just over a century ago, (2), pp. 69, 82. Singer also, (8), p. 48.

[1] 陶弘景   [2] 葛洪   [3] 魏伯陽

(iii) *The origins of the root 'Chem-'*

The first idea which has to be considered is that it was simply part of a personal name. Even though Halen in +1694 attempted to write a biography of Chemes, 'the first author of the sciences',[a] it must be evident from what has so far been said that Chemes or Chymes is far too shadowy a figure on whom to base anything certain;[b] and in any case the nature of the root which formed his name remains to be determined. However, we cannot dismiss persons so easily, for there is someone else in the field, someone whom most chemists have totally forgotten about since they were introduced to the O.T. in their early youth—Ham the son of Noah, brother of Shem and Japheth.[c] In Greek his name was spelt with a *ch* (Χαμ), hence his possible connection with 'chem-'.

But not only this; there was an ancient corpus of legend about him recalling that concerned with Enoch and the angels. According to John Cassianus, who was writing about +428, Cham was expert in all the arts and sciences of the antediluvian generations, and wished to save this accumulated natural knowledge of mankind.[d] However, Noah and his two other sons were so holy that it was not possible for Cham to bring into the Ark any handbooks on the ancient 'superstitious, wicked and profane arts', so he inscribed them on metal plates and buried them underground. After the flood waters went down he succeeded in finding them again, and thus 'transmitted to his descendants a seedbed of profanity and perpetual sin'. In Pseudo-Clement of Rome (fictional material written about +220),[e] Cham figures as the first great magician, handing down his technical knowledge to his sons, especially Mizraim, ancestor of the Egyptians,[f] Babylonians and Persians,[g] and finally being burnt to death by his own conjured star-sparks. Nor is the sexual element lacking, for it will be remembered that after the flood Cham was cursed because he had seen his father's nakedness, Noah being drunk and the weather doubtless hot.[h] That there were books about Cham, or purporting to have been written by him, is not in question, for at a still earlier date Clement of Alexandria, writing just before +200,[i] quotes the Gnostic Isidorus, son of Basilides, as saying (in order to exemplify Greek indebtedness to Hebrew origins) that 'Pherecydes drew on the 'Prophecies of Ham' (*Cham Prophēteias*, Χαμ προφητείας)'.[j] Book-titles deduced from several other patristic references were listed by J. A. Fabricius, when drawing up his censuses of biblical apocryphs and pseudepigraphs at the beginning of the +18th century;[k] but virtually nothing is known of their content.

---

[a] *De Chemo Scientiarum Auctore*, an Uppsala monograph.

[b] As Kopp also saw, (2), p. 77.   [c] See the flood story in Gen. 7. 1 ff.

[d] *Conlationes*, VIII, 21; Petschenig ed., xiii (2), p. 240; Gibson tr., p. 384. Cassianus' dates were +360 to *c.* +435.   [e] *Recognitiones*, IV, xxvii; tr. T. Smith, p. 297.

[f] Cf. the Arabic name for Egypt, al-Miṣr.

[g] Hence a confusion in later writers with Zoroaster.   [h] Gen. 9. 20–27.

[i] *Stromata*, VI, vi; Wilson tr., vol. 2, p. 335. Cf. Hilgenfeld (1), p. 215; Grant (1), p. 139.

[j] For the general background of this, see the interesting study of Pherecydes and his book by West (1), esp. pp. 3, 39, 43, 45.

[k] Apart from the 'Prophecies of Cham' (*Vet.* no. XCII, in vol. 1, pp. 291 ff.), there are also 'Treatises of Cham' on magic and astrology (the *metallorum laminis* in fact) mentioned, besides Pseudo-Clement,

It seems to have been S. Bochartus in +1692 who first launched the idea that *chēmeia* was derived from Cham;[a] he knew the Zosimus passage but believed (wrongly) that the first use of the name of the art had occurred in Firmicus Maternus (p. 340 above). The association, however, was doubtless much older. The great drawback of it is that there is no direct statement in any ancient author that the arts which Cham transmitted were chemical or metallurgical in character. Though Reuvens found the whole story an 'absurd fable'[b] and Kopp a mere 'fantasy',[c] such Victorian scholars' opinions could not discredit the fact that ancient legends are themselves historical data, and Hoffmann was perhaps nearer the mark when he opined that 'chem-' was the mother of Chymes, and Cham his father.[d] But this only leads us back to square one, faced again by the problem of the origin of 'chem-' itself.

If personal names do not solve our problem, could the root have come from the name of a country? That this country was in fact Egypt is a suggestion which has had quite a long run, ever since it was first put forward in relatively modern times by Hermann Conring in +1648.[e] The idea has always rested on one single text, the statement of Plutarch (c. +95) in his book on Isis and Osiris that 'Egypt itself, by reason of the extreme blackness of the soil, is called by them [the Egyptians, or the priests] Chēmia (Χημία), the very same name which is given to the black part or pupil of the eye.'[f] The truth of this, confirmed by modern Egyptology during the past century and a half, cannot be questioned. For example, Champollion affirmed that Chēmi, Kēmī or Kimī (dialectal variations) was 'the veritable and only Egyptian name of Egypt' also meaning black—the people were *rem-chmé* or *reman-chimi*, the black of the eye was *pichēmi ambal*, the Nile was *Ou-chamé*, and so on.[g] In other works he gave the hieroglyphic character, a pictograph of a crocodile's tail, 'emblem of obscurity and darkness', as also the versions in hieratic and demotic.[h] Conring's suggestion was popularised by von Humboldt,[i] adopted by the weighty philologist Pott in 1876, and smiled upon by Hoffmann[j] and von Lippmann,[k] but it has almost insuperable objections, one being that such a derivation of *chemeia* is given by no ancient author. Moreover, as Ruska pointed out,[l] Egypt is never known in Greek literature as

by Epiphanius, in *Haeres.* XXXIX (*Vet.* nos. XCIII and XCIV, in vol. 1, pp. 294, 297). One can ignore Fabricius' numerous late non-patristic authorities. His 'Book of Cham' on chemistry and alchemy (*Vet.* no. XCV, in vol. 1, p. 301) was simply another incarnation of the 'Book of Chemes' already discussed. It is curious that Fabricius listed a 'Book of Mirjam', sister of Moses, on chemistry (*Vet.* no. CLXV, in vol. 2, p. 869); this was a common confusion with Alexandrian Mary the Jewess, the chief reference being to George Syncellos, *Chronographia*, p. 248, who of course knew the Corpus well. The 'Domestic Chemistry of Moses' seems not to have been included. A condensed list of all these presumed texts is given in Migne (1), pp. xliv, xlv.

---

[a] *Phaleg*, IV, 1, in *Op. Omnia*, vol. 3, pp. 203 ff. The book was posthumous, Bochart having died in +1667.

[b] (1), pp. 69–70, in the 3rd letter.　　　　　　　　[c] (2), pp. 66–7.

[d] (1), p. 521, cf. pp. 517–18. He meant, presumably, that by a convergence of sound the Cham epic became accreted with the 'chem-' root.　　　　[e] *De Hermetica Medicina*, p. 19.

[f] *De Iside et Osiride*, ch. 33, Parthey ed., p. 58; cf. Squire (1), text, p. 83, tr. pp. 43–4; Gwyn Griffiths (1). Khme (Coptic) is derivative.

[g] (1), vol. 1, pp. 101 ff. Cf. Erman & Grapow (1), vol. 5, p. 123, no. 1, (2), p. 196, no. 1.

[h] (2), p. 152, (3), pp. 62, 178.　　　　　　　[i] (1), 1847 ed., vol. 2, p. 451.

[j] (1), p. 524.　　　　[k] (1), p. 295. One can add Gundel (4); Forbes (32).

[l] (11), pp. 319 ff.

Chēmia but always as *Aigyptos* (Αἴγυπτος). And this is also true specifically of the Corpus, where all mentions of 'the holy art of Egypt' or 'the techniques of the Egyptians' use the Greek name, not the Egyptian one,[a] just as we normally say Finland rather than Suomi. Favoured though it has been by popular books on the history of chemistry, this derivation is too weak to stand.

But suppose we retain the notion of 'blackness'? The Alexandrian proto-chemists of the Corpus certainly had a lot to say about 'blackening' (cf. pt. 2, p. 23), and taking up Plutarch's hint one might be disposed to think of *chēmeutēs* as 'blackener' or '*black-maker*', and *chēmeutikai bibloi* as 'books concerned with blackening'. This was Hoffmann's favourite theory,[b] but it strikes the uncomfortable fact that the normal Greek word for blackening was *melan* (μέλανσις) or *melasmos* (μελασμός).[c] Worse still, the Corpus generally speaks of *melanōsis* (μελάνωσις).[d] The word is not in what is left of Pseudo-Democritus, who has only *leucōsis* and *xanthōsis*,[e] but starts with Zosimus,[f] and when Olympiodorus comes to speak of a 'black preparation' he says *melana zōmon* (μέλανα ζωμόν).[g] The fact is that 'blackening' is never called *chēmi* or *chēmeia* in the Corpus (or the papyri either), and so in spite of fancied resemblances with the 'black art' of medieval times, there is simply no ground for deriving the 'chem-' root from the notion of blackness.[h]

But was there no other word in the Ancient Egyptian language embodying the phoneme 'km-' which could be relevant to the origin of 'chem-'? One at least has been suggested, in an ingenious and interesting but hardly convincing theory. The verb *km* meant 'to complete, achieve, attain, execute' or 'bring to a close', and in pyramid texts it was indeed applied to the making of ointments or metalwork.[i] From it was derived (so it is thought) the title of a book of the − 3rd millennium, *Kmj.t* (Book of Completion), some fragments of which have survived to this day. This was not a 'wisdom-book',[j] nor a religious text, nor a fictional work, but a compendium of excerpts from the best accepted writings intended to teach sound grammar to the youthful scribe, and a good style of composition.[k] It was probably being put together already by about − 2300, for it must have been widely current and in standard use by − 1970, in the time of Sesostris I of the XIIth Dynasty,[l] since Kheti the son of Duauf quoted it in his 'Satire on the Trades', the most famous example of what became a characteristic genre of Ancient Egyptian literature.[m] These texts exalt the

---

a See for example *Corp.* I, xiii, 1 and II, ii, 1.

b (1), pp. 525, 529.   c Ruska (11), *loc. cit.*

d The spelling varies. *Melanōsis* occurs in Comarius (*Corp.* IV, xx, 5) and Moses (*Corp.* IV, xxii, 47), but *melansis* (μέλανσις) in Zosimus and Olympiodorus, *passim*, as also *Corp.* III, xliv, 5 = VI, xv, 5 bis, anonymous texts.   e Cf. Vol. 5, pt. 2, p 23.

f Berthelot & Ruelle (1), vol. 2, pp. 107–252.

g *Corp.* II, iv, 40.

h Von Lippmann (1), pp. 301–7 notwithstanding, who supported Hoffmann in this.

i Erman & Grapow (1), vol. 5, p. 128, no. 12, (2), p. 195, no. 12.

j As some lexicographers tentatively defined it; Erman & Grapow (1), vol. 5, p. 130, no. 12.

k It was assuredly the book *Qemi* which Berthelot (1), p. 10, heard about from G. Maspero (2), p. 125. Chinese parallels will come up for discussion in Vol. 6, Sect. 38 on botany.

l *Kmj.t* is known to have been very popular also in the XIXth and XXth Dynasties (c. − 1320 to − 1160, contemporary with Shang times in China).

m See the general survey of van de Walle (1).

scribal function according to the general theme that 'a clever scholar is worthy to stand before rulers',[a] and they warn the student to keep his nose well into the book *Kmj.t* if he wishes to avoid the miserable lot of all those manual workers—potters, builders, weavers, dyers, brewers, smiths and metal-smelters, sailors and farmers—men of aching backs and stinking hands.[b] This genre is familiar to many of us because a typical excerpt from it was given by Wallis Budge in his classical introduction to Ancient Egyptian,[c] and several other translations have appeared.

Now A. Hermann (1), to whom we are indebted for much of the foregoing information, has suggested that the name of the 'Book of Chymes', so vital for us, was derived from nothing other than the *Kmj.t*.[d] It is not claimed that there was ever anything at all in that ancient book about metallurgy, chemistry or other techniques, simply that it was quoted by Kheti son of Duauf in a text which had something to do with them. Admittedly, again, his text was written precisely in disdain of those techniques,[e] nor do any extant versions of the 'Satires' actually describe any of the details of the operations.[f] Hermann however felt that by the Ptolemaic period labour conditions had eased with rising technological invention, so that chemical-metallurgical work, especially with the precious metals, became more attractive to intellectuals. But it seems to us bizarre that the Hellenistic proto-chemists should have drawn the name of their holy proto-bible (or unholy, according to your point of view), from an ancient book which never had any connection with the chemical arts, except in so far as it was quoted in satires against them. We conclude that it would be quixotic to derive the root 'chem-' from this source.

Perhaps the most obvious source of it would be a group of Greek words with the general sense of liquid and pouring—verbs like *cheein* and *chōneuein* (χέειν, χωνεύειν), to melt or pour, and *epicheein* (ἐπιχέειν), to pour on or off—nouns like *chyma, cheuma* (χύμα, χεῦμα) meaning fusion or the molten state, *chymos* (χυμός) juice or liquid (cf. chyme, p. 366),[g] finally *chytra, chytridion* (χύτρα, χυτρίδιον) and many similar forms,[h] signifying crucibles of different kinds. There is no question that these occur very frequently both in the Corpus[i] and the papyri, and the derivation is a good deal more attractive than

---

[a] Cf. Prov. 22. 29. Ecclesiasticus 38. 24 to 39. 11 reproduces closely a version of the trades-satire literature, but although it debars the manual workers from high political leadership, it is highly appreciative of their work. This is more than can be said of the classical Greek attitude to 'banausic' occupations.

[b] The descriptions, of brutal frankness or abusiveness, are often very exaggerated.

[c] (5), pp. 212 ff., where it appears as 'The Proverbs of Ṭuauu-f-se-Kharthåi'.

[d] R. J. Forbes, after having adhered to Egypt as such (32), rallied to Hermann's proposal, as we know from letters published by Mahdihassan (15), pp. 94–5.

[e] Indeed it is one of the most ancient extant statements of the differentiation of social classes, contrasting the learned scribe sitting at ease with those who sweat under the sun or by the fires of forge and furnace.

[f] This greatly weakens Hermann's suggestion that by the time of Zosimus, the *Kmj.t* was just confused with the 'Satire on Trades'.

[g] This was the preference of Mahn (1), but for a bad reason. Though he called in (Skr.) *rasāyana* (cf. pp. 352, 498) to his support, he believed that the earliest phase of Mediterranean chemistry was primarily medical and herbal, which is just the opposite of the truth.

[h] E.g. *chōnos, chōnon, chōnē, chythra, chōstra* (χῶνος, χῶνον, χώνη, χύθρα, χώστρα).

[i] See for example III, xli, 2, xlii, IV, xxii, 38, V, i, 1, 16, VI, xiv, 6, 8, 15, xv, 4, 8. Also Stephanus of Alexandria (Ideler ed.), pp. 202, 210, 212.

others so far considered. It goes back to Rolfinck's *Chimia*[a] and Vossius' *Etymologicon*,[b] both of +1662, and has had impressive support, e.g. from Hoefer,[c] Gildemeister (1) following Cl. Salmasius (+1588 to +1653),[d] and Diels (1, 4), to say nothing of Stephanides (3) and Hammer-Jensen (2); and it was the considered conclusion of Ruska (11). Ruska felt that the identification was particularly convincing when he found the words of the Philosophus Anonymus (late +7th cent.):[e] 'The present book is called the "Book of Metallurgic and Chymeutic (Art)" ($\mu\epsilon\tau\alpha\lambda\lambda\iota\kappa\dot{\eta}$ ($\kappa\alpha\dot{\iota}$) $\chi\upsilon\mu\epsilon\upsilon\tau\iota\kappa\dot{\eta}$ ($\tau\acute{\epsilon}\chi\nu\eta$)), dealing with the fabrication of gold and silver, and the fixation of mercury, with vapours, with tinctures prepared from living things,[f] with the making of green gems and shining gems and all other colours, with pearls and the dyeing red of skins (leather) fit for princes; all this with the help of brine and eggs[g] and the metallurgical art'. Thus it was certainly a chemical compendium. But even this solution is subject to a serious criticism, namely that most of the words mentioned had been in use in classical Greek long before. Why should *chymeia* and *chēmeia* have waited till the +3rd or 4th century before making their appearance? Among the Greek inscriptions *chyma* or *cheuma* in the sense of ingot can be found in −170 or −70, and *chytra* even earlier.[h] Similar forms occur in Aristotle's *Historia Animalium* (c. −350) and even in the Hippocratic Corpus, evidence for the second half of the −5th century.[i] Hoffmann added other arguments—*chēmia* or *chēmeia* never applies in Greek to a preparation or a substance, as later Arabic *kīmiyā'* certainly does; *chymoi* in the Corpus means bodily humours only; liquids as such are usually called *zōmoi* ($\zeta\omega\mu o\acute{\iota}$)[j] or *chyloi* ($\chi\upsilon\lambda o\acute{\iota}$); and the word which might most logically be expected, *chymateia* ($\chi\upsilon\mu\alpha\tau\epsilon\acute{\iota}\alpha$) is never found. In sum, the derivation from 'pouring' and 'molten' is much less convincing when closely considered than it seems at first sight.[k]

The only other well-known proposal was the paradoxical one of Lagercrantz (2) that by a process common in Indo-European linguistics *chymeia* was derived by an inversion of consonants from *moicheia* ($\mu o\iota\chi\epsilon\acute{\iota}\alpha$), i.e. adultery, falsification, counterfeiting, deception.[l] But this was completely demolished by Ruska (11), chiefly on two grounds, first that *dolos* ($\delta\acute{o}\lambda o\varsigma$) is the word always used in the papyri and the Corpus

---

[a] *Chimia in Artis Formam Redacta*, p. 19. Werner Rolfinck (+1559 to +1673) was the first professor of chemistry at Jena; cf. Partington (7), vol. 2, pp. 312ff.

[b] P. 17. Ruhland in his *Lexicon Alchemiae* (+1661) held the same: 'Chymia *apo to chuo* fundo. Unde *chymē* succus et chymia ars succum faciens, seu res solidas in succum resolvens...' (p. 149).

[c] (1), 1842 ed., vol. 1, p. 219, 1866 ed., vol. 1, pp. 226, 275.

[d] The idea can even be traced back to Ermolao Barbaro (d. +1493), *In Dioscoridem Corollariorum Libri Quinque* (Cologne, +1530). Cf. Kopp (2), p. 74, followed by Schorlemmer (1).

[e] *Corp.* III, xliv, 7.

[f] *Murex* purple for example.   [g] Cf. pt. 2, pp. 73–4, 253.

[h] See Boeckh (1), *Corpus Inscriptionum Graecarum*, no. 161 (vol. 1, p. 286), no. 1570 (vol. 1, pp. 750–3).

[i] For the *De Arte*, 12, is of genuine Hippocratic date, though perhaps Cnidian rather than Coan.

[j] Cf. *Corp. Alchem. Gr.* III, xxxix, 6.

[k] Naturally von Lippmann (1), pp. 295ff.; Pott (1); and others, on account of their different preferences, agreed with this.

[l] The idea seems to have arisen in part from what may be one of the earliest texts in the Corpus (1, xiii, 1), the 'Epistle of Isis to Horus'. Here Isis is visited by exalted angels, notably Amnael, who seek to lie with her, but in return she successfully obtains from him the great secrets of proto-chemistry (gilding, leucosis, etc.). The text (which may be of the +1st century) clearly echoes the Enoch myth; though the word *moicheia* does not in fact occur in it.

for falsification, and secondly, more important, that it is a wholly erroneous idea of Hellenistic proto-chemistry to think of it only or even primarily as falsification. Readers of this volume will be well aware from pt. 2, pp. 21 ff. that in the Hellenistic world there was not only aurifiction but also aurifaction—and very numinous at that— the 'holy and divine art'. Lastly, the Arabic authorities quoted by Lagercrantz were carefully dissected by Ruska, no man being better qualified to do it, and he showed that although the meaning of falsification does attach to some extent to *al-kīmīyā*, it is always far from the whole story.

If Egyptian and Greek do not help, what about returning to Hebrew? There is a word *chometz* (or *chāmetz*) which means leavened bread, hence leaven itself, or fermentation, though the technical term for yeast is *sĕ'or*; thus forming a parallel with *maza* and *zymē* (p. 365 below). Since there was a clear association between Hellenistic and Jewish proto-chemists earlier than the first appearance of the word *chēmeia* itself,[a] this possibility of origin must be taken seriously—especially as *maza* and *massa* came to be synonymous with alchemy in later times. The same Semitic root appears in Arabic and Aramaic as *khamir* (leaven, leavened bread). *Al-kīmiyā'* would hardly have derived direct from that, however, but rather as the offspring of *chēmeia* (pp. 355, 481) by way of Syriac. Such a derivation of 'chem-' from *chometz* derives its force from the prominence of the idea of fermentation in the process of projection by the philosopher's stone (p. 367 below). Projection was in fact a kind of fermentation. But here the difficulty is that the fermentation concept was only one among many multifarious thoughts and techniques envisaged in aurifaction and aurifiction, not universally applicable, and competing with colour-change, death-and-resurrection, dilution, distillation, reflux distillation, etc. Still, of all the proposals for explaining the prophet Chemes and his book, this one seems among the more attractive.[b]

If then it must be concluded that none of the classical and Western derivations of the root 'chem-' are entirely satisfactory, room should be left open for other suggestions, and one there is which particularly concerns us.

In 1946 I suggested that 'chem-' should be equated with Ch.*kim*,[1] i.e. *chin*,[1] as in *lien chin shu*,[2] the commonest expression for 'the art of transmuting metals (or gold)', generally implying aurifaction or aurifiction.[c] I remarked that this phrase would have been pronounced in Cantonese *lien kim shok*; but actually a more correct transcription of that kind would be *lin kêm*[d] *shut*.[e] The idea then in mind was that Canton was the terminus for so long a time of the sea route to and from China frequented by the Arabs, but since we know that the term *chēmeia* (*chimeia*, *chymeia*) was in use by the

---

[a] Cf. pt. 2, pp. 17, 19, 74, 253.

[b] We are indebted to Mr Ronald Hassett and Prof. E. Wiesenberg for suggesting and perpending this possibility. Yet another derivation could be drawn from (Heb.) *ḥokhmah*, wisdom, since the initial guttural of a root like *ḥkm* tended to disappear in transliterations (Wright (1), pp. 48 ff.; Brockelmann (3), pp. 120 ff., (4), pp. 48 ff.; Moscati (1), pp. 39 ff.). This could have given *chi* or *k* and *m* for the consonants, the vowels being, as we saw, very variable. Cf. Job 28, and obvious links with *sophia* and *logos*, but nothing about gold or anything operative, so the connection would have been very abstract.

[c] Needham (58), Fr. ed., p. 209, Eng. ed., p. 216.

[d] Near *gêm* or *gum*, with hard *g*.   [e] Or *ssu(t)*, as spoken by my friend Chhen Fei-Hua.

[1] 金   [2] 煉金術

+3rd century, an overland route such as the Old Silk Road would seem a more likely channel of communication—and transmission, if such there was. Hence it is important that the key word, chin,[1] was (and still is) pronounced kim, kin, king[a] in many other dialects and related languages, not only Hakka, Amoyese and Fukienese,[b] but also Korean, Japanese and Annamese. What is much more important is that the ancient pronunciation throughout China would have been something like lien kiem dzhiuet (lien kiam dẑ'įuĕt),[c] so that a root 'kem-' could conceivably have come Westwards.

Unknown to me at the time, a similar suggestion had been put forward earlier in the same year by Mahdihassan (14). According to present recollection my proposal had originated during the war years in many conversations with chemical friends in China, such as Huang Tzu-Chhing, Li Hsiang-Chieh and Chang Tzu-Kung. Mahdihassan on the other hand had been led to it in the course of his studies on words of possible Chinese origin in other languages such as Arabic, Persian, Turkish, Urdu, Hindustani and English,[d] pondering especially (38) the case of kincob,[e] a Hobson-Jobson word for brocade, gold damask, and cloth of gold.[f] There was already good authority[g] for tracing this to chin hua[2,3];[h] whence Persian kimkhwā, Hindi kimkhwāb and—most significantly—Byz. Greek kamchanē, kamouchas and chamouchas (καμχανή, καμουχᾶς, χαμουχᾶς),[i] current from the end of the +13th century onwards. Mahdihassan first attempted (14) to derive chēmeia, kīmiyā', from chin mi,[4] 'gold deception' or 'infatuation',[j] and then from chin mi,[5] 'gold secret';[k] but a few years later, perceiving that the middle consonants were tautologous, proposed (9)[l] chin i,[6] 'gold juice'

---

[a] Always with the hard g, approximating k but not aspirated.

[b] Fuchow in particular.

[c] K185j, 652a, 497d. Again the k definite though probably spoken as a hard g. Karlgren used a date around +600 as his 'ancient' point de répère, and −700 for his 'archaic'; the latter forms are conjectural and we need not consider them for our present purpose.

[d] See, e.g. (1) on monsoon, (2) on turquoise and jade, (3), (5) on porcelain, (6) on carboy, (11) on dijinn, (35), (36) on godown, (37) on plague, (49) on paper, (50) on kutcherry and tussore.

[e] Current in English from +1712 onwards.

[f] The meaning could include all kinds of polychrome drawloom flowered or patterned silks, but gold thread was prominent in the conception. Cf. the confusion referred to in Vol. 1, p. 6.

[g] Yule & Burnell (1), p. 368.

[h] The pronunciations of these characters vary respectively only in tone.

[i] For the first of these forms the reference is to the 'Letter of Theodorus the Hyrtacenian to Lucites, Protonotary and Protovestiary of the Trapezuntians', written in +1300; the other two are in du Cange's glossary of late Greek, defined as 'pannus sericus, sive ex bombyce confectus'. A reference of +1330 is in Yule (2), vol. 3, pp. 99, 155, vol. 4, p. 17.

[j] Apart from the fact that the phrase is practically unknown in Chinese, this suffered from the same defect as the theory of Lagercrantz (2) just discussed.

[k] Also unknown in Chinese, a pure construct. The second character in the phrase would certainly have been pronounced pi in ancient and medieval times anyway. Mahdihassan realised this, but did not think of mi,[7] the easy way out—but, as ancient Chinese phraseology, equally poor.

[l] And often afterwards asserted (13, 15, 20, etc.). We are not able to follow him in all his suggestions. For example in (15), so far as we can understand it, he made a sharp separation of (Gk.) chymeia from chēmeia, though we believe the spellings were indiscriminate. The former he took to have been a translation from (Skr.) rasāyana, 'the way of the juices', and that in turn a direct translation from (Ch.) i tao,[8] though this is an expression we have never encountered in that language (cf. (12), p. 90, (21), p. 173). Chēmeia on the other hand he derived from (Ar.) kīmiyā', rather than the reverse (cf. (25), p. 40), this being itself derived from (Ch.) chin i;[9] but he also left room for its being purely Egyptian, i.e. from the

---

[1] 金   [2] 金華   [3] 錦花   [4] 金迷   [5] 金秘   [6] 金液
[7] 密   [8] 液道   [9] 金液

or 'liquid', the ancient pronunciation of which would have been *kiem iak* (*kiem i̯äk*).[a]

Mahdihassan found this phrase in one of the translations of Wu & Davis,[b] who rendered it 'gold fluid'; but it was the same as Ware's 'potable gold' or 'gold exudate'[c] in the *Pao Phu Tzu* book.[d] Indeed Ko Hung often mentions a *Chin I Ching*[1] (Manual of the Potable Gold),[e] not now identifiable. In the present volume we have frequently used the phrase 'potable gold' (pt. 3, pp. 40, 49, 82–3), though we have also been obliged to recognise the quite different meaning of 'metallous fluid' in physiological alchemy (pt. 2, p. 90). Ware tentatively fixed the proto-chemical sense of the term as mercuric oxide,[f] though the *Shih Yao Erh Ya* gives it as a synonym of mercurous chloride (calomel, cf. pt. 2, p. 152);[g] probably it was a phrase with many meanings, varying with schools and periods. Mahdihassan liked to think of it as 'gold-making plant-juice',[h] linking this up with Ayurvedic *bhasmas* and his theory (discussed above, pt. 3, pp. 48–9) of the ancient use of organic acids in the surface-enrichment of gold-containing alloys.[i] However that may be, there is no dispute about the ubiquity of the phrase *chin i*, and it goes back quite a long way, for apart from the *Pao Phu Tzu* book and its long antecedent tradition, it occurs in those versions of the *Lieh Hsien Chuan* (Lives of Famous Immortals)[j] which contain the biography of Ma Ming-Shêng[2] (*fl. c.* + 100).[k] It is also mentioned in one of the poems of Shen Yo[3] (+ 441 to + 513). There is little doubt that it was current early enough to have travelled Westwards in Han times, though one may well feel that the word *kiem* (*kim, chin*) itself was sufficient, without any companion word.[l] Mahdihassan's equation was afterwards accepted by Dubs (34), Schneider (1) and others, while many have found it attractive.[m]

What arguments present themselves as favourable for such an unexpected, and (at first sight) unlikely, transmission? First it has been well observed that none of the

book *Kmj.t* discussed in the note on p. 349 above. Subsequently, however, Mahdihassan saw the improbability of *chēmeia* being derived from *kīmiyā'*, and boldly suggested that it might have come from *chin i*[1] directly (cf. (21), p. 174, (30), p. 43, (32), p. 340, (33), pp. 102ff. and (22), (19), p. 4), though he clung to the idea of pre-Islamic Arabs, perhaps sailors or sea-merchants, as the intermediaries who carried it to Alexandria.

---

[a] K800*n*.   [b] (2), p. 250.   [c] (5), pp. 64, 68–9, 89ff., 112.
[d] Ch. 3, p. 7*b*; ch. 4, pp. 1*a, b,* 14*a, b,* 16*a, b*; ch. 6, p. 3*a*.
[e] Ch. 4, pp. 2*a*, 15*a*; ch. 18, p. 2*a*; Ware (5), pp. 70, 91, 303. Several book titles beginning with this phrase will be found in our bibliography.
[f] We know of no justification for this (cf. RP44). Stannic sulphide is far more likely (cf. pt. 3, p. 103, pt. 2, p. 271).
[g] Ch. 1, p. 1*b*. Cf. RP45, 46.   [h] (9), p. 120.
[i] Unfortunately, so far as we know, *chin i* in Chinese never refers to the juice or extract of a plant.
[j] The material in this dates from − 35 onwards, stabilised finally by about + 400. Cf. Dubs (34), pp. 33–4.
[k] See also that in *YCCC*, ch. 106, pp. 20*b*ff. The historicity of this character is in doubt (cf. Sivin (1), p. 58) but that does not affect the argument.
[l] Mahdihassan (9), p. 115, (15), p. 82, drew support for his doublet from the lexicographer 'Abdallāh al-Ṣafadī (+ 1297 to + 1363, cf. Mieli (1), 2nd ed., p. 268) who averred that *kīmiyā'* had two syllables only. This confirmed his view that *kīm-* was indissoluble. But it did not prove that more than one word had been transmitted.
[m] E.g. Prof. H. B. Collier (priv. comm. 1952), Prof. B. Farrington (priv. comm. 1961), Prof. Guido Majno (priv. comm. 1968).

[1] 金液經   [2] 馬鳴生   [3] 沈約

usual derivations of 'chem-' from Greek, Hebrew or Egyptian origins have any connection with names for gold, though aurifiction and aurifaction constituted the central pivot of ancient proto-chemistry in so far as the elixir theme was missing from it.[a] 'Chem-' would thus have a built-in connection, as it were, with *chin i* (i.e. *kiem i̯äk*),[b] or more simply with *chin* (*kiem*) alone. This is a consideration of real weight. The fact that another *chin* (*kiem*) derivative could generate a *chi* (χ) in Byzantine Greek (p. 352 above) is also very relevant here. The Old Silk Road began to function soon after − 110 (cf. Vol. 1, p. 176), and if any one thing would have interested the merchants who handled its cargoes through Central Asia more than anything else it would have been gold. If just after the time of Li Shao-Chün (− 133) it became common knowledge that Chinese adepts and technicians had ways of fabricating gold artificially,[c] talk about 'the goldery', 'the gold art' or 'liquid gold' would almost naturally have been expected to spread over the rest of the Old World. Whoever were the real Persians who stood behind the name Ostanes (cf. p. 334), so revered by Pseudo-Democritus and his successors, might they not have brought with them among their semi-mystical impedimenta the Chinese name for a Chinese technique, whether or not in fact the manipulations of the Persian and Graeco-Egyptian thaumaturgists closely resembled it? There is much evidence of contact between China and the Eastern Mediterranean both by land and sea between the − 1st and + 3rd centuries, Parthian middlemen and Roman-Syrian traders being prominent in the picture.[d] Indeed the dates are even consistent with a seed sown by Chang Chhien[1] himself, who was away in Western Central Asia between approximately − 136 and − 126, or by one of his followers, or by some members of the other slightly later Chinese expeditions to that region of Sino-Iranian contact.[e] He and they had plenty of time to talk about their ideas and beliefs, explaining thaumaturgical Taoist metallurgy to some sympathetic Hun or Bactrian Greek, for example, and no doubt they also carried with them scrolls and other material objects. Two historical facts which make this kind of conversation very plausible come at once to mind. We know that iron-casting was introduced to Ferghana and Bactria by Chinese technicians in the neighbourhood of − 110.[f] And we have a striking example of the kind of thing that Chinese travellers used to like to talk about, in the account of the interview of the Buddhist pilgrims Hui-Shêng[2] and Sung Yün[3] with the King of Udyāna, when they expatiated upon the palaces of silver and gold of the holy immortals, and many other aspects of Taoism, medical and proto-scientific.[g]

a Mahdihassan (15), p. 81.
b Mahdihassan (14b), (21), p. 174, (22), (30), p. 43, (32), p. 340, (33), pp. 102ff.
c And this is to say nothing of the preceding aurifiction (cf. pt. 3, pp. 26ff.), going back to − 144 for certain, and probably half a century earlier than that.
d See Vol. 1, pp. 150, 155, 157ff., 177ff., 181ff., 191ff., 197ff., 199ff. and Vol. 4, pt. 3, pp. 422ff. Some scholars have not hesitated to assert their belief that scraps of Chinese alchemy, if not more, were transmitted to the eastern Roman world during the + 1st century; cf. Wang Chi-Min (1), p. 10. On Indian–Alexandrian contacts at this time see the reviews of Frend (1, 2).   e See Vol. 1, pp. 173ff.
f Vol. 1, pp. 234-5. We have discussed this further in pt. 2, p. 219 above.
g Vol. 1, pp. 207, 209. This was in + 520 or thereabouts, but the instance is a striking one, for although the two were Buddhists, they were also patriots, and Taoism (with its corollary of alchemy) was a constituent of the national culture.

¹ 張騫   ² 惠生   ³ 宋雲

If some have found an influence of *chin* (*kiem*) on *chēmeia* (*chimeia*, *chymeia*) difficult to accept, there has been less desire to question its influence on *al-kīmiyā'*.[a] No Arabic etymologist ever produced a plausible derivation of the word from Semitic roots,[b] and there is the further point that both *chin i* and *kīmiyā'* could and did mean an actual substance or elixir as well as the art of making elixirs, while *chēmeia* does not seem to have been used as a concrete noun of that kind.[c] We are left then with the possibility that the name of the Chinese 'gold art', crystallised in the syllable *chin* (*kiem*), spread over the length and breadth of the Old World, evoking first the Greek terms for chemistry and then, indirectly or directly, the Arabic one.

(iv) *Parallelisms of content*

We have now reached the point when we can ask in what sense there was a community of interest between the proto-chemistry of the Hellenistic world and the alchemy of China. Clarification of this will raise in a natural way the further question of whether or not we ought to think of transmissions at this early time, and if so in what directions. A natural division presents itself—things, methods and ideas—so let us pursue it.

Before doing this, however, we must just recall that fundamental distinction which we were obliged to make in Vol. 5, pt. 2. The reason why we wrote 'proto-chemistry' for the Graeco-Egyptian world in the preceding paragraphs, and 'alchemy' for China, was because macrobiotics (longevity or material immortality attained by the aid of chemical knowledge) played hardly any part in the former culture but was central for the latter. The papyri described aurifiction,[d] and the adepts of the Corpus visualised aurifaction,[e] but the 'drug of immortality' was primarily a metaphor in the Greek context and primarily a real material thing for the Chinese.[f] There seem to be no exceptions to this rule,[g] and it needs bearing in mind during the sub-sections that follow.

[a] The initial consonant here is definitely *kāf*, not *kh*. This alone would exclude a possible derivation from *khamir* (leaven, yeast, ferment), as my friend Dr Said Durrani has pointed out. We are grateful also to Professor R. B. Serjeant and Professor D. M. Dunlop for advice on this subject.

[b] For example, the derivations of al-Ṣafadī, and Muḥ. al-Khwārizmī al-Kātib (+976), from *kama*, to conceal, and (Heb.) *Yahwé*, God, could not today be taken seriously. Cf. Mahdihassan (9), p. 115, commenting on Wiedemann's article, *s.v.* in the 'Encyclopaedia of Islam' (30). On this al-Khwārizmī see Mieli (1), 2nd ed., p. 94, and Wiedemann (7). About the same time (+980), Bar Bahlul, in his Syriac lexicon, connected *chēmeia* with Kīmā, the Pleiades, eight mixtures being governed by eight stars (cf. Berthelot & Duval (1), p. 133; Ruska, 13). The same work also confused it with *kémélaya* (*chameleon*, χαμαιλέων), because of the colour changes in chemistry (cf. Hoffmann (1), p. 530; Berthelot & Duval, *loc. cit.*).

[c] Reiterated in his usual way by Mahdihassan (9), p. 109, (15), p. 83, (17), p. 67, (25), p. 40. Arabic has to say *ṣanʿat al-kīmiyā'*, *'ilm al-kīmiyā'*, 'the craft of...', or 'the science of...', chemy.

[d] Vol. 5, pt. 2, pp. 18ff.   [e] *Ibid.*, pp. 16ff., 21ff.

[f] *Ibid.*, pp. 72ff.

[g] Occasionally something suspicious turns up, but it tends to evaporate on analysis. Almost a century ago, Draper wrote: 'Ptolemy II, Philadelphus (r. −285 to −246) was haunted towards the close of his life by an intolerable dread of death, and spent much time in the discovery of an elixir, devoting himself to this with great assiduity. There was a chemical laboratory (in the Museum at Alexandria) to which people flocked from all countries', (1), p. 20, (2), vol. 1, pp. 189–90, conflated. Justification for this statement has never been forthcoming; there is nothing to be found in the works of the standard historians such as Mahaffy (1) and Bevan (4), though we learn that he was an enlightened monarch, interested in Egyptian religion, doing much for commerce and industry, a benefactor to the Museum and Library, and devoted to mistresses both Egyptian and Greek after the death of Arsinoe II. Only in

On the substances known and used there is not much to say, for as far as one can see they were quite similar in East and West. The two noble metals and all the ancient base metals were known to everyone, as also mercury and its sulphide, cinnabar, together with the two sulphides of arsenic naturally occurring. Besides these there were a number of crude ores and minerals, a variety of salts and alums, and the oxides of the metals.[a] In nomenclature there was of course no similarity, with a few exceptions; lead was 'black lead' and tin was 'white lead' in China as in the West,[b] while it is somewhat eyebrow-raising to find that the epithet 'male' was applied at both ends of the Old World to one of the sulphides of arsenic. However, *hsiung huang*,[1] the 'male yellow', was the disulphide, realgar, while *arsenicon* (ἀρσενικόν) was *auripigmentum*, orpiment, the trisulphide.[c] It would not be wise to build very much upon this curious fact, in the absence of any other evidence of contact, which remains to be seen. It is also noteworthy, however, that the greatest achievements of ancient chemistry (apart from metallurgy) both concerned sulphides, the calcium polysulphides among the Greeks and stannic sulphide in China. And while we are upon the subject of colours,[d] there is always that disturbing fact that both in Hellenistic Alexandria and in China purple was of such great importance. Among the Greeks it was the highest stage of the transmutation of other metals into gold, the *iōsis* in the *chrysopoia*, while among the Chinese 'purple sheen gold' was the most wonderful of the forms of gold, and purple was the colour of Taoist ineffability.[e]

Lists of names and cover-names[f] occur in both civilisations.[g] The greatest in China was probably the *Shih Yao Erh Ya* compiled just after +800 (cf. pt. 3, p. 152), corresponding quite reasonably to the *Lexicon of Chrysopoia* contained in the Corpus, which may date from the end of the +10th century.[h] Parts of this are doubtless much

Athenaeus (*Deipnosophists*, XII, 51) do we get a clue: 'Phylarchus, in the 22nd book of his Histories, tells us that "Ptolemy of Egypt, the most admirable of all princes, and the most learned and accomplished of men, was so beguiled and debased in his mind by his unseasonable luxury that he actually dreamed that he should live for ever, and said that he alone had found out how to become an immortal".' This however could have been theurgic, or something like the deification of Roman emperors—certainly it is not in itself enough foundation for attributing alchemical elixirs to Ptolemaic Alexandria. If the account had referred to the +3rd century rather than to the −3rd, it might have been tempting perhaps to see in it a remote echo of Chinese Taoist alchemy, especially in the light of what has been said in the previous pages about contacts and transmissions, but Ptolemy Philadelphus is really rather too early for this. It was Dr S. Mahdihassan who first noticed this passage in Draper and brought it to our attention.

a Further similarities in mineral materials are fairly obvious—stalactites and stalagmites for calcium carbonate (Vol. 3, p. 605), aetites with its legends (Vol. 3, p. 652), and 'thunder-axes' (Needham (56), p. 34). On all these things de Mély (6) is is still worth reading.

b Cf. Berthelot (1), p. 230, and Bailey (1), *s.v.* for Pliny. In China lead was also 'black tin'.

c Berthelot (2), p. 210. De Mély (6), p. 328, saw this coincidence (if such it is) long ago, but confused the two sulphides.

d Cf. the interesting and learned discussion of Dronke (3).

e Vol. 5, pt. 2, pp. 23, 253, 262 ff., pt. 3, pp. 173, 194.

f One of the frankest statements on the usage of cover-names occurs in the Jābirian 'Book of the Western Mercury', tr. Berthelot & Houdas (1), p. 214. Here cf. Siggel (3).

g And both tended to poetical fancies such as the blood, semen, horn or other parts, of mythical or coloured animals.

h *Corp. Alchem. Gr.*, 1, ii. See also Berthelot (1), p. 24, (2), pp. 10–11; Festugière (1), pp. 220–1; von Lippmann (1), pp. 11, 325–6.

1 雄黃

older, though some versions betray their date by giving Arabic terms in Greek,[a] but Mei Piao's work was certainly not the first in China. What is more startling, in view of the universal Chinese use of the expression *huang pai chih shu*,[1] 'the art of the yellow and the white', for aurifiction and aurifaction, is that Ps.-Democritus is said to have prepared a 'Catalogue of the Yellow and the White'. So Synesius,[b] but Zosimus also refers to his 'Books of the White and the Yellow'.[c] Of course, such expressions referring to silver and gold would be natural enough, and could have arisen, one supposes, quite independently.

Passing to the methods used, we need say little about apparatus, since this is discussed in detail elsewhere (pp. 83 ff.), and we have given attention also to the fundamental types of operation used (pp. 8 ff.), showing that there was little or nothing to choose between China and the West in this particular.[d] Of outstanding interest here, however, is the fact that 'projection', i.e. great chemical change (often from base to noble metal) brought about by the addition of only a small amount of some other chemical substance, occurs early at both ends of the Old World, with China perhaps slightly leading. *Tien*[2] in Chinese was *epiballein* (ἐπιβάλλειν) in Graeco-Egyptian circles. It seems to have been present in −1st-century China, judging from the reports concerning Mao Ying *c.* −40 and Chhêng Wei, whose *floruit* was either −95 or *c.* −20; later on, after Yin Kuei about +300, it becomes exceedingly common.[e] In spite of attempts to deny the fact,[f] there can equally be no doubt that it is fully developed in the Hellenistic Corpus, occurring not only in Olympiodorus[g] and Synesius[h] but in Zosimus[i] and even in Ps.-Democritus himself;[j] this means from the +1st century onwards.[k] So also transmutation (*hua*[3]) is represented by several words —*diabasis* (διάβασις), *strepsis* and *ecstrepsis* (στρέψις, ἔκστρεψις), *strophē* and *ecstrophē* (στροφή, ἐκστροφή), while *cataspaō* (κατασπάω) seems to have meant the pulling away of the previous nature or form (cf. pt. 2, p. 22).[l] Even the philosophers' stone is there, the 'stone that is not a stone' (*lithon ou lithon*, λίθον οὐ λίθον),[m] and the powder (*xērion*, ξήριον) of projection.[n] Presumably it is conceivable that all this should have developed

---

[a] Berthelot & Ruelle (1), vol. 3, p. 18.

[b] *Corp.* II, iii, 2; cf. Berthelot (1), pp. 155–6.   [c] *Corp.* III, xxv, 2, 3.

[d] At any rate, in general principle. Distillation and reflux distillation seem to have started earlier in the Hellenistic world than in China, where there was more skill anciently in steaming techniques; and when distillation developed in China it was with a logically antithetical design. Cf. pp. 62 ff. above.

[e] For the details see pt. 3, pp. 38–9 *et passim*.

[f] E.g. Dubs (5), p. 81, in answer to which the late Prof. J. R. Partington supplied us most kindly with Corpus references (priv. comm. 1959).

[g] *Corp.* II, iv, 12; 'project the powder', he says, very rightly, being engaged in the making of arsenical copper.   [h] *Corp.* II, iii, 2; 'the projections of the Egyptians'.

[i] *Corp.* III, vii, 1; III, x, 2; III, xiii, end; III, xxiv, 7 'the prophet Chimes exclaimed with enthusiasm: "After projection, etc."'; III, xxviii, 2, 3 quoting Mary, 'project the yellow sandarac (realgar) in cloth bags'; III, lvi, 1.

[j] *Corp.* II, i, 4, 24; perhaps also making arsenical copper.

[k] The word *epiballein* or similar forms is found also in the +3rd-century papyri.

[l] Hence again the origin of the word 'spagyrical'.

[m] The two titles, *Corp.* IV, xx, tit. (Comarius to Ps.-Cleopatra), and III, xxix, tit. (Zosimus), may be later additions, but the stone is in the text at any rate in two places, III, vi, 6 and III, xxix, 21, both Zosimus.   [n] *Corp.* III, xxix, 24 (Zosimus). Cf. p. 473 below.

[1] 黃白之術   [2] 點   [3] 化

twice over, but if one is resolutely determined to reject any influences from the East, it must give a somewhat uncomfortable feeling to know that aurifiction and aurifaction were going on in China a couple of centuries earlier than the first evidence we have of the techniques in the Mediterranean region.

As for the fundamental chemistry, mostly rather simple, that gave rise to the idea of projection in the first place, we now have a number of reasonable explanations. An idea common, it seems, to all the Old World civilisations, it has kept on cropping up throughout this Section (pt. 2, pp. 18, 195, 223, 225). Besides the mere debasement of the noble metals in *diplōsis*, Western Asia and the Mediterranean region had zinc compounds with which to make the gold-like brasses, and arsenic to turn copper silvery or golden, while to these East Asia added nickel for the production of the really silvery cupro-nickel. There were also always of course the amalgams, where 'all the mercury was turned to silver', and other possible uniform-substrate alloys which have been discussed in the metallurgical-chemical introduction (pt. 2, pp. 242 ff.). But this was only half the armamentarium of the aurifictors and aurifactors, for striking results could be achieved by mere surface-films, whether quite coarse as in the case of the classical gilding and silvering processes, or rather fine, as when base metals were treated with sulphurous and arsenical vapours. Indeed if projection was made with very small amounts of substance relative to the mass to be transmuted, it is probably only to be explained in terms of oxidation and deoxidation. Cadmium or antimony in lead gives oxide films of striking colours on cooling, and such films can even be dyed with organic colours (as in the case of the anodised aluminium oxide films so commonly used on metal objects at the present day); and this was very probably known both to the Alexandrians and the Chinese. Again, high-tin bronzes are liable to 'sweat', crystals of pure tin growing out at the surface by segregation during solidification, and this can produce a highly silvery appearance. Many effects of that kind could have been produced. And finally, though it was probably not usually thought of as projection, there was the clever—and very ancient—technique of leaching out base metals from the alloy surface, leaving a thin layer of almost pure gold or silver. A great deal must always have depended on what use for the aurifactive 'gold or 'silver' was in mind, for re-casting would destroy surface-films, and uniform-substrate alloys would therefore be required if this was to be done. With all such possibilities before us, one can feel fairly ready to diagnose any case of aurifaction in the literature, either Eastern or Western, provided always that the description is sufficiently full.

Also under the heading of methods two further aspects should not be forgotten, the importance attached to secrecy and oral instruction (*khou chüeh*[1])[a] both in East and West, and the accompaniment of chemical operations by magic, exorcisms, talismans, etc. (*fu*[2]).[b]

How now to organise the realm of ideas? As it happens, there is a quite convenient way, for the Hellenistic proto-chemists were fond of certain sayings which they called

---

[a] *Corp.* II, i, 3, 29 (Ps.-Dem.). Cf. Berthelot (1), p. 162; Bidez & Cumont (1), vol. 2, p. 317; Festugière (1), pp. 221, 332ff.

[b] Here there is a wealth of material; we refer only to Berthelot (1), pp. 15, 16, 35.

[1] 口訣　　　[2] 符

ainigmata, 'enigmas' (αἰνίγματα), or, as we might say, axioms or aphorisms. Let us then take a look at the most important of these, adding a running commentary in which the similarities and differences with Chinese thought can be described. To begin, then,

[1] The All is One, by It arises every thing, [all things tend to the One],[a] and if the All were not One, it would be nothing at all.[b]

The All is One, by It every thing is engendered; the One is the All, and if it did not contain all things, it could not engender them [or, it could never have come into existence].[c]

[1a] The Serpent is One, it possesses the Power according to the Two Symbols.[d]

Here the main affirmation would have commanded the assent of every ancient Chinese philosopher. 'The sage embraces the Oneness of the universe, making it his testing-instrument for everything under Heaven'.[e] The unity and uniformity of Nature is of course the basic assumption of natural science. 'Only the enlightened man holding on to the idea of the One can bring about changes in things and affairs'.[f] Side by side with this first Alexandrian aphorism we could reasonably set that beautiful passage from the *Lieh Tzu* book,[g] purporting to be the words of Lieh Yü-Khou's master Huchhiu Tzu-Lin, talking to his fellow-disciple Pohun Wu-Jen.

> There is an Engenderer which was not itself generated.
> There is a Changer which is itself unchanging.
> The Ungenerated can generate generation,
> The Unchanging can transform the things that change.
> What is engendered cannot but generate in turn,
> What changes cannot but undergo further change.
> Therefore there is perpetual generation and perpetual transformation,
> And there is never any time when generated things are not generating, and changing
>   things not undergoing change.

He was speaking of the Tao, as it shows itself in the Yin and Yang, the four seasons, and all the cycles of growth and dissolution. So also in their way were the Greeks and Egyptians.[h] Presumably anyone anywhere in the world meditating on Nature could have said the same kind of thing, but in considering the cosmic arrangements we must remember that there were also more detailed parallelisms. Certain symbolic correlations

---

[a] Only in Ps.-Cleopatra's Chrysopoia figure, opp. p. 64 in Berthelot (1), cf. p. 61. Also (2), pp. 133, 135.

[b] *Corp.* II, iv, 27 (Olympiodorus, quoting Chēmes); III, xx, 1 (Zosimus). Cf. Berthelot (1), pp. 61, 178, 284. Cf. 'Gospel of Thomas', log. 81, in Doresse (1).

[c] *Corp.* III, xviii, 1 (Zosimus, quoting Chymes).

[d] Again Ps.-Cleopatra only. This has to do with the Ouroboros symbol, on which see pp. 374 ff.

[e] *Tao Tê Ching*, ch. 22, cf. Vol. 2, p. 46.   [f] *Kuan Tzu*, ch. 49, cf. Vol. 2, *loc. cit.*

[g] Ch. 1, p. 1*b*, tr. auct. This book consists of material of all periods between the −5th century and +380.

[h] The Greek background of Aphorism [1] has been studied by Sheppard (4). The idea of the world as a single whole or organism (*unum esse omnia*) seems to have started with Xenophanes (*c.* −530; cf. Diels (5), vol. 1, no. 21 (11), p. 121, no. 31 (2), from Simplicius, *Phys.* 22, 22 ff.; Diels–Freeman, p. 93), unless an Orphic fragment preserved by Clement of Alexandria is earlier. Certainly Plato (*Soph.*, 242D, Fowler ed., p. 359) attributed it to the Eleatics before Xenophanes. Contemporaries of the Alexandrian proto-chemists often voiced the affirmation of cosmic unity, e.g. *Poimandres*, XII, 8 (Hermes to Thoth, in *Corp. Hermet.*, ed. Nock & Festugière, vol. 1, p. 177) and another Gnostic work mentioned by Clement—as also Hippolytus and Galen (for references see Sheppard).

in the world of the Hellenistic proto-chemists were similar to those of the Chinese, though not identical, e.g. the association of spatial directions with colours,[a] and of elements with metals, planets, and even lunar mansions.[b]

Next comes a very famous aphorism.

[2] One nature takes pleasure in another nature, one nature triumphs over another nature, one nature dominates another nature.[c]

[2a] You, O King, must know this, and all leaders, priests and prophets must know this; that he who has not learnt to recognise the substances, and has not combined them, and has not understood their forms and joined like with like, will labour in vain and his efforts will be fruitless—for the natures of things take pleasure in each other, are charmed by one another, destroy one another, transform one another, and generate again each the other.[d]

On this splendid proto-chemical saying much has been written, but perhaps it has not so far been pointed out that we almost seem to be listening to a version of the 'mutual production order' (*hsiang shêng*[1]) and the 'mutual conquest order' (*hsiang shêng*[2]) of the Chinese Five Elements as that doctrine developed after the time of Tsou Yen in the −4th century.[e] According to this there was a defined sequence in which the elements came into being, one from the other, and a second sequence in which they overcame and destroyed each other. When these doctrines were first explained we noted that there were certain hints of similar ideas among the pre-Socratics, notably Pherecydes of Syros (*c.* −550)[f] and Heraclitus of Ephesus (*c.* −500), but they are not much more.[g] To this should be added also that Plato, in the *Timaeus* (*c.* −360), envisaged an interchangeability between water, air and fire, in an ascending series, only earth, a *caput mortuum*, remaining always unchanged.[h] Nevertheless it seems doubtful whether this aphorism can be fully accounted for by Greek theories of preceding centuries, so here is a case where it might be well to leave open the possibility that some other influence, perhaps adjuvant rather than capital, came Westwards along the Old Silk Road.

As regards the second quotation, a point of interest lies in the words 'will join like with like' (*ta genē synapsei tois genesin*, τὰ γένη συνάψει τοῖς γένεσιν), for it echoes again the concern which the Chinese had with the idea of *thung lei*.[3] We have taken occasion to compare the Greek with the Chinese thought on this subject in another place (pp. 318 ff.), and found the latter more developed, since the *lei* were divisions cutting

---

[a] Cf. Berthelot (1), pp. 35, 182. Jung (8), pp. 195, 292, was much impressed by this, though not too sure on his facts.   [b] Cf. Berthelot (1), p. 49, (2), pp. 73 ff.

[c] *Corp.* I, iii, 12 (Philos. Egg), I, v, 3 (Ouroboros), I, xiii, 7 (Isis to Horus), II, i, 3 (the temple vision in Ps.-Democritus), II, iii, 1 (Synesius, attributing to Ostanes), III, i, 7 and III, xix, 2 (Zosimus). Parts of the aphorism recur as a refrain in the *Physica kai Mystica*, II, i, 4–12, 14, 16–18, 20–28. Cf. Bidez & Cumont (1), vol. 1, pp. 244 ff.; Festugière (1), pp. 229, 231, 259; Berthelot (1), p. 151.

[d] *Corp.* VI, xiv, 3 (Philos. Anonym. quoting Ps.-Democr.). Cf. 'Gospel of Philip', 109, 113, 126 and Gaertner (1).

[e] See Vol. 2, pp. 253, 255–6. The similarity was noted briefly by Tsêng Chao-Lun (2). Cf. p. 312.

[f] Cf. p. 346 above.

[g] See also Vol. 2, p. 245. In later times echoes are more frequent and more explicit. The astrological work of Nechepso-Petosiris, supposed to be of the mid −2nd century, says: 'una natura ab alia vincitur, unusque deus ab altero...' (Riess fragm. no. 28, from Firmicus Maternus, *Math.* IV, 16). Or again Ptolemy: 'the lesser cause yields always to the greater and stronger', *Tetrabiblos* I, 3 (cf. Boll, 6).

[h] 56C to 57C; cf. Cornford (7), pp. 224 ff.

[1] 相生   [2] 相勝   [3] 同類

across the basic distinction between Yin and Yang things. For this last we can find a parallel in Zosimus, where he speaks of some substances originating from water, others from fire.[a] Such are the roots of the idea of chemical affinity in both civilisations.

The next aphorism does not perhaps have such close similarities with Chinese thought.

[3] If you do not take away from bodies their corporeal estate, and if you do not transform incorporeal things into bodies, [and if you do not make two bodies into a single one],[b] you will never obtain what you are seeking [or, none of the results you hope for will be produced].[c]

On the surface there is of course community, for the interconversion of vapour (*chhi*,[1] *aithalē*, αἰθάλη) and solid substances, whether in sublimation, volatilisation, distillation, condensation, or chemical combination, was an exceedingly familiar phenomenon both in East and West; but the aphorism probably means a good deal more than that, having reference to the basic Greek proto-chemical doctrine of the deprivation and addition of forms (cf. pt. 2, pp. 22 ff.). It is a striking thing that in Chinese theory there does not seem to be any appearance of the death-and-resurrection motif, though in the Corpus this can be found explicitly referred to.[d] Nor can one find in China any parallel to the concept of *prima materia*, stripped of all forms, both those which it had before and those with which the aurifactor would endow it. The distinction between matter and spirit was always in China much less sharp (cf. pt. 2, pp. 86, 92), and at this early time, corresponding to the Han period, the later Neo-Confucian conception of *chhi*[2] and *li*,[3] matter and organisation,[e] had not developed—even when it did, it was not at all the same as Peripatetic matter and form. What substituted for these theories was the interaction of the forces represented by the trigrams and hexagrams (*kua*[4]) of the *I Ching*,[f] and of that there seems to have been no trace in Europe.

Though 'favourable times' (*kairoi*, καιροί) are mentioned in the Corpus now and then,[g] and time itself (*chronos*, χρόνος) measured and waited upon, there was nothing like so great an emphasis on time in Hellenistic proto-chemistry as in Chinese alchemy.[h] Our theoretical sub-section (pp. 242 ff.) has demonstrated how vital it there

[a] *Corp.* III, x, 2. In III, xxviii, 9, 10 we have 'similars' (*ta homoia*, τὰ ὅμοια) and even 'relationship' or affinity (*syngeneia*, συγγένεια). Cf. Berthelot (1), p. 160.

[b] Only in the quotations attributed to Mary the Jewess. Cf. 'Gospel of Thomas', log. 110 tr. Ménard; Doresse.

[c] *Corp.* III, iv, a fragment specifically attributed to Hermes. Mary is quoted in Olympiodorus (II, iv, 40) and in Zosimus or Ps.-Zosimus (III, xxix, 1). The aphorism also occurs in I, iii, 12 (Egg) and IV, i, 9 (Pelagius quoting Zosimus, but the first part only). Cf. Festugière (1), pp. 242, 253; Berthelot (1), pp. 134, 171–2.

[d] *Corp.* III, viii, 2 (Zos.). It is curious that fertility rituals figure very little in the Chinese material which Frazer collected for his 'Golden Bough'. Prof. Dubs (priv. comm. *c.* 1950) doubted whether the theme had ever played any prominent part in Chinese culture. This may be understandable if the idea of the resurrection of the body was primarily ancient Egyptian and Mesopotamian (cf. pt. 2, p. 79).

[e] See Vol. 2, pp. 472 ff.

[f] A point well made by Huang Tzu-Chhing (1), p. 723.

[g] E.g. III, xiv, xv (both Zosimus, but referring to Hermes). See especially Festugière (1), pp. 243, 264, 277, 278–9.

[h] As Sivin (2) was perhaps the first to emphasise.

[1] 氣   [2] 氣   [3] 理   [4] 卦

was, not only in connection with the experimental heating times (*huo hou*[1])[a] but also for the slow development of minerals and metals in the earth, this being artificially accelerated by the alchemist in the laboratory. Only perhaps in the Ouroboros symbol (cf. p. 375) did the Greeks and Egyptians implicitly accept the importance of cyclical processes in chemical change and the time inevitably taken by them.

On the other hand, and linking up with what has just been said on the interconversion of vapours and solid substances, there was much community of thought in China and the West about the formation of mineral and chemical substances in the earth. The two 'exhalations', dry and wet (*anathumiaseis*, ἀναθυμιάσεις), of Aristotle[b] were mirrored in the role of *chhi*[2] as the Chinese understood it. A classical statement of theirs runs as follows:[c]

> Rock and stone are the kernel of the *chhi*, and the bones of the earth. Large masses form cliffs and boulders, the smallest particles form sands and dusts. The seminal essence of the *chhi* gives gold and jade, its poisonous part produces white arsenical ore and arsenious oxide. When the *chhi* congeals it gives rise to the red and caerulean pigments; when it becomes transformed it trickles out in the form of alum (waters) and mercury. Such are its transmutations. Now it changes from something soft into something hard, as when concentrated brine crystallises into dense masses; now it changes from something moving into something still, as when petrifactions are formed from plants and trees. Even flying things and running things can be changed into stone, and the once animate pass into the inanimate. Even thunder and shooting stars can appear as stones,[d] the formless passing into that which has form. Thus do the treasures of the great earth come into existence. Though metals and minerals, formed on the great potter's wheel of Nature, and in its furnaces and bellows, may seem but dull and stupid things, the Shaping Forces have produced them in inexhaustible abundance; and human beings may rely on them as full of value in the preservation of life and health. Though they may seem but dead gewgaws their use and profit to mankind is inexhaustible.

With this de Mély compared the statements one can find in Seneca's *Quaestiones Naturales* (*c.* +60) on the breath of the earth producing metalliferous veins and all its other minerals.[e] And there are also closer parallels, such as the idea that lead is the ancestor of all the other metals.[f] And what the *Huai Nan Tzu* book says about the generation of metals in the earth (p. 225) can be paralleled closely enough by medieval European statements such as one we have given already from Vincent of Beauvais (*c.* +1246),[g] or others in Arabic sources such as Ibn Sīnā's *Kitāb al-Shifā'* (+1022).[h]

---

a Or, more precisely perhaps, the phasing of the alchemist's application of fire or other heat prescribed, on theoretical grounds.

b Discussed in Vol. 3 above, pp. 469, 636ff. One of the best accounts is that of Eichholz (1).

c *PTKM*, ch. 8, (p. 1), tr. auct. The now inadequate translation of de Mély (6), p. 317, was taken from *Wakan Sanzai Zue* (de Mély (1), text, p. 1, tr. p. 3).

d 'Thunder-axes' and meteorites.

e E.g. II, x, III, xv, VI, xiii, xvi (Clarke tr., pp. 60, 126, 240, 245).

f This is in *Corp.* III, xvii, 1 (Zosimus), cf. Berthelot (1), p. 229. Correspondingly, *PTKM*, ch. 8, (p. 12), taken from the *Thu Hsiu Chen Chün Pên Tshao, c.* +1040; thence de Mély (1), text, p. 22, tr. p. 27. Berthelot (3) noticed this. Of course Chinese texts cannot always be taken *au pied de la lettre*; apart from the symbolism of 'lead' in physiological alchemy, lead also stands for the Yin or feminine Metal element in the *Tshan Thung Chhi* type-process. Cf. p. 253.

g Vol. 3, p. 639.   h Tr. Holmyard & Mandeville (1), pp. 38ff.

¹ 火候   ² 氣

One can only say that these ideas were common property throughout the length and breadth of the Old World from the beginning of our era onwards.

Moreover there was in antiquity what Halleux (1) has called a 'biological conception of the mineral world'. Already in the *Iliad* the formation of silver is described as *genethlē* (γενέθλη), a birth word normally only used of men.[a] *Nascitur*, or *gignesthai* (γίγνεσθαι), was the standard term later on for metals and minerals, opposed in Dioscorides, for example, to *skeuazomai* (σκευάζομαι) for artificial products. Thus the mineral world was assimilated to the vegetable and animal worlds,[b] hence indeed the idea that mines should be allowed to have a fallow period.[c] In Chinese we do not see this 'biological analogy' so clearly since the verb *shêng*[1] was always indifferently used for birth and for all natural production. This brings us, however, to the role of sex in proto-chemistry.

The next aphorisms take us right into the realm of the Yin and the Yang.

[4] Above, the celestial things, below, the terrestrial; by the male and the female the work is accomplished.[d]

[5] Join the male and the female, and you will find what you are seeking.[e]

[6] If the two do not become one, and the three one, and the whole of the composition one, the result attained will be nothing.[f]

Certainly nothing could be more like the *thien ti yang yin*[2] complementarity than the first of these,[g] but we have to reckon with a widespread tendency to sexualisation in the ancient West also, doubtless because sexual union was one of the most primitive analogies for all chemical reaction.[h] According to Seneca, the Egyptians recognised a male and a female manifestation of each element, earth, fire, air and water—male crags and boulders being opposed to female cultivatable land, male burning to female glow, windy maleness to misty femaleness, and the sea being manly while fresh water was feminine.[i] The ancients also believed that certain stones were of two sexes, male (*arrēn*, ἄρρην) and female (*thēlys*, θῆλυς). Three such are in Theophrastus, the *lyngurion*, (λυγγούριον), i.e. tourmaline or fossil amber, rated according to the degree of transparency, and the sex of the lynx the urine of which was supposed to have formed it; the *cyanos* (κύανος), i.e. azurite, basic copper carbonate, rated according to the depth of

[a] *Il.* 2.857.

[b] There is much further material on this in different times and places in Daubrée (1); Sébillot (1); Cline (1) and Eliade (5).

[c] On Virgil's remarks about the iron ore of Elba Servius commented that the iron grows again there. As Bailey says (1), vol. 2, pp. 175–6, weathering would transform exposed sulphide to the sulphate, but the idea in the minds of the ancients was certainly that of vegetative growth.

[d] *Corp.* III, x, 4 (Zosimus), but also as a kind of caption to one of the MS. drawings of apparatus, Berthelot (2), fig. 37, from 2327, fol. 81 v, cf. pp. 161, 163.

[e] *Corp.* III, xxix, 13 (Zosimus, reporting a saying of Mary). Cf. 'Gospel of Thomas', log. 1, 22, in Puech (4); log. 27, in Doresse (1); cf. Pagels (1).

[f] *Corp.* I, iii, 13 (Philos. Egg, an ancient fragment).

[g] As Holgen (1) already realised in 1917 (p. 404). Yoshida Mitsukuni, too, (7), p. 210, was much impressed by the parallelism.

[h] Olympiodorus records a symbolic correlation of Chinese type, making the east male (Yang) and the west female (Yin); *Corp.* II, iv, 32, commented on by Berthelot (1), p. 64. The immense importance for later Western alchemy of sexual ideas and imagery, as in the 'marriage of contraries', etc., often elsewhere referred to in these volumes (pt. 3, pp. 69, 214, and here, pp. 121, 253, 259) was well emphasised in a rather quaint paper by Redgrove (1).   [i] *Quaest. Nat.* III, xiv (Clarke tr., p. 125).

[1] 生   [2] 天地陽陰

the blue colour, and *sardion* (σάρδιον), i.e. cornaline or sardonyx, also by the depth of its red.[a] This was handing down older traditions.[b] Moreover it was only natural that magnetite should exercise a feminine attraction on martial iron,[c] and fire be struck from male flint.[d] And 'pregnant stones', aetites, geodes, nodular concretions, etc. we have already mentioned.[e] In the Corpus the actual words for marriage and sexual union are used in connection with chemical reactions again and again, e.g. *syngamēsōsin* (συγγαμήσωσιν) in Ps.-Democritus, corresponding precisely to the Chinese terms such as *chiao kou*[1] which we constantly encounter in the other parts of this Section.

Nor can one overlook the important presence of women adepts in proto-chemistry and alchemy from the very beginning—Pseudo-Cleopatra, Mary the Jewess and Theosebeia[f] correspond most strikingly with the wife of Chhêng Wei, the wife of Ko Hung, Wei Hua-Tshun or Kêng the Teacher in China.[g] Gnostic prophetesses or women Taoists, they were always in the picture.[h] If there were ancient connections between the culinary and the chemical arts, this is not so surprising, but surely there were much deeper reasons—there were certain things which could only be accomplished by collaboration between the sexes, and for the birth of new compounds as well as for the birth of children, goddesses and women could do what gods and men alone could not. Now no ancient civilisation had this dual function so deeply embedded in the nature of all being as did the Chinese, for the Yin and Yang were the most fundamental manifestations of the Tao; *i Yin i Yang chih wei Tao*[2].[i] The extent therefore to which this profound ontological sexualisation, so characteristic of Chinese thought, had influence beyond its borders Westwards during the Hellenistic period is a question which ought to be raised, and indeed kept open, while research continues.

Further aspects of generation, including fermentation, arise in the last of the Hellenistic aphorisms which we must consider.

[7] Gold engenders gold, as wheat produces wheat, and women give birth to men.[j]

[Or:] He who sows wheat harvests wheat, he who sows gold and silver will obtain more gold and silver.[k]

    [a] *De Lapidibus*, see Caley & Richards (1), pp. 109, 122; Eichholz (2), p. 107. Pliny follows suit with the carbuncle and the sandastros, *Hist. Nat.*, XXXVII, 92, 101.

    [b] Cf. Boson (3) on the Assyrian and Babylonian texts.

    [c] Pliny, however, divided the magnet-stones themselves into male or female according to their attractive power (*Hist. Nat.* XXXVI, 128, 129).

    [d] Cf. Nonnus, *Dionysiaca*, II, 493 ff.

    [e] P. 205 above, and Vol. 3, p. 652. See also especially Bromehead (2); Bailey (1), vol. 2, p. 253.

    [f] The eminent role of women in Hellenistic proto-chemistry was underlined by Berthelot (1), p. 64.

    [g] To say nothing of all the semi-legendary or mythical instructresses like Su Nü,[3] or rather the real women of charismatic skill who stood behind them. Cf. Vol. 2, pp. 147–8 and pt. 5 below.

    [h] There were many ways in which they could have been important in chemical technology at an earlier time still. Already we have met with an Akkadian perfume-craft mistress of the −13th century—Tapputī-Bēlatēkallim; p. 83 above.

    [i] *I Ching*, Hsi Tzhu App. I, ch. 5, (ch. 2, p. 35a, Wilhelm–Baynes tr., vol. 1, p. 319).

    [j] *Corp.* I, xiii, 8 (Isis to Horus), II, iv, 32 (Olympiodorus). In both cases a husbandman is appealed to as witness (Acharantus and Achaab respectively), and in the second the saying is attributed to Hermes.   [k] *Corp.* IV, i, 8 (Pelagius). Cf. Berthelot (1), pp. 51–2, 186.

[1] 交媾   [2] 一陰一陽之謂道   [3] 素女

We are standing here at the beginning of human knowledge about the phenomena of catalytic action, but while part of it was based on a true appreciation of the working of ferments, another part originated from a metallurgical misunderstanding. Noble metals have the property of retaining many of their characteristics even with considerable 'dilution', hence the possibility of 'debasement'. As we know (pt. 2, pp. 18ff.), the papyri and the Corpus constantly speak of the imitation, or 'multiplication', of the precious metals by alloying them with others such as copper, tin and lead, but at least some of the gold and silver mixture (*asem, electrum*) had to be there, though quite a small amount might do. Hence the proto-chemical artisans and philosophers talked of the *diplōsis* or *triplōsis* (doubling or tripling) of *asem*, the necessary amount of which was often thought of as a 'seed', possibly something like what we might call a nucleus of crystallisation. Moreover, it is clear enough (p. 357 above) that the theme of 'projection', the conversion of much by the addition of little, was already in the Corpus, just as it had been in China a century or so earlier;[a] and it was most natural that the action of a small amount of substance (corresponding to the later philosophers' stone) should have been likened to the action of yeast. After all, this had been made use of, if not understood, for three or four millennia already. The idea of a pinch of leaven leavening a great mass was taken over simply from the empirical human technology of beer and bread—the 'domestication of yeasts', which must go back at least to Babylonian times.[b]

The word loosely used for ferment in the Corpus is *maza* (μάζα), meaning leavened barley-bread,[c] thought of, presumably, in the unrisen state. The ordinary word for yeast itself, *zymē* (ζύμη), was not used. Zosimus, writing to Theodore, speaks of that 'inexhaustible *maza*' that Moses obtained according to the precept of the Lord,[d] and explains that the *maza* is the copper, to be converted as the bread is. From this it is sure that Zosimus (*c.* +300) visualised projection as a kind of fermentation.[e] The 'inexhaustible stock' is referred to also in both the artisanal papyri.[f] What is more, it seems that the word *maza* came quite early to be identical with the name of the art, *chēmeia*, itself, for Zosimus describes a procedure contained in 'the *Maza* of Moses'.[g] This writing cannot be the 'Domestic Chemistry of Moses' (cf. p. 327) as we have it now, because that text[h] is assessed at a +7th-century date, but it could have been some earlier writing attributed to Moses; in any case Zosimus goes on to quote a sentence occurring 'towards the end of the *Maza* of Moses'. This by itself might not

---

[a] One recalls the theological-liturgical parallel; pt. 3, p. 38, to which add Staniloae (1).

[b] On beer and brewing in ancient Mesopotamia see Huber (3), almost but not quite replaced by Röllig (1). Modern biochemistry has enabled us to make a clear distinction between the process of fermentation as such and the growth by cell-division of the population of organisms concerned, but such knowledge was not available to the ancients.

[c] Inferior to leavened wheaten bread, *artos* (ἄρτος). Strange that a word so similar in sound, *matza*, referred in Hebrew to *un*leavened bread.

[d] *Corp.* III, xliii, 6. This is a *catalogue raisonné* of chapter headings and contents in a book now lost. Some of the subjects in the relevant chapter are of much interest, e.g. decomposition (*sēpseōs*, σήψεως), fermentation (*zymiōseōs*, ζυμιώσεως), transformation (*metabolēs*, μεταβολῆς), and regeneration (*palingenesias*, παλιγγενεσίας).

[e] Indeed he says as much explicitly, quoting Ps.-Democritus, in *Corp.* III, lii, 4.

[f] Leiden pap. no. 7, Stockholm pap. no. 8; cf. Caley (1, 2).

[g] *Corp.* III, xxiv, 4, 5. A similar reference to Chimes immediately follows.   [h] *Corp.* IV, xxii.

be very convincing were it not for the undoubted fact that later on *massa*, the Latin equivalent, came to be a synonym of chemia and alchemia. *Massa* normally means, of course, a lump, as of dough, or an inchoate mass of anything.[a]

From all this ancient literature the idea of the metallurgical 'ferment' passed into the alchemy of the Arabs, where nearly all the writers have it—the Jābirian Corpus and Ibn Umail, also the *Turba Philosophorum* and the +11th-century 'Book of Alums and Salts'. Thence to Geber about +1300, and to the Villanovan Corpus of the early +14th century, where *massa* and *azymum* have become the regular words for the philosophers' stone as ferment,[b] actual gold and silver still being thought of as necessary elements in its composition. Indeed a garbled etymology at times derived alchemia itself from *archymum* (i.e. *azymon*, ἄζυμον, unfermented dough), so that all chemistry (*archēmia*, ἀρχημία) was synonymous with the 'yeasty craft' (*maza pragma*, μάζα πρᾶγμα), as in the early +14th-century Byzantine Greek translation of Pseudo-Albertus *Semita Recta* or *De Alchemia*.[c] And so *massa* came to rest in Ruhland's lexicon (+1661): 'Kymus, id est massa.[d] Kuria vel kymia, id est, massa, heist dieselb Kunst/alchimia, al-kymia'.[e] This replication process, whereby a certain thing could make more of itself, like the widow's cruse of oil, *ad infinitum*, was surely a strange and not often recognised ancestral foreshadowing of the knowledge we now have about the self-replicating ribonucleoproteins in all cell-division. And by the same token it well exemplifies the truth that 'the alchemists took a path just the opposite of later chemistry, for while we seek to explain biological processes in terms of chemical ones, they conversely explained inorganic phenomena in terms of biological events.'[f]

In Chinese alchemical literature we have not found so much use of the terminology of fermentation, whether for the yeast or the mass of material which it transforms into something else,[g] but by contrast a rich related vocabulary, that of embryology and foetal development, is very much to the fore. To *maza* and *massa* corresponds the 'chaos' of the yolk and white, the *hung-jung*,[1] as it is called already in the *Lun Hêng*,[h]

---

a One cannot help wondering (in spite of Murray's Etymological Dictionary) whether this word, associated as it was with bread, could have generated the term 'mass' used in the Western Church for the eucharist or holy liturgy. The usual derivation is of course from the words of 'sending away' or dismissal, either of the catechumens before the anaphora or of the whole people at the end of the rite—unless 'commission' was a direct parallel of 'leitourgia'. But if *maza* accomplished (fermentative) change, so did the words of institution when pronounced by an ordained presbyter. Could *matza* also be relevant (though only the Western church preferred unleavened bread)? Strange that in Aramaic *patir*, unleavened bread, also meant 'to dismiss', 'to send home'.

  b See Ganzenmüller (2), pp. 148, 177; Darmstädter (1), pp. 122, 181. One Geber reference is *Liber Fornacum*, chs. 25–27, Russell tr., p. 255.

  c See Berthelot (2), pp. 208–9.

  d P. 272. Also p. 149, 'Chymus, id est, massa'. Cf. p. 403 below.

  e P. 271. Cit. Berthelot (2), p. 257. There is even a suggestion that *massa* may have been the origin of the German word for brass, *messing* (von Lippmann (1), p. 573). In view of what we have seen (pt. 2, pp. 195ff. above) about the imitation of gold by means of the low-zinc brases, this may have distinct plausibility.   f Ganzenmüller (2), p. 150, Cf. p. 363 above.

  g It does nevertheless from time to time occur, as in the passage from the *Tshan Thung Chhi* reproduced on p. 317 from ch. 32 (ch. 3, p. 5a) or *TT*990, ch. 3, p. 6b. Here it is closely in the context of *similia similibus...*, i.e. the *thung lei*[2] concept, categorical identity or similarity, a thought-complex probably derived in part from the biological phenomena of fermentation and generation in all the ancient civilisations.   h Vol. 2, p. 370.

  ¹ 鴻溶   ² 同類

and to the 'hermetic' vessel which came to be called in the West the 'egg of the philosophers'[a] correspond *tan*,[1,2] the hen's egg, and even more frequently *thai*,[3] the womb or matrix. Certain paragraphs in our sub-section on theory in China have had to be devoted especially to these parallelisms,[b] which even went so far as to involve the actual use of avian egg-shells in the experimental or technical set-ups. The image of the chicken egg and embryo (*chi tzu*[4]) constantly recurs, and one often finds the expression *tan phi*,[5] the 'foetus of the elixir', i.e. intermediate stages in its formation. So also in the Hellenistic Corpus there are two tractates concerning the 'egg', the first entitled 'What the ancients said about the Egg',[c] the second 'Nomenclature of the Egg, the Mystery of the Art'.[d] Both these explain figurative cover-names for many reagents drawn from the components of the hen's egg, and there is a third of the same kind (not in the Corpus): 'Elucidation of the Parts of the Egg according to Justinian'.[e] The first two could be of the +2nd or +3rd century, the third probably of the late +7th. On the other hand, a further short piece is called 'Techniques of the Emperor Justinian'[f] and this deals with the dry distillation of real eggs, a procedure which occurs commonly elsewhere in the Corpus and led, as we know (pt. 2, pp. 252–3) to the 'divine' or 'sulphurous' water (calcium polysulphides).

The two ideas, of leaven and life-germing, were closely connected. Very early the development of embryos was thought of as a kind of fermenting, morphological differentiation, with the appearance of complex organs, muscles, nerves and vessels, being analogised in a simple-minded way with the varied textures, shapes and colours which appear in maturing cheese. In the early Jewish Wisdom Literature, Job is made to say: 'Hast thou not poured me out as milk, and curdled me like cheese? Thou hast clothed me with skin and flesh, and knit me together with bones and sinews.'[g] Aristotle was saying the same thing—for him the menstrual blood was the material basis of the foetus and the semen provided the form, acting upon it just as rennet acts upon milk.[h] This idea, though little further developed, remained a commonplace throughout the Western Middle Ages, prominent for example in the visions of Hildegard of Bingen (+1098 to +1180),[i] and related to what Albertus Magnus had in mind when he said that 'eggs grow into embryos because their wetness is like the wetness of yeast'.[j]

These lines of thought were to lead in the end to our present understanding of proteins and enzyme proteins. The fascination which Alexandrian proto-chemists and Chinese alchemists alike had felt about the extraordinary potentialities of the avian egg and its contents was felt again by Sir Thomas Browne (+1605 to +1682) in his

[a] Illustration in Berthelot (2), p. 170; the *kērotakis* or aludel.
[b] Cf. pp. 292 ff. above.
[c] *Corp.* I, iii, re-titled by Berthelot & Ruelle.
[d] *Corp.* I, iv, a shorter fragment.
[e] In Berthelot (2), pp. 214–15. He suggests (pp. 176, 297) a Pseudo-Justinian II.
[f] *Corp.* v, xxiv. On the general significance of eggs cf. Berthelot (1), p. 51.
[g] Job 10, 10. The book is now dated soon after −400, written somewhere in Palestine or Arabia in the post-exilic period.
[h] For fuller discussion see Needham (2).
[i] See Singer (3, 4).   [j] *De Animalibus*, XVII, ed. Stadler (1).

[1] 蛋   [2] 蠶   [3] 胎   [4] 雞子   [5] 丹胚

chymical elaboratory at Norwich in the middle of the century, where he carried out many experiments with the apparatus then available to try and find out more about these proteinaceous substances.[a] Work paralleling this on the mammalian amniotic and allantoic fluids was done by his contemporary Walter Needham, and further efforts to unveil the secrets of the proteins of eggs were discussed at length by Hermann Boerhaave in his *Elementa Chemiae* of +1732. But no real break-through in the understanding of protein structure was possible of course until the development of organic chemistry in the nineteenth century, and the classical work of Emil Fischer stood nearer the end of that than the beginning.

This concludes what had to be said about the Hellenistic aphorisms and the light which they throw on the earliest Western proto-chemistry, sometimes very close to Chinese alchemy in its ideas and theories, sometimes further away. But the love of aphorisms and oracular sayings did not cease with the Greeks, it was handed down through all later European chemical technology and alchemy, stemming in large measure from the wisdom of its Arabic counterpart. Let us look at the most outstanding gnomic utterance of these subsequent times, the *Tabula Smaragdina* or 'Emerald Table'.[b] This was a short statement in about a dozen verses, purporting to reveal the whole secret of alchemy—to anyone who could understand it. Immensely influential, or at least highly regarded, throughout the later Middle Ages and post-Renaissance periods, it was first printed in +1541[c] and often afterwards, as in the *Musaeum Hermeticum* of +1678.[d] But it was already well known in the +13th century because it is mentioned in the *De Rebus Metallicis et Mineralibus*[e] (c. +1280) authentically of Albertus Magnus, and reproduced (in one of its many versions)[f] by Roger Bacon in his edition of Pseudo-Aristotle, *Secretum Secretorum* (+1255, with an introduction of c. +1275).[g] This was then, however, far from new, because in fact it was an Arabic work, the *Kitāb Sirr al-Asrār*, a book of advice to kings, first translated into Latin by Johannes Hispalensis c.+1140 and again by Philip of Tripoli, c.+1243. The book itself may have been compiled at some time around +800, and it may have had a Syriac original, though this is not certain.[h] The text of the *Tabula*[i] underwent several other translations into Latin, including one probably by Plato of Tivoli c. +1140 and another by Hugh of Santalla in the middle of the same century.[j] The former, studied and collated by Steele & Singer (1), was the source of the well known version of Holmyard;[k] that of Read, however,[l] derived from the early printed versions through

a On this and the rest of the paragraph further information is in Needham (2).

b Emerald was a name for any green stone, including green glass. For example, the 'sacro catino', a great dish taken by the Crusaders at the sack of Caesarea in +1101, said to have been brought to Solomon by the Queen of Sheba, and to have been used at the Last Supper, turned out, when taken from Genoa to Paris in 1809, to be of green glass.

c Anon. (101), the *De Alchemia* of Nuremberg.

d Anon. (87), tr. Waite (8), vol. 2, p. 243.   e I, i, 3.

f Several of these were collected in Tenney Davis (9).

g Steele (1), fasc. 5, pp. xlviiiff., 115ff. For English incunabula and later versions see Manzalaoui (1).

h Cf. Mieli (1), 2nd ed., pp. 69–70.

i Its Arabic title was *Kitāb al-Lauḥ al-Zumurrudi*.

j This last was published by Nau (2), and is reproduced in Ruska (8), p. 178. Cf. Haskins (1), p. 67.

k (1), p. 95.   l (1), p. 51.

chemical historians such as Thomson[a] and Rodwell.[b] For the corpus of legend which was handed down about the *Tabula*, its discovery by Alexander the Great in the tomb of Hermes,[c] etc. etc. we may refer simply to Read, and expatiate no further here.

As was pointed out by von Lippmann,[d] no Greek original was known, and none has come to light since his time. Apart from the Pseudo-Aristotle,[e] Holmyard (13) was the first to find another Arabic version, in the *Kitāb Usṭuqus al-Uss al-Thānī* (Second Book of the Elements of the Foundation), one of the treatises in the Jābirian Corpus.[f] Actually the *Tabula* text appears in at least one of the other books in this, the *Kitāb al-Ḥayy* (Book of the Living).[g] But though that would imply the middle or latter half of the +9th century, the date was pushed still further back when Ruska (8) discovered the *Tabula* at the end of the *Kitāb Sirr al-Khalīqa wa Ṣanʿat al-Ṭabīʿa* (Book of the Secret of Creation and the Art (of Reproducing) Nature),[h] otherwise entitled *Kitāb Balaniyūs al-Ḥakīm fī ʾl-ʿIlal* (Book of Apollonius the Wise on the Causes).[i] This is not in the Corpus but it was extremely influential upon it,[j] and it has no connection with the +1st-century Apollonius of Tyana (Balīnās or Balīnūs) except in so far as he retained a great name among the Arabs for natural philosophy and magic.[k] Internal evidence shows that this book was written possibly as early as +650, more probably

[a] (1), vol. 1, p. 10.

[b] (1), p. 62.

[c] The Arabs later on were particularly partial to stories of occult inscriptions discovered in underground tombs or temples. Perhaps one may be allowed the surmise that the idea of a cavern containing treasures of (chemical) wisdom may go back to the cycle of legends associated with Adam's cave. Jewish apocryphal writings which first appeared about the −3rd century related that after their expulsion from paradise Adam and Eve found a cave in which they hid the treasures of the earth and in which they were themselves finally buried. This story germinated in the Adam-Book of the Sethian Gnostics (cf. Doresse (1), vol. 1, pp. 281ff.), but it also assumed a Christian form which said that what Adam hid in the Cavern were the treasures which the Magi afterwards took with them to Bethlehem. Eventually it gained wide diffusion in many languages—Hebrew, Aramaic, Syriac, Armenian, Arabic, etc. As finally edited by a Nestorian early in the +4th century, the Syriac 'Book of the Cavern of Treasures' (tr. Budge, 7) goes on to say that when they found the cave at the top of a mountain Adam and Eve were still virgin, but that he was now consumed with passion for her. So they took from the borders of the Garden gold, frankincense and myrrh (symbols of kings, priests and physicians to come) and blessing them laid them up in the Cavern to be a house of prayer for ever; which things being done they went down to the foot of the holy mountain and there they lay together. Thus Eve became pregnant with Cain and his twin sister Lebhūdhā, and again with Abel and his twin sister Kelīmath. Later the parents of mankind were buried in the Cavern. For further information on this legend cycle see Bezold (3), pp. 7–8; Preuschen (1); Götze (1); Monneret de Villard (2); Foerster (1) and Doresse (1), vol. 1, p. 202.

Of course one need not insist on a Hebrew origin for the cave-complex in view of the prevalence of cave-tombs on each side of the Nile valley, to say nothing of the adyta of the pyramids, but it might be another case of that confluence of Hebrew and Egyptian ideas which is so evident in the Greek proto-chemical Corpus.

[d] (1), p. 58. Cf. Ullmann (1), pp. 170ff.

[e] A +12th-century Arabic MS., ostensibly dictated by the priest Sergius of Nablus, was discovered by Ruska (25). The translation of this *Tabula* version is in Ruska (8), pp. 113ff.

[f] Kraus (2), p. 12, no. 6. Holmyard gave an English translation; also in Ruska (8), pp. 120ff.

[g] Kraus (2), p. 47, no. 133; cf. (3), p. 280.

[h] The last half of the title refers to the *Tabula Smaragdina*, which was supposed to describe in veiled terms an esoteric doctrine of the making of the elixir (Kraus (3), pp. 302–3). On Ruska's discovery see Winderlich (1), pp. 15–16; Plessner (3).

[i] This was first studied by de Sacy (1) in +1799. Cf. Kraus (3), pp. 272ff.

[j] Kraus (3), p. 282.

[k] As has been well emphasised by Multhauf (5), pp. 125ff., 131ff

about +820 under al-Ma'mun.[a] Its content has made it seem close to that of the Syriac 'Book of the Treasures' written by Job of Edessa early in the +9th century,[b] and also to what we have of Nemesius of Emesa (Homs)[c] who wrote in Greek about +400, contemporary with Synesius, but there is no direct parallel or overlap with the texts of either of these.[d] Thus there is still no ancient version of the *Tabula* either in Syriac or Greek, and the possibility presents itself that a source farther in the east should be looked for.

It was the considered opinion of Ruska that the origin of the text lay in a north-eastern direction, from Further Asia.[e] It was, he said, neither Islamic, nor Persian, nor Christian; could it be 'Chaldean' or Harranian (cf. p. 426)? Ought we not to think, he went on, of 'the great culture-oases in the region of the Oxus and Jaxartes rivers, of Merv and Balkh, or Khiva, Bokhara and Samarqand, those great cities which since ancient times had seen the exchange of material and intellectual goods between West and East, and where Greek traditions endured for such a surprisingly long time?' These two rivers flowing into the Aral Sea enclose Sogdia and Bactria, with Khwarizm to the west and Ferghana to the east—but to mention such names is to evoke the spirit of Chang Chhien and to mention China also. And Ruska did not shrink from this, for he visualised the cities on the Central Asian trade-routes north and east of the Sassanian empire as filled with a mixed population of Persians, Turanians, Syrians, Indians and Chinese,[f] places, as we know, where Buddhism, Manichaeism and the Chinese cults met together with Nestorian Christianity. Here the Graeco-Egyptian culture-world had passed into a cloudy distance, but alchemy, astrological and macrobiotic, was very much alive.[g]

Resting one day in 1968 at a wayside auberge in the South of France, I read again the text of the *Tabula Smaragdina* and felt so much the Chinese flavour of it that I started translating it to see how it would look in that language. Later on, returning home, I found that Chang Tzu-Kung (1) had suggested a possible Chinese original as long ago as 1945.[h] Perhaps we may dare here to reproduce the text inserting the Chinese words which one might imagine could have been in the primary form of it.[i] It runs as follows:

---

[a] Ruska (8), pp. 125ff., 127, 166; Kraus (2), p. lviii, (3), pp. 272ff. Ruska considered that it could hardly be earlier than the +7th century.

[b] Tr. Mingana (1).   [c] Lat. tr. Matthaei (1).

[d] Kraus (3), pp. 276, 278.

[e] (8), p. 167. This is in agreement with the general view of other scholars, e.g. Ganzenmüller (2), p. 32.   [f] (8), pp. 174–5.

[g] As a living witness to this conception we may cite that leaf of paper discovered in Eastern Sinkiang which has on one side a Chinese Buddhist *sūtra* text and on the other a mystical-magical mineralogy or lapidary of Hellenistic type written in Turkish but evidently translated from Sogdian. This has been described by Thomsen (1), with comments by Andreas. On the geography of the Central Asian routes of communication the studies of Herrmann (2, 3, 5, 6) have not really been superseded. The importance of the Old Silk Road for Chinese–Hellenistic intellectual contacts was already realised by Holgen (1) in 1917 (p. 471) and later emphasised again by Huang Tzu-Chhing (2) and others. Cf. Fig. 1531a.

[h] He knew only the version of Read (1). Some thirty years later Yoshida Mitsukuni (7), p. 209, translated it into Japanese, using many expressions similar to or the same as those which occurred to us. The whole atmosphere of the alchemical creed seemed to him very reminiscent of the *Tshan Thung Chhi* (cf. Vol. 5, pt. 3, pp. 50ff.).

[i] Using the German translation of the oldest known Arabic text, in Ruska, (8) pp. 159ff.

1. True, true,[a] with no room for doubt, certain, worthy of all trust (is this).

2. See, the highest (*shang*[1]) comes from the lowest (*hsia*[2]), and the lowest from the highest; indeed a marvellous work of the One (*Tao*[3]).[b]

[3*a*. See how all things (*wan wu*[4]) originated (*shêng*[5]) from It, by a single process.][c]

[3*d*. How wonderful is Its work (*tsao hua*[6])! It is the principle (*li*[7]) of the world and its sustainer (*chu*[8]).][d]

4. The father of it (the elixir, *tan*[9])[e] is the sun (*Yang*[10]), its mother the moon (*Yin*[11]); the wind (*chhi*[12] or *fêng*[13]) bore it in its belly (*thai*[14]), and the earth (*thu*[15]) nourished it (*yang chih*[16]).

5. This is the father (*tsu*[17]) of wondrous works (*shen ming pien hua*[18]),[f] the guardian (*pao*[19]) of mysteries (*shêng jen miao yung*[20]), perfect in its powers (*tê*[21]), the animator of lights (*kuang*[22])

6. This fire (*huo*[23]) will be poured upon the earth (*thu*[15])....

7. So do thou separate the earth (*thu*[15]) from the fire (*huo*[23]), the subtle (*chhing*[24]) from the gross (*cho*[25]), acting prudently and with art (*shen shu*[26]).[g]

8. It (the Tao[3]?) ascends from the earth (*ti*[27]) to the heavens (*thien*[28]), [and orders the lights (*yao*[29]) above],[c] then descends again to the earth; and in it is the power of the highest and the lowest.[h] Thus when thou hast the light of lights, darkness will flee away from thee.

9. With this power of powers (the elixir, *tan*[9]) shalt thou be able to get the mastery of every subtle thing (*wei*[30]),[i] and be able to penetrate (*thung*[31]) everything that is gross (*cho*[25]).

10. In this way was the great world itself formed (*tsao wu*[32]).

[11. Hence thus and thus marvellous operations (*miao fa*[33]) will be achieved.][j]

12. Hence I am called Hermes, thrice great in wisdom.

[13. ....][k]

---

[a] This emphasis on truth reminds one that Jung (8), p. 348, was much impressed with the parallelism between the 'inner spiritual man', actually, the 'true man' (*alēthinos anthrōpos*, ἀληθινὸς ἄνθρωπος) of the Gnostics, and the *chen jen*[34] or 'perfected' or 'realised immortal', later 'adept', of China. It is hard to know what to make of a coincidence like this; standing by itself it may not mean very much, but who knows what words of wisdom and what same-seeming ideals were exchanged in the Central Asian cities? Cf. Leisegang (1), pp. 78ff., who points out that Philo Judaeus was one of the chief users of the expression.

[b] It would also come naturally in Chinese thought to write *Chhien*[35] and *Khun*[36] here (cf. Vol. 2, p. 315, Table 14), and the whole affirmation looks remarkably like the doctrine that the extreme of Yang (*Yang chi*[37]) generates Yin[38] and vice versa (see Vol. 4, pt. 1, p. 9, Fig. 277, as also Fig. 1515 above; and all the discussion of physiological alchemy in Vol. 5, pt. 5).

[c] Considered to be a later addition.   [d] A much later addition, thought Ruska.

[e] Or, 'the primary vitality (*yuan chhi*[39])'.

[f] We conjecture that the original of 'works' here could have been the typical 'changes and transformations'.   [g] Or possibly *chhiao kung*.[40]

[h] Is this not a palpable reference to the powers of pure Yang (*shun Yang*[41]) and pure Yin (*shun Yin*[42]) at their transient maxima?

[i] Or perhaps *chi*.[43]   [j] Only in one Arabic version.

[k] Not in the Arabic, nor are any of the other Latin codicils either.

---

[1] 上	[2] 下	[3] 道	[4] 萬 物	[5] 生	[6] 造 化
[7] 理	[8] 柱	[9] 丹	[10] 陽	[11] 陰	[12] 氣
[13] 風	[14] 胎	[15] 土	[16] 養 之	[17] 祖	[18] 神 明 變 化
[19] 保	[20] 聖人妙用		[21] 德	[22] 光	[23] 火
[24] 清	[25] 濁		[26] 神 術		[27] 地
[28] 天	[29] 耀	[30] 微	[31] 通	[32] 造 物	[33] 妙 法
[34] 眞 人	[35] 乾	[36] 坤	[37] 陽 極	[38] 陰	[39] 元 氣
[40] 巧 工	[41] 純 陽	[42] 純 陰	[43] 幾		

Whatever may be thought of this rather unusual exercise,[a] it is at least reasonable to suggest that a sharp look-out should be kept for possible primary sources in the Chinese alchemical and philosophical literature. Chang Tzu-Kung, who noted many parallels in Chinese tradition for the inscribing of gnomic utterances on slabs and steles in caves and temples,[b] was inclined to see the origin of the *Tabula* in the Nei yeh[1] chapter (49) of the *Kuan Tzu*[2] book, a text datable in the late −4th century or the early −3rd. We are not sure how convincing this is, for no exact parallelisms of wording occur; and the long chapter, in rhymed prose, is mainly concerned with Taoist ataraxy, harmony of the self with the universe, and the beginnings of physiological alchemy in diet and breath control.[c] Still, there are beautiful passages about the world of Nature, two of which may be quoted.

Always the essence (*ching*[3]) of things is what gives them birth (*shêng*[4]);
Below, it gives life to the five grains; above, it orders the stars in their ranks.
Coursing through all things between heaven and earth, it may be called daemonic and
      spiritual,
Stored within the breast of a man, it may be called the sagely.
Therefore this breath (*chhi*[5]) of life—
How bright it is! As if mounting the heavens,
How dark! As if entering an abyss of gloom,
How vast! As if filling the whole ocean,
How compact! As if held within the self. . . .[d]

Or again:[e]

What enables transformation (*hua*[6]) in unity with things (*i wu*[7]) is called spirit (*shen*[8]),
What enables change (*pien*[9]) in unity with (human) affairs (*i shih*[10]) is called wisdom (*chih*[11]).
To transform without altering one's breath of life,
To change without altering one's wisdom,
Only the enlightened man who grasps the unity of Nature (*chih i chih chün tzu*[12]) is able to do
      this!
And since he grasps this unity, and does not lose it,
He is able to reign as prince over the myriad things.
The enlightened man commands things, and is not commanded by things,
Because he has gained the principle of the One (*tê i chih li*[13]). . .

But it is not quite what we are looking for, and something much nearer would be needed to confirm our suspicion.

---

  a I had, at least (without knowing it), the inspiring example of Genzmer (1), who succeeded in reconstructing the original German of a verse or two of epic theodicy (of −500 or so) from Tacitus' Latin (*Germ.* 11). This was most kindly brought to my knowledge by Dr Peter Dronke. For its further background see Ineichen, Scindler & Bodmer (1), pp. 708 ff.
  b A typical example has been given in translation in pt. 3, p. 195 above.
  c We drew upon it now and then in Vol. 2, e.g., pp. 46, 60–1, but it is fully translated by Rickett (1), pp. 158 ff.
  d Pp. 1*a, b*, tr. Rickett, *op. cit.*, mod. auct.
  e Pp. 3*b*, tr. Rickett, *op. cit.*, mod. auct.; cf. Haloun (2) in Vol. 2, *loc. cit.*

¹ 內業	² 管子	³ 精	⁴ 生	⁵ 氣	⁶ 化
⁷ 一物	⁸ 神	⁹ 變	¹⁰ 一事	¹¹ 智	¹² 執一之君子
¹³ 得一之理					

What does so rather more is the little that we know of the content of the 'Book of the Secret of Creation' of Balīnās himself, the first in which the *Tabula* occurs, and perhaps to be put in the +7th or +8th century. It has never been printed, so it is accessible only in the Arabic and Latin MSS, but something has been told of it by Multhauf.[a] It seems to contain several of the most characteristic motifs of later Arabic alchemy—knowledge of sal ammoniac and borax, the idea that all metals are mixtures of sulphur and mercury, and even (in one version)[b] the numerological or 'quantitative' analysis of elements and qualities in chemical substances. But in the account of 'creation', or cosmic evolution, the thought grows closer to Chinese origins. Heat acted as male, cold as female, their union producing humidity and dryness, just as Yang and Yin would be expected to do. The minerals and metals are each connected with the planets, as they would be in the Five-Element symbolic correlation system, but further the metals themselves are male or female (gold, iron and lead Yang, copper, tin and silver Yin, and mercury hermaphroditic or equally balanced).[c] Just as in the *Huai Nan Tzu* passage,[d] ores are concocted in the bosom of the earth century after century, mercury 'fermenting' with sulphur there, as so often it was made to do in the 'regenerated elixir' experiments of China.[e] Although it is generally agreed that 'Balīnās' refers to Apollonius of Tyana in Syria, one is almost driven to wonder whether this consensus should be questioned (for it rests on very little), and whether we might not have to have recourse to an Apollonius of Bactria? After all, the Euthydemid and Eucratid kings left behind them there a population Greek in names, and partly in speech, for centuries afterwards; and from a country where under many rulers the coins bore images of Hindu gods with Greek inscriptions almost anything might be expected. But even if we stick to Apollonius of Tyana (+1st century) as the eponymous hero we must remember that his biography registers much Asian contact, especially with the 'magi' in Mesopotamia and the brahmins and ascetics of India.[f] There is probably much more to be discovered yet about what Ruska called the 'Apollonische Schriftenkreise' than anything we can at present say.[g]

---

[a] (5), pp. 125ff., 132–3, following de Sacy (1). Cf. Plessner (9).

[b] According, at least, to one of the Jābirian writers; Kraus (3), pp. 188, 196.

[c] Of course this assignment of sexes is quite different from the Chinese. Cf. Fig. 1533.

[d] See p. 224. [e] Cf. pt. 3, pp. 20, 73–4, etc.

[f] Cf. Meile (1); Goosens (1); Filliozat (6). The 'Life of Apollonius of Tyana' was written by Philostratus of Lemnos about +218; tr. Conybeare (1); Jones (1). He was one of the learned circle of the Syrian empress Julia Domna in Athens, and died c. +240. Most of the travels of Apollonius about which he wrote are considered quite apocryphal, but they include a visit to Taxila in India, and to Ethiopia. It remains, however, a most insipid book, and I find it very difficult to understand how Apollonius can ever have got such a reputation among the Arabs for alchemical natural philosophy.

[g] The other great *Tabula* of the Western Middle Ages was the *Tabula Chemica* attributed to Senior Zadith Filius Hamuel. There is no doubt about the identification of this writer with Muḥammad ibn Umail al-Ṣādiq al-Tamīmī, who died c. +960. The piece is a very allegorical poem in some 90 strophes, and derives directly from two works in Arabic, the *Kitāb al-Mā' al-Waraqī wa'l-Arḍ al-Najmīyah* (Book of the Silvery Water and the Starry Earth), and the *Risālat al-Shams ilā 'l-Hilāl* (Epistle of the Sun to the Cresent Moon). The Arabic texts were discovered by Rosen in 1877 and de Slane in 1883, and the identification was made by Ali, Stapleton & Husain (1), who published the Arabic and the Latin together in 1933. See also the discussions of Ruska (9, 10). The Latin versions are extremely corrupt, and the Arabic very Hellenistic in origin, quoting Hermes, Ps.-Democritus, Mary, Zosimus, etc. Although it uses their aphorisms freely, it is too long-winded to be regarded as aphorismic itself, so we do not consider it further here. Cf. also Holmyard (1), pp. 99–100.

In closing this balance-sheet, and in this connection, we should like to allude to a point which was acutely made by Berthelot[a] and which has already been referred to (pt. 2, p. 25). The word atom never once passes the lips of the Hellenistic proto-chemists. Thus the writers of the Corpus (like the artisans of the papyri) seem to stand aloof from the traditions of Greek atomism and Indian atomism alike, and though they were certainly close to some versions of Peripatetic philosophy (cf. pt. 2, p. 26), their lack of interest in atoms seems to link them, as it were, along the Old Silk Road, with those other, more northerly, regions of Iranian and Chinese culture where atoms were not in favour either.

(v) *Parallelisms of symbol*

Lastly comes the question of symbolism. As is well known, the MSS of the Hellenistic proto-chemical Corpus are illustrated by four depictions of a serpent with its tail in its mouth.[b] This (Fig. 1525) is the emblem (or representation of a mythological being) traditionally known by its Greek name as Ouroboros,[c] the 'dragon tail-eater' ($\delta\rho\acute{\alpha}\kappa\omega\nu$ Ο$\mathring{\upsilon}\rho o\beta\acute{o}\rho os$). Three of the pictures are captioned by variant forms of Aphorism [1] above; and it is not difficult to think of reasons why the Graeco-Egyptian proto-chemists should have chosen Ouroboros as their chief symbol.[d] The infinite rotation of transformations giving new things within the unity of matter and of Nature,[e] the reproduction of the original substances started with, as in the oxidation and reduction of metal and calx, the repeatability of combining and decomposition, the cycle of aurifactive or aurifictive changes produced by adding arsenic to copper and blowing it off again, even the *kērotakis* technique of reflux distillation,[f] all would justify the symbol of eternal recurrence.

In the Corpus itself there are three chief mentions, two short descriptions among the oldest fragments,[g] and one curious paragraph apparently written by Olympiodorus late in the +5th century.[h] The first begins (obviously referring to chemical change):

Here is the mystery: the serpent Ouroboros (devouring its tail) is the composition which as a whole is devoured and melted away, dissolved and transformed by fermentation (or putrefaction, *sēpsis*, σῆψις)...[1]

---

[a] (1), pp. 263–4.

[b] Two of these are pictorial and two diagrammatic. All are reproduced in Sheppard (4), perhaps the best paper on the subject, and of course in Berthelot (1), p. 59, opp. p. 64, 284; (2), figs. 11, 13, 34, pp. 193, 196. Sherwood Taylor (2) reproduces three, and all treatments of Greek proto-chemistry say something on them.

[c] A better English spelling would be Uroboros, analogous to words like 'urodele'. The etymology is from *oura* (ο$\mathring{\upsilon}$ρά), tail, and the root of *bora* (βορά), food, *boros* (βορός), voracious.

[d] Serpents as such were also numinous because the sloughing of the skin typified regeneration and rebirth; cf. Macrobius, *Saturnalia*, 1, xx. This could have been connected with the death-and-resurrection motif of *prima materia*; cf. Sheppard (5).

For a later Chinese Buddhist parallel cf. Vol. 2, p. 422 and Fig. 47.

[e] Berthelot (1), p. 284.

[f] Sherwood Taylor (2).

[g] *Corp.* I, v and vi, both titled by Berthelot & Ruelle.

[h] *Corp.* II, iv, 18.

[i] This may well have been a reference to the attacking of the base metal by the corrosive vapours in the *kērotakis*.

Fig. 1525   Fig. 1526   Fig. 1527

Fig. 1525. Hellenistic Ouroboros. A representation from MS. Marcianus 299, fol. 188v. The inscription says: 'The One is the All' (in *Chrysopoia*, ps-Cleopatra).

Fig. 1526. Hellenistic Ouroboros. Another representation, from Paris MS. 2327, fol. 196 (in Olympiodorus, late +5th century). The tail-eating serpent's three concentric rings are coloured green, yellow and red, from inside outwards respectively. The four feet are said to represent the basic elements (*tetrasōmia*), and the three ears the sublimed vapours (*aithalai*), probably sulphur, mercury and arsenic. Both these redrawings are from Sheppard (4).

Fig. 1527. Hellenistic Ouroboros on a Gnostic gem, one of the many Abraxas talismans. From King (3), redrawn by Sheppard (4). Date ± 1st century, contemporary with ps-Cleopatra.

And it goes on to explain another of the pictures (Fig. 1526) by saying that 'the four feet constitute the *tetrasōmia* (τετρασωμία)'[a] and 'the three ears are the three sublimed vapours (*aithalai*, αἰθάλαι)'.[b] For his part, Olympiodorus wrote:

Agathodaemon, having placed the original principle in the end, and the end in the original principle, affirmed it to be that serpent Ouroboros; and if he spoke thus it was not in jealousy (to hide the truth) as the uninitiated think. That is made obvious by the use of the plural— eggs.[c] But you who know everything—who was Agathodaemon? Some say he was the ancientest among the philosophers of Egypt, others that he was a mysterious messenger, the good angel of that land. Others have called him the heavens, perhaps because a serpent is the image of the world. And indeed certain Egyptian hierographists, wishing to represent the world on obelisks, or express it in sacred characters, have engraved the serpent Ouroboros, with its body studded with stars.

This was what I wanted to explain about the original principle, said Agathodaemon.[d] It was he who published the book on chēmeutics.[e]

His reference to the star-studded serpent reminds us that in fact Ouroboros was a symbol far older in time and more widespread in space than the proto-chemistry of Hellenistic Alexandria. The image of the serpent or dragon eating its own tail was very ancient both in Egypt and the Mesopotamian region. A primeval serpent was part of the cosmogonic mythology of the Pyramid Texts (c. −2300)[f] and occurs

[a] The Pb, Cu, Sn, Fe alloy supposed to be the starting-point of aurification.

[b] Certainly Hg, As and S. The second fragment is essentially a shorter version of the first. Both of them end with a passage about the ritual slaying and sacrifice of a temple-guarding serpent, which closely parallels one of the visions of Zosimus (*Corp.* III, i, 5); cf. Berthelot (1), pp. 60, 180–1. We have already encountered dragon-killing metaphors connected with alchemy in ancient China (pt. 3, pp. 7–8).

[c] Which generate fowls, and they eggs once again.

[d] Now a real person must be intended.   [e] Cf. p. 344 above.

[f] Rundle Clark (1), pp. 50ff. In the 'Book of the Dead', even older, the serpent Āpepi personifies the darkness which the rising sun (Horus), symbolised by a cat, must daily conquer (Budge (4), pp. 248, 280).

figured in Ouroboros form both in Coffin Text pictures (*c.* −1300)[a] and in the mythological papyri (*c.* −1050).[b] But two intertwined Ouroboroi are seen also in a relief of black asphalt excavated from a level of the Elamite necropolis at Susa antedating Hammurabi's time (*c.* −2000).[c] These ideas of cosmic serpents surrounding the universe were quite appropriate as symbols of the recurrences of the planetary revolutions, accurate in their several periods but incommensurable.[d] They would also have come to be connected quite naturally with celestial serpents and dragons, either in the form of constellations or of the monster which controlled solar and lunar eclipses.[e]

Ouroboros reached his apogee, so to say, in the Gnostic period (−2nd to +3rd centuries)[f] when the Greek and West Asian cultures had come together.[g] Many theologies resulted from that union.[h] The tail-eater appears on innumerable inscribed gems, seals and amulets which have come down to us (cf. Fig. 1527);[i] since the seventeenth century these have been known as Abraxas gems, because that word often occurs on them, as here, where it is accompanied by Iaō as well. The former was a Gnostic incantation or word of power,[j] the latter the name of one of the archons or evil demiurges and rulers who had created and now governed the material world,

a Rundle Clark (1), pp. 53, 81, 240ff., figs. 8, 11. Hermopolis, standing for the whole world, is surrounded by Ouroboros.

b Piankoff & Rambova (1), p. 73 and facsim. pl. 1 (the Papyrus of Her-uben A). Cf. Mahdihassan (26), fig. 3.

c *Mémoires de la Délégation en Perse*, vol. 13, pl. XXXVII, 8, in connection with the paper of Pottier (1). Cf. Deonna (3).

d 'The nous-demiurge, encompassing the circles and whirling them round with thunderous speed, set his creations in eternal revolution, so that every ending is a new beginning', *Poimandres*, 11 (*c.* +1st cent., *Corp. Hermet.*, Nock & Festugière (1), vol. 1; cf. Jonas (1), p. 150). Cf. Sir Thomas Browne: 'All things began in Order, so shall they end, and so shall they begin again; according to the Ordainer of Order, and the mysticall Mathematicks of the City of Heaven', 'Garden of Cyrus' (+1658), ch. 5.

e Cf. Vol. 3, pp. 228, 252. See also the fragment *De Dracone Coelesti* preserved in Cumont (5), vol. 8, pt. 1, pp. 194ff.; and further in Bouché–Leclercq (1), pp. 122ff.

f Perhaps before the life of Christ we should speak of proto-Gnostic ideas, cf. Grant (1), p. 14.

g The symbol had not been known in classical Greece or in the Etruscan and early Roman cultures.

h On the general relations of Gnosticism and proto-chemistry see Sheppard's interesting study (1). As Doresse (1), vol. 1, pp. 105ff. and Puech (3) have pointed out, there are very Gnostic passages in Zosimus (*Corp. Alchem. Gr.*, III, xlix, 'On the Letter Ω'), and in Olympiodorus (*Corp.* II, lv, 32) where the cosmic mythology of Adam and Eve is already sketched in purely chemical terms, foreshadowing the medieval equations: Adam = philosophical sulphur, and Eve = philosophical mercury (cf. Doresse, *op. cit.*, p. 130).

Again, there are echoes outside the Corpus. According to St Ephraim of Syria (d. +373), the Gnostics and Manichees say that the mingled constituents of good and evil 'conquer one another and are conquered by one another' (Mitchell (1), vol. 1, p. xvii). This is reminiscent of Aphorism [2].

The historian of chemistry should also be aware that Gnosticism in due time found a place within the bosom of Islam as part of the theology of the Ismailite (*Ismāʿīlīya*) movement (+8th to +10th centuries). This was closely connected both with the scientific writings of the Brethren of Sincerity (cf. Vol. 2, pp. 95–6, Vol. 3, p. 602) and with the Jābirian Corpus in Arabic alchemy (cf. pp. 396ff. below). On this connection, the full implications of which are as yet far from being fully understood, see Corbin (1); Strothmann (1).

i See King (4), pl. XII, 1*a*, descr. pp. 206–7; and (3), pls. C 5 and M 2, also pp. 103, 213ff. Sometimes Ouroboros appears as one constituent part of the design, but often, as here, it forms the frame. These religious and magical representations, sometimes very complex and even now not entirely explained, were first studied and published by de Montfaucon (1) at the beginning of the +18th century. See the English edition, vol. 2, pt. 2, bk. 3, pls. 48–53. Cf. Berthelot (1), p. 62; Preisendanz (2); Bonner (1), pp. 19, 158, 250, pls. II, 39, VII, 141, 153, VIII, 172, IX, 191.

j Derived from the figure 365 in Greek mathematical-alphabetic notation.

planetary gods with names derived from those of the Old Testament God of Israel.[a] Ouroboros has been supposed to symbolise eternity here, but as like as not it stands for all the aeons, terrifying vastnesses of time and space in which man's spirit was imprisoned if no saviour came, each cyclical and perpetual like planetary revolutions.[b] By the +1st century the symbol was penetrating Roman,[c] Gallo-Roman[d] and Scythian[e] culture.

The Gnostic literature also has many references to Ouroboros. The tail-eater, 'king of the worms of the earth', occurs, identified with the serpent of Genesis, in the +2nd-century Jewish-Syriac 'Acts of Kyriakos and Julitta';[f] while in the apocryphal 'Acts of the Holy Apostle Thomas' (+3rd cent.) he too meets the tempter serpent, son of the world-encircling tail-eater.[g] This book is particularly interesting because it incorporates the older 'Song of the Apostle Judas Thomas in the Land of the Indians',[h] a beautiful allegory in which the son of an Asian king[i] is sent to Egypt to steal a pearl of great price guarded by a coiled dragon there. The details we omit, only drawing attention to the remarkable appearance in +2nd-century Syriac of an astronomical motif characteristically Chinese.[j] This is only one more indication of that Westward current which gave Buddhism such an influence not only on

[a] The real God of Light and Life, alien to this world but man's only true home, was sharply distinguished by the Gnostics from either Yahwé or Zeus, since (like Mani after them) they regarded all material things, including food and sex, as irredeemably evil. Their creator therefore was also evil, and the good God was not a creator. For one of the best accounts of Gnosticism see Jonas (1). But there is a rich expository literature, for example Leisegang (3); Wilson (1, 2) and the older book of Burkitt (2). Burkitt was one of my greatest teachers, but his conviction that all the Gnostic systems ought to be regarded as Christian heresies is hardly, I think, acceptable in the light of modern knowledge. Even though some of their books made much use of personages and *logia* taken from the Christian Gospels, as well as parallel soteriological doctrines, they were really forms of another religion. We have no certainty that any form of Gnosticism as such ever spread as far as China, but it has long been suspected that the Taoist Trinity (San Chhing[1]) may have derived, at least in part, from the doctrines of Basilides; cf. Quispel (1).   [b] Jonas (1), pp. 51ff.; cf. Leisegang (3), p. 35.

[c] E.g. the bronze mask of Jupiter on an eagle within an Ouroboros, ascribed to the +1st century; Cumont (7). And the Mithraic monuments discussed in Cumont (6), vol. 1, p. 80, vol. 2, pp. 208 (no. 25, fig. 36), 453 (no. 15, fig. 407).   [d] E.g. the patera found near Geneva and described by Deonna (2).

[e] Here the serpent-dragon has become a lion or a horse; see Rice (1), several examples, and Rice (2), fig. 64. Cf. Mahdihassan (26), figs. 19–21.   [f] In Reitzenstein (2), p. 78.

[g] 3, 32, ed. Lipsius & Bonnet (1), vol. 2, pt. 2, p. 149; James ed. (1), p. 379; Bornkamm (1). Cf. Doresse (1), vol. 2, pp. 44ff. Cf. Fig. 1529.

[h] Called by modern editors 'Hymn of the Pearl' or 'of the Soul'; 9, 108–113, ed. Lipsius & Bonnet (1), vol. 2, pt. 2, pp. 219ff.; James ed. (1), pp. 411ff. A translation in verses by A. A. Bevan, revised by W. R. Schoedel, is given in Grant (1), pp. 116ff. For an exegesis see Jonas (1), pp. 112ff. Cf. too Doresse (1), vol. 1, p. 102; Reitzenstein (4).

[i] Seemingly Parthian. The 'Acts' and the 'Song' are considered closely related to the culture of the city of Edessa. One of the most recent discussions of the Thomas legends will be found in Dihle (2).

[j] See Vol. 3, p. 252, and *in extenso* de Visser (2). Another example of this kind can be found in a Gnostic prayer which has been deeply studied by Peterson (1). It occurs in one of a collection of magical papyri and addresses to the creator god Aeon (= Sabaōth), an appeal to save Adam and his descendants from that destiny (*heimarmenē, εἱμαρμένη*) meted out by the powers of the air (*daimōn aerios, δαίμων ἀέριος*) ruling like a high official between the Light and the Darkness. But Aeon is termed 'ruler of the Pole' (*akinokratōr, ἀκινοκράτωρ*) and is said to be 'throned on the Great Bear' (*heptameriou statheis, ἑπταμερίου σταθείς*). This is an eyebrow-raising parallelism, to say the least, for nothing more Chinese could be imagined than deities or sub-deities in that constellation (cf. Vol. 3, p. 240 and Fig. 90). See also Doresse (1), vol. 1, pp. 112ff., 186, 302.

Interesting examples of Indian–Gnostic connections were discussed by Kennedy (2) in his study of the Gospels of the Infancy, the *Lalita Vistara* and the *Vishnu Purāṇa*.

[1] 三清

Manichaeism[a] but on Gnosticism itself.[b] Naturally Ouroboros was especially important among the Ophite Gnostics, whose cosmology was discussed by Celsus (c. +178) and his opponent Origen in +248. For the Ophites Ouroboros–Leviathan[c] surrounded the seven planetary spheres; indeed it was the firmament itself, beyond which lay paradise.[d] Also in the +3rd century come mentions and depictions of Ouroboros in the magical papyri cognate to and even overlapping with the chemical-technological ones.[e] Finally, in the Egyptian Gnostic-Christian work *Pistis Sophia* (+4th cent.) purporting to record the teachings of Jesus during eleven years after his resurrection, the tail-eater appears twice, once as the outer darkness encircling the universe,[f] and again, paradoxically, as the sun's disc in glorious light.[g] These examples may suffice.[h]

We need not pursue the Ouroboros into its late medieval and post-Renaissance manifestations.[i] But it is interesting that a number of these are double, i.e. formed by two tail-biting animals, not one only. This device occurs in Arabic alchemical texts (if indeed it did not start with them), as in Ibn Umail in the first half of the +10th century.[j] It also appeared in China, as Rousselle found when in the early thirties he came to kneel in the centre of the carpet in the initiation hall of the Taoist community to which he was admitted in Peking.[k] The design figured prominently the double

---

[a] In this religion, which spread from Susiana to the shores of the Atlantic as well as those of the Pacific, and lasted a full twelve centuries (if it is not still latent though unacknowledged in some pseudo-Christian attitudes), the Gnosis came to full ecclesiastical form. This was already realised by J. C. Wolf in +1707. The founder Mani (+216 to +277) claimed to be the completer of the gospels of Zoroaster, Gautama Buddha and Jesus; among the best accounts of him and his religion are the books of Burkitt (1) and more recently Puech (1). Its absolute dualism was expressed with crystalline brevity by the Dominican Anselm of Alexandria (or whoever wrote the *Tractatus de Hereticis*) c. +1265: 'Notandum quod in Persia fuit quidam qui vocabitur Manes, qui ait primo intra se: Si deus est, unde sunt mala? Si deus non est, unde bona? Ex hoc posuit duo principia.' See Dondaine (1), p. 308; Puech (2), p. 65. The medieval tendency to treat Manichaeism as a Christian heresy was surely misguided; it was an essentially different religion. Perhaps it may be well to recall that its identification of evil with matter and darkness had nothing whatever in common with the Yin–Yang dualism, equally strong, of Chinese natural philosophy; that we explained in Vol. 2, p. 277. Cf. Bianchi (1).

[b] On this the very suggestive paper of Conze (8) should be read. Cf. also Kennedy (1) on the likenesses of the system of Basilides with Buddhism, and Przyłuski (2) on Persian in relation to Buddhist dualism and element-theories. Chavannes & Pelliot (1), 2nd pt., pp. 312–13, recall that in the early days of orientalism the Augustinian friar P. Georgi published a large work (+1762) attempting to prove that Buddhism in general and Lamaism in particular were nothing but the Manichaean 'heresy', in disguise, so to say. The boot is actually on the other foot – how much Gnosticism and Manichaeism drew from Buddhism, already by their time half a millennium old.

[c] Cf. Isaiah 27, 1; Psalms 74, 14 and 104, 26; Job 41, 1–10.

[d] *Contra Celsum*, VI, 24–26, 31, 35, with diagram and passwords for the archons of the spheres, tr. Chadwick (1), pp. 337ff. On the Ophites see esp. Leisegang (3), pp. 111ff., 160, Hilgenfeld (1), pp. 277ff.; Grant (1), pp. 52ff.; 89ff.

[e] Reuvens (1) and Berthelot (1), pp. 9, 18 on Leiden V and W; Leisegang (3), opp. p. 112 reproduces a drawing from the great British Museum magical papyrus.

[f] Tr. C. Schmidt (1), p. 207, § 319, 2nd ed. § 317; Mead (2), p. 265, § 319; McDermot (1).

[g] Tr. C. Schmidt (1), p. 233, § 359, 2nd ed. § 354; Mead (2), p. 296, § 359; McDermot (1).

[h] On the whole subject see further in Deonna (3); Preisendanz (3); Eliade (1).

[i] Reference need be made only to Jung (1), figs. 6, 13, 20, 46, 47, 92, 117, 253, 256; Burckhardt (1), p. 137.

[j] Ali, Stapleton & Husain (1), pls. 1, A, 2, B, C. Hence perhaps its appearance (though in single form) on reliefs, relatively recent, in Dahomey, where it is said, according to Bebey (1), to represent the god of the rainbow.

[k] Rousselle (4a), Eng. vers. p. 68, (4b), p. 37. At the present time a double Ouroboros design of fishes is portrayed on Thaiwanese postage-stamps.

dragon, which there was taken to symbolise the 'backward-flowing circulation of the creative force'.[a] Naturally all the double tail-eaters were recognised by Mahdihassan[b] as dualistic symbols, putting him in mind of the world-famous Chinese geometrical Yin–Yang pattern.[c] There is no reason for doubting this likeness, and every probability that it did spring from later Chinese influence on the single Hellenistic Ouroboros.[d]

Such was the Western symbol of the cyclical processes of Nature. But there was no place where these were more appreciated and emphasised than in China, that civilisation which yet basically always rejected the other-worldliness and anti-worldliness of Indian, Iranian and Hellenistic religion. As we noted long ago,[e] the Taoists were obsessed by the problem of change, and particularly by cyclical change, adaptation to which makes a man a sage. The Tao is the tranquillity at the centre of all the disturbances of birth and death, rising and setting. Already the *I Ching* says:[f] '(If we examine) the original beginnings of things and their return to their endings,[g] we shall understand their coming-into-being and their passing-away (*yuan shih fan chung ku chih ssu shêng chih shuo*[1])'. And among its aphorisms we find: 'Wherever there is an ending, there is a new beginning (*chung tsê yu shih*[2])';[h] and also 'There is no going that is not followed by a returning (*wu wang pu fu*[3])'.[i] 'Return brings good fortune (*chhi lai fu chi*[4])';[j] 'Missing the time for reversion—misfortune! (*mi fu hsiung*[5])'.[k] In fact 'the Tao is made up of returning and reverting (*fan fu chhi Tao*[6])',[l] which echoes that great sentence in the *Tao Tê Ching*:

---

[a] See pt. 5 below on physiological alchemy.   [b] (16, 26, 30, 34, 43).

[c] It is surprising that no monograph seems to have been written on the development of this, the *liang i*[7] sign of the 'two forces' or 'instruments'. Most art historians think that its origins must have been with Neo-Confucianism, but with the root rather than the flower; this points to Chhen Thuan (+10th cent.) or even Li Ao (+9th cent.); cf. Vol. 2, pp. 452, 467; and Needham (76). If so, the influence of the design would have reached Ibn Umail quite quickly.

[d] Mahdihassan's attempt (28) to interpret forms like Fig. 1525 as dualistic may be, however, less convincing; as also his efforts (29) to bring the uniped or ophidian shapes of the Chinese organiser god and goddess Fu-Hsi and Nü-Kua into the picture; cf. Vol. 1, p. 163 and Fig. 28, Vol. 2, p. 210, Vol. 3, pp. 23, 95; Przyłuski (1); Chêng Tê-Khun (7). It is interesting, however, that triple forms can occur in Tibetan and other Mahāyāna iconography. Leisegang (3) has reproduced, opp. p. 32, and discussed, pp. 18ff., a mandala the centre of which is formed by a pigeon, a snake and a pig, biting each other's tails in Ouroboros form. This he considered to represent the three cardinal vices of hate, greed and unfeeling stupidity. Leisegang again pondered on the possible historical relations between Buddhism and the Gnosis. Mani at least acknowledged Gautama as one of his great forerunners.

[e] Vol. 2, pp. 74ff.

[f] Hsi Tzhu App. I, ch. 4, (ch. 2, p. 34*b*, Wilhelm–Baynes tr., vol. 1, p. 316). This, the Great Appendix, is probably of the −2nd century. The first four words became proverbial and were often used by Chinese naturalists of later times, e.g. Wang Khuei in the *Li Hai Chi* (cf. Vol. 6).

[g] Or: '(if we realise that) the original beginnings of things by reversion become their endings....'

[h] Kua no. 32, Hêng,[8] 'constancy' (ch. 2, p. 3*a*, Wilhelm–Baynes tr., vol. 2, p. 190).

[i] Kua no. 11, Thai,[9] 'prosperity' (ch. 1, p. 29*a*, Wilhelm–Baynes tr., vol. 1, p. 52).

[j] Kua no. 40, Chieh,[10] 'unravelling' (ch. 2, p. 15*a*, Wilhelm–Baynes tr., vol. 1, p. 165).

[k] Kua no. 24, Fu,[11] 'return' (ch. 1, p. 49*b*, Wilhelm–Baynes tr., vol. 1, p. 106).

[l] The same (ch. 1, p. 48*b*, Wilhelm–Baynes tr., vol. 1, p. 103). All these are in the canonical text, which may be as old as the −7th century.

[1] 原始反終故知死生之說   [2] 終則有始   [3] 無往不復   [4] 其來復吉
[5] 迷復凶   [6] 反復其道   [7] 兩儀   [8] 恆   [9] 泰
[10] 解   [11] 復

'Returning is the characteristic motion of the Tao (*fan chê Tao chih tung*[1])'[a]. It also says:[b]

> The myriad things all do their works and acts
> But I have seen how each has its returning (*fu*[2]);
> All beings howsoever they flourish
> Return and go home to the roots that bore them
> (*ko fu kuei chhi kên*[3]).

All this was naturally applied to history too, as in the opening sentence of the *San Kuo Chih Yen I*[c] which became proverbial—'Whenever there has long been division reunion must come, but union cannot last for ever and division will assuredly occur again (*thien hsia fên chiu pi ho, ho chiu pi fên*[4])'.

That it was also applied in chemistry we have already seen most abundantly both in the practical and theoretical sub-sections, *huan tan*,[5] the 'cyclically-transformed elixir' and the 'regenerated (or regenerative) enchymoma', as the case may be.[d] No phrase or technical term is more all-pervading in the alchemical literature. One of its oldest occurrences may be in connection with Chüeh Tung Tzu[6] in the −2nd century;[e] and we can never forget that Tsou Yen[7] in the −4th began with a 'method of repeated transmutation' (*chhung tao*[8]).[f] The cyclically-transformed elixir, with its nine repetitions (*chiu chuan*[9]), is prominent in the writings of Ko Hung[10] and his contemporaries (*c.* +300);[g] and it echoes on for centuries in the operations of such men as Chang Yuan-Yu[11] (+555),[h] Liu Tao-Ho[12] (+760),[i] and Chhen Thuan[13] (+970).[j] We have generally had most in mind the successive formations and decompositions of mercuric sulphide, but this was probably not the only process involved, for the repeated purification of gold by cupellation,[k] and its isolation by amalgamation with mercury,[l] have also to be considered. Again, just as the Hellenistic protochemists had the image of Ouroboros for their reflux distillations, so also the Chinese could have thought of it for their cycles of fire-phasing[m] and their arrangements for microcosmic circulation.[n]

This being so it would be natural to ask whether the Ouroboros motif occurs in Chinese art, and the answer is yes, though with less explicit relation to cosmological

---

a Ch. 40 (Waley (4), p. 192. Cf. ch. 25 (Waley (4), p. 174, and Vol. 2, p. 50 above).

b Ch. 16, tr. auct., adjuv. Duyvendak (18), p. 49, Chhu Ta-Kao (2), p. 26, Lin Yü-Thang (1), p. 109, Blakney (1), p. 68. Cf. Waley (4), p. 162.

c The famous Yuan historical novel by Lo Kuan-Chung (cf. Vol. 1, p. 112).

d See particularly pp. 218–19, 249, 261–2, and pt. 3, pp. 86, 109, 140, 195.

e Vol. 5, pt. 3, p. 20. The authority for this alchemist, Li Hsiu,[14] is rather late.

f *Ibid.*, p. 14.

g Vol. 5, pt. 2, p. 128; pt. 3, pp. 82–3, 86, 90, 109. Cf. also Ware (5), pp. 64, 82.

h Vol. 5, pt. 3, p. 131.   i *Ibid.*, p. 140   j *Ibid.*, p. 194.

k Vol. 5, pt. 2, p. 277.   l *Ibid.*, p. 278.

m Based doubtless on their recognition of cosmic cycles in time (cf. above, p. 242). Descriptions in Vol. 5, pt. 3, pp. 60, 73–4, and also above, pp. 266 ff.

n Described above, pp. 281 ff.

[1] 反者道之動	[2] 復	[3] 各復歸其根	[4] 天下分久必合，合久必分
[5] 還丹	[6] 絕洞子	[7] 鄒衍	[8] 重道 [9] 九轉
[10] 葛洪	[11] 張遠遊	[12] 劉道合	[13] 陳搏 [14] 李修

or chemical theory than in the West. Dragons of course appear in all Chinese ornamentation everywhere, but those pursuing the moon-pearl and those coiled with their
heads in the centre[a] are not what we are looking for. Some, however, do have their
tails in or near their mouths[b]—we illustrate one in jade[c] and one in bronze[d] from the
Middle Chou period, another from the Thang, and a compact jade ring form which
may be as old as the Shang[e] (Fig. 1528a–e). At least seven other Ouroboros forms in
stone, jade and bronze, have been recovered from Shang-ling-tshun, and three or four
more from Chün-hsien (Hsin-tshun) and Chia-ko-chuang (−9th to −7th centuries),
including some double ones.[f] The motif is also found more to the West in the Tagar I
culture (−8th to −6th centuries) of the Minusinsk basin in Southern Siberia near
Krasnoyarsk north of the Altai mountains.[g] Coiled dragon-like monsters occur, too,
in Scythian tombs in the Crimea;[h] and the dating now available indicates that this
influence was travelling from east to west rather than in the opposite direction. Perhaps
the primeval sky-serpent of ancient Egypt met the tail-eating dragon of ancient
China in the lands of the Scyths.

Tail-eaters widely various in date and context keep on turning up in
Chinese culture. Hopkins showed long ago (17, 18) that the word *chhen*,[1]
meaning an asterism used as a sidereal reference mark, was originally a pictograph of a dragon or serpent coiled round almost in a circle (see cut).[i]
One of the 'great markers' (*ta chhen*[2]) is defined in the *Erh Ya*[j] as 'the
house, the heart and the tail', which being interpreted means the lunar
mansion constellations Fang,[3] Hsin[4] and Wei,[5] covering together[k] a region
from 238° to 265° R.A. and from 20° to 45° Decl. S. approximately, i.e.

just about the area taken up by our constellation Scorpio. Old copies of the *Erh Ya* and
the *Hsing Ching* often represent this as an almost circular ring of stars, but in fact the
whole body is of course quite long drawn out. Nevertheless the tail in the Chinese sky

---

[a] The *phan lung*[6] pattern. E.g. Watson (4), pl. 26a, b and p. 37 (−11th cent.); Anon. (27), pls. XVIII,
XIX (−9th cent.); Kuo Pao-Chün (2), pl. LXXXII, no. 1, p. 51, no. 87 (−9th cent.). A pair of bronze
piao[7] horse-bit ornaments with this design (−10th cent.) in the collection of Dr Chêng Tê-Khun was
exhibited in the Oriental Studies Institute at Cambridge, May 1971.

[b] E.g. Salmony (5), pls. XXII, 1, 2 (−12th cent.), XXXVIII, 3, 4 (−10th cent.), LXIII, 1 (−8th
cent.); Anon. (27), pls. XXIX, 7, LII, 4 (−8th cent.); Salmony (4), pl. XIX, 3 (−5th cent.); Salmony
(5), pls. LXXXIV, 7, LXXXV, 1, 3, 4, 6 (−5th cent.); Salmony (1), pl. LXII, 3 (−4th cent.); Pope-
Hennessy (1), pl. LV and p. 123 (+8th cent.); Gray (1) and Gure (1), pl. 99, no. 260, p. 49 (+11th
cent.). All these datings, especially the pre-era ones, are of course very approximate, fixable only plus or
minus a century or two. The animals, moreover, are not all dragons, but may be tigers, serpents, etc.

[c] Jenyns (3), pl. XXXVI B (−9th to −6th cents).

[d] Palmgren (1), pl. XXIV, 5 and p. 112 (−9th to −7th cents), described and figured also in Heine-
Geldern (4), p. 385.

[e] Buhot (1), fig. 72 and p. 95. We add an incomplete specimen from the Seligman Collection
(communicated by Mrs Brenda Seligman, 1954), cf. Hansford (2). These are theriomorphic variants of
a very old and simple ornament, the split-ring disc; cf. Salmony (5), pls. XXX, 8, 9 (−10th cent.),
LXXXIX, 1–9 (−5th cent.). Other examples are figured in Salmony (1), pl. XXII, 2, 4 (−9th cent.),
and Dr Chêng Tê-Khun has in his collection a beautiful theriomorph split-ring disc from the Shang
period (−12th or −11th cent.).

[f] Cf. Watson (6), pp. 107, 168, fig. 48, pls. 70, 72.   [g] *Ibid.*, pp. 107ff., pl. 71.

[h] *Ibid.*, pls. 73, 74. Cf. also Gryaznov (1); Artemenko *et al.* (1).

[i] Cf. K455 *f, g*. Both bone and bronze forms occur.   [j] Ch. 8, p. 12b (Shih Thien or Fêng Yü).

[k] Hsiu nos. 4, 5 and 6; see Table 24 in Vol. 3.

[1] 辰   [2] 大辰   [3] 房   [4] 心   [5] 尾   [6] 蟠龍   [7] 鑣

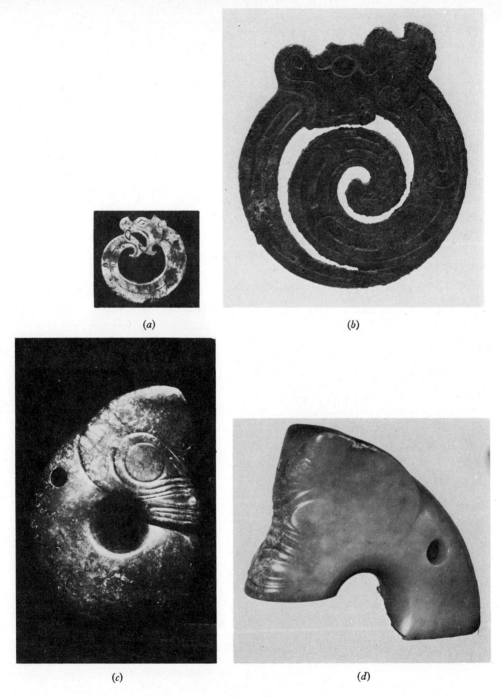

(a)

(b)

(c)

(d)

Fig. 1528

(e)

Fig. 1528. Examples of the Chinese Ouroboros.

(a) Jade ring of the Chou period (−9th to −6th century) from the Eumorphopoulos Collection in the British Museum (Soame Jenyns (3), pl. 36B). Diam. 2·3 cms.

(b) Bronze ornament of the Chou period (−9th to −7th century) in the Museum of Far Eastern Antiquities, Stockholm (Palmgren (1), pl. 24).

(c) Compact jade split ring from Shang or Chou, or possibly a Sung imitation, in the Musée Guimet, Paris (Buhot (1), fig. 72).

(d) Incomplete specimen of a similar kind, from the collection of Mrs Brenda Seligman. Cf. the catalogue of Hansford (2).

(e) Jade ornament of Thang date, +8th century, from the Eumorphopoulos Collection in the British Museum. Cf. Pope-Hennessy (1), pl. 55.

Fig. 1529. A painting by Tung Chhi-Chhang (+1555 to +1636), one of the first Chinese painters to try the European style. Laufer (28), p. 103, supposed the subject to be John the Baptist, but it must surely be St Thomas of the Indies, who in the apocryphal *Acts of Thomas* had indeed an encounter with an Ouroboros. A similar figure can be seen in the cartouche of the oriental map of Elwe (1).

does in fact coil round upon itself, so that one can see how the idea arose.[a] But this astronomical nomenclature had no close connection with cosmological or chemical symbolism, signifying rather certain patterns in the sky, and moreover there were several other ring-like constellations in Chinese astrography.

On the other hand, at the further end of history, there is a close and very strange connection. Early in this century Laufer (28) acquired an album of paintings by Tung Chhi-Chhang[1] (+1555 to +1636) who was one of the first Chinese artists to try painting in European style. The fourth of this series (Fig. 1529) represents some Christian saint holding an Ouroboros; this was identified by Laufer for no particular reason as John the Baptist, Hagios Prodromos—but surely in the light of what we read above in one of the Apocryphal Acts he must be St Thomas the Apostle of the Indies, with the very sign of the noxious serpent which in the Gnostic story he first compelled to restore to life a young man whom it had killed, and then destroyed utterly.[b] One can only conclude that some Jesuit or layman in China at the time knew this apocryphal text and suggested the theme to Tung Chhi-Chhang.

However that may be, there was something singularly prophetic, or perhaps one should say, not devoid of insight, in the cyclical aspect of chemical things which the Chinese and the Greeks both recognised. For in fact the development of modern chemical and biochemical science has revealed a wealth of cyclical processes in the organic as well as the inorganic world.[c] In the +17th century Stahl evolved the idea of a cycle of phlogiston as the vehicle of nutrition of plants and animals, and though that hypothetical entity died long ago the carbon–nitrogen cycle which he visualised is clearly acceptable today.[d] So also in the inorganic world there is the very different carbon–nitrogen cycle in stars, a series of nuclear reactions which convert hydrogen into positrons and helium with great release of energy;[e] and all the many examples of catalysis (such as the role of bromine in the oxidation of sulphur by nitric acid) have been shown to involve chain reactions with restoration of the catalyst. Then there came the recognition of the phosphate cycle in yeast fermentation by Harden & Young;[f] and in modern biochemistry the tricarboxylic acid cycle which oxidises pyruvic acid in muscle,[g] the phosphorylation cycles transferring energy in that tissue,[h] and the ornithine cycle which synthesises urea in many phyla of animals.[i] Thus in chemistry as well as in physiology[j] the intuitions of ancient men have proved justified by the growth of knowledge.

With this we can bring to an end our digression on serpentine symbolism, leaving

---

[a] See Fig. 94 in Vol. 3. Antares (Ta huo[2]), the central star of Hsin hsiu, is specifically called a Ta Chhen[3] in the *Kungyang Chuan*, ch. 23, p. 5a (Duke Chao, 17th year).

[b] 'Acts of Thomas', 3, 30–38; M. R. James tr. (1), pp. 378ff.; Bornkamm tr. (1), pp. 459ff.

[c] See the classical biological cosmography of Lotka (1).

[d] His ideas have been expounded by Strube (1).

[e] Cf. Bethe (1).   [f] This great advance was made in 1908.

[g] See Krebs (1), a paper which discusses other examples of the chains of enzyme cycles which build up metabolic cycles.   [h] Cf. D. M. Needham (2).   [i] Cf. Krebs (2).

[j] On ancient Chinese conceptions of the circulation of the blood, long antedating Europe, see Sects. 43 and 44 in Vol. 6; and meanwhile Lu Gwei-Djen & Needham (5). Understanding of the meteorological water-cycle was probably one of the most ancient appreciations of natural circulation in all cultures; cf. Vol. 3, pp. 467ff.

[1] 董其昌   [2] 火火   [3] 大辰

(very appropriately) two stings in the tail. First, that Ouroboros actually lives—in the shape of the South African armadillo lizard, which when disturbed holds the tip of its tail in its mouth in order to protect its belly by its spring scales.[a] Not impossible therefore is it that the ancients had a living pattern before them, rather than having to form one entirely out of their imaginations. Secondly, there is considerable evidence that the tail-eating serpent provided a stimulus for Kekulé's first cyclic formula in organic chemistry, that of the benzene ring (1865).[b] His own autobiographical account mentions that a year or two before this the idea came to him as he visualised in a reverie a snake seize hold of its own tail.[c] Such points as these alone, apart from the perennial interest of archetypes, might justify our disquisition, but the main lesson still remains, namely that however the streams of influence flowed back and forth, the whole of the Old World was 'in the circuit'; and in no context more so than that of the cycles of Nature and the symbols than men made for them.

From all the foregoing discussion there remains, as they say, one loose end—why was there that curious connection of Gnosticism and Proto-Manichaeism with the earliest Alexandrian proto-chemistry? How exactly it came about is not at all obvious, for if the material world was essentially evil there should surely not have been much inducement to study it—and that an inhibition such as this might be effective, to some extent at least, we have already seen reason to suspect in the case of Chinese Buddhism.[d] But Gnosticism, like its successor Manichaeism, did believe in the possibility of the ultimate salvation of the souls or sparks of light imprisoned in the darkness, so perhaps the 'ascent' of matter implicit in aurifaction was one of the ideas that inspired the Alexandrian proto-chemists.[e] Other parallels can easily be imagined—for example many a Graeco-Egyptian philosophical artisan must have pondered the model of an oil separating from an aqueous solution after shaking, first the confused mass, then the dispersed globules gathering with their likes, finally the two phases homogeneously asunder. Just so did the Gnostic and the Manichee visualise the kingdoms of God and Demiurge, Light and Darkness; 'the spirits of just men made perfect' ascending from the latter to the former.

One wonders whether the technique of distillation itself (first developed, in its Western form,[f] so far as we can see, by the Alexandrian proto-chemists of the + 1st and + 2nd centuries) did not have something to do with this mentality of modelling a

---

[a] Carr (1), pp. 166–7.

[b] Madhihassan (27, 28, 40); Read (3), p. 180. See also Schneider (2); Benfey & Fikes (1); Partington (7), vol. 4, pp. 533 ff, 553 ff.

[c] Berthelot's publications on the Corpus did not begin till 1885, but Kekulé had studied under the historian of chemistry, Kopp, who would certainly have known about Ouroboros, and in any case Kekulé could have been familiar with many of its representations in late alchemical books. He had also been involved in early life, it seems, in a legal case in which some jewellery with an interlinked design of the Susa type (p. 376) had prominently figured, and this made a deep impression upon him. As a parallel for the Ouroboros stimulus to Kekulé, Rather (1) has devoted an intriguing paper to the possibility that a kabbalistic structural-combinatorial principle (the 'creative word') was a powerful influence on the development of the idea of genetic coding in modern molecular biology, starting with Nägeli and Hertwig.   [d] Vol. 2, pp. 417 ff.

[e] Sometimes the theological thought is expressed very chemically, as in Basilides (see Hippolytus, *Ref. Omn. Haeresium*, VII, 22, 26; McMahon & Salmond tr., vol. 1, pp. 277–8, 283, 285; Legge tr., vol. 2, pp. 69–70, 74, 77). For a case of similar Buddhist thinking on 'upward' and 'downward' transformations, cf. Vol. 2, pp. 421–2.   [f] Cf. pp. 84–5 above.

particular world-conception. Burkitt used the word several times to translate the
Manichaean idea of the ascension of souls into the realm of light, e.g. in a hymn
quoted by St Ephraim the Syrian, who died in +373:

> Day by day [they say] diminishes
> The number of souls (on the earth)
> As they are distilled and mount up.[a]

So also, translating a sentence of Cumont on the Manichees:[b] 'Man knows henceforth
the way of enfranchisement; he must consecrate his life to keeping the soul from all
corporal defilement by practising continence and renunciation, so as to set free little
by little from the bonds of matter the divine substance both within him and disseminated throughout Nature, thereby joining in the great work of distillation which God
is carrying out in the universe.'[c] One can thus begin to see something of the mystical
significance which would have attached to vapours and volatile substances, whether
aqueous or oily, arsenical or sulphurous, in the minds of the Graeco-Egyptian and
Persian proto-chemists, 'spirit' rising from the 'hell' of the distilling flask to be
caught in the heaven of the receiver. An analogous idea was that of the cosmic noria
or water-raising wheel[d] with twelve buckets which lifted up the souls of the Manichaean
elect at death into the heavens, running out there like glittering water into the waxing
moon. As it waned it trans-shipped them to that other celestial vessel, the sun,
eventually to rise aloft beyond the spheres in a 'column of glory'.[e]

[a] 'Discourse to Hypatius', v; in C. W. Mitchell (1), vol. 1, pp. cix, 162, vol. 2, p. clxxxiii.

[b] (8), p. 49.

[c] It was Burkitt (1), pp. 32, 35, who inserted the idea of 'distillation'; Mitchell had 'refined and
mount up' (*badmiṣṭal-lān wsālḳān*), Cumont wrote 'épuration'. But Burkitt may have shown a deep
insight in his rendering.

[d] An extended discussion of this will be found in Vol. 4, pt. 2, pp. 356ff. We concluded that the noria
was an Indian invention which reached the Hellenistic world about the −1st century, and the illustration of the +2nd-century Apamea mosaic (our Fig. 596) is directly relevant here. But equally the
sāqīya could have suggested the idea; this we recognised as typically Hellenistic (from −200 or so),
and now a beautiful −2nd-century fresco painting of one of these machines in a tomb at Wardian near
Alexandria has been published by Riad (1).

[e] See Puech (3) and the special study of Cumont (9). Chavannes & Pelliot (1), p. 517, were the first
to realise, from a Chinese Manichaean text, one of the Tunhuang MSS, that Mani's model must have
been the noria (by which they may well have meant the *sāqīya*), both of these machines having certainly
been common in Mani's +3rd-century Mesopotamia. In this text the wheel is called *yeh lun*,[1] and
seems to be of threefold nature, or perhaps supplied by three auxiliary wheels, those of water, wind and
fire. Here one cannot help being reminded of the *ching*,[2] *chhi*[3] and *shen*[4] so prominent in physiological
alchemy (see Vol. 5, pt. 5). The other source of the noria idea must assuredly have been the zodiac,
with its twelve divisions. Greek Pseudo-Zoroaster apocrypha of the −1st and −2nd centuries have a
theory of souls traversing the zodiacal round. A similar notion has lived on among the Parsis (de
Menasce, 2). One classical exposition of the noria doctrine occurs in Hegemonius' *Acta Archelai*,
VIII, 6–7, an important +4th-century account (more than somewhat romancé, however) of a disputa-
tion between Bp. Archelaus of Khalkar in Mesopotamia and Mani himself about +262; see Migne,
Patrol. Graeca, vol. 10, col. 1439; cf. Puech (1), pp. 22ff. Another is in the *Panarios* of Epiphanius, Bp.
of Constantia in Cyprus from +367 onwards (d. +403), heresy no. 66, § xxvi; in Migne, *Patrol. Graeca*,
vol. 42, col. 74. Further on the whole subject see Burkitt (1), p. 43; Puech (1), pp. 83, 176.

The cosmic noria may even have got into book-titles. A lost *Ying Lun Ching*[5] is mentioned in one
text, and the *Thung Chih Lüeh* lists a *Ying Lun Hsin Chao*[6] by one Chiang Chhüan-Chhing,[7] seemingly
astrological. See Chavannes & Pelliot (1), 1st pt., pp. 555–6, 2nd pt., pp. 104–5.

One wonders whether the name noria (*al-nāʿūra*) could have had anything to do with the puzzling

[1] 業輪   [2] 精   [3] 氣   [4] 神   [5] 應輪經   [6] 應輪心照
[7] 蔣權卿

We do not wish to be understood as saying that Gnostic philosophy or theology was responsible in the first place for a technique like distillation, which arose much more probably from prior artisanal practice pondered by philosophical minds, but it may have given some inspiration to people like Pseudo-Democritus, Comarius, Pseudo-Cleopatra and Mary the Jewess in the conduct of their experiments. It seems clear, at all events, that the doctrine of the fundamentally evil nature of the material world, held by both Gnostics and Manichaeans, did not prevent all proto-scientific exploration of it; perhaps they felt that they ought to know more about the nature of their prison-house. It is rather striking that when the + 4th-century *Pistis Sophia* (cf. p. 378 above) enumerates in its tedious way the 89 distinctions which the mystery of Jesus will explain, it includes most of the questions which science asks—the differences of animals and minerals, what really are gold, silver, copper, iron and lead,[a] 'why the matter of glass has arisen and why the matter of wax has arisen', finally why plants are what they are.[b] Obviously it would be an elementary mistake to think of the Alexandrian proto-chemists—or their contemporaries the Taoist elixir-seekers either —as scientific workers of modern type, interested in the analysis of natural phenomena for its own sake, and the real powers which they are aware that this will bring; no, the former probably looked upon chemical operations primarily as numinous symbolic rituals, the latter more as numinous natural magic. But both were extremely interested in cosmic models,[c] these being valued partly no doubt for their demonstrative analogical significance, partly for their believed powers of sympathetic magic, and partly for the real chemical effects (as we should think of them today) which happened in them. But the cosmologies, the macrocosms of these experimental microcosms, were widely different in East and West; and so were the main objectives, aurifaction in the West, material immortality in the East. The common factor was simply the belief that chemical cosmic working models could be made, and that it was worth while to make them.

To sum it all up, there existed from ancient times a trans-Asian continuity, greatly enhanced after Alexander's conquests (− 320) and further facilitated after Chang Chhien's diplomatic and commercial expeditions (− 110). Ostanes the Mede personifies it. We may never be able to trace the exact capillary channels which connected Tsou Yen's tradition with that of Bolus and Pseudo-Democritus; all we can do is to go on deepening our understanding of both of them and contrasting them each with the other. Most likely the two foci of aurifiction and aurifaction, centered primarily on Chhang-an and Alexandria, had essentially independent origins—the question is how far there were mutual influences once they had begun to develop. Westward may have

personage Norea discussed by Pearson (1). A wicked temptress spirit in Jewish apocryphal literature, she was for the Gnostics a Hagia Sophia figure, a moving symbol of cosmic redemption. In the legendary corpus, including the Nag Hammadi texts, she gets confused, among others, with Na'amah the sister of Tubal Cain, as well as Noah's dubious wife, and appears as a beautiful seducer of the angels in the Enoch legend (cf. p. 341).

---

[a] A sixth metal seems to be mentioned in the text, but the word has not been identified.

[b] See § 206–16, esp. C. Schmidt tr. (1), 2nd ed., p. 136, § 212; Schmidt & Till (1); Mead (2), p. 176, § 210. Burkitt (2), p. 75, was particularly struck by this.

[c] Seen abundantly for China in the sub-section on theories, pp. 279ff. above.

come the root of the name *chēmeia* (that 'goldery' that would have so much interested the merchants on the Old Silk Road); the idea of the loves and wars of the elemental natures; a strengthening of the sexual Yin–Yang concept of chemical reaction at the birth of all novelty; and perhaps the belief in the possibility of 'projection'. What certainly did not come at this time was the basic idea of material immortality and the belief in an elixir of life; for that a different eschatology would have been needed; nor did the emphasis on time in Chinese alchemical theory have much echo in the West. The conviction of the value of mineral and metallic medicines did not get through; nor yet the scheme of natural forces symbolised by the trigrams and hexagrams of the 'Book of Changes', unexportable, this last, for two and a half millennia until the age of modern scholarship transcending ethnic barriers had dawned, with men like Wilhelm and Jung. Similarly the death-and-resurrection motif of Greek *prima materia* never found its way to China; but it is possible that the idea of distillation did, though it could only have been a stimulus diffusion since the design of Chinese stills was so radically different. As for biological analogies, the West emphasised fermentation, while China emphasised rather generation. Common factors, however, were the majority of the chemical reagents, and the role of breaths, *chhi, anathumiaseis*, and the like in Nature's operations. And finally neither the Graeco-Egyptians nor the Chinese cared much about atoms, leaving them to the Graeco-Roman philosophers, the Indians and the Buddhists. It was all a pattern of very imperfect communication,[a] but that there was no communication, and no will for it, could hardly be sustained in the light of modern knowledge.

(2) CHINA AND THE ARABIC WORLD

When between +635 and +660 the tribesfolk of the Arabian deserts, inspired by the new religion of the prophet Muḥammad and determined to replace their poverty by a fuller life, poured forth into the surrounding areas of age-long culture, a fresh civilisation with its own language and its own characteristic features was born. It was destined, as everyone knows, to inherit the major part of Hellenistic science and technology, and to pass it on in due course to the Latin West; a process of absorption, enrichment and transference facilitated geographically by the fact that Islam conquered not only the Near and Middle East but also North Africa and Spain. But its cultural boundaries stretched much farther eastwards, reaching to the borders of India and the bounds of Sinkiang, covering everywhere in fact as far east as the longitude of Lop Nor and all the space between the Chad and the Caspian. Hence it is

a We are not by any means the first to have raised the question of possible Chinese influences on Hellenistic proto-chemistry, or at least of mutual contacts at that early time. A case for it was eloquently put forward by Barnes (3) in 1935, and it has been supported by Huang Tzu-Chhing (1) and Ganzen-müller (2), p. 32.

In an interesting paper Sheppard (6) speaks of a 'multi-focal' origin of alchemy, but the definition of it then used by him was not the same as ours (cf. pt. 2, pp. 9 ff. above). Provided his 'alchemy' be understood in the sense of aurifiction and aurifaction only, we would be inclined to agree with him and to see these practices arising independently in Hellenistic, Persian, Indian and Chinese cultures—but macrobiotics is quite another thing. We see the *hsien* Taoism of China as its only original home, whatever the external stimuli may have been which helped it to crystallise in that milieu.

easy to understand that Hellenistic knowledge was not at all the only river which flowed into the lake of Islam—Persia and Iranian tradition was swallowed up in it, and strong currents of influence came westwards now from India and now from China.[a] Obviously when Arabic culture began to concern itself with chemical matters much would be added to the proto-chemistry of the Hellenistic world, and in what follows we must try to trace particularly the passage westwards of Chinese alchemical theory and practice.

(i) *Arabic alchemy in rise and decline*

Although at the present time relatively few scholars are devoting themselves to the study of alchemy in Arabic culture, there were giants who worked on this subject in the fairly recent past (Kraus, Ruska, Stapleton and Wiedemann, for example), and we can learn from the results of their labours.[b] The first question to be decided is whether there was any significant chemical movement under the second Caliphate, the Umayyad (+661 to +750), or whether it began rather under the third, the 'Abbāsid (+750 to +1258), a period corresponding with the golden age of alchemy in China, the Middle and Late Thang and the Northern Sung. While there are hints that something was brewing in the late +7th and early +8th centuries,[c] the main figures of this period to whom alchemical activities were traditionally attributed have been shown with fair certainty to have had no such interests. Indeed Arabic alchemy begins with a striking paradox, in that Khālid and Ja'far were quite real historical personages but not alchemists, while Jābir was among the greatest of alchemists but not real, that is to say not a single person, rather a syndicate of heterodox natural philosophers.

Khālid ibn Yazīd ibn Mu'āwiya (c. +665 to +704) was a kind of crown prince who did not obtain the Caliphate; he supposedly occupied himself with alchemy, was taught by a Byzantine named Stephanus, and wrote alchemical poems. But the tradition has been demolished by the critical analysis of Ruska (4).[d] Debate on the matter has been going on ever since the +14th century, for while Ibn abī Ya'qūb al-Nadīm al-Warrāq al-Baghdādī reported three alchemical books by Khālid in his *Fihrist al-'Ulūm* (Bibliography of the Sciences) finished in +987,[e] Ibn Khaldūn in his *Muqaddima* denied all possibility of this attribution.[f] Khālid's name appears in a kind of colophon to one of the early Arabic alchemical writings, the *Kitāb Qarāṭīs al-Ḥakīm* (Book of Crates the Wise, or the Physician)[g] but there cannot be any real

---

[a] On the focal character of Islamic science, uniting West and East as never before, much has already been said in Vol. 1, pp. 214ff., 220ff.

[b] Their greatest successor today is Ullmann (1). Convenient digests of what is known about alchemy in Arabic culture will also be found in Leicester (1), pp. 62ff.; Multhauf (5), pp. 117ff. Older, but very useful, general papers are those of Wiedemann (15, 21, 24, 25, 29, 32). See also Haschmi (6).

[c] For example the story of Bishr ibn Marwān (p. 475 below) and the embassy of 'Umāra ibn Ḥamza (p. 391 below).

[d] Its authenticity is still defended by Dunlop (6), pp. 205ff., (7), p. 3, but Ruska's rejection has been followed by Mieli (1), 2nd ed., p. 55 and by Hitti (1), 2nd ed., pp. 255, 380.

[e] Fück tr., p. 93; cf. Dodge (1), vol. 2, p. 851.　　　　[f] Rosenthal tr. (1), vol. 3, pp. 229–30.

[g] Tr. Berthelot & Houdas (1), pp. 9ff., 44ff. The personage of the title is not otherwise known. We shall refer again to the book on p. 427.

connection, for this text can be dated to the end of the +8th or the beginning of the +9th centuries, when it was put together from visionary Graeco-Egyptian and Ḥarrānian[a] materials of perhaps the +6th.[b] As for the poems, they must be later forgeries since sal ammoniac is prominent in them, and this salt was almost certainly not known to the Greek proto-chemists in +700 (see pp. 432ff. below). Finally, Ibn Khallikān about +1280 recorded a correspondence between Khālid and a Byzantine hermit named Morienus, probably modelled on a Greek dialogue between Heraclius and Stephanus of Alexandria;[c] this had a great vogue after its translation into Latin though resting on a purely fictional base.[d]

Here already at the outset something about Khālid detains us, however, because of its significance for the East Asian influence on Arabic alchemy which we shall develop in what follows. In the +11th century Qāḍī al-Rashīd ibn al-Zubair wrote a lapidary entitled *Kitāb al-Dhakhā'ir wa'l-Tuḥaf* which dealt with gems, precious metals, and minerals with strange properties. In this he averred that Khālid gained his alchemical knowledge from a book on the subject sent by the emperor of China to his Mu'āwiya grandfather, the first Umayyad Caliph.[e] This would imply a transmission between +661 and +680, the time of Thang Kao Tsung; and it is imaginable that some work of Thao Hung-Ching, Su Yuan-Ming or Sun Ssu-Mo could have been sent,[f] but what is really hard to imagine is its translation into Arabic at that early time.[g] Perhaps the story need not be taken too seriously as history, but the existence of the story some centuries later is the historical and significant thing.

The traditions about Ja'far al-Ṣādiq, the sixth Imām,[h] were similarly demolished by Ruska (5). The main claim of this religious teacher (+699 to +765) is that he is mentioned as the instructor of Jābir in the *Fihrist*,[i] as also in many of the books of the Jābirian Corpus itself,[j] but this is evidence only for the second half of the +9th century or the first half of the +10th, not for the +8th. Ja'far may well have been interested in the 'occult arts', since his name was persistently associated from the beginning of the +9th century onwards with geomancy (sand divination), prognostics from twitches and cramps, weather forecasting, physiognomy, oneiromancy, etc. but there is no real basis for his alchemy.[k] Certain alchemical texts of later times do indeed

---

[a] See pp. 426ff. below.   [b] Cf. Kraus (3), p. 35; Ruska (36).

[c] See the *Kitāb Wafayāt al-A'yān*, a collection of biographies, tr. McGuckin de Slane (2), vol. 1, pp. 481ff.; and, for the whole question, Ruska (4), pp. 31ff.

[d] There is a modern English translation by Stavenhagen (2). Cf. p. 403 below.

[e] Ullmann (1), pp. 120, 192.

[f] See Vol. 5, pt. 3, pp. 120ff., 130, 132ff., 140.

[g] It would have been early even if the language had been Syriac or Greek. There is more than one candidate for the honour of being the first translation of a secular text into Arabic—the usual view is that it was a medical text from Syriac in +684 (cf. Dunlop (7), p. 2). Ullmann (1), p. 152, seems to favour an alchemical one, a treatise of Zosimus done from the Greek and dated +659; but this depends on a very late manuscript described by Stapleton & Azo (2), a +15th-century copy of a +13th-century collection (cf. p. 415).

[h] The imāms, seven or twelve in number, according to the reckoning of diverse Shī'ah sects, were the direct descendants of 'Alī, alone invested (for Shī'ah Muslims) with the spiritual authority of the Prophet, hence infallible and impeccable. See Hitti (1), 2nd ed., pp. 255, 380, 441–2.

[i] Fück tr., p. 96; cf. Dodge (1), vol. 2, p. 853.

[j] See Kraus (2), pp. xxviff., (3), pp. 35, 77, 114, 141, 183, etc.

[k] Ruska (5), pp. 26ff.

bear the name, but it has been shown that in such connections we must speak of Pseudo-Ja'far.[a]

Arabic alchemy does not really begin until the +9th century,[b] but it may be significant that we have a circumstantial account of aurifaction seen by an Arab envoy at Byzantium towards the end of the previous one. His name was 'Umāra ibn Ḥamza, and being despatched on a mission by the Caliph al-Manṣūr in +772 he was present at a demonstration in a secret elaboratory in the imperial palace when lead was turned to silver by the projection of a white preparation, and copper to gold by the projection of a red one.[c] The story is told in a geographical work, the *Kitāb al-A'lāq al-Nafīsa*, written by Ibn al-Faqīh of Hamadan about +902.[d] At the end of his narrative 'Umāra concludes that it was this incident which awakened the interest of the Caliphs in alchemy. There is no particular reason for disbelieving the story, but whether aurifaction was really the first chemical exercise to intrigue the Arabs is doubtful, for the pursuit of macrobiotics may have been known at least as early, as we shall duly see; and that must have come from a diametrically opposite quarter.

The great days of Arabic alchemy are reached with that flood of books and tractates which go under the name of 'Jābir ibn Ḥayyān' and can be dated with certainty to the last half of the +9th century and the first half of the +10th. Understanding of this was the solution of one of the most intractable puzzles in the history of chemistry, namely the relation of the 'Geber' who wrote in Latin towards the end of the +13th century and the 'Jābir' who lived in the golden age of the 'Abbāsids. The breakthrough came in two classical papers by Ruska (2) and Kraus (1) published side by side in 1930.[e] Historians of the last century (Schmieder, Hoefer) generally confused Geber and Jābir, though Kopp first realised that the Geberian titles were not to be found in the Arabic bibliographies, while Berthelot & Houdas[f] not only recognised the great difference between the two types of texts but also knew that already in +987 the author of the *Fihrist* recorded grave doubts concerning Jābir's authorship and historicity.[g] Jābir does not know many things which are in Geber, and Geber shows no trace of having been translated from the Arabic, though Latin translations of a few of the Jābirian works have been found.[h] The fact is that the Jābirian writings form a

---

[a] So, for example, Ruska (5) was able to identify a *Kitāb Risālat Ja'far al-Ṣādiq fī 'Ilm al-Ṣanā'a wa'l-Ḥajar al-Mukarram* (Book of the Letters of Ja'far al-Ṣādiq on the Science of the Art and the Noble Stone) with a *Ta'wīdh al-Ḥākim bi-'amri'llāh fī 'Ilm al-Ṣan'a al-āliya* (Talisman of al-Ḥākim (the Ruler, by the Grace of God, Fatimid Caliph, r. +996 to +1020) on the Science of the Exalted Work), which was found in an Indian library and published by Stapleton & Azo (2), pp. 77ff. From these translations it is clear that the text must have been written between about +1050 and +1280 and has nothing whatever to do with Ja'far.

[b] The book of Balīnās, of which we have already spoken (p. 369), and to which we shall refer again (p. 457), may indeed be as old as about +750, the very beginning of the 'Abbāsid Caliphate, but though it was influential on the later Arabic writers it contains, strictly speaking, no alchemy. One may of course be inclined to speculate that there were alchemists among Balīnās' Central Asian circles, and that they wrote texts which have been lost, but here we have no solid evidence. And in any case the neighbourhood of +820 is a better date for his book. It got into Latin early, with Hugh of Santalla, about +1130.

[c] On white and red elixirs cf. p. 392. Haschmi (5) discusses them in terms of ion exchange coatings on metals.

[d] It may be read in full in the translation of Dunlop (6), pp. 217ff.

[e] See also Ruska (31, 35).   [f] (1), p. 17.   [g] Fück tr., p. 96; cf. Dodge (1), vol. 2, p. 855.

[h] Berthelot (10), pp. 320ff. (12); Ruska (3). Others are surmised; Plessner (8).

Corpus, the work of many different writers with a common philosophical outlook; none can be earlier than about +850 and the whole collection must have been completed not only before +987 but before about +930 because there are quotations in Ibn al-Waḥshīya al-Nabaṭī.[a] As for the real existence of Jābir ibn Ḥayyān himself, it has been and still is a matter of debate,[b] but if he is accepted as historical his dates cannot have been far from *c.* +720 to +815,[c] perhaps some decades later.[d] Whether he wrote any of the Corpus texts, even the earliest, remains undecided.[e]

Partial lists of the titles in the Jābirian Corpus have been given from time to time with appropriate commentaries,[f] but the main authority is still that of the two magnificent monographs of Kraus (2, 3) published in Egypt by 1943. The census of titles and MSS which he conducted gave no less than 1143 books and tractates on alchemy, 847 on magic and theurgy, sympathies and antipathies, 500 on medicine and pharmacy, 300 on philosophy, 100 on mathematics and astronomy, and another hundred on theology.[g] This invites comparison with the *Tao Tsang* itself, though there of course the range of datings is much wider.[h] All the Jābirian texts are roughly similar in style, but Kraus and Ruska were able to establish that they were produced in an order which we can still trace.[i]

The oldest book in the Corpus is probably the *Kitāb al-Raḥma al-Kabīr* (Greater Book of Pity),[j] so entitled because of the compassion which the writer felt for the common aurifictors who got into such trouble straying from the true path. This text is very Hellenistic in a way, containing developed forms of the famous aphorisms (cf. pp. 358 ff. above); but it strikes also many new and previously unheard notes, speaking of elixirs in a markedly chemo-therapeutic manner (cf. p. 479 below), and hinting already at the theory of mercury and sulphur as constituents of all the metals.[k] A decade or so before +850 would be a good guess for its date, and then during the

---

[a] See Kraus (2), p. lix. There is doubt about the historicity of this personage, but his books were real enough, if not entirely what they purported to be.

[b] Holmyard, the leading Arabist in the field outside Germany, was slow to be convinced of the great distinction between Geber and Jābir (cf. 3, 8, 16) but he agreed in the end, retaining as long as he lived, however, a belief in the historicity of the latter (cf. (1), pp. 66 ff. (17), etc.).

[c] Dunlop (6), p. 209, following Holmyard (in Richard Russell (1), mod. ed.).

[d] As Ruska (5) persuasively suggests.

[e] Recent attempts by Sezgin (1, 2) and Haschmi (4) to defend their 'authenticity' in this sense, as a whole, have been refuted by Plessner (4), cf. Ullmann (1), p. 199; Rex (1).

[f] For example Holmyard (2); Ruska (3), in (37).

[g] It must be understood that these figures are somewhat inflated because Kraus left many blank numbers out of respect for the Arabic bibliographers' rough estimates of the wealth of the literature, and in the expectation that many further MSS would come to light. If we deduct these vacancies there remain 568 alchemical books and tractates—still a goodly collection—few of which have had the study they deserve, and almost none translated. The figures for the other subjects need reduction in like manner. But the enumeration is very difficult because on some counts the individual chapters of the larger works are reckoned as independent tractates.

[h] The time of the Jābirian Corpus corresponded to the last half-century of the Thang and the whole of the Wu Tai period. Cf. Vol. 5, p. 3, pp. 141 ff., 167 ff.

[i] Using the classical philological method of notes on who is cited by whom, besides much other evidence; cf. Kraus (2), pp. xxxiv ff., lvii ff.

[j] Kr 5, i.e. No. 5 in the census of Kraus (2). This abbreviation will be taken as standard hereafter. Tr. Berthelot & Houdas (1), pp. 163 ff.

[k] Berthelot & Houdas tr., pp. 166–7, 170, 172–3, 181. The red and white elixirs also come in, pp. 180–1, 189.

next forty years came the two groups called respectively the One Hundred and Twelve Books[a] and the Seventy Books.[b] These contain the marrow of Jābirian alchemy, purely technical, concerned with substances, apparatus and processes; though the latter group is more systematic than the former. At and around the turn of the century we have to place the Books of the Balances (*Kutub al-Mawāzīn*),[c] those strange treatises in which it was sought to determine the proportions of elementary constituents in the composition of substances (cf. pp. 394, 459 below). Afterwards come the Five Hundred Books, or Epistles,[d] writings in which the alchemical-practical is subordinated to gnostic allegories and theological speculations of a Shī'ah character.[e] In these many things of interest can however be found, as for example the *Kitāb al-Ḥajar* (Book of the Stone),[f] which treats of the relations between alchemy and medicine. Towards +930 or rather later come the last books of the Corpus. The *Kitāb al-Khawāṣṣ al-Kabīr* (Greater Book of Properties)[g] and the *Kitāb al-Baḥth* (Book of the Search)[h] revive or continue the ancient lore of sympathies and antipathies in fuller medieval form;[i] but of much more chemical interest are the Books of the Seven Metals (*Kutub al-Ajsād al-Sab'a*).[j] Finally the *Kitāb al-Raḥma al-Ṣaghīr* (Lesser Book of Pity),[k] which presupposes the existence of all the others, purports to expound the essence of alchemy but does so more vaguely than ever, yet echoes the concern of the earlier work of the same name for the unfortunate aurifictors.[l]

As for the identity of the writers of the many Jābirian books, very little positive is known.[m] Some of those named in later sources as 'Jābir's' disciples may be suspected,[n] and 'commentators' may well have been actual authors.[o] One name we have of a man who was asserted by a contemporary to have written some of the books,[p] but that is all.

When we survey the actual content of Jābirian and all Arabic alchemy we find ourselves in a world quite different from that of Hellenistic proto-chemistry, even though Greek influences were manifold and went very deep.[q] Putting it epigrammatically one could say that aurification and aurifaction no longer dominate, for macrobiotics and 'chemo-therapy' have come prominently into the picture, together with biological products and substances, more pharmacological interest,[r] and a certain

---

[a] Kr 6 to 122.   [b] Kr 123 to 191.

[c] Kr 303 to 446.   [d] Kr 447 to 826.   [e] Cf. p. 396 below.   [f] Kr 553.

[g] Kr 1900.   [h] Kr 1800.   [i] Cf. pp. 311 ff. above.

[j] Kr 947 to 956.   [k] Kr 969, tr. Berthelot & Houdas (1), pp. 133 ff.

[l] It is worth recalling that the Jābirian period corresponds with the time of Tuku Thao and Hokan Chi, publisher of the first printed book on alchemy in any civilisation, the *Hsüan Chieh Lu*, as also with the time of appearance of that important book of metallurgical chemistry, the *Pao Tsang Lun*. See Vol. 5, pt. 3, pp. 158, 167, 180, 211.

[m] See especially Kraus (2), pp. lxii ff.

[n] For example 'Uthmān ibn Suwayd al-Ikhmīmī (*fl.* +890), whom we shall shortly meet again (p. 399) in another connection.

[o] E.g. Abū Qirān al-Nisibī (perhaps a significant patronymic, cf. p. 410), or Abū Bakr 'Alī ibn Muḥammad al-Khurāsānī (perhaps also significant, p. 425), or Abū Ja'far Muḥammad ibn abi al-'Azāqir al-Shalmaghānī (d. +933).

[p] This is al-Ḥasan ibn al-Nakad al-Mawṣilī (*fl.* +932). He made money on it too.

[q] Brief introductions to the subject will be found in Leicester (1), pp. 64 ff.; Multhauf (5), pp. 128 ff.; but the only extended and penetrating treatment is that of Kraus (3).

[r] Cf. the *Kitāb al-Sumūm wa-daf'Maḍārrihā* (Book of Poisons and Antidotes), a 'veritable Summa of toxicology', Kr 2145 in the Corpus; as also another *Kitāb al-Sumūm*, written by Ibn al-Waḥshīya about +930 and showing stronger East Asian influences, on which see Levey (8). Cf. p. 449.

thread of preoccupation with all life phenomena. Theory also plays a considerably greater role, and Arabic alchemy is hence much more precise and logical than Hellenistic proto-chemistry, even though the structure is often based upon the most arbitrary and (to us) implausible assumptions.

For its theory of matter Jābirian alchemy adopted the four Aristotelian principles of heat, cold, moisture and dryness,[a] looking upon these, however, not so much as qualities or accidents but as real material constituents of things.[b] Substances, such as metals, had both external (*barrānī*) and internal (*jawwānī*) characteristics, so that gold, for example, was hot and moist externally but cold and dry inside. To convert one thing into another, as in this case silver to gold, it was only necessary to bring out the internal characteristics of the less noble metal; as for chemical change in general, everything depended on the admixture or *krasis* (κρᾶσις), *mizāj* in Arabic, of the primary constituents, external and internal. The agents for changing these balances, and so converting one substance into something else by a transmutation (*qalb* or *iqlāb*), were none other than the elixirs (*al-iksīr*), among which there was a supreme elixir (*al-iksir al-a'ẓam*); these and this were capable both of neutralising constituents present in excess and also of supplying the deficiencies of others. On account of the vital importance of the elixir concept for the general history of chemistry in the whole of the Old World we shall have to examine minutely a little later on (pp. 472 ff.) both the name and the thought. Here, continuing our sketch, it may be added that the Jābirian and Arabic alchemists believed that the actual qualities or constituents themselves could be obtained pure if one went on operating long enough.[c]

All this was associated with a highly elaborate body of theory known as the Science of the Balance ('Ilm al-Mīzān).[d] This was nothing less than an attempt to determine the proportional constitution of every natural object, but although the idea was exact and quantitative the execution was based not on experimental weighings but on numerological computations. We have already come across indications of this kind of thing in Chinese contexts (p. 304), but the Arabic theory remains perhaps the greatest example of such a procedure in all the history of science. Among its various sources the rough Galenic attempt to classify all drugs in four degrees (*taxeis*, τάξεις) of pharmaco-dynamic intensity was certainly one,[e] but there were also origins of a more theurgic or magical character, and the system lost all contact with reality when it divined the composition of substances from the letters and syllables of their Arabic names.[f] Thus the heat, cold, dryness and moisture in metals or salts were solemnly 'measured' in

---

a The Arabic terms are, respectively, *ḥarāra, burūda, ruṭūba* and *yubūsa*.

b There is one very 'Jābirian' text in the Hellenistic Corpus, namely v, ii, 'On the Work of the Four Elements', tr. Berthelot & Ruelle (1), vol. 3, pp. 322ff. They were sure it was later than the +7th century, and indeed it could well be of the +9th or +10th and derivative from the Arabic.

c For a quotation exemplifying this see Leicester (1), p. 66, taken from Kraus (3), p. 10, translating from the *Kitāb al-Talkhīṣ* (Book of the Reduction), Kr 164. The mania of the Arabs for almost endlessly repeated distillations, etc. up to 700 or 800 times, is often shown, as in the only English translation of a Jābirian book, that of Steele (3), the *Kitāb Hatk al-Astār* (Book of the Rending of Veils), Kr 972. This was no great contribution.

d Cf. Kraus (3), pp. 23 ff. but especially pp. 187 ff.

e Cf. Kraus (3), pp. 189 ff.; Harig (1).

f Kraus, *op. cit.*, pp. 223 ff.

units of much precision (*qīrāṭ, dirham, dānaq*, etc.).[a] In all these computations, the numbers 1, 3, 5, 8 and 17 keep on coming in, and these have more importance than one might think, because of the light they throw on origins and transmissions (cf. p. 459 below).[b]

At the same time, Jābirian and Arabic alchemy was more advanced than Hellenistic proto-chemistry because of its clearer and more rational classifications. These have come down to us in much detail, on which the specialist works are to be consulted;[c] here perhaps we need only say that in the Corpus, there are in general five spirits or volatile substances (*arwāḥ* or *nufūs*), seven metals, malleable and sonorous (*ajsām*), and an indefinite number of pulverisable minerals (*ajsād*) which later are divided into vitriols (*zājāt*), boraxes (*būraq*), salts (*milḥ*), stones (*ḥajar*) and the like. And here the Arabs have gone beyond the Greeks because a new volatile spirit is added to the classical sulphur, mercury and arsenic, namely ammonia in the form of sal ammoniac; for all the Arabic writings are characterised from the beginning by knowledge and use of ammonium chloride ('mineral ammoniac', *nūshādir*) from natural sources in Central Asia, and ammonium carbonate ('derived ammoniac', *nūshādir mustanbaṭ*) obtained by the dry distillation of hair and other animal substances.[d] This points up what has already been said, namely that Arabic alchemy is full of animal and plant substances of every kind,[e] very often submitted to destructive distillation, and so producing gases, inflammable materials, liquids, oils and ash (a palpable demonstration, for those times, of the air, fire, earth and water elements). Or they could be treated with weak acids, alkalies and alcohol solutions, sublimed for camphor or distilled wet for essential oils. A certain development of laboratory apparatus also occurred during these Arabic centuries. Moreover, a new salt was added to those previously known, saltpetre (potassium nitrate).[f] As for the volatile substances it must also be said that in the Jābirian Corpus we find the Western beginnings of the idea that all the metals (and perhaps other substances too) are combinations of sulphur (*al-kibrīt*) and mercury (*al-zībaq*) in different proportions,[g] all having naturally developed with great slowness in the bosom of the earth (cf. pp. 224, 454, 458).[h]

One should add that Jābirian alchemy was surrounded by an aura of speculative philosophy bordering on the magical. Reference has already been made to the Science of Properties ('Ilm al-Khawāṣṣ), by which the Arabic writers meant the tradition of causes and effects, sympathies and antipathies, going back to Bolus of Mendes[i] and

[a] Kraus, *op. cit.*, pp. 230 ff.   [b] Kraus, *op. cit.*, pp. 194 ff., 219.

[c] Cf. Kraus (3), pp. 18 ff.

[d] Kraus, *op. cit.*, p. 41. A typical work on this technique is the *Kitāb al-Ḥukūma* (Book of Government), Kr 134, cf. Ruska (3).

[e] An extensive treatment of the use of these in chemistry is found in the *Kitāb al-Ḥayy* (Book of the Living), Kr 133, cf. Ruska (3).

[f] See pp. 195 ff. above, where the exceptional importance of this salt is made clear.

[g] The oldest Arabic statement of this doctrine is found in the 'Book of the Secret of Creation' of Balīnās (see p. 369 above), early in the +9th century; later (pp. 455, 459) we shall consider its possible Chinese origin and the developments to which it led. See Ruska (8), p. 151. It also comes in the 'Epistles of the Brethren of Sincerity', in Ibn Sīnā, and in most of the Arabic alchemists.

[h] Cf. statements deriving from al-Rāzī in Ruska (21), pp. 62, 64, 96, 98.

[i] And, for all we know, to Huai Nan Tzu also (cf. p. 311 above).

generally covered by the Greek term *physica* (φυσικά).[a] But besides this there was also the Science of Theurgy and Apotropaics ('Ilm al-Ṭilasmāt)[b] and the Science of Generation ('Ilm al-Takwīn).[c] This last had to do not only with the formation of all kinds of ores and minerals but with the spontaneous and artificial generation of plants, animals and men; hence the idea of the homunculus, which later was absorbed into Latin alchemy, and some very strange directions for creating life, including the incubation of an artificial foetus within a model of the celestial spheres maintained in perpetual motion. This pseudo-science is of so much interest for the origins of the elixir idea that we must presently return to it.[d] Last of all, the Jābirian Corpus includes a number of books on cosmology and cosmogony, in which the chemical properties of substances are brought into relation with the construction of the universe itself.[e]

Here we come close to theology. It is thus of great interest that Kraus and Ruska were able to establish that all the works of the Jābirian Corpus were produced by a school or group essentially Ismā'īli in character,[f] and so foreshadowing the notable scientific collection called *Rasā'il Ikhwān al-Ṣafā'* (Epistles of the Brethren of Sincerity). These were produced rather later, in the second half of the +10th century, and covered a wider spectrum of heavenly and earthly sciences than the alchemy of the Corpus; we have already had a good deal to say about them.[g]

What was the Ismā'īliya movement? It arose as part of the schism caused by the insistent belief of the Shī'ah Muslims in the transmission of the Prophet's spiritual authority through his son-in-law 'Alī, the first Imām.[h] Partly political from the beginning, this movement became even more so as it gained the support of the Iranian people, maintaining their own traditions over against purely Arab ideas. The theory of the Imāmate involved belief in a succession of either seven or twelve inspired leaders, and the Ismā'īlis took their name from the seventh, either Ismā'īl ibn Ja'far or Muḥammad ibn Ismā'īl. Both before and after their time the imāms had been, and became again, invisible, but one of the incarnate succession would one day appear in a kind of *parousia* as al-Mahdī to rule the earth in justice, peace and righteousness.[i] This ecclesiology and eschatology has evident similarities with ideas such as the apostolic succession, the Second Coming, and the rebirths of Buddhas and Bodhisattvas, but most of all with millenniarism and chiliasm, whatever religion they might be attached to. Hence the interest of the Qarmatian movement,[j] an Ismā'īli organisation deeply

---

[a] See Kraus (3), pp. 61 ff.

[b] E.g. talismans.

[c] See Kraus, *op. cit.*, pp. 97 ff., 109, 119. The principal treatise on these subjects is that entitled *Kitāb al-Tajmī'* (Book of the Concentration), Kr 398. Partial translation in Berthelot & Houdas (1), pp. 191 ff. ending with a late interpolation (p. 225).

[d] See pp. 485 ff. below.

[e] Kraus (3), pp. 135 ff. The most important treatise on these subjects is the *Kitāb al-Taṣrīf* (Book of the Transmutation, or, the Morphology), Kr 404.

[f] See especially Kraus (2), pp. xlviii ff.

[g] Vol. 2, pp. 95 ff. and *passim* in other volumes. See the fuller accounts also in Mieli (1), 2nd ed., pp. 128 ff.; Hitti (1), 2nd ed., pp. 372–3.

[h] Cf. Hitti (1), 2nd ed., pp. 441 ff.; Mieli (1), 2nd ed., pp. 59 ff.

[i] One cannot help being reminded of Maitreya, the Buddha to come.

[j] Encountered already in Vol. 2, p. 96. Chinese parallels are discussed in Needham (56).

socialist or communist in practice, which began about +885, succeeded in establishing an independent State near Bahrein on the Persian Gulf,[a] kept up continual war with the Caliphate throughout the +10th century, and even after being overthrown in Iraq bequeathed much of its equalitarian doctrine to the Fatimid dynasty of Egypt and to the Neo-Ismāʿīlites of Alamūt and Syria, groups destined to last on until the Mongol floods of +1260.[b] So what the Brethren of Sincerity in their time (c. +960 to +980), and the writers of the Jābirian Corpus in theirs (c. +890 to +920),[c] had in common with the early medieval Taoists was the simultaneous possession of mystical, scientific and political tendencies. They all acknowledged the existence of mysteries in Nature transcending *a priori* ratiocination, they all believed in the efficacy of manual operations for the development of sciences based on observation and experiment (though of course they found it very hard to distinguish between real effects and magical claims), and they all looked for the coming of an equalitarian and classless society. It is very remarkable that we can still find such clear traces of the movements of this kind which manifested themselves both in Western and Eastern Asia several centuries before the scientific revolution in Europe.[d]

Copious though it is, the Jābirian Corpus was far from exhausting the alchemical effort of the +9th and +10th centuries, and mention has to be made of other writers seemingly independent of it; three remarkable men and one strange book the authorship of which we are still not quite sure about. The three men are Dhū al-Nūn al-Miṣrī (d. +859), Ibn Isḥāq al-Kindī (+800 to +867) and the great al-Rāzī (+865 to +925). The book was the *Turba Philosophorum* (to give it its best-known name) or 'Congress of the Philosophers', datable somewhere very close to +900. It must be symptomatic of the amplitude already attained by Arabic civilisation that the Corpus never refers to any of these, nor do they refer to texts that are in it.

With the Egyptian Dhū al-Nūn ('Him of the Fish') we are in presence of the allegorical and ecstatic-visionary trend already noticed in the Corpus, which goes back to Zosimus and Olympiodorus among the Greeks. Hence it is of much interest that Dhū al-Nūn, whose full name was Abū al-Fayḍ Thawbān ibn Ibrāhīm al-Ikhmīmī al-Miṣrī, came from the same city as Zosimus—for Ikhmīm was the Arab name for Panopolis.[e] He is often regarded as one of the first of the sufis or mystical Neo-Platonic philosophers of Islam, and seems to have been close to the Ismāʿīlis, but there is nothing practical attributable to him.[f] The case of al-Kindī is altogether different. Abū Yūsuf Yaʿqūb ibn Isḥāq ibn al-Ṣabbāḥ al-Kindī was one of

---

[a] Quite reminiscent of the semi-independent theocratic Taoist State in Northern Szechuan founded by Chang Tao-Ling[1] in the +2nd century. On this one may read Maspero (13, 32).

[b] The Ismāʿīlis are by no means extinct even today, and form a number of more or less independent sects, some ten million people in all, in the Middle East and India.

[c] The Qarmatians are occasionally mentioned by name in the Corpus (Kraus (2), p. xlix), e.g. in the *Kitāb Ikhrāj mā fī'l-Quwwa ilā'l-Fiʿl* (Book of the Passage from Potentiality to Actuality), Kr 331; along with Indians, Mazdaeans and Greeks. Tr. Rex (2).

[d] On the wider significance of this association of mystical naturalism with early science and revolutionary social movements see Sect. 10(f) in Vol. 2, pp. 86ff., or Needham (77).

[e] In the Nile delta.

[f] Cf. Mieli (1), 2nd ed., p. 64; Hitti (1), 2nd ed., p. 435; Dunlop (6), p. 297.

[1] 張道陵

the greatest and most prolific philosophers of Islam, but also a great naturalist, meteorologist, mathematician and physicist.[a] We have already met with his important work on the chemistry of perfumes and distillations, *Kitāb Kīmiyā' al-'Iṭr wa'l-Taṣ'īdāt* (pp. 127 ff. above).[b] At the same time he was against aurifaction, and wrote an 'Epistle in Refutation of those who Claim the Artificial Fabrication of Gold and Silver'.

Just before al-Kindī died there was born at Rayy in Persia one of the greatest of all Arabic scientific men, Abū Bakr Muḥammad ibn Zakarīyā al-Rāzī.[c] In him chemistry and alchemy were combined not with philosophy but with medicine, for he was the leading physician of his time, and headed the great hospital in Baghdad. On his bibliography much work has been done,[d] though few of his texts have been adequately studied, apart from the *Kitāb Sirr al-Asrār* (Book of the Secret of Secrets), an integral translation of which we owe to Ruska (14).[e] The characteristic of his chemical writing is a complete matter-of-factness and freedom from all mysticism; there is none of that aura of 'nonsense' which pervades the Corpus. His classification of naturally occurring substances ('*aqāqīr*) is similar to that in the Corpus, if more elaborate and clearer;[f] in other ways also there is much similarity,[g] as in the concept of elixirs, the knowledge of sal ammoniac, the mention of East Asian things, the great use of plant and animal materials, and the preparation of caustic alkalies.[h] One of the works most influential in the Latin alchemy of the Middle Ages, the *De Aluminibus et Salibus*, has been shown to derive from parts of the 'Book of the Secret of Secrets'; it must have been translated and enlarged by some very practical Spanish alchemist of the +11th century, and the further work of translation is often attributed to Gerard of Cremona (+1114 to +1187).[i] But practical though al-Rāzī was, he never doubted the possibility of aurifaction, and one of his tractates was entitled: 'Refutation of al-Kindī with regard to his including Alchemy in the Category of the Impossible'.[j] It is to be feared that both these refutations are lost.

Next a word about the *Turba Philosophorum*.[k] One cannot refer to its proper title in Arabic because the original version has not yet been found,[l] and we depend on

---

[a] Cf. Mieli (1), 2nd ed., p. 80; Dunlop (6), pp. 178, 223, 229, 231.

[b] Tr. Garbers (1).

[c] Cf. Mieli (1), 2nd ed., pp. 89ff., 132ff.; Hitti (1), 2nd ed., pp. 365ff.; Leicester (1), pp. 68–9; Kraus & Pines (1).

[d] See Ranking (1); Ruska (15). Ruska (16) and Kraus (5) have discussed the account of al-Rāzī's life and writings given by al-Bīrūnī about +1036. A translation is given by Dunlop (6), pp. 237ff.

[e] It seems that al-Rāzī wrote two books, one 'of Secrets', the other 'of the Secret of Secrets'. Both are combined in a MS. collection in the Tashkent Library, and Karimov (1) has published a Russian translation of the latter, apparently the larger work of the two. He believes that Ruska's translation was of the former. These texts have nothing to do with the *Secretum Secretorum* that Roger Bacon was interested in (pp. 297, 368 above, pp. 494, 497 below).

[f] There is a study of this in Stapleton, Azo & Husain (1).

[g] General accounts of al-Rāzī's chemistry have been given by Partington (17) and Heym (2).

[h] Translations and textual comparisons (but no interpretations) in Ruska & Garbers (1).

[i] There is an excellent monograph on this book and its history by Ruska (21). See also Multhauf (5), pp. 160ff., especially on the part played by Gerard.　　　　[j] Ranking (1), no. 40.

[k] The most important papers on this are those of Ruska (6) and Plessner (5, 6, 7). There is an English translation, not perhaps meeting present-day scholarly standards, by Waite (13).

[l] It was probably *Muṣḥaf al-Jamā'a* (Book of the Assembly); Ullmann (1), pp. 213ff.

several Latin translations none older than the +12th century, though more or less parallel texts in Arabic are not unknown.[a] The structure of the work is intriguing; in a series of speeches reminiscent of those at a symposium or congress nine pre-Socratic Greek philosophers give each their divergent opinions, from Anaximander to Xenophanes, after which follow sixty-three other speeches all dealing directly with alchemy—these more international since they include Astanius (Ostanes), Bonellus (Balīnās) and others. The dating of this work has been difficult, but that it must have been put together in the neighbourhood of +900 appears from the following considerations. The first person to quote from it is Ibn Umail, who died about +960.[b] The *Fihrist* says that 'Uthmān ibn Suwayd al-Ikhmīmī, who was certainly living about the turn of the century, wrote a work entitled 'Book of the Controversies and Conferences of the Philosophers', and this was most probably the *Turba*.[c] It must have been put together after the appearance in Arabic of the Indian book of poisons, *Kitāb Shānāq*, c. +830, because the theme of the poison-maiden, which is used in the *Turba* as an alchemical allegory, came into Arabic literature by this means.[d] If Ruska and Berthelot were right in dating the *Kitāb al-Ḥabīb* to the middle or early +9th century, then it may well have been a precursor of the *Turba* for it certainly contains many speeches and dialogues.[e] Ruska (36) identified it with a work listed in the *Fihrist*, 'Book (of Dialogues) of Mary the Copt with the Philosophers who assembled at her House'.[f] The idea was surely very much in the air in the +9th century because similar ideas about symposia can be found in the Jābirian Corpus.[g] As for the earlier roots of the *Turba*, it has been possible to show borrowing from the *Refutatio Omnium Haeresium* written by Hippolytus[h] in +222, and a close connection with the opinions of philosophers as reported by Olympiodorus (+6th century) in the Hellenistic proto-

---

[a] A *Jamā'a Fīthāghūras* (Compendium of Pythagoras) is listed in the *Fihrist* bibliography (see Fück (1), p. 94) There was an Arabic text of it in Cairo in the +16th century, so it may still be recovered. Kraus also found a MS. entitled *Min Maṣḥaf al-Jamā'a* (From the Book of the Assembly), see Ruska (6), p. 297. [b] We shall speak further of him immediately below.

[c] Fück (1), p. 107; Dodge (1), vol. 2, p. 865. The suggestion is due to Plessner (5), and has been generally accepted, as by Nasr (1), p. 283.

[d] This is an interesting story in itself. The *Kitāb Shānāq fī 'l-Sumūm wa'l-Tariyāq* (Book of Canakya on Poisons and Theriaca) was based, as textual parallelisms show, on the *Arthaśāstra* (+3rd cent.), and Canakya was Kautilya. The *Suśruta* and *Caraka samhitas* also afforded material. About +790 it was translated from Sanskrit into Persian by one of the Indian physicians at Jundi-shāpūr, Mankah or Kankah (Kanaka), to whom there is a reference in the Jābirian Corpus (Kraus (3), p. 59). Then by +830 it was done into Arabic from Persian by one Abū Ḥātim, and revised by al-'Abbās ibn Sa'īd al-Jauharī, who added Greek materials. It is not very closely related to the *Kitāb al-Sumūm* of the Corpus (Kr 2145), on which see Ruska (38). We owe an exhaustive study of the Shānāq book to Bettina Strauss (1), afterwards the wife of Paul Kraus. Its translation under the 'Abbāsids has to be set beside the similar incorporation of Indian medical and astronomical sources into the Arabic literature described in Vol. 1, p. 216 (cf. Dunlop (7), p. 6ff.). On Kautilya see also Vol. 5, pt. 3, pp. 164–5. On the history of the poison-maiden theme see at length Penzer (2) and Hertz (1).

[e] Al-Ḥabīb ('friend') appears to be a personal name. Translation in Berthelot & Houdas (1), pp. 76ff. Though all the personages have Greek names, many Chinese and Indian notes are struck in what they say; cf. pp. 469, 470, 471, below.

[f] See Fück (1), p. 94.

[g] The details are given by Kraus (3), p. 59, citing especially the *Kitāb al-Mujarradāt* (Book of Abstractions), Kr 63, 64.

[h] Who has come into our argument before, p. 124 above. All the nine pre-Socratics appear in his book, and there are textual resemblances with the *Turba*, as Plessner (5) was the first to notice.

chemical Corpus,[a] though these are descriptive or doxographic rather than in dialogue
form.

The general view now is that the 'Mob, or Congress...' was an attempt to put
Hellenistic proto-chemistry and natural philosophy into Arabic form and adapt it to
Islamic science.[b] There can be no doubt that the writer had a remarkable knowledge
of Greek thought, but the essential points which he wanted to make (and made
through the mouth of Xenophanes) were first the importance of Islamic monotheism,

Fig. 1530. A Chinese *Turba Philosophorum*, from *Shen Hsien Thung Chien*, Hua tsang thu, p. 45 a. The
conclave consists of Yin Chhang-Shêng (cf. Vol. 5, pt. 3, pp. 75–6), Chang Kung-Chao, Ma Ming-
Shêng (ibid. p. 77), Khung Yuan-Fang, Lu Tung-Hsüan and Liu Tzu-Nan. Its date would be about
the middle of the + 2nd century, in the Later Han period.

secondly the uniformity of Nature, and thirdly the universality of the four elements as
components of all created things. Obviously Greek and Byzantine influence was
paramount in the work, yet even here traits distinctively Further Asian keep on
coming in—for example, a great emphasis on sexuality in chemical substances and

---

[a] *Corp.* ii, iv, 19–28; also Plessner's find.
[b] There is nothing quite similar in Chinese literature, though dialogues occasionally occur; we shall
meet with one on physiological alchemy in Vol. 5, p. 5. But paintings and drawings of natural philosophers
in conclave are quite frequently found, and we reproduce here a picture of six ancient alchemists in
plenary session (Fig. 1530). Ma Ming-Shêng and Yin Chhang-Shêng we know from Vol. 5, pt. 3, pp. 43,
49, 51, 75–7, and the other four are also of the later Han period, + 2nd century (*Shen Hsien Thung Chien*,
Hua tsang thu, p. 45 a).

reactions, quite reminiscent of the theory of Yin and Yang,[a] a marked imagery of the processes of animal generation,[b] and explicit references to India.[c]

Ibn Umail has just been mentioned, and he is the next of our landmarks.[d] Muḥammad ibn Umail al-Ṣādiq al-Tamīmī (*c.* +900 to +960) wrote much on alchemy, but his most renowned work was the *Kitāb al-Māʾ al-Waraqī waʾl-Arḍ al-Najmīya* (Book of the Silvery Water and the Starry Earth); this found its way into Latin as the *Tabula Chemica* (cf. p. 373 above) with his name still attached to it in the somewhat disguised form of Senior Zadith Filius Hamuel.[e] Similarly, his alchemical poem, the *Risālat al-Shams ilāʾl-Hilāl*, on which the former work was really a commentary, was translated into Latin, in this case keeping its exact title *Epistola Solis ad Lunam Crescentem*. There is something ominous about alchemical poems. They presage and preside over the decaying end of a tradition, when the hard factual side has been pushed as far as it will go within the prevailing intellectual cadre, and there is no real way further forward; this one can see very clearly both in the Hellenistic proto-chemical and the Chinese alchemical traditions as well as in the Arabic.[f] Ibn Umail's writing is by no means devoid of a practical and experimental basis, but it has much in common with that of Dhū al-Nūn, being very allegorical and mystical, with visions in the crypts of the pyramids and so forth; it also accepts the possibility of aurifaction and belongs to the mineral-metallurgical school, warning against the use of animal substances in alchemy. Then with Maslama ibn Aḥmad al-Majrīṭī a little later (he died *c.* +1007) comes the development of Arabic alchemy in Spain, for his *Rutbat al-Ḥakīm* (The Sage's Step) reveals much practical knowledge.[g] He discusses

[a] This runs all through the text. In the speech of 'Socrates' chemical reaction is compared with generation, lead is male and orpiment female; according to 'Diamedes' both male and female substances are needed, mercury being the former and sulphur the latter; 'Ostanes' makes copper female and mercury male; 'Theophilus' has allegories of nights of love between man and wife, etc. etc. See the translation of Ruska (6), pp. 200, 215–16, 229, and 247; also his summaries, sects. 54, 55, 57, 59. Many traces of the *Turba* are found in later Latin texts, such as the +14th-century *Consilium Conjugii, seu de Massa Solis et Lunae*, a very Yin–Yang production; on this see Ruska (6), pp. 333 ff., 342; Ferguson (1), vol. 1, p. 176; Berthelot (10), p. 249. On the meaning of *massa* here see p. 366 above. Ruska himself well appreciated how much non-Greek ideology the *Turba* contains, (6), p. 295.

[b] Notable e.g. in the speech of 'Bonellus' (Balīnās) analogising the alchemical work with the development of the embryo in egg or womb (Ruska (6), p. 247, summary 59; cf. p. 292 above). We have already mentioned the strange ideas of the Jābirian Corpus on artificial generation within cosmic models (p. 396 above), and we shall return to this presently because of its close connection with the idea of the life-giving elixir (pp. 485 ff. below).

[c] In his speech, 'Leucippus' says: 'What Democritus had on the science of the natures [i.e. the four elements] he got from me, and hence (in the last resort) from the philosophers of India and Babylonia; I think, however, that he excelled all others of his time in science.' Perhaps Babylonia here meant Persia and the Ostanes tradition, as opposed to the Graeco-Egyptian Hermes-Agathodaemon tradition. It will be remembered that the Hellenistic Corpus contains a 'Letter of Pseudo-Democritus to Leucippus' (*Corp.* II, ii). The only mentions of India in this Corpus both concern wootz steel (cf. Needham, 32); one fragment may be quite early (*Corp.* I, xvii, 3), but the other must be late, perhaps +10th century (*Corp.* V, v).

[d] The fullest study of him is that of Ali, Stapleton & Husain (1), but Holmyard (1), p. 99 and Leicester (1), pp. 63, 80 have something to say on him too. He quotes al-Rāzī (cf. Stapleton & Azo, 2) as well as the *Turba*. See also Ruska (9).

[e] Cf. pp. 366 above.   [f] Cf. pt. 2, p. 17, pt. 3, pp. 148, 195, and p. 327 above.

[g] See especially Holmyard (11) and Ruska (28). Mentions also in Mieli (1), 2nd ed., pp. 180–1; Holmyard (1), p. 98; Leicester (1), p. 71. But there is some discrepancy about the dating of the 'Sage's Step'; the +950 of some MSS seems too early, and the +1050 of others is certainly too late, hence, though the question is still unsettled, the book may be by Maslama al-Majrīṭī's immediate pupils.

cupellation and cementation,[a] and describes the oxidation of mercury, but he believed in transmutation by elixirs, and indeed that every alloy was a new species. Al-Majrīṭī and his school also propagated in the West the writings of the Brethren of Sincerity, and doubtless of the Jābirians too.

The +11th century yields two names of importance, one much better known than the other. In +1034 Ibn ʿAbd al-Malik al-Ṣāliḥī al-Khwārizmī al-Kaṭī produced his ʿAin al-Ṣanʿa wa-ʿAun al-Ṣanaʿa (Essence of the Art and Aid to the Workers).[b] This was basically a book of metallurgical chemistry in the Hellenistic tradition, allowing of aurifiction as well as aurifaction; 'tingeing' is prominent, calcium polysulphide used, arsenical copper known, as well as the dilution of silver with copper, and one finds also the surface-enrichment of gold–copper alloys by a sulphide method.[c] East Asian influence might also be detected, however, in the attention given to the combination and liberation of mercury and sulphur, while sal ammoniac is used to obtain the chlorides of tin from a tin–mercury amalgam. Similarly the description of corrosive sublimate if not calomel suggests knowledge of antecedents in Chinese chemistry.[d] Far more celebrated was the great physician and naturalist Ibn Sīnā—Abū ʿAlīal-Ḥusain Ibn Sīnā (+980 to +1037), who certainly occupied himself at one time or another with chemical operations, even though the totality of the books and tractates on alchemy, mostly in Latin, afterwards attributed to him, is an apocryphal literature.[e] As we know, he resolutely denied the possibility of transmutation or aurifaction in a genuine book, the Kitāb al-Shifāʾ (Book of the Remedy),[f] and the same opinion is advanced in another genuine text, the tractate Ishāra ilā Fasād ʿIlm Aḥkām al-Nujūm (Demonstration of the Futility of Astrology).[g]

By the middle of the +11th century the great days of Arabic alchemy were over, and little remained save poets and commentators.[h] Within less than a century, however, the process of translation out of Arabic into the Latin was to commence. During this period there are only two Arabic names to mention, first al-Ṭughrāʾī and then Ibn Arfaʿ Raʾs; both wrote poetical works. Abū Ismāʿīl al-Ḥusain al-Ṭughrāʾī, who was executed in a religious persecution in +1121, wrote a Kitāb al-Jauhar al-Naḍīr fī Ṣināʿat al-Iksīr (Book of the Brilliant Stone and the Preparation of the Elixir), which has been studied and expounded by Razuq (1). It is much more allegorical than practical, but he may have been the Artephius of the Latins.[i] Later in the century there was Ibn Arfa ʿRaʾs al-Andalusī (d. +1197), chiefly known for his Dīwān Shudhūr al-Dhahab (Poem on the Particles of Gold). This contains some references and information not found anywhere else.[j]

a Cf. Vol. 5, pt. 2, pp. 36ff., 51ff.
b See Stapleton & Azo (1) and Ahmad & Datta (1). Mieli (1), 2nd ed., p. 133 has a word on him.
c Cf. Vol. 5, pt. 2, pp. 39, 251.   d Cf. Vol. 5, pt. 3, pp. 123, 127ff.
e Ruska (26, 27). Nevertheless, see also Stapleton, Azo, Husain & Lewis (1).
f Holmyard & Mandeville (1). Their translation of the passage has been quoted already in Vol. 5, pt. 2, p. 30.   g Cf. Ruska (26); Ullmann (1), p. 252.
h To keep step with events in China, we may recall that this was the time of the Imperial Alchemical Elaboratory at the capital, the activities of that outstanding metallurgical adept Wang Chieh, and the publication of the military encyclopaedia Wu Ching Tsung Yao, with all its information on gunpowder weapons (see Vol. 5, pt. 3, pp. 182, 184, 187).   i See Mieli (1), 2nd ed., p. 156; and p. 495 below.
j Cf. Holmyard (1), p. 100; Mieli (1), 2nd ed., p. 289; Hopkins (1), p. 152.

These men were contemporaries of the translators, busy as bees between $+1120$ and $+1180$ mostly in Spain but also in the Near East.[a] The majority of the alchemical translations have remained anonymous, but of the many names known some must have occupied themselves in such work.[b] One thinks of Hugh of Santalla (*fl.* $+1125$), Adelard of Bath (*fl.* $+1130$) who worked in the East, Dominic Gundisalvi the arch-deacon (*fl.* $+1135$), Hermann the Carinthian or Dalmatian (*fl.* $+1141$) companion of Robert of Chester, finally Gerard of Cremona (*fl.* $+1167$). Perhaps it was Robert who produced the first exactly dated alchemical translation, the *De Compositione Alchemiae* of $+1144$, a rough précis of the Khālid–Morienus dialogue preceded and followed by other materials,[c] but this has been the subject of some controversy.[d] A translation of rather similar date, though cast in a different form, is the *De Anima in Arte Alchimiae*, compiled from various Arabic sources about $+1140$ but nothing whatever to do, in spite of what it says, with Ibn Sīnā himself.[e] Some of al-Rāzī's work was also trans-lated at this time, as in the *Liber Secretorum Bubacaris*, i.e. of Abū Bakr.[f] And in this first wave there seems to have been also a translation, very garbled, of the Seventy Books (i.e. tractates) in the Jābirian Corpus, entitled *Liber de Septuaginta Johannis* and done by another scholar from Cremona, Renaldus.[g] At all events by $+1250$ a whole flood of Latin translations, however crude, from Arabic alchemical texts had become available in the Western world.

Meanwhile Arabic alchemy continued, though the steam had mostly gone out of it. Early in the $+13$th century there was 'Abd al-Raḥīm al-Dimashqī al-Jaubarī with his *Kitāb al-Mukhtār fī Kashf al-Asrār wa-Hatk al-Astār* (Choice Book of the Revelation of Secrets and the Tearing of Veils), chiefly notable for its strong stand against aurifaction.[h] The allegorical trend was continued in the second half of the

---

[a] During this time in China Wu Wu and others were writing much on alchemical apparatus, and the development of the Northern and Southern Schools of Taoism was occurring, accompanied by a strong wave of interest in physiological alchemy. See Vol. 5, pt. 3, pp. 198, 200; and pt. 5 below.

[b] On the Latin translations see Thorndike (1), vol. 2, pp. 217 ff.; Berthelot (10), pp. 229 ff.; Dunlop (7); and on the translators Thorndike (1), vol. 2, pp. 19, 78, 83, 86–7; Steinschneider (1, 4); Burnett (2).

[c] Robert's prologue is the *locus classicus* where 'alchemia' is treated as a concrete noun, a substance, in fact the elixir. The words are: 'Alchymia est substantia corporea ex uno et per unum composita, preciosiora ad invicem per cognationem et effectum conjungens, et eadem naturali commixtione ingeniis melioribus naturaliter convertens'. Cf. p. 366 above.

[d] First Reitzenstein (5) recognised the very Hellenistic nature of Morienus' explanations, and showed a close connection with the Heraclius–Stephanus dialogue of *c.* $+630$ (cf. p. 327 above). Then Holmyard (19) demonstrated marked textual parallelisms with the *Kitāb al-'Ilm al-Muktasab* of al-'Irāqī (cf. p. 404 below) who would have used the Morienus document about $+1270$. Nevertheless Ruska (4), esp. pp. 31 ff., 35 ff. (41, 42) came to believe, for various reasons, that the whole thing, both prologue and text, was a pastiche, plausibly Italian, of the late $+13$th or $+14$th century. His scepticism was not, how-ever, supported by Thorndike (1), vol. 1, pp. 773–4, vol. 2, pp. 214 ff. Steele & Singer (1) also defended the authenticity of the work, and they have now been vindicated by Stavenhagen (1) after a thorough study of the widely scattered manuscripts. This does not of course reinstate the historicity of Morienus or the alchemical interest of Khālid. There is a modern English translation by Stavenhagen (2). Although the main emphasis of the work is on projective aurifaction, there is an interesting, perhaps significant, allusion to the longevity of Morienus (pp. 7, 66). But the original Arabic text itself can hardly be older than about $+820$ because it contains passages (pp. 20, 25, 27) implying knowledge of ammonium chloride (sal ammoniac, cf. p. 437 below). There is also of course 'the marriage of fire and water' (p. 33); and the elixir is frequently mentioned, 'alchemy' being used as a noun synonymous with it (pp. 35, 41, 43, 47). See also Burnett (1).

[e] Ruska (26); Berthelot (10), p. 293.

[f] Cf. Berthelot (10), p. 306.

[g] Kraus (2), p. 42; Berthelot (10), pp. 69, 77, 320 ff.

[h] Cf. Mieli (1), 2nd ed., p. 156.

century by Abū al-Qāsim al-Sīmāwī al-'Irāqī with his *Kitāb al-'Ilm al-Muktasab fī Zirā'at al-Dhahab* (Book of Knowledge acquired concerning the Cultivation of Gold), studied and translated by Holmyard (5).[a] This is not entirely divorced from practice,[b] but it tends to philosophy and the discreet revealing of the cover-names in the trade.[c] Al-'Irāqī drew heavily from Ibn Umail, and was in turn commented upon by 'Alī ibn al-Amīr Aidamur al-Jildakī (*fl.* +1342).[d] If there was more of quotation than of practice in his *Al-Burhān fī 'Ilm Asrār 'Ilm al-Mīzān* (Proofs of the Secret Science of the Balance), his other main work, the *Nihāyat al-Ṭalab* (End of the Search), on which Taslimi (1) and Holmyard (15) have written at length, contains many concrete instructions and shows a genuine love of experiment. Holmyard called him 'the last of the outstanding Muslim alchemists'.[e]

At the end of the day it is possible to take a retrospective survey of Arabic alchemy by means of those remarkable chapters on the subject which Ibn Khaldūn incorporated in his 'Prolegomena' or 'Introduction to History', the *Muqaddima*, written in +1377. 'Abd al-Raḥmān ibn Muḥammad ibn Khaldūn, the first historical sociologist, perhaps, in any civilisation, discussed in his book many sciences human and divine, and one of them was alchemy. Here aurifaction was dominant, but not exclusively so, for Ibn Khaldūn entertained the speculation that if only the *mizāj* (*krasis*) of the elements of man's being were made perfect by an elixir, he would live eternally; just as gold, with its *krasis* more perfect than in any other metal, persists for ever without spontaneous decay.[f] Still, he defines alchemy as

a science which studies the substance through which the generation of gold and silver may be artificially accomplished, and comments on the operations leading to it. The (alchemists) acquire knowledge of the tempers and powers of all created things, and investigate them critically. They hope that they may thus come upon the substance that is prepared to (produce gold and silver). They even study the waste matter of animals, such as bones, feathers, hair, eggs and excrement;[g] to say nothing of minerals.... The (alchemists) assume that all these techniques lead to the production of a natural substance which they call 'elixir'.[h]

The alchemists, he goes on to say,[i] in a reference to the *materia prima* of Hellenistic proto-chemistry, have been inspired by the thought that mineral substances may be changed and transformed artificially one into another because of the plain matter devoid of all qualities which is common to them all. But he also mentions fermentation.

    [a] See also Holmyard (1), p. 100; Nasr (1), p. 278; Mieli (1), 2nd ed., p. 156.
    [b] Cf. Fig. 1533.
    [c] About the same time, in +1283, the collection of Persian and Arabic alchemical texts known as the Rampur Corpus was made; on this see Stapleton & Azo (2).
    [d] Cf. Mieli (1), 2nd ed., p. 289; Wiedemann (32), pp. 21 ff.
    [e] It will be remembered that Arabic alchemy was now facing, as it were, powerful competition from Latin alchemy, for the time of Geber had come, with all its basic discoveries, especially that of the strong mineral acids. Saltpetre and gunpowder, with all that that implied, were also now reaching Western Europe. As for China, as we have seen, the late Sung and Yuan periods were a time of decline in proto-chemical alchemy, with physiological alchemy finding increasing favour.
    [f] See Rosenthal (1), vol. 3, p. 232; we shall quote the passage in full presently (p. 480).
    [g] A reference to the making of ammonium carbonate by dry distillation. Cf. pp. 432 ff.
    [h] Rosenthal tr., vol. 3, pp. 227 ff.   [i] Rosenthal tr., vol. 3, p. 267.

Competent (alchemists) think that the elixir is a substance composed of the four elements. The special (alchemical) processing and treatment give the substance a certain temper and certain natural powers. These powers are such as to assimilate to themselves everything with which they come into contact, and transform it into their own form and temper. They transmit their own qualities and powers to it just as yeast in bread assimilates the dough to its own essence and produces in the bread its own looseness and fluffiness, so that the bread will be easily digestible in the stomach and quickly transformed into nourishment.[a] In the same way the elixir of gold and silver assimilates the minerals with which it comes into contact to (gold and silver) and changes them into the forms of (gold and silver). This is in general the sum total of the theory (of the alchemists).[b]

Ibn Khaldūn was well read in the alchemical literature of his culture, from 'Jābir' through Maslama al-Majrīṭī to al-Ṭughrā'ī, and (as we saw) he vehemently denied that Khālid ibn Yazīd had started it.[c] But while he was not against the real knowledge of chemical substances which the alchemists had acquired, he believed that most of the practitioners had been rank charlatans—he did not know of one successful honest alchemist. It was an infatuation. The deceivers, especially Berber 'students' in the Maghrib, would cover silver with a gold veneer or copper with silver, diluting the noble metals with the base, or 'blanching' copper with mercury.[d] As for the theory of aurifaction itself, he could trace a debate which had been going on for centuries. Abū Naṣr al-Fārābī (d. +950)[e] and the Spanish philosophers who followed him maintained that the metals were not 'distinct species' but variations, as it were, on a single theme (the *prima materia*), therefore aurifaction was possible—and al-Fārābī wrote a special monograph to prove it, *Kitāb fī Wujūb Ṣinā'at al-Kīmiyā'* (Book on the Necessity of Alchemy).[f] Ibn Sīnā (d. +1037)[g] and the Eastern philosophers who followed him, however, maintained the opposite, therefore aurifaction was impossible. Al-Ṭughrā'ī (d. +1121) countered this by drawing attention to the phenomena of biological metamorphoses (some quite real, more based only on legend).[h] Ibn Sīnā's scepticism, Ibn Khaldūn felt, would not resist al-Ṭughrā'ī's argument, but he had thought of something better—the alchemists, it was admitted on all hands, sought to accelerate that production of gold which happens very slowly in any case within the bosom of the earth, but this 'embryology' of gold could obviously be known only to God— moreover, if instant aurifaction were possible, why should Nature take such a long time about it underground? Ibn Khaldūn went on to say that 'those who claim to have made gold with the help of alchemy are like those who might claim success in the artificial creation of man from semen'.[i] Curiously enough, that was exactly what

---

[a] The *massa* theory, on which see p. 366 above. [b] Rosenthal tr., vol. 3, p. 268.

[c] Rosenthal tr., vol. 3, pp. 228–9.

[d] Rosenthal tr., vol. 3, pp. 269, 271. Impatience with charlatanism seems a typical sign of the impending death of an alchemical tradition; cf. our comparison in Vol. 5, pt. 3, pp. 212ff. For a glimpse of the medieval Islamic underworld see the interesting study of Bosworth (1).

[e] Eminent philosopher, born at a place on the Jaxartes R., i.e. in Sogdia or Ferghana (Fig. 1531a).

[f] Dunlop (6), p. 241.

[g] Equally great philosopher and physician, born at Bokhara on the Oxus, i.e. a Khwarizmian. It might not be a coincidence that both these men were from almost the borders of China, differing in their attitude to alchemy but deeply interested in it. More will be said on this point later, p. 424.

[h] Like Petrus Bonus later on, cf. Vol. 5, pt. 2, p. 64.

[i] Rosenthal tr., vol. 3, p. 276.

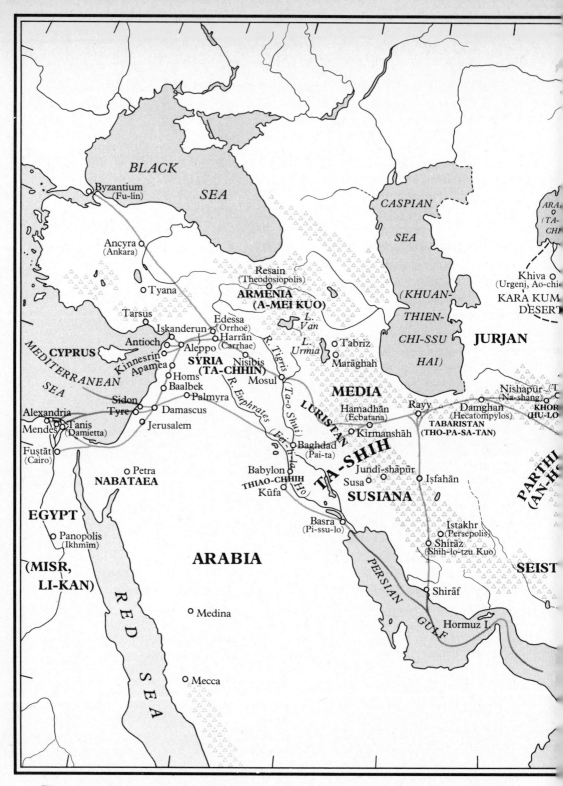

Fig. 1531*a*. Map to show the communication routes, mainly overland, between the Chinese and Arabic culture-areas, as also the relations between Chinese, Iranian and Eastern Mediterranean lands and cities. Focussed on Central Asia, the map was drawn on a Harvard–Yenching Institute physical blank, extended westwards to the same scale, 1:6,250,000, with the assistance of the Cambridge Department of Geography and Cartography. The area covered is from approximately longitude 25° East to longitude 95° East, and from approximately latitude 45° North to 20° North. The reproduction here is reduced by just over three times in longitude and twice in latitude. Mountainous regions are

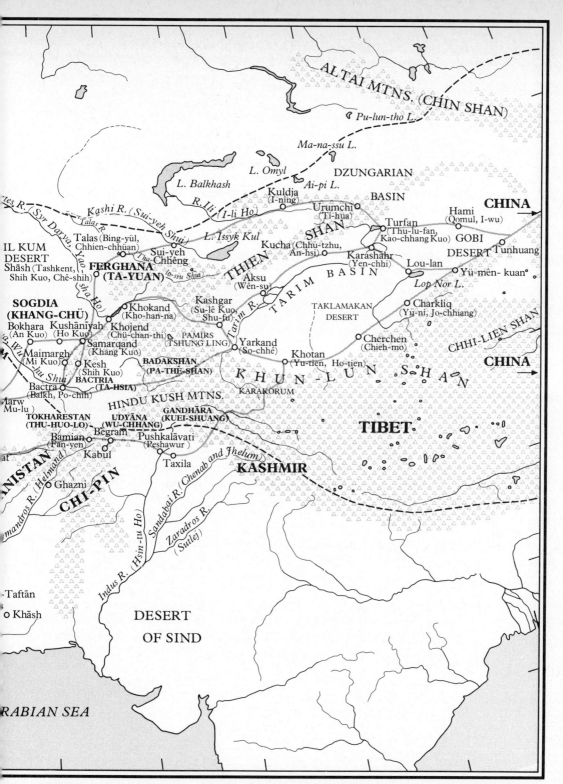

marked by a hatching of small triangles. The range is from Tunhuang in the east, at the threshold of China, to Alexandria and Byzantium (Istanbul) in the west; it goes as far north as the northern ends of the Caspian and Aral Seas, and as far south as the mouth of the Persian Gulf. Most of the Chinese names for places outside China are those current in the Thang period (+ 7th to − 10th centuries). All the area enclosed by the - - -line was in Chinese occupation or subject to Chinese suzerainty in the Thang, according to the historical atlas of Thung Shih-Hêng (1).

the Jābirians had claimed (cf. pp. 326 above and 485 below), but either Ibn Khaldūn had not read those particular books or dismissed them as the wildest nonsense. Finally he brought up a few supplementary arguments. God, he said, had provided gold and silver as the means of exchange in the daily life of the peoples, how then could he allow the making of alchemical gold to upset this pre-established harmony? Moreover, with this fermentation theory, the transformation of dough by yeast is, after all, a corruption or destruction, while the production of gold from other minerals by fermentation would be creative and constructive—therefore the whole analogy is wrong.[a] There is no such thing as aurifaction—but if alchemy cannot be a natural craft, its effects might always be produced by sorcery or miracle-working—and there Ibn Khaldūn was prepared to leave the matter.

The curtain falls with 'Abd al-Wahhāb al-Sha'rānī (d. +1565) and his *Laṭā'if al-Minan wa'l-Akhlāq* (Pleasant Gifts and Traits of Character).[b] Here deceptive aurifaction by impostors and charlatans is most prominent, but the belief still persists, for among a thousand unsuccessful practitioners, there was one who achieved his aim— yet for religious reasons declined to use any of the gold he made, Shaikh Aḥmad ibn Sulaimān al-Zāhid (d. *c.* +1420). May Allāh have mercy on his soul, for he loved not the goods of this present world.

(ii) *The meeting of the streams*

Throughout the foregoing account there have been inescapable indications that the early chemistry of the Arabs was not at all exclusively Graeco-Egyptian in origin, that this inheritance in fact was paralleled by a contribution both theoretical and practical coming westwards from China, India, Central Asia and Persia. Presently we shall briefly but carefully study what the main features of this contribution were; first let us enquire into the historical geography that provided the scene for so great a cultural movement of synthesis.

Long ago, Ruska (32), studying the sources of the material in the Jābirian Corpus, came to the conclusion that the lands along the route of the Old Silk Road[c] had been particularly important. From North and East Persia many influences now radiated, from Tabaristan along the south shore of the Caspian, from Khorasan with cities like

---

[a] This sounds many echoes of thought about upward and downward transformations in Nature; we have seen them in China in Buddhist contexts, cf. Vol. 2, pp. 420 ff. What is particularly interesting to recall here is that in +1678 there was published in London an anonymous pamphlet entitled 'Of a Degradation of Gold made by an Anti-Elixir; a strange Chymical Narative'. As we learn from Ihde (1), who has gone into the question, Robert Boyle's pen was accepted by contemporaries as the source, and indeed his name was printed on the title-page of the second edition of +1739. The anti-elixir, the nature of which was never disclosed, was a dark red powder which would convert pure gold into a brittle silvery mass and baser materials such as a yellowish-brown powder partly vitrified. The de-aurifaction purported to have been proved by the touchstone, cupellation and the hydrostatic balance. Ihde considers various possibilities—a joke, but it was not in character for Robert Boyle to jest—a fraud, but that would have been beneath him—a parable enshrining a set of beliefs, and Boyle was indeed prepared to take auri-faction seriously— or possibly an experiment 'cooked' by one of his laboratory assistants. Ihde decides for the last suggestion, but it is not a very convincing one, and the incident remains passing strange. See Anon. (104).

[b] See Dunlop (6), pp. 248–9; Hitti (1), 2nd ed., p. 742.

[c] Cf. Vol. 1, pp. 181 ff., and here Fig. 1531a.

Tus, Nishapur and Herat in its mountain ranges, from Khwarizm in the Oxus valley, with Khiva as its capital, from old Sogdia and Bactria between the Oxus and the Jaxartes, the land of Balkh, Bokhara and Samarqand, finally from Ferghana beyond the Jaxartes on the very borders of Sinkiang. 'Historical research', wrote Ruska, 'is realising more and more the role of transmission played by Central Asia after Alexander the Great between the human cultures of West, East and South. And it is also seeing more and more how Islam became the great basin into which all these streams of ideology flowed, uniting themselves in Central Asia from East and West and also from India to form remarkable (new) religious and philosophical patterns'.[a] In Arabic alchemy, he went on to say, a Western 'Egyptian-Spanish' set of ideas and techniques, infused with Hellenistic metallurgical mysticism, joined with an 'Eastern-Persian' set, characteristically chemo-therapeutic in nature. 'Both flow side by side and interpenetrate at times, finally delivering all into the alchemy of the Latin Middle Ages.'

The horizons of Kraus were rather more limited to the Hellenistic world, and more Greek than Egyptian at that, but even he could write as follows:[b] 'Jābirian alchemical theory...has few traits in common with what remains to us of ancient (Graeco-Egyptian proto-chemistry). The Graeco-Oriental tradition from which it derives was much more experimental in tendency and more systematical, it was more averse from symbolism and allegory, and it made use of animal substances and products, emphasising (a new volatile agent), "sal ammoniac" (ammonium chloride and carbonate) unknown to the Hellenistic world. Also one finds in the Jābirians a theory of (composition and) transmutation based on numerological principles hardly any trace of which is found in the Greek Corpus. The (Western) traditions on which the Jābirian system depended come more into focus in the science of "properties" (sympathies and antipathies), the Greek, not to say Neo-Pythagorean, inspiration of which can be established with greater certainty. One can thus try to define what parts of the whole complex derived from materials originating further East, in India and even perhaps in China.' Kraus also recalled the claim of 'Jābir' to have visited India;[c] and in another place pin-pointed the Ṣābians of Ḥarrān as a group which seems to have transmitted both Pythagorean, Hermetic and Gnostic ideas from the Mediterranean area as well as indigenous Chaldaean 'Nabataean' notions and certain characteristically Chinese terms, things and concepts.[d] We shall go into all this in a moment, but first we must trace the exact way in which Hellenistic proto-chemistry and the other sciences got into Arab dress.[e]

The key area was Syria and the northern half of Mesopotamia, the latter having long been bisected by a north–south line forming the frontier between the Roman (Byzantine) Empire and the Persian Empire (Sassanid from +224). But Susiana was

---

[a] (32), p. 270.   [b] (3), p. viii.
[c] E.g. in *Kitāb al-A'rāḍ* (Book of Accidents), Kr 182. Cf. Kraus (3), p. 91.
[d] (3), pp. 305–16.
[e] There are many fuller accounts of this process to which reference can be made. One may mention certain admirable summaries like Meyerhof (3) or books such as O'Leary (1); Dunlop (7); Berthelot & Duval (1); see also for authoritative surveys Hitti (1), 2nd ed., pp. 309 ff.; Mieli (1), 2nd ed., pp. 65 ff.

also important, east of the lower waters of the Tigris. By the +5th century the city of Edessa (al-Ruhā', mod. Urfa) east of the upper Euphrates and within its great bend, had become a notable academic centre of Syrian Christians where much translation from Greek into Syriac was carried out.[a] But then came the theological controversy occasioned by the views of Nestorius (*fl.* +431) and the ensuing closure of the academy at Edessa by the emperor Zeno in +489. The Nestorian scholars fled across the border, first to Nisibis[b] (near the upper Tigris, north-west of Mosul), some later to join the new academy founded about +555 by the Persian king Khosrau Anūshirwān at Jundi-shāpūr.[c] This became a very great university where Greek science and medicine were taught in Syriac and probably Persian, having also a celebrated hospital as well as facilities for translators. Throughout the +8th century it flourished much, not declining till the end of the +9th, by which time it had played a great part in transmitting Hellenistic science through Syriac into the world of Islam.[d]

Edessa, Nisibis and Jundi-shāpūr were by no means the only important centres of Syriac learning. There was also Resain (Ra's al-'Ayn, Theodosiopolis) in Mesopotamia between Edessa and Nisibis, home of a great scholar Bp. Sergius (d. +536), certainly also a physician, for he translated Galen,[e] and perhaps also interested in Hellenistic proto-chemistry.[f] Running south from Aleppo (across the Euphrates from Edessa) there was a chain of centres, Kinnesrin (Qinnasrīn),[g] Homs and Baalbek (Heliopolis),[h] all places where Greek science and philosophy were transferred into the Syriac tongue. And Ḥarrān (class. Carrhae), only a short distance south of Edessa, west of Mosul, was also in the circuit, yet standing very much by itself because from the +5th to the +11th centuries it was primarily non-Christian and non-Islamic, perpetuating older indigenous religion, and influences from far in the East. After the

---

[a] For example, Thaufīl al-Rūmī (Theophilus of Edessa, d. +785) was an important astrologer and translator in the time of the Caliph al-Mahdī. O'Leary emphasises the role of Christianity as a Hellenising force in the Near East during the first four or five centuries.

[b] This was the home town of the famous bishop and translator Severus Sebokht (*fl.* +660), already spoken of in Vol. 1, p. 220, Vol. 3, *passim*. Another bishop, the Monophysite Georgius (d. +724) working in Mesopotamia about the same time, also translated works of Aristotle into Syriac.

[c] In Susiana not far south-west of Susa. In the meantime, in +529, the academy of Athens had also been closed, by the emperor Justinian, as a measure directed against the teaching of the non-Christian philosophers there, after which they all migrated to Persia; and though some of them later found a home within the Byzantine domains at Alexandria (+533), others may have stayed, most probably at Jundi-shāpūr (cf. Sandys (1), vol. 1, p. 375; Dunlop (6), p. 172; Nasr (1), p. 189). The city itself had been founded about +270 by Shāpūr I, and it has been conjectured that the metallurgical aurifictors expelled by the Diocletian edict of +296 (cf. p. 332) also took refuge at Jundi-shāpūr if not at Edessa (Hopkins (1), pp. 8, 125). The city was damaged during the Arab occupation of Susiana (Khūzistān) in +639, but not irreparably.

[d] The date of the Hejira, +622, from which all Arabic chronology starts, may here be recalled. During the Umayyad Caliphate (+661 to +750) arms still prevailed over letters, but from the beginning of the 'Abbāsid Caliphate onwards (+750 to +1258) the Arabs were ready to conquer all the worlds of the intellect.

[e] Cf. Brunet & Mieli (1), p. 880. This was about the time of the great anonymous compendium of Galen in Syriac put into English by Wallis Budge (6).

[f] On the extant Syriac proto-chemical and alchemical MSS something is said elsewhere (p. 88 above and pp. 411, 430, 473 below). The only collection is that of Berthelot & Duval (1).

[g] Severus Sebokht's see.

[h] One of its sons was none other than Callinicus, the inventor of 'Greek Fire' at Byzantium in or about +673.

+9th century the mainstream of science was flowing in channels other than Syriac, but as late as the +13th the language could still count men of considerable importance, such as Bar Hebraeus, alias Patriarch Gregorius of the Jacobites, alias Abū al-Faraj al-Malaṭī (d. +1386), a Christian of Jewish stock who worked with the Muslim and Chinese astronomers at the Marāghah observatory,[a] and wrote not only on cosmology but also on history, theology and philosophy.

Meanwhile in +762 Baghdad had been founded as the new Arabic capital, and three years later al-Manṣūr summoned for consultation a Christian Syrian physician from Jundi-shāpūr, Jūrjis (George) al-Bakhtīshū' (d. c. +771), a man destined to found a famous medical family at the seat of the Caliphate. This was a symbol of the 'brain-drain' that the Arabs were creating, for other physicians and naturalists made the same move. By the time when al-Ma'mūn founded the academy there called Bait al-Ḥikma (+830, the House of Wisdom) the second great epoch of translation, from Syriac into Arabic, had already got under way. It seems to have started as early as +684 when a Jewish physician of Basra, Māsarjawayh,[b] translated, almost certainly from Syriac, a book of notes on medicine by the priest Aaron of Alexandria (*Kunnāsh Ahrūn al-Qass*). But the +9th century it was which saw the peak of activity, with the work of Ḥunain ibn Isḥāq al-'Ibādī (d. +877), 'the Shaikh of the Translators', and several members of his family.[c] Another Christian, Qusṭā ibn Lūqā al-Ba'labakkī (of Baalbek, d. c. +912) was outstanding in translation as well as an original writer. But perhaps significantly Ṣābians were also prominent, such as al-Ḥajjāj ibn Yūsuf ibn Maṭar (*fl.* +786 to +833), first successful translator of Euclid[d] and Ptolemy. Corresponding to Ḥunain as head of the Nestorian scholars, the head of the Ṣābian ones was assuredly Thābit ibn Qurra (+836 to +901) who organised the translation of the bulk of the Greek mathematical and astronomical works.[e] The names of the translators of the proto-chemical literature have unfortunately not been preserved. So much for the passage from Greek through Syriac to Arabic.

By contrast with all this, what can be said about the transmission of ideas into Arabic from Sanskrit, and even more interesting to us, from Chinese? Immediately we find ourselves arrested by certain remarkable texts. One of the *ḥadīths*, or 'undoubted' sayings of the Prophet himself, handed down in tradition, runs: 'Go in quest of knowledge even unto China.'[f] Then al-Nadīm, concluding the section on alchemy in his *Fihrist* of +987, wrote as follows:[g]

I, Muḥammad ibn Isḥāq, have lastly only to add that the books on this subject are too numerous and extensive to be recorded in full, and besides the authors keep on repeating the statements of their predecessors. The Egyptians especially have many alchemical writers

[a] Cf. Vol. 1, p. 218.　　　[b] Dunlop (6), pp. 38–9, 213. We shall meet him again, p. 476 below.

[c] He was a Nestorian Christian physician of Jundi-shāpūr, and known to the later Latins as Johannitius.

[d] Cf. Steinschneider (3).

[e] Biographies of some thirty eminent Ṣābian scholars will be found in Chwolson (1), vol. 1, pp. 542 ff. Cf. p. 426 below.

[f] Or, 'Seek for knowledge even though it be as far away as China'. Suhrawardy (1), no. 273. Cf. Wensinck (2); Mahdihassan (15), p. 82.

[g] Flügel ed., vol. 1, p. 360; tr. Berthelot & Houdas (1), p. 40; Fück (1), p. 109; Dodge (1), vol. 2, p. 868, mod. auct.

and scholars, and (some say) that that was the country where the science was born; the temples (*al-barābī*, with their elaboratories) were there,[a] and Mary (the Jewess, or, the Copt) worked there.[b] But others say that the discussions on the Art originated among the first Persians,[c] (while) according to others the Greeks were the first who dealt with it. And others yet again say that alchemy originated either in India or in China. But Allāh knoweth best (what is the truth)!

Thus by this time, in spite of the powerful influence of Hellenistic proto-chemistry and its traditions, it could be thought possible that not only Iran but the cultures of East Asia could have produced the spagyrical art.[d] After this it comes as less of a surprise to find that Hermes himself could be regarded as an inhabitant of China.[e] The Spanish Muslim alchemist of the +12th century, Ibn Arfa' Ra's, has already been mentioned, and in an anonymous Letter derivative from him we can read a curious passage.[f]

> The real name of Hermes was Aḥnuḥ (Enoch)[g] and also Idrīs (Adam's son).... He was a dweller in the upper land of China, as the author of the 'Particles of Gold' pointed out, where he says: 'Mining was looked after by Hermes in China, and Āris [Horus?] found out how to protect the workings from flooding by water.' Now Āris lived in lower China, and belonged to the first of the Indians. Further he says that Aḥnuḥ (on whom be peace) came down from the upper to the lower land of China, into India, and went up a river valley in Sarandīb (Ceylon) till he came to the mountain (of that island) whence Adam (peace be upon him) descended.[h] That was how he found the cavern which he called the Cave of Treasures....

This was filled with gold and gems and engraved tablets expounding the treasures of the sciences, from among which Hermes chose out the most beautiful and best— a typical example of the cave-legends already referred to in connection with the *Tabula Smaragdina*.[i] The anonymous letter, which has to be dated early in the +13th century, then goes on to some rather vague descriptions of aurifactive methods, very syncretistic in character because sal ammoniac is prominent in it, together with *diplōsis*, *leucōsis* and *xanthōsis*,[j] effected by elixirs.[k] Elsewhere, in a pre-Jābirian apocryphon entitled *Muṣḥaf al-Ḥayāt* (Book of Life) there is a discussion between a Byzantine

---

  a For an explanation of this see Fück, *loc. cit.*, p. 90.

  b Cf. pp. 32, 327 above.   c Cf. Vol. 5, pt. 2, pp. 201 ff., pp. 253 ff. above.

  d Other suspicious 'oriental' references in the *Fihrist*'s tenth discourse, that on alchemy, can be noted in Fück (1), p. 89, 91, 95, etc. Al-Nadīm knew of a book entitled: 'Conversations of Uṣtānis al-Rūmī (Ostanes) with Tauhīr, King of India', and another 'Epistle of the Indians to Iskandar (Alexander)'.

  e On the legend of the Three Hermes (like a line of reincarnated Buddhas), first formulated in Arabic literature by Abū Maꞌshar Jaꞌfar (d. *c.* +886) in his *Kitāb al-Ulūf* (Book of the Thousands), see Burnett (1). For Albumasar (as the Latins called him) and his translator Hermann of Carinthia (+1140) the second Hermes, founder of astrology and alchemy, gained his knowledge in the Indies. Perhaps the fact that Abū Maꞌshar was also named al-Balkhī may not be without significance (cf. p. 424).

  f This is the *Qabas al-Qābis fī Tadbīr Hirmis al-Harāmisa*, tr. Siggel (4), pp. 299 ff. Cf. Ullmann (1), p. 168; Plessner (2); Pingree (1), p. 10.

  g Cf. pp. 341 ff. above.

  h A reference to the well-known footprint on Adam's Peak.

  i Cf. pp. 335, 369 above.

  j Cf. Vol. 5, pt. 2, pp. 18, 23; and pp. 358, 365 above.

  k Siggel's account of this work was quoted by von Lippmann (11), and this must have been the source of Chang Tzu-Kung's suggestion (1) that Hermes was Chinese.

emperor, Theodorus, and an adept named Āras (Horus?) who came from 'Lower China'.[a]

Another interesting exchange demands mention here because a medieval Persian writer of a history of China attributed the invention of chemistry to a Chinese named Hua Jen,[1] or 'Changer'; while (at first sight) the Persian's Chinese source regarded him as a man from the Far West. Rashīd al-Dīn al-Hamdānī, in his history of China finished in +1304,[b] speaking of the time of the High King Mu of the Chou,[2] mentions the exploits of the legendary charioteer Tsao Fu,[3] and then goes on to say:

at that time there lived a man called Ḥwār.n (Hua Jen). He invented the science of chemistry and also understood the knowledge of poisons, so well that he could change his appearance in an instant of time...[c]

Here there is no suggestion that Hua Jen was anything but a Chinese.

In order to clarify Rashīd al-Dīn's source one has to know two things expounded by Jahn & Franke: first that he and his assistants were helped by two Chinese Buddhist physicians, Li Ta-Chih and K.msūn;[d] and secondly that he depended on a little-known genre of Chinese historical writing, general surveys done from a Buddhist angle and incorporating the lives of Buddhas, arhats and bodhisattvas within the framework of Confucian secular history. The oldest of these, says Franke (18), was the *Li Tai San Pao Chi*,[4] written by Fei Chhang-Fang[5] in +597; but the closest to Rashīd al-Dīn's history was the work of a monk named Nien-Chhang,[6] entitled *Fo Tsu Li Tai Thung Tsai*[7] (General Record of Buddhist and Secular History through the Ages)[e] and printed in +1341. Since this was twenty-three years after the death of the great Persian scholar it cannot have been his direct source as such, but he could have used an earlier manuscript form of it, and this was probably what happened.

The statements of Nien-Chhang[f] about 'Changer' are as follows:[g]

In King Mu's time a Changer appeared from the Furthest West. He could overturn mountains and reverse the flow of rivers, he could remove towns and cities, pass through fire and water, and pierce metal and stone—there was no end to the myriad changes and transformations (which he could effect, and undergo). The King revered him as a sage, and built for him a Tower of Middle Heaven to dwell in; indeed his appearance was like that of Mañjuśrī or Maudgalyāyana or some such bodhisattva. But what the King did not know was that he had in fact been a disciple of the Buddha.

The story echoes familiarly, for it was nothing but a condensation and Buddhist adaptation of the opening part of the third chapter of the *Lieh Tzu* book, datable

---

[a] Ullmann (1), p. 190.

[b] Part of his famous *Jāmiʿ al-Tawārīkh*, all completed by +1316. Cf. Vol. 1, p. 218.

[c] Tr. Jahn & Franke (1), p. 43, eng. auct.

[d] The characters for their names are not known. They recommended a Buddhist compilation on Chinese history by three monks, Fo-Hsien, Fei and Shan-Huan. We do not know the characters for their names either, nor can this work be identified.

[e] Still extant, and preserved in *Ta Tsang* (Taishō Tripiṭaka), vol. 49, pp. 477 ff.

[f] Born in +1282, his secular name had been Huang.[8]

[g] Comm. in ch. 3 (p. 495.3), tr. auct. The translation given on p. 43 of Jahn & Franke (1) seems not to be from the *Fo Tsu Li Tai Thung Tsai* as stated (except for the last sentence), but from the *Lieh Tzu* text itself.

[1] 化人   [2] 周穆王   [3] 造父   [4] 歷代三寶紀   [5] 費長房
[6] 念常   [7] 佛祖歷代通載   [8] 黃

therefore to any time between the -3rd and the $+4$th centuries.[a] Now, in its Buddhist incarnation, it evoked the magic powers (*siddhi*) of the Tantric saints.[b] Its original intention had probably been to suggest that the visible world was like a dream or a magician's illusion, and Changer was certainly not a historical person, but the chemical artisans of the middle ages did not appreciate such fine distinctions, so it was wholly natural that Changer should have become in due course the technic deity[c] and patron saint of the art, craft and science of chemical change.

As for the 'Furthest West' in *Lieh Tzu* and the *Fo Tsu Li Tai Thung Tsai*, it never meant Europe or the Roman Empire, but rather that legendary land of the immortals, thought of as somewhere near Tibet or Sinkiang, where reigned the Great Queen Mother of the West, Hsi Wang Mu,[1] nothing short of a goddess. King Mu of Chou paid her a celebrated visit, the main theme of the ancient book *Mu Thien Tzu Chuan*,[d] and also referred to in *Lieh Tzu*.[e] When centuries later the story came to the knowledge of real Westerners like the group around Rashīd al-Dīn all this was omitted, and they took Changer (Hua Jen) to have been a Chinese with marvellous chemical knowledge.[f] The significant fact that early in the $+14$th century they were quite ready to do this is the only justification of these paragraphs.[g]

The reality that lay behind these fabulous ideas can be approached from several different angles. First one can take a look at the known intensity of Arab–Chinese intercourse over the centuries in question. Secondly one can notice the fact that so many of the greatest scholars of the Islamic world came from countries on the borders of the Chinese culture-area, and though they made their fortunes in metropolitan Iraq or Egypt they may well have been recipients and transmitters of ideas current in their home-lands. Thirdly one may see whether we can identify any particular place at the western end of the Old Silk Road which could have been a focal entrepôt in the transfer of Chinese notions—something analogous in the realm of the intellect to Begram in that of the products of the arts.[h]

a Ch. 3, pp. 1 aff., tr. Graham (6), pp. 61 ff. Most of the phrases used by Nien-Chhang are verbal quotations from the *Lieh Tzu*, though not in full nor in the same order.

b Cf. Vol. 5, pt. 5.   c See Vol. 1, pp. 51 ff.

d On this see Vol. 3, p. 507.

e Graham tr., p. 64. Her fabled visit to the Emperor Wu of the Former Han is the nub of the *Han Wu Ti Nei Chuan*, tr. Schipper (1).

f Of course the activities of magical men, 'tricksters' or 'transformers', are found in mythologies world-wide; cf. Radin (1); F. Boas (1), pp. 407, 474. They are often beings of selfish motivation, but also often endowed with the beneficent attributes of culture-heroes and deified inventors. Boas remarked (p. 414) that the more sophisticated an Amerindian tribe was, the more sharply would the line be likely to be drawn between the trickster and the culture-hero. I am indebted to Prof. Gene Weltfish for discussions and references on this.

More recently, the trickster-figure has been recognised as an archetype in psycho-analysis; cf. Jung (15). A traditional Thai trickster-tale has been edited and translated by Brun (1).

g A somewhat parallel case concerns the transmission of the stories concerning the ancient sacrifices of girls to the god of the Yellow River (cf. Vol. 2, p. 137). Mahdihassan (57) found this in Islamic tradition, but attributed to the Nile instead, with the Caliph 'Umar replacing the humanitarian Confucian governor.

h In 1937–9 a French expedition to Begram, near Kabul on the southern slopes of the Hindu Kush and in the lands of ancient Bactria, unearthed two walled-up stores of transit trade-goods, appropriated as customs-duty, no doubt, by the Kushān kings in the years before $+250$. There were lovely carved

1 西王母

As for texts and books it may be said at once that in contrast to the flood of Greek scientific books which poured into Arabic we do not so far know of one Chinese work which was translated into that language until a very late date. Of Persian writings there were many and of Sanskrit more than a few,[a] but because Chinese books remained behind the ideographic-alphabetic barrier that is no reason whatever for thinking that Chinese ideas also did.[b] Indeed seminal concepts divested of verbiage might be all the more compelling. Unfortunately very few Persian texts on alchemy or metallurgy have so far been discovered. Almost the only one as yet reported on, and that but briefly, has been a 'Treatise of Jāmās (Jāmāsp) for Ardashīr the King on the Hidden Secret', of which an Arabic translation was found by Stapleton & Azo[c] in the Rampur MS. A colophon says that the text was copied by al-Ṭughrā'ī himself in the early +12th century, the collection was made in +1283, and the MS. is no earlier than the +15th century. The content of the 'Treatise' is very similar to that of the books of the Hellenistic Corpus, with mentions of Pseudo-Democritus and Ostanes;[d] Stapleton & Azo were inclined to consider it a genuine production of the +3rd century, in which case its date would be about +235, but this seems extremely doubtful, and pending further study it may more conservatively be regarded as an Arabic translation of some Hellenistic writing of the +4th to the +7th centuries— especially as it seems to have nothing characteristically Persian about it.[e] A parallel text exists in the Cairo Library.[f] Stapleton (4) regarded the *Risālat al-Ḥadhar* (Book of Warning), supposedly addressed to his disciples by Agathodaemon when dying,[g] as a typical Ḥarrānian text; if so, they had not got much beyond the Hellenistic Corpus.[h] Finally for Syriac alchemical and metallurgical texts not simply translations from the Greek we are better off, as may be seen from those printed and rendered into French (very indifferently, it seems)[i] by Berthelot & Duval (1); these we shall discuss below as occasion arises.[j]

plaques of bone and ivory from India, lacquer-ware, especially boxes, from China, and bronzes and glass vessels from Syria and Alexandria going eastwards in exchange for the Asian products (cf. Fig. 293 in Vol. 4, pt. 1). See the reports of Hackin & Hackin (1); Hackin, Hackin, Carl & Hamelin (1).

[a] Here the classical guide is Steinschneider (2). And we have discussed the book of Shānāq on p. 399 above. See also Ullmann (1), pp. 165, 186–7.

[b] Here we recall the tradition mentioned on p. 390 above about the gift of alchemical texts by the Chinese emperor to the first Umayyad Caliph.

[c] (2), p. 59.

[d] Cf. pp. 333 ff. above. A reference to the 'water of life' (Stapleton (4), p. 28) gives one pause, but it seems that this is nothing more than Hellenistic metaphor concerning reflux distillation (cf. pt. 2, p. 72 above). Parallels in Ibn Umail; Stapleton, Lewis & Taylor (1), pp. 72, 76.

[e] For the entry in Ḥajī Khalīfa's bibliography see Flügel (2), vol. 3, p. 384, no. 6068.

[f] This is entitled 'Book of Asfīdūs on the Wisdom of Aflārūs'. It awaits further study.

[g] There is a précis translation by Stapleton & Husain (1).

[h] Stapleton himself regarded this text as pre-Hellenistic, but here we cannot follow him, nor in his attempted chronology of the ancient Mediterranean proto-chemists either, (4), p. 36.

[i] Cf. Ruska (13).

[j] They consist mainly of two MSS in the British Museum, Egerton 709 and Or. 1593. Each is in two parts, a Syriac half and an Arabic half written in Syriac script. The former is considered to be contemporary with the early parts of the Jābirian Corpus (mid +9th cent.) but based on Syriac antecedents of the +5th and +6th centuries. Giving a 'Doctrine of (Pseudo-) Democritus', it is very Hellenistic in tone, but uses the term elixir, and ends with diagrams of apparatus. The latter is of the +9th to the +11th centuries, much of it contemporary with the later parts of the Jābirian Corpus and

The Prophet died in +632, and within forty years Khorasan fell, with all the rest of Persia, to the Arab empire. Early in the +8th century all the great cities of Western Central Asia were taken—Balkh in Bactria (Tokharistan), Bokhara and Samarqand in Sogdia, Khiva in Khwarizm. By +715 Ferghana was added nominally to Islam, though Tashkent not occupied till +751. With these conquests of Transoxiana (*mā warā' al-nahr*, that which lies beyond the River), and their consequent suppression of Zoroastrianism and Buddhism, Islam stood militarily face to face with the outer defences of the Chinese culture-area.[a] But strangely the confrontation came to no climax and brought no long conflict. For in the same year that Tashkent was taken the Arabs secured a Pyrrhic victory against a Chinese army led by a Korean general, Kao Hsien-Chih,[1] at the Battle of the Talas River;[b] the Chinese were defeated but the Arabs so mauled that they could press no further. Soon afterwards, because of the rebellion of An Lu-Shan,[2] the Chinese withdrew from the whole of Turkestan (Sinkiang) leaving a vacuum as it were between the two civilisations; and very soon afterwards al-Manṣūr was to be seen despatching (in +756) a contingent of Muslim troops to help the young emperor Su Tsung regain control after An Lu-Shan's revolt.[c] Thus it came about that no Arab army ever crossed the Chinese border in hostility. And already a closeness of cultural contact had appeared, for many Chinese artisans taken prisoner at the Talas River settled with their arts and crafts in Baghdad and other Arabic cities, some returning home in +762 but others (like the paper-makers and weavers) staying to exert permanent effects—very likely some workers with chemical knowledge were among them, especially as painters and gilders are mentioned.[d] We even know their names.

During the last half of the +7th century and the first quarter of the +8th, Chinese contacts were mostly with the ousted Iranian ruling families, who were granted asylum at the Chinese imperial court,[e] and with the not yet Islamised peoples of Central Asia. Around +660 there were Turki dancing-girls as far south as Kweilin,[f] and burials in Sinkiang and Shensi have disclosed Arab and Thang coins side by side.[g] That there was an Arabic embassy to China in +651 is very unlikely, and probably rests on the tradition that one of the Companions of the Prophet, Sa'd ibn abī Waqqāṣ, the conqueror of Persia, was sent as an envoy to Canton, where indeed his supposed tomb is still venerated. Chinese historical records mention however a real ambassador

al-Rāzī. Here sal ammoniac is prominent. Late +13th-century interpolations include mentions of gunpowder and mineral acids. The third MS. (Cambr. Mm 6. 29) is very miscellaneous in content, with some Hellenistic material as early as the +2nd to the +6th centuries.

a For the whole story of the Arab conquest of Central Asia see H. A. R. Gibb (4, 5).

b Cf. Vol. 1, pp. 125, 179, 187, 215. Detailed descriptions of the battle, which was technically a great Arab victory, will be found in Chavannes (14), pp. 142–3, 297.

c Besides the 3000 Arab troops there were many contingents from the Central Asian States, such as Tokharestan and Ferghana. See Chavannes (14), pp. 158, 299 and Mahler (1), pp. 76, 100–1. The reason for this will be more understandable from p. 423 below. The rebellion was over by +759. Mahler contests the tradition that Islam in West China dates from this time, on rather insufficient grounds.

d Cf Vol. 1, p. 236.   e Vol. 1, p. 214.   f Schafer (16), p. 76; Chavannes (14), pp. 301–3.

g See Yang Lien-Shêng (8), and Yen Chi (1) retailing Hsia Nai (3). Sassanian coins have been found in Chinese tombs several times also; cf. Hsia Nai (2).

1 高仙芝   2 安祿山

from the Umayyads, one Sulaimān, in +726, probably sent to explore the situation after the Central Asian acquisitions.[a] But from the start of the ʿAbbāsid Caliphate in +750 onwards merchants were much more important than diplomats, and since the overland routes were temporarily rather obstructed they reached Canton (Khanfu) by sea in ships.[b] There, and at Hangchow (Khinzai, Quinsay), and other ports, they set up 'factories' and everywhere an Arab quarter (*fan fang*[1]) with a headman (*qāḍī*, or *fang chang*[2]) responsible to the Chinese magistrate for law and order.[c] Sometimes this arrangement broke down, as in +758, when Arabs (possibly from a settlement in Hainan) burned and looted Canton, or in +878, when thousands of them were killed in a Chinese peasant rebellion, equalitarian turned xenophobic, led by the salt-merchant Huang Chhao.[3] But the trade and intercourse went steadily on. Described by many Arabic geographers, it was the actual experience of some writers and informants, such as Sulaimān al-Tājir (the Merchant) who told of the period centering on +851 in the book entitled *Akhbār al-Ṣīn waʾl-Hind*.[d] One such trader, Ibn Wahb al-Baṣrī, was received, it seems, by the Thang emperor Hsi Tsung in +876. And for the whole Arab–Chinese mercantile field between the +8th and the +13th centuries we have the incomparable *Chu Fan Chih*[4] (Records of Foreign Peoples)[e] written by Chao Ju-Kua[5] in +1225. The extinction of the ʿAbbāsid Caliphate by the Mongols under Hūlāgū Khan thirty-three years later made hardly any difference for Arabic–Chinese intercourse except to intensify the traffic on the overland route as compared with the sea one. And so it continued until the arrival of the Portuguese in the Indian Ocean at the end of the +15th century ushered in a modern world. But this later period (+13th to +15th centuries, Southern Sung, Yuan and Ming)[f] is much less important for our present purpose than the earlier one (+7th to +11th centuries, Thang, Wu Tai and Northern Sung) because that was when Arabic alchemy received its fertilising Chinese influences.[g]

[a] Vol. 1, p. 215, with sources. Other authorities give +716 and +729 (*Tshê Fu Yuan Kuei*, ch. 971, p. 1a, ch. 974, p. 17a, ch. 975, p. 9b; Chavannes (17), pp. 32, 50). The man in +716 declined to kowtow, but was let off; Sulaimān apparently had no such objection. Bretschneider (2), vol. 2, p. 46, put the first embassy in +713, but probably only because that was the first year of the reign-period; on the other hand he adduced Arabic sources for fixing it during the caliphate of Sulaimān, +715 to +717. There was another embassy in +798, from Harūn al-Rashīd, cf. H. A. R. Gibb (1). During the Jurchen-Chin and Sung dynasties in the +10th and +11th centuries more than twenty diplomatic or semi-diplomatic missions were exchanged with the Buwayhid Caliphate.

[b] Vol. 1, pp. 179ff., Vol. 4, pt. 3, *passim*. On this Arab–Chinese trade and the shipping used in it see Ferrand (1); Hourani (1); Huzzayin (1); Hadi Hasan (1); Richards (1).

[c] He also led the prayers in the mosque, gave the Friday sermon, and administered Islamic law and charities. It was a curious anticipation of the extraterritoriality of the later foreign concessions, though doubtless with much less pretension of immunity from Chinese law.

[d] Ed. and tr. Sauvaget (2); earlier Ferrand (2) and others.

[e] Tr. Hirth & Rockhill (1); and for Islamic lands only, Hirth (11).

[f] See Vol. 1, pp. 218ff.

[g] On the whole story of Arab–Chinese contacts there is a classical survey by Schefer (2); to say nothing of the older literature. Schefer gives an interesting account of Abūʾl ʿAbbās Aḥmad Shihāb al-Dīn al-ʿUmarī (d. +1338), who in his *Masālik al-Absār fī Mamālik al-Amsār* (Ways of the Eyes to Survey the Provinces of the Great States) speaks of the great skill of Chinese craftsmen—their saddles (cf. p. 452 below), their clothes made of plant fibre, their sugar and their paper money—and even a story about a dental prosthesis told by one Saʿīd Tāj al-Dīn Ḥasan ibn al-Khallāl al-Samarqandī.

[1] 番坊   [2] 坊長   [3] 黃巢   [4] 諸番志   [5] 趙汝适

Of course the Arabic and Persian merchants, whether themselves from Egypt, Iraq, Iran or Central Asia, were not confined to the foreign quarters in the great coastal ports, for many came in overland along the Old Silk Road; and since in Thang times foreign people and things were all the rage, there was hardly a city in China unfamiliar with the hu[1] merchants (Fig. 1531, b), as they were universally called.[a] Hu girls were also widely in service as maids and entertainers.[b]

Fig. 1531 b.

They excelled in beautiful dances with whirling gyrations to left and right 'as swift as the wind', with long hair, fluttering sleeves and gauzy scarves. A special troupe of them was attached to the imperial court, occupying a College in Thai-chhang Ssu, and their skill was celebrated by poets such as Pai Chü-I and Yuan Chen. Judging by the frequency of the figurines representing them, another numerous group of Central Asians in Thang China were the Turkic grooms and camel-drivers, some of whom must have brought, and doubt-less exchanged, knowledge of veterinary medicine. Armenian-looking wine-sellers often appear, too, a fact which might well have some significance in connection with the history of freezing-out and distillation (cf. pp. 141 ff., 151 ff. above). And for philosophical discourse, besides the Buddhists, there were Zoroastrian Persians, Manichaean Tocharis, Nestorian Christian Uighurs, Muslim Arabs, perhaps even Ṣābian Ḥārranis. At the Chinese capital there were whole streets of the shops and warehouses of foreign merchants, restaurants where they ate their strange food and spoke their strange tongues, caravanserais where they and their convoys put up, lastly the temples and churches of many religions. It could easily have been said of Chhang-an that if you stayed there long enough you could meet representatives of every country in the known world. Not only 'Parthians, Medes and Elamites, and the dwellers in Mesopotamia...'[c] all were there, but rubbing shoulders too with Koreans, Japanese, Vietnamese, Tibetans, Indians, Burmese and Sinhalese; and no one who did not have something to contribute on the

Fig. 1531 b. Statuette of a Persian or Arabic merchant, Thang in date. Buff clay, all-over crazed yellow glaze with flecks of green (photo. Royal Ontario Museum, Toronto). Many similar representations of Central and Western Asian travellers, pedlars, camel-drivers, grooms, envoys and servants, are given in the illustrations of Mahler (1).

[a] Cf. Vol. 1, p. 125 and Fig. 22, opp. p. 128. Strictly speaking, the term hu meant Sogdian, but by extension was commonly used for all Persians (Po-ssu[2]) and Arabs (Ta-shih[3]) as well. The Manichaean astronomer who came to China in +719 (Vol. 1, p. 205, Vol. 3, p. 204) was a hu; and so was the Zoroastrian princess Māsiš who died in Sian at the age of 26 in +874, and whose bilingual tomb inscription in Pehlevi and Chinese has been translated by Itō Gikyo (1).

[b] Ichida Mikinosuke (1).

[c] Acts 2, 8.

[1] 胡   [2] 波斯   [3] 大食

nature of the world, and the wonders thereof. One may well ask oneself how they looked to the eyes of their hosts.[a]

Fortunately there is a colourful source from which we can gain a good idea of what the Arabs and Persians at least meant to the ordinary Chinese, and by good luck it brings us back precisely to our main theme of alchemy and early chemistry. In +977, as part of a general programme of producing new encyclopaedias and collections,[b] the second Sung emperor, Thai Tsung, commissioned a treasury of 'rustic histories', 'biographical traditions' and 'short tales'. This was the *Thai-Phing Kuang Chi*[1] (Miscellaneous Records collected in the Thai-Phing reign-period), edited by Li Fang[2] and twelve colleagues, and finished in the following year. Exactly how much of this material might be considered based on solid historical facts it is not now possible to say, certainly a great deal of it was fictional, but in the present context that does not matter, for the texts give a clear indication of how the *hu* merchants seemed to the general run of literate Chinese in the Thang and Wu Tai periods.[c] Hence the extraordinary interest of the fact that they were often mixed up with alchemy and Taoism, skilled in recognising the gold of aurifiction or aurifaction, or engaged in studying the art itself, and not only that but concerned with life-elixirs and physiological alchemy as well. This is an element not to be overlooked in the case for transmission of Chinese alchemical ideas to Arabic culture. So, to summarise, as Schafer put it, the *hu* merchant in China was wealthy and generous, a befriender of young and indigent scholars, extremely learned in the knowledge of gems, minerals and precious metals, a dealer in wonders and not devoid of magical and mysterious powers. Let us look at a few of the stories to gain the advantage of some concrete detail.

In one report, dated between +806 and +816, a young man, Wang Ssu-Lang,[3] masters the technique of making artificial gold (*hua chin*[4]), and saves his uncle from financial difficulties by giving him an ingot of it. Of this gold it is said that 'the Arabic and Persian merchants from the Western countries particularly wanted to buy it. It had no fixed price, and he (Wang) used to ask what he liked for it'.[d] Another account, referring specifically to +746, tells of a man named Tuan Lüeh[5] who met a merchant in a shop in Wei-chün who had more than ten catties of drugs most valuable for the preservation of longevity (*yang shêng*[6]) and for helping one to avoid cereal food (*pi ku*[7]). Some of these were very difficult to get, however, and each day he used to go to the market to enquire of the Arabic and Persian merchants if they had any for sale.[a] Here then the *hu* apothecaries are directly involved in a trade related to

[a] Cf. Hsiang Ta (3). A mine of information is also contained in the book of Schafer (13) on Thang 'exotica'. Many contemporary statuettes of *hu* merchants exist, on which see the monograph of Mahler (1).

[b] One result of this was the indispensable *Thai-Phing Yü Lan* (+983), so often quoted in these volumes.

[c] Much credit is due to Schafer (2) for having seen this and discussed it in a pioneer paper. The *Thai-Phing Kuang Chi* is of peculiar value because more than half the books which it excerpted have long been lost.

[d] *TPKC*, ch. 35, pp. 3*a*–4*a* (vol. 1, pp. 189–90). Attention was drawn to this by Fêng Chia-Shêng (5), p. 135. The editors drew the story from the *Chi I Chi*[8] of Hsüeh Yung-Jo.[9]

[1] 太平廣記　　　[2] 李昉　　　[3] 王四郎　　　[4] 化金　　　[5] 段碣
[6] 養生　　　　　[7] 辟穀　　　[8] 集異記　　　[9] 薛用弱

the characteristically Chinese physiological alchemy. Relevant to this, and to the idea of bodily incorruptibility,[b] is a third tale, that of Li Kuan,[1] who having received a beautiful pearl from a dying Persian merchant in recompense for his kindness, decided to place it in the mouth of the dead man. Many years later when the tomb was opened, the body was found quite undecayed, with the pearl still in position.[c] Another alchemical story is that of Mr Lu and Mr Li,[2] both Taoist adepts doing gymnastics and breathing exercises on Thai-pai Shan. One having acquired great wealth by means of aurifaction bestowed an alchemical staff upon the other, which he said could be sold for a great sum at the 'Persian shop' at Yangchow, and so indeed it turned out.[d] Evidently the hu merchants knew something valuable when they saw it. Our last example is perhaps the longest and most impressive, the story of Tu Tzu-Chhun[3].[e]

Tu, we are told, was an idle young scholar who met a strange old man in the Persian Bazaar in the Western Market at Chhang-an, and having caught his fancy was transferred from hunger and cold to a rich life of the utmost comfort. But before long it appeared that the stranger required his services for the accomplishment of an alchemical procedure designed to make an immortality elixir. There is a graphic description of Tu Tzu-Chhun arriving at a remote palace near Hua Shan forty li or so out of the capital, and there in the great hall was the old man attired in Taoist vestments, with an alchemical furnace nine feet high pouring out purple vapours, through which could dimly be seen nine Jade Maidens bearing the insignia of the Caerulean Dragon and the White Tiger.[f] But then the story takes a curious twist, for after ingesting certain drugs and sitting down to gaze in meditation at a blank wall, Tu found himself undergoing the torments of a variety of Buddhist hells, and was eventually reincarnated in another body before breaking the spell by a burst of uncontrollable emotion. Thus having failed to master these terrifying apparitions, Tu awoke, and the experiment, which would have gained hsien immortality for both the old Persian and himself, also ended in failure.

Putting all this together, it seems evident that in the eyes of ordinary people at least during the Thang period, the merchants from Persia and the Arab countries were very

---

a This was also noted from *TPKC* by Fêng Chia-Shêng, *loc. cit.* but his reference (ch. 6) seems to be wrong, and we have not been able to find the right one. Central Asian drugs were of course coming as tribute, and being imported commercially all through this period, cf. e.g. Mahler (1), pp. 75, 79, 89. In +729 the monk Nan-Tho[4] was sent from Balkh in Tokharestan with tribute of valuable drugs (*TFYK*, ch. 971, p. 8a), and in +746 the king or emir of Tabaristan offered 'thousand-year longevity jujube-dates' (*TFYK*, ch. 971, p. 15b). In all, between +713 and +755 some two hundred different drugs, together with perfumes, precious red stones and glass, came from the Turkic countries (*CTS*, ch. 221B, p. 5a, tr. Chavannes (14), pp. 157–8; cf. (17), pp. 50, 57, 76 from *TFYK*). On this period see also Chhen Pang-Hsien (1), pp. 154ff., and our Vol. 1, pp. 204–6.

b This has been discussed already in Vol. 5, pt. 2, pp. 294ff.

c *TPKC*, ch. 402, p. 6b (vol. 3, p. 1543). The story was taken from the *Tu I Chih*[5] of Li Jung,[6] a book we have had occasion to quote from before.

d *TPKC*, ch. 17, pp. 3b–4b (vol. 1, pp. 136–7). The story was taken from the *I Shih*[7] of a writer named Lu[8] whose given name has been lost.

e *TPKC*, ch. 16, pp. 1b–4a (vol. 1, pp. 132–3). The story was taken from the *Hsüan Kuai Hsü Lu*[9] written by Li Fu-Yen.[10]   f Cf. Vol. 5, pt. 5.

| [1] 李灌 | [2] 盧李二生 | [3] 杜子春 | [4] 難陁 | [5] 獨異志 |
| [6] 李冗 | [7] 逸史 | [8] 盧氏 | [9] 玄怪續錄 | [10] 李復言 |

interested indeed both in the metallurgical and the macrobiotic aspects of Chinese alchemy.[a] Looked at from this angle it seems almost to stand to reason that Chinese ideas would have found their way westwards to join with the Hellenistic ones that had been taken up into Arabic thought. This conclusion is not affected at all by the seeming fact that no one attempted the translation of an entire book of the *Tao Tsang*,[b] for instance, into Arabic—and how difficult that would have been—because all we are looking for is a new substance here and there, a few theories which might or might not have been misunderstood, and one basic grand conception not misunderstood, namely that chemical operations could perform miracles of life-giving and life-prolonging.

If one would like to make the personal acquaintance of a group of *hu* merchant-naturalists in the China of this period, one could hardly do better than consider the Li family of Szechuan.[c] Li Hsün[1] (*fl.* +900 to +930) was of a Persian family which had settled in China during the Sui (*c.* +600) and moved to Szechuan about +880; they were wholesalers, shipowners and caravan patrons in the spice trade. Li Hsün, apart from renown as a poet, became expert in materia medica, perfumes and natural history, the author of a *Hai Yao Pên Tshao*[2] on the plant and animal drugs of the southern countries beyond the seas. This was in the same genre as the +8th-century *Hu Pên Tshao*[3] of Chêng Chhien[4],[d] but unfortunately neither work has survived entire.[e] Li Hsün's younger brother, Li Hsien,[5] was more of an alchemist, occupying himself with arsenical and other mineral drugs as well as essential oils and their distillation; he was also noted as a chess-player. This was the time of the Chhien Shu kingdom in Szechuan, and the younger sister, Li Shun-Hsien,[6] herself a poet of great elegance, became one of the ladies of that court.[f] Probably unrelated to this family, though of the same name, was the *hu* physician Li Mi-I,[7] who had sailed east to Japan in +735, and participated in the cultural renaissance of the Nara period.[g]

Let us turn now to the lands just west of the westernmost borders of Chinese Turkestan. There is no need to labour the point that they did indeed produce many of the most famous scholars of the Islamic centuries, but it is worth giving a few examples.[h] Taking a bird's-eye view of Western Central Asia let us proceed westwards

---

[a] Here we should not forget the possible role of Jewish merchants. From Vol. 3, p. 681, we know that in the +9th century a group of these called al-Rādhānīyah (perhaps from their centre at the town of al-Rayy, near mod. Teheran) travelled regularly both by land and sea between China and Provence. They spoke all the languages current along the routes, and since both Damascus and Oman were among their entrepôts, they could well have been a link between the Chinese alchemists and the Jābirians.

[b] One should perhaps remember here that in the Thang the Chinese alchemical literature was not nearly so rich as it became afterwards during the late Sung, for example.

[c] There is a special paper by Read (10) on the transmissions in natural history arising from Arabic-Chinese contacts.　　　[d] *Fl.* +742. *SIC*, p. 1366.

[e] It would be well worth while collecting all the quotations to assemble as much as is left.

[f] To illustrate this paragraph, we give a picture of a silver ewer of strikingly Arabic or Persian design and Sung date, now preserved in the Provincial Historical Museum at Chhêngtu, the home city of the Li family (Fig. 1532).　　　[g] Cf. Vol. 1, pp. 187–8.

[h] Arabic personal names, though excessively complicated, have the advantage that a man's birthplace is immediately evident in them.

[1] 李珣　　　[2] 海藥本草　　　[3] 胡本草　　　[4] 鄭虔　　　[5] 李玹
[6] 李舜絃　　　[7] 李密醫

Fig. 1532. Evidence of the close relations between China and the Islamic countries in the Sung; a silver ewer of West Asian character (orig. photo. 1972, at the Exhibition of History and Archaeology, Chhêngtu).

from the north–south mountain-range which joins the Thien Shan with the Pamirs and the Hindu Kush, that range through which run the most important of the passes on the Old Silk Road.[a] Thus we should speak first of Ferghana, with Tashkent (al-Shāsh) as its capital;[b] then of old Sogdia and Bactria between the two north-westward-flowing rivers (the Jaxartes and the Oxus) with great cities at Bokhara, Samarqand and Balkh respectively;[c] then again of Khwarizm along the Oxus, centred on Khiva;[d] and finally of Khorasan, with the towns of Merv, Tus, Nishapur and modern Meshed strung along the NW–SE rampart of the Kuh-i-Alādāgh mountains. The city of Kashgar, on the eastern side of the Thien Shan–Hindu Kush connecting range, was the most easterly place ever under Islamic sovereignty, as at times it was, so perhaps the Taklamakan Desert and the inhospitable Tarim Basin were even greater barriers than the range itself. But broadly speaking the Arabic flood washed up to that range and stopped, leaving there a permanent cultural tide-mark. It is not part of our present purpose to demonstrate traces of Chinese influence in the writings of the scholars who came from these parts; all that needs pointing out is how near they were

---

[a] Cf. the map (Fig. 32) in Vol. 1; and here, Fig. 1531a on pp. 406–7 above.

[b] It may be of interest from the chemical point of view to note that when the Arabs obtained Ferghana they got at the same time important mines of mercury, as well as one of the few centres of asbestos production in medieval times (cf. Vol. 3, pp. 655ff.).

[c] Balkh is just south of the latter river.   [d] Cf. Tolstov (1, 2, 4).

to it—and if it be urged that the shrouds of linguistic understanding were too great, all that need be said is that young and intelligent merchants, both Chinese and Arabic-Persian, did sometimes themselves personally travel in the caravans, and must have talked. We cannot impose on them a silence *ex hypothesi*.

Moreover, what is not generally understood—histories of the Arab world tend to overlook it—is that for a century or so before the Arab conquests the Turkic States of Western Central Asia had mostly been Chinese protectorates more or less self-governing.[a] After the first flooding of the whole region by the Turkic tribes (Thu-chüeh[1]) a series of campaigns by Chinese armies had led to the acceptance of the sovereignty of the Son of Heaven from +659 onwards; this applied to Ferghana, Sogdia and Bactria, Khwarizm in part but not Khorasan, though including a number of States in the mountains to the south in what is now Afghanistan, reaching down through old Gandhāra to the plains of India. Thus in their turn, for a time, the Chinese had been able to impose their rule on many lands west of the mountain ranges that plug like a cork the western end of Sinkiang. Ferghana they knew as Pa-han-na;[2] and Tashkent (al-Shāsh), Shih,[3] appropriated sometimes the ancient term for the whole country, Ta-yuan.[4] Sogdia retained its classical name, Khang-chü,[5] Samarqand being Khang[6] and Bokhara An.[7] Bactria or Tokharestan was transliterated as Tu-huo-lo,[8] with Balkh itself as Po-chih.[9] Khwarizm, though tributary only, not protected, was Huo-li-hsi-mi[10] or Huo-hsün,[11] but Khiva seems not to appear in Chinese, unless it was Chi-to-chü-chê[12] which sounds more like Korkandj. Thus everyone who was born in any of these parts after the middle of the +7th century must have had a mental background in which many typical Chinese ideas circulated—and that without any direct access, or very little, to the Chinese literature itself.

From Ferghana, then, one can name one of the most celebrated Muslim astronomers of the +9th century,[b] and somewhat surprisingly, an authority on the poetry in the Arabic language.[c] A little further West, the neighbourhood of Bokhara produced perhaps the greatest of all Islamic scientists, the 'Third Master (al-muʿallim al-thālith)',[d] Ibn Sīnā himself.[e] Samarqand was significant rather later, in the +13th

---

[a] The best account of all this is to be found in Chavannes (14), pp. 268 ff., 299 ff., together with his additional notes in (17). Cf. Gibb (4, 5). A summary set in the context of a general survey of Chinese history is given by Gernet (3), pp. 209, 221 ff., 250 ff.

[b] Abū al-ʿAbbās ibn Kaṭīr al-Farghānī, the Alfraganus of the later Latins, *fl.* +861. See Mieli (1), 2nd ed., pp. 82–3; Hitti (1), 2nd ed., p. 376.

[c] Abū Muḥriz Khalaf al-Aḥmar (d. *c.* +800). See Dunlop (6), p. 28.

[d] The first was Aristotle. But the second was also a scholar from Turkestan, namely Abū Naṣr Muḥammad ibn Muḥammad ibn Ṭarkhān ibn Uzlagh al-Fārābī (d. +950), commentator of Aristotle, Porphyry, Plotinus, etc. and interested in such matters as the classification of the sciences. See Mieli (1), 2nd ed., p. 94.

[e] Abū ʿAlī al-Ḥusain Ibn Sīnā, the Avicenna of the Latins (+980 to +1037). See Mieli (1), 2nd ed., p. 102. His *Qānūn fiʾ l-Ṭibb* is still one of the greatest of all medical works in any civilisation. And, as we shall see in Sect. 44, his extreme emphasis on sphygmology, the role of the pulse in diagnosis, demonstrates even in detail a close connection with the medical practice and writings of earlier Chinese physicians.

---

[1] 突厥   [2] 拔汗那   [3] 石   [4] 大宛   [5] 康居   [6] 康

[7] 安   [8] 覩貨羅   [9] 蒲知   [10] 貨利習彌   [11] 火尋   [12] 急多颮遮

and + 14th centuries, with an important astronomer[a] and a famous physician.[b] Balkh, however, kept on producing men of great authority, in the + 8th century the first of all the Islamic mystics (sufis),[c] in the + 9th the first of a series of great geographers[d] and a widely known astrologer-astronomer,[e] in the + 10th an Ismā'īli polymath whose writings are still read with interest today.[f] And on the political side, where much influence must have been exerted, there was the famous Barmecide family of viziers, who practically ruled the Caliphate till + 803. The founder of this line, Khālid ibn Barmak (*fl.* + 765), came from Balkh, where his father was said to have been a Buddhist lay devotee (*upasaka*).[g]

When we reach Khwarizm we find no less than four prominent al-Khwārizmī's, the proper differentiation of whom has doubtless been a pitfall to scholars inexpert in Arabic fields before now. One was perhaps the most celebrated mathematician in all Islamic history, he who gave us the very name of algebra;[h] and a second, of more immediate interest, was an alchemist whose work has already been mentioned (pp. 83, 402).[i] The other two were both encyclopaedists, one in the + 10th,[j] the other in the + 12th century,[k] and both chose for their books very similar titles meaning 'Key to the Sciences'. Though the second was almost wholly philological the former has much present relevance, for it contained a large section on alchemy which has been carefully studied and translated in modern times. And the neighbourhood of Khwarizm was also the birthplace of one who did not take his name from that locality—the great

---

[a] This was 'Aṭā ibn Aḥmad al-Samarqandī (*fl.* + 1362), whose contact with Chinese astronomy was particularly close. See Vol. 1, p. 218.

[b] Muhammad ibn 'Alī Najīb al-Dīn al-Samarqandī (d. + 1223). See Mieli (1), 2nd ed., p. 163.

[c] I.e. Ibrāhīm ibn Adham (d. c. + 777). See Hitti (1), 2nd ed., p. 434.

[d] Namely Abū Zayd Aḥmad ibn Sahl al-Balkhī (+ 849 to + 934), whose *Kitāb al-Ashkāl* (Book of the Seven Climes), written c. + 920, was the foundation of the famous *Kitāb al-Masālik wa'l-Mamālik* (Book of Roads and Provinces) after its enlargements and re-writings by al-Iṣṭakhrī about + 950 and Ibn Ḥauqal. See Dunlop (6), pp. 164–5; Mieli (1), 2nd ed., p. 117.

[e] Abū Ma'shar Ja'far al-Balkhī, the Albumasar of the Latins (d. + 886). His best known book was the *Kitāb al-Mudhākarāt*. Details in Mieli (1), 2nd ed., p. 89; Dunlop (6), pp. 174, 176, (9).

[f] Nāṣir-i-Khusraw al-Qubādiyānī (d. + 1060). Mieli (1), 2nd ed., p. 116.

[g] Ruska (32); Hitti (1), 2nd ed., p. 294. But I remember that the late Professor F. C. Burkitt used to say that judging from the 'Barmecide feast' and other evidence the Barmak family was much more likely to have been Manichaean than Buddhist. Another tradition makes them hereditary managers of a Zoroastrian fire-temple.

[h] Abū 'Abd Allāh Muhammad ibn Mūsā al-Khwārizmī, of course, c. + 750 to c. + 850. He was the Librarian of the Khizāna of the Bait al-Ḥikma (House of Wisdom) at Baghdad. For his mathematical geography, *Kitāb Ṣūrat al-Arḍ*, he drew upon sources much wider than Ptolemy; while his *Kitāb al-Sindhind* was based on the Indian astronomical work *Brāhmasphuṭa Siddhānta* which had been translated in + 771 by al-Fazārī. The book that laid the foundations of post-Diophantine algebra was the *Kitāb al-Mukhtaṣar fī Ḥisab al-Jabr wa'l-Muqābala*; see Dunlop (6), pp. 150, 152, 154, 217; Mieli (1), 2nd ed., p. 82.

[i] Ibn 'Abd al-Malik al-Ṣāliḥī al-Kāṭī al-Khwārizmī, c. + 1034. Mieli (1), 2nd ed., p. 133; Stapleton & Azo (1); Ahmad & Datta (1).

[j] Abū 'Abd Allāh Muhammad ibn Aḥmad ibn Yūsuf al-Khwārizmī, the secretary (al-Kātib), whose *Mafātīḥ al-'Ulūm* was finished in + 976. The important part on alchemy has been translated by Wiedemann (15), and many other parts examined and translated in Wiedemann (23). Another study of the alchemy is that of Stapleton, Azo & Husain (1). See also Dunlop (6), pp. 246–7; Mieli (1), 2nd ed., pp. 94, 136.

[k] Abū Ya'qūb ibn abī Bakr al-Khwārizmī, the die-engraver (al-Sakkākī), + 1160 to + 1229, whose book was entitled *Miftāḥ al-'Ulūm*. Dunlop (6), pp. 246–7.

Abū al-Raiḥan Muḥammad ibn Aḥmad al-Bīrūnī (+973 to +1048).[a] Often mentioned in our previous volumes,[b] he was one of the greatest of Arabic scientists, making important contributions in mathematics, astronomy and geography, but also in the biological and medical sciences. Besides this he was a famous traveller who stayed long in India, and expounded the arts, sciences and customs of that culture in great detail to the rest of the world; an association with Further Asia which makes him particularly relevant in the present context.

It remains to speak of Khorasan, further west in Persia proper, and as far as it would be reasonable to go geographically here. It was a region of high culture, producing eminent Koranic commentators,[c] historians,[d] literary men,[e] and the greatest theologian of Islam,[f] to say nothing of the statesman who under the first Ilkhān was to be the founder of the Marāghah observatory.[g] But what gives particular pause in the line of thought we are following is the persistent tradition that Jābir ibn Ḥayyān was a man of Tus, al-Ṭūsī, either born there or brought up there.[h] The biographical memorials concerning this great but shadowy alchemical figure which Holmyard (17) studied suggest that this was so, but even if the historicity of Jābir himself is quite dismissed, the tradition may conceivably enshrine a certain truth about the origin of Jābirian alchemy. In this connection one should recall the conclusions reached above (p. 370) concerning the probable area of origin of the Balīnās or 'Apollonius' literature, especially the *Kitāb Sirr al-Khalīqa* of *c.* +750, or more likely *c.* +820, which may be looked upon as the start of the whole Arabic alchemical literature.[i] Balkh in Bactria, or indeed any of the Transoxianic centres beyond, can be found in the forefront of the picture.

Finally, apart from the merchants already envisaged, great scholars travelled, we should remember, to Khorasan and points east. One need instance only the geographer al-Mas'ūdī (+895 to *c.* +957), who wrote about those parts in his *Murūj al-Dhahab*...[j] But there were also official expeditions like that of Sallām al-Tarjamān

[a] Cf. Mieli (1), pp. 98ff.; Hitti (1), pp. 376–7.

[b] For example, Vol. 1, pp. 216–17; Vol. 3, pp. 252, 612; Vol. 4, pt. 2, p. 534; Vol. 4, pt. 3, pp. 502, 579.

[c] E.g. Abū Isḥāq Aḥmad ibn Muḥammad al-Tha'ālibī from Nishapur (*c.* +950 to *c.* +1020), cf. Dunlop (6), p. 55.

[d] Like Ibn abī Ṭāhir Ṭayfūr (al-Khurāsānī), *fl.* +819 to +833, who wrote a good 'History of Baghdad'. See Dunlop (6), pp. 80–1. On China in Islamic historiography, see Jahn (1).

[e] Like Abū Manṣūr 'Abd al-Malik al-Tha'ālibī, also from Nishapur (+961 to +1038) whose *Laṭā'if al-Ma'ārif* (Pleasant Sorts of Knowledge) gave him an extraordinarily high reputation. Among other things it contains a list of 'firsts', technical inventors, etc.; cf. Vol. 1, pp. 51ff. for Chinese parallels. On this al-Tha'ālibī see Dunlop (6), pp. 54ff. Unlike the historian just mentioned, he was highly regarded for the perfection of his Arabic style.

[f] Abū Ḥamīd al-Ghazālī al-Ṭūsī (Algazel to the Latins, the Aquinas of Islam), +1059 to +1111, sufi and philosopher of religion. See Mieli (1), 2nd ed., p. 94; Hitti (1), 2nd ed., p. 431.

[g] Naṣīr al-Dīn al-Ṭūsī (+1201 to +1274). See Vol. 1, pp. 217–18. We know the name, though not the characters, of one Chinese astronomer at Marāghah, Fu Mêng-Chi, and one of his colleagues was an al-Andalusī who wrote on the calendrical science of the Chinese and the Uighurs. Scientific contacts between China and Persia during the Mongol period have been summarised by Jahn (2).

[h] He was also commonly termed al-ṣūfī, which again might be significant.

[i] See Ruska (8); Multhauf (5), p. 128.

[j] *Murūj al-Dhahab wa-Ma'ādin al-Jawāhir* (Meadows of Gold and Fields of Gems); cf. Mieli (1), 2nd ed., p. 114; Dunlop (6), pp. 99ff. We often quote the translation of de Meynard & de Courteille (1).

(the interpreter) and his party to the 'Dyke' (*sadd*) of Dhū'l-Qarnain ('Him of the Horns', Alexander the Great) i.e. the Great Wall, also known to the Arabs as the Wall of Gog and Magog.[a] This took place in the reign of the Caliph al-Wāthiq (+842 to +847), but whether they reached the Wall itself or only passes with suggestive names in the Thien Shan remains uncertain.[b] Nevertheless, ideas could well have travelled with such men.

The last subject needing discussion here is the possible identification of a place or places on or near the Old Silk Road which could have acted as a focal point for the spread of East Asian ideas among the Syrians and Arabs. There is one city which has attracted suspicion of this kind, namely Ḥarrān, the classical Carrhae, a short distance south-east of Edessa (Orrhoë) and within the great bend of the Euphrates, in the province known about +400 as Osrhoene Euphratensis. This city is near the top of the arch of the Fertile Crescent rather than at its eastern end, where one might expect such a transmission-point to have been, but here perhaps religion was more important than geography, and Ḥarrān was unique in that it was neither Christian nor Muslim.[c] The Ṣābians of Ḥarrān were doggedly 'pagan', not because they perpetuated the conventional worship of the Graeco-Roman pantheon, nor yet because they were given to one or other of the Hellenistic mystery-religions, but because they had a cult of their own, based apparently on ancient Babylonian or 'Chaldean' practices[d] more than anything else. There seems to be nothing wrong with the tradition that the Ṣābians were adherents of a special religion who adopted the name comparatively late in order to enjoy the privileges of being considered 'People of the Book' (*ahl al-kitāb*) like Jews and Christians.[e] Ṣābians had indeed been mentioned in the Holy *Qur'ān*,[f] though who these were remains still problematical; in any case it is suggested that the book which the men of Ḥarrān adopted as their sacred scripture consisted of the writings attributed to Hermes-Agathodaemon, which in those days people on all hands were prepared to venerate no less than the Torah or the Gospels.[g] Now Ḥarrān in the first three Arab centuries was not only a great trade centre[h] but also famous for its astrolabists, alembic makers, and producers of astrological talismans;[i]

a Holy *Qur'ān*, Sur. XVIII, 82 ff.

b See the discussion in Vol. 4, pt. 3, pp. 56–7, and Dunlop (6), pp. 167–8.

c Neither Zoroastrian ('Magian'), Mazdaean, nor Mandaean either, one might add.

d For example, the historian Ḥamza al-Iṣfahānī (+897 to +967) says in his *Ta'rīkh Sinī Mulūk al-Arḍ wa'l-Anbiyā'* (Chronology of the Kings and Prophets of the Earth) that originally all easterners were *samīniyyūn* (from *sramana*), i.e. Buddhists, while all westerners were *kaldāniyyūn*, i.e. Chaldaeans in religion, as the Ṣābians still are. Dunlop (6), p. 114.

e As late as *c.* +830, according to Chwolson (1), vol. 1, p. 470. This notable work, now some two decades more than a century old, remains to the present day the indispensable source of information about the Ṣābians. See also the *Fihrist*, tr. Dodge (1), vol. 2, pp. 745 ff.

f Sur. V, 72–3.

g As Massignon remarked, they hoped that the Hermetic connection would get them accepted as monotheists. Cf. Stapleton (4), pp. 22, 25 and Stapleton & Husain (1), discussed critically on p. 415 above. See also Stapleton, Azo & Husain (1), p. 398. The role of the Ṣābians in transmitting texts attributed to Hermes (in Persian, Hōsheng; cf. Seybold, 1) has naturally made them of particular interest to those who study that literature; see e.g. Sarton (13); Massignon (4); Plessner (2); Stapleton, Lewis & Taylor (1); Yates (1), pp. 49 ff.

h It was a cross-roads of the E–W Susa–Baghdad–Aleppo–Antioch route and the N–S route from the Hittite Pontus region down to Babylon and Basra. Cf. Meyerhof (4).

i Chwolson (1), vol. 1, p. 344.

that alchemy or proto-chemistry was much cultivated there is quite clear if only from internal evidence in the writings of al-Rāzī.[a]

The Ṣābian religion, though very syncretistic, was really an old Syrian–Mesopotamian system Hellenised—though not to the point of giving up human sacrifices. Worship was paid to the sun and moon,[b] and to the five planets, considered as demiurges, assistants of the creator-god or gods.[c] Each one of the planets was associated with a particular colour, a particular geometrical shape (embodied in the construction of its temple),[d] a particular sacrifice and liturgy, and a particular metal (used for the image), as in the table overleaf.[e]

Of course, there is something rather Chinese about a symbolic correlation system such as this in itself.[f] But our attention is then caught by the fact that mercury was excluded

[a] See especially Stapleton, Azo & Husain (1), pp. 340–2 (cf. also 317, 335, 345, 361, 398, 401); also Stapleton & Azo (2), pp. 68, 72. The authority most quoted by al-Rāzī in his historical work *Kitāb al-Shawāhid* (Book of Evidences) is one Sālim al-Ḥarrānī.

[b] The great temple of the Assyrian moon-god, Sin, at or near Ḥarrān, had already been important in −2000 and was sacked by the Persians in −610. An interesting article by Seton Lloyd describes an archaeological prospection of it (*Times*, 21 March 1951). Later archaeological discoveries on and near the site have been reported by Seton Lloyd (2); Lloyd & Gökce (1) and Gadd (1). On what remains of Ḥarrān itself see Lloyd & Brice (1). Surrounding medieval Ṣābian monuments have been investigated by Rice (1) and Segal (1).

[c] Chwolson (1), vol. 1, pp. 158ff., pp. 725ff.; Segal (2). Much information on these liturgies and their sacrifices has come down to us in the Arabic literature. Some of the fullest detail occurs in that strange book of planetary, astral and talismanic magic known as *Picatrix* in its Latin form; a translation of the *Ghāyat al-Ḥakīm* (Aim of the Wise), compiled in Spain in the mid +11th century (by +1056), and attributed to Maslama al-Majrīṭī the alchemist, but certainly not by him (cf. Plessner, 1). The peculiar Latin name was a corruption of Buqrāṭis, which in turn may have corrupted Hippocrates. There are full instructions for the adoration of the planetary gods (trs. Dozy & de Goeje (2); Ritter & Plessner (1), vol. 2, pp. 167ff., 206ff., 213ff.) and curious accounts of other Ṣābian rites such as bull sacrifice (to Saturn), child sacrifice (to Jupiter), initiation ceremonies of young men (Ritter & Plessner, *op. cit.*, pp. 237ff.), and the strange dissolution in oil of the human sacrifice to Mars, with the story of the prophesying or divining by its separated head (Ritter & Plessner, *op. cit.*, pp. 146ff., 240–1). One is tempted to regard the legend of the brazen head of Roger Bacon as derivative from this Ṣābian fable, the factual basis of which remains highly problematical. Another echo is Scandinavian. Davidson (1) tells that in Viking mythology the ancient giant Mimir, guardian of the sacred well under the World-tree Yggdrasil, was beheaded by the gods, 'but afterwards Odin embalmed his head and kept it so that he might consult with it when he was in urgent need of counsel'. The motif has lived on into our own time, as witness the novel of C. S. Lewis, 'That Hideous Strength'.

[d] The *Kitāb al-Qarāṭis al-Ḥakīm* (Book of Crates the Wise), one of the oldest Arabic alchemical books (cf. p. 389 above) has a distinctly Ṣābian flavour, for the principal vision takes place in a temple of Venus which has an 'Indian' high priest; cf. the translated text in Berthelot & Houdas (1), pp. 61ff. Furthermore, the 'Indian' temple attendants aim their arrows at Crates, and we know that shooting the sacrifice to death was precisely one of the ceremonies at the Ṣābian planetary temples (in the liturgy for the moon, cf. Stapleton, Azo & Husain (1), pp. 400, 402).

Two relevant extant books are attributed to Balīnās (Ullmann (1), pp. 173–4). One is entitled *Kitāb al-Aṣnām al-Sabʿa* (Book of the Seven Idols); the other, the *Kitāb al-Qamar al-Akbar* (Greatest Book of the Moon). Cf. p. 369.

Finally, strengthening the alchemical connection, we are told by al-Masʿūdī that al-Rāzī himself wrote a book on the Ḥarrānian religion (de Meynard & de Courteille (1), ch. 64; comm. Stapleton, Azo & Husain (1), p. 341).

[e] Modified from Stapleton, Azo & Husain (1), pp. 398ff., 403, after Chwolson (1), vol. 2, pp.380ff. The chief authority was al-Dimashqī, writing about +1300.

[f] Cf. Vol. 2, pp. 261ff. We might also remember here the considerable influence exerted in China by the Chhi Yao[1] books on astronomy and calendrical science between the +5th and the +8th centuries; see Vol. 3, pp. 204ff. These 'Seven Luminaries' texts were recognised as West Asian in origin, so perhaps they were of Ṣābian inspiration acting eastwards.

[1] 七曜

Planet	Colour	Shape	No. of steps in the temple	Metal
Saturn (Zuḥal)	black	hexagon	9	lead
Jupiter (Mushtarī)	green	triangle	8	tin
Mars (Mirrīkh)	red	oblong rectangle	7	iron
Sun (Shams)	yellow	square	6	gold
Venus (Zuharah)	blue and white	triangle	5	copper
Mercury ('Uṭārid)	brown	hexagon and square	4	*khārṣīnī*
Moon (Qamar)	white	pentagon	3	silver

from the metals, and 'Chinese arrow-head metal', *khārṣīnī*,[a] put in its place. Since all the Arabic writers agree about this, and since religious custom is notoriously so conservative, there can hardly be any other conclusion than that Chinese connections here must have gone back a long way. Exactly what *khārṣīnī* was we shall shortly discuss (p. 431)—some have suggested metallic zinc, but cupro-nickel seems really more probable—at any rate it was something which came from China in relatively small amounts and which neither Ṣābians nor Arabs knew how to make themselves. Perhaps because of its scarcity, the image of 'Uṭārid was cast, it was claimed, of an alloy of all the metals as well as *khārṣīnī*. For the rest, the Ṣābian religion had cosmic male and female forces prominently in its system,[b] much light-mysticism,[c] and a special interest in cosmic cycles similar to the Great Year;[d] other hints of East and

Chinese influence in *Picatrix* has not so far been looked for, but it seems to be very much there, and some of the numerous 'Indian' ascriptions may really refer to 'Further India'. The question is important because it bears on the East Asian contacts of the Ṣābians. Listening to Professor Plessner at the Barcelona Congress of 1959 I was much impressed not only by the prominence of the 28 *manāzil* or lunar mansions (= *hsiu*,[1] *nakshatra*) in the book (see Ritter & Plessner (1), vol. 2, pp. 14 ff.)—so characteristic a feature of Chinese and Indian astronomy—but by the fact that they and other asterisms are depicted in the typically Chinese 'ball-and-link' convention of drawing constellations, i.e. as hollow circles connected by straight lines (cf. Vol. 3, pp. 276 ff.). Examples from Ritter & Plessner (1), vol. 2, pp. 85, 111, 112, 114, 310, 320 ff.) are shown in the inset drawings. Others can be seen in the monograph

of Winkler (1) on Muslim magic seals, pp. 150 ff. and esp. p. 166. Mr Destombes has told us that a similar style of drawing occurs on Arabic astrolabes and spread to Latin astronomical MSS of the + 12th and + 13th centuries. Further on the lunar mansion lists in Arabic see Plessner (2). Other suspicious features in the *Picatrix* book include (*a*) Chinese tutty, onyx and the 'laughing-stone' (cf. p. 449 below), (*b*) the use of magic squares as childbirth charms (p. 463 below), (*c*) microcosmic–macrocosmic correlations between human anatomy and celestial patterns and numbers (as Hartner (13) has already noted), and (*d*) an apotropaic authority called Kīnās the Pneumatologist, who lived till the age of 540 years and could control the *pneumata*—just like a *hsien*[2] with powers over the *chhi*.[3] See Ritter & Plessner (1), vol. 2, pp. 40 ff., 172–3, 189, 259 ff., 405, 407. All this is relevant to the question of Chinese relations with Ḥarrān.

---

a Also known as *ḥadīd ṣīnī*, 'Chinese iron'.
b Recalling Yin and Yang.   c Doubtless with Gnostic connections.
d Perhaps the Indian influence of *kalpas* and *mahākalpas* might be descried here.

¹ 宿   ² 仙   ³ 氣

South Asian connections are also to be found.[a] At all events, Ḥarrān seems to have been a place through which ideas emanating from those parts of the world may have been channelled into the Arabic mind.

By the + 10th century the very orthodox Ashʿarīs got the upper hand in Islam, and Ḥarrān became very uncomfortable. In +933 the *muḥtasib* or police chief of Baghdad, a man named al-Iṣṭakhrī, demanded the extermination of the Ṣābians; and it seems that most of them now gradually accepted Islam, their last official head, Ḥukaim ibn ʿĪsā ibn Marwān, dying in +944.[b] But during the previous three or four centuries the role of Ḥarrān may have been truly important in transmitting both ideas and things. Let us now see what some of these were—and first, the things.

(iii) *Material influences*

As we have just been talking about a certain Chinese metal or alloy, *khārṣīnī* (barb or arrow-head metal of China), or *ḥadīd al-Ṣīn* (iron of China), which made such a great impression in Arabic culture, let us first see what more there is to be said about it, even though it was certainly not the most important of the substances which the Arabs got to know about from their friends further East. Still, it figured regularly (presumably under Ṣābian influence) in Arabic lists[c] of the seven metals (*ajsād*),[d] displacing mercury or glass (*zujāj*).[e] Other mentions of it in the Arabic alchemical texts are quite numerous.[f] Indeed, there was a special book devoted to it in the Jābirian Corpus, the *Kitāb al-Khārṣīnī*.[g] So we shall want to know when it first appeared in the West Asian lands, and what in all probability it really was.

[a] For example, Indian imagery in the temple of Saturn. Such connections were taken very seriously by Chwolson (1), vol. 1, p. 798. Though A. J. Hopkins was one of the first to appreciate the possible importance of Ḥarrān, he went much too far in saying, (1), p. 156, that the Ṣābian religion 'was derived from Iran and further back from China'.

[b] The case is slightly reminiscent of the liquidation of Aztec culture by the Spaniards centuries later, and one has similar mixed feelings about it, for very much the same reasons. Sarton (13) seems to suggest that some Ṣābians found refuge in Christendom, but that sounds rather unlikely. More probably the Hermetic strain continued within the bosom of Islam itself (cf. Massignon, 4), and found its way to Latin Europe and the Renaissance by way of Spain.

[c] On these lists in the Jābirian Corpus see Kraus (3), pp. 19, 22–3. One is in the *Kitāb al Khawāṣṣ* (Book of Properties), Kr 78, another in the *Kitāb al-Khamsīn* (Book of the Fifty), Kr 1825–1874. For al-Rāzī (c. +900) see Stapleton, Azo & Husain (1), pp. 321, 340–1, 345, 363, 370, 405. For al-Kātib al-Khwārizmī (+976) see Stapleton, Azo & Husain (1), p. 363; Wiedemann (15), p. 80. For al-Qazwīnī (c. +1275) see Stapleton, Azo & Husain (1), p. 406.

[d] The Hellenistic proto-chemists had recognised only six; cf. *Corp.* III, xvii, 1, classed by Berthelot & Ruelle as a Zosimus text. So also the *Caraka* and *Suśruta samhitas* in India, with all later tradition; Ray (1), vol. 1, pp. 25, 44, 48, 72, 127, 157.

[e] As in the Jābirian Seventy Books, towards the end, pointed out by Stapleton, Azo & Husain (1), p. 405; Kraus (3), p. 21. They were referring to the *Kitāb al-Ghasl* (Book of Washing, i.e. Purification), Kr 183.

[f] Cf. Siggel (2), p. 79, (3), p. 12. For the Jābirian Corpus see Kraus (2), p. xxxv, (3), pp. 19, 22ff. *Ḥadīd ṣīnī* comes in the *Kitāb al-Sirr al-Maknūn* (Book of the Hidden Secret), Kr 389. For al-Rāzī see Ruska (14), pp. 42, 84, 85, 134, 138, 156; Stapleton, Azo & Husain (1), pp. 405–11. For al-Bīrūnī (c. +1020), see Kraus (3), pp. 22–3. For al-Ṣafadī (d. +1363) see Wiedemann (32), p. 9. For Ibn Khaldūn (+1377) see Rosenthal (1), vol. 3, p. 271. *Khārṣīnī* seems not to occur in *Picatrix*, but instructions are given for a charm to be inscribed on a ring made of 'Chinese iron' (Ritter & Plessner (1), p. 411).

[g] Kr 953; see Kraus (2), p. 116.

If the Arabo-Syriac MS. published by Berthelot & Duval were a text of the early
+9th century it might be the first mention of 'Chinese arrow-head metal' in the West,
for the name occurs in connection with the 'filtration' of metals, or descensory
distillation, through a crucible with a perforated floor, in the *bot-bar-bot* apparatus
(cf. p. 33 above);[a] but this work is considered to include material as late as the
late+10th century.[b] Moreover, *khārṣīnī* is not mentioned in its list of metals.[c] The
Jābirian Corpus therefore seems to be the beginning.[d]

Various ideas were current about *khārṣīnī*. It was exceedingly scarce,[e] it caused
mortal wounds when actually made into arrow-heads,[f] and it resembled the metal
of which mirrors were made (though somewhat softer),[g] while pots and cauldrons
were also manufactured from it.[h] In his encyclopaedia *Kitāb 'Ajā'ib al-Makhlūqāt...*
(Marvels of Creation)[i] Abū Yaḥyā al-Qazwīnī (*c.* +1275) considered that when the

---

[a] Berthelot & Duval (1), pp. 149, 150. Actually 'Chinese iron' is the reading here rather than
'Chinese arrow-head metal'. The device was the ancestor of the Gooch crucible.

[b] Later mentions in Syriac are plentiful. For example, the 'Book of Dialogues' of Severus bar
Shakko (d. +1241), has *parzla* (iron) of China; cf. Ruska (40), p. 159. For this reference and for help in
the study of the Syriac texts in general we are much indebted to Dr Sebastian Brock.

[c] *Op. cit.*, p. 156.

[d] But we may find a reference from an unexpected source which would take back 'Chinese iron' to
a point some four centuries earlier. It occurs in Sir Harold Bailey's fascinating account of the past half-
century of Iranian studies. What he says is of such philological erudition that we cannot but quote it
verbatim.

'The Kharoṣṭhī Kroraina texts [he says, (1), p. 103] are important because of the Iranian words which
they contain; they assure a date around +300, before the bulk of the extant Khotan Śaka texts were
written. One problem they have raised is the source of the north-western Dardic Pašai word *čimä́r*,
"iron", and the related words of Dardistan and Nuristan. In a Buddhist Sanskrit manuscript of the
Saṃghāṭa-sūtra from the Gilgit monument called a *stūpa*, probably therefore about +400, there occurs
the word *cīmara-kāra*, "a worker in cīmara metal". The Chinese translation proposed the meaning
"iron", the Tibetan translation gave "copper". Modern dialects have the meaning of "iron". Now this
word probably occurs in Kroraina in the phrase *cina-cīmara*...If we render by "Chinese cīmara", that
is, "Chinese iron", one is at once reminded of the Arabic *ḥadīd-ṣīnī*, "Chinese iron", possibly meaning
nickel [or cupro-nickel]. This Buddhist Sanskrit word *cīmara-* has such a similar appearance to Turkish
timür, "iron", that the older form of that word is likely to be *čimr-* with -*ür* replacing -*r*- after consonant, as the Turks turned Persian *babr*, "tiger" into the name Babur.'

For us all this is rather strong support for the view expressed a few pages below that 'Chinese iron'
or 'Chinese arrow-head metal' was cupro-nickel and not zinc, because it takes back the export to
Chin times; cf. Vol. 5, pt. 2, pp. 225 ff. It also means that the expression 'Chinese iron' was in some
Iranian languages long before it ever got into Arabic.

[e] Cf. al-Rāzī, in Stapleton, Azo & Husain (1), p. 345; Kraus (3), p. 23.

[f] Cf. Laufer (1), p. 555, quoting Steingass' Persian dictionary, p. 438. There may have been some
substance in this notion, because from the earliest days of proto-gunpowder in China, incendiary
weapons involved toxic compositions and smokes; cf. Davis & Ware (1). The time with which we are
here concerned would have been rather early for proto-gunpowder, but not for incendiaries, and there
could have been confusion with a strange metal.

[g] See al-Rāzī (tr. Stapleton, Azo & Husain (1), p. 371), and al-Dimashqī, *c.* +1325, who speaks of
distorting mirrors (Wiedemann (33), p. 403). This suggests whiteness (as of a high tin bronze like gun-
metal), and reflectivity. Whiteness is also indicated by the alternative name of 'Chinese iron'. The
Persian terms *isfīd-rūy* and *sepīd-rūy* meant 'white copper', but they raise the same nomenclature
problems that we have faced in Chinese (pt. 2, p. 232 above), and in fact they were almost certainly used
to mean high tin bronzes rather than *pai thung*, cupro-nickel. Cf. von Lippmann (1), p. 417. For another
reference to Chinese mirror metal in al-Rāzī see Stapleton, Azo & Husain (1), p. 387. That mirrors from
China said to be made of steel were sold in Baghdad around +990 for double or several times their
weight in silver we learn from al-Tha'ālibī's *Laṭā'if al-Ma'ārif* (Dunlop (6), p. 58).

[h] Al-Ṣafadī, in Wiedemann (32); at Badakshan, says al-Qazwīnī, in Kraus (3), p. 23.

[i] Mieli (1), 2nd ed., p. 150.

proportions of mercury and sulphur were just right for silver but injured by cold before coction, *khārṣīnī* was formed.

The formation of *khārṣīnī* [he wrote] is like that of the other metals already mentioned. Its mine is in the land of China. Its colour is white, with a reddish tinge. All spear or arrow heads made from it are very injurious. It is also worked into fish-hooks by means of which large fishes are caught, for when once they have swallowed one of these they cannot escape except with the greatest difficulty. From this metal also is made a kind of mirror which is the best treatment for palsy, since a paralysed man will derive benefit if he sits and gazes at it in a darkened room. Pincers are furthermore made from it good for the pulling out of hairs, and if the place where they were is then oiled several times, the hair will not grow again.[a]

Al-Qazwīnī adds that no other metal yields a ring (of resonance) equalling that of this one, and that none is so suitable for the making of bells large and small.[b] Yet it was not regarded as very potent, for to 1 part of the elixir there corresponded the following quantities:[c]

'mineral ammoniac' (ammonium chloride)	$2\frac{1}{2}$ parts
gold	5
'derived ammoniac' (ammonium carbonate)	$6\frac{1}{4}$
silver	10
copper	$14\frac{2}{7}$
tin	20
lead	25
iron	50
khārṣīnī	100

This was a strange 'Pythagorean' numerological anticipation of modern experimental orders such as the atomic weight or the electrochemical series. In the *Kitāb al-Ḥadīd* (Tractate on Iron)[d] the Jābirian writer expressly says that the planet Mercury is to be correlated not with mercury but with *khārṣīnī*; and Kraus remarked that the Arabic alchemists who preferred to regard mercury as one of the volatile spirits needed something else to complete the set of metals.

What then was *khārṣīnī*? There have been two main opinions—de Sacy, Laufer and Ruska said cupro-nickel,[e] Humbert and Stapleton said zinc.[f] Others could not decide between them.[g] Stapleton was rather optimistic, we suspect, in believing that al-Rāzī was too good a practical chemist to have classed it among the basic metals unless he had satisfied himself that it was not an alloy.[h] Would he really have been able to distinguish about +900? Another approach is now possible on the basis of what we have learnt about the dates of first preparation of metallic zinc and cupro-

---

[a] Tr. Stapleton, Azo & Husain (1), p. 407.

[b] Wiedemann (33), pp. 403–4; Kraus (3), p. 23. Kashgar, it seems, was famous for them.

[c] Kraus (3), p. 23 ff.   [d] Kr 950.

[e] Laufer (1), p. 555; Ruska (14), pp. 42–3; de Sacy (2), vol. 3, pp. 452 ff. Or, as some of them put it, tutenag.   [f] Stapleton, Azo & Husain (1), p. 407; Humbert (1), p. 171.

[g] E.g. Bocthor (1); Dozy & Engelmann (1), pp. 252, 294.

[h] Stapleton *et al. loc. cit.* Other alloys discussed by al-Rāzī (Stapleton *et al. op. cit.*, pp. 324, 408 ff.) were *isfīd-rūy*, bronzes; *shābah* or *rūḥ-i-tūtiyā*, brasses; and *ṭālīqūn*, an alloy of copper and lead, also called *nuḥās* (or *mis*) *ṣīnī*, i.e. 'Chinese copper'. Some Chinese coin compositions were indeed rather high in lead, up to nearly 30% (cf. Table 98 and p. 215 in Vol. 5, pt. 2).

The *Kitāb al-Aḥjār* or Lapidary of Pseudo-Aristotle, a Syrian–Persian–Indian compilation of the early +9th century, attributes to *ṭālīqūn* several of the same properties as those ascribed by al-Qazwīnī to nickel elsewhere (pt. 2, pp. 212 ff., 225 ff.). From that it is clear that the Chinese could not have exported metallic zinc before this very time, about +900, while cupro-nickel could have been sent to the western countries at any period from the late Han or San Kuo onwards, say the +2nd or +3rd century.[a] Consequently if some at least of the Jābirian references are to be placed in the early or middle +9th century the time would have been rather too early for zinc, and cupro-nickel would be more probable. True, one gets the impression that *khārṣīnī* occurs in the later Arabic alchemical books rather than in the earliest ones, so the point is rather difficult to prove by datings. However, the references to the bell-like resonance of *khārṣīnī* speak almost decisively in favour of cupro-nickel and against zinc, for this was one of the characteristics of paktong most admired in +18th-century Europe. In any case, we are clearly in presence of one material substance which greatly interested the Arabs, and which they knew came to them from China.

Khārṣīnī was far from being the only substance to which the Arabs applied the epithet 'Chinese'. When they learnt of saltpetre in the +13th century they called it 'Chinese snow' (*thalj al-Ṣīn* or *thalj ṣīnī*),[b] just as rockets were known among them as 'Chinese arrows' (*sahm al-Khiṭāi*).[c] We shall come across several other examples of this nomenclature presently. But the transmission of saltpetre and gunpowder belongs to a period rather later than that on which we are now concentrating in connection with the transmission of the elixir idea. There was one new substance however which characterised Arabic alchemy almost from its beginnings, a substance of striking properties and reactivity, and one which unquestionably came (to begin with) from regions further East—this was sal ammoniac, *nūshādir* or *nūshādur*. Properly speaking, the term means ammonium chloride (NH_4Cl), but the Arabs, and indeed the later Latins, did not clearly distinguish it from ammonium carbonate (($NH_4)_2CO_3$), except that the former was considered mineral or natural (*al-ḥajar*) and the latter derived or artificial (*mustanbaṭ*). In explanation one can broadly say that the former was exported from the mountainous regions of Central Asia such as the Thien Shan, while the latter, seemingly a characteristic product of Arabic alchemy, was obtained by the dry distillation of hair. A little later it was found, also in the Near East, that the chloride itself could be obtained by sublimation from heated soot. But in China the natural product always occupied the chief place.

khārṣīnī; cf. Ruska (19); Stapleton, Azo & Husain (1), p. 409. But it would have been neither silvery nor resonant.

---

[a] The episode of the cupro-nickel coinage in ancient Bactria will also not be forgotten (cf. pt. 2, pp. 237 ff. above). Could this perhaps have decanted a continuous Central Asian tradition into the Arabic world?

[b] Cf. Partington (5), pp. 202 ff., 310, 313. This comes first in Ibn al-Baiṭār, c. +1240 (cf. p. 194 above). Ḥasan al-Rammāḥ the military pyrotechnician, writing about +1280, adds many similar expressions such as 'Chinese flowers', 'Chinese wheel', 'Chinese arsenic' and 'Chinese iron', which was probably not the same as *khārṣīnī*.

[c] Cf. Laufer (1), pp. 555–6. It may have already been suggested that the two names betray a passage westwards by the sea and land routes respectively; see Vol. 1, p. 169. The Liao dynasty of the Chhitan Tartars lasted from +907 to +1125, when the Mongols eliminated it, but the appellation 'Cathay' remained in use in various forms in Central Asia for long afterwards, and it was natural that things travelling over the Old Silk Road should receive the adjective 'Cathayan'.

That the sal ammoniac of modern times, the *nūshādir* of the Arabs and the *nao-sha* of China,[a] was quite different from the 'ammoniac salt' (*halas ammōniakon*, ἅλας ἀμμωνιακόν) of the Western ancients, Beckmann at the end of the +18th century was probably the first to realise.[b] The mentions of this natural product in Herodotus,[c] Columella,[d] Dioscorides[e] and other writers[f] clearly show that it came from the neighbourhood of the desert oasis of Siwa (near the present-day border between Egypt and Libya), with its famous temple of Amūn-Ra; later Pliny confused the etymology by bringing in the Greek word for sand (*ammos*, ἄμμος).[g] But from all descriptions the properties of this salt were quite unlike those of ammonium chloride, and by general agreement it must have been either rock-salt (sodium chloride) or natron (sodium carbonate).[h] This has to be remembered when considering, as presently we must, the names which ammonium chloride has borne in the history of the nations.

If we open one of those vintage Victorian chemical textbooks which are often so much more useful to the historian of science than contemporary expositions, which have become more than half physics; a volume, for example, in one of the many editions of Roscoe & Schorlemmer,[i] we can find a history of sal ammoniac briefly set forth. It is of great importance for it is the pre-history of the most useful volatile alkali, gaseous ammonia. This was Priestley's 'alkaline air' (+1774), the composition of which (NH_3) was determined by Berthollet (+1785) and more accurately by Austin three years later. Down to Priestley's time ammonia was known only in aqueous solution, by two names, *spiritus volatilis salis ammoniaci* and 'spirits of hartshorn'. Glauber in the +17th century already knew that a volatile alkali could be produced by the action of a fixed alkali on sal ammoniac,[j] hence the first name. The second recalled the original method, essentially Arabic, of destructive distillation of animal refuse such as hoofs, horns, bones, hair, etc.,[k] the ammonium carbonate produced ('salt of hartshorn' or *sal volatile*) now being neutralised by hydrochloric acid.[l] Most of the sal ammoniac used in Renaissance and seventeenth-century Europe was imported from Egypt, where it was obtained by sublimation from soot, especially that of camel dung fires such as those of the baths (*ḥammām*)[m] and the

---

[a] See pp. 443 ff. below.   [b] (1), vol. 2, pp. 396 ff.

[c] IV, 181.   [d] *De Re Rust.*, VI, xvii, 7, 8.   [e] V, 126.

[f] 'Ammoniac salt' is twice mentioned in the Leiden Papyrus (cf. pt. 2, p. 16), but Berthelot (2), pp. 30, 45, well realised that it was not sal ammoniac. Similarly, there is no trace of sal ammoniac under any name in the earlier parts of the Hellenistic Corpus (Ruska (13), p. 5).

[g] *Hist. Nat.*, XXXI, xxxix. He and other Latin writers used an initial aspirate form making 'hammoniacum'.

[h] On 'natron' and 'nitre' cf. pp. 179 ff. above. The properties of ammonium chloride, volatilising so completely without decrepitation and liberating so easily 'insupportable vapours' could not have been overlooked. Cf. Bailey (1), vol. 1, pp. 41, 163.

[i] (1), 3rd or 4th edition, vol. 1, pp. 452 ff., vol. 2, pp. 287 ff., 297 ff.

[j] And therefore that it consisted of an alkaline and an acid part, cf. Partington (7), vol. 2, p. 353. This was in +1647; Tachenius in his *Hippocrates Chemicus* of +1666 (II, viii) expressed the matter even more clearly. Cf. Partington, *op. cit.* pp. 293–4.

[k] Cf. Stapleton (1), p. 28, quoting al-Khwārizmī al-Kātib's *Mafātīḥ* (+976).

[l] Since Sala (+1620); Partington (10), p. 318, (7), vol. 2, p. 279. Or taken through the sulphate to react with common salt, cf. Aikin & Aikin (1), vol. 2, pp. 281 ff.

[m] Accounts of this go back a long way indeed, to Abū Isḥāq al-Iṣṭakhrī, for instance, *c.* +970 (Mieli (1), 2nd ed., p. 115), to Abū Ja'far al-Ghāfiqī, d. +1165 (Mieli, *op. cit.*, p. 205), and to Abū

ovens used for artificial incubation of hen's eggs (*ma'mal al-katākīt*).[a] Latin Geber, *c.* +1300, has an account of the preparation of sal ammoniac by the distillation of urine, sweat and wood soot with salt,[b] hence the origin, it has been supposed, of the term *spiritus salis urinae*.[c] But whether this method could ever have produced any sublimed crystals of the chloride has been gravely doubted since the beginning of the +18th century, as Multhauf (8) has shown, so it may well be that the Geberian formula was a cover-up for something else, and that the Venetians who in the previous century were supposed to be using it were really importing the chloride from Egypt.[d]

The historical importance of sal ammoniac lay in its great reactivity, providing the Arabic alchemists with a new 'volatile spirit' to set beside mercury, sulphur and arsenic; for both the chloride and the carbonate sublimed readily, and unchanged. The former will attack, or colour,[e] many metals, even silver, producing the chlorides; it reduces metal oxides as a flux giving a clean metallic surface suitable for tinning, silvering or gilding;[f] it also found employment in the dyeing craft where ammonia alum was wanted as a mordant;[g] and it has useful pharmacological properties.[h] A strongly refrigerant effect on solution attracted notice early,[i] and the 'English drops' or 'smelling-salts' familiar since the +17th century were only confections of ammonium carbonate.[j] Stapleton sought evidence of some of these uses in the +12th-

---

'Abd Allāh al-Anṣārī al-Dimashqī, *c.* +1320 (Mieli, *op. cit.*, p. 275). Such references have been collected in von Lippmann (1), pp. 403, 418. For similar accounts of fairly recent date see Aikin & Aikin (1), vol. 2, pp. 280ff.; Parkes (1), vol. 4, pp. 339ff.; Ure (1), vol. 1, pp. 140ff.; Clow & Clow (1), p. 420.

a See Needham (2), pp. 6–7. It was the report of the French Consul at Cairo, Lemere, in +1719, that drew attention to this (in Parkes, *loc. cit.*).

b *De Invent. Veritatis*, ch. 4, tr. Darmstädter (1), pp. 105, 174. Stapleton (1), p. 28, had this reference wrong, for *De Investigat. Perfectionis*, ch. 4 (Darmstädter tr., p. 97) has only a re-sublimation process for the purification of the salt; but he deserved great credit for his pioneer monograph on the whole subject. The only other studies comparable in importance are those of Ruska (13, 39), and we draw on all of these here.

c Cf. Libavius, *Syntagma*, VI, viii, 39; Partington (7), vol. 2, p. 264.

d In modern times all other sources have been ousted by the ammonium sulphate of 'gas liquor', a by-product in the distillation of coal and coal-tar. But ammonium carbonate distilled from bones and offal was still a mainstay of the industry at least as late as 1830; cf. Multhauf (8); Ure (1), vol. 1, pp. 135ff.

e Cf. Partington (7), vol. 2, p. 19.

f Cf. Agricola, *De Nat. Foss.* (+1546), IV, ix; Partington (7), vol. 2, p. 53.

g Cf. Ure (1), vol. 1, p. 147.

h An expectorative stimulant, mild cholagogue, and diaphoretic; cf. Sollmann (1), pp. 556ff. Acting as dispenser for my father, when a boy, I remember many prescriptions containing 'am. carb.' and 'am. chlor.' Sal ammoniac, we now learn, was commonly added to snuff (Parkes (1), vol. 4, pp. 339ff.).

i Cf. pt. 3, p. 225 above. There can be a temperature fall to $-10°$, or with snow or pounded ice, to $-18°$ (Ure (1), vol. 2, pp. 296–7).

j Our remark at an earlier point (pt. 2, p. 90) will be remembered, namely that Ko Hung would have considered these a perfect instantaneous elixir capable of recalling the absent souls and restoring consciousness. Indeed, he may well have used *nao-sha* in some such way (cf. p. 440 below). This was probably the application which brought sal ammoniac into the poems of Robert Burns. In his 'Death and Doctor Hornbook' (a satire on John Wilson, schoolmaster of Tarbolton, who professed medical knowledge), speaking of quack drugs and Latin patter, he wrote:

> 'Forbye some new uncommon weapons—
> *Urinus spiritus* of capons,
> Or mite-horn shavings, filings, scrapings
>   Distill'd *per se*,
> Sal alkali o' midge-tail clippings
>   And mony mae.'

century *De Anima in Arte Alchemiae* of Pseudo-Avicenna.[a] The 'hardening of mercury' and the softening of other metals by the 'water of hair' must refer to the action upon them of ammonium chloride prepared from the carbonate got by dry distillation.[b] The mention of ceration (ready melting without fumes on a hot plate) may mean that the late Arabs and later Latins succeeded in preparing one or other of the complex salts of ammonia and mercury.[c] The addition of hair (and perhaps salt) to the zinc carbonate used in the making of brass, and the 'fumes of hair turning copper yellow', probably imply the use of the ammonia salts as fluxes. And since sal ammoniac when heated in a confined space often acts like gaseous hydrochloric acid, one can see that the early medieval Arabic alchemists had come into possession of a new reagent of real importance.

That sal ammoniac (*nūshādir*) is constantly mentioned in Arabic alchemical texts needs no proof. The writings of the Jābirian Corpus often speak of it in both its forms,[d] and though no individual title embodies its name, there is, significantly, a 'Book of Hair' (*Kitāb al-Sha'ar*),[e] as well as more general treatments in the 'Book of the Living' and the 'Book of Government'.[f] They constituted but one aspect of that emphasis on organic in addition to inorganic materials which the Arabs shared with the Chinese,[g] in contrast to the Greeks. All this suggests that we can confidently place Arabic knowledge of the natural ammonium chloride as far back as about $+850$ and of the carbonate (from hair and other animal material)[h] by about $+875$. Knowledge

[a] (1), pp. 37ff., elucidating v, xx, vii, iii, etc.

[b] Presumably then either via the sulphate using blue vitriol followed by common salt, or directly using the magnesium chloride of bittern, processes later industrially current; cf. Multhauf (8).

[c] I.e. 'fusible white precipitate' ($2NH_3 . HgCl_2$, mercuri-diammonium chloride), cf. 'infusible white precipitate' ($NH_2 . HgCl$, amino-mercuric chloride), and 'sal alembroth' (($NH_4)_2 . HgCl_4 . H_2O$, ammonium chloro-mercurate); see Partington (10), pp. 399, 401–2; Durrant (1), pp. 375–6; Aikin & Aikin (1), vol. 2, pp. 82, 283. The second of these was of practical importance because it multiplied the solubility of corrosive sublimate some twenty times. The term alembroth, pointing so unmistakably to Arabic origins, has been considered a corruption of al-Rāzī's references to the axes required for breaking up certain hard minerals or preparations (von Lippmann (1), vol. 3, pp. 29–30). Sal alembroth was used in medicine, being regarded as less irritant than other mercurials (Sollmann (1), p. 638).

[d] E.g. *Kitāb al-Raḥma al-Kabīr*, Kr 5, see Berthelot & Houdas (1), pp. 167, 187, corr. by Ruska (13), p. 10. The second reference speaks of *nūshādir* purified by resublimation. *Kitāb al-Tajmī'*, Kr 398, see Berthelot & Houdas (1), p. 205, corr. by Ruska (13), p. 11. *Kitāb al-Aḥjār*, Kr 40, see Kraus (3), pp. 226–7. *Kitāb al-Khawāṣṣ al-Kabīr*, Kr 1900, see Kraus (3), pp. 18–19, *nūshādir al-sha'ar*, 'hair ammoniac'. *Kitāb al-Sumūm*, Kr 2145, see Ruska (39). Consult also Kraus (3), pp. 25, 41–2, 109, 233.

[e] Kr 34. Other titles in the Hundred and Twelve Books which tell the same story are the *Kitāb al-Bayḍ* (Book of Eggs), Kr 32; the *Kitāb al-Dam* (Book of Blood), Kr 33; the *Kitāb al-Bawl* (Book of Urine), Kr 56; the *Kitāb al-Ḥayawān* (Book of Animals), Kr 55; and perhaps the *Kitāb al-Nabāt* (Book of Plants), Kr 35.

[f] I.e. the *Kitāb al-Ḥayy*, Kr 133, and the *Kitāb al-Ḥukūma*, Kr 134, respectively, both among the Seventy Books; cf. Ruska (3), p. 43. The first twenty tractates of this group are all on animal chemistry; Kraus (2), pp. 44, 47. Apparently the different parts and excreta of animal and human bodies each had their 'partisans' as sources of *nūshādir*. The *Kitāb al-Lāhūt* (Book of Divine Grace), Kr 123, and the *Kitāb al-Balāgha* (Book of Attainment), Kr 135, name marrow, blood, hair, bones, urine and sperm as the most important; among the best animals the 'hottest', i.e. lion, viper and fox; among human beings bilious men, thin men, Yemenis, sea islanders, Sindhis and Copts; failing any of these one has to be content with cattle, gazelles, and asses wild and tame. Cf. Ruska (3), p. 41; Kraus (3), p. 4.

[g] Cf. pp. 393, 398, 401 and 404 above, and p. 497 below.

[h] Stapleton (1), pp. 30ff., gives a long excursus on the magical properties attributed to hair, etc. in Asian cultures, and adds examples of biological metamorphoses, real and supposed, which made its transformation into an important elixir chemical seem more likely (cf. pt. 2, p. 64 above on Petrus

of the chloride obtained by sublimation from dung and its soot would have followed about +900. This last can be traced in many poetical verses praising that alchemical treasure which is to be got from the lowest and most repulsive origins.[a] Of these perhaps the most famous lines are those attributed, doubtless wrongly, to Khālid ibn Yazīd (p. 389 above):

> Take talc, with *ushshaq* gum,[b] and what thou findest in the streets,
> Add a substance resembling borax, and weigh them without error;
> Then, if thou lovest God, thou shalt be master of all his works.[c]

But there are numerous other statements of the idea.[d] Naturally from the beginning of the +10th century sal ammoniac is seen in constant use.[e]

To fix the first appearance of *nūshādir* in Western Asia more exactly is a difficult matter, for the helpfulness of our Syriac documents is vitiated by the many interpolations from Arabic which they afterwards received. The two main Syriac MSS studied by Berthelot & Duval speak of *melḥē armōnīqōn*[f] (which would be the ammoniac salt or soda of the ancients),[g] and *nshādr*, *nūshādr* (which must be the ammonium chloride of the Arabs),[h] but as they are no earlier than the early parts of the Jābirian

---

Bonus). Such passages can be found in the +10th-century text translated by Berthelot & Duval (1), p. 155, and in Ibn Khaldūn's *Muqaddima* quoting al-Ṭughrā'ī (d. +1121), tr. Rosenthal (1), vol. 3, p. 272.

[a] A deeply important theme of introspective psychology, this—the 'lotus rising from the mud' in other civilisations. Or 'samsara *is* nirvana'.

[b] I.e. 'gum ammoniac' (the name a coincidence here), from the Persian umbellifer *Dorema Aucheri* or *ammoniacum*. It is a balsam or gum-resin still used in pharmaceutical flavouring; cf. Sollmann (1), p. 121.

[c] Tr. de Meynard & de Courteille (1); Wiedemann (27), p. 346, repr. (23), vol. 1, p. 52; Ruska (4), p. 28; Dunlop (6), p. 206; v. Lippmann (1), pp. 357–8, eng. et mod. auct.

[d] Both in prose and verse. For example the Jābirian *Kitāb al-Raḥma al-Ṣaghīr*, Kr 969, see Berthelot & Houdas (1), p. 136. Or the post-Jābirian *Kitāb al-Jāmi'*, see Berthelot & Houdas, *op. cit.*, p. 118, cf. p. 125; Kraus (2), p. 197.

[e] See, e.g. Partington (17) for al-Rāzī, as also Heym (2); and Stapleton, Azo & Husain (1) and Ruska (14) translating the *Kitāb Sirr al-Asrār*, *passim*. On the derivative *De Aluminibus et Salibus* see Ruska (21), pp. 82ff., 125. Cf. Stapleton (1), p. 28 on al-Khwārizmī al-Kātib (+976). Add Ruska (5), pp. 74–5, 79, 86, 106, 112, 124, translating the *Ta'wīdh al-Ḥākim*...of *c.* +1050, cf. p. 391 above. And Stapleton & Azo (2), p. 80. Further, Stapleton (1), p. 29, on Ibn al-Tilmīdh and Abū Ja'far al Ghāfiqī (both d. +1165); and al-Sharīf al-Idrīsī (d. +1166). For al-Khwārizmī al-Kātī see Stapleton & Azo (1); here sal ammoniac of Khurāsān is often mentioned (+1034). There are many references also in the tractates of the Rampur Codex, see Stapleton & Azo (2), pp. 60, 64, 79, 80, 83.

[f] Or *armenaitā* (Berthelot & Duval (1), p. 70, text, p. 39, corr. Ruska (13), p. 16).

[g] (1), pp. 8, 9, 86, text pp. 4, 48. The difficulty is that Duval may have punctuated his text wrongly in some places, so we cannot be sure that this adjectival combination is always present where he thought it was; see Ruska (13), pp. 14–17. *Armōnīqōn* may sometimes be a separate word meaning a mineral from Armenia. There could also be a confusion with the gum already referred to (Berthelot & Duval, *op. cit.*, p. 10; Ruska, *op. cit.*, p. 16). The translations of Duval were distinctly careless and did not take adequate account of the various technical terms used in the texts, hence the later corrections of Ruska are indispensable. Ruska dated the lists incorporated in these texts, and in which the words occur, about the time of Ḥunain ibn Isḥāq the great translator, i.e. *c.* +850 to +870. We cannot follow him in all his interpretations, however.

[h] (1), pp. 13, 64, 66, 70, etc. *Nūshādir* is prominent in the +11th-century Arabic MS. written in Syriac script, as would be expected, cf. (1), pp. 143, 155, 159, 160, 183, 197. Among late Syriac references one could quote Severus bar Shakko (d. +1241); see Ruska (40).

Corpus they do not take us much further.[a] The focal point remains at $+850$ or the decades just preceding that date.[b] Where, one may now ask, did the natural sal ammoniac then obtained by the Arabs come from?

There can be no doubt that Central Asia was the region and that the salt was collected at the mouths of clefts or vents in the earth which also gave forth variously gases, flames and smoke; often, according to the accounts, these vents were situated within natural caves. All the Arabic writers, from the $+10$th to the $+14$th century, agreed that mineral sal ammoniac came from these eastern lands, China, Ferghana and Persia—whether Ibn 'Alī al-Mas'ūdī $c.$ $+947$ or al-Dimashqī the cosmographer $c.$ $+1320$.[c] Parallel Chinese accounts, which we shall be considering in a moment, fully confirm the importance of Sinkiang as a source of sal ammoniac. The medieval and traditional descriptions were not of course precise about the geological nature of the phenomena, and it has long been assumed that the mountains of Central Asia, if not numbering any volcanoes still eruptive, were yet capable of sufficient activity of that kind to account for the sal ammoniac production. That ammonium chloride is deposited around the openings of volcanic vents (fumaroles, solfataras),[d] and crystallises on the surface of cooling lava in its cracks and fissures, is not to be doubted in view of the many eye-witness accounts available.[e] But all the geological evidence goes to show that the Thien Shan and Altai mountain ranges ceased to be the scene of volcanic activity many ages ago, and that the real source of sal ammoniac in these regions was the burning of underground seams of coal.[f]

[a] Nor does the *Lexicon* of Bar Bahlul ($c.$ $+980$), Berthelot & Duval (1), pp. 135, 137; Ruska (13), pp. 17–18. See p. 447 below. Nor does the Cambridge MS., which is of very miscellaneous content, hard to date; Berthelot & Duval, *op. cit.*, pp. 248, 297.

[b] According to Multhauf (5), p. 126, sal ammoniac is mentioned in the *Kitāb Sirr al-Khalīqa* (Book of the Secret of Creation, see p. 370 above), which may be dated at about $+820$. Another early mention is that in the Arabic Lapidary of Pseudo-Aristotle which belongs to the same time; Ruska (19), p. 173. It recurs of course in the enlarged Lapidary of al-Qazwīnī ($c.$ $+1250$); Ruska (24), p. 40.

[c] See the *Murūj al-Dhahab*, tr. de Meynard & de Courteille (1), and the *Nukhbat al-Dahr*, tr. Mehren (1), pp. 93, 169, 308, cit. von Lippmann (1), p. 418. A collection of the accounts of Arabic writers will be found in Ruska (39) and Laufer (1), p. 507.

[d] This is well known for the fumaroles of Hecla, Etna, Vesuvius, Pozzuoli and other European volcanic regions; Roscoe & Schlorlemmer (1), vol. 2, p. 287; Singer (8), pp. 172–4, 203; Bischof (1), Eng. ed., vol. 1, p. 345, 2nd Germ. ed., vol. 1, pp. 636ff. A Vesuvian sample analysed by M. H. Klaproth (1), vol. 2, pp. 67ff., in $+1794$ was found to be almost pure. The first account of natural volcanic sal ammoniac in Europe was given, it seems, by J. B. da Porta in $+1589$ (Partington (7), vol. 2, p. 21).

[e] See, for example, Abich (1) on Vesuvius, and Pough (1) on Parícutin in Mexico. Here 'the gas vents in the lava were lined with snowy crystals of ammonium chloride with an occasional seasoning of mustard-coloured iron ammonium chlorides'. A colour photograph is given in fig. 4.

[f] There is a special study of this problem by Ruska (39) well worth reading. As the Atlas of Chang Chhi-Yün (2), vol. 2, map 15, shows, the Thien Shan range is rich in coal deposits everywhere along its length, though especially to the north round the rim of the Dzungarian Basin. From Friederichsen's monograph on Thien Shan morphology (1) we know that there are igneous rocks in the mountains, and one dead volcano just north of Kashgar, but absolutely no volcanic activity. Since coal combustion loci would be likely to vary from time to time it is not surprising that modern travellers such as Regel (1) or v. Lecoq (1, 2) have failed to remark on them. Coal as the source was suggested long ago by Bischof (1), 2nd Germ. ed., vol. 1, pp. 636ff., and indeed by von Humboldt himself. As for ignition, lightning may have played a part as well as camp fires on outcrops, and geologists seem not to exclude the possibility of some kind of spontaneous combustion.

Still, the mountainous country north of Kucha (Khu-chhê[1]) in Chinese Turkestan, especially Pai-shan[2] or Huo-shan[3],[a] was famous century after century for its sal ammoniac caves.[b] So also was Pei-shan[4] mountain on the northern side of the Thien Shan south of Kuldja (I-ning[5]),[c] places on the southern side further east near Turfan,[d] and a field of 'solfataras' near Urumchi (Ti-hua[6]) five miles in circumference.[e] The Altai range to the north across Dzungaria is also said to have produced sal ammoniac,[f] and apparently Khotan (Ho-tien[7]) too, from the slopes of the Khun-lun Shan to the south.[g] It has been widely considered that all the regions of the earth were excelled by Central Asia for sal ammoniac deposits,[h] and this is not unlikely. But if Sinkiang was the main region there were lesser areas of importance both east and west of it.

In China proper, where almost no active volcanoes have existed in historical times, [i] vents in northern Shansi, according to von Humboldt, yielded much sal ammoniac;[j] while in the south of the province slowly-burning unworked coal-seams gave, he knew, another harvest of the salt.[k] Westward of Sinkiang, on the other hand, across

a We give the Chinese names with diffidence since we have no modern maps sufficiently detailed to mark the small places. But references to the neighbourhood of Kucha recur constantly in the Chinese sources, as we shall see; here we may mention only the geographical work *Ta Ming I Thung Chih*,[8] finished in +1461; cf. Bretschneider (2), vol. 2, p. 243. Other accounts are translated and considered in Liu Mao-Tsai (1), pp. 17–18, 160, 171, 238–9.

b See, e.g. von Humboldt (4), vol. 1, pp. 100ff.; Fuchs (1), pp. 271ff.; Ritter (1), vol. 2, pp. 333–7; J. Klaproth (6), vol. 2, pp. 357ff.; von Richthofen (2), vol. 1, p. 560; Timkovsky (1), vol. 1, pp. 389ff.; Keferstein (1), pp. 156–7, probably the authority for Porter Smith (1), p. 190.

c Cf. Ruska (39); Fuchs (1), pp. 271ff.

d Von Humboldt (4), vol. 1, pp. 118ff.; Fuchs, *op. et loc. cit.*; Ritter (1), vol. 2, p. 342. This was the subject of a correspondence between Rémusat (10) and L. Cordier (1) in 1824; the former translated a passage from the *Wakan Sanzai Zue* and believed that it proved the existence of active volcanoes in Sinkiang. Though this is unacceptable now, many Chinese sources justify the old Western statements about the production of the salt in this area, for example, *Hsin Wu Tai Shih*, ch. 74, p. 8b, *Sung Shih*, ch. 490, pp. 11b, 12a, *Ming Shih*, ch. 329, p. 19b, and *Ta Ming I Thung Chih*, cf. Bretschneider (2), vol. 2, p. 193. Individual localities with burning vents in the Turfan depression are Karakhojo (Huo-chou[9]) and Liu-chhêng[10] just east of Turfan city itself; *TMITC*, cf. Bretschneider, *loc. cit.*

e Fuchs (1), pp. 271ff.; Ritter (1), vol. 2 pp. 386–8. Validation comes from *Tu Shih Fang Yü Chi Yao*, ch. 65, p. 51a, b and *Ming Shih*, ch. 329, p. 19b, cf. Bretschneider (2), vol. 2, p. 190 (quoting Chang Khuang-Yeh's account, cf. p. 442 below).

f According to von Humboldt (4), vol. 1, pp. 120, 141. But there is little coal there.

g Von Humboldt (4), vol. 1, pp. 118ff. I could find no confirmation of this in Rémusat (7) however. Also there are no coal deposits near by. But the export is confirmed by *Sung Shih*, ch. 490, p. 7a, and *TMITC*, cf. Chang Hung-Chao (1), p. 221.

h Fuchs (1), pp. 271ff.; Bischof (1); both accepted the volcanic explanation, but the latter suspected coal-seams as well. The Uighur people were in general closely associated with the production, cf. *Hsin Wu Tai Shih*, ch. 74, p. 10a.

i See Anon. (145). The Khun-lun Shan in southern Sinkiang has long had a few active volcanoes.

j (4), vol. 1, p. 213, referring especially to Pao-tê[11] near Ho-chhü,[12] on the Yellow River just south of the Great Wall's crossing. There is much coal in this neighbourhood (Chang Chhi-Yün (2), vol. 5, map 17); and authentication of the report is forthcoming from *TSFY*, ch. 40, p. 19b.

k (4), vol. 1, p. 215. The same is true of Kansu, especially around Linthao, south of Lanchow, according to *TMITC* and *TSFY*, ch. 60, p. 4a. Working out along the Old Silk Road, at least three other places were associated with sal ammoniac, (a) Chhih-chin, between Chiayükuan and Yümen (cf. Vols. 1 and 4, pt. 3), *TMITC*, cf. Bretschneider (2), vol. 2, p. 214, (b) Anhsi city, junction for Tunhuang and Hami, cf. *Hsin Thang Shu*, ch. 40, p. 11a, (c) Tunhuang itself, the Shachow Exarchate), cf. *Sung Shih*, ch. 490, p. 23b. There is coal in these desert regions, but they could have been centres of transmission rather than of production.

¹ 庫車   ² 白山   ³ 火山   ⁴ 北山   ⁵ 伊寧   ⁶ 廸化
⁷ 和闐   ⁸ 大明一統志   ⁹ 火州   ¹⁰ 柳城   ¹¹ 保德
¹² 河曲

the mountain barrier, the whole area between Samarqand in Sogdia and Tashkent in Ferghana has been said to possess sal ammoniac vents;[a] while the region of Bokhara was renowned for the product.[b] The most westerly region of all lies further to the south, in Persian Baluchistan, where the Damindān (now Tamindan) valley in the Kūh-i-Taftān range, a relatively inactive volcanic massif, produces sal ammoniac down to this day.[c] Perhaps this is the only place where the substance was almost certainly volcanic in origin rather than sedimentary. The general upshot is that Central Asia, broadly speaking, was the first source of ammonium chloride for the Arabic alchemists; but whether they obtained knowledge of it directly or from their Chinese contacts might depend in our estimation on how long the chemical had been known in China beforehand, and this must be our next enquiry.[d]

Broadly speaking, it had been familiar there for some three and a half centuries, if not indeed for as many as seven, before the Arabs came to know about its interesting properties. Under the name of *nao sha*[1] (with a very varying orthography)[e] it appears with certainty from about +500 onwards, and it will be worth while giving a brief account of this literature, but first let us look at what may be the oldest references.[f] In the *Tshan Thung Chhi*, datable at +142, we find the following admirable passage:[g]

If the chemical substances used are not of the right sorts, if their categories (*lei*[2]) are not compatible, and if the measuring out of the mixture (of reactants) is at fault, then the natural pattern (*kang chi*[3])[h] will be lost. In such a case, even if Huang Ti were to set up the furnace and Thai-I to work the fire, even if the Eight Adepts were to take charge of the process and the Prince of Huai-Nan to moderate and harmonise it;[i] however impassioned the prayers to the spirits, however splendid the alchemical temple—failure will be inevitable. It would be like mending a cauldron with glue, or bathing a boil with sal ammoniac, or driving away cold with ice, etc. etc.....

The difficulty here is that the word used is *lu*,[4] and while *lu sha*[5] has meant sal ammoniac in medieval and modern times,[j] it is hard to be sure that this held good in

[a] Ruska (39), but the evidence is not very sure. Still, tribute of the salt from Hei-lou,[6] a country believed to be Khorasan, is recorded in *Ming Shih*, ch. 332, p. 22b, for +1453; cf. Bretschneider (2), vol. 2, p. 272. The Buttam Mts. in eastern Ferghana are regarded as an important source (Barthold (2), p. 169).

[b] See Burnes (1), vol. 2, p. 166, 'found in its native state among the hills near Juzzak'. The sal ammoniac of Bokhara was first studied by Model in +1758, and M. H. Klaproth's analysis of it in +1794 showed it to be nearly pure.

[c] See the description of Skrine (1). The region is in Seistan just west of the modern Pakistan border and north of the city of Khāsh. Damindān as an earthly hell (like Gehenna, cf. pt. 2, p. 79 above) has had centuries of renown; for the Iranian *Bundahishn* (93.3) calls it a cavern perpetually smoking, a fountain of *dōzakh* (hell). The poem *Zarārusht-nāma* has: 'may he save me from *dōzakh* and the demons of Damindān' (Rosenberg ed., 1565). We are grateful to Sir Harold Bailey for these references.

[d] Something has already been said about sal ammoniac in Vol. 3, p. 654, but its importance for Arabic–Chinese relations compels us to look at it again here from a somewhat different angle.

[e] We write here one of the commonest, and perhaps the oldest, forms. See further on p. 445 below.

[f] If it was mentioned in the *Pên Ching* (i.e. the *Shen Nung Pên Tshao Ching*) we should have to place Chinese knowledge of it in the −2nd century. But none of the reconstructions of the text justify this. Only occasionally later pharmaceutical natural histories attribute *nao sha*[7] to the *Pên Ching* (e.g. the *Shao-Hsing Chiao-Ting Ching-Shih Chêng-Lei Pei-Chi Pên Tshao* of +1159, Okanishi Tameto ed. tshê 1, ch. 5A); this was presumably a mistake.

[g] Ch. 30, p. 26b, tr. auct., adjuv. Wu & Davis (1), p. 257.

[h] Cf. Vol. 2, pp. 554ff. and Needham (50).

[i] Cf. p. 168 above.   [j] Giles (2), dict.; RP126.

[1] 硇砂   [2] 類   [3] 綱紀   [4] 磠   [5] 磠砂   [6] 黑婁   [7] 碙砂

the +2nd century. But the ancient meaning of sand, shingle or pebbles would make no sense here, while sal ammoniac would, for its refrigerant effect[a] would soothe an inflammation without doing anything to cure it. Next, *nao sha*[1] is not in the *Pao Phu Tzu* book (*c.* +300) as we have it today, but Li Shih-Chen in the *Pên Tshao Kang Mu* gives a quotation which may indicate that the substance was known to Ko Hung, since he makes *Pao Phu Tzu* say that 'there are many ways of subduing sal ammoniac (*fu nao yao*[2])'.[b] Most of these involve calcium salts, so presumably the non-volatile calcium chloride was being made. Li Shih-Chen here also quotes Lei Hsiao[3] to the effect that 'when *nao*[1] meets with "red feathery crystalline mercury" (*chhih hsü hung*[4]) it stays in the metal reaction-vessel (*liu chin ting*[5])'. The reference is an unusual one, but probably to the sulphide or the oxide,[c] though the exact salt does not matter for the chlorides of mercury, copper and tin are evidently being formed. The most likely origin of this is the *Lei Kung Yao Tui*, which is datable perhaps about +450 but certainly not later than +560.

When we reach the lifetime of Thao Hung-Ching we find a man to whom sal ammoniac was almost certainly known. There is no reference to it in what is left of the *Pên Tshao Ching Chi Chu*, but a longish description in the *Ming I Pieh Lu* was carefully copied in most of the pharmaceutical natural histories from the *Hsin Hsiu Pên Tshao* (+659)[d] onwards, through the many different editions of the *Chêng Lei Pên Tshao*,[e] and into later compendia. This would point to +510 as a rather firm date. Soon the character and the definition of the substance makes appearance in the literary dictionaries, such as the *Yü Phien*[6] of +543[f] and the *Chhieh Yün*[7] of +601.[g] At the same time reports begin to accumulate of sal ammoniac brought in tribute and trade from Central Asia, the first perhaps that of the *Wei Shu*[8] in +554 telling of its coming from Sogdia,[h] though Thao Hung-Ching or his disciples had already said that it was produced by the western barbarians (Hsi Jung[9]). In +610 Phei Chü[10] told in his *Hsi Yü Thu Chi*[11] how White Mountain in Kucha (Chiu-tzhu,[12] Pai-shan[13]) was a great source of sal ammoniac,[i] and this was confirmed by the *Sui Shu* in +636,[j] adding, together with the *Pei Shih* (+670),[k] details about Sogdia. The eighth century brings many more references. Chhen Tshang-Chhi in his *Pên Tshao Shih I* (+725) tells us that the Hu[14] people call it *nêng sha*;[15] Wang Thao in the *Wai Thai Pi Yao* (+752) describes its pharmacological properties; Hsiao Ping[16] in *Ssu Shêng Pên Tshao*[17] (*c.* +775)

---

   [a] Cf. p. 434.

   [b] Ch. 11, (p. 59). It must always be remembered that Li Shih-Chen had many writings at his disposal not available to us now.

   [c] There might be some connection here with the legendary Taoist Chhih Hsü Tzu (cf. *PPT/NP*, ch. 15, p. 11*b* and Kaltenmark (2), p. 135), but the simplest interpretation arises from the character of the salt.   [d] Ch. 5, p. 8*b*.

   [e] E.g. *Ta-Kuan Pên Tshao*[18] (+1108), ch. 5, p. 7*a*, *b* (p. 111), and *Chêng Lei Pên Tshao* (+1249), ch. 5, (pp. 125.2, 126.1).   [f] Ch. 22, p. 15*a*.

   [g] Judging from *Kuang Yün*[19] (Chiao Pên), ch. 2, p. 11*b*, and *Chi Yün*,[20] ch. 3, p. 16*a*.

   [h] Ch. 102, p. 27*b*.   [i] See Chang Hung-Chao (1), p. 222.

   [j] Ch. 83, pp. 8*b*, 11*a*.   [k] Ch. 97, p. 26*a*.

[1]	[2] 伏硇藥	[3] 雷斅	[4] 赤鬚汞	[5] 留金鼎
[6] 玉篇	[7] 切韻	[8] 魏書	[9] 西戎	[10] 裴矩
[11] 西域圖記	[12] 龜茲	[13] 白山	[14] 胡	[15] 濃沙
[16] 蕭炳	[17] 四聲本草	[18] 大觀本草	[19] 廣韻	[20] 集韻

was perhaps the first to refer to the rich sal ammoniac production of Pei-thing (mod. Urumchi or Ti-hua[1]) in the Thien Shan, hence the name *pei-thing sha*;[2] and the *Tan Fang Ching Yuan* of about +780 seems to have been the earliest text to introduce the expressive name *chin tsei*,[3] the 'thief of the metals', alluding of course to the propensity of ammonium chloride for attacking them and forming their salts. All this was a long time before the discussions of *nūshādir* in the Jābirian corpus.

It may be good, however, to follow the Chinese evidence through the +9th and +10th centuries a short way. The former opens with the *Shih Yao Erh Ya* defining sal ammoniac with further names (+806);[a] and the latter with two texts of +918, in China the *Pao Tsang Lun* of Chhing Hsia Tzu (cf. pt. 3, p. 180), in Japan the *Honzō Wamyō*.[b] Tuku Thao discusses the salt in his *Tan Fang Chien Yuan* of c. +938, and in +972 the *Jih Hua Chu Chia Pên Tshao* brings a new name, *ti yen*,[4] 'barbarian salt';[c] the geographical work *Thai-Phing Huan Yü Chi* (+980 approximately) confirms the import from Sogdia and Kucha.[d] Lastly the *Hsin Wu Tai Shih* in +1070 emphasises the contribution of sal ammoniac from Turfan and the Uighurs,[e] while the *Pên Tshao Thu Ching* of +1061 adds another name, *chhi sha*[5] or 'pneumatic salt', clearly derived from the property of volatilising completely without decrepitation.[f] This is of course far from being the end of the Chinese literature on sal ammoniac, but we have said enough to prove the point that it flourished long before that of the Arabs, hence the strong probability that this was the line of transmission.

A rather impressive amount of knowledge about ammonium chloride accumulated in China during those early centuries. The endothermic effect occurring on solution seems to have been known very early,[g] and also the formation of ammonium carbonate by exchange with lime salts. The reduction of oxides on metal surfaces to the chlorides and the consequent cleaning effect was seized upon by the metal-workers, and as an outstanding flux ammonium chloride 'can be used', says the *Ming I Pieh Lu* text, 'in soldering (*kho wei han*[6])'. This function is particularly prominent in the many processes described in the *Thai-Chhing Tan Ching Yao Chüeh* of Sun Ssu-Mo,[h] datable about +640. But the fact that under suitable conditions ammonium chloride will attack all the base metals and silver much more strongly was also known to Thao Hung-Ching or his followers: 'it softens gold and silver (*jou chin yin*[7])' says the *Ming I Pieh Lu*;[i] 'it can dissolve the five metals and the eight minerals, and it rots the guts of man', says the *Yao Hsing Lun* some fifty years later. Hence the name 'thief of the five metals (*wu chin tsei*[8])' found in the *Pao Tsang Lun*. But both this book and another

---

[a] Ch. 1, p. 2a. E.g. *niu sha*,[9] perhaps by phonetic corruption. In some, the adjective 'yellow' occurs, probably because the salt was sometimes mixed with the mustard-yellow iron ammonium chlorides.

[b] Ch. 1, p. 11a. As usual, several different ways of writing *nao* are given.

[c] Cit. *CLPT*, loc. cit.   [d] Ch. 183, p. 3b and ch. 181, p. 5b.

[e] Ch. 74, pp. 8b, 10a.   [f] *CLPT*, loc. cit., *PTKM*, ch. 11, (p. 58).   [g] Cf. pp. 434ff. above.

[h] Pp. 20b, 21a, b, 23b, 24a, b, 25a, 26b, Sivin (1) tr., pp. 194, 196–7, 201, 203–4, 207–8, 283. *Huang nao sha* (yellow sal ammoniac) comes on p. 22b (Sivin tr., pp. 198, cf. 279); if this was not a cover-name for sulphur (as in *SYEY*) it would be the mixed iron ammonium chlorides.

[i] In *HHPT* and *CLPT*, locc. citt. So also the *Yao Hsing Lun*, preserved by the *Chia-Yu Pên Tshao* of +1057.

[1] 廸化   [2] 北庭砂   [3] 金賊   [4] 狄鹽   [5] 氣砂   [6] 可爲銲
[7] 柔金銀   [8] 五金賊   [9] 狃砂

+10th-century work, Tuku Thao's *Tan Fang Ching Yuan*, warn how poisonous the metal chlorides are if taken internally; the resulting powders (*hui shuang*[1]) lead to grave illness, with boils and ulcers as in elixir poisoning.[a] By +1116 the *Pên Tshao Yen I* tells of a kind of cupellation process for distinguishing between true and false gold and silver: 'these can be detected using sal ammoniac, for if thrown in when melted, the false metal is all dissolved and dissipated (*wei wu chin hsiao san*[2])' as, of course, the chlorides.[b] Much is said in the literature about purification of the salt, generally solution and filtration or decantation (*shui fei*[3]) followed by three successive sublimations (*fei*[4]). As for its medical uses, the diaphoretic action and the stimulation of the central nervous system are referred to, but the most striking effect noted was expectorant, the relief of coughing in bronchial and other catarrhs by the secretion of thinner and less tenacious mucus. The *Wai Thai Pi Yao* even applies this to the removal of fish-bones lodged in the throat. Otherwise we hear of the healing of scorpion stings[c] and the curing of certain eye-diseases by ammoniacal vapours,[d] while one strange relation, that of Su Sung and his collaborators in the mid +11th century, avers that some of the men of Central Asia use sal ammoniac for the pickling or curing of meat.[e] Perhaps they did.

And so we come back to the Chinese accounts of the way in which these Turkic and Tartar peoples collected the salt. The oldest dates from between +981 and +984, during which years a Chinese envoy, Chang Khuang-Yeh,[5] was in those parts. After his return he wrote an account of his travels entitled *Hsing Chhêng Chi*,[6] in which the following words occur:[f]

In the Pei-thing Shan[7] (mountains) north of the prince's palace (i.e. near Urumchi) sal ammoniac (*nao sha*[8, 9]) is produced. There are places in these hills where smoke and vapours sometimes issue forth even on the clearest days. In the evening light, the flames look like torches burning, and shed a ruddy glow even on birds and rats. Those who collect (this salt round the openings) put on wooden shoes to gather it, for otherwise the soles of their feet would be scorched...

Another account of what must have been the burning coal-seams near Urumchi is found in the *Hsi Yü Chiu Wên*[10],[g] probably from the *Hsi Yü Wên Chien Lu*[11] of +1777.

Near Urumchi thirty li west of Pu-la-kho-thai, there's a place more than 100 li around with ashy dust that flies in the air. At the centre there are flames springing up. If you throw in a stone you get a sudden issue of black smoke which takes a long time to die down. In winter

---

[a] See on this subject Sect. 45 below, and meanwhile Ho Ping-Yü & Needham (4).
[b] On cupellation methods see pt. 2, pp. 36 ff. above.
[c] Cf. Wang Chia-Yin (*1*), p. 54.   [d] This is the *Pên Tshao Yen I* of Khou Tsung-Shih again.
[e] Could there have been a confusion here with saltpetre or even with borax?
[f] This passage was very often reproduced, as for example in *Hsi Chhi Tshung Hua* (c. +1150), ch. 2, p. 34b, in *Sung Shih*, ch. 490, pp. 11b, 12a, and in *PTKM*, ch. 11, (p. 58). Tr. auct. It was known to Ritter (*1*), vol. 2, p. 342, and to Schott (2). It goes on to speak about a bluish-green mud which comes out with the salt, and itself turns to a granular salt, but this is probably only a reference to the colour of the surrounding earth. The local people use the sal ammoniac for working leather.
[g] P. 14a (p. 227), in *Chou Chhê So Chih* coll., tr. auct. This passage was known to von Humboldt (4), vol. 1, pp. 100 ff.

[1] 灰霜	[2] 僞物盡消散	[3] 水飛	[4] 飛	[5] 張匡鄴
[6] 行程記	[7] 北庭山	[8] 硇砂	[9] 磠砂	[10] 西域曹聞
[11] 西域聞見錄				

when the snow can lie ten feet deep in the neighbourhood, only this place has no snow. They call it Huo-yen[1] (Blazefield).[a] Even the birds don't dare to fly over it.

In a separate place, the same geographer, the 71-year old Mr Chhun-Yuan,[2] spoke of the sal ammoniac industry in the hills north of Kucha:[b]

Nao sha[3] is produced in the mountains of that name which lie to the north of Kucha city. In spring, summer and autumn, the many caves there are full of fire. From a distance at night they look like thousands of lamps, so bright that it is hard to go near them. In winter here the cold is extreme, sometimes with heavy snow, and at this season the fires die down. Then the local people go there to collect the sal ammoniac, entering the caves naked (because of the heat). The nao sha accumulates inside the caves like stalactite drippings.

The fullest description occurs very late, from the last decade of the +18th century. In the Chu Yeh Thing Tsa Chi[4] (Miscellaneous Records of the Bamboo-Leaf Pavilion), Yao Yuan-Chih[5] wrote as follows:[c]

According to Hsü Hsing-Po,[6] the mountains where sal ammoniac is produced near Kucha have no special name now but were called in Thang times the 'Great Magpie Mountains'. They have extremely hot places which look from afar at night like so many lanterns. No one dares go near them in spring or summer. Even in very cold weather, the people take off their ordinary clothes, and wear leather bags with holes through which they can see. They enter the caves to dig up (the sal ammoniac), but come out after one or two hours and could not possibly stay longer than three; even then the leather bag is scorching hot. The nao sha sparkles on the ground with a reddish glow; they collect it and bring it out mixed with lumps of rock (on which it was deposited), and for every dozen pounds of rock spoil they do not get more than one or two tenths of an ounce of the salt. The product has to be kept in earthenware jars holding more than a picul, with their mouths tightly sealed, yet not full, and kept cool, otherwise it will all disappear. It will also disappear if subjected to wind, wetness or damp, leaving only an unchanging white residue of granular appearance. Though this is the least valuable part, it is probably the only kind which finds its way to the central provinces of China.[d] I suspect that these fire-mountains of Kucha are only one area of a whole ancient region of like activity.

Similar descriptions are to be found in Arabic authors, also from the +10th century onwards.[e]

Much—perhaps too much—has now been said of the diffusion of the wonderful new volatile salt in East and West, but the fascinating problem of the name is still unanswered: what has nao sha got to do with nūshādir and nūshādur? Stapleton, rushing in where angels feared to tread, affirmed that the oldest term was the Chinese

[a] Cf. TSFYCY, ch. 65, p. 51a, b.
[b] Hsi Yü Wên Chien Lu, in Hsi Yü Chiu Wên, p. 20b (p. 240), in Chou Chhê So Chih coll., tr. auct., corr. Vol. 3, p. 655. The passage was known to many earlier Western writers, notably von Richthofen, (2), vol. 1, p. 560; Ritter (1), vol. 2, pp. 333–7; von Humboldt (4), vol. 1, pp. 100ff.; Timkovsky (1), vol. 1, pp. 389ff.
[c] Cit. Chang Hung-Chao (1), p. 222, tr. auct., corr. Vol. 3, p. 655.
[d] This was much too pessimistic, for when Hanbury (6) analysed several samples sent from Peking in 1865 by William Lockhart, he found that two out of three were rather pure.
[e] See Ruska (39); Ouseley (1), p. 233; de Meynard & de Courteille (1), vol. 1, p. 347.

[1] 火燄 [2] 椿園七十一老人 [3] 硇砂 [4] 竹葉亭雜記 [5] 姚元之
[6] 徐星伯

and that it had given rise to a Persian form with the ending *dārū*, drug or medicine.[a] Laufer, a dozen years later, dismissed this suggestion as that of a chemist, not a philologist, and felt compelled to seek the original name in Sogdia (one of the homes of the product), since he regarded the Chinese characters as phonetic transcriptions of some foreign word and not an indigenous coinage.[b] He therefore suggested Sogdian forms such as **navša* or **nafša*. Other scholars since then have proposed various alternatives, such as Schafer's Iranian **njau-ṣa*,[c] but all of them remain purely conjectural. However, there is a real word in Sogdian, *nwš' 'tr* (= **naušātur*), which occurs, in the only instance so far found, as the name of one of the constituents in a magical-pharmaceutical recipe dating from about +600. Since it is described as broken or pounded, it is presumably our salt.[d] In Iranian the word is interpreted as derived from *anōsh*, immortal, and *ātur*, fire; but this could be a 'popular' explanation of an essentially foreign term.[e]

Although Laufer did not say it in so many words, he for his part was clearly suspicious of the autochthonous origin of *nao sha* as a name because there were so many ways of writing it. And indeed it is true that the orthography of the phrase shows a singularly wide range of alternatives. One must admit however that there are other instances of a similar phenomenon, such as the famous hundred ways of writing the character *shou*,[1] longevity, for which it would be quixotic to claim an implication of foreignness.[f] Establishing the facts here is distinctly difficult because successive reprintings may have followed the fashion of their time rather than the script of the original text, and quotations in later books cannot be taken as firm evidence for the original way of writing; yet we can also rely to some extent on the deeply-rooted conservatism of the Chinese literati. Inscriptions would of course be the best evidence, but as they do not often talk about drugs and chemicals such research would be very difficult, perhaps impossible, and in any case has not yet been done. Therefore what we shall say here is subject to all necessary reserves. One further point at the outset— there is no semantic distinction between the two forms of *sha*,[2, 3] 'sand' or granular crystalline chemical; they were used in *nao sha* indiscriminately.

With one important exception to which we shall return in a moment, there were in pre-Thang times four different ways of writing the character for sal ammoniac, *nao*.[g]

---

[a] (1), pp. 40–1. Arabic etymologists, said Stapleton, had suggested *nūsh dārū*, 'life-giving medicine' (not a bad name for smelling-salts), but he preferred to see *nao sha* as the origin of the *n-sh* part of the word. Earlier de Mély (6), p. 339, (1), p. li, had had the same thought, but could present no theory about the last syllable.

[b] (1), pp. 503 ff.   [c] (13), p. 218.

[d] Paris Sogd. text 3.173, tr. Benveniste (2). We owe our knowledge of this to the kindness of Sir Harold Bailey, who advised us on the probable date. The recipe also includes camphor, sandal-wood, costus and musk, so it was probably fumigatory in nature.

[e] Bailey, priv. comm. Ruska (13), p. 7, thought of *nōsh aḍar*, 'drinking fire', as just a possibility, but the same impression would apply.

[f] Doubtless this abundance derived from the caprices of decorative artists, yet there are very many characters which can quite correctly be written in two or three or even up to a dozen different ways. One of the real difficulties for beginners in classical Chinese is to know when one or two strokes 'make all the difference', and when they do not.

[g] *Nao*[4, 5, 6, 7]. The first of these is regarded as the primary form. The second is supposed to derive from Rad. no. 122, *wang*,[8] net. The phonetic in the third is *kang*,[9] a ridge, and would be expected to

[1] 壽   [2] 砂   [3] 沙   [4] 硇   [5] 硇   [6] 碙   [7] 砳   [8] 罔   [9] 岡

During the Thang period (+7th to +10th centuries) nine more were added, some purely phonetic with no visually relevant component in the character at all.[a] In Wu Tai and Sung times we find seven more;[b] and finally by the +15th century one last one, generally admitted to be 'incorrect', brought up the rear.[c] No wonder that Laufer, sensing this multitude of forms, believed that the original name must have been a foreign one, and that the Chinese could never make up their minds how to write it.

But all those who have worried over this problem hitherto have reckoned without the form *lu sha*[1] which we met with in the oldest reference of all, the *Tshan Thung Chhi*,[d] and this can no longer be overlooked. It may well be correct there, and no corruption. Yet the considered opinion of Chang Hung-Chao[e] was that all the forms of *nao* derived from *nao*.[2] Here the right-hand phonetic component was an old word pronounced *hsin*, and defined in the *Shuo Wên* (+121)[f] as the fontanelles of the skull where the cranial bones meet.[g] The ancient form of the character given by Hsü Shen here is shown in (*a*) in the inset cut. Chang suggested that this character was borrowed

(a)   *(b)*   *(c)*

as a substitute for *nao*,[3,4] the brain, to give the sound in *nao*.[5] He did not say why this substitution took place, but one could imagine without difficulty that the more complex phonetic was already occupied, in use for agate, *ma nao*.[6,7] Nor did he say why some connection with the brain was wanted, but here Li Shih-Chen comes to the

---

lead to the pronunciation *kang* or *khang*, but in all these cases where guidance is given by commentaries they indicate the sound *nao*. So also for the last, though derived presumably from *ka*[8] or *ko*[8], to beg.

[a] *Nao*.[9,10,11,12,13,14,15,16,17] The first four of these were close to the primary form. The fifth, which was favoured in Japan, has special significance for the argument we are developing. The expected pronunciation of the sixth and seventh would be *nêng* or *nung*, but *nao* is always indicated; as we noted on p. 440 above, one of our medieval authorities, Chhen Tshang-Chhi in +725, regarded this as an attempt to transliterate some name of *hu*[18] (Persian, Turkic, Sogdian) linguistic origin. *Niu*, the eighth, anciently meant a small gong, and *niu*, the ninth, perverse or evil; though here pronounced *nao* they had no visible connection with minerals or even water.

[b] *Nao*.[19,20,21,22,23,24,25] The first three of these were fairly close to the primary form, and the fourth not very far away from it, though obviously borrowed from the old word for a vent, flue or impluvium, *chhuang*[26] (cf. Vol. 4, pt. 3, p. 121). The fifth and sixth come again from the *wang*[27] radical (122), while the seventh brought in yet another theme, that of *chiung*,[28] waste border land (Rad. no. 13), generating *kung*,[29] the bright light on it.

[c] *Nao*.[30] This was explicitly rejected in a note at the end of the entry for sal ammoniac in the later editions of *PTKM*, ch. 11, (p. 61).

[d] And as mentioned in Vol. 3, p. 654, we have checked this in the *Tao Tsang* text as well as several other editions.   [e] (*1*), pp. 221 ff.; and the Morohashi dict. too, viii, 359.

[f] P. 216.2.

[g] This character has never died out, though its pronunciation has changed slightly, and one finds it in the current colloquial expressions for fontanelles, *hsing mên*,[31] and skull-cap, *hsing mao*.[32] In the course of time it came to be written in a more complicated way as *hsing*[33] or *hsing*,[34] and this assimilated naturally to the *tshung*[35] phonetic (meaning 'hurried') as *hsing*.[36] In this form we find it in the title of an interesting anatomical-medical tractate on the fontanelles dating from the end of the Thang or the beginning of the Sung, the *Lu Hsing Ching*.[37]

[1] 磠砂	[2] 硇	[3] 瑙	[4] 腦	[5] 硇	[6] 瑪瑙	[7] 瑪磠
[8] 㲹	[9] 硇	[10] 匘	[11] 硇	[12] 淎	[13] 硇	[14] 磄
[15] 濃	[16] 鐃	[17] 狃	[18] 胡	[19] 匘	[20] 硇	[21] 淎
[22] 硇	[23] 𥖨	[24] 硐	[25] 硐	[26] 囱	[27] 网	[28] 冂
[29] 囧	[30] 硇	[31] 囟門	[32] 囟帽	[33] 顖	[34] 顋	[35] 悤
[36] 顖	[37] 顱顖經					

rescue, saying that if much sal ammoniac is taken the brain is disturbed or irritated, hence the name.[a] This might be dismissed as a late 'fanciful etymology', but again it might not, for the effect of this and other simple salts upon the blood alkalinity and hence upon mental processes, is no fable.[b] Furthermore, what Chang Hung-Chao did not notice was that the *Shuo Wên* goes on to say that the old way of writing *hsin*[1] was *hsin*,[2] i.e. something extremely similar to *lu*.[3]

Chang was inclined to write off all instances of the use of this last word as corruptions of *nao*[4] but he had overlooked the appearance of it in the *Tshan Thung Chhi*. Although lexicographers in general tended to define it as just sand or gravel as such (without specifying what kind of sand),[c] it keeps on cropping up in situations where it can only mean sal ammoniac. These texts differ widely in date; one might mention as examples the *Thai-Phing Huan Yü Chi* of *c.* +980 (cf. p. 441 above) and the *Chu Yeh Thing Tsa Chi* at the end of the +18th century.[d] If we look into the origins of *lu*[5] itself, the *Shuo Wên* tells us that it meant 'salty soils in the West',[e] an interesting definition in the present context; adding the seal form shown in (*b*) in the inset cut. The form of the character in Chou inscriptions is simply as we see in (*c*) of the inset cut,[f] an interesting graph because it does not have the 'signal' at the top which characterises the *lu*[5] (salt) radical (no. 198) and connects it with the *pu*[6] (divination) radical (no. 25). Then in the +16th century Wei Hsiao gave, as another ancient alternative, the form *lu*,[7] with the dots but without the 'signal'.[g] And finally Tsêng Hsi-Shu also says[h] that in the Shih Chou[8] script[i] the character was simplified in yet

---

[a] *PTKM*, ch. 11, (p. 58): *Nao sha hsing tu, fu chih shih jen nao luan, ku yüeh nao sha.*[9] The second of these three *nao* is obviously intended to be read *nao*[10] (= *nao*[11]), disturbed, or more likely *nao*,[12] the brain, itself.

[b] The late J. B. S. Haldane, my predecessor as Sir Wm. Dunn Reader in Biochemistry at Cambridge, made many experiments of this kind on himself in his classical studies of induced acidosis between 1920 and 1930. I remember once meeting him in a befuddled state on the staircase of the Institute, and on offering to help was told: 'I shall be all right in an hour or two; at present I'm about 80 % sodium haldanate.' Haldane began by breathing high CO_2 concentrations and ingesting sodium bicarbonate; see Davies, Haldane & Kennaway (1). Apart from many results of physiological interest, he devoted a special paper (3) to the mild but peculiar hallucinations resulting. He then went on to take as much as 55 gms. of ammonium chloride at one time, obtaining a marked and prolonged acidosis also associated with neuropsychological abnormalities; see Haldane (2); Baird, Douglas, Haldane & Priestley (1); Haldane, Wigglesworth & Woodrow (1, 2); Haldane, Linder, Hilton & Fraser (1). The condition was regularly termed an 'intoxication', accompanied by 'air hunger' and physical exhaustion. Ammonium chloride is still used, though not on Haldane's heroic scale, for the purpose of producing experimental acidosis, as for example when it is desired to establish a low urinary pH in studying the excretion of a basic drug; or clinically to sterilise the urinary tract after a urethral infection. We are grateful to another former collaborator of J. B. S. Haldane's, Dr Martin Case, for some of this information.

[c] E.g. the +11th-century *Chi Yün*, ch. 5, p. 18*b*. Cf. Morohashi dict., viii, 397.

[d] Both Giles (2) in his dictionary and Read & Pak (1) in their glossary admit it as in wide nineteenth-century use.

[e] P. 247.1, *hsi fang hsien ti yeh.*[13] The word ultimately came to mean rock-salt.

[f] Cf. K 71*b*, and Tsêng Hsi-Shu (1), Hai chi, hsia, p. 17*a*.

[g] In the *Liu Shu Ching Yün*[14] (Collected Essentials of the Six Scripts) by Wei Hsiao[15] (+1483 to +1543). See Tsêng Hsi-Shu, *loc. cit.*

[h] (1), *loc. cit.*

[i] Also of the Chou, supposedly early in the Chhun Chhiu period. On Shih Chou and his *Phien*, see Vol. 6, Sect. 38.

[1] 囟	[2] 𠧚	[3] 硇	[4] 硇	[5] 鹵	[6] 卜
[7] 𡆙	[8] 史籀	[9] 硇砂性毒服之使人硇亂故曰硇砂			[10] 㨔
[11] 撋	[12] 腦	[13] 西方鹹地也	[14] 六書精蘊		[15] 魏校

another way, with all the dots left out, i.e. *lu*,[1] approximating again to *hsin*.[2] Thus in the end there was very little differentiation between the *lu* derivatives and the *hsin* (and *nao*) derivatives. There is no reason to talk of corruption; people borrowed because they had a need to do so.

What does all this add up to? Simply the suggestion that ammonium chloride was after all first known and studied in China before the end of the Han. One remembers the evidence that the north-western provinces produced it; it did not have to be imported from abroad. If so, and if Ko Hung about +300 also spoke of *lu*[3] rather than *nao*,[4] perhaps that was the original technical term. It would not have been defined with modern precision. But this very special medicinal salty 'sand', which could affect men's minds like strong drink, needed to be distinguished from sand in general, so at least by the time of Thao Hung-Ching (*c.* +500) a graph very similar to it but without the dots was borrowed from a word meaning the brain-cover. This brought with it the sound of the character for brain as such (*nao*), which itself embodied graphically the essential square component with its diagonal cross.[a] One could say that this was punning, but that would be neither bar nor criticism, for puns have often played a part in Chinese ideographic development. Later on, as the knowledge of the salt spread westwards, it would have been very natural for the Sogdians to speak of *naušātur* by the +7th century, though whether their penultimate consonant could have derived from *tu*[5] or *thu*,[6, 7] or some other Chinese word, we could not presume to say.[b] Thus an attentive study of the nomenclature and its history, so far as we can make it on the evidence we have, justifies Stapleton rather than Laufer and points to China as the true native land of sal ammoniac.

From Sogdian to Arabic and to *nūshādur* was no great step, and little need be said about it,[c] but we can hardly escape without a final word on the transference of the name and the knowledge to the European West. Into the languages of that region *nūshādur* entered but did not permanently stay, being soon replaced by sal ammoniac, a reincarnation of the ancient 'ammoniac salt' with a totally different meaning. Ruska saw this development as having taken place in two stages.[d] First, when Bar Bahlul was writing his Syriac encyclopaedia about +980, his Nestorian medical colleagues were so convinced of the greatness of Galen and Dioscorides that they imagined that all the drugs of Persia must have been known to the Greeks, so they looked for equivalents and equations and naturally found them. That was why Bar Bahlul wrote: '*Armōnīqōn*, i.e. the salt *anūshādōr* or *nūshādīr*', as well as two other similar entries. After that the way was open for the Latin translators of the +12th century to replace words like *almizadir*[e] or *alnuzadir*, which at first they tended to use, by sal ammoniacum.[f] Ruska thought that it was probably the Jewish physicians, as

---

[a] A certain abbreviation or simplification took place at the same time.
[b] Poison (active principle), earth, and 'to smear on' or 'bathe in' respectively.
[c] Late Skr. *navasara* or *navasadara* show the spread into Indian culture.
[d] (13), pp. 18ff.
[e] This occurs in the *De Compositione Alchemiae* (+12th century); cf. p. 403.
[f] See e.g. the *Liber Sacerdotum*, a translation from the Arabic done in the late +12th or early +13th century, Berthelot (10), pp. 81ff., 179ff., tr. 187ff., 209ff.; 'Almiçadir, id est sal ammoniacum', p. 217.

[1] 卤   [2] 囟   [3] 磠   [4] 硇   [5] 毒   [6] 土   [7] 塗

much at home in Spain as in Syria, who affirmed and authorised this change of terminology. The Byzantine Greeks adapted another word, *tzaparicon* (τζαπαρικόν),[a] but it had no future because the further development of alchemy lay in Latin rather than Greek; and even in Russian, *nashatyr*, taken directly from Arabic sources, remained permanently in possession. Thus by the time of the Geberian Latin writings[b] the term sal ammoniac had become fully accepted in its modern meaning; and with a mention of the shortened form, *salmiak*, current in German from *c.* +1600 onwards, we can conclude the whole story.[c] What it all seems to show is that the knowledge of this volatile salt with striking chemical properties, capable of liberating a rather alkaline gas (ammonia) and a strong acid (hydrochloric),[d] started in China about the beginning of the +1st millennium, then spread across the length and breadth of the Old World, reaching Western Asia by the +9th century and the Latin West by the +12th. If we have expatiated upon it rather fully that is because its peregrinations constitute a model for those of the elixir idea itself.

If time, space and patience permitted, much more could be said about the material influences of China on the alchemy and natural history of the Arabs, but we must confine ourselves to a few particularly interesting examples. An early source of value is the *Kitāb al-Aḥjār* or Lapidary, falsely ascribed to Aristotle, and done into Arabic from the Syriac some time during the first few decades of the +9th century by Luka

Cf. von Lippmann (1), p. 484. The same form occurs in the translation from the Jābirian Corpus, *Liber Septuaginta*, cf. Berthelot (10), p. 327. But a whole chapter is devoted to sal armoniacus in *De Anima in Arte Alchemiae* (cf. p. 403 above).

---

[a] This was derived from one of the Arabic cover-names (Ruska (13), p. 6). For examples, see *Corp. Alchem. Gr.* VI, xvi, 11, a fragment of Cosmas, +11th century; also V, i, 5, another late text; also the +13th-century Mount Athos icon-painters' treatise; Partington (18). In other similar texts, however, *halas ammōniacon* (ἅλας ἀμμωνιακόν) seems to have its new meaning of sal ammoniac, not its old one (*Corp.* V, xxi, 1). Von Lippmann (1) was self-contradictory here; at one moment he said that 'the Byzantine Greeks were hardly acquainted with sal ammoniac before the +13th century' (p. 107), but at another he opined that 'the Arabs got to know of sal ammoniac from the late Alexandrian chemists' (pp. 392, 398). The former dating was too late, the latter statement (perhaps due to a mistake of al-Jāḥiẓ) was the fallacy of Bar Bahlul, the idea that the Greeks knew everything.

[b] *De Invent. Veritatis* and *De Investigat. Perfectionis*, cf. p. 391 above.

[c] On the developments following, reference may again be made to Multhauf (5), pp. 333ff., (8). In tribute to friends who have passed away I should like to mention here a long and instructive correspondence on the sal ammoniac problem which took place between Dr H. E. Stapleton, Dr Dorothea Singer, Prof. H. H. Dubs, Prof. Gustav Haloun and myself in 1950–1.

[d] Sal ammoniac deserves truly to be regarded as one of those 'seminal' substances or 'limiting factors' on which the most fundamental chemical advances depended, alongside saltpetre (potassium nitrate) and copperas (green vitriol, ferrous sulphate), which we discussed in this light at an earlier place (pp. 195ff.). For the +12th-century *De Aluminibus et Salibus* (chs. 11, 81), deriving from al-Rāzī (cf. Ruska, 21), used sal ammoniac to prepare corrosive sublimate from mercury (cf. pt. 3, pp. 123ff. above). 'This preparation', wrote Multhauf (5), pp. 162–3, 'marks the beginning of the most significant period in the history of the science of matter between ancient times and the organisation of chemistry in the +18th century.' For thus began 'the systematic pursuit of synthetic chemistry', the chloride of mercury being a reactive substance capable of chlorinating other substances, as the writer of the *De Aluminibus* knew. And in the following (+13th) century, when the Geberian practitioners added sal ammoniac to the mixture of saltpetre, alum and copperas which gave them on distillation nitric acid, they got the strongest acid of all, *aqua regia* (Sherwood Taylor (4), pp. 90–1). This making of nitric and hydrochloric acids could thus be seen as a development from the earlier chlorination of mercury—and all depended on Thao Hung-Ching's *nao sha* and *hsiao shih*.

bar Serapion.[a] This was afterwards the basis for an enlargement made by Zakariyā' ibn Maḥmūd al-Qazwīnī and incorporated in his 'Cosmography' about +1275.[b] The original writer seems to have been some Syrian who knew both the Greek and the Eastern traditional lore about gems and minerals but depended much more on the latter than on Theophrastus; he used many Persian words, and the places of origin of his stones were very often Persia, Khorasan, India or China. For example, a kind of sand called *sunbādhaj*, which could be used for polishing, and if ground very fine as tooth-powder, came from the islands in the Chinese seas.[c] Similarly China was known as one of the principal sources of onyx (*jaz'*),[d] a precious stone equated by Laufer with caerulean jade (*pi yü*[1]).[e] But the most amusing item—almost too amusing —was the Chinese stone *al-bāhit*, which drove people mad when they saw it, and made them laugh themselves to death.[f] In some versions this was connected with the Alexander-Romance,[g] because it was said that the world-conqueror built a city wall of it without any city inside, so that all those who approached, being attracted as if by a magnet for men, climbed up, fell in laughing and were never seen again.[h] There must have been a lot of interest in the laughing-mad stone, for very soon after Pseudo-Aristotle a whole tractate in the Jābirian Corpus, the *Kitāb al-Bāhit* (Book of the Surprising) was consecrated to it.[i] A century later it was prominent again in the *Kitāb al-Sumūm* (Book of Poisons) attributed to Ibn al-Waḥshīya,[j] and we find it too in the Latin *Picatrix* (cf. pp. 313, 427–8 above),[k] the Arabic source of which would have been finished by +1056. It may well be that something very solid lay behind these mythological ghosts and shadows, for there was indeed in China a hallucinogenic mushroom which leads to uncontrollable laughter and may well kill in excessive doses.[l] But one could hardly say the same for another queer stone that came from China, *khuṣyat iblīs*, 'devil's testicles', good though they were said to be for defending travellers against all danger of brigands.[m]

The shape of these things might suggest that they were of fossil origin. Sometimes

[a] I.e. Lūqā ibn Sarāfyūn, cf. Mieli (1), 2nd ed., pp. 69, 71. The best study and translation of this work is by Ruska (19), who thought (p. 46) that it might have been drafted by Ḥunain ibn Isḥāq (d. +877) himself.

[b] This was studied and translated by Ruska (24).

[c] Ruska (19), p. 150, (24), p. 26.   [d] Ruska (19), p. 145, (24), p. 12.

[e] (1), p. 554, (13), p. 52.

[f] Ruska (19), pp. 8ff., (24); Kraus (3), p. 74.

[g] Cf. Vol. 4, pt. 3, pp. 56–7, 674, cf. pt. 2, p. 572.

[h] This is most fully developed by al-Qazwīnī, who has two entries for the stone. He also has a lot to say about magnets (*lāqiṭ*) for almost everything—gold, silver, lead, hair, wool, nails, bones, cotton, whetstone and brass; Ruska (19), pp. 16, 155ff. This may possibly have been a Chinese idea (see Vol. 4, pt. 1, p. 235), but so far we have not encountered any text which would support that.

[i] Kr 118, see Kraus (2), p. 39, (3), pp. 74–5. The *bāhit* stone is also discussed in the *Kitāb al-Khawāṣṣ al-Kabīr* (Greater Book of Properties), Kr 1900.

[j] Levey (6), pp. 11, 27, cf. (8). There is always uncertainty about the authenticity of this writer; according to Kraus (2), p. lix, the book should be placed about +980 rather than +930, and attributed to Ibn Aḥmad al-Zayyāt.

[k] Ritter & Plessner (1), vol. 2, p. 405.

[l] Cf. pt. 2, p. 121 above, and more fully in Sect. 45 below. *Chün Phu*, p. 3a.

[m] See Ruska (24), p. 21; Laufer (1), p. 554.

[1] 碧玉

29

other accounts in these books put us in mind of Chinese originals, though the country the stones come from is not necessarily named. For example:[a]

Mirād. A wonderful stone. Aristotle says: 'It is found in the lands of the south. If it is taken from the earth when the sun is in the south its nature is hot and dry, but if the sun is in the north then its nature is cold and wet. It is red in colour in the former case, but green in the latter.' In Greek it is called *sarūṭāṭīs*, which is to say 'flying stone'; and that is because this stone is formed in the air from fine dust or mist that rises from the earth, stirred up by the wind, so that it is driven here and there, going round in circles. In the air its colour is greenish-black, like the colour of indigo used by the dyers. If the wind blows harder, the movements of these stones become wilder, but when the sun goes down it stops so that many of them fall to the ground and can be picked up; thus they rise into the air and fall again. If anyone takes one of these stones about with him, the *shaiṭān* (satans, fallen angels) follow him and teach him everything that he wishes to know, if he allows himself to be taught by them.

This reminds one of nothing so much as the Chinese 'stone-swallows' story, i.e. the fossil brachiopods which we met with in an earlier volume,[b] their flight in the air disproved by the sceptical scholar Tu Wan in +1133. Elixir stones turning copper or mercury to silver, and silver to gold, are not unexpected in these books,[c] but it is curious to find that *shādanaj* (haematite) comes in two sorts, male and female (Yang and Yin), presumably the red and the brown.[d]

Chinese kaolin was also known to the Arabs, imported and used by them, not for the making of porcelain, which nobody outside China could accomplish until the +18th century, but for medicinal purposes, since it acts as a valuable antacid and adsorbent in the gastro-intestinal tract, besides coating the stomach wall and preventing ulceration.[e] Hence it is interesting that the *Fihrist* lists a book, presumably of the +9th or the early +10th century, entitled *Kitāb al-Suyūb wa'l-Ma'jūnāt wa'l-Ghaḍār al-Ṣīnī* (Book on Ores, Electuaries and Chinese Clay).[f] Its author was Ja'far ibn al-Ḥusain, a worthy nothing to do with the Sixth Imām.[g] Other cases there are where we listen to Arabic alchemists talking about Chinese substances no longer identifiable. For example, 'Chinese salt', which they say was extremely hard to get, remains for us incognito.[h] And Bar Bahlul was in a morass of confusion about 'Claudianos—the Chelidonium of China';[i] he was not sure whether it was the old multiple alloy of the Alexandrians[j] or a salt of arsenic or copper, or even a plant root or an animal product. All we can tell is that substances of Chinese origin were being used in alchemy in his time in Syria.

  a Ruska (19), p. 88, (24), p. 36, eng. auct. This is in al-Qazwīnī but not in Pseudo-Aristotle, as we now have it.
  b Vol. 3, pp. 614–15. The present entry concludes with an interesting echo of Enoch and the fallen angels, especially the Slavonic version, on which cf. p. 343 above.
  c Ruska (19), p. 157, (24), pp. 17, 21, 30, 33–4, etc.
  d Ruska (24), pp. 26–7. On sex in minerals see also p. 363 above.
  e See e.g. Clark (1), pp. 340, 343, 573.
  f The Persians called it *khāk-i-chīnī* (Laufer (1), p. 556). Cf. Dodge (1), vol. 2, pp. 743–4.
  g See Ruska (5), pp. 8–9. Wiedemann (31), p. 6, gave a wrong translation of the title.
  h Berthelot & Duval (1), p. 146; Stapleton, Azo & Husain (1), p. 375—references in al-Rāzī's writings or of his time.
  i Berthelot & Duval (1), p. 138.   j Cf. pt. 2, pp. 20, 195 above.

Thus far we have been speaking of natural products but there is something to say about artificial ones also, many of which were very relevant to the activities of the Arabic alchemists. First of all, the Jābirian corpus has references to the casting of iron, a technique which, as is well known, no one in the world could carry out before the end of the +14th century—except the Chinese.[a] One of the Seventy Books, the *Kitāb al-Naqd* (Book of Testing, or of Coinage)[b] has a good deal on the casting of iron, information which must have reached this group of writers from further East.[c] The time would be the second half of the +9th century. But what it says is very strange and garbled, for it envisages the melting of the iron in a crucible suspended within a cosmic model like an armillary sphere, and kept in perpetual motion above the furnace by some kind of machinery.[d] This is closely connected with the Arabic ideas on artificial spontaneous generation, mentioned already (p. 396) and to be looked at again (p. 485). The *Kitāb al-Naqd* also speaks of the transmutation of iron into silver and gold, which recalls the Chinese episode of Wang Chieh and his predecessors (cf. pt. 3, pp. 186 ff.).

Then there was glass, and porcelain, both important for chemical apparatus. Wiedemann (26) showed that glass from China (*zujāj al-Ṣīn*) was appreciated in or soon after Jābirian times,[e] though the imitation of gems in Alexandrian fashion remained a living tradition among the Arabs.[f] As for porcelain, it was just called 'china' (*ṣīnī*) as it is with us.[g] One might well ask when such vessels first became known in Arabic lands, and the answer seems to be that it was just about the same time as paper, i.e. in the course of the Central Asian conquests of the mid +8th century. The question was asked almost a century ago by Hirth (25), and Laufer (10) touched upon it, but the best studies are those of Kahle (7, 8).[h] The oldest Arabic reference so far found comes from the same year as the Battle of Talas River, +751, for when at that time Abū Dā'ūd Khālid ibn Ibrāhīm took a city near Samarqand he got great store of Chinese porcelain, some painted and (more surprisingly) gilded.[i] Subsequent mentions are numerous, as by Ibn Khurdādhbih in +846, and he is well justified by the abundant finds of Chinese porcelain contemporary with him at Samarra and elsewhere.

But sometimes the china got broken, and cement was necessary for mending it. That however is not at all the only thing which gives great interest to a whole set of

---

[a] See Needham (31, 32). [b] Kr 156. Cf. Kraus (2), p. 53, (3), pp. 57–8.

[c] Parallel passages are to be found in two other books of the Corpus, the *Kitāb al-Rāwūq* (Book of the Filter), also among the Seventy (Kr 140), and the *Kitāb al-Khawāṣṣ al-Kabīr*, ch. 9 (Kr 1900). Eng. tr. of the passage in the former by Nasr (1), p. 260.

[d] The great authority on all these methods is one Arius (Ariyūs), a name which neither Kraus (3), pp. 54–5, nor anyone else has ever succeeded in explaining.

[e] On the history of glass in Chinese culture see Vol. 4, pt. 1, pp. 99ff.

[f] He tells of a special work on this subject by Ibn Muḥammad al-Bisṭāmī, a man apparently of the +12th century, but not otherwise known.

[g] See the glossary of Siggel (2), p. 83. According to Laufer (1), p. 556, (10), p. 126, the Persian term for porcelain was *faghfūr-i-chīnī*, derived from Sogdian and a literal translation of 'Son of Heaven', the Chinese emperor's name. Thence it got into all Slavonic languages. Cf. Vol. 1, p. 169.

[h] See also Krenkow (2).

[i] The story is in Abū Ja'far al-Ṭabarī (de Goeje ed., ser. III, vol. 1, p. 79). The conqueror also got many valuable Chinese saddles, an interesting point in view of what immediately follows.

Chinese formulae or recipes for dyes, inks, varnishes and other chemical preparations occupying several chapters of the *Kitāb al-Khawāṣṣ al-Kabīr*.[a] Some of the more interesting of these may be listed as follows.

1. A waterproof and dust-repelling cream or varnish for clothes, weapons, etc.
2. A Chinese lacquer, varnish or cream (*duhn ṣīnī*) for protecting leather harness straps, belts, scabbards, bow-cases, etc.[b]
3. An absolutely fire-proof cement for glass and porcelain (*ghaḍār ṣīnī*).
4. How to make Chinese saddles (*surūj*).[c]
5. Recipes for Chinese and Indian ink.
6. A waterproof cream for impregnating silk, useful for the garments of divers.[d]
7. Other impregnating preparations for clothes, swords, wood, silk, etc.
8. Imitation of a Tibetan wood.
9. A Chinese cream for polishing mirrors.
10. Methods for making riding-whips (*miqra'a*).
11. Methods for transforming wrought iron (*narmāhan*) into steel (*fūlādh*).[e]

Thus by the latter half of the +9th century there were not only natural products from China in use by the Arabs, but also those of art, and even the transcription and adoption of some of the methods of these arts themselves. This is a significant item of evidence in one's estimate of the westward passage of chemical ideas.

In the preceding paragraphs we have occasionally mentioned plants and vegetable materials, and in concluding we ought to remind ourselves how much from the Chinese materia medica passed over to the Arabs.[f] Of the activities of the drug merchants of both cultures something was earlier said (p. 419), and the implication of the great trade in silks and tea was naturally part of the background. But there were many objects of much higher medical importance—cinnamon (*dār ṣīnī* the China tree, or *dār ṣīnī al-Ṣīn*, real China tree from China),[g] zedoary root (*jadwāre khitāi*),[h] ginger (*zanjabīl ṣīnī*),[i] bitter coptis root (*māmīrān ṣīnī*),[j] sweet-flag (*wajj* or *ighir*),[k] and

---

[a] Chs. 28 to 31 of Kr 1900. A detailed discussion of four of them was given by Ruska (7), with translation and chemical commentary. Cf. also Kraus (3), pp. 78–9.

[b] Recommended by the writer on the ground of personal experiments.

[c] It seems that a considerable export trade from China to the Arab countries developed in these articles; cf. B. Lewis (1), ch. 5.

[d] The writer claims to have got this from al-Faḍl ibn Yaḥyā al-Barmakī (+765 to +803), who said it was from an old unidentifiable MS. If the statement is acceptable it would place the coming of the recipe about the time of the Arab conquest of Central Asia.

[e] It would be interesting to know whether this was some form of the characteristic co-fusion process (see Needham, 31, 32).

[f] The point is made by Nasr (1), pp. 116ff., and was long ago expounded by Laufer (1), pp. 535ff. See also Levey & al-Khaledy (1).

[g] *Cinnamomum Cassia, chün kuei*[1] (R494). Cf. Meyerhof & Sohby (1), pp. 468ff., translating the abridgment of al-Ghāfiqī's 'Book of Simple Drugs' (late +12th century) by Bar Hebraeus (late +13th). See also Levey (6), *passim*.

[h] From *Kaempferia pandurata* (formerly called *Curcuma zedoaria*), *phêng o shu*[2] (R648).

[i] *Zingiber officinale, shêng chiang*[3] (R650). Cf. Levey (6), p. 87.

[j] *Coptis Teeta, huang lien*[4] (R534).

[k] *Acorus Calamus, chhang phu*[5] (CC1918), listed in the +9th-century *Aqrābādhīn* discussed by Levey (10). Of the items in this pharmacopoeia 31 % were of Persian or Indian origin.

[1] 菌桂   [2] 蓬莪茂   [3] 生薑   [4] 黃連   [5] 菖莆

officinal rhubarb (*rīwand ṣīnī*).[a] This is only to list a few, and to say nothing of fruits such as the peach[b] and the apricot,[c] which came westwards so early that they lost all national suffixes or epithets.[d]

There is one final point which ought to be made. Though no one has so far looked into the matter, it would be well worth examining how far the Arabic alchemists inherited some of the terms for chemical states and operations which their earlier Chinese confrères had used. Having noted one or two possible examples of this during our studies we mention them here with all due reservations. For instance, the conception of 'resistant to fire' (*qayyūm al-nār*), brought about when a fugacious or volatile substance is changed into another which no longer has this property, seems to mirror the Chinese idea of 'subduing' (*fu*[1]) or conquering (*shêng*[2]). Mentions of this sort of fixation are quite easy to find.[e] One wonders whether it derives entirely from the Alexandrians. Or again, the Arabic *mudbir*, 'regeneration', turning or returning, seems reminiscent of the Chinese *huan*,[3] so often used in the description of cyclical processes for the making of elixirs.[f] But this question must be left for further research. It is time that we turned to the ideological rather than the material influences of Chinese upon Arabic alchemy.

(iv) *Theoretical influences*

In Ben Jonson's 'The Alchemist' (+1610) we hear Subtle engage in an instructive harangue on what is meant by 'remote matter'. 'It is', he says,[g]

> of the one part
> A humid Exhalation, which we call
> *Materia liquida*, or the unctuous *Water*;

[a] *Rheum officinale, ta huang*[4] (R582). Cf. Vol. 1, p. 183; and Levey (6), p. 58.

[b] *Prunus persica, thao*[5] (R448). [c] *Prunus armeniaca, hsing*[6] (R444).

[d] Nor is it to say anything of the rich veins of Arabic toxicology, which drew upon all the lore of the known world from Spain to China. In the context of Chinese connections it is interesting to meet in the *Kitāb al-Sumūm* attributed to Ibn al-Waḥshīya (just mentioned, p. 449) with a reference to 'purging croton, that calamitous thing'. This is of course *Croton Tiglium* or *pa tou*[7] (R322), indeed a dangerously poisonous purgative. Even more characteristic of China, however, are other references to *ku*[8] poison (though not so named), made by letting several tarantula spiders sting themselves to death, and then working up the survivor. We have already had a good deal to say about this strange procedure (Vol. 2, p. 136) and shall return to it in Sect. 45; it deserves much more investigation than it has yet had, both scientific and philological. Meanwhile, something of the part which *ku* could play in national scandals and court intrigues in the Han period may be appreciated from the study of Loewe (5). Thirdly we hear much of lethal incenses, a subject very relevant to the Chinese fumigations which we have discussed above, pt. 2, pp. 148ff. In the translation of Levey (6), for croton see pp. 12, 94, for *ku* poison pp. 52, 67, and for poisonous smokes, pp. 39ff.; cf. also his summary (8). The *Kitāb al-Sumūm* of the Jābirian Corpus (Kr 2145), quite a different work and perhaps rather earlier, has been translated by Siggel (5); its overt literary sources appear to be entirely Greek, but many Persian drugs are mentioned, and there is talk of toxic smokes.

Perhaps one could also include under this rubric the Chinese doll (*timthāl* or *ṣanam*) mentioned in the Jābirian corpus used to bring about insomnia; Kraus (3), p. 85.

[e] For example in the +9th-century *Kitāb al-Ḥabīb*, tr. Berthelot & Houdas (1), pp. 79, 83, 112. Cf. Mahdihassan (12), p. 97.

[f] Attention was drawn to this first by Mahdihassan (17), p. 81. Cf. pp. 380ff. above.

[g] Pp. 382-3.

[1] 伏   [2] 勝   [3] 還   [4] 大黃   [5] 桃   [6] 杏   [7] 巴豆

[8] 蠱

On the other part, a certain crass and viscous
Portion of Earth, both which, concorporate,
Do make the Elementary Matter of Gold,
Which is not yet *propria materia*,
But commune to all Metals, and all Stones.
For where it is forsaken of that moisture,
And hath more driness, it becomes a Stone,
Where it retains more of the humid fatness,
It turns to Sulphur, or to Quicksilver,
Who are the Parents of all other Metals.
Nor can this remote Matter suddenly
Progress so from extreme unto extreme,
As to grow Gold, and leap o're all the Means.
Nature doth first beget th' imperfect, then
Proceeds she to the perfect. Of that aiery
And oily Water, *Mercury* is engendred;
Sulphur o' the fat and earthy part; the one
(Which is the last) supplying the place of Male,
The other of Female, in all Metals....

Of all the theories which the pioneers of chemistry entertained through the ages none was more important or more widespread than the belief that all metals, or all fusible bodies, were composed of mercury and sulphur in one form or another. It is generally acknowledged that this doctrine first appears in Arabic alchemy at the beginning of the +9th century, there being no antecedent for it in the writings of the Hellenistic proto-chemists.[a] But after that its ramifications continued down to the +18th century, even across the threshold, one might say, of modern chemistry. Having become a commonplace among the Arabic writers from the Jābirians to al-Jildakī it entered naturally into Latin alchemy,[b] and can be demonstrated in a thousand quotations from *c.* +1150 through the Geberian texts (*c.* +1290) onwards.[c] Petrus Bonus of Ferrara could write in +1330:[d]

Sulphur is a certain earthy fatness, thickened and hardened by well-tempered decoction, and it is related to quicksilver as the male to the female, and as the proper agent to the proper matter. Some sulphur is fusible and some is not, according as the metals to which it belongs are fusible or not. Quicksilver is coagulated in the bowels of the earth by its own proper sulphur. Hence we ought to say that these two, quicksilver and sulphur, in their joint mutual operation, are the first principles of metals.

For a time in Latin alchemy there came about a certain division, the Villanovan writers of the early +14th century stressing the role of mercury while those of the

[a] Cf. Leicester (1), pp. 65, 72; Multhauf (5), p. 134; Partington (4), p. 29; Nasr (1), p. 266.
[b] For example, the *De Aluminibus et Salibus* (cf. p. 398 above) says (in the +12th century) that gold and silver are formed in the earth by heat from mercury and sulphur during a thousand years, yet by God-given knowledge the alchemist can perform the process in a day (ch. 27); see Ruska (21) and Steele (2), cf. Multhauf (5), p. 162. How Chinese this was can be seen from p. 244 above.
[c] E.g. *De Investigat. Perfect.* chs. 3, 5, tr. Darmstädter (1), pp. 97, 99.
[d] *Pretiosa Margarita Novella*, tr. Waite (7), pp. 191 ff.

Lullian Corpus in the late + 14th emphasised rather that of sulphur.[a] But the two principles continued in explanatory vogue until the early + 16th century, when Paracelsus, adding salt, formed the celebrated Tria Prima of that chemistry which Boyle destroyed.[b] Later, van Helmont, Sylvius and Tachenius brought in two more principles, phlegm and earth, but the destiny of sulphur was far from accomplished, for Paracelsian sulphur became in the system of J. J. Becher (+ 1635 to + 1682) *terra pinguis* or 'fatty earth', something found particularly in organic materials and leaving them when they were burnt.[c] This then generated, in the hands of his pupil G. E. Stahl (+ 1660 to + 1734), the concept of phlogiston, on which historians of modern chemistry have written so many pages. It was used to explain all the phenomena which we now think of as concerned with oxidation–reduction, though in an opposite sense to ours, yet in spite of this it has been considered 'the first great unifying principle in chemistry' because it embodied a transfer principle, some kind of component being donated by one substance and received by another.[d] By the end of the eighteenth century the phlogiston theory had been killed stone dead by Lavoisier and his colleagues, but from its body sprang some of the most important ideas of the revolutionary modern chemistry.[e] Even the other member of the pair, mercury, still had a part to play, for the ancient archetype of contraries combining never ceased to haunt the minds of chemists, and doubtless facilitated the general acceptance of the electrochemical theory of affinity introduced by J. J. Berzelius (+ 1779 to 1848), most influential of the chemists of the first half of the nineteenth century.[f] The mercury–sulphur theory, therefore, had quite a career.[g]

It has occurred to many that the natural properties of sulphur and mercury uniquely fitted them for occupying such central positions in proto-chemical thought.[h] Sulphur

[a] Leicester (1), p. 87. [b] Leicester, *op. cit.*, p. 97.
[c] See Leicester's summary (1), p. 121; Multhauf (5), pp. 277–8.
[d] Leicester, *op. cit.*, pp. 122–3. Also the monographs of White (1); Metzger (1).
[e] Leicester, *op. cit.*, pp. 142 ff.
[f] Leicester, *op. cit.*, p. 168. For the background of this see Multhauf (5), pp. 299 ff.
[g] Before the exhaustive work of Dobbs (4) there was little realisation of the extent to which the alchemical work of Isaac Newton himself was dominated by this theory (cf. pp. 82, 128, 134 ff., 145, 150, 160, 181, 221). She knew of its Arabic origin (pp. 135, 220) but went no further back. It is now generally appreciated that Newton spent at least as much of his time in the 'chymical elaboratory' which he had in Trinity College as in thinking about optics and celestial mechanics (cf. Vol. 5, pt. 2, pp. 34–6). There he worked to 'open' metals and extract their 'mercury' (Dobbs (4), pp. 145, 198); in elixir iatro-chemistry he was not interested at all. Searching for a unified science of Nature, he was assuredly engaged in what Dobbs calls a 'chemicalisation' of alchemy, and although he may not have got very far with that, he favoured (like Boyle) a corpuscularian chemical philosophy which would include gravitational and magnetic forces beyond the purely mechanical impacts of Descartes (pp. xi, 88–9, 211). All this was in the period from + 1668 to + 1685 and later. With Newton, mercury and sulphur found one of their ultimate incarnations in what he called particles of 'earth' and 'acid', but so penetrating was his insight that these sound almost like protons and electrons, the sub-atomic particles out of which all sorts of matter—'one catholick matter', as Boyle put it (pp. 199 ff.)—would be built by variants of their stable configurations. This would involve many levels of size, degrees of complexity, and differences in density, as Figala (1) has described. It is exciting to find the *shui yin* and the *liu huang* of ancient Chinese alchemy coming through thus to the threshold of modern science.
[h] For example, Mahdihassan (16), pp. 19 ff., 23–4, (18), p. 42, (21), p. 196, (25), p. 42, (26), p. 19 (28), pp. 100–1, (59). The late Prof. J. R. Partington often used to emphasise in conversation the origin of the phlogiston theory from the sulphur of the sulphur–mercury theory (e.g. priv. comm. Feb. 1959); and Prof. J. D. Bernal, in a lecture of about 1937, pointed out how reasonable it had been to take sulphur as an elementary principle of burning, and mercury as an elementary principle of 'metallicity'.

is a solid, yellow like the sun, arousing thought-associations of raucous, stifling heat, therefore obviously male.[a] By contrast mercury is a white metallic liquid, cold, smooth and insinuating, therefore obviously female.[b] At the very beginning of this Section attention was drawn to the immense importance of the colour red, the colour of blood and life, in the proto-chemical thought of all the ancient civilisations (pt. 3, pp. 2–3). Hence it was of great significance that the combination of these two primary substances produced a blood-red substance, cinnabar. Furthermore each of them alone would produce a red thing, for sulphur heated to 180 °C turns to a dark orange-red liquid, while mercury heated in air to 300 °C forms the bright red oxide.[c] And it was from cinnabar that Li Shao-Chün[1] in the −2nd century set out to make that artificial gold which would confer immortality upon his emperor (pt. 3, pp. 29, 31 above). Besides, it has been pointed out that though both sulphur and mercury were powerful reagents in the vaporous state, the former was eminently combustible and therefore invited identification as the spirit of the element Fire, while the latter, eminently fusible and alone among the metals liquid at room temperatures, constituted naturally the spirit of the element Water.[d] Hence the combination of the two, and any process which could be analogised with it, was a *conjunctio oppositorum* (cf. pt. 3, pp. 69, 145 and above, pp. 121, 363).[e] 'Marry the male and the female' is the theme repeated endlessly, and passages based on it can be read in numerous ancient proto-chemical and alchemical texts. In Olympiodorus:[f]

He who knows the secret art of *chumeia* says to them: 'How can one understand transmutation?[g] How can water and fire, inimical and contrary the one to the other, be united in the same body, be made of one mind in grace and friendship? What a paradoxical *krasis*! Whence comes this unexpected amity of foes?'

And passages of very similar character occur in the Arabic Ostanes text,[h] the *Kitāb al-Ḥabīb*[i] and many other Arabic sources. After all, Lactantius in the +4th century had said that the very creation itself had been just this, the combination of *calor* and

He enlarged upon this is his Beard Lectures at Oxford (1), p. 203, and more recently, in lectures posthumously published, (2), p. 113, where he is at his most provocative in saying that 'the real origin of chemistry came from China'. He is referring to the combination of Yin mercury and Yang sulphur to make blood-red cinnabar, and adds: 'if we now turn it into our physical terms, we are dealing with a superfluity of electrons in mercury, a lack of electrons in sulphur, and a balance of electrons—the sulphide'. His conclusion is that the sulphur–mercury theory was essentially Chinese, and adopted by the Arabs.

---

    a The Yang dragon (Yang *lung*[2]) as in China we would say.

    b The Yin tiger (Yin *hu*[3]) as it would be in China.

    c Even that other partner, lead, important, as we know (pp. 254 ff. above, and pt. 5), in some Chinese systems, also gave rise on heating to a red thing, minium, red lead, Pb_3O_4.

    d Hopkins (1), p. 116.

    e Hammer-Jensen (2), pp. 17 ff. One of the Zosimus texts has an interesting passage on the reaction of sulphur and mercury, though rather obscurely worded, *Corp. Alchem. Gr.* III, xlix, 14.

    f *Corp.* II, iv, 41, tr. Berthelot & Ruelle (1), eng. et mod. auct.

    g As usual the term employed is the ancestor of our own 'metabolism'.

    h Berthelot & Houdas (1), pp. 120–1.

    i Berthelot & Houdas (1), pp. 79, 100.

    [1] 李少君   [2] 陽龍   [3] 陰虎

humor;[a] and he was echoed by many other theologians, Christian, Gnostic and Muslim alike.

Let us now look a little more closely at the Arabic point of origin of the sulphur–mercury theory. The idea that all metals are concocted from sulphur (*al-kibrīt*) and mercury (*al-zībaq*) is already found in the texts attributed to Balīnās, especially the *Kitāb Sirr al-Khalīqa...*,[b] which means the first few decades of the +9th century. It cannot be found in the West at any prior time.[c] It then runs throughout the Jābirian Corpus, expounded particularly however in the *Kitāb al-Ghasl* (Book of Washing) and the *Kitāb al-Īḍāḥ al-Maʿrūf bi-Thalāthīn Kalima* (Book of Enlightenment; commonly called, the Thirty Words).[d] After the +9th century it becomes an accepted doctrine of all Arabic writers on alchemy,[e] and is taken over directly into the Latin tongue[f] to fulfil the destiny we have already sketched.

It has been usual to maintain that this theory was derived by the Arabs from the two terrestrial exhalations of Aristotle.[g] One of these vapours (*anathumiaseis*, ἀναθυμιά-σεις) given off by the earth under the influence of the sun, was hot and fiery, dry and gaseous (*pneumatōdestera*, πνευματωδεστέρα), the other moist, cool and aqueous (*at-midōdestera*, ἀτμιδωδεστέρα). The former generated the idea of the sulphur component, the latter that of mercury, so many historians of chemistry have thought,[h] though Aristotle himself did not make any connection with these two elements.[i] Some scholars imply that men such as Ibn Sīnā[j] or the Geberian writer[k] explicitly did so, but this we have not been able to confirm by any original text, though of course they clearly stated the theory itself—'All Metallick bodies are compounded of Argentvive and Sulphur'. Other scholars have therefore contested the derivation from the *anathumiaseis*, proposing rather that the idea was originally a Chinese one, or at any

---

[a] *Div. Inst.* II, 9, 12.

[b] See Ruska (8), p. 151; Kraus (3), p. 283; Multhauf (5), p. 133.

[c] *Pace* Tenney Davis (4).

[d] Kr 183 and Kr 125 respectively. On the dominance of the theory cf. Kraus (3), p. 1. For further examples in Jābirian texts, see Berthelot & Houdas (1), p. 170; Nasr (1), p. 266.

[e] To illustrate this we reproduce (Fig. 1533) a drawing from an Arabic MS. in the British Museum (Add. 25.724) on which Ploss *et al.* (1), p. 116, have commented. It shows the six metals all held firmly under control by sulphur and mercury (or sun and moon, or Yang and Yin, as you will). The text discusses the permanence of substances with perfect *krasis* and the impermanence of those in which it is unbalanced.

[f] Cf. von Lippmann (1), vol. 1, p. 488, vol. 2, p. 180.

[g] *Meteorologica*, I, iv (341 b 6 ff.)., II, ix (369 a 13 ff.) and esp. III, vi (378 a 13 ff.); Lee ed. pp. 28, 222, 287. We discussed them in the Sections on meteorology and mineralogy in Vol. 3, pp. 469, 636. It has also, however, been proposed that Dioscorides was in part the author of the theory; Berthelot (1), p. 68, attributing to him the following words: 'Some say that mercury is a constituent of all metals.' But this seems to be only a mistranslation due to the double meaning of *metalla*, etc. in Greek, and the proper rendering ought to be: 'And some say that hydrargyrum is found by itself (i.e. native) in the mines.' See Goodyer's translation, in Gunther (1), p. 638, confirmed by Lenz (1), p. 74; *De Mat. Med.* v, 110.

[h] E.g. Sherwood Taylor (3), p. 80; Darmstädter (1), p. 137. According to von Lippmann (3), vol 2, pp. 109, 149, Pebechius said that there was mercury in all things, but we have not been able to find this in the Hellenistic Corpus.

[i] Aristotle's own view was that the dry exhalation formed all the *fossiles*, i.e. ores, minerals and rocks, while the moist vaporous one formed all the metals.

[j] For example, Leicester (1), p. 70. But the relevant passages in the *Kitāb al-Shifāʾ* (tr. Holmyard & Mandeville (1), pp. 38 ff.) do not contain this identification.

[k] For instance, Leicester (1), p. 85. But we find no supporting passage. Multhauf (5) avoided this statement.

Fig. 1533. Yin and Yang (the Moon and the Sun) as controllers of the Six Metals, an illustration from an Arabic MS. (British Museum Add. MSS 25, 724) of Abū al-Qāsim al-'Irāqī, c. + 1280 (cf. p. 404). The text reads: 'Know that (the constituents of) this compound substance (the elixir) possess precise weights which bring them into equilibrium so that the heat does not exceed the cold, nor the dryness the wetness. And whatever attains (such) equilibrium is permanent, and no more subject to change, while all things that do not attain it will be overcome by impermanence and transformation.'

rate something arising from the contact of the Arabic alchemists with the earlier Chinese tradition.[a]

The position here is a little ambiguous. While we cannot call to mind any Chinese alchemical text which explicitly states the theory in the customary way, it will have become abundantly evident to the reader of the earlier parts of this Section (pt. 2, pp. 128, 326 ff., pt. 3, pp. 14, 74, 86, 126, etc.) that mercury and sulphur played a much more prominent part in Chinese alchemy than any other substances, and that cinnabar itself, mercuric sulphide, far excelled all other chemicals for prominence in Chinese alchemical song and story. Moreover, the passage from Khung Ying-Ta (c. +640) which we quoted on p. 156 above shows clearly that the element Metal was regarded in orthodox Chinese natural philosophy as a mixture of Yin and Yang, with the former

a E. g. Huang Tzu-Chhing (2); Mahdihassan (16), pp. 21, 23–4, (26), p. 19, (28), pp. 100–1; Haschmi (5), p. 62. Especially Abrahams (1), p. xix, has stressed the sulphur–mercury theory as a derivate of the idea of Yang and Yin as constituents of all the metals. Chinese dualism, says Subbarayappa (2), had much influence on Indian alchemy also.

predominating; and from there it would have been a very short step to the idea that the metals known to man were different in properties because of the varying proportions of Yin and Yang (mercury and sulphur) which they contained. So if the doctrine just stated was not actually received by any individual Arabic *ḥakīm* from his Chinese *chen jen*,[1] it could have sprung very easily from a knowledge of how the adepts of China conceived of elixir-making and aurifaction. Indeed it hardly amounted to more than saying that all the metals were varieties of cinnabar.[a]

There is a further argument. References to *anathumiasis* are rare in the Jābirian Corpus, but when they do occur they never concern the sulphur–mercury theory. For instance, the *Kitāb al-Ḥāṣil* (Book of the Result)[b] quotes the *Placita Philosophorum* of Pseudo-Plutarch[c] as saying that the *anathumiasis* of the world-soul is damp in itself so that it is of the same kind as those which emanate from living beings on the earth—here the word is rendered *bukhār*.[d] If the Jābirian sulphur–mercury thesis had really been derived from Aristotle's *anathumiaseis* it seems hardly believable that this or some other rendering of the Greek word should never be found in association with it.

The next subject takes us further away from chemistry as ordinarily understood, but it illuminates the relations of Arabic and Chinese proto-scientific thought in a most curious way. At an earlier point, in describing the characteristics of the Jābirian Corpus (p. 394), we had occasion to refer to the Theory of the Balance ('Ilm al-Mīzān) which runs widely through the books and tractates contained in it.[e] This set out to be a science of the quantitative composition of all bodies, but it was 'computational' and numerological, not primarily empirical or experimental. The system is most fully described in the 'Books of the Balances' (*Kutub al-Mawāzīn*),[f] some 144 tractates which may be dated to the neighbourhood of +900, about the middle of the whole period through which the Corpus was written. It was based upon Greek medicine and natural history to the extent that the four qualities (*ṭabā'i'*) or natures (hot, cold, moist and dry) were regarded as the ultimate constituents of all things, whether metals, minerals, plants or animals; these qualities being of course traditionally related to the four Aristotelian elements (earth, fire, air and water), and to the Hippocratic–Galenic humours (blood, black bile, yellow bile and phlegm).[g] Galen's four degrees (*taxeis*, τάξεις) of intensity of each quality in a thing,[h] orginally a pharmacodynamic classification mounting from foods to poisons and powerful drugs, was accepted by the Jābirian

---

[a] As was acutely remarked by Mahdihassan (16). According to Pelliot (54) our word derives from Ar. *sinjafar* or *zunjafur*, and that from Sogdian **shingafr* through Persian *shangarf*. Cf. Siggel (2), p. 97. But Iranians also said *sīmshangarf*, meaning '(quick-)silver cinnabar', a phrase exactly analogous to the Chinese *yin chu*,[2] quick-silver being *shui yin*.[3] This could argue for ancient mercantile and proto-chemical contact.

[b] Kr 323.   [c] Though Plutarch's name is not mentioned.

[d] See Kraus (3), p. 333.

[e] Brief descriptions can be found in various places, e.g. Multhauf (5), p. 135; Nasr (1), pp. 263ff., but the only profound discussion is still that of Kraus (3), pp. 187ff.

[f] Kr 303 to 446.

[g] Gruman (1) has put it this way: what the Jābirians did 'was to project the Galenic system on to the problems of inorganic chemistry, seeking to identify the characteristic make-up or balance of the qualities in each metal and chemical substance', p. 60.

[h] The chief source for this is the *De Simpl. Med.*

[1] 眞人   [2] 銀硃   [3] 水銀

writers, but in arguing about relative dosages in *dirhams*, etc., they believed that they must go beyond the fallible empirical impressions of the superficial senses, and apply a theoretical system founded essentially on a numerology.[a] A simple exposé of this is seen in Fig. 1534. While the Galenic degrees were all equal, the *martabas* of the Jābirians were related to each other, starting from the lowest, in the proportion of 1, 3, 5 and 8, making a total of 17; and this number was regarded as being the base (*qā'ida*) of the whole theory.[b] Furthermore, taking the first *martaba*, we find that it

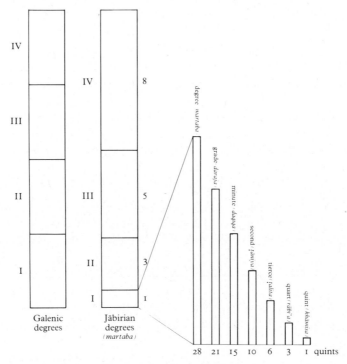

Fig. 1534. Chart of the 'Ilm al-Mīzān (Science of the Balance) of the Jābirian Corpus. Building on the Galenic estimation of pharmaceutical potencies as related to the *krasis* or relative proportions of the Greek elements and humours, the Arabic alchemists of the + 9th and + 10th centuries produced a much more sophisticated, but equally numerological, system of believed component quantities. Perfect balance, if only it could be attained, would exempt men and metals, indeed all things, from change and decay.

was divided into 28 of the smallest units, the quints (*khāmisāt*), five other named units being available for intermediate magnitudes between these two.[c] Thus since there were 17 units equivalent to the first and smallest *martaba* in the whole series, there was a total range of 476 *khāmisāt*, not just simply 4 degrees. This was clearly a more sophisticated parameter.

But it was based neither on observation nor experiment.[d] The Jābirians proceeded

[a] See Kraus (3), pp. 193–4ff.   [b] Kraus, *op. cit.*, p. 227.
[c] On all matters connected with these see Kraus (3), pp. 190, 193–4, 196, 270.
[d] Except in so far as the Jābirians embarked on their computations with a considerable previous knowledge of the properties of the substances in question.

to what we might call, coining a word, a *krasi-gnosis*, an estimate of the relative compositions of substances, by taking the numerical values of their names in Arabic; a kind of glyphomancy[a] known as the *mīzān al-ḥurūf* (balance of the letters).[b] But this *hijā'* (spelling out) gave only the exterior nature of the substance, and the latent, complementary or interior nature had to be found by conjecture (*ḥads*) to make it up to a value of 17.[c] Thus was the *krasis* or equilibrium ('*adl*) found out, and all the operations of alchemy were directed to altering these inner and outer equilibria. Elixirs did precisely this. Spontaneous change could also occur, as in the rusting or corrosion of metals, and in one place this is applied to the evaporation of ammonium carbonate and chloride (*nūshādir*). 'If the natures in a body undergo change', says the text, 'leading to excess or defect,[d] it loses its normal state, breaking down of its own accord.'[e] For all this there were some antecedents in the Hellenistic world, and kabbalistic glyphomancy with marked Jewish connections permeated Arabic thought, but there is no need for us to explore it here.[f] More important questions to ask are first, how far back did its chemical application go, and secondly, where did the numerological succession, 1, 3, 5, 8, 17 and 28 come from?

It seems clear that the theory of the Balance was not well known to the earlier Jābirians, writing the Seventy Books in the middle of the +9th century,[g] nor is there anything much about it in the present text of the book attributed to Balīnās (cf. pp. 369 ff. above), the *Kitāb Sirr al-Khalīqa wa-Ṣanʿat al-Ṭabīʿa* (Book of the Secret of Creation and the Art (of Reproducing) Nature), *c.* +820 to +830, though the term *mīzān* occurs.[h] But other parts of the Corpus, notably the *Kitāb al-Aḥjār* just mentioned (Kr 307), do attribute to Balīnās a discussion of the six basic numbers,[i] so there is no barrier to believing that the theory was taking shape in the first decade or two of the +9th century. Moreover the Corpus also contains a set of tractates called '*Ashara Kutub 'alā Ra'y Balīnās Ṣāḥib al-Tilasmāt* (Ten Books of the Opinions of Balīnās, Lord of Talismans)[j] where the *mīzān* also comes in, but the significant thing in the light of what follows is that these are very Ṣābian texts, dealing with the seven metals, planets and divine images (*aṣnām*). Hence we must suspect (cf. p. 426 above) East Asian connections.

[a] Cf. Vol. 2, p. 364.

[b] An analysis of this kind appears in the translation of the *Kitāb al-Mīzān al-Ṣaghīr* (Kr 369) by Berthelot & Houdas (1), pp. 158–9.

[c] Kraus (3), pp. 223 ff. works out a number of concrete examples. A discussion on the maximum of 17 components occurs in the *Kitāb al-Tajmīʿ* (Kr 398), tr. Berthelot & Houdas (1), pp. 191, 200, 204.

[d] Cf. Vol. 2, pp. 463, 566.

[e] This is in the *Kutub al-Aḥjār 'alā Ra'y Balīnās* (Books of Minerals according to the Opinions of Balinas), Kr 307 to 310; four in number. See Kraus (3), p. 233.

[f] See further in Kraus (3), pp. 236 ff., 266. Allegro (1) tells us in his 'Dead Sea Scrolls' that the Qumran community, very gnostic and dualistic, believed that the behaviour of each man and woman was determined by the activities of two warring spirits within. They thought that the inheriting of these spirits depended upon the astrological situation at the birth of the individual, and that their proportions could be numerically reckoned up. This he derives (p. 125) from the 'Manual of Discipline', *c.* −1st century. Here then was a much earlier Balance.

[g] Kraus (3), p. 235.

[h] Kraus, *op. cit.*, pp. 283, 289. Needless to say, the system was entirely unknown to the Greeks, notwithstanding the identification of Balīnās with some Apollonius or other.

[i] *Op. cit.*, p. 286.   [j] Kr 293 to 302.

Kraus was extremely puzzled about the origin of the numerological succession.[a] He devoted to it an immensely learned disquisition, recalling (not very convincingly) the *Timaeus* and Pythagoras, searching for some connection with the music of the spheres, and alluding to the 17 consonants of the Greek language.[b] But his work in the forties ended with no solution to the problem. During the fifties Stapleton found it in the simplest magic square, a mathematical achievement of ancient China, which Cammann during the sixties has set in its fullest perspective and context.

Stimulated by some earlier papers of Coyaji,[c] Stapleton (4a) suggested in 1950 that the mysterious Jābirian numbers could all be derived from the magic square of three.[d] A magic square is an arrangement of numbers in the form of a square or other matrix such that every column, every row, and each of the diagonals adds up to the same figure, the constant.[e]

4	9	2
3	5	7
8	1	6

In this simplest case (see inset cut) they all add up to 15. Stapleton then applying what he called gnomonic analysis made it clear that if divided in this way the gnomon's total is 28, while the numbers in the remaining four compartments are the four lowest numbers required, together with their total of 17.[f] If the Jābirians saw in these, as they certainly did, the fundamental numbers of Nature, the reason could well have been that the magic square of three had somewhere sometime been regarded as a numinous cosmic cantrap or diagram of the highest sanctity and solemnity. Stapleton knew that veneration of this kind had been paid to it for centuries in China, and correctly pointed out that it was nothing other than the Lo Shu[1] (Writing from the River Lo), often associated with the Ming Thang[2] (Bright Hall), the cosmic temple of Han times, and with the Chiu Kung[3] (Nine Palaces), a further conception, sometimes celestial, sometimes terrestrial. He therefore asserted that a cardinal influence had been exerted by Chinese cosmism upon Arabic proto-chemistry and alchemy, and indeed it seems that he was right.

In Section 19 we gave a fairly full account of the Lo Shu magic square and its history.[g] As one of the legendary diagrams bestowed by Heaven upon the engineer-emperor Yü the Great it certainly goes back to pre-Confucian times, but not until the latter part of the −4th century do we get evidence (from the *Chuang Tzu* book)[h] that an arrangement of numbers formed the essential point of it. This is reinforced by a passage in the Hsi Tzhu (Great Appendix) of the *I Ching* (Book of Changes) dating

[a] Cf. Kraus (3), p. 297.

[b] See especially Kraus (3), pp. 199, 207, 218. On the music of the spheres, p. 203, the *Timaeus* p. 220, and the Greek alphabet p. 209.   [c] (2–5) but especially (6).

[d] His chief papers on this subject are (2, 3) and the longest (4); for the gnomon see (5). All are very *touffu*, ranging over many subjects from Zoroastrianism to ziggurats, and full of arbitrary judgments, *non sequiturs* and mistakes in fields other than those of Arabic studies—but in this brilliant discovery he was right.

[e] As a general rule for odd-number squares, if n = the base number (the number of cells on one side), m = the number in the central cell, c = the constant (the sum of all the rows and columns and the two principal diagonals), and t = the total sum of all the numbers; then $nm = c$ and $n^2m = t$. Cammann (9), p. 48.

[f] Kraus (3), p. 219, had touched upon the subject of gnomons, but missed the clue that a magic square arrangement was essential.

[g] Vol. 3, pp. 56ff.   [h] Ch. 14, tr. Legge (5), vol. 1, p. 346.

[1] 洛書   [2] 明堂   [3] 九宮

from the −2nd century,[a] and clinched by a quite definite statement in the *Ta Tai Li Chi* datable about +80.[b] To its subsequent history we shall return in a moment. In spite of the criticisms of Cammann (7) we find relatively little that needs alteration in our account of Chinese magic squares, but we record with gratitude his demonstration that Theon of Smyrna (*fl. c.* +130) was a man of straw in this connection.[c] The point is important, because it can now be reliably stated that magic squares were completely unknown in any part of the Hellenistic world, and appear in the West only in the Jābirian period.[d] Cammann (12) assumed that the earliest appearance there of the magic square of three was as a pregnancy and childbirth charm in the Jābirian *Kitāb al-Mīzān al-Ṣaghīr* (Lesser Book of the Balance),[e] datable about +900. He was more uncertain about an earlier appearance in a text of Thābit ibn Qurra (al-Ḥarrānī, +826 to +901),[f] though this has the authority of Suter.[g] But there is a still earlier example in a gynaecological text in the 'Paradise of Wisdom' of 'Alī ibn Sahl Rabbān al-Ṭabarī, who died in +860, again for cosmic aid in difficult labour, discovered by Siggel (6).[h] We must thus conclude that the world-emblem of the magic square of three reached the Arabic West just about the time of the Balīnās books, in the early decades of the +9th century. This agrees strikingly enough with other examples of Chinese influence which we have been examining.

It now seems reasonably certain, wrote Cammann, that the magic square of three was invented in ancient China, first discussed there, and first put to practical use in the philosophy and religion of that culture.[i] Why was it that the Chinese were the foremost to embark on this line of mathematical development, leading ultimately to what is known as combinatorial analysis? Because, as he said,[j] 'the Shang Chinese were apparently the first people in the world who could, and did, consistently express any number, however large, with only nine digits; and they were regularly doing this some two thousand years before the Hindus learnt to do the same thing with the numerals we now call Arabic.' Later Chinese contributions included, as he says, the first magic squares of five, six, seven and nine, the first bordered magic squares, the earliest known composite magic square and probably the first augmented square. The most ingenious solutions for the squares of six and nine ever devised were, and remained, uniquely Chinese. Unquestionably most of these developments were based upon the

---

[a] Pt. 1, ch. 9, tr. Wilhelm (2), vol. 1, p. 234.   [b] Ch. 67.

[c] We were certainly misled by Sarton (1), vol. 1, p. 272, who, like Homer, nodded at times.

[d] Cammann (9), pp. 45–6, confesses to a feeling that the first magic square ought to have been Pythagorean, or rather perhaps that one would expect it first in Babylonia, spreading out both westwards and eastwards—but, as he says, there is no trace of evidence for anywhere but China.

Elsewhere, (7), p. 118, he drew attention to the Arabic acceptance in the +9th century very clearly.

[e] Kr369, see Kraus (2), p. 90, (3), p. 73; Berthelot & Houdas (1), p. 150; Hermelink (1).

[f] Cammann (7), p. 118, (9), p. 46; Noble (1); cf. Mieli (1), 2nd ed., p. 86. Note his Ṣābian origin.

[g] (1), p. 36.

[h] Later mentions are less important for us, e.g. the *Rasā'il Ikhwān al-Ṣafā'* (*c.* +990), in which the first set of magic squares of ascending complexity in the West is found, 'even the construction methods sometimes showing obvious Chinese influence'; Cammann (12), p. 190. The square of three as a childbirth charm turns up again in *Picatrix* and its Arabic original, i.e. by +1056 (Ritter & Plessner (1), p. 407). The many varieties of magic squares of higher orders are beyond the scope of our argument here, but may be followed in Cammann (7, 8, 12, 13).

[i] (8), pp. 52–3.   [j] (9), p. 40, following our Vol. 3, p. 15.

Lo Shu[1] diagram already mentioned, so they were probably made during the period when this was a very sacred and esoteric thing,[a] i.e. before the +10th century, when the tradition lost its associations with religious cosmism and came to the surface, turning into a secular commonplace. This point is typified by the work of the Taoist scholar Chhen Thuan[2] (+895 to +989).[b] In subsequent centuries major creativity in this field passed to the Arabs, Byzantines and Indians,[c] but in his book of +1275 Yang Hui[3] preserved a great collection of what had been done in China down to, and including, his own time.[d]

But all this agreed, why did the magic square of three have such a charisma? It might be fair to say that the number nine has been too much overshadowed in our considerations of ancient Chinese thought by the prominence of the classification of everything in fives, following the rise of the Five Element theory after Tsou Yen[4].[e] Nine was always a convenient cosmic number because of the way in which centrality could be surrounded by eight directions of space. Thus in ancient Chinese natural philosophy we find, above, the Nine Spaces of the heavens (Chiu Yeh[5]),[f] below, the Nine Provinces of China (Chiu Chou[6])[g] and the nine cauldrons of the Hsia associated with them (Chiu Ting[7]),[h] the 'well-field' (ching thien[8]) arrangement of land allocation in nine lots (chiu thien[9]),[i] and, on a broader scale, the Nine Continents (Chiu Chou[6]) of Tsou Yen, of which China was only one.[j] There was also the elusive expression Chiu Kung[10] (Nine Palaces, or Halls), which turns up in all sorts of contexts referring to the heavens, the rooms (shih[11]) of the cosmic Ming Thang[12] temple, and the lay-outs of divination devices (shih[13, 14]),[k] mathematical matrices and boards for chess and proto-chess.[l] In all these uses the theme of microcosm and macrocosm was vital, and the Lo

---

a This was the time of Hsü Yo[15] (c. +190) with his Chiu Kung suan fa[16] (Nine Hall computing method), as also of his commentator Chen Luan[17] (c. +560) who drew upon a lost book entitled Huang Ti Chiu Kung Ching[18] (cf. Sui Shu, ch. 34, p. 21a). See Vol. 3, pp. 58–9 and Cammann (9), pp. 42–3.

b Cf. Vol. 3, p. 59 and Cammann (9), p. 76.

c See Cammann (12, 13) and literature there cited; also Nasr (3), e.g. p. 211.

d This was the Hsü Ku Chai Chi Suan Fa[19] (Choice Mathematical Remains collected to preserve the Achievements of Old), analysed by Li Nien (4), vol. 3, pp. 59ff., (21), vol. 1, p. 175, and Cammann (8).

e See Vol. 2, p. 232, also pp. 242, 253ff., 261ff.

f Huai Nan Tzu, ch. 3, pp. 2b, 3a, tr. Chatley (1), p. 5. To get in all the 28 lunar mansions (hsiu[20]) three were allotted to each Space, including the centre, and four to the northern one. The astronomical background will be understood from Vol. 3, p. 240 and Table 24.

g The classical description is in the Yü Kung chapter of the Shu Ching, on which see Vol. 3, pp. 500–1.

h See Vol. 3, pp. 503–4. The liturgical work Shang-Chhing Ling-Pao Ta Fa[21] (TT1204–6, cf. Vol. 5, pt. 2, p. 129) depicts the nine cauldrons marked with the Lo Shu numbers (ch. 12, p. 9a).

i See Vol. 4, pt. 3, pp. 256ff.

j An account of this doctrine has been given in Vol. 2, p. 236.

k On the connection with the shih see Cammann (9), pp. 70ff. Chang Hêng,[22] the great +2nd-century mathematician, physicist and naturalist, is known to have recommended magic squares for divination purposes as a method already old in his time (Hou Han Shu, ch. 89, p. 11b). On the connection of the shih with the history of the magnetic compass, see Sect. 26i (4) in Vol. 4, pt. 1.

l Many references will be found in Vols. 2, 3 and 4, pt. 1, passim.

---

1 洛書	2 陳摶	3 楊輝	4 騶衍	5 九野	6 九州
7 九鼎	8 井田	9 九田	10 九宮	11 室	12 明堂
13 栻	14 式	15 徐岳	16 九宮算法		17 甄鸞
18 黃帝九宮經		19 續古摘奇算法		20 宿	21 上清靈寶大法
22 張衡					

Shu exemplified it to perfection, a magical expression of centrality and universal order, an *imago mundi*, a miniature emblem of the cosmos.[a]

Moreover it was for centuries associated with the worship of one of the greatest deities of Chinese cosmism, the supreme pole-star sky-god Thai I[1,2].[b] How far he was ever regarded as a creator remains obscure because that concept was always very uncharacteristic of Chinese thinking, but a passage in the *Lü Shih Chhun Chhiu* (−239) has Thai I giving rise to the two principles (*liang i*[3]) Yin and Yang, and hence to the myriad things.[c] Liturgical honours for him were suggested to Han Wu Ti in −124 by Miu Chi[4] and carried out at a special altar,[d] but by −113 Thai I had displaced Huang Ti (the Yellow Emperor) from the central position at the imperial sacrifices and was worshipped at a central altar with three concentric levels,[e] the altars of the Five Emperors all being round about. By +7, under Wang Mang, he was universally recognised as the supreme deity,[f] and he it is that we see riding in the chariot of the Great Bear in Han tomb-reliefs such as those of Wu Liang (+147).[g]

But Thai I did not stay otiose in the Purple Forbidden Enclosure of the circumpolar stars. He was believed to go round on an annual tour of visitations through the spaces of the universe, a peregrination represented on the Lo Shu by the order 5–1–2–3–4–5, then rest awhile, then 5–6–7–8–9–5. This eternal cyclical balanced rhythm is first described in Han prognostication books such as the *I Wei Chhien Tso Tu*[5] (Apocryphal Treatise on the (Book of) Changes; a Penetration of the Regularities of Chhien),[h] datable in the +1st or +2nd century.[i] But it had a long innings, for Taoist talismans of precisely this complicated array, as it traces out on the Lo Shu magic square, are depicted in the *Tao Fa Hui Yuan*,[6] a Thang encyclopaedia of Taoist liturgy and apotropaics (see Fig. 1535).[j] By +1116, when Yuan Miao-Tsung[7] finished his *Thai-*

---

[a] Since odd numbers were Yang and even ones Yin it embodied an expression of their ever-changing balance, as also of the generations and destructions of the five elements; and there were ingenious correlations with the trigrams and hexagrams of the *I Ching*. All these have been most fully worked out by Cammann (9), cf. esp. pp. 46, 50, 56, 65, 73. Although he tends to regard the Lo Shu symbolism as essentially celestial, the terrestrial aspects surely always clung to it also. For example, it was certainly understood as terrestrial by such adepts of Chinese cosmism as Hsiao Chi[8] in his *Wu Hsing Ta I*[9] (Main Principles of the Five Elements), *c.* +600, who there laid out the sacred mountains and great rivers of China in the framework of the Lo Shu cells (ch. 1, pp. 28aff.). This might, with further study, throw light on the *fên yeh*[10] system of cartography in the Thang; see Vol. 3, p. 545.

[b] Also written Thai I[11,12] like Thai-shan. Literally translated, the words mean 'Great Unity'. It was the name of an asterism which may have been one of the most ancient of Chinese pole-stars; cf. Vol. 3, pp. 260–1. [c] Ch. 22 (vol. 1, p. 44).

[d] *Shih Chi*, ch. 28, p. 23b, tr. Chavannes (1), vol. 3, p. 467. Also *Chhien Han Shu*, ch. 25A p. 20a.

[e] Like the traditional later Altar of Heaven, still to be seen at Peking (cf. Vol. 3, p. 257). Cf. Chavannes (1), vol. 3, p. 490. Purple vestments replaced the former yellow, in correlation with the Purple Palace of the circumpolar stars (cf. Vol. 3, pp. 259ff.). One can see how the ancient connection between pole-star and emperor was at work here. See also Loewe (6), p. 11.

[f] Cammann (9), p. 63. Cf. *CHS*, ch. 99B, p. 13a.

[g] Vol. 3, Fig. 90, p. 241.

[h] See Tables 13 and 14 in Vol. 2; the first of the *kua* (trigrams and hexagrams).

[i] Ch. 2, p. 3a; the passage is translated in Vol. 3, p. 58. It is quoted also in the *I Wei Ho Thu Shu*[13] (in *Ku Wei Shu*, ch. 16, p. 2a); and, with Chêng Hsüan's valuable commentary, in *Hou Han Shu*, ch. 89, p. 11b, the biography of Chang Hêng.

[j] Ch. 94, p. 7a, ch. 98, p. 15a, b.

[1] 太一　　　[2] 太乙　　　[3] 兩義　　　[4] 謬忌　　　[5] 易緯乾鑿度

[6] 道法會元　[7] 元妙宗　　[8] 蕭吉　　　[9] 五行大義　[10] 分野　　　　[11] 泰一

[12] 泰乙　　　[13] 易緯河圖數

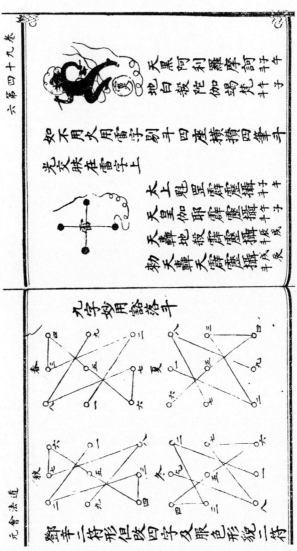

Fig. 1535. The tours of the god Thai I through the spaces of the universe represented by the nine cells of the magic square of three; from the Thang and Sung liturgical encyclopaedia *Tao Fa Hui Yuan*, ch. 94, p. 7a. The four diagrams give the routes for each of the four seasons. The caption calls it 'the mystery of the nine words of the *huo lo tou*'.

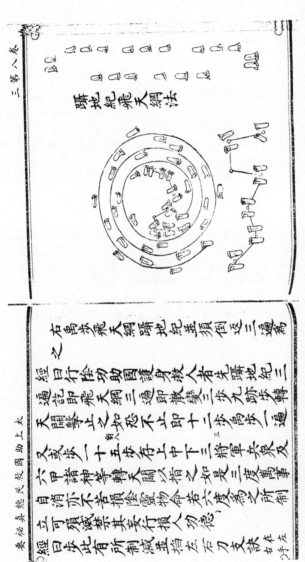

Fig. 1536. Directions for a Taoist ritual dance symbolising the circling of the Great Bear (*pei tou*) in the heavens; from the *Thai-Shang Chu Kuo Chiu Min Tsung Chen Pi Yao* (+1116), ch. 8, p. 3*b*. The caption calls it 'rules for treading the earth in memory of the flight of the (vehicle) Thien Kang' (cf. Vol. 3, Fig. 90, p. 241). An exactly similar notation of footsteps is used on p. 10*b* in 'the rule for dancing the *huo lo tou*', i.e. the route of the god Thai I through the universe during the winter quarter. The names of the nine cells of the magic square are given beside the footsteps, and the text of the appropriate chant to be sung during the ritual is provided alongside.

Shang Chu Kuo Chiu Min Tsung Chen Pi Yao,[1] a similar prayer-book or missal, the steps are marked out with drawings of footprints, thus showing that the idea of the progression had generated a liturgical dance.[a] Like the 'Pace of Yü', this was part of what came to be called *tha kang pu tou*,[2] 'treading the *kang* and stepping the *tou*', the former being the box of the Great Bear and the latter the whole constellation. These were the activities of Taoists, but the imperial court was still engaged in elaborate ritual representations of the processions of Thai I, and sacrifices to him, throughout the +8th century;[b] and the last known enacting of these did not take place until after the beginning of the Sung, in +1008.[c]

Thus to sum it all up, the scholars of ancient China saw in the numbers of the Lo Shu the two forces of Yin and Yang at work, the cycles of the Four Seasons and the Five Elements, and the deployment of the Nine Directions of space, emphasis always remaining on cosmic centrality like a kind of power-house. So cosmic a symbol was bound to be thaumaturgic, needing care lest it should get into the wrong hands— hence the secrecy which surrounded the magic square of three down to the end of the Thang. Though mentioned so much in texts, it seems never to have been seen in public.[d] Only after the fall of that dynasty, with Chhen Thuan's work in the Wu Tai period, did it enter into general knowledge and circulation.

Accordingly when the Arabic alchemists came into possession of the Lo Shu magic square about the beginning of the +9th century, they were receiving a cosmic symbol loaded with a great weight of reverent belief, eight centuries' worth in fact of numinous well thought out natural philosophy. There seems little cause for surprise that the Jābirians should have taken it as a great secret of the universe, whence they could extract those numbers which they believed would be found at the basis of the constitution of all natural substances.[e] There are still some obscurities in the whole transaction,[f] and we have no concrete evidence as to how the handing over came

---

[a] Ch. 8, p. 10*b*. See Fig. 1536.

[b] An account of these, with references, is given by Cammann (9), pp. 74–5, largely based on *Thang Hui Yao*, ch. 10B (pp. 256 ff.). The *Chiu Kung than*[3] or Altar of the Nine Halls, with nine 'flying thrones' (*fei wei*[4]), was used for the services, and Thai I and the other gods were perhaps impersonated by actors, as in 'live chess'. The connections with chess are always in the background, and need further investigation, as e.g. in the case of the curious writings mentioned in Vol. 3, p. 542.

[c] There may also have been important repercussions in Japan. During the Heian period it was considered impious and dangerous to travel in certain directions at particular times of year, since it meant opposing the 'divinités ambulantes' (*yu hsing shen*[5]), as Frank (1) calls them in his monograph on the subject. Thus there were 'direction-prohibitions' (*kata-imi*[6]) and 'direction-disobedience' (*kata-tagae*[7]).

[d] Unless it was pictured in the full versions of some of the apocryphal Han prognostication books (*Chhan Wei shu*[8]), most of which were destroyed in literary inquisitions such as that of +605, and none of which are available to us now.

[e] It may be of interest to remark that attempts at the quantification of the Galenic qualities were still going on as late as Elizabethan times, as witness the study of John Dee's grading mathematics by Clulee (1). On Dee himself (+1527 to +1608) see French (1).

[f] For example, why one gnomon should have been chosen out of the possible four. Also magic squares do not seem to have played any very significant part in Chinese alchemy, though of course invisibly represented in *kua* arrangements. They were prominent, however, in Burmese alchemy of the +15th century; Htin Aung (1), p. 54. Could this imply Arabic influence?

[1] 太上助國救民總眞秘要   [2] 踏罡步斗   [3] 九宮壇   [4] 飛位
[5] 遊行神   [6] 方忌   [7] 方違   [8] 讖緯書

about,[a] but when one remembers the opportunities for intercourse both in Central Asia and in the coastal cities of China (cf. pp. 417, 422 ff. above), one can see that it would have taken hardly more than a brace of well-educated merchants in Ferghana, or a couple of intelligent Arab physicians in Canton,[b] who happened to fall in with a milieu of Taoist literati, to imagine how the entire gift might have been conveyed. Of course the use made of it was different, but the magic square was the same, and it looks as if here the Arabs owed much more to Taoists than to Pythagoreans.

We have now seen reasons for believing that both the sulphur–mercury theory of metals and the use of the Lo Shu numbers as the fundamental 'constants' in Arabic alchemy originated from intercourse with the earlier alchemy of China. But any reader with a Chinese background who studies attentively the available translations and descriptions of Arabic texts acquires gradually an uneasy feeling of *déjà vu*. Certain ideas and expressions seem unexpectedly familiar. It is therefore greatly to be hoped that much further work may be done, preferably by scholars who are masters both of Arabic and Chinese as well as having some acquaintance with chemical science, on the similarities which thus present themselves. Here we can do no more than draw attention to some of them. What needs to be done is the converse of that which was accomplished for the Greek sources by Kraus (3), namely to find out what Chinese sources there could have been for Jābirian ideas, and try to pin-point which of these are more easily explicable from the Chinese than from the Greek angle.

For example, the structure of the implicit natural philosophy should be examined carefully. What connection could there have been between the Chinese Five Elements[c] and the continuing Arabic predilection for speaking of a pentad of principles? 'Five elements or kinds of matter', it has been said, 'are consistently differentiated by Arabic writers'[d]—and with many variations this is true from al-Kindī in the +9th century through al-Rāzī and al-Farābī in the +10th to Solomon ben Gabīrōl[e] in the +11th and beyond.[f] Moreover there are passages speaking of the mutual generation and destruction of the elements.[g] Then maleness and femaleness, recalling so strongly the Yang and Yin, is found throughout the Arabic texts;[h] even to such statements as: 'once it has arrived at its maximum of perfection, a thing can only decrease.'[i] The metals were divided into two groups, gold, iron, tin, lead and *khārṣīnī* being male; silver, copper and mercury female.[j] So also were other chemical substances; for example there were two forms of *maghnīsiyā*, one female and the other male.[k] Only

[a] There have been two obvious hints pointing to the Ṣābians and Ḥarrān in the preceding pages.
[b] One could even hazard a guess that they were gynaecologists in view of the frequency with which the Lo Shu appears in Arabic texts from the beginning as a childbirth charm.
[c] Cf. Vol. 2, pp. 242ff., 253ff.
[d] Multhauf (5), p. 147, cf. p. 121. On the five principles, found alongside the four Aristotelian elements in the Jābirian Corpus, see Kraus (3), p. 137.
[e] More often known as Ibn Gabirol or Avicebron (+1021 to c. +1058).
[f] On some of the later Arabic alchemical writers cf. Taslimi (1); Razuq (1).
[g] Berthelot & Houdas (1), pp. 78, 100 (K. al-Ḥabīb).
[h] For example, Berthelot & Houdas (1), p. 69 (K. al-Qarāṭīs), 76–7, 79, 103 (K. al-Ḥabīb), 121 (Ar. Ostanes), 131 (Kitāb al-Mulk, Book of Royalty, Kr 454). Cf. Multhauf (5), p. 132 on Balīnās.
[i] Berthelot & Houdas (1), p. 77 (K. al-Ḥabīb).
[j] Cf. Siggel (3); Multhauf (5), p. 133. Mercury was sometimes judged to be hermaphrodite.
[k] Stapleton & Azo (1), p. 57 (after al-Khwārizmī al-Kāṭī and the Rasā'il of the Brethren of Sincerity).

those with a command of the whole range of Arabic texts will be able to tell us whether this Yin–Yang mentality is wholly capable of explanation from the Mediterranean background; meanwhile we suspect at least a Chinese contribution. And from time to time there are the strangest of Taoist echoes—the superiority of movement to rest (the 'unrestingness of the sage'),[a] the cosmogonic procession of numbers (as in the *Tao Tê Ching*),[b] and similarly the overcoming of the strong by the weak.[c]

The Yin–Yang differentiation comes in again with regard to the two spiritual entities envisaged by Arabic naturalists as intrinsic to all things and substances.[d] There was the *rūḥ* 'spirit', the male 'animus' tending to return upwards to the heavens, and the *nafs* 'soul', the female 'anima' tending to sink into the earth below.[e] The former assured continuity (hence also, later, longevity), the latter moulded the visible form and the individuality; but there was much variation in usage, so that *rūḥ* could be applied, for example, to distillate vapours. In the Jābirian period *rūḥ* normally translated Greek *pneuma* ($\pi\nu\epsilon\hat{\upsilon}\mu\alpha$)[f] and *nafs* Greek *psychē* ($\psi\upsilon\chi\acute{\eta}$),[g] but chemically the former signified the group of volatile substances, sometimes five, sometimes eleven in number,[h] i.e. the *arwāḥ*. Two of these, however, sulphur and arsenic, were at times distinguished as *nafs*, together with the volatile essential oils. In the Corpus we find three tractates on *rūḥ*[i] and one on *nafs*.[j] The question is, how far can all this be accounted for purely on Hellenistic, Syriac and Persian grounds, and whether the possibility of some parallelisms with the *hun*[1] and *pho*[2] theories of Chinese naturalism would not be well worth investigating (cf. pt. 2, pp. 85 ff. above).[k]

The same considerations present themselves at the more practically chemical level. The relations of cinnabar to gold are sometimes spoken of in a very Chinese way,[l] and 'our gold is better than that of the vulgar',[m] just as Pao Phu Tzu would have said (cf. pt. 2, pp. 68 ff. above). Of natural sympathies and antipathies there is a great deal

---

[a] Berthelot & Houdas (1), p. 76 (*K. al.-Ḥabīb*). Cf. Vol. 4, pt. 1, p. 61.

[b] *Op. cit.*, p. 91, from the same book.   [c] *Op. cit.*, p. 105, again from the same book.

[d] Attention was drawn to this by Mahdihassan in a number of papers: (31), pp. 24, 31, (32), pp. 335, 339, 343, (33), p. 82, (34), p. 81. Massignon (5) made a special study of it.

[e] A typical passage showing how these ideas were used may be seen in the Jābirian *Kitāb al-Zībaq al-Sharqī* (Book of the Eastern Mercury), Kr 470, tr. Berthelot & Houdas (1), p. 208.

[f] Hence it equated also with *chhi*.[3]

[g] See Kraus (3), pp. 153, 160, 166, 285 for *rūḥ*, pp. 21, 330 for *nafs*.

[h] Sulphur, arsenic, mercury, sal ammoniac (cf. p. 435 above), and camphor were the standard five, but elsewhere sulphur was differentiated into four kinds, the red, the yellow, the black and the white; arsenic into its two sulphides, the red and the yellow; while both mercury and sal ammoniac were considered to come in two forms, the 'mineral' or natural, and the 'derived' or artificial. The former could not have differed, but the latter doubtless distinguished between the chloride and the carbonate, cf. p. 432 above.

[i] The *Kitāb al-Rūḥ* (Kr 25); the *K. al-Rūḥ fi'l-Mawāzīn* (Book of the Animus and the (Science of the) Balances), Kr 1009; and *K. Rawḥ al-Arwāḥ* (Book of the Repose of the Animi), Kr 1007, presumably on the condensation of volatile substances.

[j] *Kitāb al-Nafs wa'l-Manfūs* (Book of the Anima and the Animate), Kr 822.

[k] Both in Neo-Platonic and later Latin medieval thought there is much to be pondered on in this context. For example Pagel (18) has examined the concepts of archaeus, astral body, entelechy, ochema, sophic fire, and the like, in the Gnostic and Hermetic background of Paracelsus.

[l] Berthelot & Houdas (1), p. 87, *K. al-Ḥabīb* again.

[m] *Op. cit.*, p. 181 (*K. al-Raḥma al-Kabīr*, Kr 5)

[1] 魂   [2] 魄   [3] 氣

in the Jābirian Corpus as elsewhere,[a] but only a detailed and wide-ranging analysis would enable us to decide how much of this goes back to Bolus of Mendes and how much to the Prince of Huai-Nan.[b] This was no doubt the pseudo-science out of which came the earliest conceptions of chemical affinity and reaction, so it is interesting to find general statements[c] recalling the Chinese 'categories' which we studied at an earlier point.[d] Lastly, the usage of 'cover-names' was universal among the Arabs just as it was among the Chinese. At one place we have the practice admitted very clearly,[e] at another we find names reminiscent of Chinese imagery, such as 'clouds-and-rain' for mercury.[f]

Here we are very near the borders of magic and divination; and there is evidence that some of this came westwards from China to Arabic culture also. Ibn Khaldūn in his *Muqaddima* (+1377) was very interested in a method of divination called *zā'iraja*.[g] One form of this was performed on a chart called *zā'irajat al-'ālam* or 'table of the universe', said to have been introduced by one Aḥmad al-Sabtī of Ceuta in the +12th century. It had a matrix system 55 compartments in breadth and 131 in length superimposed upon concentric circles representing the spheres and elements of the sublunary world as well as that of the spirits. Highly complex calculations done in connection with this started out from the astrological situation at the time, but then followed principles depending on the numerical values of the letters of the Arabic alphabet; so there was clearly a certain connection with the 'science of the balance' (p. 459 above) used in Jābirian alchemy. An exposition of the system is given by Dunlop,[h] who regards it as a good example of Arabic science in decline, though the *'ilm al-mīzān* of the Jābirian flowering-time was surely not much better. Interest for us lies rather in the fact that one variant of the technique was called the *zā'iraja Khiṭā'iyya*, and said to be due to one 'Umar ibn Aḥmad ibn 'Alī al-Khiṭā'i, presumably a Chinese from the borderlands who lived his life as a Muslim.[i] One would not be at all surprised to find that complex magic squares had played a part somewhere in the development of these systems.

There may have been another contact much earlier, judging from what we learn of the divination books associated with the name of the Sixth Imām, Ja'far al-Ṣādiq (p. 390 above). Ruska gives a table of 'elegant sand divination' (*qur'at raml laṭīfa*) with long and short lines, taken from one of these traditional books;[j] the symbols are distinctly reminiscent of the trigrams and hexagrams of the *I Ching*[1] (Book of

[a] *Op. cit.*, pp. 150ff. (*K. al-Mīzān al-Ṣaghīr*, Kr 369). Cf. Kraus (3), pp. 61ff., 65ff.
[b] Cf. pp. 311ff. above.
[c] Berthelot & Houdas (1), pp. 53, 58 (*K. al-Qarāṭīs*), 78 (*K. al-Ḥabīb*).   [d] Cf. pp. 305ff. above.
[e] Berthelot & Houdas (1), p. 214 (*Kitāb al-Zībaq al-Gharbī*, Book of the Western Mercury, Kr 471). Western mercury was a cover-name for water, and eastern mercury for oil.
[f] Siggel (3), p. 27.
[g] Tr. Rosenthal (1), vol. 3, pp. 182 to 227, with MS. illustration, pp. 204–5.
[h] (6), pp. 242ff. It has been used within living memory in Egypt; Lane (1), p. 239.
[i] This we know from the great Turkish bibliographer Ḥajī Khalīfa, who in his *Kashf al-Ẓunūn* (Discovery of the Thoughts) lists four books on it; Flügel (2) ed., vol. 3, pp. 532–3. We are much indebted to Professor D. M. Dunlop, who was curious about the Chinese connection, for bringing this subject to our knowledge.
[j] (5), pp. 26ff., 28.
[1] 易經

Changes),[a] or even more, perhaps, of the series of 81 tetragrams developed by Yang Hsiung[1] in his *Thai Hsüan Ching*[2] (Manual of the Great Mystery) around $+10$.[b] There would have been plenty of time for such ideas to have passed to the Arabs by the time of Ja'far, $c.+740$.

All such possibilities are doubtless somewhat nebulous and require much further investigation, but we have said enough in the preceding pages to suggest, if not indeed to prove, that theoretical influences more than negligible were exerted upon Arabic alchemy from the Chinese culture-area. Some were proto-science, paralleling the real knowledge of actual substances which also came, and others were pseudo-science, but even these last were straws on the stream to show the way the current was flowing. We have left till the last the most important influence of all, the name and concept of elixir, and it is to this that we are now in a position to turn.

(v) *The name and concept of 'elixir'*

In Arabic alchemical thought, *al-iksīr* was a substance which when added in projection (*ṭarḥ*) to any imperfect thing brought about a change for the better in the balance or *krasis* of its qualities, i.e. a transmutation (*qalb* or *iqlāb*).[c] Even a change to the perfect equilibration seen in gold was possible. Living things also were capable of a similar perfection, which in their case meant health and longevity, so that the *iksīrs* were naturally thought of as drugs, the 'medicines of man as well as of metals'.[d] And just as *iksīrs* would powerfully work on plants, animals and human beings no less than on mineral or metallic substances, so in their turn they could be prepared by art from any of the three natural kingdoms—a Chinese rather than a Hellenistic trait. The different schools (*ṭawā'if*) which emphasised one or other of these realms as raw material starting-points were discussed at length in one of the books of the Jābirian Corpus, the *Kitāb al-Lāhūt* (Book of Divine Grace).[e] In another work, perhaps older, the 'Opinions of Balīnās on Mineral Substances' (*Kitāb al-Aḥjār 'alā Ra'y Balīnās*),[f] it is declared that there are seven types of *iksīrs*, three uncombined, three with constituents drawn from two of the realms in different combinations, and one made of substances taken from all three realms. Processes of distillation nearly always enter into the preparations.[g]

---

[a] See Vol. 2, pp. 304ff.   [b] Vol. 2, p. 329.
[c] The best account is still that of Kraus (3), pp. 2ff. It seems hardly necessary to say that the word occurs consistently from the very beginning of Arabic alchemical writing—as in the Balīnās texts (Ruska, 8) and the *Kitāb Qarāṭīs al-Ḥakīm* (Book of Crates), on which see Berthelot & Houdas (1), pp. 65, 66, 70.
[d] 'The idea of the medical use of elixirs,' wrote Kraus, 'so widespread in (later) Latin and Indian alchemy, seems to have been unknown to the Greeks'.
[e] Kr 123, cf. Berthelot (12), p. 310.   [f] Kr 307 to 310.
[g] It is interesting to see the forerunners here of the fire-decomposition processes purporting to prove Aristotelian element-theory which were so devastatingly criticised by Robert Boyle in the 'Sceptical Chymist'. Fire was represented by gases (*nār* or *ṣibgh*) from the combustible material, Air by the condensable vapours or oils (*duhn*), Water naturally by the aqueous fraction (*muhallil*), and Earth by the mineral residue or *caput mortuum* (*arḍ*). This is common to al-Rāzī and most of the other Arabic writers.

[1] 揚 雄   [2] 太 玄 經

The provenance of the word *al-iksīr* has given rise to a good deal of discussion, for it has no obvious Arabic root.[a] Little help is available from the Jābirian writers themselves, who engage in the manner of their time in fanciful etymologies. For example, the *Kitāb al-Raḥma al-Kabīr* says:[b]

Al-iksīr was thus named because it has so great a power over the substances on which it is projected, transforming them and conferring its own nature upon them.[c] Others aver that the name originated because the elixir breaks and divides itself up;[d] and others yet again say that it got its name because of its nobility and superiority.[e]

It is probable, however, that the second of these suggestions was fairly near the mark. For since the first proposal of Fleischer in 1836 it has been generally assumed that *iksīr* was taken wholly from the Greek word *xērion* (ξήριον), found quite often in the Hellenistic Corpus.[f] In one clear statement of Olympiodorus (+6th century) this is identified as the 'dry powder of projection' (*epiballeis to xērion*, ἐπιβάλλεις τὸ ξηρίον, said of adding arsenic to copper);[g] but in many other occurrences the word 'projection' was supplied, not unreasonably, by Berthelot & Ruelle.[h] There is even a fragment entitled *Peri Xēriou* (περὶ ξηρίου, On the Powder), presumably part of a lost tractate,[i] in which it is said that the truest powder (*alēthestaton xērion*, ἀληθέστατον ξηρίον) has three powers, those of penetration, tincture and fixation. And it is interesting that the word, which originally probably meant any dry powder, had slight medical undertones, for the physicians used it to signify styptic preparations suitable for strewing on open wounds.[j]

In the Syriac texts of the +7th to the +9th centuries but based largely on early material in Greek, the word is perfectly recognisable in its new forms—*ksyra, ksirin,*

[a] For example, *kasr* (reduction, breaking up) would have given rather *taksīr* (priv. comm. from Dr Said Durrani).

[b] Kr 5, tr. Berthelot & Houdas (1), p. 181 (para. 46).

[c] Here the pun seems to be on *qūwa* (power).

[d] This must be the derivation already mentioned from *kasr* (reduction, division into pieces), however unconvincing.

[e] Here several alternative words seem to be available for consideration as roots, and one cannot tell which the writer had in mind. Thus we have *qāhira* (noble, superior; as in Cairo); *qudra* (capability, competence); and *kathīr* (manifold, numerous, multipotent). The translation of Houdas here, as often elsewhere, is a mere précis of the richness of the Arabic text, as we have had occasion to appreciate in the study of this passage with Dr Muhammad N. Yacout of Caius, to whom warmest thanks are due.

[f] For example see Singer (8), p. 49. And the derivation is given as standard in various modern dictionaries.

[g] *Corp. Alchem. Gr.* II, iv, 12.

[h] For example, *Corp.* III, vi, 11 and xxix, 24 (Zosimus, c. +300, speaking of the qualities of the powder); *Corp.* IV, xix, 9 (Iamblichus, also c. +300); *Corp.* VI, xv, 9 (Philosophus Anonymus, late +7th cent.); *Corp.* VI, xvii, 1 (a fragment of obscure origin, perhaps late); *Corp.* VI, xx, 12, 13, 17 (Nicephoras Blemmydes, +13th cent.).

[i] *Corp.* III, xxxi, included by the editors among the Zosimus material.

[j] The earliest occurrence of this is not, as sometimes said, in Galen, but in the Oxyrhynchus Papyrus 1142.7 (+3rd cent.), thereafter in Aetius, 6, 65 and Alexander of Tralles, 1, 15 (both of the +6th century). Galen recommends (in *De Meth. Med.*, Kuhn ed., vol. 10, pp. 320–1) a mixture of incense gum and bitter aloes as a styptic; according to Schleifer (1), p. 350 it is probable that this passage was the model for one in the Syriac 'Book of Medicines' (see Budge (6), vol. 1 (Syriac), p. 43, vol. 2 (Eng.), p. 41). But here the word *ksyryn* is used, though *xērion* is not in Galen. Budge missed the point, and translated 'powdered aloes'. We gratefully acknowledge discussions with Dr Sebastian Brock on these matters.

ksyryn, iksirin, eksirin—and in association with projection (*arma*, from *rma*, to throw). [a] Indeed it is used even more frequently. But it seems no longer to mean a dry powder only, for it is said to be like honey, like ice, like a metal, like rust, or a distillate, or even an oil. This suggests that by the beginning of the +8th century an inflow of other ideas was entering the Arabic world from some quite different quarter, accompanied perhaps by a similar sound which was identified with the *x* or *ks* phoneme, and carried with it a powerful reinforcement of the idea (so strong as to be essentially a new thing), that the elixir partook of the nature of a medicine.

This consideration leads us to see some value in the proposals which have been made to derive *iksīr* from Chinese roots, just as in the case of 'chem-' which we examined earlier (pp. 351 ff.). Dissatisfied with the purely Arabic or Greek derivations, Mahdihassan in 1957 suggested that Chinese phrases such as *yao chi*,[1] 'medicinal dose', or *yao chih*,[2] 'medicinal mushroom',[b] should be perpended.[c] The Thang pronunciation of the former would have been something like *iäk-dziei*, and that of the latter perhaps *iäk-tsi*,[d] but unfortunately neither phrase is at all a classical one.[e] Some years later Mahdihassan changed his mind and proposed the more unlikely *i chhi*,[3] which would mean something like 'unitary pneuma' (medieval pronunciation *ik-si*), not a phrase with any very close proto-chemical connections.[f] Soon afterwards the eminent sinologist Dubs came in with yet another suggestion, namely *i chih*,[4] literally 'essence of a juice', or as he took it, 'the substance of a fluid secretion';[g] the Thang pronunciation of which would have been something like *iäk-ts'iət*. There, apart from criticisms by Mahdihassan,[h] the matter has rested. The suggestion as a whole does not carry the weight which, as we saw, can be attached to *chin*[5] (*kim*), gold, as the origin of 'chem-', but it deserves perhaps to be retained for a while if only as a possible case of an erroneous but suggestive linguistic identification. If one visualises an Arab merchant of the +8th century in Canton or Hangchow or Sinkiang discussing alchemy with agreeable Taoist contacts, one can imagine his interest at finding a phrase which sounded so like the Syriac *iksirin* or Arabic *iksīr* with which he was already familiar, and we know how easy it always is in such cases to make an unjustifiable judgment of identity. 'How extraordinary—that's just what *we* say!' But imperceptibly of course he was absorbing a number of things which had not previously been said

[a] In Berthelot & Duval (1), for the main Syriac MS. see instances on pp. 41, 42ff., 46, 48ff., 51ff., 55, 76, 95–6; and for the Cambridge Syriac MS., p. 258. In the Arabic MS. of rather later date, written in Syriac script, see pp. 142, 168, 182–3. Here it always appears as *al-iksīr*.

[b] On this aspect of Chinese alchemical symbolism cf. pt. 2, pp. 116, 121, 125 above.

[c] (9), p. 128.   [d] Cf. Karlgren (1), esp. 800n.

[e] For example neither appears in the *Pao Phu Tzu* concordances of Ware and Schipper.

[f] (17), p. 75, (20), (21), pp. 197–8. Mahdihassan picked the phrase up from a reference by Li Chhiao-Phing (1), p. 17 to the *chen i chih chhi*,[6] the *chhi* of the primary vital unity; and this was in a way a misunderstanding since Li was speaking of a book of physiological alchemy, the *Huang Pai Ching*[7] (cf. pt. 5) under the impression that it was talking about proto-chemical processes. Cf. pt. 3, pp. 216ff. above.

[g] (34), p. 35. Correct *RBS*, 1968, vol. 7, no. 755, where the situation is wrongly stated.

[h] (19) and (25), pp. 49ff. He suggests, it is not quite clear why, that *kīmiyā'* travelled westwards by sea and *iksīr* overland.

[1] 藥劑   [2] 藥芝   [3] 一氣   [4] 液汁   [5] 金
[6] 眞一之氣   [7] 黃白經

among the Greeks, Syrians and Arabs, namely e.g. that the powder of projection was also a mighty medicine, the panacea of man as well as metals.[a]

In order to prove that this was how the Arabs saw the affair it will be necessary to give a number of direct quotations. Gruman (1), to whom we owe the best monograph on the history of macrobiotics, or what he called prolongevity, in all the Old World cultures, was disappointed in what he could find about such elixir effects in the translated texts which were available to him.[b] I think we can show that the harvest is not so disappointing, and that we come into the presence of macrobiotic medicines before the end of the century which saw the death of the Prophet. Let us start with two records of this early period, one which may imply only aurifaction though the word elixir is used, and one which unquestionably refers to a longevity drug.

The latter comes from the *Kitāb al-Imāma wa'l-Siyāsa* (Book of the Religious and Civil Authority) attributed to Ibn Qutayba who died in +889, but perhaps rather by one of his contemporaries.[c] The Caliph 'Abd al-Malik (r. +685 to +705) appointed his brother Bishr ibn Marwān as Governor of Basra, with Mūsā ibn Nuṣair as his principal adviser. Now Bishr was fond of pleasure and handed over the conduct of all affairs to Mūsā. While thus withdrawn from business:

One of the men of Iraq came before him, and said: 'In God's name, is it your wish that I give you a drink which will cause you never to grow old, subject to certain conditions which I shall lay upon you?' 'What are these conditions?' asked Bishr. 'That you do not allow yourself to be angry, do not mount a horse, and have no dealings with women, nor yet take any bath, for forty (days and) nights.' Bishr accepted these conditions, and drank what was given to him, shutting himself up from all men, near and far, and remaining secluded in his palace. And so he continued, till news suddenly reached him that he had been given the Governorship of Kufa as well as of Basra. At this, his joy and delight could not be contained. He called for a horse to go to Kufa, but the same man appeared and urged him not to set forth, and not to stir by the least movement from his place. But Bishr would not listen to him. When the man saw his determination, he said: 'Bear me witness against yourself that you have disobeyed me!' And Bishr did so, testifying that the man was free of blame.

Then he rode out to Kufa, but he had not gone many miles when, having placed his hand upon his beard, lo! it fell away in his hand.[d] Seeing this he turned back to Basra, but remained there not many days until he died. When the news of the death of Bishr reached 'Abd al-Malik, he sent al-Ḥajjāj ibn Yūsuf as Governor in his room.

There may be some degree of the fictional in this story, but the fact that it was current so soon after the event suggests at the least that people were talking about elixirs of perpetual youth or life around the time of the death of Bishr ibn Marwān, which can be fixed at +694. The remarkable story is repeated in a work indubitably written by Ibn Qutayba, the *Kitāb al-Ma'ārif* (Book of Knowledge in General), where he says that Bishr died after drinking the remedy called *idhrīṭūs* or *adhrīṭūs*.

---

[a] That Arabic alchemy had a far more medical character than Hellenistic proto-chemistry was noted long ago by Temkin (3), and we agree with him rather than with Ullmann (1), p. 150; cf. Vol. 5, pt. 2, p. 15. But it is even more true of Chinese alchemy, where, as we have seen in Vol. 5, pt. 3, the great majority of adepts were physicians too.   [b] Pp. 59 ff.

[c] What follows is in the words of Dunlop (6), pp. 208–9 and (7), pp. 3–4, with minor modifications.

[d] As Professor Darbishire pointed out at Guelph University, Ontario, this would suggest the presence of arsenic, lead or thallium in the elixir.

The identification of this term was solved by Dunlop (8), who suggested that the statement: 'a night spent at Ḥīrah is more healthful and profitable than the taking of a draught of the sherbet of Theodoretus'[a] had nothing to do with the + 5th-century theologian,[b] but referred again to *ahdrīṭūs* (*thādurīṭūs*), and that this was most probably just the Greek *adōrētos* (ἀδώρητος), 'not given'. And indeed the word does occur in the Zosimus writings, applied to some kind of agent used for blanching or whitening metal surfaces—a far cry from life elixirs. The passage runs thus:[c]

> This is the uncommunicated mystery which none of the prophets dared to divulge in words but revealed only to the initiated. In their symbolical scriptures they called it the stone which is not a stone, the thing unknown yet known to everyone, the despised thing of great price, the thing given by God and yet not given (*ton adōrēton kai theodōrēton*, τὸν ἀδώρητον καὶ θεοδώρητον). For my part I shall praise it under the name of the thing given by God and yet not given, for in all our works it is the only thing which dominates matter. Such is the drug of power (*to pharmakon to tēn dynamin echon*, τὸ φάρμακον τὸ τὴν δύναμιν ἔχον), the mithriac mystery.[d]

Thus the terminology here must have been of Greek origin, taken from a passage which even ends with a poetical reference to a drug. But the idea of a medicine of eternal youth was exceedingly un-Greek, and what is more, the story includes some remarkably typical Chinese features, especially the injunctions to refrain from all the passions during the course of the treatment or training—vital, as we shall find (pt. 5 below) in physiological alchemy, and very likely to have been stipulated by Taoist adepts offering life elixirs. Of course there is no need to suppose that the physician-alchemist in this case was actually Chinese, only that he must have been in contact with Chinese culture; and this could have been true of one man whose participation in the events has been suggested by Dunlop already, namely Māsarjawayh, the Syriac-speaking Jewish physician of Basra (*fl.* + 683 to + 717), later frequently referred to by Arabic men of science, and certainly living in a great trading centre where Chinese contacts in depth may well have been likely.[e]

The other story from this early period concerns a Jacobite bishop of the + 8th century, Isaac of Ḥarrān, and comes from an anonymous Syriac fragment edited and translated by Brooks (1).[f] This Isaac was a bad character, a *budmāsh*, irregularly instituted in the first place; to him came a strange wandering monk, who performed an

---

[a] From near the end of the 6th book of *Annals* of Ḥamza al-Iṣfahānī, dealing with + 7th-century events.

[b] However, in the time of Paul of Aegina (*fl.* + 640, cf. Sarton (1), vol. 1, p. 479) there really was an antidote called Theodoretus (see Adams tr. (2), vol. 3, p. 520) made up with or without anacardia. It only meant 'God-given'. Paul stayed in Alexandria after the fall of the city, and had great influence on Arabic medicine.

[c] *Corp.* III, ii, 1.

[d] I.e. Mithraic, a *hapax legomenon* in the Corpus, but interesting for its Persian background.

[e] Hitti (1), 2nd ed., p. 255; Dunlop (6), p. 213. He was one of the earliest of the translators from Greek and Syriac into Arabic, as we saw on p. 411.

[f] Parallel texts are those of Michael the Syrian, *Chronicle*, XI, 25 (Chabot ed., vol. 2, pp. 523 ff.) and Bar Hebraeus, *Chron. Ecclesiast.* (ed. Abbeloos–Lamy, vol. 1, p. 315). Both translate the Syriac word *eksirin* wrongly as *secretum*. The significance of Isaac's birthplace and background (cf. pp. 426 ff.) will not be overlooked. We are much indebted to Dr Sebastian Brock for communicating to us knowledge of this fragmentary chronicle, in which further references to alchemy in one form or another occur.

aurifaction in his presence using an elixir. A day or two later, Isaac, accompanying him on his way, murdered him by throwing him down a well, but found in his cloak neither a supply of the *eksirin* nor instructions for making it. Isaac eventually got himself made some kind of patriarch because of his pretended art, but when he proved unable to teach it to the secular Muslim ruler, the emir had him executed in +756. Here there is nothing overt about macrobiotics save the name of the substance, but the story has an uncanny similarity to a hundred others found in Chinese texts of earlier as well as subsequent times.[a]

Hardly more than half a century later we are at the beginning of the Jābirian Corpus, which contains a good deal more about the idea of human imperishability than has sometimes been thought. Here everything resolves round the conception of *'adl* or *krasis*, that perfect heavenly equilibrium, if only it could be attained. Though the idea is in part very Greek, very Galenic, it is also in part very Chinese,[b] since perfection of equilibrium between Yin and Yang had been the highest longed-for good in that culture since the middle of the −1st millennium. Thus

I have shown you by examples [says the writer of the *Kitāb al-Mīzān al-Ṣaghīr*][c] the necessity of the equilibrium of the qualities in the performance of the Great Work; apart from that it does not often matter so much. You must know that this equilibrium is indispensable in the Science of the Balance and the practice of the Work; one way may be easier than the other but the principle is identical. Thus I have told you about waters, equilibrium, synthesis, analysis, softening and the like, I have shown you everything if you have eyes to see—but if you have eyes and see not, the fault cannot be attributed to me.

In the 'Book of the Concentration' (*Kitāb al-Tajmīʿ*)[d] I wrote that 'if we could take a man, dissect him in such a way as to balance his qualities, and then restore him to life, he would no longer be subject to death.' Admit the good sense of the argument—have I not said that living beings need an equilibration of their qualities? This once obtained, they are no more affected by change and decay, and undergo no further modifications, so that neither they nor their children can ever perish. By God's grace no longer need they fear diseases such as leprosy or elephantiasis. He who does not know this knows nothing, and there is no science in him.

Similar notes are struck in many Jābirian texts. The *Kitāb al-Raḥma al-Kabīr*, for example, says that the death of a man is always due to one of the humours (the qualities) overwhelming another, or all the others; this is why the soul has to leave the body.[e] Elsewhere in the same book we read:[f]

The least fragile things are those which have the least opposition (of qualities) within themselves. They are the best equilibrated, the best composed, they last longer than others and are less liable to dissolution. They resist best all destructive forces tending to the separation of the spiritual and corporeal elements. Now those things which suffer most from

---

[a] Cf. pt. 3, pp. 106, 212, 215 above.

[b] Cf. pp. 226, 253 above. Another Arabic term which could be used for *krasis* was *i'tidāl*, perfect proportion or equilibrium (Bürgel (1); Ullmann (1), p. 173).

[c] Kr 369, tr. Berthelot & Houdas (1), pp. 147–8, eng. auct. adjuv. Gruman (1), p. 60.

[d] Kr 398. This is the book which contains so much on the principles and practice of artificial generation; cf. p. 396 above and pp. 485 ff. below.

[e] Kr 5, tr. Berthelot & Houdas (1), p. 174. One of the earliest of the texts.

[f] Tr. Berthelot & Houdas (1), p. 173, eng. auct., adjuv. Gruman (1), pp. 60–1.

internal oppositions are the living beings, animals and especially man. So long as the qualities are balanced in their oppositions, he remains in a state of health, but if one of them gains over the others he falls ill and the gravity of his illness is proportional to the excess of one of the qualities over the others. If it dominates them too violently death will ensue and the soul will separate from the body. Thus was it that God created man; if He had wanted him to live for ever He would have planted in his being only concordant elements, not warring ones. Since He did not do this He must have intended every living thing to die. And since God did not want any living being to subsist for ever, he afflicted man with this diversity of the four qualities which leads to death...

As Gruman says, this is rather different from the confident tone of Ko Hung and the alchemists of China,[a] but something must be allowed for the conventionalities of Muslim piety.[b] Of course no other culture had the Chinese belief in a distinctively material immortality, but it was possible for that to turn into a belief in extreme longevity when it found its way into the monotheistic lands of the 'People of the Book', who had always before them the examples of the Old Testament patriarchs. This point we must return to shortly, here more is to be said about the elixir as a medicine both metallurgical and human.

The 'fevers' of metals are cured by the elixir. In the same *Kitāb al-Raḥma* we are told that red and yellow copper are patients suffering from a hot fever of the yellow bile and blood, while tin and mercury have a cold fever due to the black bile and the phlegm.[c] This is imagery for which there are some slight echoes in the Hellenistic Corpus,[d] but in other places the term 'remedy' is freely applied to reagents affecting transformations in the appearance or composition of metals and alloys.[e] In another book, the *Kitāb al-Zībaq al-Gharbī* (Book of the Western Mercury), this mysterious substance is apostrophised in words which recall the metaphorical rhapsodies of the Hellenistic Corpus (cf. pt. 2, p. 72 above) and yet suggest the presence of something newer and more concrete, certainly not much constrained by the piety we have just encountered.[f]

Know that this 'water' has been named Divine because it brings the qualities out from among the qualities and revivifies the dead; therefore it has been called the 'Water of Living Things', just as the stone has been called the 'Animate Stone'. It is the Water of Life, and he that has drunk of it can never more die...

This is phraseology which we find often repeated in later Arabic writers.

---

[a] In another place (Berthelot & Houdas tr., p. 188) it is suggested that the union of 'souls' (volatile substances) with 'bodies' (non-volatile substances) is a model for the dead whom God will resurrect at the last day. They will then be complete and immortal, but some destined for heaven and some for hell.

[b] Islamic theologians may well have doubted the legitimacy of interfering with an individual's allotted life-span. Hence perhaps the *Responsum de Longævitate* in the *Kitāb al-Ajal* of Moshe ben Maimōn (Maimonides) in the +12th century; the great Jewish philosopher opined that longevity techniques counselled by physicians were not an infringement of divine authority because God would already have known and planned both the consultation and the actions. There is a translation by Weil (1), for knowledge of which we are indebted to Prof. F. Klein-Franke. A corresponding non-theistic paradox faced by the Taoists has been discussed in pt. 2, p. 83.

[c] See Berthelot & Houdas (1), p. 172.

[d] E.g. a Zosimus text, *Corp.* III, xix, 2, where the aurifaction of copper by polysulphide films is compared with the healthful effects of balanced food and drink on men.

[e] Berthelot & Houdas (1), p. 181.   [f] Kr 471, tr. Berthelot & Houdas, p. 213.

But where the Arabs went far beyond anything earlier in the West was their actual administration of elixir preparations to desperately sick human beings—a proceeding which brought them completely into line with the lineage of Chinese alchemist-physicians. Three striking Jābirian stories have been translated out of many more, and they are worth giving entire, both for their colour and for their revolutionary nature. All these come from the *Kitāb al-Khawāṣṣ al-Kabīr*.[a]

One day [saith Jābir], when my renown as a learned man and true disciple of my Master had already become known, I found myself at the house of Yaḥyā ibn Khālid. This man had a noble slave-girl endowed with perfect beauty, intelligent, well brought up, and good at music; nobody else had anyone like her. But being afflicted by some illness she had taken a purgative which made her so sick that in view of her constitution it seemed doubtful whether she could recover. She vomited so much that she could hardly breathe or speak.

Yaḥyā having been informed of her state by a messenger asked me what I thought of the case. As I could not see her I recommended cold water treatment, for at that time I knew nothing better for use against poisons. However, it did no good, nor hot treatment either, for I had counselled warming her abdomen with hot salt and bathing her feet with hot water.

As she continued to get worse, Yaḥyā took me to see her, and I found her half dead and greatly exhausted. Now I had with me a little of this elixir, and I made her drink two grains of it with three ounces of pure *sakanjabīn*.[b] By God and by my Master, it was not long before I had to cover my face before this girl, for in less than half an hour she regained all her beauty.

Then Yaḥyā prostrated himself before me and embraced my feet, at which I begged him, as a brother, to give over. So he asked me about this medicinal elixir, and I offered him the rest of what I had with me. However, he would not take it, but from that time began to study and practise the sciences until he had acquired much knowledge. Yet his son Ja'far went beyond him in intelligence and learning.[c]

On the exact historicity of this account there is no need to insist, what matters is the conviction that chemicals could be used in this way. 'Jābir' had a similar experience with a slave-girl of his own.[d]

According to what she said, she had taken unwittingly as much as an ounce of yellow arsenic. I could not find any remedy for her condition though I tried all the antidotes I knew. Finally I made her drink a grain of this elixir in honey and water. No sooner had it entered into her body than she vomited the arsenic and was restored to health.

And thirdly, there was a case of snake-bite poisoning.[e]

As I went out one morning to go to the house of my Master Ja'far (may the blessing of God be upon him), I came upon a man whose whole right side was dreadfully swollen, without exaggeration as green as a beetroot, and in some places already blue. I asked what the matter was, and he answered that this had come on after he had been bitten by a viper. I therefore obliged him to take two grains of this elixir dissolved in cold water, for I believed

[a] Kr 1900. What follows is from the translation of Kraus (2), p. xxxviii, eng. auct., adjuv. Gruman (1), p. 61; Temkin (3), p. 145.
[b] Oxymel, vinegar and honey evaporated to a syrup; cf. Browne (1), pp. 41, 87.
[c] This is the putative teacher of Jābir ibn Ḥayyān discussed on p. 390 above.
[d] Tr. Kraus (2), p. xxxix, eng. auct., adjuv. Gruman (1), p. 61.
[e] Tr. Kraus, *loc. cit.*, eng. auct., adjuv. Gruman, *loc. cit.*

him to be on the point of death. By God, the green and blue discolorations disappeared and were replaced by the natural colour of the body; and after some time the swelling went down and his side became normal. Having recovered his speech he got up and went home, entirely cured.

One cannot help being reminded of the indiscriminate use of the term *tan*[1] in Chinese for elixir and compounded medicine.

Another feature of the Jābirian writers is that from time to time they actually have to do with adepts of incredible longevity. Ḥarbī the Ḥimyarite is a case in point. In the *Kitāb al-Ḥāṣil* (Book of the Result),[a] it is said, in connection with the glyphomantic part of the Balance Theory, that 'Jābir' learnt the names of the metals in the Ḥimyaritic language from this shaikh who was aged 463 years. They could then take their place in a table of such names along with Arabic, Greek, 'Alexandrian'[b] and Persian. Ḥarbī, who appears again in a number of other Jābirian books,[c] is claimed by Jābir as his master, and actually appears in the title of one of them,[d] so that he must have been, or was credited with having been, an alchemical adept himself. A macrobiotic shaikh of this kind ranks almost as a *hsien*.[2] And have not certain scholars,[e] greatly daring, ventured to suggest that *djinn* (*jinn*) was a loan-word derived from *hsien*? We would not like to be committed to following them, for the accepted etymology is from *djanna*, 'covered, veiled', a term in Semitic languages for poetic and prophetic possession as well as madness, and hence for the spirits so possessing.[f] Nevertheless the idea is suggestive.

After the end of the Jābirian Corpus period, i.e. towards the end of the +10th century, the doctrine of '*adl* (*krasis*) in relation to longevity and immortality was again very clearly stated by Ibn Bishrūn. His 'Epistle on Alchemy' has survived only because it was incorporated by Ibn Khaldūn in his *Muqaddima* or 'Introduction to History' (+1377).[g] What Ibn Bishrūn wrote was this:

It should be understood and realised that all philosophers have praised the soul and have thought that it governs, sustains and defends the body in which it is active. For when the soul leaves the body it dies and grows cold, unable any more to move or defend itself because there is no life in it and no light. I have mentioned the body and soul only because this craft (alchemy) is similar to the body of man, which is built up by regular meals and which persists and is perfected by the living luminous soul, that soul which enables the body to do those great and mutually contradictory things that only its informing presence can authorise. Man

---

[a] Kr 323, one of the *Kutub al-Mawāzīn*. See Kraus (3), p. 261.

[b] Presumably demotic Egyptian, Coptic.

[c] For example, *Kitāb Usṭuqus al-Uss al-Thālith* (Third Book of the Elements of the Foundation), Kr 8; *Kitāb al-Rāhib* (Book of the Hermit), Kr 630; *Kitāb al-Tajmī'* (Book of the Concentration), Kr 398; *Kitāb al-Dhahab* (Book of Gold), Kr 947.

[d] The *Kitāb Muṣaḥḥaḥāt Ḥarbī* (Book of the Rectifications of Harbi'), Kr 211. In other places he is called Ḥarbī al-Yemenī (Stapleton & Azo (2), p. 72). Ḥimyar was part of the Yemen.

[e] Tenney Davis (4); Mahdihassan (9, 11). The latter draws attention also to the parallelism between *yü nü*[3] and *houris*.

[f] See Wensinck (3). The old derivation from Lat. *genius* is now discredited.

[g] Tr. Rosenthal (1), vol. 3, pp. 230 ff. The letter was written, somewhere about +1000, by Abū Bakr ibn Bishrūn to Ibn al-Samḥ, both having been pupils of Maslama al-Majrīṭī (cf. p. 401) in Spain. The passage quoted is on p. 232 of Rosenthal's translation, mod. auct.

[1] 丹   [2] 仙   [3] 玉女

suffers from the disharmony of his component elements. If his elements were in complete harmony, and thus not affected by accidents and inner contradictions, the soul would never be able to leave the body. Man would then live eternally. Praised be He who governs all things—exalted is He.

Here the last conclusion is essentially identical with that of the Jābirian passages with which we began. But there is something curious about the third sentence, for why should Ibn Bishrūn compare the alchemical craft with the human body if he was not implying that the latter could be so improved, so equilibrated, as to fix and retain the soul in perpetuity—or even, conceivably, so perfected in its constitution as to need the aid of the soul no more.[a] The 'regular meals' would have been a way of referring to the contagious transmission of the elixir's perfect equilibration to whatever it was made to work upon,[b] whether base metal or flesh and blood. Just as the equilibration of the qualities in the metals was the essence of the aurifactive process, so once again we see the 'imperishable' metal a model for the everlasting man.[c]

It was about this time (+1010) that the great Iranian poet Firdawsī was completing Persia's national epic, the *Shāhnāma*. In this there are clear references to life and health elixirs, as has been pointed out by Sarwar & Mahdihassan (1).[d] In Firdawsī's writing, the word *al-kīmiyā'* (cf. pp. 351, 355 above) has a variety of meanings, including (a) sagacity, stratagem, plot, plan, (b) fraud, deceit, (c) the art of alchemy, (d) an actual substance, the elixir or philosophers' stone,[e] and (e) a plant of immortality. At one place an urgent request is sent for an elixir made from sal ammoniac which restores the wounded to health, at another there is an account of how the dead on a battlefield were not resuscitated by a vegetable elixir brought from India. It would be worth while pursuing further the pioneer work of Sarwar & Mahdihassan. A few decades later there was produced in Egypt the *Ta'widh al-Ḥākim bi-'amri'llāh* already referred to (pp. 391, 436), an alchemical work which contains a good deal about the 'Water of Life' or 'Divine Water' bringing resurrection to everlasting life.[f] But as this kind of language occurs in the close context of chemical operations it may be a late echo of Hellenistic allegory rather than Chinese materialism and concreteness.

[a] This point was overlooked by Gruman (1), p. 60, who quoted only the penultimate three sentences.
[b] This is enlarged upon in a later passage of the same text, Rosenthal tr., p. 268.
[c] It may be worth pointing out that if the idea of *i'tidāl* (*krasis*) was near to modern physiology and endocrinology, it was singularly far distant from the modern classical concept of chemical purity. Probably this could never have arisen before the age of atomic chemistry, with its populations of molecules all chemically identical and free from 'interlopers'. Nor would it have seemed strange to ancient and medieval proto-chemists that their ethically 'purest' substance, the Philosophers' Stone or Elixir (for there is always a psychological undertone about 'purity'), should be a chemically 'impure' mixture. We remember Ko Hung thinking that aurifactive gold was in some sense a compound body (Vol. 5, pt. 2, p. 2). Of course in modern times definitions of chemical purity have all had to be revised in the light of our new knowledge of large molecules of biological origin; as Pirie (2) has said: 'purity is a concept that has no meaning except with reference to the methods and assumptions used in studying the substance under discussion'. See also Pirie (3). Thanks are due to Dr Brian Cragg of Toronto for raising this question.
[d] Cf. also Mahdihassan (15), p. 84.
[e] This is the sense in which we find the word in the *Kīmiyā' al-Sa'āda* (Elixir of Blessedness), title of a notable book of ethical and devotional character by the religious philosopher al-Ghazālī al-Ṭūsī (+1059 to +1111), tr. Ritter (3).
[f] See Ruska (5), pp. 84, 119–20. This is one of the Pseudo-Ja'far texts.

The way people's minds were working can sometimes be appreciated better from poetry than from expository prose. For this reason the writings of Ibn 'Alī al-Ḥusain al-Ṭughā'ī (p. 402) are of considerable interest.[a] Besides his treatises, this alchemist (d. +1121) has a poem in the *Maṣābīḥ al-Ḥikma* about a king who was childless and who sought successfully upon an island (as it might be Phêng-Lai)[b] the Water of Life which cured him. This is also in the *dīwān* of al-Ṭughrā'ī (*al-Maqāṭī' fī'l-Ṣan'a*), where we find poems too on the curing of an impotent king, and the restoration of youth to an old one. There is further an alchemical dream experienced by the author himself—a vision of a bottle of the Water of Life and a silver pot containing the Soil of Paradise; al-Ṭughrā'ī drank some of the former, but was forbidden by the Prophet in person to eat any of the latter.

Among the representatives of late Arabic alchemy we may take Ibn Aidamur al-Jildakī, who died in +1342. His remarkable work *Nihāyat al-Ṭalab* (The End of the Search)[c] was essentially a commentary on the *Kitāb al-'Ilm al-Muktasab...* (Book of Acquired Knowledge in the Cultivation of Gold) written by Abū'l-Qāsim al-Sīmawī al-'Irāqī about +1270, always referred to by al-Jildakī as 'the Shaikh'.[d] But although so respected, it has been shown that al-'Irāqī lifted much of his material bodily from Ibn Umail's +10th-century *Kitāb al-Mā' al-Waraqī* (cf. p. 401 above)— at any rate this assures us of being in the presence of a single long and continuous tradition. In al-Jildakī's works it is easy to find traces of life-elixir ideas. He comments, for instance, on passages from the Jābirian 'Book of the Western Mercury' such as the following:[e]

Western mercury is considered by the sages to be the soul, for it is cold and moist. It is also the Divine Water, since it liquefies the parts and prevents fire from burning. Its coldness is due to its whiteness and to moisture, for it is a water, and every water is cold and moist. Some of the sages have said that it is dry, for it does not respond to the smelting-fire, and in comparison with oil it actually is dry. Some other sages have stated that it is moist, and what they meant was that the tincture will not penetrate unless it is dissolved in it....

Mercury is soul,[f] and there is nothing of the same status in the world. It is the living soul which on mixing with a body animates it and transforms it from one state to another and from one colour to the next.[g] It is the Water of Life and the spring of vitality from which whoever drinks never dies....

[al-J.: Understand the words of this learned teacher, whose eminence, in theory as in practice, has been matched by no one, either among those who preceded him or those who have come after him.][h]

---

[a] See the monograph by Razuq (1), especially pp. 161, 185, 245, 255 and 268 for the points referred to.

[b] As regards Eastern influences, all the names of 58 previous alchemists given by al-Ṭughrā'ī are Greek, Jewish, Syriac, Arabic or Persian, but he does mention one Asfīdiyūs who was the first of five philosophical ministers of an Indian king incongruously named Adriyānūs. And a work existed with the title *Kitāb al-Wuzarā' al-Khamsa li-Malik al-Hind*.

[c] On this there is a detailed monographic study by Taslimi (1), pp. 134–5, 189, 202, 374, 540–1 and 547 of which document the points made immediately below.

[d] On al-'Irāqī see Mieli (1), 2nd ed., p. 156.

[e] Cf. pp. 404, 471 above. Parts of this quotation were omitted by Berthelot & Houdas (1), pp. 212 ff.

[f] Surely a reference to its volatility, or an idea derived from that.

[g] This sounds rather like the Hellenistic colour succession, cf. pt. 2, pp. 22 ff. above.

[h] We were led to this passage in Taslimi, pp. 134–5 by a reference in Mahdihassan (32), p. 346.

Elsewhere al-Jildakī speaks of the preparation of a 'foodstuff' from sulphur and mercury as the secret of success in the Art, reminiscent of the 'regular meals' of life elixir to which Ibn Bishrūn referred (p. 480). And he goes on to say that 'the difficulty lies in combining "mercury" and "sulphur" in such a way as to form a simple homogeneous substance.' When this is done it resembles milk, and is called 'the Divine Water, preserving all those who taste of it from death, the Water of Life, the First Child, Viper's Saliva, Red Blood, and Birds' Milk.' This is the Water which helped Shaiṭān to expel Adam and Eve from the 'Middle Sphere', it 'kills the living and reanimates the dead, blackens the white and whitens the black.' Al-Jildakī often repeats such phrases and one begins with him to hear the apocalyptic tone of medieval Latin alchemy. But he emphasises the role of the elixir in medicine.

The philosophers' gold, when applied three times to the eyes of a person suffering from continuous flow of tears, cures him; if an eyelash is plucked with a pair of tweezers made of this gold it will grow no more;[a] if a plate of this gold is placed on the heart of someone suffering from palpitations he is sure to recover; and if this gold is dissolved and taken, it will cure all atrabilious diseases. Common gold exhibits none of these properties....

The philosophers' silver cures hot fevers, and in solution in date-wine constitutes a remedy for atrabilious diseases, while common silver does none of these things....

The elixir cures patients suffering from leprosy if it is applied to his sores and given him to drink as a potion. The sores burst and effuse a yellow water, after which new skin develops and no mark is left on the body...[b]

Finally, al-Jildakī warns his reader never to taste or smell the elixir, and to avoid its vapour at the time of projection, for it is a very dangerous poison, and that is the reason why it can subdue the poison of leprosy and other bodily diseases. But one can dilute the toxic nature of the elixir by mixing it with other drugs, and in this way it can be put to many good uses.

Our last witness can come from the +14th century, like al-Jildakī after the time of Latin Geber—Muḥammad ibn al-Akfānī al-Sakhāwī (d. +1348) who wrote in Egypt. The entry for alchemy in his encyclopaedia *Irshād al-Qāṣid ilā Asnā al-Maqāṣid* (Guide for the Struggling, on the Highest Questions) has been studied by Wiedemann. There he says that 'the elixir changes substances just as a poison does in a living body, but it changes them to health.'[c] And he also says:

The elixir is furthermore used in medical practice, bringing results beyond those of all ordinary drugs. It heals epilepsy and leprosy and suchlike diseases, just as Ḥunain ibn Isḥāq [+809 to +877] says it does in his 'Discussion' (*Maqāla*) on this question.[d]

As we have read through these various excerpts one particular point has become more and more noticeable, namely the conception of the elixir as poison. When Berthelot noticed this in the early *Kitāb Qarāṭīs* (Book of Crates),[e] he was reminded of the *ios* (ἰός) or *virus* often referred to in the Corpus and other writings of Hellenistic

---

[a] This was an old attribute of *khārṣīnī*, cf. p. 431 above.
[b] Al-Jildakī gives his own eye-witness account of a case of this kind.
[c] (21), p. 106. [d] *Op. cit.*, p. 107.
[e] In Berthelot & Houdas (1), pp. 54–5, 65, 67.

times, and was tempted to relate the 'fiery poison' of 'Crates'[a] to the 'fiery drug' (*to pyrinon pharmakon*, τὸ πύρινον φάρμακον) of Mary the Jewess.[b] But *ios* was an extremely vague and confused word in Greek; it could mean smells, odours and 'virtues', magnetic attraction, pharmacological active property, but also rusts and oxides, as well as violet or purple colorations, and even refining processes in general.[c] We have no need to deny that the idea of poison was present in Hellenistic proto-chemistry— how could anyone work with the vapours of mercury and arsenic and not have it? But the conception of the elixir as supremely beneficial both to inorganic and to organic things, and at the same time supremely poisonous, has an especially good parallel in Chinese thought, where *tu*,[1] so often wrongly translated 'poison', had for the pharmaceutical naturalists throughout the ages the meaning rather of 'active principle', active for good or evil according to the conditions. Many of them, indeed, would have warmly appreciated the great dictum of Paracelsus already referred to (pt. 3, p. 135) that whether or not a thing is a dangerous poison depends entirely on the dose.[d] The poison principle runs down through all Arabic alchemy, as has been seen, and thence directly into that of the Latins. At one end the Jābirian writer says:[e]

> The result we have sought to obtain is that the elixir when fully prepared should be a light, subtle, spiritual and corporeal poison.... It is called poison because of its subtlety and penetration (like the wafting of perfumes), and fiery because it resists the fire.[f]

At the other end Petrus Bonus is saying in +1330: 'sic et hic lapis efficit in metallis leprosis, et ideo quandoque venenum quandoque theriaca dicitur (thus also this Stone works in leprous metals, and therefore it is sometimes called 'poison' and sometimes "universal antidote")'.[g] And he quotes from Arabic and quasi-Arabic sources. Haly speaks of the 'fiery poison' (*toxicum igneum*);[h] and Morienus says: 'unde a quibusdam venenum appellari solet, quia sicut venenum in corpore humano ita elixir in corpore metallino...(hence (the Stone) is often called "poison" by some, because in the body of metals it works just as poisons do in the body of man).'[i] Thus did the tradition come down until the end of alchemy itself.

[a] It comes again in *Kitāb al-Ḥabīb*, *op. cit.*, p. 93. And in the MS. written in Syriac script; Berthelot & Duval (1), p. 182.

[b] See *Corp.* II, iv, 54 and III, xxviii, 8. The identification seems rather doubtful.

[c] Berthelot (1), pp. 61, 178, (2), pp. 13–14, 133, 254–5.

[d] Dritte Defension in 'Sieben Defensiones' (+1537 or a year later), ed. Sudhoff, vol. 11, p. 138. Our thanks are due to Dr Walter Pagel for the exact location of the statement.

[e] *K. al-Raḥma al-Kabīr*, tr. Berthelot & Houdas (1), pp. 174–5, 181.

[f] Similar statements are in the two *Kutub al-Zībaq*, Berthelot & Houdas, *op. cit.*, pp. 208, 213.

[g] *Pretiosa Margarita Novella*, in Manget, *Bibliotheca Chemica Curiosa*, +1702 ed., vol. 2, p. 49. The passage was noted by Stapleton (1), pp. 36–7 and App., p. ii.

[h] 'Et Haly in suis Secretis:...' Presumably this was a confusion between the advice to monarchs given in the *Secretum Secretorum* of Pseudo-Aristotle (cf. p. 368), or something in the alchemical work of al-Rāzī, *Kitāb Sirr al-Asrār* (Book of the Secret of Secrets), already often mentioned (pp. 195, 398), and some text of Haly Abbas, i.e. 'Alī ibn al-'Abbās al-Majūsī (the Mage) who died in +994, one of the greatest physicians of Islam (see Mieli (1), 2nd ed., pp. 120ff.; Sarton (1), vol. 1, pp. 677–8). Al-Rāzī certainly shared the ideas about the poisonous nature of the elixir (Ruska (14), p. 75). As for 'Alī ibn al-'Abbās, he wrote no 'Secrets' so far as we know, but one of the titles of his canon of medicine was *Kitāb al-Malakī*, in Latin *Liber Regius*, the Royal Book, so the mistaken attribution is understandable.

[i] On the Morienus book see pp. 390, 403 above.

[1] 毒

We have now seen that there is in fact a great deal in the Arabic alchemical literature on elixirs of life and everlasting life. Of course it is different in general character and also in details from anything which we find in the Chinese texts—that would be expected—but evidently the atmosphere in the Arab world from +700 onward is radically different from that of Hellenistic proto-chemistry. If this can be sensed only on the basis of texts which have been studied in modern research and translated into Western languages, what may we expect when the literally thousands of Arabic alchemical books not yet examined are placed at the disposal of the world republic of learning? But there is one final point to be made. Immortality or longevity elixir ideas did not have to reach Europe only through Islamic culture. If Gruman's pessimism had proved to be more justified than in fact it seems to have been this would have been even more important, but it always has to be remembered. Nestorian contacts and transmissions sometimes took place directly,[a] the Armenian kingdoms could sometimes be foci for ideas,[b] and in the travels of the magnetic compass we have already seen one vivid possibility of transmission through the +12th-century Western Liao kingdom, the Qarā-Khiṭāi.[c] The mid +13th century was not at all too late for direct influences, and that was just the time when Franciscans like William de Rubruquis (Ruysbroeck) were discussing sphygmology in China,[d] and Odoric of Pordenone disputing with Mahāyanist ho-shang (monks) about reincarnation.[e] The Italian merchants at Yangchow in the +14th century might have been a little on the late side, and even Marco Polo and his contemporaries too, but that there were channels short-circuiting both Islam and India we need be in no doubt. How far they carried the ideas with which we are here concerned remains to be seen.

There is one very important theme of Arabic alchemy which seems never before to have been set properly in the context of elixir doctrine, though Kraus gave it close and learned study.[f] This was the so-called Science of Generation ('Ilm al-Takwīn), concerned not only with the production of ores and minerals in Nature and in the laboratory, including the generation of the noble metals from the base, but also with the artificial asexual in vitro generation of plants, animals and even human beings.[g] It will not do to dismiss such ideas as merely 'medieval nonsense'. They often give deep insight into the minds of the men of that age, and they also illuminate what passed from one lot of men to another.

Let us therefore take a closer look at this extraordinary development, as we find it in the most explicit source, the Kitāb al-Tajmī' in the Jābirian Corpus.[h] The artificial

---

[a] There are many examples from the +5th to the +9th century, and later there was the epic of Rabbān Bar Sauma described in Vol. 1, pp. 221, 225. Cf. Budge (2).

[b] Cf. Vol. 1, p. 224. Also the Radhanite Jews and the Khazars (Vol. 3, pp. 681 ff.).

[c] Vol. 4, pt. 1, p. 332.

[d] Vol. 1, p. 224; we shall be returning to him in Sect. 44.

[e] Vol. 1, p. 190. A particularly delicate subject in the present context. Odoric might have talked with Taoists as well.

[f] (3), pp. 97 ff.

[g] We have already referred to it in the sketch of Jābirian alchemy on p. 396 above.

[h] 'Book of the Concentration' (Kr 398), cf. p. 435 above. There is also some discussion of artificial generation in the Kitāb al-Ikhrāj mā fī'l-Qūwa ilā'l-Fi'l (Book of the Passage from Potentiality to Actuality), Kr 331, tr. Rex (2); as well as many references in other books of the Corpus.

creation of minerals (*takwīn al-aḥjār*), of plants (*takwīn al-nabāt*), of animals (*takwīn al-ḥayawān*) and even of men and prophets (*takwīn aṣḥāb al-nawāmīs*),[a] by human artisanal action (*ṣāniʻ*), imitating the demiurge (*bāriʼ*) or creator of the world,[b] was a cardinal belief of the +9th century. These were the two sorts of generation (*kawn*) or creation (*khalq*) distinguished in the Balīnās texts, the first (*al-kawn al-awwal*), by God, the second (*al-kawn al-thānī*), by man.[c] A Jābirian writer, speaking of the elixir, says:[d]

If you can succeed in composing (or organising) the isolated things, you will assume the very place of the (World-) Soul in relation to Substance,[e] the isolated things occupying in relation to yourself the place of the (four) qualities (or natures)—thus you will be able to transform them into anything you wish.

And aurifaction was only one special case of this general principle. In Ibn Khaldūn's definition,[f] alchemy

is a science that studies the substance (the elixir) through which the generation of gold and silver may be artificially accomplished, and comments on the operation leading to it.

Moreover, the possibility of an artificial generation (*takwīn*) of plants and animals was not confined to Jābirian circles, itwas widely believed and discussed. Ibn al-Waḥshīya's *Kitāb al-Taʻfīn* (Book of Putrefaction),[g] *c.* +930, has much on it, and it was well known at the farther end of the Mediterranean in Muslim Spain, as is shown by the *Kitāb Ghāyat al-Ḥakīm* of Maslama al-Majrīṭī (or Pseudo-Majrīṭī), *c.* +1050 or a few years later. It was of course connected with the idea of natural spontaneous generation, prominent in the *Kitāb Sirr al-Khalīqa* (*c.* +820). Perhaps significantly, the *Rasāʼil Ikhwān al-Ṣafāʼ* and many other texts attribute that idea to India (or the Further Indies) and even place the creation of the first man by this means in India or Ceylon.[h] One has therefore to take the whole matter seriously. And the practical directions include some fascinating detail.

What sort of thing did they involve? In one procedure, in the *Kitāb al-Tajmīʻ*, a theromorphic glass vessel, shaped according to the animal intended,[i] contained the semen, blood, and samples of many parts of the organism[j] to be reproduced,[k] together

---

[a] Or legislators.

[b] Cf. Kraus (3), pp. 99, 104, 126.

[c] *Kitāb al-Aḥjār* (Book of Minerals), Kr 40. See also *Kitāb al-Mīzān al-Ṣaghīr* (Lesser Book of the Balance), Kr 369.

[d] In *Kitāb Maydān al-ʻAql* (Book of the Arena of the Intelligence), Kr 362. Tr. Kraus (3), eng. auct.

[e] Note how well this justifies our interpretation of the passage from Ibn Bishrūn quoted on p. 480 above.

[f] *Muqaddima*, Rosenthal tr., vol. 3, p. 227.

[g] Also entitled *Kitāb Asrār al-Shams waʼl-Qamar* (Book of the Secrets of the Sun and the Moon).

[h] Cf. Kraus (3), p. 121.

[i] This was the 'form' (*ṣura*), 'mould' (*miṭāl*) or 'effigy' (*ṣanam*).

[j] For birds the egg-white of the species was to be used, and *nūshādir* (sal ammoniac) combined with added dyeing materials to give to the feathers any colour desired. Cf. Kraus (3), p. 109.

[k] It was even thought possible to produce animals not existing in Nature. For example the sperm of a bird in a human mould would give rise to a winged man (Kraus (3), p. 116). How strange it is to reflect that in modern experimental embryology it has become possible to do this kind of thing, as also to mix a variety of tissue-rudiments and have them sort themselves out into a considerable measure of individuated organisation.

with drugs and chemicals[a] chosen in kind and quantity according to the method of the Balance;[b] all this enclosed at the centre of a cosmic model, a celestial sphere (*kura*), globular, latticed, or armillary,[c] set in continuous perpetual motion by a mechanical device.[d] Meanwhile a fire of the first, or unit, intensity (i.e. a mild one) was kept burning underneath.[e] If the exactly correct time was not reached, or if it was exceeded, no success whatever would be achieved. Other schools were partisans of 'putrefaction' (*sēpsis*, σῆψις, *ta'fīn*), or stressed the importance of aeration and stronger heat, or considered that blood was more essential than the chemicals;[f] some said that semen was indispensable if the new being were to have the power of speech, and parts of the brain if it were to be endowed with thought, memory and imagination. It was even averred that higher beings would come forth from the apparatus equipped already with the knowledge of all the sciences.[g] There can be no question that the origin of the famous homunculus of Paracelsus[h] lies here, but how far Aldous Huxley would have been surprised to find his 'Brave New World' of separated totipotent blastomeres and artificially incubated 'test-tube babies' anticipated in the dreams of these Arabic alchemists we would not undertake to say.[i]

A parallel passage about a perpetually rotating spherical cosmic model within which the transmutation of all the base metals into gold was performed, occurs in the *Kitāb al-Rāwūq* (Book of the Filter),[j] and may be read in the translations of Said Husain Nasr[k] and Kraus.[l]

All these constructions seem very un-Hellenistic, but they do signally recall the Chinese armillary spheres and celestial globes kept in continuous rotation by water-power, instruments which derived from polar-equatorial, not ecliptic-planetary, astronomy, and came into use much earlier than anything of the kind in the West.[m]

[a] Some constituents specified had an Eastern (Iranian) provenance.

[b] N.B. an elixir mixture.

[c] Instructions for making this are said to have been given in the *Ta'ālīm al-Handasa* (Teachings of Geometry), Kr 2805.

[d] In some instructions the central reaction-vessel also had to be made to rotate.

[e] For variant processes see Kraus (3), pp. 109, 110, 111 ff., 115, 117.

[f] It will be remembered that in Arabic alchemy elixirs were prepared not only from inorganic but also from plant and animal substances.

[g] Cf. Kraus (3), pp. 115–19.

[h] The main passage is in *De Nat. Rerum*, I, vi, Sudhoff ed., vol. 11, pp. 316 ff. The procedure is clearly derivative, for human semen was to be allowed to 'putrefy' in a cucurbit for forty days, then 'fed' cautiously with the arcanum of human blood for forty weeks (cf. Needham (2), p. 65). The theme recurs in Goethe's 'Faust', Pt. II, Act ii, Sc. 2, the laboratory. Earlier Latin allusions of intermediate date occur in the writings of William of Auvergne (d. +1249), a theologian much given to the study of magic, and in the *De Essentiis* of Pseudo-Thomas Aquinas (perhaps c. +1310) which attributes artificial generation processes to al-Rāzī (Thorndike (1), vol. 2, p. 353, vol. 3, p. 139). The origin of the homunculus has been sought in the 'little man' or *anthroparion* (ἀνθρωπάριον) who appears, with a 'silver man' and a 'gold man', in the visions of Zosimus (*Corp.* III, i, 2, 5); but as usual the term seems purely allegorical, no presage of the Jābirian cosmic incubators—in spite of Berthelot (1), pp. 60, 180; von Lippmann (20), pp. 35–6; Kraus (3), pp. 120–1.

[i] Two interesting papers have been consecrated by von Schwarzenfeld (1, 2) to the magical and alchemical activities of Rudolf II at Prague, and especially the objects in his collections still preserved in the Hradschin Castle. We did not expect to be able to illustrate a homunculus, but one of Rudolf's, actually a human figure of blown glass enclosed in a prismatic block of glass, perhaps only a model of what one should expect, appears in Fig. 1537.

[j] Kr 140.   [k] (1), p. 260.   [l] (3), p. 57 in French.

[m] See Vol. 4, pt. 2, pp. 481 ff. and its background in Vol. 3, Sect. 20.

Similar Indian ideas, especially concerning perpetual motion, are also recalled.[a] On alchemical cosmic models as such there are plenty of Chinese analogues and predecessors, as we have duly seen.[b] So much for the rotating cosmic shell.

As for the central vivification, Kraus' ingenuity was much exercised to find Hellenistic antecedents, but little was available save spontaneous generation, automata, and rituals for the animation of religious images, none of which is very much to the point. Artificial generation in the Arabic sense was, Kraus admitted, unknown in Greek writings.[c] Spontaneous generation on the other hand was of course widely believed, as of bees from the corpses of lions, and so on throughout the European centuries, faith in it dying out only with the growth of modern biology in the Enlightenment period.[d] It was equally widespread in Chinese culture.[e] But it was uncontrollable by men. As for moving and singing automata or puppets there is surely no need to refer to the works of the Alexandrian mechanicians,[f] but there were other more uncanny Graeco-Egyptian stories of speaking statues[g] and ever-rotating columns, which the Arabs inherited.[h] However, honours are about even here again, for Chinese culture also had a wealth of legends concerning automata, some of which, like the Taoist robot of King Mu of the Chou,[i] came very near indeed to being artificial flesh and blood. On the third point 'Jābir' connects the artificial generation schools with the image-makers (eidōlopoioi, εἰδωλοποιοί, muṣawwirūn),[j] raising therefore the matter of theurgic animation techniques. It was not a question necessarily of causing statues of the gods to move, but rather preparing them in such a way as to serve as the real abodes of the spirits which were to be worshipped through them, to assure the real

a See Vol. 4, pt. 2, p. 539 and Needham, Wang, & Price (1); also Lynn White (14), p. 70, (15).

b Pp. 279ff. above.

c (3), pp. 119, 123. It is true that 'Jābir' often claims to be only commenting on a 'Book of (Artificial) Generation' (Kitāb al-Tawlīd = Peri Genneseōs, περὶ γεννήσεως) by Porphyry of Tyre, the Neo-Platonist (b. +223), but it must be apocryphal as there is no such title among the well-authenticated books of that philosopher.

d See Needham (2), passim, and the classical monograph of von Lippmann (20). In the present context Kraus (3), pp. 106ff.

e Cf. Vol. 2, p. 421 and passim, also Sect. 39 in Vol. 6.

f See Vol. 4, pt. 2, pp. 156ff. and the translation of Woodcroft (1); with Diels (1), pp. 62ff., the basic contributions of A. G. Drachmann, and much other well-known literature. On Daedalus' wooden Aphrodite, moved, it was said, by mercury, see Aristotle, De Anima, I, 3 (406 b 12).

g One sometimes wonders whether there could have been any Ṣābian influence on the Arabic system of artificial generation; remembering especially the Martian sacrifice of the speaking head (p. 427 above). Here might be relevant also the later Jewish tradition of the golem, on which the chief study is a rare monograph by K. Mueller (1).

h In Egypt, it was reported, there were two statues borne by a column of iron which spontaneously and perpetually rotated on a mirror; the Jābirians said that this was discussed with other similar things in the Kitāb al-Ashkāl al-Ṭabī'iyya (Book of Natural Figures), Kr 2655. Other stories spoke of a leaning column perpetually rotating. Kraus (3), p. 113, surmised confusions with the famous Memnon statues whistling in the dawn wind, the fabled mirror of the Pharos at Alexandria, and the concave sundial or scaphe (cf. Vol. 3, pp. 301–2) with its inclined gnomon and ever-moving shadow. See also Reitzenstein (4); and Carra de Vaux (5), translating the Mukhtaṣar al-'Ajā'ib (Breviary of Marvels) of Pseudo-Mas'ūdī (c. +970), notably pp. 161, 198ff., 272, 278. Further in Dodds (2), pp. 194–5.

i See Vol. 2, pp. 53–4, translating a passage from the Lieh Tzu book well worth re-reading.

j It was strange that all this should have arisen in Islam, for the Prophet was severe, in one of his ḥadīth saying: 'On the Day of Resurrection the makers of images will receive the heaviest of punishments, and they will be told, "Give life to that which you have created".' See Suhrawardy (1) and other collections; Kraus (3), pp. 123–4, 134.

Fig. 1537. **Esoteric objects still** extant from the collections of Rudolf II in Prague (von Schwarzenfeld, 1, 2). On the left, a homunculus, or 'devil in a glass', actually a glass-blown figure enclosed in a prismatic block of glass. In the centre two mandrakes or mandrake-like objects; on the right a bell of Tibetan flavour. From the Curiosa of the Hradschin Castle, Prague.

presence, as it were, of these gods and spirits. The Neo-Platonists accepted the idea and wrote much on the practice;[a] from one source we learn that the liturgists observed the heavens to get the right time, and then placed the appropriate herbs, gems and perfumes in the statue, which itself had been moulded from clay mixed with holy water, aromatic plant and other material powdered and sieved, together with comminuted metals and precious stones.[b] But once again there was not much to choose between Hellenistic and East Asian practices, for in China and Japan there was the readying of images for the presence of gods, lokapalas, bodhisattvas, etc., even to the insertion of model viscera to make them complete,[c] then their formal consecration by the dotting in of the pupil of the eye.[d] One can only conclude that the Arabs did not have to rely exclusively on Hellenistic culture for what they knew (or thought they knew) about spontaneous generation, mechanically operated simulacra, or the animation of religious images. All this may have a certain relevance, yet it does not get to the root of the matter.[e]

[a] See Kraus (3), pp. 127ff.

[b] Porphyry, in his works: 'Philosophy of the Oracles' and 'On Statues', discussed by Bidez (2); as also his 'Letter to Anebo', an Egyptian priest, ed. and tr. (It.) by Sodano (1). Many references will be found in Kraus (3), p. 123, including the interesting book of Weynants-Ronday (1) and the revealing paper of I. Levy (1), esp. p. 129.

[c] A remarkable example of this is still extant at the Seiryo-ji[1] Temple (the Shakadō)[2] at Saga on the outskirts of Kyoto. The statue of Shakyamuni was made at Khaifêng in +985 for the Japanese monk Chōnen[3] and taken back by him to Kyoto. The viscera, in appropriately different colours and shapes, are made of textile materials stuffed, and have much importance for the study of medieval East Asian anatomy. They have been closely studied by Morita Kōmon (1); Watanabe Kōzō (1, 2); and Ishihara Akira (1, 2). In 1964 Dr Lu Gwei-Djen and I had the opportunity of examining them personally at the Shakadō, for which our best thanks are due to the Abbot. We shall have more to say about these remarkable objects in Sect. 43 in Vol. 6. On the statue and its history in general see Henderson & Hurvitz (1).

[d] This 'eye-opening ceremony' paralleled a Confucian custom, the adding of one missing character stroke at the dedication of an ancestral tablet, but we do not know how far this goes back.

[e] Kraus suggested, rather awkwardly, (3), p. 134, that the artificial generation system was mainly the

[1] 清涼寺　　　[2] 釋迦堂　　　[3] 奝然

No, the fundamental feature of the Arabic creation of the rabbit out of the hat lay, as we see it, in those chemical substances which were added to the animal materials in the central container, for they represented nothing other than the *al-iksīr* of life, and the entire pattern of pseudo-scientific operations—how far ever tried out in practice remains somewhat obscure—was simply a new and original Arabic exercise using the powers of the life-giving *tan*.[1] The Chinese elixir idea was at the centre, and the Chinese perpetual-motion cosmic model surrounded it;[a] beyond this some part was doubtless played by earlier Mediterranean ideas on the subjects just discussed. In general, therefore, this giving of life to the lifeless, by chemical means, was, we conclude, a particular Arabic application of a characteristically East Asian conception, the giving of eternal life to the living, by chemical means. It reminds one of Kungsun Cho[2] in the −4th century, saying with typical Chinese optimism: 'I can heal hemiplegia. If I were to give a double dose of the same medicine I could probably raise the dead!'[b]

Summing it all up, we think one could say that Arabic alchemical theory was a marriage between the Taoist idea of longevity or immortality brought about by the ingestion of chemical substances and the Galenic rating of pharmacal potency in accordance with the *krasis* or balance of the four primary qualities (the natures). Gruman was quite right in remarking[c] that Arabic alchemists generally emphasised their ties with Hellenistic literature and traditions; that is indeed the dominant impression one gets in studying their writings—but perhaps if those were the books they read, the Persian, Indian and especially Chinese ideas and practices were what they talked about, few or no texts from those lands being available in Arabic translation at any time. The macrobiotics of China seems to have come westwards through a filter, as it were, leaving behind inevitably the concept of material immortality on earth or among the clouds and stars; after all, Paradise for Muslims was quite similar to the Heaven of the Christians, irretrievably subject to 'ethical polarisation' (cf. pt. 2, p. 80 above). Nevertheless some vital smaller molecules filtered through—(i) the conviction of the possibility of a chemically induced longevity, validated always by the example of the Old Testament patriarchs, (ii) hope in a similar conservation or restoration of youth, (iii) speculation on what the achievement of a perfect balance of qualities might be able to accomplish, (iv) the enlargement of the life-extension idea to life-donation in artificial generation systems, and (v) the uninhibited application of elixir chemicals in the medical treatment of disease. This last new development was the subject of a classical paper by Temkin (3), who perceived that the whole course of Hellenistic proto-chemistry was primarily metallurgical (aurifictive and aurifactive as we should

---

old theory of the animation of statues transformed by a strictly monotheist environment. This does not seem very convincing.

---

[a] It is curious that in the *Picatrix* of the mid +11th century (cf. pp. 313, 427–8) the artificial generation of monsters is ascribed to the 'Indians' (Ritter & Plessner (1), p. liv, pp. 288–90). This may always mean the Further Indies, but perpetual motion ideas were prominent in India proper, as already noted.

[b] Cf. Vol. 2, p. 72.   [c] (1), p. 59.

[1] 丹   [2] 公孫緯

say), while Arabic joined with Chinese alchemy in the profoundly medical nature of its preoccupations. Ko Hung, Thao Hung-Ching and Sun Ssu-Mo had glorious successors of the same cast of mind in al-Kindī, the Jābirians, al-Rāzī and Ibn Sīnā. Temkin found no link between chemistry and medicine in Greek until the poems of Theophrastes (c. +620) and Heliodorus (c. +716),[a] for although Dioscorides and Paul of Aegina of course knew of mineral medicines, Gnostic philosophy was as oil and water with the Hippocratic tradition, and chemical macrobiotics quite foreign to the Hellenistic world (cf. pt. 2, pp. 71 ff. above). Then eventually the first two of the ideas just listed, together with the fifth, passed through into the Latin culture of Western Europe at the time of the translations in the +12th century. If nothing living was ever seen to step forth from Jābir ibn Ḥayyān's cosmic incubators, chemo-therapy with all its marvellous achievements of today was assuredly born from the Chinese–Arabic tradition, with Paracelsus as its midwife.[b]

(3) MACROBIOTICS IN THE WESTERN WORLD

Returning now at last to our own European home, we are not in duty bound, we feel, in view of our responsibilities to the civilisation of China, to document in detail the attitudes of the Latin alchemists and the Renaissance hygienists and Paracelsians to longevity and material immortality. There is too great a cloud of witnesses available to everyone with knowledge of the European tongues. Suffice it to say, with Gruman,[c] that prolongevity had remained a neglected theme, hardly indeed perceptible, in the West throughout antiquity and far into the Middle Ages, till suddenly in the +13th century there appears full-fledged a macrobiotic alchemy. It must have been brewing from the middle of the previous century onwards, as the translations from the Arabic multiplied,[d] but after about +1230 the idea in one form or another was generally accepted.

Take Albertus Magnus for example, Albert of Bollstadt (+1206 to +1280), that

---

[a] It is possible that these two may have been only one person, at the later date; their background is rather obscure. Cf. p. 327 above.

[b] These words are reminiscent of those with which Wilson (2e), p. 619, concluded his one-man symposium in 1940. Much credit is due to the past pioneers in the history of chemistry who have reached similar conclusions, often with more insight than evidence at their disposal. An eloquent statement of the case was made by Edkins (17) already in 1855, and elaborated by Martin (8) in 1871. Thirty years later Martin's convictions were still the same, (2), pp. 24, 44ff., 52, 61, 63, 69. Other remarkable pioneers were Hjortdahl (1) in 1909 who got the essence of the pattern right in a few pages, and Holgen (1) in 1917. The twenties brought the classical statement of Campbell (1), pp. 53–4, and the weak but epoch-making book of Johnson (1). Waley (14) and Partington (8c) added their weight in the thirties, and since then there have been the studies of Dubs (5), pp. 84–5; Sherwood Taylor (3), p. 71, (7), pp. 32–3; Tenney Davis (2, 3, 4); Figurovsky (1); Chang Tzu-Kung (1); Haschmi (6); Bernal (1), p. 203; Arntz (1), pp. 203, 208; and Mahdihassan in many papers. We may also mention the interesting reviews of Debus (4), pp. 44–5, (24). Among Chinese scholars of high repute the same convictions have not infrequently obtained, as witness Li Thao (12), p. 212, (14), p. 112; Fêng Chia-Shêng (5), p. 120; Wang Chi-Min (1), p. 11 and Hsing Tê-Kang (1), p. 252. See also Anon. (83), p. 50; (167), p. 455. Miki Sakae (2), p. 20, traces the line, just as we should, from Ko Hung through Jābir to Paracelsus; and Florkin (1), p. 58, in his general history of biochemistry, recognises that the macrobiotic theme of Chinese alchemy was transmitted through the Arabs to join the transmutative theme of Hellenistic proto-chemistry. So did Seligman (2) and Kroeber (3).   [c] (1), p. 49.

[d] Cf. Dronke (2) on the School of Chartres, so open to Arabic learning. And the classic book of Haskins (2).

outstanding medieval naturalist so fortunate as to be beatified by historians of science as well as by the Church. He does not speak much of artificial longevity, but he knows the medical value of elixirs. In the *De Mineralibus* he says:[a]

> We do not intend to show how a certain *istorum* is transmitted into another, or how by means of a medical antidote, which the alchemists call elixir, diseases are healed, and men's secrets revealed, or conversely their open knowledge is concealed, but rather to show that the stones (i.e. solid chemical substances) are compounded from the elements, and how each one is constituted in its own species...

In other words, he intends to talk about natural minerals and gems, not about alchemy, but in the meantime lets slip that elixirs do something more than aurifaction. As for *istorum*, Dunlop thought it might be a corruption of an Arabic form, perhaps *'unṣur* (pl. *'anāṣir*) which means element, quality or nature in Jābir, matter in Balīnās; perhaps *iss*, i.e. principle.[b]

Far more daringly does Roger Bacon (+1214 to +1292) affirm time after time that when men have unravelled all the secrets of alchemy there is almost no limit to the longevity that they will be able to attain.[c] It was but a part, of course, of his general scientific and technological optimism that makes him seem so modern a figure, so far ahead of his time. Towards the end of his *Opus Majus*, addressed to Clement IV in +1266 or +1267, there is a section entitled 'Capitulum de secunda praerogativa scientiae experimentalis'.[d] Here, in the second 'Example', he says:[e]

> Another example can be given in the field of medicine, and it concerns the prolongation of human life, for which the medical art has nothing to offer except regimens of healthy living. In fact, there are possibilities for a far greater extension of the span of life. In the beginning of the world the lives of men were much longer than now, when life has been unduly shortened...

Bacon goes on to say that many believe that this has been according to the will of heaven, adding dubious astrological arguments about the senescence of the world, but he will have none of this, and recommends not only hygienic regimen but also marvellous medicines, some already known and some yet to be found out.

> Although the regimen of health [he says] should be observed from infancy onwards, in food and drink, in sleeping and waking, in motion and rest, in evacuation and retention, in the disposition of the airs and the control of the passions, no one wishes to give thought to these things, not even physicians, among whom hardly one in a thousand can order such matters gently and surely. Very rarely does it happen that anyone pays sufficient heed to the rules of health. No one does so in his youth, but sometimes one in three thousand thinks of these matters when he is old and approaching death....
> Sins also weaken the powers of the soul, so that it becomes incompetent for the natural control of the body; therefore the powers of the body are enfeebled and life is shortened.

---

[a] 1, i (*Opera*, Jammy ed., vol. 2, p. 210).

[b] (7), p. 72, cf. Kraus (3), pp. 110, 165, 285.

[c] On this in general see Thorndike (1), vol. 2, pp. 655 ff.; Ganzenmüller (2), pp. 80, 181 ff.; Frankowska (1), pp. 43, 88–9, 107.

[d] This follows upon the 12th chapter of Part VI.

[e] Jebb ed., p. 466, tr. Burke (1), vol. 2, pp. 617–18; this and the following passages will be found in the edition of Bridges (1), vol. 2, pp. 205 ff., 210–12. On the genuine (and spurious) alchemical writings of Roger Bacon see D. W. Singer (2).

Thus a weakened constitution is passed down from fathers to sons, and so transmitted further.

These are the two natural causes on account of which the longevity of man does not follow the natural order established in the beginning. It is praeternaturally abbreviated.

But now it is proved by certain experiments that many things can retard this hastening and decline; and secrets found out by experiment show that longevity can be prolonged by many years. Many authors have written on these things, and the possibility of remedies against this ruin ought to be known.[a]

The whole macrobiotic system of Roger Bacon is enshrined on this page.[b] He did not disparage the Hippocratic and Galenic systems of regimen which had come down from antiquity,[c] and he added a reference to the effects of sin, possibly out of respect for his cloth though not devoid of psychological validity; yet what was uppermost in his mind was the actual prolongation of the human life-span by material and chemical means. The traditional hygiene had aimed simply at fulfilling the 'natural' span of life; what Bacon offered was, as Gruman says, something radically new in the Western world, a methodical rationale for the prolonging of human life beyond its 'natural' span.

After all, it was agreed throughout Christendom that the soul was immortal. Why should it not be retained by art a good while longer in its mortal husk? As Bacon wrote elsewhere:[d]

The possibility of the prolongation of life is confirmed by the consideration that the soul is naturally immortal and not capable of dying. So, after the Fall, a man might live for a thousand years; only since then has the length of life gradually shortened. Therefore it follows that this shortening is accidental and may be remedied wholly or in part.

Here the reference is to Methuselah's 969 years, but there is no doubt from other passages that Roger Bacon took heart from the examples of all the Old Testament patriarchs, just as the Arabic alchemists had done before him.[e] In this way could the material immortality of China find a foothold in Europe. In that Western part of the world there were, as Gruman has worked out in detail,[f] three main types of legend which helped the acculturation; the antediluvian type, the hyperborean type and the fountain type. The patriarchs naturally belonged to the first of these. The second pictured certain very far parts of the earth as peopled by extremely long-lived races.[g] The third, analogous to certain Chinese Taoist paradise descriptions,[h] spoke of rivers

[a] Jebb ed., pp. 467–8, tr. Burke (1), vol. 2, *loc. cit.*, mod. auct.

[b] Other passages of much interest are to be found in his *De Retardatione Accidentium Senectutis*, tr. Browne (1), pp. 53 ff., 136. His criticism of reliance upon hygienic regimen alone was that no one can be protected against all environmental hazards (Browne tr., pp. 13–14, cit. Gruman (1), p. 65). Something more was needed—the 'admirable virtues' in things, not yet fully discovered (Browne tr., pp. 46–7).

[c] Mem. the translation of Galen's *De Sanitate Tuenda* by R. M. Green (1).

[d] *Epistola de Mirab. Potest.*, tr. Davis (16), p. 35.

[e] Gen. 9. One of the best studies of patriarchal longevity and the puzzlement of Renaissance thinkers as to how to take it is that of Egerton (1). Explanations were mythical, denying all validity to the tradition, or metaphorical, taking the patriarchs as symbols of tribes or dynasties, or literal—even as early as Josephus changes of diet were invoked. We should look upon it now simply as a variety of golden-age primitivism (Vol. 2, pp. 127 ff.), cf. Boas (1); Lovejoy & Boas (1). On the other hand Roger Bacon could—and probably did—cite the prophecies of Isaiah, e.g. 65. 17, 20, where in the new heaven and the new earth no one will be less than centenarian.　　　　[f] (1), pp. 20 ff.

[g] This goes back as far as Pindar, *Pyth. Odes*, 10, and has marked Indian connections.

[h] Cf. Vol. 2, p. 142.

or fountains of life or youth, powerfully restorative and preservative if anyone could find them and bathe in them.[a] The role of all these in Baconian and Paracelsian optimism remains to be determined more fully.

A few pages further on Roger Bacon takes up the powers of alchemy.[b] A paragraph full of burning enthusiasm ends as follows:[c]

And the experimental science (of the future) will know, from the 'Secret of Secrets' of Aristotle,[d] how to produce gold not only of twenty-four degrees but of thirty or forty or however many desired. This was why Aristotle said to Alexander 'I wish to show you the greatest of secrets', and indeed it is the greatest. For not only will it conduce to the well-being of the State, and provide everything desirable that can be bought for abundant supplies of gold, but what is infinitely more important, it will give the prolongation of human life. For that medicine which would remove all the impurities and corruptions of baser metal so that it should become silver and the purest gold, is considered by the wise to be able to remove the corruptions of the human body to such an extent that it will prolong life for many centuries. And this is the body composed with an equal temperament of the elements, about which I spoke previously.

Here then is Ko Hung (and Jābir too) in Latin dress at last. The final sentence strikes a note familiar to us, and indeed Bacon explicitly reproduces the Arabic doctrine of perfect equilibration, which must have reached him through the translators in Spain and Catalonia. What he had said a page or two earlier was this:[e]

Now if truly the elements should be prepared and purified in some sort of mixture so that there should be no nocive action (*infectio*) of one element on another, but that all should be reduced to a pure simplicity, then the wisest men have judged that they would have the highest and most perfect medicine. For in this way the elements would be on an equality....

This condition will exist in our bodies after the resurrection. For an equalisation of the elements in these bodies excludes corruption for all eternity. For equality is the ultimate end-in-view or final cause of all natural matter in mixed bodies, since it is the most noble of states, soothing and quietening all appetite in matter so that it desires no other thing.

The body of Adam did not possess a full equality of the elements, for there were in him the actions and passions of contradictory elements, so that waste occurred and food was necessary to make it good. And this was why it was told him that he should not eat of the fruit of life. But since the elements in him approached equality there was very little waste in him; and hence he was fit for immortality, which he could have secured if he had eaten always of the fruit of the tree of life. For that fruit is considered to have the elements in a condition approaching equality, and it could have continued the incorruptibility of Adam—which would have happened, if he had not sinned.

The wise have therefore laboured to reduce the elements in various forms of food and drink to an equality, or nearly so, and have taught means to this end.

From this it is evident that the ideas of Ibn Bishrūn and his colleagues were fully at

---

[a] This may have originated in part from Hebrew legend, cf. Gen. 2. 10, Psalms, 36. 9 continued in Rev. 22. 1. It certainly comes in several versions of the Alexander-Romance (cf. Cary (1); P. Meyer (1), pp. 174 ff.). General studies of the 'Fountain of Youth' motif are those of Hopkins (4); Masson (1). Apparently it was what Juan Ponce de León, the conqueror of Florida, was looking for in +1513; Beauvois (1). Cf. Wünsche (1).   [b] In the third Exemplum.

[c] *Opus Majus*, Jebb ed., p. 472, Burke tr., vol. 2, p. 627, mod. auct.

[d] This is Pseudo-Aristotle, *Kitāb Sirr al-Asrār*, the book of advice to kings, edited by Roger Bacon himself (cf. Steele, 1). It originated probably about +800. Cf. p. 297 above, and p. 497 below.

[e] *Opus Majus*, Jebb, ed., pp. 470–1, Burke tr. vol. 2, pp. 624–5.

work among the Latin alchemists, even though there might have been some raised eyebrows among the theologians studying Roger Bacon's interpretations of Genesis.

This comes out in another way in other books.[a] Thus in the *Opus Tertium* (+ 1267) there is an interesting passage[b] on speculative and operational alchemy[c] which treats explicitly 'of the generation of things from their elements', not only inanimate minerals and metals but also plants and animals.[d] This is the very idea of the Arabic *takwīn*, and now and then we can even catch Roger Bacon in the use of Arabic phrases so typical as 'if God wills'.[e] There was nothing very new in the belief that Art could produce in a single day what Nature takes a thousand years to accomplish,[f] but we ought not to miss the point that Roger was also extremely interested in the possibility of perpetual motion machines,[g] probably to be achieved with magnets, as indeed his friend Pierre de Maricourt was constantly occupied in attempting.[h] Here then were the two components of the Arabic artificial generation system, though Bacon probably never knew its full details; he would have been very excited if he had, and would certainly have found some ingenious way of reconciling it with Christian theology.

Finally, he adduced a number of case histories to demonstrate the possibility of extraordinary longevity, and if they sound very unconvincing to us they may have carried more weight with his contemporaries. The 'oriental' reference is significant— Artephius, for instance, wandered all over the east seeking knowledge, much of which he got from Tantalus the teacher of a King of India, so that he was enabled to live for 1025 years, by 'secret experiments on the nature of things'.[i] Bacon was always quoting the story of the Sicilian farmer:[j]

In the time of King William of Sicily a man was found who renewed the period of his youth in strength and sagacity beyond all human calculation for about sixty years, and from

[a] Statements about elixirs, and gold elixirs (*chin tan*[1] !) in particular, are scattered everywhere in Bacon's books. Cf. *Opus Minus*, Brewer ed., pp. 314–15, 375, where we hear that gold *per magisterium* is better than natural gold; or again *De Secretis Operibus Naturae...*, Brewer ed., pp. 538ff. A further source on elixirs is the rather muddled group of tractates attributed to Roger Bacon and published as *De Arte Chymiae* in + 1603, the title beginning *Sanioris Medicinae Magistri...* (cf. Multhauf (5), p. 190). In this see especially pp. 285–291. Here elixirs restore sight, and potable gold brings back youthfulness even when a man is at the point of death. The authenticity is disputable, but many ideas close to Bacon's are present in the book.

[b] Ch. XII, text in Brewer (1), pp. 39ff., partial tr., p. lxxx.

[c] He deplores at the conclusion that there are so few skilled practical men who understand how to conduct chemical operations.

[d] The passage repeats the valuation of alchemy as so beneficial for the State treasury, but better still as a means of prolonging life, which is unduly short because of lack of regimen and the inheritance of corrupt constitutions.

[e] E.g. *Opus Minus*, text in Brewer (1), pp. 314–15. Here again we read of the equilibration of the qualities, which is what, if accomplished, will prolong life beyond the single century.

[f] *Opus Minus, loc. cit.*

[g] See the *De Secretis Operibus Naturae...*, (before + 1250), Brewer ed., p. 537. This is the place which refers to Pierre de Maricourt as 'Experimentator tamen fidelis et magnificus'. See also *Opus Majus*, Jebb ed., pp. 465–6, the first Exemplum.

[h] Cf. Vol. 4, pt. 2, pp. 540–1. There are many warm mentions of Peter de Maharn-Curia the Picard in Roger Bacon; cf. Brewer (1), pp. xxxvii, lxxv, 35, 43, 46, e.g.

[i] *Opus Majus*, Jebb ed., p. 469; *De Mirab. Potest.*, Davis tr., pp. 34–5. Thorndike (1), vol. 2, p. 354, identifies Artephius with al-Tughrā'ī, on whom see p. 402 above.

[j] *Opus Majus*, Jebb ed., p. 469, Burke tr., vol. 2, p. 622; *De Retardat. Accid. Senectut.*, Browne tr. p. 75; *De Mirab. Potest.*, Davis tr., pp. 33–4.

[1] 金丹

a rustic ploughman became a messenger of the king. While ploughing in the fields he found hidden in the earth a golden vessel which contained an excellent liquor. Thinking that this liquor was dew from the sky he drank of it, and washed his face, and was restored in mind and body beyond all measure.

There are many other reports of the same kind in Bacon's writings,[a] further study of which we may omit, but this one deserves particular mention because he used it explicitly to recommend potable gold.[b]

And a good experimenter says in his book on the regimen of the aged that if one took that which is tempered (or equilibrated) in the fourth degree (i.e. the highest), and what swims in the sea, and what grows in the air, and what is thrown up by the sea, and a certain plant of the Indies, and that which is found in the viscera of long-lived animals, and the two creeping things that live in the lands of the Tyrians and the Ethiopians, all prepared and elaborated as they ought to be with the treasure of a noble animal, then it would be possible to prolong the duration of human life by many times, delaying the onset of old age and mitigating the affections of senility.

Now truly that which is tempered in the fourth degree is gold, as is said in the book *De Spiritibus et Corporibus*, the greatest friend of nature above all others. And if by a certain experiment this could be made as good as possible (which would be far better than natural gold),[c] as the alchemical art has power to do, like the vessel which the farmer found, and if it could be dissolved into a liquid like that which the ploughman drank—then a marvellous operation would take place in the body of man...

If one compares all this with the hopes of Han Wu Ti and his Taoist advisers (pt. 3, pp. 29 ff. above) the coincidence is striking.

This last passage contains some rather mysterious allusions, but they are cleared up in other Baconian writings, notably the tractate *De Retardatione Accidentium Senectutis*, which can be dated between $+1236$ and $+1245$.[d] There the seven *occulta* turn out to be as follows, first gold (as just stated), second ambergris or spermaceti (that which swims on the sea or is cast up by it),[e] third the flesh of vipers or lizards, 'dragons', from Ethiopia, fourth rosemary, sixth a bone believed to come from the heart of the stag, and seventh lign-aloes[f] (the 'plant' from India).[g] The fifth proves to be something more remarkable than any of these, namely *fumus juventutis*, i.e. the exhalations

---

 [a] See, for example, *Opus Minus*, Brewer ed., pp. 373 ff. William of Sicily reappears, together with many other cases, in *De Secretis Operibus Naturae...*, Brewer ed., pp. 538 ff.

 [b] *Opus Majus*, Jebb ed., pp. 469–70, Burke tr., vol. 2, p. 623, mod. auct. Gruman (1), p. 63 remarks, quite rightly, that the thought here is strikingly similar to that of Li Shao-Chün in the -2nd century.

 [c] What did Ko Hung say in $+315$? See pt. 3, p. 2.

 [d] I shall always remember the astonishment I experienced when first reading the text and exposé of this work by Little & Withington (1), so Taoist in its implications. What follows is based upon pp. 15, 57 ff., 140 ff. of their study.

 [e] On ambergris see pt. 2, p. 142 above. Spermaceti wax is not an intestinal concretion but a product from the heads of a number of whales and dolphins; Bacon seems not to have distinguished between them. Ambergris had been used as an aphrodisiac traditionally among the Arabs and perhaps also in China (cf. Davenport (1), pp. 37 ff.).

 [f] On the perfumed aloes-wood see Vol. 5, pt. 2, p. 141 above.

 [g] Nearly four hundred years later quite a similar list of medicines conducive to longevity was given by Francis Bacon in his *Historia Vitae et Mortis* ($+1623$). They include gold in all its forms, pearls, emerald, hyacinth, bezoar, ambergris and lign-aloes (Montagu ed., vol. 10, pp. 178 ff.). Frequent blood-letting was to be avoided. Cf. Walker (3).

or effluvia of healthy young persons. As the *Secretum Secretorum* says:[a] 'Si sentis dolorem in stomacho...tunc medicina necessaria tibi est amplecti puellam calidam et speciosam'. This was a kind of contagion, for Bacon also says: 'Infirmitas hominis in hominem transit, ita est sanitas.' The geriatric benefit supposedly derived from proximity to a healthy and beautiful young girl, with the absorption of her breath, is an idea presumably as old as King David,[b] and it was certainly still current in the +16th, +17th and +18th centuries;[c] but when we go on to read that coitus entirely destroys the effect we can no longer forbear from recalling that Chinese physiological alchemy (*nei tan*[1]) which will be the subject of the following sub-sections in pt. 5, and it looks as if Bacon was recommending the transfer of *chhi*,[2] for what else could *fumus juventutis* mean? Strangely also, if this happy solution was unattainable, Roger Bacon recommended as a substitute some kind of arcanum prepared from human blood.[d] For more reasons than one, therefore, he probably felt it necessary to be as discreet as possible in discussing elixirs with the Pope or with his Franciscan colleagues. But his texts remain for us the supreme and first great example of *hsien*[3] medicines and *hsien*[3] hagiography in the Western world.

Any intimations of chemical macrobiotics which one can find in Europe reinforcing Roger Bacon's convictions during the following half century or so are obviously of great interest for the theme of transfer from the East. Hence we should not overlook a striking passage in Marco Polo which occurs in his account of India (Malabar).[e] Speaking of men whom we might think of as *sadhus* he says:[f]

And these Braaman (Brahmins) live more (i.e. longer) than any other people in the world, and this comes about through little eating and drinking and through great abstinence which they practise more than any other people....

Moreover they have among them regulars and orders of monks according to their faith, who serve the churches where their idols are; who are called 'ciugui'[g] (and) who certainly

---

[a] Steele ed. (1), p. 73, tr., p. 198. This was the Arabic 'Book of the Secret of Secrets' again, i.e. advice to kings, Pseudo-Aristotle addressing Alexander, edited by Roger Bacon about +1255 with an introduction added by him some twenty years later. Cf. pp. 297, 368, 494 above.

[b] I Kings, 1. 1–4, Abishag the Shunamite.

[c] In +1573 our own second founder, Dr John Caius, adopted some such technique in his last illness, though it may also have been connected with a diet of human milk for his disordered stomach, perhaps due to a carcinoma. In the following century the same was reported of our William Harvey by John Aubrey in his 'Brief Lives' (Dick ed., p. 213). So also Thomas Sydenham praised the balsamic exhalations 'ex sano et athletico corpore'; and Francis Bacon had said 'Neque negligenda sunt fomenta ex corporibus vivis', going on to speak of King David (*Hist. Vit. et Mort.*, Montagu ed., vol. 10, p. 244; cf. Grmek (2), pp. 44–5). And in the +18th century came Cohausen's book on the subject, which we shall discuss in pt. 5 below. One can hardly forbear from adding a reference to Nabokov's famous novel, 'Lolita'.

[d] This recalls the conjecture of Multhauf (5), pp. 190, 192, that Roger Bacon belonged to a school of Latin alchemists which firmly believed in what they had learnt from Arabic sources, the preparing of elixirs from organic materials. Multhauf suggests that the majority, more faithful perhaps to Hellenistic traditions, accepted only mineral and metallic magisteries, abhorring the others. Further research may be expected to throw more light on this.

[e] Attention was drawn to it by Berthelot (10), p. 201.

[f] Ch. 177, Moule & Pelliot tr., vol. 1, pp. 403–4, text, vol. 2, p. lxxxii. The omitted sentences enlarge on the remarkably good teeth of these sages, who do not bleed themselves or others, and whose food is mainly bread, rice and milk. [g] I.e. yogi (Pelliot (47), p. 391).

[1] 內丹　　[2] 氣　　[3] 仙

live more than all others in the world, for they commonly live from 150 years to 200. And yet they are all quite capable in their bodies so that they are well able to go and come wherever they wish, and they do well all the service which is needed for their monastery and for their idols, and though they are so old they render it as well as if they were younger....

And again I tell you that these ciugui who live so long time...eat also what I shall explain, and it will seem indeed a great thing to you, very strange to hear. I tell you that they take quicksilver and sulphur and mix them together with water and make a drink of them; and they drink it and say that it increases their life, and they live longer by it. They do it twice in the week, and sometimes twice each month, and you may know that those people use this drink from their infancy (so as) to live longer, and without mistake those who live so long use this drink of sulphur and of quicksilver....[a]

And he goes on to expatiate on the gymnosophists. The passage is particularly interesting because the dietetic-hygienic element and the elixir-pharmaceutic element are both so prominently present; Li Shao-Chün's cinnabar is living again in Rusticianus' Latin. Marco Polo was a contemporary of Roger Bacon's; he reached China in +1275 and left for India in the year of Bacon's death, +1292, returning to Italy by +1295, so that the dictating of his reminiscences belongs to the ensuing decade. Of course Marco's information did not spread with the rapidity of a mass-produced paperback of the present day, but it attained diffusion in a considerable number of manuscripts which were widely read,[b] and what he reported of the chemically-induced longevity of Asian saints and sages must at least have chimed in with those other notes which emanated from specifically Arabic sources.[c]

After this time the theme of elixirs goes continuously on. In the following century the Villanovan Corpus has a *Liber de Conservatione Juventutis et Retardatione Senectutis*.[d] About +1320 John Dastin wrote a letter on alchemy to John XXII which opens as follows:[e]

[a] Whatever the true facts may have been, Marco Polo is borne out by a text two hundred years older than himself, the *Rasārṇavakalpa*. There we read that 'half a *pala* of sulphur and one *pala* of mercury, taken for a year, bestows a longevity of three hundred years' (Roy & Subbarayappa tr., p. 85).

[b] For example, there is an evident echo in Francis Bacon's *Hist. Vit. et. Mort.* (+1623): 'Etiam Seres, Indorum populus, cum vino suo ex palmis, longaevi habiti sunt, usque ad annum centesimum tricesimum' (Montagu ed., vol. 10, pp. 164–5).

[c] Some of the ideas in the foregoing passage had, to be sure, been current in Europe a long time before Marco Polo, for they can be found in the Alexander-Romance, that corpus of legend about Alexander the Great which first took form in the +3rd or +4th century. We have had occasion to refer to it before, in connection with aerial cars (Vol. 4, pt. 2, p. 572), diving-bells (Vol. 4, pt. 3, p. 674) and the Great Wall (Vol. 4, pt. 3, pp. 56–7). Its Sino-Indian element occurs especially in the *Commonitorium Palladii* (ed. Pfister (1); cf. Cary (1), pp. 12ff.; Derrett (1); Ross (1), pp. 30ff.; Coedès (1), pp. 98ff.). There we find mention of (*a*) Serica, the land of the silk-producing Seres, (*b*) the Isles of the Blest (like Phêng-Lai), (*c*) the capture of ships with iron nails by magnetic rocks (cf. Vol. 4, pt. 1, p. 235), (*d*) the longevity of the brahmins, never less than 150 years, (*e*) their simplicity, piety and vegetarian diet. Palladius, who must have written his account by +375, did not claim to have gone to India or the Further Indies himself, but got much information from one Thebeus Scholasticus, who had been there from +356 to +362. Those who collect pepper, he says, get it from troglodytic dwarfs in the uttermost islands, little men who share the virtues and the longevity of the brahmins. Among the echoes of this story is the dwarf-motif in the +11th-century German poem 'Ruodlieb' (Werner Braun). But the mercury and sulphur are in Marco Polo alone. Cf. p. 483 above.

We have to thank Dr Peter Dronke for illuminating discussions on these subjects.

[d] Cf. Little & Withington (1). Arnold died in +1311, but this would be rather later.

[e] Josten (1), p. 43; cf. Ferguson (1), vol. 1, p. 199.

This is the secret of secrets, the priceless treasure, the very true and infallible work concerning the composition of the most noble matter (the philosophers' stone) which, according to the tradition of all philosophers, transforms any metallic body into very pure gold and silver, which conserves (bodies in their) essence, and fortifies (them) in (their) virtue, which makes an old man young, and drives out all sickness from the body.

And after a disquisition on equilibration of temperament, sulphur and mercury, ferment and the like, he ends by saying that lastly

it is inccrated so that the spirit may be incorporated and fixed in the body, until it becomes one with it, standing, penetrating and perfusing, tingeing and remaining—of which, according to the philosopher, one part converts a million parts of any body you may choose into the most genuine gold and silver respectively, depending on which of the two elixirs you have prepared. And it has effective virtue over all other medicines of the philosophers to cure all infirmity, because, if it were an illness of one month it cures it within one day, but if it were an illness of a year it cures it in twelve days. But if it were an inveterate illness (like old age) it cures it in a month. And therefore this medicine ought to be sought for by all men everywhere, and before all other medicines in this world.[a]

By this time alcohol had become widely known and used.[b] John of Rupescissa (*fl.* +1345) was perhaps the first to identify it with the quintessence or missing fifth element,[c] and though gold leaf suspended in alcohol was more impressive as an elixir symbolically than effectually, the new solvent did give access to higher concentrations of many active substances from the plant and animal world.[d]

Henceforward the elixir idea becomes a universal commonplace. Thomas Norton, speaking of the Ruby Stone of the Philosophers in his *Ordinall of Alchemy*, c. +1440, wrote:[e]

> Whereof said Mary, sister of Aaron
> 'Life is short, and Science is full long',
> Natheless it greatly retardeth Age
> When it is ended (accomplished) by strong Courage...

And we may end this phase of the story by the inevitable quotation from Ben Jonson.[f]

> *Mamm*: Ha!
> Do you think I fable with you? I assure you,
> He that has once the Flower of the Sun,
> The perfect Ruby, which we call Elixir,
> Not only can do that, but by its Vertue,
> Can confer Honour, Love, Respect, Long Life,
> Give Safety, Valour, yea, and Victory,
> To whom he will. In eight and twenty days

---

[a] *Op. cit.*, p. 51.
[b] For an account of the discovery of alcohol and its spread cf. pp. 122ff. above.
[c] Cf. Leicester (1), p. 89; Multhauf (5), p. 211.
[d] See on the history of the quintessence Sherwood Taylor (6).
[e] Holmyard (12), p. 87, facsimile of 1652.
[f] *The Alchemist*, 1610, p. 372. Cf. pt. 3, p. 214. After all this, and with the background we now have, it hardly comes as any surprise to find that a work on cinnabar as a drug was published by Gabriel Clauder at Jena in +1684.

> I'll make an old Man, of Fourscore, a Child.
> *Surly*:   No doubt, he's that already. *Mamm*: Nay, I mean
> Restore his Years, renew him, like an Eagle,
> To the fifth Age; make him get Sons and Daughters,
> Young Giants; as our Philosophers have done
> (The antient Patriarchs afore the Flood)
> But taking, once a week, on a Knives Point,
> The quantity of a Grain of Mustard of it:
> Become stout Marses, and beget young Cupids...

If the general picture so far outlined is approximately correct, namely that there was a passage of the elixir idea from the Arabic alchemists to the Latins, reaching full acceptance by them, according to their lights, in the time of Roger Bacon; then it might be expected that similar macrobiotic hopes would have become known in Byzantine culture a couple of centuries earlier. This is exactly what we find. If we open the history of fourteen Byzantine rulers written by Michael Psellus about +1063, his *Chronographia*, we can read a very peculiar passage about the reign of the Empress Theodora (+1055 to +1056). Psellus wrote:[a]

The extremely generous persons [installed by her in positions of authority in the church][b] who surpassed all bounds of liberality with their munificent gifts, were not angels carrying messages to her from God, but men, who imitated the angelic beings in outward appearance, yet at heart were hypocrites. I am referring to the Naziraeans of our time.[c] These men model themselves on the Divine, or rather they have a code of laws which is, superficially, based on the imitation of the Divine. While still subject to the limitations of human nature, they behave as though they were demigods among us. For the other attributes of Divinity they affect utter contempt. There is no effort to harmonise the soul with heavenly things, no repression of the human desires, no attempt by the use of oratory to hold in check some men and goad on others. These things they regard as of minor importance. Some of them utter prophecies with the assurance of an oracle, solemnly declaring the will of God. Others profess to change natural laws, cancelling some altogether and extending the scope of others; they claim to make immortal the dissoluble human body and to arrest the natural changes which affect it. To prove these assertions they say that they always wear armour, like the ancient Acarnanians;[d] and for long periods of time walk in the air—descending very rapidly, however, when they smell savoury meat on earth! I know their kind and I have often seen them. Well, these were the men who led the empress astray, telling her that she would live for ever; and through their deceit she very nearly came to grief herself and brought ruin on the Empire as well.

They predicted for her a life going on centuries without end. Yet in fact she was already nearing the day which Fate had decreed should be her last. I ought not to use such an expression—what I mean is that she had nearly finished her life and the end was at hand. As a matter of fact she was assailed by a very terrible illness....

And indeed she died in the summer of the second year of her reign, aged 76.

a Theodora sect., paras. XVIII, XIX, cf. also XV, Sathas ed., pp. 186–7; Renauld ed., vol. 2, pp. 80–1; Sewter tr. p. 269.

b The grammar is faulty here but the reference seems to be to ecclesiastical promotions mentioned in a just preceding paragraph.

c I.e. monks, from Heb. *nazir*, separated.

d An allusion to Thucydides, I, 5, speaking of a semi-civilised people.

From this it seems clear that Theodora was under the influence of a group of monks who claimed to be in possession of macrobiotic techniques.[a] Though these are not described, they could well have been psycho-physiological as well as chemo-therapeutic, and the whole passage has a very Taoist, or perhaps one should say rather at such a time and place, a Sufi, or even Siddhi, character. Walking on air is just what one expects of a Taoist *hsien*, and the remark about the failure to repress human desires might be an obscure reference to something like that physiological alchemy which will be the subject of the remainder of this Section.[b] Unfortunately none of the commentators has anything whatever to say about this strange group of Christian monks, so we can only record their existence.[c]

The name of Michael Psellus ought to strike a familiar note in the mind of anyone who has patiently followed our exposition from the beginning of Vol. 5.[d] For he was indeed none other than that Psellus who addressed an 'Epistle on the Chrysopoia' to the Patriarch of Byzantium in +1045 or +1046.[e] He wrote a preface to the Greek proto-chemical Corpus, and may indeed have been its first collector.[f] He was in touch with Arab scholars, and had Arabs among his pupils, this at a time when many Arabic writings were being translated into Greek.[g] In another place in the *Chronographia* he has an interesting passage on the chemical interests of the Empress Zoe, who died in +1050 aged 72 under Constantine IX; she turned her apartments into a veritable laboratory and never tired of investigating the properties of perfumes and their combinations.[h] Thus here we end, as we began, with Michael Psellus, a polymathic man whose life and times would repay, it seems, much further study by historians of science.

After this there is little more for us to say by way of conclusion. In the field of macrobiotics with which we are concerned there were two great movements during the scientific revolution. First, the ancient Greek tradition of medical hygiene, which had by no means been repudiated by Roger Bacon and the alchemists who followed him, gained from their elixir beliefs a new impetus and a new lease of life. In +1550 Luigi Cornaro published his *Discorsi della Vita Sobria*;[i] this, though largely dietetic, laid much emphasis on the avoidance of psychological strains and submission to the passions.[j] In these ways the innate moisture could be conserved.[k] Widely translated

---

[a] Of course it was customary for Byzantine ecclesiastics favoured by the emperor to predict long life for him, and length of days, just as they threatened an ill-disposed one with an early death through the assured wrath of God—but here there seems to be something more than these usual reactions.

[b] In pt. 5.

[c] The Hesychasts, of course, borrowed meditation techniques from Buddhism or Hinduism, but their movement was much later, in the first half of the +14th century.

[d] Pt. 2, p. 17, and p. 328 above. There is now a biography by Pingree (2).

[e] See Bidez (1), which includes an Italian translation. The Patriarch in question was apparently Michael Cerularius, not Psellus' friend Joannes Xiphilinus, as has often been stated.

[f] Cf. Berthelot (1), pp. 102, 248–9, 279.

[g] Bidez (1), p. 23.   [h] Sewter tr., pp. 186–7.

[i] Cornaro (+1467 to +1565) was a friend of Fracastoro. On him and his work see Sigerist (2); Walker (1). No less than nine English translations had appeared by 1825.

[j] This was strikingly similar to Chinese physiological alchemy; cf. pt. 5 below.

[k] This was Aristotelian and Galenic orthodoxy (cf. Gruman (1), pp. 15ff.) yet it reminds one of the necessity for the conservation of the *ching*.[1]

[1] 精

and approved, Cornaro's book had many successors, notably Lessius' *Hygiasticon* of
+1614[a] and Sir William Temple's essay on health and longevity (+1770).[b] In +1796
came Christopher Hufeland's *Art of Prolonging Life*, in which the term macrobiotics
was first used, appearing indeed in the original German version of the title. The
influence of Hufeland, who was a friend of Goethe, Schiller and Herder, extended all
over the world, and his prescriptions for longevity, in themselves very reasonable,
passed into Japanese literature in the translations of Ogata Kōan[1],[c] as has been shown
by Achiwa Gorō (1) in his interesting study of the theory of nature-healing in the
Rangaku period. Hufeland also exerted a great effect on many nineteenth-century
writers on medical hygiene and macrobiotics,[d] following the ideas of William Godwin
and A. N. de Condorcet.

The other great movement just mentioned was of course that of iatro-chemistry,
especially as it developed to the fullness of the Paracelsian form. This was the great
empirical phase of chemistry developing in opposition to Galilean–Newtonian
mechanicism, along with movements of lesser scope such as that of the biologically-
minded Cambridge Platonists.[e] Necessarily it too had Pythagorean and Neo-Platonic,
not to say Gnostic and Hermetic, roots.[f] How far it could have had certain East Asian
roots, transmitted either through the Arabs or by way of more direct contacts in the
+13th century and later, would be very hard to say, yet it really is the case that much
of the Paracelsian thought-world has a strangely Chinese air.[g] For example, the very
idea of an organic universe, with an interconnectedness of all things,[h] the prominence
of the macrocosm–microcosm analogy,[i] and the readiness to conceive of action at a
distance, based on resonance and 'magneticall' phenomena[j]—in all these things one
has to speak at least of a parallelism with traditional Chinese world-views. But there
are more detailed and disturbing similarities. The Paracelsians spoke of two kinds of

[a] English translation (Cambridge) by +1634.   [b] *Works*, vol. 3, p. 266.
[c] For example, *Byogaku Tsūron*[2] (Survey of Pathology), and *Hushi Keiken Ikun*[3] (Mr Hu's Well-
Tested Advice to Posterity).
[d] Notably Sweetser (1), who was the first to speak of 'mental hygiene' (1867), Jacques (1), who
emphasised the will of the individual in determining his own fate, and Thoms (1), who collected as much
evidence as possible about human longevity from historical records. The term 'folk-lore' was coined
by him. 'Gerontology' was introduced by Metchnikov in 1903, and 'geriatrics' by Nascher in 1909.
[e] Cf. Vol. 2, pp. 296, 503–4, where we have touched on the Chinese parallelisms before.
[f] Particular attention has been paid to these by Pagel (28) and Pagel & Winder (1, 2). Gnostics and
Paracelsians both had a predilection for classifying celestial–terrestrial beings, processes and elements
in groups of eight. This recalls the eight trigrams (*pa kua*[4]) of the *I Ching* (on which see Vol. 2, p. 313).
Hearsay about these might have strengthened the ogdoad tendencies in the West, but it would have to
have happened quite early.
[g] The fact that the Paracelsian pharmaceutical revolution, lucidly sketched by Debus (25), used
mineral drugs in defiance of the herbal *idée fixe* of the Galenical Colleges, was alone enough to unite
it with the age-old tradition of Chinese pharmacy (cf. pt. 3, p. 46 above, and Needham (64), p. 284), and
cannot but suggest some trains of influence. Lach (5), vol. 2, pt. 3, pp. 422 ff., considers the tradition that
Paracelsus stayed some time with the Tartars in Russia, and ponders on how their name entered chemical
terminology.
[h] Cf. Debus (2), p. xxxiv on Elias Ashmole, and (18), pp. 19, 86, (26) on the Paracelsians.
[i] Cf. Debus (6), p. 391, (7), p. 47, for comparison with Vol. 2, pp. 294 ff. and *passim*. Cf. Zippert (1);
Pagel (28), pp. 38, 124.
[j] Cf. Debus (6), p. 390 and 400–1 on Kenelm Digby, as also Dobbs (1); and Gelbart (1) on Walter
Charleton.

[1] 緒方洪庵   [2] 病學通論   [3] 扶氏經驗遺訓   [4] 八卦

fire,[a] strangely echoing the Chinese division of that element into 'princely fire' (*chün huo*[1]) and 'ministerial fire' (*hsiang huo*[2]).[b] Sexuality was very prominent in their thinking,[c] as it had been in that of all proto-chemists and alchemists from the beginning.[d] Robert Fludd coined the words 'voluny' and 'noluny',[e] the former to express sympathy, light, warmth, life and expansion, the latter to express antipathy, dark, cold, death and contraction—can they have been anything other than Yang[3] and Yin[4] respectively? By this time Jesuit-transmitted knowledge could have been coming in, a phase of contact which might also have been responsible for the play which he made of the 'light' and 'heavy' antithesis (*chhing*,[5] *cho*[6]) in cosmogony.[f] After such parallelisms it is hardly surprising to find Fludd engaging in symbolic correlations between spatial directions and the viscera of the body;[g] while every Paracelsian wrote on sympathies and antipathies,[h] categories of reactivity,[i] and numerology rather than mathematics.[j] Pervading all was their characteristic empiricism[k] and their emphasis on the medical and macrobiotic side of alchemy.[l] We are not saying that all these traits were marks of the future that modern science had before it, obviously in many ways the exact reverse was the case, but among them certain great convictions stand out, notably that chemo-therapy in unimagined power was a realisable goal for man;[m] and if indeed there were East Asian contributions, however indirect, to these ideas, then some invaluable sense came through along with the nonsense.[n] About the intermediation of the Arabs enough has already been said; for this period one should perhaps look for more direct contacts (Fig. 1538).

[a] See Pagel (28), p. 70; Pagel & Winder (1), (2), pp. 102 ff. This was gnostic, rabbinic and kabbalistic doctrine, so if there was any connection it must have been much earlier than the +16th century.

[b] Cf. Vol. 4, pt. 1, p. 65, and for a full explanation, Vol. 6. The doubling came about in China because of the necessity of meeting the need of medical philosophy for a sixfold rather than the fivefold classification.   [c] See Pagel (28), pp. 62 ff.

[d] Cf. pp. 363 ff. above.

[e] See Debus (6), p. 405. This was in Fludd's *Philosophia Moysaica*, published posthumously in +1638. The idea of positive and negative here was applied by him to magnetism, following William Gilbert; and as time went on to electricity by +18th-century physicists. Polarity also played a great part in the Naturphilosophie school (cf. Pagel (1), pp. 291–2).

[f] Cf. Debus (8), p. 266.   [g] Cf. Debus (8), p. 272, (18), p. 116.

[h] Cf. Debus (6), p. 391, (18), p. 90.   [i] Cf. Debus (6), p. 407.

[j] Cf. Debus (7), p. 49.   [k] Debus, *op. cit.*, p. 43.

[l] Cf. Debus (18), pp. 23, 146, (21, 22, 23).

[m] The fact that I am sitting here writing these words is in itself an indication of what we all owe to the elixir alchemists and the Paracelsians. Without public hygiene, sulpha-drugs, immunology and antibiotics I should have been carried away thirty years ago or more. In Roger Bacon's time old age began at 45.

[n] For those interested in 'nonsense' (and who ever knows what may come out of it?) an interesting study might be the comparison of the cosmic-chemical charts of different cultures. Singer, Anderson & Addis reproduced one of these from a +15th-century alchemical MS. (BM, Egerton 845, (1), their no. 440), and Heym (1) gave another from an early +18th-century work on mystical alchemy, the *Aurea Catena Homeri*. It is very likely that the Kabbalah literature had something to do with this, for it delighted in charts of creation (the 'Sephirothic Tree'), as may be seen in such works as Athanasius Kircher's *Oedipus Aegyptiacus* as well as the Hebrew originals (cf. e.g. Hall (1), pl. CXXIII); to this we drew attention in Vol. 2, p. 297 in connection with the influence of Chinese organic and correlative thinking on Europe. Indeed the first model for all subsequent cosmic-chemical charts could conceivably have been the Neo-Confucian *Thai Chi Thu*, described and discussed in Vol. 2, p. 461. Thanks are due to Mrs Alice Howell of Westbury, L.I., for raising this point.

[1] 君火   [2] 相火   [3] 陽   [4] 陰   [5] 清   [6] 濁

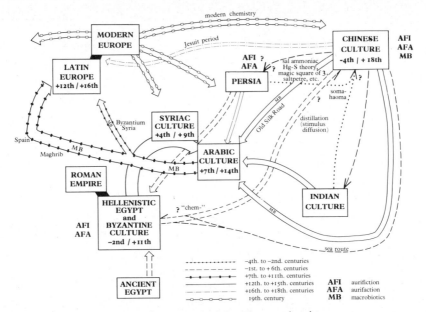

Fig. 1538. Chart to illustrate the multi-focal origins of proto-chemistry.

This chart is to be seen as superimposed roughly upon a map of the Old World, allowance being made for diagrammatic displacements occasioned by the different time-periods of the cultures. Single or double arrow-heads mark the arrival of well-established influences; less certain ones are qualified by interrogation marks. The pattern itself may also be visualised in depth, as the older lines of influence are drawn underneath the later ones at places where they cross.

The phrase 'multi-focal' is that of Sheppard (6), but since we define alchemy as macrobiotics+ aurifaction we find it applicable not to alchemy but to proto-chemistry. It is now unquestionably clear that there were two foci of aurifiction and aurifaction, Hellenistic Egypt and China; perhaps there were four if Persia and India should be included as well, their independence remaining still in some doubt. But there was only one focus of chemical macrobiotics, China, hence the home of all alchemy *sensu stricto*; and one can see this influence spreading westwards when the time came through Arabic and Byzantine culture to the Latin West, and therefore into Paracelsian iatro-chemistry and modern chemotherapy.

The stimuli which may have acted upon Chinese culture in ancient times with the generation of the *hsien tan* complex are very obscure, but two dotted lines are shown to indicate the possible role of *soma haoma* ideas and practices. Gilgamesh's herb of immortality, back in the − 3rd millennium, might have been the ancestor of these, but Ancient Mesopotamia has been excluded from the chart for the sake of simplicity, since geographically it would underlie the Arabic culture. Moreover, we still have very little sure information about the chemical knowledge of the Babylonians and Assyrians. Presumably the technology of the precious metals, glass, ceramics, fermentations, bitumens, perfumes and the like, should be shown by a series of dashes influencing the Hellenistic, Byzantine and Syriac cultures, as has been done to indicate the influence of the traditional technology of Ancient Egypt; but none of this demonstrates either aurifiction or aurifaction, still less chemical macrobiotics.

Lastly, it will be seen that Indian culture is drawn in a rather isolated position. This is not because we believe that it really was so, but because Indian philology and archaeology are as yet so undeveloped that the dating of texts (and therefore of ideas and practices) presents grave difficulties, while at the same time the study of ancient artifacts has so much more yet to tell us. Almost the only things that can be said for certain are that Indian medicine, mineralogy and other sciences exerted a great influence on the Arabic world from the + 7th century onwards, and that Chinese macrobiotics influenced India to some extent at a rather earlier date.

It was not common for the +17th-century mystical chymists to make any direct reference to the ideas of the Chinese, but one such allusion does occur in Thomas Vaughan's *Magia Adamica* (+1650).[a] Here, however, the justification claimed was not because of any priority of theirs, but rather by their recognition and confirmation of a universally valid *philosophia aeterna*. Vaughan was arguing that natural magic was the way in which God himself, as the Holy Spirit, worked and had worked in creation, that it was essentially chemical in nature, a vivification of matter, that man can use the same creative forces if he is in tune with the divine Word or Logos, and that the Kabbalah as well as Greek and Egyptian magic were only imperfect anticipations of Christian magic. If the Christian cosmology thus superseded all others, then all peoples should accept it once it was called to their attention, in proof of which he adduced the Nestorian Stone, that famous stele erected near Sian in +781 recording the development of the Church in China since Bishop Alopên's coming in +635.[b] The imperial favour which the religion received, and its approbation by many of the learned of the empire, demonstrated, Vaughan thought, the universal truth of 'Christian chemical creation'.

Thomas Vaughan, no doubt, was rather on the fringes of the iatro-chemical movement. Before leaving it we ought to take one more look at the central doctrine taught by Paracelsus. From the opening of Vol. 5, pt. 2 onwards we have emphasised his great watchword that the business of alchemy was not to make gold but to prepare medicines, showing that this was in a direct line of descent from the elixir ideas of Li Shao-Chün and Ko Hung. What then were the actual words of Paracelsus? The first statement,[c] that alchemy is not aurifaction, occurs in the *Paragranum* (+1530):[d]

It is not as the praters maintain, that alchemy is to make gold, and to make silver; the grand principle is that it is to make arcana,[e] and to direct them against diseases—that is the aim and target of all true Alchemy.[f]

In another passage, he says that alchemy is purificatory chemistry, its true task the liberation of the medicinal substances contained in the crude materials. So in his *Labyrinthus Medicorum Errantium* (|1538) he wrote:[g]

If you see a herb, a stone or a tree, you see only the husk, the gangue or the slag, and underneath that lies the drug. You must take out the drug and separate it from the dross;

[a] Vaughan (2), repr. in Waite (4, 5). The passage occurs on pp. 176–7 of the 1968 edition of (4). Thanks are due to Mr David Hallam for calling this to our attention.

[b] Cf. Vol. 1, p. 128. For a full account see Saeki (1, 2). The exact nature of the sources used by Thomas Vaughan presents a somewhat puzzling problem, but knowledge of the Stone and its inscription had been circulating in Europe since Nicholas Trigault's first translation in +1625.

[c] An excellent study of the main views of Paracelsus is that of Walden (4).

[d] Sudhoff ed., vol. 8, p. 185; Strebel ed., vol. 5, p. 114.

[e] Compound or simple medicines endowed with the divine power of the Creator, embodying astral virtues, and prepared by chemical means. The concept was complex; further exposition will be found in Pagel (10).

[f] The temptation is irresistible to give one or two of these passages in the extremely colourful +16th-century German that Paracelsus wrote. So here: '...entgegen den Schwätzern, die sagen: die Alchimia mache Gold, mache Silber; hie ist das fürnemen: mach Arcana und richte dieselbigen gegen den Krankheiten. Das ist Zweck und Ziel der wahren Alchimie.' Even with the varying degrees of modernisation that editors have provided, the vitality of Paracelsus always impresses.

[g] Sudhoff ed., vol. 11, pp. 187–8; Strebel ed., vol. 1, p. 192.

only then will you have it. That is Alchemy. And the task of the apothecary and the operator in the elaboratory is the same as that of Vulcan.[a]

And a page or two further on he wrote:[b]

Alchemy is what brings to its perfection that which is not yet perfect.[c] Those who draw lead from its ore, and work it up, are alchemists of metals. So also there are alchemists of minerals, sulphur, vitriol and salt. If you ask what Alchemy is, know that it is simply the art of purifying the impure by fire. Alchemists of wood are like carpenters who make wood into a house; so also the wood-carver chips out from wood, and throws away, what does not belong to it, and thus creates a figure. Just so there are alchemists of drugs, who purge away from the drug all that which is not drug. See then what kind of an art is Alchemy—that which separates the useless from the useful, and brings this last to its final matter and its final perfection.[d]

And again, in the *Paramirum*:[e]

Whoever takes what Nature has generated that can be useful to man, and brings it to that place and estate that Nature has appointed for it, that man is an Alchemist.[f]

And finally:[g]

No physician can do without alchemy; if he ignores it he will be but a sluttish scullion compared with the master-cook of a princely palace.[h]

In such sayings Paracelsus linked the Chinese medieval alchemists with the biochemical pharmacologists of modern times, extracting a few milligrams of a substance from a ton or a gallon of raw material.[i] Yet alchemy still embraced, in Paracelsus' thought, the chemistry and metallurgy of gold, in so far as it contributed to that remedy in which he still believed, the *aurum potabile*.[j] In +1526 he had written, in his book *Von den natürlichen Dingen*:[k]

Aqua salis (hydrochloric acid) is distilled from the calcination to a spirit which dissolves gold to an oil, whereby potable gold can be prepared.

a '...sie sehen nur die Schlacke, innen aber unter der Schlacke da liegt die Arznei. Nun muss zuerst die Schlacke der Arznei genommen werden. Dann ist die Arznei da. Das ist Alchimie....'

b Sudhoff ed., vol. 11, pp. 188–9; Strebel ed., vol. 1, p. 194.

c Paracelsus often compared the alchemist to the archaeus operating in bodily chemical processes; cf. Ganzenmüller (5), p. 430. For example: 'Der Archeus fabriziert als innerer Alchymisten-Geist' (*Paragranum*, Strebel ed., vol. 5, p. 106).

d 'Das ist Alchimie, das zum Ende zu bringen, was nicht zu seinem Ende gekommen ist.... So gibt es Alchimisten der Arznei, die von der Arznei das entfernen was nit Arznei ist. Jetzt sehet, welche Kunst die Alchimie ist. Sie ist die Kunst, die das Nutzlose vom Nützlichen entfernt, und es zu der letzten Materie und zum letzten Wesen bringt.'

e Sudhoff ed., vol. 8, p. 181; Strebel ed., vol. 5, p. 110.

f 'Wer also das, was in der Natur dem Menschen zunutze wächst, dahin bringt, wozu Natur es bestimmt hat, ist ein Alchimist.'

g Sudhoff ed., vol. 1, p. 125.

h 'Kein Arzt kann ohne die Alchimie sein, sonst ist er wie ein Saukoch gegenüber einem Fürstenkoch.' Cf. the opening of *Labyrinthus*, ch. 5, in Strebel ed., vol. 1, p. 190.

i Indeed the Chinese iatro-chemists had done just this from the +11th century onwards when they prepared the steroid hormones from urine (cf. Vol. 5, pt. 5).

j Direct descendant of that ancient Chinese conception, chin i.[1]

k Sudhoff ed., vol. 2, p. 106; Strebel ed., vol. 8, p. 246.

[1] 金液

To sum it up, therefore, Paracelsus was the vital pivot linking Chinese and then Arabic elixir alchemy through iatro-chemistry with the pharmacology and medicine based on modern chemistry. Of course the emphasis on therapy was not entirely new with him because it had come down from Roger Bacon's macrobiotics through such writers as Arnold of Villanova, John of Rupescissa and Michael Savonarola,[a] who all exalted the preparation of medicines for the conservation and prolongation of human life above any dubious aurifaction;[b] in a line of inheritance which (as we have seen) can be traced back to Islam and ultimately to China. But Paracelsus was the definitive figure who broke with aurifaction for ever as the main aim of alchemy, and pointed the way to all later chemistry and pharmacy.[c]

Having come now to the term of our long discussion of laboratory proto-chemistry and alchemy it may be just worth while to look at a few aspects of modern gerontology.[d] Prolongevity, after all, was the main objective of these ancient and medieval sciences.

The greatest ages to which human beings can attain have long been a matter of interest both in East and West.[e] William Harvey himself performed an autopsy on Thomas Parr, who died in +1635 at the supposed age of 152.[f] In recent times there have been some scientific studies of the super-centenarians of the Andes, especially the Vilcabamba Valley in Ecuador, and the conditions of their lives.[g] Here the two oldest men were 123 and 143 respectively, but there were several women of 103 and 105. Another region known for super-centenarians is Abkhasia in Russian Georgia, where the climate of the mountain valleys of the Caucasus may be similar to that of the Andes.[h] Here life-spans of 700 centenarians recently examined ranged up to 141 years. A third well-established longevity area is Hunza in the Karakorum range of the Himalayas.[i] Common to all these regions is a combination of mountain environment and a primarily vegetable diet, often low in calorific value, factors strangely justifying the ideas of the Taoists of old. The oldest human record confirmed by documents in Europe is 130, but lives ending between 100 and 111 years have not been extremely uncommon in England. The conservation of youthful vigour into such seniorities is still a task for the future, but one can easily see how a few super-centenarians in ancient China could have given colour to the conception of the *hsien*.

Perhaps it is more interesting to look at the unquestionable demographic fact that the life-length expectation of men and women has been rising continuously in the

---

[a] At the castle of Olsztyn in Warmia (Poland), where Nicholas Copernicus was Treasurer between +1516 and +1522, there is preserved among the books of the great astronomer a collection of Villanovan medical tractates bound up with some of the writings of Savonarola. Dr Lu Gwei-Djen and I had the pleasure of studying this in September, 1973.

[b] A good study of this line of succession has been made by Ganzenmüller (5).

[c] I am much indebted to my old friend Dr Walter Pagel for guidance through the *labyrinthus tractatorum Paracelsianorum*.

[d] On the biology of ageing and death in general there are excellent accounts by Grmek (2, 3) and Comfort (1), with references to an extensive literature. The more recent book of Rosenfeld (1) covers what is known about the biochemistry of ageing. On the historical background of gerontology see Burstein (1); Gruman (1, 2); Veith (6).

[e] An interesting book on the subject was published in 1907 by Nakamura Mokukō (1).

[f] Cf. Keynes (2), pp. 219 ff. [g] Davies (1, 2); Halsell (1).

[h] Cf. Benet (1).

[i] A close study of all these three parts of the world has been reported on by Leaf & Launois (1).

Western world since the Middle Ages, with the first rapid increase coming in the +18th century.[a] The following rough table, derived from several well-defined statistical analyses, shows the size of the change.

Date	Life-expectancy at birth, years	
	men	women
+1300	24	33
+1400	24	33
+1500	27	35
+1600	28	36
+1700	32	37
+1800	39	43
+1900	52	60
+1950	65	72

For other parts of the world we have no figures, but the same process must be occurring wherever modern science, medicine and technology, with the fuller understanding of nutrition and hygiene, is penetrating.[b] Though many other factors, such as food supplies, communications, housing and sanitation, have also had leading parts to play, the conviction of those thousands of pioneers, both Chinese, Arabic and European, that greater chemical knowledge could really lead to a lengthening of human life has surely proved true beyond dispute. Seen from Ko Hung's point of view, all hygiene and bacteriology, all pharmacy and nutritional science, would have been but extensions of the chemical knowledge needed for preparing the *tan*. The only failing of the early pioneers was the idea that there was one single substance alone which would be the universal medicine of man as well as metals; yet the elixir conception, from Tsou Yen through Jābir to Roger Bacon, was a veritably great creative dream. The kernel of truth in it was that the human body has a chemistry of its own, like all other compounded bodies whether inorganic or organic, and that if man could gain deep knowledge of that he would be able to prolong his life beyond belief.[c] If *hsien* immortality still eludes us, one begins to wonder whether it will always do so. But what unimaginable changes in human society centuries hence will have to come about to control such knowledge, if we ever attain it!

As a concluding epilogue, let us read the exquisitely Taoist words[d] of one of the

---

[a] See Hollingsworth (1a), p. 358, (1b), pp. 56–7, (2); Peller (1), p. 98; Wrigley (1), p. 171; Cipolla (3), p. 101; Russell (1), p. 47; Mols (1), p. 69; Armengaud (1), p. 48.

[b] There has been much debate on the role of medicine in the great +18th-century rise of population. While Griffith (1) attributed much importance to it, McKeown & Brown (1), agreeing with a standpoint of our own (Needham (59), repr. in (64), pp. 406 ff.) find that at that stage it was not a leading factor.

[c] We shall pursue this in Vol. 5, pt. 5, on Chinese physiological alchemy and iatrochemistry.

[d] P. 39, tr. Debus (7), p. 46, (18), p. 20, mod. auct.; also quoted by Partington (7), vol. 2, p. 164. As Norpoth (1) has reminded us, this exhortation was completely in accord with the convictions of Paracelsus himself. In the *Sieben Defensiones* he justified his wanderings: 'Die Schrift wird erforscht durch ihre Buchstaben, die Natur aber durch Land zu Land, als oft ein Land, als oft ein Blatt. Also ist Codex Naturae, also muss man ihre Blätter umkehren' (Strebel ed., vol. 1, p. 118).

great Paracelsian physicians, Peter Severinus, archiater to the King of Denmark. In his *Idea Medicinae Philosophicae* (+1571) he wrote of the necessity of replacing book-learning and scholastic philosophy by practical experience of natural phenomena, and practical experimentation. Only so could the inspiring Paracelsian aim be achieved, that alchemists should make not gold but medicines. So to his readers he said:

Sell your lands, your houses, your clothes and your jewellery; burn up your books. Instead of those things, buy yourselves stout shoes and travel to the mountains, search the valleys, the deserts, the shores of the sea and the deepest depressions of the earth; note with care the distinctions between animals, the differences of plants, the various kinds of minerals, and the properties and mode of origin of everything that exists. Be not ashamed to study diligently the astronomy and terrestrial philosophy of the country people. Lastly purchase coal, build furnaces, watch and operate with the fire never wearying. In this way, and in no other, will you arrive at a knowledge of things and their properties.

BIBLIOGRAPHIES

A   CHINESE AND JAPANESE BOOKS BEFORE +1800
B   CHINESE AND JAPANESE BOOKS AND JOURNAL ARTICLES SINCE +1800
C   BOOKS AND JOURNAL ARTICLES IN WESTERN LANGUAGES

In Bibliographies A and B there are two modifications of the Roman alphabetical sequence: transliterated *Chh-* comes after all other entries under *Ch-*, and transliterated *Hs-* comes after all other entries under *H-*. Thus *Chhen* comes after *Chung* and *Hsi* comes after *Huai*. This system applies only to the first words of the titles. Moreover, where *Chh-* and *Hs-* occur in words used in Bibliography C, i.e. in a Western language context, the normal sequence of the Roman alphabet is observed.

When obsolete or unusual romanisations of Chinese words occur in entries in Bibliography C, they are followed, wherever possible, by the romanisations adopted as standard in the present work. If inserted in the title, these are enclosed in square brackets; if they follow it, in round brackets. When Chinese words or phrases occur romanised according to the Wade–Giles system or related systems, they are assimilated to the system here adopted (cf. Vol. 1, p. 26) without indication of any change. Additional notes are added in round brackets. The reference numbers do not necessarily begin with (1), nor are they necessarily consecutive, because only those references required for this volume of the series are given.

Korean and Vietnamese books and papers are included in Bibliographies A and B. As explained in Vol. 1, pp. 21 ff., reference numbers in italics imply that the work is in one or other of the East Asian languages.

ABBREVIATIONS

See also p. xxv

A	*Archeion*	*AJPA*	*Amer. Journ. Physical Anthropology*
AA	*Artibus Asiae*		
AAA	*Archaeologia*	*AJSC*	*American Journ. Science and Arts* (Silliman's)
AAAA	*Archaeology*		
A/AIHS	*Archives Internationales d'Histoire des Sciences* (continuation of *Archeion*)	*AM*	*Asia Major*
		AMA	*American Antiquity*
		AMH	*Annals of Medical History*
AAN	*American Anthropologist*	*AMS*	*American Scholar*
AAPWM	*Archiv. f. Anat., Physiol., and Wiss. Med.* (Joh. Müller's)	*AMY*	*Archaeometry* (Oxford)
		AN	*Anthropos*
ABAW/PH	*Abhandlungen d. bayr. Akad. Wiss. München* (Phil.-Hist. Klasse)	*ANATS*	*Anatolian Studies* (British School of Archaeol. Ankara)
ACASA	*Archives of the Chinese Art Soc. of America*	*ANS*	*Annals of Science*
		ANT	*Antaios* (Stuttgart)
ACF	*Annuaire du Collège de France*	*ANTJ*	*Antiquaries Journal*
ADVC	*Advances in Chemistry*	*AP*	*Aryan Path.*
ADVS	*Advancement of Science* (British Assoc., London)	*APH*	*Actualités Pharmacologiques*
		AP/HJ	*Historical Journal, National Peiping Academy*
AEM	*Anuario de Estudios Medievales* (Barcelona)	*APAW/PH*	*Abhandlungen d. preuss. Akad. Wiss. Berlin* (Phil.-Hist. Klasse)
AEPHE/SHP	*Annuaire de l'Ecole Pratique des Hautes Études* (Sect. Sci. Hist. et Philol.)	*APHL*	*Acta Pharmaceutica Helvetica*
		APNP	*Archives de Physiol. normale et pathologique*
AEPHE/SSR	*Annuaire de l'Ecole Pratique des Hautes Études* (Sect. des Sci. Religieuses)	*AQ*	*Antiquity*
		AR	*Archiv. f. Religionswissenschaft*
AESC	*Aesculape* (Paris)	*ARB*	*Annual Review of Biochemistry*
AEST	*Annales de l'Est* (Fac. des Lettres, Univ. Nancy)	*ARLC/DO*	*Annual Reports of the Librarian of Congress* (Division of Orientalia)
AF	*Ärztliche Forschung*	*ARMC*	*Ann. Reports in Medical Chemistry*
AFG	*Archiv. f. Gynäkologie*		
AFGR/CINO	*Atti della Fondazione Giorgio Ronchi e Contributi dell'Istituto Nazionale di Ottica* (Arcetri)	*ARO*	*Archiv Orientalni* (Prague)
		ARQ	*Art Quarterly*
AFP	*Archivum Fratrum Praedicatorum*	*ARSI*	*Annual Reports of the Smithsonian Institution* (Washington, D.C.)
AFRA	*Afrasian* (student Journal of London Inst. Oriental & African Studies)	*AS/BIHP*	*Bulletin of the Institute of History and Philology, Academia Sinica*
		AS/CJA	*Chinese Journal of Archaeology, Academia Sinica*
AGMN	*Archiv. f. d. Gesch. d. Medizin u. d. Naturwissenschaften* (Sudhoff's)	*ASEA*	*Asiatische Studien; Études Asiatiques*
AGMW	*Abhandlungen z. Geschichte d. Math. Wissenschaft*	*ASN/Z*	*Annales des Sciences Naturelles; Zoologie* (Paris)
AGNT	*Archiv. f. d. Gesch. d. Naturwiss. u. d. Technik* (cont. as *AGMNT*)	*ASSF*	*Acta Societatis Scientiarum Fennicae* (Helsingfors)
		AT	*Atlantis*
AGP	*Archiv. f. d. Gesch. d. Philosophie*	*ATOM*	*Atomes* (Paris)
AGR	*Asahigraph*	*AX*	*Ambix*
AGWG/PH	*Abhdl. d. Gesell. d. Wiss. Z. Göttingen* (Phil.-Hist. Kl.)	*BABEL*	*Babel; Revue Internationale de la Traduction*
AHES/AHS	*Annales d'Hist. Sociale*		
AHOR	*Antiquarian Horology*	*BCGS*	*Bull. Chinese Geological Soc.*
AIENZ	*Advances in Enzymology*	*BCP*	*Bulletin Catholique de Pékin*
AIP	*Archives Internationales de Physiologie*	*BCS*	*Bulletin of Chinese Studies* (Chhêngtu)
AJA	*American Journ. Archaeology*	*BDCG*	*Ber. d. deutsch. chem. Gesellschaft.*
AJOP	*Amer. Journ. Physiol.*	*BDP*	*Blätter f. deutschen Philosophie*

BE/AMG	*Bibliographie d'Études (Annales du Musée Guimet)*	*CHIM*	*Chimica* (Italy)
BEC	*Bulletin de l'École des Chartes* (Paris)	*CHIND*	*Chemistry and Industry* (Journ. Soc. Chem. Ind. London)
BEFED	*Bulletin de l'Ecole Française de l'Extrême Orient* (Hanoi)	*CHJ*	*Chhing-Hua Hsüeh Pao (Chhing-Hua (Ts'ing-Hua) University Journal of Chinese Studies)*
BGSC	*Bulletin of the Chinese Geological Survey*	*CHJ/T*	*Chhing-Hua (T'sing-Hua) Journal of Chinese Studies* (New Series, publ. Thaiwan)
BGTI	*Beiträge z. Gesch. d. Technik u. Industrie* (continued as *Technik Geschichte—see BGTI/TG*)		
BGTI/TG	*Technik Geschichte*	*CHWSLT*	*Chung-Hua Wên-Shih Lun Tshung (Collected Studies in the History of Chinese Literature)*
BHMZ	*Berg und Hüttenmännische Zeitung*	*CHYM*	*Chymia*
BIHM	*Bulletin of the (Johns Hopkins) Institute of the History of Medicine* (cont. as *Bulletin of the History of Medicine*)	*CHZ*	*Chemiker Zeitung*
		CIBA/M	*Ciba Review* (Medical History)
		CIBA/MZ	*Ciba Zeitschrift* (Medical History)
		CIBA/S	*Ciba Symposia*
BJ	*Biochemical Journal*	*CIBA/T*	*Ciba Review* (Textile Technology)
BJRL	*Bull. John Rylands Library* (Manchester)	*CIMC/MR*	*Chinese Imperial Maritime Customs (Medical Report Series)*
BK	*Bunka (Culture)*, Sendai	*CIT*	*Chemie Ingenieur Technik*
BLSOAS	*Bulletin of the London School of Oriental and African Studies*	*CJ*	*China Journal of Science and Arts*
		CJFC	*Chin Jih Fo Chiao (Buddhism Today)*, Thaiwan
BM	*Bibliotheca Mathematica*		
BMFEA	*Bulletin of the Museum of Far Eastern Antiquities* (Stockholm)	*CLINR*	*Clinical Radiology*
		CLR	*Classical Review*
BMFJ	*Bulletin de la Maison Franco–Japonaise* (Tokyo)	*CMJ*	*Chinese Medical Journal*
		CN	*Chemical News*
BMJ	*British Medical Journal*	*CNRS*	Centre National de la Recherche Scientifique
BNJ	*British Numismatic Journ.*		
BOE	*Boethius; Texte und Abhandlungen d. exakte Naturwissenschaften* (Frankfurt)	*COCJ*	*Coin Collectors' Journal*
		COPS	*Confines of Psychiatry*
		CP	*Classical Philology*
BR	*Biological Reviews*	*CQ*	*Classical Quarterly*
BS	*Behavioural Science*	*CR*	*China Review* (Hongkong and Shanghai)
BSAA	*Bull. Soc. Archéologique d'Alexandrie*		
		CRAS	*Comptes Rendus hebdomadaires de l'Acad. des Sciences* (Paris)
BSAB	*Bull. Soc. d'Anthropologie de Bruxelles*		
		CREC	*China Reconstructs*
BSCF	*Bull. de la Société Chimique de France*	*CRESC*	*Crescent* (Surat)
		CRR	*Chinese Recorder*
BSGF	*Bull. de la Société Géologique de France*	*CRRR*	*Chinese Repository*
		CS	*Current Science*
BSJR	*Bureau of Standards Journ. of Research*	*CUNOB*	*Cunobelin; Yearbook of the British Association of Numismatic Societies*
BSPB	*Bull. Soc. Pharm. Bordeaux*		
BUA	*Bulletin de l'Université de l'Aurore* (Shanghai)	*CUP*	Cambridge University Press
		CUQ	*Columbia University Quarterly*
BV	*Bharatiya Vidya* (Bombay)	*CURRA*	*Current Anthropology*
		CVS	*Christiania Videnskabsselskabet Skrifter*
CA	*Chemical Abstracts*		
CALM	*California Medicine*	*CW*	*Chemische Weekblad*
CBH	*Chūgoku Bungaku-hō (Journ. Chinese Literature)*	*CWR*	*China Weekly Review*
CCJ	*Chung-Chi Journal* (Chhung-Chi Univ. Coll. Hongkong)	*DAZ*	*Deutscher Apotheke Zeitung*
		DB	*The Double Bond*
CDA	*Chinesisch-Deutschen Almanach* (Frankfort a/M)	*DI*	*Die Islam*
		DK	*Dōkyō Kenkyū (Researches in the Taoist Religion)*
CEM	*Chinese Economic Monthly* (Shanghai)		
		DMAB	*Abhandlungen u. Berichte d. Deutsches Museum* (München)
CEN	*Centaurus*		
CHA	*Chemische Apparatur*	*DS*	*Desalination (International Journ. Water Desalting)* (Amsterdam and Jerusalem, Israel)
CHEMC	*Chemistry in Canada*		
CHI	*Cambridge History of India*		

DV	*Deutsche Vierteljahrschrift*
DVN	*Dan Viet Nam*
DZZ	*Deutsche Zahnärztlichen Zeit.*
EARLH	*Earlham Review*
EECN	*Electroencephalography and Clinical Neurophysiology*
EG	*Economic Geology*
EHOR	*Eastern Horizon* (Hongkong)
EHR	*Economic History Review*
EI	*Encyclopaedia of Islam*
EMJ	*Engineering and Mining Journal*
END	*Endeavour*
EPJ	*Edinburgh Philosophical Journal* (continued as *ENPJ*)
ERE	*Encyclopaedia of Religion and Ethics*
ERJB	*Eranos Jahrbuch*
ERYB	*Eranos Yearbook*
ETH	*Ethnos*
EURR	*Europaïsche Revue* (Berlin)
EXPED	*Expedition* (*Magazine of Archaeology and Anthropology*), Philadelphia
FCON	*Fortschritte d. chemie d. organischen Naturstoffe*
FER	*Far Eastern Review* (London)
FF	*Forschungen und Fortschritte*
FMNHP/AS	*Field Museum of Natural History* (Chicago) *Publications*; Anthropological Series
FP	*Federation Proceedings* (USA)
FPNJ	*Folia Psychologica et Neurologica Japonica*
FRS	*Franziskanischen Studien*
GBA	*Gazette des Beaux-Arts*
GBT	*Global Technology*
GEW	*Geloof en Wetenschap*
GJ	*Geographical Journal*
GR	*Geographical Review*
GRM	*Germanisch–Romanische Monatsschrift*
GUJ	*Gutenberg Jahrbuch*
HCA	*Helvetica Chimica Acta*
HE	*Hesperia* (*Journ. Amer. Sch. Class. Stud. Athens*)
HEJ	*Health Education Journal*
HERM	*Hermes; Zeitschr. f. Klass. Philol.*
HF	*Med Hammare och Fackla* (Sweden)
HHS	*Hua Hsüeh* (*Chemistry*), Ch. Chem. Soc.
HHSTH	*Hua Hsüeh Thung Hsün* (*Chemical Correspondent*), Chekiang Univ.
HITC	*Hsüeh I Tsa Chih* (*Wissen und Wissenschaft*), Shanghai
HJAS	*Harvard Journal of Asiatic Studies*
HMSO	Her Majesty's Stationery Office
HOR	*History of Religion* (Chicago)
HOSC	*History of Science* (annual)

HRASP	*Histoire de l'Acad. Roy. des Sciences, Paris*
HSS	*Hsüeh Ssu* (*Thought and Learning*), Chhêngtu
HU/BML	*Harvard University Botanical Museum Leaflets*
HUM	*Humanist* (RPA, London)
IA	*Iron Age*
IBK	*Indogaku Bukkyōgaku Kenkyū* (*Indian and Buddhist Studies*)
IC	*Islamic Culture* (Hyderabad)
ID	*Idan* (Medical Discussions), Japan
IEC/AE	*Industrial and Engineering Chemistry; Analytical Edition*
IEC/I	*Industrial and Engineering Chemistry; Industrial Edition*
IHQ	*Indian Historical Quarterly*
IJE	*Indian Journ. Entomol.*
IJHM	*Indian Journ. History of Medicine*
IJHS	*Indian Journ. History of Science*
IJMR	*Indian Journ. Med. Research*
IMIN	*Industria Mineraria*
IMW	*India Medical World*
INDQ	*Industria y Quimica* (Buenos Aires)
INM	*International Nickel Magazine*
IPEK	*Ipek; Jahrb. f. prähistorische u. ethnographische Kunst* (Leipzig)
IQB	*Iqbal* (Lahore), later *Iqbal Review* (*Journ. of the Iqbal Academy* or *Bazm-i Iqbal*)
IRAQ	*Iraq* (British Sch. Archaeol. in Iraq)
ISIS	*Isis*
ISTC	*I Shih Tsa Chih* (*Chinese Journal of the History of Medicine*)
IVS	*Ingeniörvidenskabelje Skrifter* (Copenhagen)
JA	*Journal Asiatique*
JAC	*Jahrb. f. Antike u. Christentum*
JACS	*Journ. Amer. Chem. Soc.*
JAHIST	*Journ. Asian History* (*International*)
JAIMH	*Pratibha; Journ. All-India Instit. of Mental Health*
JALCHS	*Journal of the Alchemical Society* (London)
JAN	*Janus*
JAOS	*Journal of the American Oriental Society*
JAP	*Journ. Applied Physiol.*
JAS	*Journal of Asian Studies* (continuation of *Far Eastern Quarterly, FEQ*)
JATBA	*Journal d'Agriculture tropicale et de Botanique appliqué*
JBC	*Journ. Biol. Chem.*
JBFIGN	*Jahresber. d. Forschungsinstitut f. Gesch. d. Naturwiss.* (Berlin)
JC	*Jimnin Chūgoku* (*People's China*), Tokyo
JCE	*Journal of Chemical Education*
JCP	*Jahrb. f. class. Philologie*

JCS	Journal of the Chemical Society	JUB	Journ. Univ. Bombay
JEA	Journal of Egyptian Archaeology	JUS	Journ. Unified Science (continuation of Erkenntnis)
JEGP	Journal of English and Germanic Philology		Journal of the West China Border
		JWCBRS	Journal of the West China Border Research Society
JEH	Journal of Economic History		
JEM	Journ. Exper. Med.	JWCI	Journal of the Warburg and Courtauld Institutes
JFI	Journ. Franklin Institute		
JGGBB	Jahrbuch d. Gesellschaft f. d. Gesch. u. Bibliographie des Brauwesens	JWH	Journal of World History (UNESCO)
JGMB	Journ. Gen. Microbiol.		
JHI	Journal of the History of Ideas	KHS	Kho Hsüeh (Science)
JHMAS	Journal of the History of Medicine and Allied Sciences	KHSC	Kho-Hsüeh Shih Chi-Khan (Ch. Journ. Hist. of Sci.)
JHS	Journal of Hellenic Studies	KHTP	Kho Hsüeh Thung Pao (Science Correspondent)
JI	Jissen Igaku (Practical Medicine)		
JIM	Journ. Institute of Metals (UK)	KHVL	Kungliga Humanistiska Vetenskapsamfundet i Lund Årskerättelse (Bull. de la Soc. Roy. de Lettres de Lund)
JIMA	Journ. Indian Med. Assoc.		
JKHRS	Journ. Kalinga Historical Research Soc. (Orissa)		
JMBA	Journ. of the Marine Biological Association (Plymouth)	KKD	Kiuki Daigaku Sekai Keizai Kenkyūjo Hōkoku (Reports of the Institute of World Economics at Kiuki Univ.)
JNMD	Journ. Nervous & Mental Diseases		
JMS	Journ. Mental Science		
JNPS	Journ. Neuropsychiatr.	KKTH	Khao Ku Thung Hsün (Archaeological Correspondent), cont. as Khao Ku
JOP	Journ. Physiol.		
JOSHK	Journal of Oriental Studies (Hongkong Univ.)		
		KKTS	Ku Kung Thu Shu Chi Khan (Journal of the Imperial Palace Museum and Library), Thaiwan
JP	Journal of Philology		
JPB	Journ. Pathol. and Bacteriol.		
JPC	Journ. f. prakt. Chem.	KSVA/H	Kungl. Svenske Vetenskapsakad. Handlingar
JPCH	Journ. Physical Chem.		
JPH	Journal de Physique	KVSUA	Kungl. Vetenskaps Soc. i Uppsala Årsbok (Mem. Roy. Acad. Sci. Uppsala)
JPHS	Journ. Pakistan Historical Society		
JPHST	Journ. Philos. Studies		
JPOS	Journal of the Peking Oriental Society	KW	Klinische Wochenschrift
JRAI	Journal of the Royal Anthropological Institute	LA	Annalen d. Chemie (Liebig's)
		LCHIND	La Chimica e l'Industria (Milan)
JRAS	Journal of the Royal Asiatic Society	LEC	Lettres Édifiantes et Curieuses écrites des Missions Étrangères (Paris, 1702–1776)
JRAS/B	Journal of the (Royal) Asiatic Society of Bengal		
		LH	l'Homme; Revue Française d'Anthropologie
JRAS/BOM	Journ. Roy. Asiatic Soc., Bombay Branch		
		LIN	L'Institut (Journal Universel des Sciences et des Sociétés Savantes en France et à l'Étranger)
JRAS/KB	Journal (or Transactions) of the Korea Branch of the Royal Asiatic Society		
		LN	La Nature
JRAS/M	Journal of the Malayan Branch of the Royal Asiatic Society	LP	La Pensée
		LSYC	Li Shih Yen Chiu (Journal of Historical Research), Peking
JRAS/NCB	Journal (or Transactions) of the Royal Asiatic Society (North China Branch)		
		LSYKK	Li Shih yü Khao Ku (History and Archaeology; Bulletin of the Shenyang Museum), Shenyang
JRAS/P	Journ. of the (Royal) Asiatic Soc. of Pakistan		
		LT	Lancet
JRIBA	Journ. Royal Institute of British Architects	LYCH	Lychnos (Annual of the Swedish Hist. of Sci. Society)
JRSA	Journal of the Royal Society of Arts		
JS	Journal des Sçavans (1665–1778) and Journal des Savants (1816–)	MAAA	Memoirs Amer. Anthropological Association
JSA	Journal de la Société des Americanistes	MAI/NEM	Mémoires de l'Académie des Inscriptions et Belles-Lettres, Paris (Notices et Extraits des MSS)
JSCI	Journ. Soc. Chem. Industry		
JSHS	Japanese Studies in the History of Science (Tokyo)	MAIS/SP	Mémoires de l'Acad. Impériale des Sciences, St Pétersbourg

MAS/B	Memoirs of the Asiatic Society of Bengal	MS	Monumenta Serica
MB	Monographiae Biologicae	MSAF	Mémoires de la Société (Nat.) des Antiquaires de France
MBLB	May and Baker Laboratory Bulletin	MSGVK	Mitt. d. Schlesische Gesellschaft f. Volkskunde
MBPB	May and Baker Pharmaceutical Bulletin	MSIV/MF	Memoire di Mat. e. Fis della Soc. Ital. (Verona)
MCB	Mélanges Chinois et Bouddhiques	MSOS	Mitteilungen d. Seminar f. orientalischen Sprachen (Berlin)
MCE	Metallurgical and Chemical Engineering	MSP	Mining and Scientific Press
MCHSAMUC	Mémoires concernant l'Histoire, les Sciences, les Arts, les Mœurs et les Usages, des Chinois, par les Missionnaires de Pékin (Paris 1776–)	MUJ	Museum Journal (Philadelphia)
		MUSEON	Le Muséon (Louvain)
		N	Nature
		NAGE	New Age (New Delhi)
MDGNVO	Mitteilungen d. deutsch. Gesellsch. f. Natur. u. Volkskunde Ostasiens	NAR	Nutrition Abstracts and Reviews
		NARSU	Nova Acta Reg. Soc. Sci. Upsaliensis
MDP	Mémoires de la Délégation en Perse	NC	Numismatic Chronicle (and Journ. Roy. Numismatic Soc.)
MED	Medicus (Karachi)		
MEDA	Medica (Paris)	NCDN	North China Daily News
METL	Metallen (Sweden)	NCGH	Nihon Chūgoku Gakkai-hō (Bulletin of the Japanese Sinological Society)
MGG	Monatsschrift f. Geburtshilfe u. Gynäkologie		
MGGW	Mitteilungen d. geographische Gesellschaft Wien	NCH	North China Herald
		NCR	New China Review
MGSC	Memoirs of the Chinese Geological Survey	NDI	Niigata Daigaku Igakubu Gakushikai Kaihō (Bulletin of the Medical Graduate Society of Niigata University)
MH	Medical History		
MI	Metal Industry		
MIE	Mémoires de l'Institut d'Egypte (Cairo)	NFR	Nat. Fireworks Review
		NHK	Nihon Heibon Keisha (publisher)
MIFC	Mémoires de l'Institut Français d'Archéol. Orientale (Cairo)	NIZ	Nihon Ishigaku Zasshi (Jap. Journ. Hist. Med.)
MIK	Mikrochemie		
MIMG	Mining Magazine	NN	Nation
MIT	Massachusetts Institute of Technology	NQ	Notes and Queries
		NR	Numismatic Review
MJ	Mining Journal, Railway and Commercial Gazette	NRRS	Notes and Records of the Royal Society
MJA	Med. Journ. Australia	NS	New Scientist
MJPGA	Mitteilungen aus Justus Perthes Geogr. Anstalt (Petermann's)	NSN	New Statesman and Nation (London)
MKDUS/HF	Meddelelser d. Kgl. Danske Videnskabernes Selskab (Hist.-Filol.)	NU	The Nucleus
		NUM/SHR	Studies in the History of Religions (Supplements to Numen)
MM	Mining and Metallurgy (New York, contd. as Mining Engineering)	NW	Naturwissenschaften
MMN	Materia Medica Nordmark		
MMVKH	Mitteilungen d. Museum f. Völkerkunde (Hamburg)	OAZ	Ostasiatische Zeitschrift
		ODVS	Oversigt over det k. Danske Videnskabernes Selskabs Forhandlinger
MMW	Münchener Medizinische Wochenschrift	OE	Oriens Extremus (Hamburg)
MOULA	Memoirs of the Osaka University of Liberal Arts and Education	OLZ	Orientalische Literatur-Zeitung
		ORA	Oriental Art
MP	Il Marco Polo	ORCH	Orientalia Christiana
MPMH	Memoirs of the Peabody Museum of American Archaeology and Ethnology, Harvard University	ORD	Ordnance
		ORG	Organon (Warsaw)
		ORR	Orientalia (Rome)
MRASP	Mémoires de l'Acad. Royale des Sciences (Paris)	ORS	Orientalia Suecana
		OSIS	Osiris
MRDTB	Memoirs of the Research Dept. of Tōyō Bunko (Tokyo)	OUP	Oxford University Press
		OUSS	Ochanomizu University Studies
MRS	Mediaeval and Renaissance Studies	OX	Oxoniensia

PAAAS	Proceeding of the British Academy	RBS	Revue Bibliographique de Sinologie
PAAQS	Proceedings of the American Anti-quarian Society	RDM	Revue des Mines (later Revue Universelle des Mines)
PAI	Paideuma	RGVV	Religionsgeschichtliche Versuche und Vorarbeiten
PAKJS	Pakistan Journ. Sci.		
PAKPJ	Pakistan Philos. Journ.	RHR/AMG	Revue de l'Histoire des Religions (Annales du Musée Guimet, Paris)
PAPS	Proc. Amer. Philos. Soc.		
PCASC	Proc. Cambridge Antiquarian Soc.		
PEW	Philosophy East and West (Univ. Hawaii)	RHS	Revue d'Histoire des Sciences
		RHSID	Revue d'Histoire de la Sidérurgie (Nancy)
PF	Psychologische Forschung		
PHI	Die Pharmazeutische Industrie	RIN	Rivista Italiana di Numismatica
PHREV	Pharmacological Reviews	RKW	Repertorium f. Kunst. wissenschaft
PHY	Physis (Florence)		
PJ	Pharmaceut. Journal (and Trans. Pharmaceut. Soc.)	RMY	Revue de Mycologie
		ROC	Revue de l'Orient Chrétien
PKAWA	Proc. Kon. Akad. Wetensch. Amsterdam	RP	Revue Philosophique
		RPA	Rationalist Press Association (London)
PKR	Peking Review		
PM	Presse Medicale	RPCHG	Revue de Pathologie comparée et d'Hygiène générale (Paris)
PMG	Philosophical Magazine		
PMLA	Publications of the Modern Language Association of America	RPLHA	Revue de Philol., Litt. et Hist. Ancienne
PNHB	Peking Natural History Bulletin	RR	Review of Religion
POLYJ	Polytechnisches Journal (Dingler's)	RSCI	Revue Scientifique (Paris)
PPHS	Proceedings of the Prehistoric Society	RSH	Revue de Synthèse Historique
		RSI	Reviews of Scientific Instruments
PRGS	Proceedings of the Royal Geographical Society	RSO	Rivista di Studi Orientali
		RUB	Revue de l'Univ. de Bruxelles
PRIA	Proceedings of the Royal Irish Academy		
PRPH	Produits Pharmaceutiques	S	Sinologica (Basel)
PRSA	Proceedings of the Royal Society (Series A)	SA	Sinica (originally Chinesische Blätter f. Wissenschaft u. Kunst)
PRSB	Proceedings of the Royal Society (Series B)	SAEC	Supplemento Annuale all'Enciclopedia di Chimica
PRSM	Proceedings of the Royal Society of Medicine	SAEP	Soc. Anonyme des Études et Pub. (publisher)
PSEBM	Proc. Soc. Exp. Biol and Med.	SAM	Scientific American
PTRS	Philosophical Transactions of the Royal Society	SB	Shizen to Bunka (Nature and Culture)
		SBE	Sacred Books of the East series
QSGNM	Quellen u. Studien z. Gesch. d. Naturwiss. u. d. Medizin (continuation of Archiv. f. Gesch. d. Math., d. Naturwiss. u. d. Technik, AGMNT, formerly Archiv. f. d. Gesch. d. Naturwiss. u. d. Technik, AGNT)	SBK	Seikatsu Bunka Kenkyū (Journ. Econ. Cult.)
		SBM	Svenska Bryggareföreningens Månadsblad
		SC	Science
		SCI	Scientia
		SCIS	Sciences; Revue de la Civilisation Scientifique (Paris)
QSKMR	Quellenschriften f. Kunstgeschichte und Kunsttechnik des Mittelalters u. d. Renaissance (Vienna)	SCISA	Scientia Sinica (Peking)
		SCK	Smithsonian Contributions to Knowledge
		SCM	Student Christian Movement (Press)
RA	Revue Archéologique		
RAA/AMG	Revue des Arts Asiatiques (Annales du Musée Guimet)	SCON	Studies in Conservation (Journ. Internat. Instit. for the Conservation of Museum objects)
RAAAS	Reports, Australasian Assoc. Adv. of Sci.	SET	Structure et Evolution des Techniques
RAAO	Revue d'Assyriologie et d'Archéologie Orientale	SGZ	Shigaku Zasshi (Historical Journ. of Japan)
RALUM	Revue de l'Aluminium		
RB	Revue Biblique	SHA	Shukan Asahi
RBPH	Revue Belge de Philol. et d'Histoire	SHAW/PH	Sitzungsber. d. Heidelberg. Akad. d. Wissensch. (Phil.-Hist. Kl.)

SHST/T	Studies in the History of Science and Technol. (Tokyo Univ. Inst. Technol.)	TIMM	Transactions of the Institution of Mining and Metallurgy
SI	Studia Islamica (Paris)	TJSL	Transactions (and Proceedings) of the Japan Society of London
SIB	Sibrium (Collana di Studi e Documentazioni, Centro di Studi Preistorici e Archeologici Varese)	TLTC	Ta Lu Tsa Chih (Continent Magazine), Thaipei
SILL	Sweden Illustrated	TMIE	Travaux et Mémoires de l'Inst. d'Ethnologie (Paris)
SK	Seminarium Kondakovianum (Recueil d'Études de l'Institut Kondakov)	TNS	Transactions of the Newcomen Society
SM	Scientific Monthly (formerly Popular Science Monthly)	TOCS	Transactions of the Oriental Ceramic Society
SN	Shirin (Journal of History), Kyoto	TP	T'oung Pao (Archives concernant l'Histoire, les Langues, la Géographie, l'Ethnographie et les Arts de l'Asie Orientale), Leiden
SNM	Sbornik Nauknych Materialov (Erivan, Armenia)		
SOS	Semitic and Oriental Studies (Univ. of Calif. Publ. in Semitic Philol.)		
SP	Speculum	TQ	Tel Quel (Paris)
SPAW/PH	Sitzungsber. d. preuss. Akad. d. Wissenschaften (Phil.-Hist. Kl.)	TR	Technology Review
		TRAD	Tradition (Zeitschr. f. Firmengeschichte und Unternehmerbiographie)
SPCK	Society for the Promotion of Christian Knowledge	TRSC	Trans. Roy. Soc. Canada
SPMSE	Sitzungsberichte d. physik. med. Soc. Erlangen	TS	Tōhō Shūkyō (Journal of East Asian Religions)
SPR	Science Progress	TSFFA	Techn. Studies in the Field of the Fine Arts
SSIP	Shanghai Science Institute Publications	TTT	Theoria to Theory (Cambridge)
STM	Studi Medievali	TYG	Tōyō Gakuhō (Reports of the Oriental Society of Tokyo)
SWAW/PH	Sitzungsberichte d. k. Akad. d. Wissenschaften Wien (Phil.-Hist. Klasse), Vienna	TYGK	Tōyōgaku (Oriental Studies), Sendai
		TYKK	Thien Yeh Khao Ku Pao Kao (Archaeological Reports)
TAFA	Transactions of the American Foundrymen's Association	UCC	University of California Chronicle
TAIME	Trans. Amer. Inst. Mining Engineers (continued as TAIMME)	UCR	University of Ceylon Review
		UNASIA	United Asia (India)
TAIMME	Transactions of the American Institute of Mining and Metallurgical Engineers	UNESC	Unesco Courier
		UNESCO	United Nations Educational, Scientific and Cultural Organisation
TAPS	Transactions of the American Philosophical Society (cf. MAPS)	UUA	Uppsala Univ. Årsskrift (Acta Univ. Upsaliensis)
TAS/J	Transactions of the Asiatic Society of Japan	VBA	Visva-Bharati Annals
TBKK	Tōhoku Bunka Kenkyūshitsu Kiyō (Record of the North-Eastern Research Institute of Humanistic Studies), Sendai	VBW	Vorträge d. Bibliothek Warburg
		VK	Vijnan Karmee
		VKAWA/L	Verhandelingen d. Koninklijke Akad. v. Wetenschappen te Amsterdam (Afd. Letterkunde)
TCS	Trans. Ceramic Society (formerly Trans. Engl. Cer. Soc., contd as Trans. Brit. Cer. Soc.)	VMAWA	Verslagen en Meded. d. Koninklijke Akad. v. Wetenschappen te Amsterdam
TCULT	Technology and Culture		
TFTC	Tung Fang Tsa Chih (Eastern Miscellany)	VVBGP	Verhandlingen d. Verein z. Beförderung des Gewerbefleisses in Preussen
TGAS	Transactions of the Glasgow Archaeological Society		
TG/T	Tōhō Gakuhō, Tōkyō (Tokyo Journal of Oriental Studies)	WA	Wissenschaftliche Annalen
		WKW	Wiener klinische Wochenschrift
TH	Thien Hsia Monthly (Shanghai)	WS	Wên Shih (History of Literature), Peking
THG	Tōhōgaku (Eastern Studies), Tokyo		
TICE	Transactions of the Institute of Chemical Engineers	WWTK	Wên Wu (formerly Wên Wu Tshan Khao Tzu Liao, Refer-

	ence Materials for History and Archaeology)	ZAC	Zeitschr. f. angewandte chemie
		ZAC/AC	Angewandte Chemie
WZNHK	Wiener Zeitschr. f. Nervenheil-kunde	ZAES	Zeitschrift f. Aegyptische Sprache u. Altertumskunde
		ZASS	Zeitschr. f. Assyriologie
YCHP	Yenching Hsüeh Pao (Yenching University Journal of Chinese Studies)	ZDMG	Zeitschrift d. deutsch. Morgen-ländischen Gesellschaft
		ZGEB	Zeitschr. d. Gesellsch. f. Erdkunde (Berlin)
YJBM	Yale Journal of Biology and Medicine	ZMP	Zeitschrift f. Math. u. Physik
YJSS	Yenching Journal of Social Studies	ZPC	Zeitschr. f. physiologischen Chemie
		ZS	Zeitschr. f. Semitistik
Z	Zalmoxis; Revue des Études Reli-gieuses	ZVSF	Zeitschr. f. vergl. Sprachforschung

ADDENDA TO ABBREVIATIONS

This list is Conflated with that on p. 271 of *SCC*, Vol. 5, part 3. The items which appeared in that list are indicated here by an asterisk.

AAS	Arts Asiatiques		Health and Tibbi Research, Karachi)
*ACTAS	Acta Asiatica (Bull. of Eastern Culture, Tōhō Gakkai, Tokyo)	JARCHS	Journ. Archaeol. Science
ADR	American Dyestuff Reporter	JJHS	Japanese Journ. History of Science
AGMNT	Archiv f. d. Geschichte d. Mathem-atik, d. Naturwiss. u. d. Technik	JPMA	Journ. Pakistan Med. Assoc.
		MAGW	Mitt. d. Anthropol. Gesellschaft in Wien
AIND	Ancient India (Bull. Archaeol. Survey of India)	MARCH	Mediaeval Archaeology
AOAW/PH	Anzeiger d. Österr. Aka d. d. Wiss. (Vienna, Phil.-Hist. Klasse)	MLJ	Mittel-Lateinisches Jahrbuch
		MMLPS	Memoirs of the Manchester Liter-ary and Philosophical Soc.
BCED	Biochemical Education		
*BILCA	Boletim do Instituto Luis de Camoes (Macao)	NAMSL	Nouvelles Archives des Missions Scientifiques et Littéraires
BIOL	The Biologist	*NGM	National Geographic Magazine
BJHOS	Brit. Journ. History of Science	NT	Novum Testamentum
BSAC	Bull. de la Soc. d'Acupuncture	NTS	New Testament Studies
*CFC	Cahiers Franco-Chinois (Paris)	PAKARCH	Pakistan Archaeology
*CHEM	Chemistry (Easton, Pa.)	PAR	Parabola
CLMED	Classica et Mediaevalia	PBM	Perspectives in Biol. and Med.
*COMP	Comprendre (Soc. Eu. de Culture, Venice)	PHYR	Physical Review
		PIH	Pharmacy in History
*CR/MSU	Centennial Review of Arts and Science (Michigan State Univ-ersity)	*POLREC	Polar Record
		POPST	Population Studies
		PRPSG	Proc. Roy. Philos. Soc. Glasgow
DZA	Deutsche Zeitschr. f. Akupunktur	*PV	Pacific Viewpoint (New Zealand)
EB	Encyclopaedia Britannica	RIAC	Revue Internationale d'Acupunc-ture
*ECB	Economic Botany		
ENZ	Enzymologia	RTS	Religious Tract Society
EPI	Episteme	SCRM	Scriptorium
ESSOM	Esso Magazine	SHM	Studies in the History of Medicine
GERI	Geriatrics	SOB	Sobornost
GESN	Gesnerus	TCPP	Transactions and Studies of the College of Physicians of Phila-delphia
HAHR	Hispanic American Historical Re-view		
HAM	Hamdard Voice of Eastern Medi-cine (Organ of the Inst. of	ZGNTM	Zeitschr. f. Gesch. d. Naturwiss., Technik u. Med.

A. CHINESE AND JAPANESE BOOKS BEFORE +1800

Each entry gives particulars in the following order:

(a) title, alphabetically arranged, with characters;
(b) alternative title, if any;
(c) translation of title;
(d) cross-reference to closely related book, if any;
(e) dynasty;
(f) date as accurate as possible;
(g) name of author or editor, with characters;
(h) title of other book, if the text of the work now exists only incorporated therein; or, in special cases, references to sinological studies of it;
(i) references to translations, if any, given by the name of the translator in Bibliography C;
(j) notice of any index or concordance to the book if such a work exists;
(k) reference to the number of the book in the *Tao Tsang* catalogue of Wieger (6), if applicable;
(l) reference to the number of the book in the *San Tsang* (Tripiṭaka) catalogues of Nanjio (1) and Takakusu & Watanabe, if applicable.

Words which assist in the translation of titles are added in round brackets.

Alternative titles or explanatory additions to the titles are added in square brackets.

It will be remembered (p. 305 above) that in Chinese indexes words beginning *Chh-* are all listed together after *Ch-*, and *Hs-* after *H-*, but that this applies to initial words of titles only.

Where there are any differences between the entries in these bibliographies and those in Vols. 1–4, the information here given is to be taken as more correct.

An interim list of references to the editions used in the present work, and to the *tshung-shu* collections in which books are available, has been given in Vol. 4, pt. 3, pp. 913 ff., and is available as a separate brochure.

ABBREVIATIONS

C/Han	Former Han.
E/Wei	Eastern Wei.
H/Han	Later Han.
H/Shu	Later Shu (Wu Tai).
H/Thang	Later Thang (Wu Tai).
H/Chin	Later Chin (Wu Tai).
S/Han	Southern Han (Wu Tai).
S/Phing	Southern Phing (Wu Tai).
J/Chin	Jurchen Chin.
L/Sung	Liu Sung.
N/Chou	Northern Chou.
N/Chhi	Northern Chhi.
N/Sung	Northern Sung (before the removal of the capital to Hangchow).
N/Wei	Northern Wei.
S/Chhi	Southern Chhi.
S/Sung	Southern Sung (after the removal of the capital to Hangchow).
W/Wei	Western Wei.

A-Nan Ssu Shih Ching 阿難四事經.
　Sūtra on the Four Practices spoken to Ānanda.
　India.
　Tr. San Kuo, betw. +222 and +230 by Chih-Chhien 支謙.
　N/696; TW/493.

A-Phi-Than-Phi Po-Sha Lun 阿毘曇毘婆沙論.
　Abhidharma Mahāvibhāsha.
　India (this recension not much before +600).
　Tr. Hsüan-Chuang, +659 玄奘.
　N/1263; TW/1546.

Chang Chen-Jen Chin Shih Ling Sha Lun.
　See *Chin Shih Ling Sha Lun.*

Chao Fei-Yen Pieh Chuan 趙飛燕別傳.
　[= *Chao Hou I Shih.*]
　Another Biography of Chao Fei-Yen [historical novelette].
　Sung.
　Chhin Shun 秦醇.

Chao Fei-Yen Wai Chuan 趙飛燕外傳.
　Unofficial Biography of Chao Fei-Yen (d. −6, celebrated dancing-girl, consort and empress of Han Chhêng Ti).
　Ascr. Han, +1st.
　Attrib. Ling Hsüan 伶玄.

Chao Hou I Shih 趙后遺事.
　A Record of the Affairs of the Empress Chao (−1st century).
　See *Chao Fei-Yen Pieh Chuan.*

Chao Hun 招魂.
　The Summons of the Soul [ode].
　Chou (Chhu), c. −240.
　Prob. Ching Chhai 景差.
　Tr. Hawkes (1), p. 103.

Chen Chhi Huan Yuan Ming 眞氣還元銘.
　The Inscription on the Regeneration of the Primary Chhi.
　Thang or Sung, must be before the mid +13th century.
　Writer unknown.
　TT/261.

Chen Chung Chi 枕中記.
　[= *Ko Hung Chen Chung Shu.*]
　Pillow-Book (of Ko Hung).
　Ascr. Chin, c. +320, but actually not earlier than the +7th century.
　Attrib. Ko Hung 葛洪.
　TT/830.

Chen Chung Chi 枕中記.
　See *Shê Yang Chen Chung Chi.*

Chen-Chung Hung-Pao Yuan-Pi Shu 枕中鴻寶苑祕書.
　The Infinite Treasure of the Garden of Secrets; (Confidential) Pillow-Book (of the Prince of Huai-Nan).
　See *Huai-Nan Wang Wan Pi Shu.*
　Cf. Kaltenmark (2), p. 32.

Chen Hsi 眞系.
　The Legitimate Succession of Perfected, or Realised, (Immortals).
　Thang, +805.
　Li Po 李渤.
　In *YCCC*, ch. 5, pp. 1a ff.

Chen Kao 眞誥.
> Declarations of Perfected, or Realised,
> (Immortals) [visitations and revelations of
> the Taoist pantheon].
> Chin and S/Chhi. Original material from
> +364 to +370, collected from +484 to
> +492 by Thao Hung-Ching (+456 to
> +536), who provided commentary and
> postface by +493 to +498; finished
> +499.
> Original writers unknown.
> Ed. Thao Hung-Ching 陶弘景.
> *TT*/1004.

Chen Yuan Miao Tao Hsiu Tan Li Yen Chhao
    眞元妙道修丹歷驗抄.
> [= *Hsiu Chen Li Yen Chhao Thu.*]
> A Document concerning the Tried and
> Tested (Methods for Preparing the)
> Restorative Enchymoma of the Mysterious
> Tao of the Primary (Vitalities) [physio-
> logical alchemy].
> Thang or Sung, before +1019.
> Tung Chen Tzu (ps.) 洞眞子.
> In *YCCC*, ch. 72, pp. 17*b* ff.

Chen Yuan Miao Tao Yao Lüeh 眞元妙道要畧.
> Classified Essentials of the Mysterious Tao
> of the True Origin (of Things) [alchemy
> and chemistry].
> Ascr. Chin, +3rd, but probably mostly
> Thang, +8th and +9th, at any rate
> after +7th as it quotes Li Chi.
> Attrib. Chêng Ssu-Yuan 鄭思遠.
> *TT*/917.

Chêng I Fa Wên (Thai-Shang) Wai Lu I 正一法
    文太上外籙儀.
> The System of the Outer Certificates, a Thai-
> Shang Scripture.
> Date unknown, but pre-Thang.
> Writer unknown.
> *TT*/1225.

Chêng Lei Pên Tshao 證類本草.
> See *Ching-Shih Chêng Lei Pei-Chi Pên Tshao*
> and *Chhung-Hsiu Chêng-Ho Ching-Shih*
>   *Chêng Lei Pei-Yung Pên Tshao*

Chêng Tao Pi Shu Shih Chung 證道祕書十種.
> Ten Types of Secret Books on the Verifica-
> tion of the Tao.
> See Fu Chin-Chhüan (6)

Chi Hsiao Hsin Shu 紀效新書.
> A New Treatise on Military and Naval
> Efficiency.
> Ming, *c.* +1575.
> Chhi Chi-Kuang 戚繼光.

Chi Hsien Chuan 集仙傳.
> Biographies of the Company of the Immortals.
> Sung, *c.* +1140.
> Tsêng Tshao 曾慥.

Chi I Chi 集異記.
> A Collection of Assorted Stories of Strange
> Events.
> Thang.
> Hsüeh Yung-Jo 薛用弱.

Chi Ni Tzu 計倪子.
> [= *Fan Tzu Chi Jan* 范子計然.]
> The Book of Master Chi Ni.
> Chou (Yüeh), −4th century.
> Attrib. Fan Li 范蠡, recording the
> philosophy of his master Chi Jan 計然.

Chi Shêng Fang 濟生方.
> Prescriptions for the Preservation of Health.
> Sung, *c.* +1267.
> Yen Yung-Ho 嚴用和.

Chi Than Lu 劇談錄.
> Records of Entertaining Conversations.
> Thang, *c.* +885.
> Khang Phien 康駢 or 軿.

Chi Yün 集韻.
> Complete Dictionary of the Sounds of
> Characters [cf. *Chhieh Yün* and *Kuang
> Yün*].
> Sung, +1037.
> Compiled by Ting Tu 丁度 *et al.*
> Possibly completed in +1067 by Ssuma
> Kuang 司馬光.

Chia-Yu Pên Tshao 嘉祐本草.
> See *Chia-Yu Pu-Chu Shen Nung Pên Tshao.*

Chia-Yu Pu-Chu Shen Nung Pên Tshao 嘉祐補
    註神農本草.
> Supplementary Commentary on the *Pharma-
> copoeia of the Heavenly Husbandman*,
> commissioned in the Chia-Yu reign-
> period.
> Sung, commissioned +1057, finished
> +1060.
> Chang Yü-Hsi 掌禹錫,
> Lin I 林億,
> & Chang Tung 張洞.

Chiang Huai I Jen Lu 江淮異人錄.
> Records of (Twenty-five) Strange Magician-
> Technicians between the Yangtze and the
> Huai River (during the Thang, Wu and
> Nan Thang Dynasties, *c.* +850 to +950).
> Sung, *c.* +975.
> Wu Shu 吳淑.

Chiang Wên-Thung Chi 江文通集.
> Literary Collection of Chiang Wên-Thung
> (Chiang Yen).
> S/Chhi, *c.* +500.
> Chiang Yen 江淹.

Chiao Chhuang Chiu Lu 蕉窗九錄.
> Nine Dissertations from the (Desk at the)
> Banana-Grove Window.
> Ming, *c.* +1575.
> Hsiang Yuan-Pien 項元汴.

Chien Wu Chi 漸悟集.
> On the Gradual Understanding (of the
> Tao).
> Sung, mid +12th century.
> Ma Yü 馬鈺.
> *TT*/1128.

Chih Chen Tzu Lung Hu Ta Tan Shih 至眞子
    龍虎大丹詩.
> Song of the Great Dragon-and-Tiger En-
> chymoma of the Perfected-Truth Master.

Chi Chen Tzu Lung Hu Ta Tan Shih (cont.)
Sung, +1026.
Chou Fang (Chih Chen Tzu) 周方.
Presented to the throne by Lu Thien[-Chi]
盧天驥, c. +1115.
TT/266.
Chih-Chhuan Chen-Jen Chiao Chêng Shu 稚川
眞人校證術.
Technical Methods of the Adept (Ko) Chih-
Chhuan (i.e. Ko Hung), with Critical
Annotations [and illustrations of al-
chemical apparatus].
Ascr. Chin, c. +320, but probably later.
Attrib. Ko Hung 葛洪.
TT/895.
Chih Chih Hsiang Shuo San Chhêng Pi Yao 直
指祥說三乘秘要.
See *Wu Chen Phien Chih Chih Hsiang Shuo
San Chhêng Pi Yao.*
Cf. Davis & Chao Yün-Tshung (6).
Chih-Chou hsien-sêng Chin Tan Chih Chih 紙舟
先生金丹直指.
Straightforward Indications about the
Metallous Enchymoma by the Paper-
Boat Teacher.
Sung, prob. +12th.
Chin Yüeh-Yen 金月巖.
TT/239.
Chih Hsüan Phien 指玄篇.
A Pointer to the Mysteries [psycho-physio-
logical alchemy].
Sung, c. +1215.
Pai Yü-Chhan 白玉蟾.
In *Hsiu Chen Shih Shu* (*TT*/260), chs. 1–8.
Chih Kuei Chi 指歸集.
Pointing the Way Home (to Life Eternal); a
Collection.
Sung, c. +1165.
Wu Wu 吳悞.
TT/914.
Cf. Chhen Kuo-Fu (1), vol. 2, pp. 389,
390.
Chih Tao Phien 旨道篇 (or 編).
A Demonstration of the Tao.
Sui or just before, c. +580.
Su Yuan-Ming (or -Lang) 蘇元明(朗)
= Chhing Hsia Tzu 青霞子.
Now extant only in quotations.
Chih Tshao Thu 芝草圖.
See *Thai-Shang Ling-Pao Chih Tshao Thu.*
Chin Hua Chhung Pi Tan Ching Pi Chih 金華
冲碧丹經祕旨.
Confidential Instructions on the Manual of
the Heaven-Piercing Golden Flower
Elixir [with illustrations of alchemical
apparatus].
Sung, +1225.
Phêng Ssu 彭耜 & Mêng Hsü 孟煦
(pref. and ed. Mêng Hsü).
Received from Pai Yü-Chhan 白玉蟾 and
Lan Yuan-Lao 蘭元老.
TT/907.

The authorship of this important work is
obscure. In his preface Mêng Hsü says
that in +1218 he met in the mountains
Phêng Ssu, who transmitted to him a
short work which Phêng himself had received from Pai Yü-Chhan. This is ch. 1
of the present book. Two years later Mêng
met an adept named Lan Yuan-Lao, who
claimed to be an avatar of Pai Yü-Chhan
and transmitted to Mêng a longer text;
this is the part which contains descriptions
of the complicated alchemical apparatus
and appears as ch. 2 of the present work.
The name of the book is taken from that
of the alchemical elaboratory of Lan Yuan-
Lao, which was called Chin Hua Chhung
Pi Tan Shih 金華冲碧丹室.
Chin Hua Tsung Chih 金華宗旨
[= *Thai-I Chin Hua Tsung Chih*, also entitled
Chhang Shêng Shu; former title: *Lü
Tsu Chhuan Shou Tsung Chih.*]
Principles of the (Inner) Radiance of the
Metallous (Enchymoma) [a Taoist *nei tan*
treatise on meditation and sexual tech-
niques, with Buddhist influence].
Ming and Chhing, c. +1403, finalised
+1663, but may have been transmitted
orally from an earlier date. Present title
from +1668.
Writer unknown. Attrib. Lü Yen 呂喦
(Lü Tung-Pin) and his school, late
+8th.
Commentary by Tan Jan-Hui 澹然慧
(1921).
Prefaces by Chang San-Fêng 張三峯
(c. +1410) and several others, some per-
haps apocryphal.
See also *Lü Tsu Shih Hsien-Thien Hsü Wu
Thai-I Chin Hua Tsung Chih.*
Cf. Wilhelm & Jung (1).
Chin Hua Yü I Ta Tan 金華玉液大丹.
The Great Elixir of the Golden Flower (or,
Metallous Radiance) and the Juice of
Jade.
Date unknown, probably Thang.
Writer unknown.
TT/903.
Chin Hua Yü Nü Shuo Tan Ching 金華玉女
說丹經.
Sermon of the Jade Girl of the Golden
Flower about Elixirs and Enchymomas.
Wu Tai or Sung.
Writer unknown.
In *YCCC*, ch. 64, pp. 1 a ff.
Chin I Huan Tan Pai Wên Chüeh 金液還丹百
問訣.
Questions and Answers on Potable Gold
(Metallous Fluid) and Cyclically-
Transformed Elixirs and Enchymomas.
Sung.
Li Kuang-Hsüan 李光玄.
TT/263.

Chin I Huan Tan Yin Chêng Thu 金液還丹印
證圖.
  Illustrations and Evidential Signs of the
  Regenerative Enchymoma (constituted
  by, or elaborated from) the Metallous
  Fluid.
  Sung, prob. +12th, perhaps *c.* +1218,
  date of preface.
  Lung Mei Tzu (ps.) 龍眉子.
  TT/148.
Chin Ku Chhi Kuan 今古奇觀.
  Strange Tales New and Old.
  Ming, *c.* +1620; pr. betw. +1632 and
  +1644.
  Fêng Mêng-Lung 馮夢龍.
  Cf. Pelliot (57).
Chin Mu Wan Ling Lun 金木萬靈論.
  Essay on the Tens of Thousands of
  Efficacious (Substances) among Metals
  and Plants.
  Ascr. Chin, *c.* +320. Actually prob. late
  Sung or Yuan.
  Attrib. Ko Hung 葛洪.
  TT/933.
Chin Pi Wu Hsiang Lei Tshan Thung Chhi 金碧
五相類參同契.
  Gold and Caerulean Jade Treatise on the
  Similarities and Categories of the Five
  (Substances) and the *Kinship of the Three*
  [a poem on physiological alchemy].
  Ascr. H/Han, *c.* +200.
  Attrib. Yin Chhang-Shêng 陰長生.
  TT/897.
  Cf. Ho Ping-Yü (12).
  Not to be confused with the *Tshan Thung
  Chhi Wu Hsiang Lei Pi Yao*, q.v.
Chin Shih Ling Sha Lun 金石靈砂論.
  A Discourse on Metals, Minerals and
  Cinnabar (by the Adept Chang).
  Thang, between +713 and +741.
  Chang Yin-Chü 張隱居.
  TT/880.
Chin Shih Pu Wu Chiu Shu Chüeh 金石簿五
九數訣.
  Explanation of the Inventory of Metals and
  Minerals according to the Numbers Five
  (Earth) and Nine (Metal) [catalogue of
  substances with provenances, including
  some from foreign countries].
  Thang, perhaps *c.* +670 (contains a story
  relating to +664).
  Writer unknown.
  TT/900.
Chin Shih Wu Hsiang Lei 金石五相類.
  [= *Yin Chen Chün Chin Shih Wu Hsiang
  Lei.*]
  The Similarities and Categories of the Five
  (Substances) among Metals and Minerals
  (sulphur, realgar, orpiment, mercury and
  lead) (by the Deified Adept Yin).
  Date unknown (ascr. +2nd or +3rd
  century).

Attrib. Yin Chen-Chün 陰眞君 (Yin
  Chhang-Shêng).
  TT/899.
Chin Tan Chen Chuan 金丹眞傳.
  A Record of the Primary (Vitalities, re-
  gained by) the Metallous Enchymoma.
  Ming, +1615.
  Sun Ju-Chung 孫汝忠.
Chin Tan Chêng Li Ta Chhüan 金丹正理大全
  Comprehensive Collection of Writings on
  the True Principles of the Metallous
  Enchymoma [a florilegium].
  Ming, *c.* +1440.
  Ed. Han Chhan Tzu 涵蟾子.
  Cf. Davis & Chao Yün-Tshung (6).
Chin Tan Chieh Yao 金丹節要.
  Important Sections on the Metallous
  Enchymoma.
  Part of *San-Fêng Tan Chüeh* (q.v.).
Chin Tan Chih Chih 金丹直指.
  Straightforward Explanation of the Metal-
  lous Enchymoma.
  Sung, prob. +12th.
  Chou Wu-So 周無所.
  TT/1058.
  Cf. *Chih-Chou hsien-sêng Chin Tan Chih
  Chih.*
  See Chhen Kuo-Fu (1), vol. 2, pp. 447 ff.
Chin Tan Chin Pi Chhien Thung Chüeh 金丹金
碧潛通訣.
  Oral Instructions explaining the Abscondite
  Truths of the Gold and Caerulean Jade
  (Components of the) Metallous Enchym-
  oma.
  Date unknown, not earlier than Wu Tai.
  Writer unknown.
  Incomplete in *YCCC*, ch. 73, pp. 7a ff.
Chin Tan Fu 金丹賦.
  Rhapsodical Ode on the Metallous Enchy-
  moma.
  Sung, +13th.
  Writer unknown.
  Comm. by Ma Li-Chao 馬涖昭.
  TT/258.
  Cf. *Nei Tan Fu*, the text of which is very
  similar.
Chin Tan Lung Hu Ching 金丹龍虎經.
  Gold Elixir Dragon and Tiger Manual.
  Thang or early Sung.
  Writer unknown.
  Extant only in quotations, as in *Chu Chia
  Shen Phin Tan Fa*, q.v.
Chin Tan Pi Yao Tshan Thung Lu 金丹秘要
參同錄.
  Essentials of the Gold Elixir; a Record of the
  Concordance (or Kinship) of the Three.
  Sung.
  Mêng Yao-Fu 孟要甫.
  In *Chu Chia Shen Phin Tan Fa*, q.v.
Chin Tan Ssu Pai Tzu 金丹四百字.
  The Four-Hundred Word Epitome of the
  Metallous Enchymoma.

Chin Tan Ssu Pai Tzu (*cont.*)
Sung, *c.* +1065.
Chang Po-Tuan 張伯端.
In *Hsiu Chen Shih Shu* (*TT*/260), ch. 5,
pp. 1*a* ff.
TT/1067.
Comms. by Phêng Hao-Ku and Min I-Tê
in *Tao Tsang Hsü Pien* (*Chhu chi*), 21.
Tr. Davis & Chao Yün-Tshung (2).

Chin Tan Ta Chhêng 金丹大成.
Compendium of the Metallous Enchymoma.
Sung, just before +1250.
Hsiao Thing-Chih 蕭廷芝.
In *TTCY* (*mao chi*, 4), and in *TT*/260,
Hsiu Chen Shih Shu, chs. 9–13 incl.

Chin Tan Ta Yao 金丹大要.
[= *Shang Yang Tzu Chin Tan Ta Yao*.]
Main Essentials of the Metallous Enchy-
moma; the true Gold Elixir.
Yuan, +1331 (pref. +1335).
Chhen Chih-Hsü 陳致虛
(Shang Yang Tzu 上陽子).
In *TTCY* (*mao chi*, 1, 2, 3).
TT/1053.

Chin Tan Ta Yao Hsien Phai (*Yuan Liu*) 金丹
大要仙派源流.
[= *Shang Yang Tzu Chin Tan Ta Yao
Hsien Phai*.]
A History of the Schools of Immortals
mentioned in the *Main Essentials of the
Metallous Enchymoma; the true Gold Elixir*.
Yuan, *c.* +1333.
Chhen Chih-Hsü 陳致虛
(Shang Yang Tzu 上陽子).
In *TTCY*, *Chin Tan Ta Yao*, ch. 3, pp.
40 ff.
TT/1056.

Chin Tan Ta Yao Lieh Hsien Chih 金丹大要
列仙誌.
[= *Shang Yang Tzu Chin Tan Ta Yao Lieh
Hsien Chih*.]
Records of the Immortals mentioned in the
*Main Essentials of the Metallous Enchy-
moma; the true Gold Elixir*.
Yuan, *c.* +1333.
Chhen Chih-Hsü 陳致虛
(Shang Yang Tzu 上陽子).
TT/1055.

Chn Tan Ta Yao Pao Chüeh 金丹大藥寶訣.
Precious Instructions on the Great Medi-
cines of the Golden Elixir (Type).
Sung, *c.* +1045.
Tshui Fang 崔昉.
Preface preserved in *Kêng Tao Chi*, ch. 1,
p. 8*b*, but otherwise only extant in
occasional quotations.
Perhaps the same book as the *Wai Tan
Pên Tshao* (q. v.).

Chin Tan Ta Yao Thu 金丹大要圖.
[= *Shang Yang Tzu Chin Tan Ta Yao Thu*.]
Illustrations for the *Main Essentials of the
Metallous Enchymoma; the true Gold Elixir*.

Yuan, +1333.
Chhen Chih-Hsü 陳致虛
(Shang Yang Tzu 上陽子).
Based on drawings and tables of the +10th
century onwards by Phêng Hsiao 彭曉,
Chang Po-Tuan 張伯端 (hence the
name *Tzu Yang Tan Fang Pao Chien
Thu*), Lin Shen-Fêng 林神鳳 and
others.
In *TTCY* (*Chin Tan Ta Yao*, ch. 3,
pp. 26*a* ff.).
TT/1054.
Cf. Ho Ping-Yü & Needham (2).

Ching Chhu Sui Shih Chi 荊楚歲時記.
Annual Folk Customs of the States of
Ching and Chhu [i.e. of the districts cor-
responding to those ancient States;
Hupei, Hunan and Chiangsi].
Prob. Liang, *c.* +550, but perhaps partly
Sui, *c.* +610.
Tsung Lin 宗懍.
See des Rotours (1), p. cii.

Ching-Shih Chêng Lei Pei-Chi Pên Tshao 經史
證類備急本草.
The Classified and Consolidated Armament-
arium of Pharmaceutical Natural History.
Sung, +1083, repr. +1090.
Thang Shen-Wei 唐慎微.

Ching Shih Thung Yen 警世通言.
Stories to Warn Men.
Ming, *c.* +1640.
Fêng Mêng-Lung 馮夢龍.

Ching Tien Shih Wên 經典釋文.
Textual Criticism of the Classics.
Sui, *c.* +600.
Lu Tê-Ming 陸德明.

Ching Yen Fang 經驗方.
Tried and Tested Prescriptions.
Sung, +1025.
Chang Shêng-Tao 張聲道.
Now extant only in quotations.

Ching Yen Liang Fang 經驗良方.
Valuable Tried and Tested Prescriptions.
Yuan.
Writer unknown.

Chiu Chêng Lu 就正錄.
Drawing near to the Right Way; a Guide
[to physiological alchemy].
Chhing, prefs. +1678, +1697.
Lu Shih-Chhen 陸世忱.
In *Tao Tsang Hsü Pien* (*Chhu chi*), 8.

Chiu Chuan Chhing Chin Ling Sha Tan 九轉青
金靈砂丹.
The Ninefold Cyclically Transformed
Caerulean Golden Numinous Cinnabar
Elixir.
Date unknown.
Writer unknown, but much overlap with
TT/886.
TT/887.

Chiu Chuan Ling Sha Ta Tan 九轉靈砂大
丹.

Chiu Chuan Ling Sha Ta Tan (*cont.*)
   The Great Ninefold Cyclically Transformed
      Numinous Cinnabar Elixir.
   Date unknown.
   Writer unknown.
   TT/886.
*Chiu Chuan Ling Sha Ta Tan Tzu Shêng Hsüan
   Ching* 九轉靈砂大丹資聖玄經.
   Mysterious (or Esoteric) Sagehood-
      Enhancing Canon of the Great Ninefold
      Cyclically Transformed Numinous Cinna-
      bar Elixir (or Enchymoma).
   Date unknown, probably Thang; the text is
      in sūtra form.
   Writer unknown.
   TT/879.
Chiu Chuan Liu Chu Shen Hsien Chiu Tan Ching
   九轉流珠神仙九丹經.
   Manual of the Nine Elixirs of the Holy
      Immortals and of the Ninefold Cyclically
      Transformed Mercury.
   Not later than Sung, but contains material
      from much earlier dates.
   Thai-Chhing Chen Jen 太清眞人.
   TT/945.
Chiu Huan Chin Tan Erh Chang 九還金丹二章.
   Two Chapters on the Ninefold Cyclically
      Transformed Gold Elixir.
   Alternative title of *Ta-Tung Lien Chen Pao
      Ching, Chin Huan Chin Tan Miao Chüeh*
      (q.v.).
   In *YCCC*, ch. 68, pp. 8*a* ff.
Chiu Phu 酒譜.
   A Treatise on Wine.
   Sung, +1020.
   Tou Phing 竇苹.
Chiu Shih 酒史.
   A History of Wine.
   Ming, +16th (but first pr. +1750).
   Fêng Shih-Hua 馮時化.
Chiu Thang Shu 舊唐書.
   Old History of the Thang Dynasty [+618
      to +906].
   Wu Tai (H/Chin), +945.
   Liu Hsü 劉昫.
   Cf. des Rotours (2), p. 64.
   For translations of passages see the index of
      Frankel (1).
Chiu Ting Shen Tan Ching Chüeh
   See *Huang Ti Chiu Ting Shen Tan Ching
      Chüeh.*
Cho Kêng Lu 輟耕錄.
   [Sometimes *Nan Tshun Cho Kêng Lu.*]
   Talks (at South Village) while the Plough is
      Resting.
   Yuan, +1366.
   Thao Tsung-I 陶宗儀.
Chou Hou Pei Chi Fang 肘後備急方.
   [= *Chou Hou Tsu Chiu Fang*
      or *Chou Hou Pai I Fang*
      or *Ko Hsien Ong Chou Hou Pei Chi Fang.*]
   Handbook of Medicines for Emergencies.

Chin, *c.* +340.
   Ko Hung 葛洪.
Chou Hou Pai I Fang 肘後百一方
   See *Chou Hou Pei Chi Fang.*
Chou Hou Tsu Chiu Fang 肘後卒救方
   See *Chou Hou Pei Chi Fang.*
Chou I Tshan Thung Chhi 周易參同契.
   See also titles under *Tshan Thung Chhi.*
Chou I Tshan Thung Chhi Chieh 周易參同契解.
   The *Kinship of the Three and the Book of
      Changes,* with Explanation.
   Text, H/Han, *c.* +140.
   Comm., Sung, +1234.
   Ed. & comm. Chhen Hsien-Wei 陳顯微.
   TT/998.
Chou I Tshan Thung Chhi Chu 周易參同契註.
   The *Kinship of the Three and the Book of
      Changes,* with Commentary.
   Text, H/Han, *c.* +140.
   Comm. ascr. H/Han, *c.* +160, but prob-
      ably Sung.
   Attrib., ed. and comm. Yin Chhang-Shêng
      陰長生.
   TT/990.
Chou I Tshan Thung Chhi Chu 周易參同契註.
   The *Kinship of the Three and the Book of
      Changes,* with Commentary.
   Text, H/Han, *c.* +140.
   Comm. probably Sung.
   Ed. and comm. unknown.
   TT/991.
Chou I Tshan Thung Chhi Chu 周易參同契註.
   The *Kinship of the Three and the Book of
      Changes,* with Commentary.
   Text, H/Han, *c.* +140.
   Comm. probably Sung.
   Ed. and comm. unknown.
   TT/995.
Chou I Tshan Thung Chhi Chu 周易參同契註.
   The *Kinship of the Three and the Book of
      Changes,* with Commentary.
   Text, H/Han, *c.* +140.
   Comm., Sung, *c.* +1230.
   Ed. & comm. Chhu Hua-Ku 儲華谷.
   TT/999.
Chou I Tshan Thung Chhi Chu (*TT*/992).
   Alternative title for *Tshan Thung Chhi
      Khao I* (Chu Hsi's) q.v.
Chou I Tshan Thung Chhi Fa Hui 周易參同契
   發揮.
   Elucidations of the *Kinship of the Three and
      the Book of Changes* [alchemy].
   Text, H/Han, *c.* +140.
   Comm., Yuan, +1284.
   Ed. & comm. Yü Yen 兪琰.
   Tr. Wu & Davis (1).
   TT/996.
Chou I Tshan Thung Chhi Fên Chang Chu (*Chieh*)
   周易參同契分章註(解).
   The *Kinship of the Three and the Book of
      Changes* divided into (short) chapters,
      with Commentary and Analysis.

Chou I Tshan Thung Chhi Fên Chang Chu (Chieh)
(*cont.*)
Text, Han, *c.* +140.
Comm., Yuan, *c.* +1330.
Comm. Chhen Chih-Hsü 陳致虛
(Shang Yang Tzu 上陽子).
TTCY pên 93.

Chou I Tshan Thung Chhi Fên Chang Thung
Chen I 周易參同契分章通真義.
The *Kinship of the Three and the Book of*
Changes divided into (short) chapters for
the Understanding of its Real Meanings.
Text, H/Han, *c.* +140.
Comm., Wu Tai +947.
Ed. & comm. Phêng Hsiao 彭曉.
Tr. Wu & Davis (1).
TT/993.

Chou I Tshan Thung Chhi Shih I 周易參同契
釋疑.
Clarification of Doubtful Matters in the
Kinship of the Three and the Book of
Changes.
Yuan, +1284.
Ed. & comm. Yü Yen 俞琰.
TT/997.

Chou I Tshan Thung Chhi Su Lüeh 周易參同
契疏略.
Brief Explanation of the *Kinship of the Three*
and the Book of Changes.
Ming, +1564.
Ed. & comm. Wang Wên-Lu 王文祿.

Chou I Tshan Thung Chhi Ting Chhi Ko Ming
Ching Thu 易周參同契鼎器歌明鏡
圖.
An Illuminating Chart for the Mnemonic
Rhymes about Reaction-Vessels in the
Kinship of the Three and the Book of
Changes.
Text, H/Han, *c.* +140 (*Ting Chhi Ko*
portion only).
Comm., Wu Tai, +947.
Ed. & comm. Phêng Hsiao 彭曉.
TT/994.

Chu Chêng Pien I 諸證辨疑.
Resolution of Diagnostic Doubts.
Ming, late +15th.
Wu Chhiu 吳球.

Chu Chhüan Chi 竹泉集.
The Bamboo Springs Collection [poems
and personal testimonies on physiological
alchemy].
Ming, +1465.
Tung Chhung-Li *et al.* 蘁重理.
In *Wai Chin Tan* (q.v.), ch. 3.

Chu Chia Shen Phin Tan Fa 諸家神品丹法.
Methods of the Various Schools for Magical
Elixir Preparations (an alchemical an-
thology).
Sung.
Mêng Yao-Fu 孟要甫
(Hsüan Chen Tzu 女真子) *et al.*
TT/911.

Chu Fan Chih 諸蕃志.
Records of Foreign Peoples (and their Trade).
Sung, *c.* +1225. (This is Pelliot's dating;
Hirth & Rockhill favoured between
+1242 and +1258.)
Chao Ju-Kua 趙汝适.
Tr. Hirth & Rockhill (1).

Chu Yeh Thing Tsa Chi 竹葉亭雜記.
Miscellaneous Records of the Bamboo Leaf
Pavilion.
Chhing, begun *c.* +1790 but not finished
till *c.* 1820.
Yao Yuan-Chih 姚元之.

Chuan Hsi Wang Mu Wo Ku Fa 傳西王母握
固法.
[= *Thai-Shang Chuan Hsi Wang Mu Wo*
Ku Fa.]
A Recording of the Method of Grasping
the Firmness (taught by) the Mother
Goddess of the West.
[Taoist heliotherapy and meditation. 'Grasp-
ing the firmness' was a technical term for
a way of clenching the hands during
meditation.]
Thang or earlier.
Writer unknown.
Fragment in *Hsiu Chen Shih Shu* (*TT*/260),
ch. 24, p. 1a ff.
Cf. Maspero (7), p. 376.

Chuang Lou Chi 妝樓記.
Records of the Ornamental Pavilion.
Wu Tai or Sung, *c.* +960.
Chang Mi 張泌.

Chün-Chai Tu Shu Chih 郡齋讀書志.
Memoir on the Authenticities of Ancient
Books, by (Chhao) Chün-Chai.
Sung, +1151.
Chhao Kung-Wu 晁公武.

Chün-Chai Tu Shu Fu Chih 郡齋讀書附志.
Supplement to Chün-Chai's (Chhao Kung-
Wu's) *Memoir on the Authenticities of*
Ancient Books.
Sung, *c.* +1200.
Chao Hsi-Pien 趙希弁.

Chün-Chai Tu Shu Hou Chih 郡齋讀書後志.
Further Supplement to Chün-Chai's (Chhao
Kung-Wu's) *Memoir on the Authenticities*
of Ancient Books.
Sung, pref. +1151, pr. +1250.
Chhao Kung-Wu 晁公武, re-compiled by
Chao Hsi-Pien 趙希弁, from the edi-
tion of Yao Ying-Chi 姚應績.

Chün Phu 菌譜.
A Treatise on Fungi.
Sung, +1245.
Chhen Jen-Yü 陳仁玉.

Chung Hua Ku Chin Chu 中華古今注.
Commentary on Things Old and New in
China.
Wu Tai (H/Thang), +923 to +926.
Ma Kao 馬縞.
See des Rotours (1), p. xcix.

Chung Huang Chen Ching 中黃眞經
    [= *Thai-Chhing Chung Huang Chen Ching*
    or *Thai Tsang Lun*.]
    True Manual of the Middle (Radiance) of
    the Yellow (Courts), (central regions of the
    three parts of the body) [Taoist anatomy
    and physiology with Buddhist influence].
    Prob. Sung, +12th or +13th.
    Chiu Hsien Chün (ps.) 九仙君.
    Comm. Chung Huang Chen Jen (ps.) 中
    黃眞人.
    TT/810.
    Completing *TT*/328 and 329 (Wieger).
    Cf. Maspero (7), p. 364.

Chung Lü Chuan Tao Chi 鍾呂傳道集.
    Dialogue between Chungli (Chhüan) and
    Lü (Tung-Pin) on the Transmission of
    the Tao (and the Art of Longevity, by
    Rejuvenation).
    Thang, +8th or +9th.
    Attrib. Chungli Chhüan 鍾離權 and Lü
    Yen 呂嵒.
    Ed. Shih Chien-Wu 施肩吾.
    In *Hsiu Chen Shih Shu* (*TT*/260), chs.14–16
    incl.

Chung Shan Yü Kuei Fu Chhi Ching 中山玉
    櫃服氣經.
    Manual of the Absorption of the Chhi,
    found in the Jade Casket on Chung-
    Shan (Mtn). [Taoist breathing exercises.]
    Thang or Sung, +9th or +10th.
    Attrib. Chang Tao-Ling (Han) 張道陵 or
    Pi-Yen Chang Tao-chê 碧巖張道者
    or Pi-Yen hsien-sêng 碧巖先生.
    Comm. by Huang Yuan-Chün 黃元君.
    In *YCCC*, ch. 60, pp. 1 *a* ff.
    Cf. Maspero (7), pp. 204, 215, 353.

Chungli Pa Tuan Chin Fa 鍾離八段錦法.
    The Eight Elegant (Gymnastic) Exercises of
    Chungli (Chhüan).
    Thang, late +8th.
    Chungli Chhüan 鍾離權.
    In *Hsiu Chen Shih Shu* (*TT*/260), ch. 19.
    Tr. Maspero (7), pp. 418 ff.
    Cf. Notice by Tsêng Tshao in *Lin Chiang
    Hsien* (*TT*/260, ch. 23, pp. 1 *b*, 2 *a*) dated
    +1151. This says that the text was in-
    scribed by Lü Tung-Pin himself on stone
    and so handed down.

Chhang Chhun Tzu Phan-Hsi Chi 長春子磻溪
    集.
    Chhiu Chhang-Chhun's Collected (Poems)
    at Phan-Hsi.
    Sung, *c.* +1200.
    Chhiu Chhu-Chi 邱處機.
    TT/1145.

Chhang Shêng Shu 長生術.
    The Art and Mystery of Longevity and
    Immortality.
    Alternative title of *Chin Hua Tsung Chih* (q.v.).

Chhen Wai Hsia Chü Chien 塵外退舉牋.
    Examples of Men who Renounced Official
Careers and Shook off the Dust of the
    World [the eighth and last part (ch. 19)
    of *Tsun Shêng Pa Chien*, q.v.].
    Ming, +1591.
    Kao Lien 高濂.

Chhi Chü An Lo Chien 起居安樂牋.
    On (Health-giving) Rest and Recreations in
    a Retired Abode [the third part (Chs. 7,
    8) of *Tsun Shêng Pa Chien*, q.v.].
    Ming, +1591.
    Kao Lien 高濂.

Chhi Fan Ling Sha Ko 七返靈砂歌.
    Song of the Sevenfold Cyclically Trans-
    formed Numinous Cinnabar (Elixir).
    See *Chhi Fan Tan Sha Chüeh*.

Chhi Fan Ling Sha Lun 七返靈砂論.
    On Numinous Cinnabar Seven Times
    Cyclically Transformed.
    Alternative title for *Ta-Tung Lien Chen Pao
    Ching, Hsiu Fu Ling Sha Miao Chüeh*
    (q.v.).
    In *YCCC*, ch. 69, pp. 1 *a* ff.

Chhi Fan Tan Sha Chüeh 七返丹砂訣.
    [= *Wei Po-Yang Chhi Fan Tan Sha Chüeh*
    or *Chhi Fan Ling Sha Ko*.]
    Explanation of the Sevenfold Cyclically
    Transformed Cinnabar (Elixir), (of Wei
    Po-Yang).
    Date unknown (ascr. H/Han).
    Writer unknown (attrib. Wei Po-Yang).
    Comm. by Huang Thung-Chün 黃童君.
    Thang or pre-Thang, before +806.
    TT/881.

Chhi Hsiao Liang Fang 奇效良方.
    Effective Therapeutics.
    Ming, *c.* +1436, pr. +1470.
    Fang Hsien 方賢.

Chhi Kuo Khao 七國考.
    Investigations of the Seven (Warring) States.
    Chhing, *c.* +1660.
    Tung Yüeh 董說.

Chhi Lu 七錄.
    Bibliography of the Seven Classes of Books.
    Liang, +523.
    Juan Hsiao-Hsü 阮孝緒.

Chhi Min Yao Shu 齊民要術.
    Important Arts for the People's Welfare
    [lit. Equality].
    N/Wei (and E/Wei or W/Wei), between
    +533 and +544.
    Chia Ssu-Hsieh 賈思勰.
    See des Rotours (1), p.c; Shih Shêng-Han (1).

*Chhi Yün Shan Wu Yuan Tzu Hsiu Chen Pien
    Nan (Tshan Chêng)* 棲雲山悟元子修
    眞辨難參證.
    See *Hsiu Chen Pien Nan (Tshan Chêng)*.

Chhieh Yün 切韻.
    Dictionary of the Sounds of Characters
    [rhyming dictionary].
    Sui, +601.
    Lu Fa-Yen 陸法言.
    See *Kuang Yün*.

Chhien Chin Fang Yen I 千金方衍義.
Dilations upon the *Thousand Golden Remedies*.
Chhing, +1698.
Chang Lu 張璐.

Chhien Chin I Fang 千金翼方.
Supplement to the *Thousand Golden Remedies* [i.e. Revised Prescriptions saving lives worth a Thousand Ounces of Gold].
Thang, between +660 and +680.
Sun Ssu-Mo 孫思邈.

Chhien Chin Shih Chih 千金食治.
A Thousand Golden Rules for Nutrition and the Preservation of Health [i.e. Diet and Personal Hygiene saving lives worth a Thousand Ounces of Gold], (included as a chapter in the *Thousand Golden Remedies*).
Thang, +7th (c. +625, certainly before +659).
Sun Ssu-Mo 孫思邈.

Chhien Chin Yao Fang 千金要方.
A Thousand Golden Remedies [i.e. Essential Prescriptions saving lives worth a Thousand Ounces of Gold].
Thang, between +650 and +659.
Sun Ssu-Mo 孫思邈.

Chhien Han Shu 前漢書.
History of the Former Han Dynasty [−206 to +24].
H/Han (begun about +65), c. +100.
Pan Ku 班固, and (after his death in +92) his sister Pan Chao 班昭.
Partial trs. Dubs (2), Pfizmaier (32–34, 37–51), Wylie (2, 3, 10), Swann (1).
Yin-Tê Index, no. 36.

Chhien Hung Chia Kêng Chih Pao Chi Chhêng 鉛汞甲庚至寶集成.
Complete Compendium on the Perfected Treasure of Lead, Mercury, Wood and Metal [with illustrations of alchemical apparatus].
On the translation of this title, cf. Vol. 5, pt. 3.
Has been considered Thang, +808; but perhaps more probably Wu Tai or Sung. Cf. p. 276.
Chao Nai-An 趙耐菴.
TT/912.

Chhien Khun Pi Yün 乾坤秘韞.
The Hidden Casket of Chhien and Khun (kua, i.e. Yang and Yin) Open'd.
Ming, c. +1430.
Chu Chhüan 朱權.
(Ning Hsien Wang 寧獻王, prince of the Ming.)

Chhien Khun Shêng I 乾坤生意.
Principles of the Coming into Being of Chhien and Khun (kua, i.e. Yang and Yin).
Ming, c. +1430.
Chu Chhüan 朱權.

(Ning Hsien Wang 寧獻王, prince of the Ming.)

Chhih Shui Hsüan Chu 赤水玄珠.
The Mysterious Pearl of the Red River [a system of medicine and iatro-chemistry].
Ming, +1596.
Sun I-Khuei 孫一奎.

Chhih Shui Hsüan Chu Chhüan Chi 赤水玄珠全集.
The Mysterious Pearl of the Red River; a Complete (Medical) Collection.
See *Chhih Shui Hsüan Chu*.

Chhih Shui Yin 赤水吟.
Chants of the Red River.
See Fu Chin-Chhüan (1).

Chhih Sung Tzu Chou Hou Yao Chüeh 赤松子肘後藥訣.
Oral Instructions of the Red-Pine Master on Handy (Macrobiotic) Prescriptions.
Pre-Thang.
Writer unknown.
Part of the *Thai-Chhing Ching Thien-Shih Khou Chüeh*.
TT/876.

Chhih Sung Tzu Hsüan Chi 赤松子玄記.
Arcane Memorandum of the Red-Pine Master.
Thang or earlier, before +9th.
Writer unknown.
Quoted in *TT*/928 and elsewhere.

Chhin Hsüan Fu 擒玄賦.
Rhapsodical Ode on Grappling with the Mystery.
Sung, +13th.
Writer unknown.
TT/257.

Chhing Hsiang Tsa Chi 青箱雜記.
Miscellaneous Records on Green Bamboo Tablets.
Sung, c. +1070.
Wu Chhu-Hou 吳處厚.

Chhing Hsiu Miao Lun Chien 清修妙論牋.
Subtile Discourses on the Unsullied Restoration (of the Primary Vitalities) [the first part (chs. 1, 2) of *Tsun Shêng Pa Chien*, q.v.].
Ming, +1591.
Kao Lien 高濂.

Chhing I Lu 清異錄.
Records of the Unworldly and the Strange.
Wu Tai, c. +950.
Thao Ku 陶穀.

Chhing-Ling Chen-Jen Phei Chün (Nei) Chuan 清靈眞人裴君內傳.
Biography of the Chhing-Ling Adept, Master Phei.
L/Sung or S/Chhi, +5th, but with early Thang additions.
Têng Yün Tzu 鄧雲子
(Phei Hsüan-Jen 裴玄仁 was a semi-legendary immortal said to have been born in −178).

Chhing-Ling Chen-Jen Phei Chün (Nei) Chuan
(*cont.*)
    In *YCCC*, ch. 105.
    Cf. Maspero (7), pp. 386 ff.
Chhing Po Tsa Chih 清波雜志.
    Green-Waves Memories.
    Sung, +1193.
    Chou Hui 周煇.
Chhing Wei Tan Chüeh (or *Fa*) 清微丹訣 (法).
    Instructions for Making the Enchymoma in
    Calmness and Purity [physiological
    alchemy].
    Date unknown, perhaps Thang.
    Writer unknown.
    TT/275.
Chhiu Chhang-Chhun Chhing Thien Ko 邱長春
    青天歌.
    Chhiu Chhang-Chhun's Song of the Blue
    Heavens.
    Sung, *c.* +1200.
    Chhiu Chhu-Chi 邱處機.
    TT/134.
Chhu Chhêng I Shu 褚澄遺書.
    Remaining Writings of Chhu Chhêng.
    Chhi, *c.* +500, probably greatly remodelled
    in Sung.
    Chhu Chhêng 褚澄.
Chhü Hsien Shen Yin Shu 臞仙神隱書.
    Book of Daily Occupations for Scholars in
    Rural Retirement, by the Emaciated
    Immortal.
    Ming, *c.* +1430.
    Chu Chhüan 朱權.
    (Ning Hsien Wang 寧獻王, prince of
    the Ming.)
Chhu Hsüeh Chi 初學記.
    Entry into Learning [encyclopaedia].
    Thang, +700.
    Hsü Chien 徐堅.
Chhü I Shuo Tsuan 祛疑說纂.
    Discussions on the Dispersal of Doubts.
    Sung, *c.* +1230.
    Chhu Yung 儲泳.
Chhüan-Chen Chi Hsüan Pi Yao 全眞集玄祕要.
    Esoteric Essentials of the Mysteries (of the
    Tao), according to the Chhüan-Chen
    (Perfect Truth) School [the Northern
    School of Taoism in Sung and Yuan times].
    Yuan, *c.* +1320.
    Li Tao-Shun 李道純.
    TT/248.
Chhüan-Chen Tso Po Chieh Fa 全眞坐鉢捷法.
    Ingenious Method of the Chhüan-Chen
    School for Timing Meditation (and other
    Exercises) by a (Sinking-) Bowl Clepsydra.
    Sung or Yuan.
    Writer unknown.
    TT/1212.
Chhüan Ching 拳經.
    Manual of Boxing.
    Chhing, +18th.
    Chang Khung-Chao 張孔昭.

Chhun Chhiu Fan Lu 春秋繁露.
    String of Pearls on the *Spring and Autumn
    Annals*.
    C/Han, *c.* −135.
    Tung Chung-Shu 董仲舒.
    See Wu Khang (1).
    Partial trs. Wieger (2); Hughes (1);
    d'Hormon (1) (ed.).
    Chung-Fa Index no. 4.
Chhun Chhiu Wei Yuan Ming Pao 春秋緯元
    命苞.
    Apocryphal Treatise on the *Spring and
    Autumn Annals*; the Mystical Diagrams
    of Cosmic Destiny [astrological-
    astronomical].
    C/Han, *c.* −1st.
    Writer unknown.
    In *Ku Wei Shu*, ch. 7.
Chhun Chhiu Wei Yün Tou Shu 春秋緯運斗樞.
    Apocryphal Treatise on the *Spring and
    Autumn Annals*; the Axis of the Turning
    of the Ladle (i.e. the Great Bear).
    C/Han, −1st or later.
    Writer unknown.
    In *Ku Wei Shu*, ch. 9, pp. 4*b* ff. and
    YHSF, ch. 55, pp. 22*a* ff.
Chhun Chu Chi Wên 春渚紀聞.
    Record of Things Heard at Spring Island.
    Sung, *c.* +1095.
    Ho Wei 何薳.
Chhun-yang etc.
    See *Shun-yang*.
*Chhung-Hsiu Chêng-Ho Ching-Shih Chêng Lei
    Pei-Yung Pên Tshao* 重修政和經史證
    類備用本草.
    New Revision of the Pharmacopoeia of the
    Chêng-Ho reign-period; the Classified
    and Consolidated Armamentarium.
    (A Combination of the *Chêng-Ho...Chêng
    Lei...Pên Tshao* with the *Pên Tshao Yen I*.)
    Yuan, +1249; reprinted many times after-
    wards, esp. in the Ming, +1468, with at
    least seven Ming editions, the last in
    +1624 or +1625.
    Thang Shen-Wei 唐愼微.
    Khou Tsung-Shih 寇宗奭.
    Pr. (or ed.) Chang Tshun-Hui 張存惠.
Chhung-Yang Chhüan Chen Chi 重陽全
    眞集.
    (Wang) Chhung-Yang's [Wang Chê's]
    Records of the Perfect Truth (School).
    Sung, mid +12th cent.
    Wang Chê 王嚞.
    TT/1139.
Chhung-Yang Chiao Hua Chi 重陽教化集.
    Memorials of (Wang) Chhung-Yang's
    [Wang Chê's] Preaching.
    Sung, mid +12th cent.
    Wang Chê 王嚞.
    TT/1140.
Chhung-Yang Chin-Kuan Yü-Suo Chüeh 重陽
    金關玉鎖訣.

Chhung-Yang Chin-Kuan Yü-Suo Chüeh (*cont.*)
(Wang) Chhung-Yang's [Wang's Chê's] Instructions on the Golden Gate and the Lock of Jade.
Sung, mid +12th cent.
Wang Chê 王嚞.
TT/1142.

Chhung-Yang Fên-Li Shih-Hua Chi 重陽分梨十化集.
Writings of (Wang) Chhung-Yang [Wang Chê] (to commemorate the time when he received a daily) Ration of Pears, and the Ten Precepts of his Teacher.
Sung, mid +12th cent.
Wang Chê 王嚞.
TT/1141.

Chhung-Yang Li-Chiao Shih-Wu Lun 重陽立教十五論.
Fifteen Discourses of (Wang) Chhung-Yang [Wang Chê] on the Establishment of his School.
Sung, mid +12th cent.
Wang Chê 王嚞.
TT/1216.

Đai-Viêt Sú-ký Toàn-thú 大越史記全書.
The Complete Book of the History of Great Annam.
Vietnam, *c.* +1479.
Ngô Si-Liên 吳士連.

Fa Yen 法言.
Admonitory Sayings [in admiration, and imitation, of the *Lun Yü*].
Hsin, +5.
Yang Hsiung 揚雄.
Tr. von Zach (5).

Fa Yuan Chu Lin 法苑珠林.
Forest of Pearls from the Garden of the [Buddhist] Law.
Thang, +668, +688.
Tao-Shih 道世.

Fun Tzu Chi Jan 范子計然.
See *Chi Ni Tzu.*

Fang Hu Wai Shih 方壺外史.
Unofficial History of the Land of the Immortals, Fang-hu. (Contains two *nei tan* commentaries on the *Tshan Thung Chhi*, +1569 and +1573.)
Ming, *c.* +1590.
Lu Hsi-Hsing 陸西星.
Cf. Liu Tshun-Jen (1, 2).

Fang Yü Chi 方興記.
General Geography.
Chin, or at least pre-Sung.
Hsü Chiai 徐鍇.

Fei Lu Hui Ta 斐錄彙答.
Questions and Answers on Things Material and Moral.
Ming, +1636.
Kao I-Chih (Alfonso Vagnoni) 高一志.
Bernard-Maître (18), no. 272.

Fên Thu 粉圖.
See *Hu Kang Tzu Fên Thu.*

Fêng Su Thung I 風俗通義.
The Meaning of Popular Traditions and Customs.
H/Han, +175.
Ying Shao 應劭.
Chung-Fa Index, no. 3.

Fo Shuo Fo I Wang Ching 佛說佛醫王經
Buddha Vaidyarāja Sūtra; or *Buddha-prokta Buddha-bhaiṣajyarāja Sūtra* (Sūtra of the Buddha of Healing, spoken by Buddha).
India.
Tr. San Kuo (Wu) +230.
Trs. Liu Yen (Vinayātapa) & Chih-Chhien. 支謙.
N/1327; TW/793.

Fo Tsu Li Tai Thung Tsai 佛祖歷代通載.
General Record of Buddhist and Secular History through the Ages.
Yuan, +1341.
Nien-Chhang (monk) 念常.

Fu Chhi Ching I Lun 服氣精義論.
Dissertation on the Meaning of 'Absorbing the Chhi and the Ching' (for Longevity and Immortality), [Taoist hygienic, respiratory, pharmaceutical, medical and (originally) sexual procedures].
Thang, *c.* +715.
Ssuma Chhêng-Chên 司馬承貞.
In *YCCC*, ch. 57.
Cf. Maspero (7), pp. 364 ff.

Fu Hung Thu 伏汞圖.
Illustrated Manual on the Subduing of Mercury.
Sui, Thang, J/Chin or possibly Ming.
Shêng Hsüan Tzu 昇女子.
Survives now only in quotations.

Fu Nei Yuan Chhi Ching 服內元氣經.
Manual of Absorbing the Internal Chhi of Primary (Vitality).
Thang, +8th, probably *c.* +755.
Huan Chen hsien-sêng (Mr Truth-and-Illusion) 幻眞先生.
TT/821, and in *YCCC*, ch. 60, pp. 10b ff.
Cf. Maspero (7), p. 199.

Fu Shih Lun 服石論.
Treatise on the Consumption of Mineral Drugs.
Thang, perhaps Sui.
Writer unknown.
Extant only in excerpts preserved in the *I Hsin Fang* (+982).

Fu Shou Tan Shu 福壽丹書.
A Book of Elixir-Enchymoma Techniques for Happiness and Longevity.
Ming, +1621.
Chêng Chih-Chhiao 鄭之僑 (at least in part).
Partial tr. of the gymnastic material, Dudgeon (1).

34

Fusō Ryakuki 扶桑畧記.
　Classified Historical Matters concerning the
　　Land of Fu-Sang (Japan) [from +898 to
　　+1197].
　Japan (Kamakura) +1198.
　Kōen (monk).

Genji Monogatari 源氏物語.
　The Tale of (Prince) Genji.
　Japan, +1021.
　Murasaki Shikibu 紫式部.

Hai Yao Pên Tshao 海藥本草.
　[= *Nan Hai Yao Phu*.]
　Materia Medica of the Countries Beyond
　　the Seas.
　Wu Tai (C/Shu), *c*. +923.
　Li Hsün 李珣.
　Preserved only in numerous quotations in
　　Chêng Lei Pên Tshao and later pandects.

Han Fei Tzu 韓非子.
　The Book of Master Han Fei.
　Chou, early —3rd century.
　Han Fei 韓非.
　Tr. Liao Wên-Kuei (1).

Han Kuan I 漢官儀.
　The Civil Service of the Han Dynasty and
　　its Regulations.
　H/Han +197.
　Ying Shao 應劭.
　Ed. Chang Tsung-Yuan 張宗源 (+1752
　　to 1800).
　Cf. Hummel (2), p. 57.

Han Kung Hsiang Fang 漢宮香方.
　On the Blending of Perfumes in the Palaces
　　of the Han.
　H/Han, +1st or +2nd.
　Genuine parts preserved *c*. +1131 by
　　Chang Pang-Chi 張邦基.
　Attrib. Tung Hsia-Chou 童渡周.
　Comm. by Chêng Hsüan 鄭玄.
　'Restored', *c*. +1590, by Kao Lien 高濂.

Han Thien Shih Shih Chia 漢天師世家.
　Genealogy of the Family of the Han
　　Heavenly Teacher.
　Date uncertain.
　Writers unknown.
　With Pu Appendix, 1918, by Chang Yuan-
　　Hsü 張元旭 (the 62nd Taoist Patriarch,
　　Thien Shih).
　TT/1442.

Han Wei Tshung-Shu 漢魏叢書.
　Collection of Books of the Han and Wei Dyn-
　　asties [first only 38, later increased to
　　96].
　Ming, +1592.
　Ed. Thu Lung 屠隆.

Han Wu (Ti) Ku Shih 漢武(帝)故事.
　Tales of (the Emperor) Wu of the Han
　　(r. —140 to —87).
　L/Sung and Chhi, late +5th.
　Wang Chien 王儉.

Perhaps based on an earlier work of the
　same kind by Ko Hung 葛洪.
　Tr. d'Hormon (1).

Han Wu (Ti) Nei Chuan 漢武(帝)內傳.
　The Inside Story of (Emperor) Wu of the
　　Han (r. —140 to —87).
　Material of Chin, L/Sung, Chhi, Liang and
　　perhaps Chhen date, +320 to +580,
　　probably stabilised about +580.
　Attrib. Pan Ku, Ko Hung, etc.
　Actual writer unknown.
　TT/289.
　Tr. Schipper (1).

Han Wu (Ti) Nei Chuan Fu Lu 漢武(帝)內傳
　附錄.
　See *Han Wu (Ti) Wai Chuan*.

Han Wu (Ti) Wai Chuan 漢武(帝)外傳.
　[=*Han Wu (Ti) Nei Chuan Fu Lu*.]
　Extraordinary Particulars of (Emperor) Wu
　　of the Han (and his collaborators), [largely
　　biographies of the magician-technicians
　　at Han Wu Ti's court].
　Material of partly earlier date collected and
　　stabilised in Sui or Thang, early +7th
　　century.
　Writers and editor unknown.
　Introductory paragraphs added by Wang
　　Yu-Yen 王游巖 (+746).
　TT/290.
　Cf. Maspero (7), p. 234, and Schipper (1).

Hei Chhien Shui Hu Lun 黑鉛水虎論.
　Discourse on the Black Lead and the Water
　　Tiger.
　Alternative title of *Huan Tan Nei Hsiang
　　Chin Yo Shih*, q.v.

Ho Chi Chü Fang 和劑局方.
　Standard Formularies of the (Government)
　　Pharmacies [based on the *Thai-Phing
　　Shêng Hui Fang* and other collections].
　Sung, *c*. +1109.
　Ed. Chhen Chhêng 陳承, Phei Tsung-
　　Yuan 裴宗元, & Chhen Shih-Wên
　　陳師文.
　Cf. *SIC*, p. 974.

Honan Chhen Shih Hsiang Phu 河南陳氏香譜.
　See *Hsiang Phu* by Chhen Ching.

Honan Chhêng Shih I Shu 河南程氏遺書.
　Remaining Records of Discourses of the
　　Chhêng brothers of Honan [Chhêng I and
　　Chhêng Hao, +11th-century Neo-
　　Confucian philosophers].
　Sung, +1168, pr. *c*. +1250.
　Chu Hsi (ed.) 朱熹.
　In *Erh Chhêng Chhüan Shu*, q.v.
　Cf. Graham (1), p. 141.

Honan Chhêng Shih Tshui Yen 河南程氏粹言.
　Authentic Statements of the Chhêng brothers
　　of Honan [Chhîng I and Chhêng Hao,
　　+11th-century Neo-Confucian philo-
　　sophers. In fact more altered and abridged
　　than the other sources, which are therefore
　　to be preferred.]

Honan Chhêng Shih Tshui Yen (cont.)
> Sung, first collected *c.* +1150, supposedly
> ed. +1166, in its present form by
> *c.* +1340.
> Coll. Hu Yin   胡寅.
> Supposed ed. Chang Shih   張栻.
> In *Erh Chhêng Chhüan Shu*, q.v., since
> +1606.
> Cf. Graham (1), p. 145.

Honzō-Wamyō   本草和名.
> Synonymic Materia Medica with Japanese
> Equivalents.
> Japan, +918.
> Fukane no Sukehito   深根輔仁.
> Cf. Karow (1).

Hou Han Shu   後漢書.
> History of the Later Han Dynasty [+25 to
> +220].
> L/Sung, +450.
> Fan Yeh   范曄.
> The monograph chapters by Ssuma Piao
> 司馬彪 (d. +305), with commentary by
> Liu Chao   劉昭 (*c.* +510), who first in-
> corporated them in the work.
> A few chs. tr. Chavannes (6, 16); Pfizmaier
> (52, 53).
> Yin-Tê Index, no. 41.

Hou Tê Lu   厚德錄.
> Stories of Eminent Virtue.
> Sung, early +12th.
> Li Yuan-Kang   李元綱.

Hu Kang Tzu Fên Thu   狐剛子粉圖.
> Illustrated Manual of Powders [Salts], by
> the Fox-Hard Master.
> Sui or Thang.
> Hu Kang Tzu   狐剛子.
> Survives now only in quotations; originally
> in *TT* but lost. Cf. Vol. 4, pt. 1, p. 308.

Hua Tho Nei Chao Thu   佗佗內照圖.
> Hua Tho's Illustrations of Visceral Anatomy.
> See *Hsüan Mên Mo Chüeh Nei Chao Thu.*
> Cf. Miyashita Saburo (1).

Hua-Yang Thao Yin-Chü Chuan   華陽陶隱居傳.
> A Biography of Thao Yin-Chü (Thao
> Hung-Ching) of Huayang [the great
> alchemist, naturalist and physician].
> Thang.
> Chia Sung   賈嵩.
> *TT*/297.

Hua Yen Ching   華嚴經.
> *Buddha-avataṃsaka Sūtra*; The Adorn-
> ment of Buddha.
> India.
> Tr. into Chinese, +6th century.
> TW/278, 279.

Huai Nan Hung Lieh Chieh   淮南鴻烈解.
> See *Huai Nan Tzu.*

Huai Nan Tzu   淮南子.
> [= *Huai Han Hung Lieh Chieh*   淮南鴻烈
> 解.]
> The Book of (the Prince of) Huai-Nan
> [compendium of natural philosophy].

C/Han, *c.* −120.
> Written by the group of scholars gathered
> by Liu An (prince of Huai-Nan)   劉安.
> Partial trs. Morgan (1); Erkes (1); Hughes
> (1); Chatley (1); Wieger (2).
> Chung-Fa Index, no. 5.
> *TT*/1170.

Huai-Nan (Wang) Wan Pi Shu   淮南(王)萬畢
> 術.
> [Prob. = *Chen-Chung Hung-Pao Yuan-Pi
> Shu* and variants.]
> The Ten Thousand Infallible Arts of (the
> Prince of) Huai-Nan [Taoist magical and
> technical recipes].
> C/Han, −2nd century.
> No longer a separate book but fragments
> contained in *TPYL*, ch. 736 and elsewhere
> Reconstituted texts by Yeh Tê-Hui in
> *Kuan Ku Thang So Chu Shu*, and Sun
> Fêng-I in *Wên Ching Thang Tshung-Shu.*
> Attrib. Liu An   劉安.
> See Kaltenmark (2), p. 32.
> It is probable that the terms *Chen-Chung*
> 枕中 Confidential Pillow-Book; *Hung-
> Pao* 鴻寶 Infinite Treasure; *Wan-Pi*
> 萬畢 Ten Thousand Infallible; and
> *Yuan-Pi*   苑祕 Garden of Secrets; were
> originally titles of parts of a *Huai-Nan
> Wang Shu*   淮南王書 (Writings of the
> Prince of Huai-Nan) forming the Chung
> Phien   中篇 (and perhaps also the
> Wai Shu   外書) of which the present
> *Huai Nan Tzu* book (q.v.) was the Nei
> Shu   內書.

Huan Chen hsien-sêng, etc.   幻真先生.
> See *Thai Hsi Ching* and *Fu Nei Yuan Chhi
> Ching.*

Huan Chin Shu   還金述.
> An Account of the Regenerative Metallous
> Enchymoma.
> Thang, probably +9th.
> Thao Chih   陶植.
> *TT*/915, also excerpted, in *YCCC*, ch. 70,
> pp. 13 *a* ff.

Huan Tan Chou Hou Chüeh   還丹肘後訣.
> Oral Instructions on Handy Formulae for
> Cyclically Transformed Elixirs [with
> illustrations of alchemical apparatus].
> Ascr. Chin, *c.* +320.
> Actually Thang, including a memorandum
> of +875 by Wu Ta-Ling   仵達靈, and
> the rest probably by other hands within a
> few years of this date.
> Attrib. Ko Hung   葛洪.
> *TT*/908.

Huan Tan Chung Hsien Lun   還丹眾仙論.
> Pronouncements of the Company of the
> Immortals on Cyclically Transformed
> Elixirs.
> Sung, +1052.
> Yang Tsai   楊在.
> *TT*/230.

Huan Tan Fu Ming Phien 還丹復命篇.
 Book on the Restoration of Life by the
 Cyclically Transformed Elixir.
 Sung, +12th cent., *c*. +1175.
 Hsüeh Tao-Kuang 薛道光.
 TT/1074.
Huan Tan Nei Hsiang Chin Yo Shih 還丹內象
 金鑰匙.
 [= *Hei Chhien Shui Hu Lun* and *Hung
 Chhien Huo Lung Lun*.]
 A Golden Key to the Physiological Aspects
 of the Regenerative Enchymoma.
 Wu Tai, *c*. +950.
 Phêng Hsiao 彭曉.
 Now but half a chapter in *YCCC*, ch. 70,
 pp. 1 *a* ff., though formerly contained in
 the *Tao Tsang*.
*Huan Tan Pi Chüeh Yang Chhih-Tzu Shen
 Fang* 還丹祕訣養赤子神方.
 The Wondrous Art of Nourishing the
 (Divine) Embryo (lit. the Naked Babe) by
 the use of the secret Formula of the Re-
 generative Enchymoma [physiological
 alchemy].
 Sung, probably late +12th.
 Hsü Ming-Tao 許明道.
 TT/229.
Huan Yü Shih Mo 寰宇始末.
 On the Beginning and End of the World
 [the Hebrew-Christian account of crea-
 tion, the Four Aristotelian Causes,
 Elements, etc.].
 Ming, +1637.
 Kao I-Chih (Alfonso Vagnoni) 高一志.
 Bernard-Maître (18), no. 283.
Huan Yuan Phien 還原篇.
 Book of the Return to the Origin [poems on
 the regaining of the primary vitalities in
 physiological alchemy].
 Sung, *c*. +1140.
 Shih Thai 石泰.
 TT/1077. Also in *Hsiu Chen Shih Shu*
 (*TT*/260), ch. 2.
Huang Chi Ching Shih Shu 皇極經世書.
 Book of the Sublime Principle which
 governs All Things within the World.
 Sung, *c*. +1060.
 Shao Yung 邵雍]
 TT/1028. Abridged in *Hsing Li Ta Chhüan*
 and *Hsing Li Ching I*.
Huang Chi Ho Pi Hsien Ching 皇極闔闢仙經.
 [= *Yin Chen Jen Tung-Hua Chêng Mo Huang
 Chi Ho Pi Chêng Tao Hsien Ching*.]
 The Height of Perfection (attained by)
 Opening and Closing (the Orifices of the
 Body); a Manual of the Immortals [phys-
 iological alchemy, *nei tan* techniques].
 Ming or Chhing.
 Attrib. Yin chen jen (Phêng-Thou).
 尹眞人 (蓬頭).
 Ed. Min I-Tê 閔一得, *c*. 1830.
 In *Tao Tsang Hsü Pien* (*Chhu chi*), 2, from

a MS. preserved at the Blue Goat Temple
 青羊宮 (Chhêngtu).
Huang Pai Ching 黃白鏡.
 Mirror of (the Art of) the Yellow and the
 White [physiological alchemy].
 Ming, +1598.
 Li Wên-Chu 李文燭.
 Comm. Wang Chhing-Chêng 王清正.
 In *Wai Chin Tan* coll., ch. 2 (*CTPS, pên*
 7).
*Huang-Thien Shang-Chhing Chin Chhüeh Ti
 Chün Ling Shu Tzu-Wên Shang Ching*
 皇天上清金闕帝君靈書紫文上經.
 Exalted Canon of the Imperial Lord of the
 Golden Gates, Divinely Written in Purple
 Script; a Huang-Thien Shang-Chhing
 Scripture.
 Chin, late +4th, with later revisions.
 Writer unknown.
 TT/634.
Huang Thing Chung Ching Ching 黃庭中景經.
 [= *Thai-Shang Huang Thing Chung Ching
 Ching*.]
 Manual of the Middle Radiance of the
 Yellow Courts (central regions of the
 three parts of the body) [Taoist anatomy
 and physiology].
 Sui.
 Li Chhien-Chhêng 李千乘.
 TT/1382, completing *TT*/398–400.
 Cf. Maspero (7), pp. 195, 203.
*Huang Thing Nei Ching Wu Tsang Liu Fu Pu
 Hsieh Thu* 黃庭內景五臟六府補瀉圖
 Diagrams of the Strengthening and Weaken-
 ing of the Five Yin-viscera and the Six
 Yang-viscera (in accordance with) the
 (*Jade Manual of the*) *Internal Radiance of
 the Yellow Courts*.
 Thang, *c*. +850.
 Hu An 胡愔.
 TT/429.
Huang Thing Nei Ching Wu Tsang Liu Fu Thu
 黃庭內景五臟六府圖.
 Diagrams of the Five Yin-viscera and the
 Six Yang-viscera (discussed in the *Jade
 Manual of the*) *Internal Radiance of the
 Yellow Courts* [Taoist anatomy and physi-
 ology; no illustrations surviving, but much
 therapy and pharmacy].
 Thang, +848.
 Hu An 胡愔 (title: Thai-pai Shan Chien
 Su Nü) 太白山見素女.
 In *Hsiu Chen Shih Shu* (*TT*/260), ch. 54.
 Illustrations preserved only in Japan, MS. of
 before +985.
 SIC, p. 223; Watanabe Kozo (*1*), pp. 112 ff.
Huang Thing Nei Ching Yü Ching 黃庭內景
 玉經.
 [= *Thai-Shang Huang Thing Nei Ching Yü
 Ching*.]
 Jade Manual of the Internal Radiance of the
 Yellow Courts (central regions of the

Huang Thing Nei Ching Yü Ching (cont.)
three parts of the body) [Taoist anatomy
and physiology]. In 36 *chang*.
L/Sung, Chhi, Liang or Chhen, +5th or
+6th. The oldest parts date probably
from Chin, about +365.
Writer unknown. Allegedly transmitted by
immortals to the Lady Wei (Wei Fu Jen),
i.e. Wei Hua-Tshun 魏華存.
TT/328.
Paraphrase by Liu Chhang-Shêng 劉長生
(Sui), *TT*/398.
Comms. by Liang Chhiu Tzu 梁丘子
(Thang), *TT*/399, and Chiang Shen-Hsiu
蔣愼修 (Sung), *TT*/400.
Cf. Maspero (7), p. 239.
Huang Thing Nei Ching Yü Ching Chu 黃庭內
景玉經注.
Commentary on (and paraphrased text of)
the *Jade Manual of the Internal Radiance
of the Yellow Courts*.
Sui.
Liu Chhang-Shêng 劉長生.
TT/398.
Huang Thing Nei Ching (Yü) Ching Chu 黃庭
內景(玉)經注.
Commentary on the *Jade Manual of the
Internal Radiance of the Yellow Courts*.
Thang, +8th or +9th.
Liang Chhiu Tzu (ps.) 梁丘子.
TT/399, and in *Hsiu Chen Shih Shu*
(*TT*/260), chs. 55-57; and in *YCCC*,
chs. 11, 12 (where the first 3 *chang* (30
verses) have the otherwise lost commentary
of Wu Chhêng Tzu 務成子).
Cf. Maspero (7), pp. 239 ff.
Huang Thing Nei Wai Ching Yü Ching Chieh
黃庭內外景玉經解.
Explanation of the *Jade Manuals of the
Internal and External Radiances of the
Yellow Courts*.
Sung.
Chiang Shen-Hsiu 蔣愼修.
TT/400.
Huang Thing Wai Ching Yü Ching 黃庭外景
玉經.
[= *Thai-Shang Huang Thing Wai Ching Yü
Ching*.]
Jade Manual of the External Radiance of
the Yellow Courts (central regions of the
three parts of the body) [Taoist anatomy
and physiology]. In 3 *chüan*.
H/Han, San Kuo or Chin, +2nd or +3rd.
Not later than +300.
Writer unknown.
TT/329.
Comms. by Wu Chhêng Tzu 務成子 (early
Thang) *YCCC*, ch. 12; Liang Chhiu Tzu
梁丘子 (late Thang), *TT*/260, chs. 58-60;
Chiang Shen-Hsiu 蔣愼修 (Sung),
TT/400.
Cf. Maspero (7), pp. 195 ff., 428 ff.

Huang Thing Wai Ching Yü Ching Chu 黃庭外
景玉經註.
Commentary on the *Jade Manual of the
External Radiance of the Yellow Courts*.
Sui or early Thang, +7th.
Wu Chhêng Tzu (ps.) 務成子.
In *YCCC*, ch. 12, pp. 30*a* ff.
Cf. Maspero (7), p. 239.
Huang Thing Wai Ching Yü Ching Chu 黃庭外
景玉經註.
Commentary on the *Jade Manual of the
External Radiance of the Yellow Courts*.
Thang, +8th or +9th.
Liang Chhiu Tzu (ps.) 梁丘子.
In *Hsiu Chen Shih Shu* (*TT*/260), chs. 58-
60.
Cf. Maspero (7), pp. 239 ff.
Huang Ti Chiu Ting Shen Tan Ching Chüeh
黃帝九鼎神丹經訣.
The Yellow Emperor's Canon of the Nine-
Vessel Spiritual Elixir, with Explanations.
Early Thang or early Sung, but incorpo-
rating as ch. 1 a canonical work probably
of the +2nd cent.
Writer unknown.
TT/878. Also, abridged, in *YCCC*, ch. 67,
pp. 1*a* ff.
Huang Ti Nei Ching, Ling Shu 黃帝內經靈樞.
The Yellow Emperor's Manual of Corporeal
(Medicine), the Vital Axis [medical
physiology and anatomy].
Probably C/Han, *c.* −1st century.
Writers unknown.
Edited Thang, +762, by Wang Ping 王冰.
Analysis by Huang Wên (1).
Tr. Chamfrault & Ung Kang-Sam (1).
Commentaries by Ma Shih 馬蒔 (Ming)
and Chang Chih-Tshung 張志聰
(Chhing) in *TSCC*, *I shu tien*, chs. 67 to 88.
Huang Ti Nei Ching, Ling Shu, Pai Hua Chieh
See Chhen Pi-Liu & Chêng Cho-Jen (1).
Huang Ti Nei Ching, Su Wên 黃帝內經素問.
The Yellow Emperor's Manual of Corpor-
eal (Medicine); Questions (and Answers)
about Living Matter [clinical medicine].
Chou, remodelled in Chhin and Han,
reaching final form *c.* −2nd century.
Writers unknown.
Ed. & comm., Thang (+762), Wang Ping
王冰; Sung (*c.* +1050), Lin I 林億.
Partial trs. Hübotter (1), chs. 4, 5, 10, 11,
21; Veith (1); complete, Chamfrault &
Ung Kang-Sam (1).
See Wang & Wu (1), pp. 28 ff.; Huang Wên
(1).
Huang Ti Nei Ching Su Wên I Phien 黃帝內
經素問遺篇.
The Missing Chapters from the *Questions
and Answers* of the *Yellow Emperor's
Manual of Corporeal (Medicine)*.
Ascr. pre-Han.
Sung, preface, +1099.

Huang Ti Nei Ching Su Wên I Phien (cont.)
    Ed. (perhaps written by) Liu Wên-Shu
    劉溫舒.
    Often appended to his *Su Wên Ju Shih Yün*
    Chhi Ao Lun (q.v.). 素問入式運氣奧論.
Huang Ti Nei Ching Su Wên, Pai Hua Chieh
    See Chou Fêng-Wu, Wang Wan-Chieh &
    Hsü Kuo-Chhien (1).
Huang Ti Pa-shih-i Nan Ching Tsuan Thu Chü
    Chieh 黃帝八十一難經纂圖句解.
    Diagrams and a Running Commentary for
    the *Manual of (Explanations Concerning)*
    Eighty-one Difficult (Passages) in the Yellow
    Emperor's (Manual of Corporeal Medicine).
    Sung, +1270 (text H/Han, +1st).
    Li Kung 李駉.
    TT/1012.
Huang Ti Pao Tsang Ching 黃帝寶藏經.
    Perhaps an alternative name for *Hsien-*
    Yuan Pao Tsang (Chhang Wei) Lun, q.v.
Huang Ti Yin Fu Ching 黃帝陰符經.
    See *Yin Fu Ching*.
Huang Ti Yin Fu Ching Chu 黃帝陰符經註.
    Commentary on the *Yellow Emperor's Book*
    on the Harmony of the Seen and the Unseen.
    Sung.
    Liu Chhu-Hsüan 劉處玄.
    TT/119.
Huang Yeh Fu 黃冶賦.
    Rhapsodic Ode on 'Smelting the Yellow'
    [alchemy].
    Thang, *c.* +840.
    Li Tê-Yü 李德裕.
    In *Li Wên-Jao Pieh Chi*, ch. 1.
Huang Yeh Lun 黃冶論.
    Essay on the 'Smelting of the Yellow'
    [alchemy].
    Thang, *c.* +830.
    Li Tê-Yü 李德裕.
    In *Wên Yuan Ying Hua*, ch. 739, p. 15*a*,
    and *Li Wên-Jao Wai Chi*, ch. 4.
Hui Ming Ching 慧命經.
    [= *Tsui-Shang I Chhêng Hui Ming Ching*,
    also entitled *Hsü Ming Fang*.]
    Manual of the (Achievement of) Wisdom
    and the (Lengthening of the) Life-Span.
    Chhing, +1794.
    Liu Hua-Yang 柳華陽.
    Cf. Wilhelm & Jung (1), editions after 1957.
Hung Chhien Huo Lung Lun 紅鉛火龍論.
    Discourse on the Red Lead and the Fire
    Dragon.
    Alternative title of *Huan Tan Nei Hsiang*
    Chin Yo Shih, q.v.
Hung Chhien Ju Hei Chhien Chüeh 紅鉛入黑
    鉛訣.
    Oral Instructions on the Entry of the Red
    Lead into the Black Lead.
    Probably Sung, but some of the material
    perhaps older.
    Compiler unknown.
    TT/934.

Huo Kung Chhieh Yao 火攻挈要.
    Essentials of Gunnery.
    Ming, +1643.
    Chiao Hsü 焦勗.
    With the collaboration of Thang Jo-Wang
    (J. A. Schall von Bell) 湯若望.
    Bernard-Maître (18), no. 334.
Huo Lien Ching 火蓮經.
    Manual of the Lotus of Fire [physiological
    alchemy].
    Ming or Chhing.
    Attrib. Liu An, 劉安 (Han).
    In *Wai Chin Tan*, coll., ch. 1 (*CTPS, pên* 6).
Huo Lung Ching 火龍經.
    The Fire-Drake (Artillery) Manual.
    Ming, +1412.
    Chiao Yü 焦玉.
    The first part of this book, in three sections,
    is attributed fancifully to Chuko Wu-Hou
    (i.e. Chuko Liang), and Liu Chi 劉基
    (+1311 to +1375) appears as co-editor,
    really perhaps co-author.
    The second part, also in three sections, is
    attributed to Liu Chi alone, but edited,
    probably written, by Mac Hsi-Ping
    毛希秉 in +1632.
    The third part, in two sections, is by Mao
    Yuan-I 毛元儀 (*fl.* +1628) and edited
    by Chuko Kuang-Jung 諸葛光榮
    whose preface is of +1644, Fang Yuan-
    Chuang 方元壯 & Chung Fu-Wu 鍾伏武.
Huo Lung Chüeh 火龍訣.
    Oral Instructions on the Fiery Dragon
    [proto-chemical and physiological alchemy].
    Date uncertain, ascr. Yuan, +14th.
    Attrib. Shang Yang Tsu Shih 上陽祖師.
    In *Wai Chin Tan* (coll.), ch. 3 (*CTPS, pên* 8).
Hupei Thung Chih 湖北通志.
    Historical Geography of Hupei Province.
    Min Kuo, 1921, but based on much older
    records.
    See Yang Chhêng-Hsi (ed.) (1) 楊承禧.
Hsi Chhi Tshung Hua 西溪叢話
    (*SKCS* has *Yü* 語).
    Western Pool Collected Remarks.
    Sung, *c.* +1150.
    Yao Khuan 姚寬.
Hsi Chhing Ku Chien 西清古鑑.
    Hsi Chhing Catalogue of Ancient Mirrors
    (and Bronzes) in the Imperial Collection.
    (The collection was housed in the Library
    of Western Serenity, a building in the
    southern part of the Imperial Palace).
    Chhing, +1751.
    Liang Shih-Chêng 梁詩正.
Hsi Shan Chhun Hsien Hui Chen Chi 西山羣
    仙會眞記.
    A True Account of the Proceedings of the Com-
    pany of Immortals in the Western Mountains.
    Thang, *c.* +800.
    Shih Chien-Wu 施肩吾.
    TT/243.

Hsi Shang Fu Than 席上腐談.
Old-Fashioned Table Talk.
Yuan, *c.* +1290.
Yü Yen 俞琰.

Hsi Wang Mu Nü Hsiu Chêng Thu Shih Tsê
西王母女修正途十則.
The Ten Rules of the Mother (Goddess)
Queen of the West to Guide Women
(Taoists) along the Right Road of
Restoring (the Primary Vitalities) [phy-
siological alchemy].
Ming or Chhing.
Attrib. Lü Yen 呂喦 (+8th century).
Shen I-Ping *et al.* 沈一炳.
Comm. Min I-Tê 閔一得 (*c.* 1830).
In *Tao Tsang Hsü Pien* (*Chhu chi*), 19.

Hsi-Yang Huo Kung Thu Shuo 西洋火攻圖說.
Illustrated Treatise on European Gunnery.
Ming, before +1625.
Chang Tao 張燾 & Sun Hsüeh-Shih
孫學詩.

Hsi Yo Hua-Shan Chih 西嶽華山誌.
Records of Hua-Shan, the Great Western
Mountain.
Sung, *c.* +1170.
Wang Chhu-I 王處一.
TT/304.

Hsi Yo Tou hsien-sêng Hsiu Chen Chih Nan
西嶽竇先生修眞指南.
Teacher Tou's South-Pointer for the
Regeneration of the Primary (Vitalities),
from the Western Sacred Mountain.
Sung, probably early +13th.
Tou hsien-sêng 竇先生.
In *Hsiu Chen Shih Shu* (*TT*/260), ch. 21,
pp. 1*a* to 6*b*.

Hsi Yu Chi 西遊記.
A Pilgrimage to the West [novel].
Ming, *c.* +1570.
Wu Chhêng-Ên 吳承恩.
Tr. Waley (17).

Hsi Yu Chi.
See *Chhang-Chhun Chen Jen Hsi Yu Chi*.

Hsi Yü Chiu Wên 西域舊聞.
Old Traditions of the Western Countries [a
conflation, with abbreviations, of the
Hsi Yü Wên Chien Lu and the *Shêng Wu
Chi*, q.v.].
Chhing, +1777 and 1842.
Chhun Yuan Chhi-shih-i Lao-jen 椿園七
十一老人 & Wei Yuan 魏源.
Arr. Chêng Kuang-Tsu (1843) 鄭光祖.

Hsi Yü Thu Chi 西域圖記.
Illustrated Record of Western Countries.
Sui, +610.
Phei Chü 裴矩.

Hsi Yü Wên Chien Lu 西域聞見錄.
Things Seen and Heard in the Western
Countries.
Chhing, +1777.
Chhun Yuan Chhi-shih-i Lao-jen
椿園七十一老人.

[The 71-year-old Gentleman of the Cedar
Garden.]
Bretschneider (2), vol. 1, p. 128.

Hsi Yuan Lu 洗冤錄.
The Washing Away of Wrongs (i.e. False
Charges) [treatise on forensic medicine].
Sung, +1247.
Sung Tzhu 宋慈.
Partial tr., H. A. Giles (7).

Hsiang Chhêng 香乘.
Records of Perfumes and Incense [in-
cluding combustion-clocks].
Ming, betw. +1618 and +1641.
Chou Chia-Chou 周嘉胄.

Hsiang Chien 香牋.
Notes on Perfumes and Incense.
Ming, *c.* +1560.
Thu Lung 屠隆.

Huang Kuo 香國.
The Realm of Incense and Perfumes.
Ming.
Mao Chin, 毛晉.

Hsiang Lu 香錄.
[= *Nan Fan Hsiang Lu*.]
A Catalogue of Incense.
Sung, +1151.
Yeh Thing-Kuei 葉廷珪.

Hsiang Phu 香譜.
A Treatise on Aromatics and Incense
[-Clocks].
Sung, *c.* +1073.
Shen Li 沈立.
Now extant only in the form of quotations
in later works.

Hsiang Phu 香譜.
A Treatise on Perfumes and Incense.
Sung, *c.* +1115.
Hung Chhu 洪芻.

Hsiang Phu 香譜.
[= *Hsin Tsuan Hsiang Phu*
or *Honan Chhen shih Hsiang Phu*.]
A Treatise on Perfumes and Aromatic Substances [including incense and combust-
ion-clocks].
Sung, late +12th or +13th; may be as late
as +1330.
Chhen Ching 陳敬.

Hsiang Phu 香譜.
A Treatise on Incense and Perfumes.
Yuan, +1322.
Hsiung Phêng-Lai 熊朋來.

Hsiang Yao Chhao 香藥抄.
Memoir on Aromatic Plants and Incense.
Japan, *c.* +1163.
Kuan-Yu (Kanyu) 觀祐.MS. preserved at
the 滋賀石山寺 Temple. Facsim. re-
prod. in Suppl. to the Japanese Tripiṭaka,
vol. 11.

Hsieh Thien Chi 泄天機.
A Divulgation of the Machinery of Nature
(in the Human Body, permitting the
Formation of the Enchymoma).

Hsieh Thien Chi (cont.)
Chhing, *c.* +1795.
Li Ong (Ni-Wan shih) 李翁 (Mr Ni-Wan).
Written down in 1833 by Min Hsiao-Kên
閔小艮.
In *Tao Tsang Hsü Pien (Chhu chi)*, 4.
Hsien Lo Chi 仙樂集.
(Collected Poems) on the Happiness of the
Holy Immortals.
Sung, late +12th cent.
Liu Chhu-Hsüan 劉處玄.
TT/1127.
Hsien-Yuan Huang Ti Shui Ching Yao Fa 軒轅
黃帝水經藥法.
(Thirty-two) Medicinal Methods from the
Aqueous (Solutions) Manual of Hsien-
Yuan the Yellow Emperor.
Date uncertain.
Writer unknown.
TT/922.
Hsien-Yuan Pao Tsang Chhang Wei Lun 軒轅
寶藏暢微論.
The Yellow Emperor's Expansive yet
Detailed Discourse on the (Contents of
the) Precious Treasury (of the Earth)
[mineralogy and metallurgy].
Alternative title of *Pao Tsang Lun*, q.v.
Hsien-Yuan Pao Tsang Lun 軒轅寶藏論.
The Yellow Emperor's Discourse on the
Contents of the Precious Treasury (of the
Earth).
See *Pao Tsang Lun*.
Hsin Hsiu Pên Tshao 新修本草.
The New (lit. Newly Improved) Pharma-
copoeia.
Thang, +659.
Ed. Su Ching (= Su Kung) 蘇敬(蘇恭)
and a commission of 22 collaborators
under the direction first of Li Chi 李勣
& Yü Chih-Ning 于志寧, then of
Chhangsun Wu-Chi 長孫無忌. This
work was afterwards commonly but in-
correctly known as *Thang Pên Tshao*. It
was lost in China, apart from MS. frag-
ments at Tunhuang, but copied by a
Japanese in +731 and preserved in Japan
though incompletely.
Hsin Lun 新論.
New Discussions.
H/Han, *c.* +10 to +20, presented +25.
Huan Than 桓譚.
Cf. Pokora (9).
Hsin Lun 新論.
New Discourses.
Liang, *c.* +530.
Liu Hsieh 劉勰.
Hsin Thang Shu 新唐書.
New History of the Thang Dynasty
[+618 to +906].
Sung, +1061.
Ouyang Hsiu 歐陽修 & Sung Chhi
宋祁.

Cf. des Rotours (2), p. 56.
Partial trs. des Rotours (1, 2); Pfizmaier (66–
74). For translations of passages see the
index of Frankel (1).
Yin-Tê Index, no. 16.
Hsin Tsuan Hsiang Phu 新纂香譜.
See *Hsiang Phu* by Chhen Ching.
Hsin Wu Tai Shih 新五代史.
New History of the Five Dynasties [+907
to +959].
Sung, *c.* +1070.
Ouyang Hsiu 歐陽修.
For translations of passages see the index of
Frankel (1).
Hsin Yü 新語.
New Discourses.
C/Han, *c.* −196.
Lu Chia 陸賈.
Tr. v. Gabain (1).
Hsing Li Ching I 性理精義.
Essential Ideas of the Hsing-Li (Neo-
Confucian) School of Philosophers [a con-
densation of the *Hsing Li Ta Chhüan*, q.v.].
Chhing, +1715.
Li Kuang-Ti 李光地.
Hsing Li Ta Chhüan (Shu) 性理大全(書).
Collected Works of (120) Philosophers of
the Hsing-Li (Neo-Confucian) School
[*Hsing* = human nature; *Li* =
principle of organisation in all Nature].
Ming, +1415.
Ed. Hu Kuang *et al.* 胡廣.
Hsing Ming Kuei Chih 性命圭旨.
A Pointer to the Meaning of (Human)
Nature and the Life-Span [physiological
alchemy; the *kuei* is a pun on the two
kinds of *thu*, central earth where the
enchymoma is formed].
Ascr. Sung, pr. Ming and Chhing, +1615,
repr. +1670.
Attrib. Yin Chen Jen 尹眞人.
Written out by Kao Ti 高第.
Prefs. by Yü Yung-Ning *et al.* 余永寧.
Hsing Shih Hêng Yen 醒世恆言.
Stories to Awaken Men.
Ming, *c.* +1640.
Fêng Mêng-Lung 馮夢龍.
Hsiu Chen Chih Nan 修眞指南.
South-Pointer for the Regeneration of the
Primary (Vitalities).
See *Hsi Yo Tou hsien-sêng Hsiu Chen Chih
Nan*.
Hsiu Chen Li Yen Chhao Thu 修眞歷驗鈔圖.
[= *Chen Yuan Miao Tao Hsiu Tan Li Yen
Chhao.*]
Transmitted Diagrams illustrating Tried and
Tested (Methods of) Regenerating the
Primary Vitalities [physiological alchemy].
Thang or Sung, before +1019.
No writer named but the version in *YCCC*,
ch. 72, has Tung Chen Tzu (ps.) 洞眞子.
TT/149.

Hsiu Chen Nei Lien Pi Miao Chu Chüeh 修眞
內煉秘妙諸訣.
Collected Instructions on the Esoteric
Mysteries of Regenerating the Primary
(Vitalities) by Internal Transmutation.
Sung or pre-Sung.
Writer unknown.
Perhaps identical with *Hsiu Chen Pi
Chüeh* (q.v.); now extant only in
quotations.
Hsiu Chen Pi Chüeh 脩眞秘訣.
Esoteric Instructions on the Regeneration of
the Primary (Vitalities).
Sung or pre-Sung, before +1136.
Writer uncertain.
In *Lei Shuo*, ch. 49, pp. 5*a* ff.
Hsiu Chen Pien Nan (Tshan Chêng) 修眞辯難
參證.
[*Chhi Yün Shan Wu Yuan Yzu Hsiu Chen
Pien Nan Tshan Chêng.*]
A Discussion of the Difficulties encountered
in the Regeneration of the Primary
(Vitalities) [physiological alchemy]; with
Supporting Evidence.
Chhing, +1798.
Liu I-Ming 劉一明 (Wu Yuan Tzu
悟元子).
Comm., Min I-Tê 閔一得 (*c.* 1830).
In *Tao Tsang Hsü Pien (Chhu chi)*, 23.
Hsiu Chen Shih Shu 修眞十書.
A Collection of Ten Tractates and Treat-
ises on the Regeneration of the Primary
(Vitalities) [in fact, many more than
ten].
Sung, *c.* +1250.
Editor unknown.
TT/260.
Cf. Maspero (7), pp. 239, 357.
Hsiu Chen Thai Chi Hun Yuan Thu 修眞太
極混元圖.
Illustrated Treatise on the (Analogy of the)
Regeneration of the Primary (Vitalities)
(with the Cosmogony of) the Supreme
Pole and Primitive Chaos.
Sung, *c.* +1100.
Hsiao Tao-Tshun 蕭道存.
TT/146.
Hsiu Chen Thai Chi Hun Yuan Chih Hsüan Thu
修眞太極混元指玄圖.
Illustrated Treatise Expounding the Mystery
of the (Analogy of the) Regeneration of
the Primary (Vitalities) (with the Cos-
mogony of) the Supreme Pole and
Primitive Chaos.
Thang, *c.* +830.
Chin Chhüan Tzu 金全子.
TT/147.
Hsiu Chen Yen I 修眞演義.
A Popular Exposition of (the Methods of)
Regenerating the Primary (Vitalities)
[Taoist sexual techniques].
Ming, *c.* +1560.

Têng Hsi-Hsien 鄧希賢 (Tzu Chin
Kuang Yao Ta Hsien 紫金光耀大仙 .
See van Gulik (3, 8).
Hsiu Hsien Pien Huo Lun 修仙辨惑論.
Resolution of Doubts concerning the
Restoration to Immortality.
Sung, *c.* +1220.
Ko Chhang-Kêng 葛長庚
(Pai Yü-Chhan 白玉蟾).
In *TSCC, Shen i tien*, ch. 300, *i wên*, pp.
11 *a* ff.
Hsiu Lien Ta Tan Yao Chih 修鍊大丹要旨.
Essential Instructions for the Preparation of
the Great Elixir [with illustrations of
alchemical apparatus].
Probably Sung or later.
Writer unknown.
TT/905.
Hsiu Tan Miao Yung Chih Li Lun 修丹妙用
至理論.
A Discussion of the Marvellous Functions
and Perfect Principles of the Practice of
the Enchymoma.
Late Sung or later.
Writer unknown.
TT/231.
Refers to the Sung adept Hai-Chhan hsien-
sêng 海蟾先生 (Liu Tshao 劉操).
Hsü Chen-Chün Pa-shih-wu Hua Lu 許眞君
八十五化錄.
Record of the Transfiguration of the Adept
Hsü (Hsün) at the Age of Eighty-five.
Chin, +4th cent.
Shih Tshên 施岑.
TT/445.
Hsü Chen-Chün Shih Han Chi 許眞君石函記.
The Adept Hsü (Sun's) Treatise, found in a
Stone Coffer.
Ascr. Chin, +4th cent., perhaps *c.* +370.
Attrib. Hsü Hsün 許遜.
TT/944.
Cf. Davis & Chao Yün-Tshung (6).
Hsü Hsien Chuan 續仙傳.
Further Biographies of the Immortals.
Wu Tai (H/Chou), between +923 and
+936.
Shen Fên 沈汾.
In *YCCC*, ch. 113.
Hsü Ku Chai Chi Suan Fa 續古摘奇算法.
Choice Mathematical Remains Collected to
Preserve the Achievements of Old [magic
squares and other computational examples].
Sung, +1275.
Yang Hui 楊輝.
(In *Yang Hui Suan Fa*.)
Hsü Kuang-Chhi Shou Chi 徐光啓手跡.
Manuscript Remains of Hsü Kuang-Chhi
[facsimile reproductions].
Shanghai, 1962.
Hsü Ming Fang 續命方.
Precepts for Lengthening the Life-span.
Alternative title of *Hui Ming Ching* (q.v.).

Hsü Po Wu Chih 續博物志.
　　Supplement to the *Record of the Investiga-tion of Things* (cf. *Po Wu Chih*).
　　Sung, mid +12th century.
　　Li Shih 李石.
Hsü Shen Hsien Chuan 續神仙傳.
　　Supplementary Lives of the Hsien (cf. *Shen Hsien Chuan*).
　　Thang.
　　Shen Fên 沈汾.
Hsü Shih Shih 續事始.
　　Supplement to the *Beginnings of All Affairs* (cf. *Shih Shih*).
　　H/Shu, *c.* +960.
　　Ma Chien 馬鑑.
Hsü Yen-Chou Shih Hua 許彥周詩話.
　　Hsü Yen-Chou's Talks on Poetry.
　　Sung, early +12th, prob. *c.* +1111.
　　Hsü Yen-Chou 許彥周.
Hsüan Chieh Lu 懸解錄.
　　See *Hsüan Chieh Lu* 玄解錄.
Hsüan Chieh Lu 玄解錄.
　　The Mysterious Antidotarium [warnings against elixir poisoning, and remedies for it].
　　Thang, anonymous preface of +855, prob. first pr. between +847 and +850.
　　Writer unknown, perhaps Hokan Chi 紇干臮.
　　The first printed book in any civilisation on a scientific subject.
　　TT/921, and in *YCCC*, ch. 64, pp. 5*a* ff.
Hsüan Fêng Chhing Hui Lu 玄風慶會錄.
　　Record of the Auspicious Meeting of the Mysterious Winds [answers given by Chhiu Chhu-Chi (Chhang-Chhun Chen Jen) to Chingiz Khan at their interviews at Samarqand in +1222].
　　Sung, +1225.
　　Chhiu Chhu-Chi 邱處機.
　　TT/173.
Hsüan-Ho Po Ku Thu Lu 宣和博古圖錄.
　　[= *Po Ku Thu Lu*.]
　　Hsüan-Ho reign-period Illustrated Record of Ancient Objects [catalogue of the archaeological museum of the emperor Hui Tsung].
　　Sung, +1111 to +1125.
　　Wang Fu 王黼 or 戫 *et al.*
Hsüan Kuai Hsü Lu 玄怪續錄.
　　The *Record of Things Dark and Strange*, continued.
　　Thang.
　　Li Fu-Yen 李復言.
Hsüan Mên Mo Chüeh Nei Chao Thu 玄門脈訣內照圖.
　　[= *Hua Tho Nei Chao Thu.*]
　　Illustrations of Visceral Anatomy, for the Taoist *Sphygmological Instructions*.
　　Sung, +1095, repr. +1273 by Sun Huan 孫煥 with the inclusion of Yang Chieh's illustrations.

Attrib. Hua Tho 華佗.
　　First pub. Shen Chu 沈銖.
　　Cf. Ma Chi-Hsing (2).
Hsüan Ming Fên Chuan 玄明粉傳.
　　On the 'Mysterious Bright Powder' (puri-fied sodium sulphate, Glauber's salt).
　　Thang, *c.* +730.
　　Liu Hsüan-Chen 劉玄眞.
Hsüan Nü Ching 玄女經.
　　Canon of the Mysterious Girl [or, the Dark Girl].
　　Han.
　　Writer unknown.
　　Only as fragment in *Shuang Mei Ching An Tshung Shu*, now conflated with *Su Nü Ching*, q.v.
　　Partial trs., van Gulik (3, 8).
Hsüan Phin Lu 玄品錄.
　　Record of the (Different) Grades of Im-mortals.
　　Yuan.
　　Chang Thien-Yü 張天雨.
　　TT/773.
　　Cf. Chhen Kuo-Fu (*1*), 1st ed., p. 260.
Hsüan Shih Chih 宣室志.
　　Records of Hsüan Shih.
　　Thang, *c.* +860.
　　Chang Tu 張讀.
Hsüan Shuang Chang Shang Lu 玄霜掌上錄.
　　Mysterious Frost on the Palm of the Hand; or, Handy Record of the Mysterious Frost [preparation of lead acetate].
　　Date unknown.
　　Writer unknown.
　　TT/938.

I Chen Thang Ching Yen Fang 頤眞堂經驗方.
　　Tried and Tested Prescriptions of the True-Centenarian Hall (a surgery or pharmacy).
　　Ming, prob. +15th, *c.* +1450.
　　Yang shih 楊氏.
I Chi Khao 醫籍考.
　　Comprehensive Annotated Bibliography of Chinese Medical Literature.
　　See Taki Mototane (*1*).
I Chai Ta Fa 醫家大法.
　　See *I Yin Thang I Chung Ching Kuang Wei Ta Fa.*
I Chien Chih 夷堅志.
　　Strange Stories fom I-Chien.
　　Sung, *c.* +1185.
　　Hung Mai 洪邁.
I Chin Ching 易筋經.
　　Manual of Exercising the Muscles and Tendons [Buddhist].
　　Ascr. N/Wei.
　　Chhing, perhaps +17th.
　　Attrib. Ta-Mo (Bodhidharma) 達摩
　　Author unknown.
　　Reproduced in Wang Tsu-Yuan (*1*).

I Ching 易經.
The Classic of Changes [Book of Changes].
Chou with C/Han additions.
Compilers unknown.
See Li Ching-Chih (*1, 2*); Wu Shih-Chhang (1).
Tr. R. Wilhelm (2); Legge (9); de Harlez (1).
Yin-Tê Index, no. (suppl.) 10.

I Hsin Fang (Ishinhō) 醫心方.
The Heart of Medicine [partly a collection of ancient Chinese and Japanese books].
Japan, +982 (not printed till 1854).
Tamba no Yasuyori 丹波康頼.

I Hsüeh Ju Mên 醫學入門.
Janua Medicinae [a general system of medicine].
Ming, +1575.
Li Chhan 李梴.

I Hsüeh Yuan Liu Lun 醫學源流論.
On the Origins and Progress of Medical Science.
Chhing, +1757.
Hsü Ta-Chhun 徐大椿.
(In *Hsü Ling-Thai I Shu Chhüan Chi*.)

Mên Pi Chih 醫門秘旨.
Confidential Guide to Medicine.
Ming, +1578.
Chang Ssu-Wei 張四維.

I Shan Tsa Tsuan 義山雜纂.
Collected Miscellany of (Li) I-Shan [Li Shang-Yin, epigrams].
Thang, *c.* +850.
Li Shang-Yin 李商隱.
Tr. Bonmarchand (1).

I Shih 逸史.
Leisurely Histories.
Thang.
Lu Shih 盧氏.

I Su Chi 夷俗記.
Records of Barbarian Customs.
Alternative title of *Pei Lu Fêng Su*, q.v.

I Thu Ming Pien 易圖明辨.
Clarification of the Diagrams in the (*Book of*) *Changes* [historical analysis].
Chhing, +1706.
Hu Wei 胡渭.

I Wei Chhien Tso Tu 易緯乾鑿度.
Apocryphal Treatise on the (*Book of*) *Changes*; a Penetration of the Regularities of Chhien (the first *kua*).
C/Han; −1st or +1st century.
Writer unknown.

I Wei Ho Thu Shu 易緯河圖數.
Apocryphal Treatise on the (*Book of*) *Changes*; the Numbers of the Ho Thu (Diagram).
H/Han.
Writer unknown.

I Yin Thang I Chung Ching Kuang Wei Ta Fa 伊尹湯液仲景廣為大法.
[= *I Chia Ta Fa* or *Kuang Wei Ta Fa*.]
The Great Tradition (of Internal Medicine) going back to I Yin (legendary minister) and his Pharmacal Potions, and to (Chang) Chung-Ching (famous Han physician).
Yuan, +1294.
Wang Hao-Ku 王好古.
ICK, p. 863.

Ishinhō
See *I Hsin Fang*.

Jih Chih Lu 日知錄.
Daily Additions to Knowledge.
Chhing, +1673.
Ku Yen-Wu 顧炎武.

Jih Hua Chu Chia Pên Tshao 日華諸家本草.
The Sun-Rays Master's Pharmaceutical Natural History, collected from Many Authorities.
Wu Tai and Sung, *c.* +972.
Often ascribed by later writers to the Thang, but the correct dating was recognised by Thao Tsung-I in his *Cho Kêng Lu* (+1366) ch. 24, p. 17*b*.
Ta Ming 大明.
(Jih Hua Tzu 日華子 the Sun-Rays Master.)
(Perhaps Thien Ta-Ming 田大明).

Jih Yüeh Hsüan Shu Lun 日月玄樞論.
Discourse on the Mysterious Axis of the Sun and Moon [i.e. Yang and Yin in natural phenomena; the earliest interpretation (or recognition) of the *Chou I Tshan Thung Chhi* (q.v.) as a physiological rather than (or, as well as) a proto-chemical text].
Thang, *c.* +740.
Liu Chih-Ku 劉知古.
Now extant only as quotations in the *Tao Shu* (q.v.), though at one time contained in the *Tao Tsang* separately.

Ju Yao Ching 入藥鏡.
Mirror of the All-Penetrating Medicine (the enchymoma), [rhyming verses].
Wu Tai, *c.* +940.
Tshui Hsi-Fan 崔希範.
TT/132, and in *TTCY* (*hsü chi*, 5).
With commentaries by Wang Tao-Yuan 王道淵 (Yuan); Li Phan-Lung 李攀龍 (Ming) & Phêng Hao-Ku 彭好古 (Ming).
Also in *Hsiu Chen Shih Shu* (*TT*/260), ch. 13, pp. 1*a* ff. with commentary by Hsiao Thing-Chih 蕭廷芝 (Ming).
Also in *Tao Hai Chin Liang*, pp. 35*a* ff., with comm. by Fu Chin-Chhüan 傅金銓 (Chhing).
See also *Thien Yuan Ju Yao Ching*.
Cf. van Gulik (8), pp. 224 ff.

Kan Chhi Shih-liu Chuan Chin Tan 感氣十六轉金丹.
The Sixteen-fold Cyclically Transformed Gold Elixir prepared by the 'Responding

Kan Chhi Shih-liu Chuan Chin Tan (*cont.*)
    to the Chhi' Method [with illustrations of
    alchemical apparatus].
Sung.
Writer unknown.
TT/904.

Kan Ying Ching　感應經.
    On Stimulus and Response (the Resonance
    of Phenomena in Nature).
Thang, *c.* +640.
Li-Shun-Fêng　李淳風.
See Ho & Needham (2).

Kan Ying Lei Tshung Chih　感應類從志.
    Record of the Mutual Resonances of
    Things according to their Categories.
Chin, *c.* +295.
Chang Hua　張華.
See Ho & Needham (2).

Kao Shih Chuan　高士傳.
    Lives of Men of Lofty Attainments.
Chin, *c.* +275.
Huangfu Mi　皇甫謐.

Kêng Hsin Yü Tshê　庚辛玉冊.
    Precious Secrets of the Realm of Kêng and
    Hsin (i.e. all things connected with
    metals and minerals, symbolised by these
    two cyclical characters) [on alchemy and
    pharmaceutics. Kêng-Hsin is also an
    alchemical synonym for gold].
Ming, +1421.
Chu Chhüan　朱權, (Ning Hsien Wang
    寧獻王, prince of the Ming).
Extant only in quotations.

Kêng Tao Chi　庚道集.
    Collection of Procedures of the Golden Art
    (Alchemy).
Sung or Yuan, date unknown but after +1144
Writers unknown.
Compiler, Mêng Hsien chü shih　蒙軒居士.
TT/946.

Khai-Pao Hsin Hsiang-Ting Pên Tshao　開寶新
    詳定本草.
    New and More Detailed Pharmacopoeia of
    the Khai-Pao reign-period.
Sung, +973.
Liu Han　劉翰, Ma Chih　馬志, and
    7 other naturalists, under the direction of
    Lu To-Hsün　盧多遜.

Khai-Pao Pên Tshao　開寶本草.
    See *Khai-Pao Hsin Hsiang-Ting Pên Tshao*.

Khun Yü Ko Chih　坤輿格致.
    Investigation of the Earth [Western min-
    ing methods based on Agricola's *De Re
    Metallica*].
Ming, +1639 to 1640, perhaps never printed.
Têng Yü-Han (Johann Schreck)　鄧玉函
    & (or) Thang Jo-Wang　湯若望 (John
    Adam Schall von Bell).

Khung Chi Ko Chih　空際格致.
    A Treatise on the Material Composition of
    the Universe [the Aristotelian Four
    Elements, etc.].

Ming, +1633.
Kao I-Chih (Alfonso Vagnoni)　高一志.
Bernard-Maître (18), no. 227.

Khung shih Tsa Shuo　孔氏雜說.
    Mr Khung's Miscellany.
Sung, *c.* +1082.
Khung Phing-Chung　孔平仲.

Ko Chih Ching Yuan　格致鏡原.
    Mirror of Scientific and Technological
    Origins.
Chhing, +1735.
Chhen Yuan-Lung　陳元龍.

Ko Chih Tshao　格致草.
    Scientific Sketches [astronomy and cos-
    mology; part of *Han Yü Thung*, q.v.].
Ming, +1620, pr. +1648.
Hsiung Ming-Yü　熊明遇.

Ko Hsien Ong Chou Hou Pei Chi Fang　葛仙翁
    肘後備急方.
    The Elder-Immortal Ko (Hung's) Hand-
    book of Medicines for Emergencies.
Alt. title of *Chou Hou Pei Chi Fang* (q.v.).
TT/1287.

Ko Hung Chen Chung Shu　葛洪枕中書.
    Alt. title of *Chen Chung Chi* (q.v.).

Ko Ku Yao Lun　格古要論.
    Handbook of Archaeology, Art and Anti-
    quarianisn.
Ming, +1387, enlarged and reissued +1459.
Tshao Chao　曹昭.

Ko Wu Tshu Than　格物麤談.
    Simple Discourses on the Investigation of
    Things.
Sung, *c.* +980.
Attrib. wrongly to Su Tung-Pho　蘇東坡.
Actual writer (Lu) Tsan-Ning　(錄)贊寧
    (Tung-Pho hsien-sêng). With later addi-
    tions, some concerning Su Tung-Pho.

Konjaku Monogatari　今昔物語.
    Tales of Today and Long Ago (in three
    collections: Indian, 187 stories and tradi-
    tions, Chinese, 180, and Japanese, 736).
Japan (Heian), +1107.
Compilers unknown.
Cf. Anon. (103), pp. 97 ff.

Konjaku Monogatarishū　今昔物語集.
    See *Konjaku Monogatari*.

Ku Chin I Thung (*Ta Chhüan*)　古今醫統(大全).
    Complete System of Medical Practice, New
    and Old.
Ming, +1556.
Hsü Chhun-Fu　徐春甫.

Ku Thung Thu Lu　鼓銅圖錄.
    Illustrated Account of the (Mining), Smelt-
    ing and Refining of Copper (and other Non-
    Ferrous Metals).
See Masuda Tsuna (1).

Ku Wei Shu　古微書.
    Old Mysterious Books [a collection of the
    apocryphal Chhan-Wei treatises].
Date uncertain, in part C/Han.
Ed. Sun Chio　孫彀 (Ming).

Ku Wên Lung Hu Ching Chu Su 古文龍虎經
註疏 and *Ku Wên Lung Hu Shang Ching
Chu* 古文龍虎上經註.
  See *Lung Hu Shang Ching Chu.*
Ku Wên Tshan Thung Chhi Chi Chieh 古文參
同契集解.
  See *Ku Wên Chou I Tshan Thung Chhi Chu.*
Ku Wên Tshan Thung Chhi Chien Chu Chi Chieh
古文參同契箋註集解.
  See *Ku Wên Chou I Tshan Thung Chhi Chu.*
Ku Wên Chou I Tshan Thung Chhi Chu 古文
周易參同契註.
  Commentary on the Ancient Script Version
    of the *Kinship of the Three.*
  Chhing, +1732.
  Ed. and comm. Yuan Jen-Lin 袁仁林.
  See Vol. 5, pt. 3.
*Ku Wên Tshan Thung Chhi San Hsiang Lei Chi
Chieh* 古文參同契三相類集解.
  See *Ku Wên Chou I Tshan Thung Chhi Chu.*
Kuan Khuei Pien 管窺編.
  An Optick Glass (for the Enchymoma).
  See Min I-Tê (1).
Kuan Yin Tzu 關尹子.
  [= *Wên Shih Chen Ching.*]
  The Book of Master Kuan Yin.
  Thang, +742 (may be Later Thang or Wu
    Tai). A work with this title existed in the
    Han, but the text is lost.
  Prob. Thien Thung-Hsiu 田同秀.
Kuang Chhêng Chi 廣成集.
  The Kuang-chhêng Collection [Taoist writ-
    ings of every kind; a florilegium].
  Thang, late +9th; or early Wu Tai, before
    +933.
  Tu Kuang-Thing 杜光庭.
  TT/611.
Kuang Wei Ta Fa 廣爲大法.
  See *I Yin Thang I Chung Ching Kuang Wei
    Ta Fa.*
Kuang Ya 廣雅.
  Enlargement of the *Erh Ya*; *Literary
    Expositor* [dictionary].
  San Kuo (Wei) +230.
  Chang I 張揖.
Kuang Yün 廣韻.
  Enlargement of the *Chhieh Yün*; *Dictionary
    of the Sounds of Characters.*
  Sung.
  (A completion by later Thang and Sung
    scholars, given its present name in +1011.)
  Lu Fa-Yen *et al.* 陸法言.
Kuei Chung Chih Nan 規中指南.
  A Compass for the Internal Compasses; or,
    Orientations concerning the Rules and
    Measures of the Inner (World) [i.e. the
    preparation of the enchymoma in the
    microcosm of man's body].
  Sung or Yuan, +13th and +14th.
  Chhen Chhung-Su 陳沖素 (Hsü Pai
    Tzu 虛白子).
  TT/240, and in *TTCY* (*shang mao chi*, 5).

Kungyang Chuan 公羊傳.
  Master Kungyang's Tradition (or Com-
    mentary) on the *Spring and Autumn
    Annals.*
  Chou (with Chhin and Han additions),
    late −3rd and early −2nd centuries.
  Attrib. Kungyang Kao 公羊高 but more
    probably Kungyang Shou 公羊壽.
  See Wu Khang (1); van der Loon (1).
Kuo Shih Pu 國史補.
  Emendations to the National Histories.
  Thang, *c.* +820.
  Li Chao 李肇.
Kuo Yü 國語.
  Discourses of the (ancient feudal) States.
  Late Chou, Chhin and C/Han, containing
    much material from ancient written
    records.
  Writers unknown.

Lao Hsüeh An Pi Chi 老學庵筆記.
  Notes from the Hall of Learned Old Age.
  Sung, *c.* +1190.
  Lu Yu 陸游.
Lao Tzu Chung Ching 老子中經.
  The Median Canon of Lao Tzu [on
    physiological micro-cosmography].
  Writer unknown.
  Pre-Thang.
  In *YCCC*, ch. 18.
Lao Tzu Shuo Wu Chhu Ching 老子說五廚經.
  Canon of the Five Kitchens [the five
    viscera] Revealed by Lao Tzu [respiratory
    techniques].
  Thang or pre-Thang.
  Writer unknown.
  In *YCCC*, ch. 61, pp. 5b ff.
Lei Chen Chin Tan 雷震金丹.
  Lei Chen's Book of the Metallous Encyh-
    moma.
  Ming, after +1420.
  Lei Chen (ps. ?) 雷震.
  In *Wai Chin Tan*, ch. 5 (*CTPS, pên* 10).
Lei Chen Tan Ching 雷震丹經.
  Alternative title of *Lei Chen Chin Tan*
    (q.v.).
Lei Chêng Phu Chi Pên Shih Fang 類證普濟本
事方.
  Classified Fundamental Prescriptions of
    Universal Benefit.
  Sung, +1253.
  Attrib. Hsü Shu-Wei 許叔微 (*fl.* +1132)
Lei Ching Fu I 類經附翼.
  Supplement to the Classics Classified; (the
    Institutes of Medicine).
  Ming, +1624.
  Chang Chieh-Pin 張介賓.
Lei Kung Phao Chih 雷公炮製.
  (Handbook based on the)*Venerable Master
    Lei's (Treatise on) the Preparation (of
    Drugs).*
  L/Sung, *c.* +470.

Lei Kung Phao Chi (cont.)
Lei Hsiao 雷斅.
Ed. Chang Kuang-Tou 張光斗 (Chhing),
1871.

Lei Kung Phao Chih Lun 雷公炮炙論.
The Venerable Master Lei's Treatise on
the Decoction and Preparation (of Drugs).
L/Sung, c. +470.
Lei Hsiao 雷斅.
Preserved only in quotations in *Chêng Lei
Pên Tshao* and elsewhere, and reconsti-
tuted by Chang Chi 張驥.
LPC, p. 116.

Lei Kung Phao Chih Yao Hsing (*Fu*) *Chieh*
雷公炮製藥性(賦)解.
(Essays and) Studies on the *Venerable
Master Lei's* (*Treatise on*) *the Natures of
Drugs and their Preparation.*
First four chapters J/Chin, c. +1220.
Li Kao 李杲.
Last six chapters Chhing, c. 1650.
Li Chung-Tzu 李中梓.
(Contains many quotations from earlier
Lei Kung books, +5th century onwards.)

Lei Kung Yao Tui 雷公藥對.
Answers of the Venerable Master Lei (to
Questions) concerning Drugs.
Perhaps L/Sung, at any rate before N/Chhi.
Attrib. Lei Hsiao 雷斅.
Later attrib. a legendary minister of Huang
Ti.
Comm. by Hsü Chih-Tshai 徐之才,
N/Chhi +565.
Now extant only in quotations.

Lei Shuo 類說.
A Classified Commonplace-Book [a great
florilegium of excerpts from Sung and
pre-Sung books, many of which are
otherwise lost].
Sung, +1136.
Ed. Tsêng Tshao 曾慥.

Li Chi 禮記.
[= *Hsiao Tai Li Chi.*]
Record of Rites [compiled by Tai the
Younger].
(Cf. *Ta Tai Li Chi.*)
Ascr. C/Han, c. −70/−50, but really
H/Han, between +80 and +105, though
the earliest pieces included may date from
the time of the *Analects* (c. −465 to −450).
Attrib. ed. Tai Shêng 戴聖.
Actual ed. Tshao Pao 曹褒.
Trs. Legge (7); Couvreur (3); R. Wilhelm
(6).
Yin-Tê Index, no. 27.

Li Hai Chi 蠡海集.
The Beetle and the Sea [title taken from the
proverb that the beetle's eye view cannot encompass the wide sea—a biological
book].
Ming, late +14th century.
Wang Khuei 王逵.

Li Sao 離騷.
Elegy on Encountering Sorrow [ode].
Chou (Chhu), c. −295, perhaps just before
−300. Some scholars place it as late as
−269.
Chhü Yuan 屈原.
Tr. Hawkes (1).

Li Shih Chen Hsien Thi Tao Thung Chien 歷世
真仙體道通鑑.
Comprehensive Mirror of the Embodiment
of the Tao by Adepts and Immortals
throughout History.
Prob. Yuan.
Chao Tao-I 趙道一.
TT/293.

Li Tai Ming I Mêng Chhiu 歷代名醫蒙求.
Brief Lives of the Famous Physicians in
All Ages.
Sung, +1040.
Chou Shou-Chung 周守忠.

(*Li Tai*) *Shen Hsien* (*Thung*) *Chien* (歷代)神仙
(通)鑑.
(Cf. *Shen Hsien Thung Chien.*)
General Survey of the Lives of the Holy
Immortals (in all Ages).
Chhing, +1712.
Hsü Tao 徐道 (assisted by Li Li 李理) &
Chhêng Yü-Chhi 程毓奇 (assisted by
Wang Thai-Su 王太素).

Li Wei Tou Wei I 禮緯斗威儀.
Apocryphal Treatise on the *Record of Rites*;
System of the Majesty of the Ladle [the
Great Bear].
C/Han, −1st or later.
Writer unknown.

Li Wên-Jao Chi 李文饒集.
Collected Literary Works of Li Tê-Yü
(Wên-Jao), (+787 to +849).
Thang, c. +855.
Li Tê-Yü 李德裕.

Liang Chhiu Tzu (*Nei* or *Wai*) 梁丘子.
See *Huang Thing Nei Ching* (*Yü*) *Ching Chu*
and *Huang Thing Wai Ching* (*Yü*) *Ching
Chu.*

Liang Ssu Kung Chi 梁四公記.
Tales of the Four Lords of Liang.
Thang, c. +695.
Chang Yüeh 張說.

Liao Yang Tien Wên Ta Pien 寥陽殿問答編.
[= *Yin Chen Jen Liao Yang Tien Wên Ta Pien.*]
Questions and Answers in the (Eastern
Cloister of the) Liao-yang Hall (of the
White Clouds Temple at Chhing-
chhêng Shan in Szechuan) [on physio-
logical alchemy, *nei tan*].
Ming or Chhing.
Attrib. Yin Chen Jen 尹眞人 (Phêng-
Thou 蓬頭).
Ed. Min I-Tê 閔一得, c. 1830.
In *Tao Tsang Hsü Pien* (*Chhu chi*), 3, from
a MS. preserved at the Blue Goat Temple
青羊宮 (Chhêngtu).

Lieh Hsien Chhüan Chuan 列仙全傳.
Complete Collection of the Biographies of the Immortals.
Ming, *c.* +1580.
Wang Shih-Chên 王世貞.
Collated and corrected by Wang Yün-Phêng 汪雲鵬.

Lieh Hsien Chuan 列仙傳.
Lives of Famous Immortals (cf. *Shen Hsien Chuan*).
Chin, +3rd or +4th century, though certain parts date from about −35 and shortly after +167.
Attrib. Liu Hsiang 劉向.
Tr. Kaltenmark (2).

Lin Chiang Hsien 臨江仙.
The Immortal of Lin-chiang.
Sung, +1151.
Tsêng Tshao 曾慥.
In *Hsiu Chen Shih Shu* (*TT*/260), ch. 23, pp. 1 a ff.

Ling-Pao Chiu Yu Chhang Yeh Chhi Shih Tu Wang Hsüan Chang 靈寶九幽長夜起尸度亡玄章.
Mysterious Cantrap for the Resurrection of the Body and Salvation from Nothingness during the Long Night in the Nine Underworlds; a Ling-Pao Scripture.
Date uncertain.
Writer unknown.
TT/605.

Ling-Pao Chung Chen Tan Chüeh 靈寶衆眞丹訣.
Supplementary Elixir Instructions of the Company of the Realised Immortals, a Ling-Pao Scripture.
Sung, after +1101.
Writer unknown.
TT/416.
On the term Ling-Pao see Kaltenmark (4).

Ling-Pao Wu Fu (Hsü) 靈寶五符(序).
See *Thai-Shang Ling-Pao Wu Fu* (*Ching*).

Ling-Pao Wu Liang Tu Jen Shang Phin Miao Ching 靈寶無量度人上品妙經.
Wonderful Immeasurable Highly Exalted Manual of Salvation; a Ling-Pao Scripture.
Liu Chhao, perhaps late +5th, probably finalised in Thang, +7th.
Writers unknown.
TT/1.

Ling Pi Tan Yao Chien 靈祕丹藥牋.
On Numinous and Secret Elixirs and Medicines [the seventh part (chs. 16–18) of *Tsun Shêng Pa Chien*, q.v.].
Ming, +1591.
Kao Lien 高濂.

Ling Piao Lu I 嶺表錄異.
Strange Things Noted in the South.
Thang, *c.* +890.
Liu Hsün 劉恂.

Ling Sha Ta Tan Pi Chüeh 靈砂大丹祕訣.
Secret Doctrine of the Numinous Cinnabar and the Great Elixir.
Sung, after +1101, when the text was received by Chang Shih-Chung 張侍中.
Writer unknown, but edited by a Chhan abbot Kuei-Yen Chhan-shih 鬼眼禪師.
TT/890.

Ling Shu Ching
See *Huang Ti Nei Ching, Ling Shu*.

Ling Wai Tai Ta 嶺外代答.
Information on What is Beyond the Passes (lit. a book in lieu of individual replies to questions from friends).
Sung, +1178.
Chou Chhü-Fei 周去非.

Liu Shu Ching Yün 六書精蘊.
Collected Essentials of the Six Scripts.
Ming, *c.* +1530.
Wei Hsiao 魏校.

Liu Tzu Hsin Lun 劉子新論.
See *Hsin Lun*.

Lo-Fou Shan Chih 羅浮山志.
History and Topography of the Lo-fou Mountains (north of Canton).
Chhing, +1716 (but based on older histories).
Thao Ching-I 陶敬益.

Lu Hsing Ching 顱顖經.
A Tractate on the Fontanelles of the Skull [anatomical-medical].
Late Thang or early Sung, +9th or +10th.
Writer unknown.

Lu Huo Chien Chieh Lu 爐火監戒錄.
Warnings against Inadvisable Practices in the Work of the Stove [alchemical].
Sung, *c.* +1285.
Yü Yen 兪琰.

Lu Huo Pên Tshao 爐火本草.
Spagyrical Natural History.
Possible alternative title of *Wai Tan Pên Tshao* (q.v.).

Lü Tsu Chhin Yuan Chhun 呂祖沁園春.
The (Taoist) Patriarch Lü (Yen's) 'Spring in the Prince's Gardens' [a brief epigrammatic text on physiological alchemy]
Thang, +8th (if genuine).
Attrib. Lü Yen 呂嵒.
TT/133.
Comm. by Fu Chin-Chhüan 傅金銓 (*c.* 1822).
In *Tao Hai Chin Liang*, p. 45 a, and appended to *Shih Chin Shih* (*Wu Chen Ssu Chu Phien* ed.).

Lü Tsu Chhuan Shou Tsung Chih 呂祖傳授宗旨.
Principles (of Macrobiotics) Transmitted and Handed Down by the (Taoist) Patriarch Lü (Yen, Tung-Pin).
Orig. title of *Chin Hua Tsung Chih* (q.v.).

Lü Tsu Shih Hsien-Thien Hsü Wu Thai-I Chin Hua Tsung Chih 呂祖師先天虛無太一金華宗旨.
Principles of the (Inner) Radiance of the Metallous (Enchymoma) (explained in terms of the) Undifferentiated Universe, and of all the All-Embracing Potentiality of the Endowment of Primary Vitality, taught by the (Taoist) Patriarch Lü (Yen, Tung-Pin).
Alternative name for *Chin Hua Tsung Chih* (q.v.), but with considerable textual divergences, especially in ch. 1.
Ming and Chhing.
Writers unknown.
Attrib. Lü Yen 呂喦 (Lü Tung-Pin) and his school, late +8th.
Ed. and comm. Chiang Yuan-Thing 蔣元庭 and Min I-Tê 閔一得, c. 1830.
In *TTCY* and in *Tao Tsang Hsü Pien* (*Chhu chi*), 1.

Lü Tsu Shih San Ni I Shih Shuo Shu 呂祖師三尼醫世說述.
A Record of the Lecture by the (Taoist) Patriarch Lü (Yen, Tung-Pin) on the Healing of Humanity by the Three Ni Doctrines (Taoism, Confucianism and Buddhism) [physiological alchemy in mutationist terms].
Chhing, +1664.
Attrib. Lü Yen 呂喦 (+8th cent.).
Pref. by Thao Thai-Ting 陶太定.
Followed by an appendix by Min I-Tê 閔一得.
In *Tao Tsang Hsü Pien* (*Chhu chi*), 10, 11.

Lun Hêng 論衡.
Discourses Weighed in the Balance.
H/Han, +82 or +83.
Wang Chhung 王充.
Tr. Forke (4); cf. Leslie (3).
Chung-Fa Index, no. 1.

Lung Hu Chhien Hung Shuo 龍虎鉛汞說.
A Discourse on the Dragon and Tiger, (Physiological) Lead and Mercury, (addressed to his younger brother Su Tzu-Yu).
Sung, c. +1100.
Su Tung-Pho 蘇東坡.
In *TSCC*, *Shen i tien*, ch. 300, *i wên*, pp. 6 b ff.

Lung Hu Huan Tan Chüeh 龍虎還丹訣.
Explanation of the Dragon-and-Tiger Cyclically Transformed Elixir.
Wu Tai, Sung, or later.
Chin Ling Tzu 金陵子.
TT/902.

Lung Hu Huan Tan Chüeh Sung 龍虎還丹訣頌.
A Eulogy of the Instructions for (preparing) the Regenerative Enchymoma of the Dragon and the Tiger (Yang and Yin), [physiological alchemy].

Sung, c. +985.
Lin Ta-Ku 林大古 (Ku Shen Tzu 谷神子).
TT/1068.

Lung Hu Shang Ching Chu 龍虎上經註.
Commentary on the *Exalted Dragon-and-Tiger Manual*.
Sung.
Wang Tao 王道.
TT/988, 989.
Cf. Davis & Chao Yün-Tshung (6).

Lung Hu Ta Tan Shih 龍虎大丹詩.
Song of the Great Dragon-and-Tiger Enchymoma.
See *Chih Chen Tzu Lung Hu Ta Tan Shih*.

Lung-Shu Phu-Sa Chuan 龍樹菩薩傳.
Biography of the Bodhisattva Nāgārjuna (+2nd-century Buddhist patriarch).
Prob. Sui or Thang.
Writer unknown.
TW/2047.

Man-Anpō 萬安方.
A Myriad Healing Prescriptions.
Japan, +1315.
Kajiwara Shozen 梶原性全.

Manyōshū 萬葉集.
Anthology of a Myriad Leaves.
Japan (Nara), +759.
Ed. Tachibana no Moroe 橘諸兄.
or Ōtomo no Yakamochi 大伴家持.
Cf. Anon. (103), pp. 14 ff.

Mao Shan Hsien Chê Fu Na Chhi Chüeh 茅山賢者服內氣訣.
Oral Instructions of the Adepts of Mao Shan for Absorbing the Chhi [Taoist breathing exercises for longevity and immortality].
Thang or Sung.
Writer unknown.
In *YCCC*, ch. 58, pp. 3 b ff.
Cf. Maspero (7), p. 205.

Mao Thing Kho Hua 茅亭客話.
Discourses with Guests in the Thatched Pavilion.
Sung, before +1136.
Huang Hsiu-Fu 黃休復.

Mei-Chhi Shih Chu 梅溪詩注.
(Wang) Mei-Chhi's Commentaries on Poetry.
Short title for *Tung-Pho Shih Chi Chu* (q.v.).

Mêng Chhi Pi Than 夢溪筆談.
Dream Pool Essays.
Sung, +1086; last supplement dated +1091.
Shen Kua 沈括.
Ed. Hu Tao-Ching (1); cf. Holzman (1).

Miao Chieh Lu 妙解錄.
See *Yen Mên Kung Miao Chieh Lu*.

Miao Fa Lien Hua Ching 妙法蓮花經.
Sūtra on the Lotus of the Wonderful Law

Miao Fa Lien Hua Ching (cont.)
India.
Tr. Chin, betw. +397 and +400 by Ku-mārajīva (Chiu-Mo-Lo-Shih 鳩摩羅什).
N/134; TW/262.

Ming I Pieh Lu 名醫別錄.
Informal (or Additional) Records of Famous Physicians (on Materia Medica).
Ascr. Liang, *c.* +510.
Attrib. Thao Hung-Ching 陶弘景.
Now extant only in quotations in the pharmaceutical natural histories, and a reconstitution by Huang Yü (*1*).
This work was a disentanglement, made by other hands between +523 and +618 or +656, of the contributions of Li Tang-Chih (*c.* +225) and Wu Phu (*c.* +235) and the commentaries of Thao Hung-Ching (+492) from the text of the *Shen Nung Pên Tshao Ching* itself. In other words it was the non-*Pên-Ching* part of the *Pên Tshao Ching Chi Chu* (q.v.). It may or may not have included some or all of Thao Hung-Ching's commentaries.

Ming Shih 明史.
History of the Ming Dynasty [+1368 to +1643].
Chhing, begun +1646, completed +1736, first pr. +1739.
Chang Thing-Yü 張廷玉 *et al.*

Ming Thang Hsüan Chen Ching Chüeh 明堂女真經訣.
[= *Shang-Chhing Ming Thang Hsüan Chen Ching Chüeh.*]
Explanation of the Manual of (Recovering the) Mysterious Primary (Vitalities of the) Cosmic Temple (i.e. the Human Body) [respiration and heliotherapy].
S/Chhi or Liang, late +5th or early +6th (but much altered).
Attrib. to the Mother Goddess of the West, Hsi Wang Mu 西王母.
Writer unknown.
TT/421.
Cf. Maspero (7), p. 376.

Ming Thang Yuan Chen Ching Chüeh 明堂元真經訣.
See *Ming Thang Hsüan Chen Ching Chüeh.*

Ming Thung Chi 冥通記.
Record of Communication with the Hidden Ones (the Perfected Immortals).
Liang, +516.
Chou Tzu-Liang 周子良.
Ed. Thao Hung-Ching 陶弘景.

Mo Chuang Man Lu 墨莊漫錄.
Recollections from the Estate of Literary Learning.
Sung, *c.* +1131.
Chang Pang-Chi 張邦基.

Mo O Hsiao Lu 墨娥小錄.
A Secretary's Commonplace-Book [popular encyclopaedia].

Yuan or Ming, +14th, pr. +1571.
Compiler unknown.

Mo Tzu (incl. *Mo Ching*) 墨子.
The Book of Master Mo.
Chou, −4th century.
Mo Ti (and disciples) 墨翟.
Tr. Mei Yi-Pao (*1*); Forke (3).
Yin-Tê Index, no. (suppl.) 21.
TT/1162.

Montoku-Jitsuroku 文德實錄.
Veritable Records of the Reign of the Emperor Montoku [from +851 to +858].
Japan (Heian) +879.
Fujiwara Mototsune 藤原基經.

Nan Fan Hsiang Lu 南蕃香錄.
Catalogue of the Incense of the Southern Barbarians.
See *Hsiang Lu.*

Nan Hai Yao Phu 南海藥譜.
A Treatise on the Materia Medica of the South Seas (Indo-China, Malayo-Indonesia, the East Indies, etc.).
Alternative title of *Hai Yao Pên Tshao,* q.v. (according to Li Shih-Chen).

Nan Tshun Cho Kêng Lu 南村輟耕錄.
See *Cho Kêng Lu.*

Nan Yo Ssu Ta Chhan-Shih Li Shih Yuan Wên 南嶽思大禪師立誓願文.
Text of the Vows (of Aranyaka Austerities) taken by the Great Chhan Master (Hui-)Ssu of the Southern Sacred Mountain.
Chhen, *c.* +565.
Hui-Ssu 慧思.
TW/1933, N/1576.

Nei Chin Tan 內金丹.
[= *Nei Tan Pi Chih* or *Thien Hsien Chih Lun Chhang Shêng Tu Shih Nei Lien Chin Tan Fa.*]
The Metallous Enchymoma Within (the Body), [physiological alchemy].
Ming, +1622, part dated +1615.
Perhaps Chhen Ni-Wan 陳泥丸 (Mr Ni-Wan, Chhen), or Wu Chhung-Hsü 伍冲虛.
Contains a system of symbols included in the text.
CTPS, pên 12.

Nei Ching.
See *Huang Ti Nei Ching, Su Wên* and *Huang Ti Nei Ching, Ling Shu.*

Nei Ching Su Wên.
See *Huang Ti Nei Ching, Su Wên.*

Nei Kung Thu Shuo 內功圖說.
See *Wang Tsu-Yuan (1).*

Nei Tan Chüeh Fa 內丹訣法.
See *Huan Tan Nei Hsiang Chin Yo Shih.*

Nei Tan Fu 內丹賦.
[= *Thao Chen Jen Nai Tan Fu.*]
Rhapsodical Ode on the Physiological Enchymoma.

Nei Tan Fu (cont.)
  Sung, +13th.
  Thao Chih 陶植.
  With commentary by an unknown writer.
  TT/256.
  Cf. *Chin Tan Fu*, the text of which is very
    similar.

Nei Tan Pi Chih 內丹秘指.
  Confidential Directions on the Enchymoma.
  Alternative title for *Nei Chin Tan* (q.v.).

Nei Wai Erh Ching Thu 內外二景圖.
  Illustrations of Internal and Superficial
    Anatomy.
  Sung, +1118.
  Chu Hung 朱肱.
  Original text lost, and replaced later;
    drawings taken from Yang Chieh's *Tshun*
    Chen Huan Chung Thu.

Nêng Kai Chai Man Lu 能改齋漫錄.
  Miscellaneous Records of the Ability-to-
    Improve-Oneself Studio.
  Sung, mid +12th century.
  Wu Tshêng 吳曾.

Ni-Wan Li Tsu Shih Nü Tsung Shuang Hsiu Pao
  Fa 泥丸李祖師女宗雙修寶筏.
  See *Nü Tsung Shuang Hsiu Pao Fa*.

Nihon-Koki 日本後記.
  Chronicles of Japan, further continued
    [from +792 to +833].
  Japan (Heian), +840.
  Fujiwara Otsugu 藤原緒嗣.

Nihon-Koku Ganzai-sho Mokuroku 日本國
  見在書目錄.
  Bibliography of Extant Books in Japan.
  Japan (Heian), c. +895.
  Fujiwara no Sukeyo 藤原佐世.
  Cf. Yoshida Mitsukuni (6), p. 196.

Nihon Sankai Meibutsu Zue 日本山海各物
  圖會.
  Illustrations of Japanese Processes and
    Manufactures (lit., of the Famous Pro-
    ducts of Japan).
  Japan (Tokugawa), Osaka, +1754.
  Hirase Tessai 平瀨徹齋.
  Ills. by Hasegawa Mitsunobu 長谷川光
    & Chigusa Shinemon 千種屋新右衛
    門.
  Facsim. repr. with introd. notes, Meicho
    Kankokai, Tokyo, 1969.

Nihon-shoki 日本書記.
  See *Nihongi*.

Nihon Ryo-iki 日本靈異記.
  Record of Strange and Mysterious Things in
    Japan.
  Japan (Heian), +823.
  Writer unknown.

Nihongi 日本記.
  [= *Nihon-shoki*.]
  Chronicles of Japan [from the earliest
    times to +696].
  Japan (Nara), +720.
  Toneri-shinnō (prince), 舍人親王,

Ōno Yasumaro, 大安萬呂,
  Ki no Kiyobito *et al.*
  Tr. Aston (1).
  Cf. Anon. (103), pp. 1 ff.

Nihongi Ryaku 日本記畧.
  Classified Matters from the *Chronicles of*
    Japan.
  Japan.

Nittō-Guhō Junrei Gyōki 入唐求法巡禮行記
  Record of a Pilgrimage to China in Search
    of the (Buddhist) Law.
  Thang, +838 to +847.
  Ennin 圓仁.
  Tr. Reischauer (2).

Nü Kung Chih Nan 女功指南.
  A Direction-Finder for (Inner) Achieve-
    ment by Women (Taoists).
  [Physiological alchemy, *nei tan* gymnastic
    techniques, etc.]
  See *Nü Tsung Shuang Hsiu Pao Fa*.

Nü Tsung Shuang Hsiu Pao Fa 女宗雙修寶筏.
  [= *Ni-Wan Li Tsu Shih Nü Tsung Shuang*
    Hsiu Pao Fa, or *Nü Kung-Chih Nan*.]
  A Precious Raft (of Salvation) for Women
    (Taoists) Practising the Double Re-
    generation (of the primary vitalities, for
    their nature and their life-span, *hsing*
    ming), [physiological alchemy].
  Chhing, c. +1795.
  Ni-Wan shih 泥丸氏, Li Ong (late +16th),
    李翁, Mr Ni-Wan, the Taoist Patriarch
    Li.
  Written down by Thai-Hsü Ong 太虛翁,
    Shen I-Ping 沈一炳, Ta-Shih (Taoist
    abbot), c. 1820.
  In *Tao Tsang Hsü Pien* (*Chhu chi*), 20.
  Cf. *Tao Hai Chin Liang*, p. 34a, *Shih Chin*
    Shih, p. 12a.

Pai hsien-sêng Chin Tan Huo Hou Thu 白先生
  金丹火候圖.
  Master Pai's Illustrated Tractate on the
    'Fire-Times' of the Metallous Enchy-
    moma.
  Sung, c. +1210.
  Pai Yü-Chhan 白玉蟾.
  In *Hsiu Chen Shih Shu* (*TT*/260), ch. 1.

Pao Phu Tzu 抱樸 (or 朴)子.
  Book of the Preservation-of-Solidarity
    Master.
  Chin, early +4th century, probably c. +320.
  Ko Hung 葛洪.
  Partial trs. Feifel (1, 2); Wu & Davis (2)
  Full tr. Ware (5), *Nei Phien* chs. only.
  TT/1171–1173.

Pao Phu Tzu Shen Hsien Chin Shuo Ching
  抱朴子神仙金汋經.
  The Preservation-of-Solidarity Master's
    Manual of the Bubbling Gold (Potion) of
    the Holy Immortals.
  Ascr. Chin c. +320. Perhaps pre-Thang,
    more probably Thang.

Pao Phu Tzu Shen Hsien Chin Shuo Ching (cont.)
Attrib. Ko Hung 葛洪.
TT/910.
Cf. Ho Ping-Yü (11).

Pao Phu Tzu Yang Shêng Lun 抱朴子養
生論.
The Preservation-of-Solidarity Master's
Essay on Hygiene.
Ascr. Chin c. +320.
Attrib. Ko Hung 葛洪.
TT/835.

Pao Shêng Hsin Chien 保生心鑑.
Mental Mirror of the Preservation of Life
[gymnastics and other longevity tech-
niques].
Ming, +1506.
Thieh Fêng chü-shih 鐵峰居士
(The Recluse of Iron Mountain, ps.).
Ed. c. +1596 by Hu Wên-Huan 胡文煥.

Pao Shou Thang Ching Yen Fang 保壽堂經
驗方.
Tried and Tested Prescriptions of the Pro-
tection-of-Longevity Hall (a surgery or
pharmacy).
Ming, c. +1450.
Liu Sung-shih 劉松石.

Pao Tsang Lun 寶藏論.
[=*Hsien-Yuan Pao Tsang Chhang Wei Lun.*]
(The Yellow Emperor's) Discourse on the
(Contents of the) Precious Treasury (of
the Earth), [mineralogy and metallurgy].
Perhaps in part Thang or pre-Thang; com-
pleted in Wu Tai (S/Han). Tsêng Yuan-
Jung (1) notes Chhao Kung-Wu's dating
of it at +918 in his *Chhun Chai Tu Shu
Chih.* Chang Tzu-Kao (2), p. 118, also
considers it mainly a Wu Tai work.
Attrib. Chhing Hsia Tzu 青霞子.
If Su Yuan-Ming 蘇元明 and not
another writer of the same pseudonym,
the earliest parts may have been of the
Chin time (+3rd or +4th); cf Yang
Lieh-Yü (1).
Now only extant in quotations.
Cf. *Lo-fou Shan Chih*, ch. 4, p. 13a.

Pao Yen Thang Pi Chi 寶顏堂祕笈.
Private Collection of the Pao-Yen Library.
Ming, six collections printed between
+1606 and +1620.
Ed. Chhen Chi-Ju 陳繼儒

Pei Lu Fêng Su 北虜風俗.
[=*I Su Chi.*]
Customs of the Northern Barbarians (i.e.
the Mongols).
Ming, +1594.
Hsiao Ta-Hêng 蕭大亨.

Pei Mêng So Yen 北夢瑣言.
Fragmentary Notes Indited North of
(Lake) Mêng.
Wu Tai (S/Phing), c. +950.
Sun Kuang-Hsien 孫光憲.
See des Rotours (4), p. 38.

Pei Shan Chiu Ching 北山酒經.
Northern Mountain Wine Manual.
Sung, +1117.
Chu Hung 朱肱.

Pei Shih 北史.
History of the Northern Dynasties [Nan
Pei Chhao period, +386 to +581].
Thang, c. +670.
Li Yen-Shou 李延壽.
For translations of passages see the index of
Frankel (1).

Pên Ching Fêng Yuan 本經逢原.
(Additions to Natural History) aiming at
the Original Perfection of the *Classical
Pharmacopoeia (of the Heavenly
Husbandman).*
Chhing, +1695, pr. +1705.
Chang Lu 張璐.
LPC, no. 93.

Pên Tshao Chhiu Chen 本草求眞.
Truth Searched out in Pharmaceutical
Natural History.
Chhing, +1773.
Huang Kung-Hsiu 黄宮繡.

Pên Tshao Ching Chi Chu 本草經集注.
Collected Commentaries on the *Classical
Pharmacopoeia (of the Heavenly Husband-
man).*
S/Chhi, +492.
Thao Hung-Ching 陶弘景.
Now extant only in fragmentary form as a
Tunhuang or Turfan MS., apart from
the many quotations in the pharma-
ceutical natural histories, under Thao
Hung-Ching's name.

Pên Tshao Hui 本草滙.
Needles from the Haystack; Selected Essen-
tials of Materia Medica.
Chhing, +1666, pr. +1668.
Kuo Phei-Lan 郭佩蘭.
LPC, no. 84.
Cf. Swingle (4).

Pên Tshao Hui Chien 本草彙箋.
Classified Notes on Pharmaceutical Natural
History.
Chhing, begun +1660, pr. +1666.
Ku Yuan-Chiao 顧元交.
LPC, no. 83.
Cf. Swingle (8).

Pên Tshao Kang Mu 本草綱目.
The Great Pharmacopoeia; or, The Pan-
dects of Natural History (Mineralogy,
Metallurgy, Botany, Zoology etc.),
Arrayed in their Headings and Sub-
headings.
Ming, +1596.
Li Shih-Chen 李時珍.
Paraphrased and abridged tr. Read &
collaborators (2–7) and Read & Pak (1)
with indexes. Tabulation of plants in
Read (1) (with Liu Ju-Chhiang).
Cf. Swingle (7).

Pên Tshao Kang Mu Shih I   本草綱目拾遺.
Supplementary Amplifications for the
Pandects of Natural History (of Li Shih-
Chen).
Chhing, begun *c.* +1760, first prefaced
+1765, prolegomena added +1780, last
date in text 1803.
Chhing, first pr. 1871.
Chao Hsüeh-Min   趙學敏.
LPC, no. 101.
Cf. Swingle (11).
Pên Tshao Mêng Chhüan   本草蒙筌.
Enlightenment on Pharmaceutical Natural
History.
Ming, +1565.
Chhen Chia-Mo   陳嘉謨.
Pên Tshao Pei Yao   本草備要.
Practical Aspects of Materia Medica.
Chhing, *c.* +1690, second ed. +1694.
Wang Ang   汪昂.
LPC, no. 90; ICK, pp. 215 ff.
Cf. Swingle (4).
Pên Tshao Phin Hui Ching Yao   本草品彙精要.
Essentials of the Pharmacopoeia Ranked
according to Nature and Efficacity (Im-
perially Commissioned).
Ming, +1505.
Liu Wên-Thai   劉文泰, Wang Phan 王槃
& Kao Thing-Ho   高廷和.
Pên Tshao Shih I   本草拾遺.
A Supplement for the Pharmaceutical
Natural Histories.
Thang, *c.* +725.
Chhen Tshang-Chhi   陳藏器.
Now extant only in numerous quotations.
Pên Tshao Shu   本草述.
Explanations of Materia Medica.
Chhing, before +1665, first pr. +1700.
Liu Jo-Chin   劉若金.
LPC, no. 79.
Cf. Swingle (6).
Pên Tshao Shu Kou Yuan   本草述鉤元.
Essentials Extracted from the *Explanations
of Materia Medica.*
See Yang Shih-Thai (1).
Pên Tshao Thu Ching   本草圖經.
Illustrated Pharmacopoeia; or, Illustrated
Treatise of Pharmaceutical Natural
History.
Sung, +1061.
Su Sung   蘇頌 *et al.*
Now preserved only in numerous quota-
tions in the later pandects of pharma-
ceutical natural history.
Pên Tshao Thung Hsüan   本草通玄.
The Mysteries of Materia Medica Un-
veiled.
Chhing, begun before +1655, pr. just
before +1667.
Li Chung-Tzu   李中梓.
LPC, no. 75.
Cf. Swingle (4).

Pên Tshao Tshung Hsin   本草從新.
New Additions to Pharmaceutical Natural
History.
Chhing, +1757.
Wu I-Lo   吳儀洛.
LPC, no. 99.
Pên Tshao Yao Hsing   本草藥性.
The Natures of the Vegetable and Other
Drugs in the Pharmaceutical Treatises.
Thang, *c.* +620.
Chen Li-Yen   甄立言 & (perhaps) Chen
Chhüan   甄權.
Now extant only in quotations.
Pên Tshao Yen I   本草衍義.
Dilations upon Pharmaceutical Natural
History.
Sung, pref. +1116, pr. +1119, repr. +1185,
+1195.
Khou Tsung-Shih   寇宗奭.
See also *Thu Ching Yen I Pên Tshao*
(*TT*/761).
Pên Tshao Yen I Pu I   本草衍義補遺.
Revision and Amplification of the *Dilations
upon Pharmaceutical Natural History.*
Yuan, *c.* +1330.
Chu Chen-Hêng   朱震亨.
LPC, no. 47.
Cf. Swingle (12).
Pên Tshao Yuan Shih   本草原始.
Objective Natural History of Materia
Medica; a True-to-Life Study.
Chhing, begun +1578, pr. +1612.
Li Chung-Li   李中立.
LPC, no. 60.
Phan Shan Yü Lu   盤山語錄.
Record of Discussions at Phan Mountain
[dialogues of pronouncedly medical
character on physiological alchemy].
Sung, prob. early +13th.
Writer unknown.
In *Hsiu Chen Shih Shu* (*TT*/260), ch. 53.
Phêng-Lai Shan Hsi Tsao Huan Tan Ko   蓬萊
山西竈還丹歌.
Mnemonic Rhymes of the Cyclically
Transformed Elixir from the Western
Furnace on Phêng-lai Island.
Ascr. *c.* −98. Probably Thang.
Huang Hsüan-Chung   黃玄鍾.
TT/909.
Phêng Tsu Ching   彭祖經.
Manual of Phêng Tsu [Taoist sexual tech-
niques and their natural philosophy].
Late Chou or C/Han, −4th to −1st.
Attrib. Phêng Tsu   彭祖.
Only extant as fragments in *CSHK*
(Shang Ku Sect.), ch. 16, pp. 5*b* ff.
Phu Chi Fang   普濟方.
Practical Prescriptions for Everyman.
Ming, *c.* +1418.
Chu Hsiao   朱橚 (Chou Ting Wang 周定王,
prince of the Ming).
ICK, p. 914.

Pi Yü Chu Sha Han Lin Yü Shu Kuei 碧玉朱砂寒林玉樹匱.
On the Caerulean Jade and Cinnabar Jade-Tree-in-a-Cold-Forest Casing Process.
Sung, early +11th cent.
Chhen Ching-Yuan 陳景元.
TT/891.

Pien Huo Pien 辯惑編.
Disputations on Doubtful Matters.
Yuan, +1348.
Hsieh Ying-Fang 謝應芳.

Pien Tao Lun 辨道論.
On Taoism, True and False.
San Kuo (Wei), c. +230.
Tshao Chih (prince of the Wei), 曹植.
Now extant only in quotations.

Po Wu Chi 博物記.
Notes on the Investigation of Things.
H/Han, c. +190.
Thang Mêng (b) 唐蒙.

Po Wu Chih 博物志.
Records of the Investigation of Things (cf. *Hsü Po Wu Chih*).
Chin, c. +290 (begun about +270).
Chang Hua 張華.

Pu Wu Yao Lan 博物要覽.
The Principal Points about Objects of Art and Nature.
Ming, c. +1560.
Ku Thai 谷泰.

Rokubutsu Shinshi 六物新志.
New Record of Six Things [including the drug mumia]. (In part a translation from Dutch texts.)
Japan, +1786.
Ōtsuki Gentaku 大槻玄澤.

San Chen Chih Yao Yü Chüeh 三眞旨要玉訣.
Precious Instructions concerning the Message of the Three Perfected (Immortals), [i.e. Yang Hsi (*fl.* +370) 楊羲; Hsü Mi (*fl.* +345) 許謐; and Hsü Hui (d. c. +370) 許翽].
Taoist heliotherapy, respiration and meditation.
Chin, c. +365, edited probably in the Thang.
TT/419.
Cf. Maspero (7), p. 376.

San-Fêng Chen Jen Hsüan Than Chhüan Chi 三峯眞人玄譚全集.
Complete Collection of the Mysterious Discourses of the Adept (Chang) San-Fêng [physiological alchemy].
Ming, from c. +1410 (if genuine).
Attrib. Chang San-Fêng 張三峯.
Ed. Min I-Tê (1834) 閔一得.
In *Tao Tsang Hsü Pien (Chhu chi)*, 17.

San-Fêng Tan Chüeh 三峯丹訣 (includes *Chin Tan Chieh Yao* and *Tshai Chen Chi Yao*,

with the *Wu Kên Shu* series of poems, and some inscriptions).
Oral Instructions of (Chang) San-Fêng on the Enchymoma [physiological alchemy].
Ming, from c. +1410 (if genuine).
Attrib. Chang San-Fêng 張三峯.
Ed., with biography, by Fu Chin-Chhüan 傅金銓 (Chi I Tzu 濟一子) c. 1820.

San Phin I Shen Pao Ming Shen Tan Fang 三品頤神保命神丹方.
Efficacious Elixir Prescriptions of Three Grades Inducing the Appropriate Mentality for the Enterprise of Longevity.
Thang, Wu Tai & Sung.
Writers unknown.
YCCC, ch. 78, pp. 1 a ff.

San-shih-liu Shui Fa 三十六水法.
Thirty-six Methods for Bringing Solids into Aqueous Solution.
Pre-Thang.
Writer unknown.
TT/923.

San Tshai Thu Hui 三才圖會.
Universal Encyclopaedia.
Ming, +1609.
Wang Chhi 王圻.

San Tung Chu Nang 三洞珠囊.
Bag of Pearls from the Three (Collections that) Penetrate the Mystery [a Taoist florilegium].
Thang, +7th.
Wang Hsüan-Ho (ed.) 王懸河.
TT/1125.
Cf. Maspero (13), p. 77; Schipper (1), p. 11.

San Yen 三言.
See *Hsing Shih Hêng Yen*, *Yü Shih Ming Yen*, *Ching Shih Thung Yen*.

Setsuyō Yoketsu.
See *Shê Yang Yao Chüeh*.

Shan Hai Ching 山海經.
Classic of the Mountains and Rivers.
Chou and C/Han, −8th to −1st.
Writers unknown.
Partial tr. de Rosny (1).
Chung-Fa Index, no. 9.

Shang-Chhing Chi 上清集.
A Literary Collection (inspired by) the Shang-Chhing Scriptures [prose and poems on physiological alchemy].
Sung, c. +1220.
Ko Chhang-Kêng 葛長庚 (Pai Yü-Chhan 白玉蟾).
In *Hsiu Chen Shih Shu TT*/260), chs. 37 to 44

Shang-Chhing Ching 上清經.
[Part of *Thai Shang San-shih-liu Pu Tsun Ching*.]
The Shang-Chhing (Heavenly Purity) Scripture.
Chin, oldest parts date from about +316.
Attrib. Wei Hua-Tshun 魏華存, dictated to Yang Hsi 楊羲.
In *TT*/8.

Shang-Chhing Chiu Chen Chung Ching Nei Chüeh 上清九眞中經內訣.
Confidential Explanation of the Interior Manual of the Nine (Adepts); a Shang-Chhing Scripture.
Ascr. Chin, +4th, probably pre-Thang.
Attrib. Chhih Sung Tzu 赤松子 (Huang Chhu-Phing 黃初平).
TT/901.

Shang Chhing Han Hsiang Chien Chien Thu 上清含象劍鑑圖.
The Image and Sword Mirror Diagram; a Shang-chhing Scripture.
Thang, c. +700.
Ssuma Chhêng-Chên 司馬承貞.
TT/428.

Shang-Chhing Hou Shêng Tao Chün Lieh Chi 上清後聖道君列紀.
Annals of the Latter-Day Sage, the Lord of the Tao; a Shang-Chhing Scripture.
Chin, late +4th.
Revealed to Yang Hsi 楊羲.
TT/439.

Shang-Chhing Huang Shu Kuo Tu I 上清黃書過度儀.
The System of the Yellow Book for Attaining Salvation; a Shang-Chhing Scripture [the rituale of the communal Taoist liturgical sexual ceremonies, +2nd to +7th centuries].
Date unknown, but pre-Thang.
Writer unknown.
TT/1276.

Shang-Chhing Ling-Pao Ta Fa 上清靈寶大法.
The Great Liturgies; a Shang-Chhing Ling-Pao Scripture.
Sung, +13th.
Chin Yün-Chung 金允中.
TT/1204, 1205, 1206.

Shang-Chhing Ming Thang Hsüan Chen Ching Chüeh 上清明堂玄眞經訣.
See *Ming Thang Hsüan Chen Ching Chüeh.*

Shang-Chhing San Chen Chih Yao Yü Chüeh 上清三眞旨要玉訣.
See *San Chen Chih Yao Yü Chüeh.*

Shang-Chhing Thai-Shang Pa Su Chen Ching 上清太上八素眞經.
Realisation Canon of the Eight Purifications (or Eightfold Simplicity); a Shang-Chhing Thai-Shang Scripture.
Date uncertain, but pre-Thang.
Writer unknown.
TT/423.

Shang-Chhing Thai-Shang Ti Chün Chiu Chen Chung Ching 上清太上帝君九眞中經.
Ninefold Realised Median Canon of the Imperial Lord; a Shang-Chhing Thai-Shang Scripture.
Compiled from materials probably of Chin period, late +4th.
Writers and editor unknown.
TT/1357.

Shang-Chhing Tung-Chen Chiu Kung Tzu Fang Thu 上清洞眞九宮紫房圖.
Description of the Purple Chambers of the Nine Palaces; a Tung-Chen Scripture of the Shang-Chhing Heavens [parts of the microcosmic body corresponding to stars in the macrocosm].
Sung, probably +12th century.
Writer unknown.
TT/153.

Shang-Chhing Wo Chung Chüeh 上清握中訣.
Explanation of (the Method of) Grasping the Central (Luminary); a Shang-Chhing Scripture [Taoist meditation and heliotherapy].
Date unknown, Liang or perhaps Thang.
Writer unknown.
Based on the procedures of Fan Yu-Chhung 范幼沖 (H/Han).
TT/137.
Cf. Maspero (7), p. 373.

Shang Phin Tan Fa Chieh Tzhu 上品丹法節次.
Expositions of the Techniques for Making the Best Quality Enchymoma [physiological alchemy].
Chhing.
Li Tê-Hsia 李德洽.
Comm. Min I-Tê 閔一得, c. 1830.
In *Tao Tsang Hsü Pien (Chhu chi)*, 6.

Shang Shu Ta Chuan 尚書大傳.
Great Commentary on the *Shang Shu* chapters of the *Historical Classic.*
C/Han, c. −185.
Fu Shêng 伏勝.
Cf. Wu Khang (1), p. 230.

Shang-Tung Hsin Tan Ching Chüeh 上洞心丹經訣.
An Explanation of the Heart Elixir and Enchymoma Canon; a Shang-Tung Scripture.
Date unknown, perhaps Sung.
Writer unknown.
TT/943.
Cf. Chhen Kuo-Fu (1), vol. 2, pp. 389, 435.

Shang Yang Tzu Chin Tan Ta Yao 上陽子金丹大要.
See *Chin Tan Ta Yao.*

Shang Yang Tzu Chin Tan Ta Yao Hsien Phai (Yuan Liu) 上陽子金丹大要仙派 (源流).
See *Chin Tan Ta Yao Hsien Phai (Yuan Liu).*

Shang Yang Tza Chin Tan Ta Yao Lieh Hsien Chih 上陽子金丹大要列仙誌.
See *Chin Tan Ta Yao Lieh Hsien Chih.*

Shang Yang Tzu Chin Tan Ta Yao Thu 上陽子金丹大要圖.
See *Chin Tan Ta Yao Thu.*

Shao-Hsing Chiao-Ting Ching-Shih Chêng Lei Pei-Chi Pên Tshao 紹興校定經史證類備急本草.

Shao-Hsing Chiao-Ting Ching-Shih Chêng Lei Pei-Chi Pên Tshao (cont.)|
The Corrected Classified and Consolidated Armamentarium; Pharmacopoeia of the Shao-Hsing Reign-Period.
S/Sung, pres. +1157, pr. +1159, often copied and repr. especially in Japan.
Thang Shen-Wei 唐愼微 ed. Wang Chi-Hsien 王繼先 *et al.*
Cf. Nakao Manzō (*1*, *1*); Swingle (*11*).
Illustrations reproduced in facsimile by Wada (*1*); Karow (*2*).
Facsimile edition of a MS. in the Library of Ryokoku University, Kyoto 龍谷大學 圖書館.
Ed. with an analytical and historical introduction, including contents table and indexes (別冊) by Okanishi Tameto 岡西 爲人 (Shunyōdō, Tokyo, 1971).

Shê Ta Chhêng Lun Shih 攝大乘論釋.
Mahāyāna-samgraha-bhāshya (Explanatory Discourse to assist the Understanding of the Great Vehicle).
India, betw. +300 and +500.
Tr. Hsüan-Chuang 玄奘, *c.* +650.
N/1171 (4); TW/1597.

(Shê Yang) Chen Chung Chi (or Fang) (攝養)枕 中記(方).
Pillow-Book on Assisting the Nourishment (of the Life-Force).
Thang, early +7th.
Attrib. Sun Ssu-Mo 孫思邈.
TT/830, and in *YCCC*, ch. 33.

Shê Yang Yao Chüeh (*Setsuyō Yoketsu*) 攝養要訣.
Important Instructions for the Preservation of Health conducive to Longevity.
Japan (Heian), *c.* +820.
Mononobe Kōsen (imperial physician) 物部廣泉.

Shen Hsien Chin Shuo Ching 神仙金汋經.
See *Pao Phu Tzu Shen Hsien Chin Shuo Ching.*

Shen Hsien Chuan 神仙傳.
Lives of the Holy Immortals.
(Cf. *Lieh Hsien Chuan* and *Hsü Shen Hsien Chuan.*)
Chin, +4th century.
Attrib. Ko Hung 葛洪.

Shen Hsien Fu Erh Tan Shih Hsing Yao Fa 神仙服餌丹石行藥法.
The Methods of the Holy Immortals for Ingesting Cinnabar and (Other) Minerals, and Using them Medicinally.
Date unknown.
Attrib. Ching-Li hsien-sêng 京里先生.
TT/417.

Shen Hsien Fu Shih Ling-Chih Chhang-Phu Wan Fang 神仙服食靈芝菖蒲丸方.
Prescriptions for Making Pills from Numinous Mushrooms and Sweet Flag (*Calamus*), as taken by the Holy Immortals.
Date unknown

Writer unknown.
TT/837.

Shen Hsien Lien Tan Tien Chu San Yuan Pao Ching Fa 神仙鍊丹點鑄三元寶鏡法.
Methods used by the Holy Immortals to Prepare the Elixir, Project it, and Cast the Precious Mirrors of the Three Powers (or the Three Primary Vitalities), [magical].
Thang, +902.
Writer unknown.
TT/856.

Shen Hsien Thung Chien 神仙通鑑.
(Cf. (*Li Tai*) *Shen Hsien* (*Thung*) *Chien.*)
General Survey of the Lives of the Holy Immortals.
Ming, +1640.
Hsüeh Ta-Hsün 薛大訓.

Shen I Chi 神異記.
(Probably an alternative title of *Shen I Ching*, q.v.)
Records of the Spiritual and the Strange.
Chin, *c.* +290.
Wang Fou 王浮.

Shen I Ching 神異經.
Book of the Spiritual and the Strange.
Ascr. Han, but prob. +3rd, +4th or +5th century.
Attrib. Tungfang Shuo 東方朔.
Probable author, Wang Fou 王浮.

Shen Nung Pên Tshao Ching 神農本草經.
Classical Pharmacopoeia of the Heavenly Husbandman.
C/Han, based on Chou and Chhin material, but not reaching final form before the +2nd century.
Writers unknown.
Lost as a separate work, but the basis of all subsequent compendia of pharmaceutical natural history, in which it is constantly quoted.
Reconstituted and annotated by many scholars; see Lung Po-Chien (*1*), pp. 2 ff., 12 ff.
Best reconstructions by Mori Tateyuki 森立之 (1845), Liu Fu 劉復 (1942).

Shen shih Liang Fang 沈氏良方.
Original title of *Su Shen Liang Fang* (q.v.).

Shen Thien-Shih Fu Chhi Yao Chüeh 申天師 服氣要訣.
Important Oral Instructions of the Heavenly Teacher (or Patriarch) Shen on the Absorption of the Chhi [Taoist breathing exercises].
Thang, *c.* +730.
Shen Yuan-Chih 申元之.
Now extant only as a short passage in *YCCC*, ch. 59, pp. 16*b* ff.

Shêng Chi Tsung Lu 聖濟總錄.
Imperial Medical Encyclopaedia [issued by authority].
Sung, *c.* +1111 to +1118.
Ed. by twelve physicians.

Shêng Shih Miao Ching 生尸妙經.
  See *Thai-Shang   Tung-Hsüan   Ling-Pao
    Mieh Tu* (or *San Yuan*) *Wu Lien Shêng
    Shih Miao Ching.*
Shêng Shui Yen Than Lu 澠水燕談錄.
  Fleeting Gossip by the River Shêng [in
    Shantung].
  Sung, late +11th century (before +1094).
  Wang Phi-Chih 王闢之.
Shih Chin Shih 試金石.
  On the Testing of (what is meant by)
    'Metal' and 'Mineral'.
  See Fu Chin-Chhüan (5).
Shih Han Chi 石函記.
  See *Hsü Chen Chün Shih Han Chi.*
Shih I Chi 拾遺記.
  Memoirs on Neglected Matters.
  Chin, c. +370.
  Wang Chia 王嘉.
  Cf. Eichhorn (5).
Shih I Tê Hsiao Fang 世醫得効方.
  Efficacious Prescriptions of a Family of
    Physicians.
  Yuan, +1337.
  Wei I-Lin 危亦林.
Shih Liao Pên Tshao 食療本草.
  Nutritional Therapy; a Pharmaceutical
    Natural History.
  Thang, c. +670.
  Mêng Shen 孟詵.
Shih Lin Kuang Chi 事林廣記.
  Guide through the Forest of Affairs
    [encyclopaedia].
  Sung, between +1100 and +1250; first
    pr. +1325.
  Chhen Yuan-Ching 陳元靚.
  (A unique copy of a Ming edition of
    +1478 is in the Cambridge University
    Library.)
Shih Ming 釋名.
  Explanation of Names [dictionary].
  H/Han, c. +100.
  Liu Hsi 劉熙.
Shih Pien Liang Fang 十便良方.
  Excellent Prescriptions of Perfect
    Convenience.
  Sung, +1196.
  Kuo Than 郭坦.
  Cf. SIC, p. 1119; ICK, p. 813.
Shih Wu Chi Yuan 事物紀原.
  Records of the Origins of Affairs and
    Things.
  Sung, c. +1085.
  Kao Chhêng 高承.
Shih Wu Pên Tshao 食物本草.
  Nutritional Natural History.
  Ming, +1571 (repr. from a slightly earlier
    edition).
  Attrib. Li Kao 李杲 (J/Chin) or Wang
    Ying 汪穎 (Ming) in various editions;
    actual writer Lu Ho 盧和.
  The bibliography of this work in its several

different forms, together with the ques-
  tions of authorship and editorship, are
  complex.
  See Lung Po-Chien (1), pp. 104, 105, 106;
    Wang Yü-Hu (1), 2nd ed. p. 194;
    Swingle (1, 10).
Shih Yao Erh Ya 石藥爾雅.
  The Literary Expositor of Chemical Physic;
    or, Synonymic Dictionary of Minerals
    and Drugs.
  Thang, +806.
  Mei Piao 梅彪.
  TT/894.
Shih Yuan 事原.
  On the Origins of Things.
  Sung.
  Chu Hui 朱繪.
Shoku-Nihongi 續日本記.
  Chronicles of Japan, continued [from +697
    to +791].
  Japan (Nara), +797.
  Ishikawa Natari 石川,
    Fujiwara Tsuginawa 藤原繼繩,
    Sugeno Sanemichi 菅野眞道 *et al.*
Shoku-Nihonkoki 續日本後記.
  Chronicles of Japan, still further continued
    [from +834 to +850].
  Japan (Heian), +869.
  Fujiwara Yoshifusa 藤原良房.
Shou Yü Shen Fang 壽域神方.
  Magical Prescriptions of the Land of the
    Old.
  Ming, c. +1430.
  Chu Chhüan 朱權 (Ning Hsien Wang
    寧獻王, prince of the Ming).
Shu Shu Chi I 數術記遺.
  Memoir on some Traditions of Math-
    ematical Art.
  H/Han, +190, but generally suspected of
    having been written by its commentator
    Chen Luan 甄鸞, c. +570. Some place
    the text as late as the Wu Tai period
    (+10th. cent.), e.g. Hu Shih; and others
    such as Li Shu-Hua (2) prefer a Thang
    dating.
  Hsü Yo 徐岳.
Shu Yuan Tsa Chi 菽園雜記.
  The Bean-Garden Miscellany.
  Ming, +1475.
  Lu Jung 陸容.
Shuang Mei Ching An Tshung Shu 雙梅景闇
  叢書.
  Double Plum-Tree Collection [of ancient
    and medieval books and fragments on
    Taoist sexual techniques].
  See Yeh Tê-Hui (1) 葉德輝 in Bib. B.
Shui Yün Lu 水雲錄.
  Record of Clouds and Waters [iatro-
    chemical].
  Sung, c. +1125.
  Yeh Mêng-Tê 葉夢得.
  Extant now only in quotations.

Shun Yang Lü Chen-Jen Yao Shih Chih 純陽
呂眞人藥石製.
The Adept Lü Shun-Yang's (i.e. Lü
Tung-Pin's) Book on Preparations of
Drugs and Minerals [in verses].
Late Thang.
Attrib. Lü Tung-Pin 呂洞賓.
TT/896.
Tr. Ho Ping-Yü, Lim & Morsingh (1).

Shuo Wên.
See *Shuo Wên Chieh Tzu*.

Shuo Wên Chieh Tzu 說文解字.
Analytical Dictionary of Characters (lit.
Explanations of Simple Characters and
Analyses of Composite Ones).
H/Han, +121.
Hsü Shen 許慎.

So Sui Lu 瑣碎錄.
Sherds, Orts and Unconsidered Fragments
[iatro-chemical].
Sung, prob. late +11th.
Writer unknown.
Now extant only in quotations. Cf. *Winter's
Tale*, iv, iii, *Timon of Athens*, iv, iii, and
Julius Caesar, iv, i.

Sou Shen Chi 搜神記.
Reports on Spiritual Manifestations.
Chin, *c.* +348.
Kan Pao 干寶.
Partial tr. Bodde (9).

Sou Shen Hou Chi 搜神後記.
Supplementary Reports on Spiritual
Manifestations.
Chin, late +4th or early +5th century.
Thao Chhien 陶潛.

· *Ssu Khu Thi Yao Pien Chêng* 四庫提要辨證.
See Yü Chia-Hsi (1).

Ssu Shêng Pên Tshao 四聲本草.
Materia Medica Classified according to the
Four Tones (and the Standard Rhymes),
[the entries arranged in the order of the
pronunciation of the first character of
their names].
Thang, *c.* +775.
Hsiao Ping 蕭炳.

Ssu Shih Thiao Shê Chien 四時調攝牋.
Directions for Harmonising and Strengthen-
ing (the Vitalities) according to the Four
Seasons of the Year [the second part
(chs. 3–6) of *Tsun Shêng Pa Chien*, q.v.].
Ming, +1591.
Kao Lien 高濂.
Partial tr. of the gymnastic material,
Dudgeon (1).

Ssu Shih Tsuan Yao 四時纂要.
Important Rules for the Four Seasons
[agriculture and horticulture, family
hygiene and pharmacy, etc.].
Thang, *c.* +750.
Han O 韓鄂.

Su Nü Ching 素女經.
Canon of the Immaculate Girl.

Han.
Writer unknown.
Only as fragment in *Shuang Mei Ching An
Tshung Shu*, now containing the *Hsüan Nü
Ching* (q.v.).
Partial trs. van Gulik (3, 8).

Su Nü Miao Lun 素女妙論.
Mysterious Discourses of the Immaculate
Girl.
Ming, *c.* +1500.
Writer unknown.
Partial tr. van Gulik (3).

Su Shen Liang Fang 蘇沈良方.
Beneficial Prescriptions collected by Su
(Tung-Pho) and Shen (Kua).
Sung, *c.* +1120. Some of the data go back
as far as +1060. Preface by Lin Ling-Su
林靈素.
Shen Kua 沈括 and Su Tung-Pho
蘇東坡 (posthumous).
The collection was at first called *Shen
shih Liang Fang*, so that most of the
entries are Shen Kua's, but as some certainly stem from Su Tung-Pho, the latter
were probably added by editors at the
beginning of the new century.
Cf. *ICK*, pp. 737, 732.

Su Wên Ling Shu Ching.
See *Huang Ti Nei Ching, Su Wên* and
Huang Ti Nei Ching, Ling Shu.

Su Wên Nei Ching.
See *Huang Ti Nei Ching, Su Wên*.

Sui Shu 隋書.
History of the Sui Dynasty [+581 to
+617].
Thang, +636 (annals and biographies);
+656 (monographs and bibliography).
Wei Chêng 魏徵 *et al.*
Partial trs. Pfizmaier (61–65); Balazs (7, 8);
Ware (1).
For translations of passages see the index of
Frankel (1).

Sun Kung Than Phu 孫公談圃.
The Venerable Mr Sung's Conversation
Garden.
Sung, *c.* +1085.
Sun Shêng 孫升.

Sung Chhao Shih Shih 宋朝事實.
Records of Affairs of the Sung Dynasty.
Yuan, +13th.
Li Yu 李攸.

Sung Shan Thai-Wu hsien-sêng Chhi Ching
嵩山太无先生氣經.
Manual of the (Circulation of the) Chhi,
by Mr Grand-Nothingness of Sung
Mountain.
Thang, +766 to +779.
Prob. Li Fêng-Shih 李奉時 (Thai-Wu
hsien-sêng).
TT/817, and in *YCCC*, ch. 59 (partially),
pp. 7a ff.
Cf. Maspero (7), p. 199.

Sung Shih 宋史.
    History of the Sung Dynasty [+960 to +1279].
    Yuan, *c.* +1345.
    Tho-Tho (Toktaga) 脫脫 & Ouyang Hsüan 歐陽玄.
    Yin-Tê Index, no. 34.
Szechuan Thung Chih 四川通志.
    General History and Topography of Szechuan Province.
    Chhing, +18th century (pr. 1816).
    Ed. Chhang Ming 常明, Yang Fang-Tshan 楊芳燦 *et al.*

Ta Chao 大招.
    The Great Summons (of the Soul), [ode].
    Chhu (between Chhin and Han), −206 or −205.
    Writer unknown.
    Tr. Hawkes (1), p. 109.
Ta Chih Tu Lun 大智度論.
    Mahā-prajñapāramito-padeśa Śāstra (Commentary on the Great Sūtra of the Perfection of Wisdom).
    India.
    Attrib. Nāgārjuna, +2nd.
    Mostly prob. of Central Asian origin.
    Tr. Kumārajīva, +406.
    N/1169; TW/1509.
Ta Chün Ku Thung 大鈞皷銅.
    (Illustrated Account of the Mining), Smelting and Refining of Copper [and other Non-Ferrous Metals], according to the Principles of Nature (lit. the Great Potter's Wheel).
    See Masuda Tsuna (1).
Ta Fang Kuang Fo Hua Yen Ching 大方廣佛華嚴經.
    Avataṁsaka Sūtra.
    India.
    Tr. Śikshānanda, +699.
    N/88; TW/279.
Ta Huan Tan Chao Chien 大還丹照鑑.
    An Elucidation of the Great Cyclically Transformed Elixir [in verses].
    Wu Tai (Shu), +962.
    Writer unknown.
    TT/919.
Ta Huan Tan Chhi Pi Thu 大還丹契祕圖.
    Esoteric Illustrations of the Concordance of the Great Regenerative Enchymoma.
    Thang or Sung.
    Writer unknown.
    In *YCCC*, ch. 72, pp. 1*a* ff.
    Cf. *Hsiu Chen Li Yen Chhao Thu* and *Chin I Huan Tan Yin Chêng Thu.*
Ta-Kuan Ching-Shih Chêng Lei Pei-Chi Pên Tshao 大觀經史證類備急本草.
    The Classified and Consolidated Armamentarium; Pharmacopoeia of the Ta-Kuan reign-period.
    Sung, +1108; repr. +1211, +1214 (J/Chin), +1302 (Yuan).

Thang Shen-Wei 唐愼微.
    Ed. Ai Shêng 艾晟.
Ta Ming I Thung Chih 大明一統志.
    Comprehensive Geography of the (Chinese) Empire (under the Ming dynasty).
    Ming, commissioned +1450, completed +1461.
    Ed. Li Hsien 李賢.
Ta Tai Li Chi 大戴禮記.
    Record of Rites [compiled by Tai the Elder] (cf. *Hsiao Tai Li Chi*; *Li Chi*).
    Ascr. C/Han, *c.* −70 to −50, but really H/Han, between +80 and +105.
    Attrib. ed. Tai Tê 戴德, in fact probably ed. Tshao Pao 曹褒.
    See Legge (7).
    Trs. Douglas (1); R. Wilhelm (6).
Ta Tan Chhien Hung Lun 大丹鉛汞論.
    Discourse on the Great Elixir [or Enchymoma] of Lead and Mercury.
    If Thang, +9th, more probably Sung.
    Chin Chu-Pho 金竹坡.
    TT/916.
    Cf. Yoshida Mitsukuni (5), pp. 230–2.
Ta Tan Chi 大丹記.
    Record of the Great Enchymoma.
    Ascr. +2nd cent., but probably Sung, +13th.
    Attrib. Wei Po-Yang 魏伯陽.
    TT/892.
Ta Tan Chih Chih 大丹直指.
    Direct Hints on the Great Elixir.
    Sung, *c.* +1200.
    Chhiu Chhu-Chi 邱處機.
    TT/241.
Ta Tan Wên Ta 大丹問答.
    Questions and Answers on the Great Elixir (or Enchymoma) [dialogues between Chêng Yin and Ko Hung].
    Date unknown, prob. late Sung or Yuan.
    Writer unknown.
    TT/932.
Ta Tan Yao Chüeh Pên Tshao 大丹藥訣本草.
    Pharmaceutical Natural History in the form of Instructions about Medicines of the Great Elixir (Type), [iatro-chemical].
    Possible alternative title of *Wai Tan Pên Tshao* (q.v.).
Ta-Tung Lien Chen Pao Ching, Chiu Huan Chin Tan Miao Chüeh 大洞鍊眞寶經九還金丹妙訣.
    Mysterious Teachings on the Ninefold Cyclically Transformed Gold Elixir, supplementary to the Manual of the Making of the Perfected Treasure; a Ta-Tung Scripture.
    Thang, +8th, perhaps *c.* +712.
    Chhen Shao-Wei 陳少微.
    TT/884. A sequel to *TT*/883, and in *YCCC*, ch. 68, pp. 8*a* ff.
    Tr. Sivin (4).

Ta-Tung Lien Chen Pao Ching, Hsiu Fu Ling Sha Miao Chüeh 大洞鍊眞寶經修伏靈砂妙訣.
Mysterious Teachings on the Alchemical Preparation of Numinous Cinnabar, supplementary to the Manual of the Making of the Perfected Treasure; a Ta-Tung Scripture.
Thang, +8th, perhaps *c.* +712.
Chhen Shao-Wei 陳少微.
TT/883. Alt. title: *Chhi Fan Ling Sha Lun*, as in *YCCC*, ch. 69, pp. 1*a* ff.
Tr. Sivin (4).

Ta Yu Miao Ching 大有妙經.
[= *Tung-Chen Thai-Shang Su-Ling Tung-Yuan Ta Yu Miao Ching*.]
Book of the Great Mystery of Existence [Taoist anatomy and physiology; describes the *shang tan thien*, upper region of vital heat, in the brain].
Chin, +4th.
Writer unknown.
TT/1295.
Cf. Maspero (7), p. 192.

Tai I Phien 代疑篇.
On Replacing Doubts by Certainties.
Ming, +1621.
Yang Thing-Yün 楊廷筠.
Preface by Wang Chêng 王徵.

Taketori Monogatari 竹取物語.
The Tale of the Bamboo-Gatherer.
Japan (Heian), *c.* +865. Cannot be earlier than *c.* +810 or later than *c.* +955.
Writer unknown.
Cf. Matsubara Hisako (1, 2).

Tan Ching Shih Tu 丹經示讀.
A Guide to the Reading of the Enchymoma Manuals.
See Fu Chin-Chhüan (3).

Tan Ching Yao Chüeh.
See *Thai-Chhing Tan Ching Yao Chüeh.*

Tan Fang Ao Lun 丹房奧論.
Subtle Discourse on the (Alchemical) Elaboratory (of the Human Body, for making the Enchymoma).
Sung, +1020.
Chhêng Liao-I 程了一.
TT/913, and in *TTCY* (*chung mao chi*, 5).

Tan Fang Chien Yuan 丹方鑑源.
The Mirror of Alchemical Processes (and Reagents); a Source-book.
Wu Tai (H/Shu), *c.* +938 to +965.
Tuku Thao 獨孤滔.
Descr. Fêng Chia-Lo & Collier (1).
See Ho Ping-Yü & Su Ying-Hui (1).
TT/918.

Tan Fang Ching Yuan 丹房鏡源.
The Mirror of the Alchemical Elaboratory; a Source-book.
Early Thang, not later than +800.
Writer unknown.

Survives only incorporated in *TT*/912 and in *CLPT*.
See Ho Ping-Yü & Su Ying-Hui (1).

Tan Fang Hsü Chih 丹房須知.
Indispensable Knowledge for the Chymical Elaboratory [with illustrations of apparatus].
Sung, +1163.
Wu Wu 吳悞.
TT/893.

Tan Fang Pao Chien Chih Thu 丹房寶鑑之圖.
[= *Tzu Yang Tan Fang Pao Chien Chih Thu*.]
Precious Mirror of the Elixir and Enchymoma Laboratory; Tables and Pictures (to illustrate the Principles).
Sung, *c.* +1075.
Chang Po-Tuan 張伯端 (Tzu Yang Tzu 紫陽子 or Tzu Yang Chen Jen).
Incorporated later in *Chin Tan Ta Yao Thu* (q.v.)
In *Chin Tan Ta Yao* (*TTCY* ed.), ch. 3, pp. 34*a* ff. Also in *Wu Chen Phien* (in *Hsiu Chen Shih Shu*, *TT*/260, ch. 26, pp. 5*a* ff.).
Cf. Ho Ping-Yü & Needham (2).

Tan I San Chüan 丹擬三卷.
See Pa Tzu-Yuan (1).

Tan Lun Chüeh Chih Hsin Ching 丹論訣旨心鏡 (*Chien* or *Chao* 鑑, 照 occur as tabu forms in the titles of some versions.)
Mental Mirror Reflecting the Essentials of Oral Instruction about the Discourses on the Elixir and the Enchymoma.
Thang, probably +9th.
Chang Hsüan-Tê 張玄德, criticising the teachings of Ssuma Hsi-I 司馬希夷.
TT/928, and in *YCCC*, ch. 66, pp. 1*a* ff.
Tr. Sivin (5).

Tan Thai Hsin Lu 丹臺新錄.
New Discourse on the Alchemical Laboratory.
Early Sung or pre-Sung.
Attrib. Chhing Hsia Tzu 青霞子 or Hsia Yu-Chang 夏有章.
Extant only in quotations.

Tan-Yang Chen Jen Yü Lu 丹陽眞人玉錄.
Precious Records of the Adept Tan-Yang.
Sung, mid +12th cent.
Ma Yü 馬鈺.
TT/1044.

Tan-Yang Shen Kuang Tshan 丹陽神光燦.
Tan Yang (Tzu's Book) on the Resplendent Glow of the Numinous Light.
Sung, mid +12th cent.
Ma Yü 馬鈺.
TT/1136.

Tan Yao Pi Chüeh 丹藥祕訣.
Confidential Oral Instructions on Elixirs and Drugs.
Prob. Yuan or early Ming.
Hu Yen 胡演.
Now only extant as quotations in the pharmaceutical natural histories.

Tao Fa Hsin Chhuan 道法心傳.
  Transmission of (a Lifetime of) Thought on
    Taoist Techniques [physiological al-
    chemy with special reference to micro-
    cosm and macrocosm; many poems and a
    long exposition].
  Yuan, +1294.
  Wang Wei-I 王惟一.
  TT/1235, and *TTCY* (*hsia mao chi*, 5).

Tao Fa Hui Yuan 道法會元.
  Liturgical and Apotropaic Encyclopaedia of
    Taoism.
  Thang and Sung.
  Writers and compiler unknown.
  TT/1203.

Tao Hai Chin Liang 道海津梁.
  A Catena (of Words) to Bridge the Ocean
    of the Tao.
  See Fu Chin-Chhüan (*4*).

Tao Shu 道樞.
  Axial Principles of the Tao [doctrinal
    treatise, mainly on the techniques of
    physiological alchemy].
  Sung, early +12th; finished by 1145.
  Tsêng Tshao 曾慥.
  TT/1005.

Tao Su Fu 擣素賦.
  Ode on a Girl of Matchless Beauty [Chao
    nü, probably Chao Fei-Yen]; or, Of
    What does Spotless Beauty Consist?
  C/Han, *c.* −20.
  Pan chieh-yü 班婕妤.
  In *CSHK*, Chhien Han Sect., ch. 11,
    p. 7*a* ff.

Tao Tê Ching 道德經.
  Canon of the Tao and its Virtue.
  Chou, before −300.
  Attrib. Li Erh (Lao Tzu) 李耳(老子).
  Tr. Waley (4); Chhu Ta-Kao (2); Lin Yü-
    Thang (1); Wieger (7); Duyvendak (18);
    and very many others.

Tao Tsang 道藏.
  The Taoist Patrology [containing 1464
    Taoist works].
  All periods, but first collected in the Thang
    about +730, then again about +870 and
    definitively in +1019. First printed in
    the Sung (+1111 to +1117). Also printed in
    J/Chin (+1168 to +1191), Yuan (+1244,
    +1607). and Ming (+1445, +1598 and
  Writers numerous.
  Indexes by Wieger (6), on which see Pelliot's
    review (58); and Ong Tu-Chien (Yin-Tê
    Index, no. 25).

Tao Tsang Chi Yao 道藏輯要.
  Essentials of the Taoist Patrology [con-
    taining 287 books, 173 works from the
    Taoist Patrology and 114 Taoist works
    from other sources].
  All periods, pr. 1906 at Erh-hsien-ssu
    二仙寺, Chhêngtu.
  Writers numerous.

Ed. Ho Lung-Hsiang 賀龍驤 & Phêng
  Han-Jan 彭瀚然 (Chhing).

Tao Tsang Hsü Phien Chhu Chi 道藏續篇初集.
  First Series of a Supplement to the Taoist
    Patrology.
  Chhing, early 19th cent.
  Edited by Min I-Tê 閔一得.

Tao Yin Yang Shêng Ching 導引養生經.
  [= *Thai-Chhing Tao Yin Yang Shêng Ching*.]
  Manual of Nourishing the Life-Force (or,
    Attaining Longevity and Immortality) by
    Gymnastics.
  Late Thang, Wu Tai, or early Sung.
  Writer unknown.
  TT/811, and in *YCCC*, ch. 34.
  Cf. Maspero (7), pp. 415 ff.

Têng Chen Yin Chüeh 登眞隱訣.
  Confidential Instructions for the Ascent to
    Perfected (Immortality).
  Chin and S/Chhi. Original material from
    the neighbourhood of +365 to +366;
    commentary (the 'Confidential Instruct-
    ions' of the title) by Thao Hung-Ching
    (+456 to +536) written between +493
    and +498.
  Original writer unknown.
  Ed. Thao Hung-Ching 陶弘景.
  TT/418, but conservation fragmentary.
  Cf. Maspero (7), pp. 192, 374.

Thai-Chhing Chen Jen Ta Tan 太清眞人大丹.
  [Alternative later name of *Thai-Chhing
    Tan Ching Yao Chüeh*.]
  The Great Elixirs of the Adepts; a Thai-
    Chhing Scripture.
  Thang, mid +7th (*c.* +640).
  Prob. Sun Ssu-Mo 孫思邈.
  In *YCCC*, ch. 71.
  Tr. Sivin (1), pp. 145 ff.

Thai-Chhing Chin I Shen Chhi Ching 太清金
  液神氣經.
  Manual of the Numinous Chhi of Potable
    Gold; a Thai-Chhing Scripture.
  Ch. 3 records visitations by the Lady Wei
    Hua-Tshun and her companion divinities
    mostly paralleling texts in the *Chen Kao*.
    They were taken down by Hsü Mi's great-
    grandson Hsü Jung-Ti (d. +435), *c.* +430.
    Chs 1 and 2 are Thang or Sung, before
    +1150. If pre-Thang, cannot be earlier
    than +6th.
  Writers mainly unknown.
  TT/875.

Thai-Chhing Chin I Shen Tan Ching 太清金液
  神丹經.
  Manual of the Potable Gold (or Metallous
    Fluid), and the Magical Elixir (or
    Enchymoma); a Thai-Chhing Scripture.
  Date unknown, but must be pre-Liang
    (Chhen Kuo-Fu (*1*), vol. 2, p. 419). Con-
    tains dates between +320 and +330, but
    most of the prose is more probably of the
    early +5th century.

Thai-Chhing Chin I Shen Tan Ching (*cont.*)
Preface and main texts of *nei tan* character,
all the rest *wai tan*, including laboratory
instructions.
Writer unknown; chs. variously attributed.
The third chapter, devoted to descriptions
of foreign countries which produced
cinnabar and other chemical substances,
may be of the second half of the +7th
century (see Maspero (14), pp. 95 ff.).
Most were based on Wan Chen's *Nan
Chou I Wu Chih* (+3rd cent.), but not
the one on the Roman Orient (Ta-Chhin)
translated by Maspero. Stein (5) has
pointed out however that the term *Fu-
Lin* for Byzantium occurs as early as
+500 to +520, so the third chapter may
well be of the early +6th century.
TT/873.
Abridged in *YCCC* ch. 65, pp. 1 *a* ff.
Cf. Ho Ping-Yü (10).

Thai-Chhing Ching Thien-Shih Khou Chüeh
太清經天師口訣.
Oral Instructions from the Heavenly Masters
[Taoist Patriarchs] on the Thai-Chhing
Scriptures.
Date unknown, but must be after the mid
+5th cent. and before Yuan.
Writer unknown.
TT/876.

Thai-Chhing Chung Huang Chen Ching 太清中
黃眞經.
See *Chung Huang Chen Ching*.

Thai-Chhing Shih Pi Chi 太清石壁記.
The Records in the Rock Chamber (lit.
Wall); a Thai-Chhing Scripture.
Liang, early +6th, but includes earlier work
of Chin time as old as the late +3rd,
attributed to Su Yuan-Ming.
Edited by Chhu Tsê hsien-sêng 楚澤先生.
Original writer, Su Yuan-Ming 蘇元明
(Chhing Hsia Tzu 青霞子),
TT/874.
Tr. Ho Ping-Yü (8).
Cf. *Lo-fou Shan Chih*, ch. 4, p. 13 *a*.

Thai-Chhing Tan Ching Yao Chüeh 太清丹經
要訣.
[= *Thai-Chhing Chen Jen Ta Tan*.]
Essentials of the Elixir Manuals, for Oral
Transmission; a Thai-Chhing Scripture.
Thang, mid +7th (c. +640).
Prob. Sun Ssu-Mo 孫思邈.
In *YCCC*, ch. 71.
Tr. Sivin (1), pp. 145 ff.

Thai-Chhing Tao Yin Yang Shêng Ching 太清
導引養生經.
See *Tao Yin Yang Shêng Ching*.

Thai-Chhing Thiao Chhi Ching 太清調氣經.
Manual of the Harmonising of the Chhi; a
Thai-Chhing Scripture [breathing exer-
cises for longevity and immortality].
Thang or Sung, +9th or +10th.

Writer unknown.
TT/813.
Cf. Maspero (7), p. 202.

Thai-Chhing (*Wang Lao*) (*Fu Chhi*) *Khou Chüeh*
(or *Chhuan Fa*) 太清王老服氣口訣
(傳法).
The Venerable Wang's Instructions for
Absorbing the Chhi; a Thai-Chhing
Scripture [Taoist breathing exercises].
Thang or Wu Tai (the name of Wang added
in the +11th).
Writer unknown.
Part due to a woman Taoist, Li I 李液.
TT/815, and in *YCCC*, ch. 62, pp. 1 *a* ff.
and ch. 59, pp. 10 *a* ff.
Cf. Maspero (7), p. 209.

Thai-Chhing Yü Pei Tzu 太清玉碑子.
The Jade Stele (Inscription); a Thai-
Chhing Scripture [dialogues between
Chêng Yin and Ko Hung].
Date unknown, prob. late Sung or Yuan.
Writer unknown.
TT/920.
Cf. *Ta Tan Wên Ta* and *Chin Mu Wan Ling
Lun*, which incorporate parallel passages.

*Thai-Chi Chen-Jen Chiu Chuan Huan Tan
Ching Yao Chüeh* 太極眞人九轉還丹
經要訣.
Essential Teachings of the Manual of the
Supreme-Pole Adept on the Ninefold
Cyclically Transformed Elixir.
Date unknown, perhaps Sung on account
of the pseudonym, but the Manual
(*Ching*) itself may be pre-Sui because its
title is in the *Sui Shu* bibliography. Mao
Shan influence is revealed by an account
of five kinds of magic plants or mush-
rooms that grow on Mt Mao, and in-
structions of Lord Mao for ingesting
them.
Writer unknown.
TT/882.
Partial tr. Ho Ping-Yü (9).

Thai-Chi Chen-Jen Tsa Tan Yao Fang 太極眞
人雜丹藥方.
Tractate of the Supreme-Pole Adept on
Miscellaneous Elixir Recipes [with illus-
trations of alchemical apparatus].
Date unknown, but probably Sung on
account of the philosophical significance
of the pseudonym.
Writer unknown.
TT/939.

Thai-Chi Ko Hsien-Ong Chuan 太極葛仙翁傳.
Biography of the Supreme-Pole Elder-
Immortal Ko (Hsüan).
Prob. Ming.
Than Ssu-Hsien 譚嗣先.
TT/447.

Thai Hsi Ching 胎息經.
Manual of Embryonic Respiration.
Thang, +8th, c. +755.

Thai Hsi Ching (cont.)
　　Huan Chen hsien-sêng　幻眞先生
　　　　(Mr Truth-and-Illusion).
　　TT/127, and *YCCC*, ch. 60, pp. 22*b* ff.
　　Tr. Balfour (1).
　　Cf. Maspero (7), p. 211.
Thai Hsi Ching Wei Lun　胎息精微論.
　　Discourse on Embryonic Respiration and
　　　　the Subtlety of the Seminal Essence.
　　Thang or Sung.
　　Writer unknown.
　　In *YCCC*, ch. 58, pp. 1*a* ff.
　　Cf. Maspero (7), p. 210.
Thai Hsi Kên Chih Yao Chüeh　胎息根旨要訣.
　　Instruction on the Essentials of (Under-
　　　　standing) Embryonic Respiration [Taoist
　　　　respiratory and sexual techniques].
　　Thang or Sung.
　　Writer unknown.
　　In *YCCC*, ch. 58, pp. 4*b* ff.
　　Cf. Maspero (7), p. 380.
Thai Hsi Khou Chüeh　胎息口訣.
　　Oral Explanation of Embryonic Respiration.
　　Thang or Sung.
　　Writer unknown.
　　In *YCCC*, ch. 58, pp. 12*a* ff.
　　Cf Maspero (7), p. 198.
Thai Hsi Shui Fa　泰西水法.
　　Hydraulic Machinery of the West.
　　Ming, +1612.
　　Hsiung San-Pa (Sabatino de Ursis)　熊三拔
　　　　& Hsü Kuang-Chhi　徐光啓.
Thai Hsüan Pao Tien　太玄寶典.
　　Precious Records of the Great Mystery [of
　　　　attaining longevity and immortality by
　　　　physiological alchemy, *nei tan*].
　　Sung or Yuan, +13th or +14th.
　　Writer unknown.
　　TT/1022, and in *TTCY* (*shang mao chi*, 5).
Thai-I Chin Hua Tsung Chih　太一(or 乙)金華
　　宗旨.
　　Principles of the (Inner) Radiance of the
　　　　Metallous (Enchymoma), (explained in
　　　　terms of the) Undifferentiated Universe.
　　See *Chin Hua Tsung Chih*.
Thai-Ku Chi　太古集.
　　Collected Works of (Ho) Thai-Ku [Ho Ta-
　　　　Thung].
　　Sung, *c.* +1200.
　　Ho Ta-Thung　郝大通.
　　TT/1147.
Thai Ku Thu Tui Ching　太古土兌經.
　　Most Ancient Canon of the Joy of the Earth;
　　　　or, of the Element Earth and the Kua
　　　　Tui [mainly on the alchemical sub-
　　　　duing of metals and minerals].
　　Date unknown, perhaps Thang or slightly
　　　　earlier.
　　Attrib. Chang hsien-sêng　張先生.
　　TT/942.
Thai Pai Ching　太白經.
　　The Venus Canon.

Thang, *c.* +800.
　　Shih Chien-Wu　施肩吾.
　　TT/927.
Thai Phing Ching　太平經.
　　[= *Thai Phing Chhing Ling Shu*.]
　　Canon of the Great Peace (and Equality).
　　Ascr. H/Han, *c.* +150 (first mentioned
　　　　+166) but with later additions and inter-
　　　　polations.
　　Part attrib. Yü Chi　于吉.
　　Perhaps based on the *Thien Kuan Li Pao
　　　　Yuan Thai Phing Ching* (*c.* −35) of Kan
　　　　Chung-Kho　甘忠可.
　　TT/1087. Reconstructed text, ed. Wang
　　　　Ming (2).
　　Cf. Yü Ying-Shih (2), p. 84.
　　According to Hsiung Tê-Chi (1) the parts
　　　　which consist of dialogue between a
　　　　Heavenly Teacher and a disciple corre-
　　　　spond with what the *Pao Phu Tzu*
　　　　bibliography lists as *Thai Phing Ching*
　　　　and were composed by Hsiang Khai
　　　　襄楷.
　　The other parts would be for the most part
　　　　fragments of the *Chia I Ching*　甲乙經,
　　　　also mentioned in *Pao Phu Tzu*, and due
　　　　to Yü Chi and his disciple Kung Chhung
　　　　宮崇 between +125 and +145.
Thai Phing Chhing Ling Shu　太平清領書.
　　Received Book of the Great Peace and
　　　　Purity.
　　See *Thai Phing Ching*.
Thai-Phing Huan Yü Chi　太平寰宇記.
　　Thai-Phing reign-period General Descrip-
　　　　tion of the World [geographical record].
　　Sung, +976 to +983.
　　Yüeh Shih　樂史.
Thai-Phing Hui Min Ho Chi Chü Fang　太平惠
　　民和劑局方.
　　Standard Formularies of the (Government)
　　　　Great Peace People's Welfare Pharmacies
　　　　[based on the *Ho Chi Chü Fang*, etc.].
　　Sung, +1151.
　　Ed. Chhen Shih-Wên　陳師文, Phei
　　　　Tsung-Yuan　裴完元, and Chhen
　　　　Chhêng　陳承.
　　Cf. Li Thao (1, 6); SIC, p. 973.
Thai-Phing Kuang Chi　太平廣記.
　　Copious Records collected in the Thai-
　　　　Phing reign-period [anecdotes, stories,
　　　　mirabilia and memorabilia].
　　Sung, +978.
　　Ed. Li Fang　李昉.
Thai-Phing Shêng Hui Fang　太平聖惠方.
　　Prescriptions Collected by Imperial
　　　　Benevolence during the Thai-Phing
　　　　reign-period.
　　Sung, commissioned +982; completed
　　　　+992.
　　Ed. Wang Huai-Yin　王懷隱, Chêng Yen
　　　　鄭彥　*et al.*
　　SIC, p. 921; *Yü Hai*, ch. 63.

Thai-Phing Yü Lan 太平御覽.
Thai-Phing reign-period Imperial Encyclo-
paedia (lit. the Emperor's Daily Readings).
Sung, +983.
Ed. Li Fang 李昉.
Some chs. tr. Pfizmaier (84–106).
Yin-Tê Index, no. 23.

*Thai-Shang Chu Kuo Chiu Min Tsung Chen Pi
Yao* 太上助國救民總眞秘要.
Arcane Essentials of the Mainstream of
Taoism, for the Help of the Nation and
the Saving of the People; a Thai-Shang
Scripture [apotropaics and liturgy].
Sung +1116.
Yuan Miao-Tsung 元妙宗.
TT/1210.

Thai-Shang Chuan Hsi Wang Mu Wo Ku Fa
太上傳西王母握固法.
See *Chuan Hsi Wang Mu Wo Ku Fa*.

Thai-Shang Huang Thing Nei (or *Wai* or *Chung*)
Ching (*Yü*) *Ching* 太上黃庭內(外,中)
景(玉)經.
See *Huang Thing*, etc.

Thai-Shang Lao Chün Yang Shêng Chüeh 太上
老君養生訣.
Oral Instructions of Lao Tzu on Nourishing
the Life-Force; a Thai-Shang Scripture
[Taoist respiratory and gymnastic exer-
cises].
Thang.
Attrib. Hua Tho 華佗 and Wu Phu
吳普.
Actual writer unknown.
TT/814.

Thai-Shang Ling-Pao Chih Tshao Thu 太上靈
寶芝草圖.
Illustrations of the Numinous Mushrooms;
a Thai-Shang Ling-Pao Scripture.
Sui or pre-Sui.
Writer unknown.
TT/1387.

Thai-Shang Ling-Pao Wu Fu (*Ching*) 太上靈
寶五符(經).
(Manual of) the Five Categories of For-
mulae (for achieving Material and
Celestial Immortality); a Thai-Shang
Ling-Pao Scripture [liturgical].
San Kuo, mid +3rd.
Writers unknown.
TT/385.
On the term Ling-Pao see Kaltenmark (4).

Thai-Shang Pa-Ching Ssu-Jui Tzu-Chiang (*Wu-
Chu*) *Chiang-Shêng Shen Tan Fang* 太上
八景四蘂紫漿(五珠)降生神丹方.
Method for making the Eight-Radiances
Four-Stamens Purple-Fluid (Five-Pearl)
Incarnate Numinous Elixir; a Thai-
Shang Scripture.
Chin, probably late +4th.
Putatively dictated to Yang Hsi 楊羲.
In *YCCC*, ch. 68; another version in
TT/1357.

Thai-Shang Pa Ti Yuan (*Hsüan*) *Pien Ching*
太上八帝元(玄)變經.
See *Tung-Shen Pa Ti Yuan* (*Hsüan*) *Pien
Ching*.

Thai Shang-San-shih-liu pu Tsun Ching 太上
三十六部尊經.
The Venerable Scripture in 36 Sections.
TT/8.
See *Shang Chhing Ching*.

Thai-Shang Tung Fang Nei Ching Chu 太上洞
房內經注.
Esoteric Manual of the Innermost Chamber,
a Thai-Shang Scripture; with Commen-
tary.
Ascr. −1st cent.
Attrib. Chou Chi-Thung 周季通.
TT/130.

Thai-Shang Tung-Hsüan Ling-Pao Mieh Tu (or
San Yuan) *Wu Lien Shêng Shih Miao
Ching* 太上洞玄靈寶滅度 (or 三元)
五鍊生尸妙經.
Marvellous Manual of the Resurrection (or
Preservation) of the Body, giving Salvation
from Dispersal, by means of (the Three
Primary Vitalities and) the Five Trans-
mutations; a Ling-Pao Thai-Shang Tung-
Hsüan Scripture.
Date uncertain.
Writer unknown.
TT/366.

Thai-Shang Tung-Hsüan Ling-Pao Shou Tu I
太上洞玄靈寶授度儀.
Formulae for the Reception of Salvation; a
Thai-Shang Tung-Hsüan Ling-Pao
Scripture [liturgical].
L/Sung, *c.* +450.
Lu Hsiu-Ching 陸修靜.
TT/524.

*Thai-Shang Wei Ling Shen Hua Chiu Chuan
Tan Sha Fa* 太上衛靈神化九轉丹砂
法.
Methods of the Guardian of the Mysteries
for the Marvellous Thaumaturgical
Transmutation of Ninefold Cyclically
Transformed Cinnabar; a Thai-Shang
Scripture.
Sung, if not earlier.
Writer unknown.
TT/885.
Tr. Spooner & Wang (1); Sivin (3).

Thai-Shang Yang Shêng Thai Hsi Chhi Ching
太上養生胎息氣經.
See *Yang Shêng Thai Hsi Chhi Ching*.

Thai Tsang Lun 胎臟論.
Discourse on the Foetalisation of the
Viscera (the Restoration of the Embry-
onic Condition of Youth and Health).
Alternative title of *Chung Huang Chen
Ching* (q.v.).

*Thai-Wei Ling Shu Tzu-Wên Lang-Kan Hua
Tan Shen Chen Shang Ching* 太微靈書
紫文琅玕華丹神眞上經.

Thai-Wei Ling Shu Tzu-Wên Lang-Kan Hua Tan Shon Chen Shang Ching (*cont.*)
Divinely Written Exalted Spiritual Realisation Manual in Purple Script on the Lang-Kan (Gem) Radiant Elixir; a Thai-Wei Scripture.
Chin, late +4th century, possibly altered later.
Dictated to Yang Hsi 楊羲.
TT/252.

Thai-Wu hsien-sêng Fu Chhi Fa 太无先生服氣法.
See *Sung Shan Thai-Wu hsien-sêng Chhi Ching*.

Than hsien-sêng Shui Yün Chi 譚先生水雲集.
Mr Than's Records of Life among the Mountain Clouds and Waterfalls.
Sung, mid +12th cent.
Than Chhu-Tuan 譚處端.
TT/1146.

Thang Hui Yao 唐會要.
History of the Administrative Statutes of the Thang Dynasty.
Sung, +961.
Wang Phu 王溥.
Cf. des Rotours (2), p. 92.

Thang Liu Tien 唐六典.
Institutes of the Thang Dynasty (lit. Administrative Regulations of the Six Ministries of the Thang).
Thang, +738 or +739.
Ed. Li Lin-Fu 李林甫.
Cf. des Rotours (2), p. 99.

Thang Pên Tshao 唐本草.
Pharmacopoeia of the Thang Dynasty.
= *Hsin Hsiu Pên Tshao*, (q.v.).

Thang Yü Lin 唐語林.
Miscellanea of the Thang Dynasty.
Sung, collected c. +1107.
Wang Tang 王讜.
Cf. des Rotours (2), p. 109.

Thao Chen Jen Nei Tan Fu 陶眞人內丹賦.
See *Nei Tan Fu*.

Thi Kho Ko 體殼歌.
Song of the Bodily Husk (and the Deliverance from its Ageing).
Wu Tai or Sung, in any case before +1040
Yen Lo Tzu (ps.) 煙蘿子.
In *Hsiu Chen Shih Shu* (*TT*/260), ch. 18.

Thiao Chhi Ching 調氣經.
See *Thai-Chhing Thiao Chhi Ching*.

Thieh Wei Shan Tshung Than 鐵圍山叢談.
Collected Conversations at Iron-Fence Mountain.
Sung, c. +1115.
Tshai Thao 蔡絛.

Thien-Hsia Chün Kuo Li Ping Shu 天下郡國利病書.
Merits and Drawbacks of all the Countries in the World [geography].
Chhing, +1662.
Ku Yen-Wu 顧炎武.

Thien Hsien Chêng Li Tu 'Fa Tien Ching' 天仙正理讀法點睛.
The Right Pattern of the Celestial Immortals; Thoughts on Reading the *Consecration of the Law*.
See Fu Chin-Chhüan (2).

Thien Hsien Chih Lun Chhang Shêng Tu Shih Nei Lien Chin Tan (Chüeh Hsin) Fa 天仙直論長生度世內煉金丹(訣心)法.
(Confidential) Methods for Processing the Metallous Encymhoma; a Plain Discourse on Longevity and Immortality (according to the Principles of the) Celestial Immortals for the Salvation of the World.
Alternative title for *Nei Chin Tan* (q.v.).

Thien Kung Khai Wu 天工開物.
The Exploitation of the Works of Nature.
Ming, +1637.
Sung Ying-Hsing 宋應星.
Tr. Sun Jen I-Tu & Sun Hsüeh-Chuan (1).

Thien-thai Shan Fang Wai Chih 天臺山方外志.
Supplementary Historical Topography of Thien-thai Shan.
Ming.
Chhuan-Têng (monk) 傳燈.

Thien Ti Yin-Yang Ta Lo Fu 天地陰陽大樂賦.
Poetical Essay on the Supreme Joy.
Thang, c. +800.
Pai Hsing-Chien 白行簡.

Thien Yuan Ju Yao Ching 天元入藥鏡.
Mirror of the All-Penetrating Medicine (the Enchymoma; restoring the Endowment) of the Primary Vitalities.
Wu Tai, +940.
Tshui Hsi-Fan 崔希範.
In *Hsiu Chen Shih Shu* (*TT*/260), ch. 21, pp. 6b to 9b; a prose text without commentary, not the same as the *Ju Yao Ching* (q.v.) and ending with a diagram absent from the latter.
Cf. van Gulik (8), pp. 224 ff.

Tho Yo Tzu 橐籥子.
Book of the Bellows-and-Tuyère Master [physiological alchemy in mutationist terms].
Sung or Yuan.
Writer unknown.
TT/1174, and *TTCY* (hsin mao chi, 5).

Thou Huang Tsa Lu 投荒雜錄.
Miscellaneous Jottings far from Home.
Thang, c. +835.
Fang Chhien-Li 房千里.

Thu Ching (Pên Tshao) 圖經(本草).
Illustrated Treatise (of Pharmaceutical Natural History). See *Pên Tshao Thu Ching*.
The term *Thu Ching* applied originally to one of the two illustrated parts (the other being a *Yao Thu*) of the *Hsin Hsiu Pên*

Thu Ching (Pên Tshao) (cont.)
Tshao of +659 (q.v.); cf. *Hsin Thang Shu*, ch. 59, p. 21 *a* or *TSCCIW*, p. 273. By the middle of the +11th century these had become lost, so Su Sung's *Pên Tshao Thu Ching* was prepared as a replacement. The name *Thu Ching Pên Tshao* was often afterwards applied to Su Sung's work, but (according to the evidence of the *Sung Shih* bibliographies, *SSIW*, pp. 179, 529) wrongly.

Thu Ching Chi-Chu Yen I Pên Tshao 圖經集注衍義本草.
Illustrations and Collected Commentaries for the *Dilations upon Pharmaceutical Natural History.*
TT/761 (Ong index, no. 767).
See also *Thu Ching Yen I Pên Tshao.*
The *Tao Tsang* contains two separately catalogued books, but the *Thu Ching Chi-Chu Yen I Pên Tshao* is in fact the introductory 5 chapters, and the *Thu Ching Yen I Pên Tshao* the remaining 42 chapters of a single work.

Thu Ching Yen I Pên Tshao 圖經衍義本草.
Illustrations (and Commentary) for the *Dilations upon Pharmaceutical Natural History.* (An abridged conflation of the *Chêng-Ho...Chêng Lei...Pên Tshao* with the *Pên Tshao Yen I.*)
Sung, *c.* +1223.
Thang Shen-Wei 唐慎微, Khou Tsung-Shih 寇宗奭, ed. Hsü Hung 許洪.
TT/761 (Ong index, no. 768).
See also *Thu Ching Chi-Chu Yen I Pên Tshao.*
Cf. Chang Tsan-Chhen (2); Lung Po-Chien (1), nos. 38, 39.

Thu Hsiu Chen Chün Tsao-Hua Chih Nan 土宿眞君造化指南.
Guide to the Creation, by the Earth's Mansions Immortal.
See *Tsao-Hua Chih Nan.*

Thu Hsiu Pên Tshao 土宿本草.
The Earth's Mansions Pharmacopoeia.
See *Tsao-Hua Chih Nan.*

Thung Hsüan Pi Shu 通玄秘術.
The Secret Art of Penetrating the Mystery [alchemy].
Thang, soon after +864.
Shen Chih-Yen 沈知言.
TT/935.

Thung Su Pien 通俗編.
Thesaurus of Popular Terms, Ideas and Customs.
Chhing, +1751.
Tsê Hao 翟灝.

Thung Ya 通雅.
Helps to the Understanding of the *Literary Expositor* [general encyclopaedia with much of scientific and technological interest].
36

Ming and Chhing, finished +1636, pr. +1666.
Fang I-Chih 方以智.

Thung Yu Chüeh 通幽訣.
Lectures on the Understanding of the Obscurity (of Nature) [alchemy, protochemical and physiological].
Not earlier than Thang.
Writer unknown.
TT/906.
Cf. Chhen Kuo-Fu (1), vol. 2, p. 390.

Tien Hai Yü Hêng Chih 滇海虞衡志.
A Guide to the Region of the Kunming Lake (Yunnan).
Chhing, *c.* +1770, pr. +1799.
Than Tshui 檀萃.

Tien Shu 典術.
Book of Arts.
L/Sung.
Wang Chien-Phing 王建平.

Ting Chhi Ko 鼎器歌.
Song (or, Mnemonic Rhymes) on the (Alchemical) Reaction-Vessel.
Han, if indeed originally, as it is now, a chapter of the *Chou I Tshan Thung Chhi* (q.v.).
It has sometimes circulated separately.
In *Chou I Tshan Thung Chhi Fên Chang Chu Chieh*, ch. 33 (ch. 3, pp. 7 *a* ff.).
Cf. *Chou I Tshan Thung Chhi Ting Chhi Ko Ming Ching Thu* (*TT*/994).

Ton Isho 頓醫抄.
Medical Excerpts Urgently Copied.
Japan, +1304.
Kajiwara Shozen 梶原性全.

Tongŭi Pogam 東醫寶鑑.
See *Tung I Pao Chien.*

Tou hsien-sêng Hsiu Chen Chih Nan 竇先生修眞指南.
See *Hsi Yo Tou hsien-sêng Hsiu Chen Chih Nan.*

Tsao Hua Chhien Chhui 造化鉗鎚.
The Hammer and Tongs of Creation (i.e. Nature).
Ming, *c.* +1430.
Chu Chhüan 朱權.
(Ning Hsien Wang 寧獻王, prince of the Ming.)

Tsao-Hua Chih Nan 造化指南.
[= *Thu Hsiu Pên Tshao.*]
Guide to the Creation (i.e. Nature).
Thang, Sung or possibly Ming. A date about +1040 may be the best guess, as there are similarities with the *Wai Tan Pên Tshao* (q.v.).
Thu Hsiu Chen Chün 土宿眞君 (the Earth's Mansions Immortal).
Preserved only in quotation, as in *PTKM.*

Tsê Ko Lu 則克錄.
Methods of Victory.
Title, in certain editions, of the *Huo Kung Chieh Yao* (q.v.).

Tsêng Kuang Chih Nang Pu 增廣智囊補.
Additions to the *Enlarged Bag of Wisdom*
Supplemented.
Ming, c. +1620.
Fêng Mêng-Lung 馮夢龍.

Tshai Chen Chi Yao 採眞機要.
Important (Information on the) Means (by
which one can) Attain (the Regeneration
of the) Primary (Vitalities) [physiological
alchemy, poems and commentary].
Part of *San-Fêng Tan Chüeh* (q.v.).

Tshan Thung Chhi 參同契.
The Kinship of the Three; or, The Accor-
dance (of the *Book of Changes*) with the
Phenomena of Composite Things
[alchemy].
H/Han, +142.
Wei Po-Yang 魏伯陽.

Tshan Thung Chhi.
See also titles under *Chou I Tshan Thung
Chhi.*

Tshan Thung Chhi Chang Chü 參同契章句.
The *Kinship of the Three* (arranged in)
Chapters and Sections.
Chhing, +1717.
Ed. Li Kuang-Ti 李光地.

Tshan Thung Chhi Khao I 參同契考異.
[= *Chou I Tshan Thung Chhi Chu.*]
A Study of the *Kinship of the Three.*
Sung, +1197.
Chu Hsi 朱熹 (originally using pseudonym
Tsou Hsin 鄒訢).
TT/992.

Tshan Thung Chhi Shan Yu 參同契闡幽.
Explanation of the Obscurities in the *Kin-
ship of the Three.*
Chhing, +1669, pref. +1729, pr. +1735.
Ed and comm. Chu Yuan-Yü 朱元育.
TTCY.

Tshan Thung Chhi Wu Hsiang Lei Pi Yao 參同
契五相類祕要.
Arcane Essentials of the Similarities and
Categories of the Five (Substances) in the
Kinship of the Three (sulphur, realgar,
orpiment, mercury and lead).
Liu Chhao, possibly Thang; prob. between
+3rd and +7th cents., must be before
the beginning of the +9th cent., though
ascr. +2nd.
Writer unknown (attrib. Wei Po-Yang).
Comm. by Lu Thien-Chi 盧天驥, wr.
Sung, +1111 to +1117, probably +1114.
TT/898.
Tr. Ho Ping-Yü & Needham (2).

Tshao Mu Tzu 草木子.
The Book of the Fading-like-Grass Master.
Ming, +1378.
Yeh Tzu-Chhi 葉子奇.

Tshê Fu Yuan Kuei 册府元龜.
Collection of Material on the Lives of
Emperors and Ministers, (lit. (Lessons of)
the Archives, (the True) Scapulimancy);

[a governmental ethical and political
encyclopaedia.]
Sung, commissioned +1005, pr. +1013.
Ed. Wang Chhin-Jo 王欽若 & Yang I
楊億.
Cf. des Rotours (2), p. 91.

Tshui Hsü Phien 翠虛篇.
Book of the Emerald Heaven.
Sung, c. +1200.
Chhen Nan 陳楠.
TT/1076.

Tshui Kung Ju Yao Ching Chu (or Ho) Chieh
崔公入藥鏡註(合)解.
See *Ju Yao Ching* and *Thien Yuan Ju Yao
Ching.*

Tshun Chen Huan Chung Thu 存眞環中圖.
Illustrations of the True Form (of the Body)
and of the (Tracts of) Circulation (of the
Chhi).
Sung, +1113.
Yang Chieh 楊介.
Now partially preserved only in the *Ton-
Isho* and the *Man-Anpō* (q.v.). Some of
the drawings are in Chu Hung's *Nei
Wai Erh Ching Thu*, also in *Hua Tho
Nei Chao Thu* and *Kuang Wei Ta Fa*
(q.v.).

Tshun Fu Chai Wên Chi 存復齋文集.
Literary Collection of the Preservation-and-
Return Studio.
Yuan, +1349.
Chu Tê-Jun 朱德潤.

Tso Chuan 左傳.
Master Tso chhiu's Tradition (or Enlargement) of the *Chhun Chhiu* (*Spring and
Autumn Annals*), [dealing with the period
−722 to −453].
Late Chou, compiled from ancient written
and oral traditions of several States be-
tween −430 and −250, but with addi-
tions and changes by Confucian scholars
of the Chhin and Han, especially Liu
Hsin. Greatest of the three commen-
taries on the *Chhun Chhiu*, the others
being the *Kungyang Chuan* and the
Kuliang Chuan, but unlike them, prob-
ably originally itself an independent book
of history.
Attrib. Tsochhiu Ming 左邱明.
See Karlgren (8); Maspero (1); Chhi Ssu-
Ho (1); Wu Khang (1); Wu Shih-
Chhang (1); van der Loon (1), Eberhard,
Müller & Henseling (1).
Tr. Couvreur (1); Legge (11); Pfizmaier
(1–12).
Index by Fraser & Lockhart (1).

Tso Wang Lun 坐忘論.
Discourse on (Taoist) Meditation.
Thang, c. +715.
Ssuma Chhêng-Chên 司馬承貞.
TT/1024, and in TTCY (*shang mao chi*,
5).

Tsui Shang I Chhêng Hui Ming Ching 最上一乘慧命經.
Exalted Single-Vehicle Manual of the Sagacious (Lengthening of the) Life-Span.
See *Hui Ming Ching*.

Tsun Shêng Pa Chien 遵生八牋.
Eight Disquisitions on Putting Oneself in Accord with the Life-Force [a collection of works].
Ming, +1591.
Kao Lien 高濂.
For the separate parts see:
1. *Chhing Hsiu Miao Lun Chien* (chs. 1, 2).
2. *Ssu Shih Thiao Shê Chien* (chs. 3-6).
3. *Chhi Chü An Lo Chien* (chs. 7, 8).
4. *Yen Nien Chhio Ping Chien* (chs. 9, 10).
5. *Yin Chuan Fu Shih Chien* (chs. 11-13).
6. *Yen Hsien Chhing Shang Chien* (chs. 14, 15).
7. *Ling Pi Tan Yao Chien* (chs. 16-18).
8. *Lu Wai Hsia Chü Chien* (ch. 19).

Tsurezuregusa 徒然草.
Gleanings of Leisure Moments [miscellanea, with much on Confucianism, Buddhism and Taoist philosophy].
Japan, c. +1330.
Kenkō hōshi 兼好法師 (Yoshida no Kaneyoshi 吉田兼好).
Cf. Anon. (103), pp. 197 ff.

Tu Hsing Tsa Chih 獨醒雜志.
Miscellaneous Records of the Lone Watcher.
Sung, +1176.
Tsêng Min-Hsing 曾敏行.

Tu I Chih 獨異志.
Things Uniquely Strange.
Thang.
Li Jung 李冗 (or 冗).

Tu Jen Ching 度人經.
See *Ling-Pao Wu Liang Tu Jon Shang Phin Miao Ching*.

Tu Shih Fang Yü Chi Yao 讀史方輿紀要.
Essentials of Historical Geography.
Chhing, first pr. +1667, greatly enlarged before the author's death in +1692, and pr. c. +1799.
Ku Tsu-Yü 顧祖禹.

Tung-Chen Ling Shu Tzu-Wên Lang-Kan Hua Tan Shang Ching 洞眞靈書紫文琅玕華丹上經.
Divinely Written Exalted Manual in Purple Script on the Lang-Kan (Gem) Radiant Elixir; a Tung-Chen Scripture.
Alternative name of *Thai-Wei Ling Shu Tzu-Wên Lang-Kan Hua Tan Shen Chen Shang Ching* (q.v.).

Tung-Chen Thai-Shang Su-Ling Tung-Yuan Ta Yu Miao Ching 洞眞太上素靈洞元大有妙經.
See *Ta Yu Miao Ching*.

Tung-Chen Thai-Wei Ling Shu Tzu-Wên Shang Ching 洞眞太微靈書紫文上經.
Divinely Written Exalted Canon in Purple Script; a Tung-Chen Thai-Wei Scripture.
See *Thai-Wei Ling Shu Tzu-Wên Lang-Kan Hua Tan Shen Chen Shang Ching*, which it formerly contained.

Tung Hsien Pi Lu 東軒筆錄.
Jottings from the Eastern Side-Hall.
Sung, end +11th.
Wei Thai 魏泰.

Tung-Hsüan Chin Yü Chi 洞玄金玉集.
Collections of Gold and Jade; a Tung-Hsüan Scripture.
Sung, mid +12th cent.
Ma Yü 馬鈺.
TT/1135.

Tung-Hsüan Ling-Pao Chen Ling Wei Yeh Thu 洞玄靈寶眞靈位業圖.
Charts of the Ranks, Positions and Attributes of the Perfected (Immortals); a Tung-Hsüan Ling-Pao Scripture.
Ascr. Liang, early +6th.
Attrib. Thao Hung-Ching 陶弘景.
TT/164.

Tung Hsüan Tzu 洞玄子.
Book of the Mystery-Penetrating Master.
Pre-Thang, perhaps +5th century.
Writer unknown.
In *Shuang Mei Ching An Tshung Shu*.
Tr van Gulik (3).

Tung I Pao Chien 東醫寶鑑.
Precious Mirror of Eastern Medicine [system of medicine].
Korea, commissioned in +1596, presented +1610, printed +1613.
Hǒ Chun 許浚.

Tung-Pho Shih Chi Chu 東坡詩集注.
[= *Mei-Chhi Shih Chu*.]
Collected Commentaries on the Poems of (Su) Tung-Pho.
Sung, c. +1140.
Wang Shih-Phêng 王十朋 (i.e. Wang Mei-Chhi 王梅溪).

Tung Shen Ching 洞神經.
See *Tung Shen Pa Ti Miao Ching Ching* and *Tung Shen Pa Ti Yuan Pien Ching*.

Tung Shen Pa Ti Miao Ching Ching 洞神八帝妙精經.
Mysterious Canon of Revelation of the Eight (Celestial) Emperors; a Tung-Shen Scripture.
Date uncertain, perhaps Thang but more probably earlier.
Writer unknown.
TT/635.

Tung Shen Pa Ti Yuan (Hsüan) Pien Ching 洞神八帝元(玄)變經.
Manual of the Mysterious Transformations of the Eight (Celestial) Emperors; a Tung-Shen Scripture [nomenclature of

Tung Shen Pa Ti Yuan (Hsüan) Pien Ching
    (*cont.*)
    spiritual beings, invocations, exorcisms,
    techniques of rapport].
    Date uncertain, perhaps Thang but more
    probably earlier.
    Writer unknown.
    TT/1187.
Tzu Chin Kuang Yao Ta Hsien Hsiu Chen Yen I
    紫金光耀大仙修眞演義.
    See *Hsiu Chen Yen I*.
Tzu-Jan Chi 自然集.
    Collected (Poems) on the Spontaneity of
    Nature.
    Sung, mid +12th cent.
    Ma Yü 馬鈺.
    TT/1130.
Tzu-Yang Chen Jen Nei Chuan 紫陽眞人內傳.
    Biography of the Adept of the Purple Yang.
    H/Han, San Kuo or Chin, before +399.
    Writer unknown.
    This Tzu-Yang Chen Jen was Chou I-Shan
    周義山 (not to be confused with Chang
    Po-Tuan).
    Cf. Maspero (7), p. 201; (13), pp. 78, 103.
    TT/300.
Tzu-Yang Chen Jen Wu Chen Phien 紫陽眞人
    悟眞篇.
    See *Wu Chen Phien*.
Tzu Yang Tan Fang Pao Chien Chih Thu 紫陽
    丹房寶鑑之圖.
    See *Tan Fang Pao Chien Chih Thu*.

Wai Chin Tan 外金丹.
    Disclosures (of the Nature of) the Metallous
    Enchymoma [a collection of some thirty
    tractates on *nei tan* physiological alchemy,
    ranging in date from Sung to Chhing and
    of varying authenticity].
    Sung to Chhing.
    Ed. Fu Chin-Chhüan 傅金銓, *c.* 1830.
    In *CTPS*, *pên* 6-10 incl.
Wai Kho Chêng Tsung 外科正宗.
    An Orthodox Manual of External Medicine.
    Ming, +1617.
    Chhen Shih-Kung 陳實功.
Wai Kuo Chuan 外國傳.
    See *Wu Shih Wai Kuo Chuan*.
Wai Tan Pên Tshao 外丹本草.
    Iatrochemical Natural History.
    Early Sung, *c.* +1045.
    Tshui Fang 崔昉.
    Now extant only in quotations.
    Cf. *Chin Tan Ta Yao Pao Chüeh* and *Ta
    Tan Yao Chüeh Pên Tshao*.
Wai Thai Pi Yao (Fang) 外臺秘要(方).
    Important (Medical) Formulae and Pre-
    scriptions now revealed by the Governor
    of a Distant Province.
    Thang, +752.
    Wang Thao 王燾.
    On the title see des Rotours (1), pp. 294,

721. Wang Thao had had access to the
    books in the Imperial Library as an
    Academician before his posting as a high
    official to the provinces.
Wakan Sanzai Zue 和漢三才圖會.
    The Chinese and Japanese Universal
    Encyclopaedia (based on the *San Tshai
    Thu Hui*).
    Japan, +1712.
    Terashima Ryōan 寺島良安.
Wamyō-Honzō. See *Honzō-Wamyō.*
Wamyō Ruijūshō 和 (or 倭) 名類聚抄.
    General Encyclopaedic Dictionary.
    Japan (Heian), +934.
    Minamoto no Shitagau 源順.
Wamyōshō 和名抄.
    See *Wamyō Ruijushō.*
Wan Hsing Thung Phu 萬姓統譜.
    General Dictionary of Biography.
    Ming, +1579.
    Ling Ti-Chih 凌迪知.
Wan Ping Hui Chhun 萬病回春.
    The Restoration of Well-Being from a
    Myriad Diseases.
    Ming, +1587, pr. +1615.
    Kung Thing-Hsien 龔廷賢.
Wan Shou Hsien Shu 萬壽仙書.
    A Book on the Longevity of the Immortals
    [longevity techniques, especially gym-
    nastics and respiratory exercises].
    Chhing, +18th.
    Tshao Wu-Chi 曹無極.
    Included in Pa Tzu-Yuan (*1*).
Wang Hsien Fu 望仙賦.
    Contemplating the Immortals; a Hymn of
    Praise [ode on Wangtzu Chhiao and
    Chhih Sung Tzu].
    C/Han, −14 or −13.
    Huan Than 桓譚.
    In *CSHK* (Hou Han sect.), ch. 12, p. 7*b*;
    and several encyclopaedias.
Wang Lao Fu Chhi Khou Chüeh 王老服氣口
    訣.
    See *Thai-Chhing Wang Lao Fu Chhi Khou
    Chüeh.*
*Wang-Wu Chen-Jen Khou Shou Yin Tan Pi
    Chüeh Ling Phien* 王屋眞人口授陰丹
    秘訣靈篇.
    Numinous Record of the Confidential Oral
    Instructions on the Yin Enchymoma
    handed down by the Adept of Wang-Wu
    (Shan).
    Thang, perhaps *c.* +765; certainly between
    +8th and late +10th.
    Probably Liu Shou 劉守.
    In *YCCC*, ch. 64, pp. 13*a* ff.
*Wang-Wu Chen-Jen Liu Shou I Chen-Jen Khou
    Chüeh Chin Shang* 王屋眞人劉守依眞
    人口訣進上.
    Confidential Oral Instructions of the Adept
    of Wang-Wu (Shan) presented to the
    Court by Liu Shou.

Wang-Wu Chen-Jen Liu Shou I Chen-Jen Khou
Chüeh Chin Shang (cont.)
Thang, *c.* +785 (after +780); certainly
between +8th and late +10th.
Liu Shou 劉守.
In *YCCC*, ch. 64, pp. 14*a* ff.
Wei Lüeh 緯畧.
Compendium of Non-Classical Matters.
Sung, +12th century (end), *c.* +1190.
Kao Ssu-Sun 高似孫.
Wei Po-Yang Chhi Fan Tan Sha Chüeh.
See *Chhi Fan Tan Sha Chüeh.*
Wei Shêng I Chin Ching 衛生易筋經.
See *I Chin Ching.*
Wei Shu 魏書.
History of the (Northern) Wei Dynasty
[+386 to +550, including the Eastern
Wei successor State].
N/Chhi, +554, revised +572.
Wei Shou 魏收.
See Ware (3).
One ch. tr. Ware (1, 4).
For translations of passages, see the index
of Frankel (1).
Wên Shih Chen Ching 文始眞經.
True Classic of the Original Word (of Lao
Chün, third person of the Taoist
Trinity).
Alternative title of *Kuan Yin Tzu* (q.v.).
Wên Yuan Ying Hua 文苑英華.
The Brightest Flowers in the Garden of
Literature [imperially commissioned
collection, intended as a continuation of
the *Wên Hsüan* (q.v.) and containing
therefore compositions written between
+500 and +960].
Sung, +987; first pr. +1567.
Ed. Li Fang 李昉, Sung Pai 宋白 *et
al.*
Cf des Rotours (2), p. 93.
Wu Chen Phien 悟眞篇.
[= *Tzu-Yang Chen Jen Wu Chen Phien.*]
Poetical Essay on Realising (the Necessity
of Regenerating the) Primary (Vitalities)
[Taoist physiological alchemy].
Sung, +1075.
Chang Po-Tuan 張伯端.
In, e.g., *Hsiu Chen Shih Shu* (*TT*/260), chs.
26–30 incl.
TT/138. Cf. *TT*/139-43.
Tr. Davis & Chao Yün-Tshung (7).
*Wu Chen Phien Chih Chih Hsiang Shuo San
Chhêng Pi Yao* 悟眞篇直指祥說三乘
祕要.
Precise Explanation of the Difficult Essen-
tials of the *Essay on Realising the Neces-
sity of Regenerating the Primary Vitalities*,
in accordance with the Three Classes of
(Taoist) Scriptures.
Sung, *c.* +1170.
Ong Pao-Kuang 翁葆光.
TT/140.

Wu Chen Phien San Chu 悟眞篇三註.
Three Commentaries on the *Essay on
Realising the Necessity of Regenerating the
Primary Vitalities* [Taoist physiological
alchemy].
Sung and Yuan, completed *c.* +1331.
Hsüeh Tao-Kuang 薛道光 (or Ong Pao-
Kuang 翁葆先), Lu Shu 陸墅 &
Tai Chhi-Tsung 戴起宗 (or Chhen
Chih-Hsü 陳致虛).
TT/139.
Cf. Davis & Chao Yün-Tshung (7).
Wu Chhêng Tzu 務成子.
See *Huang Thing Wai Ching Yü Ching
Chu.*
Wu Chhu Ching 五厨經.
See *Lao Tzu Shuo Wu Chhu Ching.*
Wu Hsiang Lei Pi Yao 五相類祕要.
See *Tshan Thung Chhi Wu Hsiang Lei Pi
Yao.*
Wu Hsing Ta I 五行大義.
Main Principles of the Five Elements.
Sui, *c.* +600.
Hsiao Chi 蕭吉.
Wu Hsüan Phien 悟玄篇.
Essay on Understanding the Mystery (of
the Enchymoma), [Taoist physiological
alchemy].
Sung, +1109 or +1169.
Yü Tung-Chen 余洞眞.
TT/1034, and in *TTCY* (*shang mao chi*,
5).
Wu I Chi 武夷集.
The Wu-I Mountains Literary Collection
[prose and poems on physiological
alchemy].
Sung, *c.* +1220.
Ko Chhang-Kêng 葛長庚 (Pai Yü-Chhan
白玉蟾).
In *Hsiu Chen Shih Shu* (*TT*/260), chs. 45–52.
Wu Kên Shu 無根樹.
The Rootless Tree [poems on physiological
alchemy].
Ming, *c.* +1410 (if genuine).
Attrib. Chang San-Fêng 張三峯.
In *San-Fêng Tan Chüeh* (q.v.).
Wu Lei Hsiang Kan Chih 物類相感志.
On the Mutual Responses of Things accord-
ing to their Categories.
Sung, *c.* +980.
Attrib. wrongly to Su Tung-Pho 蘇東
坡.
Actual writer (Lu) Tsan-Ning (monk)
錄贊寧.
See Su Ying-Hui (1, 2).
Wu Li Hsiao Shih 物理小識.
Small Encyclopaedia of the Principles of
Things.
Ming and Chhing, finished by +1643, pr.
+1664.
Fang I-Chih 方以智.
Cf. Hou Wai-Lu (3, 4).

Wu Lu 吳錄.
 Record of the Kingdom of Wu.
 San Kuo, +3rd century.
 Chang Pho 張勃.
Wu Shang Pi Yao 無上秘要.
 Essentials of the Matchless Books (of
  Taoism), [a florilegium].
 N/Chou, between +561 and +578.
 Compiler unknown.
 TT/1124.
 Cf. Maspero (13), p. 77; Schipper (1), p. 11.
Wu shih Pên Tshao 吳氏本草.
 Mr Wu's Pharmaceutical Natural History.
 San Kuo (Wei), c. +235.
 Wu Phu 吳普.
 Extant only in quotations in later literature.
Wu Shih Wai Kuo Chuan 吳時外國傳.
 Records of the Foreign Countries in the
  Time of the State of Wu.
 San Kuo, c. +260.
 Khang Thai 康泰.
 Only in fragments in *TPYL* and other
  sources.
Wu Tai Shih Chi.
 See *Hsin Wu Tai Shih.*
Wu Yuan 物原.
 The Origins of Things.
 Ming, +15th.
 Lo Chhi 羅頎.

Yang Hsing Yen Ming Lu 養性延命錄.
 On Delaying Destiny by Nourishing the
  Natural Forces (or, Achieving Longevity
  and Immortality by Regaining the Vitality
  of Youth), [Taoist sexual and respiratory
  techniques].
 Sung, betw. +1013 and +1161 (acc. to
  Maspero), but as it appears in *YCCC* it
  must be earlier than +1020, very prob-
  ably pre-Sung.
 Attrib. Thao Hung-Ching or Sun Ssu-Mo.
 Actual writer unknown.
 TT/831, abridged version in *YCCC*, ch. 32,
  pp. 1 a ff.
 Cf. Maspero (7), p. 232.
Yang Hui Suan Fa 楊輝算法.
 Yang Hui's Methods of Computation.
 Sung, +1275.
 Yang Hui 楊輝.
Yang Shêng Shih Chi 養生食忌.
 Nutritional Recommendations and Pro-
  hibitions for Health [appended to *Pao
  Shêng Hsin Chien*, q.v.].
 Ming, c. +1506.
 Thieh Fêng Chü-Shih 鐵峰居士.
  (The Recluse of Iron Mountain, ps.).
 Ed. Hu Wên-Huan (c. +1596) 胡文煥.
Yang Shêng Tao Yin Fa 養生導引法.
 Methods of Nourishing the Vitality by
  Gymnastics (and Massage), [appended to
  Pao Shêng Hsin Chien, q.v.].
 Ming, c. +1506.

Thieh Fêng Chü-Shih 鐵峰居士.
 (The Recluse of Iron Mountain, ps.)
 Ed. Hu Wên-Huan (c. +1596) 胡文煥.
Yang Shêng Thai Hsi Chhi Ching 養生胎息氣
  經.
 [= *Thai-Shang Yang Shêng Thai Hsi Chhi
  Ching.*]
 Manual of Nourishing the Life-Force (or,
  Attaining Longevity and Immortality) by
  Embryonic Respiration.
 Late Thang or Sung.
 Writer unknown.
 TT/812.
 Cf. Maspero (7), pp. 358, 365.
Yang Shêng Yen Ming Lu 養生延命錄.
 On Delaying Destiny by Nourishing the
  Natural Forces.
 Alternative title for *Yang Hsing Yen Ming
  Lu* (q.v.).
Yao Chung Chhao 藥種抄.
 Memoir on Several Varieties of Drug Plants.
 Japan, c. +1163.
 Kuan-Yu (Kanyu) 觀祐. MS. preserved
  at the 滋賀石山寺 Temple. Facsim.
  reprod. in Suppl. to the Japanese Tripiṭaka,
  vol. 11.
Yao Hsing Lun 藥性論.
 Discourse on the Natures and Properties of
  Drugs.
 Liang (or Thang, if identical with *Pên
  Tshao Yao Hsing*, q.v.).
 Attrib. Thao Hung-Ching 陶弘景.
 Only extant in quotations in books on
  pharmaceutical natural history.
 ICK, p. 169.
Yao Hsing Pên Tshao 藥性本草.
 See *Pên Tshao Yao Hsing.*
Yao Ming Yin Chüeh 藥名隱訣.
 Secret Instructions on the Names of Drugs
  and Chemicals.
 Perhaps an alternative title for the *Thai-
  Chhing Shih Pi Chi* (q.v.).
Yeh Chung Chi 鄴中記.
 Record of Affairs at the Capital of the Later
  Chao Dynasty.
 Chin.
 Lu Hui 陸翽.
 Cf. Hirth (17).
Yen Fan Lu 演繁露.
 Extension of the *String of Pearls* (on the
  Spring and Autumn Annals), [on the
  meaning of many Thang and Sung
  expressions].
 Sung, +1180.
 Chhêng Ta-Chhang 程大昌.
 See des Rotours (1), p. cix.
Yen Hsien Chhing Shang Chien 燕閒清賞牋.
 The Use of Leisure and Innocent Enjoy-
  ments in a Retired Life [the sixth part
  (chs. 14, 15) of *Tsun Shêng Pa Chien*, q.v.].
 Ming, +1591.
 Kao Lien 高濂.

Yen I I Mou Lu 燕翼詒謀錄.
Handing Down Good Plans for Posterity
from the Wings of Yen.
Sung, +1227.
Wang Yung 王栐.

*Yen-Ling hsien-sêng Chi Hsin Chiu Fu Chhi
Ching* 延陵先生集新舊服氣經.
New and Old Manuals of Absorbing the Chhi,
Collected by the Teacher of Yen-Ling.
Thang, early +8th, *c.* +745.
Writer unidentified.
Comm. by Sang Yü Tzu (+9th or +10th)
桑榆子.
TT/818, and (partially) in *YCCC*, ch. 58,
p. 2*a et passim*, ch. 59, pp. 1*a* ff., 18*b* ff.,
ch. 61, pp. 19*a* ff.
Cf. Maspero (7), pp. 220, 222.

Yen Mên Kung Miao Chieh Lu 鴈門公妙解錄.
The Venerable Yen Mên's Record of Marvellous Antidotes [alchemy and elixir
poisoning].
Thang, probably in the neighbourhood of
+847 since the text is substantially
identical with the *Hsüan Chieh Lu*
(q.v.) of this date.
Yen Mên 鴈門 (perhaps a ps. taken
from the pass and fortress on the
Great Wall, cf. Vol. 4, pt. 3, pp. 11,
48 and Fig. 711).
TT/937.

Yen Nien Chhio Ping Chien 延年却病牋.
How to Lengthen one's Years and Ward off
all Diseases [the fourth part (chs. 9, 10)
of *Tsun Shêng Pa Chien*, q.v.].
Ming, +1591.
Kao Lien 高濂.
Partial tr. of the gymnastic material,
Dudgeon (1).

Yen Shou Chhih Shu 延壽赤書.
Red Book on the Promotion of Longevity.
Thang, perhaps Sui.
Phei Yü (or Hsüan) 裴煜 (玄).
Extant only in excerpts preserved in the
I Hsin Fang (+982), SIC, p. 465.

Yen Thieh Lun 鹽鐵論.
Discourses on Salt and Iron [record of the
debate of −81 on State control of com-
merce and industry].
C/Han, *c.* −80 to −60.
Huan Khuan 桓寬.
Partial tr. Gale (1); Gale, Boodberg & Lin.

Yin Chen Chün Chin Shih Wu Hsiang Lei 陰眞
君金石五相類.
Alternative title of *Chin Shih Wu Hsiang
Lei* (q.v.).

Yin Chen Jen Liao Yang Tien Wên Ta Pien
尹眞人寥陽殿問答編.
See *Liao Yang Tien Wên Ta Pien*.

*Yin Chen Jen Tung-Hua Chêng Mo Huang Chi
Ho Pi Chêng Tao Hsien Ching* 尹眞人
東華正脈皇極闔闢證道仙經.
See *Huang Chi Ho Pi Hsien Ching*.

Yin Chuan Fu Shih Chien 飲饌服食牋.
Explanations on Diet, Nutrition and
Clothing [the fifth part (chs. 11–13) of
Tsun Shêng Pa Chien, q.v.].
Ming, +1591.
Kao Lien 高濂.

Yin Fu Ching 陰符經.
The Harmony of the Seen and the Unseen.
Thang, *c.* +735 (unless in essence a pre-
served late Warring States document).
Li Chhüan 李筌.
TT/30.
Cf. *TT*/105–124. Also in *TTCY* (*tou chi*, 6).
Tr. Legge (5).
Cf. Maspero (7), p. 222.

Yin Shan Chêng Yao 飲膳正要.
Principles of Correct Diet [on deficiency
diseases, with the aphorism 'many
diseases can be cured by diet alone'].
Yuan, +1330, re-issued by imperial order
in +1456.
Hu Ssu-Hui 忽思慧.
See Lu & Needham (1).

Yin Tan Nei Phien 陰丹內篇.
Esoteric Essay on the Yin Enchymoma.
Appendix to the *Tho Yo Tzu* (q.v.).

*Yin-Yang Chiu Chuan Chhêng Tzu-Chin Tien-
Hua Huan Tan Chüeh* 陰陽九轉成紫
金點化還丹訣.
Secret of the Cyclically Transformed Elixir,
Treated through Nine Yin-Yang Cycles
to Form Purple Gold and Projected to
Bring about Transformation.
Date unknown.
Writer unknown, but someone with Mao
Shan affiliations.
TT/888.

Ying Chhan Tzu Yü Lu 瑩蟾子語錄.
Collected Discourses of the Luminous-
Toad Master.
Yuan, *c.* +1320.
Li Tao-Shun 李道純 (Ying Chhan Tzu
瑩蟾子).
TT/1047.

Ying Yai Shêng Lan 瀛涯勝覽.
Triumphant Visions of the Ocean Shores
[relative to the voyages of Chêng Ho].
Ming, +1451. (Begun +1416 and com-
pleted about +1435.)
Ma Huan 馬歡.
Tr. Mills (11); Groeneveldt (1); Phillips (1);
Duyvendak (10).

Ying Yai Shêng Lan Chi 瀛涯勝覽集.
Abstract of the *Triumphant Visions of the
Ocean Shores* [a refacimento of Ma Huan's
book].
Ming, +1522.
Chang Shêng (b) 張昇.
Passages cit. in *TSCC*, *Pien i tien*, chs. 58,
73, 78, 85, 86, 96, 97, 98, 99, 101, 103,
106.
Tr. Rockhill (1).

Yōjōkun 養生訓.
Instructions on Hygiene and the Prolongation of Life.
Japan (Tokugawa), *c.* +1700.
Kaibara Ekiken 貝原益軒 (ed. Sugiyasu Saburō 杉靖三郎).

Yü-Chhing Chin-Ssu Chhing-Hua Pi-Wên Chin-Pao Nei-Lien Tan Chüeh 玉清金笥青華祕文金寶內鍊丹訣.
The Green-and-Elegant Secret Papers in the Jade-Purity Golden Box on the Essentials of the Internal Refining of the Golden Treasure, the Enchymoma.
Sung, late +11th century.
Chang Po-Tuan 張伯端.
TT/237.
Cf. Davis & Chao Yün-Tshung (5).

Yü-Chhing Nei Shu 玉清內書.
Inner Writings of the Jade-Purity (Heaven).
Probably Sung, but present version incomplete, and some of the material may be, or may have been, older.
Compiler unknown.
TT/940.

Yü Fang Chih Yao 玉房指要.
Important Matters of the Jade Chamber.
Pre-Sui, perhaps +4th century.
Writer unknown.
In *I Hsin Fang* (*Ishinhō*) and *Shuang Mei Ching An Tshung Shu*.
Partial trs. van Gulik (3, 8).

Yü Fang Pi Chüeh 玉房祕訣.
Secret Instructions concerning the Jade Chamber.
Pre-Sui, perhaps +4th century.
Writer unknown.
Partial tr. van Gulik (3).
Only as fragment in *Shuang Mei Ching An Tshung Shu* (q.v.).

Yu Huan Chi Wên 游宦紀聞.
Things Seen and Heard on my official Travels.
Sung, +1233.
Chang Shih-Nan 張世南.

Yü Phien 玉篇.
Jade Page Dictionary.
Liang, +543.
Ku Yeh-Wang 顧野王.
Extended and edited in the Thang (+674) by Sun Chhiang 孫強.

Yü Shih Ming Yen 喻世明言.
Stories to Enlighten Men.
Ming, *c.* +1640.
Fêng Mêng-Lung 馮夢龍.

Yü Tung Ta Shen Tan Sha Chen Yao Chüeh 玉洞大神丹砂眞要訣.
True and Essential Teachings about the Great Magical Cinnabar of the Jade Heaven [paraphrase of +8th-century materials].
Thang, not before +8th.
Attrib. Chang Kuo 張果.
TT/889.

Yu-Yang Tsa Tsu 酉陽雜俎.
Miscellany of the Yu-yang Mountain (Cave) [in S.E. Szechuan].
Thang, +863.
Tuan Chhêng-Shih 段成式.
See des Rotours (1), p. civ.

Yuan Chhi Lun 元氣論.
Discourse on the Primary Vitality (and the Cosmogonic Chhi).
Thang, late +8th or perhaps +9th.
Writer unknown.
In *YCCC*, ch. 56.
Cf. Maspero (7), p. 207.

Yuan-Shih Shang Chen Chung Hsien Chi 元始上眞衆仙記.
Record of the Assemblies of the Perfected Immortals; a Yuan-Shih Scripture.
Ascr. Chin, *c.* +320, more probably +5th or +6th.
Attrib. Ko Hung 葛洪.
TT/163.

Yuan Yang Ching 元陽經.
Manual of the Primary Yang (Vitality).
Chin, L/Sung, Chhi or Liang, before +550.
Writer unknown.
Extant only in quotations, in *Yang Hsing Yen Ming Lu*, etc.
Cf. Maspero (7), p. 232.

Yuan Yu 遠遊.
Roaming the Universe; or, The Journey into Remoteness [ode].
C/Han, *c.* −110.
Writer's name unknown, but a Taoist.
Tr. Hawkes (1).

Yüeh Wei Tshao Thang Pi Chi 閱微草堂筆記.
Jottings from the Yüeh-wei Cottage.
Chhing, 1800.
Chi Yün 紀昀.

Yün Chai Kuang Lu 雲齋廣錄.
Extended Records of the Cloudy Studio.
Sung.
Li Hsien-Min 李獻民.

Yün Chhi Yu I 雲溪友議.
Discussions with Friends at Cloudy Pool
Thang, *c.* +870.
Fan Shu 范攄.

Yün Chi Chhi Chhien 雲笈七籤.
The Seven Bamboo Tablets of the Cloudy Satchel [an important collection of Taoist material made by the editor of the first definitive form of the *Tao Tsang* (+1019), and including much material which is not in the Patrology as we now have it].
Sung, *c.* +1022.
Chang Chün-Fang 張君房.
TT/1020.

Yün Hsien Tsa Chi 雲仙雜記.
Miscellaneous Records of the Cloudy Immortals.
Thang or Wu Tai, *c.* +904.
Fêng Chih 馮贄.

Yün Hsien San Lu 雲仙散錄.
    Scattered Remains on the Cloudy Im-
      mortals.
    Ascr. Thang or Wu Tai, *c.* +904, actually
      probably Sung.
    Attrib. Fêng Chih 馮贄, but probably by
      Wang Chih 王銍.

Yün Kuang Chi 雲光集.
    Collected (Poems) of Light (through the)
      Clouds.
    Sung, *c.* +1170.
    Wang Chhu-I 王處一.
    TT/1138.

ADDENDA TO BIBLIOGRAPHY A

Hai Kho Lun 海客論
    Guests from Overseas [descriptions of
      alchemical exotica] Sung.
    Li Kuang-Hsüan 李光玄.
    TT/1033.

Ju Lin Wai Shih 儒林外史.
    Unofficial History of the World of Learning
      [satirical novel on the life of the literati
      in the Ming period].
    Chhing, begun before +1736, completed
      +1749.
    Wu Ching-Tzu 吳敬梓.

    Tr. Yang & Yang (1); Tomkinson (2).
    Cf. Chang Hsin-Tshang (2).

Shih Shuo Hsin Yu 世說新語.
    New Discourse on the Talk of the Times
      [notes of minor incidents from Han to
      Chin].
    Cf. *Hsü Shih Shuo.*
    L/Sung, +5th.
    Liu I-Chhing 劉義慶.
    Commentary by Liu Hsün
      劉峻 (Liang).

CONCORDANCE FOR
TAO TSANG BOOKS AND TRACTATES

Wieger nos.		Ong nos.	Wieger nos.		Ong nos.
634	*Huang-Thien Shang-Chhing Chin Chhüeh Ti Chün Ling Shu Tzu-Wên Shang Ching*	639	890	*Ling Sha Ta Tan Pi Chüeh*	896
635	*Tung Shen Pa Ti Miao Ching Ching*	640	891	*Pi Yü Chu Sha Han Lin Yü Shu Kuei*	897
761	*Thu Ching (Chi Chu) Yen I Pên Tshao*	{767 768	892	*Ta Tan Chi*	898
773	*Hsüan Phin Lu*	780	893	*Tan Fang Hsü Chih*	899
810	*(Thai-Chhing) Chung Huang Chen Ching*	816	894	*Shih Yao Erh Ya*	900
811	*(Thai-Chhing) Tao Yin Yang-Shêng Ching*	817	895	*(Chih-Chhuan Chen Jen) Chiao Chêng Shu*	901
812	*(Thai-Shang) Yang-Shêng Thai-Hsi Chhi Ching*	818	896	*Shun-Yang Lü Chen Jen Yao Shih Chih*	902
813	*Thai-Chhing Thiao Chhi Ching*	819	897	*Chin Pi Wu Hsiang Lei Tshan Thung Chhi*	903
814	*Thai-Shang Lao Chün Yang-Shêng Chüeh*	820	898	*Tshan Thung Chhi Wu Hsiang Lei Pi Yao*	904
815	*Thai-Chhing (Wang Lao) Fu Chhi Khou Chüeh (or Chhuan Fa)*	821	899	*(Yin Chen Chün) Chin Shih Wu Hsiang Lei*	905
817	*Sung Shan Thai-Wu hsien-sêng Chhi Ching*	823	900	*Chin Shih Pu Wu Chiu Shu Chüeh*	906
818	*(Yen-Ling hsien-sêng Chi) Hsin Chiu Fu Chhi Ching*	824	901	*Shang-Chhing Chiu Chen Chung Ching Nei Chüeh*	907
821	*(Huan-Chen hsien-sêng) Fu Nei Yuan Chhi Chüeh*	827	902	*Lung Hu Huan Tan Chüeh*	908
830	*Chen Chung Chi*	836	903	*Chin Hua Yü I Ta Tan*	909
(830)	*(Shê Yang) Chen Chung Chi (or Fang)*	(836)	904	*Kan Chhi Shih-liu Chuan Chin Tan*	910
831	*Yang Hsing Yen Ming Lu*	837	905	*Hsiu Lien Ta Tan Yao Chih (or Chüeh)*	911
835	*(Pao Phu Tzu) Yang Shêng Lun*	841	906	*Thung Yu Chüeh*	912
838	*Shang-Chhing Ching Chen Tan Pi Chüeh*	844	907	*Chin Hua Chhung Pi Tan Ching Pi Chih*	913
856	*Shen Hsien Lien Tan Tien Chu San Yuan Pao Chao Fa*	862	908	*Huan Tan Chou Hou Chüeh*	914
873	*Thai-Chhing (or Shang-Chhing) Chin I Shen Tan Ching*	879	909	*Phêng-Lai Shan Hsi Tsao Huan Tan Ko*	915
874	*Thai-Chhing Shih Pi Chi*	880	910	*(Pao Phu Tzu) Shen Hsien Chin Shuo Ching*	916
875	*Thai-Chhing Chin I Shen Chhi Ching*	881	911	*Chu Chia Shen Phin Tan Fa*	917
876	*(Thai-Chhing Ching) Thien-Shih Khou Chüeh*	882	912	*Chhien Hung Chia Kêng Chih Pao Chi Chhêng*	918
878	*(Huang Ti) Chiu Ting Shen Tan Ching Chüeh*	884	913	*Tan Fang Ao Lun*	919
879	*Chiu Chuan Ling Sha Ta Tan Tzu Shêng Hsüan Ching*	885	914	*Chih Kuei Chi*	920
880	*(Chang Chen Jen) Chin Shih Ling Sha Lun*	886	915	*Huan Chin Shu*	921
881	*(Wei Po-Yang) Chhi Fan Tan Sha Chüeh*	887	916	*Ta Tan Chhien Hung Lun*	922
882	*(Thai-Chi Chen Jen) Chiu Chuan Huan Tan Ching Yao Chüeh*	888	917	*Chen Yuan Miao Tao Yao Lüeh*	923
883	*(Ta-Tung Lien Chen Pao Ching) Hsiu Fu Ling Sha Miao Chüeh*	889	918	*Tan Fang Chien Yuan*	924
884	*(Ta-Tung Lien Chen Pao Ching) Chiu Huan Chin Tan Miao Chüeh*	890	919	*Ta Huan Tan Chao Chien*	925
885	*(Thai-Shang Wei Ling Shen Hua) Chiu Chuan Tan Sha Fa*	891	920	*Thai-Chhing Yü Pei Tzu*	926
886	*Chiu Chuan Ling Sha Ta Tan*	892	921	*Hsüan Chieh Lu*	927
887	*Chiu Chuan Chhing Chin Ling Sha Tan*	893	922	*(Hsien-Yuan Huang Ti) Shui Ching Yao Fa*	928
888	*Yin-Yang Chiu Chuan Chhêng Tzu Chin Tien Hua Huan Tan Chüeh*	894	923	*San-shih-liu Shui Fa*	929
889	*Yü-Tung Ta Shen Tan Sha Chen Yao Chüeh*	895	927	*Thai Pai Ching*	933
			928	*Tan Lun Chüeh Chih Hsin Chien*	934
			932	*Ta Tan Wên Ta*	938
			933	*Chin Mu Wan Ling Lun*	939
			934	*Hung Chhien Ju Hei Chhien Chüeh*	940
			935	*Thung Hsüan Pi Shu*	941
			937	*Yen Mên Kung Miao Chieh Lu (= 921)*	943
			938	*Hsüan Shuang Chang Shang Lu*	944
			939	*Thai-Chi Chen Jen Tsa Tan Yao Fang*	945
			940	*Yü Chhing Nei Shu*	946
			942	*Thai-Ku Thu Tui Ching*	948
			943	*Shang-Tung Hsin Tan Ching Chüeh*	949
			944	*(Hsü Chen-Chün) Shih Han Chi*	950

Wieger nos.		Ong nos.
945	*Chiu Chuan Liu* (or *Ling*) *Chu Shen Hsien Chiu Tan Ching*	951
946	*Kêng Tao Chi*	952
988	(*Ku Wên*) *Lung Hu Ching Chu Su*	994
989	(*Ku Wên*) *Lung Hu Shang Ching Chu*	995
990	*Chou I Tshan Thung Chhi* (*Chu*) comm. by Yin Chhang-Shêng	996
991	*Chou I Tshan Thung Chhi Chu* comm. anon.	997
992	*Tshan Thung Chhi Khao I* (or *Chou I Tshan Thung Chhi Chu*) comm. by Chu Hsi	998
993	*Chou I Tshan Thung Chhi Fên Chang Thung Chen I* comm. by Phêng Hsiao	999
994	*Chou I Tshan Thung Chhi Ting Chhi Ko Ming Ching Thu* comm. by Phêng Hsiao	1000
995	*Chou I Tshan Thung Chhi Chu* comm. anon.	1001
996	*Chou I Tshan Thung Chhi Fa Hui* comm. by Yü Yen	1002
997	*Chou I Tshan Thung Chhi Shih I* comm. by Yü Yen	1003
998	*Chou I Tshan Thung Chhi Chieh* comm. by Chhen Hsien-Wei	1004
999	*Chou I Tshan Thung Chhi Chu* comm. by Chhu Hua-Ku	1005
1004	*Chen Kao*	1010
1005	*Tao Shu*	1011
1020	*Yün Chi Chhi Chhien*	1026
1022	*Thai Hsüan Pao Tien*	1028
1024	*Tso Wang Lun*	1030
1028	*Huang Chi Ching Shih* (*Shu*)	1034
1034	*Wu Hsüan Phien*	1040
1044	*Tan-Yang Chen Jen Yü Lu*	1050
1047	*Ying Chhan Tzu Yü Lu*	1053
1053	(*Shang Yang Tzu*) *Chin Tan Ta Yao*	1059
1054	(*Shang Yang Tzu*) *Chin Tan Ta Yao Thu*	1060
1055	(*Shang Yang Tzu*) *Chin Tan Ta Yao Lieh Hsien Chih*	1061
1056	(*Sheng Yang Tzu*) *Chin Tan Ta Yao Hsien Phai* (*Yuan Liu*)	1062
1058	*Chin Tan Chih Chih*	1064
1067	*Chin Tan Ssu Pai Tzu* (*Chu*)	1073
1068	*Lung Hu Huan Tan Chüeh Sung*	1074
1074	*Huan Tan Fu Ming Phien*	1080
1076	*Tshui Hsü Phien*	1082

Wieger nos.		Ong nos.
1077	*Huan Yuan Phien*	1083
1087	*Thai Phing Ching*	1093
1124	*Wu Shang Pi Yao*	1130
1125	*San Tung Chu Nang*	1131
1127	*Hsien Lo Chi*	1133
1128	*Chien Wu Chi*	1134
1130	*Tzu-Jan Chi*	1136
1135	*Tung-Hsüan Chin Yü Chi*	1141
1136	*Tan-Yang Shen Kuang Tshan*	1142
1138	*Yün Kuang Chi*	1144
1139	(*Wang*) *Chhung-Yang Chhüan Chen Chi*	1145
1140	(*Wang*) *Chhung-Yang Chiao Hua Chi*	1146
1141	(*Wang*) *Chhung-Yang Fên-Li Shih-Hua Chi*	1147
1142	(*Wang*) *Chhung-Yang* (*Chen Jen*) *Chin-Kuan* (or *Chhüeh*) *Yü-So Chüeh*	1148
1145	*Chhang-Chhun Tzu Phan-Chhi Chi*	1151
1146	*Than hsien-sêng Shui Yün Chi*	1152
1147	*Thai-Ku Chi*	1153
1162	*Mo Tzu*	1168
1170	*Huai Nan* (*Tzu*) *Hung Lieh Chieh*	1176
1171	*Pao Phu Tzu, Nei Phien*	1177
1172	*Pao Phu Tzu, Pieh Chih*	1178
1173	*Pao Phu Tzu, Wai Phien*	1179
1174	*Tho Yo Tzu*	1180
1187	*Tung Shen Pa Ti Yuan* (*Hsüan*) *Pien Ching*	1193
1204		1211
1205	*Shang-Chhing Ling-Pao Ta Fa*	1212
1206		1213
1212	*Chhüan-Chen Tso Po Chieh Fa*	1219
1216	(*Wang*) *Chhung-Yang Li-Chiao Shih-Wu Lun*	1223
1225	*Chêng I Fa Wên* (*Thai-Shang*) *Wai Lu I*	1233
1235	*Tao Fa Hsin Chhuan*	1243
1273	*Shang-Chhing Ching Pi Chüeh*	1281
1276	*Shang-Chhing Huang Shu Kuo Tu I*	1284
1287	*Ko Hsien-Ong* (*Ko Hung*) *Chou Hou Pei Chi Fang*	1295
1295	(*Tung-Chen Thai-Shang Su-Ling Tung-Yuan*) *Ta Yu Miao Ching*	1303
1357	*Shang-Chhing Thai-Shang Ti Chün Chiu Chen Chung Ching*	1365
1382	*Huang Thing Chung Ching Ching*	1390
1405	*Thai-Shang Lao Chün Thai Su Ching* (see index s.v. *Thai Su Chuan*)	1413
1442	*Han Thien Shih Shih Chia*	1451

B. CHINESE AND JAPANESE BOOKS AND JOURNAL ARTICLES SINCE +1800

Achiwa Gorō (1) 阿知波五郎.
   Rangaku-ki no Shizen Ryū-nō-setsu Kenkyū
   蘭學期の自然良能說研究.
   A Study of the Theory of Nature-Healing
      in the Period of Dutch Learning in Japan.
   ID, 1965 No. 31, 2223.
Akitsuki Kanei (1) 秋月觀瑛.
   Kōrō Kannen no Shiroku 黃老觀念の
   系譜.
   On the Genealogy of the Huang-Lao Con-
      cept (in Taoism).
   THG, 1955, **10**, 69.
Andō Kōsei (1) 安藤更生.
   Kanshin 鑑眞.
   Life of Chien-Chen (+688 to +763),
      [outstanding Buddhist missionary to
      Japan, skilled also in medicine and archi-
      tecture].
   Bijutsu Shuppansha, Tokyo 1958, repr.
      1963.
   Abstr. *RBS*, 1964, **4**, no. 889.
Andō Kōsei (2) 安藤更生.
   Nihon no Miira 日本のミイラ.
   Mummification in Japan.
   Mainichi Shimbunsha, Tokyo, 1961.
   Abstr. *RBS*, 1968, **7**, no. 575.
Anon. (10).
   Tunhuang Pi Hua Chi 敦煌壁畫集.
   Album of Coloured Reproductions of the
      fresco-paintings at the Tunhuang cave-
      temples.
   Peking, 1957.
Anon. (11).
   Changsha Fa Chüeh Pao-Kao 長沙發掘
   報告.
   Report on the Excavations (of Tombs of
      the Chhu State, of the Warring States
      period, and of the Han Dynasties) at
      Chhangsha.
   Acad. Sinica Archaeol. Inst., Kho-Hsüeh,
      Peking, 1957.
Anon. (17).
   Shou-hsien Tshai Hou Mu Chhu Thu I Wu
   壽縣蔡侯墓出土遺物.
   Objects Excavated from the Tomb of the
      Duke of Tshai at Shou-hsien.
   Acad. Sinica. Archaeol. Inst., Peking, 1956.
Anon. (27).
   Shang-Tshun-Ling Kuo Kuo Mu Ti
   上村嶺虢國墓地.
   The Cemetery (and Princely Tombs) of the
      State of (Northern) Kuo at Shang-
      tshun-ling (near Shen-hsien in the San-
      mên Gorge Dam Area of the Yellow
      River).

Institute of Archaeology, Academia Sinica,
      Peking, 1959 (Field Expedition Reports,
      Ting Series, no. 10), (Yellow River
      Excavations Report no. 3).
Anon. (28).
   *Yünnan Chin-Ning Shih-Chai Shan Ku Mu
   Chhün Fa-Chüeh Pao-Kao* 雲南晉寧
   石寨山古墓羣發掘報告.
   Report on the Excavation of a Group of
      Tombs (of the Tien Culture) at Shih-chai
      Shan near Chin-ning in Yunnan.
   2 vols.
   Yunnan Provincial Museum.
   Wên-Wu, Peking, 1959.
Anon. (57).
   Chung Yao Chih 中藥志.
   Repertorium of Chinese Materia Medica
      (Drug Plants and their Parts, Animals
      and Minerals).
   4 vols.
   Jen-min Wei-shêng, Peking, 1961.
Anon. (73) (Anhui Medical College Physio-
      therapy Dept.).
   Chung I An-Mo Hsüeh Chien Phien 中醫
   按摩學簡編.
   Introduction to the Massage Techniques in
      Chinese Medicine.
   Jen-min Wei-shêng, Peking, 1960, repr. 1963
Anon. (74) (National Physical Education
      Council).
   Thai Chi Chhüan Yün Tung 太極拳運動.
   The Chinese Boxing Movements [instruct-
      ions for the exercises].
   Jen-min Thi-yü, Peking, 1962.
Anon. (77).
   Chhi Kung Liao Fa Chiang I 氣功療法講義.
   Lectures on Respiratory Physiotherapy.
   Kho-hsüeh Chi-shu, Shanghai, 1958.
Anon. (78).
   *Chung-Kuo Chih Chhien chih Ting Liang
   Fên-Hsi* 中國制錢之定量分析.
   Analyses of Chinese Coins (of different
      Dynasties).
   KHS, 1921, **6** (no. 11), 1173.
   Table reprinted in Wang Chin (2), p. 88.
Anon. (100).
   Shao-Hsing Chiu Niang Tsao 紹興酒釀造.
   Methods of Fermentation (and Distillation)
      of Wine used at Shao-hsing (Chekiang).
   Chhing Kung Yeh, Peking, 1958.
Anon. (101).
   Chung-Kuo Ming Tshai Phu 中國名菜譜.
   Famous Dishes of Chinese Cookery.
   12 vols.
   Chhing Kung Yeh, Peking, 1965.

Anon. (*103*).
　Nihon Miira no Kenkyū　日本ミイラの
　　研究.
　Researches on Mummies (and Self-Mummi-
　　fication) in Japan.
　Heibonsha, Tokyo, 1971.
Anon. (*104*).
　*Chhangsha Ma Wang Tui i Hao Han Mu
　　Fa-Chüeh Chien-Pao*　長沙馬王堆一號
　　漢墓發掘簡報.
　Preliminary Report on the Excavation of Han
　　Tomb No. 1 at Ma-wang-tui (Hayagriva
　　Hill) near Chhangsha [the Lady of Tai,
　　c. − 180].
　Wên Wu, Peking, 1972.
Anon. (*105*).
　*Kōkogaku-shō no Shin-Hakken; Nisen-yonen
　　mae no Kinue Orimono Sono-hoka*　考古
　　學上の新發見；二千余年まえの緝
　　繪織物その他.
　A New Discovery in Archaeology; Painted
　　Silks, Textiles and other Things more than
　　Two Thousand Years old.
　JC, 1972 (no. 9), 68, with colour-plates.
Anon. (*106*).
　*Wên-Hua Ta Ko-Ming Chhi Chien Chhu Thu
　　Wên Wu*　文化大革命期間出土
　　文物.
　Cultural Relics Unearthed during the period
　　of the Great Cultural Revolution (1965–71),
　　vol. 1 [album].
　Wên Wu, Peking, 1972.
Anon. (*109*).
　Chung-Kuo Kao Têng Chih-Wu Thu Chien
　　中國高等植物圖鑑.
　Iconographia Cormophytorum Sinicorum
　　(Flora of Chinese Higher Plants).
　2 vols. Kho-Hsüeh, Peking, 1972 (for Nat.
　　Inst. of Botany).
Anon. (*110*).
　Chhang Yung Chung Tshao Yao Thu Phu
　　常用中草藥圖譜.
　Illustrated Flora of the Most Commonly
　　Used Drug Plants in Chinese Medicine.
　Jen-min Wei-shêng, Peking, 1970.
Anon. (*111*).
　Man-chhêng Han Mu Fa-Chüeh Chi Yao
　　滿城漢墓發掘紀要.
　The Essential Findings of the Excavations of
　　the (Two) Han Tombs at Man-chhêng
　　(Hopei), [Liu Shêng, Prince Ching of
　　Chung-shan, and his consort Tou Wan].
　KKTH, 1972, (no. 1), 8.
Anon. (*112*).
　*Man-chhêng Han Mu 'Chin Lou Yü-I' ti
　　Chhing-Li ho Fu-Yuan*　滿城漢墓「金縷
　　玉衣」的清理和復原.
　On the Origin and Detailed Structure of
　　the Jade Body-cases Sewn with Gold
　　Thread found in the Han Tombs at Man-
　　chhêng.
　KKTH, 1972, (no. 2), 39.

Anon. (*113*).
　*Shih Than Chi-nan Wu-ying Shan Chhu-
　　Thu-ti Hsi Han, Lo Wu, Tsa Chi, Yen Huan
　　Thao Yung*　試談濟南无影山出土的
　　西漢榮舞雜技宴歡陶俑.
　A Discourse on the Early Han pottery models
　　of musicians, dancers, acrobats and mis-
　　cellaneous artists performing at a banquet,
　　discovered in a Tomb at Wu-ying Shan
　　(Shadowless Hill) near Chinan (in Shan-
　　tung province).
　WWTK, 1972, (no. 5), 19.
Anon. (*115*).
　Tzhu-Hang Ta Shih Chuan　慈航大師傳.
　A Biography of the Great Buddhist Teacher,
　　Tzhu-Hang (d., self-mummified, 1954).
　Thaipei, 1959 (Kan Lu Tshung Shu ser.
　　no. 11).
Aoki Masaru (*1*)　青水正兒.
　Chūka Meibutsu Kō　中華名物考.
　Studies on Things of Renown in (Ancient
　　and Medieval) China, [including aro-
　　matics, incense and spices].
　Shunjūsha, Tokyo, 1959.
　Abstr. *RBS*, 1965, **5**, no. 836.
Asahina Yasuhiko (*1*) (ed.)　朝比奈泰彥 with
　　16 collaborators.
　Shōsōin Yakubutsu　正倉院藥物.
　The Shōsōin Medicinals; a Report on
　　Scientific Researches.
　With an English abstract by Obata Shige-
　　yoshi.
　Shokubutsu Bunken Kankō-kai, Osaka, 1955.

Chan Jan-Hui (*1*)　湛然慧.
　Alternative orthography of Tan Jan-Hui (*1*).
Chang Chhang-Shao (*1*)　張昌紹.
　Hsien-tai-ti Chung Yao Yen-Chiu　現代的
　　中藥研究.
　Modern Researches on Chinese Drugs.
　Kho-hsüeh Chi-shu, Shanghai, 1956.
Chang Chhi-Yün (*2*) (ed.)　張其昀.
　Chung-Hua Min Kuo Ti-Thu Chi　中華民
　　國地圖集.
　Atlas of the Chinese Republic (5 vols.):
　　vol. 1 Thaiwan, vol. 2, Central Asia,
　　vol. 3 North China, vol. 4 South China,
　　vol. 5 General maps.
　National Defence College⎫
　National Geographical　　⎬ Thaipei, 1962–3.
　　Institute　　　　　　　　⎭
Chang Ching-Lu (*1*).
　*Chung-Kuo Chin-Tai Chhu-Pan Shih-Liao
　　Chhu Phien*　中國近代出版史料初編.
　Materials for a History of Modern Book-
　　Publishing in China, Pt. 1.
Chang Hsin-Chhêng (*1*)　張心澂.
　Wei Shu Thung Khao　僞書通考.
　A Complete Investigation of the (Ancient
　　and Medieval) Books of Doubtful
　　Authenticity.
　2 vols., Com. Press, 1939, repr. 1957.

Chang Hsing-Yün (1) 章杏雲.
    Yin Shih Pien 飲食辯.
    A Discussion of Foods and Beverages.
    1814, repr. 1824.
    Cf. Dudgeon (1).
Chang Hsüan (1) 張瑄.
    Chung Wên Chhang Yung San Chhien Tzu
        Hsing I Shih 中文常用三千字形義釋.
    Etymologies of Three Thousand Chinese
        Characters in Common Use.
    Hongkong Univ. Press, 1968.
Chang Hung-Chao (1) 章鴻釗.
    Shih Ya 石雅.
    Lapidarium Sinicum; a Study of the Rocks,
        Fossils and Minerals as known in
        Chinese Literature.
    Chinese Geol. Survey, Peiping: 1st ed. 1921,
        2nd ed. 1927.
    MGSC (ser. B), no. 2, 1–432 (with Engl.
        summary).
    Crit. P. Demiéville, BEFEO, 1924, 24,
        276.
Chang Hung-Chao (3) 章鴻釗.
    Chung-Kuo Yung Hsin ti Chhi-Yuan
        中國用鋅的起源.
    Origins and Development of Zinc Tech-
        nology in China.
    KHS, 1923, 8 (no. 3), 233, repr. in Wang
        Chin (2), p. 21.
    Cf. Chang Hung-Chao (2).
Chang Hung-Chao (6) 章鴻釗.
    Tsai Shu Chung-Kuo Yung Hsin ti Chhi-
        Yuan 再述中國用鋅的起源.
    Further Remarks on the Origins and
        Development of Zinc Technology in
        China.
    KHS, 1925, 9 (no. 9), 1116, repr. in Wang
        Chin (2), p. 29.
    Cf. Chang Hung-Chao (3).
Chang Hung-Chao (8) 章鴻釗.
    Lo shih 'Chung-Kuo I-Lan' Chüan Chin
        Shih I Chêng 洛氏「中國伊蘭」卷金
        石譯證.
    Metals and Minerals as Treated in Laufer's
        'Sino-Iranica', translated with Commen-
        taries.
    MGSC 1925 (Ser. B), no. 3, 1–119.
    With English preface by Ong Wên-Hao.
Chang Tzu-Kao (1) 張子高.
    Kho-Hsüeh Fa Ta Lüeh Shih 科學發達
        略史.
    A Classified History of the Natural Sciences.
    Com. Press, Shanghai, 1923, repr. 1936.
Chang Tzu-Kao (2) 張子高.
    Chung-Kuo Hua-Hsüeh Shih Kao (Ku-Tai
        chih Pu) 中國化學史稿(古代之部).
    A Draft History of Chemistry in China
        (Section on Antiquity).
    Kho-Hsüeh, Peking, 1964.
Chang Tzu-Kao (3) 張子高.
    Lien Tan Shu Fa-Shêng yü Fa-Chan
        鍊丹術發生與發展.

On the Origin and Development of Chinese
    Alchemy.
    CHJ, 1960, 7 (no. 2), 35.
Chang Tzu-Kao (4) 張子高.
    Tshung Tu Hsi Thung Chhi Than Tao 'Wu'
        Tzu Pên I 從鍍錫銅器談到「鋈」字
        本義.
    Tin-Plated Bronzes and the Possible
        Original Meaning of the Character wu.
    AS/CJA, 1958 (no. 3), 73.
Chang Tzu-Kao (5) 張子高.
    Chao Hsüeh-Min 'Pên Tshao Kang Mu
        Shih I' Chu Shu Nien-Tai, Chien-Lun
        Wo-Kuo Shou-Tzhu Yung Chhiang-Shiu
        Kho Thung Pan Shih 趙學敏「本草綱
        目拾遺」著述年代彙論我國首次用强
        水刻銅版事.
    On the Date of Publication of Chao Hsüeh-
        Min's Supplement to the Great Pharma-
        copoeia, and the Earliest Use of Acids for
        Etching Copper Plates in China.
    KHSC, 1962, 1 (no. 4), 106.
Chang Tzu-Kao (6) 張子高.
    Lun Wo Kuo Niang Chiu Chhi-Yuan ti
        Shih-Tai Wên-Thi 論我國釀酒起源的
        時代問題.
    On the Question of the Origin of Wine in
        China.
    CHJ, 1960, 17 (7), no. 2, 31.
Chang Tzu-Kung (1) 張資珙.
    Lüeh Lun Chung-Kuo ti Nieh Chih Pai-
        Thung ho tha tsai Li-Shih shang yü Ou-
        Ya Ko Kuo ti Kuan-Hsi 略論中國的
        鎳質白銅和他在歷史上與歐亞各國
        的關係.
    On Chinese Nickel and Paktong, and on
        their Role in the Historical Relations
        between Asia and Europe.
    KHS, 1957, 33 (no. 2), 91.
Chang Tzu-Kung (2) 張資珙.
    Yuan Su Fa-Hsien Shih 元素發現史.
    The Discovery of the Chemical Elements
        (a translation of Weeks (1), with some
        40% of original material added).
    Shanghai, 1941.
Chang Wên-Yuan (1).
    Thai Chi Chhüan Chhang Shih Wên Ta
        太極拳常識問答.
    Explanation of the Standard Principles of
        Chinese Boxing.
    Jen-min Thi-yü, Peking, 1962.
Chao Pi-Chhen (1) 趙避塵.
    Hsing Ming Fa Chüeh Ming Chih 性命法
        訣明指.
    A Clear Explanation of the Oral Instruct-
        ions concerning the Techniques of the
        Nature and the Life-Span.
    Chhi-shan-mei, Thaipei, Thaiwan, 1963.
    Tr. Lu Khuan-Yü (4).
Chi Yün (1).
    Yüeh Wei Tshao Thang Pi Chi 閱微草堂
        筆記.

Chi Yün (1) (cont.)

Jottings from the Yüeh-wei Cottage.
1800.

Chia Tsu-Chang & Chia Tsu-Shan (1) 買祖璋
買祖珊.

Chung-Kuo Chih-Wu Thu Chien 中國植物
圖鑑.

Illustrated Dictionary of Chinese Flora
[arranged on the Engler system; 2602
entries].

Chung-hua, Peking, 1936, repr. 1955, 1958.

Chia Yo-Han (Kerr, J. G.) 嘉約翰 & Ho
Liao-Jan 何了然 (1).

Hua Hsüeh Chhu Chiai 化學初階.

First Steps in Chemistry.

Canton, 1870.

Chiang Thien-Shu (1) 蔣天樞.

'*Chhu Tzhu Hsin Chu*' *Tao Lun* 「楚辭新
注」導論.

A Critique of the *New Commentary on the
Odes of Chhu*.

CHWSLT, 1962, **1**, 81.

Abstr. *RBS*, 1969, **8**, no. 558.

Chiang Wei-Chhiao (1) [Yin Shih Tzu] 蔣維喬.
Yin Shih Tzu Ching Tso Fa 因是子靜坐
法.

Yin Shih Tzu's Methods of Meditation
[Taoist].

Shih-yung, Hongkong, 1914, repr. 1960,
1969.

With Buddhist addendum, Hsü Phien
續編.

Cf. Lu Khuan-Yü (1), pp. 167, 193.

Chi ang Wei-Chhiao (2) [Yin Shih Tzu] 蔣維喬
Ching Tso Fa Chi Yao 靜坐法輯要.

The Important Essentials of Meditation
Practice.

Repr. Thaiwan Yin Ching Chhu, Thaipei,
1962.

[Chiang Wei-Chhiao] (3) Yin Shih Tzu 蔣維喬.
Hu Hsi Hsi Ching Yang Shêng Fa 呼吸習
靜養生法.

Methods of Nourishing the Life-Force by
Respiratory Physiotherapy and Meditation Technique.

Repr. Thai-Phing, Hongkong, 1963.

[Chiang Wei-Chhiao] (4) Yin Shih Tzu 蔣維喬.
*Yin Shih Tzu Ching Tso Wei Shêng Shih
Yen Than* 因是子靜坐衛生實驗談.

Talks on the Preservation of Health by
Experiments in Meditation.

Printed together with the Hsü Phien of (1).

Tzu-Yu, Thaichung, Thaiwan⎱ 1957.
           Hongkong⎰

Cf. Lu Khuan-Yü, (1), pp. 157, 160, 193.

Chiang Wei-Chhiao (5) 蔣維喬.
*Chung-Kuo-ti Hu Hsi Hsi Ching Yang
Shêng Fa (Chhi Kung Fang Chih Fa)*
中國的呼吸習靜養生法(氣功防
治法).

The Chinese Methods of Prolongevity by
Respiratory and Meditational Technique

(Hygiene and Health due to the Circulation of the Chhi).

Wei-Shêng, Shanghai, 1956, repr. 1957.

Chiang Wei-Chhiao 蔣維喬 & Liu Kuei-Chen
(1).

Chung I Than Chhi Kung Liao Fa 中醫談
氣功療法.

Respiratory Physiotherapy in Chinese
Medicine.

Thai-Phing, Hongkong, 1964.

Chieh Hsi-Kung (1) 解希恭.

*Thai-yuan Tung-thai-pao Chhu Thu ti Han
Tai Thung Chhi* 太原東太堡出土的
漢代銅器.

Bronze Objects of Han Date Excavated at
Tung-thai-pao Village near Thaiyuan
(Shansi), [including five unicorn-foot
horse-hoof gold pieces, about 140 gms. wt.,
with almost illegible inscriptions].

WWTK, 1962 (no. 4/5), no. 138-9, 66
(71), ill. p. 11.

Abstr. *RBS*, 1969, **8**, no. 360 (p. 196).

Chikashige Masumi (1) = (1) 近重眞澄.

*Tōyō Renkinjutsu; Kagakujō yori mitaru
Tōyōjōdai no Bunka*
東洋鍊金術；化學上より見たる東
洋上代の文化.

East Asian Alchemy; the Culture of
East Asia in Early Times seen from the
Chemical Point of View.

Rokakuho Uchida, Tokyo, 1929, repr.
1936.

Based partly on (4) and on papers in
SN 1918, **3** (no. 2) and 1919, **4** (no. 2).

Chikashige Masumi (2) 近重眞澄.
Tōyō Kodōki no Kagaku-teki Kenkyū
東洋古銅器の化學的研究.

A Chemical Investigation of Ancient
Chinese Bronze [and Brass] Vessels.

SN, 1918, **3** (no. 2), 177.

Chikashige Masumi (3) 近重眞澄.
Kagaku yori mitaru Tōyōjōdai no Bunka
化學より觀たる東洋上代の文化.

The Culture of Ancient East Asia seen
from the Viewpoint of Chemistry.

SN, 1919, **4** (no. 2), 169.

Chikashige Masumi (4) 近重眞澄.
Tōyō Kodai Bunka no Kagakukan
東洋古代文化の化學觀.

A Chemical View of Ancient East Asian
Culture.

Pr. pr. Tokyo, 1920.

Chojiya Heibei (1).
Shoseki Seirenho 硝石製煉法.

The Manufacture of Saltpetre.

Yedo, 1863.

Chŏn Sangun (2) 全相運.
Han'guk kwahak kisul sa
韓國科學技術史.

A Brief History of Science and Technology
in Korea.

World Science Co. Seoul, 1966.

Chou Fêng-Wu 周鳳梧, Wang Wan-Chieh
王萬杰 & Hsü Kuo-Chhien 徐國仟 (1).
Huang Ti Nei Ching Su Wên, Pai Hua Chieh
黃帝內經素問白話解.
The Yellow Emperor's Manual of Corporeal
(Medicine); Questions (and Answers) about
Living Matter; done into Colloquial
Language.
Jen-min Wei-shêng, Peking, 1963.

Chou Shao-Hsien (1) 周紹賢.
Tao Chia yü Shen Hsien 道家與神仙.
The Holy Immortals of Taoism; the
Development of a Religion.
Chung-Hua, Thaipei (Thaiwan), 1970.

Chu Chi-Hai (1) 朱季海.
'*Chhu Tzhu' Chieh Ku Shih I* 「楚辭」解故
識遺.
Commentary on Parts of the Odes of Chhu
(especially Li Sao and Chiu Pien), [with
special attention to botanical identifi-
cations].
CHWSLT, 1962, **2**, 77.
Abstr. *RBS*, 1969, **8**, no. 557.

Chu Lien (1) 朱璉.
Hsin Chen Chiu Hsüeh 新針灸學.
New Treatise on Acupuncture and Moxi-
bustion.
Jen-min Wei-shêng, Peking, 1954.

Chhang Pi-Tê (1) 昌彼得.
Shuo Fu Khao 說郛考.
A Study of the Shuo Fu Florilegium.
Chinese Planning Commission for East
Asian Studies, Thaipei (Thaiwan), 1962.

Chhen Ching (1) 陳經.
Chhiu Ku Ching Shê Chin Shih Thu 求古
精舍金石圖.
Illustrations of Antiques in Bronze and
Stone from the Spirit-of-Searching-Out-
Antiquity Cottage.

Chhen Kung-Jou (3) 陳公柔.
Pai-Sha Thang Mu Chien Pao 白沙唐墓
簡報.
Preliminary Report on (the Excavation of) a
Thang Tomb at the Pai-sha (Reservoir),
(in Yü-hsien, Honan).
KKTH, 1955 (no. 1), 22.

Chhen Kuo-Fu (1) 陳國符.
'*Tao Tsang' Yuan Liu Khao* 「道藏」源流考.
A Study on the Evolution of the Taoist
Patrology.
1st ed. Chung-Hua, Shanghai, 1949.
2nd ed. in 2 vols., Chung-Hua, Peking, 1963.

Chhen Mêng-Chia (4) 陳夢家.
Yin Hsü Pu Tzhu Tsung Shu 殷虛卜辭綜
述.
A study of the Characters on the Shang
Oracle-Bones.
Kho-Hsüeh, Peking, 1956.

Chhen Pang-Hsien (1) 陳邦賢.
Chung Kuo I-Hsüeh Shih 中國醫學史.
History of Chinese Medicine.
Com. Press, Shanghai, 1937, 1957.

Chhen Phan (7) 陳槃.
*Chan-Kuo Chhin Han Chien Fang-Shih
Khao Lun* 戰國秦漢間方士考論.
Investigations on the Magicians of the
Warring States, Chhin and Han periods.
AS/BIHP, 1948, **17**, 7.

Chhen Pi-Liu 陳璧琉 & Chêng Cho-Jen (1)
鄭卓人.
Ling Shu Ching, Pai Hua Chieh 靈樞經
白話解.
The Yellow Emperor's Manual of Corporeal
(Medicine); the Vital Axis; done into
Colloquial Language.
Jen-min Wei-shêng, Peking, 1963.

Chhen Thao (1).
Chhi Kung Kho-Hsüeh Chhang Shih 氣功
科學常識.
A General Introduction to the Science of
Respiratory Physiotherapy.
Kho-hsüeh Chi-shu, Shanghai, 1958.

Chhen Wên-Hsi (1) 陳文熙.
Lu-kan-shih 'Tutty' Thou-shih Thang-Thi
爐甘石 Tutty 鍮石鐺銻.
A Study of the Designations of Zinc Ores,
lu-kan-shih, tutty and brass.
HITC 1933, **12**, 839; 1934, **13**, 401.

Chhen Yin-Kho (3) 陳寅恪.
*Thien Shih Tao yü Pin-Hai Ti-Yü chih
Kuan-Hsi* 天師道與濱海地域之關係
On the Taoist Church and its Relation to
the Coastal Regions of China (c. +126 to
+536).
AS/BIHP, 1934 (no. 3/4), 439.

Chhen Yuan (4) 陳垣.
Shih Hui Chü Li 史諱舉例.
On the Tabu Changes of Personal Names in
History; Some Examples.
Chung-Hua, Peking, 1962, repr. 1963.

Dohi Keizō (1) 土肥慶藏.
Shōsōin Yakushi no Shiteki Kōsatsu 正倉
院藥種の史的考察.
Historical Investigation of the Drugs preserved in the Imperial Treasury at
Nara.
In *Zoku Shōsōin Shiron* 續正倉院史論.
1932, No. 15, Neiyaku 寧藥.
1st pagination, p. 133.

Dōno Tsurumatsu (1) 道野鶴松.
*Kodai no Shina ni okeru Kagakushisō toku
ni Genzoshisō ni tsuite* 古代の支那に
於ける化學思想特に元素思想に就
いて.
On Ancient Chemical Ideas in China, with
Special Reference to the Idea of Ele-
ments [comparison with the Four
Aristotelian Elements and the Spagyrical
Tria Prima].
TG/T, 1931, **1**, 159.

Fan Hsing-Chun (6) 范行準.
Chung Hua I-Hsüeh Shih 中華醫學史.

Fan Hsing-Chun (6) (cont.)
Chinese Medical History.
ISTC, 1947, **1** (no. 1), 37, (no. 2), 21;
1948, **1** (no. 3/4), 17.
Fan Hsing-Chun (12) 范行準.
Liang Han San Kuo Nan Pei Chhao Sui Thang I Fang Chien Lu 兩漢三國南北朝隋唐醫方簡錄.
A Brief Bibliography of (Lost) Books on Medicine and Pharmacy written during the Han, Three Kingdoms, Northern and Southern Dynasties and Sui and Thang Periods.
CHWSLT, 1965, **6**, 295.
Fêng Chhêng-Chün (1) 馮承鈞.
Chung-Kuo Nan-Yang Chiao-Thung Shih 中國南洋交通史.
History of the Contacts of China with the South Sea Regions.
Com. Press, Shanghai, 1937, repr. Thai-Phing, Hongkong, 1963.
Fêng Chia-Shêng (1) 馮家昇.
Huo-Yao ti Fa-Hsien chi chhi Chhuan Pu 火藥的發現及其傳佈.
The Discovery of Gunpowder and its Diffusion.
AP/HJ, 1947, **5**, 29.
Fêng Chia-Shêng (2) 馮家昇.
Hui Chiao Kuo wei Huo-Yao yu Chung-Kuo Chhuan Ju Ou-Chou ti Chhiao Liang 回教國為火藥由中國傳入歐州的橋梁.
The Muslims as the Transmitters of Gunpowder from China to Europe.
AP/HJ, 1949, 1.
Fêng Chia-Shêng (3) 馮家昇.
Tu Hsi-Yang ti Chi Chung Huo-Chhi Shih Hou 讀西洋的幾種火器史後.
Notes on reading some of the Western Histories of Firearms.
AP/HJ, 1947, **5**, 279.
Fêng Chia-Shêng (4) 馮家昇.
Huo-Yao ti Yu Lai chi chhi Chhuan Ju Ou-chou ti Ching Kuo 火藥的由來及其傳入歐洲的經過.
On the Origin of Gunpowder and its Transmission to Europe.
Essay in Li Kuang-Pi & Chhien Chün-Yeh (q.v.), p. 33.
Peking, 1955.
Fêng Chia-Shêng (5) 馮家昇.
Lien Tan Shu ti Chhêng Chhang chi chhi Hsi Chhuan 煉丹術的成長及其西傳.
Achievements of (ancient Chinese) Alchemy and its Transmission to the West.
Essay in Li Kuang-Pi & Chhien Chün-Yeh (q.v.), p. 120.
Peking 1955.
Fêng Chia-Shêng (6) 馮家昇.
Huo-Yao ti Fa-Ming ho Hsi Chhuan 火藥的發明和西傳.
The Discovery of Gunpowder and its Transmission to the West.

Hua-Tung, Shanghai, 1954.
Revised ed. Jen-Min, Shanghai, 1962.
Fu Chhin-Chia (1) 傅勤家.
Chung-Kuo Tao Chiao Shih 中國道教史.
A History of Taoism in China.
Com. Press, Shanghai, 1937.
Fu Chin-Chhüan (1) 傅金銓.
Chhih Shui Yin 赤水吟.
Chants of the Red River [physiological alchemy].
1823.
In *CTPS, pên* 4.
Fu Chin-Chhüan (2) 傅金銓.
Thien Hsien Chêng Li Tu 'Fa Tien Ching' 天仙正理讀法點睛.
The Right Pattern of the Celestial Immortals; Thoughts on Reading the *Consecration of the Law* [physiological alchemy]. *Tien ching* refers to the ceremony of painting in the pupils of the eyes in an image or other representation].
1820.
In *CTPS, pên* 5.
Fu Chin-Chhüan (3) 傅金銓.
Tan Ching Shih Tu 丹經示讀.
A Guide to the Reading of the Enchymoma Manuals [dialogue of pupil and teacher on physiological alchemy]
c. 1825.
In *CTPS, pên* 11.
Fu Chin-Chhüan (4) 傅金銓.
Tao Hai Chin Liang 道海津梁.
A Catena (of Words) to Bridge the Ocean of the Tao [mutationism, Taoist–Buddhist–Confucian syncretism, and physiological alchemy].
1822.
In *CTPS, pên* 11.
Fu Chin-Chhüan (5) 傅金銓.
Shih Chin Shih 試金石.
On the Testing of (what is meant by) 'Metal' and 'Mineral'.
c. 1820.
In *Wu Chen Phien Ssu Chu* ed.
Fu Chin-Chhüan (6) (ed.) 傅金銓.
Chêng Tao Pi Shu Shih Chung 證道秘書十種.
Ten Types of Secret Books on the Verification of the Tao.
Early 19th.
Fu Lan-Ya (Fryer, John) 傅蘭雅 & Hsü Shou 徐壽 (1) (tr.).
Hua-Hsüeh Chien Yuan 化學鹽原.
Authentic Mirror of Chemical Science (translation of Wells, 1).
Chiangnan Arsenal Transl. Bureau, Shanghai, 1871.
Fukui Kōjun (1) 福井康順.
Tōyō Shisō no Kenkyū 東洋思想の研究.
Studies in the History of East Asian Philosophy.
Risōsha, Tokyo, 1956.
Abstr. *RBS*, 1959, **2**, no. 564.

Fukunaga Mitsuji (1)　福永光司.
　Hōzensetsu no Keisei　封禪說の形成.
　The Evolution of the Theory of the Fêng and
　　Shan Sacrifices (in Chhin and Han Times).
　TS, 1954, **1** (no. 6), 28, (no. 7), 45.

Harada Yoshito　原田淑人 & Tazawa Kingo (1)
　田澤金吾.
　Rakurō Gokan-en Ō Ku no Fumbo　樂浪五
　　官掾王旴の墳墓.
　Lo-Lang; a Report on the Excavation of
　　Wang Hsü's Tomb in the Lo-Lang Pro-
　　vince (an ancient Chinese Colony in Korea).
　Tokyo Univ. Tokyo, 1930.

Hasegawa Usaburo (1)　長谷川卯三郎.
　Shin Igaku Zen　新醫學禪.
　New Applications of Zen Buddhist Tech-
　　niques in Medicine.
　So Gensha, Tokyo, 1970. (In the Hara-o-
　　tsukuruzen Series.)

Hiraoka Teikichi (2)　平岡楨吉.
　'Enanji' ni arawareta Ki no Kenkyū　「淮南
　　子」に現われた氣の研究.
　Studies on the Meaning and the Conception
　　of 'chhi' in the *Huai Nan Tzu* book.
　Kan Gi Bunka Gakkai, Tokyo, 1961.
　Abstr. *RBS*, 1968, **7**, no. 620.

Ho Han-Nan (1)　何漢南.
　*Sian Shih Hsi-yao-shih Tshun Thang Mu
　　Chhing-Li Chi*　西安市西窯實村唐
　　墓清理記.
　A Summary Account of the Thang Tomb at
　　Hsi-yao-shih Village near Sian [the tomb
　　which yielded early Arabic coins].
　Cf. Hsia Nai (3).
　KKTH, 1965, no. 8 (no. 108), pp. 383, 388.

Ho Hsin (1) (Hobson, Benjamin)　合信.
　Po Wu Hsin Phien　博物新編.
　New Treatise on Natural Philosophy and
　　Natural History [the first book on modern
　　chemistry in Chinese].
　Shanghai, 1855.

Ho Ping-Yü　何丙郁 & Chhen Thieh-Fan (1)
　陳鐵凡.
　*Lun 'Shun Yang Lü Chen Jen Yao Shih
　　Chih' ti Chu Chhing Shih-Tai*　論「純陽
　　呂眞人藥石製」的著成時代.
　On the Dating of the 'Manipulations of
　　Drugs and Minerals, by the Adept Lü
　　Shun-Yang', a Taoist Pharmaceutical and
　　Alchemical Manual.
　JOSHK, 1971, **9**, 181–228.

Ho-Ping-Yü　何丙郁 & Su Ying-Hui　蘇瑩輝
　(1).
　'Tan Fang Ching Yuan' Khao　「丹房鏡源」
　　考.
　On the *Mirror of the Alchemical
　　Elaboratory*, (a Thang Manual of
　　Practical Experimentation).
　JOSHK,1970, **8** (no. 1), 1, 23.

Hori Ichirō (1)　堀一郎.
　*Yudono-san Kei no Sokushimbutsu (Miira) to
　　sono Haikei*　湯殿山系の即身佛(ミイ
　　ラ)とその背景.
　The Preserved Buddhas (Mummies) at the
　　Temples on Yudono Mountain.
　TBKK, 1961, no. 35 (no. 3).
　Repr. in Hori Ichirō (2), p. 191.

Hori Ichirō (2)　堀一郎.
　Shūkyō Shūzoku no Seikatsu Kisei　宗教習
　　俗の生活規制.
　Life and Customs of the Religious Sects (in
　　Buddhism).
　Miraisha, Tokyo, 1963.

Hou Pao-Chang (1)　侯寶璋.
　Chung-Kuo Chieh-Phou Shih　中國解剖史
　A History of Anatomy in China.
　ISTC, 1957, **8** (no. 1), 64.

Hou Wai-Lu (3)　侯外廬.
　*Fang I-Chih—Chung-Kuo ti Pai Kho
　　Chhüan Shu Phai Ta Chê-Hsüeh Chia*
　　方以智—中國的百科全書派大哲
　　學家.
　Fang I-Chih—China's Great Encyclopaedist
　　Philosopher.
　LSYC, 1957 (no. 6), 1; 1957 (no. 7), 1.

Hou Wai-Lu (4)　侯外廬.
　*Shih-liu Shih-Chi Chung-Kuo ti Chin-Pu
　　ti Chê-Hsüeh Ssu-Chhao Kai-Shu*　十六
　　世紀中國的進步的哲學思潮概述.
　Progressive Philosophical Thinking in
　　+16th-century China.
　LSYC, 1959 (no. 10), 39.

Hou Wai-Lu　侯外廬, Chao Chi-Pin　趙紀彬,
　Tu Kuo-Hsiang　杜國庠 & Chhiu
　Han-Shêng (1)　邱漢生.
　Chung-Kuo Ssu-Hsiang Thung Shih　中國
　　思想通史.
　General History of Chinese Thought.
　5 vols.
　Jen-Min, Peking, 1957.

Hu Shih (7)　胡適.
　Lun Hsüeh Chin Chu, ti-i Chi　論學近著
　　第一集.
　Recent Studies on Literature (first series).

Hu Yao-Chen (1).
　Chhi Kung Chien Shen Fa　氣功健身法.
　Respiratory Exercises and the Strengthen-
　　ing of the Body.
　Thai-Phing, Hongkong, 1963.

Huang Chu-Hsün (1)　黃著勳.
　Chung-Kuo Khuang Chhan　中國鑛產.
　The Mineral Wealth and Productivity of
　　China.
　2nd ed., Com. Press, Shanghai, 1930.

Hung Huan-Chhun (1)　洪煥椿.
　*Shih chih Shih-san Shih-Chi Chung-Kuo
　　Kho-Hsüeh-ti Chu-Yao Chhêng-Chiu.*
　　十至十三世紀中國科學的主要成就.
　The Principal Scientific (and Techno-
　　logical) Achievements in China from the
　　+10th to the +13th centuries (inclusive),
　　[the Sung period].
　LSYC, 1959, **5** (no. 3), 27.

Hung Yeh (2) 洪業.
　Tsai Shuo 'Hsi Ching Tsa Chi' 再說「西
　　京雜記」.
　Further Notes on the Miscellaneous Records
　　of the Western Capital [with a study of
　　the dates of Ko Hung].
　AS/BIHP, 1963, 34 (no. 2), 397.
Hsia Nai (2) 夏鼐.
　Khao-Ku-Hsüeh Lun Wên Chi 考古學論
　　文集.
　Collected Papers on Archaeological Subjects.
　Academia Sinica, Peking, 1961.
Hsia Nai (3) 夏鼐.
　Sian Thang Mu Chhu Thu A-la-pa Chin
　　Pi 西安唐墓出土阿拉伯金幣.
　Arab Gold Coins unearthed from a Thang
　　Dynasty Tomb (at Hsi-yao-thou Village)
　　near Sian, Shensi (gold dīnārs of the
　　Umayyad Caliphs 'Abd al-Malik, +702,
　　'Umar ibn 'Abd al-'Azīz, +718, and
　　Marwān II, +746).
　Cf. Ho Han-Nan (1).
　KKTH, 1965, no. 8 (no. 108), 420, with
　　figs 1–6 on pl. 1.
Hsiang Ta (3) 向達.
　Thang Tai Chhang-An yü Hsi Yü Wên Ming
　　唐代長安與西域文明.
　Western Cultures at the Chinese Capital
　　(Chhang-an) during the Thang Dynasty.
　YCHP Monograph series, no. 2, Peiping,
　　1933.
Hsieh Sung-Mu (1) 謝誦穆.
　Chung-Kuo Li-Tai I-Hsüeh Wei Shu Khao
　　中國歷代醫學僞書考.
　A Study of the Authenticity of (Ancient
　　and Medieval) Chinese Medical Books.
　ISTC, 1947, 1 (no. 1), 53.
Hsiung Tê-Chi (1) 熊德基.
　'Thai Phing Ching' ti Tso-Chê ho Ssu-
　　Hsiang chi chhi yü Huang Chin ho Thien
　　Shih Tao ti Kuan-Hsi 「太平經」的作
　　者和思想及其與黃巾和天師道的
　　關係.
　The Authorship and Ideology of the Canon
　　of the Great Peace; and its Relation with
　　the Yellow Turbans (Rebellion) and the
　　Taoist Church (Tao of the Heavenly
　　Teacher).
　LSYC 1962 (no. 4), 8.
　Abstr. RBS, 1969, 8, no. 737.
Hsü Chien-Yin (1) 徐建寅.
　Ko Chih Tshung Shu 格致叢書.
　A General Treatise on the Natural Sciences.
　Shanghai, 1901.
Hsü Chih-I (1) 徐致一.
　Wu Chia Thai Chi Chhüan 吳家太極拳.
　Chinese Boxing Calisthenics according to
　　the Wu Tradition.
　Hsin-Wên, Hongkong, 1969.
Hsü Chung-Shu (7) 徐中舒.
　Chin Wên Chia Tzhu Shih Li 金文嘏辭
　　釋例.

Terms and Forms of the Prayers for Bles-
　　sings in the Bronze Inscriptions.
　AS/BIHP, 1936, (no. 4), 15.
Hsü Chung-Shu (8) 徐中舒.
　Chhen Hou Ssu Chhi Khao Shih 陳侯四
　　器考釋.
　Researches on Four Bronze Vessels of the
　　Marquis Chhen [i.e. Prince Wei of Chhi
　　State, r. −378 to −342].
　AS/BIHP, 1934, no. 3/4), 499.
Hsü Ti-Shan (1) 許地山.
　Tao Chiao Shih 道教史.
　History of Taoism.
　Com. Press, Shanghai, 1934.
Hsü Ti-Shan (2) 許地山.
　Tao Chia Ssu-Hsiang yü Tao Chiao 道家
　　思想與道教.
　Taoist Philosophy and Taoist Religion.
　YCHP, 1927, 2, 249.
Hsüeh Yü (1) 薛愚.
　Tao-Chia Hsien Yao chih Hua-Hsüeh Kuan
　　道家仙藥之化學觀.
　A Look at the Chemical Reactions in-
　　volved in the Elixir-making of the Taoists.
　HSS, 1942, 1 (no. 5), 126.
Huang Lan-Sun (1) (ed.) 黃蘭孫.
　Chung-Kuo Yao-Wu-ti Kho-Hsüeh Yen-
　　Chiu 中國藥物的科學研究.
　Scientific Researches on Chinese Materia
　　Medica.
　Chhien-Chhing Thang, Shanghai, 1952.

Imai Usaburō (1) 今井宇三郎.
　'Goshinpen' no Seisho to Shisō 悟眞篇の
　　成書と思想.
　The Poetical Essay on Realising the...
　　Primary Vitalities [by Chang Po-Tuan,
　　+1075]; its System of Thought and how
　　it came to be written.
　TS, 1962, 19, 1.
　Abstr. RBS, 1969, 8, no. 799.
Ishihara Akira (1) 石原明.
　Gozōnyūtai no Igi ni tsuite 五臟入胎の意
　　義について.
　The Buddhist Meaning of the Visceral
　　Models (in the Sakyamuni Statue at the
　　Seiryōji Temple).
　NIZ, 1956, 7 (nos. 1–3), 5.
Ishihara Akira (2) 石原明.
　Indo Kaibōgaku no Seiritsu to sono Ryūden
　　印度解剖學の成立とその流傳.
　On the Introduction of Indian Anatomical
　　Knowledge (to China and Japan).
　NIZ, 1956, 7 (nos. 1–3), 64.
Ishii Masako (1) 石井昌子.
　Kōhon 'Chen Kao' 稿本「眞誥.」
　Draft of an Edition of the Declarations of
　　Perfected Immortals, (with Notes on
　　Variant Readings).
　In several volumes.
　Toyoshima Shobō, Tokyo, (for Dōkyō
　　Kankōkai 道教刊行會), 1966–.

Ishii Masako (2) 石井昌子.
'*Chen Kao' no Seiritsu o Meguru Shiryō-teki
Kentō*; '*Têng Chen Yin Chüeh*', '*Chen
Ling Wei Yeh Thu*' *oyobi* '*Wu Shang Pi
Yao*' *tono Kankei wo Chūshin-ni* 「眞誥」
の成立をみぐる資料的檢討;「登眞
隱訣」,「眞靈位業圖」及び「無上秘要」
との關係を中心に.
Documents for the Study of the Formation
of the *Declarations of Perfected Immortals*....
DK, 1968, **3**, 79–195 (with French summary
on p. iv).
Ishii Masako (3) 石井昌子.
'*Chen Kao' no Seiritsu ni Kansuru Kōsatsu*
「眞誥」の成立に關する一考察.
A Study of the Formation of the *Declara-
tions of Perfected Immortals*.
DK, 1965, **1**, 215 (French summary, p. x).
Ishii Masako (4) 石井昌子.
Thao Hung-Ching Denkikō 陶弘景傳記考.
A Biography of Thao Hung-Ching.
DK, 1971, **4**, 29–113 (with French sum-
mary, p. iv).
Ishijima Yasutaka (1) 石島快隆.
Hōbokushi Insho Kō 抱朴子引書考.
A Study of the Books quoted in the *Pao
Phu Tzu* and its Bibliography.
BK, 1956, **20**, 877.
Abstr. *RBS*, 1959, **2**, no. 565.
Itō Kenkichi (1) 伊藤堅吉.
Sei no Mihotoke 性のみほとけ.
Sexual Buddhas (Japanese Tantric images
etc.).
Zufushinsha, Tokyo, 1965.
Itō Mitsutōshi (1) 伊藤光遠.
Yang Shêng Nei Kung Pi Chüeh 養生內
功祕訣.
Confidential Instructions on Nourishing the
Life Force by Gymnastics (and other
physiological techniques).
Tr. from the Japanese by Tuan Chu-
Chün 段竹君.
Thaipei (Thaiwan), 1966. ¦

Jen Ying-Chhiu (1) 任應秋.
Thung Su Chung-Kuo I-Hsüeh Shih Hua
通俗中國醫學史話.
Popular Talks on the History of Medicine.
Chungking, 1957.
Jung Kêng (3) 容庚.
Chin Wên Pien 金文編.
Bronze Forms of Characters.
Peking, 1925, repr. 1959.

Kao Chih-Hsi (1) 高至喜.
Niu Têng 牛鐙.
An 'Ox Lamp' (bronze vessel of Chhien
Han date, probably for sublimation,
with the boiler below formed in the shape
of an ox, and the rising tubes a continua-
tion of its horns).
WWTK, 1959 (no. 7), 66.

Kao Hsien (1) *et al.* 高銛.
Hua-Hsüeh Yao Phin Tzhu-Tien 化學藥
品辭典.
Dictionary of Chemistry and Pharmacy
(based on T. C. Gregory (1), with the
supplement by A. Rose & E. Rose).
Shanghai Sci & Tech. Pub., Shanghai,
1960.
Kawabata Otakeshi 川端男勇 & Yoneda
Yūtarō 米田祐太郎 (1)
Tōsei Biyaku-kō 東西媚藥考.
Die Liebestränke in Europa und Orient.
Bunkiūsha, Tokyo.
Kawakubo Teirō (1) 川久保悌郎.
*Shindai Manshū ni okeru Shōka no Zokusei
ni tsuite* 清代滿洲における燒鍋の簇
生について.
On the (Kao-liang) Spirits Distilleries in
Manchuria in the Chhing Period and
their Economic Role in Rural Colonisa-
tion.
Art. in *Wada Hakase Koki Kinen Tōyōshi
Ronsō* (Wada Festschrift) 和田博士古
稀記念東洋史論叢, Kōdansha, Tokyo,
1961, p. 303.
Abstr. *RBS*, 1968, **7**, no. 758.
Khung Chhing-Lai *et al.* (1) 孔慶萊 (13 col-
laborators).
Chih-Wu-Hsüeh Ta Tzhu Tien 植物學大
辭典.
General Dictionary of Chinese Flora.
Com. Press, Shanghai and Hongkong,
1918, repr. 1933 and often subsequently.
Kimiya Yasuhiko (1) 木宮泰彥.
Nikka Bunka Kōryūshi 日華文化交流史.
A History of Cultural Relations between
Japan and China.
Fuzambō, Tokyo, 1955.
Abstr *RBS*, 1959, **2**, no. 37.
Kobayashi Katsuhito (1) 小林勝人.
Yō Shu Gakuha no hitobite 楊朱學派の
人々.
On the Disciples and Representatives of the
(Hedonist) School of Yang Chu.
TYGK, 1961, **5**, 29.
Abstr. *RBS*, 1968, **7**, no. 606.
Koyanagi Shikita (1) 小柳司氣太.
Tao Chiao Kai Shuo 道教概說.
A Brief Survey of Taoism.
Tr. Chhen Pin-Ho 陳斌和.
Com. Press, Shanghai, 1926.
Repr. Com. Press, Thaipei, 1966.
Kuo Mo-Jo (8) 郭沫若.
Chhu Thu Wên Wu Erh San Shih 出土文
物二三事.
One or two Points about Cultural Relics
recently Excavated (including Japanese
coin inscriptions).
WWTK, 1972 (no. 3), 2.
Kuo Pao-Chün (1) 郭寶鈞.
*Hsün-hsien Hsin-tshun Ku Tshan Mu chih
Chhing Li* 濬縣辛村古殘墓之清理.

Kuo Pao-Chün (1) (cont.)
Preliminary Report on the Excavations at
the Ancient Cemetery of Hsin-tshun
village, Hsün-hsien (Honan).
TYKK, 1936, **1**, 167.
Kuo Pao-Chün (2) 郭寶鈞.
Hsün-hsien Hsin-tshun 濬縣辛村.
(Archaeological Discoveries at) Hsin-tshun
Village in Hsün-hsien (Honan).
Inst. of Archaeology, Academia Sinica,
Peking, 1964 (Field Expedition Reports,
I series, no. 13).
Kurihara Keisuke (1) 栗原圭介.
Gusai no Gireiteki Igi 虞祭の「儀禮」的
意義.
The Meaning and Practice of the Yü
Sacrifice, as seen in the *Personal
Conduct Ritual*.
NCGH, 1961, **13**, 19.
Abstr. *RBS*, 1968, **7**, no. 615.
Kuroda Genji (1) 黑田源次.
Ki 氣.
On the Concept of Chhi (*pneuma*; in
ancient Chinese thought).
TS, 1954 (no. 4/5), 1; 1955 (no. 7), 16.

Lai Chia-Tu (1) 賴家度.
'*Thien Kung Khai Wu*' *chi chhi Chu chê*;
Sung Ying-Hsing 「天工開物」及其著者
朱應星.
The *Exploitation of the Works of Nature* and
its Author; Sung Ying-Hsing.
Essay in Li Kuang-Pi & Chhien Chün-
Yeh (q.v.), p. 338.
Peking, 1955.
Lai Tou-Yen (1) 賴斗岩.
I Shih Sui Chin 醫史碎錦.
Medico-historical Gleanings.
ISTC, 1948, **2** (nos 3/4), 41.
Lao Kan (6) 勞榦.
*Chung-Kuo Tan-Sha chih Ying-Yung chi
chhi Thui-Yen* 中國丹砂之應用及其
推演.
The Utilisation of Cinnabar in China and
its Historical Implications.
AS/BIHP, 1936, **7** (no. 4), 519.
Li Chhiao-Phing (1) 李喬苹.
Chung-Kuo Hua-Hsüeh Shih 中國化學
史.
History of Chemistry in China.
Com. Press, Chhangsha, 1940, 2nd (en-
larged) ed. Thaipei, 1955.
Li Kuang-Pi 李光璧 & Chhien Chün-Yeh (1)
錢君曄.
*Chung-Kuo Kho-Hsüeh Chi-Shu Fa-Ming ho
Kho-Hsüeh Chi-Shu Jen Wu Lun Chi*
中國科學技術發明和科學技術人物
論集.
Essays on Chinese Discoveries and Inven-
tions in Science and Technology, and on
the Men who made them.
San-lien Shu-tien, Peking, 1955.

Li Nien (4) 李儼.
Chung Suan Shih Lun Tshung 中算史論
叢.
Gesammelte Abhandlungen ü. die Ge-
schichte d. chinesischen Mathematik.
3 vols. 1933–5; 4th vol. (in 2 parts), 1947.
Com. Press, Shanghai.
Li Nien (21) 李儼.
Chung Suan Shih Lun Tshung (second
series) 中算史論叢.
Collected Essays on the History of Chinese
Mathematics—vol. 1, 1954; vol. 2,
1954; vol. 3, 1955; vol. 4, 1955; vol. 5,
1955.
Kho-Hsüeh, Peking.
Li Shu-Hua (3) 李書華.
Li Shu-Hua Yu Chi 李書華遊記.
Travel Diaries of Li Shu-Hua [recording
visits to temples and other notable places
around Huang Shan, Fang Shan, Thien-
thai Shan, Yen-tang Shan etc. in 1935
and 1936].
Chhuan-chi Wên-hsüeh, Thaipei, 1969.
Li Shu-Huan (1) 李叔還.
Tao Chiao Yao I Wên Ta Chi Chhêng 道教
要義問答集成.
A Catechism of the Most Important Ideas
and Doctrines of the Taoist Religion.
Pr. Kao-hsiung and Thaipei, Thaiwan, 1970.
Distributed by the Chhing Sung Kuan
(Caerulean Pine-tree Taoist Abbey),
Chhing Shan (Castle Peak), N.T.
Hongkong.
Liang Chin (1) 梁津.
Chou Tai Ho Chin Chhêng Fên Khao 周代
合金成分考.
A Study of the Analysis of Alloys of the
Chou period.
KHS, 1925, **9** (no. 3), 1261; repr. in Wang
Chin (2), p. 52.
Lin Thien-Wei (1) 林天蔚.
Sung-Tai Hsiang Yao Mou-I Shih Kao
宋代香藥貿易史稿.
A History of the Perfume and Drug
Trade during the Sung Dynasty.
Chung-kuo Hsüeh-shê, Hongkong, 1960.
Ling Shun-Shêng (6) 凌純聲.
Chung-Kuo Chiu chih Chhi Yuan 中國酒
之起源.
On the Origin of Wine in China.
AS/BIHP, 1958, **29**, 883 (Chao Yuan-Jen
Presentation Volume).
Liu Kuei-Chen (1).
Chhi Kung Liao Fa Shih Chien 氣功療法
實踐.
The Practice of Respiratory Physiotherapy.
Hopei Jen-min, Paoting, 1957.
Also published as: *Shih Yen Chhi Kung Liao
Fa* 試驗氣功療法.
Experimental Tests of Respiratory Physio-
therapy.
Thai-Phing, Hongkong, 1965.

Liu Po (*1*) 劉波.
Mo-Ku chi chhi Tsai-Phei 蘑菇及其栽培.
Mushrooms, Toadstools, and their Culti-
vation.
Kho-Hsüeh, Peking, 1959, repr, 1960, 2nd
ed. enlarged, 1964.

Liu Shih-Chi (*1*) 劉仕驥.
Chung-Kuo Tsang Su Sou Chhi 中國葬俗
搜奇.
A Study of the Curiosities of Chinese Burial
Customs.
Shanghai Shu-chü, Hongkong, 1957.

Liu Shou-Shan *et al.* (*1*) 劉壽山.
Chung Yao Yen-Chiu Wên-Hsien Tsê-Yao
1820-1961 中藥研究文獻摘要.
A Selection of the Most Important Findings
in the Literature on Chinese Drugs from
1820 to 1961.
Kho-Hsüch, Peking, 1963.

Liu Wên-Tien (*2*) 劉文典.
Huai Nan Hung Lieh Chi Chieh 淮南鴻
烈集解.
Collected Commentaries on the Huai Nan
Tzu Book.
Com. Press, Shanghai, 1923, 1926.

Liu Yu-Liang (*1*) 劉友樑.
Khuang Wu Yao yü Tan Yao 礦物藥與
丹藥.
The Compounding of Mineral and In-
organic Drugs in Chinese Medicine.
Sci. & Tech. Press, Shanghai, 1962.

Lo Hsiang-Lin (*3*) 羅香林.
Thang Tai Kuang-chou Kuang-Hsiao Ssu
yü Chung-Yin Chiao-Thung chih Kuan-
Hsi 唐代廣州光孝寺與中印交通之
關係.
The Kuang-Hsiao Temple at Canton during
the Thang period, with reference to Sino-
Indian Relations.
Chung-kuo Hsüeh-shê, Hongkong, 1960.

Lo Tsung-Chen (*1*) 羅宗眞.
Chiangsu I-Hsing Chin Mu Fa-Chüeh Pao-
Kao 江蘇宜興晉墓發掘報告
(with a postscript by Hsia Nai 夏鼐).
Report of an Excavation of a Chin Tomb at
I-hsing in Chiangsu [that of Chou Chhu,
d. +297, which yielded the belt-orna-
ments containing aluminium; see p. 192).
AS/CJA, 1957 (no. 4), no. 18, 83.
Cf. Shen Shih-Ying (*1*); Yang Kên (*1*).

Lo Tsung-Chen (*2*) 羅宗眞.
Rejoinder to Shen Shih-Ying (*1*).
KKTH, 1963 (no. 3), 165.

Lu Khuei-Shêng (*1*) (ed.) 陸奎生.
Chung Yao Kho-Hsüeh Ta Tzhu-Tien
中藥科學大辭典.
Dictionary of Scientific Studies of Chinese
Drugs.
Shanghai Pub. Co., Hongkong, 1957.

Lung Po-Chien (*1*) 龍伯堅.
Hsien Tshun Pên-Tshao Shu Lu 現存本草
書錄.

Bibliographical Study of Extant Pharmacopoeias and Treatises on Natural History
(from all periods).
Jên-min Wei-shêng, Peking, 1957.

Ma Chi-Hsing (*2*) 馬繼興.
Sung-Tai-ti Jen Thi Chieh Phou Thu
宋代的人體解剖圖.
On the Anatomical Illustrations of the Sung
Period.
ISTC, 1957, 8 (no. 2), 125.

Maeno Naoaki (*1*) 前野直彬.
Meikai Yūkō 冥界游行.
On the Journey into Hell [critique of
Duyvendak (20) continued; a study of the
growth of Chinese conceptions of hell].
CBH, 1961, 14, 38; 15, 33.
Abstr. RBS, 1968, 7, no. 636.

Mao Phan-Lin (*1*) 茆泮林.
Ed. & comm. Huai Nan Wan Pi Shu (q.v.).
In Lung Chhi Ching Shih Tshung-Shu 龍溪
精舍叢書.
Collection from the Dragon Pool Studio.
Ed. Chêng Kuo-Hsün 鄭國勳 (1917).
c. 1821.

Masuda Tsuna (*1*) 增田綱謹. Master-
Craftsman to the Sumitomo Family.
Kodō Zuroku 鼓銅圖錄.
Illustrated Account of the (Mining,) Smelting
and Refining of Copper (and other Non-
Ferrous Metals).
Kyoto, 1801.
Tr. in CRRR, 1840, 9, 86.

Masutomi Kazunosuke (*1*) 益富壽之助.
Shōsōin Yakubutsu o Chūshin to suru Kodai
Sekiyaku no Kenkyū 正倉院藥物を中
心とする古代石藥の研究.
A study of Ancient Mineral Drugs based on
the chemicals preserved in the Shōsōin
(Treasury, at Nara).
Nihon Kōbutsu shumi no Kai, Kyoto, 1957.

Matsuda Hisao (*1*) 松田壽男.
Jūen to Ninjin to Chōbi 戎鹽と人參と貂皮.
On Turkestan salt, Ginseng and Sable Furs.
SGZ, 1957, 66, 49.

Mei Jung-Chao (*1*) 梅榮照.
Wo Kuo ti-i pên Wei-chi-fên Hsüeh ti i-pên;
'Tai Wei Chi Shih Chi' Chhu Pan I Pai
Chou Nien 我國第一本微積分學的
譯本；「代微積拾級」出版一百周年.
The Centenary of the First Translation into
Chinese of a book on Analytical Geo-
metry and Calculus; (Li Shan-Lan's
translation of Elias Loomis).
KHSC, 1960, 3, 59.
Abstr. RBS, 1968, 7, no. 747.

Mêng Nai-Chhang (*1*) 孟乃昌.
Kuan-yü Chung-Kuo Lien-Tan-Shu Chung
Hsiao-Suan-ti Ying Yung 關於中國煉
丹術中硝酸的應用.
On the (Possible) Applications of Nitric
Acid in (Mediaeval) Chinese Alchemy.
KHSC, 1966, 9, 24.

Michihata Ryōshū (1) 道端良秀.
Chūgoku Bukkyō no Kishin 中國佛教の
鬼神.
The 'Gods and Spirits' in Chinese Buddhism.
IBK, 1962, 10, 486.
Abstr. RBS, 1969, 8, no. 700.
Mikami, Yoshio (16) 三上義夫.
Shina no Muki Sanrui ni kansuru Chishiki
no Hajime 支那の無機酸類に關する
知識の始め.
Le Premier Savoir des Acides Inorganiques
en Chine.
JI, 1931, 1 (no. 1), 95.
Miki Sakae (1) 三木榮.
Chōsen Igakushi oyobi Shippeishi 朝鮮醫
學史及疾病史.
A History of Korean Medicine and of
Diseases in Korea.
Sakai, Osaka, 1962.
Miki Sakae (2) 三木榮.
Taikei Sekai Igakushi; Shoshi Teki Kenkyū
體系世界醫學史;書誌的研究.
A Systematic History of World Medicine;
Bibliographical Researches.
Tokyo, 1972.
Min I-Tê (1) 閔一得.
Kuan Khuei Pien 管窺編.
An Optick Glass (for the Enchymoma).
c. 1830.
In Tao Tsang Hsü Pien (Chhu chi), 7.
Miyagawa Torao et al. (1) 宮川寅雄.
Chhangsha Kanbo no Kiseki; Yomigaeru Tai
Hou Fu Jen no Sekai 長沙漢墓の奇
跡;よみがえる軑侯夫人の世界
Marvellous Relics from a Han Tomb; the
World of the Resurrected Lady of Tai.
SHA 1972 (增刊) no. 9–10.
Other important picture also in AGR 1972,
25 Aug.
Miyashita Saburō (1) 宮下三郎.
Kanyakū, Shūseki no Yakushi-gaku teki
Kenkyū 漢藥;秋石の藥史學的研究.
A Historical-Pharmaceutical Study of the
Chinese Drug 'Autumn Mineral' (chhiu
shih).
Priv. pr. Osaka, 1969.
Miyashita Saburō (2) 宮下三郎.
Senroku-jūichi-ren ni Chin Katsu ga Seizō
shita Sei-horumonzai ni tsuite 一〇六一
年に沈括が製造した性ホルモン劑に
ついて.
On the Preparation of 'Autumn Mineral'
[Steroid Sex Hormones from Urine] by
Shen Kua in +1061.
NIZ, 1965, 11 (no. 2), 1.
Mizuno Seiichi (3) 水野清一.
Indai Seidō Bunka no Kenkyū 殷代青銅
文化の研究.
Researches on the Bronze Culture of the
Shang (Yin) Period.
Kyoto, 1953.
Morita Kōmon (1) 森田幸門.

Josetsu 序説.
Introduction to the Special Number of
Nihon Ishigaku Zasshi (Journ. Jap. Soc.
Hist. of Med.) on the Model Human
Viscera in the Cavity of the Statue of
Sakyamuni (Buddha) at the Seiriyōji
Temple at Saga (near Kyoto).
NIZ, 1956, 7 (nos. 1–3), 1.
Murakami Yoshimi (3) 村上嘉實.
Chūgoku no Sennin; Hōbokushi no Shisō
中國の仙人;抱朴子の思想.
On the Immortals of Chinese (Taoism); a Study
of the Thought of Pao Phu Tzu (Ko Hung).
Heirakuji Shoten, Tokyo, 1956; repr. Se-u
Sōshō, Kyoto, 1957.
Abstr. RBS, 1959, 2, nos. 566, 567.

Nakajima Satoshi (1) 中島敏.
Shina ni okeru Shisshiki Shūdōhō no Kigen
支那に於ける濕式收銅法の起源.
The Origins and Development of the Wet
Method for Copper Production in China.
In Miscellany of Oriental Studies presented
to Prof. Katō Gen'ichi, Tokyo 1950,
加藤博士還曆記念東洋史集説.
Also TYG, 1945, 27 (no. 3).
Nakao, Manzō (1) 中尾万三.
Shokuryō-honsō no Kōsatsu 「食療本草」の
考察.
A Study of the [Tunhuang MS. of the] Shih
Liao Pên Tshao (Nutritional Therapy;
a Pharmaceutical Natural History), [by
Mêng Shen, c. +670].
SSIP, 1930, 1 (no. 3).
Nakaseko Rokuro (1).
Sekai Kwagakushi 世界化學史.
General History of Chemistry.
Kaniya Shoten, Kyoto, 1927.
Rev. M. Muccioli, A, 1928, 9, 379.
Ōbuchi Ninji (1) 大淵忍爾.
Dōkyō-shi no Kenkyū 道教史の研究.
Researches on the History of Taoism and
the Taoist Church.
Okayama, 1964.
Ogata Kōan (1) 緒方洪庵.
Byōgaku Tsūron 病學通論.
Survey of Pathology (after Christopher
Hufeland's theories).
Tokyo, 1849.
Ogata Kōan (2) 緒方洪庵.
Hushi Keiken Ikun 扶氏經驗遺訓.
Mr Hu's (Christopher Hufeland's) Well-tested
Advice to Posterity [medical macrobiotics].
Tokyo, 1857.
Ogata Tamotsu (1) 小片保.
Waga Kuni Sokushinbutsu seiritsu ni Kansuru
Shomondai 我國即身佛成立に關す
る諸問題.
The Self-Mummified Buddhas of Japan, and
Several (Anatomical) Questions concern-
ing them.
NDI (Spec. No.), 1962 (no. 15), 16, with 8 pls.

Okanishi Tameto (2) 岡西爲人.
Sung I-chhien I Chi Khao 宋以前醫籍考.
Comprehensive Annotated Bibliography of
Chinese Medical Literature in and before
the Sung Period.
Jên-min Wei-shêng, Peking, 1958.
Okanishi Tameto (4) 岡西爲人.
Tan Fang chih Yen-Chiu 丹方之研究.
Index to the 'Tan' Prescriptions in Chinese
Medical Works.
In Huang Han I-Hsüeh Tshung-Shu, 1936,
vol. 11.
Okanishi Tameto (5) 岡西爲人.
Chhung Chi 'Hsin Hsiu Pên Tshao' 重輯
「新修本草」.
Newly Reconstituted Version of the New
and Improved Pharmacopoeia (of +659).
National Pharmaceutical Research Insti-
tute, Thaipei, 1964.
Ong Tu-Chien (1) 翁獨健.
'Tao Tsang' Tzu Mu Yin Tê 道藏子目
引得.
An Index to the Taoist Patrology.
Harvard-Yenching, Peiping, 1935.
Ong Wên-Hao (1) 翁文灝.
Chung-Kuo Kung Chhan Chih Lüeh 中國
鑛產誌畧.
The Mineral Resources of China (Metals
and Non-Metals except Coal).
MGSC (Ser. B), 1919, no. 1, 1–270.
With English contents-table.
Ōya Shin'ichi (1) 大矢眞一.
Nihon no Sangyō Gijutsu 日本の產業技
術.
Industrial Arts and Technology in (Old)
Japan.
Sanseido, Tokyo, 1970.

Pa Tzu-Yuan (1) (ed.) 巴子園.
Tan I San Chüan 丹擬三卷.
Three Books of Draft Memoranda on
Elixirs and Enchymomas.
1801.
Phan Wei (1) 潘霨.
Wei Shêng Yao Shu 衛生要術.
Essential Techniques for the Preservation
of Health [based on earlier material on
breathing exercises, physical culture and
massage etc. collected by Hsü Ming-
Fêng 徐鳴峰].
1848, repr. 1857.
Pi Li-Kan (Billequin, M. A.) 畢利干,
Chhêng Lin 承霖 & Wang Chung-
Hsiang 王鍾祥 (1).
Hua-Hsüeh Shan Yuan 化學闡原.
Explanation of the Fundamental Principles
of Chemistry.
Thung Wên Kuan, Peking, 1882.

Richie, Donald ドナルド・リチー & Itō
Kenkichi (1) = (1) 伊藤堅吉.
Danjozō 男女像.
Images of the Male and Female Sexes
[= The Erotic Gods].
Zufushinsha, Tokyo, 1967.

Sanaka Sō (1) 佐中壯.
Tō Inkyo Shōden; Sono Senjutsu o tsujite
mita Honzōgaku to Senyaku to no Kankei
陶隱居小傳; その撰述を通じて見
た本草學と仙藥との關係.
A Biography of Thao Hung-Ching; his
Knowledge of Botany and Medicines of
Immortality.
Art. in Wada Hakase Koki Kinen Tōyōshi
Ronsō (Wada Festschrift) 和田博士古
稀記念東洋史論叢. Kōdansha, Tokyo,
1961, p. 447.
Abstr. RBS, 1968, 7, no. 756.
Sawa Ryūken (1) 左和隆研.
Nihon Mikkyō, sono Tenkai to Bijutsu
日本密教その展開と美術.
Esoteric (Tantric) Buddhism in Japan; its
Development and (Influence on the) Arts.
NHK, Tokyo, 1966, repr. 1971.
Shen Shih-Ying (1) 沈時英.
Kuan-yü Chiangsu I-Hsing Hsi Chin Chou
Chhu Mu Chhu-Thu Tai-Shih Chhêng-
Fên Wên-Thi 關于江蘇宜興西晉周處
墓出土帶飾成分問題.
Notes on the Chemical Composition of the
Belt Ornaments from the Western
Chin Period (+265 to +316) found in
the Tomb of Chou Chhu at I-hsing in
Chiangsu.
KKTH, 1962 (no. 9), 503.
Eng. tr. by N. Sivin (unpub.).
Cf. Lo Tsung-Chen (1); Yang Kên (1).
Shih Shu-Chhing (2) 史樹青.
Ku-Tai Kho-Chi Shih Wu Ssu Khao 古代
科技事物四考.
Four Notes on Ancient Scientific Tech-
nology; (a) Ceramic objects for medical
heat-treatment; (b) Mercury silvering of
bronze mirrors; (c) Cardan Suspension
perfume burners; (d) Dyeing stoves.
WWTK, 1962 (no. 3), 47.
Shima Kunio (1) 島邦男.
Inkyo Bokuji Kenkyū 殷墟卜辭研究.
Researches on the Shang Oracle-Bones and
their Inscriptions.
Chūgokugaku Kenkyūkai, Hirosaki, 1958.
Abstr. in RBS, 1964, 4, no. 520.
Shinoda Osamu (1) 篠田統.
Daki Hyō Shōkō 噯氣樽小考.
A Brief Study of the 'Daki' [Nuan Chhi]
Temperature Stabiliser (used in breweries
for the saccharification vats, cooling them
in summer and warming them in winter).
MOULA, ser. B, 1963 (no. 12), 217.
Shinoda Osamu (2) 篠田統.
Chūsei no Sake 中世の酒.
Wine-Making in Medieval (China and
Japan).

Shinoda Osamu (2) (cont.)
Art. in Yabuuchi Kiyoshi (25), p. 321.

Su Fên 蘇芬, Chu Chia-Hsüan 朱稼軒 et
al. (1).
Tzhu-Hang Fa Shih Phu-Sa Ssu Pu Hsiu
航慈法師菩薩四不朽.
The Self-Mummification of the Abbot and
Bodhisattva, Tzhu-Hang (d. 1954).
CJFC, 1959 (no. 27), pp. 15, 21, etc.

Su Ying-Hui (1) 蘇瑩輝.
Lun 'Wu Lei Hsiang Kan Chih' chih Tso-
Chhêng Shih-Tai 論「物類相感志」之作
成時代.
On the Time of Completion of the Mutual
Responses of Things according to their
Categories.
TLTC 1970, 40 (no. 10).

Su Ying-Hui (2) 蘇瑩輝.
'Wu Lei Hsiang Kan Chih' Fên Chüan Yen-
Ko Khao-Lüeh 「物類相感志」分卷沿
革考略.
A Study of the Transmission of the Mutual
Responses of Things according to their
Categories and the Vicissitudes in the
Numbering of its Chapters.
KKTS, 1970, 1 (no. 2), 23.

Sun Fêng-I (1) 孫馮翼.
Ed. & comm. Huai Nan Wan Pi Shu (q.v.).
In Wên Ching Thang Tshung-Shu 問經堂
叢書.
Collection from the Hall of Questioning the
Classics.
+1797 to 1802.

Sun Tso-Yün (1) 孫作雲.
Shuo Yü Jen 說羽人.
On the Feathered and Winged Immortals
(of early Taoism).
LSYKK, 1948.

Takeuchi Yoshio (1) 武內義雄.
Shinsen Setsu 神僊說.
The Holy Immortals (a study of ancient
Taoism).
Tokyo, 1935.

Taki Mototane (1) 多紀元胤.
I Chi Khao (Iseki-kō) 醫籍考.
Comprehensive Annotated Bibliography of
Chinese Medical Literature (Lost or
Still Existing).
c. +1825, pr. 1831.
Repr. Tokyo, 1933, and Chinese-Western
Medical Research Society, Shanghai, 1936,
with introdn. by Wang Chi-Min.

Takizawa Bakin (1) 瀧澤馬琴.
Kinsei-setsu Bishōnen-roku 近世說美少
年錄.
Modern Stories of Youth and Beauty.
Japan (Yedo), c. 1820.

Tan Jan-Hui (1) (ed.) 澹然慧.
'Chhang Shêng Shu', 'Hsü Ming Fang' Ho
Khan 「長生術」「續命方」合刊.
A Joint Edition of the Art and Mystery of

Longevity and Immortality and the Precepts for Lengthening the Life-Span. [The
former work is that previously entitled
Thai-I Chin Hua Tsung Chih (q.v.) and
the latter is that previously entitled
Tsui-Shang I Chhêng Hui Ming Ching
(q.v.).]
Peiping, 1921 (the edition used by Wilhelm
& Jung, 1).

Thang Yung-Thung 湯用彤 & Thang I-Chieh
(1) 湯一介.
Khou Chhien-Chih ti Chu-Tso yü Ssu-
Hsiang 冦謙之的著作與思想.
On the Doctrines and Writings of (the
Taoist reformer) Khou Chhien-Chih (in
the Northern Wei period).
LSYC, 1961, 8 (no. 5), 64.
Abstr. RBS, 1968, 7, no. 659.

Ting Hsü-Hsien (1) 丁緒賢.
Hua Hsüeh Shih Thung Khao 化學史
通考.
A General Account of the History of
Chemistry.
2 vols., Com. Press, Shanghai, 1936, repr.
1951.

Ting Wei-Liang (Martin, W. A. P.) (1)
丁韙良.
Ko Wu Ju Mên 格物入門.
An Introduction to Natural Philosophy.
Thung Wên Kuan, Peking, 1868.

Ting Wên-Chiang (1) 丁文江.
Biography of Sung Ying-Hsing 宋應星
(author of the Exploitation of the Works of
Nature).
In the Hsi Yung Hsüan Tshung-Shu 喜詠
軒叢書, ed. Thao Hsiang 陶湘.
Peiping, 1929.

Tōdō Kyōshun (1) 藤堂恭俊.
Shina Jōdokyō ni okeru Zuichiku Yōgo
Setsu no seiritsu Katei ni tsuite シナ淨土
教における隋逐擁護說の成立過程
について,
On the Origin of the Invocation to the
25 Bodhisattvas for Protection against
severe Judgments; a Practice of
the Chinese Pure Land (Amidist)
School.
Art. in Tsukamoto Hakase Shōju Kinen
Bukkyōshigaku Ronshū (Tsukamoto
Festschrift) 塚本博士頌壽記念佛敎
史學論集, Kyoto, 1961, p. 502.
Abstr. RBS, 1968, 7, no. 664.

Tokiwa, Daijō (1) 常盤大定.
Dōkyō Gaisetsu 道敎概說.
Outline of Taoism.
TYG, 1920, 10 (no. 3), 305.

Tokiwa, Daijō (2) 常盤大定.
Dōkyō Hattatsu-shi Gaisetsu 道敎發達史
概說.
General Sketch of the Development of
Taoism.
TYG, 1921, 11 (no. 2), 243.

Tsêng Chao-Lun (1) 曾昭掄
The Translations of the Chiangnan Arsenal Bureau.
TFTC, 1951, **38** (no. 1), 56.

Tsêng Chao-Lun (2) 曾昭掄.
Chung Wai Hua-Hsüeh Fa-Chan Kai Shu 中外化學發展概述.
Chinese and Western Chemical Discoveries; an Outline.
TFTC, 1953, **40** (no. 18), 33.

Tsêng Chao-Lun (3) 曾昭掄.
Erh-shih Nien Lai Chung-Kuo Hua-Hsüeh chih Chin-Chan 二十年來中國化學之進展
Advances in Chemistry in China during the past Twenty Years.
KHS, 1936, **19** (no. 10), 1514.

Tsêng Hsi-Shu (1) 曾熙署.
Ssu Thi Ta Tzu Tien 四體大字典.
Dictionary of the Four Scripts.
Shanghai, 1929.

Tsêng Yuan-Jung (1) 曾遠榮.
Chung-Kuo Yung Hsin chih Chhi-Yuan 中國用鋅之起源.
Origins and Development of Zinc Technology in China [with a dating of the *Pao Tsang Lun*].
Letter of Oct. 1925 to Wang Chin.
Pr. in Wang Chin (2), p. 92.

Tshai Lung-Yün (1) 蔡龍雲.
Ssu Lu Hua Chhüan 四路華拳.
Chinese Boxing Calisthenics on the Four Directions System.
Jen-min Thi-yu, Peking, 1959, repr. 1964.

Tshao Yuan-Yü (1) 曹元宇.
Chung-Kuo Ku-Tai Chin-Tan-Chia ti Shê-Pei ho Fang-Fa 中國古代金丹家的設備和方法.
Apparatus and Methods of the Ancient Chinese Alchemists.
KHS, 1933, **17** (no. 1), 31.
Reprinted in Wang Chin (2), p. 67.
Engl. précis by Barnes (1).
Engl. abstr. by H. D. C[ollier], *ISIS*, 1935, **23**, 570.

Tshao Yuan-Yü (2) 曹元宇.
Chung-Kuo Tso Chiu Hua-Hsüeh Shih-Liao 中國作酒化學史料.
Materials for the History of Fermentation (Wine-making) Chemistry in China.
HITC, 1922, **6** (no. 6), 1.

Tshao Yuan-Yü (3) 曹元宇.
Kuan-yü Thang-Tai mei-yu Chêng-Liu Chiu ti Wên-thi 關于唐代沒有蒸餾酒的問題.
On the Question of whether Distilled Alcoholic Liquors were known in the Thang Period.
KHSC, 1963, no. 6, 24.

Tsuda Sōkichi (2) 津田左右吉.
Shinsen Shisō ni kansuru ni-san no Kōsatsu 神仙思想に關する二三の考察.

Some Considerations and Researches on the Holy Immortals (and the Immortality Cult in Ancient Taoism.
In *Man-Sen Chiri Rekishi Kenkyū Hōkoku* 滿鮮地理歷史研究報告 (Research Reports on the Historical Geography of Manchuria and Korea), 1924, no. 10, 235

Tsumaki, Naoyoshi (1) 妻木直良.
Dōkyō no Kenkyū 道教の研究.
Studies in Taoism.
TYG, 1911, **1** (no. 1), 1; (no. 2), 20; 1912, **2** (no. 1), 58.

Tuan Wên-Chieh (1) (ed.) 段文杰.
Yü Lin Khu 楡林窟.
The Frescoes of Yü-lin-khu [i.e. Wan-fo-hsia, a series of cave-temples in Kansu].
Tunhuang Research Institute, Chung-kuo Ku-tien I-shu, Peking, 1957.

Tzu Chhi (1) 梓溪.
Chhing Thung Chhi Ming Tzhu Chieh-Shuo 青銅器名詞解說.
An Explanation of the Terminology of (Ancient) Bronze Vessels.
WWTK, 1958 (no. 1), 1; (no. 2), 55; (no. 3), 1; (no. 4), 1; (no. 5), 1; (no. 6), 1; (no. 7), 68.

Udagawa Yōan (1) 宇田川榕庵.
Seimi Kaisō 舍密開宗.
Treatise on Chemistry [largely a translation of W. Henry (1), but with added material from other books, and some experiments of his own].
Tokyo, 1837-46.
Cf. Tanaka Minoru (3).

Umehara Sueji (3) 梅原末治.
Senoku Seishō Shinshūhen 泉屋清賞;新收編.
New Acquisitions of the Sumitomo Collection of Ancient Bronzes (Kyoto); a Catalogue.
Kyoto, 1961.
With English contents-table.

Wada Hisanori (1) 和田久德.
'*Namban Kōroku*' to '*Shohanshi*' to no *Kankei* 「南蕃香錄」と「諸蕃志」との關係.
On the *Records of Perfumes and Incense of the Southern Barbarians* [by Yeh Thing-Kuei, c. +1150] and the *Records of Foreign Peoples* [by Chao Ju-Kua, c. +1250, for whom it was an important source].
OUSS, 1962, **15**, 133.
Abstr. *RBS*, 1969, **8**, no. 183.

Wang Chi-Liang 王季梁 & Chi Jen-Jung (1) 紀紉容.
Chung-Kuo Hua-Hsüeh Chieh chih Kuo-Chhü yü Wei-Lai 中國化學界之過去與未來.
The Past and Future of Chemistry [and Chemical Industry] in China.
HHSTH, 1942, 3.

Wang Chi-Wu (*1*) 王輯五.
Chung-Kuo Jih-Pên Chiao Thung Shih
中國日本交通史.
A History of the Relations and Connections
between China and Japan.
Com. Pr., Thaiwan, 1965 (*Chung-Kuo
Wên-Hua Shih Tshung-Shu* ser.).

Wang Chin (*1*) 王璡.
Chung-Kuo chih Kho-Hsüeh Ssu-Hsiang
中國之科學思想.
On (the History of) Scientific Thought in
China.
Art. in *Kho-Hsüeh Thung Lun* (科學通
論).
Sci. Soc. of China, Shanghai, 1934.
Orig. pub. *KHS*, 1922, **7** (no. 10), 1022.

Wang Chin (*2*) (ed.) 王璡.
Chung-Kuo Ku-Tai Chin-Shu Hua-Hsüeh chi
Chin Tan Shu 中國古代金屬化學及
金丹術.
Alchemy and the Development of Metal-
lurgical Chemistry in Ancient and Medi-
eval China [collective work].
Chung-kuo Kho-hsüeh Thu-shu I-chhi
Kung-ssu, Shanghai, 1955.

Wang Chin (*3*) 王璡.
Chung-Kuo Ku-Tai Chin Shu Yuan Chih
chih Hua-Hsüeh 中國古代金屬原質
之化學.
The Chemistry of Metallurgical Operations
in Ancient and Medieval China [smelt-
ing and alloying].
KHS, 1919, **5** (no. 6), 555; repr. in Wang
Chin (*2*), p. 1.

Wang Chin (*4*) 王璡.
Chung-Kuo Ku-Tai Chin Shu Hua Ho Wu
chih Hua Hsüeh 中國古代金屬化合物
之化學.
The Chemistry of Compounds containing
Metal Elements in Ancient and Medieval
China.
KHS, 1920, **5** (no. 7), 672; repr. in Wang
Chin (*2*), p. 10.

Wang Chin (*5*) 王璡.
Wu Shu Chhien Hua-Hsüeh Chhêng-Fên chi
Ku-Tai Ying-Yung Chhien Hsi Hsin La
Khao 五銖錢化學成分及古代應用
鉛錫鋅鑞考.
An Investigation of the Ancient Tech-
nology of Lead, Tin, Zinc and *la*, together
with Chemical Analyses of the Five-Shu
Coins [of the Han and subsequent
periods].
KHS, 1923, **8** (no. 9), 839; repr. in Wang
Chin (*2*), p. 39.

Wang Chin (*6*) 王璡.
Chung-Kuo Thung Ho Chin nei chih Nieh
中國銅合金內之鎳.
On the Chinese Copper Alloys containing
Nickel [paktong] etc.
KHS, 1929, **13**, 1418; abstr. in Wang Chin
(*2*), p. 91.

Wang Chin (*7*) 王璡.
Chung-Kuo Ku-Tai Chiu-Ching Fa-Chiao
Yeh chih i pan 中國古代酒精發酵業
之一斑.
A Brief Study of the Alcoholic Fermentation
Industry in Ancient (and Medieval)
China.
KHS, 1921, **6** (no. 3), 270.

Wang Chin (*8*) 王璡.
Chung-Kuo Ku-Tai Thao Yeh chih Kho-
Hsüeh Kuan 中國古代陶業之科學
觀.
Scientific Aspects of the Ceramics Industry
in Ancient China.
KHS, 1921, **6** (no. 9), 869.

Wang Chin (*9*) 王璡.
Chung-Kuo Huang Thung Yeh chih Chhüan
Shêng Shih-Chhi 中國黃銅業之全盛
時期.
On the Date of Full Development of the
Chinese Brass Industry.
KHS, 1925, **10**, 495.

Wang Chin (*10*) 王璡.
I-Hsing Thao Yeh Yuan Liao chih Kho-
Hsüeh Kuan 宜興陶業原料之科學觀
Scientific Aspects of the Raw Materials of
the I-hsing Ceramics Industry.
KHS, 1932, **16** (no. 2), 163.

Wang Chin (*11*) 王璡.
Chung-Kuo Ku-Tai Hua-Hsüeh ti Chhêng-
Chiu 中國古代化學的成就.
Achievements of Chemical Science in
Ancient and Medieval China.
KHTP, 1951, **2** (no. 11), 1142.

Wang Chin (*12*) 王璡.
Ko Hung i-chhien chih Chin Tan Shih Lüeh
葛洪以前之金丹史略.
A Historical Survey of Alchemy before Ko
Hung (*c.* +300).
HITC, 1935, **14**, 145, 283.

Wang Hsien-Chhien (*3*) 王先謙.
Shih Ming Su Chhông Pu 釋名疏證補.
Revised and Annotated Edition of the
[Han] *Explanation of Names* [dictionary].
Peking, 1895.

Wang Khuei-Kho (*1*) (tr.) 王奎克.
San-shih-liu Shui-Fa—Chung-Kuo Ku Tai
Kuan-yü Shui Jung I ti I Chung Tsao
Chhi Lien Tan Wên Hsien 三十六水
法—中國古代關於水溶液的一種早
期煉丹文獻.
The *Thirty-six Methods of Bringing Solids
into Aqueous Solution*—an Early Chinese
Alchemical Contribution to the Problem
of Dissolving (Mineral Substances), [a
partial translation of Tshao Thien-Chhin,
Ho Ping-Yü & Needham, J. (*1*).]
KHSC, 1963, **5**, 67.

Wang Khuei-Kho (*2*) 王奎克.
Chung-Kuo Lien-Tan-Shu Chung ti Chin-I
ho Hua-Chhih 中國煉丹術中的金液
和華池.

Wang Khuei-Kho (2) (cont.)
'Potable Gold' and Solvents (for Mineral Substances) in (Medieval) Chinese Alchemy.
KHSC, 1964, 7, 53.

Wang Ming (2) 王明.
'Thai Phing Ching' Ho Chiao 「太平經」合校.
A Reconstructed Edition of the Canon of the Great Peace (and Equality).
Chung-Hua, Peking and Shanghai, 1960.

Wang Ming (3) 王明.
'Chou I Tshan Thung Chhi' Khao Chêng 「周易參同契」考證.
A Critical Study of the Kinship of the Three.
AS/BIHP, 1948, 19, 325.

Wang Ming (4) 王明.
'Huang Thing Ching' Khao 「黃庭經」考.
A Study on the Manuals of the Yellow Courts.
AS/BIHP, 1948, 20A.

Wang Ming (5) 王明.
'Thai Phing Ching' Mu Lu Khao 「太平經」目錄考.
A Study of the Contents Tables of the Canon of the Great Peace (and Equality).
WS, 1965, no. 4, 19.

Wang Tsu-Yuan (1) 王祖源.
Nei Kung Thu Shuo 內功圖說.
Illustrations and Explanations of Gymnastic Exercises [based on an earlier presentation by Phan Wei (q.v.) using still older material from Hsü Ming-Fêng].
1881.
Modern reprs. Jen-min Wei-shêng, Peking, 1956; Thai-Phing, Hongkong, 1962.

Wang Yeh-Chhiu, 王治秋 Wang Chung-Shu 王仲殊 & Hsia Nai (1) 夏鼐.
Bunka Dai-Kakumei-Kikan Shutsudo Bumbutsu Tenran 文化大革命期間出土文物展覽.
Articles to accompany the Exhibition of Cultural Relics Excavated (in Ten Provinces of China) during the Period of the Great Cultural Revolution.
JC, 1971 (no. 10), 31, with colour-plates.

Watanabe Kōzō (1) 渡邊幸三.
Genzai suru Chūgoku Kinsei made no Gozō Rokufu Zu no Gaisetsu 現存する中國近世までの五藏六府圖の概說
General Remarks on (the History of) Dissection and Anatomical Illustration in China.
NIZ, 1956, 7 (nos. 1–3), 88.

Watanabe Kōzō (2) 渡邊幸三.
Seiryōji Shaka Tainai Gozō no Kaibō-gakuteki Kenkyū 清涼寺釋迦胎內五藏の解剖學的研究.
An Anatomical Study (of Traditional

Chinese Medicine) in relation to the Visceral Models in the Sakyamuni Statue at the Seiryōji Temple (at Saga, near Kyoto).
NIZ, 1956, 7 (nos. 1–3), 30.

Wei Yuan (1) 魏源.
Shêng Wu Chi 聖武記.
Records of the Warrior Sages [a history of the military operations of the Chhing emperors].
1842.

Wên I-To (3) 聞一多.
Shen Hua yü Shih 神話與詩.
Religion and Poetry (in Ancient Times), [contains a study of the Taoist immortality cult and a theory of its origins].
Peking, 1956 (posthumous).

Wu Chhêng-Lo (2) 吳承洛.
Chung-Kuo Tu Liang Hêng Shih 中國度量衡史.
History of Chinese Metrology [weights and measures].
Com. Press, Shanghai, 1937; 2nd ed. Shanghai, 1957.

Wu Shih-Chhang (1) 吳世昌.
Mi Tsung Su Hsiang Shuo Lüeh 密宗塑像說略.
A Brief Discussion of Tantric (Buddhist) Images.
AP/HJ, 1935, 1.

Wu Tê-To (1) 吳德鐸.
Thang Sung Wên-Hsien chung Kuan-yü Chêng-Liu Chiu yü Chêng-Liu Chhi Wên-thi 唐宋文獻中關于蒸餾酒與蒸餾器問題.
On the Question of Liquor Distillation and Stills in the Literature of the Thang and Sung Periods.
KHSC, 1966, no. 9, 53.

Yabuuchi Kiyoshi (11) (ed.) 藪內淸.
'Tenkō Kaibutsu' no Kenkyū 「天工開物」の研究.
A Study of the Thien Kung Khai Wu (Exploitation of the Works of Nature, +1637).
Japanese translation of the text, with annotative essays by several hands.
Kōseisha, Tokyo, 1953.
Rev. Yang Lien-Shêng, HJAS, 1954, 17, 307.
English translation of the text (sparingly annotated). See Sun & Sun (1).
Chinese translations of the eleven essays:
(a) 'Thien Kung Khai Wu' chih Yen-Chiu by Su Hsiang-Yü 蘇薌雨 et al.
Tshung-Shu Wei Yuan Hui, Thaiwan, and Chi-Shêng, Hongkong, 1956
(b) 'Thien Kung Khai Wu' Yen-Chiu Wên Chi by Chang Hsiung 章熊 & Wu Chieh 吳傑
Com. Press, Peking, 1961.

Yabuuchi Kiyoshi (25) (ed.) 藪內清.
Chūgoku Chūsei Kagaku Gijutsushi no
 Kenkyū 中國中世科學技術史の研究.
Studies in the History of Science and Tech-
 nology in Medieval China [a collective
 work].
Kadokawa Shoten (for the Jimbun Kagaku
 Kenkyusō), Tokyo, 1963.

Yamada Keiji (1) 山田慶兒.
'Butsurui sōkan shi' no seiritsu 「物類相感
 志」の成立.
The Organisation of the Book Wu Lei
 Hsiang Kan Chih (Mutual Responses of
 Things according to their Categories).
SBK, 1965, 13, 305.

Yamada Keiji (2) 山田慶兒.
Chūsei no Shizen-kan 中世の自然觀.
The Naturalism of the (Chinese) Middle
 Ages [with special reference to Taoism,
 alchemy, magic and apotropaics].
Art. in Yabuuchi (25), pp. 55–110.

Yamada Kentarō (1) 山田憲太郎.
Tōzai Kōyakushi 東西香藥史.
A History of Perfumes, Incense, Aromatics
 and Spices in East and West.
Tokyo, 1958.

Yamada Kentarō (2) 山田憲太郎.
Kōryō no Rekishi 香料の歷史.
History of Perfumes, Incense and Aro-
 matics.
Tokyo, 1964 (Kiino Kuniya Shinshō, ser. B,
 14).

Yamada Kentarō (3) 山田憲太郎.
Ogawa Kō-ryō Jihō 小川香料時報.
News from the Ogawa Company; A History
 of the Incense, Spice and Perfume
 Industry (in Japan).
Ogawa & Co. (pr. pr.), Osaka, 1948.

Yamada Kentarō (4) 山田憲太郎.
Tōa Kō-ryōshi 東亞香料史.
A History of Incense, Aromatics and Per-
 fumes in East Asia.
Toyōten, Tokyo, 1942.

Yamada Kentarō (5) 山田憲太郎.
Chūgoku no Ansoku-kō to Seiyō no Benzoin
 to no Genryū. 中國の安息香と西洋
 のベンゾインとの源.
A Study of the Introduction of an-hsi
 hsiang (gum guggul, bdellium) into China,
 and that of gum benzoin into Europe.
SB, 1951 (no. 2), 1–36.

Yamada Kentarō (6) 山田憲太郎.
Chūsei no Chūgokujin to Arabiajin ga Shitte
 ita Ryūnō no Sanshutsuchi toku ni
 Baritsu (no) Kuni ni tsuite 中世の中國
 人とアラビア人が知っていた龍腦
 の產出地とくに婆律國について
On the knowledge which the Medieval
 Chinese and Arabs possessed of Baros
 camphor (from Dryobalanops aromatica)
 and its Place of Production, Borneo.
NYGDR, 1966 (no. 5), 1.

Yamada Kentarō (7) 山田憲太郎.
Ryūnō-kō (Sono Shōhinshi-teki Kōsatsu)
 龍腦考(その商品史的考察)
A Study of Borneo or Baros camphor
 (from Dryobalanops aromatica), and the
 History of the Trade in it.
NYGDR, 1967 (no. 10), 19.

Yamada Kentarō (8) 山田憲太郎.
Chin sunawachi Kō 沈すなわ香.
On the 'Sinking Aromatic' (garroo wood,
 Aquilaria agallocha).
NYGDR, 1970, 7 (no. 1), 1.

Yang Chhêng-Hsi (1) et al. 楊承禧.
Hu-Pei Thung Chih 湖北通志.
Historical Geography of Hupei Province.
1921.

Yang Kên (1) 楊根.
Chin Tai Lü Thung ho-chin-ti Chien-Ting
 chi chhi Yeh-Lien Chi-Shu-ti Chhu-Pu
 Than-Thao 晉代鋁銅合金的鑑定及
 其冶煉技術的初步探討.
An Aluminium-Copper Alloy of the Chin
 Dynasty (+265 to +420); its Determina-
 tion and a Preliminary Study of the
 Metallurgical Technology (which it
 Implies).
AS/CJA, 1959 (no. 4), no. 26, 91.
Eng. tr. by D. Bryan (unpub.) for the
 Aluminium Development Association,
 1962.
Cf. Lo Tsung-Chen (1); Shen Shih-
 Ying (1).

Yang Lieh-Yü (1) 楊烈宇.
Chung-Kuo Ku-Tai Lao-Tung Jen-Min tsai
 Chin-Shu chi Ho-Chin Ying-Yung-shang-
 ti Chhêng-Chiu 中國古代勞動人民在
 金屬及合金應用上的成就.
Ancient Chinese Achievements in Practical
 Metal and Alloy Technology.
KHTP, 1955, 5 (no. 10), 77.

Yang Lien-Shêng (2) 楊聯陞.
Tao Chiao chih Tzu Po yü Fo Chiao chih
 Tzu-Phu 道教之自搏與佛教之自撲
Penitential Self-Flagellation, Violent Pros-
 tration and similar practices in Taoist and
 Buddhist Religion.
Art. in Tsukamoto Hakase Shōju Kinen
 Bukkyōshigaku Ronshū (Tsukamoto
 Festschrift), 塚本博士頌壽記念佛教
 史學論集. Kyoto, 1961, p. 962.
Abstr. RBS, 1968, 7, no. 642.
Also AS/BIHP, 1962, 34, 275; abstr. in
 RBS, 1969, 8, no. 740.

Yang Ming-Chao (1) 楊明照.
Critical Notes on the Pao Phu Tzu
 book.
BCS, 1944, 4.

Yang Po-Chün (1) 楊伯峻.
Lüeh Than Wo-Kuo Shih-Chi shang kuan-
 yü Shih Thi Fang-Fu-ti Chi-Tsai ho
 Ma-wang-tui I-hao Han-Mu Mu-Chu
 Wén-Thi 略談我國史籍上關於尸體

Yang Po-Chun (1) (cont.)
防腐的記載和馬王堆一號漢墓墓主
問題.
A Brief Discussion of Some Historical Texts
concerning the Preservation of Human
Bodies in an Incorrupt State, especially in
connection with the Han Burial in Tomb
no. 1 at Ma-wang-tui.
WWTK, 1972 (no. 9), 36.

Yang Shih-Thai (1) 楊時泰.
Pên Tshao Shu Kou Yuan 本草述鉤元.
Essentials Extracted from the Explanations
of Materia Medica.
Pref. 1833, first pr. 1842.
LPC, no. 108.

Yeh Tê-Hui (1) (ed.) 葉德輝.
Shuang Mei Ching An Tshung Shu 雙梅景
闇叢書.
Double Plum-Tree Collection [of ancient
and medieval books and fragments on
Taoist sexual techniques].
Contains Su Nü Ching (incl. Hsüan Nü
Ching), Tung Hsüan Tzu, Yü Fang Chih
Yao, Yü Fang Pi Chüeh, Thien Ti Yin
Yang Ta Lo Fu, etc. (qq.v.).
Chhangsha, 1903 and 1914.

Yeh Tê-Hui (2) 葉德輝.
Ed. & comm. Huai Nan Wan Pi Shu (q.v.).
In Kuan Ku Thang So Chu Shu 觀古堂
所著書.
Writings from the Hall of Pondering
Antiquity.
Chhangsha, 1919.

Yen Tun-Chieh (20) 嚴敦傑.
Chung-Kuo Ku-Tai Tzu-Jan Kho-Hsüeh
ti Fa-Chan chi chhi Chhêng-Chiu 中國
古代自然科學的發展及其成就.
The Development and Achievements of the
Chinese Natural Sciences (down to 1840).
KHSC, 1969, 1 (no. 3), 6.

Yen Tun-Chieh (21) 嚴敦傑.
Hsü Kuang-Chhi 徐光啓.
A Biography of Hsü Kuang-Chhi.
Art. in Chung-Kuo Ku-Tai Kho-Hsüeh Chia,
ed. Li Nien (27), 2nd ed. p. 181.

Yin Shih Tzu 丙是子.
See Chiang Wei-Chhiao.

Yoshida Mitsukuni (2) 吉田光邦
'Tenkō Kaibutsu' no Seiren Shuzō Gijutsu
「天工開物」の製錬鑄造技術.
Metallurgy in the Thien Kung Khai Wu
(Exploitation of the Works of Nature,
+1637).
Art. in Yabuuchi Kiyoshi (11), p. 137.

Yoshida Mitsukuni (5) 吉田光邦.
Chūsei no Kagaku (Rentan-jitsu) to Senjitsu
中世の化學(煉丹術)と仙術.
Chemistry and Alchemy in Medieval China.
Art. in Yabuuchi (25), p. 200.

Yoshida Mitsukuni (6) 吉田光邦.
Renkinjutsu 鍊金術.
(An Introduction to the History of)
Alchemy (and Early Chemistry in China
and Japan).
Chūō Kōronsha, Tokyo, 1963.

Yoshida Mitsukuni (7) 吉田光邦.
Chūgoku Kagaku-gijutsu-shi Ronshū 中國科
學技術史論集.
Collected Essays on the History of Science
and Technology in China.
Tokyo, 1972.

Yoshioka Yoshitoyo (1) 吉岡義豐.
Dōkyō Keiten Shiron 道教經典史論.
Studies on the History of the Canonical
Taoist Literature.
Dōkyō Kankōkai, Tokyo, 1955, repr. 1966.
Abstr. RBS, 1957, 1, no. 415.

Yoshioka Yoshitoyo (2) 吉岡義豐.
Sho Tō ni okeru Butsu-Dō Ronsō no ichi
Shiryō, 'Dōkyō Gisū' no seiritsu ni tsuite
初唐における佛道論爭の一資料
「道教義樞」の成立について.
The Tao Chiao I Shu (Basic Principles of
Taoism, by Mêng An-Phai 孟安排,
c. +660) and its Background; a Contri-
bution to the Study of the Polemics
between Buddhism and Taoism at the
Beginning of the Thang Period.
IBK, 1956, 4, 58.
Cf. RBS, 1959, 2, no. 590.

Yoshioka Yoshitoyo (3) 吉岡義豐
Eisei e no Nagai Dōkyō 永生への願い道
教.
Taoism; the Quest for Material Immortality
and its Origins.
Tankōsha, Tokyo, 1972 (Sekai no Shūkyō,
no. 9).

Yü Chia-Hsi (1) 余嘉錫.
Ssu Khu Thi Yao Pien Chêng 四庫提要
辨證.
A Critical Study of the Annotations in the
'Analytical Catalogue of the Complete
Library of the Four Caetgories (of
Literature)'.
1937.

Yü Fei-An (1) 于非闇.
Chung-Kuo Hua Yen-Sê-ti Yen-Chiu 中國
畫顏色的研究.
A Study of the Pigments Used by Chinese
Painters.
Chao-hua Mei-shu, Peking, 1955, 1957.

Yü Yün-Hsiu (1) 余雲岫.
Ku-tai Chi-Ping Ming Hou Su I 古代疾
病名候疏義.
Explanations of the Nomenclature of
Diseases in Ancient Times.
Jen-min Wei-shêng, Shanghai, 1953.
Rev. Nguyen Tran-Huan, RHS, 1956, 9, 275.

Yuan Han-Chhing (1) 袁翰青.
Chung-Kuo Hua-Hsüeh Shih Lun Wên Chi
中國化學史論文集.
Collected Papers in the History of Chemistry
in China.
San-Lien, Peking, 1956.

ADDENDA TO BIBLIOGRAPHY B

An Chih-Min (*1*) 安志敏.
  *Chhangsha Hsin Fa-hsien-ti Han Po Hua
  Shih-Than* 長沙新發現的西漢帛畫
  試探.
  A Tentative Interpretation of the Western
  (Former) Han Silk Painting recently
  discovered at Chhangsha (Ma-wang-tui,
  No. 1 Tomb).
  KKTH, 1973, no. 1 (no. 124), 43.
An Chih-Min (*2*) 安志敏.
  *Chin Pan yü Chin Ping; Chhu Han Chin
  Pi chi chhi yu Kuan Wên-Thi* 金版與
  金餅, 楚漢金幣及其有關問題.
  The 'chin pan' and 'chin ping' currency;
  a study of the Gold Coins of the Chhu
  State and the Han dynasty, and Some
  Related Problems.
  AS/CJA, 1973, No. 39 (no. 2), 61.
Anon (*116*).
  *Hsien-yang Shih Chin Nien Fa-hsien-ti
  I-Phi Chhin Han I Wu* 咸陽市近年
  發現的一批秦漢遺物.
  Recent Finds of the Chhin and Han
  Dynasties at the City of Hsienyang.
  KKTH, 1973 (no. 3), 167 and pl. 10, fig. 2.
Anon. (*117*).
  *Kuan-yü Chhangsha Ma-Wang-Tui
  I-hao Han Mu ti Tso Than Chi-Yao*
  關於長沙馬王堆一號漢墓的座談
  紀要.
  Report of an Editorial Round-Table
  Discussion on the Han Tomb Ma-
  wang-tui No. 1 at Chhangsha.
  KKTH, 1972, no. 5 (no. 122), 37.
Anon. (*118*).
  Hsi Han Po Hua 西漢帛畫.
  A Silk Painting of the Western (Former)
  Han, [the T-shaped Banner in the
  Tomb of the Lady of Tai, *c*.–166],
  (album, with explanatory introduction).
  Wên-Wu, Peking, 1972.
Anon. (*124*).
  *Tshung Sian Nan Chiao Chhu-thu-ti
  I-Yao Wên-Wu Khan Thang-Tai I-Yao-
  ti Fa-Chan* 從西安南郊出土的醫
  藥文物看唐代醫藥的發展.
  The Development of Pharmaceutical
  Chemistry in the Thang Period as seen
  from the specimens of (inorganic) drugs
  recovered from the Hoard in the
  Southern Suburbs of Sian [the silver
  boxes with the labelled chemicals,
  c.+756].
  WWTK, 1972 (no. 6), 52.

Anon. (*125*).
  *Chhangsha Ma-Wang-Tui I-hao Han Mu
  Fa-Chüeh Pao-Kao yü Wên-Tzu Thu
  Lu* 長沙馬王堆一號漢墓發掘
  報告與文資圖錄.
  The Excavation of Han Tomb No. 1 at
  Ma-wang-tui (Hayagriva Hill) near
  Chhangsha; Illustrated (Full) Report
  including the Literary Evidence.
  2 Vols. (text and album).
  Wên-Wu, Peking, 1973, for the Hunan
  Provincial Museum and the Archaeo-
  logical Institute of Academia Sinica.
  Rev. *EHOR*, 1974, **13** (no. 3), 66.
Anon. (*126*).
  *Ma-Wang-Tui-I hao Han Mu Nü Shih
  Yen-Chiu-ti Chi-ko Wên-Thi*
  馬王堆一號漢墓女屍研究的幾
  個問題.
  Some problems raised by the Researches
  into the (Preservation of the) Female
  Corpse in the Han Tomb No. 1 at Ma-
  wang-tui [the Lady of Tai; anatomy,
  pathology and chemistry].
  WWTK, 1973, no. 7 (no. 206), 74.
  Preceded by a one-page press statement on
  the same subject by the Hsinhua News
  Agency.
Anon. (*127*).
  Yung-Lo Kung 永樂宮.
  The Yung-lo Taoist Temple [at Jui-chhêng
  (Yung-chi), Shansi, and its frescoes].
  Album of 206 illustrations—the full
  publication.
  Jen-Min Mei-Shu, Peking, 1964.

Chhen Shun-Hua (*1*) 陳舜華.
  Ma-Wang-Tui Han Mu 馬王堆漢墓.
  On the Han Tomb at Ma-wang-tui [the
  tomb of the Lady of Tai, with particular
  reference to the causes of the conservation
  of her body from decay].
  Chung-hua, Hongkong, 1973.
Chhin Ling-Yün (*1*) 秦嶺雲.
  Pei-ching Fa-Hai Ssu Ming Tai Pi Hua
  北京法海寺明代壁畫.
  The Ming Frescoes of the Fa-Hai Ssu
  Temple at Peking.
  Chung-Kuo Ku Tien I Shu, Peking, 1958.
  Cf. Fig. 1312 in Vol. 5. pt. 2.

Fu Tsêng-Hsiang (*1*) 傅增湘.
  Nung Hsüeh Tsuan Yao 農學纂要.
  Essentials of Agricultural Science.
  1901.

Hashimoto Sōkichi (1) 橋本宗吉.
Ranka Naigai Sanbō Hōten 闌科內外三
法方典.
Handbook of the Three Aspects of Dutch
Internal and External Medicine.
Naniwa (Osaka), 1805.
Cf. Shimao Eikoh (1).

I Ping (1) 一冰.
Thang-Tai Yeh Yin Shu Chhu Than
唐代冶銀術初探.
A Preliminary Study of Silver Smelting
during the Thang Period.
WWTK, 1972 (no. 6), 40.

Itō Gikyo (1) 伊藤義敎.
Sian Chhu Thu Han-Po Ho Pi Mu Chih Po
Wên Yü Yen Hsüeh-ti Shih-Shih 西安
出土漢婆合璧墓志婆文語言學的
試釋.
A Linguistic Interpretation of the Pahlevi
Text of the Sino-Pahlevi Tomb Inscrip-
tion Unearthed at Sian (+874).
AS/CJA, 1964 (no. 2), 195.

Kao Yao-Thing (1) 高耀亭.
Ma-Wang-Tui I-hao Han Mu Sui Tsang
Phin chung Kung Shih Yung ti Shou
Lei 馬王堆一號漢墓隨葬品中
供食用的獸類.
On the kinds of Animals used for the
Sacrificial Food Offerings in the Han
Tomb No. 1 at Ma-wang-tui, at the
burial (of the Lady of Tai).
WWTK, 1973, no. 9 (no. 208), 76.

Kêng Chien-Thing (1) 耿鑒庭.
Sian Nan Chiao Thang-Tai Chiao Tsang
li ti I-Yao Wên-Wu 西安南郊唐代
窖藏裡的醫藥文物.
Cultural Objects of Pharmaceutical
Interest found in a Hoard in the Southern
Suburbs of Sian [the silver boxes with
labelled inorganic chemicals, c. +756].
WWTK, 1972 (no. 6), 56.

Ku Thieh-Fu (1) 顧鐵符.
Shih Lun Chhangsha Han Mu ti Pao-
Tshun Thiao Chien 試論長沙漢墓的
保存條件.
A Tentative Discussion of the Various
Preservative Conditions in the Tomb at
Chhangsha (of the Lady of Tai).
KKTH, 1972, no. 6 (no. 123), 53.

Li Thao (14) 李濤.
Chung-Kuo Tui-Yü Chin-Tai Chi Chung
Chi Chhu I-Hsüeh-ti Kung-Hsien 中國
對於近代幾種基礎醫學的貢獻.
Chinese Contributions to some of the
Fundamental Principles of Modern
Medicine.
ISTC, 1955, 7 (no. 2), 110.

Ma Yung (1) 馬雍.
Lun Chhangsha Ma-Wang-Tui I-hao
Han Mu Chhu Thu Po Hua ti Ming
Chhêng Ho Tso Yung 論長沙馬王
堆一號漢墓出土帛畫的名稱和作用.
A Discussion of the Name and Function
of the Silk Painting Unearthed from the
Han Tomb No. 1 at Ma-wang-tui,
Chhangsha.
KKTH, 1973 (no. 2), 118.

Miyashita Saburō (3) 宮下三郎.
Tonkōbon 'Chō Chūkei Gozōron' Kōyakuchū
敦煌本張仲景「五藏論」校譯注.
Critical Edition and Annotated Japanese
Translation of three Tunhuang MSS of
Chang Chung-Ching's Wu Tsang Lun
(Treatise on the Five Viscera) [S 5614,
P 2755, 2378].
TG/K, 1964, 35, 289.
Abstr. RBS, 1973, 10, no. 897.
Rev. P. Demiéville, TP, 1970, 56, 18.

Nakamura Mokuko (1) 中村木公.
Meika Chōjujitsu Rekidan 名家長壽實
歷談.
A Discussion of Some Celebrated Cases
of Great Longevity.
Tokyo, 1907.

Ōtani Shō (1) 大谷彰.
Chūgoku no Shu 中國の酒.
(A History of) Wine in China.
Shibata Shoten, Tokyo, 1974.

Shang Ping-Ho (1) 尚秉和.
Chung-Kuo Shê-Hui Fêng Su Shih 中國
社會風俗史.
A History of Manners and Customs in
Chinese Society.
Peking
Japanese tr. by Akita Seimei 秋田成明.
Tokyo, 1966.

Shih Shu-Chhing (4) 史樹青.
Wo Kuo Ku-Tai-ti Chin Tsho Kung I
我國古代的金錯工藝.
The Art of Inlaying (Gold and Silver) in
Ancient China.
WWTK, 1973, no. 6 (no. 205), 66.

Shih Wei (1) 史爲.
Chhangsha Ma-Wang-Tui I-hao Han Mu-ti
Kuan Kuo Chih-Tu 長沙馬王堆
一號漢墓的棺槨制度.
On the Design and Construction of the
Inner and Outer Coffins in the Ma-
wang-tui Tomb No. 1 at Chhangsha.
KKTH, 1972, no 6 (no. 123), 48.

Shinoda Osamu (2) 篠田統.
Chūsei no Shu 中世の酒.
Wine-Making in Medieval (China and
Japan).
Art. in Yabuuchi Kiyoshi (25), p. 321.
Repr. in Shinoda & Tanaka (1), p. 541.

Shinoda Osamu (3) 篠田統.
Sōgen Shuzōshi 宋元酒造史.
A History of Wine-making in the Sung and Yuan Periods.
Art. in Yabuuchi Kiyoshi (26), p. 279.
Repr. in Shinoda & Tanaka (1), p. 591.

Shinoda Osamu 篠田統, & Tanaka Sei-ichi (1) (ed.) 田中靜一.
Chung-Kuo Shih Ching Tshung Shu 中國食經叢書.
A Collection of Chinese Manuals of Nutrition (and Fermentations).
Tokyo, 1973.

Sun Tso-Yün (2) 孫作雲.
Ma-Wang-Tui I-hao Han Mu Chhi Kuan Hua Khao Shih 馬王堆一號漢墓漆棺畫考釋.
The Mythological Paintings on the Lacquer Coffins of the Han Tomb at Ma-wang-tui, No. 1 (Chhangsha).
KKTH, 1973, no. 4 (no. 127), 247.

Sun Tso-Yun (3) 孫作雲.
Chhangsha Ma-Wang-Tui I-hao Han Mu Chhu Thu Hua Fan Khao Shih 長沙馬王堆一號漢墓出土畫幡考釋.
Studies on the Silk Painting Unearthed from the Han Tomb Ma-wang-tui No. 1 at Chhangsha.
KKTH, 1973, no. 1 (no. 124), 54.

Têng Pai (1) (ed.) 鄧白.
Yung-Lo Kung Pi Hua 永樂宮壁畫.
Album of (Twenty) Coloured Plates selected from the Frescoes of the Yung-lo (Taoist) Temple (in Shansi), *c.* +1350.
Jen-Min I-Shu, Shanghai, 1959.

Thung Shih-Hêng (1) 童世亨.
Li Tai Chiang Yü Hsing Shih I Lan Thu 歷代疆域形勢一覽圖.
Historical Atlas of China.
Com. Press, Shanghai, 1922.

Yabuuchi Kiyoshi (4) 藪內清.
Chūgoku no Tokei 中國の時計.
[Ancient] Chinese Time-Keepers.
JJHS, 1951 (no. 19), 19.

Yü Hsing-Wu (1) 于省吾.
Kuan-yü Chhangsha Ma-Wang-Tui I-hao Han Mu Nei Kuan Kuan Shih Chieh Shuo 關于長沙馬王堆一號漢墓內棺棺飾的解說.
An Interpretation of the Decorations on the Inner Coffin of Han Tomb Ma-wang-tui No. 1 at Chhangsha.
KKTH, 1973 (no. 2), 126.

Yü Ying-Shih (1) 余英時.
Fang I-Chih Wan Chieh Khao 方以智晚節考.
Fang I-Chih; his last Years and his Death.
Inst. of Adv. Chinese Studies and Research, New Asia College, Chinese University, Hongkong, 1972.

596

C. BOOKS AND JOURNAL ARTICLES IN WESTERN LANGUAGES

ABBOTT, B. C. & BALLENTINE, D. (1).'The "Red Tide" Alga, a toxin from *Gymnodinium veneficum*. *JMBA*, 1957, **36**, 169.

ABEGG, E., JENNY, J. J. & BING, M. (1). 'Yoga'. *CIBA/M*, 1949, **7** (no. 74), 2578. *CIBA/MZ*, 1948, **10**, (no. 121), 4122.

> Includes: 'Die Anfänge des Yoga' and 'Der klassische Yoga' by E. Abegg; 'Der Kundalinī-Yoga' by J. J. Jenny; and 'Über medizinisches und psychologisches in Yoga' by M. Bing & J. J. Jenny.

ABICH, M. (1). 'Note sur la Formation de l'Hydrochlorate d'Ammoniaque à la Suite des Éruptions Volcaniques et surtout de celles du Vésuve.' *BSGF*, 1836, **7**, 98.

ABRAHAMS, H. J. (1). Introduction to the Facsimile Reprint of the 1530 Edition of the English Translation of H. Brunschwyk's *Vertuose Boke of Distillacyon*. Johnson, New York and London, 1971 (Sources of Science Ser., no. 79).

ABRAHAMSOHN, J. A. G. (1). 'Berättelse om *Kien* [*chien*], elt Nativt Alkali Minerale från China...' *KSVA/H*, 1772, **33**, 170. Cf. von Engeström (2).

ABRAMI, M., WALLICH, R. & BERNAL, P. (1). 'Hypertension Artériclle Volontaire.' *PM*, 1936, **44** (no. 17), 1 (26 Feb.).

ACHELIS, J. D. (1). 'Über den Begriff Alchemie in der Paracelsischen Philosophie.' *BDP*, 1929–30, **3**, 99.

ADAMS, F. D. (1). *The Birth and Development of the Geological Sciences*. Baillière, Tindall & Cox, London, 1938; repr. Dover, New York, 1954.

ADNAN ADIVAR (1). 'On the *Tanksuq-nāmah-i Ilkhān dar Funūn-i 'Ulūm-i Khiṭāi*.' *ISIS*, 1940 (appeared 1947), **32**, 44.

ADOLPH, W. H. (1). 'The Beginnings of Chemical Research in China.' *PNHB*, 1950, **18** (no. 3), 145.

ADOLPH, W. H. (2). 'Observations on the Early Development of Chemical Education in China.' *JCE*, 1927, **4**, 1233, 1488.

ADOLPH, W. H. (3). 'The Beginnings of Chemistry in China.' *SM*, 1922, **14**, 441. Abstr. *MCE*, 1922, **26**, 914.

AGASSI, J. (1). 'Towards an Historiography of Science.' Mouton, 's-Gravenhage, 1963. (History and Theory; Studies in the Philosophy of History, Beiheft no. 2.)

AHMAD, M. & DATTA, B. B. (1). 'A Persian Translation of the +11th-Century Arabic Alchemical Treatise 'Ain al-Ṣan'ah wa 'Aun al-Ṣana'ah (Essence of the Art and Aid to the Workers) [by 'Abd al-Malik al-Ṣāliḥī al-Khwārizmī al-Kathī, +1034].' *MAS/B*, 1927, **8**, 417. Cf. Stapleton & Azo (1).

AIGREMONT, Dr [ps. S. Schultze] (1). *Volkserotik und Pflanzenwelt; eine Darstellung alter wie moderner erotischer und sexuelle Gebräuche, Vergleiche, Benennungen, Sprichwörter, Redewendungen, Rätsel, Volkslieder, erotischer Zaubers und Aberglaubens, sexuelle Heilkunde die sich auf Pflanzen beziehen.* 2 vols. Trensinger, Halle, 1908. Re-issued as 2 vols. bound in one, Bläschke, Darmstadt, n.d. (1972).

AIKIN, A. & AIKIN, C. R. (1). *A Dictionary of Chemistry and Mineralogy.* 2 vols. Phillips, London, 1807.

AINSLIE, W. (1). *Materia Indica; or, some Account of those Articles which are employed by the Hindoos and other Eastern Nations in their Medicine, Arts and Agriculture; comprising also Formulae, with Practical Observations, Names of Diseases in various Eastern Languages, and a copious List of Oriental Books immediately connected with General Science, etc. etc.* 2 vols. Longman, Rees, Orme, Brown & Green, London, 1826.

AITCHISON, L. (1). *A History of Metals.* 2 vols. McDonald & Evans, London, 1960.

ALEXANDER, GUSTAV (1). *Herrengrunder Kupfergefässe.* Vienna, 1927.

ALEXANDER, W. & STREET, A. (1). *Metals in the Service of Man.* Pelican Books, London, 1956 (revised edition).

ALI, M. T., STAPLETON, H. E. & HUSAIN, M. H. (1). 'Three Arabic Treatises on Alchemy by Muḥammad ibn Umail [al-Ṣādiq al-Tamīnī] (d. c. +960); the *Kitāb al-Mā' al-Waraqī wa'l Arḍ al-Najmīyah* (Book of the Silvery Water and the Starry Earth), the *Risālat al-Shams Ila'l Hilāli* (Epistle of the Sun to the Crescent Moon), and the *al-Qaṣīdat al-Nūnīyah* (Poem rhyming in Nūn) —edition of the texts by M.T.A.; with an Excursus (with relevant Appendices) on the Date, Writings and Place in Alchemical History of Ibn Umail, an Edition (with glossary) of an early mediaeval Latin rendering of the first half of the *Mā' al-Waraqī*, and a Descriptive Index, chiefly of the alchemical authorities quoted by Ibn Umail [Senior Zadith Filius Hamuel], by H.E.S. & M.H.H.' *MAS/B*, 1933, **12** (no. 1), 1–213.

ALLEN, E. (ed.) (1). *Sex and Internal Secretions; a Survey of Recent Research.* Williams & Wilkins, Baltimore, 1932.

ALLEN, H. WARNER (1). *A History of Wine; Great Vintage Wines from the Homeric Age to the Present Day*. Faber & Faber, London, 1961.

ALLETON, V. & ALLETON, J. C. (1). *Terminologie de la Chimie en Chinois Moderne*. Mouton, Paris and The Hague, 1966. (Centre de Documentation Chinois de la Maison des Sciences de l'Homme, and VIe Section de l'École Pratique des Hautes Études, etc.; Matériaux pour l'Étude de l'Extrême-Orient Moderne et Contemporain; Études Linguistiques, no. 1.)

AMIOT, J. J. M. (7). 'Extrait d'une Lettre...' *MCHSAMUC*, 1791, **15**, v.

AMIOT, J. J. M. (9). 'Extrait d'une Lettre sur la Secte des Tao-sée [Tao shih].' *MCHSAMUC*, 1791, **15**, 208–59.

ANAND, B. K. & CHHINA, G. S. (1). 'Investigations of Yogis claiming to stop their Heart Beats.' *IJMR*, 1961, **49**, 90.

ANAND, B. K., CHHINA, G. S. & BALDEV SINGH (1). 'Studies on Shri Ramanand Yogi during his Stay in an Air-tight Box.' *IJMR*, 1961, **49**, 82.

ANAND, MULK RAJ (1). *Kama-Kala; Some Notes on the Philosophical Basis of Hindu Erotic Sculpture*. Nagel; Geneva, Paris, New York and Karlsruhe, 1958.

ANAND, MULK RAJ & KRAMRISCH, S. (1). *Homage to Khajuraho*. With a brief historical note by A. Cunningham. Marg, Bombay, n.d. (*c*. 1960).

ANDERSON, J. G. (8). 'The Goldsmith in Ancient China.' *BMFEA*, 1935, **7**, 1.

ANDŌ KŌSEI (1). 'Des Momies au Japon et de leur Culte.' *LH*, 1968, 8 (no. 2), 5.

ANIANE, M. (1). 'Notes sur l'Alchimie, "Yoga" Cosmologique de la Chrétienté Mediévale'; art. in *Yoga, Science de l'Homme Intégrale*. Cahiers du Sud, Loga, Paris, 1953.

ANON. (83). 'Préparation de l'Albumine d'Oeuf en Chine.' *TP*, 1897 (1e sér.), **8**, 452.

ANON. (84). *Beytrag zur Geschichte der höhern Chemie*. 1785. Cf. Ferguson (1), vol. 1, p. 111.

ANON. (85). *Aurora Consurgens* (first half of the +14th cent.). In ANON. (86). *Artis Auriferae*. Germ. tr. 'Aufsteigung der Morgenröthe' in Morgenstern (1).

ANON. (86). *Artis Auriferae, quam Chemiam vocant, Volumina Duo, quae continent 'Turbam Philosophorum', aliosą antiquiss. auctores, quae versa pagina indicat; Accessit noviter Volumen Tertium...* Waldkirch, Basel, 1610. One of the chief collections of standard alchemical authors' (Ferguson (1), vol. 1, p. 51).

ANON. (87). *Musaeum Hermeticum Reformatum et Amplificatum* (twenty-two chemical tracts). à Sande, Frankfurt, 1678 (the original edition, much smaller, containing only ten tracts, had appeared at Frankfurt, in 1625; see Ferguson (1), vol. 2, p. 119). Tr. Waite (8).

ANON. (88). *Probierbüchlein, auff Golt, Silber, Kupffer und Bley, Auch allerley Metall, wie man die Zunutz arbeyten und Probieren Soll. c.* 1515 or some years earlier; first extant pr. ed., Knappe, Magdeburg, 1524. Cf. Partington (7), vol. 2, p. 66. Tr. Sisco & Smith (2).

ANON. (89). (in Swedish) *METL*, 1960 (no. 3), 95.

ANON. (90). 'Les Chinois de la Dynastie Tsin [Chin] Connaissaient-ils déjà l'Alliage Aluminium–Cuivre?' *RALUM*, 1961, 108. Eng. tr. 'Did the Ancient Chinese discover the First Aluminium–Copper Alloy?' *GBT*, 1961, 41.

ANON. [initialled Y.M.] (91). 'Surprenante Découverte; un Alliage Aluminium–Cuivre réalisé en Chine à l'Époque Tsin [Chin].' *LN*, 1961 (no. 3316), 333.

ANON. (92). *British Encyclopaedia of Medical Practice; Pharmacopoeia Supplement* [proprietary medicines]. 2nd ed. Butterworth, London, 1967.

ANON. (93). *Gehes Codex d. pharmakologische und organotherapeutische Spezial-präparate...* [proprietary medicines]. 7th ed. Schwarzeck, Dresden, 1937.

ANON. (94). *Loan Exhibition of the Arts of the Sung Dynasty* (Catalogue). Arts Council of Great Britain and Oriental Ceramic Society, London, 1960.

ANON. (95). Annual Reports, Messrs Schimmel & Co., Distillers, Miltitz, near Leipzig, 1893 to 1896.

ANON. (96). Annual Report, Messrs Schimmel & Co., Distillers, Miltitz, near Leipzig, 1911.

ANON. (97). *Decennial Reports on Trade etc. in China and Korea* (Statistical Series, no. 6), 1882–1891. Inspectorate-General of Customs, Shanghai, 1893.

ANON. (98). 'Saltpetre Production in China.' *CEM*, 1925, **2** (no. 8), 8.

ANON. (99). 'Alkali Lands in North China [and the sodium carbonate (*chien*) produced there].' *JSCI*, 1894, **13**, 910.

ANON. (100). *A Guide to Peiping [Peking] and its Environs*. Catholic University (Fu-Jen) Press, for Peking Bookshop (Vetch), Peking, 1946.

ANON. (101) (ed.). *De Alchemia: In hoc Volumine de Alchemia continentur haec: Geber Arabis, philosophi solertissimi rerumque naturalium, praecipue metallicarum peritissimi...* (4 books); *Speculum Alchemiae* (Roger Baeon); *Correctorium Alchemiae* (Richard Anglici); *Rosarius Minor; Liber Secretorum Alchemicae* (Calid = Khalid); *Tabula Smaragdina* (with commentary of Hortulanus)...etc. Petreius, Nuremberg, 1541. Cf. Ferguson (1), vol. 1, p. 18.

ANON. (103). *Introduction to Classical Japanese Literature*. Kokusai Bunka Shinkokai (Soc. for Internat. Cultural Relations), Tokyo, 1948.

ANON. (104). *Of a Degradation of Gold made by an Anti-Elixir; a Strange Chymical Narative*. Herringman, London, 1678. 2nd ed. *An Historical Account of a Degradation of Gold made by an Anti-Elixir; a Strange Chymical Narrative. By the Hon. Robert Boyle, Esq*. Montagu, London, 1739.

ANON. (105). 'Some Observations concerning Japan, made by an Ingenious Person that hath many years resided in that Country...' *PTRS*, 1669, **4** (no. 49), 983.

ANON. (113). 'A 2100-year-old Tomb Excavated; the Contents Well Preserved.' *PKR*, 1972, no. 32 (11 Aug.), 10. *EHOR*, 1972, **11** (no. 4), 16 (with colour-plates). [The Lady of Tai (d. *c.* −186), incorrupted body, with rich tomb furnishings.] The article also distributed as an offprint at showings of the relevant colour film, e.g. in Hongkong, Sept. 1972.

ANON. (114). 'A 2100-year-old Tomb Excavated.' *CREC*, 1972, **21** (no. 9), 20 (with colour-plates). [The Lady of Tai, see previous entry.]

ANON. (115). *Antiquities Unearthed during the Great Proletarian Cultural Revolution*. n.d. [Foreign Languages Press, Peking, 1972]. With colour-plates. Arranged according to provinces of origin.

ANON. (116). *Historical Relics Unearthed in New China* (album). Foreign Languages Press, Peking, 1972.

ANTENORID, J. (1). 'Die Kenntnisse der Chinesen in der Chemie.' *CHZ*, 1902, **26** (no. 55), 627.

ANTZE, G. (1). 'Metallarbeiten aus Peru.' *MMVKH*, 1930, **15**, 1.

APOLLONIUS OF TYANA. *See* Conybeare (1); Jones (1).

ARDAILLON, E. (1). *Les Mines de Laurion dans l'Antiquité*. Inaug. Diss. Paris. Fontemoing, Paris, 1897.

ARLINGTON, L. C. & LEWISOHN, W. (1). *In Search of Old Peking*. Vetch, Peiping, 1935.

ARMSTRONG, E. F. (1). 'Alcohol through the Ages.' *CHIND*, 1933, **52** (no. 12), 251, (no. 13), 279. (Jubilee Memorial Lecture of the Society of Chemical Industry.)

ARNOLD, P. (1). *Histoire des Rose-Croix et les Origines de la Franc-Maçonnerie*. Paris, 1955.

AROUX, E. (1). *Dante, Hérétique, Révolutionnaire et Socialiste; Révélations d'un Catholique sur le Moyen Age*. 1854.

AROUX, E. (2). *Les Mystères de la Chevalerie et de l'Amour Platonique au Moyen Age*. 1858.

ARSENDAUX, H. & RIVET, P. (1). 'L'Orfèvrerie du Chiriqui et de Colombie'. *JSA*, 1923, **15**, 1.

ASCHHEIM, S. (1). 'Weitere Untersuchungen über Hormone und Schwangerschaft; das Vorkmmen der Hormone im Harn der Schwangeren.' *AFG*, 1927, **132**, 179.

ASCHHEIM, S. & ZONDEK, B. (1). 'Hypophysenvorderlappen Hormon und Ovarialhormon im Harn von Schwangeren.' *KW*, 1927, **6**, 1322.

ASHBEE, C. R. (1). *The Treatises of Benvenuto Cellini on Goldsmithing and Sculpture; made into English from the Italian of the Marcian Codex...* Essex House Press, London, 1898.

ASHMOLE, ELIAS (1). *Theatrum Chemicum Britannicum; Containing Severall Poeticall Pieces of our Famous English Philosophers, who have written the Hermetique Mysteries in their owne Ancient Language, Faithfully Collected into one Volume, with Annotations thereon by E. A. Esq*. London, 1652. Facsim. repr. ed. A. G. Debus, Johnson, New York and London, 1967 (Sources of Science ser. no. 39).

ASTON, W. G. (tr.) (1). '*Nihongi*', *Chronicles of Japan from the Earliest Times to* +697. Kegan Paul, London, 1896; repr. Allen & Unwin, London, 1956.

ATKINSON, R. W. (2). '[The Chemical Industries of Japan; I,] Notes on the Manufacture of *oshiroi* (White Lead).' *TAS/J*, 1878, **6**, 277.

ATKINSON, R. W. (3). 'The Chemical Industries of Japan; II, *Ame* [dextrin and maltose].' *TAS/J*, 1879, **7**, 313.

[ATWOOD, MARY ANNE] (1) (Mary Anne South, Mrs Atwood). *A Suggestive Enquiry into the Hermetic Mystery; with a Dissertation on the more Celebrated of the Alchemical Philosophers, being an Attempt towards the Recovery of the Ancient Experiment of Nature*. Trelawney Saunders, London, 1850. Repr. with introduction by W. L. Wilmhurst, Tait, Belfast, 1918, repr. 1920. *Hermetic Philosophy and Alchemy; a Suggestive Enquiry*. Repr. New York, 1960.

AVALON, A. (ps.). *See* Woodroffe, Sir J.

AYRES, LEW (1). *Altars of the East*. New York, 1956.

BACON, J. R. (1). *The Voyage of the Argonauts*. London, 1925.

BACON, ROGER
  Compendium Studii Philosophiae, +1271. *See* Brewer (1).
  De Mirabili Potestatis Artis et Naturae et de Nullitate Magiae, bef. +1250. *See* de Tournus (1); T. M [oufet]? (1); Tenney Davis (16).
  De Retardatione Accidentium Senectutis etc., +1236 to +1245. *See* R. Browne (1); Little & Withington (1).
  De Secretis operibus Artis et Naturae et de Nullitate Magiae, bef. +1250. *See* Brewer (1).
  Opus Majus, +1266. *See* Bridges (1); Burke (1); Jebb (1).
  Opus Minus, +1266 or +1267. *See* Brewer (1).

Opus Tertium, +1267. *See* Little (1); Brewer (1).

Sanioris Medicinae etc., pr. +1603. *See* Bacon (1).

Secretum Secretorum (ed.), *c.* +1255, introd. *c.* +1275. *See* Steele (1).

BACON, ROGER (1). *Sanioris Medicinae Magistri D. Rogeri Baconis Angli De Arte Chymiae Scripta.* Schönvetter, Frankfurt, 1603. Cf. Ferguson (1), vol. 1, p. 63.

BAGCHI, B. K. & WENGER, M. A. (1). 'Electrophysiological Correlates of some Yogi Exercises.' *EECN*, 1957, **7** (suppl.), 132.

BAIKIE, J. (1). 'The Creed [of Ancient Egypt].' *ERE*, vol. iv, p. 243.

BAILEY, CYRIL (1). *Epicurus; the Extant Remains.* Oxford, 1926.

BAILEY, SIR HAROLD (1). 'A Half-Century of Irano-Indian Studies.' *JRAS*, 1972 (no. 2), 99.

BAILEY, K. C. (1). *The Elder Pliny's Chapters on Chemical Subjects.* 2 vols. Arnold, London, 1929 and 1932.

BAIN, H. FOSTER (1). *Ores and Industry in the Far East; the Influence of Key Mineral Resources on the Development of Oriental Civilisation.* With a chapter on Petroleum by W. B. Heroy. Council on Foreign Relations, New York, 1933.

BAIRD, M. M., DOUGLAS, C. G., HALDANE, J. B. S. & PRIESTLEY, J. G. (1). 'Ammonium Chloride Acidosis.' *JOP*, 1923, **57**, xli.

BALAZS, E. (= S.) (1). 'La Crise Sociale et la Philosophie Politique à la Fin des Han.' *TP*, 1949, **39**, 83.

BANKS, M. S. & MERRICK, J. M. (1). 'Further Analyses of Chinese Blue-and-White [Porcelain and Pottery].' *AMY*, 1967, **10**, 101.

BARNES, W. H. (1). 'The Apparatus, Preparations and Methods of the Ancient Chinese Alchemists.' *JCE*, 1934, **11**, 655. 'Diagrams of Chinese Alchemical Apparatus' (an abridged translation of Tshao Yuan-Yü, 1). *JCE*, 1936, **13**, 453.

BARNES, W. H. (2). 'Possible References to Chinese Alchemy in the −4th or −3rd Century.' *CJ*, 1935, **23**, 75.

BARNES, W. H. (3). 'Chinese Influence on Western Alchemy.' *N*, 1935, **135**, 824.

BARNES, W. H. & YUAN, H. B. (1). 'Thao the Recluse (+452 to +536); Chinese Alchemist.' *AX*, 1946, **2**, 138. Mainly a translation of a short biographical paper by Tshao Yuan-Yü.

LA BARRE, W. (1). 'Twenty Years of Peyote Studies.' *CURRA*, 1960, **1**, 45.

LA BARRE, W. (2). *The Peyote Cult.* Yale Univ. Press, New Haven, Conn., repr. Shoestring Press, Hamden, Conn. 1960. (Yale Univ. Publications in Anthropology, no. 19.)

BARTHOLD, W. (2). *Turkestan down to the Mongol Invasions.* 2nd ed. London, 1958.

BARTHOLINUS, THOMAS (1). *De Nivis Usu Medico Observationes Variae.* Copenhagen, 1661.

BARTON, G. A. (1). '[The "Abode of the Blest" in] Semitic [including Babylonian, Jewish and ancient Egyptian, Belief].' *ERE*, ii, 706.

DE BARY, W. T. (3) (ed.). *Self and Society in Ming Thought.* Columbia Univ. Press, New York and London, 1970.

BASU, B. N. (1) (tr.). *The 'Kāmasūtra' of Vātsyāyana* [prob. +4th century]. Rev. by S. L. Ghosh. Pref. by P. C. Bagchi. Med. Book Co., Calcutta, 1951 (10th ed.).

BAUDIN, L. (1). 'L'Empire Socialiste des Inka [Incas].' *TMIE*, 1928, no. 5.

BAUER, W. (3). 'The Encyclopaedia in China.' *JWH*, 1966, **9**, 665.

BAUER, W. (4). *China und die Hoffnung auf Glück; Paradiese, Utopien, Idealvorstellungen.* Hanser, München, 1971.

BAUMÉ, A. (1). *Éléments de Pharmacie.* 1777.

BAWDEN, F. C. & PIRIE, N. W. (1). 'The Isolation and Some Properties of Liquid Crystalline Substances from Solanaceous Plants infected with Three Strains of Tobacco Mosaic Virus.' *PRSB*, 1937, **123**, 274.

BAWDEN, F. C. & PIRIE, N. W. (2). 'Some Factors affecting the Activation of Virus Preparations made from Tobacco Leaves infected with a Tobacco Necrosis Virus.' *JGMB*, 1950, **4**, 464.

BAYES, W. (1). *The Triple Aspect of Chronic Disease, having especial reference to the Treatment of Intractable Disorders affecting the Nervous and Muscular System.* Churchill, London, 1854.

BAYLISS, W. M. (1). *Principles of General Physiology.* 4th ed. Longmans Green, London, 1924.

BEAL, S. (2) (tr.). *Si Yu Ki* [*Hsi Yü Chi*], *Buddhist Records of the Western World, transl. from the Chinese of Hiuen Tsiang* [*Hsüan-Chuang*]. 2 vols. Trübner, London, 1881, 1884, 2nd ed. 1906. Repr. in 4 vols. with new title; *Chinese Accounts of India, translated from the Chinese of Hiuen Tsiang.* Susil Gupta, Calcutta, 1957.

BEAUVOIS, E. (1). 'La Fontaine de Jouvence et le Jourdain dans les Traditions des Antilles et de la Floride.' *MUSEON*, 1884, **3**, 404.

BEBEY, F. (1). 'The Vibrant Intensity of Traditional African Music.' *UNESC*, 1972, **25** (no. 10), 15. (On p. 19, a photograph of a relief of Ouroboros in Dahomey.)

BEDINI, S. A. (5). 'The Scent of Time; a Study of the Use of Fire and Incense for Time Measurement in Oriental Countries.' *TAPS*, 1963 (N.S.), **53**, pt. 5, 1–51. Rev. G. J. Whitrow, *A/AIHS*, 1964, **17**, 184.

BEDINI, S. A. (6). 'Holy Smoke; Oriental Fire Clocks.' *NS*, 1964, **21** (no. 380), 537.

VAN BEEK, G. W. (1). 'The Rise and Fall of Arabia Felix.' *SAM*, 1969, **221** (no. 6), 36.

BEER, G. (1) (ed. & tr.). 'Das Buch Henoch [Enoch]' in *Die Apokryphen und Pseudepigraphien des alten Testaments*, ed. E. Kautzsch, 2 vols. Mohr (Siebeck), Tübingen, Leipzig ¦and Freiburg i/B, 1900, vol. 2 (Pseudepigraphien), pp. 217 ff.

LE BEGUE, JEAN (1). *Tabula de Vocabulis Synonymis et Equivocis Colorum* and *Experimenta de Coloribus* (MS. BM. 6741 of +1431). Eng. tr. Merrifield (1), vol. 1, pp. 1–321.

BEH, Y. T. *See* Kung, S. C., Chao, S. W., Bei, Y. T. & Chang, C. (1).

BEHANAN, KOVOOR T. (1). *Yoga; a Scientific Evaluation*. Secker & Warburg, London, 1937. Paperback repr. Dover, New York and Constable, London. n.d. (*c.* 1960).

BEHMEN, JACOB. *See* Boehme, Jacob.

BELL, SAM HANNA (1). *Erin's Orange Lily*. Dobson, London, 1956.

BELPAIRE, B. (3). 'Note sur un Traité Taoiste.' *MUSEON*, 1946, **59**, 655.

BENDALL, C. (1) (ed.). *Subhāṣita-saṃgraha*. Istas, Louvain, 1905. (Muséon Ser. nos. 4 and 5.)

BENEDETTI-PICHLER, A. A. (1). 'Micro-chemical Analysis of Pigments used in the Fossae of the Incisions of Chinese Oracle-Bones.' *IEC/AE*, 1937, **9**, 149. Abstr. *CA*, 1938, **31**, 3350.

BENFEY, O. T. (1). 'Dimensional Analysis of Chemical Laws and Theories.' *JCE*, 1957, **34**, 286.

BENFEY, O. T. (2) (ed.). *Classics in the Theory of Chemical Combination*. Dover, New York, 1963. (Classics of Science, no. 1.)

BENFEY, O. T. & FIKES, L. (1). 'The Chemical Prehistory of the Tetrahedron, Octahedron, Icosahedron and Hexagon.' *ADVC*, 1966, **61**, 111. (Kekulé Centennial Volume.)

BENNETT, A. A. (1). *John Fryer; the Introduction of Western Science and Technology into Nineteenth-Century China*. Harvard Univ. Press, Cambridge, Mass. 1967. (Harvard East Asian Monographs, no. 24.)

BENSON, H., WALLACE, R. K., DAHL, E. C. & COOKE, D. F. (1). 'Decreased Drug Abuse with Transcendental Meditation; a Study of 1862 Subjects.' In 'Hearings before the Select Committee on Crime of the House of Representatives (92nd Congress)', U.S. Govt. Washington, D.C. 1971, p. 681 (Serial no. 92-1).

BENTHAM, G. & HOOKER, J. D. (1). *Handbook of the British Flora; a Description of the Flowering Plants and Ferns indigenous to, or naturalised in, the British Isles*. 6th ed. 2 vols. (1 vol. text, 1 vol. dra wings). Reeve, London, 1892. repr. 1920.

BENVENISTE, E. (1). 'Le Terme *obryza* et la Métallurgie de l'Or.' *RPLHA*, 1953, **27**, 122.

BENVENISTE, E. (2). *Textes Sogdiens* (facsimile reproduction, transliteration, and translation with glossary). Paris, 1940. Rev. W. B. Hemming, *BLSOAS*, **11**.

BERENDES, J. (1). *Die Pharmacie bei den alten Culturvölkern; historisch-kritische Studien*. 2 vols. Tausch & Grosse, Halle, 1891.

BERGMAN, FOLKE (1). *Archaeological Researches in Sinkiang*. Reports of the Sino-Swedish [scientific] Expedition [to Northwest China]. 1939, vol. 7 (pt. 1).

BERGMAN, TORBERN (1). *Opuscula Physica et Chemica, pleraque antea seorsim edita, jam ab Auctore collecta, revisa et aucta*. 3 vols. Edman, Upsala, 1779–83. Eng. tr. by E. Cullen, *Physical and Chemical Essays*, 2 vols. London, 1784, 1788; the 3rd vol. Edinburgh, 1791.

BERGSØE, P. (1). 'The Metallurgy and Technology of Gold and Platinum among the Pre-Columbian Indians.' *IVS*, 1937, no. A44, 1–45. Prelim. pub. *N*, 1936, **137**, 29.

BERGSØE, P. (2). 'The Gilding Process and the Metallurgy of Copper and Lead among the Pre-Columbian Indians.' *IVS*, 1938, no. A46. Prelim. pub. 'Gilding of Copper among the Pre-Columbian Indians.' *N*, 1938, **141**, 829.

BERKELEY, GEORGE, BP. (1). *Siris; Philosophical Reflections and Enquiries concerning the Virtues of Tar-Water*. London, 1744.

BERNAL, J. D. (1). *Science in History*. Watts, London, 1954. (Beard Lectures at Ruskin College, Oxford.) Repr. 4 vols. Penguin, London, 1969.

BERNAL, J. D. (2). *The Extension of Man; a History of Physics before 1900*. Weidenfeld & Nicolson, London, 1972. (Lectures at Birkbeck College, London, posthumously published.)

BERNARD, THEOS (1). *Haṭhayoga; the Report of a Personal Experience*. Columbia Univ. Press, New York, 1944; Rider, London, 1950. Repr. 1968.

BERNARD-MAÎTRE, H. (3). 'Un Correspondant de Bernard de Jussieu en China; le Père le Chéron d'Incarville, missionaire français de Pékin, d'après de nombreux documents inédits.' *A/AIHS*, 1949, **28**, 333, 692.

BERNARD-MAÎTRE, H. (4). 'Notes on the Introduction of the Natural Sciences into the Chinese Empire.' *YJSS*, 1941, **3**, 220.

BERNARD-MAÎTRE, H. (9). 'Deux Chinois du 18e siècle à l'École des Physiocrates Français.' *BUA*, 1949 (3e sér.), **10**, 151.

BERNARD-MAÎTRE, H. (17). 'La Première Académie des Lincei et la Chine.' *MP*, 1941, 65.

BERNARD-MAÎTRE, H. (18). 'Les Adaptations Chinoises d'Ouvrages Européens; Bibliographie chronologique depuis la venue des Portugais à Canton jusqu'à la Mission française de Pékin (+1514 à +1688).' *MS*, 1945, **10**, 1–57, 309–88.

BERNAREGGI, E. (1). 'Nummi Pelliculati' (silver-clad copper coins of the Roman Republic). *RIN*, 1965, **67** (5th. ser., **13**), 5.

BERNOULLI, R. (1). 'Seelische Entwicklung im Spiegel der Alchemie u. verwandte Disciplinen.' *ERJB*, 1935, 3, 231–87. Eng. tr. 'Spiritual Development as reflected in Alchemy and related Disciplines.' *ERYB*, 1960, **4**, 305. Repr. 1970.

BERNTHSEN, A. *See* Sudborough, J. J. (1).

BERRIMAN, A. E. (2). 'A Sumerian Weight-Standard in Chinese Metrology during the Former Han Dynasty (−206 to −23).' *RAAO*, 1958, **52**, 203.

BERRIMAN, A. E. (3). 'A New Approach to the Study of Ancient Metrology.' *RAAO*, 1955, **49**, 193.

BERTHELOT, M. (1). *Les Origines de l'Alchimie*. Steinheil, Paris, 1885. Repr. Libr. Sci. et Arts, Paris, 1938.

BERTHELOT, M. (2). *Introduction à l'Étude de la Chimie des Anciens et du Moyen-Age*. First published at the beginning of vol. 1 of the *Collection des Anciens Alchimistes Grecs* (see Berthelot & Ruelle), 1888. Repr. sep. Libr. Sci. et Arts, Paris, 1938. The 'Avant-propos' is contained only in Berthelot & Ruelle; there being a special Preface in Berthelot (2).

BERTHELOT, M. (3). Review of de Mély (1), *Lapidaires Chinois. JS*, 1896, 573.

BERTHELOT, M. (9). Les Compositons Incendiaires dans l'Antiquité et Moyen Ages.' *RDM*, 1891, **106**, 786.

BERTHELOT, M. (10). *La Chimie au Moyen Age;* vol. 1, *Essai sur la Transmission de la Science Antique au Moyen Age* (Latin texts). Impr. Nat. Paris, 1893. Photo. repr. Zeller, Osnabrück; Philo, Amsterdam, 1967. Rev. W. P[agel], *AX*, 1967, **14**, 203.

BERTHELOT, M. (12). 'Archéologie et Histoire des Sciences; avec Publication nouvelle du Papyrus Grec chimique de Leyde, et Impression originale du *Liber de Septuaginta* de Geber.' *MRASP*, 1906, **49**, 1–377. Sep. pub. Philo, Amsterdam, 1968.

BERTHELOT, M. [P. E. M.]. *See* Tenney L. Davis' biography (obituary), with portrait. *JCE*, 1934, **11** (585) and Boutaric (1).

BERTHELOT, M. & DUVAL, R. (1). *La Chimie au Moyen Age*; vol. 2, *l'Alchimie Syriaque*. Impr. Nat. Paris, 1893. Photo. repr. Zeller, Osnabrück; Philo, Amsterdam, 1967. Rev. W. P[agel], *AX*, 1967, **14**, 203.

BERTHELOT, M. & HOUDAS, M. O. (1). *La Chimie au Moyen Age*; vol. 3, *l'Alchimie Arabe*. Impr. Nat. Paris, 1893. Photo repr. Zeller, Osnabrück; Philo, Amsterdam, 1967. Rev. W. P[agel], *AX*, 1967, **14**, 203.

BERTHELOT, M. & RUELLE, C. E. (1). *Collection des Anciens Alchimistes Grecs*. 3 vols. Steinheil, Paris, 1888. Photo. repr. Zeller, Osnabrück, 1967.

BERTHOLD, A. A. (1). 'Transplantation der Hoden.' *AAPWM*, 1849, **16**, 42. Engl. tr. by D. P. Quiring. *BIHM*, 1944, **16**, 399.

BERTRAND, G. (1). Papers on laccase. *CRAS*, 1894, **118**, 1215; 1896, **122**, 1215; *BSCF*, 1894, **11**, 717; 1896, **15**, 793.

BERTUCCIOLI, G. (2). 'A Note on Two Ming Manuscripts of the *Pên Tshao Phin Hui Ching Yao*.' *JOSHK*, 1956, **2**, 63. Abstr. *RBS*, 1959, **2**, 228.

BETTENDORF, G. & INSLER, V. (1) (ed.). *The Clinical Application of Human Gonadotrophins*. Thieme, Stuttgart, 1970.

BEURDELEY, M. (1) (ed.). *The Clouds and the Rain; the Art of Love in China*. With contributions by K. Schipper on Taoism and sexuality, Chang Fu-Jui on literature and poetry, and J. Pimpaneau on perversions. Office du Livre, Fribourg and Hammond & Hammond, London, 1969.

BEVAN, E. R. (1). 'India in Early Greek and Latin Literature.' *CHI*, Cambridge, 1935, vol. 1, ch. 16, p. 391.

BEVAN, E. R. (2). *Stoics and Sceptics*. Oxford, 1913.

BEVAN, E. R. (3). *Later Greek Religion*. Oxford, 1927.

BEZOLD, C. (3). *Die 'Schatzhöhle'; aus dem Syrische Texte dreier unedirten Handschriften in's Deutsche übersetzt und mit Anmerkungen versehen... nebst einer Arabischen Version nach den Handschriften zu Rom, Paris und Oxford*. 2 vols. Hinrichs, Leipzig, 1883, 1888.

BHAGVAT, K. & RICHTER, D. (1). 'Animal Phenolases and Adrenaline.' *BJ*, 1938, **32**, 1397.

BHAGVAT SINGHJI, H. H. (Maharajah of Gondal) (1). *A Short History of Aryan Medical Science*. Gondal, Kathiawar, 1927.

BHATTACHARYA, B. (1) (ed.). *Guhya-samāja Tantra, or Tathāgata-guhyaka*. Orient. Instit., Baroda, 1931. (Gaekwad Orient. Ser. no. 53.)

BHATTACHARYA, B. (2). *Introduction to Buddhist Esoterism*. Oxford, 1932.

BHISHAGRATNA, (KAVIRAJ) KUNJA LAL SHARMA (1) (tr.). *An English Translation of the 'Sushruta Samhita',* based on the original Sanskrit Text. 3 vols. with an index volume, pr. pr. Calcutta, 1907–18. Reissued, Chowkhamba Sanskrit Series Office, Varanasi, 1963. Rev. M. D. Grmek, *A/AIHS*, 1965, **18**, 130.

BIDEZ, J. (1). '*l'Épître sur la Chrysopée' de Michel Psellus* [with Italian translation]; [also] *Opuscules et Extraits sur l'Alchimie, la Météorologie et la Démonologie...* (Pt. VI of *Catalogue des Manuscrits Alchimiques Grecques*). Lamertin, for Union Académique Internationale, Brussels, 1928.

BIDEZ, J. (2). *Vie de Porphyre le Philosophe Neo-Platonicien avec les Fragments des Traités περὶ ἀγαλμάτων et 'De Regressu Animae'.* 2 pts. Univ. Gand, Leipzig, 1913. (Receuil des Trav. pub. Fac. Philos. Lettres, Univ. Gand.)

BIDEZ, J. & CUMONT, F. (1). *Les Mages Hellenisés; Zoroastre, Ostanès et Hytaspe d'après la Tradition Grecque.* 2 vols. Belles Lettres, Paris, 1938.

BIDEZ, J., CUMONT, F., DELATTE, A. HEIBERG, J. L., LAGERCRANTZ, O., KENYON, F., RUSKA, J. & DE FALCO, V. (1) (ed.). *Catalogue des Manuscrits Alchimiques Grecs.* 8 vols. Lamertin, Brussels, 1924–32 (for the Union Académique Internationale).

BIDEZ, J., CUMONT, F., DELATTE, A., SARTON, G., KENYON, F. & DE FALCO, V. (1) (ed.). *Catalogue des Manuscrits Alchimiques Latins.* 2 vols. Union Acad. Int., Brussels, 1939–51.

BIOT, E. (1) (tr.). *Le Tcheou-Li ou Rites des Tcheou* [Chou]. 3 vols. Imp. Nat., Paris, 1851. (Photographically reproduced, Wênticnko, Pciping, 1930.)

BIOT, E. (17). 'Notice sur Quelques Procédés Industriels connus en Chine au XVIe siècle.' *JA*, 1835 (2ᵉ sér.), **16**, 130.

BIOT, E. (22). 'Mémoires sur Divers Minéraux Chinois appartenant à la Collection du Jardin du Roi.' *JA*, 1839 (3ᵉ sér.), **8**, 206.

BIRKENMAIER, A. (1). 'Simeon von Köln oder Roger Bacon?' *FRS*, 1924, **2**, 307.

AL-BĪRŪNĪ, ABŪ AL-RAIḤĀN MUḤAMMAD IBN-AḤMAD. *Ta'rīkh al-Hind* (History of India). *See* Sachau (1).

BISCHOF, K. G. (1). *Elements of Chemical and Physical Geology,* tr. B. H. Paul & J. Drummond from the 1st German edn. (3 vols., Marcus, Bonn, 1847-54), Harrison, London, 1854 (for the Cavendish Society). 2nd German ed. 3 vols. Marcus, Bonn, 1863, with supplementary volume, 1871.

BLACK, J. DAVIDSON (1). 'The Prehistoric Kansu Race.' *MGSC* (Ser. A.), 1925, no. 5.

BLAKNEY, R. B. (1). *The Way of Life; Lao Tzu—a new Translation of the 'Tao Tê Ching'.* Mentor, New York, 1955.

BLANCO-FREIJEIRO, A. & LUZÓN, J. M. (1). 'Pre-Roman Silver Miners at Rio Tinto.' *AQ*, 1969, **43**, 124.

DE BLANCOURT, HAUDICQUER (1). *L'Art de la Verrerie...* Paris, 1697. Eng. tr. *The Art of Glass...with an Appendix containing Exact Instructions for making Glass Eyes of all Colours.* London, 1699.

BLAU, J. L. (1). *The Christian Interpretation of the Cabala in the Renaissance.* Columbia Univ. Press, New York, 1944. (Inaug. Diss. Columbia, 1944.)

BLOCHMANN, H. F. (1) (tr.). *The 'Ā'īn-i Akbarī'* (*Administration of the Mogul Emperor Akbar*) of Abū'l Faẓl 'Allāmī. Rouse, Calcutta, 1873. (Bibliotheca Indica, N.S., nos. 149, 158, 163, 194, 227, 247 and 287.)

BLOFELD, J. (3). *The Wheel of Life; the Autobiography of a Western Buddhist.* Rider, London, 1959.

BLOOM, ANDRÉ [METROPOLITAN ANTHONY] (1). 'Contemplation et Ascèse; Contribution Orthodoxe', art. in *Technique et Contemplation.* Études Carmelitaines, Paris, 1948, p. 49.

BLOOM, ANDRÉ [METROPOLITAN ANTHONY] (2). 'l'Hésychasme, Yoga Chrétien?', art. in *Yoga*, ed. J. Masui, Paris, 1953.

BLOOMFIELD, M. (1) (tr.). *Hymns of the Atharva-veda, together with Extracts from the Ritual Books and the Commentaries.* Oxford, 1897 (*SBE*, no. 42). Repr. Motilal Banarsidass, Delhi, 1964.

BLUNDELL, J. W. F. (2). *Medicina Mechanica.* London.

BOAS, G. (1). *Essays on Primitivism and Related Ideas in the Middle Ages.* Johns Hopkins Univ. Press, Baltimore, 1948.

BOAS, MARIE (2). *Robert Boyle and Seventeenth-Century Chemistry.* Cambridge, 1958.

BOAS, MARIE & HALL, A. R. (2). 'Newton's Chemical Experiments.' *A/AIHS*, 1958, **37**, 113.

BOCHARTUS, S. (1). *Opera Omnia, hoc est Phaleg, Canaan, et Hierozoicon.* Boutesteyn & Luchtmans, Leiden and van de Water, Utrecht, 1692. [The first two books are on the geography of the Bible and the third on the animals mentioned in it.]

BOCTHOR, E. (1). *Dictionnaire Français–Arabe,* enl. and ed. A. Caussin de Perceval. Didot, Paris, 1828-9. 3rd ed. Didot, Paris, 1864.

BODDE, D. (5). 'Types of Chinese Categorical Thinking.' *JAOS*, 1939, **59**, 200.

BODDE, D. (9). 'Some Chinese Tales of the Supernatural; Kan Pao and his *Sou Shen Chi*.' *HJAS*, 1942, **6**, 338.

BODDE, D. (10). 'Again Some Chinese Tales of the Supernatural; Further Remarks on Kan Pao and his *Sou Shen Chi*.' *JAOS*, 1942, **62**, 305.

BOECKH, A. (1) (ed.). *Corpus Inscriptionum Graecorum.* 4 vols. Berlin, 1828-77.

BOEHME, JACOB (1). *The Works of Jacob Behmen, the Teutonic Theosopher . . . To which is prefixed, the Life of the Author, with Figures illustrating his Principles, left by the Rev. W. Law.* Richardson, 4 vols. London, 1764–81. See Ferguson (1), vol. 1, p. 111. Based partly upon: *Idea Chemiae Böhmianae Adeptae; das ist, ein Kurtzer Abriss der Bereitung deß Steins der Weisen, nach Anleitung deß Jacobi Böhm . . .* Amsterdam, 1680, 1690; and: *Jacob Böhms kurtze und deutliche Beschreibung des Steins der Weisen, nach seiner Materia, aus welcher er gemachet, nach seiner Zeichen und Farbe, welche im Werck erscheinen, nach seiner Kraft und Würckung, und wie lange Zeit darzu erfordert wird, und was insgemein bey dem Werck in acht zu nehmen . . .* Amsterdam, 1747.

BOEHME, JACOB (2). *The Epistles of Jacob Behmen, aliter Teutonicus Philosophus, translated out of the German Language.* London, 1649.

BOERHAAVE, H. (1). *Elementa Chemiae, quae anniversario labore docuit, in publicis, privatisque, Scholis.* 2 vols. Severinus and Imhoff, Leiden, 1732. Eng. tr. by P. Shaw: *A New Method of Chemistry, including the History, Theory and Practice of the Art.* 2 vols. Longman, London, 1741, 1753.

BOERHAAVE, HERMANN. See Lindeboom (1).

BOERSCHMANN, E. (11). 'Peking, eine Weltstadt der Baukunst.' *AT*, 1931 (no. 2), 74.

BOLL, F. (6). 'Studien zu Claudius Ptolemäus.' *JCP*, 1894, **21** (Suppl.), 155.

BOLLE, K. W. (1). *The Persistence of Religion; an Essay on Tantrism and Sri Aurobindo's Philosophy.* Brill, Leiden, 1965. (Supplements to *Numen*, no. 8.)

BONI, B. (3). 'Oro e Formiche Erodotee.' *CHIM*, 1950 (no. 3).

BONMARCHAND, G. (1) (tr.). 'Les Notes de Li Yi-Chan [Li I-shan], (Yi-Chan Tsa Tsouan [*I-Shan Tsa Tsuan*]), traduit du Chinois; Étude de Littérature Comparée.' *BMFJ*, 1955 (N.S.) **4** (no. 3), 1–84.

BONNER, C. (1). 'Studies in Magical Amulets, chiefly Graeco-Egyptian.' Ann Arbor, Michigan, 1950. (Univ. Michigan Studies in Humanities Ser., no. 49.)

BONNIN, A. (1). *Tutenag and Paktong; with Notes on other Alloys in Domestic Use during the Eighteenth Century.* Oxford, 1924.

BONUS, PETRUS, of Ferrara (1). *M. Petri Boni Lombardi Ferrariensis Physici et Chemici Excellentiss. Introductio in Artem Chemiae Integra, ab ipso authore inscripta Margarita Preciosa Novella; composita ante annos plus minus ducentos septuaginta, Nune multis mendis sublatis, comodiore, quam antehâc, forma edita, et indice revum ad calcem adornata.* Foillet, Montbeliard, 1602. 1st ed. Lacinius ed. Aldus, Venice, 1546. Tr. Waite (7). See Leicester (1), p. 86. Cf. Ferguson (1), vol. 1, p. 115.

BORNET, P. (2). 'Au Service de la Chine; Schall et Verbiest, maîtres-fondeurs, I. les Canons.' *BCP*, 1946 (no. 389), 160.

BORNET, P. (3) (tr.). 'Relation Historique' [de Johann Adam Schall von Bell, S.J.]; Texte Latin avec Traduction française.' Hautes Études, Tientsin, 1942 (part of *Lettres et Mémoires d'Adam Schall S.J.* ed H. Bernard[-Maître]).

BORRICHIUS, O. (1). *De Ortu et Progressu Chemiae.* Copenhagen, 1668.

BOSE, D. M., SEN, S.-N., SUBBARAYAPPA, B. V. et al. (1). *A Concise History of Science in India.* Baptist Mission Press, Calcutta, for the Indian National Science Academy, New Delhi, 1971.

BOSON, G. (1). 'Alcuni Nomi di Pietri nelle Inscrizioni Assiro-Babilonesi.' *RSO*, 1914, **6**, 969.

BOSON, G. (2). 'I Metalli e le Pietri nelle Inscrizioni Assiro-Babilonesi.' *RSO*, 1917, **7**, 379.

BOSON, G. (3). *Les Métaux et les Pierres dans les Inscriptions Assyro-Babyloniennes.* Munich, 1914.

BOETOCKE, R. (1). *The Difference between the Ancient Physicke, first taught by the godly Forefathers, insisting in unity, peace and concord, and the Latter Physicke . . .* London, 1585. Cf. Debus (12).

BOUCHÉ-LECLERCQ, A. (1). *L'Astrologie Grecque.* Leroux, Paris, 1899.

BOURKE, J. G. (1). 'Primitive Distillation among the Tarascoes.' *AAN*, 1893, **6**, 65.

BOURKE, J. G. (2). 'Distillation by Early American Indians.' *AAN*, 1894, **7**, 297.

BOURNE, F. S. A. (2). *The Lo-fou Mountains; an Excursion.* Kelly & Walsh, Shanghai, 1895.

BOUTARIC, A. (1). *Marcellin Berthelot (1827 à 1907).* Payot, Paris, 1927.

BOVILL, E. W. (1). 'Musk and Amber[gris].' *NQ*, 1954.

BOWERS, J. Z. & CARUBBA, R. W. (1). 'The Doctoral Thesis of Engelbert Kaempfer: "On Tropical Diseases, Oriental Medicine and Exotic Natural Phenomena".' *JHMAS*, 1970, **25**, 270.

BOYLE, ROBERT (1). *The Sceptical Chymist; or, Chymico-Physical Doubts and Paradoxes, touching the Experiments whereby Vulgar Spagyrists are wont to endeavour to evince their Salt, Sulphur and Mercury to be the True Principles of Things.* Crooke, London, 1661.

BOYLE, ROBERT (4). 'A New Frigoric Experiment.' *PTRS*, 1666, **1**, 255.

BOYLE, ROBERT (5). *New Experiments and Observations touching Cold.* London, 1665. Repr. 1772.

BOYLE, ROBERT. See Anon. (104).

BRADLEY, J. E. S. & BARNES, A. C. (1). *Chinese–English Glossary of Mineral Names.* Consultants' Bureau, New York, 1963.

BRASAVOLA, A. (1). *Examen Omnium Sinplicium Medicamentorum.* Rome, 1536.

BRELICH, H. (1). 'Chinese Methods of Mining Quicksilver.' *TIMM*, 1905, **14**, 483.

BRELICH, H. (2). 'Chinese Methods of Mining Quicksilver.' *MJ*, 1905 (27 May), 578, 595.

BRETSCHNEIDER, E. (1). *Botanicon Sinicum; Notes on Chinese Botany from Native and Western Sources*, 3 vols.
   Vol. 1 (Pt. 1, no special sub-title) contains
      ch. 1. Contribution towards a History of the Development of Botanical Knowledge among Eastern
         Asiatic Nations.
      ch. 2. On the Scientific Determination of the Plants Mentioned in Chinese Books.
      ch. 3. Alphabetical List of Chinese Works, with Index of Chinese Authors.
      app. Celebrated Mountains of China (list)
      Trübner, London, 1882 (printed in Japan); also pub. *JRAS/NCB*, 1881 (n.s.), **16**, 18–230 (in
      smaller format).
   Vol. 2, Pt. II, *The Botany of the Chinese Classics*, with Annotations, Appendixes and Indexes by
      E. Faber, contains
      Corrigenda and Addenda to Pt. 1
      ch. 1. Plants mentioned in the *Erh Ya*.
      ch. 2. Plants mentioned in the *Shih Ching*, the *Shu Ching*, the *Li Chi*, the *Chou Li* and other Chinese
         classical works.
      Kelly & Walsh, Shanghai etc. 1892; also pub. *JRAS/NCB*, 1893 (n.s.), **25**, 1–468.
   Vol. 3, Pt. III, *Botanical Investigations into the Materia Medica of the Ancient Chinese*, contains
      ch. 1. Medicinal Plants of the *Shen Nung Pên Tshao Ching* and the [*Ming I*] *Pieh Lu*
      with indexes of geographical names, Chinese plant names and Latin generic names.
      Kelly & Walsh, Shanghai etc., 1895; also pub. *JRAS/NCB*, 1895 (n.s.), **29**, 1–623.
BRETSCHNEIDER, E. (2). *Mediaeval Researches from Eastern Asiatic Sources; Fragments towards the Know-
   ledge of the Geography and History of Central and Western Asia from the +13th to the +17th
   century*. 2 vols. Trübner, London, 1888. New ed. Routledge & Kegan Paul, 1937. Photo-reprint, 1967.
BREUER, H. & KASSAU, E. (1). *Eine einfache Methode zur Isolierung von Steroiden aus biologischen Medien
   durch Mikrosublimation*. Proc. 1st International Congress of Endocrinology, Copenhagen, 1960,
   Session XI (*d*), no. 561.
BREUER, H. & NOCKE, L. (1). 'Stoffwechsel der Oestrogene in der menschlichen Leber'; art. in VIter
   Symposium d. Deutschen Gesellschaft f. Endokrinologie, *Moderne Entwicklungen auf dem Gesta-
   gengebiet Hormone in der Veterinärmedizin*. Kiel, 1959, p. 410.
BREWER, J. S. (1) (ed.). *Fr. Rogeri Bacon Opera quaedam hactenus inedita*. Longman, Green, Longman &
   Roberts, London, 1859 (Rolls Series, no. 15). Contains *Opus Tertium* (*c.* +1268), part of *Opus
   Minus* (*c.* +1267), part of *Compendium Studii Philosophiae* (+1272), and the *Epistola de Secretis
   Operibus Artis et Naturae et de Nullitate Magiae* (*c.* +1270).
BRIDGES, J. H. (1) (ed.). *The 'Opus Maius'* [*c.* +1266] *of Roger Bacon*. 3 vols. Oxford, 1897–1900.
BRIDGMAN, E. C. (1). *A Chinese Chrestomathy, in the Canton Dialect*. S. Wells Williams, Macao, 1841.
BRIDGMAN, E. C. & WILLIAMS, S. WELLS (1). 'Mineralogy, Botany, Zoology and Medicine' [sections of
   a Chinese Chrestomathy], in Bridgman (1), pp. 429, 436, 460 and 497.
BRIGHTMAN, F. E. (1). *Liturgies, Eastern and Western*. Oxford, 1896.
BROMEHEAD, C. E. N. (2). 'Aetites, or the Eagle-Stone.' *AQ*, 1947, **21**, 16.
BROOKS, CHANDLER McC., GILBERT, J. L., LEVEY, H. A. & CURTIS, D. R. (1). *Humors, Hormones and
   Neurosecretions; the Origins and Development of Man's present Knowlege of the Humoral Control of
   Body Function*. New York State Univ. N.Y. 1962.
BROOKS, E. W. (1). 'A Syriac Fragment [a chronicle extending from +754 to +813].' *ZDMG*, 1900, **54**,
   195.
BROOKS, G. (1). *Recherches sur le Latex de l'Arbre à Laque d'Indochine; le Laccol et ses Derivés*. Jouve,
   Paris, 1932.
BROOKS, G. (2). 'La Laque Végétale d'Indochine.' *LN*, 1937 (no. 3011), 359.
BROOMHALL, M. (1). *Islam in China*. Morgan & Scott, London, 1910.
BROSSE, T. (1). *Études instrumentales des Techniques du Yoga; Expérimentation psychosomatique* ... with
   an Introduction 'La Nature du Yoga dans sa Tradition' by J. Filliozat, École Française d'Extrême-
   Orient, Paris, 1963 (Monograph series, no. 52).
BROUGH, J. (1). 'Soma and *Amanita muscaria*.' *BLSOAS*, 1971, **34**, 331.
BROWN-SÉQUARD, C. E. (1). 'Du Rôle physiologique d'un thérapeutique d'un Suc extrait de Test-
   icules d'Animaux, d'après nombre de faits observés chez l'Homme.' *APNP*, 1889, **21**, 651.
BROWNE, C. A. (1). 'Rhetorical and Religious Aspects of Greek Alchemy; including a Commentary and
   Translation of the Poem of the Philosopher Archelaos upon the Sacred Art.' *AX*, 1938, **2**, 129; 1948,
   3, 15.
BROWNE, E. G. (1). *Arabian Medicine*. Cambridge, 1921. Repr. 1962. (French tr. H. J. P. Renaud;
   Larose, Paris, 1933.)
BROWNE, RICHARD (1) (tr.). *The Cure of Old Age and the Preservation of Youth* (tr. of Roger Bacon's
   De Retardatione Accidentium Senectutis...). London, 1683.
BROWNE, SIR THOMAS (1). *Religio Medici*. 1642.

BRÜCK, R. (1) (tr.). 'Der Traktat des Meisters Antonio von Pisa.' *RKW*, 1902, **25**, 240. A +14th-century treatise on glass-making.

BRUNET, P. & MIELI, A. (1). *L'Histoire des Sciences (Antiquité)*. Payot, Paris, 1935. Rev. G. Sarton, *ISIS*, 1935, **24**, 444.

BRUNSCHWYK, H. (1). '*Liber de arte Distillandi de Compositis': Das Buch der waren Kunst zu distillieren die Composita und Simplicia; und das Buch 'Thesaurus Pauperum', Ein schatz der armen genannt Micarium die brösamlin gefallen von den büchern d'Artzny und durch Experiment von mir Jheronimo Brunschwick uff geclubt und geoffenbart zu trost denen die es begehren*. Grüninger, Strassburg, 1512. (This is the so-called 'Large Book of Distillation'.) Eng. tr. *The Vertuose Boke of the Distillacyon...*, Andrewe, or Treveris, London, 1527, 1528 and 1530. The last reproduced in facsimile, with an introduction by H. J. Abrahams, Johnson, New York and London, 1971 (Sources of Science Ser., no. 79).

BRUNSCHWYK, H. (2). '*Liber de arte distillandi de simplicibus*' *oder Buch der rechten Kunst zu distillieren die eintzigen Dinge*. Grüninger, Strassburg, 1500. (The so-called 'Small Book of Distillation'.)

BRUNTON, T. LAUDER (1). *A Textbook of Pharmacology, Therapeutics and Materia Medica*. Adpated to the United States Pharmacopoeia by F. H. Williams. Macmillan, London, 1888.

BRYANT, P. L. (1). 'Chinese Camphor and Camphor Oil.' *CJ*, 1925, **3**, 228.

BUCH, M. (1). 'Die Wotjäken, eine ethnologische Studie.' *ASSF*, 1883, **12**, 465.

BUCK, J. LOSSING (1). *Land Utilisation in China; a Study of 16,786 Farms in 168 Localities, and 38,256 Farm Families in Twenty-two Provinces in China, 1929 to 1933*. Univ. of Nanking, Nanking and Commercial Press, Shanghai, 1937. (Report in the International Research Series of the Institute of Pacific Relations.)

BUCKLAND, A. W. (1). 'Ethnological Hints afforded by the Stimulants in Use among Savages and among the Ancients.' *JRAI*, 1879, **8**, 239.

BUDGE, E. A. WALLIS (4) (tr.). *The Book of the Dead; the Papyrus of Ani in the British Museum*. Brit. Mus., London, 1895.

BUDGE, E. A. WALLIS (5). *First Steps in [the Ancient] Egyptian [Language and Literature]; a Book for Beginners*. Kegan Paul, Trench & Trübner, London, 1923.

BUDGE, E. A. WALLIS (6) (tr.). *Syrian Anatomy, Pathology and Therapeutics; or, 'The Book of Medicines'—the Syriac Text, edited from a Rare Manuscript, with an English Translation...* 2 vols. Oxford, 1913.

BUDGE, E. A. WALLIS (7) (tr.). *The 'Book of the Cave of Treasures'; a History of the Patriarchs and the Kings and their Successors from the Creation to the Crucifixion of Christ, translated from the Syriac text of BM Add. MS. 25875*. Religious Tract Soc. London, 1927.

BUHOT, J. (1). *Arts de la Chine*. Editions du Chène, Paris, 1951.

BÜLFFINGER, G. B. (1). *Specimen Doctrinae Veterum Sinarum Moralis et Politicae; tanquam Exemplum Philosophiae Gentium ad Rem Publicam applicatae; Excerptum Libellis Sinicae Genti Classicis, Confucii sive Dicta sive Facta Complexis*. Frankfurt a/M, 1724.

BULLING, A. (14). 'Archaeological Excavations in China, 1949 to 1971.' *EXPED*, 1972, **14** (no. 4), 2; **15** (no. 1), 22.

BURCKHARDT, T. (1). *Alchemie*. Walter, Freiburg i/B, 1960. Eng. tr. by W. Stoddart: *Alchemy; Science of the Cosmos, Science of the Soul*. Stuart & Watkins, London, 1967.

BURKE, R. B. (1) (tr.). *The 'Opus Majus' of Roger Bacon*. 2 vols. Philadelphia and London, 1928.

BURKILL, I. H. (1). *A Dictionary of the Economic Products of the Malay Peninsula* (with contributions by W. Birtwhistle, F. W. Foxworthy, J. B. Scrivener & J. G. Watson). 2 vols. Crown Agents for the Colonies, London, 1935.

BURKITT, F. C. (1). *The Religion of the Manichees*. Cambridge, 1925.

BURKITT, F. C. (2). *Church and Gnosis*. Cambridge, 1932.

BURNAM, J. M. (1). *A Classical Technology edited from Codex Lucensis 490*. Boston, 1920.

BURNES, A. (1). *Travels into Bokhara...* 3 vols. Murray, London, 1834.

BURTON, A. (1). *Rush-bearing; an Account of the Old Customs of Strewing Rushes, Carrying Rushes to Church, the Rush-cart; Garlands in Churches, Morris-Dancers, the Wakes, and the Rush*. Brook & Chrystal, Manchester, 1891.

BUSHELL, S. W. (2). *Chinese Art*. 2 vols. For Victoria and Albert Museum, HMSO, London, 1909; 2nd ed. 1914.

CABANÈS, A. (1). *Remèdes d'Autrefois*. 2nd ed. Maloine, Paris, 1910.

CALEY, E. R. (1). 'The Leyden Papyrus X; an English Translation with Brief Notes.' *JCE*, 1926, **3**, 1149.

CALEY, E. R. (2). 'The Stockholm Papyrus; an English Translation with Brief Notes.' *JCE*, 1927, **4**, 979.

CALEY, E. R. (3). 'On the Prehistoric Use of Arsenical Copper in the Aegean Region.' *HE*, 1949, **8** (Suppl.), 60 (Commemorative Studies in Honour of Theodore Leslie Shear).

CALEY, E. R. (4). 'The Earliest Use of Nickel Alloys in Coinage.' *NR*, 1943, **1**, 17.

CALEY, E. R. (5). 'Ancient Greek Pigments.' *JCE*, 1946, **23**, 314.

CALEY, E. R. (6). 'Investigations on the Origin and Manufacture of Orichalcum', art. in *Archaeological Chemistry*, ed. M. Levey. Pennsylvania University Press, Philadelphia, Pennsylvania, 1967, p. 59.

CALEY, E. R. & RICHARDS, J. C. (1). *Theophrastus on the Stones*. Columbus, Ohio, 1956.

CALLOWAY, D. H. (1). 'Gas in the Alimentary Canal.' Ch. 137 in *Handbook of Physiology*, sect. 6, 'Alimentary Canal', vol. 5, 'Bile; Digestion; Ruminal Physiology'. Ed. C. F. Code & W. Heidel. Williams & Wilkins, for the American Physiological Society, Washington, D.C. 1968.

CALMET, AUGUSTIN (1). *Dissertations upon the Appearances of Angels, Daemons and Ghosts, and concerning the Vampires of Hungary, Bohemia, Moravia and Silesia*. Cooper, London, 1759, tr. from the French ed. of 1745. Repr. with little change, under the title: *The Phantom World, or the Philosophy of Spirits, Apparitions, etc....*, ed. H. Christmas, 2 vols. London, 1850.

CAMMANN, S. VAN R. (4). 'Archaeological Evidence for Chinese Contacts with India during the Han Dynasty.' *S*, 1956, **5**, 1; abstr. *RBS*, 1959, **2**, no. 320.

CAMMANN, S. VAN R. (5). 'The "Bactrian Nickel Theory".' *AJA*, 1958, **62**, 409. (Commentary on Chêng & Schwitter, 1.)

CAMMANN, S. VAN R. (7). 'The Evolution of Magic Squares in China.' *JAOS*, 1960, **80**, 116.

CAMMANN, S. VAN R. (8). 'Old Chinese Magic Squares.' *S*, 1962, **7**, 14. Abstr. L. Lanciotti, *RBS*, 1969, **8**, no. 837.

CAMMANN, S. VAN R. (9). 'The Magic Square of Three in Old Chinese Philosophy and Religion.' *HOR*, 1961, **1** (no. 1), 37. Crit. J. Needham, *RBS*, 1968, **7**, no. 581.

CAMMANN, S. VAN R. (10). 'A Suggested Origin of the Tibetan Maṇḍala Paintings.' *ARQ*, 1950, **13**, 107.

CAMMANN, S. VAN R. (11). 'On the Renewed Attempt to Revive the "Bactrian Nickel Theory".' *AJA*, 1962, **66**, 92 (rejoinder to Chêng & Schwitter, 2).

CAMMANN, S. VAN R. (12). 'Islamic and Indian Magic Squares.' *HOR*, 1968, **8**, 181, 271.

CAMMANN, S. VAN R. (13). Art. 'Magic Squares' in *EB* 1957 ed., vol. XIV, p. 573.

CAMPBELL, D. (1). *Arabian Medicine and its Influence on the Middle Ages*. 2 vols. (the second a bibliography of Latin MSS translations from Arabic). Kegan Paul, London, 1926.

CARATINI, R. (1). 'Quadrature du Cercle et Quadrature des Lunules en Mésopotamie.' *RAAO*, 1957, **51**, 11.

CARBONELLI, G. (1). *Sulle Fonti Storiche della Chimica e dell'Alchimia in Italia*. Rome, 1925.

CARDEW, S. (1). 'Mining in China in 1952.' *MJ*, 1953, **240**, 390.

CARLID, G. & NORDSTRÖM, J. (1). *Torbern Bergman's Foreign Correspondence* (with brief biography by H. Olsson). Almqvist & Wiksell, Stockholm, 1965.

CARLSON, C. S. (1). 'Extractive and Azeotropic Distillation.' Art. in *Distillation*, ed. A. Weissberger (*Technique of Organic Chemistry*, vol. 4), p. 317. Interscience, New York, 1951.

CARR, A. (1). *The Reptiles*. Time-Life International, Holland, 1963.

CARTER, G. F. (1). 'The Preparation of Ancient Coins for Accurate X-Ray Fluorescence Analysis.' *AMY*, 1964, **7**, 106.

CARTER, T. F. (1). *The Invention of Printing in China and its Spread Westward*. Columbia Univ. Press, New York, 1925, revised ed. 1931. 2nd ed. revised by L. Carrington Goodrich. Ronald, New York, 1955.

CARY, G. (1). *The Medieval Alexander*. Ed. D. J. A. Ross. Cambridge, 1956. (A study of the origins and versions of the Alexander-Romance; important for medieval ideas on flying-machine and diving-bell or bathyscaphe.)

CASAL, U. A. (1). 'The Yamabushi.' *MDGNVO*, 1965, **46**, 1.

CASAL, U. A. (2). 'Incense.' *TAS/J*, 1954 (3rd ser.), **3**, 46.

CASARTELLI, L. C. (1). '[The State of the Dead in] Iranian [and Persian Belief].' *ERE*, vol. XI, p. 847.

CASE, R. E. (1). 'Nickel-containing Coins of Bactria, −235 to −170.' *COCJ*, 1934, **102**, 117.

CASSIANUS, JOHANNES. *Conlationes*, ed. Petschenig. Cf. E. C. S. Gibson tr. (1).

CASSIUS, ANDREAS (the younger) (1). *De Extremo illo et Perfectissimo Naturae Opificio ac Principe Terraenorum Sidere Auro de admiranda ejus Natura...Cogitata Nobilioribus Experimentis Illustrata*. Hamburg, 1685. Cf. Partington (7), vol. 2, p. 371; Ferguson (1), vol. 1, p. 148.

CEDRENUS, GEORGIUS (1). *Historiōn Archomenē* (c. +1059), ed. Bekker (in *Corp. Script. Hist. Byz.* series).

CENNINI, CENNINO (1). *Il Libro dell'Arte*. MS on dyeing and painting, 1437. Eng. trs. C. J. Herringham, Allen & Unwin, London, 1897; D. V. Thompson, Yale Univ. Press. New Haven, Conn. 1933.

CERNY, J. (1). *Egyptian Religion*.

CHADWICK, H. (1) (tr.). *Origen 'Contra Celsum'; Translated with an Introduction and Notes*. Cambridge, 1953.

CHAMBERLAIN, B. H. (1). *Things Japanese*. Murray, London, 2nd ed. 1891; 3rd ed. 1898.

CHAMPOLLION, J. F. (1). 'L'Égypte sous les Pharaons; ou Recherches sur la Geographie, la Religion, la Langue, les Écritures, et l'Histoire de l'Égypte avant l'invasion de Cambyse*. De Bure, Paris, 1814.

CHAMPOLLION, J. F. (2). *Grammaire Égyptien en Écriture Hieroglyphique*. Didot, Paris, 1841.

CHAMPOLLION, J. F. (3). *Dictionnaire Égyptien en Écriture Hieroglyphique.* Didot, Paris, 1841.

CHANG, C. *See* Kung, S. C., Chao, S. W., Pei, Y. T. & Chang, C. (1).

CHANG CHUNG-YUAN (1). 'An Introduction to Taoist Yoga.' *RR*, 1956, **20**, 131.

CHANG HUNG-CHAO (1). *Lapidarium Sinicum; a Study of the Rocks, Fossils and Minerals as known in Chinese Literature* (in Chinese with English summary). Chinese Geological Survey, Peiping, 1927. *MGSC* (ser. B), no. 2.

CHANG HUNG-CHAO (2). 'The Beginning of the Use of Zinc in China.' *BCGS*, 1922, **2** (no. 1/2), 17. Cf. Chang Hung-Chao (3).

CHANG HUNG-CHAO (3). 'New Researches on the Beginning of the Use of Zinc in China.' *BCGS*, 1925, **4** (no. 1), 125. Cf. Chang Hung-Chao (6).

CHANG HUNG-CHAO (4). 'The Origins of the Western Lake at Hangchow.' *BCGS*, 1924, **3** (no. 1), 26. Cf. Chang Hung-Chao (5).

CHANG HSIEN-FÊNG (1). 'A Communist Grows in Struggle.' *CREC*, 1969, **18** (no. 4), 17.

CHANG KUANG-YU & CHANG CHÊNG-YU (1). *Peking Opera Make-up; an Album of Cut-outs.* Foreign Languages Press, Peking, 1959.

CHANG TZU-KUNG (1). 'Taoist Thought and the Development of Science; a Missing Chapter in the History of Science and Culture-Relations.' Unpub. MS., 1945. Now in *MBPB*, 1972, **21** (no. 1), **7** (no. 2), 20.

CHARLES, J. A. (1). 'Early Arsenical Bronzes—a Metallurgical View.' *AJA*, 1967, **71**, 21. A discussion arising from the data in Renfrew (1).

CHARLES, J. A. (2). 'The First Sheffield Plate.' *AQ*, 1968, **42**, 278. With an appendix on the dating of the Minoan bronze dagger with silver-capped copper rivet-heads, by F. H. Stubbings.

CHARLES, J. A. (3). 'Heterogeneity in Metals.' *AMY*, 1973, **15**, 105.

CHARLES, R. H. (1) (tr.). *The 'Book of Enoch', or 'I Enoch', translated from the Editor's Ethiopic Text, and edited with the Introduction, Notes and Indexes of the First Edition, wholly recast, enlarged and re-written, together with a Reprint from the Editor's Text of the Greek Fragments.* Oxford, 1912 (first ed., Oxford, 1893).

CHARLES, R. H. (2) (ed.). *The Ethiopic Version of the 'Book of Enoch', edited from 23 MSS, together with the Fragmentary Greek and Latin Versions.* Oxford, 1906.

CHARLES, R. H. (3). *A Critical History of the Doctrine of a Future Life in Israel, in Judaism, and in Christianity; or, Hebrew, Jewish and Christian Eschatology from pre-Prophetic Times till the Close of the New Testament Canon.* Black, London, 1899. Repr. 1913 (Jowett Lectures, 1898–9).

CHARLES, R. H. (4) 'Gehenna', art. in Hastings, *Dictionary of the Bible*, Clark, Edinburgh, 1899, vol. 2, p. 119.

CHARLES, R. H. (5) (ed.). *The Apocrypha and Pseudepigrapha of the Old Testament in English; with Introductions, and Critical and Explanatory Notes, to the Several Books...* 2 vols. Oxford, 1913 (1 Enoch is in vol. 2).

CHATLEY, H. (1). MS. translation of the astronomical chapter (ch. 3, Thien Wên) of *Huai Nan Tzu.* Unpublished. (Cf. note in *O*, 1952, **72**, 84.)

CHATLEY, H. (37). 'Alchemy in China.' *JALCHS*, 1913, **2**, 33.

CHATTERJI, S. K. (1). 'India and China; Ancient Contacts—What India received from China.' *JRAS/B*, 1959 (n.s.), **1**, 89.

CHATTOPADHYAYA, D. (1). 'Needham on Tantrism and Taoism.' *NAGE*, 1957, **6** (no. 12), 43; 1958, **7** (no. 1), 32.

CHATTOPADHYAYA, D. (2). 'The Material Basis of Idealism.' *NAGE*, 1958, **7** (no. 8), 30.

CHATTOPADHYAYA, D. (3) 'Brahman and Maya.' *ENQ*, 1959, **1** (no. 1), 25.

CHATTOPADHYAYA, D. (4). *Lokāyata, a Study in Ancient Indian Materialism.* People's Publishing House, New Delhi, 1959.

CHAVANNES, E. (14). *Documents sur les Tou-Kiue [Thu-Chüeh] (Turcs) Occidentaux, receuillis et commentés par E. C....* Imp. Acad. Sci., St Petersburg, 1903. Repr. Paris, with the inclusion of the 'Notes Additionelles', n. d.

CHAVANNES, E. (17). 'Notes Additionelles sur les Tou-Kiue [Thu-Chüeh] (Turcs) Occidentaux.' *TP*, 1904, **5**, 1–110, with index and errata for Chavannes (14).

CHAVANNES, E. (19). 'Inscriptions et Pièces de Chancellerie Chinoises de l'Époque Mongole.' *TP*, 1904, **5**, 357–447; 1905, **6**, 1–42; 1908, **9**, 297–428.

CHAVANNES, E. & PELLIOT, P. (1). 'Un Traité Manichéen retrouvé en Chine, traduit et annoté.' *JA*, 1911 (10e sér), **18**, 499; 1913 (11e sér), **1**, 99, 261.

CH'ÊN, JEROME. *See* Chhen Chih-Jang.

CHÊNG, C. F. & SCHWITTER, C. M. (1). 'Nickel in Ancient Bronzes.' *AJA*, 1957, **61**, 351. With an appendix on chemical analysis by X-ray fluorescence by K. G. Carroll.

CHÊNG, C. F. & SCHWITTER, C. M. (2). 'Bactrian Nickel and [the] Chinese [Square] Bamboos.' *AJA*, 1962, **66**, 87 (reply to Cammann, 5).

CHÊNG MAN-CHHING & SMITH, R. W. (1). *Thai-Chi; the 'Supreme Ultimate' Exercise for Health, Sport and Self-Defence*. Weatherhill, Tokyo, 1966.

CHÊNG TÊ-KHUN (2) (tr.). 'Travels of the Emperor Mu.' *JRAS/NCB*, 1933, **64**, 142; 1934, **65**, 128.

CHÊNG TÊ-KHUN (7). 'Yin Yang, Wu Hsing and Han Art.' *HJAS*, 1957, **20**, 162.

CHÊNG TÊ-KHUN (9). *Archaeology in China*.
    Vol. 1, *Prehistoric China*. Heffer, Cambridge, 1959.
    Vol. 2, *Shang China*. Heffer, Cambridge, 1960.
    Vol. 3, *Chou China*, Heffer, Cambridge, and Univ. Press, Toronto, 1963.
    Vol. 4, *Han China* (in the press).

CHENG WOU-CHAN. See Shêng Wu-Shan.

CHEO, S. W. *See* Kung, S. C., Chao, S. W., Pei, Y. T. & Chang, C. (1).

CHEYNE, T. K. (1). *The Origin and Religious Content of the Psalter*. Kegan Paul, London, 1891. (Bampton Lectures.)

CHHEN CHIH-JANG (1). *Mao and the Chinese Revolution*. Oxford, 1965. With 37 Poems by Mao Tsê-Tung, translated by Michael Bullock & Chhen Chih-Jang.

CHHEN SHOU-YI (3). *Chinese Literature; a Historical Introduction*. Ronald, New York, 1961.

CHHU TA-KAO (2) (tr.). *Tao Tê Ching, a new translation*. Buddhist Lodge, London, 1937.

CHIKASHIGE, MASUMI (1). *Alchemy and other Chemical Achievements of the Ancient Orient; the Civilisation of Japan and China in Early Times as seen from the Chemical (and Metallurgical) Point of View*. Rokakuho Uchida, Tokyo, 1936. Rev. Tenney L. Davis, *JACS*, 1937, **59**, 952. Cf. Chinese résumé of Chakashige's lectures by Chhen Mêng-Yen, *KHS*, 1920, **5** (no. 3), 262.

CHIU YAN TSZ. See Yang Tzu-Chiu (1).

CHOISY, M. (1). *La Métaphysique des Yogas*. Ed. Mont. Blanc, Geneva, 1948. With an introduction by P. Masson-Oursel.

CHŎN SANGŬN (1). *Science and Technology in Korea; Traditional Instruments and Techniques*. M.I.T. Press, Cambridge, Mass. 1972.

CHOU I-LIANG (1). 'Tantrism in China.' *HJAS*, 1945, **8**, 241.

CHOULANT, L. (1). *History and Bibliography of Anatomic Illustration*. Schuman, New York, 1945, tr. from the German (Weigel, Leipzig, 1852) by M. Frank, with essays by F. H. Garrison, M. Frank, E. C. Streeter & Charles Singer, and a Bibliography of M. Frank by J. C. Bay.

CHOU YI-LIANG. See Chou I-Liang.

CHU HSI-THAO (1). 'The Use of Amalgam as Filling Material in Dentistry in Ancient China.' *CMJ*, 1958, **76**, 553.

CHWOLSON, D. (1). *Die Ssabier und der Ssabismus*. 2 vols. Imp. Acad. Sci., St Petersburg, 1856. (On the culture and religion of the Sabians, Ṣābi, of Harrān, 'pagans' till the +10th century, a people important for the transmission of the Hermetica, and for the history of alchemy, Harrān being a cross-roads of influences from the East and West of the Old World.)

[CIBOT, P. M.] (3). 'Notice du Cong-Fou [*Kung fu*], des Bonzes Tao-sée [Tao Shih].' *MCHSAMUC*, 1779, **4**, 441. Often ascribed, as by Dudgeon (1) and others, to J. J. M. Amiot, but considered Cibot's by Pfister (1), p. 896.

[CIBOT, P. M.] (5). 'Notices sur différens Objets; (1) Vin, Eau-de-Vie et Vinaigre de Chine, (2) Raisins secs de Hami, (3) Notices du Royaume de Hami, (4) Rémèdes [*pao-hsing shih, khu chiu*], (5) Teinture chinoise, (6) Abricotier [selection, care of seedlings, and grafting], (7) Armoise.' *MCHSAMUC*, 1780, **5**, 467–518.

CIBOT, P. M. (11). (posthumous). 'Notice sur le Cinabre, le Vif-Argent et le *Ling sha*.' *MCHSAMUC*, 1786, **11**, 304.

CIBOT, P. M. (12) (posthumous). 'Notice sur le Borax.' *MCHSAMUC*, 1786, **11**, 343.

CIBOT, P. M. (13) (posthumous). 'Diverses Remarques sur les Arts-Pratiques en Chine; Ouvrages de Fer, Art de peindre sur les Glaces et sur les Pierres.' *MCHSAMUC*, 1786, **11**, 361.

CIBOT, P. M. (14). 'Notice sur le Lieou-li [*Liu-li*], ou Tuiles Vernissées.' *MCHSAMUC*, 1787, **13**, 396.

[CIBOT, P. M.] (16). 'Notice du Ché-hiang [*Shê hsiang*, musk and the musk deer].' *MCHSAMUC*, 1779, **4**, 493.

[CIBOT, P. M.] (17). 'Quelques Compositions et Recettes pratiquées chez les Chinois ou consignées dans leurs Livres, et que l'Auteur a crues utiles ou inconnues en Europe [on felt, wax, conservation of oranges, bronzing of copper, etc. etc.].' *MCHSAMUC*, 1779, **4**, 484.

CLAPHAM, A. R., TUTIN, T. G. & WARBURG, E. F. (1). *Flora of the British Isles*. 2nd ed. Cambridge, 1962.

CLARK, A. J. (1). *Applied Pharmacology*. 7th ed. Churchill, London, 1942.

CLARK, E. (1). 'Notes on the Progress of Mining in China.' *TAIME*, 1891, **19**, 571. (Contains an account (pp. 587 ff.) of the recovery of silver from argentiferous lead ore, and cupellation by traditional methods, at the mines of Yen-tang Shan.)

CLARK, R. T. RUNDLE (1). *Myth and Symbol in Ancient Egypt*. Thames & Hudson, London, 1959.

CLARK, W. G. & DEL GIUDICE, J. (1) (ed.). *Principles of Psychopharmacology*. Academic Press, New York and London, 1971.

CLARKE, J. & GEIKIE, A. (1). *Physical Science in the Time of Nero, being a Translation of the 'Quaestiones Naturales' of Seneca, with notes by Sir Archibald Geikie*. Macmillan, London, 1910.

CLAUDER, GABRIEL (1). *Inventum Cinnabarinum, hoc est Dissertatio Cinnabari Nativa Hungarica, longa circulatione in majorem efficaciam fixata et exalta*. Jena, 1684.

CLEAVES, F. W. (1). 'The Sino-Mongolian Inscription of +1240 [edict of the empress Törgene, wife of Ogatai Khan (+1186 to +1241) on the cutting of the blocks for the Yuan edition of the *Tao Tsang*].' *HJAS*, 1960, **23**, 62.

CLINE, W. (1). *Mining and Metallurgy in Negro Africa*. Banta, Menasha, Wisconsin, 1937 (mimeographed). (General Studies in Anthropology, no. 5, Iron.)

CLOW, A. & CLOW, NAN L. (1). *The Chemical Revolution; a Contribution to Social Technology*. Batchworth, London, 1952.

CLOW, A. & CLOW, NAN L. (2). 'Vitriol in the Industrial Revolution.' *EHR*, 1945, **15**, 44.

CLULEE, N. H. (1). 'John Dee's Mathematics and the Grading of Compound Qualities.' *AX*, 1971, **18**, 178.

CLYMER, R. SWINBURNE (1). *Alchemy and the Alchemists; giving the Secret of the Philosopher's Stone, the Elixir of Youth, and the Universal Solvent; Also showing that the* True *Alchemists did not seek to transmute base metals into Gold, but sought the Highest Initiation or the Development of the Spiritual Nature in Man*... 4 vols. Philosophical Publishing Co. Allentown, Pennsylvania, 1907. The first two contain the text of Hitchcock (1), but 'considerably re-written and with much additional information, mis-information and miscellaneous nonsense interpolated' (Cohen, 1).

COGHLAN, H. H. (1). 'Metal Implements and Weapons [in Early Times before the Fall of the Ancient Empires].' Art. in *History of Technology*, ed. C. Singer, E. J. Holmyard & A. R. Hall. Oxford, 1954, vol. 1, p. 600.

COGHLAN, H. H. (3). 'Etruscan and Spanish Swords of Iron.' *SIB*, 1957, **3**, 167.

COGHLAN, H. H. (4). 'A Note upon Iron as a Material for the Celtic Sword.' *SIB*, 1957, **3**, 129.

COGHLAN, H. H. (5). *Notes on Prehistoric and Early Iron in the Old World; including a Metallographic and Metallurgical Examination of specimens selected by the Pitt Rivers Museum, and contributions by I. M. Allen*. Oxford, 1956. (Pitt Rivers Museum Occasional Papers on Technology, no. 8.)

COGHLAN, H. H. (6). 'The Prehistorical Working of Bronze and Arsenical Copper.' *SIB*, 1960, **5**, 145.

COHAUSEN, J. H. (1). *Lebensverlängerung bis auf 115 Jahre durch den Hauch junger Mädchen*. Orig. title: *Der wieder lebende Hermippus, oder curieuse physikalisch-medizinische Abhandlung von der seltener Art, sein Leben durch das Anhauchen Junger- Mägdchen bis auf 115 Jahr zu verlängern, aus einem römischen Denkmal genommen, nun aber mit medicinischen Gründen befestiget, und durch Beweise und Exempel, wie auch mit einer wunderbaren Erfindung aus der philosophischen Scheidekunst erläutert und bestätiget von J. H. C*...Alten Knaben, (Stuttgart?), 1753. Latin ed. Andreae, Frankfurt, 1742. Reprinted in *Der Schatzgräber in den literarischen und bildlichen Seltenheiten, Sonderbarkeiten, etc., hauptsächlich des deutschen Mittelalters*, ed. J. Scheible, vol. 2. Scheible, Stuttgart and Leipzig, 1847. Eng. tr. by J. Campbell, *Hermippus Redivivus; or, the Sage's Triumph over Old Age and the Grave, wherein a Method is laid down for prolonging the Life and Vigour of Man*. London, 1748, repr. 1749, 3rd ed. London, 1771. Cf. Ferguson (1), vol. 1, pp. 168 ff.; Paal (1).

COHEN, I. BERNARD (1). 'Ethan Allen Hitchcock; Soldier–Humanitarian–Scholar; Discoverer of the "True Subject" of the Hermetic Art.' *PAAQS*, 1952, 29.

COLEBY, L. J. M. (1). *The Chemical Studies of P. J. Macquer*. Allen & Unwin, London, 1938.

COLLAS, J. P. L. (3) (posthumous). 'Sur un Sel appellé par les Chinois *Kièn*.' *MCHSAMUC*, 1786, **11**, 315.

COLLAS, J. P. L. (4) (posthumous). 'Extrait d'une Lettre de Feu M. Collas, Missionnaire à Péking, 1e Sur la Chaux Noire de Chine, 2e Sur une Matière appellée Lieou-li [*Liu-li*], qui approche du Verre, 3e Sur une Espèce de Mottes à Brûler.' *MCHSAMUC*, 1786, **11**, 321.

COLLAS, J. P. L. (5) (posthumous). 'Sur le Hoang-fan [*Huang fan*] ou vitriol, le Nao-cha [*Nao sha*] ou Sel ammoniac, et le Hoang-pé-mou [*Huang po mu*].' *MCHSAMUC*, 1786, **11**, 329.

COLLAS, J. P. L. (6) (posthumous). 'Notice sur le Charbon de Terre.' *MCHSAMUC*, 1786, **11**, 334.

COLLAS, J. P. L. (7) (posthumous). 'Notice sur le Cuivre blanc de Chine, sur le Minium et l'Amadou.' *MCHSAMUC*, 1786, **11**, 347.

COLLAS, J. P. L. (8) (posthumous). 'Notice sur un Papier doré sans Or.' *MCHSAMUC*, 1786, **11**, 351.

COLLAS, J. P. L. (9) (posthumous). 'Sur la Quintessence Minérale de M. le Comte de la Garaye.' *MCHSAMUC*, 1786, **11**, 298.

COLLIER, H. B. (1). 'Alchemy in Ancient China.' *CHEMC*, 1952, 41 (101).

COLLINS, W. F. (1). *Mineral Enterprise in China*. Revised edition, Tientsin Press, Tientsin, 1922. With an appendix chapter on 'Mining Legislation and Development' by Ting Wên-Chiang (V. K. Ting), and a memorandum on 'Mining Taxation' by G. G. S. Lindsey.

CONDAMIN, J. & PICON, M. (1). 'The Influence of Corrosion and Diffusion on the Percentage of Silver in Roman Denarii.' *AMY*, 1964, **7**, 98.

DE CONDORCET, A. N. (1). *Esquisse d'un Tableau Historique des Progrès de l'Esprit Humain*. Paris, 1795. Eng. tr. by J. Barraclough, *Sketch for a Historical Picture of the Human Mind*. London, 1955.

CONNELL, K. H. (1). *Irish Peasant Society; Four Historical Essays*. Oxford, 1968. ('Illicit Distillation', pp. 1–50.)

CONRADY, A. (1). 'Indischer Einfluss in China in 4-jahrh. v. Chr.' *ZDMG*, 1906, **60**, 335.

CONRADY, A. (3). 'Zu *Lao-Tze*, cap. 6' (The valley spirit). *AM*, 1932, **7**, 150.

CONRING, H. (1). *De Hermetica Aegyptiorum Vetere et Paracelsicorum Nova Medicina*. Muller & Richter, Helmstadt, 1648.

CONYBEARE, F. C. (1) (tr.). *Philostratus [of Lemnos]; the 'Life of Apollonius of Tyana'*. 2 vols. Heinemann, London, 1912, repr. 1948. (Loeb Classics series.)

CONZE, E. (8). 'Buddhism and Gnosis.' *NUM/SHR*, 1967, **12**, 651 (in *Le Origini dello Gnosticismo*).

COOPER, W. C. & SIVIN, N. (1). 'Man as a Medicine; Pharmacological and Ritual Aspects of Traditional Therapy using Drugs derived from the Human Body.' Art. in Nakayama & Sivin (1), p. 203.

CORBIN, H. (1). 'De la Gnose antique à la Gnose Ismaelienne.' Art. in *Atti dello Convegno di Scienze Morali, Storiche e Filologiche*—'Oriente ed Occidente nel Medio Evo'. Acc. Naz. dei Lincei, Rome, 1956 (Atti dei Convegni Alessandro Volta, no. 12), p. 105.

CORDIER, H. (1). *Histoire Générale de la Chine*. 4 vols. Geuthner, Paris, 1920.

CORDIER, H. (13). 'La Suppression de la Compagnie de Jésus et de la Mission de Péking' (1774). *TP* 1916, **17**, 271, 561.

CORDIER, L. (1). 'Observations sur la Lettre de Mons. Abel Rémusat...sur l'Existence de deux Volcans brûlans dans la Tartarie Centrale.' *JA*, 1824, **5**, 47.

CORDIER, V. (1). *Die chemischen Zeichensprache Einst und Jetzt*. Leykam, Graz, 1928.

CORNARO, LUIGI (1). *Discorsi della Vita Sobria*. 1558. Milan, 1627. Eng. tr. by J. Burdell, *The Discourses and Letters of Luigi Cornaro on a Sober and Temperate Life*. New York, 1842. Also nine English translations before 1825, incl. Dublin, 1740.

CORNER, G. W. (1). *The Hormones in Human Reproduction*. Univ. Press, Princeton, N.J. 1946.

CORNFORD, F. M. (2). *The Laws of Motion in Ancient Thought*. Inaug. Lect. Cambridge, 1931.

CORNFORD, F. M. (7). *Plato's Cosmology; the 'Timaeus' translated, with a running commentary*. Routledge & Kegan Paul, London, 1937, repr. 1956.

COVARRUBIAS, M. (2). *The Eagle, the Jaguar, and the Serpent; Indian Art of the Americas—North America (Alaska, Canada, the United States)*. Knopf, New York, 1954.

COWDRY, E. V. (1). 'Taoist Ideas of Human Anatomy.' *AMH*, 1925, **3**, 301.

COWIE, A. T. & FOLLEY, S. J. (1). 'Physiology of the Gonadotrophins and the Lactogenic Hormone.' Art. in *The Hormones...*, ed. G. Pincus, K. V. Thimann & E. B. Astwood. Acad. Press, New York, 1948–64, vol. 3, p. 309.

COYAJI, J. C. (2). 'Some Shahnamah Legends and their Chinese Parallels.' *JRAS/B*, 1928 (n.s.), **24**, 177.

COYAJI, J. C. (3). '*Bahram Yasht*; Analogues and Origins.' *JRAS/B*, 1928 (n.s.), **24**, 203.

COYAJI, J. C. (4). 'Astronomy and Astrology in the *Bahram Yasht*.' *JRAS/B*, 1928 (n.s.), **24**, 223.

COYAJI, J. C. (5). 'The *Shahnamah* and the *Fêng Shen Yen I*.' *JRAS/B*, 1930 (n.s.), **26**, 491.

COYAJI, J. C. (6). 'The *Sraosha Yasht* and its Place in the History of Mysticism.' *JRAS/B*, 1932 (n.s.), **28**, 225.

CRAIG, SIR JOHN (1). 'Isaac Newton and the Counterfeiters.' *NRRS*, 1963, **18**, 136.

CRAIG, SIR JOHN (2). 'The Royal Society and the Mint.' *NRRS*, 1964, **19**, 156.

CRAIGIE, W. A. (1). '[The State of the Dead in] Teutonic [Scandinavian, Belief].' *ERE*, vol. xi, p. 851.

CRAVEN, J. B. (1). *Count Michael Maier, Doctor of Philosophy and of Medicine, Alchemist, Rosicrucian, Mystic (+1568 to +1622); his Life and Writings*. Peace, Kirkwall, 1910.

CRAWLEY, A. E. (1). *Dress, Drinks and Drums; Further Studies of Savages and Sex*, ed. T. Besterman. Methuen, London, 1931.

CREEL, H. G. (7). 'What is Taoism?' *JAOS*, 1956, **76**, 139.

CREEL, H. G. (11). *What is Taoism?, and other Studies in Chinese Cultural History*. Univ. Chicago Press, Chicago, 1970.

CRESSEY, G. B. (1). *China's Geographic Foundations; a Survey of the Land and its People*. McGraw-Hill, New York, 1934.

CROCKET, R., SANDISON, R. A. & WALK, A. (1) (ed.). *Hallucinogenic Drugs and their Psychotherapeutic Use*. Lewis, London, 1961. (Proceedings of a Quarterly Meeting of the Royal Medico-Psychological Association.) Contributions by A. Cerletti and others.

CROFFUT, W. A. (1). *Fifty Years in Camp and Field; the Diary of Major-General Ethan Allen Hitchcock, U.S. Army*. Putnam, New York and London, 1909. 'A biography including copious extracts from the diaries but relatively little from the correspondence' (Cohen, 1).

CROLL, OSWALD (1). *Basilica Chymica*. Frankfurt, 1609.

CRONSTEDT, A. F. (1). *An Essay towards a System of Mineralogy*. London, 1770, 2nd ed. 1788 (greatly enlarged and improved by J. H. de Magellan). Tr. by G. von Engeström fom the Swedish *Försök till Mineralogie eller Mineral-Rikets Upställning*. Stockholm, 1758.

CROSLAND, M. P. (1). *Historical Studies in the Language of Chemistry*. Heinemann, London, 1962.

CUMONT, F. (4). *L'Égypte des Astrologues*. Fondation Égyptologique de la Reine Elisabeth, Brussels, 1937.

CUMONT, F. (5) (ed.). *Catalogus Codic. Astrolog. Graecorum*. 12 vols. Lamertin, Brussels, 1929–.

CUMONT, F. (6) (ed.). *Textes et Monuments Figurés relatifs aux Mystères de Mithra*. 2 vols. Lamertin, Brussels, 1899.

CUMONT, F. (7). 'Masque de Jupiter sur un Aigle Éployé [et perché sur le Corps d'un Ouroboros]; Bronze du Musée de Bruxelles.' Art. in *Festschrift f. Otto Benndorf*. Hölder, Vienna, 1898, p. 291.

CUMONT, F. (8). '*La Cosmogonie Manichéenne d'après Théodore bar Khōni* [Bp. of Khalkar in Mesopotamia, *c.* +600]. Lamertin, Brussels, 1908 (Recherches sur le Manichéisme, no. 1). Cf. Kugener & Cumont, (1, 2).

CUMONT, F. (9). 'La Roue à Puiser les Âmes du Manichéisme.' *RHR/AMG*, 1915, **72**, 384.

CUNNINGHAM, A. (1). 'Coins of Alexander's Successors in the East.' *NC*, 1873 (n.s.), **13**, 186.

CURWEN, M. D. (1) (ed.). *Chemistry and Commerce*. 4 vols. Newnes, London, 1935.

CURZON, G. N. (1). *Persia and the Persian Question*. London, 1892.

CYRIAX, E. F. (1). 'Concerning the Early Literature on Ling's Medical Gymnastics.' *JAN*, 1926, **30**, 225.

DALLY, N. (1). *Cinésiologie, ou Science du Mouvement dans ses Rapports avec l'Éducation, l'Hygiène et la Thérapie; Études Historiques, Théoriques et Pratiques*. Librairie Centrale des Sciences, Paris, 1857.

DALMAN, G. (1). *Arbeit und Sitte in Palästina*.
Vol. 1 *Jahreslauf und Tageslauf* (in two parts).
Vol. 2 *Der Ackerbau*.
Vol. 3 *Von der Ernte zum Mehl* (*Ernten, Dreschen, Worfeln, Sieben, Verwahren, Mahlen*).
Vol. 4 *Brot, Öl und Wein*.
Vol. 5. *Webstoff, Spinnen, Weben, Kleidung*.
Bertelsmann, Gütensloh, 1928– . (Schriften d. deutschen Palästina-Institut, nos. 3, 5, 6, 7, 8; Beiträge z. Forderung christlicher Theologie, ser. 2, Sammlung Wissenschaftlichen Monographien, nos. 14, 17, 27, 29, 33, 36.)
Vol. 6 *Zeltleben, Vieh- und Milch-wirtschaft, Jagd, Fischfang*.
Vol 7. *Das Haus, Hühnerzucht, Taubenzucht, Bienenzucht*.
Olms, Hildesheim, 1964. (Schriften d. deutschen Palästina-Institut, nos. 9, 10; Beiträge z. Forderung Christlicher Theologie. ser. 2, Sammlung Wissenschaftlichen Monographien, nos. 41, 48.)

DANA, E. S. (1). *A Textbook of Mineralogy, with an Extended Treatise on Crystallography and Physical Mineralogy*. 4th ed. rev. & enlarged by W. E. Ford. Wiley, New York, 1949.

DARMSTÄDTER, E. (1) (tr.). *Die Alchemie des Geber* [containing *Summa Perfectionis, Liber de Investigatione Perfectionis, Liber de Inventione Veritatis, Liber Fornacum*, and *Testamentum Geberis*, in German translation]. Springer, Berlin, 1922. Rev. J. Ruska, *ISIS*, 1923, **5**, 451.

DAS, M. N. & GASTAUT, H. (1). 'Variations de l'Activité électrique du Cerveau, du Coeur et des Muscles Squelettiques au cours de la Méditation et de l'Extase Yogique. Art. in *Conditionnement et Reactivité en Électro-encéphalographie*, ed. Fischgold & Gastaut (1). Masson, Paris, 1957, pp. 211 ff.

DASGUPTA, S. N. (3). *A Study of Patañjali*. University Press, Calcutta, 1920.

DASGUPTA, S. N. (4). *Yoga as Philosophy and Religion*. London, 1924.

DAUBRÉE, A. (1). 'La Génération des Minéraux dans la Pratique des Mineurs du Moyen Age d'après le "Bergbüchlein".' *JS*, 1890, 379, 441.

DAUMAS, M. (5). 'La Naissance et le Developpement de la Chimie en Chine.' *SET*, 1949, **6**, 11.

DAVENPORT, JOHN (1), ed. A. H. Walton. *Aphrodisiacs and Love Stimulants, with other chapters on the Secrets of Venus; being the two books by John Davenport entitled 'Aphrodisiacs and Anti-Aphrodisiacs'* [London, pr. pr. 1869, but not issued till 1873] *and 'Curiositates Eroticae Physiologiae; or, Tabooed Subjects Freely Treated'* [London, pr. pr. 1875]; *now for the first time edited, with Introduction and Notes* [and the omission of the essays 'On Generation' and 'On Death' from the second work] *by A.H.W....* Lyle Stuart, New York, 1966.

DAVID, SIR PERCIVAL (3). *Chinese Connoisseurship; the 'Ko Ku Yao Lun'* [+1388], (*Essential Criteria of Antiquities)—a Translation made and edited by Sir P. D....with a Facsimile of the Chinese Text*. Faber & Faber, London, 1971.

DAVIDSON, J. W. (1). *The Island of Formosa, past and present*. Macmillan, London, 1903.

DAVIES, D. (1). 'A Shangri-La in Ecuador.' *NS*, 1973, **57**, 236. On super-centenarians, especially in the Vilcabamba Valley in the Andes.

DAVIES, H. W., HALDANE, J. B. S. & KENNAWAY, E. L. (1). 'Experiments on the Regulation of the Blood's Alkalinity.' *JOP*, 1920, **54**, 32.

DAVIS, TENNEY L. (1). 'Count Michael Maier's Use of the Symbolism of Alchemy.' *JCE*, 1938, **15**, 403.

DAVIS, TENNEY L. (2). 'The Dualistic Cosmogony of Huai Nan Tzu and its Relations to the Background of Chinese and of European Alchemy.' *ISIS*, 1936, **25**, 327.

DAVIS, TENNEY L. (3). 'The Problem of the Origins of Alchemy.' *SM*, 1936, **43**, 551.

DAVIS, TENNEY L. (4). 'The Chinese Beginnings of Alchemy.' *END*, 1943, **2**, 154.

DAVIS, TENNEY L. (5). 'Pictorial Representations of Alchemical Theory.' *ISIS*, 1938, **28**, 73.

DAVIS, TENNEY L. (6). 'The Identity of Chinese and European Alchemical Theory.' *JUS*, 1929, **9**, 7. This paper has not been traceable by us. The reference is given in precise form by Davis & Chhen Kuo-Fu (2), but the journal in question seems to have ceased publication after the end of vol. 8.

DAVIS, TENNEY L. (7). 'Ko Hung (Pao Phu Tzu), Chinese Alchemist of the +4th Century.' *JCE*, 1934, **11**, 517.

DAVIS, TENNEY L. (8). 'The "Mirror of Alchemy" [*Speculum Alchemiae*] of Roger Bacon, translated into English.' *JCE*, 1931, **8**, 1945.

DAVIS, TENNEY L. (9). 'The Emerald Table of Hermes Trismegistus; Three Latin Versions Current among Later Alchemists.' *JCE*, 1926, **3**, 863.

DAVIS, TENNEY L. (10). 'Early Chinese Rockets.' *TR*, 1948, **51**, 101, 120, 122.

DAVIS, TENNEY L. (11). 'Early Pyrotechnics; I, Fire for the Wars of China, II, Evolution of the Gun, III, Early Warfare in Ancient China.' *ORD*, 1948, **33**, 52, 180, 396.

DAVIS, TENNEY L. (12). 'Huang Ti, Legendary Founder of Alchemy.' *JCE*, 1934, **11**, (635).

DAVIS, TENNEY L. (13). 'Liu An, Prince of Huai-Nan.' *JCE*, 1935, **12**, (1).

DAVIS, TENNEY L. (14). 'Wei Po-Yang, Father of Alchemy.' *JCE*, 1935, **12**, (51).

DAVIS, TENNEY L. (15). 'The Cultural Relationships of Explosives.' *NFR*, 1944, **1**, 11.

DAVIS, TENNEY L. (16) (tr.). *Roger Bacon's Letter concerning the Marvellous Power of Art and Nature, and concerning the Nullity of Magic...with Notes and an Account of Bacon's Life and Work*. Chem. Pub. Co., Easton, Pa. 1923. Cf. T. M[oufet] (1659).

DAVIS, TENNEY L. See Wu Lu-Chhiang & Davis.

DAVIS, TENNEY L. & CHAO YÜN-TSHUNG (1). 'An Alchemical Poem by Kao Hsiang-Hsien [+14th cent.]'. *ISIS*, 1939, **30**, 236.

DAVIS, TENNEY L. & CHAO YÜN-TSHUNG (2). 'The Four-hundred Word *Chin Tan* of Chang Po-Tuan [+11th cent.].' *PAAAS*, 1940, **73**, 371.

DAVIS, TENNEY L. & CHAO YÜN-TSHUNG (3). 'Three Alchemical Poems by Chang Po-Tuan.' *PAAAS*, 1940, **73**, 377.

DAVIS, TENNEY L. & CHAO YÜN-TSHUNG (4). 'Shih Hsing-Lin, disciple of Chang Po-Tuan [+11th cent.] and Hsieh Tao-Kuang, disciple of Shih Hsing-Lin.' *PAAAS*, 1940, **73**, 381.

DAVIS, TENNEY L. & CHAO YÜN-TSHUNG (5). 'The Secret Papers in the Jade Box of Chhing-Hua.' *PAAAS*, 1940, **73**, 385.

DAVIS, TENNEY L. & CHAO YÜN-TSHUNG (6). 'A Fifteenth-century Chinese Encyclopaedia of Alchemy.' *PAAAS*, 1940, **73**, 391.

DAVIS, TENNEY L. & CHAO YÜN-TSHUNG (7). 'Chang Po-Tuan of Thien-Thai; his *Wu Chen Phien* (Essay on the Understanding of the Truth); a Contribution to the Study of Chinese Alchemy.' *PAAAS*, 1939, **73**, 97.

DAVIS, TENNEY L. & CHAO YÜN-TSHUNG (8). 'Chang Po-Tuan, Chinese Alchemist of the +11th Century.' *JCE*, 1939 **16**, 53.

DAVIS, TENNEY L. & CHAO YÜN-TSHUNG (9). 'Chao Hsüeh-Min's Outline of Pyrotechnics [*Huo Hsi Lüeh*]; a Contribution to the History of Fireworks.' *PAAAS*, 1943, **75**, 95.

DAVIS, TENNEY, L. & CHHEN KUO-FU (1) (tr.). 'The Inner Chapters of *Pao Phu Tzu*.' *PAAAS*, 1941, **74**, 297. [Transl. of chs. 8 and 11; précis of the remainder.]

DAVIS, TENNEY L. & CHHEN KUO-FU (2). 'Shang Yang Tzu, Taoist writer and commentator on Alchemy.' *HJAS*, 1942, **7**, 126.

DAVIS, TENNEY L. & NAKASEKO ROKURO (1). 'The Tomb of Jofuku [Hsü Fu] or Joshi [Hsü Shih]; the Earliest Alchemist of Historical Record.' *AX*, 1937, **1**, 109, ill. *JCE*, 1947, **24**, (415).

DAVIS, TENENY L. & NAKASEKO ROKURO (2). 'The Jofuku [Hsü Fu] Shrine at Shingu; a Monument of Earliest Alchemy.' *NU*, 1937, **15** (no. 3), 60. 67.

DAVIS, TENNEY L. & WARE, J. R. (1). 'Early Chinese Military Pyrotechnics.' *JCE*, 1947, **24**, 522.

DAVIS, TENNEY L. & WU LU-CHHIANG (1). 'Ko Hung on the Yellow and the White.' *JCE*, 1936, **13**, 215.

DAVIS, TENNEY L. & WU LU-CHHIANG (2). 'Ko Hung on the Gold Medicine.' *JCE*, 1936, **13**, 103.

DAVIS, TENNEY L. & WU LU-CHHIANG (3). 'Thao Hung-Chhing.' *JCE*, 1932, **9**, 859.

DAVIS, TENNEY L. & WU LU-CHHIANG (4). 'Chinese Alchemy.' *SM*, 1930, **31**, 225. Chinese tr. by Chhen Kuo-Fu in *HHS*, 1936, **3**, 771.

DAVIS, TENNEY L. & WU LU-CHHIANG (5). 'The Advice of Wei Po-Yang to the Worker in Alchemy.'
NU, 1931, **8**, 115, 117. Repr. *DB*, 1935, **8**, 13.

DAVIS, TENNEY L. & WU LU-CHHIANG (6). 'The Pill of Immortality.' *TR*, 1931, **33**, 383.

DAWKINS, J. M. (1). *Zinc and Spelter; Notes on the Early History of Zinc from Babylon to the +18th
Century, compiled for the Curious.* Zinc Development Association, London, 1950. Repr. 1956.

DEANE, D. V. (1). 'The Selection of Metals for Modern Coinages.' *CUNOB*, 1969, no. 15. 29.

DEBUS, A. G. (1). *The Chemical Dream of the Renaissance.* Heffer, Cambridge,1968. (Churchill College
Overseas Fellowship Lectures, no. 3.)

DEBUS, A. G. (2). Introduction to the facsimile edition of Elias Ashmole's *Theatrum Chemicum Britannicum* (1652). Johnson, New York and London, 1967. (Sources of Science ser., no. 39.)

DEBUS, A. G. (3). 'Alchemy and the Historian of Science.' (An essay-review of C. H. Josten's *Elias
Ashmole*.) *HOSC*, 1967, **6**, 128.

DEBUS, A. G. (4). 'The Significance of the History of Early Chemistry.' *JWH*, 1965, **9**, 39.

DEBUS, A. G. (5). 'Robert Fludd and the Circulation of the Blood.' *JHMAS*, 1961, **16**, 374.

DEBUS, A. G. (6). 'Robert Fludd and the Use of Gilbert's *De Magnete* in the Weapon-Salve Controversy.'
JHMAS, 1964, **19**, 389.

DEBUS, A. G. (7). 'Renaissance Chemistry and the Work of Robert Fludd.' *AX*, 1967, **14**, 42.

DEBUS, A. G. (8). 'The Sun in the Universe of Robert Fludd.' Art. in *Le Soleil à la Renaissance; Sciences
et Mythes*, Colloque International, April 1963. Brussels, 1965, p. 261.

DEBUS, A. G. (9). 'The Aerial Nitre in the +16th and early +17th Centuries.' Communication to the
Xth International Congress of the History of Science, Ithaca, N.Y. 1962. In *Communications*,
p. 835.

DEBUS, A. G. (10). 'The Paracelsian Aerial Nitre.' *ISIS*, 1964, **55**, 43.

DEBUS, A. G. 11). 'Mathematics and Nature in the Chemical Texts of the Renaissance.' *AX*, 1968,
15, 1.

DEBUS, A. G. (12). 'An Elizabethan History of Medical Chemistry' [R. Bostocke's *Difference between the
Auncient Phisicke...and the Latter Phisicke*, +1585]. *ANS*, 1962, **18**, 1.

DEBUS, A. G. (13). 'Solution Analyses Prior to Robert Boyle.' *CHYM*, 1962, **8**, 41.

DEBUS, A. G. (14). 'Fire Analysis and the Elements in the Sixteenth and Seventeenth Centuries.' *ANS*,
1967, **23**, 127.

DEBUS, A. G. (15). 'Sir Thomas Browne and the Study of Colour Indicators.' *AX*, 1962, **10**, 29.

DEBUS, A. G. (16). 'Palissy, Plat, and English Agricultural Chemistry in the Sixteenth and Seventeenth
Centuries.' *A/AIHS*, 1968, **21** (nos. 82–3), 67.

DEBUS, A. G. (17). 'Gabriel Plattes and his Chemical Theory of the Formation of the Earth's Crust.'
AX, 1961, **9**, 162.

DEBUS, A. G. (18). *The English Paracelsians.* Oldbourne, London, 1965; Watts, New York, 1966. Rev. W.
Pagel, *HOSC*, 1966, **5**, 100.

DEBUS, A. G. (19). 'The Paracelsian Compromise in Elizabethan England.' *AX*, 1960, **8**, 71.

DEBUS, A. G. (20) (ed.). *Science, Medicine and Society in the Renaissance; Essays to honour Walter Pagel.*
2 vols. Science History Pubs (Neale Watson), New York, 1972.

DEBUS, A. G. (21). 'The Medico-Chemical World of the Paracelsians.' Art. in *Changing Perspectives in
the History of Science*, ed. M. Teich & R. Young (1), p. 85.

DEDEKIND, A. (1). *Ein Beitrag zur Purpurkunde.* 1898.

DEGERING, H. (1). 'Ein Alkoholrezept aus dem 8. Jahrhundert.' [The earliest version of the *Mappae
Clavicula*, now considered c. +820.] *SPAW/PH*, 1917, **36**, 503.

DELZA, S. (1). *Body and Mind in Harmony; Thai Chi Chhüan (Wu Style), an Ancient Chinese Way of
Exercise.* McKay, New York, 1961.

DEMIÉVILLE, P. (2). Review of Chang Hung-Chao (1), *Lapidarium Sinicum.* BEFEO, 1924, **24**, 276.

DEMIÉVILLE, P. (8). 'Momies d'Extrême-Orient.' *JS*, 1965, 144.

DENIEL, P. L. (1). *Les Boissons Alcooliques Sino-Vietnamiennes.* Inaug. Diss. Bordeaux, 1954. (Printed
Dong-nam-a, Saigon).

DENNELL, R. (1). 'The Hardening of Insect Cuticles.' *BR*, 1958, **33**, 178.

DEONNA, W. (2). 'Le Trésor des Fins d'Annecy.' *RA*, 1920 (5e sér), **11**, 112.

DEONNA, W. (3). 'Ouroboros.' *AA*, 1952, **15**, 163.

DEVASTHALI, G. V. (1). *The Religion and Mythology of the Brāhmaṇas with particular reference to the
'Satapatha-brāhmaṇa'.* Univ. of Poona, Poona, 1965. (Bhau Vishnu Ashtekar Vedic Research series,
no. 1.)

DEVÉRIA, G. (1). 'Origine de l'Islamisme en Chine; deux Légendes Mussulmanes Chinoises; Pelérinages
de Ma Fou-Tch'ou.' In *Volume Centenaire de l'Ecole des Langues Orientales Vivantes, 1795–1895.*
Leroux, Paris, 1895, p. 305.

DEY, K. L. (1). *Indigenous Drugs of India.* 2nd ed. Thacker & Spink, Calcutta, 1896.

DEYSSON, G. (1). 'Hallucinogenic Mushrooms and Psilocybine.' *PRPH*, 1960, **15**, 27.

DIELS, H. (1). *Antike Technik*. Teubner, Leipzig and Berlin, 1914; enlarged 2nd ed. 1920 (rev. B. Laufer, *AAN*, 1917, **19**, 71). Photolitho reproducton, Zeller, Osnabrück, 1965.

DIELS, H. (3). 'Die Entdeckung des Alkohols.' *APAW/PH*, 1913, no. 3, 1–35.

DIELS, H. (4). 'Etymologica' (incl. 2. χυμεία). *ZVSF*, 1916 (NF), **47**, 193.

DIELS, H. (5). *Fragmente der Vorsokratiker*. 7th ed., ed. W. Kranz. 3 vols.

DIERGART, P. (1)(ed.). *Beiträge aus der Geschichte der Chemie dem Gedächtnis v. Georg W. A. Kahlbaum...* Deuticke, Leipzig and Vienna, 1909.

DIHLE, A. (2). 'Neues zur Thomas-Tradition.' *JAC*, 1963, **6**, 54.

DILLENBERGER, J. (1). *Protestant Thought and Natural Science; a Historical Interpretation*. Collins, London, 1961.

DIMIER, L. (1). *L'Art d'Enluminure*. Paris, 1927.

DINDORF, W. *See* John Malala and Syncellos, Georgius.

DIVERS, E. (1). 'The Manufacture of Calomel in Japan.' *JSCI*, 1894, **13**, 108. Errata, p. 473.

DIXON, H. B. F. (1). 'The Chemistry of the Pituitary Hormones.' Art. in *The Hormones...* ed. G. Pincus, K. V. Thimann & E. B. Astwood. Academic Press, New York, 1948–64, vol. 5, p. 1.

DOBBS, B. J. (1). 'Studies in the Natural Philosophy of Sir Kenelm Digby.' *AX*, 1971, **18**, 1.

DODWELL, C. R. (1) (ed. & tr.). *Theophilus [Presbyter]; De Diversis Artibus (The Various Arts)* [probably by Roger of Helmarshausen, *c.* +1130]. Nelson, London, 1961.

DOHI KEIZO (1). 'Medicine in Ancient Japan; A Study of Some Drugs preserved in the Imperial Treasure House at Nara.' In *Zoku Shōsōin Shiron*, 1932, no. 15, Neiyaku. 1st pagination, p. 113.

DONDAINE, A. (1). 'La Hierarchie Cathare en Italie.' *AFP*, 1950, **20**, 234.

DOOLITTLE, J. (1). *A Vocabulary and Handbook of the Chinese Language*. 2 vols. Rozario & Marcal, Fuchow, 1872.

DORESSE, J. (1). *Les Livres Secrets des Gnostiques d'Égypte*. Plon, Paris, 1958–.
   Vol. 1. *Introduction aux Écrits Gnostiques Coptes découverts à Khénoboskion*.
   Vol. 2. '*L'Évangile selon Thomas', ou 'Les Paroles Secrètes de Jésus*'.
   Vol. 3. '*Le Livre Secret de Jean'; 'l'Hypostase des Archontes' ou 'Livre de Nōréa*'.
   Vol. 4. '*Le Livre Sacré du Grand Esprit Invisible' ou 'Évangile des Égyptiens'; 'l'Épître d'Eugnoste le Bienheureux'; 'La Sagesse de Jésus*'.
   Vol. 5 '*L'Évangile selon Philippe*.'

DORFMAN, R. I. & SHIPLEY, R. A. (1). *Androgens; their Biochemistry, Physiology and Clinical Significance*. Wiley, New York and Chapman & Hall, London, 1956.

DOUGLAS, R. K. (2). *Chinese Stories*. Blackwood, Edinburgh and London, 1883. (Collection of translations previously published in *Blackwood's Magazine*.)

DOUTHWAITE, A. W. (1). 'Analyses of Chinese Inorganic Drugs.' *CMJ*, 1890, **3**, 53.

DOZY, R. P. A. & ENGELMANN, W. H. (1). *Glossaire des Mots Espagnols et Portugais dérivés de l'Arabe*. 2nd ed. Brill, Leiden, 1869.

DOZY, R. P. A. & DE GOEJE, M. J. (2). *Nouveaux Documents pour l'Étude de la Religion des Ḥarrāniens*. Actes du 6e Congr. Internat. des Orientalistes, Leiden, 1883. 1885, vol. 2, pp. 281ff., 341 ff.

DRAKE, N. F. (1). 'The Coal Fields of North-East China.' *TAIME*, 1901, **31**, 492, 1008.

DRAKE, N. F. (2). 'The Coal Fields around Tsê-Chou, Shansi.' *TAIME*, 1900, **30**, 261.

DRONKE, P. (1). 'L'Amor che Move il Sole e l'Altre Stelle.' *STM*, 1965 (3ᵃ ser.), **6**, 389.

DRONKE, P. (2). 'New Approaches to the School of Chartres.' *AEM*, 1969, **6**, 117.

DRUCE, G. C. (1). 'The Ant-Lion.' *ANTJ*, 1923, **3**, 347.

DU, Y., JIANG, R. & TSOU, C. (1). See Tu Yü-Tshang, Chiang Jung-Chhing & Tsou Chhêng-Lu (1).

DUBLER, C. E. (1). *La 'Materia Medica' de Dioscorides; Transmission Medieval y Renacentista*. 5 vols. Barcelona, 1955.

DUBS, H. H. (4). 'An Ancient Chinese Stock of Gold [Wang Mang's Treasury].' *JEH*, 1942, **2**, 36.

DUBS, H. H. (5). 'The Beginnings of Alchemy.' *ISIS*, 1947, **38**, 62.

DUBS, H. H. (34). 'The Origin of Alchemy.' *AX*, 1961, **9**, 23. Crit. abstr. J. Needham, *RBS*, 1968, **7**, no. 755.

DUCKWORTH, C. W. (1). 'The Discovery of Oxygen.' *CN*, 1886, **53**, 250.

DUDGEON, J. (1). 'Kung-Fu, or Medical Gymnastics.' *JPOS*, 1895, **3** (no. 4), 341–565.

DUDGEON, J. (2). 'The Beverages of the Chinese' (on tea and wine). *JPOS*, 1895, **3**, 275.

DUDGEON, J. (4). '[Glossary of Chinese] Photographic Terms', in Doolittle (1), vol. 2, p. 518.

DUHR, J. (1). *Un Jésuite en Chine, Adam Schall*. Desclée de Brouwer, Paris, 1936. Engl. adaptation by R. Attwater, *Adam Schall, a Jesuit at the Court of China, 1592 to 1666*. Geoffrey Chapman, London, 1963. Not very reliable sinologically.

DUNCAN, A. M. (1). 'The Functions of Affinity Tables and Lavoisier's List of Elements.' *AX*, 1970, **17**, 28.

DUNCAN, A. M. (2). 'Some Theoretical Aspects of Eighteenth-Century Tables of Affinity.' *ANS*, 1962, **18**, 177, 217.

DUNCAN, E. H. (1). 'Jonson's "Alchemist" and the Literature of Alchemy.' *PMLA*, 1946, **61**, 699.

DUNLOP, D. M. (5). 'Sources of Silver and Gold in Islam according to al-Hamdānī (+10th century).' *SI*, 1957, **8**, 29.

DUNLOP, D. M. (6). *Arab Civilisation to A.D. 1500*. Longman, London and Librairie du Liban, Beirut, 1971.

DUNLOP, D. M. (7). *Arabic Science in the West*. Pakistan Historical Soc., Karachi, 1966. (Pakistan Historical Society Pubs. no. 35.)

DUNLOP, D. M. (8). 'Theodoretus-Adhrīṭūs.' Communication to the 26th International Congress of Orientalists, New Delhi, 1964. Summaries of papers, p. 328.

DÜNTZER, H. (1). *Life of Goethe*. 2 vols. Macmillan, London, 1883.

DÜRING, H. I. (1). 'Aristotle's Chemical Treatise, *Meteorologica* Bk. IV, with Introduction and Commentary.' *GHA*, 1944 (no. 2), 1–112. Sep. pub., Elander, Goteborg, 1944.

DURRANT, P. J. (1). *General and Inorganic Chemistry*. 2nd ed. repr. Longmans Green, London, 1956.

DUVEEN, D. I. & WILLEMART, A. (1). 'Some +17th-Century Chemists and Alchemists of Lorraine.' *CHYM*, 1949, **2**, 111.

DUYVENDAK, J. J. L. (18) (tr.). '*Tao Tê Ching*', the Book of the Way and its Virtue. Murray, London, 1954 (Wisdom of the East Series). Crit. revs. P. Demiéville, *TP*, 1954, **43**, 95; D. Bodde, *JAOS*, 1954, **74**, 211.

DUYVENDAK, J. J. L. (20). 'A Chinese *Divina Commedia*.' *TP*, 1952, **41**, 255. (Also sep. pub. Brill, Leiden, 1952.)

DYSON, G. M. (1). 'Antimony in Pharmacy and Chemistry; I, History and Occurrence of the Element; II, The Metal and its Inorganic Compounds; III, The Organic Antimony Compounds in Therapy. *PJ*, 1928, **121** (4th ser. **67**), 397, 520.

EBELING, E. (1). 'Mittelassyrische Rezepte zur Bereitung (Herstellung) von wohlriechenden Salben.' *ORR*, 1948 (n.s.) **17**, 129, 299; 1949, **18**, 404; 1950, **19**, 265.

ECKERMANN, J. P. (1). *Gespräche mit Goethe*. 3 vols. Vols. 1 and 2, Leipzig, 1836. Vol. 3, Magdeburg, 1848. Eng. tr. 2 vols. by J. Oxenford, London, 1850. Abridged ed. *Conversations of Goethe with Eckermann*, Dent, London, 1930. Ed. J. K. Moorhead, with introduction by Havelock Ellis.

EDKINS, J. (17). 'Phases in the Development of Taoism.' *JRAS/NCB*, 1855 (1st ser.), **5**, 83.

EDKINS, J. (18). 'Distillation in China.' *CR*, 1877, **6**, 211.

EFRON, D. H., HOLMSTEDT, BO & KLINE, N. S. (1) (ed.). *The Ethno-pharmacological Search for Psycho-active Drugs*. Washington, D.C. 1967. (Public Health Service Pub. no. 1645.) Proceedings of a Symposium, San Francisco, 1967.

EGERTON, F. N. (1). 'The Longevity of the Patriarchs; a Topic in the History of Demography.' *JHI*, 1966, **27**, 575.

EGGELING, J. (1) (tr.). *The 'Satapatha-brāhmaṇa' according to the Text of the Mādhyandina School*. 5 vols. Oxford, 1882–1900 (*SBE*, nos. 12, 26, 41, 43, 44). Vol. 1 repr. Motilal Banarsidass, Delhi, 1963.

EGLOFF, G. & LOWRY, C. D. (1). 'Distillation as an Alchemical Art.' *JCE*, 1930, **7**, 2063.

EICHHOLZ, D. E. (1). 'Aristotle's Theory of the Formation of Metals and Minerals.' *CQ*, 1949, **43**, 141.

EICHHOLZ, D. E. (2). *Theophrastus 'De Lapidibus'*. Oxford, 1964.

EICHHORN, W. (6). 'Bemerkung z. Einführung des Zolibats für Taoisten.' *RSO*, 1955, **30**, 297.

EICHHORN, W. (11) (tr.). The *Fei-Yen Wai Chuan*, with some notes on the *Fei-Yen Pieh Chuan*. Art. in *Eduard Erkes in Memoriam 1891–1958*, ed. J. Schubert. Leipzig, 1962. Abstr. *TP*, 1963, **50**, 285.

EISLER, R. (4). 'l'Origine Babylonienne de l'Alchimie; à propos de la Découverte Récente de Récettes Chimiques sur Tablettes Cunéiformes.' *RSH*, 1926, **41**, 5. Also *CHZ*, 1926 (nos. 83 and 86); *ZASS*, 1926, 1.

ELIADE, MIRCEA (1). *Le Mythe de l'Eternel Retour; Archétypes et Répétition*. Gallimard, Paris, 1949. Eng. tr. by W. R. Trask, *The Myth of the Eternal Return*. Routledge & Kegan Paul, London, 1955.

ELIADE, MIRCEA (4). 'Metallurgy, Magic and Alchemy.' *Z*, 1938, **1**, 85.

ELIADE, MIRCEA (5). *Forgerons et Alchimistes*. Flammarion, Paris, 1956. Eng.tr. S. Corrin, *The Forge and the Crucible*. Harper, New York, 1962. Rev. G. H[eym], *AX*, 1957, **6**, 109.

ELIADE, MIRCEA (6). *Le Yoga, Immortalité et Liberté*. Payot, Paris, 1954. Eng. tr. by W. R. Trask. Pantheon, New York, 1958.

ELIADE, MIRCEA (7). *Imgaes and Symbols; Studies in Religious Symbolism*. Tr. from the French (Gallimard, Paris, 1952) by P. Mairet. Harvill, London, 1961.

ELIADE, M. (8). 'The Forge and the Crucible: a Postscript.' *HOR*, 1968, **8**, 74–88.

ELLINGER, T. U. H. (1). *Hippocrates on Intercourse and Pregnancy; an English Translation of 'On Semen' and 'On the Development of the Child'*. With introd. and notes by A. F. Guttmacher. Schuman, New York, 1952.

ELLIS, G. W. (1). 'A Vacuum Distillation Apparatus.' *CHIND*, 1934, **12**, 77 (*JSCI*, **53**).

ELLIS, W. (1). *History of Madagascar*. 2 vols. Fisher, London and Paris, 1838.

VON ENGESTRÖM, G. (1). 'Pak-fong, a White Chinese Metal' (in Swedish). *KSVA/H*, 1776, **37**, 35.

VON ENGESTRÖM, G. (2). 'Försök på Fôrnt omtalle Salt eller *Kien* [*chien*].' *KSVA/H*, 1772, **33**, 172. Cf. Abrahamsohn (1).

D'ENTRECOLLES, F. X. (1). *Lettre au Père Duhalde* (on alchemy and various Chinese discoveries in the arts and sciences, porcelain, artificial pearls and magnetic phenomena) dated 4 Nov. 1734. *LEC*, 1781, vol. 22, pp. 91 ff.

D'ENTRECOLLES, F. X. (2). *Lettre au Père Duhalde* (on botanical subjects, fruits and trees, including the persimmon and the lichi; on medicinal preparations isolated from human urine; on the use of the magnet in medicine; on the feathery substance of willow seeds; on camphor and its sublimation; and on remedies for night-blindness) dated 8 Oct. 1736. *LEC*, 1781, vol. 22, pp. 193 ff.

ST EPHRAIM OF SYRIA [d. +373]. *Discourses to Hypatius* [against the Theology of Mani, Marcion and Bardaisan]. See Mitchell, C. W. (1).

EPHRAIM, F. (1). *A Textbook of Inorganic Chemistry*. Eng. tr. P. C. L. Thorne. Gurney & Jackson, London, 1926.

ERCKER, L. (1). *Beschreibung Allefürnemsten Mineralischem Ertzt und Berckwercks Arten...* Prague, 1574. 2nd ed. Frankfurt, 1580. Eng. tr. by Sir John Pettus, as *Fleta Minor, or, the Laws of Art and Nature, in Knowing, Judging, Assaying, Fining, Refining and Inlarging the Bodies of confin'd Metals...* Dawks, London, 1683. See Sisco & Smith (1); Partington (7), vol. 2, pp. 104 ff.

ERKES, E. (1) (tr.). 'Das Weltbild d. *Huai Nan Tzu*.' (Transl. of ch. 4.) *OAZ*, 1918, **5**, 27.

ERMAN, A. & GRAPOW, H. (1). *Wörterbuch d. Aegyptische Sprache*. 7 vols. (With *Belegstellen*, 5 vols. as supplement.) Hinrichs, Leipzig, 1926–.

ERMAN, A. & GRAPOW, H. (2). *Aegyptisches Handwörterbuch*. Reuther & Reichard, Berlin, 1921.

ERMAN, A. & RANKE, H. (1). *Aegypten und aegyptisches Leben in Altertum*. Tübingen, 1923.

ESSIG, E. O. (1). *A College Entomology*. Macmillan, New York, 1942.

ESTIENNE, H. (1) (Henricus Stephanus). *Thesaurus Graecae Linguae*. Geneva, 1572; re-ed. Hase, de Sinner & Fix, 8 vols. Didot, Paris, 1831–65.

ETHÉ, H. (1) (tr.). *Zakarīya ibn Muḥ. ibn Maḥmūd al-Qazwīnī's Kosmographie; Die Wunder der Schöpfung* [*c.* +1275]. Fues (Reisland), Leipzig, 1868. With notes by H. L. Fleischer. Part I only; no more published.

EUGSTER, C. H. (1). 'Brève Revue d'Ensemble sur la Chimie de la Muscarine.' *RMY*, 1959, **24**, 1.

EUONYMUS PHILIATER. See Gesner, Conrad.

EVOLA, J. (G. C. E.) (1). *La Tradizione Ermetica*. Bari, 1931. 2nd ed. 1948.

EVOLA, J. (G. C. E.) (2). *Lo Yoga della Potenza, saggio sui Tantra*. Bocca, Milan, 1949. Orig. pub. as *l'Uomo come Potenza*.

EVOLA, J. (G. C. E.) (3). *Metafisica del Sesso*. Atanòr, Rome, 1958.

EWING, A. H. (1). *The Hindu Conception of the Functions of Breath; a Study in early Indian Psychophysics*. Inaug. Diss. Johns Hopkins University. Baltimore, 1901; and in *JAOS*, 1901, **22** (no. 2).

FABRE, M. (1). *Pékin, ses Palais, ses Temples, et ses Environs*. Librairie Française, Tientsin, 1937.

FABRICIUS, J. A. (1). *Bibliotheca Graeca...* Edition of G. C. Harles, 12 vols. Bohn, Hamburg, 1808.

FABRICIUS, J. A. (2). *Codex Pseudepigraphicus Veteris Testamenti, Collectus, Castigatus, Testimoniisque Censuris et Animadversionibus Illustratus*. 3 vols. Felginer & Bohn, Hamburg, 1722–41.

FABRICIUS, J. A. (3). *Codex Apocryphus Novi Testamenti, Collectus, Castigatus, Testimoniisque Censuris et Animadversionibus Illustratus*. 3 vols. in 4. Schiller & Kisner, Hamburg, 1703–19.

FARABEE, W. C. (1). 'A Golden Hoard from Ecuador.' *MUJ*, 1912.

FARABEE, W. C. (2). 'The Use of Metals in Prehistoric America.' *MUJ*, 1921.

FARNWORTH, M., SMITH, C. S. & RODDA, J. L. (1). 'Metallographic Examination of a Sample of Metallic Zinc from Ancient Athens.' *HE*, 1949, **8** (Suppl.) 126. (Commemorative Studies in Honour of Theodore Leslie Shear.)

FEDCHINA, V. N. (1). 'The +13th-century Chinese Traveller, [Chhiu] Chhang-Chhun' (in Russian), in *Iz Istorii Nauki i Tekhniki Kitaya* (Essays in the History of Science and Technology in China), p. 172. Acad. Sci. Moscow, 1955.

FEHL, N. E. (1). 'Notes on the Lü Hsing [chapter of the *Shu Ching*]; proposing a Documentary Theory.' *CCJ*, 1969, **9** (no. 1), 10.

FEIFEL, E. (1) (tr.). '*Pao Phu Tzu (Nei Phien)*, chs. 1–3.' *MS*, 1941, **6**, 113.

FEIFEL, E. (2) (tr.). '*Pao Phu Tzu (Nei Phien)*, ch. 4.' *MS*, 1944, **9**, 1.

FEIFEL, E. (3) (tr.). '*Pao Phu Tzu (Nei Phien)*, ch. 11, Translated and Annotated.' *MS*, 1946, **11** (no. 1), 1.

FEISENBERGER, H. A. (1). 'The [Personal] Libraries of Newton, Hooke and Boyle.' *NRRS*, 1966, **21**, 42.

FÊNG CHIA-LO & COLLIER, H. B. (1). 'A Sung-Dynasty Alchemical Treatise; the "Outline of Alchemical Preparations" [*Tan Fang Chien Yuan*], by Tuku Thao [+10th cent.].' *JWCBRS*, 1937, **9**, 199.

FÊNG HAN-CHI (H. Y. Fêng) & SHRYOCK, J. K. (2). 'The Black Magic in China known as *Ku*.' *JAOS*, 1935, **65**, 1. Sep. pub. Amer. Oriental Soc. Offprint Ser. no. 5.

FERCHL, F. & SÜSSENGUTH, A. (1). *A Pictorial History of Chemistry*. Heinemann, London, 1939.

FERDY, H. (1). *Zur Verhütung der Conception*. 1900.

FERGUSON, JOHN (1). *Bibliotheca Chemica; a Catalogue of the Alchemical, Chemical and Pharmaceutical Books in the Collection of the late James Young of Kelly and Durris...* 2 vols. Maclehose, Glasgow, 1906.

FERGUSON, JOHN (2). *Bibliographical Notes on Histories of Inventions and Books of Secrets*. 2 vols. Glasgow, 1898; repr. Holland Press, London, 1959. (Papers collected from *TGAS*.)

FERGUSON, JOHN (3). 'The "Marrow of Alchemy" [1654-5].' *JALCHS*, 1915, **3**, 106.

FERRAND, G. (1). *Relations de Voyages et Textes Géographiques Arabes, Persans et Turcs relatifs à l'Extrême Orient, du 8ᵉ au 18ᵉ siècles, traduits, revus et annotés etc.* 2 vols. Leroux, Paris, 1913.

FERRAND, G. (2) (tr.). *Voyage du marchand Sulaymān en Inde et en Chine redigé en +851; suivi de remarques par Abū Zayd Ḥaṣan (vers +916)*. Bossard, Paris, 1922.

FESTER, G. (1). *Die Entwicklung der chemischen Technik, bis zu den Anfängen der Grossindustrie*. Berlin, 1923. Repr. Sändig, Wiesbaden, 1969.

FESTUGIÈRE, A. J. (1). *La Révélation d'Hermès Trismégiste*, I. *L'Astrologie et les Sciences Occultes*. Gabalda, Paris, 1944. Rev. J. Filliozat, *JA*, 1944, **234**, 349.

FESTUGIÈRE, A. J. (2). 'L'Hermétisme.' *KHVL*, 1948, no. 1, 1-58.

FIERZ-DAVID, H. E. (1). *Die Entwicklungsgeschichte der Chemie*. Birkhauser, Basel, 1945. (Wissenschaft und Kultur ser., no. 2.) Crit. E. J. Holmyard, *N*, 1946, **158**, 643.

FIESER, L. F. & FIESER, M. (1). *Organic Chemistry*. Reinhold, New York; Chapman & Hall, London, 1956.

FIGUIER, L. (1). *l'Alchimie et les Alchimistes; ou, Essai Historique et Critique sur la Philosophie Hermétique*. Lecou, Paris, 1854. 2nd ed. Hachette, Paris, 1856. 3rd ed. 1860.

FIGUROVSKY, N. A. (1). 'Chemistry in Ancient China, and its Influence on the Progress of Chemical Knowledge in other Countries' (in Russian). Art. in *Iz Istorii Nauki i Tekhniki Kitaya*. Moscow, 1955, p. 110.

FILLIOZAT, J. (1). *La Doctrine Classique de la Médécine Indienne*. Imp. Nat., CNRS and Geuthner, Paris, 1949.

FILLIOZAT, J. (2). 'Les Origines d'une Technique Mystique Indienne.' *RP*, 1946, **136**, 208.

FILLIOZAT, J. (3). 'Taoisme et Yoga.' *DVN*, 1949, **3**, 1.

FILLIOZAT, J. (5). Review of Festugière (1). *JA*, 1944, **234**, 349.

FILLIOZAT, J. (6). 'La Doctrine des Brahmanes d'après St Hippolyte.' *JA*, 1945, **234**, 451; *RHR/AMG*, 1945, **128**, 59.

FILLIOZAT, J. (7). 'L'Inde et les Échanges Scientifiques dans l'Antiquité.' *JWH*, 1953, **1**, 353.

FILLIOZAT, J. (10). 'Al-Bīrūnī et l'Alchimie Indienne.' Art. in *Al-Bīrūnī Commemoration Volume*. Iran Society, Calcutta, 1958, p. 101.

FILLIOZAT, J. (11). Review of P. C. Ray (1) revised edition. *ISIS*, 1958, **49**, 362.

FILLIOZAT, J. (13). 'Les Limites des Pouvoirs Humains dans l'Inde.' Art. in *Les Limites de l'Humain*. Études Carmelitaines, Paris, 1953, p. 23.

FISCHER, OTTO (1). *Die Kunst Indiens, Chinas und Japans*. Propylaea, Berlin, 1928.

FISCHGOLD, H. & GASTAUT, H. (1) (ed.). *Conditionnement et Reactivité en Électro-encephalographie*. Masson, Paris, 1957. For the Fédération Internationale d'Électro-encéphalographie et de Neurophysiologie Clinique, Report of 5th Colloquium, Marseilles, 1955 (*Electro-encephalography and Clinical Neurophysiology*, Supplement no. 6).

FLIGHT, W. (1). 'On the Chemical Compositon of a Bactrian Coin.' *NC*, 1868 (n.s.), **8**, 305.

FLIGHT, W. (2). 'Contributions to our Knowledge of the Composition of Alloys and Metal-Work, for the most part Ancient.' *JCS*, 1882, **41**, 134.

FLORKIN, M. (1). *A History of Biochemistry. Pt. I, Proto-Biochemistry; Pt. II, From Proto-Biochemistry to Biochemistry*. Vol. 30 of *Comprehensive Biochemistry*, ed. M. Florkin & E. H. Stotz. Elsevier, Amsterdam, London and New York, 1972.

FLUDD, ROBERT (3). *Tractatus Theologo-Philosophicus, in Libros Tres distributus; quorum I, De Vita, II, De Morte, III, De Resurrectione; Cui inseruntur nonnulla Sapientiae Veteris...Fragmenta;...collecta Fratribusq a Cruce Rosea dictis dedicata à Rudolfo Otreb Brittano*. Oppenheim, 1617.

FLÜGEL, G. (1) (ed. & tr.). *The 'Fihrist al-'Ulūm' (Index of the Sciences)* [by Abū'l-Faraj ibn abū-Ya'qūb al-Nadīm]. 2 vols. Leipzig, 1871-2.

FLÜGEL, G. (2) (tr.). *Lexicon Bibliographicum et Encyclopaedicum, a Mustafa ben Abdallah Katib Jelebi dicto et nomine Haji Khalfa celebrato compositum...*(the *Kashf al-Ẓunūn* (Discovery of the Thoughts of Muṣṭafā ibn 'Abdallāh Haji Khalfa, or Ḥajji Khalīfa, +17th-century Turkish (bibliographer)). 7 vols. Bentley (for the Or. Tr. Fund Gt. Br. & Ireland), London and Leipzig, 1835-58.

FOHNAHN, A. (1). 'New Chemical Terminology in Chinese.' *JAN*, 1927, **31**, 395.

FOLEY, M. G. (1) (tr.). *Luigi Galvani: 'Commentary on the Effects of Electricity on Muscular Motion'*, translated into English...[from *De Viribus Electricitatis in Motu Musculari Commentarius*, Bologna, +1791]; *with Notes and a Critical Introduction by I. B. Cohen, together with a Facsimile...and a Bibliography of the Editions and Translations of Galvani's Book prepared by J. F. Fulton & M. E. Stanton.* Burndy Library, Norwalk, Conn. U.S.A. 1954. (Burndy Library Publications, no.10.)

FORBES, R. J. (3). *Metallurgy in Antiquity; a Notebook for Archaeologists and Technologists.* Brill, Leiden, 1950 (in press since 1942). Rev. V. G. Childe, *A/AIHS*, 1951, **4**, 829.

[FORBES, R. J.] (4a). *Histoire des Bitumes, des Époques les plus Reculées jusqu'à l'an 1800.* Shell, Leiden, n.d.

FORBES, R. J. (4b). *Bitumen and Petroleum in Antiquity.* Brill, Leiden, 1936.

FORBES, R. J. (7). 'Extracting, Smelting and Alloying [in Early Times before the Fall of the Ancient Empires].' Art. in A *History of Technology*, ed. C. Singer, E. J. Holmyard & A. R. Hall. vol. 1, p.572. Oxford, 1954.

FORBES, R. J. (8). 'Metallurgy [in the Mediterranean Civilisations and the Middle Ages].' In *A History of Technology*, ed. C. Singer *et al.* vol. 2, p. 41. Oxford, 1956.

FORBES, R. J. (9). *A Short History of the Art of Distillation.* Brill, Leiden, 1948.

FORBES, R. J. (10). *Studies in Ancient Technology.* Vol. 1, *Bitumen and Petroleum in Antiquity; The Origin of Alchemy; Water Supply.* Brill, Leiden, 1955. (Crit. Lynn White, *ISIS*, 1957, **48**, 77.)

FORBES, R. J. (16). 'Chemical, Culinary and Cosmetic Arts' [in early times to the Fall of the Ancient Empires]. Art. in *A History of Technology*, ed. C. Singer *et al.* Vol. 1, p. 238. Oxford, 1954.

FORBES, R. J. (20). *Studies in Early Petroleum History.* Brill, Leiden, 1958.

FORBES, R. J. (21). *More Studies in Early Petroleum History.* Brill, Leiden, 1959.

FORBES, R. J. (26). 'Was Newton an Alchemist?' *CHYM*, 1949, **2**, 27.

FORBES, R. J. (27). *Studies in Ancient Technology.* Vol. 7, *Ancient Geology; Ancient Mining and Quarrying; Ancient Mining Techniques.* Brill, Leiden, 1963.

FORBES, R. J. (28). *Studies in Ancient Technology.* Vol. 8, *Synopsis of Early Metallurgy; Physico-Chemical Archaeological Techniques; Tools and Methods; Evolution of the Smith (Social and Sacred Status); Gold; Silver and Lead; Zinc and Brass.* Brill, Leiden, 1964. A revised version of Forbes (3).

FORBES, R. J. (29). *Studies in Ancient Technology.* Vol. 9, *Copper; Tin; Bronze; Antimony; Arsenic; Early Story of Iron.* Brill, Leiden, 1964. A revised version of Forbes (3).

FORBES, R. J. (30). *La Destillation à travers les Ages.* Soc. Belge pour l'Étude du Pétrole, Brussels, 1947.

FORBES, R. J. (31). 'On the Origin of Alchemy.' *CHYM*, 1953, **4**, 1.

FORBES, R. J. (32). Art. 'Chemie' in *Real-Lexikon f. Antike und Christentum*, ed. T. Klauser, 1950–3, vol. 2, p. 1061.

FORBES, T. R. (1). 'A[rnold] A[dolf] Berthold [1803–61] and the First Endocrine Experiment; some Speculations as to its Origin.' *BIHM*, 1949, **23**, 263.

FORKE, A. (3) (tr.). *Me Ti [Mo Ti] des Sozialethikers und seiner Schüler philosophische Werke.* Berlin, 1922. (*MSOS*, Beibände, **23–25**).

FORKE, A. (4) (tr.). '*Lun-Hêng*', *Philosophical Essays of Wang Chhung.* Vol. 1, 1907. Kelly & Walsh, Shanghai; Luzac, London; Harrassowitz, Leipzig. Vol. 2, 1911 (with the addition of Reimer, Berlin). (*MSOS*, Beibände, **10** and **14**.) Photolitho Re-issue, Paragon, New York, 1962. Crit. P. Pelliot, *JA*, 1912 (10e sér.), **20**, 156.

FORKE, A. (9). *Geschichte d. neueren chinesischen Philosophie* (i.e. from the beginning of the Sung to modern times). De Gruyter, Hamburg, 1938. (Hansische Univ. Abhdl. a. d. Geb. d. Auslandskunde, no. 46 (ser. B, no. 25).)

FORKE, A. (12). *Geschichte d. mittelälterlichen chinesischen Philosophie* (i.e. from the beginning of the Former Han to the end of the Wu Tai). De Gruyter, Hamburg, 1934. (Hamburg. Univ. Abhdl. a. d. Geb. d. Auslandskunde, no. 41 (ser. B, no. 21).)

FORKE, A. (13). *Geschichte d. alten chinesischen Philosophie* (i.e. from antiquity to the beginning of the Former Han). De Gruyter, Hamburg, 1927. (Hamburg. Univ. Abhdl. a. d. Geb. d. Auslandskunde, no. 25 (ser. B, no. 14).)

FORKE, A. (15). 'On Some Implements mentioned by Wang Chhung' (1. Fans, 2. Chopsticks, 3. Burning Glasses and Moon Mirrors). Appendix III to Forke (4).

FORKE, A. (20). 'Ko Hung der Philosoph und Alchymist.' *AGP*, 1932, **41**, 115. Largely incorporated in (12), pp. 204 ff.

FÖRSTER, E. (1). *Roger Bacon's 'De Retardandis Senectutis Accidentibus et de Sensibus Conservandis' und Arnald von Villanova's 'De Conservanda Juventutis et Retardanda Senectute'.* Inaug. Diss. Leipzig, 1924.

FOWLER, A. M. (1). 'A Note on ἄμβροτος.' *CP*, 1942, **37**, 77.

FRÄNGER, W. (1). *The Millennium of Hieronymus Bosch.* Faber, London, 1952.

FRANCKE, A. H. (1). 'Two Ant Stories from the Territory of the Ancient Kingdom of Western Tibet; a Contribution to the Question of Gold-Digging Ants.' *AM*, 1924, **1**, 67.

FRANK, B. (1). '*Kata-imi* et *Kata-tagae;* Étude sur les Interdits de Direction à l'Époque Heian.' *BMFJ*, 1958 (n.s.), **5** (no. 2–4), 1–246.

FRANKE, H. (17). 'Das chinesische Wort für "Mumie" [mummy].' *OR*, 1957, **10**, 253.

FRANKE, H. (18). 'Some Sinological Remarks on Rashīd al-Dīn's "History of China".' *OR*, 1951, **4**, 21.

FRANKE, W. (4). *An Introduction to the Sources of Ming History.* Univ. Malaya Press, Kuala Lumpur and Singapore, 1968.

FRANKFORT, H. (4). *Ancient Egyptian Religion; an Interpretation.* Harper & Row, New York, 1948. Paperback ed. 1961.

FRANTZ, A. (1). 'Zink und Messing im Alterthum.' *BHMZ*, 1881, **40**, 231, 251, 337, 377, 387.

FRASER, SIR J. G. (1). *The Golden Bough.* 3-vol. ed. Macmillan, London, 1900; superseded by 12-vol. ed. (here used), Macmillan, London, 1913–20. Abridged 1-vol. ed. Macmillan, London, 1923.

FRENCH, J. (1). *Art of Distillation.* 4th ed. London, 1667.

FRENCH, P. J. (1). *John Dee; the World of an Elizabethan Magus.* Routledge & Kegan Paul, London, 1971.

FREUDENBERG, K., FRIEDRICH, K. & BUMANN, I. (1). 'Über Cellulose und Stärke [incl. description of a molecular still].' *LA*, 1932, **494**, 41 (57).

FREUND, IDA (1). *The Study of Chemical Composition; an Account of its Method and Historical Development, with illustrative quotations.* Cambridge, 1904. Repr. Dover, New York, 1968, with a foreword by L. E. Strong and a brief biography by O. T. Benfey.

FRIEDERICHSEN, M. (1). 'Morphologie des Tien-schan [Thien Shan].' *ZGEB*, 1899, **34**, 1–62, 193–271. Sep. pub. Pormetter, Berlin, 1900.

FRIEDLÄNDER, P. (1). 'Über den Farbstoff des antiken Purpurs aus *Murex brandaris.*' *BDCG*, 1909, **42**, pt. 1, 765.

FRIEND, J. NEWTON (1). *Iron in Antiquity.* Griffin, London, 1926.

FRIEND, J. NEWTON (2). *Man and the Chemical Elements.* London, 1927.

FRIEND, J. NEWTON & THORNEYCROFT, W. E. (1). 'The Silver Content of Specimens of Ancient and Mediaeval Lead.' *JIM*, 1929, **41**, 105.

FRITZE, M. (1) (tr.). *Pancatantra.* Leipzig, 1884.

FRODSHAM, J. D. (1) (tr.). *The Poems of Li Ho* (+791 to +817). Oxford, 1970.

FROST, D. V. (1). 'Arsenicals in Biology; Retrospect and Prospect.' *FP*, 1967, **26** (no. 1), 194.

FRYER, J. (1). *An Account of the Department for the Translation of Foreign Books of the Kiangnan Arsenal.* NCH, 28 Jan. 1880, and offprinted.

FRYER, J. (2). 'Scientific Terminology; Present Discrepancies and Means of Securing Uniformity.' *CRR*, 1872, **4**, 26, and sep. pub.

FRYER, J. (3). *The Translator's Vade-Mecum.* Shanghai, 1888.

FRYER, J. (4). 'Western Knowledge and the Chinese.' *JRAS/NCB*, 1886, **21**, 9.

FRYER, J. (5). 'Our Relations with the Reform Movement.' Unpublished essay, 1909. See Bennett (1), p. 151.

FUCHS, K. W. C. (1). *Die vulkanische Erscheinungen der Erde.* Winter, Leipzig and Heidelberg, 1865.

FUCHS, W. (7). 'Ein Gesandschaftsbericht ü. Fu-Lin in chinesischer Wiedergabe aus den Jahren +1314 bis +1320.' *OE*, 1959, **6**, 123.

FÜCK, J. W. (1). 'The Arabic Literature on Alchemy according to al-Nadīm (+987); a Translation of the Tenth Discourse of the Book of the Catalogue (*al-Fihrist*), with Introduction and Commentary.' *AX*, 1951, **4**, 81.

DE LA FUENTE, J. (1). *Yalalag; una Villa Zapoteca Serrana.* Museo Nac. de Antropol. Mexico City, 1949. (Ser. Cientifica, no. 1.)

FYFE, A. (1). 'An Analysis of Tutenag or the White Copper of China.' *EPJ*, 1822, **7**, 69.

GADD, C. J. (1). 'The Ḥarrān Inscriptions of Nabonidus [of Babylon, −555 to −539].' *ANATS*, 1958, **8**, 35.

GADOLIN, J. (1). *Observationes de Cupro Albo Chinensium Pe-Tong vel Pack-Tong. NARSU*, 1827, **9** 137.

GALLAGHER, L. J. (1) (tr.). *China in the 16th Century; the Journals of Matthew Ricci, 1583–1610.* Random House, New York, 1953. (A complete translation, preceded by inadequate bibliographical details, of Nicholas Trigault's *De Christiana Expeditione apud Sinas* (1615). Based on an earlier publication: *The China that Was; China as discovered by the Jesuits at the close of the 16th Century: from the Latin of Nicholas Trigault.* Milwaukee, 1942.) Identifications of Chinese names in Yang Lien-Shêng (4). Crit. J. R. Ware, *ISIS*, 1954, **45**, 395.

GANZENMÜLLER, W. (1). *Beiträge zur Geschichte der Technologie und der Alchemie.* Verlag Chemie, Weinheim, 1956. Rev. W. Pagel, *ISIS*, 1958, **49**, 84.

GANZENMÜLLER, W. (2). *Die Alchemie im Mittelalter.* Bonifacius, Paderborn, 1938. Repr. Olms, Hildesheim, 1967. French, tr. by Petit-Dutailles, Paris, n.d. (c. 1940).

GANZENMÜLLER, W. (3). '*Liber Florum Geberti;* alchemistischen Öfen und Geräte in einer Handschrift des 15. Jahrhunderts.' *QSGNM*, 1942, **8**, 273. Repr. in (1), p. 272.

GANZENMÜLLER, W. (4). 'Zukunftsaufgaben der Geschichte der Alchemie.' *CHYM*, 1953, **4**, 31.

GANZENMÜLLER, W. (5). 'Paracelsus und die Alchemie des Mittelalters.' *ZAC/AC*, 1941, **54**, 427.

VON GARBE, R. K. (3) (tr.). *Die Indischen Mineralien, ihre Namen und die ihnen zugeschriebenen Kräfte; Narahari's 'Rāja-nighaṇṭu' [King of Dictionaries], varga XIII, Sanskrit und Deutsch, mit kritischen und erläuternden Anmerkungen herausgegeben...* Hirzel, Leipzig, 1882.

GARBERS, K. (1) (tr.). *'Kitāb Kimiya al-Itr wa'l-Tas'idat'; Buch über die Chemie des Parfüms und die Destillationen von Ya'qub ibn Ishaq al-Kindī; ein Beitrag zur Geschichte der arabischen Parfümchemie und Drogenkunde aus dem 9tr Jahrh. A.D., übersetzt...* Brockhaus, Leipzig, 1948. (Abhdl. f.d. Kunde des Morgenlandes, no. 30.) Rev. A. Mazaheri, *A/AIHS*, 1951, **4** (no. 15), 521.

GARNER, SIR HARRY (1). 'The Composition of Chinese Bronzes.' *ORA*, 1960, **6** (no. 4), 3.

GARNER, SIR HARRY (2). *Chinese and Japanese Cloisonné Enamels.* Faber & Faber, London, 1962.

GARNER, SIR HARRY (3). 'The Origins of "Famille Rose" [polychrome decoration of Chinese Porcelain].' *TOCS*, 1969.

GEBER (ps. of a Latin alchemist *c.* +1290). *The Works of Geber, the most famous Arabian Prince and Philosopher, faithfully Englished by R. R., a Lover of Chymistry [Richard Russell].* James, London, 1678. Repr. and ed. E. J. Holmyard. Dent, London, 1928.

GEERTS, A. J. C. (1). *Les Produits de la Nature Japonaise et Chinoise, Comprenant la Dénomination, l'Histoire et les Applications aux Arts, à l'Industrie, à l'Economie, à la Médécine, etc. des Substances qui dérivent des Trois Régnes de la Nature et qui sont employées par les Japonais et les Chinois: Partie Inorganique et Minéralogique...[only part published].* 2 vols. Levy, Yokohama; Nijhoff, 's Gravenhage, 1878, 1883. (A paraphrase and commentary on the mineralogical chapters of the *Pên Tshao Kang Mu*, based on Ono Ranzan's commentary in Japanese.)

GEERTS, A. J. C. (2). 'Useful Minerals and Metallurgy of the Japanese; [Introduction and] A, Iron.' *TAS/J*, 1875, **3**, 1, 6.

GEERTS, A. J. C. (3). 'Useful Minerals and Metallurgy of the Japanese; [B], Copper.' *TAS/J*, 1875, **3**, 26.

GEERTS, A. J. C. (4). 'Useful Minerals and Metallurgy of the Japanese; C, Lead and Silver.' *TAS/J*, 1875, **3**, 85.

GEERTS, A. J. C. (5). 'Useful Minerals and Metallurgy of the Japanese; D, Quicksilver.' *TAS/J*, 1876, **4**, 34.

GEERTS, A. J. C. (6). 'Useful Minerals and Metallurgy of the Japanese; E, Gold' (with twelve excellent pictures on thin paper of gold mining, smelting and cupellation from a traditional Japanese mining book). *TAS/J*, 1876, **4**, 89.

GEERTS, A. J. C. (7). 'Useful Minerals and Metallurgy of the Japanese; F, Arsenic' (reproducing the picture from *Thien Kung Khai Wu*). *TAS/J*, 1877, **5**, 25.

GEHES CODEX. See Anon. (93).

GEISLER, K. W. (1). 'Zur Geschichte d. Spirituserzeugung.' *BGTI*, 1926, **16**, 94.

GELBART, N. R. (1). 'The Intellectual Development of Walter Charleton.' *AX*, 1971, **18**, 149.

GELLHORN, E. & KIELY, W. F. (1). 'Mystical States of Consciousness; Neurophysiological and Clinical Aspects.' *JNMD*, 1972, **154**, 399.

GENZMER, F. (1). 'Ein germanisches Gedicht aus der Hallstattzeit.' *GRM*, 1936, **24**, 14.

GEOGHEGAN, D. (1). 'Some Indications of Newton's Attitude towards Alchemy.' *AX*, 1957, **6**, 102.

GEORGII, A. (1). *Kinésithérapie, ou Traitement des Maladies par le Mouvement selon la Méthode de Ling... suivi d'un Abrégé des Applications de Ling à l'Éducation Physique.* Baillière, Paris, 1847.

GERNET, J. (3). *Le Monde Chinois.* Colin, Paris, 1972. (Coll. Destins du Monde.)

GESNER, CONRAD (1). *De Remediis secretis, Liber Physicus, Medicus et partiam Chymicus et Oeconomicus in vinorum diversi apparatu, Medicis & Pharmacopoiis omnibus praecipi necessarius nunc primum in lucem editus.* Zürich, 1552, 1557; second book edited by C. Wolff, Zürich, 1569; Frankfurt, 1578.

GESNER, CONRAD (2). *Thesaurus Euonymus Philiatri, Ein köstlicher Schatz....* Zürich, 1555. Eng. tr. Daye, London, 1559, 1565. French tr. Lyon, 1557.

GESSMANN, G. W. (1). *Die Geheimsymbole der Chemie und Medizin des Mittelalters; eine Zusammenstellung der von den Mystikern und Alchymisten gebrauchten geheimen Zeichenschrift, nebst einen Kurzgefassten geheimwissenschaftlichen Lexikon.* Pr. pr. Graz, 1899, then Mickl, München, 1900.

GETTENS, R. J., FITZHUGH, E. W., BENE, I. V. & CHASE, W. T. (1). *The Freer Chinese Bronzes. Vol. 2. Technical Studies.* Smithsonian Institution, Washington, D.C. 1969 (Freer Gallery of Art Oriental Studies, no. 7). See also Pope, Gettens, Cahill & Barnard (1).

GHOSH, HARINATH (1). 'Observations on the Solubility *in vitro* and *in vivo* of Sulphide of Mercury, and also on its Assimilation, probable Pharmacological Action and Therapeutic Utility.' *IMW*, 1 Apr. 1931.

GIBB, H. A. R. (1). 'The Embassy of Hārūn al-Rashīd to Chhang-An.' *BLSOAS*, 1922, **2**, 619.

GIBB, H. A. R. (4). *The Arab Conquests in Central Asia.* Roy. Asiat. Soc., London, 1923. (Royal Asiatic Society, James G. Forlong Fund Pubs. no. 2.)

GIBB, H. A. R. (5). 'Chinese Records of the Arabs in Central Asia.' *BLSOAS*, 1922, **2**, 613.

GIBBS, F. W. (1). 'Invention in Chemical Industries [+1500 to +1700].' Art. in *A History of Technology*, ed. C. Singer *et al.* Vol. 3, p. 676. Oxford, 1957.

GIBSON, E. C. S. (1) (tr.). *Johannes Cassianus*' *'Conlationes'* in 'Select Library of Nicene and Post-Nicene Fathers of the Christian Church'. Parker, Oxford, 1894, vol. 11, pp. 382 ff.

GICHNER, L. E. (1). *Erotic Aspects of Hindu Sculpture*. Pr. pr., U.S.A. (no place of publication stated), 1949.

GICHNER, L. E. (2). *Erotic Aspects of Chinese Culture*. Pr. pr., U.S.A. (no place of publication stated), *c.* 1957.

GIDE, C. (1). *Les Colonies Communistes et Coopératives*. Paris, 1928. Eng. tr. by E. F. Row. *Communist and Cooperative Colonies*. Harrap, London, 1930.

GILDEMEISTER, E. & HOFFMANN, F. (1). *The Volatile Oils*. Tr. E. Kremers. 2nd ed., 3 vols. Longmans Green, London, 1916. (Written under the auspices of Schimmel & Co., Distillers, Miltitz near Leipzig.)

GILDEMEISTER, J. (1). 'Alchymie.' *ZDMG*, 1876, **30**, 534.

GILES, H. A. (2). *Chinese–English Dictionary*. Quaritch, London, 1892, 2nd ed. 1912.

GILES, H. A. (7) (tr.). 'The *Hsi Yüan Lu* or "Instructions to Coroners"; (Translated from the Chinese).' *PRSM*, 1924, **17**, 59.

GILES, H. A. (14). *A Glossary of Reference on Subjects connected with the Far East*. 3rd ed. Kelly & Walsh, Shanghai, 1900.

GILES, L. (6). *A Gallery of Chinese Immortals ('hsien'), selected biographies translated from Chinese sources (Lieh Hsien Chuan, Shen Hsien Chuan, etc.)*. Murray, London, 1948.

GILES, L. (7). 'Wizardry in Ancient China.' *AP*, 1942, **13**, 484.

GILES, L. (14). 'A Thang Manuscript of the *Sou Shen Chi*.' *NCR*, 1921, **3**, 378, 460.

GILLAN, H. (1). *Observations on the State of Medicine, Surgery and Chemistry in China* (+1794), ed. J. L. Cranmer-Byng (2). Longmans, London, 1962.

GLAISTER, JOHN (1). *A Textbook of Medical Jurisprudence, Toxicology and Public Health*. Livingstone, Edinburgh, 1902. 5th ed. by J. Glaister the elder and J. Glaister the younger, Edinburgh, 1931. 6th ed. title changed to *Medical Jurisprudence and Toxicology*, J. Glaister the younger, Edinburgh, 1938. 7th ed. Edinburgh, 1942, 9th ed. Edinburgh, 1950. 10th to 12th eds. (same title), Edinburgh, 1957 to 1966 by J. Glaister the younger & E. Rentoul.

GLISSON, FRANCIS (1). *Tractatus de Natura Substantiae Energetica, seu de Vita Naturae ejusque Tribus Primus Facultatibus; I, Perceptiva; II, Appetitiva; III, Motiva, Naturalibus*. Flesher, Brome & Hooke, London, 1672. Cf. Pagel (16, 17); Temkin (4).

GLOB, P. V. (1). *Iron-Age Man Preserved*. Faber & Faber, London; Cornell Univ. Press, Ithaca, N.Y. 1969. Tr. R. Bruce-Mitford from the Danish *Mosefolket; Jernalderens Mennesker bevaret i 2000 År*.

GLOVER, A. S. B. (1) (tr.). 'The Visions of Zosimus', in Jung (3).

GMELIN, J. G. (1). *Reise durch Russland*. 3 vols. Berlin, 1830.

GOAR, P. J. See Syncellos, Georgius.

GODWIN, WM. (1). *An Enquiry concerning Political Justice, and its Influence on General Virtue and Happiness*. London, 1793.

GOH THEAN-CHYE. See Ho Ping-Yü, Ko Thien-Chi *et al.*

GOLDBRUNNER, J. (1). *Individuation; a study of the Depth Psychology of Carl Gustav Jung*. Tr. from Germ. by S. Godman. Hollis & Carter, London, 1955.

GOLTZ, D. (1). *Studien zur Geschichte der Mineralnamen in Pharmazie, Chemie und Medizin von den Anfängen bis Paracelsus*. Steiner, Wiesbaden, 1972. (Sudhoffs Archiv. Beiheft, no. 14.)

GONDAL, MAHARAJAH OF. See Bhagvat Singhji.

GOODFIELD, J. & TOULMIN, S. (1). 'The Qaṭṭāra; a Primitive Distillation and Extraction Apparatus still in Use.' *ISIS*, 1964, **55**, 339.

GOODMAN, L. S. & GILMAN, A. (ed.) (1). *The Pharmacological Basis of Therapeutics*. Macmillan, New York, 1965.

GOODRICH, L. CARRINGTON (1). *Short History of the Chinese People*. Harper, New York, 1943.

GOODWIN, B. (1). 'Science and Alchemy', art. in *The Rules of the Game*... ed. T. Shanin (1), p. 360.

GOOSENS, R. (1). 'Un Texte Grec relatif à l'aśvamedha' [in the Life of Apollonius of Tyana by Philostratos]. *JA*, 1930, **217**, 280.

GÖTZE, A. (1). 'Die "Schatzhöhle"; Überlieferung und Quelle.' *SHAW/PH*, 1922, no. 4.

GOULD, S. J. (1). 'History *versus* Prophecy; Discussion with J. W. Harrington.' *AJSC*, 1970, **268**, 187. With reply by J. W. Harrington, p. 189.

GOWLAND, W. (1). 'Copper and its Alloys in Prehistoric Times.' *JRAI*, 1906, **36**, 11.

GOWLAND, W. (2). 'The Metals in Antiquity.' *JRAI*, 1912, **42**, 235. (Huxley Memorial Lecture 1912.)

GOWLAND, W. (3). 'The Early Metallurgy of Silver and Lead.' Pt. I, 'Lead' (no more published). *AAA* 1901, **57**, 359.

GOWLAND, W. (4). 'The Art of Casting Bronze in Japan.' *JRSA*, 1895, **43**. Repr. *ARSI*, 1895, 609.

GOWLAND, W. (5). 'The Early Metallurgy of Copper, Tin and Iron in Europe as illustrated by ancient Remains, and primitive Processes surviving in Japan.' *AAA*, 1899, **56**, 267.

GOWLAND, W. (6). 'Metals and Metal-Working in Old Japan.' *TJSL*, 1915, **13**, 20.

GOWLAND, W. (7). 'Silver in Roman and earlier Times.' Pt. I, 'Prehistoric and Protohistoric Times' (no more published). *AAA*, 1920, **69**, 121.

GOWLAND, W. (8). 'Remains of a Roman Silver Refinery at Silchester' (comparisons with Japanese technique). *AAA*, 1903, **57**, 113.

GOWLAND, W. (9). *The Metallurgy of the Non-Ferrous Metals*. Griffin, London, 1914. (Copper, Lead, Gold, Silver, Platinum, Mercury, Zinc, Cadmium, Tin, Nickel, Cobalt, Antimony, Arsenic, Bismuth, Aluminium.)

GOWLAND, W. (10). 'Copper and its Alloys in Early Times.' *JIM*, 1912, **7**, 42.

GOWLAND, W. (11). 'A Japanese Pseudo-Speiss (*Shirome*), and its Relation to the Purity of Japanese Copper and the Presence of Arsenic in Japanese Bronze.' *JSCI*, 1894, **13**, 463.

GOWLAND, W. (12). 'Japanese Metallurgy; I, Gold and Silver and their Alloys.' *JCSI*, 1896, **15**, 404. No more published.

GRACE, V. R. (1). *Amphoras and the Ancient Wine Trade*. Amer. School of Classical Studies, Athens and Princeton, N.J., 1961.

GRADY, M. C. (1). 'Préparation Electrolytique du Rouge au Japan.' *TP*, 1897 (1e sér.), **8**, 456.

GRAHAM, A. C. (5). '"Being" in Western Philosophy compared with *shih/fei* and *yu/wu* in Chinese Philosophy.' With an appendix on 'The Supposed Vagueness of Chinese'. *AM*, 1959, **7**, 79.

GRAHAM, A. C. (6) (tr.). *The Book of Lieh Tzu*. Murray, London, 1960.

GRAHAM, A. C. (7). 'Chuang Tzu's "Essay on Making Things Equal".' Communication to the First International Conference of Taoist Studies, Villa Serbelloni, Bellagio, 1968.

GRAHAM, D. C. (4). 'Notes on the Han Dynasty Grave Collection in the West China Union University Museum of Archaeology [at Chhêngtu].' *JWCBRS*, 1937, **9**, 213.

GRANET, M. (5). *La Pensée Chinoise*. Albin Michel, Paris, 1934. (Evol. de l'Hum. series, no. 25 bis.)

GRANT, R. McQ. (1). *Gnosticism; an Anthology*. Collins, London, 1961.

GRASSMANN, H. (1). 'Der Campherbaum.' *MDGNVO*, 1895, **6**, 277.

GRAY, B. (1). 'Arts of the Sung Dynasty.' *TOCS*, 1960, 13.

GRAY, J. H. (1). '[The "Abode of the Blest" in] Persian [Iranian, Thought].' *ERE*, vol. ii, p. 702.

GRAY, J. H. (1). *China: a History of the Laws, Manners and Customs of the People*. Ed. W. G. Gregor. 2 vols. Macmillan, London, 1878.

GRAY, W. D. (1). *The Relation of Fungi to Human Affairs*. Holt, New York, 1959.

GREEN, F. H. K. (1). 'The Clinical Evaluation of Remedies.' *LT*, 1954, 1085.

GREEN, R. M. (1) (tr.). *Galen's Hygiene; 'De Sanitate Tuenda'*. Springfield, Ill. 1951.

GREENAWAY, F. (1). 'Studies in the Early History of Analytical Chemistry.' Inaug. Diss. London, 1957.

GREENAWAY, F. (2). *The Historical Continuity of the Tradition of Assaying*. Proc. Xth Int. Congr. Hist. of Sci., Ithaca, N.Y., 1962, vol. 2, p. 819.

GREENAWAY, F. (3). 'The Early Development of Analytical Chemistry.' *END*, 1962, **21**, 91.

GREENAWAY, F. (4). *John Dalton and the Atom*. Heinemann, London, 1966.

GREENAWAY, F. (5). 'Johann Rudolph Glauber and the Beginnings of Industrial Chemistry.' *END*, 1970, **29**, 67.

GREGORY, E. (1). *Metallurgy*. Blackie, London and Glasgow, 1943.

GREGORY, J. C. (1). *A Short History of Atomism*. Black, London, 1931.

GREGORY, J. C. (2). 'The Animate and Mechanical Models of Reality.' *JPHST*, 1927, **2**, 301. Abridged in 'The Animate Model of Physical Process'. *SPR*, 1925.

GREGORY, J. C. (3). 'Chemistry and Alchemy in the Natural Philosophy of Sir Francis Bacon (+1561 to +1626).' *AX*, 1938, **2**, 93.

GREGORY, J. C. (4). 'An Aspect of the History of Atomism.' *SPR*, 1927, **22**, 293.

GREGORY, T. C. (1). *Condensed Chemical Dictionary*. 1950. Continuation by A. Rose & E. Rose. Chinese tr. by Kao Hsien (1).

GRIERSON, SIR G. A. (1). *Bihar Peasant Life*. Patna, 1888; reprinted Bihar Govt., Patna, 1926.

GRIERSON, P. (2). 'The Roman Law of Counterfeiting.' Art. in *Essays in Roman Coinage*, Mattingley Presentation Volume, Oxford, 1965, p. 240.

GRIFFITH, E. F. (1). *Modern Marriage*. Methuen, London, 1946.

GRIFFITH, F. LL. & THOMPSON, H. (1). '*The Demotic Magical Papyrus of London and Leiden* [+3rd Cent.]. 3 vols. Grevel, London, 1904–9.

GRIFFITH, R. T. H. (1) (tr.). *The Hymns of the 'Atharva-veda'*. 2 vols. Lazarus, Benares, 1896. Repr. Chowkhamba Sanskrit Series Office, Varanasi, 1968.

GRIFFITHS, J. GWYN (1) (tr.). *Plutarch's 'De Iside et Osiride'*. University of Wales Press, Cardiff, 1970.

GRINSPOON, L. (1). 'Marihuana.' *SAM*, 1969, **221** (no. 6), 17.

GRMEK, M. D. (2). 'On Ageing and Old Age; Basic Problems and Historical Aspects of Gerontology and Geriatrics.' *MB*, 1958, **5** (no. 2).

DE GROOT, J. J. M. (2). *The Religious System of China*. Brill, Leiden, 1892.
Vol. 1, Funeral rites and ideas of resurrection.
2 and 3, Graves, tombs, and *fêng-shui*.
4, The soul, and nature-spirits.
5, Demonology and sorcery.
6, The animistic priesthood (*wu*).

GRÖSCHEL-STEWART, U. (1). 'Plazentahormone.' *MMN*, 1970, **22**, 469.

GROSIER, J. B. G. A. (1). *De la Chine; ou, Description Générale de cet Empire*, etc. 7 vols. Pillet & Bertrand, Paris, 1818–20.

GRUMAN, G. J. (1). 'A History of Ideas about the Prolongation of Life; the Evolution of Prolongevity Hypotheses to 1800.' Inaug. Diss., Harvard University, 1965. *TAPS*, 1966 (n.s.), **56** (no. 9), 1–102.

GRUMAN, G. J. (2). 'An Introduction to the Literature on the History of Gerontology.' *BIHM*, 1957, **31**, 78.

GUARESCHI, S. (1). Tr. of Klaproth (5). *SAEC*, 1904, **20**, 449.

GUERLAC, H. (1). 'The Poets' Nitre.' *ISIS*, 1954, **45**, 243.

GUICHARD, F. (1). 'Properties of saponins of *Gleditschia*.' *BSPB*, 1936, **74**, 168.

VAN GULIK, R. H. (3). '*Pi Hsi Thu Khao*'; *Erotic Colour-Prints of the Ming Period, with an Essay on Chinese Sex Life from the Han to the Chhing Dynasty* (−206 *to* +1644). 3 vols. in case. Privately printed. Tokyo, 1951 (50 copies only, distributed to the most important Libraries of the world). Crit. W. L. Hsü, *MN*, 1952, **8**, 455; H. Franke, *ZDMG*, 1955 (NF) **30**, 380.

VAN GULIK, R. H. (4). 'The Mango "Trick" in China; an essay on Taoist Magic.' *TAS/J*, 1952 (3rd ser.), **3**, 1.

VAN GULIK, R. H. (8). *Sexual Life in Ancient China; a Preliminary Survey of Chinese Sex and Society from c. −1500 to +1644*. Brill, Leiden, 1961. Rev. R. A. Stein, *JA*, 1962, **250**, 640.

GUNAWARDANA, R. A. LESLIE H. (1). 'Ceylon and Malaysia; a Study of Professor S. Paranavitana's Research on the Relations between the Two Regions.' *UCR*, 1967, **25**, 1–64.

GUNDEL, W. (4). Art. 'Alchemie' in *Real-Lexikon f. Antike und Christentum*, ed. T. Klauser, 1950–3, vol. 1, p. 239.

GUNDEL, W. & GUNDEL, H. G. (1). *Astrologumena; das astrologische Literatur in der Antike und ihre Geschichte*. Steiner, Wiesbaden, 1966. (*AGMW* Beiheft, no. 6, pp. 1–382.)

GUNTHER, R. T. (3) (ed.). *The Greek Herbal of Dioscorides, illustrated by a Byzantine in +512, englished by John Goodyer in +1655, edited and first printed, 1933*. Pr. pr. Oxford, 1934, photolitho repr. Hafner, New York, 1959.

GUNTHER, R. T. (4). *Early Science in Cambridge*. Pr. pr. Oxford, 1937.

GUPPY, H. B. (1). 'Samshu-brewing in North China.' *JRAS/NCB*, 1884, **18**, 63.

GURE, D. (1). 'Jades of the Sung Group.' *TOCS*, 1960, 39.

GUTZLAFF, C. (1). 'On the Mines of the Chinese Empire.' *JRAS/NCB*, 1847, 43.

GYLLENSVÅRD, Bo (1). *Chinese Gold and Silver [-Work] in the Carl Kempe Collection*. Stockholm, 1953; Smithsonian Institution, Washington, D.C., 1954.

GYLLENSVÅRD, Bo (2). 'Thang Gold and Silver.' *BMFEA*, 1957, **29**, 1–230.

HACKIN, J. & HACKIN, J. R. (1). *Recherches archéologiques à Begram, 1937*. Mémoires de la Délégation Archéologique Française en Afghanistan, vol. 9. Paris, 1939.

HACKIN, J., HACKIN, J. R., CARL, J. & HAMELIN, P. (with the collaboration of J. Auboyer, V. Elisséeff, O. Kurz & P. Stern) (1). *Nouvelles Recherches archéologiques à Begram (ancienne Kāpiśi), 1939–1940*. Mémoires de la Délégation Archéologique Française en Afghanistan, vol. 11. Paris, 1954. (Rev. P. S. Rawson, *JRAS*, 1957, 139.)

HADD, H. E. & BLICKENSTAFF, R. T. (1). *Conjugates of Steroid Hormones*. Academic Press, New York and London, 1969.

HADI HASAN (1). *A History of Persian Navigation*. Methuen, London, 1928.

HAJI KHALFA (or Ḥajji Khalīfa). See Flügel (2).

HALBAN, J. (1). 'Über den Einfluss der Ovarien auf die Entwicklung der Genitales.' *MGG*, 1900, **12**, 496.

HALDANE, J. B. S. (2). 'Experiments on the Regulation of the Blood's Alkalinity.' *JOP*, 1921, **55**, 265.

HALDANE, J. B. S. (3). 'Über Halluzinationen infolge von Änderungen des Kohlensäuredrucks.' *PF*, 1924, **5**, 356.

HALDANE, J. B. S. See also Baird, Douglas, Haldane & Priestley (1); Davies, Haldane & Kennaway (1).

HALDANE, J. B. S., LINDER, G. C., HILTON, R. & FRASER, F. R. (1). 'The Arterial Blood in Ammonium Chloride Acidosis.' *JOP*, 1928, **65**, 412.

HALDANE, J. B. S., WIGGLESWORTH, V. B. & WOODROW, C. E. (1). 'Effect of Reaction Changes on Human Inorganic Metabolism.' *PRSB*, 1924, **96**, 1.

HALDANE, J. B. S., WIGGLESWORTH, V. B. & WOODROW, C. E. (2). 'Effect of Reaction Changes on Human Carbohydrate and Oxygen Metabolism.' *PRSB*, 1924, **96**, 15.

HALEN, G. E. (1). *De Chemo Scientiarum Auctore*. Upsala, 1694.

HALES, STEPHEN (2). *Philosophical Experiments; containing Useful and Necessary Instructions for such as undertake Long Voyages at Sea, shewing how Sea Water may be made Fresh and Wholsome*. London, 1739.

HALL, E. T. (1). 'Surface Enrichment of Buried [Noble] Metal [Alloys].' *AMY*, 1961, **4**, 62.

HALL, E. T. & ROBERTS, G. (1). 'Analysis of the Moulsford Torc.' *AMY*, 1962, **5**, 28.

HALL, F. W. (1). '[The "Abode of the Blest" in] Greek and Roman [Culture].' *ERE*, vol. ii, p. 696.

HALL, H. R. (1). 'Death and the Disposal of the Dead [in Ancient Egypt].' Art. in *ERE*, vol. iv, p. 458.

HALL, MANLY P. (1). *The Secret Teachings of All Ages*. San Francisco, 1928.

HALLEUX, R. (1). 'Fécondité des Mines et Sexualité des Pierres dans l'Antiquité Gréco-Romaine.' *RBPH*, 1970, **48**, 16.

HALOUN, G. (2). Translations of *Kuan Tzu* and other ancient texts made with the present writer, unpub. MSS.

HAMARNEH, SAMI, K. & SONNEDECKER, G. (1). *A Pharmaceutical View of Albucasis (al-Zahrāwī) in Moorish Spain*. Brill, Leiden, 1963.

HAMMER-JENSEN, I. (1). 'Deux Papyrus à Contenu d'Ordre Chimique.' *ODVS*, 1916 (no. 4), 279.

HAMMER-JENSEN, I. (2). 'Die ältesten Alchemie.' *MKDVS/IIF*, 1921, **4** (no. 2), 1–159.

HANBURY, DANIEL (1). *Science Papers, chiefly Pharmacological and Botanical*. Macmillan, London, 1876.

HANBURY, DANIEL (2). 'Notes on Chinese Materia Medica.' *PJ*, 1861, **2**, 15, 109, 553; 1862, **3**, 6, 204, 260, 315, 420. German tr. by W. C. Martius (without Chinese characters), *Beiträge z. Materia Medica Chinas*. Kranzbühler, Speyer, 1863. Revised version, with additional notes, references and map, in Hanbury (1), pp. 211 ff.

HANBURY, DANIEL (6). 'Note on Chinese Sal Ammoniac.' *PJ*, 1865, **6**, 514. Repr. in Hanbury (1), p. 276.

HANBURY, DANIEL (7). 'A Peculiar Camphor from China [Ngai Camphor from *Blumea balsamifera*]. *PJ*, 1874, **4**, 709. Repr. in Hanbury (1), pp. 393 ff.

HANBURY, DANIEL (8). 'Some Notes on the Manufactures of Grasse and Cannes [and Enfleurage].' *PJ*, 1857, **17**, 161. Repr. in Hanbury (1), pp. 150 ff.

HANBURY, DANIEL (9). 'On Otto of Rose.' *PJ*, 1859, **18**, 504. Repr. in Hanbury (1), pp. 164 ff.

HANSFORD, S. H. (1). *Chinese Jade Carving*. Lund Humphries, London, 1950.

HANSFORD, S. H. (2) (ed.). *The Seligman Collection of Oriental Art; Vol. 1, Chinese, Central Asian and Luristan Bronzes and Chinese Jades and Sculptures*. Arts Council G. B., London, 1955.

HANSON, D. (1). *The Constitution of Binary Alloys*. McGraw-Hill, New York, 1958.

HARADA, YOSHITO & TAZAWA, KINGO (1). *Lo-Lang; a Report on the Excavation of Wang Hsü's Tomb in the Lo-Lang Province, an ancient Chinese Colony in Korea*. Tokyo University, Tokyo, 1930.

HARBORD, F. W. & HALL, J. W. (1). *The Metallurgy of Steel*. 2 vols. 7th ed. Griffin, London, 1923.

HARDING, M. ESTHER (1). *Psychic Energy; its Source and Goal*. With a foreword by C. G. Jung. Pantheon, New York, 1947. (Bollingen series, no. 10.)

VON HARLESS, G. C. A. (1). *Jakob Böhme und die Alchymisten; ein Beitrag zum Verständnis J. B.'s...* Berlin, 1870. 2nd ed. Hinrichs, Leipzig, 1882.

HARRINGTON, J. W. (1). 'The First "First Principles of Geology".' *AJSC*, 1967, **265**, 449.

HARRINGTON, J. W. (2). 'The Prenatal Roots of Geology; a Study in the History of Ideas.' *AJSC*, 1969, **267**, 592.

HARRINGTON, J. W. (3). 'The Ontology of Geological Reasoning; with a Rationale for evaluating Historical Contributions.' *AJSC*, 1970, **269**, 295.

HARRIS, C. (1). '[The State of the Dead in] Christian [Thought].' *ERE*, vol. xi, p. 833.

HARRISON, F. C. (1). 'The Miraculous Micro-Organism' (*B. prodigiosus* as the causative agent of 'bleeding hosts'). *TRSC*, 1924, **18**, 1.

HARRISSON, T. (8). 'The *palang*; its History and Proto-history in West Borneo and the Philippines.' *JRAS/M*, 1964, **37**, 162.

HARTLEY, SIR HAROLD (1). 'John Dalton, F.R.S. (1766 to 1844) and the Atomic Theory; a Lecture to commemorate his Bicentenary.' *PRSA*, 1967, **300**, 291.

HARTNER, W. (12). *Oriens-Occidens; ausgewählte Schriften zur Wissenschafts- und Kultur-geschichte (Festschrift zum 60. Geburtstag)*. Olms, Hildesheim, 1968. (Collectanea, no. 3.)

HARTNER, W. (13). 'Notes on *Picatrix*.' *ISIS*, 1965, **56**, 438. Repr. in (12), p. 415.

HASCHMI, M. Y. (1). 'The Beginnings of Arab Alchemy.' *AX*, 1961, **9**, 155.

HASCHMI, M. Y. (2). '*The Propagation of Rays*; the Oldest Arabic Manuscript about Optics (the Burning-Mirror), [a text written by] *Ya'kub ibn Ishaq al-Kindī, Arab Philosopher and Scholar of the +9th Century. Photocopy, Arabic text and Commentary*. Aleppo, 1967.

HASCHMI, M. Y. (3). 'Sur l'Histoire de l'Alcool.' Résumés des Communications, XIIth International Congress of the History of Science, Paris, 1968, p. 91.

HASCHMI, M. Y. (4). 'Die Anfänge der arabischen Alchemie.' Actes du XIe Congrès International d'Histoire des Sciences, Warsaw, 1965, p. 290.

HASCHMI, M. Y. (5). 'Ion Exchange in Arabic Alchemy.' Proc. Xth Internat. Congr. Hist. of Sci., Ithaca, N.Y. 1962, p. 541. Summaries of Communications, p. 56.

HASCHMI, M. Y. (6). 'Die Geschichte der arabischen Alchemie.' *DMAB*, 1967, **35**, 60.

HATCHETT, C. (1). 'Experiments and Observations on the Various Alloys, on the Specific Gravity, and on the Comparative Wear of Gold...' *PTRS*, 1803, **93**, 43.

HAUSHERR, I. (1). 'La Méthode d'Oraison Hésychaste.' *ORCH*, 1927, **9**, (no. 2), 102.

HÄUSSLER, E. P. (1). 'Über das Vorkommen von a-Follikelhormon (3-oxy-17 Keto-1, 3, 5-oestratriën) im Hengsturin.' *HCA*, 1934, **17**, 531.

HAWKES, D. (1) (tr.). '*Chhu Tzhu'; the Songs of the South—an Ancient Chinese Anthology*. Oxford, 1959. Rev. J. Needham, *NSN* (18 Jul. 1959).

HAWKES, D. (2). 'The Quest of the Goddess.' *AM*, 1967, **13**, 71.

HAWTHORNE, J. G. & SMITH, C. S. (1) (tr.). '*On Divers Arts'; the Treatise of Theophilus* [Presbyter], *translated from the Mediaeval Latin with Introduction and Notes*...[probably by Roger of Helmarshausen, *c.* +1130]. Univ. of Chicago Press, Chicago, 1963.

HAY, M. (1). *Failure in the Far East; Why and How the Breach between the Western World and China First Began* (on the dismantling of the Jesuit Mission in China in the late +18th century). Spearman, London; Scaldis, Wetteren (Belgium), 1956.

HAYS, E. E. & STEELMAN, S. L. (1). 'The Chemistry of the Anterior Pituitary Hormones.' Art. in *The Hormones*..., ed. G. Pincus, K. V. Thimann & E. B. Astwood. Academic Press, New York, 1948–64. vol. 3, p. 201.

HEDBLOM, C. A. (1). 'Disease Incidence in China [16,000 cases].' *CMJ*, 1917, **31**, 271.

HEDFORS, H. (1) (ed. & tr.). *The 'Compositiones ad Tingenda Musiva'*... Uppsala, 1932.

HEDIN, SVEN A., BERGMAN, F. *et al.* (1). *History of the Expedition in Asia, 1927/1935.* 4 vols. Reports of the Sino-Swedish [Scientific] Expedition [to NW China]. 1936. Nos. 23, 24, 25, 26.

HEIM, R. (1). 'Old and New Investigations on Hallucinogenic Mushrooms from Mexico.' *APH*, 1959, **12**, 171.

HEIM, R. (2). *Champignons Toxiques et Hallucinogènes*. Boubée, Paris, 1963.

HEIM, R. & HOFMANN, A. (1). 'Psilocybine.' *CRAS*, 1958, **247**, 557.

HEIM, R., WASSON, R. G. *et al.* (1). *Les Champignons Hallucinogènes du Mexique; Études Ethnologiques, Taxonomiques, Biologiques, Physiologiques et Chimiques.* Mus. Nat. d'Hist. Nat. Paris, 1958.

VON HEINE-GELDERN, R. (4). 'Die asiatische Herkunft d. südamerikanische Metalltechnik.' *PAI*, 1954, **5**, 347.

HEMNETER, E. (1). 'The Influence of the Caste-System on Indian Trades and Crafts.' *CIBA/T*, 1937, **1** (no. 2), 46.

H[EMSLEY], W. B. (1). 'Camphor.' *N*, 1896, **54**, 116.

HENDERSON, G. & HURVITZ, L. (1). 'The Buddha of Seiryō-ji [Temple at Saga, Kyoto]; New Finds and New Theory.' *AA*, 1956, **19**, 5.

HENDY, M. F. & CHARLES, J. A. (1). 'The Production Techniques, Silver Content and Circulation History of the +12th-Century Byzantine Trachy.' *AMY*, 1970, **12**, 13.

HENROTTE, J. G. (1). 'Yoga et Biologie.' *ATOM*, 1969, **24** (no. 265), 283.

HENRY, W. (1). *Elements of Experimental Chemistry*. London, 1810. German. tr. by F. Wolff, Berlin, 1812. Another by J. B. Trommsdorf.

HERMANN, A. (1). 'Das Buch *Kmj.t* und die Chemie.' *ZAES*, 1954, **79**, 99.

HERMANN, P. (1). *Een constelijk Distileerboec inhoudende de rechte ende waerachtige conste der distilatiën om alderhande wateren der cruyden, bloemen ende wortelen ende voorts alle andere dinge te leeren distileren opt alder constelijcste, alsoo dat dies gelyke noyt en is gheprint geweest in geen derley sprake...* Antwerp, 1552.

HERMANNS, M. (1). *Die Nomaden von Tibet.* Vienna, 1949. Rev. W. Eberhard, *AN*, 1950, **45**, 942.

HERRINGHAM, C. J. (1). *The 'Libro dell'Arte' of Cennino Cennini* [+1437]. Allen & Unwin, London, 1897.

HERRMANN, A. (2). *Die Alten Seidenstrassen zw. China u. Syrien; Beitr. z. alten Geographie Asiens, I* (with excellent maps). Berlin, 1910. (Quellen u. Forschungen z. alten Gesch. u. Geographie, no. 21; photographically reproduced, Tientsin, 1941).

HERRMANN, A. (3). 'Die Alten Verkehrswege zw. Indien u. Süd-China nach Ptolemäus.' *ZGEB*, 1913, 771.

HERRMANN, A. (5). 'Die Seidenstrassen vom alten China nach dem Romischen Reich.' *MGGW*, 1915, **58**, 472.

HERRMANN, A. (6). *Die Verkehrswege zw. China, Indien und Rom um etwa 100 nach Chr.* Leipzig, 1922 (Veröffentlichungen d. Forschungs-instituts f. vergleich. Religionsgeschichte a.d. Univ. Leipzig, no. 7.)

HERTZ, W. (1). 'Die Sage vom Giftmädchen.' *ABAW/PH*, 1893, **20**, no. 1. Repr. in *Gesammelte Abhandlungen*, ed. v. F. von der Leyen, 1905, pp. 156–277.

D'HERVEY ST DENYS, M. J. L. (3). *Trois Nouvelles Chinoises, traduites pour la première fois.* Leroux, Paris, 1885. 2nd ed. Dentu, Paris, 1889.

HEYM, G. (1). 'The *Aurea Catena Homeri* [by Anton Joseph Kirchweger, +1723].' *AX*, 1937, **1**, 78. Cf. Ferguson (1), vol. 1, p. 470.

HEYM, G. (2). 'Al-Rāzī and Alchemy.' *AX*, 1938, **1**, 184.

HICKMAN, K. C. D. (1). 'A Vacuum Technique for the Chemist' (molecular distillation). *JFI*, 1932, **213**, 119.

HICKMAN, K. C. D. (2). 'Apparatus and Methods [for Molecular Distillation].' *IEC/I*, 1937, **29**, 968.

HICKMAN, K. C. D. (3). 'Surface Behaviour in the Pot Still.' *IEC/I*, 1952, **44**, 1892.

HICKMAN, K. C. D. & SANFORD, C. R. (1). 'The Purification, Properties and Uses of Certain High-Boiling Organic Liquids.' *JPCH*, 1930, **34**, 637.

HICKMAN, K. C. D. & SANFORD, C. R. (2). 'Molecular stills.' *RSI*, 1930, **1**, 140.

HICKMAN, K. C. D. & TREVOY, D. J. (1). 'A Comparison of High Vacuum Stills and Tensimeters.' *IEC/I*, 1952, **44**, 1903.

HICKMAN, K. C. D. & WEYERTS, W. (1). 'The Vacuum Fractionation of Phlegmatic Liquids.' *JACS*, 1930, **52**, 4714.

HIGHMORE, NATHANIEL (1). *The History of Generation, examining the several Opinions of divers Authors, especially that of Sir Kenelm Digby, in his Discourse of Bodies.* Martin, London, 1651.

HILGENFELD, A. (1). *Die Ketzergeschichte des Urchristenthums.* Fues (Reisland), Leipzig, 1884.

HILLEBRANDT, A. (1). *Vedische Mythologie.* Breslau, 1891–1902.

HILTON-SIMPSON, M. W. (1). *Arab Medicine and Surgery; a Study of the Healing Art in Algeria.* Oxford, 1922.

HIORDTHAL, T. See Hjortdahl, T.

HIORNS, A. H. (1). *Metal-Colouring and Bronzing.* Macmillan, London and New York, 1892. 2nd ed. 1902.

HIORNS, A. H. (2). *Mixed Metals or Metallic Alloys.* 3rd ed. Macmillan, London and New York, 1912.

HIORNS, A. H. (3). *Principles of Metallurgy.* 2nd ed. Macmillan, London, 1914.

HIRTH, F. (2) (tr.). 'The Story of Chang Chhien, China's Pioneer in West Asia.' *JAOS*, 1917, **37**, 89. (Translation of ch. 123 of the *Shih Chi*, containing Chang Chhien's Report; from §18–52 inclusive and 101 to 103. §98 runs on to §104, 99 and 100 being a separate interpolation. Also tr. of ch. 111 containing the biogr. of Chang Chhien.)

HIRTH, F. (7). *Chinesische Studien.* Hirth, München and Leipzig, 1890.

HIRTH, F. (9). *Über fremde Einflüsse in der chinesischen Kunst.* G. Hirth, München and Leipzig. 1896.

HIRTH, F. (11). 'Die Länder des Islam nach Chinesischen Quellen.' *TP*, 1894, **5** (Suppl.). (Translation of, and notes on, the relevant parts of the *Chu Fan Chih* of Chao Ju-Kua; subsequently incorporated in Hirth & Rockhill.)

HIRTH, F. (25). 'Ancient Porcelain; a study in Chinese Mediaeval Industry and Trade.' G. Hirth, Leipzig and Munich; Kelly & Walsh, Shanghai, Hongkong, Yokohama and Singapore, 1888.

HIRTH, F. & ROCKHILL, W. W. (1) (tr.). *Chau Ju-Kua; His work on the Chinese and Arab Trade in the 12th and 13th centuries, entitled 'Chu-Fan-Chi'.* Imp. Acad. Sci, St Petersburg, 1911. (Crit. G. Vacca *RSO*, 1913, **6**, 209; P. Pelliot, *TP*, 1912, **13**, 446; E. Schaer, *AGNT*, 1913, **6**, 329; O. Franke, *OAZ*, 1913, **2**, 98; A. Vissière, *JA*, 1914 (11ᵉ sér.), **3**, 196.)

HISCOX, G. D. (1) (ed.). *The Twentieth Century Book of Recipes, Formulas and Processes; containing nearly 10,000 selected scientific, chemical, technical and household recipes, formulas and processes for use in the laboratory, the office, the workshop and in the home.* Lockwood, London; Henley, New York, 1907. Lexicographically arranged. 4th ed., Lockwood, London; Henley, New York, 1914. Retitled *Henley's Twentieth Century Formulas, Recipes and Processes; containing 10,000 selected household and workshop formulas, recipes, proceesses and money-saving methods for the practical use of manufacturers, mechanics, housekeepers and home workers.* Spine title unchanged; index of contents added and 2 entries omitted.

HITCHCOCK, E. A. (1). *Remarks upon Alchemy and the Alchemists, indicating a Method of discovering the True Nature of Hermetic Philosophy; and showing that the Search after the Philosopher's Stone had not for its Object the Discovery of an Agent for the Transmutation of Metals—Being also an attempt to rescue from undeserved opprobrium the reputation of a class of extraordinary thinkers in past ages.* Crosby & Nichols, Boston, 1857. 2nd ed. 1865 or 1866. See also Clymer (1); Croffut (1).

HITCHCOCK, E. A. (2). *Remarks upon Alchymists, and the supposed Object of their Pursuit; showing that the Philosopher's Stone is a mere Symbol, signifying Something which could not be expressed openly without incurring the Danger of an Auto-da-Fé. By an Officer of the United States Army*. Pr. pr. Herald, Carlisle, Pennsylvania, 1855. This pamphlet was the first form of publication of the material enlarged in Hitchcock (1).

HJORTDAHL, T. (1). 'Chinesische Alchemie', art. in Kahlbaum Festschrift (1909), ed. Diergart (1): *Beiträge aus der Geschichte der Chemie*, pp. 215–24. Comm. by E. Chavannes, *TP*, 1909 (2e sér.), **10**, 389.

HJORTDAHL, T. (2). 'Fremstilling af Kemiens Historie' (in Norwegian). *CVS*, 1905, **1** (no. 7).

HO JU (1). *Poèmes de Mao Tsê-Tung* (French translation). Foreign Languages Press, Peking, 1960. 2nd ed., enlarged, 1961.

HO PENG YOKE. See Ho Ping-Yü.

HO PING-YÜ (5). 'The Alchemical Work of Sun Ssu-Mo.' Communication to the American Chemical Society's Symposium on Ancient and Archaeological Chemistry, at the 142nd Meeting, Atlantic City, 1962.

HO PING-YÜ (7). 'Astronomical Data in the Annamese *Đại Việt Sú-Ký Toàn-thủ*; an early Annamese Historical Text.' *JAOS*, 1964, **84**, 127.

HO PING-YÜ (8). 'Draft translation of the *Thai-Chhing Shih Pi Chi* (Records in the Rock Chamber); an alchemical book (*TT/874*) of the Liang period (early +6th Century, but including earlier work as old as the late +3rd).' Unpublished.

HO PING-YÜ (9). Précis and part draft translation of the *Thai Chi Chen-Jen Chiu Chuan Huan Tan Ching Yao Chüeh* (Essential Teachings of the Manual of the Supreme-Pole Adept on the Ninefold Cyclically Transformed Elixir); an alchemical book (*TT/882*) of uncertain date, perhaps Sung but containing much earlier metarial.' Unpublished.

HO PING-YÜ (10). 'Précis and part draft translation of the *Thai-Chhing Chin I Shen Tan Ching* (Manual of the Potable Gold and Magical Elixir; a Thai-Chhing Scripture); an alchemical book (*TT/873*) of unknown date and authorship but prior to +1022 when it was incorporated in the *Yün Chi Chhi Chhien*.' Unpublished.

HO PING-YÜ (11). 'Notes on the *Pao Phu Tzu Shen Hsien Chin Shuo Ching* (The Preservation-of-Solidarity Master's Manual of the Bubbling Gold (Potion) of the Holy Immortals); an alchemical book (*TT/910*) attributed to Ko Hung (c. +320).' Unpublished.

HO PING-YÜ (12). 'Notes on the *Chin Pi Wu Hsiang Lei Tshan Thung Chhi* (Gold and Caerulean Jade Treatise on the Similarities and Categories of the Five (Substances) and the *Kinship of the Three*); an alchemical book (*TT/897*) attributed to Yin Chhang-Shêng (H/Han, c. +200), but probably of somewhat later date.' Unpublished.

HO PING-YÜ (13). 'Alchemy in Ming China (+1368 to +1644).' Communication to the XIIth International Congress of the History of Science, Paris, 1968. Abstract Vol. p. 174. Communications, Vol. 3A, p. 119.

HO PING-YÜ (14). 'Taoism in Sung and Yuan China.' Communication to the First International Conference of Taoist Studies, Villa Serbelloni, Bellagio, 1968.

HO PING-YÜ (15). 'The Alchemy of Stones and Minerals in the Chinese Pharmacopoeias.' *CCJ*, 1968, **7**, 155.

HO PING-YÜ (16). 'The System of the *Book of Changes* and Chinese Science.' *JSHS*, 1972, No. 11, 23.

HO PING-YÜ & CHHEN THIEH-FAN (1) = (1). 'On the Dating of the *Shun-Yang Lü Chen-Jen Yao Shih Chih*, a Taoist Pharmaceutical and Alchemical Manual.' *JOSHK*, 1971, **9**, 181 (229).

HO PING-YÜ, KO THIEN-CHI & LIM, BEDA (1). 'Lu Yu (+1125 to 1209), Poet-Alchemist.' *AM*, 1972, 163.

HO PING-YÜ, KO THIEN-CHI & PARKER, D. (1). 'Pai Chü-I's Poems on Immortality.' *HJAS*, 1974, **34**, 163.

HO PING-YÜ & LIM, BEDA (1). 'Tshui Fang, a Forgotten +11th-Century Alchemist [with assembly of citations, mostly from *Pên Tshao Kang Mu*, probably transmitted by *Kêng Hsin Yü Tshê*].' *JSHS*, 1972, No. 11, 103.

HO PING-YÜ, LIM, BEDA & MORSINGH, FRANCIS (1) (tr.). 'Elixir Plants: the *Shun-Yang Lü Chen-Jen Yao Shih Chih* (Pharmaceutical Manual of the Adept Lü Shun-Yang)' [in verses]. Art. in Nakayama & Sivin (1), p. 153.

HO PING-YÜ & NEEDHAM, JOSEPH (1). 'Ancient Chinese Observations of Solar Haloes and Parhelia.' *W*, 1959, **14**, 124.

HO PING-YÜ & NEEDHAM, JOSEPH (2). 'Theories of Categories in Early Mediaeval Chinese Alchemy' (with transl. of the *Tshan Thung Chhi Wu Hsiang Lei Pi Yao*, c. +6th to +8th cent.). *JWCI*, 1959, **22**, 173.

HO PING-YÜ & NEEDHAM, JOSEPH (3). 'The Laboratory Equipment of the Early Mediaeval Chinese Alchemists.' *AX*, 1959, **7**, 57.

HO PING-YÜ & NEEDHAM, JOSEPH (4). 'Elixir Poisoning in Mediaeval China.' *JAN*, 1959, **48**, 221.

HO PING-YÜ & NEEDHAM, JOSEPH. See Tshao Thien-Chhin, Ho Ping-Yü & Needham, J.

HOEFER, F. (1). *Histoire de la Chimie.* 2 vols. Paris, 1842–3. 2nd ed. 2 vols. Paris, 1866–9.

HOENIG, J. (1). 'Medical Research on Yoga.' *COPS,* 1968, **11**, 69.

HOERNES, M. (1). *Natur- und Ur-geschichte der Menschen.* Vienna and Leipzig, 1909.

HOERNLE, A. F. R. (1) (ed. & tr.). *The Bower Manuscript; Facsimile Leaves, Nagari Transcript, Romanised Transliteration and English Translation with Notes.* 2 vols. Govt. Printing office, Calcutta, 1893–1912. (Archaeol. Survey of India, New Imperial Series, no. 22.) Mainly pharmacological text of late +4th cent. but with some chemistry also.

HOFF, H. H., GUILLEMIN, L. & GUILLEMIN, R. (1) (tr. and ed.). *The 'Cahier Rouge' of Claude Bernard.* Schenkman, Cambridge, Mass. 1967.

HOFFER, A. & OSMOND, H. (1). *The Hallucinogens.* Academic Press, New York and London, 1968. With a chapter by T. Weckowicz.

HOFFMANN, G. (1). Art. 'Chemie' in A. Ladenburg (ed.), *Handwörterbuch der Chemie,* Trewendt, Breslau, 1884, vol. 2, p. 516. This work forms Division 2, Part 3 of W. Förster (ed.), *Encyklopaedie der Naturwissenschaften* (same publisher).

HOFMANN, K. B. (1). 'Zur Geschichte des Zinkes bei den Alten.' *BHMZ,* 1882, **41**, 492, 503.

HOLGEN, H. J. (1). 'Iets over de Chineesche Alchemie.' *CW,* 1917, **24**, 400.

HOLGEN, H. J. (2). 'Iets uit de Geschiedenis van de Chineesche Mineralogie en Chemische Technologie.' *CW,* 1917, **24**, 468.

HOLLOWAY, M. (1). *Heavens on Earth; Utopian Communities in America, +1680 to 1880.* Turnstile, London, 1951. 2nd ed. Dover, New York, 1966.

HOLMYARD, E. J. (1). *Alchemy.* Penguin, London, 1957.

HOLMYARD, E. J. (2). 'Jābir ibn Ḥayyān [including a bibliography of the Jābirian corpus].' *PRSM,* 1923, **16** (Hist. Med. Sect.), 46.

HOLMYARD, E. J. (3). 'Some Chemists of Islam.' *SPR,* 1923, **18**, 66.

HOLMYARD, E. J. (4). 'Arabic Chemistry [and Cupellation].' *N,* 1922, **109**, 778.

HOLMYARD, E. J. (5). '*Kitāb al-'Ilm al-Muktasab fī Zirā'at al-Dhahab*' (*Book of Knowledge acquired concerning the Cultivation of Gold*), by Abū'l Qāsim Muḥammad ibn Aḥmad al-Irāqī [d. c. +1300]; the Arabic text edited with a translation and introduction. Geuthner, Paris, 1923.

HOLMYARD, E. J. (7). 'A Critical Examination of Berthelot's Work on Arabic Chemistry.' *ISIS,* 1924, **6**, 479.

HOLMYARD, E. J. (8). 'The Identity of Geber.' *N,* 1923, **111**, 191.

HOLMYARD, E. J. (9). 'Chemistry in Mediaeval Islam.' *CHIND,* 1923, **42**, 387. *SCI,* 1926, 287.

H[OLMYARD], E. J. (10). 'The Accuracy of Weighing in the +8th Century.' *N,* 1925, **115**, 963.

HOLMYARD, E. J. (11). 'Maslama al-Majrīṭī and the *Rutbat al-Ḥakīm* [(The Sage's Step)].' *ISIS,* 1924, **6**, 293.

HOLMYARD, E. J. (12) (ed.). *The 'Ordinall of Alchimy' by Thomas Norton of Bristoll* (c. +1440; facsimile reproduction from the *Theatrum Chemicum Brittannicum* (+1652) with annotations by Elias Ashmole). Arnold, London, 1928.

HOLMYARD, E. J. (13). 'The Emerald Table.' *N,* 1923, **112**, 525.

HOLMYARD, E. J. (14). 'Alchemy in China.' *AP,* 1932, **3**, 745.

HOLMYARD, E. J. (15). 'Aidamir al-Jildakī [+14th-century alchemist].' *IRAQ,* 1937, **4**, 47.

HOLMYARD, E. J. (16). 'The Present Position of the Geber Problem.' *SPR,* 1925, **19**, 415.

HOLMYARD, E. J. (17). 'An Essay on Jābir ibn Ḥayyān.' Art. in *Studien z. Gesch. d. Chemie; Festgabe f. E. O. von Lippmann zum 70. Geburtstage,* ed. J. Ruska (37). Springer, Berlin, 1927, p. 28.

HOLMYARD, E. J. & MANDEVILLE, D. C. (1). '*Avicennae De Congelatione et Conglutinatione Lapidum*', being Sections of the '*Kitāb al-Shifā*'; the Latin and Arabic texts edited with an English translation of the latter and with critical notes. Geuthner, Paris, 1927. Rev. G. Sarton, *ISIS,* 1928, **11**, 134.

HOLTORF, G. W. (1). *Hongkong—World of Contrasts.* Books for Asia, Hongkong, 1970.

HOMANN, R. (1). *Die wichtigsten Körpergottheiten im 'Huang Thing Ching'* (Inaug. Diss. Tübingen). Kümmerle, Göppingen, 1971. (Göppinger Akademische Beiträge, no. 27.)

HOMBERG, W. (1). Chemical identification of a carved realgar cup brought from China by the ambassador of Siam. *HRASP,* 1703, 51.

HOMMEL, R. P. (1). *China at Work; an illustrated Record of the Primitive Industries of China's Masses, whose Life is Toil, and thus an Account of Chinese Civilisation.* Bucks County Historical Society, Doylestown, Pa.; John Day, New York, 1937.

HOMMEL, W. (1). 'The Origin of Zinc Smelting.' *EMJ,* 1912, **93**, 1185.

HOMMEL, W. (2). 'Über indisches und chinesisches Zink.' *ZAC,* 1912, **25**, 97.

HOMMEL, W. (3). 'Chinesisches Zink.' *CHZ,* 1912, **36**, 905, 918.

HOOVER, H. C. & HOOVER, L. H. (1) (tr.). *Georgius Agricola 'De Re Metallica' translated from the 1st Latin edition of 1556, with biographical introduction, annotations and appendices upon the development of mining methods, metallurgical processes, geology, mineralogy and mining law from the earliest times to the 16th century.* 1st ed. Mining Magazine, London, 1912; 2nd ed. Dover, New York, 1950.

HOOYKAAS, R. (1). 'The Experimental Origin of Chemical Atomic and Molecular Theory before Boyle.' *CHYM*, 1949, **2**, 65.

HOOYKAAS, R. (2). 'The Discrimination between "Natural" and "Artificial" Substances and the Development of Corpuscular Theory.' *A/AIHS*, 1947, **1**, 640.

HOPFNER, T. (1). *Griechisch-Aegyptischer Offenbarungszauber*. 2 vols. photolitho script. (Studien z. Palaeogr. u. Papyruskunde, ed. C. Wessely, nos. 21, 23.)

HOPKINS, A. J. (1). *Alchemy, Child of Greek Philosophy*. Columbia Univ. Press, New York, 1934. Rev. D. W. Singer, *A*, 1936, **18**, 94; W. J. Wilson, *ISIS*, 1935, **24**, 174.

HOPKINS, A. J. (2). 'A Defence of Egyptian Alchemy.' *ISIS*, 1938, **28**, 424.

HOPKINS, A. J. (3). 'Bronzing Methods in the Alchemical Leiden Papyri.' *CN*, 1902, **85**, 49.

HOPKINS, A. J. (4). 'Transmutation by Colour; a Study of the Earliest Alchemy.' Art. in *Studien z. Gesch. d. Chemie* (von Lippmann Festschrift), ed. J. Ruska. Springer, Berlin, 1927, p. 9.

HOPKINS, E. W. (3). 'Soma.' Art. in *ERE*, vol. xi, p. 685.

HOPKINS, E. W. (4). 'The Fountain of Youth.' *JAOS*, 1905, **26**, 1–67.

HOPKINS, L. C. (17). 'The Dragon Terrestrial and the Dragon Celestial; I, A Study of the *Lung* (terrestrial).' *JRAS*, 1931, 791.

HOPKINS, L. C. (18). 'The Dragon Terrestial and the Dragon Celestial; II, A Study of the *Chhen* (celestial).' *JRAS*, 1932, 91.

HOPKINS, L. C. (25). 'Metamorphic Stylisation and the Sabotage of Significance; a Study in Ancient and Modern Chinese Writing.' *JRAS*, 1925, 451.

HOPKINS, L. C. (26). 'Where the Rainbow Ends.' *JRAS*, 1931, 603.

HORI ICHIRO (1). 'Self-Mummified Buddhas in Japan; an Aspect of the Shugen-dō ('Mountain Asceticism') Cult.' *HOR*, 1961, **1** (no. 2), 222.

HORI ICHIRO (2). *Folk Religion in Japan; Continuity and Change*, ed. J. M. Kitagawa & A. L. Miller. Univ. of Tokyo Press, Tokyo and Univ. of Chicago Press, Chicago, 1968. (Haskell Lectures on the History of Religions, new series, no. 1.)

D'HORME, E. & DUSSAUD, R. (1). *Les Religions de Babylonie et d'Assyrie, des Hittites et des Hourrites, des Phéniciens et des Syriens*. Presses Univ. de France, Paris, 1945. (Mana, Introd. à l'Histoire des Religions, no. 1, pt. 2.)

D'HORMON, A. *et al.* (1) (ed. & tr.). '*Han Wu Ti Ku Shih;* Histoire Anecdotique et Fabuleuse de l'Empereur Wou [Wu] des Han' in *Lectures Chinoises*. École Franco-Chinoise, Peiping, 1945 (no. 1), p. 28.

D'HORMON, A. (2) (ed.). *Lectures Chinoises*. École Franco-Chinoise, Peiping, 1945–. No. 1 contains text and tr. of the *Han Wu Ti Ku Shih*, p. 28.

HOURANI, G. F. (1). *Arab Seafaring in the Indian Ocean in Ancient and Early Mediaeval Times*. Princeton Univ. Press, Princeton, N.J. 1951. (Princeton Oriental Studies, no. 13.)

HOWARD-WHITE, F. B. (1). *Nickel, an Historical Review*. Methuen, London, 1963.

HOWELL, E. B. (1) (tr.). '*Chin Ku Chhi Kuan;* story no. XIII, the Persecution of Shen Lien.' *CJ*, 1925, **3**, 10.

HOWELL, E. B. (2) (tr.). *The Inconstancy of Madam Chuang, and other Stories from the Chinese. . .*(from the *Chin Ku Chhi Kuan, c.* +1635). Laurie, London, n.d. (1925).

HRISTOV, H., STOJKOV, G. & MIJATER, K. (1). *The Rila Monastery [in Bulgaria]; History, Architecture, Frescoes, Wood-Carvings*. Bulgarian Acad. of Sci., Sofia, 1959. (Studies in Bulgaria's Architectural Heritage, no. 6,)

HSIA NAI (6). 'Archaeological Work during the Cultural Revolution.' *CREC*, 1971, **20** (no. 10), 31.

HSIA NAI, KU YEN-WEN, LAN HSIN-WÊN *et al.* (1). *New Archaeological Finds in China*. Foreign Languages Press, Peking, 1972. With colour-plates, and Chinese characters in footnotes.

HSIAO WÊN (1). 'China's New Discoveries of Ancient Treasures.' *UNESC*, 1972, **25** (no. 12), 12.

HTIN AUNG, MAUNG (1). *Folk Elements in Burmese Buddhism*. Oxford, 1962. Rev. P. M. R[attansi], *AX*, 1962, **10**, 142.

HUANG TZU-CHHING (1). 'Über die alte chinesische Alchemie und Chemie.' *WA*, 1957, **6**, 721.

HUANG TZU-CHHING (2). 'The Origin and Development of Chinese Alchemy.' Unpub. MS. of a lecture in the Physiological Institute of Chhinghua University, *c.* 1942 (dated 1944). A preliminary form of Huang Tzu-Chhing (1) but with some material which was omitted from the German version, though that was considerably enlarged.

HUANG TZU-CHHING & CHAO YÜN-TSHUNG (1) (tr.). 'The Preparation of Ferments and Wines [as described in the *Chhi Min Yao Shu* of] Chia Ssu-Hsieh of the Later Wei Dynasty [*c.* +540]; with an introduction by T. L. Davis.' *HJAS*, 1945, **9**, 24. Corrigenda by Yang Lien-Shêng, 1946, **10**, 186.

HUANG WÊN (1). '*Nei Ching*, the Chinese Canon of Medicine.' *CMJ*, 1950, **68**, 17 (originally M.D. Thesis, Cambridge, 1947).

HUANG WÊN (2). *Poems of Mao Tsê-Tung, translated and annotated*. Eastern Horizon Press, Hongkong, 1966.

HUARD, P. & HUANG KUANG-MING (M. WONG) (1). 'La Notion de Cercle et la Science Chinoise.' *A/AIHS*, 1956, **9**, 111. (Mainly physiological and medical.)

HUARD, P. & HUANG KUANG-MING (M. WONG) (2). *La Médecine Chinoise au Cours des Siècles*. Dacosta, Paris, 1959.

HUARD, P. & HUANG KUANG-MING (M. WONG) (3). 'Évolution de la Matière Médicale Chinoise.' *JAN*, 1958, **47**. Sep. pub. Brill, Leiden, 1958.

HUARD, P. & HUANG KUANG-MING (M. WONG) (5). 'Les Enquêtes Françaises sur la Science et la Technologie Chinoises au 18e Siècle.' *BEFEO*, 1966, **53**, 137–226.

HUARD, P. & HUANG KUANG-MING (M. WONG) (7). *Soins et Techniques du Corps en Chine, au Japon et en Inde; Ouvrage précédé d'une Étude des Conceptions et des Techniques de l'Éducation Physique, des Sports et de la Kinésithérapie en Occident dépuis l'Antiquité jusquà l'Époque contemporaine*. Berg International, Paris, 1971.

HUARD, P., SONOLET, J. & HUANG KUANG-MING (M. WONG) (1). 'Mesmer en Chine; Trois Lettres Médicales [MSS] du R. P. Amiot; rédigées à Pékin, de +1783 à +1790. *RSH*, 1960, **81**, 61.

HUBER, E. (1). 'Die mongolischen Destillierapparate.' *CHA*, 1928, **15**, 145.

HUBER, E. (2). *Der Kampf um den Alkohol im Wandel der Kulteren*. Trowitsch, Berlin, 1930.

HUBER, E. (3). *Bier und Bierbereitung bei den Völkern der Urzeit,*

Vol. 1. *Babylonien und Ägypten.*

Vol. 2. *Die Völker unter babylonischen Kultureinfluss; Auftreten des gehopften Bieres.*

Vol. 3. *Der ferne Osten und Äthiopien.*

Gesellschaft f. d. Geschichte und Bibliographie des Brauwesens, Institut f. Gärungsgewerbe, Berlin, 1926–8.

HUBICKI, W. (1). 'The Religious Background of the Development of Alchemy and Chemistry at the Turn of the +16th and +17th Centuries.' Communication to the XIIth Internat. Congr. Hist. of Sci. Paris, 1968. Résumés, p. 102. Actes, vol. 3A, p. 81.

HUFELAND, C. (1). *Makrobiotik; oder die Kunst das menschliche Leben zu verlängern*. Berlin, 1823. *The Art of Prolonging Life*. 2 vols. Tr. from the first German ed. London, 1797. Hebrew tr. Lemberg (Lwów), 1831.

HUGHES, A. W. McKENNY (1). 'Insect Infestation of Churches.' *JRIBA*, 1954.

HUGHES, E. R. (1). *Chinese Philosophy in Classical Times*. Dent, London, 1942. (Everyman Library, no. 973.)

HUGHES, M. J. & ODDY, W. A. (1). 'A Reappraisal of the Specific Gravity Method for the Analysis of Gold Alloys.' *AMY*, 1970, **12**, 1.

HUMBERT, J. P. L. (1). *Guide de la Conversation Arabe*. Paris, Bonn and Geneva, 1838.

VON HUMBOLDT, ALEXANDER (1). *Cosmos; a Sketch of a Physical Description of the Universe*. 5 vols. Tr. E. Cotté, B. H. Paul & W. S. Dallas. Bohn, London, 1849–58.

VON HUMBOLDT, ALEXANDER (3). *Examen Critique de l'Histoire de la Géographie du Nouveau Continent, et des Progrès de l'Astronomie Nautique au 15e et 16e Siècles*. 2 vols. Paris, 1837.

VON HUMBOLDT, ALEXANDER (4). *Fragmens de Géologie et de Climatologie Asiatique*. 2 vols. Gide, de la Forest & Delaunay, Paris, 1831.

HUMMEL, A. W. (6). 'Astronomy and Geography in the Seventeenth Century [in China].' (On Hsiung Ming-Yü's work.) *ARLC/DO*, 1938, 226.

HUNGER, H., STEGMÜLLER, O., ERBSE, H. *et al.* (1). *Geschichte der Textüberlieferung der antiken und mittelälterlichen Literatur*. 2 vols. Vol. 1, *Antiken Literatur*. Atlantis, Zürich, 1964. See Ineichen, Schindler, Bodmer *et al.* (1).

HUSAIN, YUSUF (1) (ed. & tr.). 'Ḥauḍ al-Ḥayāt [= Baḥr al-Ḥayāt (The Ocean, or Water, of Life)], la Version Arabe de l'Amritkunḍa [text and French précis transl.].' *JA*, 1928, **213**, 291.

HUTTEN, E. H. (1). 'Culture, One and Indivisible.' *HUM*, 1971, **86** (no. 5), 137.

HUZZAYIN, S. A. (1). *Arabia and the Far East; their commercial and cultural relations in Graeco-Roman and Irano-Arabian times*. Soc. Royale de Géogr. Cairo, 1942.

ICHIDA, MIKINOSUKE (1). 'The Hu Chi, mainly Iranian Girls, found in China during the Thang Period.' *MRDTB*, 1961, **20**, 35.

IDELER, J. L. (1) (ed.). *Physici et Medici Graeci Minores*. 2 vols. Reimer, Berlin, 1841.

IHDE, A. J. (1). 'Alchemy in Reverse; Robert Boyle on the Degradation of Gold.' *CHYM*, 1964, **9**, 47. Abstr. in Proc. Xth Internat. Congr. Hist. of Sci., Ithaca, N.Y., 1962, p. 907.

ILG, A. (1). 'Theophilus Presbyter *Schedula Diversarum Artium*; I, Revidierter Text, Übersetzung und Appendix.' *QSKMR*, 1874, **7**, 1–374.

IMBAULT-HUART, C. (1). 'La Légende du premier Pape des Taoistes, et l'Histoire de la Famille Pontificale des Tchang [Chang], d'après des Documents Chinois, traduits pour la première fois.' *JA*, 1884 (8e sér.), **4**, 389. Sep. pub. Impr. Nat. Paris, 1885.

IMBAULT-HUART, C. (2). 'Miscellanées Chinois.' *JA*, 1881 (7e sér.), **18**, 255, 534.

INEICHEN, G., SCHINDLER, A., BODMER, D. *et al.* (1). *Geschichte der Textüberlieferung der antiken und mittelälterlichen Literatur.* 2 vols. Vol. 2, *Mittelälterlichen Literatur.* Atlantis, Zürich, 1964. See Hunger, Stegmüller, Erbse *et al.* (1).

INTORCETTA, P., HERDTRICH, C., [DE] ROUGEMONT, F. & COUPLET, P. (1) (tr.). '*Confucius Sinarum Philosophus, sive Scientia Sinensis, latine exposita*'...; *Adjecta est: Tabula Chronologica Monarchiae Sinicae juxta cyclos annorum LX, ab anno post Christum primo, usque ad annum praesentis Saeculi 1683* [by P. Couplet, pr. 1686]. Horthemels, Paris, 1687. Rev. in *PTRS*, 1687, **16** (no. 189), 376.

IYENGAR, B. K. S. (1). *Light on Yoga* ('*Yoga Bīpika*'). Allen & Unwin, London, 2nd ed. 1968, 2nd imp. 1970.

IYER, K. C. VIRARAGHAVA (1). 'The Study of Alchemy [in Tamilnad, South India].' Art. in *Acarya* [*P.C.*] *Ray Commemoration Volume*, ed. H. N. Datta, Meghned Saha, J. C. Ghosh *et al.* Calcutta, 1932, p. 460.

JACKSON, R. D. & VAN BAVEL, C. H. M. (1). 'Solar distillation of water from Soil and Plant materials; a simple Desert Survival technique.' *S*, 1965, **149**, 1377.

JACOB, E. F. (1). 'John of Roquetaillade.' *BJRL*, 1956, **39**, 75.

JACOBI, HERMANN (3). '[The "Abode of the Blest" in] Hinduism.' *ERE*, vol. ii, p. 698.

JACOBI, JOLANDE (1). *The Psychology of C. G. Jung; an Introduction with Illustrations.* Tr. from Germ. by R. Manheim. Routledge & Kegan Paul, London, 1942. 6th ed. (revised), 1962.

JACQUES, D. H. (1). *Physical Perfection.* New York, 1859.

JAGNAUX, R. (1). *Histoire de la Chimie.* 2 vols. Baudry, Paris, 1891.

JAHN, K. & FRANKE, H. (1). *Die China-Geschichte des Rašīd ad-Dīn* [*Rashīd al-Dīn*]; *Übersetzung, Kommentar, Facsimiletafeln.* Böhlaus, Vienna, 1971. (Österreiche Akademie der Wissenschaften, Phil.-Hist. Kl., Denkschriften, no. 105; Veröffentl. d. Kommission für Gesch. Mittelasiens, no. 1.) This is the Chinese section of the *Jāmiʿ al-Tawārīkh*, finished in +1304, the whole by +1316. See Meredith-Owens (1).

JAMES, MONTAGUE R. (1) (ed. & tr.). *The Apocryphal New Testament; being the Apocryphal Gospels, Acts, Epistles and Apocalypses, with other Narratives and Fragments, newly translated by....* Oxford, 1924, repr. 1926 and subsequently.

JAMES, WILLIAM (1). *Varieties of Religious Experience; a Study in Human Nature.* Longmans Green, London, 1904. (Gifford Lectures, 1901–2.)

JAMSHED BAKHT, HAKIM, S. & MAHDIHASSAN, S. (1). 'Calcined Metals or *kushtas*; a Class of Alchemical Preparations used in Unani-Ayurvedic Medicine.' *MED*, 1962, **24**, 117.

JAMSHED BAKHT, HAKIM, S. & MAHDIHASSAN, S. (2). 'Essences [(*araqiath*)]; a Class of Alchemical Preparations [used in Unani-Ayurvedic Medicine].' *MED*, 1962, **24**, 257.

JANSE, O. R. T. (6). 'Rapport Préliminaire d'une Mission archéologique en Indochine.' *RAA/AMG*, 1935, **9**, 144, 209; 1936, **10**, 42.

JEBB, S. (1). *Fratris Rogeri Bacon Ordinis Minorum 'Opus Majus' ad Clementum Quartum Pontificem Romanum* [r. +1265 to +1268] *ex MS. Codice Dublinensi, cum aliis quibusdam collato, nunc primum edidit...* Bowyer, London, 1733.

JEFFERYS, W. H. & MAXWELL, J. L. (1). *The Diseases of China, including Formosa and Korea.* Bale & Danielsson, London, 1910. 2nd ed., re-written by Maxwell alone. ABC Press, Shanghai, 1929.

JENYNS, R. SOAME (3). *Archaic* [*Chinese*] *Jades in the British Museum.* Brit. Mus. Trustees, London, 1951.

JOACHIM, H. H. (1). 'Aristotle's Conception of Chemical Combination.' *JP*, 1904, **29**, 72.

JOHN OF ANTIOCH (fl. +610) (1). *Historias Chronikēs apo Adam.* See Valesius, Henricus (1).

JOHN MALALA (prob. = Joh. Scholasticus, Patriarch of Byzantium, d. +577). *Chronographia*, ed. W. Dindorf. Weber, Bonn, 1831 (in *Corp. Script. Hist. Byz.* series).

JOHNSON, A. CHANDRAHASAN & JOHNSON, SATYABAMA (1). 'A Demonstration of Oesophageal Reflux using Live Snakes.' *CLINR*, 1969, **20**, 107.

JOHNSON, C. (1) (ed.) (tr.). '*De Necessariis Observantiis Scaccarii Dialogus (Dialogus de Scaccario)*', '*Discourse on the Exchequer*', by Richard Fitznigel, Bishop of London and Treasurer of England [c. +1180], text and translation, with introduction. London, 1950.

JOHNSON, OBED S. (1). *A Study of Chinese Alchemy.* Commercial Press, Shanghai, 1928. Ch. tr. by Huang Su-Fêng: *Chung-Kuo Ku-Tai Lien-Tan Shu.* Com. Press, Shanghai, 1936. Rev. B. Laufer, *ISIS*, 1929, **12**, 330; H. Chatley, *JRAS/NCB*, 1928, *NCDN*, 9 May 1928. Cf. Waley (14).

JOHNSON, R. P. (1). 'Note on some Manuscripts of the *Mappae Clavicula*.' *SP*, 1935, **10**, 72.

JOHNSON, R. P. (2). '*Compositiones Variae*'... *an Introductory Study.* Urbana, Ill. 1939. (Illinois Studies in Language and Literature, vol. 23, no. 3.)

JONAS, H. (1). *The Gnostic Religion.* Beacon, Boston, 1958.

JONES, B. E. (1). *The Freemason's Guide and Compendium.* London, 1950.

JONES, C. P. (1) (tr.). *Philostratus' 'Life of Apollonius'*, with an introduction by G. W. Bowersock, Penguin, London, 1970.

DE JONG, H. M. E. (1). *Michael Maier's ' Atalanta Fugiens'; Sources of an Alchemical Book of Emblems.* Brill, Leiden, 1969. (Janus Supplements, no. 8.)

JOPE, E. M. (3). 'The Tinning of Iron Spurs; a Continuous Practice from the +10th to the +17th Century.' *OX*, 1956, **21**, 35.

JOSEPH, L. (1). 'Gymnastics from the Middle Ages to the Eighteenth Century.' *CIBA/S*, 1949, **10**, 1030.

JOSTEN, C. H. (1). 'The Text of John Dastin's "Letter to Pope John XXII".' *AX*, 1951, **4**, 34.

JOURDAIN, M. & JENYNS, R. SOAME (1). *Chinese Export Art.* London, 1950.

JOYCE, C. R. B. & CURRY, S. H. (1) (ed.). *The Botany and Chemistry of* Cannabis. Williams & Wilkins, Baltimore, 1970. Rev. *SAM*, 1971, **224** (no. 3), 238.

JUAN WEI-CHOU. See Wei Chou-Yuan.

JULIEN, STANISLAS (1) (tr.). *Voyages des Pélerins Bouddhistes.* Impr. Imp., Paris, 1853–8. 3 vols. (Vol. 1 contains Hui Li's Life of Hsüan-Chuang; Vols. 2 and 3 contain Hsüan-Chuang's *Hsi Yu Chi*.)

JULIEN, STANISLAS (11). 'Substance anaesthésique employée en Chine dans le Commencement du 3e Siècle de notre ére pour paralyser momentanément la Sensibilité.' *CRAS*, 1849, **28**, 195.

JULIEN, STANISLAS & CHAMPION, P. (1). *Industries Anciennes et Modernes de l'Empire Chinois, d'après des Notices traduites du Chinois....* (paraphrased précis accounts based largely on *Thien Kung Khai Wu*; and eye-witness descriptions from a visit in 1867). Lacroix, Paris, 1869.

JULIUS AFRICANUS. *Kestoi.* See Thevenot, D. (1).

JUNG, C. G. (1). *Psychologie und Alchemie.* Rascher, Zürich, 1944. 2nd ed. revised, 1952. Eng. tr. R. F. C. Hull [& B. Hannah], *Psychology and Alchemy.* Routledge & Kegan Paul, London, 1953 (Collected Works, vol. 12). Rev. W. Pagel, *ISIS*, 1948, **39**, 44; G. II[eym], *AX*, 1948, **3**, 64.

JUNG, C. G. (2). 'Synchronicity; an Acausal Connecting Principle' [on extra-sensory perception]; essay in the collection *The Structure and Dynamics of the Psyche.* Routledge & Kegan Paul, London, 1960 (Collected Works, vol. 8). Rev. C. Allen, *N*, 1961, **191**, 1235.

JUNG, C. G. (3). *Alchemical Studies.* Eng. tr. from the Germ., R. F. C. Hull. Routledge & Kegan Paul, London, 1968 (Collected Works, vol. 13). Contains the 'European commentary' on the *Thai-I Chin Hua Tsung Chih*, pp. 1–55, and the 'Interpretation of the Visions of Zosimos', pp. 57–108.

JUNG, C. G. (4). *Aion; Researches into the Phenomenology of the Self.* Eng. tr. from the Germ., R. F. C. Hull. Routledge & Kegan Paul, London, 1959 (Collected Works, vol. 9, pt. 2).

JUNG, C. G. (5). *Paracelsica.* Rascher, Zürich and Leipzig, 1942. Eng. tr. from the Germ., R. F. C. Hull.

JUNG, C. G. (6). *Psychology and Religion; West and East.* Eng. tr. from the Germ., R. F. C. Hull. Routledge & Kegan Paul, London, 1958 (63 corr.) (Collected Works, vol. 11). Contains the essay 'Transformation Symbolism in the Mass'.

JUNG, C. G. (7). *Memories, Dreams and Reflections.* Recorded by A. Jaffé, tr. R. & C. Winston. New York and London, 1963.

JUNG, C. G. (8). *Mysterium Conjunctionis; an Enquiry into the Separation and Synthesis of Psychic Opposites in Alchemy.* Eng. tr. from the Germ., R. F. C. Hull. Routledge & Kegan Paul, London, 1963 (Collected Works, vol. 14). Orig. ed. *Mysterium Conjunctionis; Untersuchung ü. die Trennung u. Zusammensetzung der seelische Gegensätze in der Alchemie*, 2 vols. Rascher, Zürich, 1955, 1956 (Psychol. Abhandlungen, ed. C. G. J., nos. 10, 11).

JUNG, C. G. (9). 'Die Erlösungsvorstellungen in der Alchemie.' *ERJB*, 1936, 13–111.

JUNG, C. G. (10). *The Integration of the Personality.* Eng. tr. S. Dell. Farrar & Rinehart, New York and Toronto, 1939, Kegan Paul, Trench & Trübner, London, 1940, repr. 1941. Ch. 5, 'The Idea of Redemption in Alchemy' is the translation of Jung (9).

JUNG, C. G. (11). 'Über Synchronizität.' *ERJB*, 1952, **20**, 271.

JUNG, C. G. (12). *Analytical Psychology; its Theory and Practice.* Routledge, London, 1968.

JUNG, C. G. (13). *The Archetypes and the Collective Unconscious.* Eng. tr. by R. F. C. Hull. Routledge & Kegan Paul, London, 1959 (Collected Works, vol. 9, pt. 1).

JUNG, C. G. (14). 'Einige Bemerkungen zu den Visionen des Zosimos.' *ERJB*, 1938. Revised and expanded as 'Die Visionen des Zosimos' in *Von der Wurzeln des Bewussteins; Studien ü. d. Archetypus.* In *Psychologische Abhandlungen.* Zürich, 1954, vol. 9.

JUNG, C. G. & PAULI, W. (1). *The Interpretation of Nature and the Psyche.*
 (*a*) 'Synchronicity; an Acausal Connecting Principle', by C. G. Jung.
 (*b*) 'The Influence of Archetypal Ideas on the Scientific Theories of Kepler', by W. Pauli.
 Tr. R. F. C. Hull. Routledge & Kegan Paul, London, 1955.
 Orig. pub. in German as *Naturerklärung und Psyche*, Rascher, Zürich, 1952 (Studien aus dem C. G. Jung Institut, no. 4).

KAHLBAUM, G. W. A. See Diergart, P. (Kahlbaum Festschrift).

KAHLE, P. (7). 'Chinese Porcelain in the Lands of Islam.' *TOCS*, 1942, 27. Reprinted in Kahle (3), p. 326, with Supplement, p. 351 (originally published in *WA*, 1953, **2**, 179 and *JPHS*, 1953, **1**, 1).

KAHLE, P. (8). 'Islamische Quellen über chinesischen Porzellan.' *ZDMG*, 1934, **88**, 1, *OAZ*, 1934, **19** (N.F.), 69.

KALTENMARK, M. (2) (tr.). *Le 'Lie Sien Tchouan' [Lieh Hsien Chuan]; Biographies Légendaires des Immortels Taoistes de l'Antiquité*. Centre d'Etudes Sinologiques Franco-Chinois (Univ. Paris), Peking, 1953. Crit. P. Demiéville, *TP*, 1954, **43**, 104.

KALTENMARK, M. (4). 'Ling Pao; Note sur un Terme du Taoisme Religieux', in *Mélanges publiés par l'Inst. des Htes. Etudes Chin*. Paris, 1960, vol. 2, p. 559 (Bib. de l'Inst. des Htes. Et. Chin. vol. 14).

KANGRO, H. (1). *Joachim Jungius' [+1587 to +1657] Experimente und Gedanken zur Begründung der Chemie als Wissenschaft; ein Beitrag zur Geistesgeschichte des 17. Jahrhunderts*. Steiner, Wiesbaden, 1968 (Boethius; *Texte und Abhandlungen z. Gesch. d. exakten Naturwissenschaften*, no. 7). Rev. R. Hooykaas, *A/AIHS*, 1970, **23**, 299.

KAO LEI-SSU (1) (Aloysius Ko, S.J.). 'Remarques sur un Écrit de M. P[auw] intitulé "Recherches sur les Égyptiens et les Chinois" (1775).' *MCHSAMUC*, 1777, **2**, 365–574 (in some editions, 2nd pagination, 1–174).

KAO, Y. L. (1). 'Chemical Analysis of some old Chinese Coins.' *JWCBRS*, 1935, **7**, 124.

KAPFERER, R. (1). 'Der Blutkreislauf im altchinesischen Lehrbuch *Huang Ti Nei Ching*.' *MMW*, 1939 (no. 18), 718.

KARIMOV, U. I. (1) (tr.). *Neizvestnoe Sovrineniye al-Rāzī 'Kniga Taishnvi Taishi' (A Hitherto Unknown Work of al-Rāzī, 'Book of the Secret of Scerets')*. Acad. Sci. Uzbek SSR, Tashkent, 1957. Rev. N. A. Figurovsky, tr. P. L. Wyvill, *AX*, 1962, **10**, 146.

KARLGREN, B. (18). 'Early Chinese Mirror Inscriptions.' *BMFEA*, 1934, **6**, 1.

KAROW, O. (2) (ed.). *Die Illustrationen des Arzneibuches der Periode Shao-Hsing* (Shao-Hsing Pên Tshao Hua Thu) *vom Jahre +1159, ausgewählt und eingeleitet*. Farbenfabriken Bayer Aktiengesellschaft (Pharmazeutisch-Wissenschaftliche Abteilung), Leverkusen, 1956. Album selected from the *Shao-Hsing Chiao-Ting Pên Tshao Chieh-Thi* published by Wada Toshihiko, Tokyo, 1933.

KASAMATSU, A. & HIRAI, T. (1). 'An Electro-encephalographic Study of Zen Meditation (*zazen*).' *FPNJ*, 1966, **20** (no. 4), 315.

KASSAU, E. (1). 'Charakterisierung einiger Steroidhormone durch Mikrosublimation.' *DAZ*, 1960, **100**, 1102.

KAZANCHIAN, T. (1). *Laboratornaja Technika i Apparatura v Srednevekovoj Armenii po drevnim Armjanskim Alchimicheskim Rukopisjam* (in Armenian with Russian summary). *SNM*, 1949, **2**, 1–28.

KEFERSTEIN, C. (1). *Mineralogia Polyglotta*. Anton, Halle, 1849.

KEILIN, D. & MANN, T. (1). 'Laccase, a blue Copper-Protein Oxidase from the Latex of *Rhus succedanea*.' *N*, 1939, **143**, 23.

KEILIN, D. & MANN, T. (2). 'Some Properties of Laccase from the Latex of Lacquer-Trees.' *N*, 1940, **145**, 304.

KEITH, A. BERRIEDALE (5). *The Religion and Philosophy of the Vedas and Upanishads*. 2 vols. Harvard Univ. Press, Cambridge (Mass.), 1925. (Harvard Oriental Series, nos. 31, 32.)

KEITH, A. BERRIEDALE (7). '[The State of the Dead in] Hindu [Belief].' *ERE*, vol. xi, p. 843.

KELLING, R. (1). *Das chinesische Wohnhaus; mit einem II Teil über das frühchinesische Haus unter Verwendung von Ergebnissen aus Übungen von Conrady im Ostasiatischen Seminar der Universität Leipzig, von Rudolf Keller und Bruno Schindler*. Deutsche Gesellsch. für Nat. u. Völkerkunde Ostasiens, Tokyo, 1935 (*MDGNVO*, Supplementband no. 13). Crit. P. Pelliot, *TP*, 1936, **32**, 372.

KENNEDY, J. (1). 'Buddhist Gnosticism, the System of Basilides.' *JRAS*, 1902, 377.

KENNEDY, J. (2). 'The Gospels of the Infancy, the *Lalita Vistara*, and the *Vishnu Purana*; or, the Transmission of Religious Ideas between India and the West.' *JRAS*, 1917, 209, 469.

KENT, A. (1). 'Sugar of Lead.' *MBLB*, 1961, **4** (no. 6), 85.

KERNEIZ, C. (1). *Les 'Asanas', Gymnastique immobile du Hathayoga*. Tallandier, Paris, 1946.

KERNEIZ, C. (2). *Le Yoga*. Tallandier, Paris, 1956. 2nd ed. 1960.

KERR, J. G. (1). '[Glossary of Chinese] Chemical Terms', in Doolittle (1), vol. 2, p. 542.

KEUP, W. (1) (ed.). *The Origin and Mechanisms of Hallucinations*. Plenum, New York and London, 1970.

KEYNES, J. M. (Lord Keynes) (1) (posthumous). 'Newton the Man.' Essay in *Newton Tercentenary Celebrations* (July 1946). Royal Society, London, 1947, p. 27. Reprinted in *Essays in Biography*.

KHORY, RUSTOMJEE NASERWANJEE & KATRAK, NANABHAI NAVROSJI (1). *Materia Medica of India and their Therapeutics*. Times of India, Bombay, 1903.

KHUNRATH, HEINRICH (1). *Amphitheatrum Sapientiae Aeternae Solius Verae, Christiano-Kabalisticum, Divino-Magicum, necnon Physico-Chymicum, Tetriunum, Catholicon*...Prague, 1598; Magdeburg, 1602; Frankfurt, 1608, and many other editions.

KIDDER, J. E. (1). *Japan before Buddhism*. Praeger, New York; Thames & Hudson, London, 1959.

KINCH, E. (1). 'Contributions to the Agricultural Chemistry of Japan.' *TAS/J*, 1880, **8**, 369.

KING, C. W. (1). *The Natural History of Precious Stones and of the Precious Metals*. Bell & Daldy, London, 1867.

KING, C. W. (2). *The Natural History of Gems or Decorative Stones*. Bell & Daldy, London, 1867.

KING, C. W. (3). *The Gnostics and their Remains*. 2nd ed. Nutt, London, 1887.

KING, C. W. (4). *Handbook of Engraved Gems*. 2nd ed. Bell, London, 1885.

KLAPROTH, J. (5). 'Sur les Connaissances Chimiques des Chinois dans le 8ème Siècle.' *MAIS/SP*, 1810, **2**, 476. Ital. tr., S. Guareschi, *SAEC*, 1904, **20**, 449.

KLAPROTH, J. (6). *Mémoires relatifs à l'Asie...* 3 vols. Dondey Dupré, Paris, 1826.

KLAPROTH, M. H. (1). *Analytical Essays towards Promoting the Chemical Knowledge of Mineral Substances*. 2 vols. Cadell & Davies, London, 1801.

KNAUER, E. (1). 'Die Ovarientransplantation.' *AFG*, 1900, **60**, 322.

KNOX, R. A. (1). *Enthusiasm; a Chapter in the History of Religion, with special reference to the +17th and +18th Centuries*. Oxford, 1950.

KO, ALOYSIUS. See Kao Lei-Ssu.

KOBERT, R. (1). 'Chronische Bleivergiftung in klassischen Altertume.' Art. in Kahlbaum Festschrift (1909), ed. Diergart (1), pp. 103–19.

KOPP, H. (1). *Geschichte d. Chemie*. 4 vols. 1843–7.

KOPP, H. (2). *Beiträge zur Geschichte der Chemie*. Vieweg, Braunschweig, 1869.

KRAMRISCH, S. (1). *The Art of India; Traditions of Indian Sculpture, Painting and Architecture*. Phaidon, London, 1954.

KRAUS, P. (1). 'Der Zusammenbruch der Dschābir-Legende; II, Dschābir ibn Ḥajjān und die Ismaʿilijja.' *JBFIGN*, 1930, **3**, 23. Cf. Ruska (1).

KRAUS, P. (2). 'Jābir ibn Ḥayyān; Contributions à l'Histoire des Idées Scientifiques dans l'Islam; I, Le Corpus des Écrits Jābiriens.' *MIE*, 1943, **44**, 1–214. Rev. M. Meyerhof, *ISIS*, 1944, **35**, 213.

KRAUS, P. (3). 'Jābir ibn Ḥayyān; Contributions à l'Histoire des Idées Scientifiques dans l'Islam; II, Jābir et la Science Grecque.' *MIE*, 1942, **45**, 1–406. Rev. M. Meyerhof, *ISIS*, 1944, **35**, 213.

KRAUS, P. (4) (ed.). *Jābir ibn Ḥayyān; Essai sur l'Histoire des Idées Scientifiques dans l'Islam*. Vol. 1. *Textes Choisis*. Maisonneuve, Paris and El-Kandgi, Cairo, 1935. No more appeared.

KRAUS, P. (5). *L'Épître de Beruni sur al-Rāzī (Risālat al-Bīrūnī fī Fihrist Kutub Muḥammad ibn Zakarīyā al-Rāzī) [c. +1036]*. Paris, 1936.

KRAUS, P. & PINES, S. (1). 'Al-Rāzī.' Art. in *EI*, vol. iii, pp. 1134 ff.

KREBS, M. (1). *Der menschlichen Harn als Heilmittel; Geschichte, Grundlagen, Entwicklung, Praxis*. Marquardt, Stuttgart, 1942.

KRENKOW, F. (2). 'The Oldest Western Accounts of Chinese Porcelain.' *IC*, 1933, **7**, 464.

KROLL, J. (1). *Die Lehren des Hermes Trismegistos*. Aschendorff, Münster i.W., 1914. (Beiträge z. Gesch. d. Philosophie des Mittelalters, vol. 12, no. 2.)

KROLL, W. (1). 'Bolos und Demokritos.' *HERM*, 1934, **69**, 228.

KRÜNITZ, J. G. (1). *Ökonomisch-Technologische Enzyklopädie*. Berlin, 1773–81.

KUBO NORITADA (1). 'The Introduction of Taoism to Japan.' In *Religious Studies in Japan*, no. 11 (no. 105), 457. See Soymié (5), p. 281 (10).

KUBO NORITADA (2). 'The Transmission of Taoism to Japan, with particular reference to the *san shih* (three corpses theory).' *Proc. IXth Internat. Congress of the History of Religions*, Tokyo, 1958, p. 335.

KUGENER, M. A. & CUMONT, F. (1). *Extrait de la CXXIII ème 'Homélie' de Sévère d'Antioch*. Lamertin, Brussels, 1912. (Recherches sur le Manichéisme, no. 2.)

KUGENER, M. A. & CUMONT, F. (2). *L'Inscription Manichéenne de Salone [Dalmatia]*. (A tombstone or consecration memorial of the Manichaean Virgin Bassa.) Lamertin, Brussels, 1912. (Recherches sur le Manichéisme, no. 3.)

KÜHN, F. (3). *Die Dreizehnstöckige Pagode* (Stories translated from the Chinese). Steiniger, Berlin, 1940.

KÜHNEL, P. (1). *Chinesische Novellen*. Müller, München, 1914.

KUNCKEL, J. (1). '*Ars Vitraria Experimentalis*', oder *Vollkommene Glasmacher-Kunst...* Frankfurt and Leipzig; Amsterdam and Danzig, 1679. 2nd ed. Frankfurt and Leipzig, 1689. 3rd ed. Nuremberg, 1743, 1756. French tr. by the Baron d'Holbach, Paris, 1752.

KUNCKEL, J. (2). '*Collegium Physico-Chemicum Experimentale*', oder *Laboratorium Chymicum; in welchem deutlich und gründlich von den wahren Principiis in der Natur und denen gewürckten Dingen so wohl über als in der Erden, als Vegetabilien, Animalien, Mineralien, Metallen..., nebst der Transmutation und Verbesserung der Metallen gehandelt wird...* Heyl, Hamburg and Leipzig, 1716.

KUNG, S. C., CHAO, S. W., PEI, Y. T. & CHANG, C. (1). 'Some Mummies Found in West China.' *JWCBRS*, 1939, **11**, 105.

KUNG YO-THING, TU YÜ-TSHANG, HUANG WEI-TÊ, CHHEN CHHANG-CHHING & seventeen other collaborators (1). 'Total Synthesis of Crystalline Insulin.' *SCISA*, 1966, **15**, 544.

LACAZE-DUTHIERS, H. (1). 'Tyrian purple.' *ASN/Z*, 1859 (4e sér.), **12**, 5.

LACH, D. F. (5). *Asia in the Making of Europe*. 2 vols. in 5 parts. Univ. Chicago Press, Chicago and London, 1965–.

LAGERCRANTZ, O. (1). *Papyrus Graecus Holmiensis*. Almquist & Wiksells, Upsala, 1913. (The first publication of the +3rd-cent. technical and chemical Stockholm papyrus.) Cf. Caley (2).

LAGERCRANTZ, O. (2). 'Über das Wort Chemie.' *KVSUA*, 1937–8, 25.

LAMOTTE, E. (1) (tr.). *Le Traité de la Grande Vertu de Sagesse de Nāgārjuna (Mahāprajñāpāramitā-śāstra)*. 3 vols. Muséon, Louvain, 1944 (Bibl. Muséon, no. 18). Rev. P. Demiéville, *JA*, 1950, **238**, 375.

LANDUR, N. (1). 'Compte Rendu de la Séance de l'Académie des Sciences [de France] du 24 Août 1868.' *LIN* (1e section), 1868, **36** (no. 1808), 273. Contains an account of a communication by M. Chevreul on the history of alchemy, tracing it to the *Timaeus;* with a critical paragraph by Landur himself maintaining that in his view much (though not all) of ancient and mediaeval alchemy was disguised moral and mystical philosophy.

LANE, E. W. (1). *An Account of the Manners and Customs of the Modern Egyptians (1833 to 1835)*. Ward Lock, London, 3rd ed. 1842; repr. 1890.

LANGE, E. F. (1). 'Alchemy and the Sixteenth-Century Metallurgists.' *AX*, 1966, **13**, 92.

LATTIMORE, O. & LATTIMORE, E. (1) (ed.). *Silks, Spices and Empire; Asia seen through the Eyes of its Discoverers*. Delacorte, New York, 1968. (Great Explorers Series, no. 3.)

LAUBRY, C. & BROSSE, T. (1). 'Documents recueillis aux Indes sur les "Yoguis" par l'enregistrement simultané du pouls, de la respiration et de l'electrocardiogramme.' *PM*, 1936, **44** (no. 83), 1601 (14 Oct.). Rev. J. Filliozat, *JA*, 1937, 521.

LAUBRY, C. & BROSSE, T. (2). 'Interférence de l'Activité Corticale sur le Système Végétatif Neuro-vasculaire.' *PM*, 1935, **43** (no. 84). (19 Oct.)

LAUFER, B. (1). *Sino-Iranica; Chinese Contributions to the History of Civilisation in Ancient Iran*. *FMNHP/AS*, 1919, **15**, no. 3 (Pub. no. 201). Rev. and crit. Chang Hung-Chao, *MGSC*, 1925 (ser. B), no. 5.

LAUFER, B. (8). *Jade; a Study in Chinese Archaeology and Religion*. *FMNHP/AS*, 1912, **10**, 1–370. Repub. in book form, Perkins, Westwood & Hawley, South Pasadena, 1946. Rev. P. Pelliot, *TP*, 1912, **13**, 434.

LAUFER, B. (10). 'The Beginnings of Porcelain in China.' *FMNHP/AS*, 1917, **15**, no. 2 (Pub. no. 192), (includes description of +2nd-century cast-iron funerary cooking-stove).

LAUFER, B. (12). 'The Diamond; a study in Chinese and Hellenistic Folk-Lore.' *FMNHP/AS*, 1915, **15**, no. 1 (Pub. no. 184).

LAUFER, B. (13). 'Notes on Turquois in the East.' *FMNHP/AS*, 1913, **13**, no. 1 (Pub. no. 169).

LAUFER, B. (15). 'Chinese Clay Figures, Pt. I; Prolegomena on the History of Defensive Armor.' *FMNHP/AS*, 1914, **13**, no. 2 (Pub. no. 177).

LAUFER, B. (17). 'Historical Jottings on Amber in Asia.' *MAAA*, 1906, **1**, 211.

LAUFER, B. (24). 'The Early History of Felt.' *AAN*, 1930, **32**, 1.

LAUFER, B. (28). 'Christian Art in China.' *MSOS*, 1910, **13**, 100.

LAUFER, B. (40). 'Sex Transformation and Hermaphrodites in Ancient China.' *AJPA*, 1920, **3**, 259.

LAUFER, B. (41). 'Die Sage von der goldgrabenden Ameisen.' *TP*, 1908, **9**, 429.

LAUFER, B. (42). *Tobacco and its Use in Asia*. Field Mus. Nat. Hist., Chicago, 1924. (Anthropology Leaflet, no. 18.)

LEADBEATER, C. W. (1). *The Chakras, a Monograph*. London, n.d.

LECLERC, L. (1) (tr.). 'Le Traité des Simples par Ibn al-Beithar.' *MAI/NEM*, 1877, **23**, 25; 1883, **26**.

LECOMTE, LOUIS (1). *Nouveaux Mémoires sur l'État présent de la Chine*. Anisson, Paris, 1696. (Eng. tr. *Memoirs and Observations Topographical, Physical, Mathematical, Mechanical, Natural, Civil and Ecclesiastical, made in a late journey through the Empire of China, and published in several letters, particularly upon the Chinese Pottery and Varnishing, the Silk and other Manufactures, the Pearl Fishing, the History of Plants and Animals, etc.* translated from the Paris edition, etc., 2nd ed. London, 1698. Germ. tr. Frankfurt, 1699–1700. Dutch tr. 's Graavenhage, 1698.)

VON LECOQ, A. (1). *Buried Treasures of Chinese Turkestan; an Account of the Activities and Adventures of the 2nd and 3rd German Turfan Expeditions*. Allen & Unwin, London, 1928. Eng. tr. by A. Barwell of *Auf Hellas Spuren in Ost-turkestan*. Berlin, 1926.

VON LECOQ, A. (2). *Von Land und Leuten in Ost-Turkestan...* Hinrichs, Leipzig, 1928.

LEDERER, E. (1). 'Odeurs et Parfums des Animaux.' *FCON*, 1950, **6**, 87.

LEDERER, E. & LEDERER, M. (1). *Chromatography; a Review of Principles and Applications*. Elsevier, Amsterdam and London, 1957.

LEEDS, E. T. (1). 'Zinc Coins in Mediaeval China.' *NC*, 1955 (6th ser.), **14**, 177.

LEEMANS, C. (1) (ed. & tr.). *Papyri Graeci Musei Antiquarii Publici Lugduni Batavi...* Leiden, 1885. (Contains the first publication of the +3rd-cent. chemical papyrus Leiden X.) Cf. Caley (1).

VAN LEERSUM, E. C. (1). *Préparation du Calomel chez les anciens Hindous.* Art. in Kahlbaum Festschrift (1909), ed. Diergart (1), pp. 120–6.

LEFÉVRE, NICOLAS (1). *Traicté de la Chymie.* Paris, 1660, 2nd ed. 1674. Eng. tr. *A Compleat Body of Chymistry.* Pulleyn & Wright, London, 1664. repr. 1670.

LEICESTER, H. M. (1). *The Historical Background of Chemistry.* Wiley, New York, 1965.

LEICESTER, H. M. & KLICKSTEIN, H. S. (1). 'Tenney Lombard Davis and the History of Chemistry.' *CHYM*, 1950, **3**, 1.

LEICESTER, H. M. & KLICKSTEIN, H. S. (2) (ed.). *A Source-Book in Chemistry, +1400 to 1900.* McGraw-Hill, New York, 1952.

LEISEGANG, H. (1). *Der Heilige Geist; das Wesen und Werden der mystisch-intuitiven Erkenntnis in der Philosophie und Religion der Griechen.* Teubner, Leipzig and Berlin, 1919; photolitho reprint, Wissenschaftliche Buchgesellschaft, Darmstadt, 1967. This constitutes vol. 1 of Leisegang (2).

LEISEGANG, H. (2). '*Pneuma Hagion*'; *der Ursprung des Geistbegriffs der synoptischen Evangelien aus d. griechischen Mystik.* Hinrichs, Leipzig, 1922. (Veröffentlichungen des Forschungsinstituts f. vergl. Religionsgeschichte an d. Univ. Leipzig, no. 4.) This constitutes vol. 2 of Leisegang (1).

LEISEGANG, H. (3). *Die Gnosis.* 3rd ed. Kröner, Stuttgart, 1941. (Kröners Taschenausgabe, no. 32.) French tr.: '*La Gnose*'. Paris, 1951.

LEISEGANG, H. (4). 'The Mystery of the Serpent.' *ERYB*, 1955, 218.

LENZ, H. O. (1). *Mineralogie der alten Griechen und Römer deutsch in Auszügen aus deren Schriften.* Thienemann, Gotha, 1861. Photo reprint, Sändig, Wiesbaden, 1966.

LEPESME, P. (1). 'Les Coléoptères des Denrées alimentaires et des Produits industriels entreposés.' Art. in *Encyclopédie Entomologique*, vol. xxii, pp. 1–335. Lechevalier, Paris, 1944.

LESSIUS, L. (1). *Hygiasticon; seu Vera Ratio Valetudinis Bonae et Vitae...ad extremam Senectute Conservandae.* Antwerp, 1614. Eng. tr. Cambridge, 1634; and two subsequent translations.

LEVEY, M. (1). 'Evidences of Ancient Distillation, Sublimation and Extraction in Mesopotamia.' *CEN*, 1955, **4**, 23.

LEVEY, M. (2). *Chemistry and Chemical Technology in Ancient Mesopotamia.* Elsevier, Amsterdam and London, 1959.

LEVEY, M. (3). 'The Earliest Stages in the Evolution of the Still.' *ISIS*, 1960, **51**, 31.

LEVEY, M. (4). 'Babylonian Chemistry; a Study of Arabic and −2nd Millennium Perfumery.' *OSIS*, 1956, **12**, 376.

LEVEY, M. (5). 'Some Chemical Apparatus of Ancient Mesopotamia.' *JCE*, 1955, **32**, 180.

LEVEY, M. (6). 'Mediaeval Arabic Toxicology; the "Book of Poisons" of Ibn Waḥshīya [+10th cent.] and its Relation to Early Indian and Greek Texts.' *TAPS*, 1966, **56** (no. 7), 1–130.

LEVEY, M. (7). 'Some Objective Factors in Babylonian Medicine in the Light of New Evidence.' *BIHM*, 1961, **35**, 61.

LEVEY, M. (8). 'Chemistry in the *Kitāb al-Sumum* (Book of Poisons) by Ibn al-Waḥshīya [al-Nabaṭī, *fl.* +912].' *CHYM*, 1964, **9**, 33.

LEVEY, M. (9). 'Chemical Aspects of Medieval Arabic Minting in a Treatise by Manṣūr ibn Ba'ra [*c.* +1230].' *JSHS*, 1971, Suppl. no. 1.

LEVEY, M. & AL-KHALEDY, NOURY (1). *The Medical Formulary [Aqrābādhīn] of [Muḥ. ibn 'Alī ibn 'Umar] al-Samarqandī [c. +1210], and the Relation of Early Arabic Simples to those found in the indigenous Medicine of the Near East and India.* Univ. Pennsylvania Press, Philadelphia, 1967.

LÉVI, S. (2). 'Ceylan et la Chine.' *JA*, 1900 (9e sér.), **15**, 411. Part of Lévi (1).

LÉVI, S. (4). 'On a Tantric Fragment from Kucha.' *IHQ*, 1936, **12**, 207.

LÉVI, S. (6). *Le Népal; Étude Historique d'un Royaume Hindou.* 3 vols. Paris, 1905–8. (Annales du Musée Guimet, Bib. d'Études, nos. 17–19.)

LÉVI, S. (8). 'Un Nouveau Document sur le Bouddhisme de Basse Époque dans l'Inde.' *BLSOAS*, 1931, **6**, 417. (Nāgārjuna and gold refining.)

LÉVI, S. (9). 'Notes Chinoises sur l'Inde; V, Quelques Documents sur le Bouddhisme Indien dans l'Asie Centrale, pt. 1.' *BEFEO*, 1905, **5**, 253.

LÉVI, S. (10). 'Vajrabodhi à Ceylan.' *JA*, 1900, (9e sér.) **15**, 418. Part of Lévi (1).

LEVOL, A. (1). 'Analyse d'un Échantillon de Cuivre Blanc de la Chine.' *RCA*, 1862, **4**, 24.

LEVY, ISIDORE (1). 'Sarapis; V, la Statue Mystérieuse.' *RHR/AMG*, 1911, **63**, 124.

LEWIS, BERNARD (1). *The Arabs in History.* London.

LEWIS, M. D. S. (1). *Antique Paste Jewellery.* Faber, London, 1970. Rev. G. B. Hughes, *JRSA*, 1972, **120**, 263.

LEWIS, NORMAN (1). *A Dragon Apparent; Travels in Indo-China.* Cape, London, 1951.

LI CHHIAO-PHING (1) = (1). *The Chemical Arts of Old China* (tr. from the 1st, unrevised, edition, Chhangsha, 1940, but with additional material). J. Chem. Ed., Easton, Pa. 1948. Revs. W. Willetts, *ORA*, 1949, **2**, 126; J. R. Partington, *ISIS*, 1949, **40**, 280; Li Cho-Hao, *JCE*, 1949, **26**, 574. The Thaipei ed. of 1955 (Chinese text) was again revised and enlarged.

Li Cho-Hao (1). 'Les Hormones de l'Adénohypophyse.' *SCIS*, 1971, nos. 74–5, 69.

Li Cho-Hao & Evans, H. M. (1). 'Chemistry of the Anterior Pituitary Hormones.' Art. in *The Hormones*.... Ed. G. Pincus, K. V. Thimann & E. B. Astwood. Academic Press, New York, 1948–64, vol. 1, p. 633.

Li Hui-Lin (1). *The Garden Flowers of China*. Ronald, New York, 1959. (Chronica Botanica series, no. 19.)

Li Kuo-Chhin & Wang Chhung-Yu (1). *Tungsten, its History, Geology, Ore-Dressing, Metallurgy, Chemistry, Analysis, Applications and Economics*. Amer. Chem. Soc., New York, 1943 (Amer. Chem. Soc. Monographs, no. 94). 3rd ed. 1955 (A. C. S. Monographs, no. 130).

Liang, H. Y. (1). 'The Wah Chang [Hua-Chhang, Antimony] Mines.' *MSP*, 1915, **III**, 53. (The initials are given in the original as H. T. Liang, but this is believed to be a misprint.)

Liang, H. Y. (2). 'The Shui-khou Shan [Lead and Zinc] Mine in Hunan.' *MSP*, 1915, **110**, 914.

Liang Po-Chhiang (1). 'Überblick ü. d. seltenste chinesische Lehrbuch d. Medizin *Huang Ti Nei Ching*.' *AGMN*, 1933, **26**, 121.

Libavius, Andreas (1). *Alchemia. Andr. Libavii, Med. D[oct.], Poet. Physici Rotemburg. Operâ e Dispersis passim Optimorum Autorum, Veterum et Recentium exemplis potissimum, tum etian praeceptis quibusdam operosè collecta, adhibitâ; ratione et experientia, quanta potuit esse, methodo accuratâ explicata, et In Integrum Corpus Redacta*... Saur & Kopff, Frankfurt, 1597. Germ. tr. by F. Rex *et al.* Verlag Chemie, Weinheim, 1964.

Libavius, Andreas (2). *Singularium Pars Prima: in qua de abstrusioribus difficilioribusque nonnullis in Philosophia, Medicina, Chymia etc. Quaestionibus; utpote de Metallorum, Succinique Natura, de Carne fossili, ut credita est, de gestatione cacodaemonum, Veneno, aliisque rarioribus, quae versa indicat pagina, plurimis accuratè disseritur*. Frankfurt, 1599. Part II also 1599. Parts III and IV, 1601.

Licht, S. (1). 'The History [of Therapeutic Exercise].' Art. in *Therapeutic Exercise*, ed. S. Licht, Licht, New Haven, Conn. 1958, p. 380. (Physical Medicine Library, no. 3.)

Lieben, F. (1). *Geschichte d. physiologische Chemie*. Deuticke, Leipzig and Vienna, 1935.

Lin Yü-Thang (1) (tr.). *The Wisdom of Laotse [and Chuang Tzu] translated, edited and with an introduction and notes*. Random House, New York, 1948.

Lin Yü-Thang (7). *Imperial Peking; Seven Centuries of China* (with an essay on the Art of Peking, by P. C. Swann). Elek, London, 1961.

Lin Yü-Thang (8). *The Wisdom of China*. Joseph, London (limited edition) 1944; (general circulation edition) 1949.

Lindberg, D. C. & Steneck, N. H. (1). 'The Sense of Vision and the Origins of Modern Science', art. in *Science, Medicine and Society in the Renaissance* (Pagel Presentation Volume), ed. Debus (20), vol. 1, p. 29.

Lindeboom, G. A. (1). *Hermann Boerhaave; the Man and his Work*. Methuen, London, 1968.

Ling, P. H. (1). *Gymnastikens Allmänna Grunder*...(in Swedish). Leffler & Sebell, Upsala and Stockholm, 1st part, 1834, 2nd part, 1840 (based on observations and practice from 1813 onwards). Germ. tr.: *P. H. Ling's Schriften über Leibesübungen* (with posthumous additions), by H. F. Massmann, Heinrichshofen, Magdeburg, 1847. Cf. Cyriax (1).

Link, Arthur E. (1). 'The Taoist Antecedents in Tao-An's [+312 to +385] Prajñā Ontology.' Communication to the First International Conference of Taoist Studies, Villa Serbelloni, Bellagio, 1968.

von Lippmann, E. O. (1). *Entstehung und Ausbreitung der Alchemie, mit einem Anhange, Zur älteren Geschichte der Metalle; ein Beitrag zur Kulturgeschichte*. 3 vols. Vol. 1, Springer, Berlin, 1919. Vol. 2, Springer, Berlin, 1931. Vol. 3, Verlag Chemie, Weinheim, 1954 (posthumous, finished in 1940, ed. R. von Lippmann).

von Lippmann, E. O. (3). *Abhandlungen und Vorträge zur Geschichte d. Naturwissenschaften*. 2 vols. Vol. 1, Veit, Leipzig, 1906. Vol. 2, Veit, Leipzig, 1913.

von Lippmann, E. O. (4). *Geschichte des Zuckers, seiner Darstellung und Verwendung, seit den ältesten Zeiten bis zum Beginne der Rübenzuckerfabrikation; ein Beitrag zur Kulturgeschichte*. Hesse, Leipzig, 1890.

von Lippmann, E. O. (5). 'Chemisches bei Marco Polo.' *ZAC*, **21**, 1778. Repr. in (3), vol. 2, p. 258.

von Lippmann, E. O. (6). 'Die spezifische Gewichtsbestimmung bei Archimedes.' Repr. in (3), vol. 2, p. 168.

von Lippmann, E. O. (7). 'Zur Geschichte d. Saccharometers u. d. Senkspindel.' Repr. in (3), vol. 2, pp. 171, 177, 183.

von Lippmann, E. O. (8). 'Zur Geschichte der Kältemischungen.' Address to the General Meeting of the Verein Deutscher Chemiker, 1898. Repr. in (3). vol. 1, p. 110.

von Lippmann, E. O. (9). *Beiträge z. Geschichte d. Naturwissenschaften u. d. Technik*. 2 vols. Vol. 1, Springer, Berlin, 1925. Vol. 2, Verlag Chemie, Weinheim, 1953 (posthumous, ed. R. von Lippmann). Both vols. photographically reproduced, Sändig, Niederwalluf, 1971.

von Lippmann, E. O. (10). 'J. Ruska's Neue Untersuchungen ü. die Anfänge der Arabischen Alchemie.' *CHZ*, 1925, 2, 27.

von Lippmann, E. O. (11). 'Some Remarks on Hermes and Hermetica.' *AX*, 1938, 2, 21.

von Lippmann, E. O. (12). 'Chemisches u. Alchemisches aus Aristoteles.' *AGNT*, 1910, 2, 233–300.

von Lippmann, E. O. (13). 'Beiträge zur Geschichte des Alkohols.' *CHZ*, 1913, 37, 1313, 1348, 1358, 1419, 1428. Repr. in (9), vol. 1, p. 60.

von Lippmann, E. O. (14). 'Neue Beiträge zur Geschichte dez Alkohols.' *CHZ*, 1917, 41, 865, 883, 911. Repr. in (9), vol. 1, p. 107.

von Lippmann, E. O. (15). 'Zur Geschichte des Alkohols.' *CHZ*, 1920, 44, 625. Repr. in (9), vol. 1, p. 123.

von Lippmann, E. O. (16). 'Kleine Beiträge zur Geschichte d. Chemie.' *CHZ*, 1933, 57, 433. 1. Zur Geschichte des Alkohols. 2. Der Essig des Hannibal. 3. Künstliche Perlen und Edelsteine. 4. Chinesische Ursprung der Alchemie.

von Lippmann, E. O. (17). 'Zur Geschichte des Alkohols und seines Namens.' *ZAC*, 1912, 25, 1179, 2061.

von Lippmann, E. O. (18). 'Einige Bemerkungen zur Geschichte der Destillation und des Alkohols.' *ZAC*, 1912, 25, 1680.

von Lippmann, E. O. (19). 'Zur Geschichte des Wasserbades vom Altertum bis ins 13. Jahrhundert.' Art. in Kahlbaum Festschrift (1909), ed. Diergart (1), pp. 143–57.

von Lippmann, E. O. (20). *Urzeugung und Lebenskraft; Zur Geschichte dieser Problem von den ältesten Zeiten an bis zu den Anfängen des 20. Jahrhunderts*. Springer, Berlin, 1933.

von Lippmann, E. O. Biography, see Partington (19).

von Lippmann, E. O. & Sudhoff, K. (1). 'Thaddäus Florentinus (Taddeo Alderotti) über den Weingeist.' *AGMW*, 1914, 7, 379. (Latin text, and comm. only.)

Lipsius, A. & Bonnet, M. (1). *Acta Apostolorum Apocrypha*. 2 vols. in 3 parts. Mendelssohn, Leipzig, 1891–1903.

Little, A. G. (1) (ed.). *Part of the 'Opus Tertium'* [c. +1268] *of Roger Bacon*. Aberdeen, 1912.

Little, A. G. & Withington, E. (1) (ed.). *Roger Bacon's 'De Retardatione Accidentium Senectutis', cum aliis Opusculis de Rebus Medicinalibus*. Oxford, 1928. (Pubs. Brit. Soc. Franciscan Studies, no. 14.) Also printed as Fasc. 9 of Steele (1). Cf. the Engl. tr. of the *De Retardatione* by R. Browne, London, 1683.

Liu Mao-Tsai (1). *Kutscha und seine Beziehungen zu China vom 2 Jahrhundert v. bis zum 6 Jh. n. Chr.* 2 vols. Harrassowitz, Wiesbaden, 1969. (Asiatische Forschungen [Bonn], no. 27.)

Liu Mau-Tsai. See Liu Mao-Tsai.

Liu Pên-Li, Hsing Shu-Chieh, Li Chhêng-Chhiu & Chang Tao-Chung (1). 'True Hermaphroditism; a Case Report.' *CMJ*, 1959, 78, 449.

Liu Tshun-Jen (1). 'Lu Hsi-Hsing and his Commentaries on the *Tshan Thung Chhi*.' *CHJ/T*, 1968, (n.s.) 7, (no. 1), 71.

Liu Tshun-Jen (2). 'Lu Hsi-Hsing [+1520 to c. +1601]; a Confucian Scholar, Taoist Priest and Buddhist Devotee of the +16th Century.' *ASEA*, 1965, 18–19, 115.

Liu Tshun-Jen (3). 'Taoist Self-Cultivation in Ming Thought.' Art. in *Self and Society in Ming Thought*, ed. W. T. de Bary. Columbia Univ. Press, New York, 1970, p. 291.

Liu Ts'un-Yan. See Liu Tshun-Jen.

Lloyd, G. E. R. (1). *Polarity and Analogy; Two Types of Argumentation in Greek Thought*. Cambridge, 1971.

Lloyd, Seton (2). 'Sultantepe, II.' *ANATS*, 1954, 4, 101.

Lloyd, Seton & Brice, W. (1), with a note by C. J. Gadd. 'Ḥarrān.' *ANATS*, 1951, 1, 77–111.

Lloyd, Seton & Gökçe, Nuri (1), with notes by R. D. Barnett. 'Sultantepe, I.' *ANATS*, 1953, 3, 27.

Lo, L. C. (1) (tr.). 'Liu Hua-Yang; *Hui Ming Ching*, Das Buch von Bewusstsein und Leben.' In *Chinesische Blätter*, vol. 3, no. 1, ed. R. Wilhelm.

Loehr, G. (1). 'Missionary Artists at the Manchu Court.' *TOCS*, 1962, 34, 51.

Loewe, M. (5). 'The Case of Witchcraft in −91; its Historical Setting and Effect on Han Dynastic History' (*ku* poisoning). *AM*, 1970, 15, 159.

Loewe, M. (6). 'Khuang Hêng and the Reform of Religious Practices (−31).' *AM*, 1971, 17, 1.

Loewe, M. (7). 'Spices and Silk; Aspects of World Trade in the First Seven Centuries of the Christian Era.' *JRAS*, 1971, 166.

Loewenstein, P. J. (1). *Swastika and Yin-Yang*. China Society Occasional Papers (n. s.), China Society, London, 1942.

von Löhneyss, G. E. (1). *Bericht vom Bergwerck, wie man diselben bawen und in güten Wolstande bringen sol, sampt allen dazu gehörigen Arbeiten, Ordnung und Rechtlichen Processen beschrieben durch G.E.L.* Zellerfeld, 1617. 2nd ed. Leipzig, 1690.

Lonicerus, Adam (1). *Kräuterbuch*. Frankfort, 1578.

LORGNA, A. M. (1). 'Nuove Sperienze intorno alla Dolcificazione dell'Acqua del Mare.' *MSIV/MF*, 1786, **3**, 375. 'Appendice alla Memoria intorno alla Dolcificazione dell'Acqua del Mare.' *MSIV/MF*, 1790, **5**, 8.

LOTHROP, S. (1). 'Coclé; an Archaeological Study of Central Panama.' *MPMH*, 1937, **7**.

LOUIS, H. (1). 'A Chinese System of Gold Milling.' *EMJ*, 1891, 640.

LOUIS, H. (2). 'A Chinese System of Gold Mining.' *EMJ*, 1892, 629.

LOVEJOY, A. O. & BOAS, G. (1). *A Documentary History of Primitivism and Related Ideas*. Vol. 1. *Primitivism and Related Ideas in Antiquity*. Johns Hopkins Univ. Press, Baltimore, 1935.

LOWRY, T. M. (1). *Historical Introduction to Chemistry*. Macmillan, London, 1936.

LU GWEI-DJEN (1). 'China's Greatest Naturalist; a Brief Biography of Li Shih-Chen.' *PHY*, 1966, **8**, 383. Abridgment in Proc. XIth Internat. Congress of the History of Science, Warsaw, 1965, Summaries, vol. 2, p. 364; Actes, vol. 5, p. 50.

LU GWEI-DJEN (2). 'The Inner Elixir (*Nei Tan*); Chinese Physiological Alchemy.' Art. in *Changing Perspectives in the History of Science*, ed. M. Teich & R. Young. Heinemann, London, 1973, p. 68.

LU GWEI-DJEN & NEEDHAM, JOSEPH (1). 'A Contribution to the History of Chinese Dietetics.' *ISIS*, 1951, **42**, 13 (submitted 1939, lost by enemy action; again submitted 1942 and 1948).

LU GWEI-DJEN & NEEDHAM, JOSEPH (3). 'Mediaeval Preparations of Urinary Steroid Hormones.' *MH*, 1964, **8**, 101. Prelim. pub. *N*, 1963, **200**, 1047. Abridged account, *END*, 1968, **27** (no. 102), 130.

LU GWEI-DJEN & NEEDHAM, JOSEPH (4). 'Records of Diseases in Ancient China', art. in *Diseases in Antiquity*, ed. D. R. Brothwell & A. T. Sandison. Thomas, Springfield, Ill. 1967, p. 222.

LU GWEI-DJEN, NEEDHAM, JOSEPH & NEEDHAM, D. M. (1). 'The Coming of Ardent Water.' *AX*, 1972, **19**, 69.

LU KHUAN-YÜ (1). *The Secrets of Chinese Meditation; Self-Cultivation by Mind Control as taught in the Chhan, Mahāyāna and Taoist Schools in China*. Rider, London, 1964.

LU KHUAN-YÜ (2). *Chhan and Zen Teaching* (Series Two). Rider, London, 1961.

LU KHUAN-YÜ (3). *Chhan and Zen Teaching* (Series Three). Rider, London, 1962.

LU KHUAN-YÜ (4) (tr.). *Taoist Yoga; Alchemy and Immortality—a Translation, with Introduction and Notes, of 'The Secrets of Cultivating Essential Nature and Eternal Life' (Hsing Ming Fa Chüeh Ming Chih) by the Taoist Master Chao Pi-Chhen, b. 1860*. Rider, London, 1970.

LUCAS, A. (1). *Ancient Egyptian Materials and Industries*. Arnold, London (3rd ed.), 1948.

LUCAS, A. (2). 'Silver in Ancient Times.' *JEA*, 1928, **14**, 315.

LUCAS, A. (3). 'The Occurrence of Natron in Ancient Egypt.' *JEA*, 1932, **18**, 62.

LUCAS, A. (4). 'The Use of Natron in Mummification.' *JEA*, 1932, **18**, 125.

LÜDY-TENGER, F. (1). *Alchemistische und chemische Zeichen*. Berlin, 1928. Repr. Lisbing, Würzburg, 1972.

LUK, CHARLES. See LU KHUAN-YÜ.

LUMHOLTZ, C. S. (1). *Unknown Mexico; a Record of Five Years' Exploration among the Tribes of the Western Sierra Madre; in the Tierra Caliente of Tepic and Jalisco; and among the Tarascos of Michoacan*, 2 vols. Macmillan, London, 1903.

LUTHER, MARTIN (1). *Werke*. Weimarer Ausgabe.

MACALISTER, R. A. S. (2). *The Excavation of [Tel] Gezer, 1902–05 and 1907–09*. 3 vols. Murray, London, 1912.

MCAULIFFE, L. (1). *La Thérapeutique Physique d'Autrefois*. Paris, 1904.

MCCLURE, C. M. (1). 'Cardiac Arrest through Volition.' *CALM*, 1959, **90**, 440.

MCCONNELL, R. G. (1). *Report on Gold Values in the Klondike High-Level Gravels*. Canadian Geol. Survey Reports, 1907, 34.

MCCULLOCH, J. A. (2). '[The State of the Dead in] Primitive and Savage [Cultures].' *ERE*, vol. xi, p. 817.

MCCULLOCH, J. A. (3). '[The "Abode of the Blest" in] Primitive and Savage [Cultures].' *ERE*, vol. ii, p. 680.

MCCULLOCH, J. A. (4). '[The "Abode of the Blest" in] Celtic [Legend].' *ERE*, vol. ii, p. 688.

MCCULLOCH, J. A. (5). '[The "Abode of the Blest" in] Japanese [Thought].' *ERE*, vol. ii, p. 700.

MCCULLOCH, J. A. (6). '[The "Abode of the Blest" in] Slavonic [Lore and Legend].' *ERE*, vol. ii, p. 706.

MCCULLOCH, J. A. (7). '[The "Abode of the Blest" in] Teutonic [Scandinavian, Belief].' *ERE*, vol. ii, p. 707.

MCCULLOCH, J. A. (8). 'Incense.' Art. in *ERE*, vol. vii, p. 201.

MCCULLOCH, J. A. (9). 'Eschatology.' Art. in *ERE*, vol. v, p. 373.

MCCULLOCH, J. A. (10). 'Vampires.' *ERE*, vol. xii, p. 589.

MCDONALD, D. (1). *A History of Platinum*. London, 1960.

MACDONELL, A. A. (1). 'Vedic Religion.' *ERE*, vol. xii, p. 601.

McGovern, W. M. (1). *Early Empires of Central Asia.* Univ. of North Carolina Press, Chapel Hill, 1939.

McGowan, D. J. (2). 'The Movement Cure in China' (Taoist medical gymnastics). *CIMC/MR,* 1885 (no. 29), 42.

McGuire, J. E. (1). 'Transmutation and Immutability; Newton's Doctrine of Physical Qualities.' *AX,* 1967, **14,** 69.

McGuire, J. E. (2). 'Force, Active Principles, and Newton's Invisible Realm.' *AX,* 1968, **15,** 154.

McGuire, J. E. & Rattansi, P. M. (1). 'Newton and the "Pipes of Pan".' *NRRS,* 1966, **21,** 108.

McKenzie, R. Tait (1). *Exercise in Education and Medicine.* Saunders, Philadelphia and London, 1923.

McKie, D. (1). 'Some Notes on Newton's Chemical Philosophy, written upon the Occasion of the Tercentenary of his Birth.' *PMG,* 1942 (7th ser.), **33,** 847.

McKie, D. (2). 'Some Early Chemical Symbols.' *AX,* 1937, **1,** 75.

McLachlan, H. (1). *Newton; the Theological Manuscripts.* Liverpool, 1950.

Macquer, P. J. (1). *Élémens de la Théorie et de la Pratique de la Chimie.* 2 vols, Paris, 1775. (The first editions, uncombined, had been in 1749 and 1751 respectively, but this contained accounts of the new discoveries.) Eng. trs. London, 1775, Edinburgh, 1777. Cf. Coleby (1).

Madan, M. (1) (tr.). *A New and Literal Translation of Juvenal and Persius, with Copious Explanatory Notes by which these difficult Satirists are rendered easy and familiar to the Reader.* 2 vols. Becket, London, 1789.

Maenchen-Helfen, O. (4). *Reise ins asiatische Tuwa.* Berlin, 1931.

de Magalhaens, Gabriel (1). *Nouvelle Relation de la Chine.* Barbin, Paris, 1688 (a work written in 1668). Eng. tr. *A New History of China, containing a Description of the Most Considerable Particulars of that Vast Empire.* Newborough, London, 1688.

Magendie, F. (1). *Mémoire sur la Déglutition de l'Air atmosphérique.* Paris, 1813.

Mahdihassan, S. (2). 'Cultural Words of Chinese Origin' [*firoza* (Pers) = turquoise, *yashb* (Ar) = jade, *chamcha* (Pers) = spoon, *top* (Pers, Tk, Hind) = cannon, *silafchi* (Tk) = metal basin]. *BV,* 1950, **11,** 31.

Mahdihassan, S. (3). 'Ten Cultural Words of Chinese Origin' [*huqqa* (Tk), *qaliyan* (Tk) = tobacco-pipe, *sunduq* (Ar) = box, *piali* (Pers), *findjan* (Ar) = cup, *jaushan* (Ar) = armlet, *safa* (Ar) = turban, *qasai, qasab* (Hind) = butcher, *kah-kashan* (Pers) = Milky Way, *tugra* (Tk) = seal]. *JUB,* 1949, **18,** 110.

Mahdihassan, S. (5). 'The Chinese Origin of the Words Porcelain and Polish.' *JUB,* 1948, **17,** 89.

Mahdihassan, S. (6). 'Carboy as a Chinese Word.' *CS,* 1948, **17,** 301.

Mahdihassan, S. (7). 'The First Illustrations of Stick-Lac and their probable origin.' *PKAWA,* 1947, **50,** 793.

Mahdihassan, S. (8). 'The Earliest Reference to Lac in Chinese Literature.' *CS,* 1950, **19,** 289.

Mahdihassan, S. (9). 'The Chinese Origin of Three Cognate Words: Chemistry, Elixir, and Genii.' *JUB,* 1951, **20,** 107.

Mahdihassan, S. (11). 'Alchemy in its Proper Setting, with Jinn, Sufi, and Suffa as Loan-Words from the Chinese.' *IQB,* 1959, **7** (no. 3), 1.

Mahdihassan, S. (12). 'Alchemy and its Connection with Astrology, Pharmacy, Magic and Metallurgy.' *JAN,* 1957, **46,** 81.

Mahdihassan, S. (13). 'The Chinese Origin of Alchemy.' *UNASIA,* 1953, **5** (no. 4), 241.

Mahdihassan, S. (14). 'The Chinese Origin of the Word Chemistry.' *CS,* 1946, **15,** 136. 'Another Probable Origin of the Word Chemistry from the Chinese.' *CS,* 1946, **15,** 234.

Mahdihassan, S. (15). 'Alchemy in the Light of its Names in Arabic, Sanskrit and Greek.' *JAN,* 1961, **49,** 79.

Mahdihassan, S. (16). 'Alchemy a Child of Chinese Dualism as illustrated by its Symbolism.' *IQB,* 1959, **8,** 15.

Mahdihassan, S. (17). 'On Alchemy, Kimiya and Iksir.' *PAKPJ,* 1959, **3,** 67.

Mahdihassan, S. (18). 'The Genesis of Alchemy.' *IJHM,* 1960, **5** (no. 2), 41.

Mahdihassan, S. (19). 'Landmarks in the History of Alchemy.' *SCI,* 1963, **57,** 1.

Mahdihassan, S. (20). 'Kimiya and Iksir; Notes on the Two Fundamental Concepts of Alchemy.' *MBLB,* 1962, **5** (no. 3), 38. *MBPB,* 1963, **12** (no. 5), 56.

Mahdihassan, S. (21). 'The Early History of Alchemy.' *JUB,* 1960, **29,** 173.

Mahdihassan, S. (22). 'Alchemy; its Three Important Terms and their Significance.' *MJA,* 1961, 227.

Mahdihassan, S. (23). 'Der Chino-Arabische Ursprung des Wortes Chemikalie.' *PHI,* 1961, **23,** 515.

Mahdihassan, S. (24). 'Das Hermetische Siegel in China.' *PHI,* 1960, **22,** 92.

Mahdihassan, S. (25). 'Elixir; its Significance and Origin.' *JRAS/P,* 1961, **6,** 39.

Mahdihassan, S. (26). 'Ouroboros as the Earliest Symbol of Greek Alchemy.' *IQB,* 1961, **9,** 1.

Mahdihassan, S. (27). 'The Probable Origin of Kekulé's Symbol of the Benzene Ring.' *SCI,* 1960, **54,** 1.

Mahdihassan, S., (28). 'Alchemy in the Light of Jung's Psychology and of Dualism.' *PAKPJ*, 1962, **5**, 95.

Mahdihassan, S. (29). 'Dualistic Symbolism; Alchemical and Masonic.' *IQB*, 1963, 55.

Mahdihassan, S. (30). 'The Significance of Ouroboros in Alchemy and Primitive Symbolism.' *IQB*, 1963, 18.

Mahdihassan, S. (31). 'Alchemy and its Chinese Origin as revealed by its Etymology, Doctrines and Symbols.' *IQB*, 1966, 22.

Mahdihassan, S. (32). 'Stages in the Development of Practical Alchemy.' *JRAS/P*, 1968, **13**, 329.

Mahdihassan, S. (33). 'Creation, its Nature and Imitation in Alchemy.' *IQB*, 1968, 80.

Mahdihassan, S. (34). 'A Positive Conception of the Divinity emanating from a Study of Alchemy.' *IQB*, 1969, **10**, 77.

Mahdihassan, S. (35). '*Kursi* or throne; a Chinese word in the *Koran*.' *JRAS/BOM*, 1953, **28**, 19.

Mahdihassan, S. (36). '*Khazana*, a Chinese word in the *Koran*, and the associated word "Godown".' *JRAS/BOM*, 1953, **28**, 22.

Mahdihassan, S. (37). 'A Cultural Word of Chinese Origin; *ta'un* meaning Plague in Arabic.' *JUB*, 1953, **22**, 97. *CRESC*, 1950, 31.

Mahdihassan, S. (38). 'Cultural Words of Chinese Origin; *qaba, aba, diba, kimkhwab* (kincob).' *JKHRS*, 1950, **5**, 203.

Mahdihassan, S. (39). 'The Chinese Origin of the Words Kimiya, Sufi, Dervish and Qalander, in the Light of Mysticism.' *JUB*, 1956, **25**, 124.

Mahdihassan, S. (40). 'Chemistry a Product of Chinese Culture.' *PAKJS*, 1957, **9**, 26.

Mahdihassan, S. (41). 'Lemnian Tablets of Chinese Origin.' *IQB*, 1960, **9**, 49.

Mahdihassan, S. (42). 'Über einige Symbole der Alchemie.' *PHI*, 1962, **24**, 41.

Mahdihassan, S. (43). 'Symbolism in Alchemy; Islamic and other.' *IC*, 1962, **36** (no. 1), 20.

Mahdihassan, S. (44). 'The Philosopher's Stone in its Original Conception.' *JRAS/P*, 1962, **7** (no. 2), 263.

Mahdihassan, S. (45). 'Alchemie im Spiegel hellenistisch-buddhistische Kunst d. 2. Jahrhunderts.' *PHI*, 1965, **27**, 726.

Mahdihassan, S. (46). 'The Nature and Role of Two Souls in Alchemy.' *JRAS/P*, 1965, **10**, 67.

Mahdihassan, S. (47). 'Kekulé's Dream of the Ouroboros, and the Significance of this Symbol.' *SCI*, 1961, **55**, 187.

Mahdihassan, S. (48). 'The Natural History of Lac as known to the Chinese; Li Shih-Chen's Contribution to our Knowledge of Lac.' *IJE*, 1954, **16**, 309.

Mahdihassan, S. (49). 'Chinese Words in the Holy Koran; *qirtas* (paper) and its Synonym *kagaz*.' *JUB*, 1955, **24**, 148.

Mahdihassan, S. (50). 'Cultural Words of Chinese Origin; *kutcherry* (government office), *tusser* (silk).' Art. in Karmarker Commemoration Volume, Poona, 1947–8, p. 97.

Mahdihassan, S. (51). 'Union of Opposites; a Basic Theory in Alchemy and its Interpretation.' Art. in *Beiträge z. alten Geschichte und deren Nachleben*, Festschrift f. Franz Altheim, ed. R. Stiehl & H. E. Stier, vol. 2, p. 251. De Gruyter, Berlin, 1970.

Mahdihassan, S. (52). 'The Genesis of the Four Elements, Air, Water, Earth and Fire.' Art. in Gulam Yazdani Commemoration Volume, Hyderabad, Andhra, 1966, p. 251.

Mahdihassan, S. (53). 'Die frühen Bezeichnungen des Alchemisten, seiner Kunst und seiner Wunderdroge.' *PHI*, 1967, .

Mahdihassan, S. (54). 'The *Soma* of the Aryans and the *Chih* of the Chinese.' *MBPB*, 1972, **21** (no. 3), 30.

Mahdihassan, S. (55). 'Colloidal Gold as an Alchemical Preparation.' *JAN*, 1972, **58**, 112.

Mahler, J. G. (1). *The Westerners among the Figurines of the Thang Dynasty of China.* Ist. Ital. per il Med. ed Estremo Or., Rome, 1959. (Ser. Orientale Rom, no. 20.)

Mahn, C. A. F. (1). *Etymologische Untersuchung auf dem Gebiete der Romanischen Sprachen.* Dümmler, Berlin, 1858, repr. 1863.

Maier, Michael (1). *Atalanta Fugiens*, 1618. Cf. Tenney Davis (1); J. Read (1); de Jong (1).

Malhotra, J. C. (1). 'Yoga and Psychiatry; a Review.' *JNPS*, 1963, **4**, 375.

Manuel, F. E. (1). *Isaac Newton, Historian.* Cambridge, 1963.

Manuel, F. E. (2). *The Eighteenth Century Confronts the Gods.* Harvard Univ. Press, Cambridge, Mass. 1959.

Maqsood Ali, S. Asad & Mahdihassan, S. (4). 'Bazaar Medicines of Karachi; [IV], Inorganic Drugs.' *MED*, 1961, **23**, 125.

de la Marche, Lecoy (1). 'L'Art d'Enluminer; Traité Italien du XVe Siecle' (*De Arte Illuminandi*, Latin text with introduction). *MSAF*, 1888, **47** (5e sér.), **7**, 248.

Maréchal, J. R. (3). *Reflections upon Prehistoric Metallurgy; a Research based upon Scientific Methods.* Brimberg, Aachen (for Junker, Lammersdorf), 1963. French and German editions appeared in 1962.

MARSHALL, SIR JOHN (1). *Taxila*; *An Illustrated Account of Archaeological Excavations carried out at Taxila under the orders of the Government of India between the years 1913 and 1934.* 3 vols. Cambridge, 1951.

MARTIN, W. A. P. (2). *The Lore of Cathay.* Revell, New York and Chicago, 1901.

MARTIN, W. A. P. (3). *Hanlin Papers.* 2 vols. Vol. 1. Trübner, London; Harper, New York, 1880; Vol. 2. Kelly & Walsh, Shanghai, 1894.

MARTIN, W. A. P. (8). 'Alchemy in China.' A paper read before the Amer. Or. Soc. 1868; abstract in *JAOS*, 1871, **9**, xlvi. *CR*, 1879, **7**, 242. Repr. in (3), vol. 1, p. 221; (2), pp. 44 ff.

MARTIN, W. A. P. (9). *A Cycle of Cathay.* Oliphant, Anderson & Ferrier, Edinburgh and London; Revell New York, 1900.

MARTINDALE, W. (1). *The Extra Pharmacopoeia; incorporating Squire's 'Companion to the Pharmacopoeia'.* 1st edn. 1883. 25th edn., ed. R. G. Todd, Pharmaceutical Press, London, 1967.

MARX, E. (2). Japanese peppermint oil still. *MDGNVO*, 1896, **6**, 355.

MARYON, H. (3). 'Soldering and Welding in the Bronze and Early Iron Ages.' *TSFFA*, 1936, **5** (no. 2).

MARYON, H. (4). 'Prehistoric Soldering and Welding' (a précis of Maryon, 3). *AQ*, 1937, **11**, 208.

MARYON, H. (5). 'Technical Methods of the Irish Smiths.' *PRIA*, 1938, **44**c, no. 7.

MARYON, H. (6). *Metalworking and Enamelling; a Practical Treatise.* 3rd ed. London, 1954.

MASON, G. H. (1). *The Costume of China.* Miller, London, 1800.

MASON, H. S. (1). 'Comparative Biochemistry of the Phenolase Complex.' *AIENZ*, 1955, **16**, 105.

MASON, S. F. (2). 'The Scientific Revolution and the Protestant Reformation; I, Calvin and Servetus in relation to the New Astronomy and the Theory of the Circulation of the Blood.' *ANS*, 1953, **9** (no. 1).

MASON, S. F. (3). 'The Scientific Revolution and the Protestant Reformation; II, Lutheranism in relation to Iatro-chemistry and the German Nature-philosophy.' *ANS*, 1953, **9** (no. 2).

MASPERO, G. (2). *Histoire ancienne des Peuples d'Orient.* Paris, 1875.

MASPERO, H. (7). 'Procédés de 'nourrir le principe vital' dans la Religion Taoiste Ancienne.' *JA*, 1937, **229**, 177 and 353.

MASPERO, H. (9). 'Notes sur la Logique de Mo-Tseu [Mo Tzu] et de son École.' *TP*, 1928, **25**, 1.

MASPERO, H. (13). *Le Taoisme.* In *Mélanges Posthumes sur les Religions et l'Histoire de la Chine*, vol. 2, ed. P. Demiéville, SAEP, Paris, 1950. (Publ. du Mus. Guimet, Biblioth. de Diffusion, no 58.) Rev. J. J. L. Duyvendak, *TP*, 1951, **40**, 372.

MASPERO, H. (14). *Études Historiques.* In *Mélanges Posthumes sur les Religions et l'Histoire de la Chine*, vol. 3, ed. P. Demiéville. Civilisations du Sud, Paris, 1950. [Publ. du Mus. Guimet, Biblioth. de Diffusion, no. 59.) Rev. J. J. L. Duyvendak, *TP*, 1951, **40**, 366.

MASPERO, H. (19). 'Communautés et Moines Bouddhistes Chinois au 2e et 3e Siècles.' *BEFEO*, 1910, **10**, 222.

MASPERO, H. (20)., 'Les Origines de la Communauté Bouddhiste de Loyang.' *JA*, 1934, **225**, 87.

MASPERO, H. (22). 'Un Texte Taoiste sur l'Orient Romain.' *MIFC*, 1937, **17**, 377 (*Mélanges G. Maspero*, vol. 2). Reprinted in Maspero (14), pp. 95 ff.

MASPERO, H. (31). Review of R. F. Johnston's *Buddhist China* (London, 1913). *BEFEO*, 1914, **14** (no. 9), 74.

MASPERO, H. (32). *Le Taoïsme et les Religions Chinoises.* (Collected posthumous papers, partly from (12) and (13) reprinted, partly from elsewhere, with a preface by M. Kaltenmark.) Gallimard, Paris, 1971. (Bibliothèque des Histoires, no. 3.)

MASSÉ, H. (1). *Le Livre des Merveilles du Monde.* Chêne, Paris, 1944. (Album of colour-plates from al-Qazwīnī's Cosmography, c. +1275, with introduction, taken from Bib. Nat. Suppl. Pers. MS. 332.)

MASSIGNON, L. (3). 'The Qarmatians.' *EI*, vol. ii, pt. 2, p. 767.

MASSIGNON, L. (4). 'Inventaire de la Littérature Hermétique Arabe.' App. iii in Festugière (1), 1944. (On the role of the Sabians of Ḥarrān, who adopted the Hermetica as their Scriptures.)

MASSIGNON, L. (5). 'The Idea of the Spirit in Islam.' *ERYB*, 1969, **6**, 319 (*The Mystic Vision*, ed. J. Campbell). Tr. from the German in *ERJB*, 1945, **13**, 1.

MASSON, L. (1). 'La Fontaine de Jouvence.' *AESC*, 1937, **27**, 244; 1938, **28**, 16.

MASSON-OURSEL, P. (4). *Le Yoga.* Presses Univ. de France, Paris, 1954. (Que Sais-je? ser. no. 643.)

AL-MASʿŪDĪ. See de Meynard & de Courteille.

MATCHETT, J. R. & LEVINE, J. (1). 'A Molecular Still designed for Small Charges.' *IEC/AE*, 1943, **15**, 296.

MATHIEU, F. F. (1). *La Géologie et les Richesses Miniéres de la Chine.* Impr. Comm. et Industr., la Louvière, n.d. (1924), paginated 283–529, with 4 maps (from Pub. de l'Assoc. des Ingénieurs de l'École des Mines de Mons).

MATSUBARA, HISAKO (1) (tr.). *Die Geschichte von Bambus-sammler und dem Mädchen Kaguya* [the *Taketori Monogatari*, c. +866], with illustrations by Mastubara Naoko. Langewiesche-Brandt, Ebenhausen bei München, 1968.

MATSUBARA, HISAKO (2). 'Dies-seitigkeit und Transzendenz im *Taketori Monogatari*.' Inaug. Diss. Ruhr Universität, Bochum, 1970.

MATTHAEI, C. F. (1) (tr.). *Nemesius Emesenus 'De Natura Hominis' Graece et Latine* (c. +400). Halae Magdeburgicae, 1802.

MATTIOLI, PIERANDREA (2). *Commentarii in libros sex Pedacii Dioscoridis Anazarbei de materia medica....* Valgrisi, Venice, 1554, repr. 1565.

MAUL, J. P. (1). 'Experiments in Chinese Alchemy.' Inaug. Diss., Massachusetts Institute of Technology, 1967.

MAURIZIO, A. (1). *Geschichte der gegorenen Getränke*. Berlin and Leipzig, 1933.

MAXWELL, J. PRESTON (1). 'Osteomalacia and Diet.' *NAR*, 1934, **4** (no. 1), 1.

MAXWELL, J. PRESTON, HU, C. H. & TURNBULL, H. M. (1). 'Foetal Rickets [in China].' *JPB*, 1932, **35**, 419.

MAYERS, W. F. (1). *Chinese Reader's Manual*. Presbyterian Press, Shanghai, 1874; reprinted 1924.

MAZZEO, J. A. (1). 'Notes on John Donne's Alchemical Imagery.' *ISIS*, 1957, **48**, 103.

MEAD, G. R. S. (1). *Thrice-Greatest Hermes; Studies in Hellenistic Theosophy and Gnosis—Being a Translation of the Extant Sermons and Fragments of the Trismegistic Literature, with Prolegomena, Commentaries and Notes*. 3 vols. Theosophical Pub. Soc., London and Benares, 1906.

MEAD, G. R. S. (2) (tr.). '*Pistis Sophia*'; *a [Christian] Gnostic Miscellany; being for the most part Extracts from the 'Books of the Saviour', to which are added Excerpts from a Cognate Literature*. 2nd ed. Watkins, London, 1921.

MECHOULAM, R. & GAONI, Y. (1). 'Recent Advances in the Chemistry of Hashish.' *FCON*, 1967, **25**, 175.

MEHREN, A. F. M. (1) (ed. & tr.). *Manuel de la Cosmographie du Moyen-Âge, traduit de l'Arabe;* '*Nokhbet ed-Dahr fi Adjaib-il-birr wal-Bahr* [*Nukhbat al-Dahr fi 'Ajāib al-Birr wa'l Bahr*]' *de Shems ed-Din Abou-Abdallah Mohammed de Damas* [Shams al-Dīn Abū 'Abd-Allāh al-Anṣarī al-Ṣūfī al-Dimashqī; *The Choice of the Times and the Marvels of Land and Sea*, c. +1310]... St Petersburg, 1866 (text), Copenhagen, 1874 (translation).

MEILE, P. (1). 'Apollonius de Tyane et les Rites Védiques.' *JA*, 1945, **234**, 451.

MEISSNER, B. (1). *Babylonien und Assyrien*. Winter, Heidelberg, 1920, Leipzig, 1925.

MELLANBY, J. (1). 'Diphtheria Antitoxin.' *PRSB*, 1908, **80**, 399.

MELLOR, J. W. (1). *Modern Inorganic Chemistry*. Longmans Green, London, 1916; often reprinted.

MELLOR, J. W. (2). *Comprehensive Treatise on Inorganic and Theoretical Chemistry*. 15 vols. Longmans Green, London, 1923.

MELLOR, J. W. (3). 'The Chemistry of the Chinese Copper-red Glazes.' *TCS*, 1936, **35**.

DE MÉLY, F. (1) (with the collaborationof M. H. Courel). *Les Lapidaires Chinois*. Vol. 1 of *Les Lapidaires de l'Antiquité et du Moyen Age*. Leroux, Paris, 1896. (Contains facsimile reproduction of the mineralogical section of *Wakan Sanzai Zue*, chs. 59, 60, and 61.) Crit. rev. M. Berthelot, *JS*, 1896, 573).

DE MÉLY, F. (6). 'L'Alchimie chez les Chinois et l'Alchimie Grecque.' *JA*, 1895 (9e sér.), **6**, 314.

DE MENASCE, P. J. (2). 'The Cosmic Noria (Zodiac) in Parsi Thought.' *AN*, 1940, **35–6**, 451.

MEREDITH-OWENS, G. M. (1). 'Some Remarks on the Miniatures in the [Royal Asiatic] Society's *Jāmi' al-Tawārīkh* (MS. A27 of +1314) [by Rashīd al-Dīn, finished +1316].' *JRAS*, 1970 (no. 2, Wheeler Presentation Volume), 195. Includes a brief account of the section on the History of China; cf. Jahn & Franke (1).

MERRIFIELD, M. P. (1). *Original Treatises dating from the +12th to the +18th Centuries on the Arts of Painting in Oil, Miniature, and the Preparation of Colour and Artificial Gems*. 2 vols. London, 1847, London, 1849.

MERRIFIELD, M. P. (2). *A Treatise on Painting* [Cennino Cennini's], *translated from Tambroni's Italian text of 1821*. London, 1844.

MERSENNE, MARIN (3). *La Verité des Sciences, contre les Sceptiques on Pyrrhoniens*. Paris, 1625. Facsimile repr. Frommann, Stuttgart and Bad Cannstatt, 1969. Rev. W. Pagel, *AX*, 1970, **17**, 64.

MERZ, J. T. (1). *A History of European Thought in the Nineteenth Century*. 2 vols. Blackwood, Edinburgh and London, 1896.

METCHNIKOV, E. (= I. I.) (1). *The Nature of Man; Studies in Optimistic Philosophy*. Tr. P. C. Mitchell, New York, 1903, London, 1908; rev. ed. by C. M. Beadnell, London, 1938.

METTLER, CECILIA C. (1). *A History of Medicine*. Blakiston, Toronto, 1947.

METZGER, H. (1). *Newton, Stahl, Boerhaave et la Doctrine Chimique*. Alcan, Paris, 1930.

DE MEURON, M. (1). 'Yoga et Médecine; propos du Dr J. G. Henrotte recueillis par...' *MEDA*, 1968 (no. 69), 2.

MEYER, A. W. (1). *The Rise of Embryology*. Stanford Univ. Press, Palo Alto, Calif. 1939.

VON MEYER, ERNST (1). *A History of Chemistry, from earliest Times to the Present Day; being also an Introduction to the Study of the Science*. 2nd ed., tr. from the 2nd Germ. ed. by G. McGowan. Macmillan, London, 1898.

MEYER, H. H. & GOTTLIEB, R. (1). *Die experimentelle Pharmakologie als Grundlage der Arzneibehandlung.* 9th ed. Urban & Schwarzenberg, Berlin and Vienna, 1936.

MEYER, P. (1). *Alexandre le Grand dans la Litterature Française du Moyen Age.* 2 vols. Paris, 1886.

MEYER, R. M. (1). *Goethe.* 3 vols. Hofmann, Berlin, 1905.

MEYER-STEINEG, T. & SUDHOFF, K. (1). *Illustrierte Geschichte der Medizin.* 5th ed. revised and enlarged, ed. R. Herrlinger & F. Kudlien. Fischer, Stuttgart, 1965.

MEYERHOF, M. (3). 'On the Transmission of Greek and Indian Science to the Arabs.' *IC*, 1937, **11**, 17.

MEYERHOF, M. & SOBKHY, G. P. (1) (ed. & tr.). *The Abridged Version of the 'Book of Simple Drugs' of Aḥmad ibn Muḥammad al-Ghāfiqī of Andalusia by Gregorius Abu'l-Faraj (Bar Hebraeus).* Govt. Press, Cairo, 1938. (Egyptian University Faculty of Med. Pubs. no. 4.)

DE MEYNARD, C. BARBIER (3). '"L'Alchimiste", Comédie en Dialecte Turc Azeri [Azerbaidjani].' *JA*, 1886 (8ᵉ sér.), **7**, 1.

DE MEYNARD, C. BARBIER & DE COURTEILLE, P. (1) (tr.). *Les Prairies d'Or* (the *Murūj al-Dhahab* of al-Masʿūdī, +947). 9 vols. Paris, 1861–77.

MIALL, L. C. (1). *The Early Naturalists, their Lives and Work* (+1530 to +1789). Macmillan, London, 1912.

MICHELL, H. (1). *The Economics of Ancient Greece.* Cambridge, 1940. 2nd ed. 1957.

MICHELL, H. (2). 'Oreichalcos.' *CLR*, 1955, **69** (n.s. **5**), 21.

MIELI, A. (1). *La Science Arabe, et son Rôle dans l'Evolution Scientifique Mondiale.* Brill, Leiden, 1938. Repr. 1966, with additional bibliography and analytic index by A. Mazaheri.

MIELI, A. (3). *Pagine di Storia della Chimica.* Rome, 1922.

MIGNE, J. P. (1) (ed.). *Dictionnaire des Apocryphes; ou, Collection de tous les Livres Apocryphes relatifs à l'Ancien et au Nouveau Testament, pour la plupart, traduits en Français pour la première fois sur les textes originaux; et enrichie de préfaces, dissertations critiques, notes historiques, bibliographiques, géographiques et theologiques...* 2 vols. Migne, Paris, 1856. Vols. 23 and 24 of his *Troisième et Dernière Encyclopédie Théologique*, 60 vols.

MILES, L. M. & FÊNG, C. T. (1). 'Osteomalacia in Shansi.' *JEM*, 1925, **41**, 137.

MILES, W. (1). 'Oxygen-consumption during Three Yoga-type Breathing Patterns.' *JAP*, 1964, **19**, 75.

MILLER, J. INNES (1). *The Spice Trade of the Roman Empire*, −29 to +641. Oxford, 1969.

MILLS, J. V. (11). *Ma Huan['s] 'Ying Yai Shêng Lan', 'The Overall Survey of the Ocean's Shores' [1433]; translated from the Chinese text edited by Fêng Chhêng-Chün, with Introduction, Notes and Appendices...* Cambridge, 1970. (Hakluyt Society Extra Series, no. 42.)

MINGANA, A. (1) (tr.). *An Encyclopaedia of the Philosophical and Natural Sciences, as taught in Baghdad about +817; or, the 'Book of Treasures' by Job of Edessa: the Syriac Text Edited and Translated...* Cambridge, 1935.

MITCHELL, C. W. (1). *St Ephraim's Prose Refutations of Mani, Marcion and Bardaisan;...from the Palimpsest MS. Brit. Mus. Add. 14623...Vol. 1. The Discourses addressed to Hypatius. Vol. 2. The Discourse called 'Of Domnus', and Six other Writings.* Williams & Norgate, London, 1912–21. (Text and Translation Society Series.)

MITRA, RAJENDRALALA (1). 'Spirituous Drinks in Ancient India.' *JRAS/B*, 1873, **42**, 1–23.

MIYASHITA SABURŌ (1). 'A Link in the Westward Transmission of Chinese Anatomy in the Later Middle Ages.' *ISIS*, 1968, **58**, 486.

MIYUKI MOKUSEN (1). 'Taoist Zen Presented in the *Hui Ming Ching*.' Communication to the First International Conference of Taoist Studies, Villa Serbelloni, Bellagio, 1968.

MIYUKI MOKUSEN (2). 'The "Secret of the Golden Flower", Studies and [a New] Translation.' Inaug. Diss., Jung Institute, Zürich, 1967.

MODEL, J. G. (1). *Versuche und Gedanken über ein natürliches oder gewachsenes Salmiak.* Leipzig, 1758.

MODI, J. J. (1). 'Haoma.' Art. in *ERE*, vol. vi, p. 506.

MOISSAN, H. (1). *Traité de Chimie Minérale.* 5 vols. Masson, Paris, 1904.

MONTAGU, B. (1) (ed.). *The Works of Lord Bacon.* 16 vols. in 17 parts. Pickering, London, 1825–34.

MONTELL, G. (2). 'Distilling in Mongolia.' *ETH*, 1937 (no. 5), **2**, 321.

DE MONTFAUCON, B. (1). *L'Antiquité Expliquée et Representée en Figures.* 5 vols. with 5-vol. supplement. Paris, 1719. Eng. tr. by D. Humphreys, *Antiquity Explained, and Represented in Sculptures, by the Learned Father Montfaucon.* 5 vols. Tonson & Watts, London, 1721–2.

MONTGOMERY, J. W. (1). 'Cross, Constellation and Crucible; Lutheran Astrology and Alchemy in the Age of the Reformation.' *AX*, 1964, **11**, 65.

MOODY, E. A. & CLAGETT, MARSHALL (1) (ed. and tr.). *The Mediaeval Science of Weights ('Scientia de Ponderibus'); Treatises ascribed to Euclid, Archimedes, Thabit ibn Qurra, Jordanus de Nemore, and Blasius of Parma.* Univ. of Wisconsin Press, Madison, Wis., 1952. Revs. E. J. Dijksterhuis, *A/AIHS*, 1953, **6**, 504; O. Neugebauer, *SP*, 1953, **28**, 596.

MOORE-BENNETT, A. J. (1). 'The Mineral Areas of Western China.' *FER*, 1915, 225.

MORAN, S. F. (1). 'The Gilding of Ancient Bronze Statues in Japan.' *AA*, 1969, **30**, 55.

DE MORANT, G. SOULIÉ (2). *L'Acuponcture Chinoise*. 4 vols.
I. *l'Énergie* (*Points, Méridiens, Circulation*).
II. *Le Maniement de l'Energie*.
III. *Les Points et leurs Symptômes*.
IV. *Les Maladies et leurs Traitements*.
Mercure de France, Paris, 1939–. Re-issued as 5 vols. in one, with 1 vol. of plates, Maloine, Paris, 1972.

MORERY, L. (1). *Grand Dictionnaire Historique; ou le Mélange Curieux de l'Histoire Sacrée et Profane...* 1688, Supplement 1689. Later editions revised by J. Leclerc. 9th ed. Amsterdam and The Hague, 1702. Eng. tr. revised by Jeremy Collier, London, 1701.

MORET, A. (1). 'Mysteries, Egyptian.' *ERE*, vol. ix, pp. 74–5.

MORET, A. (2). *Kings and Gods in Egypt*. London, 1912.

MORET, A. (3). 'Du Caractère Religieux de la Royauté Pharaonique.' *BE/AMG*, 1902, **15**, 1–344.

MORFILL, W. R. & CHARLES, R. H. (1) (tr.). *The 'Book of the Secrets of Enoch'* [2 Enoch], *translated from the Slavonic...* Oxford, 1896.

MORGENSTERN, P. (1) (ed.). '*Turba Philosophorum'; Das ist, Das Buch von der güldenen Kunst, neben andern Authoribus, welche mit einander 36 Bücher in sich haben. Darinn die besten vrältesten Philosophi zusamen getragen, welche tractiren alle einhellig von der Universal Medicin, in zwey Bücher abgetheilt, unnd mit Schönen Figuren gezieret. Jetzundt newlich zu Nutz und Dienst allen waren Kunstliebenden der Natur* (so der Lateinischen Sprach unerfahren) *mit besondern Fleiß, mühe unnd Arbeit trewlich an tag geben...* König, Basel, 1613. 2nd ed. Krauss, Vienna, 1750. Cf. Ferguson (1), vol. 2, pp. 106 ff.

MORRIS, IVAN I. (1). *The World of the Shining Prince; Court Life in Ancient Japan* [in the Heian Period, +782 to +1167, here particularly referring to Late Heian, +967 to +1068]. Oxford, 1964.

MORRISON, P. & MORRISON, E. (1). 'High Vacuum.' *SAM*, 1950, **182** (no. 5), 20.

MORTIER, F. (1). 'Les Procédés Taoistes en Chine pour la Prolongation de la Vie Humaine.' *BSAB*, 1930, **45**, 118.

MORTON, A. A. (1). *Laboratory Technique in Organic Chemistry*. McGraw-Hill, New York and London, 1938.

MOSS, A. A. (1). 'Niello.' *SCON*, 1955, **1**, 49.

M[OUFET], T[HOMAS] (of Caius, d. +1605)? (1). '*Letter* [of Roger Bacon] *concerning the Marvellous Power of Art and Nature.* London, 1659. (Tr. of *De Mirabili Potestate Artis et Naturae, et de Nullitate Magiae*.) French. tr. of the same work by J. Girard de Tournus, Lyons, 1557, Billaine, Paris, 1628. Cf. Ferguson (1), vol. 1, pp. 52, 63-4, 318, vol. 2, pp. 114, 438.

MOULE, A. C. & PELLIOT, P. (1) (tr. & annot.). *Marco Polo* (+1254 to +1325); *The Description of the World*. 2 vols. Routledge, London, 1938. Further notes by P. Pelliot (posthumously pub.). 2 vols. Impr. Nat. Paris, 1960.

MUCCIOLI, M. (1). 'Intorno ad una Memoria di Giulio Klaproth sulle "Conoscenze Chimiche dei Cinesi nell 8 Secolo".' *A*, 1926, **7**, 382.

MUELLER, K. (1). 'Die Golemsage und die sprechenden Statuen.' *MSGVK*, 1918, **20**, 1–40.

MUIR, J. (1). *Original Sanskrit Texts*. 5 vols. London, 1858–72.

MUIR, M. M. PATTISON (1). *The Story of Alchemy and the Beginnings of Chemistry*. Hodder & Stoughton, London, 1902. 2nd ed. 1913.

MUIRHEAD, W. (1). '[Glossary of Chinese] Mineralogical and Geological Terms. In Doolittle (1), vol. 2, p. 256.

MUKAND SINGH, THAKUR (1). *Ilajul Awham* (*On the Treatment of Superstitions*). Jagat, Aligarh, 1893 (in Urdu).

MUKERJI, KAVIRAJ B. (1). *Rasa-jala-nidhi; or, Ocean of Indian Alchemy*. 2 vols. Calcutta, 1927.

MULTHAUF, R. P. (1). 'John of Rupescissa and the Origin of Medical Chemistry.' *ISIS*, 1954, **45**, 359.

MULTHAUF, R. P. (2). 'The Significance of Distillation in Renaissance Medical Chemistry.' *BIHM*, 1956, **30**, 329.

MULTHAUF, R. P. (3). 'Medical Chemistry and "the Paracelsians".' *BIHM*, 1954, **28**, 101.

MULTHAUF, R. P. (5). *The Origins of Chemistry*. Oldbourne, London, 1967.

MULTHAUF, R. P. (6). 'The Relationship between Technology and Natural Philosophy, *c.* +1250 to +1650, as illustrated by the Technology of the Mineral Acids.' Inaug. Diss., Univ. California, 1953.

MULTHAUF, R. P. (7). 'The Beginnings of Mineralogical Chemistry.' *ISIS*, 1958, **49**, 50.

MULTHAUF, R. P. (8). 'Sal Ammoniac; a Case History in Industrialisation.' *TCULT*, 1965, **6**, 569.

MUS, P. (1). 'La Notion de Temps Réversible dans la Mythologie Bouddhique.' *AEPHE/SSR*, 1939, 1.

AL-NADĪM, ABŪ'L-FARAJ IBN ABŪ YA'QŪB. See Flügel, G. (1).

NADKARNI, A. D. (1). *Indian Materia Medica*. 2 vols. Popular, Bombay, 1954.

NAGEL, A. (1). 'Die Chinesischen Küchengott.' *AR*, 1908, **11**, 23.

NAKAYAMA SHIGERU & SIVIN, N. (1) (ed.). *Chinese Science; Explorations of an Ancient Tradition*. M.I.T. Press, Cambridge, Mass., 1973. (M.I.T. East Asian Science Ser. no. 2.)

NANJIO, B. (1). *A Catalogue of the Chinese Translations of the Buddhist Tripiṭaka*. Oxford, 1883. (See Ross, E. D, 3.)

NARDI, S. (1) (ed.). *Taddeo Alderotti's Consilia Medicinalia'*, c. +1280. Turin, 1937.

NASR, SEYYED HOSSEIN. See Said Husain Nasr.

NAU, F. (2). 'The translation of the *Tabula Smaragdina* by Hugo of Santalla (mid +12th century).' *ROC*, 1907 (2e sér.), **2**, 105.

NEAL, J. B. (1). 'Analyses of Chinese Inorganic Drugs.' *CMJ*, 1889, **2**, 116; 1891, **5**, 193,

NEBBIA, G. & NEBBIA-MENOZZI, G. (1). 'A Short History of Water Desalination.' Art. from *Acqua Dolce dal Mare*. IIª Inchiesta Internazionale, Milan, Fed. delle Associazioni Sci. e Tecniche, 1966, pp. 129-172.

NEBBIA, G. & NEBBIA-MENOZZI, G. (2). 'Early Experiments on Water Desalination by Freezing.' *DS*, 1968, **5**, 49.

NEEDHAM, DOROTHY M. (1). *Machina Carnis; the Biochemistry of Muscle Contraction in its Historical Development*. Cambridge, 1971.

NEEDHAM, JOSEPH (2). *A History of Embryology*. Cambridge, 1934. 2nd ed., revised with the assistance of A. Hughes. Cambridge, 1959; Abelard-Schuman, New York, 1959.

NEEDHAM, JOSEPH (25). 'Science and Technology in China's Far South-East.' *N*, 1946, **157**, 175. Reprinted in Needham & Needham (1).

NEEDHAM, JOSEPH (27). 'Limiting Factors in the Advancement of Science as observed in the History of Embryology.' *YJBM*, 1935, **8**, 1. (Carmalt Memorial Lecture of the Beaumont Medical Club of Yale University.)

NEEDHAM, JOSEPH (30). 'Prospection Géobotanique en Chine Médiévale.' *JATBA*, 1954, **1**, 143.

NEEDHAM, JOSEPH (31). 'Remarks on the History of Iron and Steel Technology in China (with French translation; 'Remarques relatives à l'Histoire de la Sidérurgie Chinoise'). In *Actes du Colloque International 'Le Fer à travers les Ages'*, pp. 93, 103. Nancy, Oct. 1955. (*AEST*, 1956, Mémoire no. 16.)

NEEDHAM, JOSEPH (32). *The Development of Iron and Steel Technology in China*. Newcomen Soc. London, 1958. (Second Biennial Dickinson Memorial Lecture, Newcomen Society.) Précis in *TNS*, 1960, **30**, 141; rev. L. C. Goodrich, *ISIS*, 1960, **51**, 108. Repr. Heffer, Cambridge, 1964. French tr. (unrevised, with some illustrations omitted and others added by the editors), *RHSID*, 1961, **2**, 187, 235; 1962, **3**, 1, 62.

NEEDHAM, JOSEPH (34). 'The Translation of Old Chinese Scientific and Technical Texts.' Art. in *Aspects of Translation*, ed. A. H. Smith, Secker & Warburg, London, 1958. p. 65. (Studies in Communication, no. 22.) Also in *BABEL*, 1958, **4** (no. 1), 8.

NEEDHAM JOSEPH (36). *Human Law and the Laws of Nature in China and the West*. Oxford Univ. Press, London, 1951. (Hobhouse Memorial Lectures at Bedford College, London, no. 20.) Abridgement of (37).

NEEDHAM, JOSEPH (37). 'Natural Law in China and Europe.' *JHI*, 1951, **12**, 3 & 194 (corrigenda, 628).

NEEDHAM, JOSEPH (45). 'Poverties and Triumphs of the Chinese Scientific Tradition.' Art. in *Scientific Change; Historical Studies in the Intellectual, Social and Technical Conditions for Scientific Discovery and Technical Invention from Antiquity to the Present*, ed. A. C. Crombie, p. 117. Heinemann, London, 1963. With discussion by W. Hartner, P. Huard, Huang Kuang-Ming, B. L. van der Waerden and S. E. Toulmin (Symposium on the History of Science, Oxford, 1961). Also, in modified form: 'Glories and Defects...' in *Neue Beiträge z. Geschichte d. alten Welt*, vol. 1, *Alter Orient und Griechenland*, ed. E. C. Welskopf, Akad. Verl. Berlin, 1964. French tr. (of paper only) by M. Charlot, 'Grandeurs et Faiblesses de la Tradition Scientifique Chinoise', *LP*, 1963, no. 111. Abridged version; 'Science and Society in China and the West', *SPR*, 1964, **52**, 50.

NEEDHAM, JOSEPH (47). 'Science and China's Influence on the West.' Art. in *The Legacy of China*, e R. N. Dawson. Oxford, 1964, p. 234.

NEEDHAM, JOSEPH (48). 'The Prenatal History of the Steam-Engine.' (Newcomen Centenary Lecture.) *TNS*, 1963, **35**, 3-58.

NEEDHAM, JOSEPH (50). 'Human Law and the Laws of Nature.' Art. in *Technology, Science and Art; Common Ground*. Hatfield Coll. of Technol., Hatfield, 1961, p. 3. A lecture based upon (36) and (37), revised from Vol. 2, pp. 518 ff. Repr. in *Social and Economic Change* (Essays in Honour of Prof. D. P. Mukerji), ed. B. Singh & V. B. Singh. Allied Pubs. Bombay, Delhi etc., 1967, p. 1.

NEEDHAM, JOSEPH (55). 'Time and Knowledge in China and the West.' Art. in *The Voices of Time; a Cooperative Survey of Man's Views of Time as expressed by the Sciences and the Humanities*, ed. J. T. Fraser. Braziller, New York, 1966, p. 92.

NEEDHAM, JOSEPH (56). *Time and Eastern Man.* (Henry Myers Lecture, Royal Anthropological Institute, 1964.) Royal Anthropological Institute, London, 1965.

NEEDHAM, JOSEPH (58). 'The Chinese Contribution to Science and Technology.' Art. in *Reflections on our Age* (Lectures delivered at the Opening Session of UNESCO at the Sorbonne, Paris, 1946), ed. D. Hardman & S. Spender. Wingate, London, 1948, p. 211. Tr. from the French *Conférences de l'Unesco*. Fontaine, Paris, 1947, p. 203.

NEEDHAM, JOSEPH (59). 'The Roles of Europe and China in the Evolution of Oecumenical Science.' *JAHIST*, 1966, **1**, 1. As Presidential Address to Section X, British Association, Leeds, 1967, in *ADVS*, 1967, **24**, 83.

NEEDHAM, JOSEPH (60). 'Chinese Priorities in Cast Iron Metallurgy.' *TCULT*, 1964, **5**, 398.

NEEDHAM, JOSEPH (64). *Clerks and Craftsmen in China and the West* (Collected Lectures and Addresses). Cambridge, 1970.

NEEDHAM, JOSEPH (65). *The Grand Titration; Science and Society in China and the West.* (Collected Addresses.) Allen & Unwin, London, 1969.

NEEDHAM, JOSEPH (67). *Order and Life* (Terry Lectures). Yale Univ. Press, New Haven, Conn.; Cambridge, 1936. Paperback edition (with new foreword), M.I.T. Press, Cambridge, Mass. 1968. Italian tr. by M. Aloisi, *Ordine e Vita*, Einaudi, Turin, 1946 (Biblioteca di Cultura Scientifica, no. 14).

NEEDHAM, JOSEPH (68). 'Do the Rivers Pay Court to the Sea? The Unity of Science in East and West.' *TTT*, 1971, **5** (no. 2), 68.

NEEDHAM, JOSEPH (70). 'The Refiner's Fire; the Enigma of Alchemy in East and West.' Ruddock, for Birkbeck College, London, 1971 (Bernal Lecture). French tr. (with some additions and differences), 'Artisans et Alchimistes en Chine et dans le Monde Hellénistique.' *LP*, 1970, no. 152, 3 (Rapkine Lecture, Institut Pasteur, Paris).

NEEDHAM, JOSEPH (71). 'A Chinese Puzzle—Eighth or Eighteenth?', art. in *Science, Medicine and Society in the Renaissance* (Pagel Presentation Volume), ed. Debus (20), vol. 2, p. 251.

NEEDHAM, JOSEPH & LU GWEI-DJEN (1). 'Hygiene and Preventive Medicine in Ancient China.' *JHMAS*, 1962, **17**, 429; abridged in *HEJ*, 1959, **17**, 170.

NEEDHAM, JOSEPH & LU GWEI-DJEN (3). 'Proto-Endocrinology in Mediaeval China.' *JSHS*, 1966, **5**, 150.

NEEDHAM, JOSEPH & NEEDHAM, DOROTHY M. (1) (ed.). *Science Outpost.* Pilot Press, London, 1948.

NEEDHAM, JOSEPH & ROBINSON, K. (1). 'Ondes et Particules dans la Pensée Scientifique Chinoise.' *SCIS*, 1960, **1** (no. 4), 65.

NEEDHAM, JOSEPH, WANG LING & PRICE, D. J. DE S. (1). *Heavenly Clockwork; the Great Astronomical Clocks of Mediaeval China.* Cambridge, 1960. (Antiquarian Horological Society Monographs, no. 1.) Prelim. pub. *AHOR*, 1956, **1**, 153.

NEEF, H. (1). *Die im 'Tao Tsang' enthaltenen Kommentare zu 'Tao-Tê-Ching' Kap. VI.* Inaug. Diss. Bonn, 1938.

NEOGI, P. (1). *Copper in Ancient India.* Sarat Chandra Roy (Anglo-Sanskrit Press), Calcutta, 1918. (Indian Assoc. for the Cultivation of Science, Special Pubs. no. 1.)

NEOGI, P. & ADHIKARI, B. B. (1). 'Chemical Examination of Ayurvedic Metallic Preparations; I, *Shata-puta lauha* and *Shahashra-puta lauha* (Iron roasted a hundred or a thousand times).' *JRAS/B*, 1910 (n.s.), **6**, 385.

NERI, ANTONIO (1). *L'Arte Vetraria distinta in libri sette...* Giunti, Florence, 1612. 2nd ed. Rabbuiati, Florence, 1661, Batti, Venice, 1663. Latin tr. *De Arte Vitraria Libri Septem, et in eosdem Christoph. Merretti...Observationes et Notae.* Amsterdam, 1668. German tr. by F. Geissler, Frankfurt and Leipzig, 1678. English tr. by C. Merrett, London, 1662. Cf. Ferguson (1), vol. 2, pp. 134 ff.

NEUBAUER, C. & VOGEL, H. (1). *Handbuch d. Analyse d. Harns.* 1860, and later editions, including a revision by A. Huppert, 1910.

NEUBURGER, A. (1). *The Technical Arts and Sciences of the Ancients.* Methuen, London, 1930. Tr. by H. L. Brose from *Die Technik d. Altertums.* Voigtländer, Leipzig, 1919. (With a drastically abbreviated index and the total omission of the bibliographies appended to each chapter, the general bibliography, and the table of sources of the illustrations).

NEUBURGER, M. (1). 'Théophile de Bordeu (1722 bis 1776) als Vorläufer d. Lehre von der inneren Sekretion.' *WKW*, 1911 (pt. 2), 1367.

NEUMANN, B. (1). 'Messing.' *ZAC*, 1902, **15**, 511.

NEUMANN, B. & KOTYGA, G. (1) (with the assistance of M. Rupprecht & H. Hoffmann). 'Antike Gläser.' *ZAC*, 1925, **38**, 776, 857; 1927, **40**, 963; 1928, **41**, 203; 1929, **42**, 835.

NEWALL, L. C. (1). 'Newton's Work in Alchemy and Chemistry.' Art. in *Sir Isaac Newton, 1727 to 1927*, Hist. Sci. Soc. London, 1928, pp. 203–55.

NGUYEN DANG TÂM (1). 'Sur les Bokétonosides, Saponosides du Boket ou *Gleditschia fera* Merr. (*australis* Hemsl.; *sinensis* Lam.).' *CRAS*, 1967, **264**, 121.

Niu Ching-I, Kung Yo-Thing, Huang Wei-Tê, Ko Liu-Chün & eight other collaborators (1). 'Synthesis of Crystalline Insulin from its Natural A-Chain and the Synthetic B-Chain.' *SCISA*, 1966, **15**, 231.

Noble, S. B. (1). 'The Magical Appearance of Double-Entry Book-keeping' (derivation from the mathematics of magic squares). Unpublished MS., priv. comm.

Nock, A. D. & Festugière, A. J. (1). *Corpus Hermeticum* [Texts and French translation]. Belles lettres, Paris, 1945–54.
  Vol. 1, Texts I to XII; text established by Nock, tr. Festugière.
  Vol. 2, Texts XIII to XVIII, Asclepius; text established by Nock, tr. Festugière.
  Vol. 3, Fragments from Stobaeus I to XXII; text estab. and tr. Festugière.
  Vol. 4, Fragments from Stobaeus XXIII to XXIX (text estab. and tr. Festugière) and Miscellaneous Fragments (text estab. Nock, tr. Festugière).
  (Coll. Universités de France, Assoc. G. Budé.)

Noel, Francis (2). *Philosophia Sinica; Tribus Tractatibus primo Cognitionem primi Entis, secundo Ceremonias erga Defunctos, tertio Ethicam juxta Sinarum mentem complectens.* Univ. Press, Prague, 1711. Cf. Pinot (2), p. 116.

Noel, Francis (3) (tr.). *Sinensis Imperii Libri Classici Sex; nimirum: Adultorum Schola [Ta Hsüeh], Immutabile Medium [Chung Yung], Liber Sententiarum [Lun Yü], Mencius [Mêng Tzu], Filialis Observantia [Hsiao Ching], Parvulorum Schola [San Tzu Ching], e Sinico Idiomate in Latinum traducti....* Univ. Press, Prague, 1711. French tr. by Pluquet, *Les Livres Classiques de l'Empire de la Chine, précédés d'observations sur l'Origine, la Nature et les Effets de la Philosophie Morale et Politique dans cet Empire.* 7 vols. De Bure & Barrois, Didot, Paris, 1783–86. The first three books had been contained in Intorcetta *et al.* (1) *Confucius Sinarum Philosophus...*, the last three were now for the first time translated.

Noel, Francis (5) (tr.). MS. translation of the *Tao Tê Ching*, sent to Europe between +1690 and +1702. Present location unknown. See Pfister (1), p. 418.

Nordhoff, C. (1). *The Communistic Societies of the United States, from Personal Visit and Observation.* Harper, New York, 1875. 2nd ed. Dover, New York, 1966, with an introduction by M. Holloway.

Norin, E. (1). 'Tzu Chin Shan, an Alkali-Syenite Area in Western Shansi; Preliminary Notes.' *BGSC*, 1921, no. 3, 45–70.

Norpoth, L. (1). 'Paracelsus—a Mannerist?', art. in *Science, Medicine and Society in the Renaissance* (Pagel Presentation Volume), ed. Debus (20), vol. 1, p. 127.

Norton, T. (1). *The Ordinall of Alchimy (c. 1+440).* See Holmyard (12).

Noyes, J. H. [of Oneida] (1). *A History of American Socialisms.* Lippincott, Philadelphia, 1870. 2nd ed. Dover, New York, 1966, with an introduction by M. Holloway.

O'Flaherty, W. D. (1). 'The Submarine Mare in the Mythology of Siva.' *JRAS*, 1971, 9.

O'Leary, de Lacy (1). *How Greek Science passed to the Arabs.* Routledge & Kegan Paul, London, 1948.

Oakley, K. P. (2). 'The Date of the "Red Lady" of Paviland.' *AQ*, 1968, **42**, 306.

Ōbuchi, Ninji (1). 'How the *Tao Tsang* Took Shape.' Contribution to the First International Conference of Taoist Studies, Villa Serbelloni, Bellagio, 1958.

Oesterley, W. O. E. & Robinson, T. H. (1). *Hebrew Religion; its Origin and Development.* SPCK, London, 2nd ed. 1937, repr. 1966.

Ogden, W. S. (1). 'The Roman Mint and Early Britain.' *BNJ*, 1908, **5**, 1–50.

Ohsawa, G. See Sakurazawa, Nyoiti (1).

d'Ollone, H., Vissière, A., Blochet, E. *et al.* (1). *Recherches sur les Mussulmans Chinois.* Leroux, Paris, 1911. (Mission d'Ollone, 1906–1909: cf. d'Ollone, H.: *In Forbidden China*, tr. B. Miall, London, 1912.)

Olschki, L. (4). *Guillaume Boucher; a French Artist at the Court of the Khans.* Johns Hopkins Univ. Press, Baltimore, 1946 (rev. H. Franke, *OR*, 1950, **3**, 135).

Olschki, L. (7). *The Myth of Felt.* Univ. of California Press, Los Angeles, Calif., 1949.

Ong Wên-Hao (1). 'Les Provinces Métallogéniques de la Chine.' *BGSC*, 1920, no. 2, 37–59.

Ong Wên-Hao (2). 'On Historical Records of Earthquakes in Kansu.' *BGSC*, 1921, no. 3, 27–44.

Oppert, G. (2). 'Mitteilungen zur chemisch-technischen Terminologie im alten Indien; (1) Über die Metalle, besonders das Messing, (2) der Indische Ursprung der Kadmia (Calaminaris) und der Tutia.' Art. in Kahlbaum Festschrift (1909), ed. Diergart (1), pp. 127–42.

Orschall, J. C. (1). '*Sol sine Veste*'; *Oder dreyssig Experimenta dem Gold seinen Purpur auszuziehen...* Augsburg, 1684. Cf. Partington (7), vol. 2, p. 371; Ferguson (1), vol. 2, pp. 156 ff.

da Orta, Garcia (1). *Coloquios dos Simples e Drogas he cousas medicinais da India compostos pello Doutor Garcia da Orta.* de Endem, Goa, 1563. Latin epitome by Charles de l'Escluze, Plantin, Antwerp, 1567. Eng. tr. *Colloquies on the Simples and Drugs of India* with the annotations of the Conde de Ficalho, 1895, by Sir Clements Markham. Sotheran, London, 1913.

OSMOND, H. (1) 'Ololiuqui; the Ancient Aztec Narcotic.' *JMS*, 1955, **101**, 526.

OSMOND, H. (2). 'Hallucinogenic Drugs in Psychiatric Research.' *MBLB*, 1964, **6** (no. 1), 2.

OST, H. (1). *Lehrbuch der chemischen Technologie.* 11th ed. Jänecke, Leipzig, 1920.

OTA, K. (1). 'The Manufacture of Sugar in Japan.' *TAS/J*, 1880, **8**, 462.

OU YUN-JOEI. See Wu Yün-Jui in Roi & Wu (1).

OUSELEY, SIR WILLIAM (1) (tr.). The '*Oriental Geography*' *of Ebn Haukal, an Arabian Traveller of the Tenth Century* [Abū al-Qāsim Muḥammad Ibn Ḥawqal, *fl.* +943 to +977]. London, 1800. (This translation, done from a Persian MS., is in fact an abridgement of the *Kitāb al-Masālik wa'l-Mamālik*, 'Book of the Roads and the Countries', of Ibn Ḥawqal's contemporary, Abū Isḥāq Ibrāhīm ibn Muḥammad al-Fārisī al-Iṣṭakhrī.)

PAAL, H. (1). *Johann Heinrich Cohausen, +1665 bis +1750; Leben und Schriften eines bedeutenden Arztes aus der Blütezeit des Hochstiftes Münster, mit kulturhistorischen Betrachtungen.* Fischer, Jena, 1931. (Arbeiten z. Kenntnis d. Gesch. d. Medizin im Rheinland und Westfalen, no. 6.)

PAGEL, W. (1). 'Religious Motives in the Medical Biology of the Seventeenth Century.' *BIHM*, 1935, **3**, 97.

PAGEL, W. (2). 'The Religious and Philosophical Aspects of van Helmont's Science and Medicine.' *BIHM*, Suppl. no. 2, 1944.

PAGEL, W. (10). *Paracelsus; an Introduction to Philosophical Medicine in the Era of the Renaissance.* Karger, Basel and New York, 1958. Rev. D. G[eoghegan], *AX*, 1959, **7**, 169.

PAGEL, W. (11). 'Jung's Views on Alchemy.' *ISIS*, 1948, **39**, 44.

PAGEL, W. (12). 'Paracelsus; Traditionalism and Mediaeval Sources.' Art. in *Medicine, Science and Culture*, O. Temkin Presentation Volume, ed. L. G. Stevenson & R. P. Multhauf. Johns Hopkins Press. Baltimore, Md. 1968, p. 51.

PAGEL, W. (13). 'The Prime Matter of Paracelsus.' *AX*, 1961, **9**, 117.

PAGEL, W. (14). 'The "Wild Spirit" (Gas) of John-Baptist van Helmont (+1579 to +1644), and Paracelsus.' *AX*, 1962, **10**, 2.

PAGEL, W. (15). 'Chemistry at the Cross-Roads; the Ideas of Joachim Jungius.' *AX*, 1969, **16**, 100. (Essay-review of Kangro, 1.)

PAGEL, W. (16). 'Harvey and Glisson on Irritability, with a Note on van Helmont.' *BIHM*, 1967, **41**, 497.

PAGEL, W. (17). 'The Reaction to Aristotle in Seventeenth-Century Biological Thought.' Art. in *Singer Commemoration Volume, Science, Medicine and History*, ed. E. A. Underwood. Oxford, 1953, vol. 1, p. 489.

PAGEL, W. (18). 'Paracelsus and the Neo-Platonic and Gnostic Tradition.' *AX*, 1960, **8**, 125.

PALÉOLOGUE, M. G. (1). *L'Art Chinois.* Quantin, Paris, 1887.

PALLAS, P. S. (1). *Sammlungen historischen Nachrichten ü. d. mongolischen Völkerschaften.* St Petersburg, 1776. Fleischer, Frankfurt and Leipzig, 1779.

PALMER, A. H. (1). 'The Preparation of a Crystalline Globulin from the Albumin fraction of Cow's Milk.' *JBC*, 1934, **104**, 359.

[PALMGREN, N.] (1). 'Exhibition of Early Chinese Bronzes arranged on the Occasion of the 13th International Congress of the History of Art.' *BMFEA*, 1934, **6**, 81.

PÁLOS, S. (2). *Atem und Meditation; Moderne chinesische Atemtherapie als Vorschule der Meditation—Theorie, Praxis, Originaltexte.* Barth, Weilheim, 1968.

PARANAVITANA, S. (4). *Ceylon and Malaysia.* Lake House, Colombo, 1966.

DE PAREDES, J. (1). *Recopilacion de Leyes de los Reynos de las Indias.* Madrid, 1681.

PARENNIN, D. (1). 'Lettre à Mons. [J. J.] Dortous de Mairan, de l'Académie Royale des Sciences (on demonstrations to Chinese scholars of freezing-point depression, fulminate explosions and chemical precipitation, without explanations but as a guarantee of theological veracity; on causes of the alleged backwardness of Chinese astronomy, including imperial displeasure at ominous celestial phenomena; on the pretended origin of the Chinese from the ancient Egyptians; on famines and scarcities in China; and on the aurora borealis)'. *LEC*, 1781, vol. 22, pp. 132 ff., dated 28 Sep. 1735.

PARKES, S. (1). *Chemical Essays, principally relating to the Arts and Manufactures of the British Dominions.* 5 vols. Baldwin, Cradock & Joy, London, 1815.

PARTINGTON, J. R. (1). *Origins and Development of Applied Chemistry.* Longmans Green, London, 1935.

PARTINGTON, J. R. (2). 'The Origins of the Atomic Theory.' *ANS*, 1939, **4**, 245.

PARTINGTON, J. R. (3). 'Albertus Magnus on Alchemy.' *AX*, 1937, **1**, 3.

PARTINGTON, J. R. (4). *A Short History of Chemistry.* Macmillan, London, 1937, 3rd ed. 1957.

PARTINGTON, J. R. (5). *A History of Greek Fire and Gunpowder.* Heffer, Cambridge, 1960.

PARTINGTON, J. R. (6). 'The Origins of the Planetary Symbols for Metals.' *AX*, 1937, **1**, 61.

PARTINGTON, J. R. (7). *A History of Chemistry.*

Vol. 1, pt. 1. *Theoretical Background* [Greek, Persian and Jewish].

Vol. 2. *+1500* to *+1700*.

Vol. 3. *+1700* to *1800*.

Vol. 4. *1800 to the Present Time.*

Macmillan, London, 1961– . Rev. W. Pagel, *MH*, 1971, **15**, 406.

PARTINGTON, J. R. (8). 'Chinese Alchemy.'

(*a*) *N*, 1927, **119**, 11.

(*b*) *N*, 1927, **120**, 878; comment on B. E. Read (11).

(*c*) *N*, 1931, **128**, 1074; dissent from von Lippmann (1).

PARTINGTON, J. R. (9). 'The Relationship between Chinese and Arabic Alchemy.' *N*, 1928, **120**, 158.

PARTINGTON, J. R. (10). *General and Inorganic Chemistry...* 2nd ed. Macmillan, London, 1951.

PARTINGTON, J. R. (11). 'Trithemius and Alchemy.' *AX*, 1938, **2**, 53.

PARTINGTON, J. R. (12). 'The Discovery of Mosaic Gold.' *ISIS*, 1934, **21**, 203.

PARTINGTON, J. R. (13). 'Bygone Chemical Technology.' *CHIND*, 1923 (n.s.), **42** (no. 26), 636.

PARTINGTON, J. R. (14). 'The Kerotakis Apparatus.' *N*, 1947, **159**, 784.

PARTINGTON, J. R. (15). 'Chemistry in the Ancient World.' Art. in *Science, Medicine and History*, Singer Presentation Volume, ed. E. A. Underwood. Oxford, 1953, vol. 1, p. 35. Repr. with slight changes, 1959, 241.

PARTINGTON, J. R. (16). 'An Ancient Chinese Treatise on Alchemy [the *Tshan Thung Chhi* of Wei Po-Yang].' *N*, 1935, **136**, 287.

PARTINGTON, J. R. (17). 'The Chemistry of al-Razi.' *AX*, 1938, **1**, 192.

PARTINGTON, J. R. (18). 'Chemical Arts in the Mount Athos Manual of Christian Iconography [prob. +13th cent., MSS of +16th to +18th centuries].' *ISIS*, 1934, **22**, 136.

PARTINGTON, J. R. (19). 'E. O. von Lippmann [biography].' *OSIS*, 1937, **3**, 5.

PASSOW, H., ROTHSTEIN, A. & CLARKSON, T. W. (1). 'The General Pharmacology of the Heavy Metals.' *PHREV*, 1961, **13**, 185

PASTAN, I. (1). 'Biochemistry of the Nitrogen-containing Hormones.' *ARB*, 1966, **35** (pt. 1), 367,

DE PAUW, C. (1). *Recherches Philosophiques sur les Égyptiens et les Chinois...* (vols. IV and V of *Oeuvres Philosophiques*), Cailler, Geneva, 1774. 2nd ed. Bastien, Paris, Rep. An. III (1795). Crit. Kao Lei-Ssu [Aloysius Ko, S.J.], *MCHSAMUC*, 1777, **2**, 365, (2nd pagination) 1–174.

PECK, E. S. (1). 'John Francis Vigani, first Professor of Chemistry in the University of Cambridge, +1703 to +1712, and his Cabinet of Materia Medica in the Library of Queens' College.' *PCASC*, 1934, **34**, 34.

PELLIOT, P. (1). Critical Notes on the Earliest Reference to Tea. *TP*, 1922, **21**, 436.

PELLIOT, P. (3). 'Notes sur Quelques Artistes des Six Dynasties et des Thang.' *TP*, 1923, **22**, 214. (On the Bodhidharma legend and the founding of Shao-lin Ssu on Sung Shan, pp. 248 ff., 252 ff.)

PELLIOT, P. (8). 'Autour d'une Traduction Sanskrite du *Tao-tö-king* [*Tao Tê Ching*].' *TP*, 1912, **13**, 350.

PELLIOT, P. (10). 'Les Mongols et la Papauté.'

Pt. 1 'La Lettre du Grand Khan Güyük à Innocent IV [+1246].'

Pt. 2*a* 'Le Nestorien Siméon Rabban-Ata.'

Pt. 2*b* 'Ascelin [Azelino of Lombardy, a Dominican, leader of the first diplomatic mission to the Mongols, +1245 to +1248].'

Pt. 2*c* 'André de Longjumeau [Dominican envoy, +1245 to +1247].'

ROC, 1922, **23** (sér. 3, **3**), 3–30; 1924, **24** (sér. 3, **4**), 225–335; 1931, **28** (sér. 3, **8**), 3–84.

PELLIOT, P. (47). *Notes on Marco Polo; Ouvrage Posthume.* 2 vols. Impr. Nat. and Maisonneuve, Paris, 1959.

PELLIOT, P. (54). 'Le Nom Persan du Cinabre dans les Langues "Altaiques".' *TP*, 1925, **24**, 253.

PELLIOT, P. (55). 'Henri Bosmans, S.J.' *TP*, 1928, **26**, 190.

PELLIOT, P. (56). Review of Cordier (12), *l'Imprimerie Sino-Européenne en Chine. BEFEO*, 1903, **3**, 108.

PELLIOT, P. (57). 'Le *Kin Kou K'i Kouan* [*Chin Ku Chhi Kuan*, Strange Tales New and Old, *c.* +1635]' (review of E. B. Howell, 2). *TP*, 1925, **24**, 54.

PELLIOT, P. (58). Critique of L. Wieger's *Taoisme. JA*, 1912 (10ᵉ sér.), **20**, 141.

PELSENEER, J. (3). 'La Réforme et l'Origine de la Science Moderne.' *RUB*, 1954, **5**, 406.

PELSENEER, J. (4). 'L'Origine Protestante de la Science Moderne.' *LYCH*, 1947, 246. Repr. *GEW*, 47.

PELSENEER, J. (5). 'La Réforme et le Progrès des Sciences en Belgique au 16ᵉ Siècle.' Art. in *Science, Medicine and Hisory*, Charles Singer Presentation Volume, ed. E. A. Underwood, Oxford, 1953, vol. 1, p. 280.

PENZER, N. M. (2). *Poison-Damsels; and other Essays in Folklore and Anthropology.* Pr. pr. Sawyer, London, 1952.

PERCY, J. (1). *Metallurgy; Fuel, Fire-Clays, Copper, Zinc and Brass.* Murray, London, 1861.

PERCY, J. (2). *Metallurgy; Iron and Steel.* Murray, London, 1864.

PERCY, J. (3). *Metallurgy; Introduction, Refractories, Fuel.* Murray, London, 1875.

PERCY, J. (4). *Metallurgy; Silver and Gold.* Murray, London, 1880.

PEREIRA, J. (1). *Elements of Materia Medica and Therapeutics.* 2 vols. Longman, Brown, Green & Longman, London, 1842.

PERKIN, W. H. & KIPPING, F. S. (1). *Organic Chemistry,* rev. ed. Chambers, London and Edinburgh, 1917.

PERRY, E. S. & HECKER, J. C. (1). 'Distillation under High Vacuum.' Art. in *Distillation,* ed. A. Weissberger (*Technique of Organic Chemistry,* vol. 4), p. 495. Interscience, New York, 1951.

PERTOLD, O. (1). 'The Liturgical Use of *mahuḍa* liquor among the Bhīls.' *ARO,* 1931, **3,** 406.

PETERSON, E. (1). 'La Libération d'Adam de l'Ἀνάγκη.' *RB,* 1948, **55,** 199.

PETRIE, W. M. FLINDERS (5). 'Egyptian Religion.' Art. in *ERE,* vol. v, p. 236.

PETTUS, SIR JOHN (1). *Fleta Minor; the Laws of Art and Nature, in Knowing, Judging, Assaying, Fining, Refining and Inlarging the Bodies of confin'd Metals...* The first part is a translation of Ercker (1), the second contains: *Essays on Metallic Words, as a Dictionary to many Pleasing Discourses.* Dawkes, London, 1683; reissued 1686. See Sisco & Smith (1); Partington (7), vol. 2, pp. 104 ff.

PETTUS, SIR JOHN (2). *Fodinae Regales; or, the History, Laws and Places of the Chief Mines and Mineral Works in England, Wales and the English Pale in Ireland; as also of the Mint and Mony; with a Clavis explaining some difficult Words relating to Mines, Etc.* London, 1670. See Partington (7), vol. 2, p. 106.

PFISTER, R. (1). 'Teinture et Alchimie dans l'Orient Hellénistique.' *SK,* 1935, **7,** 1–59.

PFIZMAIER, A. (95) (tr.). 'Beiträge z. Geschichte d. Edelsteine u. des Goldes.' *SWAW/PH,* 1867, **58,** 181, 194, 211, 217, 218, 223, 237. (Tr. chs. 807 (coral), 808 (amber), 809 (gems), 810, 811 (gold), 813 (in part), *Thai-Phing Yü Lan.*)

PHARRIS, B. B., WYNGARDEN, L. J. & GUTKNECHT, G. D. (1). Art. in *Gonadotrophins, 1968,* ed. E. Rosenberg, p. 121.

PHILALETHA, EIRENAEUS (or IRENAEUS PHILOPONUS). Probably pseudonym of George Starkey (*c.* +1622 to +1665, *q.v.*). See Ferguson (1), vol. 2, pp. 194, 403.

PHILALETHES, EUGENIUS. See Vaughan, Thomas (+1621 to +1665), and Ferguson (1), vol. 2, p. 197.

PHILIPPE, M. (1). 'Die Braukunst der alten Babylonier im Vergleich zu den heutigen Braumethoden.' In Huber, E. (3), *Bier und Bierbereitung bei den Völkern d. Urzeit,* vol. 1, p. 29.

PHILIPPE, M. (2). 'Die Braukunst der alten Ägypter im Lichte heutiger Brautechnik.' In Huber, E. (3), *Bier und Bierbereitung bei den Völkern d. Urzeit,* vol. 1, p. 55.

PHILLIPPS, T. (SIR THOMAS) (1). 'Letter...communicating a Transcript of a MS. Treatise on the Preparation of Pigments, and on Various Processes of the Decorative Arts practised in the Middle Ages, written in the +12th Century and entitled *Mappae Clavicula.*' *AAA,* 1847, **32,** 183.

PHILOSTRATUS OF LEMNOS. See Conybeare (1); Jones (1).

PIANKOFF, A. & RAMBOVA, N. (1). *Egyptian Mythological Papyri.* 2 vols. Pantheon, New York, 1957 (Bollingen Series, no. 40).

PINCHES, T. G. (1). 'Tammuz.' *ERE,* vol. xii, p. 187. 'Heroes and Hero-Gods (Babylonian).' *ERE,* vol. vi, p. 642.

PINCUS, G., THIMANN, K. V. & ASTWOOD, E. B. (1) (ed.). *The Hormones; Physiology, Chemistry and Applications.* 5 vols. Academic Press, New York, 1948–64.

PINOT, V. (2). *Documents Inédits relatifs à la Connaissance de la Chine en France de 1685 à 1740.* Geuthner, Paris, 1932.

PITTS, F. N. (1). 'The Biochemistry of Anxiety.' *SAM,* 1969, **220** (no. 2), 69.

PIZZIMENTI, D. (1) (ed. & tr.). *Democritus 'De Arte Magna' sive 'De Rebus Naturalibus', necnon Synesii et Pelagii et Stephani Alexandrini et Michaelis Pselli in eundem Commentaria.* Padua, 1572, 1573, Cologne, 1572, 1574 (cf. Ferguson (1), vol. 1, p. 205). Repr. J. D. Tauber: *Democritus Abderyta Graecus 'De Rebus Sacris Naturalibus et Mysticis', cum Notis Synesii et Pelagii...* Nuremberg, 1717.

PLESSNER, M. (1). 'Picatrix' Book on Magic and its Place in the History of Spanish Civilisation.' Communication to the IXth International Congress of the History of Science, Barcelona and Madrid, 1959. Abstract in *Guiones de las Communicaciones,* p. 78. A longer German version appears in the subsequent *Actes* of the Congress, p. 312.

PLESSNER, M. (2). 'Hermes Trismegistus and Arab Science.' *SI,* 1954, **2,** 45.

PLESSNER, M. (3). 'Neue Materialen z. Geschichte d. *Tabula Smaragdina.*' *DI,* 1927, **16,** 77. (A critique of Ruska, 8.)

PLESSNER, M. (4). 'Jābir ibn Ḥayyān und die Zeit der Entstehung der arabischen Jābir-schriften.' *ZDMG,* 1965, **115,** 23.

PLESSNER, M. (5). 'The Place of the *Turba Philosophorum* in the Development of Alchemy.' *ISIS,* 1954, **45,** 331.

PLESSNER, M. (7). 'The *Turba Philosophorum*; a Preliminary Report on Three Cambridge Manuscripts.' *AX,* 1959, **7,** 159. (These MSS are longer than that used by Ruska (6) in his translation, but the authenticity of the additional parts has not yet been established.)

PLESSNER, M. (8). 'Geber and Jābir ibn Ḥayyān; an Authentic +16th-Century Quotation from Jābir.' *AX*, 1969, **16**, 113.

PLOSS, E. E., ROOSEN-RUNGE, H., SCHIPPERGES, H. & BUNTZ, H. (1). *Alchimia; Ideologie und Technologie*. Moos, München, 1970.

POISSON, A. (1). *Théories et Symboles des Alchimistes, le Grand Oeuvre; suivi d'un Essai sur la Bibliographie Alchimique du XIXe Siècle*. Paris, 1891. Repr. 1972.

POISSONNIER, P. J. (1). *Appareil Distillatoire présenté au Ministre de la Marine*. Paris, 1779.

POKORA, T. (4). 'An Important Crossroad of Chinese Thought' (Huan Than, the first coming of Buddhism; and Yogistic trends in ancient Taoism). *ARO*, 1961, **29**, 64.

POLLARD, A. W. (2) (ed.). *The Travels of Sir John Mandeville; with Three Narratives in illustration of it—The Voyage of Johannes of Plano Carpini, the Journal of Friar William de Rubruquis, the Journal of Friar Odoric*. Macmillan, London, 1900. Repr. Dover, New York; Constable, London, 1964.

POMET, P. (1). *Histoire Générale des Drogues*. Paris, 1694. Eng. tr. *A Compleat History of Druggs*. 2 vols. London, 1735.

DE PONCINS, GONTRAN (1). *From a Chinese City*. New York, 1957.

POPE, J. A., GETTENS, R. J., CAHILL, J. & BARNARD, N. (1). *The Freer Chinese Bronzes*. Vol. 1, Catalogue. Smithsonian Institution, Washington, D.C. 1967. (Freer Gallery of Art Oriental Studies, no. 7; Smithsonian Publication, no. 4706.) See also Gettens, Fitzhugh, Bene & Chase (1).

POPE-HENNESSY, U. (1). *Early Chinese Jades*. Benn, London, 1923.

PORKERT, MANFRED (1). *The Theoretical Foundations of Chinese Medicine*. M.I.T. Press, Cambridge, Mass. 1973. (M.I.T. East Asian Science and Technology Series, no. 3.)

PORKERT, MANFRED (2). 'Untersuchungen einiger philosophisch-wissenschaftlicher Grundbegriffe und Beziehungen in Chinesischen.' *ZDMG*, 1961, **110**, 422.

PORKERT, MANFRED (3). 'Wissenschaftliches Denken im alten China—das System der energetischen Beziehungen.' *ANT*, 1961, **2**, 532.

DELLA PORTA, G. B. (3). *De distillatione libri IX; Quibus certa methodo, multiplici artificii: penitioribus naturae arcanis detectis cuius libet mixti, in propria elementa resolutio perfectur et docetur*. Rome and Strassburg, 1609.

POSTLETHWAYT, MALACHY (1). *The Universal Dictionary of Trade and Commerce; translated from the French of Mons. [Jacques] Savary [des Bruslons], with large additions*. 2 vols. London, 1751–5. 4th ed. London, 1774.

POTT, A. F. (1). 'Chemie oder Chymie?' *ZDMG*, 1876, **30**, 6.

POTTIER, E. (1). 'Observations sur les Couches profondes de l'Acropole [& Nécropole] à Suse.' *MDP*, 1912, **13**, 1, and pl. xxxvii, 8.

POUGH, F. H. (1). 'The Birth and Death of a Volcano [Parícutin in Mexico].' *END*, 1951, **10**, 50.

[VON PRANTL, K.] (1). 'Die Keime d. Alchemie bei den Alten.' *DV*, 1856 (no. 1), no. 73, 135.

PREISENDANZ, K. (1). 'Ostanes.' Art. in Pauly–Wissowa, *Real-Encyklop. d. class. Altertumswiss.* Vol. xviii, pt. 2, cols. 1609 ff.

PREISENDANZ, K. (2). 'Ein altes Ewigkeitsymbol als Signet und Druckermarke.' *GUJ*, 1935, 143.

PREISENDANZ, K. (3). 'Aus der Geschichte des Uroboros; Brauch und Sinnbild.' Art. in E. Fehrle Festschrift, Karlsruhe, 1940, p. 194.

PREUSCHEN, E. (1). 'Die Apocryphen Gnostichen Adamschriften aus dem Armenischen übersetzt und untersucht.' Art. in Festschrift f. Bernhard Stade, sep. pub. Ricker (Töpelmann), Giessen, 1900.

PRYOR, M. G. M. (1). 'On the Hardening of the Ootheca of *Blatta orientalis* (and the cuticle of insects in general).' *PRSB*, 1940, **128**, 378, 393.

PRZYŁUSKI, J. (1). 'Les Unipédes.' *MCB*, 1933, **2**, 307.

PRZYŁUSKI, J. (2). (*a*) 'Une Cosmogonie Commune à l'Iran et à l'Inde.' *JA*, 1937, **229**, 481. (*b*) 'La Théorie des Eléments.' *SCI*, 1933.

PUECH, H. C. (1). *Le Manichéisme; son Fondateur, sa Doctrine*. Civilisations du Sud, SAEP, Paris, 1949. (Musée Guimet, Bibliothèque de Diffusion, no. 56.)

PUECH, H. C. (2). 'Catharisme Médiéval et Bogomilisme.' Art. in *Atti dello Convegno di Scienze Morali, Storiche e Filologiche*—'Oriente ed Occidente nel Medio Evo'. Accad. Naz. di Lincei, Rome, 1956 (Atti dei Convegni Alessandro Volta, no. 12), p. 56.

PUECH, H. C. (3). 'The Concept of Redemption in Manichaeism.' *ERYB*, 1969, **6**, 247 (*The Mystic Vision*, ed. J. Campbell). Tr. from the German in *ERJB*, 1936, **4**, 1.

PUFF VON SCHRICK, MICHAEL. See von Schrick.

PULLEYBLANK, E. G. (11). 'The Consonantal System of Old Chinese.' *AM*, 1964, **9**, 206.

PULSIFER, W. H. (1). *Notes for a History of Lead; and an Enquiry into the Development of the Manufacture of White Lead and Lead Oxides*. New York, 1888.

PUMPELLY, R. (1). 'Geological Researches in China, Mongolia and Japan, during the years 1862 to 1865.' *SCK*, 1866, **202**, 77.

PUMPELLY, R. (2). 'An Account of Geological Researches in China, Mongolia and Japan during the Years 1862 to 1865.' *ARSI*, 1866, **15**, 36.

PURKINJE, J. E. (PURKYNĚ). See Teich (1).

DU PUY-SANIÈRES, G. (1). 'La Modification Volontaire du Rhythme Respiratoire et les Phenomènes qui s'y rattachent.' *RPCHG*, 1937 (no. 486).

QUIRING, H. (1). *Geschichte des Goldes; die goldenen Zeitalter in ihrer kulturellen und wirtschaftlichen Bedeutung*. Enke, Stuttgart, 1948.

QUISPEL, G. (1). 'Gnostic Man; the Doctrine of Basilides.' *ERYB*, 1969, **6**, 210 (*The Mystic Vision*, ed. J. Campbell). Tr. from the German in *ERJB*, 1948, **16**, 1.

RAMAMURTHI, B. (1). 'Yoga; an Explanation and Probable Neurophysiology.' *JIMA*, 1967, **48**, 167.

RANKING, G. S. A. (1). 'The Life and Works of Rhazes.' (Biography and Bibliography of al-Rāzī.) Proc. XVIIth Internat. Congress of Medicine, London, 1913. Sect. 23, pp. 237–68.

RAO, GUNDU H. V., KRISHNASWAMY, M., NARASIMHAIYA, R. L., HOENIG, J. & GOVINDASWAMY, M. V. (1). 'Some Experiments on a Yogi in Controlled States.' *JAIMH*, 1958, **1**, 99.

RAO, SHANKAR (1). 'The Metabolic Cost of the (Yogi) Head-stand Posture.' *JAP*, 1962, **17**, 117.

RAO, SHANKAR (2). 'Oxygen-consumption during Yoga-type Breathing at Altitudes of 520 m. and 3800 m.' *IJMR*, 1968, **56**, 701.

RATLEDGE, C. (1). 'Cooling Cells for Smashing.' *NS*, 1964, **22**, 693.

RATTANSI, P. M. (1). 'The Literary Attack on Science in the Late Seventeenth and Eighteenth Centuries.' Inaug. Diss. London, 1961.

RATTANSI, P. M. (2). 'The Intellectual Origins of the Royal Society.' *NRRS*, 1968, **23**, 129.

RATTANSI, P. M. (3). 'Newton's Alchemical Studies', art. in *Science, Medicine and Society in the Renaissance* (Pagel Presentation Volume), ed. Debus (20), vol. 2, p. 167.

RATTANSI, P. M. (4). 'Some Evaluations of Reason in +16th- and +17th-Century Natural Philosophy', art. in *Changing Perspectives in the History of Science*, ed. M. Teich & R. Young. Heinemann, London, 1973, p. 148.

RATZEL, F. (1). *History of Mankind*. Tr. A. J. Butler, with introduction by E. B. Tylor. 3 vols. London, 1896–8.

RAWSON, P. S. (1). *Tantra*. (Catalogue of an Exhibition of Indian Religious Art, Hayward Gallery, London, 1971.) Arts Council of Great Britain, London, 1971.

RAY, P. (1). 'The Theory of Chemical Combination in Ancient Indian Philosophies.' *IJHS*, 1966, **1**, 1.

RAY, P. C. (1). *A History of Hindu Chemistry, from the Earliest Times to the middle of the 16th cent. A.D., with Sanskrit Texts, Variants, Translation and Illustrations*. 2 vols. Chuckerverty & Chatterjee, Calcutta, 1902, 1904, repr. 1925. New enlarged and revised edition in one volume, ed. P. Ray, retitled *History of Chemistry in Ancient and Medieval India*, Indian Chemical Society, Calcutta, 1956. Revs. J. Filliozat, *ISIS*, 1958, **49**, 362; A. Rahman, *VK*, 1957, 18.

RAY, P. C. See Tenney L. Davis' biography (obituary), with portrait. *JCE*, 1934, **11** (535).

RAY, T. (1) (tr.). The '*Ananga Ranga*' [written by Kalyana Malla, for Lad Khan, a son of Ahmed Khan Lodi, *c.* +1500], pref. by G. Bose. Med. Book Co. Calcutta, 1951 (3rd ed.).

RAZDAN, R. K. (1). 'The Hallucinogens.' *ARMC*, 1970 (1971), **6**.

RAZOOK. See Razuq.

RAZUQ, FARAJ RAZUQ (1). 'Studies on the Works of al-Ţughrā'ī.' Inaug. Diss., London, 1963.

READ, BERNARD E. (with LIU JU-CHHIANG) (1). *Chinese Medicinal Plants from the 'Pên Tshao Kang Mu', A.D. 1596 ... a Botanical, Chemical and Pharmacological Reference List*. (Publication of the Peking Nat. Hist. Bull.) French Bookstore, Peiping, 1936 (chs. 12–37 of *PTKM*). Rev. W. T. Swingle, *ARLC/DO*, 1937, 191. Originally published as *Flora Sinensis*, Ser. A, vol. 1, *Plantae Medicinalis Sinensis*, 2nd ed., *Bibliography of Chinese Medicinal Plants from the Pên Tshao Kang Mu, A.D. 1596*, by B. E. Read & Liu Ju-Chhiang. Dept. of Pharmacol. Peking Union Med. Coll. & Peking Lab. of Nat. Hist. Peking, 1927. First ed. Peking Union Med. Coll. 1923.

READ, BERNARD E. (2) (with LI YÜ-THIEN). *Chinese Materia Medica; Animal Drugs*.

	Serial nos.	Corresp. with chaps. of *Pên Tshao Kang Mu*
Pt. I Domestic Animals	322–349	50
II Wild Animals	350–387	51*A* & *B*
III Rodentia	388–399	51*B*
IV Monkeys and Supernatural Beings	400–407	51*B*
V Man as a Medicine	408–444	52

PNHB, **5** (no. 4), 37–80; **6** (no. 1), 1–102. (Sep. issued, French Bookstore, Peiping, 1931.)

	Serial nos.	Corresp. with chaps. of *Pên Tshao Kang Mu*

READ, BERNARD E. (3) (with LI YÜ-THIEN). *Chinese Materia Medica; Avian Drugs.*
    Pt. VI   Birds — 245–321 — 47, 48, 49
    PNHB, 1932, **6** (no. 4), 1–101. (Sep. issued, French Bookstore, Peiping, 1932.)

READ, BERNARD E. (4) (with LI YÜ-THIEN). *Chinese Materia Medica; Dragon and Snake Drugs.*
    Pt. VII   Reptiles — 102–127 — 43
    PNHB, 1934, **8** (no. 4), 297–357. (Sep. issued, French Bookstore, Peiping, 1934.)

READ, BERNARD E. (5) (with YU CHING-MEI). *Chinese Materia Medica; Turtle and Shellfish Drugs.*
    Pt. VIII   Reptiles and Invertebrates — 199–244 — 45, 46
    PNHB, (Suppl.) 1939, 1–136. (Sep. issued, French Bookstore, Peiping, 1937.)

READ, BERNARD E. (6) (with YU CHING-MEI). *Chinese Materia Medica; Fish Drugs.*
    Pt. IX   Fishes (incl. some amphibia, octopoda and crustacea) — 128–199 — 44
    PNHB (Suppl.), 1939. (Sep. issued, French Bookstore, Peiping, n.d. prob. 1939.)

READ, BERNARD E. (7) (with YU CHING-MEI). *Chinese Materia Medica; Insect Drugs.*
    Pt. X   Insects (incl. arachnidae etc.) — 1–101 — 39, 40, 41, 42
    PNHB (Suppl.), 1941. (Sep. issued, Lynn, Peiping, 1941.)

READ, BERNARD E. (10). 'Contributions to Natural History from the Cultural Contacts of East and West.' *PNHB*, 1929, **4** (no. 1), 57.

READ, BERNARD E. (11). 'Chinese Alchemy.' *N*, 1927, **120**, 877.

READ, BERNARD E. (12). 'Inner Mongolia; China's Northern Flowery Kingdom.' (This title is a reference to the abundance of wild flowers on the northern steppes, but the article also contains an account of the saltpetre industry and other things noteworthy at Hochien in S.W. Hopei.) *PJ*, 1926, **61**, 570.

READ, BERNARD E. & LI, C. O. (1). 'Chinese Inorganic Materia Medica.' *CMJ*, 1925, **39**, 23.

READ, BERNARD E. & PAK, C. (PAK KYEBYÖNG) (1). *A Compendium of Minerals and Stones used in Chinese Medicine, from the 'Pên Tshao Kang Mu'.* *PNHB*, 1928, **3** (no. 2), i–vii, 1–120. Revised and enlarged, issued separately, French Bookstore, Peiping, 1936 (2nd ed.). Serial nos. 1–135, corresp. with chaps. of *Pên Tshao Kang Mu*, 8, 9, 10, 11.

READ, J. (1). *Prelude to Chemistry; an Outline of Alchemy, its Literature and Relationships.* Bell, London, 1936.

READ, J. (2). 'A Musical Alchemist [Count Michael Maier].' Abstract of Lecture, Royal Institution, London, 22 Nov. 1935.

READ, J. (3). *Through Alchemy to Chemistry.* London, 1957.

READ, T .T. (1). 'The Mineral Production and Resources of China' (metallurgical notes on tours in China, with analyses by C. F. Wang, C. H. Wang & F. N. Lu). *TAIMME*, 1912, **43**, 1–53.

READ, T. T. (2). 'Chinese Iron castings.' *CWR*, 1931 (16 May).

READ, T. T. (3). 'Metallurgical Fallacies in Archaeological Literature.' *AJA*, 1934, **38**, 382.

READ, T. T. (4). 'The Early Casting of Iron; a Stage in Iron Age Civilisation.' *GR*, 1934, **24**, 544.

READ, T. T. (5). 'Iron, Men and Governments.' *CUQ*, 1935, **27**, 141.

READ, T. T. (6). 'Early Chinese Metallurgy.' *MI*, 1936 (6 March), p. 308.

READ, T. T. (7). 'The Largest and the Oldest Iron Castings.' *IA*, 1936, **136** (no. 18, 30 Apr.), 18 (the lion of Tshang-chou, +954, the largest).

READ, T. T. (8). 'China's Civilisation Simultaneous, not Osmotic' (letter). *AMS*, 1937, **6**, 249.

READ, T. T. (9). 'Ancient Chinese Castings.' *TAFA*, 1937 (Preprint no. 37–29 of June), 30.

READ, T. T. (10). 'Chinese Iron—A Puzzle.' *HJAS*, 1937, **2**, 398.

READ, T. T. (11). Letter on 'Pure Iron—Ancient and Modern'. *MM*, 1940 (June), p. 294.

READ, T. T. (12). 'The Earliest Industrial Use of Coal.' *TNS*, 1939, **20**, 119.

READ, T. T. (13). 'Primitive Iron-Smelting in China.' *IA*, 1921, **108**, 451.

REDGROVE, H. STANLEY (1). 'The Phallic Element in Alchemical Tradition.' *JALCHS*, 1915, **3**, 65. Discussion, pp. 88 ff.

REGEL, A. (1). 'Reisen in Central-Asien, 1876–9.' *MJPGA*, 1879, **25**, 376, 408. 'Turfan.' *MJPGA*, 1880, **26**, 205. 'Meine Expedition nach Turfan.' *MJPGA*, 1881, **27**, 380. Eng. tr. *PRGS*, 1881, 340.

REID, J. S. (1). '[The State of the Dead in] Greek [Thought].' *ERE*, vol. xi, p. 838.

REID, J. S. (2). '[The State of the Dead in] Roman [Culture].' *ERE*, vol. xi, p. 839.

REINAUD, J. T. & FAVÉ, I. (1). *Du Feu Grégeois, des Feux de Guerre, et des Origines de la Poudre à Canon, d'après des Textes Nouveaux.* Dumaine, Paris, 1845. Crit. rev. by D[efrémer]y, *JA*, 1846 (4ᵉ sér.), **7**, 572; E. Chevreul, *JS*, 1847, 87, 140, 209.

REINAUD, J. T. & FAVÉ, I. (2). 'Du Feu Grégeois, des Feux de Guerre, et des Origines de la Poudre à Canon chez les Arabes, les Persans et les Chinois.' *JA*, 1849 (4ᵉ sér.), **14**, 257.

REINAUD, J. T. & FAVÉ, I. (3). Controverse à propos du Feu Grégeois; Réponse aux Objections de M. Ludovic Lalanne.' *BEC*, 1847 (2ᵉ sér.), **3**, 427.

REITZENSTEIN, R. (1). *Die Hellenistischen Mysterienreligionen, nach ihren Grundgedanken und Wirkungen.* Leipzig, 1910. 3rd, enlarged and revised ed. Teubner, Berlin and Leipzig, 1927.

REITZENSTEIN, R. (2). *Das iranische Erlösungsmysterium; religionsgeschichtliche Untersuchungen.* Marcus & Weber, Bonn, 1921.

REITZENSTEIN, R. (3). '*Poimandres'; Studien zur griechisch-ägyptischen und frühchristlichen Literatur.* Teubner, Leipzig, 1904.

REITZENSTEIN, R. (4). *Hellenistische Wundererzähhungen.*
Pt. I *Die Aretalogie* [Thaumaturgical Fabulists]; *Ursprung, Begriff, Umbildung ins Weltliche.*
Pt. II *Die sogenannte Hymnus der Seele in den Thomas-Akten.*
Teubner, Leipzig, 1906.

RÉMUSAT, J. P. A. (7) (tr.). *Histoire de la Ville de Khotan, tirée des Annales de la Chine et traduite du Chinois; suivie de Recherches sur la Substance Minérale appelée par les Chinois Pierre de Iu [Jade] et sur le Jaspe des Anciens.* [Tr. of *TSCC, Pien i tien,* ch. 55.] Doublet, Paris, 1820. Crit. rev. J. Klaproth (6), vol. 2, p. 281.

RÉMUSAT, J. P. A. (9). 'Notice sur l'Encyclopédie Japonoise et sur Quelques Ouvrages du Même Genre' (mostly on the *Wakan Sanzai Zue*). *MAI/NEM*, 1827, **11**, 123. Botanical lists, with Linnaean Latin identifications, pp. 269–305; list of metals, p. 231, precious stones, p. 232; ores, minerals and chemical substances, pp. 233–5.

RÉMUSAT, J. P. A. (10). 'Lettre de Mons. A. R....à Mons. L. Cordier...sur l'Existence de deux Volcans brûlans dans la Tartarie Centrale [a translation of passages from *Wakan Sanzai Zue*].' *JA*, 1824, **5**, 44. Repr. in (11), vol. 1, p. 209.

RÉMUSAT, J. P. A. (11). *Mélanges Asiatiques; ou, Choix de Morceaux de Critique et de Mémoires relatifs aux Réligions, aux Sciences, aux Coutumes, à l'Histoire et à la Géographie des Nations Orientales.* 2 vols. Dondey-Dupré, Paris, 1825–6.

RÉMUSAT, J. P. A. (12). *Nouveaux Mélanges Asiatiques; ou, Recueil de Morceaux de Critique et de Mémoires relatifs aux Religions, aux Sciences, aux Coutumes, à l'Histoire et à la Géographie des Nations Orientales.* 2 vols. Schubart & Heideloff and Dondey-Dupré, Paris, 1829.

RÉMUSAT, J. P. A. (13). *Mélanges Posthumes d'Histoire et de Littérature Orientales.* Imp. Roy., Paris, 1843.

RENAULD, E. (1) (tr.). *Michel Psellus' 'Chronographie', ou Histoire d'un Siècle de Byzance, +976 à +1077; Texte établi et traduit...* 2 vols. Paris, 1938. (Collection Byzantine Budé, nos. 1, 2.)

RENFREW, C. (1). 'Cycladic Metallurgy and the Aegean Early Bronze Age.' *AJA*, 1967, **71**, 1. See Charles (1).

RENOU, L. (1). *Anthologie Sanskrite.* Payot, Paris, 1947.

RENOU, L. & FILLIOZAT, J. (1). *L'Inde Classique; Manuel des Études Indiennes.* Vol. 1, with the collaboration of P. Meile, A. M. Esnoul & L. Silburn. Payot, Paris, 1947. Vol. 2, with the collaboration of P. Demiéville, O. Lacombe & P. Meile. École Française d'Extrême Orient, Hanoi; Impr. Nationale, Paris, 1953.

RETI, LADISLAO (6). *Van Helmont, Boyle, and the Alkahest.* Clark Memorial Library, Univ. of California, Los Angeles, 1969. (In *Some Aspects of Seventeenth-Century Medicine and Science*, Clark Library Seminar, no. 27, 1968.)

RETI, LADISLAO (7). 'Le Arte Chimiche di Leonardo da Vinci.' *LCHIND*, 1952, **34**, 655, 721.

RETI, LADISLAO (8). 'Taddeo Alderotti and the Early History of Fractional Distillation' (in Spanish). MS. of a Lecture in Buenos Aires, 1960.

RETI, LADISLAO (10). 'Historia del Atanor desde Leonardo da Vinci hasta "l'Encyclopédie" de Diderot.' *INDQ*, 1952, **14** (no. 10), 1.

RETI, LADISLAO (11). 'How Old is Hydrochloric Acid?' *CHYM*, 1965, **10**, 11.

REUVENS, C. J. C. (1). *Lettres à Mons. Letronne...sur les Papyrus Bilingues et Grecs et sur Quelques Autres Monumens Gréco-Égyptiens du Musée d'Antiquités de l'Université de Leide.* Luchtmans, Leiden, 1830. Pagination separate for each of the three letters.

REX, FRIEDEMANN, ATTERER, M., DEICHGRÄBER, K. & RUMPF, K. (1). *Die 'Alchemie' des Andreas Libavius, ein Lehrbuch der Chemie aus dem Jahre 1597, zum ersten mal in deutscher Übersetzung...* herausgegeben... Verlag Chemie, Weinheim, 1964.

REY, ABEL (1). *La Science dans l'Antiquité.* Vol. 1 *La Science Orientale avant les Grecs,* 1930, 2nd ed. 1942; Vol. 2 *La Jeunesse de la Science Grecque,* 1933; Vol. 3 *La Maturité de la Pensée Scientifique en Grèce,* 1939; Vol. 4 *L'Apogée de la Science Technique Grecque (Les Sciences de la Nature et de l'Homme, les Mathematiques, d'Hippocrate à Platon),* 1946. Albin Michel, Paris. (Evol. de l'Hum. Ser. Complementaire.)

RHENANUS, JOH. (1). *Harmoniae Imperscrutabilis Chymico-Philosophicae Decades duae.* Frankfurt, 1625. See Ferguson (1), vol. 2, p. 264.

RIAD, H. (1). 'Quatre Tombeaux de la Nécropole ouest d'Alexandrie.' (Report of the −2nd-century sāqīya fresco at Wardian.) *BSAA,* 1967, **42**, 89. Prelim. pub., with cover colour photograph, *AAAA,* 1964, **17** (no. 3).

RIBÉREAU-GAYON, J. & PEYNARD, E. (1). *Analyse et Contrôle des Vins...* 2nd ed. Paris, 1958.

RICE, D. S. (1). 'Mediaeval Ḥarrān; Studies on its Topography and Monuments.' *ANATS,* 1952, **2**, 36–83.

RICE, TAMARA T. (1). *The Scythians.* 3rd ed. London, 1961.

RICE, TAMARA T. (2). *Ancient Arts of Central Asia.* Thames & Hudson, London, 1965.

RICHET, C. (1) (ed.). *Dictionnaire de Physiologie.* 6 vols. Alcan, Paris, 1895–1904.

RICHIE, D. & ITO KENKICHI (1) = (1). *The Erotic Gods; Phallicism in Japan* (English and Japanese text and captions). Zufushinsha, Tokyo, 1967.

VON RICHTHOFEN, F. (2). *China; Ergebnisse eigener Reisen und darauf gegründeter Studien.* 5 vols. and Atlas. Reimer, Berlin, 1877–1911.
  Vol. 1 Einleitender Teil
  Vol. 2 Das nördliche China
  Vol. 3 Das südliche China (ed. E. Tiessen)
  Vol. 4 Palaeontologischer Teil (with contributions by W. Dames *et al.*)
  Vol. 5 Abschliessende palaeontologischer Bearbeitung der Sammlung... (by F. French).
  (Teggart Bibliography says 5 vols. +2 Atlas Vols.)

VON RICHTHOFEN, F. (6). *Letters on Different Provinces of China.* 6 parts, Shanghai, 1871–2.

RICKARD, T. A. (2). *Man and Metals.* Fr. tr. by F. V. Laparra, *L'Homme et les Métaux.* Gallimard, Paris, 1938. Rev. L. Febvre, *AHES/AHS,* 1940, **2**, 243.

RICKARD, T. A. (3). *The Story of the Gold-Digging Ants.* UCC, 1930.

RICKETT, W. A. (1) (tr.). *The 'Kuan Tzu' Book.* Hongkong Univ. Press, Hong Kong, 1965. Rev. T. Pokora, *ARO,* 1967, **35**, 169.

RIDDELL, W. H. (2). Earliest representations of dragon and tiger. *AQ,* 1945, **19**, 27.

RIECKERT, H. (1). 'Plethysmographische Untersuchungen bei Konzentrations- und Meditations-Übungen.' *AF,* 1967, **21**, 61.

RIEGEL, BEISWANGER & LANZL (1). Molecular stills. *IEC/A,* 1943, **15**, 417.

RIETHE, P. (1). 'Amalgamfüllung Anno Domini 1528' [A MS. of therapy and pharmacy drawn from the practice of Johannes Stocker, d. +1513]. *DZZ,* 1966, **21**, 301.

RITTER, H. (1) (ed.). *Pseudo-al-Majrīṭī 'Das Ziel des Weisen'* [*Ghāyat al-Ḥakīm*]. Teubner, Leipzig, 1933. (Studien d. Bibliothek Warburg, no. 12.)

RITTER, H. (2) '*Picatrix,* ein arabisches Handbuch hellenistischer Magie.' *VBW,* 1923, **1**, 94. A much enlarged and revised form of this lecture appears as the introduction to vol. 2 of Ritter & Plessner (1), pp. xx ff.

RITTER, H. (3) (tr.). *Al-Ghazzālī's* (al-Ṭusī, +1058 to +1112] '*Das Elixir der Glückseligkeit*' [*Kīmiyā al-Sa'āda*]. Diederichs, Jena, 1923. (Religiöse Stimmen der Völkers; die Religion der Islam, no. 3.)

RITTER, H. & PLESSNER, M. (1). '*Picatrix'; das 'Ziel des Weisen'* [*Ghāyat al Ḥakīm*] *von Pseudo-Majrīṭī* 2 vols. Vol. 1, Arabic text, ed. H. Ritter. Teubner, Leipzig and Berlin, 1933 (Studien der Bibliothek Warburg, no. 12). Vol. 2, German translation, with English summary (pp. lix–lxxv), by H. Ritter & M. Plessner. Warburg Inst. London, 1962 (Studies of the Warburg Institute, no. 27). Crit. rev. W. Hartner, *DI,* 1966, **41**, 175, repr. Hartner (12), p. 429.

RITTER, K. (1). *Die Erdkunde im Verhältnis z. Natur und z. Gesch. d. Menschen; oder, Allgemeine Vergleichende Geographie.* Reimer, Berlin, 1822–59. 19 vols., the first on Africa, all the rest on Asia. Indexes after vols. 5, 13, 16 and 17.

RITTER, K. (2). *Die Erdkunde von Asien.* 5 vols. Reimer, Berlin, 1837 (part of Ritter, 1).

RIVET, P. (2). 'Le Travail de l'Or en Colombie.' *IPEK,* 1926, **2**, 128.

RIVET, P. (3). 'L'Orfèvrerie Colombienne; Technique, Aire du Dispersion, Origines.' Communication to the XXIst International Congress of Americanists, The Hague, 1924.

RIVET, P. & ARSENDAUX, H. (1). 'La Métallurgie en Amérique pre-Colombienne.' *TMIE,* 1946, no. 39.

ROBERTS [-AUSTEN], W. C. (1). 'Alloys used for Coinage' (Cantor Lectures). *JRSA*, 1884, **32**, 804, 835, 881.

ROBERTS-AUSTEN, W. C. (2). 'Alloys' (Cantor Lectures). *JRSA*, 1888, **36**, 1111, 1125, 1137.

ROBERTSON, T. BRAILSFORD & RAY, L. A. (1). 'An Apparatus for the Continuous Extraction of Solids at the Boiling Temperature of the Solvent.' *RAAAS*, 1924, **17**, 264.

ROBINSON, B. W. (1). 'Royal Asiatic Society MS. no. 178; an unrecorded Persian Painter.' *JRAS*, 1970 (no. 2, Wheeler Presentation Volume), 203. ('Abd al Karīm, active *c.* +1475, who illustrated some East Asian subjects.)

ROBINSON, G. R. & DEAKERS, T. W. (1). 'Apparatus for sublimation of anthracene.' *JCE*, 1932, **9**, 1717.

DE ROCHAS D'AIGLUN, A. (1). *La Science des Philosophes et l'Art des Thaumaturges dans l'Antiquité.* Dorbon, Paris. 1st ed. n.d. (1882), 2nd ed. 1912.

DE ROCHEMONTEIX, C. (1). *Joseph Amiot et les Derniers Survivants de la Mission Française à Pékin (1750 à 1795); Nombreux Documents inédits, avec Carte.* Picard, Paris, 1915.

ROCKHILL, W. W. (1). 'Notes on the Relations and Trade of China with the Eastern Archipelago and the Coast of the Indian Ocean during the +15th Century.' *TP*, 1914, **15**, 419; 1915, **16**, 61, 236, 374, 435, 604.

ROCKHILL, W. W. (5) (tr. & ed.). *The Journey of William of Rubruck to the Eastern Parts of the World* (+1253 to +1255) *as narrated by himself; with Two Accounts of the earlier Journey of John of Pian de Carpine.* Hakluyt Soc., London, 1900 (second series, no. 4).

RODWELL, G. F. (1). *The Birth of Chemistry.* London, 1874.

RODWELL, J. M. (1). *Aethiopic and Coptic Liturgies and Prayers.* Pr. pr. betw. 1870 and 1886.

ROGERS, R. W. (1). '[The State of the Dead in] Babylonian [and Assyrian Culture].' *ERE*, vol. xi, p. 828.

ROI, J. (1). *Traité des Plantes Médicinales Chinoises.* Lechevalier, Paris, 1955. (Encyclopédie Biologique ser. no. 47.) No Chinese characters, but a photocopy of those required is obtainable from Dr Claude Michon, 8 bis, Rue Desilles, Nancy, Meurthe & Moselle, France.

ROI, J. & WU YÜN-JUI (OU YUN-JOEI) (1). 'Le Taoisme et les Plantes d'Immortalité.' *BUA*, 1941 (3e sér.), **2**, 535.

ROLANDI, G. & SCACCIATI, G. (1). 'Ottone e Zinco presso gli Antichi' (Brass and Zinc in the Ancient World). *IMIN*, 1956, **7** (no. 11), 759.

ROLFINCK, WERNER (1). *Chimia in Artis Formam Redacta.* Geneva, 1661, 1671, Jena, 1662, and later editions.

ROLLESTON, SIR HUMPHREY (1). *The Endocrine Organs in Health and Disease, with an Historical Review.* London, 1936.

RÖLLIG, W. (1). 'Das Bier im alten Mesopotamien.' *JGGBB*, 1970 (for 1971), 9–104.

RONCHI, V. (5). 'Scritti di Ottica; Tito Lucrezio Caro, Leonardo da Vinci, G. Rucellai, G. Fracastoro, G. Cardano, D. Barbaro, F. Maurolico, G. B. della Porta, G. Galilei, F. Sizi, E. Torricelli, F. M. Grimaldi, G. B. Amici [a review].' *AFGR/CINO*, 1969, **24** (no. 3), 1.

RONCHI, V. (6). 'Philosophy, Science and Technology.' *AFGR/CINO*, 1969, **24** (no. 2), 168.

RONCHI, V. (7). 'A New History of the Optical Microscope.' *IJHS*, 1966, **1**, 46.

RONCHI, V. (8). 'The New History of Optical Microscopy.' *ORG*, 1968, **5**, 191.

RORET, N. E. (1) (ed.). *Manuel de l'Orfévre*, part of *Encyclopédie Roret* (or *Manuels Roret*). Roret, Paris, 1825– . Berthelot (1, 2) used the ed. of 1832.

ROSCOE, H. E. & SCHORLEMMER, C. (1). *A Treatise on Chemistry.* Macmillan, London, 1923.

ROSENBERG, E. (1) (ed.). *Gonadotrophins, 1968.* 1969.

ROSENBERG, M. (1). *Geschichte der Goldschmiedekunst auf technische Grundlage.* Frankfurt-am-Main.
Vol. 1 *Einführung*, 1910.
Vol. 2 *Niello*, 1908.
Vol. 3 (in 3 parts) *Zellenschmelz*, 1921, 1922, 1925.
Re-issued in one vol., 1972.

VON ROSENROTH, K. & VAN HELMONT, F. M. (1) (actually anon.). *Kabbala Denudata, seu Doctrina Hebraeorum Transcendentalis et Metaphysica*, etc. Lichtenthaler, Sulzbach, 1677.

ROSENTHAL, F. (1) (tr.). *The 'Muqaddimah' [of Ibn Khaldun]; an Introduction to History.* Bollingen, New York, 1958. Abridgement by N. J. Dawood, London, 1967.

ROSS, E. D. (3). *Alphabetical List of the Titles of Works in the Chinese Buddhist Tripitaka.* Indian Govt. Calcutta, 1910. (See Nanjio, B.)

ROSSETTI, GABRIELE (1). *Disquisitions on the Anti-Papal spirit which produced the Reformation; its Secret Influence on the Literature of Europe in General and of Italy in Particular.* Tr. C. Ward from the Italian. 2 vols. Smith & Elder, London, 1834.

ROSSI, P. (1). *Francesco Bacone; dalla Magia alla Scienza.* Laterza, Bari, 1957. Eng. tr. by Sacha Rabino-vitch, *Francis Bacon; from Magic to Science.* Routledge & Kegan Paul, London, 1968.

ROTH, H. LING (1). *Oriental Silverwork, Malay and Chinese; a Handbook for Connoisseurs, Collectors, Students and Silversmiths.* Truslove & Hanson, London, 1910, repr. Univ. Malaya Press, Kuala Lumpur, 1966.

ROTH, MATHIAS (1). *The Prevention and Cure of Many Chronic Diseases by Movements.* London, 1851.

ROTHSCHUH, K. E. (1). *Physiologie; der Wandel ihrer Konzepte, Probleme und Methoden vom 16. bis 19. Jahrhundert.* Alber, Freiburg and München, 1968. (Orbis Academicus, Bd. 2, no. 15.)

DES ROTOURS, R. (3). 'Quelques Notes sur l'Anthropophagie en Chine.' *TP*, 1963, **50**, 386. 'Encore Quelques Notes...' *TP*, 1968, **54**, 1.

ROUSSELLE, E. (1). 'Der lebendige Taoismus im heutigen China.' *SA*, 1933, **8**, 122.

ROUSSELLE, E. (2). 'Yin und Yang vor ihrem Auftreten in der Philosophie.' *SA*, 1933, **8**, 41.

ROUSSELLE, E. (3). 'Das Primat des Weibes im alten China.' *SA*, 1941, **16**, 130.

ROUSSELLE, E. (4a). 'Seelische Führung im lebenden Taoismus.' *ERYB*, 1933, **1** (a reprint of (6), with (5) intercalated). Eng. tr., 'Spiritual Guidance in Contemporary Taoism.' *ERYB*, 1961, **4**, 59 ('Spiritual Disciplines', ed. J. Campbell). Includes footnotes but no Chinese characters.

ROUSSELLE, E. (4b). *Zur Seelischen Führung im Taoismus; Ausgewählte Aufsätze.* Wissenschaftl. Buchgesellsch., Darmstadt, 1962. (A collection of three reprinted articles (7), (5) and (6), including footnotes, and superscript references to Chinese characters, but omitting the characters themselves.)

ROUSSELLE, E. (5). '*Ne Ging Tu [Nei Ching Thu]*, "Die Tafel des inneren Gewebes"; ein Taoistisches Meditationsbild mit Beschriftung.' *SA*, 1933, **8**, 207.

ROUSSELLE, E. (6). 'Seelische Führung im lebenden Taoismus.' *CDA*, 1934, 21.

ROUSSELLE, E. (7). 'Die Achse des Lebens.' *CDA*, 1933, 25.

ROUSSELLE, E. (8). 'Dragon and Mare; Figures of Primordial Chinese Mythology' (personifications and symbols of Yang and Yin, and the *kua* Chhien and Khun), *ERYB*, 1969, **6**, 103 (*The Mystic Vision*, ed. J. Campbell). Tr. from the German in *ERYB*, 1934, **2**, 1.

RUDDY, J. (1). 'The Big Bang at Sudbury.' *INM*, 1971 (no. 4), 22.

RUDELSBERGER, H. (1). *Chinesische Novellen aus dem Urtext übertragen.* Insel Verlag, Leipzig, 1914. 2nd ed., with two tales omitted, Schroll, Vienna, 1924.

RUFUS, W. C. (2). 'Astronomy in Korea.' *JRAS/KB*, 1936, **26**, 1. Sep. pub. as *Korean Astronomy.* Literary Department, Chosen Christian College, Seoul (Eng. Pub. no. 3), 1936.

RUHLAND, MARTIN (RULAND) (1). *Lexicon Alchemiae, sive Dictionarium Alchemisticum, cum obscuriorum Verborum et rerum Hermeticarum, tum Theophrast-Paracelsicarum Phrasium, Planam Explicationem Continens.* Palthenius, Frankfurt, 1612; 2nd ed. Frankfurt, 1661. Photolitho repr., Olms, Hildesheim, 1964. Cf. Ferguson (1), vol. 2, p. 303.

RULAND, M. See Ruhland, Martin.

RUSH, H. P. (1) Biography of A. A. Berthold. *AMH*, 1929, **1**, 208.

RUSKA, J. For bibliography see Winderlich (1).

RUSKA, J. (1). 'Die Mineralogie in d. arabischen Litteratur.' *ISIS*, 1913, **1**, 341.

RUSKA, J. (2). 'Der Zusammenbruch der Dschābir-Legende; I, die bisherigen Versuche das Dschābirproblem zu lösen.' *JBFIGN*, 1930, **3**, 9. Cf. Kraus (1).

RUSKA, J. (3). 'Die Siebzig Bücher des Ǧābir ibn Ḥajjān.' Art. in *Studien z. Gesch. d. Chemie; Festgabe f. E. O. von Lippmann zum 70. Geburtstage*, ed. J. Ruska. Springer, Berlin, 1927, p. 38.

RUSKA, J. (4). *Arabische Alchemisten.* Vol. 1, *Chālid* [Khālid] *ibn Jazīd ibn Mu'āwija* [Mu'awiya]. Winter, Heidelberg, 1924 (Heidelberger Akten d. von Portheim Stiftung, no. 6). Rev. von Lippmann (10); *ISIS*, 1925, **7**, 183. Repr. with Ruska (5), Sändig, Wiesbaden, 1967.

RUSKA, J. (5). *Arabische Alchemisten.* Vol. 2, *Ǧa'far* [Ja'far] *al-Ṣādiq, der sechste Imām.* Winter, Heidelberg, 1924 (Heidelberger Akten d. von Portheim Stiftung, no. 10). Rev. von Lippmann (10). Repr. with Ruska (4), Sändig, Wiesbaden, 1967.

RUSKA, J. (6). '*Turba Philosophorum*; ein Beitrag z. Gesch. d. Alchemie.' *QSGNM*, 1931, **1**, 1–368.

RUSKA, J. (7). 'Chinesisch-arabische technische Rezepte aus der Zeit der Karolinger.' *CHZ*, 1931, **55**, 297.

RUSKA, J. (8). '*Tabula Smaragdina*'; ein Beitrag z. Gesch. d. Hermetischen Literatur.* Winter, Heidelberg, 1926 (Heidelberger Akten d. von Portheim Stiftung, no. 16).

RUSKA, J. (9). 'Studien zu Muḥammad ibn 'Umail al-Tamīnī's *Kitāb al-Mā'al al-Waraqī wa'l-Arḍ al-Najmīyah*.' *ISIS*, 1936, **24**, 310.

RUSKA, J. (10). 'Der Urtext der *Tabula Chemica*.' *A*, 1934, **16**, 273.

RUSKA, J. (11). 'Neue Beiträge z. Gesch. d. Chemie (1. Die Namen der Goldmacherkunst, 2. Die Zeichen der griechischen Alchemie, 3. Griechischen Zeichen in Syrischer Überlieferung, 4. Ü. d. Ursprung der neueren chemischen Zeichen, 5. Kataloge der Decknamen, 6. Die metallurgischen Künste). *QSGNM*, 1942, **8**, 305.

RUSKA, J. (12). 'Über das Schriftenverzeichniss des Ǧābir ibn Ḥajjān [Jābir ibn Ḥayyān] und die Unechtheit einiger ihm zugeschriebenen Abhandlungen.' *AGMN*, 1923, **15**, 53.

RUSKA, J. (13). 'Sal Ammoniacus, Nušādir und Salmiak.' *SHAW/PH*, 1923 (no. 5), 1–23.

RUSKA, J. (14). 'Übersetzung und Bearbeitungen von al-Rāzī's Buch "Geheimnis der Geheimnisse" [*Kitāb Sirr al-Asrār*].' *QSGNM*, 1935, **4**, 153–238; 1937, **6**, 1–246.

RUSKA, J. (15). 'Die Alchemie al-Rāzī's.' *DI*, 1935, **22**, 281.

RUSKA, J. (16). 'Al-Bīrūnī als Quelle für das Leben und die Schriften al-Rāzī's.' *ISIS*, 1923, **5**, 26.

RUSKA, J. (17). 'Ein neuer Beitrag zur Geschichte des Alkohols.' *DI*, 1913, **4**, 320.

RUSKA, J. (18). 'Über die von Abulqāsim al-Zuhrāwī beschriebene Apparatur zur Destillation des Rosenwassers.' *CHA*, 1937, **24**, 313.

RUSKA, J. (19) (tr.). *Das Steinbuch des Aristoteles; mit literargeschichtlichen Untersuchungen nach der arabischen Handschrift der Bibliothèque Nationale herausgegeben und ubersetzt.* Winter, Heidelberg, 1912. (This early +9th-century text, the earliest of the Arabic lapidaries and widely known later as (Lat.) *Lapidarium Aristotelis*, must be termed Pseudo-Aristotle; it was written by some Syrian who knew both Greek and Eastern traditions, and was translated from Syriac into Arabic by Luka bar Serapion, or Lūqā ibn Sarāfyūn.)

RUSKA, J. (20). 'Über Nachahmung von Edelsteinen.' *QSGNM*, 1933, **3**, 316.

RUSKA, J. (21). *Das 'Buch der Alaune und Salze'; ein Grundwerk der spät-lateinischen Alchemie* [Spanish origin, +11th cent.]. Verlag Chemie, Berlin, 1935.

RUSKA, J. (22). 'Wem verdankt Man die erste Darstellung des Weingeists?' *DI*, 1913, **4**, 162.

RUSKA, J. (23). 'Weinbau und Wein in den arabischen Bearbeitungen der Geoponika.' *AGNT*, 1913, **6**, 305.

RUSKA, J. (24). *Das Steinbuch aus der 'Kosmographie' des Zakariya ibn Maḥmūd al-Qazwīnī* [*c.* +1250] *übersetzt und mit Anmerkungen versehen*... Schmersow (Zahn & Baendel), Kirchhain N-L, 1897. (Beilage zum Jahresbericht 1895–6 der prov. Oberrealschule Heidelberg.)

RUSKA, J. (25). 'Der Urtext d. *Tabula Smaragdina*.' *OLZ*, 1925, **28**, 349.

RUSKA, J. (26). 'Die Alchemie des Avicenna.' *ISIS*, 1934, **21**, 14.

RUSKA, J. (27). 'Über die dem Avicenna zugeschriebenen alchemistischen Abhandlungen.' *FF*, 1934, **10**, 293.

RUSKA, J. (28). 'Alchemie in Spanien.' *ZAC/AC*, 1933, **46**, 337; *CHZ*, 1933, **57**, 523.

RUSKA, J. (29). 'Al-Rāzī (Rhazes) als Chemiker.' *ZAC*, 1922, **35**, 719.

RUSKA, J. (30). 'Über die Anfänge der wissenschaftlichen Chemie.' *FF*, 1937, **13**.

RUSKA, J. (31). 'Die Aufklärung des Jābir-Problems.' *FF*, 1930, **6**, 265.

RUSKA, J. (32). 'Über die Quellen von Jābir's Chemische Wissen.' *A*, 1926, **7**, 267.

RUSKA, J. (33). 'Über die Quellen des [Geber's] *Liber Claritatis*.' *A*, 1934, **16**, 145.

RUSKA, J. (34). 'Studien zu den chemisch-technischen Rezeptsammlungen des *Liber Sacerdotum* [one of the texts related to *Mappae Clavicula*, etc.].' *QSGNM*, 1936, **5**, 275 (83–125).

RUSKA, J. (35). 'The History and Present Status of the Jābir Problem.' *JCE*, 1929, **6**, 1266 (tr. R. E. Oesper); *IC*, 1937, **11**, 303.

RUSKA, J. (36). 'Alchemy in Islam.' *IC*, 1937, **11**, 30.

RUSKA, J. (37) (ed.). *Studien z. Geschichte d. Chemie; Festgabe E. O. von Lippmann zum 70. Geburtstage*... Springer, Berlin, 1927.

RUSKA, J. (38). 'Das Giftbuch des Ġābir ibn Ḥajjān.' *OLZ*, 1928, **31**, 453.

RUSKA, J. (39). 'Der Salmiak in der Geschichte der Alchemie.' *ZAC*, 1928, **41**, 1321; *FF*, 1928, **4**, 232.

RUSKA, J. (40). 'Studien zu Severus [or Jacob] bar Shakko's "Buch der Dialoge".' *ZASS*, 1897, **12**, 8, 145.

RUSKA, J. & GARBERS, K. (1). 'Vorschriften z. Herstellung von scharfen Wässern bei Jābir und Rāzī.' *DI*, 1939, **25**, 1.

RUSKA, J. & WIEDEMANN, E. (1). 'Beiträge z. Geschichte d. Naturwissenschaften, LXVII; Alchemistische Decknamen. *SPMSE*, 1924, **56**, 17. Repr. in Wiedemann (23), vol. 2, p. 596.

RUSSELL, E. S. (1). *Form and Function; a Contribution to the History of Animal Morphology.* Murray, London, 1916.

RUSSELL, E. S. (2). *The Interpretation of Development and Heredity; a Study in Biological Method.* Clarendon Press, Oxford, 1930.

RUSSELL, RICHARD (1) (tr.). *The Works of Geber, the Most Famous Arabian Prince and Philosopher*... [containing *De Investigatione, Summa Perfectionis, De Inventione* and *Liber Fornacum*]. James, London, 1678. Repr., with an introduction by E. J. Holmyard, Dent, London, 1928.

RYCAUT, SIR PAUL (1). *The Present State of the Greek Church.* Starkey, London, 1679.

SACHAU, E. (1) (tr.). *Alberuni's India.* 2 vols. London, 1888; repr. 1910.

DE SACY, A. I. SILVESTRE (1). 'Le "Livre du Secret de la Création", par le Sage Bélinous [Balīnās; Apollonius of Tyana, attrib.].' *MAI/NEM*, 1799, **4**, 107–58.

DE SACY, A. I. SILVESTRE (2). *Chrestomathie Arabe; ou, Extraits de Divers Écrivains Arabes, tant en Prose qu'en Vers*... 3 vols. Impr. Imp. Paris, 1806. 2nd ed. Impr. Roy. Paris, 1826–7.

SAEKI, P. Y. (1). *The Nestorian Monument in China.* With an introductory note by Lord William Gascoyne-Cecil and a pref. by Rev. Prof. A. H. Sayce. SPCK, London, 1916.

SAEKI, P. Y. (2). *The Nestorian Documents and Relics in China.* Maruzen, for the Toho Bunkwa Gakuin, Tokyo, 1937, second (enlarged) edn. Tokyo, 1951.

SAGE, B. M. (1). 'De l'Emploi du Zinc en Chine pour la Monnaie.' *JPH*, 1804, **59**, 216. Eng. tr. in Leeds (1) from *PMG*, 1805, **21**, 242.

SAHLIN, C. (1). 'Cementkopper, en historiske Översikt.' *HF*, 1938, **9**, 100. Résumé in Lindroth (1) and *SILL*, 1954.

SAID HUSAIN NASR (1). *Science and Civilisation in Islam* (with a preface by Giorgio di Santillana). Harvard University Press, Cambridge, Mass. 1968.

SAÏD HUSAIN NASR (2). *The Encounter of Man and Nature; the Spiritual Crisis of Modern Man.* Allen & Unwin, London, 1968.

SAID HUSAIN NASR (3). *An Introduction to Islamic Cosmological Doctrines.* Cambridge, Mass. 1964.

SAKURAZAWA, NYOITI [OHSAWA, G.] (1). *La Philosophie de la Médecine d'Extrême-Orient; le Livre du Jugement Suprême.* Vrin, Paris, 1967.

SALAZARO, D. (1). *L'Arte della Miniatura nel Secolo XIV, Codice della Biblioteca Nazionale di Napoli . . .* Naples, 1877. The MS. Anonymus, *De Arte Illuminandi* (so entitled in the Neapolitan Library Catalogue, for it has no title itself). Cf. Partington (12).

SALMONY, A. (1). *Carved Jade of Ancient China.* Gillick, Berkeley, Calif., 1938.

SALMONY, A. (2). 'The Human Pair in China and South Russia.' *GBA*, 1943 (6e sér.), **24**, 321.

SALMONY, A. (4). *Chinese Jade through* [i.e. until the end of] *the* [Northern] *Wei Dynasty.* Ronald, New York, 1963.

SALMONY, A. (5). *Archaic Chinese Jades from the Edward and Louise B. Sonnenschein Collection.* Chicago Art Institute, Chicago, 1952.

SAMBURSKY, S. (1). *The Physical World of the Greeks.* Tr. from the Hebrew edition by M. Dagut. Routledge & Kegan Paul, London, 1956.

SAMBURSKY, S. (2). *The Physics of the Stoics.* Routledge & Kegan Paul, London, 1959.

SAMBURSKY, S. (3). *The Physical World of Late Antiquity.* Routledge & Kegan Paul, London, 1962. Rev. G. J. Whitrow, *A/AIHS*, 1964, **17**, 178.

SANDYS, J. E. (1). *A History of Classical Scholarship.* 3 vols. Cambridge, 1908. Repr. New York, 1964.

DI SANTILLANA, G. (2). *The Origins of Scientific Thought.* University of Chicago Press, Chicago, 1961.

DI SANTILLANA, G. & VON DECHEND, H. (1). *Hamlet's Mill; an Essay on Myth and the Frame of Time.* Gambit, Boston, 1969.

SARLET, H., FAIDHERBE, J. & FRENCK, G. 'Mise en evidence chez différents Arthropodes d'un Inhibiteur de la D-acidaminoxydase.' *AIP*, 1950, **58**, 356.

SARTON, GEORGE (1). *Introduction to the History of Science.* Vol. 1, 1927; Vol. 2, 1931 (2 parts); Vol. 3, 1947 (2 parts). Williams & Wilkins, Baltimore, (Carnegie Institution Pub. no. 376.)

SARTON, GEORGE (13). Review of W. Scott's 'Hermetica' (1). *ISIS*, 1926, **8**, 342.

SARWAR, G. & MAHDIHASSAN, S. (1). 'The Word *Kimiya* as used by Firdousi.' *IQB*, 1961, **9**, 21.

SASO, M. R. (1). 'The Taoists who did not Die.' *AFRA*, 1970, no. 3, 13.

SASO, M. R. (2). *Taoism and the Rite of Cosmic Renewal.* Washington State University Press, Seattle, 1972.

SASO, M. R. (3). 'The Classification of Taoist Sects and Ranks observed in Hsinchu and other parts of Northern Thaiwan.' *AS/BIE* 1971, **30** (vol. 2 of the Presentation Volume for Ling Shun-Shêng).

SASO, M. R. (4). 'Lu Shan, Ling Shan (Lung-hu Shan) and Mao Shan; Taoist Fraternities and Rivalries in Northern Thaiwan.' Unpubl. MS. 1973.

SASTRI, S. S. SURYANARAYANA (1). The '*Sāṃkhya Kārikā*' of Iśvarakrsna. University Press, Madras, 1930.

SATYANARAYANAMURTHI, G. G. & SHASTRY, B. P. (1). 'A Preliminary Scientific Investigation into some of the unusual physiological manifestations acquired as a result of Yogic Practices in India.' *WZNHK*, 1958, **15**, 239.

SAURBIER, B. (1). *Geschichte der Leibesübungen.* Frankfurt, 1961.

SAUVAGET, J. (2) (tr.). *Relation de la Chine et de l'Inde, redigée en* +857 *(Akhbār al-Ṣīn wa'l-Hind).* Belles Lettres, Paris, 1948. (Budé Association, Arab Series.)

SAVILLE, M. H. (1). *The Antiquities of Manabi, Ecuador.* 2 vols. New York, 1907, 1910.

SAVILLE, M. H. (2). *Indian Notes.* New York, 1920.

SCHAEFER, H. (1). *Die Mysterien des Osiris in Abydos.* Leipzig, 1901.

SCHAEFER, H. W. (1). *Die Alchemie; ihr ägyptisch-griechischer Ursprung und ihre weitere historische Entwicklung.* Programm-Nummer 260, Flensburg, 1887; phot. reprod. Sändig, Wiesbaden, 1967.

SCHAFER, E. H. (1). 'Ritual Exposure [Nudity, etc.] in Ancient China.' *HJAS*, 1951, **14**, 130.

SCHAFER, E. H. (2). 'Iranian Merchants in Thang Dynasty Tales.' *SOS*, 1951, **11**, 403.

SCHAFER, E. H. (5). 'Notes on Mica in Medieval China.' *TP*, 1955, **43**, 265.

SCHAFER, E. H. (6). 'Orpiment and Realgar in Chinese Technology and Tradition.' *JAOS*, 1955, **75**, 73.

SCHAFER, E. H. (8). 'Rosewood, Dragon's-Blood, and Lac.' *JAOS*, 1957, **77**, 129.

SCHAFER, E. H. (9). 'The Early History of Lead Pigments and Cosmetics in China.' *TP*, 1956, **44**, 413.

SCHAFER, E. H. (13). *The Golden Peaches of Samarkand; a Study of Thang Exotics.* Univ. of Calif. Press, Berkeley and Los Angeles, 1963. Rev. J. Chmielewski, *OLZ*, 1966, **61**, 497.

SCHAFER, E. H. (16). *The Vermilion Bird; Thang Images of the South.* Univ. of Calif. Press, Berkeley and Los Angeles, 1967. Rev. D. Holzman, *TP*, 1969, **55**, 157.

SCHAFER, E. H. (17). 'The Idea of Created Nature in Thang Literature' (on the phrases *tsao wu chê* and *tsao hua chê*). *PEW*, 1965, **15**, 153.

SCHAFER, E. H. & WALLACKER, B. E. (1). 'Local Tribute Products of the Thang Dynasty.' *JOSHK*, 1957, **4**, 213.

SCHEFER, C. (2). 'Notice sur les Relations des Peuples Mussulmans avec les Chinois dépuis l'Extension de l'Islamisme jusqu'à la fin du 15e Siècle.' In *Volume Centenaire de l'École des Langues Orientales Vivantes, 1795–1895.* Leroux, Paris, 1895, pp. 1–43.

SCHELENZ, H. (1). *Geschichte der Pharmazie.* Berlin, 1904; photographic reprint, Olms, Hildesheim, 1962.

SCHELENZ, H. (2). *Zur Geschichte der pharmazeutisch-chemischen Destilliergeräte.* Miltitz, 1911. Reproduced photographically, Olms, Hildesheim, 1964. (Publication supported by Schimmel & Co., essential oil distillers, Miltitz.)

SCHIERN, F. (1). *Über den Ursprung der Sage von den goldgrabenden Ameisen.* Copenhagen and Leipzig, 1873.

SCHIPPER, K. M. (1) (tr.). *L'Empereur Wou des Han dans la Légende Taoiste; le 'Han Wou-Ti Nei-Tchouan [Han Wu Ti Nei Chuan]'.* Maisonneuve, Paris, 1965. (Pub. de l'École Française d'Extrême Orient, no. 58.)

SCHIPPER, K. M. (2). 'Priest and Liturgy; the Live Tradition of Chinese Religion.' MS. of a Lecture at Cambridge University, 1967.

SCHIPPER, K. M. (3). 'Taoism; the Liturgical Tradition.' Communication to the First International Conference of Taoist Studies, Villa Serbelloni, Bellagio, 1968.

SCHIPPER, K. M. (4). 'Remarks on the Functions of "Inspector of Merits" [in Taoist ecclesiastical organisation; with a description of the Ordination ceremony in Thaiwan Chêng-I Taoism].' Communication to the Second International Conference of Taoist Studies, Chino (Tateshina), Japan, 1972.

SCHLEGEL, G. (10). 'Scientific Confectionery' (a criticism of modern chemical terminology in Chinese). *TP*, 1894 (1e sér.), **5**, 147.

SCHLEGEL, G. (11). 'Le Tchien [Chien] en Chine.' *TP*, 1897 (1e sér.), **8**, 455.

SCHLEIFER, J. (1). 'Zum Syrischen Medizinbuch; II, Der therapeutische Teil.' *RSO*, 1939, **18**, 341. (For Pt I see *ZS*, 1938 (n.s.), **4**, 70.)

SCHMAUDERER, E. (1). 'Kenntnisse ü. das Ultramarin bis zur ersten künstlichen Darstellung um 1827.' *BGTI/TG*, 1969, **36**, 147.

SCHMAUDERER, E. (2). 'Künstliches Ultramarin im Spiegel von Preisaufgaben und der Entwicklung der Mineralanalyse im 19. Jahrhundert.' *BGTI/TG*, 1969, **36**, 314.

SCHMAUDERER, E. (3). 'Die Entwicklung der Ultramarin-fabrikation im 19. Jahrhundert.' *TRAD*, 1969, **3–4**, 127.

SCHMAUDERER, E. (4). 'J. R. Glaubers Einfluss auf die Frühformen der chemischen Technik.' *CIT*, 1970, **42**, 687.

SCHMAUDERER, E. (5). 'Glaubers Alkahest; ein Beispiel für die Fruchtbarkeit alchemischer Denkansätze im 17. Jahrhundert.'; in the press.

SCHMIDT, C. (1) (ed.). *Koptisch-Gnostische Schriften* [including *Pistis Sophia*]. Hinrichs, Leipzig, 1905 (Griech. Christliche Schriftsteller, vol. 13). 2nd. ed. Akad. Verlag, Berlin, 1954.

SCHMIDT, R. (1) (tr.). *Das 'Kāmasūtram' des Vātsyāyana; die indische Ars Amatoria nebst dem vollständigen Kommentare (Jayamangalā) des Yasodhara—aus dem Sanskrit übersetzt und herausgegeben...* Berlin, 1912. 7th ed. Barsdorf, Berlin, 1922.

SCHMIDT, R. (2). *Beitäage z. Indischen Erotik; das Liebesleben des Sanskritvolkes, nach den Quellen dargestellt von R. S...* 2nd ed. Barsdorf, Berlin, 1911. Reissued under the imprint of Linser, in the same year.

SCHMIDT, R. (3) (tr.). *The 'Rati Rahasyam' of Kokkoka* [said to be +11th cent. under Rājā Bhōja]. Med. Book Co. Calcutta, 1949. (Bound with Tatojaya (1), *q.v.*)

SCHMIDT, W. A. (1). *Die Griechischen Papyruskunden der K. Bibliothek Berlin; III, Die Purpurfärberei und der Purpurhandel in Altertum.* Berlin, 1842.

SCHMIEDER, K. C. (1). *Geschichte der Alchemie.* Halle, 1832.

SCHRIMPF, R. (1). 'Bibliographie Sommaire des Ouvrages publiés en Chine durant la Période 1950–60 sur l'Histoire du Développement des Sciences et des Techniques Chinoises.' *BEFEO*, 1963, **51**, 615. Includes chemistry and chemical industry.

SCHNEIDER, W. (1). 'Über den Ursprung des Wortes "Chemie".' *PHI*, 1959, **21**, 79.

SCHNEIDER, W. (2). 'Kekule und die organische Strukturchemie.' *PHI*, 1958, **20**, 379.

SCHOLEM, G. (3). *Jewish Gnosticism, Merkabah* [apocalyptic or Messianic] *Mysticism, and the Talmudic Tradition.* New York, 1960.

SCHOLEM, G. (4). 'Zur Geschichte der Anfänge der Christlichen Kabbala.' Art. in L. Baeck Presentation Volume, London, 1954.

SCHOTT, W. (2). 'Ueber ein chinesisches Mengwerk, nebst einem Anhang linguistischer Verbesserungen zu zwei Bänden der Erdkunde Ritters' [the *Yeh Huo Pien* of Shen Tê-Fu (Ming)]. *APAW/PH*, 1880, no. 3.

SCHRAMM, M. (1). 'Aristotelianism; Basis of, and Obstacle to, Scientific Progress in the Middle Ages— Some Remarks on A. C. Crombie's "From Augustine to Galileo".' *HOSC*, 1963, **2**, 91; 1965, **4**, 70.

VON SCHRICK, MICHAEL PUFF (1). *Hienach volget ein nüczliche Materi von manigerley ausgeprânten Wasser, wie Man die nüczen und pruchen sol zu Gesuntheyt der Menschen; Ûn das Puchlein hat Meiyster Michel Schrick, Doctor der Erczney durch lijebe und gepet willen erberen Personen ausz den Pûchern zu sammen colligiert un beschrieben.* Augsburg, 1478, 1479, 1483, etc.

SCHUBARTH, DR (1). 'Ueber das chinesisches Weisskupfer und die vom Vereine angestellten Versuche dasselbe darzustellen.' *VVBGP*, 1824, **3**, 134. (The Verein in question was the Verein z. Beförderung des Gewerbefleisses in Preussen.)

SCHULTES, R. E. (1). *A Contribution to our Knowledge of Rivea corymbosa, the narcotic Ololiuqui of the Aztecs.* Botanical Museum, Harvard Univ. Cambridge, Mass. 1941.

SCHULTZE, S. See Aigremont, Dr.

SCHURHAMMER, G. (2). 'Die Yamabushis nach gedrückten und ungedrückten Berichten d. 16. und 17. Jahrhunderts.' *MDGNVO*, 1965, **46**, 47.

SCOTT, HUGH (1). *The Golden Age of Chinese Art; the Lively Thang Dynasty.* Tuttle, Rutland, Vt. and Tokyo, 1966.

SCOTT, W. (1) (ed.). *Hermetica.* 4 vols. Oxford, 1924–36.
    Vol 1, Introduction, Texts and Translation, 1924.
    Vol 2, Notes on the Corpus Hermeticum, 1925.
    Vol 3, Commentary; Latin Asclepius and the Hermetic Excerpts of Stobaeus, 1926 (posthumous ed. A. S. Ferguson.)
    Vol 4, Testimonia, Addenda, and Indexes (posthumous, with A. S. Ferguson).
    Repr. Dawson, London, 1968. Rev. G. Sarton, *ISIS*, 1926, **8**, 342.

SÉBILLOT, P. (1). *Les Travaux Publics et les Mines dans les Traditions et les Superstitions de tous les Peuples.* Paris, 1894.

SEGAL, J. B. (1). 'Pagan Syrian Monuments in the Vilayet of Urfa [Edessa].' *ANATS*, 1953, **3**, 97.

SEGAL, J. B. (2). 'The Ṣābian Mysteries; the Planet Cult of Ancient Ḥarrān.' Art. in *Vanished Civilisations*, ed. E. Bacon. 1963.

SEIDEL, A. (1). 'A Taoist Immortal of the Ming Dynasty; Chang San-Fêng.' Art. in *Self and Society in Ming Thought*, ed. W. T. de Bary. Columbia Univ. Press, New York, 1970, p. 483.

SEIDEL, A. (2). *La Divinisation de Lao Tseu [Lao Tzu] dans le Taoisme des Han.* École Française de l'Extrême Orient, Paris, 1969. (Pub. de l'Éc. Fr. de l'Extr. Or., no. 71.) A Japanese version is in *DK*, 1968, **3**, 5–77, with French summary, p. ii.

SELIMKHANOV, I. R. (1). 'Spectral Analysis of Metal Articles from Archaeological Monuments of the Caucasus.' *PPHS*, 1962, **38**, 68.

SELYE, H. (1). *Textbook of Endocrinology.* Univ. Press and *Acta Endocrinologica*, Montreal, 1947.

SEN, SATIRANJAN (1). 'Two Medical Texts in Chinese Translation.' *VBA*, 1945, **1**, 70.

SENCOURT, ROBERT (1). *Outflying Philosophy; a Literary Study of the Religious Element in the Poems and Letters of John Donne and in the Works of Sir Thomas Browne and Henry Vaughan the Silurist, together with an Account of the Interest of these Writers in Scholastic Philosophy, in Platonism and in Hermetic Physic; with also some Notes on Witchcraft.* Simpkin, Marshall, Hamilton & Kent, London, n.d. (1923).

SENGUPTA, KAVIRAJ N. N. (1). *The Ayurvedic System of Medicine.* 2 vols. Calcutta, 1925.

SERRUYS, H. (1) (tr.). '*Pei Lu Fêng Su*; Les Coutumes des Esclaves Septentrionaux [Hsiao Ta-Hêng's book on the Mongols, +1594].' *MS*, 1945, **10**, 117–208.

SEVERINUS, PETRUS (1). *Idea Medicinae Philosophicae*, 1571. 3rd ed. The Hague, 1660.

SEWTER, E. R. A. (1) (tr.). *Fourteen Byzantine Rulers; the 'Chronographia' of Michael Psellus [+1063, the last part by +1078].* Routledge & Kegan Paul, London; Yale Univ. Press, New Haven, Conn., 1953. 2nd revised ed. Penguin, Baltimore, and London, 1966.

SEYBOLD, C. F. (1). Review of J. Lippert's 'Ibn al-Qifṭi's Ta'rīkh al-Ḥukamā', auf Grund der Vorar-beiten Aug. Müller (Dieter, Leipzig, 1903).' *ZDMG*, 1903, **57**, 805.

SEYYED HOSSEIN NASR. See Said Husain Nasr.

SEZGIN, F. (1). 'Das Problem des Jābir ibn Ḥayyān im Lichte neu gefundener Handschriften.' *ZDMG*, 1964, **114**, 255.

SHANIN, T. (1) (ed.). *The Rules of the Game; Cross-Disciplinary Essays on Models in Scholarly Thought.* Tavistock, London, 1972.

SHAPIRO, J. (1). 'Freezing-out, a Safe Technique for Concentration of Dilute Solutions.' *S*, 1961, **133**, 2063.

SHASTRI, KAVIRAJ KALIDAS (1). *Catalogue of the Rasashala Aushadhashram Gondal* (Ayurvedic Pharma-ceutical Works of Gondal), [founded by the Maharajah of Gondal, H. H. Bhagvat Singhji]. 22nd ed. Gondal, Kathiawar, 1936. 40th ed. 1952.

SHAW, THOMAS (1). *Travels or Observations relating to sereral parts of Barbary and the Levant.* Oxford, 1738; London, 1757; Edinburgh, 1808. *Voyages dans la Régence d'Alger.* Paris, 1830.

SHEA, D. & FRAZER, A. (1) (tr.). *The 'Dabistan', or School of Manners* [by Mobed Shah, +17th Cent.], *translated from the original Persian, with notes and illustrations . . .* 2 vols. Paris, 1843.

SHEAR, T. L. (1). 'The Campaign of 1939 [excavating the ancient Athenian agora].' *HE*, 1940, **9**, 261.

SHEN TSUNG-HAN (1). *Agricultural Resources of China.* Cornell Univ. Press, Ithaca, N.Y., 1951.

SHÊNG WU-SHAN (1). *Érotologie de la Chine; Tradition Chinoise de l'Érotisme.* Pauvert, Paris, 1963. (Bibliothèque Internationale d'Érotologie, no. 11.) Germ. tr. *Die Erotik in China*, ed. Lo Duca. Desch, Basel, 1966. (Welt des Eros, no. 5.)

SHEPPARD, H. J. (1). 'Gnosticism and Alchemy.' *AX*, 1957, **6**, 86. 'The Origin of the Gnostic-Alchemical Relationship.' *SCI*, 1962, **56**, 1.

SHEPPARD, H. J. (2). 'Egg Symbolism in Alchemy.' *AX*, 1958, **6**, 140.

SHEPPARD, H. J. (3). 'A Survey of Alchemical and Hermetic Symbolism.' *AX*, 1960, **8**, 35.

SHEPPARD, H. J. (4). 'Ouroboros and the Unity of Matter in Alchemy; a Study in Origins.' *AX*, 1962, **10**, 83. 'Serpent Symbolism in Alchemy.' *SCI*, 1966, **60**, 1.

SHEPPARD, H. J. (5). 'The Redemption Theme and Hellenistic Alchemy.' *AX*, 1959, **7**, 42.

SHEPPARD, H. J. (6). 'Alchemy; Origin or Origins?' *AX*, 1970, **17**, 69.

SHEPPARD, H. J. (7). 'Egg Symbolism in the History of the Sciences.' *SCI*, 1960, **54**, 1.

SHEPPARD, H. J. (8). 'Colour Symbolism in the Alchemical *Opus*.' *SCI*, 1964, **58**, 1.

SHERLOCK, T. P. (1). 'The Chemical Work of Paracelsus.' *AX*, 1948, **3**, 33.

SHIH YU-CHUNG (1). 'Some Chinese Rebel Ideologies.' *TP*, 1956, **44**, 150.

SHIMAO EIKOH (1). 'The Reception of Lavoisier's Chemistry in Japan.' *ISIS*, 1972, **63**, 311.

SHIRAI, MITSUTARŌ (1). 'A Brief History of Botany in Old Japan.' Art. in *Scientific Japan, Past and Present*, ed. Shinjo Shinzo. Kyoto, 1926. (Commemoration Volume of the 3rd Pan-Pacific Science Congress.)

SIGERIST, HENRY E. (1). *A History of Medicine.* 2 vols. Oxford, 1951. Vol. 1, *Primitive and Archaic Medicine.* Vol. 2, *Early Greek, Hindu and Persian Medicine.* Rev. (vol. 2), J. Filliozat, *JAOS*, 1926, **82**, 575.

SIGERIST, HENRY E. (2). *Landmarks in the History of Hygiene.* London, 1956.

SIGGEL, A. (1). *Die Indischen Bücher aus dem 'Paradies d. Weisheit über d. Medizin' des 'Alī Ibn Sahl Rabban al-Ṭabarī.* Steiner, Wiesbaden, 1950. (Akad. d. Wiss. u. d. Lit. in Mainz; Abhdl. d. geistes- und sozial-wissenschaftlichen Klasse, no. 14.) Crit. O. Temkin, *BIHM*, 1953, **27**, 489.

SIGGEL, A. (2). *Arabisch-Deutsches Wörterbuch der Stoffe aus den drei Natur-reichen die in arabischen alchemistischen Handschriften vorkommen; nebst Anhang, Verzeichnis chemischer Geräte.* Akad. Verlag. Berlin, 1950. (Deutsche Akad. der Wissenchaften zu Berlin; Institut f. Orientforschung, Veröffentl. no. 1.)

SIGGEL, A. (3). *Decknamen in der arabischen Alchemistischen Literatur.* Akad. Verlag. Berlin, 1951. (Deutsche Akad. der Wissenschaften zu Berlin; Institut f. Orientforschung, Veröffentl. no. 5.) Rev. M. Plessner, *OR*, **7**, 368.

SIGGEL, A. (4). 'Das Sendschreiben "Das Licht über das Verfahren des Hermes der Hermesse dem, der es begehrt".' (*Qabas al-Qabīs fī Tadbīr Harmas al-Harāmis*, early +13th cent.) *DI*, 1937, **24**, 287.

SIGGEL, A. (5) (tr.). '*Das Buch der Gifte*' [*Kitāb al-Sumūm wa daf 'maḍārrihā*] des *Jābir ibn Ḥayyān* [Kr/2145]; *Arabische Text in Faksimile . . . übers. u. erläutert . . .* Steiner, Wiesbaden, 1958. (Ver-öffentl. d. Orientalischen Komm. d. Akad. d. Wiss. u. d. Lit. no. 12.) Cf Kraus (2), pp. 156 ff.

SIGGEL, A. (6). 'Gynäkologie, Embryologie und Frauenhygiene aus dem "Paradies der Weisheit [*Firdaws al-Ḥikma*] über die Medizin" des Abū Ḥasan 'Alī ibn Sahl Rabban al-Ṭabarī [d. *c.* +860], nach der Ausgabe von Dr. M. Zubair al-Ṣiddīqī (Sonne, Berlin-Charlottenberg, 1928).' *QSGNM*, 1942, **8**, 217.

SILBERER, H. (1). *Probleme der Mystik und ihrer Symbolik.* Vienna, 1914. Eng. tr. S. E. Jelliffe, *Problems of Mysticism, and its Symbolism.* Moffat & Yard, New York, 1917.

SINGER, C. (1). *A Short History of Biology.* Oxford, 1931.

SINGER, C. (3). 'The Scientific Views and Visions of St. Hildegard.' Art. in Singer (13), vol. 1, p. 1. Cf. Singer (16), a parallel account.

SINGER, C. (4). *From Magic to Science; Essays on the Scientific Twilight.* Benn, London, 1928.

SINGER, C. (8). *The Earliest Chemical Industry; an Essay in the Historical Relations of Economics and Technology, illustrated from the Alum Trade.* Folio Society, London, 1948.

SINGER, C. (13) (ed.). *Studies in the History and Method of Science.* Oxford, vol. 1, 1917; vol. 2, 1921. Photolitho reproduction, Dawson, London, 1955.

SINGER, C. (16). 'The Visions of Hildegard of Bingen.' Art. in Singer (4), p. 199.

SINGER, C. (23). 'Alchemy' (art. in *Oxford Classical Dictionary*). Oxford.

SINGER, CHARLES (25). *A Short History of Anatomy and Physiology from the Greeks to Harvey.* Dover, New York, 1957. Revised from *The Evolution of Anatomy.* Kegan Paul, Trench, & Trubner, London, 1925.

SINGER, D. W. (1). *Giordano Bruno; His Life and Thought, with an annotated Translation of his Work 'On the Infinite Universe and Worlds'.* Schuman, New York, 1950.

SINGER, D. W. (2). 'The Alchemical Writings attributed to Roger Bacon.' *SP*, 1932, **7**, 80.

SINGER, D. W. (3). 'The Alchemical Testament attributed to Raymund Lull.' *A*, 1928, **9**, 43. (On the pseudepigraphic nature of the Lullian corpus.)

SINGER, D. W. (4). 'l'Alchimie.' Communiction to the IVth International Congress of the History of Medicine, Brussels, 1923. Sep. pub. de Vlijt, Antwerp, 1927.

SINGER, D. W., ANDERSON, A. & ADDIS, R. (1). *Catalogue of Latin and Vernacular Alchemical Manuscripts in Great Britain and Ireland before the 16th Century.* 3 vols. Lamertin, Brussels, 1928–31 (for the Union Académique Internationale).

SINGLETON, C. S. (1) (ed.). *Art, Science and History in the Renaissance.* Johns Hopkins, Baltimore, 1968.

SISCO, A. G. & SMITH, C. S. (1) (tr.). *Lazarus Ercker's Treatise on Ores and Assaying, translated from the German edition of +1580.* Univ. Chicago Press, Chicago, 1951.

SISCO, A. G. & SMITH, C. S. (2). '*Bergwerk- und Probier-büchlein*'; *a Translation from the German of the '*Berg-büchlein*', a Sixteenth-Century Book on Mining Geology, by A. G. Sisco, and of the '*Probierbüchlein*', a Sixteenth-Century Work on Assaying, by A. G. Sisco & C. S. Smith; with technical annotations and historical notes.* Amer. Institute of Mining and Metallurgical Engineers, New York, 1949.

SIVIN, N. (1). 'Preliminary Studies in Chinese Alchemy; the *Tan Ching Yao Chüeh* attributed to Sun Ssu-Mo (+581? to after +674).' Inaug. Diss., Harvard University, 1965. Published as: *Chinese Alchemy; Preliminary Studies.* Harvard Univ. Press, Cambridge, Mass. 1968. (Harvard Monographs in the History of Science, no. 1.) Ch. 1 sep. pub. *JSHS*, 1967, **6**, 60. Revs. J. Needham, *JAS*, 1969, 850; Ho Ping-Yü, *HJAS*, 1969, **29**, 297; M. Eliade, *HOR*, 1970, **10**, 178.

SIVIN, N. (1a). 'On the Reconstruction of Chinese Alchemy.' *JSHS*, 1967, **6**, 60 (essentially ch. 1 of Sivin, 1).

SIVIN, N. (2). 'Quality and Quantity in Chinese Alchemy.' Priv. circ. 1966; expanded as: 'Reflections on Theory and Practice in Chinese Alchemy.' Contribution to the First International Conference of Taoist Studies, Villa Serbelloni, Bellagio, 1968.

SIVIN, N. (3). Draft Translation of *Thai-Shang Wei Ling Shen Hua Chiu Chuan Tan Sha Fa* (*TT*/885). Unpublished MS., copy deposited in Harvard-Yenching Library for circulation.

SIVIN, N. (4). Critical Editions and Draft Translations of the Writings of Chhen Shao-Wei (*TT*/883 and 884, and YCCC, chs. 68–9). Unpublished MS.

SIVIN, N. (5). Critical Edition and Draft Translation of *Tan Lun Chüeh Chih Hsin Ching* (*TT*/928 and YCCC, ch. 66). Unpublished MS.

SIVIN, N. (6). 'William Lewis as a Chemist.' *CHYM*, 1962, **8**, 63.

SIVIN, N. (7). 'On the *Pao Phu Tzu* (*Nei Phien*) and the Life of Ko Hung (+283 to +343).' *ISIS*, 1969. **60**, 388.

SIVIN, N. (8). 'Chinese Concepts of Time.' *EARLH*, 1966, **1**, 82.

SIVIN, N. (9). *Cosmos and Computation in Chinese Mathematical Astronomy.* Brill, Leiden, 1969. Reprinted from *TP*, 1969, **55**.

SIVIN, N. (10). 'Chinese Alchemy as a Science.' Contrib. to '*Nothing Concealed*' (Wu Yin Lu); *Essays in Honour of Liu (Aisin-Gioro) Yü-Yün*, ed. F. Wakeman, Chinese Materials and Research Aids Service Centre, Thaipei, Thaiwan, 1970, p. 35.

SKRINE, C. P. (1). 'The Highlands of Persian Baluchistan.' *GJ*, 1931, **78**, 321.

DE SLANE, BARON McGUCKIN (2) (tr.). *Ibn Khallikan's Dictionary* (translation of Ibn Khallikān's *Kitāb Wafayāt al-A'yān*, a collection of 865 biographies, +1278). 4 vols. Paris, 1842–71.

SMEATON, W. A. (1). 'Guyton de Morveau and Chemical Affinity.' *AX*, 1963, **11**, 55.

SMITH, ALEXANDER (1). *Introduction to Inorganic Chemistry.* Bell, London, 1912.

SMITH, C. S. (4). 'Matter versus Materials; a Historical View.' *SC*, 1968, **162**, 637.

SMITH, C. S. (5). 'A Historical View of One Area of Applied Science—Metallurgy.' Art. in *Applied Science and Technological Progress*. A Report to the Committee on Science and Astronautics of the United States House of Representatives by the National Academy of Sciences, Washington, D.C. 1967.

SMITH, C. S. (6). 'Art, Technology and Science; Notes on their Historical Interaction.' *TCULT*, 1970, **11**, 493.

SMITH, C. S. (7). 'Metallurgical Footnotes to the History of Art.' *PAPS*, 1972, **116**, 97. (Penrose Memorial Lecture, Amer. Philos. Soc.)

SMITH, C. S. & GNUDI, M. T. (1) (tr. & ed.). *Biringuccio's 'De La Pirotechnia' of +1540, translated with an introduction and notes*. Amer. Inst. of Mining and Metallurgical Engineers, New York, 1942, repr. 1943. Reissued, with new introductory material. Basic Books, New York, 1959.

SMITH, F. PORTER (1). *Contributions towards the Materia Medica and Natural History of China, for the use of Medical Missionaries and Native Medical Students*. Amer. Presbyt. Miss. Press, Shanghai; Trübner, London, 1871.

SMITH, F. PORTER (2). 'Chinese Chemical Manufactures.' *JRAS/NCB*, 1870 (n.s.), **6**, 139.

SMITH, R. W. (1). 'Secrets of Shao-Lin Temple Boxing.' Tuttle, Rutland, Vt. and Tokyo, 1964.

SMITH, T. (1) (tr.). The *'Recognitiones' of Pseudo-Clement of Rome* [c. +220]. In Ante-Nicene Christian Library, ed. A. Roberts & J. Donaldson, Clark, Edinburgh, 1867. vol. 3, p. 297.

SMITH, T., PETERSON, P. & DONALDSON, J. (1) (tr.). The *Pseudo-Clementine Homilies* [c. +190, attrib. Clement of Rome, *fl.* +96]. In Ante-Nicene Christian Library, ed. A. Roberts & J. Donaldson, Clark, Edinburgh, 1867. vol. 17.

SMITHELLS, C. J. (1). 'A New Alloy of High Density.' *N*, 1937, **139**, 490.

SMYTHE, J. A. (1). *Lead; its Occurrence in Nature, the Modes of its Extraction, its Properties and Uses, with Some Account of its Principal Compounds*. London and New York, 1923.

SNAPPER, I. (1). *Chinese Lessons to Western Medicine; a Contribution to Geographical Medicine from the Clinics of Peiping Union Medical College*. Interscience, New York, 1941.

SNELLGROVE, D. (1). *Buddhist Himalaya; Travels and Studies in Quest of the Origins and Nature of Tibetan Religion*. Oxford, 1957.

SNELLGROVE, D. (2). The *'Hevajra Tantra', a Critical Study*. Oxford, 1959.

SODANO, A. R. (1) (ed. & tr.). *Porfirio* [Porphyry of Tyre]; *Lettera ad Anebo* (Greek text and Italian tr.). Arte Tip., Naples, 1958.

SOLLERS, P. (1). 'Traduction et Presentation de quelques Poèmes de Mao Tsê-Tung.' *TQ*, 1970, no. 40, 38.

SOLLMANN, T. (1). *A Textbook of Pharmacology and some Allied Sciences*. Saunders, 1st ed. Philadelphia and London, 1901. 8th ed., extensively revised and enlarged, Saunders, Philadelphia and London, 1957.

SOLOMON, D. (1) (ed.). *LSD, the Consciousness-Expanding Drug*. Putnam, New York, 1964. Rev. W. H. McGlothlin, *NN*, 1964, **199** (no. 15), 360.

SOYMIÉ, M. (4). 'Le Lo-feou Chan (Lo-fou Shan]; Étude de Géographie Religieuse.' *BEFEO*, 1956, **48**, 1–139.

SOYMIÉ, M. (5). 'Bibliographie du Taoisme; Études dans les Langues Occidentales' (pt. 2). *DK*, 1971, **4**, 290–225 (1–66); with Japanese introduction, p. 288 (3).

SOYMIÉ, M. (6). 'Histoire et Philologie de la Chine Médiévale et Moderne; Rapport sur les Conférences' (on the date of *Pao Phu Tzu*). *AEPHE/SHP*, 1971, 759.

SOYMIÉ, M. & LITSCH, F. (1). 'Bibliographie du Taoisme; Études dans les Langues Occidentales' (pt. 1). *DK*, 1968, **3**, 318–247 (1–72); with Japanese introduction, p. 316 (3).

SPEISER, E. A. (1). *Excavations at Tepe Gawra*. 2 vols. Philadelphia, 1935.

SPENCER, J. E. (3). 'Salt in China.' *GR*, 1935, **25**, 353.

SPENGLER, O. (1). *The Decline of the West*, tr. from the German, *Die Untergang des Abendlandes*, by C. F. Atkinson. 2 vols. Vol. 1, *Form and Actuality;* vol. 2, *Perspectives of World History*. Allen & Unwin, London, 1926, 1928.

SPERBER, D. (1). 'New Light on the Problem of Demonetisation in the Roman Empire.' *NC*, 1970 (7th ser.), **10**, 112.

SPETER, M. (1). 'Zur Geschichte der Wasserbad-destillation; das "Berchile" Abul Kasims.' *APHL*, 1930, **5** (no. 8), 116.

SPIZEL, THEOPHILUS (1). *De Re Literaria Sinensium Commentarius*... Leiden, 1660 (frontispiece, 1661).

SPOONER, R. C. (1). 'Chang Tao-Ling, the first Taoist Pope.' *JCE*, 1938, **15**, 503.

SPOONER, R. C. (2). 'Chinese Alchemy.' *JWCBRS*, 1940 (A), **12**, 82.

SPOONER, R. C. & WANG, C. H. (1). 'The Divine Nine-Turn Tan-Sha Method, a Chinese Alchemical Recipe.' *ISIS*, 1948, **28**, 235.

VAN DER SPRENKEL, O. (1). 'Chronology, Dynastic Legitimacy, and Chinese Historiography' (mimeographed). Paper contributed to the Study Conference at the London School of Oriental Studies

1956, but not included with the rest in *Historians of China and Japan*, ed. W. G. Beasley & E. G. Pulleybank, 1961.

SQUIRE, S. (1) (tr.). *Plutarch 'De Iside et Osiride', translated into English* (sep. pagination, text and tr.). Cambridge, 1744.

STADLER, H. (1) (ed.). *Albertus Magnus 'De Animalibus, libri XXVI.'* 2 vols. Münster i./W., 1916–21.

STANLEY, R. C. (1). *Nickel, Past and Present.* Proc. IInd Empire Mining and Metallurgical Congress, 1928, pt. 5, Non-Ferrous Metallurgy, 1–34.

STANNUS, H. S. (1). 'Notes on Some Tribes of British Central Africa [esp. the Anyanja of Nyasaland]. *JRAI*, 1912, **40**, 285.

STAPLETON, H. E. (1). 'Sal-Ammoniac; a Study in Primitive Chemistry.' *MAS/B*, 1905, **1**, 25.

STAPLETON, H. E. (2). 'The Probable Sources of the Numbers on which Jābirian Alchemy was based.' *A/AIHS*, 1953, **6**, 44.

STAPLETON, H. E. (3). 'The Gnomon as a possible link between one type of Mesopotamian *Ziggurat* and the Magic Square Numbers on which Jābirian Alchemy was based.' *AX*, 1957, **6**, 1.

STAPLETON, H. E. (4). 'The Antiquity of Alchemy.' *AX*, 1953, **5**, 1. The Summary also printed in *A/AIHS*, 1951, **4** (no. 14), 35.

STAPLETON, H. E. (5). 'Ancient and Modern Aspects of Pythagoreanism; I, The Babylonian Sources of Pythagoras' Mathematical Knowledge; II, The Part Played by the Human Hand with its Five Fingers in the Development of Mathematics; III, Sumerian Music as a possible intermediate Source of the Emphasis on Harmony that characterises the 6th-century Teaching of both Pythagoras and Confucius; IV, The Belief of Pythagoras in the Immaterial, and its Co-existence with Natural Phenomena.' *OSIS*, 1958, **13**, 12.

STAPLETON, H. E. & AZO, R. F. (1). 'Alchemical Equipment in the +11th Century.' *MAS/B*, 1905, **1**, 47. (Account of the '*Ainu al-San'ah wa-l 'Aunu al-Sana'ah* (Essence of the Art and Aid to the Workers) by Abū-l Ḥakīm al-Sālihī al-Kāthī, +1034.) Cf. Ahmad & Datta (1).

STAPLETON, H. E. & AZO, R. F. (2). 'An Alchemical Compilation of the +13th Century.' *MAS/B*, 1910, **3**, 57. (A florilegium of extracts gathered by an alchemical copyist travelling in Asia Minor and Mesopotamia about +1283.)

STAPLETON, H. E., AZO, R. F. & HUSAIN, M. H. (1). 'Chemistry in Iraq and Persia in the +10th Century.' *MAS/B*, 1927, **8**, 315–417. (Study of the *Madkhal al-Ta'līmī* and the *Kitāb al-Asrār* of al-Rāzī (d. +925), the relation of Arabic alchemy with the Sabians of Ḥarrān, and the role of influences from Hellenistic culture, China and India upon it.) Revs. G. Sarton, *ISIS*, 1928, **11**, 129; J. R. Partington, *N*, 1927, **120**, 243.

STAPLETON, H. E., AZO, R. F., HUSAIN, M. H. & LEWIS, G. L. (1). 'Two Alchemical Treatises attributed to Avicenna.' *AX*, 1962, **10**, 41.

STAPLETON, H. E. & HUSAIN, H. (1) (tr.). 'Summary of the Cairo Arabic MS. of the "Treatise of Warning (*Risālat al-Ḥaḍar*)" of Agathodaimon, his Discourse to his Disciples when he was about to die.' Published as Appendix B in Stapleton (4), pp. 40 ff.

STAPLETON, H. E., LEWIS, G. L. & TAYLOR, F. SHERWOOD (1). 'The Sayings of Hermes as quoted in the *Mā al-Waraqī* of Ibn Umail' (c. +950). *AX*, 1949, **3**, 69.

STARKEY, G. [Eirenaeus Philaletha] (1). *Secrets Reveal'd; or, an Open Entrance to the Shut-Palace of the King; Containing the Greatest Treasure in Chymistry, Never yet so plainly Discovered. Composed by a most famous English-man styling himself Anonymus, or Eyrenaeus Philaletha Cosmopolita, who by Inspiration and Reading attained to the Philosophers Stone at his Age of Twenty-three Years, A.D. 1645...* Godbid for Cooper, London, 1669. Eng. tr. of first Latin ed. *Introitus Apertus...* Jansson & Weyerstraet, Amsterdam, 1667. See Ferguson (1), vol. 2, p. 192.

STARKEY, G. [Eirenaeus Philaletha] (2). *Arcanum Liquoris Immortalis, Ignis-Aquae Seu Alkehest.* London, 1683, Hamburg, 1688. Eng. tr. 1684.

STAUDENMEIER, LUDWIG (1). *Die Magie als experimentelle Wissenschaft.* Leipzig, 1912.

STEELE, J. (1) (tr.). *The 'I Li', or Book of Etiquette and Ceremonial.* 2 vols. London, 1917.

STEELE, R. (1) (ed.). *Opera Hactenus Inedita Rogeri Baconi.* 9 fascicles in 3 vols. Oxford, 1914–.

STEELE, R. (2). 'Practical Chemistry in the +12th Century; Rasis *De Aluminibus et Salibus*, the [text of the] Latin translation by Gerard of Cremona, [with an English précis].' *ISIS*, 1929, **12**, 10.

STEELE, R. (3) (tr.). *The Discovery of Secrets* [a Jābirian Corpus text]. Luzac (for the Geber Society), London, 1892.

STEELE, R. & SINGER, D. W. (1). 'The Emerald Table [*Tabula Smaragdina*].' *PRSM*, 1928, **21**, 41.

STEIN, O. (1). 'References to Alchemy in Buddhist Scriptures.' *BLSOAS*, 1933, **7**, 263.

STEIN, R. A. (5). 'Remarques sur les Mouvements du Taoisme Politico-Religieux au 2e Siècle ap. J. C.' *TP*, 1963, **50**, 1–78. Japanese version revised by the author, with French summary of the alterations. *DK*, 1967, **2**.

STEIN, R. A. (6). 'Spéculations Mystiques et Thèmes relatifs aux "Cuisines" [*chhu*] du Taoisme.' *ACF*, 1972, **72**, 489.

STEINGASS, F. J. (1). *A Comprehensive Persian–English Dictionary*. Routledge & Kegan Paul, London, 1892, repr. 1957.

STEININGER, H. (1). *Hauch- und Körper-seele, und der Dämon, bei 'Kuan Yin Tzu'*. Harrassowitz, Leipzig, 1953. (Sammlung orientalistischer Arbeiten, no. 20.)

STEINSCHNEIDER, M. (1). 'Die Europäischen Übersetzungen aus dem Arabischen bis mitte d. 17. Jahrhunderts. A. Schriften bekannter Übersetzer; B, Übersetzungen von Werken bekannter Autoren deren Übersetzer unbekannt oder unsicher sind.' *SWAW /PH*, 1904, **149** (no. 4), 1–84; 1905, **151** (no. 1), 1–108. Also sep. issued. Repr. Graz, 1956.

STEINSCHNEIDER, M. (2). 'Über die Mondstationen (Naxatra) und das Buch Arcandam.' *ZDMG*, 1864, **18**, 118. 'Zur Geschichte d. Übersetzungen ans dem Indischen in Arabische und ihres Einflusses auf die Arabische Literatur, insbesondere über die Mondstationen (Naxatra) und daraufbezüglicher Loosbücher.' *ZDMG*, 1870, **24**, 325; 1871, **25**, 378. (The last of the three papers has an index for all three.)

STEINSCHNEIDER, M. (3). 'Euklid bei den Arabern.' *ZMP*, 1886, **31** (Hist. Lit. Abt.), 82.

STEINSCHNEIDER, M. (4) *Gesammelte Schriften*, ed. H. Malter & A. Marx. Poppelauer, Berlin, 1925.

STENRING, K. (1) (tr.). *The Book of Formation, 'Sefer Yetzirah', by R. Akiba ben Joseph*...With introd. by A. E. Waite, Rider, London, 1923.

STEPHANIDES, M. K. (1). *Symbolai eis tēn Historikē tōn Physikōn Epistēmōn kai Idiōs tēs Chymeias* (in Greek). Athens, 1914. See Zacharias (1).

STEPHANIDES, M. K. (2). Study of Aristotle's views on chemical affinity and reaction. *RSCI*, 1924, **62**, 626.

STEPHANIDES, M. K. (3). *Psammourgikē kai Chymeia* (Ψαμμουργική καί Χυμεία) [in Greek]. Mytilene, 1909.

STEPHANIDES, M. K. (4). 'Chymeutische Miszellen.' *AGNWT*, 1912, **3**, 180.

STEPHANUS OF ALEXANDRIA. *Megalēs kai Hieras Technēs* [*Chymeia*]. Not in the *Corpus Alchem. Gr.* (Berthelot & Ruelle) but in Ideler (1), vol. 2.

STEPHANUS, HENRICUS. See Estienne, H. (1).

STILLMAN, J. M. (1). *The Story of Alchemy and Early Chemistry*. Constable, London and New York, 1924. Repr. Dover, New York, 1960.

STRASSMEIER, J. N. (1). *Inschriften von Nabuchodonosor* [−6th cent.]. Leipzig, 1889.

STRASSMEIER, J. N. (2). *Inschriften von Nabonidus* [r. −555 to −538]. Leipzig. 1889.

STRAUSS, BETTINA (1). 'Das Giftbuch des Shānāq; eine literaturgeschichtliche Untersuchung.' *QSGNM*, 1934, **4**, 89–152.

VON STRAUSS-&-TORNEY, V. (1). 'Bezeichnung der Farben Blau und Grün in Chinesischen Alterthum.' *ZDMG*, 1879, **33**, 502.

STRICKMANN, M. (1). 'Notes on Mushroom Cults in Ancient China.' Rijksuniversiteit Gent (Gand), 1966. (Paper to the 4e Journée des Orientalistes Belges, Brussels, 1966.)

STRICKMANN, M. (2). 'On the Alchemy of Thao Hung-Ching.' Unpub. MS. Revised version contributed to the 2nd International Conference of Taoist Studies, Tateshina, Japan, 1972.

STRICKMANN, M. (3). 'Taoism in the Lettered Society of the Six Dynasties.' Contribution to the 2nd International Conference of Taoist Studies, Chino (Tateshina), Japan, 1972.

STROTHMANN, R. (1). 'Gnosis Texte der Ismailiten; Arabische Handschrift Ambrosiana H 75.' *AGWG/PH*, 1943 (3rd ser.), no. 28.

STRZODA, W. (1). *Die gelben Orangen der Prinzessin Dschau, aus dem chinesischen Urtext*. Hyperion Verlag, München, 1922.

STUART, G. A. (1). *Chinese Materia Medica; Vegetable Kingdom, extensively revised from Dr F. Porter Smith's work*. Amer. Presbyt. Mission Press, Shanghai, 1911. An expansion of Smith, F.P. (1).

STUART, G. A. (2). 'Chemical Nomenclature.' *CRR*, 1891; 1894, **25**, 88; 1901, **32**, 305.

STUHLMANN, C. C. (1). 'Chinese Soda.' *JPOS*, 1895, **3**, 566.

SUBBARAYAPPA, B. V. (1). 'The Indian Doctrine of Five Elements.' *IJHS*, 1966, **1**, 60.

SUDBOROUGH, J. J. (1). *A Textbook of Organic Chemistry; translated from the German of A. Bernthsen, edited and revised*. Blackie, London, 1906.

SUDHOFF, K. (1). 'Eine alchemistische Schrift des 13. Jahrhunderts betitelt *Speculum Alkimie Minus*, eines bisher unbekannten Mönches Simeon von Köln.' *AGNT*, 1922, **9**, 53.

SUDHOFF, K. (2). 'Alkoholrezept aus dem 8. Jahrhundert?' [The earliest version of the *Mappae Clavicula*, now considered *c.* +820.] *NW*, 1917, **16**, 681.

SUDHOFF, K. (3). 'Weiteres zur Geschichte der Destillationstechnik.' *AGNT*, 1915, **5**, 282.

SUDHOFF, K. (4). 'Eine Herstellungsanweisung für "Aurum Potabile" und "Quinta Essentia" von dem herzogliche Leibarzt Albini di Moncalieri (14ter Jahrh.).' *AGNT*, 1915, **5**, 198.

SÜHEYL ÜNVER, A. (1). *Tanksuknamei Ilhan der Fününu Ulumu Hatai Mukaddinesi* (Turkish tr.) T. C. Istanbul Universitesi Tib Tarihi Enstitusu Adet 14. Istanbul, 1939.

SÜHEYL ÜNVER, A. (2). *Wang Shu-ho eseri hakkinda* (Turkish with Eng. summary). Tib. Fak. Mecmuasi. Yil 7, Sayr 2, Umumi no. 28. Istanbul, 1944.

AL-SUHRAWARDY, ALLAMA SIR ABDULLAH AL-MAMUN (1) (ed.). *The Sayings of Muḥammad* [ḥadith]. With foreword by M. K. (Mahatma) Gandhi. Murray, London, 1941. (Wisdom of the East series.)

SUIDAS (1). *Lexicon Graece et Latine...(c.* +1000), ed. Aemilius Portus & Ludolph Kuster, 3 vols. Cambridge, 1705.

SULLIVAN, M. (8). 'Kendi' (drinking vessels, Skr. *kundika*, with neck and side-spout). *ACASA*, 1957, **11**, 40.

SUN JEN I-TU & SUN HSÜEH-CHUAN (1) (tr.). '*Thien Kung Khai Wu*', *Chinese Technology in the Seventeenth Century, by Sung Ying-Hsing.* Pennsylvania State Univ. Press; University Park & London, Penn. 1966.

SUTER, H. (1). *Die Mathematiker und Astronomen der Araber und ihre Werke.* Teubner, Leipzig, 1900. (Abhdl. z. Gesch. d. Math. Wiss. mit Einschluss ihrer Anwendungen, no. 10; supplement to *ZMP*, **45**.) Additions and corrections in *AGMW*, 1902, no. 14.

SUZUKI SHIGEAKI (1). 'Milk and Milk Products in the Ancient World.' *JSHS*, 1965, **4**, 135.

SWEETSER, WM. (1). *Human Life.* New York, 1867.

SWINGLE, W. T. (12). 'Notes on Chinese Accessions; chiefly Medicine, Materia Medica and Horticulture.' *ARLC/DO*, 1928/1929, 311. (On the *Pên Tshao Yen I Pu I*, the *Yeh Tshai Phu*, etc.; including translations by M. J. Hagerty.)

SYNCELLOS, GEORGIOS (1). *Chronographia* (*c.* +800), ed. W. Dindorf. Weber, Bonn, 1829 (in *Corp. Script. Hist. Byz.* series). Ed. P. J. Goar, Paris, 1652.

TANAKA, M. (1). *The Development of Chemistry in Modern Japan.* Proc. XIIth Internat. Congr. Hist. of Sci., Paris, 1968. Abstracts & Summaries, p. 232; Actes, vol. 6, p. 107.

TANAKA, M. (2). 'A Note to the History of Chemistry in Modern Japan, [with a Select List of the most important Contributions of Japanese Scientists to Modern Chemistry].' *SHST/T*, Special Issue for the XIIth Internat. Congress of the Hist. of Sci., Paris, 1968.

TANAKA, M. (3). 'Einige Probleme der Vorgeschichte der Chemie in Japan; Einführung und Aufnahme der modernen Materienbegriffe.' *JSHS*, 1967, **6**, 96.

TANAKA, M. (4). 'Ein Hundert Jahre der Chemie in Japan.' *JSHS*, 1964, **3**, 89.

TARANZANO, C. (1). *Vocabulaire des Sciences Mathématiques, Physiques et Naturelles.* 2 vols. Hsien-hsien, 1936.

TARN, W. W. (1). *The Greeks in Bactria and India.* Cambridge, 1951.

TASLIMI, MANUCHECHR (1). 'An Examination of the *Nihāyat al-Ṭalab* (The End of the Search) [by 'Izz al-Dīn Aidamur ibn 'Ali ibn Aidamur al-Jildakī, *c.* +1342] and the Determination of its Place and Value in the History of Islamic Chemistry.' Inaug. Diss. London, 1954.

TATARINOV, A. (2). 'Bemerkungen ü. d. Anwendung schmerzstillender Mittel bei den Operationen, und die Hydropathie, in China.' Art. in *Arbeiten d. k. Russischen Gesandschaft in Peking über China, sein Volk, seine Religion, seine Institutionen, socialen Verhältnisse, etc.*, ed. C. Abel & F. A. Mecklenburg. Heinicke, Berlin, 1858. Vol. 2, p. 467.

TATOJAYA, YATODHARMA (1) (tr.). *The 'Kokkokam' of Ativira Rama Pandian* [a Tamil prince at Madura, late +16th cent.]. Med. Book Co., Calcutta, 1949. Bound with R. Schmidt (3).

TAYLOR, F. SHERWOOD (2). 'A Survey of Greek Alchemy.' *JHS*, 1930, **50**, 109.

TAYLOR, F. SHERWOOD (3). *The Alchemists.* Heinemann, London, 1951.

TAYLOR, F. SHERWOOD (4). *A History of Industrial Chemistry.* Heinemann, London, 1957.

TAYLOR, F. SHERWOOD (5). 'The Evolution of the Still.' *ANS*, 1945, **5**, 185.

TAYLOR, F. SHERWOOD (6). 'The Idea of the Quintessence.' Art. in *Science, Medicine and History* (Charles Singer Presentation Volume), ed. E. A. Underwood, Oxford, 1953. Vol. 1, p. 247.

TAYLOR, F. SHERWOOD (7). 'The Origins of Greek Alchemy.' *AX*, 1937, **1**, 30.

TAYLOR, F. SHERWOOD (8) (tr. and comm.). 'The Visions of Zosimos [of Panopolis].' *AX*, 1937, **1**, 88.

TAYLOR, F. SHERWOOD (9) (tr. and comm.). 'The Alchemical Works of Stephanos of Alexandria.' *AX*, 1937, **1**, 116; 1938, **2**, 38.

TAYLOR, F. SHERWOOD (10). 'An Alchemical Work of Sir Isaac Newton.' *AX*, 1956, **5**, 59.

TAYLOR, F. SHERWOOD (11). 'Symbols in Greek Alchemical Writings.' *AX*, 1937, **1**, 64.

TAYLOR, F. SHERWOOD & SINGER, CHARLES (1). 'Pre-scientific Industrial Chemistry [in the Mediterranean Civilisations and the Middle Ages].' Art. in *A History of Technology*, ed. C. Singer *et al.* Oxford, 1956. Vol. 2, p. 347.

TAYLOR, J. V. (1). *The Primal Vision; Christian Presence amid African Religion.* SCM Press, London, 1963.

TEGENGREN, F. R. (1). 'The Iron Ores and Iron Industry of China; including a summary of the Iron situation of the Circum-Pacific Region.' *MGSC*, 1921 (Ser. A), no. 2, pt. I, pp. 1–180, with Chinese abridgement of 120 pp. 1923 (Ser. A), no. 2, pt. II, pp. 181–457, with Chinese abridgement of 190 pp. The section on the Iron Industry starts from p. 297: 'General Survey; Historical Sketch' [based mainly on Chang Hung-Chao (*1*)], pp. 297–314; 'Account of the Industry [traditional] in different Provinces', pp. 315–64; 'The Modern Industry', pp. 365–404; 'Circum-Pacific Region', pp. 405-end.

TEGENGREN, F. R. (2). 'The Hsi-khuang Shan Antimony Mining Fields in Hsin-hua District, Hunan.' *BGSC*, 1921, no. 3, 1–25.

TEGENGREN, F. R. (3). 'The Quicksilver Deposits of China.' *BGSC*, 1920, no. 2, 1–36.

TEGGART, F. J. (1). *Rome and China; a Study of Correlations in Historical Events.* Univ. of California Press, Berkeley, Calif. 1939.

TEICH, MIKULÁŠ (1) (ed.). *J. E. Purkyně, 'Opera Selecta'.* Prague, 1948.

TEICH, MIKULÁŠ (2). 'From "Enchyme" to "Cyto-Skeleton"; the Development of Ideas on the Chemical Organisation of Living Matter.' Art. in *Changing Perspectives in the History of Science...,* ed. M. Teich & R. Young, Heinemann, London, 1973, p. 439.

TEICH, MIKULÁŠ & YOUNG, R. (1) (ed.). *Changing Perspectives in the History of Science...* Heinemann, London, 1973.

TEMKIN, O. (3). 'Medicine and Graeco-Arabic Alchemy.' *BIHM*, 1955, **29**, 134.

TEMKIN, O. (4). 'The Classical Roots of Glisson's Doctrine of Irritation.' *BIHM*, 1964, **38**, 297.

TEMPLE, SIR WM. (3). 'On Health and Long Life.' In *Works*, 1770 ed. vol. 3, p. 266.

TESTE, A. (1). *Homoeopathic Materia Medica, arranged Systematically and Practically.* Eng. tr. from the French, by C. J. Hempel. Rademacher & Shelk, Philadelphia, 1854.

TESTI, G. (1). *Dizionario di Alchimia e di Chimica Antiquaria.* Mediterranea, Rome, 1950. Rev. F. S[herwood] T[aylor], *AX*, 1953, **5**, 55.

THACKRAY, A. (1). '"Matter in a Nut-shell"; Newton's "Opticks" and Eighteenth-Century Chemistry.' *AX*, 1968, **15**, 29.

THELWALL, S. & HOLMES, P. (1) (tr.). *The Writings of Tertullian* [c. +200]. In Ante-Nicene Christian Library, ed. A. Roberts & J. Donaldson. Clark, Edinburgh, 1867, vols. 11, 15 and 18.

THEOBALD, W. (1). 'Der Herstelling der Bronzefarbe in Vergangenheit und Gegenwart.' *POLYJ*, 1913, **328**, 163.

THEOPHANES (+758 to +818) (1). *Chronographia*, ed. Classen (in *Corp. Script. Hist. Byz.* series).

[THEVENOT, D.] (1) (ed.). *Scriptores Graeci Mathematici, Veterum Mathematicorum Athenaei, Bitonis, Apollodori, Heronis et aliorum Opera Gr. et Lat. pleraque nunc primum edita* [including the *Kestoi* of Julius Africanus]. Paris, 1693.

THOMAS, E. J. (2). '[The State of the Dead in] Buddhist [Belief].' *ERE*, vol. xi, p. 829.

THOMAS, SIR HENRY (1). 'The Society of Chymical Physitians; an Echo of the Great Plague of London, +1665.' Art. in Singer Presentation Volume, *Science, Medicine and History*, ed. E. A. Underwood. 2 vols. Oxford, 1953. Vol. 2, p. 56.

THOMPSON, D. V. (1). *The Materials of Mediaeval Painting.* London, 1936.

THOMPSON, D. V. (2) (tr.). *The 'Libro dell Arte' of Cennino Cennini* [+1437]. Yale Univ. Press, New Haven, Conn. 1933.

THOMPSON, NANCY (1). 'The Evolution of the Thang Lion-and-Grapevine Mirror.' *AA*, 1967, **29**. Sep. pub. Ascona, 1968; with an addendum on the Jen Shou Mirrors by A. C. Soper.

THOMPSON, R. CAMPBELL (5). *On the Chemistry of the Ancient Assyrians* (mimeographed, with plates of Assyrian cuneiform tablets, romanised transcriptions and translations). Luzac, London, 1925.

THOMS, W. J. (1). *Human Longevity; its Facts and Fictions.* London, 1873.

THOMSEN, V. (1). 'Ein Blatt in türkische "Runen"-schrift aus Turfan.' *SPAW/PH*, 1910, 296. Followed by F. C. Andreas: 'Zwei Soghdische Exkurse zu V. Thomsen's "Ein Blatt...".' 307.

THOMSON, JOHN (2). '[Glossary of Chinese Terms for] Photographic Chemicals and Apparatus.' In Doolittle (1), vol. 2, p. 319.

THOMSON, T. (1). *A History of Chemistry.* 2 vols. Colburn & Bentley, London, 1830.

THORNDIKE, LYNN (1). *A History of Magic and Experimental Science.* 8 vols. Columbia Univ. Press, New York:

    Vols. 1 & 2 (The First Thirteen Centuries), 1923, repr. 1947;
    Vols. 3 and 4 (Fourteenth and Fifteenth Centuries), 1934;
    Vols. 5 and 6, (Sixteenth Century), 1941;
    Vols. 7 and 8 (Seventeenth Century), 1958.
    Rev. W. Pagel, *BIHM*, 1959, **33**, 84.

THORNDIKE, LYNN (6). 'The *cursus philosophicus* before Descartes.' *A/AIHS*, 1951, **4** (**30**), 16.

THORPE, SIR EDWARD (1). *History of Chemistry.* 2 vols. in one. Watts, London, 1921.

THURSTON, H. (1). *The Physical Phenomena of Mysticism*, ed. J. H. Crehan. Burns & Oates, London, 1952. French tr. by M. Weill, *Les Phenomènes Physiques du Mysticisme aux Frontières de la Science.* Gallimard, Paris, 1961.

TIEFENSEE, F. (1). *Wegweiser durch die chinesischen Höflichkeits-Formen.* Deutschen Gesellsch. f. Natur- u. Völkerkunde Ostasiens, Tokyo, 1924 (*MDGNVO*, **18**), and Behrend, Berlin, 1924.

TIMKOVSKY, G. (1). *Travels of the Russian Mission through Mongolia to China, and Residence in Peking in the Years 1820–1*, with corrections and notes by J. von Klaproth. Longmans, Rees, Orme, Brown & Green, London, 1827.

TIMMINS, S. (1). 'Nickel German Silver Manufacture', art. in *The Resources, Products and Industrial History of Birmingham and the Midland Hardware District*, ed. S. Timmins. London, 1866, p. 671.

TOBLER, A. J. (1). *Excavations at Tepe Gawra*. 2 vols. Philadelphia, 1950.

TOLL, C. (1). *Al-Hamdānī, 'Kitāb al-Jauharatain' etc., 'Die beiden Edelmetalle Gold und Silber', herausgegeben u. übersetzt*...University Press, Uppsala, 1968 (*UUA*, Studia Semitica, no. 1).

TOLL, C. (2). 'Minting Technique according to Arabic Literary Sources.' *ORS*, 1970, **19–20**, 125.

TORGASHEV, B. P. (1). *The Mineral Industry of the Far East*. Chali, Shanghai, 1930.

DE TOURNUS, J. GIRARD (1) (tr.). *Roger Bachon de l' Admirable Pouvoir et Puissance de l' Art et de Nature, ou est traicté de la pierre Philosophale*. Lyons, 1557, Billaine, Paris, 1628. Tr. of *De Mirabili Potestate Artis et Naturae, et de Nullitate Magiae*.

TRIGAULT, NICHOLAS (1). *De Christiana Expeditione apud Sinas*. Vienna, 1615; Augsburg, 1615. Fr. tr.: *Histoire de l'Expédition Chrétienne au Royaume de la Chine, entrepris par les PP. de la Compagnie de Jésus, comprise en cinq livres...tirée des Commentaires du P. Matthieu Riccius, etc.* Lyon, 1616; Lille, 1617; Paris, 1618. Eng. tr. (partial): *A Discourse of the Kingdome of China, taken out of Ricius and Trigautius*, In *Purchas his Pilgrimes*. London, 1625, vol. 3, p. 380. Eng. tr. (full): see Gallagher (1). Trigault's book was based on Ricci's *I Commentarj della Cina* which it follows very closely, even verbally, by chapter and paragraph, introducing some changes and amplifications, however. Ricci's book remained unprinted until 1911, when it was edited by Venturi (1) with Ricci's letters; it has since been more elaborately and sumptuously edited alone by d'Elia (2).

TSHAO THIEN-CHHIN, HO PING-YÜ & NEEDHAM, JOSEPH (1). 'An Early Mediaeval Chinese Alchemical Text on Aqueous Solutions' (the *San-shih-liu Shui Fa*, early +6th century). *AX*, 1959, **7**, 122. Chinese tr. by Wang Khuei-Kho (1), *KHSC*, 1963, no. 5, 67.

TSO, E. (1). 'Incidence of Rickets in Peking; Efficacy of Treatment with Cod-liver Oil.' *CMJ*, 1924, **38**, 112.

TSUKAHARA, T. & TANAKA, M. (1). 'Edward Divers; his Work and Contribution to the Foundation of [Modern] Chemistry in Japan.' *SHST/T*, 1965, 4.

TU YÜ-TSHANG, CHIANG JUNG-CHHING & TSOU CHHÊNG-LU (1). 'Conditions for the Successful Resynthesis of Insulin from its Glycyl and Phenylalanyl Chains.' *SCISA*, 1965, **14**, 229.

TUCCI, G. (4). 'Animadversiones Indicae; VI, A Sanskrit Biography of the Siddhas, and some Questions connected with Nāgārjuna.' *JRAS/B*, 1930, **26**, 138.

TUCCI, G. (5). *Teoria e Practica del Maṇḍala*. Rome, 1949. Eng. tr. London, 1961.

ULSTADT, PHILIP (1). *Coelum Philosophorum seu de Secretis Naturae Liber*. Strassburg, 1526 and many subsequent eds.

UNDERWOOD, A. J. V. (1). 'The Historical Development of Distilling Plant.' *TICE*, 1935, **13**, 34.

URDANG, G. (1). 'How Chemicals entered the Official Pharmacopoeias.' *A/AIHS*, 1954, **7**, 303.

URE, A. (1). A *Dictionary of Arts, Manufactures and Mines*. 1st ed., 2 vols, London, 1839. 5th ed. 3 vols. ed. R. Hunt, Longman, Green, Longman & Roberts, London, 1860.

VACCA, G. (2). 'Nota Cinesi.' *RSO*, 1915, **6**, 131. (1) A silkworm legend from the *Sou Shen Chi*. (2) The fall of a meteorite described in *Mêng Chhi Pi Than*. (3) Invention of movable type printing (*Mêng Chhi Pi Than*). (4) A problem of the mathematician I-Hsing (chess permutations and combinations) in *Mêng Chhi Pi Than*. (5) An alchemist of the +11th century (*Mêng Chhi Pi Than*).

VAILLANT, A. (1) (tr.). *Le Livre des Secrets d'Hénoch; Texte Slave et Traduction Française*. Inst. d'Études Slaves, Paris, 1952. (Textes Publiés par l'Inst. d'Ét. Slaves, no. 4.)

VALESIUS, HENRICUS (1). *Polybii, Diodori Siculi, Nicolai Damasceni, Dionysii Halicar[nassi], Appiani, Alexand[ri] Dionis[ii] et Joannis Antiocheni, Excerpta et Collectaneis Constantini Augusti [VII] Porphyrogenetae...nunc primum Graece edidit, Latine vertit, Notisque illustravit*. Du Puis, Paris, 1634.

DE LA VALLÉE-POUSSIN, L. (9). '[The "Abode of the Blest" in] Buddhist [Belief].' *ERE*, vol. ii, p. 686.

VANDERMONDE, J. F. (1). 'Eaux, Feu (et Cautères), Terres etc., Métaux, Minéraux et Sels, du *Pên Ts'ao Kang Mou*.' MS., accompanied by 80 (now 72) specimens of inorganic substances collected and studied at Macao or on Poulo Condor Island in +1732, then presented to Bernard de Jussieu, who deposited them in the Musée d'Histoire Naturelle at Paris. The samples were analysed for E. Biot (22) by Alexandre Brongniart (in 1835 to 1840), and the MS. text (which had been acquired from the de Jussieu family by the Museum in 1857) printed in excerpt form by de Mély (1), pp. 156–248. Between 1840 and 1895 the collection was lost, but found again by Lacroix, and the MS. text, not catalogued at the time of acquisition, was also lost, but found again by Deniker; both in time for the work of de Mély.

VARENIUS, BERNARD (1). *Descriptio Regni Japoniae et Siam; item de Japoniorum Religione et Siamensium; de Diversis Omnium Gentium Religionibus...* Hayes, Cambridge, 1673.

VARENIUS, BERNARD (2). *Geographiae Generalis, in qua Affectiones Generales Telluris explicantur summa cura quam plurimus in locis Emendata, et XXXIII Schematibus Novis, aere incisis, una cum Tabb. aliquot quae desiderabantur Aucta et Illustrata, ab Isaaco Newton, Math. Prof. Lucasiano apud Cantabrigiensis.* Hayes, Cambridge, 1672. 2nd ed. (*Auctior et Emendatior*), 1681.

VĀTH, A. (1) (with the collaboration of L. van Hée). *Johann Adam Schall von Bell, S. J., Missionar in China, Kaiserlicher Astronom und Ratgeber am Hofe von Peking; ein Lebens- und Zeit-bild.* Bachem, Köln, 1933. (Veröffentlichungen des Rheinischen Museums in Köln, no. 2.) Crit. P. Pelliot, *TP*, 1934, **31**, 178.

VAUGHAN, T. [Eugenius Philalethes] (1), (attrib.), *A Brief Natural History, intermixed with a Variety of Philosophical Discourses, and Observations upon the Burning of Mount Aetna; with Refutations of such vulgar Errours as our Modern Authors have omitted.* Smelt, London, 1669. See Ferguson (1), vol. 2, p. 197; Waite (4), p. 492.

VAUGHAN, T. (Eugenius Philalethes] (2). *Magia Adamica; or, the Antiquitie of Magic, and the Descent thereof from Adam downwards proved; Whereunto is added, A Perfect and True Discoverie of the True Coelum Terrae, or the Magician's Heavenly Chaos, and First Matter of All Things.* London, 1650. Repr. in Waite (5). Germ. ed. Amsterdam, 1704. See Ferguson (1), vol. 2, p. 196.

DE VAUX, B. CARRA (5). '*L'Abrégé des Merveilles*' (*Mukhtaṣaru'l-'Ajā'ib*) *traduit de l'Arabe*...(A work attributed to al-Mas'ūdī.) Klincksieck, Paris, 1898.

VAVILOV, S. I. (1). 'Newton and the Atomic Theory.' Essay in *Newton Tercentenary Celebrations Volume* (July 1946). Royal Society, London, 1947, p. 43.

DE VEER, GERARD (1). 'The Third Voyage Northward to the Kingdoms of Cathaia, and China, Anno 1596.' In *Purchas his Pilgrimes*, 1625 ed., vol. 3, pt. 2, bk. iii. p. 482; ed. of McLehose, Glasgow, 1906, vol. 13, p. 91.

VEI CHOW JUAN. See Wei Chou-Yuan.

VELER, C. D. & DOISY, E. A. (1). 'Extraction of Ovarian Hormone from Urine.' *PSEBM*, 1928, **25**, 806.

VON VELTHEIM, COUNT (1). *Von den goldgrabenden Ameisen und Greiffen der Alten; eine Vermuthung.* Helmstadt, 1799.

VERHAEREN, H. (1). *L'Ancienne Bibliothèque du Pé-T'ang.* Lazaristes Press, Peking, 1940.

DI VILLA, E. M. (1). *The Examination of Mines in China.* North China Daily Mail, Tientsin, 1919.

DE VILLARD, UGO MONNERET (2). *Le Leggende Orientali sui Magi Evangelici.* Vatican City, 1952. (Studie Testi, no. 163.)

DE VISSER, M. W. (2). *The Dragon in China and Japan.* Müller, Amsterdam, 1913. Orig. in *VKAWA/L*, 1912 (n. r.), **13** (no. 2.).

V[OGT], E. (1). 'The Red Colour Used in [Palaeolithic and Neolithic] Graves.' *CIBA/T*, 1947, **5** (no. 54), 1968.

VOSSIUS, G. J. (1). *Etymologicon Linguae Latinae.* Martin & Allestry, London, 1662; also Amsterdam, 1695, etc.

WADDELL, L. A. (4). '[The State of the Dead in] Tibetan [Religion].' *ERE*, vol. xi, p. 853.

WAITE, A. E. (1). *Lives of Alchemystical Philosophers, based on Materials collected in 1815 and supplemented by recent Researches; with a Philosophical Demonstration of the True Principles of the Magnum Opus or Great Work of Alchemical Re-construction, and some Account of the Spiritual Chemistry...; to Which is added, a Bibliography of Alchemy and Hermetic Philosophy.* Redway, London, 1888. Based on: [Barrett, Francis], (attrib.). *The Lives of Alchemystical Philosophers; with a Critical Catalogue of Books in Occult Chemistry, and a Selection of the most Celebrated Treatises on the Theory and Practice of the Hermetic Art.* Lackington & Allen, London, 1814, with title-page slightly changed, 1815. See Ferguson (1), vol. 2, p. 41. The historical material in both these works is now totally unreliable and outdated; two-thirds of it concerns the 17th century and later periods, even as enlarged and re-written by Waite. The catalogue is 'about the least critical compilation of the kind extant'.

WAITE, A. E. (2). *The Secret Tradition in Alchemy; its Development and Records.* Kegan Paul, Trench & Trübner, London; Knopf, New York, 1926.

WAITE, A. E. (3). *The Hidden Church of the Holy Graal* [Grail]; *its Legends and Symbolism considered in their Affinity with certain Mysteries of Initiation and Other Traces of a Secret Tradition in Christian Times.* Rebman, London, 1909.

WAITE, A. E. (4) (ed.). *The Works of Thomas Vaughan; Eugenius Philalethes*...Theosophical Society, London, 1919.

WAITE, A. E. (5) (ed.). *The Magical Writings of Thomas Vaughan* (*Eugenius Philalethes*); *a verbatim reprint of his first four treatises;* '*Anthroposophia Theomagica*', '*Anima Magica Abscondita*', '*Magia Adamica*' *and the* '*Coelum Terrae*'. Redway, London, 1888.

WAITE, A. E. (6) (tr.). *The Hermetic and Alchemical Writings of Aureolus Philippus Theophrastus Bombast of Hohenheim, called Paracelsus the Great*...2 vols. Elliott, London, 1894. A translation of the Latin Works, Geneva, 1658.

WAITE, A. E. (7) (tr.). *The 'New Pearl of Great Price', a Treatise concerning the Treasure and most precious Stone of the Philosophers* [by P. Bonus of Ferrara, *c.* +1330]. Elliott, London, 1894. Tr. from the Aldine edition (1546).

WAITE, A. E. (8) (tr.). *The Hermetic Museum Restored and Enlarged; most faithfully instructing all Disciples of the Sopho-Spagyric Art how that Greatest and Truest Medicine of the Philosophers' Stone may be found and held; containing Twenty-two most celebrated Chemical Tracts.* 2 vols. Elliott, London, 1893, later repr. A translation of Anon. (87).

WAITE, A. E. (9). *The Brotherhood of the Rosy Cross; being Records of the House of the Holy Spirit in its Inward and Outward History.* Rider, London, 1924.

WAITE, A. E. (10). *The Real History of the Rosicrucians.* London, 1887.

WAITE, A. E. (11) (tr.). *The 'Triumphal Chariot of Antimony', by Basilius Valentinus, with the Commentary of Theodore Kerckringius.* London, 1893. A translation of the Latin *Currus Triumphalis Antimonii,* Amsterdam, 1685.

WAITE, A. E. (12). *The Holy Kabbalah; a Study of the Secret Tradition in Israel as unfolded by Sons of the Doctrine for the Benefit and Consolation of the Elect dispersed through the Lands and Ages of the Greater Exile.* Williams & Norgate, London, 1929.

WAITE, A. E. (13) (tr.). *The 'Turba Philosophorum', or, 'Assembly of the Sages'; called also the 'Book of Truth in the Art' and the Third Pythagorical Synod; an Ancient Alchemical Treatise translated from the Latin, [together with] the Chief Readings of the Shorter Codex, Parallels from the Greek Alchemists, and Explanations of Obscure Terms.* Redway, London, 1896.

WAITE, A. E. (14). 'The Canon of Criticism in respect of Alchemical Literature.' *JALCHS*, 1913, **1**, 17. His reply to the discussion, p. 32.

WAITE, A. E. (15). 'The Beginnings of Alchemy.' *JALCHS*, 1915, **3**, 90. Discussion, pp. 101 ff.

WAITE, A. E. See also Stenring (1).

WAKEMAN, F. (1) (ed.). *Wu Yin Lu, 'Nothing Concealed'; Essays in Honour of Liu (Aisin-Gioro) Yü-Yün.* Chinese Materials and Research Aids Service Centre, Thaipei, Thaiwan, 1970.

WALAAS, O. (1) (ed.). *The Molecular Basis of Some Aspects of Mental Activity.* 2 vols. Academic Press, London and New York, 1966–7.

WALDEN, P. (1). *Mass, Zahl und Gewicht in der Chemie der Vergangenheit; ein Kapitel aus der Vorgeschichte des Sogenannten quantitative Zeitalters der Chemie.* Enke, Stuttgart, 1931. Repr. Liebing, Würzburg, 1970. (Samml. chem. u. chem. techn. Vorträge, N.F. no. 8.)

WALDEN, P. (2). 'Zur Entwicklungsgeschichte d. chemischen Zeichen.' Art. in *Studien z. Gesch. d. Chemie* (von Lippmann Festschrift), ed. J. Ruska. Springer, Berlin, 1927, p. 80.

WALDEN, P. (3). *Geschichte der Chemie.* Universitätsdruckerei, Bonn, 1947. 2nd ed. Athenäum, Bonn, 1950.

WALDEN, P. (4). 'Paracelsus und seine Bedeutung für die Chemie.' *ZAC/AC*, 1941, **54**, 421.

WALEY, A. (1) (tr.). *The Book of Songs.* Allen & Unwin, London, 1937.

WALEY, A. (4) (tr.). *The Way and its Power; a Study of the 'Tao Tê Ching' and its Place in Chinese Thought.* Allen & Unwin, London, 1934. Crit. Wu Ching-Hsiang, *TH*, 1935, **1**, 225.

WALEY, A. (10) (tr.). *The Travels of an Alchemist; the Journey of the Taoist [Chhiu] Chhang-Chhun from China to the Hindu-Kush at the summons of Chingiz Khan, recorded by his disciple Li Chih-Chhang.* Routledge, London, 1931. (Broadway Travellers Series.) Crit. P. Pelliot, *TP*, 1931, **28**, 413.

WALEY, A. (14). 'Notes on Chinese Alchemy, supplementary to Johnson's "Study of Chinese Alchemy".' *BLSOAS*, 1930, **6**, 1. Revs. P. Pelliot, *TP*, 1931, **28**, 233; Tenney L. Davis, *ISIS*, 1932, **17**, 440.

WALEY, A. (17). *Monkey, by Wu Chhêng-Ên.* Allen & Unwin, London, 1942.

WALEY, A. (23). *The Nine Songs; a study of Shamanism in Ancient China* [the *Chiu Ko* attributed traditionally to Chhü Yuan]. Allen & Unwin, London, 1955.

WALEY, A. (24). 'References to Alchemy in Buddhist Scriptures.' *BLSOAS*, 1932, **6**, 1102.

WALEY, A. (27) (tr.). *The Tale of Genji.* 6 vols. Allen & Unwin, London; Houghton Mifflin, New York, 1925–33.
    Vol. 1   *The Tale of Genji.*
    Vol. 2   *The Sacred Tree.*
    Vol. 3   *A Wreath of Cloud.*
    Vol. 4   *Blue Trousers.*
    Vol. 5   *The Lady of the Boat.*
    Vol. 6   *The Bridge of Dreams.*

WALKER, D. P. (1). 'The Survival of the "Ancient Theology" in France, and the French Jesuit Missionaries in China in the late Seventeenth Century.' MS. of Lecture at the Cambridge History of Science Symposium, Oct. 1969. Pr. in Walker (2) pp. 194 ff.

WALKER, D. P. (2). *The Ancient Theology; Studies in Christian Platonism from the +15th to the +18th Century.* Duckworth, London, 1972.

WALKER, D. P. (3). 'Francis Bacon and *Spiritus*', art. in *Science, Medicine and Society in the Renaissance* (Pagel Presentation Volume), ed. Debus (20), vol. 2, p. 121.

WALKER, W. B. (1). 'Luigi Cornaro; a Renaissance Writer on Personal Hygiene.' *BIHM*, 1954, **28**, 525.

WALLACE, R. K. (1). 'Physiological Effects of Transcendental Meditation.' *SC*, 1970, **167**, 1751.

WALLACE, R. K. & BENSON, H. (1). 'The Physiology of Meditation.' *SAM*, 1972, **226** (no. 2), 84.

WALLACE, R. K., BENSON, H. & WILSON, A. F. (1). 'A Wakeful Hypometabolic Physiological State.' *AJOP*, 1971, **221**, 795.

WALLACKER, B. E. (1) (tr.). *The 'Huai Nan Tzu' Book*, [Ch.] 11; *Behaviour, Culture and the Cosmos.* Amer. Oriental Soc., New Haven, Conn. 1962. (Amer. Oriental Series, no. 48.)

VAN DE WALLE, B. (1). 'Le Thème de la Satire des Métiers dans la Littérature Egyptienne.' *CEG*, 1947, **43**, 50.

WALLESER, M. (3). 'The Life of Nāgārjuna from Tibetan and Chinese Sources.' *AM* (Hirth Anniversary Volume), **1**, 1.

WALSHE, W. G. (1). '[Communion with the Dead in] Chinese [Thought and Liturgy].' *ERE*, vol. iii, p. 728.

WALTON, A. HULL. See Davenport, John.

WANG, CHHUNG-YU (1). *Bibliography of the Mineral Wealth and Geology of China.* Griffin, London, 1912.

WANG, CHHUNG-YU (2). *Antimony; its History, Chemistry, Mineralogy, Geology, Metallurgy, Uses, Preparations, Analysis, Production and Valuation; with Complete Bibliographies.* Griffin, London, 1909.

WANG CHHUNG-YU (3). *Antimony; its Geology, Metallurgy, Industrial Uses and Economics.* Griffin, London, 1952. ('3rd edition' of Wang Chhung-Yu (2), but it omits the chapters on the history, chemistry, mineralogy and analysis of antimony, while improving those that are retained.)

WANG CHI-MIN & WU LIEN-TÊ (1). *History of Chinese Medicine.* Nat. Quarantine Service, Shanghai, 1932, 2nd ed. 1936.

WANG CHIUNG-MING (1). 'The Bronze Culture of Ancient Yunnan.' *PKR*, 1960 (no. 2), 18. Reprinted in mimeographed form, Collet's Chinese Bookshop, London, 1960.

WANG LING (1). 'On the Invention and Use of Gunpowder and Firearms in China.' *ISIS*, 1947, **37**, 160.

WARE, J. R. (1). 'The *Wei Shu* and the *Sui Shu* on Taoism.' *JAOS*, 1933, **53**, 215. Corrections and emendations in *JAOS*, 1934, **54**, 290. Emendations by H. Maspero, *JA*, 1935, **226**, 313.

WARE, J. R. (5) (tr.). *Alchemy, Medicine and Religion in the China of +320; the 'Nei Phien' of Ko Hung ('Pao Phu Tzu').* M.I.T. Press, Cambridge, Mass. and London, 1966. Revs. Ho Ping-Yü, *JAS*, 1967, **27**, 144; J. Needham, *TCULT*, 1969, **10**, 90.

WARREN, W. F. (1). *The Earliest Cosmologies; the Universe as pictured in Thought by the Ancient Hebrews, Babylonians, Egyptians, Greeks, Iranians and Indo-Aryans—a Guidebook for Beginners in the Study of Ancient Literatures and Religions.* Eaton & Mains, New York; Jennings & Graham, Cincinnati, 1909.

WASHBURN, E. W. (1). 'Molecular Stills.' *BSJR*, 1929, **2** (no. 3), 476. Part of a collective work by E. W. Washburn, J. H. Bruun & M. M. Hicks. *Apparatus and Methods for the Separation, Identification and Determination of the Chemical Constituents of Petroleum*, p. 467.

WASITZKY, A. (1). 'Ein einfacher Mikro-extraktionsapparat nach dem Soxhlet-Prinzip.' *MIK*, 1932, **11**, 1.

WASSON, R. G. (1). 'The Hallucinogenic Fungi of Mexico; an Enquiry into the Origins of the Religious Idea among Primitive Peoples.' *HU/BML*, 1961, **19**, no. 7. (Ann. Lecture, Mycol. Soc. of America.)

WASSON, R. G. (2). '*Ling Chih* [the Numinous Mushroom]; Some Observations on the Origins of a Chinese Conception.' Unpub. MS. Memorandum, 1962.

WASSON, R. G. (3). *Soma; Divine Mushroom of Immortality.* Harcourt, Brace & World, New York; Mouton, The Hague, 1968. (Ethno-Mycological Studies, no. 1.) With extensive contributions by W. D. O'Flaherty. Rev. F. B. J. Kuiper, *IIJ*, 1970, **12** (no. 4), 279; followed by comments by R. G. Wasson, 286.

WASSON, R. G. (4). 'Soma and the Fly-Agaric; Mr Wasson's Rejoinder to Prof. Brough.' Bot. Mus. Harvard Univ. Cambridge, Mass. 1972. (Ethno-Mycological Studies, no. 2.)

WASSON, R. G. & INGALLS, D. H. H. (1). 'The Soma of the *Rig Veda*; what was it?' (Summary of his argument, followed by critical remarks by Ingalls.) *JAOS*, 1971, **91** (no. 2). Separately issued as: *R. Gordon Wasson on Soma and Daniel H. H. Ingalls' Response.* Amer. Oriental Soc. New Haven, Conn. 1971. (Essays of the Amer. Orient. Soc. no. 7.)

WASSON, R. G. & WASSON, V. P. (1). *Mushrooms, Russia and History.* 2 vols. Pantheon, New York, 1957.

WATERMANN, H. I. & ELSBACH, E. B. (1). 'Molecular stills.' *CW*, 1929, **26**, 469.

WATSON, BURTON (1) (tr.). '*Records of the Grand Historian of China*', translated from the '*Shih Chi*' of *Ssuma Chhien*. 2 vols. Columbia University Press, New York, 1961.

WATSON, R., Bp of Llandaff (1). *Chemical Essays*. 2 vols. Cambridge, 1781; vol. 3, 1782; vol. 4, 1786; vol. 5, 1787. 2nd ed. 3 vols. Dublin, 1783. 5th ed. 5 vols., Evans, London, 1789. 3rd ed. Evans, London, 1788. 6th ed. London, 1793-6.

WATSON, WM. (4). *Ancient Chinese Bronzes*. Faber & Faber, London, 1962.

WATTS, A. W. (2). *Nature, Man and Woman; a New Approach to Sexual Experience*. Thames & Hudson, London; Pantheon, New York, 1958.

WAYMAN, A. (1). 'Female Energy and Symbolism in the Buddhist Tantras.' *HOR*, 1962, **2**, 73.

WESBTER, C. (1). 'English Medical Reformers of the Puritan Revolution; a Background to the "Society of Chymical Physitians".' *AX*, 1967, **14**, 16.

WEEKS, M. E. (1). *The Discovery of the Elements; Collected Reprints of a series of articles published in the Journal of Chemical Education; with Illustrations collected by F. B. Dains*. Mack, Easton, Pa. 1933. Chinese tr. *Yuan Su Fa-Hsien Shih* by Chang Tzu-Kung, with additional material. Shanghai, 1941.

WEI CHOU-YUAN (VEI CHOU JUAN) (1). 'The Mineral Resources of China.' *EG*, 1946, **41**, 399-474

VON WEIGEL, C. E. (1). *Observationes Chemicae et Mineralogicae*. Pt. 1, Göttingen, 1771; pt. 2, Gryphiae, 1773.

WEISS, H. B. & CARRUTHERS, R. H. (1). *Insect Enemies of Books* (63 pp. with extensive bibliography). New York Public Library, New York, 1937.

WELCH, HOLMES, H. (1). *The Parting of the Way; Lao Tzu and the Taoist Movement*. Beacon Press, Boston, Mass. 1957.

WELCH, HOLMES H. (2). 'The Chang Thien Shih ["Taoist Pope"] and Taoism in China.' *JOSHK*, 1958, **4**, 188.

WELCH, HOLMES H. (3). 'The Bellagio Conference on Taoist Studies.' *HOR*, 1970, **9**, 107.

WELLMANN, M. (1). 'Die Stein- u. Gemmen-Bücher d. Antike.' *QSGNM*, 1935, **4**, 86.

WELLMANN, M. (2). 'Die Φυσικὰ des Bolos Democritos und der Magier Anaxilaos aus Larissa.' *APAW/PH*, 1928 (no. 7).

WELLMANN, M. (3). 'Die "Georgika" des [Bolus] Demokritos.' *APAW/PH*, 1921 (no. 4), 1-.

WELLS, D. A. (1). *Principles and Applications of Chemistry*. Ivison, Blakeman & Taylor, New York and Chicago, 1858. Chinese tr. by J. Fryer & Hsü Shou, Shanghai, 1871.

WELTON, J. (1). *A Manual of Logic*. London, 1896.

WENDTNER, K. (1). 'Assaying in the Metallurgical Books of the +16th Century.' Inaug. Diss. London, 1952.

WENGER, M. A. & BAGCHI, B. K. (1). 'Studies of Autonomic Functions in Practitioners of Yoga in India.' *BS*, 1961, **6**, 312.

WENGER, M. A., BAGCHI, B. K. & ANAND, B. K. (1). 'Experiments in India on the "Voluntary" Control of the Heart and Pulse.' *CIRC*, 1961, **24**, 1319.

WENSINCK, A. J. (2). *A Handbook of Early Muhammadan Tradition, Alphabetically Arranged*. Brill, Leiden, 1927.

WENSINCK, A. J. (3). 'The Etymology of the Arabic Word *djinn*.' *VMAWA*, 1920, 506.

WERTHEIMER, E. (1). Art. 'Arsenic' in *Dictionnaire de Physiologie*, ed. C. Richet, vol. i. Paris.

WERTIME, T. A. (1). 'Man's First Encounters with Metallurgy.' *SC*, 1964, **146**, 1257.

WEST, M. (1). 'Notes on the Importance of Alchemy to Modern Science in the Writings of Francis Bacon and Robert Boyle.' *AX*, 1961, **9**, 102.

WEST, M. L. (1). *Early Greek Philosophy and the Orient*. Oxford, 1971.

WESTBERG, F. (1). *Die Fragmente des Toparcha Goticus* (*Anonymus Tauricus*, '*Zapisk gotskogo toparcha*'); *Nachdruck der Ausgabe St. Petersburg, 1901, mit einem wissenchafts-geschichtlichen Vorwort in englischer Sprache von Ihor Ševčenko* (*Washington*). Zentralantiquariat der D. D. R., Leipzig, 1971. (Subsidia Byzantina, no. 18.)

WESTBROOK, J. H. (1). 'Historical Sketch [of Intermetallic Compounds].' Xerocopy of art. without indication of place or date of pub., comm. by the author, General Electric Co., Schenectady, N.Y.

WESTERBLAD, C. A. (1). *Pehr Henrik Ling; en Lefnadsteckning och några Sympunkter* [in Swedish]. Norstedt, Stockholm, 1904. *Ling, the Founder of the Swedish Gymnastics*. London, 1909.

WESTERBLAD, C. A. (2). *Ling; Tidshistoriska Undersökningar* [in Swedish]. Norstedt, Stockholm. Vol. 1, *Den Lingska Gymnastiken i dess Upphofsmans Dagar*, 1913. Vol. 2, *Personlig och allmän Karakteristik samt Litterär Analys*, 1916.

WESTFALL, R. S. (1). 'Newton and the Hermetic Tradition', art. in *Science, Medicine and Society in the Renaissance* (Pagel Presentation Volume), ed. Debus (20), vol. 2. p. 183.

WEULE, K. (1). *Chemische Technologie der Naturvölker*. Stuttgart, 1922.

WEYNANTS-RONDAY, M. (1). *Les Statues Vivantes; Introduction à l'Étude des Statues Égyptiennes*... Fond. Egyptol. Reine Elis:, Brussels, 1926.

WHELER, A. S. (1). 'Antimony Production in Hunan Province.' *TIMM*, 1916, **25**, 366.

WHELER, A. S. & LI, S. Y. (1). 'The Shui-ko-shan [Shui-khou Shan] Zinc and Lead Mine in Hunan [Province].' *MIMG*, 1917, **16**, 91.

WHITE, J. H. (1). *The History of the Phlogiston Theory*. Arnold, London, 1932.

WHITE, LYNN (14). *Machina ex Deo; Essays in the Dynamism of Western Culture*. M.I.T. Press, Cambridge, Mass. 1968.

WHITE, LYNN (15). 'Mediaeval Borrowings from Further Asia.' *MRS*, 1971, **5**, 1.

WHITE, W. C., Bp. of Honan (3). *Bronze Culture of Ancient China; an archaeological Study of Bronze Objects from Northern Honan dating from about −1400 to −771*. Univ. of Toronto Press, Toronto, 1956 (Royal Ontario Museum Studies, no. 5).

WHITFORD, J. (1). 'Preservation of bodies after arsenic poisoning.' *BMJ*, 1884, pt. 1, 504.

WHITLA, W. (1). *Elements of Pharmacy, Materia Medica and Therapeutics*. Renshaw, London, 1903.

WHITNEY, W. D. & LANMAN, C. R. (1) (tr.). *Atharva-veda Saṃhitā*. 2 vols. Harvard Univ. Press, Cambridge, Mass. 1905. (Harvard Oriental Series, nos. 7, 8.)

WIBERG, A. (1). 'Till Frågan om Destilleringsförfarandets Genesis; en Etnologisk-Historisk Studie' [in Swedish]. *SBM*, 1937 (nos. 2-3), 67, 105.

WIDENGREN, GEO. (1). 'The King and the Tree of Life in Ancient Near Eastern Religion.' *UUA*, 1951, **4**, 21.

WIEDEMANN, E. (7). 'Beiträge z. Gesch. d. Naturwiss.; VI, Zur Mechanik und Technik bei d. Arabern.' *SPMSE*, 1906, **38**, 1. Repr. in (23), vol. 1, p. 173.

WIEDEMANN, E. (11). 'Beiträge z. Gesch. d. Naturwiss.; XV, Über die Bestimmung der Zusammensetzung von Legierungen.' *SPMSE*, 1908, **40**, 105. Repr. in (23), vol. 1, p. 464.

WIEDEMANN, E. (14). 'Beiträge z. Gesch. d. Naturwiss.; XXV, Über Stahl und Eisen bei d. muslimischen Völkern.' *SPMSE*, 1911, **43**, 114. Repr. in (23), vol. 1, p. 731.

WIEDEMANN, E. (15). 'Beiträge z. Gesch. d. Naturwiss.; XXIV, Zur Chemie bei den Arabern' (including a translation of the chemical section of the *Mafātīḥ al-'Ulūm* by Abū 'Abdallah al-Khwārizmī al-Kātib, c. +976). *SPMSE*, 1911, **43**, 72. Repr. in (23), vol. 1, p. 689.

WIEDEMANN, E. (21). 'Zur Alchemie bei den Arabern.' *JPC*, 1907, **184** (N.F. **76**), 105.

WIEDEMANN, E. (22). 'Über chemische Apparate bei den Arabern.' Art. in the Kahlbaum Gedächtnisschrift: *Beiträge aus d. Gesch. d. Chemie*...ed. P. Diergart (1), 1909, p. 234.

WIEDEMANN, E. (23). *Aufsätze zur arabischen Wissenschaftsgeschichte* (a reprint of his 79 contributions in the series 'Beiträge z. Geschichte d. Naturwissenschaften' in *SPMSE*), ed. W. Fischer, with full indexes. 2 vols. Olm, Hildesheim and New York, 1970.

WIEDEMANN, E. (24). 'Beiträge z. Gesch. d. Naturwiss.; I, Beiträge z. Geschichte der Chemie bei den Arabern.' *SPMSE*, 1902, **31**, 15. Repr. in (23), vol. 1, p. 1.

WIEDEMANN, E. (25). 'Beiträge z. Gesch. d. Naturwiss.; LXIII, Zur Geschichte der Alchemie.' *SPMSE*, 1921, **53**, 97. Repr. in (23), vol. 2, p. 545.

WIEDEMANN, E. (26). 'Beiträge z. Mineralogie u.s.w. bei den Arabern.' Art. in *Studien z. Gesch. d. Chemie* (von Lippmann Festschrift), ed. J. Ruska. Springer, Berlin, 1927, p. 48.

WIEDEMANN, E. (27). 'Beitrage z. Gesch. d. Naturwiss.; II, 1. Einleitung, 2. Ü. elektrische Erscheinungen, 3. Ü. Magnetismus, 4. Optische Beobachtungen, 5. Ü. einige physikalische usf. Eigenschaften des Goldes, 6. Zur Geschichte d. Chemie (a) Die Darstellung der Schwefelsäure durch Erhitzen von Vitriolen, die Wärme-entwicklung beim Mischen derselben mit Wasser, und ü. arabische chemische Bezeichnungen, (b) Astrologie and Alchemie, (c) Anschauungen der Araber ü. die Metallverwandlung und die Bedeutung des Wortes al-Kimiya.' *SPMSE*, 1904, **36**, 309. Repr. in (23), vol. 1, p. 15.

WIEDEMANN, E. (28). 'Beiträge z. Gesch. d. Naturwiss.; XL, Über Verfälschungen von Drogen usw. nach Ibn Bassām und Nabarāwī.' *SPMSE*, 1914, **46**, 172. Repr. in (23), vol. 2, p. 102.

WIEDEMANN, E. (29). 'Zur Chemie d. Araber.' *ZDMG*, 1878, **32**, 575.

WIEDEMANN, E. (30). 'Al-Kīmīyā.' Art. in *Encyclopaedia of Islam*, vol. ii, p. 1010.

WIEDEMANN, E. (31). 'Beiträge zur Gesch. der Naturwiss.; LVII, Definition verschiedener Wissenschaften und über diese verfasste Werke.' *SPMSE*, 1919, **50-51**, 1. Repr. in (23), vol. 2, p. 431.

WIEDEMANN, E. (32). *Zur Alchemie bei den Arabern*. Mencke, Erlangen, 1922. (Abhandlungen zur Gesch. d. Naturwiss. u. d. Med., no. 5.) Translation of the entry on alchemy in Haji Khalfa's Bibliography and of excerpts from al-Jildakī, with a biographical glossary of Arabic alchemists.

WIEDEMANN, E. (33). 'Beiträge zur Gesch. der Naturwiss.; V, Auszüge aus arabischen Enzyklopädien und anderes.' *SPMSE*, 1905, **37**, 392. Repr. in (23), vol. 1, p. 109.

WIEGER, L. (2). *Textes Philosophiques*. (Ch and Fr.) Mission Press, Hsien-hsien, 1930.

WIEGER, L. (3). *La Chine à travers les Ages; Précis, Index Biographique et Index Bibliographique*. Mission Press, Hsien-hsien, 1924. Eng. tr. E. T. C. Werner.

WIEGER, L. (6) *Taoisme*. Vol. 1. *Bibliographie Générale*: (1) Le Canon (Patrologie); (2) Les Index Officiels et Privés. Mission Press. Hsien-hsien, 1911. Crit. P. Pelliot, *JA*, 1912 (10e Sér.) **20**, 141.

WIEGER, L. (7). *Taoisme*. Vol. 2. *Les Pères du Système Taoiste* (tr. selections of Lao Tzu, Chuang Tzu, Lieh Tzu). Mission Press, Hsien-hsien, 1913.

WIEGLEB, J. C. (1). *Historisch-kritische Untersuchung der Alchemie, oder den eingebildeten Goldmacher-kunst; von ihrem Ursprunge sowohl als Fortgange, und was nun von ihr zu halten sey.* Hoffmanns Wittwe und Erben, Weimar, 1777. 2nd ed. 1793. Photolitho repr. of the original ed., Zentral-Antiquariat D.D.R. Leipzig, 1965. Cf. Ferguson (1), vol. 2, p. 546.

WIGGLESWORTH, V. B. (1). 'The Insect Cuticle.' *BR*, 1948, **23**, 408.

WILHELM, HELLMUT (6). 'Eine Chou-Inschrift über Atemtechnik.' *MS*, 1948, **13**, 385.

WILHELM, RICHARD & JUNG, C. G. (1). *The Secret of the Golden Flower; a Chinese Book of Life* (including a partial translation of the *Thai-I Chin Hua Tsung Chih* by R. W. with notes, and a 'European commentary' by C. G. J.).

Eng. ed. tr. C. F. Baynes, (with C. G. J.'s memorial address for R. W.). Kegan Paul, London and New York, 1931. From the Germ. ed. *Das Geheimnis d. goldenen Blute; ein chinesisches Lebensbuch.* Munich, 1929.

Abbreviated preliminary version: '*Tschang Scheng Shu [Chhang Shêng Shu]*; die Kunst das mensch-lichen Leben zu verlängern.' *EURR*, 1929, **5**, 530.

Revised Germ. ed. with new foreword by C. G. J., Rascher, Zürich, 1938. Repr. twice, 1944.

New Germ. ed. entirely reset, with new foreword by Salome Wilhelm, and the partial translation of a Buddhist but related text, the *Hui Ming Ching*, from R. W.'s posthumous papers, Zürich, 1957.

New Eng. ed. including all the new material, tr. C. F. Baynes. Harcourt, New York and Routledge, London, 1962, repr. 1965, 1967, 1969. Her revised tr. of the 'European commentary' alone had appeared in an anthology: *Psyche und Symbol*, ed. V. S. de Laszlo. Anchor, New York, 1958. Also tr. R. F. C. Hull for C. G. J.'s *Collected Works*, vol. 13, pp. 1–55, i.e. Jung (3).

WILLETTS, W. Y. (1). *Chinese Art*. 2 vols. Penguin, London, 1958.

WILLETTS, W. Y. (3). *Foundations of Chinese Art; from Neolithic Pottery to Modern Architecture.* Thames & Hudson, London, 1965. Revised, abridged and re-written version of (1), with many illustrations in colour.

WILLIAMSON, G. C. (1). *The Book of 'Famille Rose'* [polychrome decoration of Chinese Porcelain]. London, 1927.

WILSON, R. McLACHLAN (1). *The Gnostic Problem; a Study of the Relations between Hellenistic Judaism and the Gnostic Heresy*. Mowbray, London, 1958.

WILSON, R. McLACHLAN (2). *Gnosis and the New Testament*. Blackwell, Oxford, 1968.

WILSON, R. McLACHLAN (3) (ed. & tr.). *New Testament Apocrypha* (ed. E. Hennecke & W. Schnee-melcher). 2 vols. Lutterworth, London, 1965.

WILSON, W. (1) (tr.). *The Writings of Clement of Alexandria* (b. c. +150] (including *Stromata, c.* +200). In Ante-Nicene Christian Library, ed. A. Roberts & J. Donaldson. Clark, Edinburgh, 1867, vols 4 and 12.

WILSON, W. J. (1). 'The Origin and Development of Graeco-Egyptian Alchemy.' *CIBA/S*, 1941, **3**, 926.

WILSON, W. J. (2) (ed.). 'Alchemy in China.' *CIBA/S*, 1940, **2** (no. 7), 594.

WILSON, W. J. (2a). 'The Background of Chinese Alchemy.' *CIBA/S*, 1940, **2** (no. 7), 595.

WILSON, W. J. (2b). 'Leading Ideas of Early Chinese Alchemy.' *CIBA/S*, 1940, **2** (no. 7), 600.

WILSON, W. J. (2c). 'Biographies of Early Chinese Alchemists.' *CIBA/S*, 1940, **2** (no. 7), 605.

WILSON, W. J. (2d). 'Later Developments of Chinese Alchemy.' *CIBA/S*, 1940, **2** (no. 7), 610.

WILSON, W. J. (2e). 'The Relation of Chinese Alchemy to that of other Countries.' *CIBA/S*, 1940, **2** (no. 7), 618.

WILSON, W. J. (3). 'An Alchemical Manuscript by Arnaldus [de Lishout] de Bruxella [written from +1473 to +1490].' *OSIS*, 1936, **2**, 220.

WINDAUS, A. (1). 'Über d. Entgiftung der Saponine durch Cholesterin.' *BDCG*, 1909, **42**, 238.

WINDAUS, A. (2). 'Über d. quantitative Bestimmung des Cholesterins und der Cholesterinester in einigen normalen und pathologischen Nieren.' *ZPC*, 1910, **65**, 110.

WINDERLICH, R. (1) (ed.). *Julius Ruska und die Geschichte d. Alchemie, mit einem Völlstandigen Verzeichnis seiner Schriften; Festgabe zu seinem 70. Geburtstage...* Ebering, Berlin, 1937. (Abhdl. z. Gesch. d. Med. u. d. Naturwiss., no. 19).

WINDERLICH, R. (2). 'Verschüttete und wieder aufgegrabene Quellen der Alchemie des Abendlandes' (a biography of J. Ruska and an account of his work). Art. in Winderlich (1), the Ruska Presentation Volume.

WINKLER, H. A. (1). *Siegel und Charaktere in der Mohammedanische Zauberei*. De Gruyter, Berlin and Leipzig, 1930. (*DI* Beiheft, no. 7.)

WISE, T. A. (1). *Commentary on the Hindu System of Medicine.* Thacker, Ostell & Lepage, Calcutta; Smith Elder, London, 1845.

WISE, T. A. (2). *Review of the History of Medicine [among the Asiatic Nations].* 2 vols. Churchill, London, 1867.

WOLF, A. (1) (with the co-operation of F. Dannemann & A. Armitage). *A History of Science, Technology, and Philosophy in the 16th and 17th Centuries.* Allen & Unwin, London, 1935; 2nd ed., revised by D. McKie, London, 1950. Rev. G. Sarton, *ISIS*, 1935, **24**, 164.

WOLF, A. (2). *A History of Science, Technology and Philosophy in the 18th Century.* Allen & Unwin, London, 1938; 2nd ed. revised by D. McKie, London, 1952.

WOLF, JOH. CHRISTOPH (1). *Manichaeismus ante Manichaeos, et in Christianismo Redivivus; Tractatus Historico-Philosophicus...* Liebezeit & Stromer, Hamburg, 1707. Repr. Zentralantiquariat D. D. R., Leipzig, 1970.

WOLF, T. (1). *Viajes Cientificos.* 3 vols. Guayaquil, Ecuador, 1879.

WOLFF, CHRISTIAN (1). 'Rede über die Sittenlehre der Sineser', pub. as *Oratio de Sinarum Philosophia Practica* [that morality is independent of revelation]. Frankfurt a/M, 1726. The lecture given in July 1721 on handing over the office of Pro-Rector, for which Christian Wolff was expelled from Halle and from his professorship there. See Lach (6). The German version did not appear until 1740 in vol. 6 of Wolff's *Kleine Philosophische Schriften*, Halle.

WOLTERS, O. W. (1). 'The "Po-Ssu" Pine-Trees.' *BLSOAS*, 1960, **23**, 323.

WONG K. CHIMIN. See Wang Chi-Min.

WONG, M. or MING. See Huang Kuang-Ming, Huard & Huang Kuang-Ming.

WONG MAN. See Huang Wên.

WONG WÊN-HAO. See Ong Wên-Hao.

WOOD, I. F. (1). '[The State of the Dead in] Hebrew [Thought].' *ERE*, vol. xi, p. 841.

WOOD, I. F. (2). '[The State of the Dead in] Muhammadan [Muslim, Thought].' *ERE*, vol. xi, p. 849.

WOOD, R. W. (1). 'The Purple Gold of Tut'ankhamēn.' *JEA*, 1934, **20**, 62.

WOODCROFT, B. (1) (tr.). *The 'Pneumatics' of Heron of Alexandria.* Whittingham, London, 1851.

WOODROFFE, SIR J. G. (ps. A. Avalon) (1). *Śakti and Śakta; Essays and Addresses on the Śakta Tantra-śāstra.* 3rd ed. Ganesh, Madras; Luzac, London, 1929.

WOODROFFE, SIR J. G. (ps. A. Avalon) (2). *The Serpent Power* [Kuṇḍalinī Yoga], *being the Ṣat-cakra-nirūpana* [i.e. ch. 6 of Pūrnānanda's *Tattva-chintāmaṇi*] *and 'Pādukā-panchaka', two works on Laya Yoga...* Ganesh, Madras; Luzac, London, 1931.

WOODROFFE, SIR J. G. (ps. A. Avalon) (3) (tr.). *The Tantra of the Great Liberation, 'Mahā-nirvāna Tantra', a translation from the Sanskrit.* London, 1913. Ganesh, Madras, 1929 (text only).

WOODS, J. H. (1). *The Yoga System of Patañjali; or, the Ancient Hindu Doctrine of Concentration of Mind...* Harvard Univ. Press, Cambridge, Mass. 1914. (Harvard Oriental Series, no. 17.)

WOODWARD, J. & BURNETT, G. (1). *A Treatise on Heraldry, British and Foreign...* 2 vols. Johnston, Edinburgh and London, 1892.

WOOLLEY, C. L. (4). 'Excavations at Ur, 1926–7, Part II.' *ANTJ*, 1928, **8**, 1 (24), pl. viii, 2.

WOULFE, P. (1). 'Experiments to show the Nature of *Aurum Mosaicum*.' *PTRS*, 1771, **61**, 114.

WRIGHT, SAMSON, (1). *Applied Physiology.* 7th ed. Oxford, 1942.

WU KHANG (1). *Les Trois Politiques du Tchounn Tsieou [Chhun Chhiu] interpretées par Tong Tchong-Chou [Tung Chung-Shu] d'après les principes de l'école de Kong-Yang [Kungyang].* Leroux, Paris, 1932. (Includes tr. of ch. 121 of *Shih Chi*, the biography of Tung Chung-Shu.)

WU LU-CHHIANG. See Tenney L. Davis' biography (obituary). *JCE*, 1936, **13**, 218.

WU LU-CHHIANG & DAVIS, T. L. (1) (tr.). 'An Ancient Chinese Treatise on Alchemy entitled *Tshan Thung Chhi*, written by Wei Po-Yang about +142...' *ISIS*, 1932, **18**, 210. Critique by J. R. Partington, *N*, 1935, **136**, 287.

WU LU-CHHIANG & DAVIS, T. L. (2) (tr.). 'An Ancient Chinese Alchemical Classic; Ko Hung on the Gold Medicine, and on the Yellow and the White; being the 4th and 16th chapters of *Pao Phu Tzu...*' *PAAAS*, 1935, **70**, 221.

WU YANG-TSANG (1). 'Silver Mining and Smelting in Mongolia.' *TAIME*, 1903, **33**, 755. With a discussion by B. S. Lyman, pp. 1038 ff. (Contains an account of the recovery of silver from argentiferous lead ore, and cupellation by traditional methods, at the mines of Ku-shan-tzu and Yen-tung Shan in Jehol province. The discussion adds a comparison with traditional Japanese methods observed at Hosokura). Abridged version in *EMJ*, 1903, **75**, 147.

WULFF, H. E. (1). *The Traditional [Arts and] Crafts of Persia; their Development, Technology and Influence on Eastern and Western Civilisations.* M.I.T. Press, Cambridge, Mass. 1966. Inaug. Diss. Univ. of New South Wales, 1964.

WUNDERLICH, E. (1). 'Die Bedeutung der roten Farbe im Kultus der Griechern und Römer.' *RGVV* 1925, **20**, 1.

YABUUCHI KIYOSHI (9). 'Astronomical Tables in China, from the Han to the Thang Dynasties.' Eng. art. in Yabuuchi Kiyoshi (25) (ed.), *Chūgoku Chūsei Kagaku Gijutsushi no Kenkyū* (Studies in the History of Science and Technology in Mediaeval China). Jimbun Kagaku Kenkyusō, Tokyo, 1963.

YAMADA KENTARO (1). *A Short History of Ambergris [and its Trading] by the Arabs and the Chinese in the Indian Ocean.* Kinki University, 1955, 1956. (Reports of the Institute of World Economics, *KKD*, nos. 8 and 11.)

YAMADA KENTARO (2). *A Study of the Introduction of 'An-hsi-hsiang' into China and of Gum Benzoin into Europe.* Kinki University, 1954, 1955. (Reports of the Institute of World Economics, *KKD*, nos. 5 and 7.)

YAMASHITA, A. (1). 'Wilhelm Nagayoshi Nagai [Nakai Nakayoshi], Discoverer of Ephredrin; his Contributions to the Foundation of Organic Chemistry in Japan.' *SHST/T*, 1965, 11.

YAMAZAKI, T. (1). 'The Characteristic Development of Chemical Technology in Modern Japan, chiefly in the Years between the two World Wars.' *SHST/T*, 1965, 7.

YAN TSZ CHIU. See Yang Tzu-Chiu (1).

YANG LIEN-SHÊNG (8). 'Notes on Maspero's "Les Documents Chinois de la Troisième Expédition de Sir Aurel Stein en Asie Centrale".' *HJAS*, 1955, **18**, 142.

YANG TZU-CHIU (1). 'Chemical Industry in Kuangtung Province.' *JRAS/NCB*, 1919, **50**, 133.

YATES, FRANCES A. (1). *Giordano Bruno and the Hermetic Tradition.* Routledge & Kegan Paul, London, 1964. Rev. W. P[agel], *AX*, 1964, **12**, 72.

YATES, FRANCES A. (2). 'The Hermetic Tradition in Renaissance Science.' Art. in *Art, Science and History in the Renaissance*, ed. C. S. Singleton. Johns Hopkins Univ. Press, Baltimore, 1968, p. 255.

YATES, FRANCES A. (3). *The Rosicrucian Enlightenment.* Routledge & Kegan Paul, London, 1972.

YEN CHI (1). 'Ancient Arab Coins in North-West China.' *AQ*, 1966, **40**, 223.

YETTS, W. P. (4). 'Taoist Tales; III, Chhin Shih Huang's Ti's Expeditions to Japan.' *NCR*, 1920, **2**, 290.

YOUNG, S. & GARNER, SIR HARRY M. (1). 'An Analysis of Chinese Blue-and-White [Porcelain]', with 'The Use of Imported and Native Cobalt in Chinese Blue-and-White [Porcelain].' *ORA*, 1956 (n. s.), **2** (no. 2).

YOUNG, W. C. (1) (ed.). *Sex and Internal Secretions.* 2 vols. Williams & Wilkins, Baltimore, 1961.

YÜ YING-SHIH (2). 'Life and Immortality in the Mind of Han China.' *HJAS*, 1965, **25**, 80.

YUAN WEI-CHOU. See Wei Chou-Yuan.

YULE, SIR HENRY (1) (ed.). *The Book of Ser Marco Polo the Venetian, concerning the Kingdoms and Marvels of the East, translated and edited, with Notes, by H. Y....*, 1st ed. 1871, repr. 1875. 2 vols. ed. H. Cordier. Murray, London, 1903 (reprinted 1921). 3rd ed. also issued Scribners, New York, 1929. With a third volume, *Notes and Addenda to Sir Henry Yule's Edition of Ser Marco Polo*, by H. Cordier. Murray, London, 1920.

YULE, SIR HENRY (2). *Cathay and the Way Thither; being a Collection of Mediaeval Notices of China.* 2 vols. Hakluyt Society Pubs. (2nd ser.) London, 1913–15. (1st ed. 1866.) Revised by H. Cordier, 4 vols. Vol. 1 (no. 38), *Introduction; Preliminary Essay on the Intercourse between China and the Western Nations previous to the Discovery of the Cape Route.* Vol. 2 (no. 33), *Odoric of Pordenone.* Vol. 3 (no. 37), *John of Monte Corvino and others.* Vol. 4 (no. 41), *Ibn Baṭṭūṭa and Benedict of Goes.* (Photo-litho reprint, Peiping, 1942.)

YULE, H. & BURNELL, A. C. (1). *Hobson-Jobson; being a Glossary of Anglo-Indian Colloquial Words and Phrases....* Murray, London, 1886.

YULE & CORDIER. See Yule (1).

ZACHARIAS, P. D. (1). 'Chymeutike, the real Hellenic Chemistry.' *AX*, 1956, **5**, 116. Based on Stephanides (1), which it expounds.

ZIMMER, H. (1). *Myths and Symbols in Indian Art and Civilisation*, ed. J. Campbell. Pantheon (Bollingen), Washington, D.C., 1947.

ZIMMER, H. (3). 'On the Significance of the Indian Tantric Yoga.' *ERYB*, 1961, **4**, 3, tr. from German in *ERJB*, 1933, **1**.

ZIMMER, H. (4). 'The Indian World Mother.' *ERYB*, 1969, **6**, 70 (*The Mystic Vision*, ed. J. Campbell). Tr. from the German in *ERJB*, 1938, **6**, 1.

ZIMMERN, H. (1). 'Assyrische Chemische-Technische Rezepte; insbesondere f. Herstellung farbiger glasierter Ziegel, im Umschrift und Übersetzung.' *ZASS*, 1925, **36** (N.F. **2**), 177.

ZIMMERN, H. (2). 'Babylonian and Assyrian [Religion].' *ERE*, vol. ii, p. 309.

ZONDEK, B. & ASCHHEIM, S. (1). 'Hypophysenvorderlappen und Ovarium; Beziehungen der endokrinen Drüsen zur Ovarialfunktion.' *AFG*, 1927, **130**, 1.

ZURETTI, C. O. (1). *Alchemistica Signa; Glossary of Greek Alchemical Symbols.* Vol. 8 of Bidez, Cumont, Delatte, Heiberg et al. (1).

ZURETTI, C. O. (2). *Anonymus 'De Arte Metallica seu de Metallorum Conversione in Aurum et Argentum'* [early +14th cent. Byzantine]. Vol. 7 of Bidez, Cumont, Delatte, Heiberg et al. (1), 1926.

ADDENDA TO BIBLIOGRAPHY C

ABRAMS, S. I. (1). 'Synchronicity and Alchemical Psychology.' Unpub. book.

ADAMS, F. (2) (tr.). *The Seven Books of Pauius Aegineta, translated from the Greek, with a Commentary embracing a Complete View of the Knowledge of the Greeks, Romans and Arabians on all subjects connected with Medicine and Surgery.* 3 vols. Sydenham Society, London, 1844–7.

ADLER, JEREMY (1). *A Study in the Chemistry of Goethe's 'Die Wahlverwandtschaften'.* Inaug. Diss. London (Westfield College), 1974.

ALI, MAULANA MUHAMMAD (1). *A Manual of Hadith; Arabic Text and English Translation.* Curzon, London, 1977.

ALLCHIN, F. R. (1). 'Stamped *tangas* and Condensers (Receivers); Evidence of Distillation at Sheikhān Dherī [near Charsala in the Vale of Peshawar, the site of the ancient city of Pushkalāvatī, sister capital to Taxila in Gandhāra].' MS. in the press.

ALLEAU, RENÉ (1) (ed.). *L'Alchémie des Philosophes* [a florilegium]. (Passages from Arabic translated by Y. Marquet & V. Monteil, from Indian languages by G. Mazars, and from Chinese by M. Kaltenmark & K. Schipper.) Pr. pr., *Art et Valeur*, Paris, 1978.

ALLEGRO, J. M. (1). *The Dead Sea Scrolls.* Penguin (Pelican), London, 1956, 2nd ed. 1958, 1959. Repr. 1961.

ANDERSSON, J. G. (6). 'Prehistoric Sites in Honan.' *BMFEA*, 1947, **19**, 1.

ANDRIEU, M. (1). '*Immixtio et Consecratio*'; la Consécration par Contact dans les Documents Liturgiques du Moyen Âge.* Picard, Paris, 1924. (Univ. de Strasbourg, Bibl. de l'Inst. de Droit Canonique, no. 2.)

ANON. (117). 'The Study of a Body Two Thousand Years Old [the Lady of Tai, d. *c.* − 166].' *CREC*, 1973, **22** (no. 10), 32. (Cf. pt. 2, p. 304.)

ANON. (145). 'Volcanoes [in Sinkiang, Inner Mongolia, Chilin, Heilungchiang, Shansi and Yunnan].' *CREC*, 1974, **23** (no. 8), 40.

ANON. (146). 'Pottery Warriors [life-size figurines near the tomb of Chhin Shih Huang Ti] and another Soft Body [incorrupt corpse of a magistrate buried in the Early Han period, − 167; tomb no. 168 at Feng-huang Shan].' *EHOR*, 1975, **14** (no. 4), 52.

ARDAILLON, E. (1). *Les Mines de Laurion dans l'Antiquité* (Inaug. Diss., Paris). Thorin, Paris, 1897. Repr. in Kounas (1). (Cf. pt. 2, pp. 41, 218.)

ARMENGAUD, A. (1). 'Population in Europe, +1700 to 1914.' Art. in *Fontana Economic History of Europe*, ed. C. M. Cipolla, vol. 3, p. 22.

ARNTZ, H. (1). *Weinbrenner; die Geschichte vom Geist des Weines.* Seewald, Stuttgart, 1975 (for the Asbach Distillery, Rüdesheim-am-Rhein).

ARTEMENKO, I., BIDZILIA, V., MOZOLEVSKY, B. & OTROSHCHENKO, V. (1). 'Excavations of Scythian *kurgans* (burial-places); the Golden Cup of Gaimanov; Idyll on a Royal Breastplate; a Horse's Finery capped by a Goddess of the Chase, etc.' With colour plate of panther Ouroboros. UNESC, 1976, **29** (no. 12), 17.

BAILEY, SIR HAROLD (2). 'Trends in Iranian Studies.' *MRDTB*, 1971, **29**, 1. (Cf. pt. 2, p. 116 (g), another statement on the etymology of *soma-haoma*.)

BAILEY, SIR HAROLD (3). 'The Range of the Colour *zar-* in Khotan Saka Texts.' Art. in *Mémorial Jean de Menasce*, 1974, ed. Gignoux & Tafazzoli, p. 369. (On colour names in relation to *soma-haoma*.)

DE BARRIOS, V. B. (1). 'A Guide to Tequila, Mezcal and Pulque.' Minutiae Mexicana, Mexico City, 1971.

BECKMANN, J. (1). *A History of Inventions, Discoveries and Origins.* German ed. 5 vols. 1786 to 1805. 4th ed., 2 vols. tr. by W. Johnston, Bohn, London, 1846. Enlarged ed., 2 vols. Bell & Daldy, London, 1872. Bibliography in John Ferguson (2).

BENET, SULA (1). *The Abkhazians, Long-living People of the Caucasus.* Holt, Rinehart & Winston, New York, 1974.

BENFEY, O. T. (3). 'How not to Die from the Elixir of Life; or, were the Chinese first again? Kidney Stones *vs.* Mercury Poisoning; Eat Spinach with your Tuna; Popeye to the Rescue; and Another Research Problem for *Chemistry*'s Readers.' *CHEM*, 1974, **47** (no. 7), 2 (Editorial).

BENVENISTE, E. (3). *Les Mages dans l'ancien Iran.* Maisonneuve, Paris, 1938. (Publications de la Soc. des Études Iraniennes, no. 15.)

BESSMERTNY-HEIMANN, B. (1) 'L'Esprit Moderne chez Bernard Palissy.' *A*, 1936, **18**, 166. (The attack on aurifaction in his 'Discours Admirables...', +1580.)

BETHE, H. A. (1). 'The Carbon-Nitrogen Cycle as Energy-Source of Stars.' *PHYR*, 1939, **55**, 103, 434.

BEVAN, E. R. (4). 'The Ptolemaic Dynasty.' Contribution to *A History of Egypt*, vol. 4, ed. W. M. Flinders Petrie. Methuen, London, 1927.

BEYER, S. (1). *The Cult of Tārā; Magic and Ritual in Tibet.* Univ. California Press, Berkeley and Los Angeles, 1973. (Includes material on alchemy in Tibet.)

BIANCHI, UGO (1). *Selectrd Essays on Gnosticism, Dualism and Mysteriosophy*. Brill, Leiden, 1978.

BISHOP, E. (1) (ed.). *Indicators*. Pergamon, Oxford, 1973. Contains an introductory chapter by E. Rancke-Madsen on the history of acid–base indicators since the 17th century.

BLACKER, C. & LOEWE, M. (1) (ed.). *Ancient Cosmologies*. Allen & Unwin, London, 1975.

BLOCH, M. R. (5). 'Two Ancient Saltpetre Plants at the Dead Sea.' Introduction to Report on the Excavations at Um Baraque [on the Western Coast of the Dead Sea], by Gichon. In the press.

BLYTH, R. H. (1). 'Mushrooms in Japanese Verse.' *TAS/J*, 1973 (3rd ser.), **11**, 93.

BOAS, FRANZ (1). *Race, Language and Culture*. Macmillan, New York, 1940, 1948. Collier-Macmillan, London; Free Press, New York, 1966.

BOECKH, AUGUST (1). *A Dissertation on the Silver Mines of Laurion in Attica* (1815). Repr. in Kounas (1). (Originally part of Boeckh's *The Public Economy of Athens*, tr. from the German by G. C. Lewis, 2nd ed., Parker, London, 1842, pp. 615–74.)

BONELLI, M. L. RIGHINI & SHEA, W. R. (1), (ed.). *Reason, Experiment and Mysticism in the Scientific Revolution*. Science History Publications, New York, 1975.

BORNKAMM, G. (1) (tr.). 'The "Acts of Thomas"'. In Hennecke & Schneemelcher (1), Eng. ed., vol. 2, p. 425.

BOSWORTH, C. E. (1). *The Mediaeval Islamic Underworld; the Banū Sāsān in Arabic Society and Litera-ture. Pt. 1, The Banū Sāsān in Arabic Life and Lore. Pt. 2, The Arabic Jargon Texts, the Qaṣīda Sāsāniyyas of Abū Dulaf and Ṣāfī al-Dīn*. Brill, Leiden, 1976.

BOSWORTH, C. E. (3) (ed.). *Iran and Islam* (Memorial Volume for Vladimir Minorsky). Edinburgh Univ. Press, Edinburgh, 1971.

BOXER, C. R. (5). 'A Note on the Interaction of Portuguese and Chinese Medicine at Macao and Peking in the +16th to +18th Centuries.' *BILCA*, 1974, **8**, 33. (Contains biographical material on J. F. Vandermonde, cf. pt. 2, pp. 160–1. He was appointed medical officer to the city of Macao in 1723 and returned to France in 1732.)

BRAUN, WERNER (1). *Studien zum Ruodlieb; Ritterideal, Erzählstruktur und Darstellungs-stil*. de Gruyter, Berlin, 1962. (Quellen und Forschungen z. Sprach- u. Kulturgesch. d. germanischen Völker, no. 131; N.F., no. 7.)

BREHM, E. (1). 'Roger Bacon's Place in the History of Alchemy.' *AX*, 1976, **23**, 53.

BROCKELMANN, C. (3). *Grundriss d. vergleichenden Grammatik der Semitischen Sprachen*. Berlin, 1908.

BROCKELMANN, C. (4). *Kurzgefasste vergleichenden Grammatik der Semitischen Sprachen*. Berlin, 1908.

BRUMAN, H. J. (1) 'The Asiatic Origin of the Huichol Still.' *GR*, 1944, **34**, 418.

BRUMAN, H. J. (2). 'Early Coconut Culture in Western Mexico.' *HAHR*, 1945, **25**, 212.

BRUMAN, H. J. (3). 'Aboriginal Drink Areas in New Spain.' Inaug. Diss., Berkeley, Calif. 1940.

BRUN, VIGGO (1). *Sug, the Trickster who Fooled the Monk; a Northern Thai Tale with Vocabulary*. Studentlitteratur, Lund; Curzon, London, 1976. (Scandinavian Institute of Asian Studies Monograph Ser. no. 27.)

BUDGE, E. A. WALLIS (2) (tr.). *The Monks of Kublai Khan; or, the History of the Life and Travels of Rabban Sawma and Markos* [Marqos Bayniel]. (Tr. from Syriac.), RTS, London, 1928.

BULLING, A. (15). 'The "Guide of the Souls" Picture in the Western Han Tomb in Ma Wang Tui near Chhangsha.' *ORA*, 1974, **20**, 1.

BÜRGEL, C. (1). '*Adab* und i'*Kidtal* in al-Ruhāwī's *Abab al-Tabīb*.' *ZDMG*, 1967, **117**, 90.

BURNETT, C. S. F. (1). 'The Legend of the Three Hermes and Abū Ma'shar's *Kitāb al-Ulūf* (Book of the Thousands) in the Latin Middle Ages.' *JWCI*, 1976, **39**, 231.

BURNETT, C. S. F. (2). 'Arabic into Latin in +12th-Century Spain; the Works of Hermann of Carinthia.' *MLJ*, 1978, **13**, 100.

BURSTEIN, S. R. (1). 'Papers on the Historical Background of Gerontology.' *GERI*, 1955, **10**, 189, 328, 536.

CAMPBELL, DONALD (1). *Arabian Medicine and its Influence on the Middle Ages*. 2 vols. Trübner, London, 1926.

CAPON, E. & McQUITTY, W. (1). *Princes of Jade*. Cardinal (Nelson), London, 1973. (Cf. pt. 2, p. 303, the tombs and jade body cases of Liu Shêng and Tou Wan; also the tomb of the Lady of Tai, pp. 8 ff., 162 ff.)

CASTANEDA, CARLOS (1). *The Teachings of Don Juan; a Yaqui Way of Knowledge*. Ballantine Books, New York, 1974, orig. pub. California Univ. Press, 1968.

CASTANEDA, CARLOS (2). *A Separate Reality; Further Conversations with Don Juan*. Simon & Schuster (Touchstone), New York, 1971; Pocket Books, New York, 1973.

CASTANEDA, CARLOS (3). *Journey to Ixtlan; the Lessons of Don Juan*. Simon & Schuster, New York, 1972; Pocket Books, New York, 1974.

CHANG, H. C. (CHANG HSIN-CHANG). *See* Chang Hsin-Tshang.

CHANG HSIN-TSHANG (1). *Allegory and Courtesy in [Edmund] Spenser; a Chinese View.* Edinburgh Univ. Press, Edinburgh, 1956. Rev. Liu Jung-Ju (J. J. Y. Liu), *JRAS*, 1956, 87.

CHANG HSIN-TSHANG (2) (tr. and comm.). *Chinese Literature; Popular Fiction and Drama.* University Press, Edinburgh, 1973.

CHANG HSIN-TSHANG (4), (tr. and comm.). *Chinese Literature; Nature Poetry.* Edinburgh University Press, Edinburgh, 1977.

DE LA CHARME, A. (1). *Confucii Chi-King sive Liber Carminum. Ex Latina P. Lacharmi Interpretatione editit J. Mohl* (The Latin translation of the *Shih Ching* (Book of Odes) originally made in 1733). Stuttgart and Tübingen, 1830.

CHŎN SANGUN (1). *Science and Technology in Korea; Traditional Instruments and Techniques.* M.I.T. Press, Cambridge, Mass., 1974. (M.I.T. East Asian Science Series, no. 4.) Based largely on Chŏn Sangun (2).

CHURCHILL, S. J. A. (1). 'The Alchemist [comments on Mirza Fath-ali Akhunzade's play "The History of Mullah Ibrāhīm Khalil the Alchemist", written in 1851].' *JRAS*, 1886, **18** (N.S.), 463. See also Barbier de Meynard (3) and G. le Strange (5).

CIPOLLA, C. M. (3). *The Economic History of World Population.* Penguin (Pelican), London, 1962; 7th ed., 1978.

CIPOLLA, C. M. (4) (ed.). *The Fontana Economic History of Europe.* Collins–Fontana, London, 1972, repr. 1976.
    Vol. 1.   *The Middle Ages.*
    Vol. 2.   *The 16th and 17th centuries.*
    Vol. 3.   *The Industrial Revolution.*
    Vol. 4.   *The Emergence of Industrial Societies.*
    Vol. 5.   *The 20th Century.*
    Vol. 6.   *Contemporary Economies.*

COLLIER, H. B. (2). 'The Black Copper of Yunnan [*wu thung*].' *JCE*, 1940, vol. 19. 'X-Ray Fluorescence Analysis of Black Copper of Yunnan.' *NW*, 1977, **64**, 484.

COMFORT, ALEX (1). *The Biology of Senescence.* Routledge & Kegan Paul, London, 1956.

CRISCIANI, C. (1). 'The Conception of Alchemy as expressed in the *Pretiosa Margarita* of Petrus Bonus of Ferrara.' *AX*, 1973, **20**, 165.

CRISCIANI, C. (2) (ed.). *The Preziosa Margarita Novella (New Pearl of Great Price) by Pietro Bono da Ferrara* [Italian translation of the +15th century of Petrus Bonus' book of +1330]. Nuova Italia, Firenze, 1976. (Centro di Studi del Pensiero Filosofico del Cinquecento e del Seicento in Relazione ai Problemi della Scienza Pubs.) Rev. A. G. Keller, *AX*, 1978, **25**, 148.

DANI, A. H. (1). 'Shaikhān Dherī Excavations, 1963 and 1964.' *APAK*, 1966, **2**, 17–214.

DAVIDSON, H. R. ELLIS (1). 'Scandinavian Cosmologies.' Art. in *Ancient Cosmologies*, ed. C. Blacker & M. Loewe. Allen & Unwin, London, 1975, p. 172.

DAVIES, D. (2). *The Centenarians of the Andes.* Barrie & Jenkins, London 1975. Rev. A. Comfort, *N*, 1975 **258** Rev. Suppl., 41.

DAVIES, D. (3). 'Rediscovering Islamic Science.' *N*, 1976, **260**, 474.

DAVIES, NIGEL (1). *The Aztecs; a History.* Abacus (Sphere), London, 1977.

DAY, JOAN (1). *Bristol Brass; the History of the Industry.* David & Charles, Newton Abbot, 1973.

DEBUS, A. (22). 'The Chemical Debates of the +17th Century; the Reaction to Robert Fludd and Jean Baptiste van Helmont.' Art. in *Reason, Experiment and Mysticism in the Scientific Revolution*, ed. Bonelli & Shea (1), p. 19.

DEBUS, A. (23). 'The Chemical Philosophers; Chemical Medicine from Paracelsus to van Helmont.' *HOSC*, 1974, **12**, 235.

DEBUS, A. (24). 'Alchemy.' Art. in *Dictionary of the History of Ideas*, ed. P. P. Wiener. Scribner, New York, 1973, vol. 1, p. 27.

DEBUS, A. G. (25). 'The Pharmaceutical Revolution of the Renaissance.' *CMED*, 1976, **11**, 307.

DEBUS, A. G. (26). 'Chemistry, Pharmacy and Cosmology; a Renaissance Union.' *PIH*, 1978, **20** (no. 4), 127.

DECAISNE, J. (1). 'Note sur les Deux Espèces de Nerprun [buckthorn] qui fournissent le Vert de Chine.' *CRAS*, 1857, **44**. Repr. in Rondot, Persoz & Michel (1), pp. 139ff.

DEMACHY, J. F. (1). 'L'Art du Distillateur des Eaux-Fortes.' Art. in *Description des Arts et Métiers...*, vol. 14. Académie des Sciences, Paris, 1773.
    Also in *Recueil des Dissertations Physico-Chimiques présentées à différentes Académies.* Amsterdam and Paris, 1774.

DERRETT, J. D. M. (1). 'The History of "Palladius on the Races of India and the Brahmins".' *CLMED*, 1960, **21**, 64.

DIGBY, SIR KENELM (1). *A Choice Collection of Rare Chymical Secrets and Experiments in Philosophy. As also Rare and unheard-of Medicines, Menstruums, and Alkahests; with the True Secret of Volatilising* the fixt Salt of Tartar. Collected and Experimented by the Honourable and truly Learned Sir Kenelm Digby, Kt., Chancellour to Her Majesty the Queen-Mother. Hitherto kept Secret since his Decease but now Published for the good and benefit of the Publick, by George Hartman, London: Printed for the Publisher, and are to be Sold by the Book-Sellars of London, and at his own House in Hewes Court in Black-Fryers. London, 1682. See Dobbs (3).

DIGUET, L. (1). 'La Sierra du Nayarit et ses Indigènes.' *NAMSL*, 1899, **9**, 571.

DOBBS, B. J. T. (2). 'Studies in the Natural Philosophy of Sir Kenelm Digby; II, Digby and Alchemy.' *AX*, 1973, **20**, 143.

DOBBS, B. J. T. (3). 'Studies in the Natural Philosophy of Sir Kenelm Digby; III, Digby's Experimental Alchemy—the Book of Secrets.' *AX*, 1974, **21**, 1.

DOBBS, B. J. T. (4). *The Foundations of Newton's Alchemy; or, 'The Hunting of the Greene Lyon'.* Cambridge, 1975. Cf. Figala (1). Revs. D. T. Whiteside, *ISIS*, 1977, **68**, 116; M. B. Hall, *BJHOS*, 1977, **10** (no. 3), 262.

DODDS, E. R. (2). *The Ancient Concept of Progress; and other Essays on Greek Literature and Belief.* Oxford, 1973.

DODGE, BAYARD (1) (tr.). *The* Fihrist al-'Ulūm (*Bibliography of the Sciences*) [by Ibn abī Ya'qub al-Nadīm al-Warrāq al-Baghdā. (+987)]; a +10th-Century Survey of Muslim Culture. 2 vols. Columbia University Press, New York, 1968.

DŌKE TATSUMASA (1). 'Udagawa Yōan; a Pioneer Scientist of Early Nineteenth-Century Feudal Japan.' *JSHS*, 1973, **12**, 99.

DORESSE, J. (1). *Les Livres Secrets des Gnostiques d'Égypte.* Plon, Paris, 1958–.
Vol. 1. *Introduction aux Écrits Gnostiques Coptes découverts à Khénoboskion.*
Vol. 2. *'L'Évangile selon Thomas', ou 'les Paroles Secrètes de Jésus'.*
Vol. 3. *'Le Livre secret de Jean'; 'l'Hypostase des Archontes' ou 'Livre de Nōréa'.*
Vol. 4. *'Le Livre Sacré du Grand Espirt Invisible' ou 'Evangile des Égyptiens'; 'l'Épître d'Eugnoste le Bienheureux'; 'La Sagesse de Jésus'.*
Vol. 5. *'L'Évangile selon Philippe'.*
Eng. tr. of the first two volumes, by P. Mairet: *The Secret Books of the Egyptian Gnostics; an Introduction to the Gnostic Coptic Manuscripts discovered at Chenoboskion; with an English Translation and Critical Evaluation of the 'Gospel according to Thomas'.* Hollis & Carter, London, 1960.

DOSHI, S. L. (1). *The Bhīls; between Society Self-awareness and Cultural Synthesis.* New Delhi, 1971.

DRAPER, J. W. (1). *A History of the Conflict between Religion and Science.* 19th ed. London, 1888.

DRAPER, J. W. (2). *The Intellectual Development of Europe.* 2 vols. Bell, London, 1891.

DRONKE, P. (3). 'Tradition and Innovation in Mediaeval Western Colour-Imagery.' *ERYB*, 1972, **41**, 51.

DUNCAN, A. M. (3). 'William Keir's *De Attractione Chemica* (1778), and the Concepts of Chemical Saturation, Attraction and Repulsion.' *ANS*, 1967, **23**, 149.

DUNCKER, H. G. L. & SCHEIDEWIN, F. G. (1). *Sancti Hippolyti Episcopi et Martyris 'Refutatione Omnium Haeresium' Librorum Decem quae Supersunt...* (Greek and Latin texts). Dieterich, Göttingen, 1859.

DUNLOP, D. M. (9). 'The *Mudhākarāt fi 'Ilm al-Nujūm* (Dialogues on Astrology) attributed to Abū Ma'shar al-Balkhī (Albumasar).' Art. in *Iran and Islam* (Memorial Volume for Vladimir Minorsky), ed. C. E. Bosworth, p. 229.

DURÁN, DIEGO (1). *Historia de las Indias de Nueva España e Islas de la Tierra Firme.* 2 vols. ed. A. M. Garibay. Porrúa, Mexico City, 1967 (published from the autograph MS. in Madrid). Eng. tr. *The Aztecs...* by D. Heyden & F. Horcasitas Orion, New York, 1964.

EDELSTEIN, S. M. (1). 'The Origins of Dry Cleaning.' *ADR*, 1957, **46**, (no. 1), 1. Repr. in Edelstein (2), p. 7.

EDELSTEIN, S. M. (2). 'Historical Notes on the Wet-Processing Industry, II' (eleven collected papers). *Amer. Dyestuff Reporter*, New York, 1964.

EDELSTEIN, S. M. (3). 'Lo-Kao—the story of Chinese Green.' *ADR*, 1957, **46** (no. 12), 433. Repr. in Edelstein (2), p. 15. (*Rhamnus utilis* and *Rhamnus chlorophorus*.)

EICHHORN, W. (12). 'Die Wiedereinrichtung der Staatsreligion im Anfang der Sung-Zeit.' *MS*, 1964, **23**, 205–63.

EISSFELDT, O. (1). *The Old Testament; an Introduction—including the Apocrypha and Pseudepigrapha, and also Works of similar type from Qumran. The History of the Formation of the Old Testament.* tr. from the German by P. R. Ackroyd. Blackwell, Oxford, 1965.

ELIVE, I. B. (1). *Carte de la Perse, de l'Armémie, de la Natolie et de l'Arabie.* Elive, Amsterdam, 1792.

ELWIN, VERRIER (1). *The Baiga.* London, 1939.

EVANS, R. J. W. (1). *Rudolf II and his World; a Study in Intellectual History, +1576 to +1612.* Oxford, 1973.

FARRAND, W. R. (1). 'Frozen Mammoths and Modern Geology.' *SC*, 1961, **133**, 729. (Cf. pt. 2, p. 304.)

FELICIANO, R. T. (1). 'Illicit Beverages.' *PHJS*, 1926, **29**, 465.

FERGUSON, JOHN (4). 'Some Early Treatises on Technological Chemistry.' *PRPSG*, 1888, **19**, 126. With five supplements: *PRPSG*, 1894, **25**, 232; 1909, **41**, 113; 1911, **43**, 232; 1912, **44**, 1; 1916, **48**, 1.

FIGALA, K. (1). 'Newton as Alchemist.' (Essay-review of B. J. T. Dobbs' *Foundations of Newton's Alchemy . . .*) *HOSC*, 1977, **15**, 102.

FOERSTER, W. (1), assisted by E. Haenchen, M. Krause & K. Rudolph. *Die Gnosis.* 2 vols. Artemis, Zürich, 1969–71. (Bibl. d. alten Welt, Reibe Antike und Christentum.) Eng. tr. by P. W. Coxon & K. H. Kuhn, ed. R. McLachlan Wilson. *Gnosis; a Selection of Gnostic Texts*, 2 vols. Oxford, 1972–4.

FRANKOWSKA, M. (1). Scientia *as interpreted by Roger Bacon.* tr. from Polish. U.S. Dept. of Commerce, Environmental Science Services Administration, and National Science Foundation, Washington, D.C.; with the Sci. Pubs. Foreign Cooperation Centre of the Central Institute for Scientific, Technical and Economic Information, Warsaw, 1971.

FREND, W. H. C. (1). Some Cultural Links between India and the West in the Early Christian Centuries.' *TTT*, 1968, **2**, 306. Repr. in Frend (2), no. 7.

FREND, W. H. C. (2). 'Religion Popular and Unpopular in the Early Christian Centuries.' (Collected essays.) Variorum Reprints, London, 1976.

FUKUI KŌJUN (1). 'A Study of the *Chou I Tshan Thung Chhi.*' *ACTAS*, 1974, no. 27, 19.

VON FÜRER-HAIMENDORF, G. (1). *The Chenchus, Jungle Folk of the Deccan [Eastern India, in Andhra Pradesh].* London, 1943.

GAERTNER, BERTIL E. (1). 'The Pauline and Johannine Idea of "to know God" against the Hellenistic Background; the Greek philosophical Principle "Like by Like" in Paul and John.' *NTS*, 1968, **14**, 209.

GAINES, A. M. & HANDY, J. L. (1). 'The Decomposition of Jade by ammonia from decaying Human Bodies.' *N*, 1975, **253**, 433. Abstract in *Times*, 11 Feb. 1975.

GASTER, M. (1). 'Alchemy.' Art. in the *Jewish Encyclopaedia*. 12 vols. ed. Funk & Wagnall, New York, 1901–6, repr. 1907, vol. 1, p. 328.

GERSHEVITCH, I. (2). 'An Iranianist's View of the Soma Controversy.' Art. in *Mémorial Jean de Menasce*, 1974, ed. Gignoux & Tafazzoli, p. 45.

GESNER, CONRAD (3). *The Treasure of Euonymus, conteyning the Wonderfull hid Secretes of Nature touching the most apt Formes to prepare and destyl Medicines, for the Conservation of Helth: as Quintessences, Aurum Potabile, Hippocras, Aromaticall Wynes, Balmes, Oyles, Perfumes, garnishing Waters, and other manifold excellent Confections. Whereunto are joyned the formes of Sundry apt fornaces and vessels, required in this art.* Translated (with great diligence and laboure) out of Latin by Peter Morwyng, Fellow of Magdalene Colledge in Oxford. Daye, London, 1559, 1565. 3rd ed., tr. George Baker, London, 1570, 1576. 4th ed., tr. George Baker, London, 1599.

GETTENS, R. J., FELLER, R. L. & CHASE, W. T. (1). 'Vermilion and Cinnabar.' *SCON*, 1972, **17**, 45.

GHIRSHMAN, R. (2). 'Essai de Recherche Historico-archéologique' [digest of the work of S. P. Tolstov]. *AA*, 1952, **16**, 209 and 292.

GHOSH, A. (1) '(Further Excavations at) Taxila (Sirkap), 1944–5.' *AIND*, 1947, No. 4, 41–84.

GIBERT, L. (1). *Dictionnaire Historique et Géographique de la Mandchourie.* Missions Étrangères, Hongkong, 1934.

GIGNOUX, P. & TAFAZZOLI, A. (1) (ed.). *Mémorial Jean de Menasce.* Imp. Orientaliste, Louvain, 1974. (Pub. Fondation Culturelle Iranienne, no. 185.)

GLASS, D. V. & EVERSLEY, D. E. C. (1), (ed.). *Population in History; Essays in Historical Demography.* Arnold, London, 1965.

GRAHAM, A. C. (8) (tr.). *Poems of the Late Thang, translated with an introduction.* Penguin, London, 1965.

GRANET, M. (1). *Danses et légendes de la Chine Ancienne.* 2 vols. Alcan, Paris, 1926.

GRANT, EDWARD (1) (ed. and tr.). *A Source-Book of Mediaeval Science.* Harvard University Press, Cambridge, Mass., 1973. (Contains translations of essential passages on alchemy from Ibn Sīnā (Avicenna), Petrus Bonus, Albertus Magnus, Thomas Aquinas and Albert of Saxony.)

GRIFFITH, G. TALBOT (1). *Population Problems of the Age of Malthus.* Cambridge, 1926.

GRMEK, M. D. (3). 'Le Vieillissement et la Mort.' Art. in *Biologie*. Encyclopédie de la Pléiade, Paris, 1965, p. 779.

GRYAZNOV, M. P. (1). 'Horses for the Hereafter.' With photograph of a bronze Ouruboros of mammalian type found at Arzhan. *UNESC*, 1976, **29** (no. 12), 38.

GUERLAC, HENRY (2). *John Mayow and the Aerial Nitre.* Actes du VIIe Congrès International d'Histoire des Sciences, Jerusalem, 1953, p. 332.

HALOUN, G. & HENNING, W. B. (1). 'The "Compendium of the Doctrines and Styles of the Teaching of Mani, the Buddha of Light" [*Mo-Ni Kuang Fo Chiao Fa I Lüeh*, +731, S/3969].' *AM*, 1953, (N.S.) **3**, 184.

HALSELL, GRACE (1). *Los Viejos*. (A study of the centenarians and super-centenarians of Ecuador.) Rodale, Emmaus, Pa. 1974.

HARDEN, A. & YOUNG, W. J. (1). 'The Alcoholic Ferment of Yeast-Juice; III, the Function of Phosphates in the Fermentation of Glucose by Yeast-Juice.' *PRSB*, 1908, **80**, 299.

HARIG, G. (1). *Bestimmung der Intensität im medizinischen System Galens*. Berlin, 1974.

HARNER, M. J. (1). *Hallucinogens and Shamanism*. Oxford, 1973. Rev. *SAM*, 1973, **229** (no. 4), 129. (Cf. pt. 2, p. 116.)

HARRIS, J. R. (1) *Lexicographical Studies on Ancient Egyptian Minerals*. Berlin, 1961. (Veröffentl. d. Inst. f. Orientforsch. d. deutsche Akad. d. Wiss. Berlin, no. 54.)

HASKINS, C. H. (1). *Studies in the History of Mediaeval Science*. Harvard Univ. Press, Cambridge, Mass., 1927.

HASKINS, C. H. (2). *The Renaissance of the Twelfth Century*. Harvard Univ. Press, Cambridge, Mass., 1927, revised 1955. 6th reprint (paperback), 1976.

HEFFERN, R. (1). *Secrets of the Mind-Altering Plants of Mexico*. Pyramid, New York, 1974. (Cf. pt. 2, p. 116.)

HENNECKE, E. & SCHNEEMELCHER, H. (1) (ed.). *Neutestamentliche Apokryphen in deutscher Übersetzung*, 3rd revised edition, 2 vols. Mohr (Siebeck), Tübingen, 1959–64. Eng. tr., ed. R. McLachlan Wilson (3), 2 vols. Lutterworth, London, 1963–5.

HENNING, W. B. (2). 'The "Book of the Giants" [the version of the "Book of Enoch" in the Manichaean Canonical Scriptures].' *BLSOAS*, 1943, **11**, 52.

HENNING, W. B. (3). 'Ein Manichäische Henochbuch.' *SPAW/PH*, 1934 (no. 4), 27.

HENNING, W. B. (4). 'Neue Materialen z. Gesch. d. Manichäismus.' *ZDMG*, 1936, **90** (N.F. **15**), 1.

HERMELINK, H. (1). 'Die ältesten magischen Quadrate höherer Ordnung und ihre Bildungsweise.' *AGMN*, 1953, **42**, 199.

HO PING-YÜ (18). 'Chinese Alchemical and Medical Prescriptions; a Preliminary Study'. *Proc. 14th Internat. Congress of the History of Science, Tokyo, 1974. Abstracts*, p. 40. *Proceedings*, vol. 3, p. 295.

HOLLINGSWORTH, T. H. (1). (*a*) 'A Demographic Study of the British Ducal Families.' *POPST*, 1957, **11**, 4. Repr. in '*Population in History* . . .', ed. Glass & Eversley (1), p. 354. (*b*) Extended as: 'The Demography of the British Peerage', Suppl. to *POPST*, 1964, **18**, 1–108.

HOLLINGSWORTH, T. H. (2). *Historical Demography*. Hodder & Stoughton, London, 1969.

HOLMYARD, E. J. (5). '*Kitāb al- 'Ilm al-Muktasab fī Zirā'at al-Dhabab*' (*Book of Knowledge acquired concerning the Cultivation of Gold*), by *Abū'l Qāsim Muḥammad ibn Aḥmad al-Irāqī* [d. *c.* +1300]; *the Arabic text edited with a translation and introduction*. Geuthner, Paris, 1923. Repr. in *HAM*, 1977, **20**, 7–68, but reset, so that the indexes no longer correspond; add 6 or 7 to each entry.

HOLMYARD, E. J. (18). 'Alchemical Equipment [in the Mediterranean Civilisations and the Middle Ages]. Art. in *A History of Technology*, ed. C. Singer *et al*. Oxford, 1956, vol. 2, p. 731.

HOLMYARD, E. J. (19). 'A Romance of Chemistry.' (The *De Compositione Alchemiae* of +1144.) *CHIND*, 1925, **3**, 75, 105, 136, 272, 300, 327 (*JSCI*, **44**).

HUMMEL, A. W. (2) (ed.). *Eminent Chinese of the Chhing Period*. 2 vols. Library of Congress, Washington, 1944.

HUNG HSIN (1). 'A Treasure Lake [the Hungtsê Hu, in the Lower Huai River Valley in Chiangsu province].' *CREC*, 1974, **23** (no. 8), 46. (Cf. pt. 2, p. 121 on *Ganoderma lucidum*, now cultivated for pharmaceutical use in China; nephritis, neurasthenia, general tonic properties, no hallucinogens.)

HUSAIN, MIAN MOHD SIDDIQ (1). 'The Islamic Contribution to [Chemical] Medicine.' *JPMA*, 1953, **3** (no. 4). Chinese tr. by Chao Shih-Chhiu in *ISTC*, 1954, **6** (no. 3), 214.

HUXLEY, ALDOUS (1). *The Doors of Perception; and, Heaven and Hell*. Penguin, London, 1959; many times reprinted. The first orig. pub. Chatto & Windus, London, 1954, the second, Chatto & Windus, London, 1956. (Cf. pt. 2, p. 116.)

HYDE, MARGARET O. (1). *Mind Drugs*. McGraw-Hill, New York, 1972; 2nd ed., revised and enlarged, Simon & Schuster (Pocket Books), New York, 1973. (Cf. pt. 2, pp. 116, 150.)

IMAZEKI, ROKUYA (1). 'Japanese Mushroom Names.' *TAS/J*, 1973 (3rd ser.), **11**, 26.

IMAZEKI, ROKUYA & WASSON, R. G. (1). 'Kinpu; Mushroom Books of the Tokugawa Period.' *TAS/J*, 1973 (3rd ser.), **11**, 81.

JAHN, K. (1). 'China in der islamischen Geschichts-schreibung.' *AOAW/PH*, 1972, **108**, 63.

JAHN, K. (2). 'Wissenschaftliche Kontakte zwischen Iran und China in der Mongolenzeit.' *AOAW/PH*, 1969, **106**, 204.

JEON SANG-WOON. *See* Chŏn Sangun.

JUNG, C. G. (15). 'On the Psychology of the Trickster-Figure', in *Four Archetypes*. Routledge & Kegan Paul, London, 1972 (from vol. 9, pt. 1 of the Collected Works).

KALTENMARK, M. (5). 'Hygiène et Mystique en Chine.' *BSAC*, 1959 (no. 33), 21.

KALTENMARK, M. (6). 'L'Alchimie en Chine.' *BSAC*, 1960 (no. 37) 21.

KARLGREN, B. (1). 'Grammata Serica; Script and Phonetics in Chinese and Sino-Japanese.' *BMFEA*, 1940, **12**, 1. (Photographically reproduced as separate volume, Peiping, 1941.) Revised edition, *Grammata Serica Recensa*, Stockholm, 1957.

KELLY, ISABEL (1). 'Vasijas de Colima con Boca de Estribo.' *BINAH*, 1970, no. 42, 26.

KELLY, ISABEL (2). 'On the bifid and trifid vessels of the Capacha complex of Western Mexico (Colima region).' Art. in *Archaeology of West Mexico*, ed. B. Bell. Ajijic, Jalisco, Mexico, 1974.

KENNY, E. J. A. (1). 'Quaestiones Hydraulicae.' Inaug. Diss., Cambridge, 1936.

KERR, G. H. (1). *Okinawa; the History of an Island People*. Tokyo, 1958. (States on p. 40 that the voyages of exploration which discovered the Liu-Chhiu Islands in +607 and later were motivated by an interest in Phêng-Lai and the other isles of the holy immortals (cf. pt. 3, p. 18) on the part of the Sui emperor, Yang Ti (cf. pt. 3, p. 132). But *Sui Shu*, ch. 81, p. 10a, does not bear this out. It records that the first sighting was reported in +605, by a sea-captain, Ho Man, and that the first successful reconnaissance was made in +607 by a military officer, Chu Kuan, who was accompanied by the sailor. This was followed by an expeditionary force under Chhen Lêng in +609.)

KHAWAM, R. R. (1). 'Les Statues Animées dans les Mille et Une Nuits.' *AHES/AESC*, 1975, **30** (no. 5), 1084.

KIMURA EIICHI (1). 'Taoism and Chinese Thought.' *ACTAS*, 1974, no. 27, 1.

KOUNAS, DIONYSIOS A. (1) (ed.). *Studies on the Ancient Silver Mines at Laurion*. Coronado Press, Lawrence, Kansas, 1972. (Reprints by litho-offset Boeckh (1), Ardaillon (1) and Xenophon 'On the Revenues' (c. −355), tr. H. G. Dakyns; with an introduction by Kounas.)

KREBS, H. A. (1). 'Cyclic Processes in Living Matter.' *ENZ*, 1948, **12**, 88.

KREBS, H. A. (2). 'The Discovery of the Ornithine Cycle of Urea Synthesis.' *BCED*, 1973, **1** (no. 2), 19.

KROEBER, A. L. (3). 'The Ancient *Oikoumenē* as an Historic Culture Aggregate' (Huxley Memorial Lecture). *JRAI*, 1945, **75**, 9.

LACH, D. F. (5). *Asia in the Making of Europe*. 2 vols. in 5 parts. Univ. Chicago Press, Chicago and London, 1965–.
Vol. 1. The Century of Discovery.
Vol. 2. A Century of Wonder: pt. 1, The Visual Arts; pt. 2, The Literary Arts; pt. 3, The Scholarly Disciplines.

LAL, B. B. (1). 'Excavation at Hastināpura [near Meerut], and other Explorations in the Upper Gangā and Sutlej Basins, 1950–2; New Light on the Dark Age between the End of the Harappa Culture and the Early Historical Period.' *AIND*, 1954, No. 10/11, 4–151.

LEAF, A. & LAUNOIS, J. (1). 'Every Day is a Gift when you are over 100' (report on the centenarians of Abkhasia in the Caucasus, Hunza in the Himalayas, and Vilcabamba in the Andes). *NGM*, 1973, **143** (no. 1), 93.

LEGGE, F. (1) (tr.). '*Philosophumena; or, the Refutation of all Heresies*', formerly attributed to Origen but now to Hippolytus, Bishop and Martyr, who flourished about +220. 2 vols. SPCK, London, 1921. (Translations of Christian Literature; I, Greek Texts, no. 14.) For text see Duncker & Scheidewin (1).

LEICESTER, H. M. (2). *The Development of Biochemical Concepts from Ancient to Modern Times*. Harvard Univ. Press, Cambridge, Mass., 1974. (Harvard Monographs in the History of Science, no. 7.)

LEISS, W. (1). *The Domination of Nature*. Braziller, New York, 1972.

LESLIE, D. (2). 'The Problem of Action at a Distance in Early Chinese Thought' (discussion on lecture by J. Needham). *Actes du VIIe Congrès International d'Histoire des Sciences, Jerusalem 1953* (1954), p. 186.

LESLIE, D. (7). 'Les Théories de Wang Tch'ong [Wang Chhung] sur la Causalité.' Art. in *Mélanges offertes à Monsieur Paul Demiéville*. Paris, 1974, p. 179.

LEVEY, M. (10). 'Some Facets of Mediaeval Arabic Pharmacology.' *TCPP*, 1963, **30**, 157.

LEWIN, GÜNTER (1). *Die ersten Fünfzig Jahre der Sung-Dynastie in China; Beitrag zu einer Analyse des sozial-ökonomischen Formation...* Akademie, Berlin, 1973.

LI CHI (5). 'The Tuan-Fang Altar Set Re-examined.' *MMJ*, 1970, **3**, 51.

LI HUI-LIN (6). 'An Archaeological and Historical Account of *Cannabis* in China.' *ECB*, 1974, **28**, 437.

LI HUI-LIN (7). 'The Origin and Use of *Cannabis* in Eastern Asia; Linguistic-Cultural Implications.' *ECB*, 1974, **28**, 293.

LI HUI-LIN (8). *The Garden Flowers of China*. Ronald, New York, 1959. (Chronica Botanica series, no. 19.) [Correct from (1) in Vol. 5, pts. 2, 3.]

LINDROTH, S. H. (2). *Christopher Polhem och Stora Kopparberget; ett Bidrag till Bergsmekanikens Historia*. Almqvist & Wiksell, Upsala, 1951 (for Stora Kopparbergs Bergslags Aktiebolag).

LINS, P. A. & ODDY, W. A. (1). 'The Origins of Mercury Gilding.' *JARCHS*, 1975, **2**, 365.

LONGO, V. G. (1). *Neuropharmacology and Behaviour*. Freeman, New York, 1973. Rev. *SAM*, 1973, **229** (no. 4), 129. (Cf. pt. 2. p. 116.)

LONICERUS, ADAM (1). *Kreuterbuch. Kunstliche Conterfeytinge der Bäume, Stauden, Hecken, Kräuter, Getreydt, Gewürtze, etc., mit eygent liche Beschreibung derselben Namen...und derselben Gestalt, natürlicher Kraft und Wirckung. Sampt vorher gesetztem und Gantz auszführlich beschriebenem Bericht der schönen und nützlichen Kunst zu Destillieren...Nunmehr durch Petrum Uffenbach auf das allerfleissigst übersehen, corrigiert und verbessert.* Frankfurt, 1630.

LOTKA, A. J. (1). *Elements of Physical Biology.* Williams & Wilkins, Baltimore, 1925.

LU GWEI-DJEN & NEEDHAM, JOSEPH (5). *Celestial Lancets; a History and Rationale of Acupuncture and Moxibustion.* Cambridge, in the press.

LYONS, SIR HENRY (1). *The Royal Society, 1660–1940; a History of its Administration under its Charters.* Cambridge, 1944.

McDERMOT, VIOLET (1) (tr.). Pistis Sophia; *an English Rendering.* Brill, Leiden, 1977. (Nag Hammadi Studies, no. 9.)

McGUIRE, E. B. (1). *Irish Whiskey; a History of Distilling, the Spirit Trade and Excise Controls in Ireland.* Gill & McMillan, Dublin; Barnes & Noble, New York, 1973.

McINTYRE, L., SEIDLER, N. & SEIDLER, R. (1). 'The Lost Empire of the Incas.' *NGM*, 1973, **144** (no. 6), 729–87. (Includes an account of the perfect preservation by freezing, over five centuries, of the bodies of men and boys sacrificed at shrines from 17,000 to 20,000 ft altitude in the Andean range of mountains. Cf. pt. 2, p. 304.)

McKEOWN, T. & BROWN, R. G. (1). 'Medical Evidence related to English Population Changes in the +18th Century.' *POPST*, 1955, **9**, 119. Repr. in *Population in History...*, ed. Glass & Eversley (1), p. 285.

McMAHON, J. H. & SALMOND, S. D. F. (1) (tr.). *The 'Refutation of All Heresies' by Hippolytus; with Fragments from his Commentaries on Various Books of Scripture.* 2 vols. Clark, Edinburgh, 1868. (Ante-Nicene Christian Library, nos. 6 and 9.) For text see Duncker & Scheidewin (1).

MAHAFFY, J. P. (1). 'The Ptolemaic Dynasty.' Contribution to *A History of Egypt*, vol. 4, ed. W. M. Flinders Petrie. Methuen, London, 1899.

MAHDIHASSAN, S. (56). 'The Earliest Distillation Units of Pottery in Indo-Pakistan.' *PAKARCH*, 1972 (1975), **8**, 159.

MAHDIHASSAN, S. (57). 'A Legend attributed to the Caliph 'Umar and its Chinese Basis.' *ABRN*, 1976, **16**, 115.

MAHDIHASSAN, S. (58). 'Alchemy with the Egg as its Symbol.' *JAN*, 1976, **63**, 133.

MAHDIHASSAN, S. (59). 'The Bases of Alchemy.' *SHM*, 1977, 49.

MAHDIHASSAN, S. (60). *Indian Alchemy, or Rasayana, in the light of Asceticism and Geriatrics.* Institute of the History of Medicine and Medical Research, Tughlaqabad, New Delhi, 1977.

MAJOR, JOHN (1). 'The Cosmological Chapter in the *Huai Nan Tzu* Book (ch. 3).' Inaug. Diss. Harvard, 1975.

MAJUMDAR, R. C. (2). '[Indian] Medicine.' Art. in *Concise History of Science in India*, ed. Bose, Sen & Subbarayappa, New Delhi, 1971.

MANZALAOUI, M. (1). *The 'Secretum Secretorum' (of Pseudo-Aristotle) [incunabula and other later English versions].* 2 vols. Oxford, 1977.

DEL MAR, ALEXANDER (1). *A History of the Precious Metals from the Earliest Times to the Present.* Begun, 1858; laid aside, 1862; completed, 1879. Bell, London, 1880; 2nd ed., revised, 1902. Repr. Burt Franklin, New York, 1968.

MATHER, R. B. (1). 'The Fine Art of Conversation; the Yen Yü Phien of the *Shih Shuo Hsin Yü.*' *JAOS*, 1971, **91**, 222. (The story of Ho Yen and the Five-Mineral Powder (pt. 3, p. 45) is translated and discussed on p. 232.)

MATTIOLI, PIERANDREA (2). *Commentarii in libros sex Pedacii Dioscoridis Anazarbei de materia medica.* With appendix: 'De Ratione Distillandi aquas ex omnibus plantis et quomodo genuini odores in ipsos aquis conservari possint.' Valgrisi, Venice, 1554. Repr. 1558, 1559, 1560, 1563, 1674. French tr., Lyon, 1562. German tr., Prague, 1563.

MÉNARD, J. É. (1) (ed. & tr.). *L'Évangile selon Thomas.* Brill, Leiden, 1975. (Nag Hammadi Studies, no. 5.)

MEYERHOF, M. (4). 'Von Alexandrien nach Baghdad.' *SPAW/PH*, 1930, **23**, 389–429.

MICHEL, A. F. (1). 'Sur la Matière Colorante des Nerpruns [buckthorns] Indigènes.' In Rondot, Persoz & Michel (1), pp. 183ff.

MILIK, J. T. (1). *The Books of Enoch.* Oxford, 1976. (Aramaic and Hebrew fragments among the Qumran Scrolls.)

MISH, J. L. (1). 'Creating an Image of Europe for China; [Giulio] Aleni's [+1582 to +1649] *Hsi Fang Ta Wên* (Questions and Answers concerning the Western World), with Introduction, Translation and Notes.' *MS*, 1964, **23**, 1–87. Abstr. *RBS*, 1973, **10**, no. 866. (Contains discussions of geography, astronomy and geomancy in China and the West, with mention of alchemy on p. 76. This interview with the Jesuit was edited by Chiang Tê-Ching, later Minister of Rites under the last Ming emperor, in +1637.)

MIYAKAWA HISAYUKI (1). 'The Legate Kao Phien [d. +887] and a Taoist Magician, Lü Yung-Chih, in the Time of Huang Chhao's Rebellion [+875 to +884].' *ACTAS*, 1974, no. 27, 75. (The Taoist entourage of the general who suppressed it, including several alchemists, Chuko Yin, Tshai Thien and Shenthu shêng = Pieh-Chia.) (Cf. pt. 3, pp. 173–4.)

MOLESWORTH, J. T. (1). *A Dictionary of Marathi and English.* 2nd ed. Bombay, 1857.

MOLS, R. (1). 'Population in Europe, +1500 to +1700.' Art. in *Fontana Economic History of Europe* (ed. C. M. Cipolla), vol. 2, p. 15.

MONGAIT, A. L. (1). *Archaeology in the U.S.S.R.* Penguin (Pelican), London, 1961.

MOORHOUSE, S., GREENAWAY, F., MOORE, C. C., BELLAMY, C. V., NICOLSON, W. E. & BIEK, L. (1). 'Mediaeval Distilling-Apparatus of Glass and Pottery' (fragments found at British sites). *MARCH*, 1972, **16**, 79.

MOSCATI, SABATINO (1) (ed.). *An Introduction to the Comparative Grammar of the Semitic Languages.* Wiesbaden, 1964.

MULTHAUF, R. P. (9) 'An Enquiry into Saltpetre Supply and the Early Use of Firearms.' Abstracts of Scientific Section Papers. 15th. Internat. Congress Hist. of Sci. Edinburgh, 1977, p. 37.

MURAKAMI YOSHIMI (1). 'The Affirmation of Desire in Taoism.' *ACTAS*, 1974, no. 27, 57.

NEEDHAM, D. M. (2). 'Chemical Cycles in Muscle Contraction.' Art. in *Perspectives in Biochemistry* (Hopkins Presentation Volume), ed. J. Needham & D. E. Green. Cambridge, 1938 (several times reprinted), p. 201.

NEEDHAM, JOSEPH (43). 'The Past in China's Present.' *CR/MSU*, 1960, **4**, 145 and 281; repr. with some omissions, *PV*, 1963, **4**, 115. French tr.: 'Du Passé Culturel, Social et Philosophique Chinois dans ses Rapports avec la Chine Contemporaine', by G. M. Merkle-Hunziker. *COMP*, 1960, No. 21–2, 261; 1962, No. 23–4, 113; repr. in CFC, 1960, No. 8, 26; 1962, No. 15–16, 1.

NEEDHAM, JOSEPH (73) (ed.). *The Chemistry of Life; Eight Lectures on the History of Biochemistry, edited, with an introduction....* Cambridge, 1970.

NEEDHAM, JOSEPH (75). 'The Cosmology of Early China.' Art. in *Ancient Cosmologies*, ed. C. Blacker & M. Loewe, 1975, p. 87.

NEEDHAM, JOSEPH (76). 'The Institute's Symbol.' (A note on the origins of the traditional *liang i* symbol of Yin and Yang, incorporated in the device of the Institute of Biology, London.) *BIOL*, 1977.

NEEDHAM, JOSEPH (77). 'The Pattern of Nature-Mysticism and Empiricism in the Philosophy of Science; −3rd-Century China, +10th-Century Arabia, and +17th-Century Europe.' Art. in *Science, Medicine and History.* (Singer Presentation Volume.) Ed. E. A. Underwood. Oxford, 1953.

NEEDHAM, JOSEPH & LU GWEI-DJEN (9). 'Manfred Porkert's Interpretations of Terms in Mediaeval Chinese Natural and Medical Philosophy.' *ANS*, 1975, **32**, 491.

NEEDHAM, JOSEPH & LU GWEI-DJEN (11). 'Trans-Pacific Echoes and Resonances; Listening Once Again.' Art. in *The Smoking Mirror; Early Asian and Amerindian Civilisations*, ed. A. Kehoe, 1979.

PAGEL, W. (27). 'Paracelsus' Ätherähnliche Substanzen und ihre Pharmakologische Auswertung an Hühnern'. *GESN*, 1964, **21**, 113.

PAGEL, W. (28). *Das medizinische Weltbild des Paracelsus, sein Zusammenhang mit Neuplatonismus und Gnosis.* Steiner, Wiesbaden, 1962. (Kosmosophie; Forschungen und Texte zur Geschichte des Weltbildes, der Naturphilosophie, der Mystik und des Spiritualismus von Spätmittelalter bis zur Romantik, no. 1.)

PAGEL, W. & WINDER, M. (1). 'The Eightness of Adam and related "Gnostic" Ideas in the Paracelsian Corpus.' *AX*, 1969, **16**, 119.

PAGEL, W. & WINDER, M. (2). 'The Higher Elements and Prime Matter in Renaissance Naturalism and in Paracelsus.' *AX*, 1974, **21**, 93.

PAGELS, ELAINE H. (1). 'The Gnostic Vision; Varieties of Androgyny illustrated by Texts from the Nag Hammadi Library.' *PAR*, 1978, **3** (no. 4), 6.

PANTHEO, GIOVANNI AGOSTINO (1). *Voarchadumia contra Alchimiam; Ars Distincta ab Archimia & Sophia; cum Additionibus, Proportionibus, Numeris et Figuris opportunis...* [an assayer's counterblast to aurifactive claims]. Venice, 1530; 2nd ed. Paris, 1550. (The opening word, meaning gold thoroughly refined, is said to be formed from a 'Chaldaean' word and a Hebrew phrase.) Ferguson (1), vol. 2, pp. 166–7. (Cf. pt. 2. p. 32.)

PEARSON, BIRGER A. (1). 'The Figure of Norea in Gnostic Literature.' Art. in *Proc. Internat. Colloquium on Gnosticism*, Stockholm, 1973, ed. G. Widengren & D. Hellholm, p. 143 (Kungl. Vitterhets Historie och Antikvitets Akademiens Handlingar, Filol.-Filos. Ser., no. 17).

PEARSON, R. J. (1). *The Archaeology of the Ryukyu [Liu-Chhiu] Islands; a Regional Chronology from −3000 to the Historic Period.* Univ. Hawaii Press, Honolulu, 1969.

PELLER, S. (1). 'Births and Deaths among Europe's Ruling Families since +1500.' Art. in *Population in History...*, ed. Glass & Eversley (1), p. 87.

PERSOZ, J. (1). 'Sur une Matière Colorante Verte qui vient de Chine' (*lü kao*, from buckthorn bark and twigs). *CRAS*, 1852, **35**, 558. Repr. in Rondot, Persoz & Michel (1), pp. 129 ff.

PERSOZ, J. (2). 'Des Propriétés Chimiques et Tinctoriales du Vert de Chine' (*lü kao, Rhamnus* spp.). In Rondot, Persoz & Michel (1), pp. 151 ff.

PERSOZ, J. (3). 'Sur la Teinture en Jaune avec le *hoang-tchi* [*huang-chih*]' (*Gardenia* spp.). In Rondot, Persoz & Michel (1), pp. 199 ff., cf. pp. 86 ff.

PETERSON, W. J. (1). 'Western Natural Philosophy published in Late Ming China.' *PAPS*, 1973, **117**, 295.

PFISTER, F. (1). *Kleine Texte zum Alexander-roman.* Heidelberg, 1910. (Sammlung Vulgärlat. Texte, no. 4.)

PINGREE, DAVID (1). *The Thousands of Abū Ma'shar.* Warburg Inst. London, 1968.

PINGREE, DAVID (2). 'Michael Psellus' (biography), in *Dictionary of Scientific Biography*, ed. C. C. Gillespie, Scribner, New York, 1975, Vol. 11, p. 182.

PINTO, FERNÃO MENDES (1). *The Voyages and Adventures of Fernand Mendez Pinto, a Portugal: During his Travels for the Space of one and twenty years in the Kingdoms of Ethiopia, China, Tartaria, Cauchin-china, Calaminham, Siam, Pegu, Japan, and a great part of the East-Indiaes. With a Relation and Description of most of the Places thereof; their Religion, Laws, Riches, Customs and Government in time of Peace and War. Where he five times suffered Shipwrack, was sixteen times sold, and thirteen times made a Slave. Written Originally by himself in the Portugal Tongue, and Dedicated to the Majesty of Philip King of Spain. Done into English by H. C[ogan], Gent.* Macock, Cripps & Lloyd, London, 1653. Facsimile edition, Dawson, London, 1969.

PIRIE, N. W. (2). 'The Criteria of Purity used in the Study of Large Molecules of Biological Origin.' *BR*, 1940, **15**, 377.

PIRIE, N. W. (3). 'Principles of "Mini-Life".' Introduction to Ciba Symposium on *Pathogenic Mycoplasmas.* Elsevier, Amsterdam, 1972.

PIRSIG, R. M. (1). *Zen and the Art of Motorcycle Maintenance; an Inquiry into Values.* Bodley Head, London, 1974.

PLESSNER, M. (6). *Vorsokratische Philosophie und Griechische Alchemic; Studien zu Text und Inhalt der Turba Philosophorum.* Steiner, Wiesbaden, 1975. Ed. posthumously F. Klein-Franke. (Boethuis; Texte und Abhandlungen z. Gesch. d. exakten Naturwissenschaften, no. 4.)

PLESSNER, M. (9). 'Balīnūs.' Art. in *EI* (2nd ed.), vol. 1, p. 995.

POPHAM, H. (1). 'Softest Clothing, Woolly, Bright.' *ESSOM*, 1972, **21** (no. 3), 13. (On the dry-cleaning industry, the title taken from William Blake.)

PORTER, W. N. (1) (tr.). *The Miscellany of a Japanese Priest, being a Translation of the 'Tsurezuregusa'* [+1338], *by Kenkō* [*Hōshi*], *Yoshida* [*no Kaneyoshi*]. With an introduction by Ichikawa Sanki. Clarendon (Milford), London, 1914. Repr. Tuttle, Rutland, Vt. and Tokyo, 1974.

PUECH, H. C. (4) (tr.). 'Gnostic Gospels and Related Documents' in Hennecke & Schneemelcher (1). Eng. ed. vol. 1, p. 231.

PUECH, H. C. (5). 'La Gnose et le Temps.' *ERJB*, 1952, **20**, 57.

RADIN, PAUL (1). *The Trickster, a Study in American Indian Mythology.* Routledge & Kegan Paul, London, 1956. Schocken, New York, 1972. (With commentaries by K. Kerényi and C. G. Jung.)

RATHER, L. J. (1). 'Alchemistry, the Kabbala, the Analogy of the "Creative Word" and the Origins of Molecular Biology.' *EPI*, 1972, **6**, 83.

REGMI, D. R. (1). *Mediaeval Nepal.* 2 vols. Calcutta, 1966.

REITZENSTEIN, R. (5). 'Alchemistische Lehrschriften und Märchen bei den Arabern.' *RGVV*, 1923, **19**, 66.

RENONDEAU, G. (1). *Le 'Shugendō'; Histoire, Doctrine et Rites des Yamabushi.* Paris, 1965. (Cahiers de la Société Asiatique, no. 18.) (Cf. pt. 2, p. 299.)

REX, FRIEDEMANN (1). Review of F. Sezgin's *Geschichte des arabischen Schrifttums*, vol. 4 (Alchemie, Chemie, Botanik, Agrikultur bis *c*. 430 H. [+1052]). *DI*, 1972, **49**, 305.

REX, FRIEDEMANN (2), (tr). *Zur Theorie der Naturprozesse in der früharabischen Wissenschaft; das 'Kitāb al-Ikhrāj mā fi'l-Qūwa ila'l-Fi'l'* [Buch vom Überführen dessen, was in der Potenz ist, in den Akt, Kr 331] *übersetzt und erklärt; ein Beitrag zum alchemistischen Weltbild der Jābir-Schriften* (8–10 *Jahrhundert nach Chr.*). Inaug. Diss. (Habilitationsschrift), Tübingen, 1975. Steiner, Wiesbaden, 1975 (Collection des Travaux de l'Acad. Int. d'Hist. des Sci., no. 22).

RICHARDS, D. S. (1), (ed.). *Islam and the Trade of Asia; a Colloquium.* Cassirer, Oxford; Univ. of Pennsylvania Press, Philadelphia 1971. (Papers on Islamic History, no. 2.) Contains: M. Rogers: 'China and Islam; the Archaeological Evidence in the Mashriq.' G. T. Scanlon: 'Egypt and China; Trade and Imitation.' N. Chittick: 'East African Trade with the Orient.' G. F. Hudson: 'The Mediaeval Trade of China.'

ROBINSON, J. M. (1) (ed.). *The Nag Hammadi Library in English; translated by members of the Coptic Gnostic Library Project of the Institute for Antiquity and Christianity* [Claremont, California]. Brill, Leiden; Harper & Row, San Francisco, 1977.

ROM, H. (1). 'Alchemy.' Art. in *Universal Jewish Encyclopaedia*, ed. I. Landman. 11 vols. Ktav, New York, 1969, vol. 1, p. 164.

RONDOT, N. (3). 'Notice du Vert de Chine et de la Teinture en Vert chez les Chinois...' (*Rhamnus* spp.). In Rondot, Persoz & Michel (1), pp. 1 ff.

RONDOT, N., PERSOZ, J. & MICHEL, A. F. (1). *Notice du Vert de Chine et de la Teinture en Vert chez les Chinois, suivie d'une Étude des Propriétés Chimiques et Tinctoriales du 'lo-kao [lü kao]' par Mons. J. P...., et de Recherches sur la Matière Colorante des Nerpruns Indigènes par Mons. A. F. M...'*. Lahure (for the Chambre de Commerce de Lyon), Paris, 1858.

ROSENFELD, ALBERT (1). *Prolongevity; a Report on the Scientific Discoveries now being made about Ageing and Dying, and their Promise of an Extended Human Life Span—without Old Age*. Knopf, New York, 1976.

ROSS, D. J. A. (1). *Alexander Historiatus; a Guide to Mediaeval Illustrated Alexander Literature*. Warburg Institute, London, 1963. (Warburg Institute Surveys, no. 1.)

ROSSI, P. (2). 'The Equivalence of Intellects.' Symposium contribution at the 15th International Congress of the History of Science, Edinburgh, 1977. Synopses of Symposia Papers, p. 14.

ROSZAK, T. (1). *The Making of a Counter-Culture; Reflections on the Technocratic Society and its Youthful Opposition*. New York, 1968, repr. 1969. Faber & Faber, London, 1970, repr. 1971.

ROSZAK, T. (2). *Where the Wasteland Ends; Politics and Transcendence in Post-Industrial Society*. New York and London, 1972–3.

ROY, MIRA & SUBBARAYAPPA, B. V. (1) '*Rasārṇavakalpa*' (*The Manifold Powers of the Ocean of Rasa*); an [*alchemical*] *Text* [*of the + 11th Century*] *edited and translated into English...* Indian National Science Academy, New Delhi, 1976. (Monograph series, no. 5.) Rev. S. Mahdihassan, *IJHS*, 1977, **12** (no. 1), 76.

RUDENKO, S. I. (2). *Frozen Tombs of Siberia; the Pazyryk [perma-frost] Burials of Iron-Age Horsemen*. Dent, London, 1970. First pub. in Russian, 1953. Dent, London, 1970. With an explanatory introduction by the translator, M. W. Thompson. (Cf. pt. 2, p. 304.)

RUDOLPH, K. (1) *et al*. *Gnosis und Gnostizismus*. Wissenschaftliche Buchgesellschaft, Darmstadt, 1975. (Wege der Forschung, no. 262.)

RUDOLPH, R. C. (8). 'Two Recently Discovered Han Tombs.' *AAAA*, 1973, **26** (no. 2), 106.

RUSKA, J. (41). 'Zwei Bücher *De Compositione Alchemiae* und ihre Vorreden.' *AGMNT*, 1928, **11**, 28.

RUSKA, J. (42). 'Methods of Research in the History of Chemistry.' *AX*, 1937, **1**, 21.

RUSSELL, J. C. (1). 'Population in Europe, +500 to +1500.' Art. in *Fontana Economic History of Europe* (ed. C. M. Cipolla), vol. 1, p. 25.

SANFORD, J. H. (1). 'Japan's "Laughing Mushrooms".' *ECB*, 1972, **26**, 174. (Cf. pt. 2, p. 121.)

SCARISBRICK, J. (1). *Spirit Manual; Historical and Technical*. 2nd ed. Whitehead, Wolverhampton, 1894. (Revenue series, no. 2.)

SCHENK, MARTA (1). 'Alkoholische Getränke in Chinas früher Geschichte.' Inaug. Diss. (Magisterarbeit), Berlin, 1962.

SCHILDKNECHT, H. (1). *Zone Melting*. Academic Press, New York, 1966.

SCHILDKNECHT, H. & SCHLEGELMILCH, F. (1). 'Normales Erstarren zur Anreicherung und Reinigung organischer und anorganischer Verbindungen.' *CIT*, 1963, **35**, 637.

SCHMIDT, C. & TILL, W. K. (1). '*Pistis Sophia*', ein gnostisches Originalwerk des 3. Jahrhunderts aus dem Koptischen übersetzt. Leipzig, 1925.

SCHOLER, D. M. (1) (ed.). *Nag Hammadi Bibliography, 1948–1969*. Brill, Leiden, 1971. (Nag Hammadi Studies, no. 1.) Supplemented annually subsequently in *NT*, 1971, **13**, 322; 1972, **14**, 312; 1973, **15**, 327; 1974, **16**, 316; 1975, **17**, 305; 1977, **19**, 293.

SCHORLEMMER, C. (1). 'On the Origin of the Word Chemistry.' *MMLPS*, 1882 (3rd ser.), **7**, 75.

SCHULTES, R. E. & HOFMANN, A. (1). *The Botany and Chemistry of Hallucinogens*. Thomas, New York, 1973. Rev. *SAM*, 1973, **229** (no. 4), 129. (Cf. pt. 2, p. 116.)

VON SCHWARZENFELD, G. (1). 'Prag als Esoterikerzentrum von Rudolf II bis Kafka.' *ANT*, 1963, **4**, 341.

VON SCHWARZENFELD, G. (2). 'Magica aus der Zeit Rudolfs II.' *ANT*, 1963, **4**, 478.

SELER, E. (1). 'Die Huichol-Indianer des Staates Jalisco in Mexico.' *MAGW*, 1901, **31**, 138.

SELIGMAN, C. G. (2). 'The Roman Orient and the Far East.' *AQ*, 1937, **11**, 5; *ARSI* 1938, 547.

SEZGIN, F. (2). *Geschichte des arabischen Schrifttums*, vol. 4: Alchimie, Chemie, Botanik, Agrikultur bis ca. 430 H. [+1052]. Brill, Leiden, 1971.

SHAMASASTRY, R. (1) (tr.). *Kauṭilya's 'Arthaśāstra.'* With introduction by F. J. Fleet. Wesleyan Mission Press, Mysore, 1929.

SIGISMUND, R. (1). *Die Aromata in ihrer Bedeutung für Religion, Sitten, Gebräuche, Handel und Geographie des Alterthums bis zu den ersten Jahrhunderten unserer Zeitrechnung*. Winter, Leipzig, 1884; photolitho reprint, Zentralantiquariat d. D.D.R., Leipzig, 1974. (Cf. pt. 2, pp. 136 ff.)

N S C

SIRAISI, N. G. (1). 'Taddeo Alderotti and Bartolomeo da Varignani on the Nature of Medical Learning.' *ISIS*, 1977, **68**, 27.

SIVIN, N. (13) (ed.). *Science and Technology in East Asia*. (Nineteen articles by sixteen writers reprinted from *Isis*, 1914–1976.) Science History Publications, New York, 1977.

SIVIN, N. (14). 'Chinese Alchemy and the Manipulation of Time.' *ISIS*, 1976, **67**, 513. Repr. in *Science and Technology in East Asia*, ed. Sivin (13), p. 108.

SIVIN, N. (15). 'On the Word "Taoist" as a Source of Perplexity with special reference to the Relations of Science and Religion in Traditional China.' *HOR*, 1978, **17**, 303.

SMITH, C. S. (8). 'An Examination of the Arsenic-Rich Coating on a Bronze Bull from Horoztepe.' Art. in *Application of Science...Works of Art*, 1973, ed. W. J. Young (1); p. 96.

SMITH, C. S. & HAWTHORNE, J. G. (1) (ed. & tr.). '*Mappae Clavicula*; A Little Key to the World of Medieval Techniques.' *TAPS*, 1974 (N.S.), **64** (pt. 4), 1–128. (Annotated translation based on a collation of the Sélestat and Phillipps-Corning MSS, with reproductions of both.) Rev. H. Silvestre, *SCRM*, 1977, **31**, 319.

SNYDER, A. E. (1). 'Desalting Water by Freezing.' *SAM*, 1962, **207** (no. 6), 41.

SOLOMON, D. (2) (ed.). *The Marijuana Papers*. New York, 1966; 2nd ed., revised, Granada (Panther Books), 1969; repr. 1970, 1972. (Cf. pt. 2, p. 150.)

STANILOAE, DUMITRU (1). 'The Cross in Orthodox Theology and Worship.' *SOB*, 1977, **7** (no. 4), 233.

STAVENHAGEN, L. (1). 'The Original Text of the Latin *Morienus*.' *AX*, 1970, **17**, 1.

STAVENHAGEN, L. (2), (tr.). *A Testament of Alchemy, being the Revelations of Morienus, Ancient Adept and Hermit of Jerusalem, to Khālid ibn Yazīd ibn Muʿāwiyya, King of the Arabs, on the Divine Secrets of the Magisterium and the Accomplishment of the Alchemical Art*. Brandeis Univ. Press and University Press of New England, Hanover, New Hampshire, 1974.

LE STRANGE, G. (5) (tr.). 'The Alchemist; a Persian Play' (written by Mirza Fath-ali Akhunzade of Derbend, 1851, first pub. at Tiflis, 1861, repr. by Mīrzā Jaʿafar of Karājeh-Dāgh, 1874). *JRAS*, 1886, **18** (N.S.), 103.

STREBEL, J. (1), (ed.). *Theophrastus von Hohenheim, gen. Paracelsus, sämtliche Werke in zeitgemässer kurzer Auswahl*. 8 vols. Zollikofer, St Gallen, 1944–1949.

STRICKMANN, M. (4). 'Taoist Literature.' Art. in EB, 15th ed. 1974, p. 1051.

STRICKMANN, M. (5). 'A Taoist Confirmation of Liang Wu Ti's Proscription of Taoism.' Unpub. MS., 1975.

STRICKMANN, M. (6). 'The Longest Taoist Scripture [*Tu Jen Ching*].' *HOR*, 1978, **17**, 331.

STRUBE, I. (1). 'Die Phlogistonlehre Georg Ernest Stahl (1660 bis 1734).' *ZGNTM*, 1961, **1**, 27.

SUBBARAYAPPA, B. V. (2). 'Some Trends in Indian Alchemy.' Proc. 15th. Internat. Congr. History of Science. Edinburgh, 1977. Abstracts, p. 48.

SUDHOFF, K. (5), (ed.). *Theophrast von Hohenheim, gen. Paracelsus, sämtliche Werke*. 14 vols, Oldenbourg, München and Berlin, 1922–1933. Abstr. and crit. rev. P. Walden, *ZAC/AC*, 1941. **54**, 431.

SULER, B. (1). 'Alchemy.' Art. in *Encyclopaedia Judaica* (German), ed. J. Klatzkim, 10 vols. Eshkol, Berlin, 1928–34. vol. 2, cols. 137ff.

SULER, B. (2). 'Alchemy.' Art. in *Encyclopaedia Judaica* (Israel), ed. C. Roth, 16 vols. Keter, Jerusualem, 1971–2. Vol. 2, cols. 542ff. (enlarged by the editorial staff).

TAMBURELLO, A. (1). '"Taoismus" in Japan' (tr. from Ital. by G. Glaesser). *ANT*, 1970, **12**, 125. (Cf. pt. 2, p. 300.)

TAUBE, E. (1). 'Knock and Nickel; a Name Study.' *GR*, 1944, **34**, 428.

TEICH, MIKULÁŠ (3). 'Born's Amalgamation Process and the International Metallurgical Gathering at Skleno in +1786.' *ANS*, 1975, **32**, 305.

TÊNG SSU-YÜ & BIGGERSTAFF, K. (1). *An Annotated Bibliography of Selected Chinese Reference Works*. Harvard-Yenching Inst. Peiping, 1936. (Yenching Journ. Chin. Studies, monograph no. 12.)

THURM, H. G. (1). Shao Chui, *Gebrannter Wein im alten China*. Team-Fachverlag, Karlstein a/M., 1978. (Sonderausgabe der Alkohol-Industrie, Edition 78.)

DE TIZAC, J. H. D'ARDENNE (1). *Les Hautes Époques de l'Art Chinois, d'après les Collections du Musée Cernuschi*.' Nilsson, Paris, n.d. [1930].

DE TIZAC, J. H. D'ARDENNE (2). *L'Art Chinois Classique*. Laurens, Paris, 1926.

TOLSTOV, S. P. (1). *Drevniy Choresm* (Ancient Chorasmia) (in Russian). University Press, Moscow, 1948. Germ. tr. *Auf den Spuren d. altchoresmischen Kultur*, by O. Mehlitz. Kultur & Vorschritt, Berlin, 1953, rev. A. D. H. Bivar, *ORA*, 1955 (N.S.) **1**, 129. See also Ghirshman (2) and Mongait (1), pp. 235 ff.

TOLSTOV, S. P. (2). 'Les Résultats des Travaux de l'Expédition archéologique et ethnographique de l'Académie des Sciences de l'U.R.S.S. au Khorezin en 1951–1955.' *AAS*, 1957, **4**, 83 and 187.

TOLSTOV, S. P. (4). 'Le Kharezm Ancien.' Art. in *Ourarton, Neapolis des Scythes, Kharezm*, ed. C. Virolleaud. Maisonneuve, Paris, 1954, p. 107.

TOMKINSON, L. (2) (tr.). *Tales of Ming Scholars (the Ju Lin Wai Shih)*. Unpub. MS. in the East Asian History of Science Library, Cambridge.

TSUNODA RYUSAKU & GOODRICH, L. CARRINGTON (1). *Japan in the Chinese Dynastic Histories*. Perkins, South Pasadena, 1951. (Perkins Asiatic Monographs, no. 2.)

TURNER, G. l'E. & LEVERE, T. H. (1). *Van Marum's Scientific Instruments in [Pieter] Teyler [van der Hulst's] Museum [at Haarlem]*. Vol. 4 of *Martinus van Marum; Life and Work* (6 vols., 1969–76), ed. R. J. Forbes, E. Lefebvre & J. G. de Bruyn. Noordhoff, Leiden, 1973.

ULLMANN, M. (1). 'Die Natur- und Geheim-wissenschaften in Islam.' Art. in *Handbuch d. Orientalistik*, 1. Abt. *Der Nahe und der mittler Osten*. Ergänzungsband VI, 2er Abschnitt, 1–500. Brill, Leiden and Köln, 1972. Rev. D. M. Dunlop, *JRAS*, 1974 (no. 1), 64.

ULLMANN, M. (2). 'Die Medizin in Islam.' Art. in *Handbuch d. Orientalistik*, 1. Abt. *Der Nahe und der mittlere Osten*. Ergänzungsband VI, 1er Abschnitt. Brill, Leiden and Köln, 1970.

VEITH, I. (6). 'Historical Reflections on Longevity.' *PBM*, 1970, **13**, 255.

VERESHCHAGIN, N. K. (1). 'The Mammoth (Woolly Elephant) "Cemeteries" of North-East Siberia.' *POLREC*, 1974, **17** (no. 106), 3. (Cf. pt. 2, p. 304.)

VERHAEREN, H. (2) (ed.). *Catalogue de la Bibliothèque du Pé-T'ang* (the Pei Thang Jesuit Library in Peking). Lazaristes Press, Peking, 1949. Photographically reproduced, Belles Lettres (Cathasia). Paris, 1969.

DE VOCHT, H. (1). *Ben Jonson's 'The Alchemist', edited from the Quarto of +1612, with Comments on its Text*. Uystpruyst, Louvain, 1950. (Materials for the Study of the Old English Drama, no. 22.)

WADDINGTON, C. H. (5). 'The New Atlantis Revisited.' *PRSB*, 1975, **190**, 301. (Bernal Lecture to the Royal Society.)

WAGNER, R. G. (1). 'Lebens-stil und Drogen im chinesischen Mittelalter.' *TP*, 1973, **59**, 79–178. (An exhaustive study of the tonic Han Shih San, a powder of four inorganic substances (Ca, Mg, Si) and nine plant substances containing alkaloids.)

WALLACKER, B. E. (2). 'Liu An, Second Prince of Huai-Nan (c. −180 to −122).' *JAOS*, 1972, **92**, 36.

WASSON, R. G. (5). 'The Hallucinogenic Mushrooms of Mexico, and Psilocybin; a Bibliography.' *HV/BML*, 1962, **20**, no. 2.

WASSON, R. G. (6). 'Mushrooms and Japanese Culture.' *TAS/J*, 1973 (3rd ser.), **11**, 5.

WATSON, WM. (5). *The Genius of China*. (Catalogue of an Exhibition of Archaeological Finds of the People's Republic of China at the Royal Academy, London, Sep. 73 to Jan. 74.) Rainbird, Bell, Westerham & Dorstel, for *The Times* and *Sunday Times* in association with the Royal Academy and the Great Britain/China Committee, London, 1973.

WATSON, WM. (6). *Cultural Frontiers in Ancient East Asia*. University Press, Edinburgh, 1971.

WATSON, WM. & WILLETTS, W. (1). *Archaeology in Modern China; Descriptive Catalogue of the Sites and Photographs [shown at the Chinese Archaeological Exhibition, London, Oxford, etc.]* (mimeographed). Britain–China Friendship Association, London, 1959.

WEIL, GOTTHOLD (1) (tr. & ed.). *Moses ben Maimon: 'Responsum' on the Duration of Life (Über die Lebensdauer)—ein unediertes 'Responsum' herausgegeben, übersetzt und erklärt von G. W. Basel and New York, 1958.

WELCH, HOLMES H. (4). *The Practice of Chinese Buddhism, 1900 to 1950*. Harvard Univ. Press, Cambridge, Mass., 1967.

WÊN PIEN (1). 'The World's Oldest Painting on Silk; Visions of Heaven, Earth and the Underworld in a Two-Thousand-Year-Old Chinese Tomb [the Lady of Tai, d. c. −166].' *UNESC*, 1974, **27** (no. 4), 18.

WESTMAN, R. S. (1). 'Magical Reform and Astronomical Reform; the Yates Thesis Reconsidered.' Art. in *Hermeticism and the Scientific Revolution*. Clark Library, Los Angeles, 1977.

WHEELER, SIR R. E. M. (8). '(Excavations at) Brahmagiri and Chandravalla, 1947; Megalithic and other Cultures in the Chitaldrug District, Mysore State.' *AIND*, 1947, No. 4, 180–310.

WILHELM, RICHARD (2) (tr.). '*I Ging' [I Ching]; Das Buch der Wandlungen*. 2 vols. (3 books, pagination of 1 and 2 continuous in first volume). Diederichs, Jena, 1924. (Eng. tr. C. F. Baynes (2 vols.). Bollingen-Pantheon, New York, 1950.) (See Vol. 2, p. 308.)

WRIGHT, W. (1). *Lectures on the Comparative Grammar of the Semitic Languages*. Cambridge, 1890.

WRIGLEY, E. A. (1). *Population and History*. Weidenfeld & Nicolson (World University Library), London, 1969.

WÜNSCHE, A. (1). *Die Sagen vom Lebensbaum und Lebenswasser; Altorientalische Mythen*. Leipzig, 1905. (Ex Oriente Lux 1, no. 2/3.)

YANG HSIEN-YI & YANG, GLADYS (1) (tr.). *The Scholars ('Ju Liu Wai Shih')*. Foreign Languages Press, Peking, 1957. Repr. 1973. (With an appendix on official rank and on the examination system.)

YASUDA YURI (1) (tr.). *Old Tales of Japan.* With illustrations by Sakakura Yoshinobu and Mitsui Eiichi. Tuttle, Tokyo, and Rutland, Vt., 1947, repr. 1953. Revised ed. 1956, many times reprinted. (Contains (pp. 133 ff.) a very abridged version in English of the *Taketori Monogatari* (+9th cent.), 'The Luminous Princess'.) Cf. pt. 3, p. 176.

YATES, ROBIN (1). Translations of the alchemical poems of Lu Kuei-Mêng (unpub. MS.).

YOUNG, W. J. (1) (ed.). *The Application of Science in the Examination of Works of Art.* Museum of Fine Arts, Boston, 1973.

YÜ YING-SHIH (1). *Trade and Expansion in Han China; a Study in the Structure of Sino-Barbarian Economic Relations.* Univ. Calif. Press, Berkeley and Los Angeles, 1967.

ZIPPERT, E. (1). 'Leib–Seele, Körper–Geist in Ostasien und bei Paracelsus.' *DZA*, 1955, **4** (no. 3–4), 23.

GENERAL INDEX

by MURIEL MOYLE

NOTES

(1) Articles (such as 'the', 'al-', etc.) occurring at the beginning of an entry, and prefixes (such as 'de', 'van', etc.) are ignored in the alphabetical sequence. Saints appear among all letters of the alphabet according to their proper names. Styles such as Mr, Dr, if occurring in book titles or phrases, are ignored; if with proper names, printed following them.

(2) The various parts of hyphenated words are treated as separate words in the alphabetical sequence. It should be remembered that, in accordance with the conventions adopted, some Chinese proper names are written as separate syllables while others are written as one word.

(3) In the arrangement of Chinese words, Chh- and Hs- follow normal alphabetical sequence, and *ü* is treated as equivalent to *u*.

(4) References to footnotes are not given except for certain special subjects with which the text does not deal. They are indicated by brackets containing the superscript letter of the footnote.

(5) Explanatory words in brackets indicating fields of work are added for Chinese scientific and technological persons (and occasionally for some of other cultures), but not for political or military figures (except kings and princes).

夏 Hsia kingdom (legendary?)		c. −2000 to c. −1520
商 Shang (Yin) kingdom		c. −1520 to c. −1030
周 Chou dynasty (Feudal Age)	Early Chou period	c. −1030 to −722
	Chhun Chhiu period 春秋	−722 to −480
	Warring States (Chan Kuo) period 戰國	−480 to −221
First Unification 秦 Chhin dynasty		−221 to −207
漢 Han dynasty	Chhien Han (Earlier or Western)	−202 to +9
	Hsin interregnum	+9 to +23
	Hou Han (Later or Eastern)	+25 to +220
三國 San Kuo (Three Kingdoms period)		+221 to +265
First Partition	蜀 Shu (Han)	+221 to +264
	魏 Wei	+220 to +265
	吳 Wu	+222 to +280
Second Unification 晉 Chin dynasty: Western		+265 to +317
Eastern		+317 to +420
劉宋 (Liu) Sung dynasty		+420 to +479
Second Partition Northern and Southern Dynasties (Nan Pei chhao)		
齊 Chhi dynasty		+479 to +502
梁 Liang dynasty		+502 to +557
陳 Chhen dynasty		+557 to +589
魏 Northern (Thopa) Wei dynasty		+386 to +535
Western (Thopa) Wei dynasty		+535 to +556
Eastern (Thopa) Wei dynasty		+534 to +550
北齊 Northern Chhi dynasty		+550 to +577
北周 Northern Chou (Hsienpi) dynasty		+557 to +581
Third Unification 隋 Sui dynasty		+581 to +618
唐 Thang dynasty		+618 to +906
Third Partition 五代 Wu Tai (Five Dynasty period) (Later Liang, Later Thang (Turkic), Later Chin (Turkic), Later Han (Turkic) and Later Chou)		+907 to +960
遼 Liao (Chhitan Tartar) dynasty		+907 to +1124
West Liao dynasty (Qarā-Khiṭāi)		+1124 to +1211
西夏 Hsi Hsia (Tangut Tibetan) state		+986 to +1227
Fourth Unification 宋 Northern Sung dynasty		+960 to +1126
宋 Southern Sung dynasty		+1127 to +1279
金 Chin (Jurchen Tartar) dynasty		+1115 to +1234
元 Yuan (Mongol) dynasty		+1260 to +1368
明 Ming dynasty		+1368 to +1644
清 Chhing (Manchu) dynasty		+1644 to +1911
民國 Republic		+1912

N.B. When no modifying term in brackets is given, the dynasty was purely Chinese. Where the overlapping of dynasties and independent states becomes particularly confused, the tables of Wieger (1) will be found useful. For such periods, especially the Second and Third Partitions, the best guide is Eberhard (9). During the Eastern Chin period there were no less than eighteen independent States (Hunnish, Tibetan, Hsienpi, Turkic, etc.) in the north. The term 'Liu chhao' (Six Dynasties) is often used by historians of literature. It refers to the south and covers the period from the beginning of the +3rd to the end of the +6th centuries, including (San Kuo) Wu, Chin, (Liu) Sung, Chhi, Liang and Chhen. For all details of reigns and rulers see Moule & Yetts (1).

SUMMARY OF THE CONTENTS OF VOLUME 5

CHEMISTRY AND CHEMICAL TECHNOLOGY

Part 2, Spagyrical Discovery and Invention:
Magisteries of Gold and Immortality

Part 4, Spagyrical Discovery and Invention:

Apparatus and Theory

Laboratory apparatus and equipment

The laboratory bench
The stoves *lu* and *tsao*
The reaction-vessels *ting* (tripod, container, cauldron) and
    kuei (box, casing, container, aludel)
The sealed reaction-vessels *shen shih* (aludel, lit. magical
    reaction-chamber) and *yao fu* (chemical pyx)
Steaming apparatus, water-baths, cooling jackets, con-
    denser tubes and temperature stabilisers
Sublimation apparatus
Distillation and extraction apparatus
    Destillatio per descensum
    The distillation of sea-water
    East Asian types of still
    The stills of the Chinese alchemists
    The evolution of the still
    The geographical distribution of still types
The coming of Ardent Water
    The Salernitan quintessence
    Ming naturalists and Thang 'burnt-wine'
    Liang 'frozen-out wine'
    From icy mountain to torrid still
    Oils in stills; the rose and the flame-thrower
Laboratory instruments and accessory equipment

Reactions in aqueous medium

The formation and use of a mineral acid
'Nitre' and *hsiao*; the recognition and separation of
    soluble salts
Saltpetre and copperas as limiting factors in East and West
The precipitation of metallic copper from its salts by
    iron
The role of bacterial enzyme actions
Geodes and fertility potions
Stabilised lacquer latex and perpetual youth

The theoretical background of elixir alchemy [with Nathan Sivin]

Introduction
    Areas of uncertainty
    Alchemical ideas and Taoist revelations
The spectrum of alchemy
The role of time
    The organic development of minerals and metals
    Planetary correspondences, the First Law of Chinese
        physics, and inductive causation
    Time as the essential parameter of mineral growth
    The subterranean evolution of the natural elixir
The alchemist as accelerator of cosmic process
    Emphasis on process in theoretical alchemy
    Prototypal two-element processes
    Correspondences in duration
    Fire phasing
Cosmic correspondences embodied in apparatus
    Arrangements for microcosmic circulation
    Spatially oriented systems
    Chaos and the egg
Proto-chemical anticipations
    Numerology and gravimetry
    Theories of categories

Comparative survey

China and the Hellenistic world
    Parallelisms of dating
    The first occurrence of the term 'chemistry'
    The origins of the root 'chem-'
    Parallelisms of content
    Parallelisms of symbol
China and the Arabic world
    Arabic alchemy in rise and decline
    The meeting of the streams
    Material influences
    Theoretical influences
    The name and concept of 'elixir'
Macrobiotics in the Western world

Part 5, Spagyrical Discovery and Invention:

Physiological Alchemy

The outer and the inner macrobiogens; the elixir and the enchymoma

Esoteric traditions in European alchemy
Chinese physiological alchemy; the theory of the enchym-
    oma (*nei tan*) and the three primary vitalities
    The quest for material immortality
    Rejuvenation by the union of opposites; an *in vivo*
        reaction
    The *Hsiu Chen* books and the *Huang Thing* canons
The historical development of physiological alchemy
The techniques of macrobiogenesis
    Respiration control, aerophagy, salivary deglutition and
        the circulation of the *chhi*
    Gymnastics, massage and physiotherapeutic exercise
    Meditation and mental concentration
    Phototherapeutic procedures
    Sexuality and the role of theories of generation
The borderline between proto-chemical (*wai tan*) and
    physiological (*nei tan*) alchemy

Late enchymoma literature of Ming and Chhing
    The 'Secret of the Golden Flower' unveil'd
Chinese physiological alchemy (*nei tan*) and the Indian
    Yoga, Tantric and Hathayoga systems
    Originalities and influences; similarities and differences
Conclusions; *nei tan* as proto-biochemistry

The enchymoma in the test-tube; medieval preparations of urinary steroid and protein hormones

Introduction
The sexual organs in Chinese medicine
Proto-endocrinology in Chinese medical theory
The empirical background
The main iatro-chemical preparations
Comments and variant processes
The history of the technique

ROMANISATION CONVERSION TABLES

by Robin Brilliant

PINYIN/MODIFIED WADE–GILES

Pinyin	Modified Wade–Giles	Pinyin	Modified Wade–Giles
a	a	chou	chhou
ai	ai	chu	chhu
an	an	chuai	chhuai
ang	ang	chuan	chhuan
ao	ao	chuang	chhuang
ba	pa	chui	chhui
bai	pai	chun	chhun
ban	pan	chuo	chho
bang	pang	ci	tzhu
bao	pao	cong	tshung
bei	pei	cou	tshou
ben	pên	cu	tshu
beng	pêng	cuan	tshuan
bi	pi	cui	tshui
bian	pien	cun	tshun
biao	piao	cuo	tsho
bie	pieh	da	ta
bin	pin	dai	tai
bing	ping	dan	tan
bo	po	dang	tang
bu	pu	dao	tao
ca	tsha	de	tê
cai	tshai	dei	tei
can	tshan	den	tên
cang	tshang	deng	têng
cao	tshao	di	ti
ce	tshê	dian	tien
cen	tshên	diao	tiao
ceng	tshêng	die	dieh
cha	chha	ding	ting
chai	chhai	diu	tiu
chan	chhan	dong	tung
chang	chhang	dou	tou
chao	chhao	du	tu
che	chhê	duan	tuan
chen	chhên	dui	tui
cheng	chhêng	dun	tun
chi	chhih	duo	to
chong	chhung	e	ê, o

Pinyin	Modified Wade–Giles	Pinyin	Modified Wade–Giles
en	ên	jia	chia
eng	êng	jian	chien
er	êrh	jiang	chiang
fa	fa	jiao	chiao
fan	fan	jie	chieh
fang	fang	jin	chin
fei	fei	jing	ching
fen	fên	jiong	chiung
feng	fêng	jiu	chiu
fo	fo	ju	chü
fou	fou	juan	chüan
fu	fu	jue	chüeh, chio
ga	ka	jun	chün
gai	kai	ka	kha
gan	kan	kai	khai
gang	kang	kan	khan
gao	kao	kang	khang
ge	ko	kao	khao
gei	kei	ke	kho
gen	kên	kei	khei
geng	kêng	ken	khên
gong	kung	keng	khêng
gou	kou	kong	khung
gu	ku	kou	khou
gua	kua	ku	khu
guai	kuai	kua	khua
guan	kuan	kuai	khuai
guang	kuang	kuan	khuan
gui	kuei	kuang	khuang
gun	kun	kui	khuei
guo	kuo	kun	khun
ha	ha	kuo	khuo
hai	hai	la	la
han	han	lai	lai
hang	hang	lan	lan
hao	hao	lang	lang
he	ho	lao	lao
hei	hei	le	lê
hen	hên	lei	lei
heng	hêng	leng	lêng
hong	hung	li	li
hou	hou	lia	lia
hu	hu	lian	lien
hua	hua	liang	liang
huai	huai	liao	liao
huan	huan	lie	lieh
huang	huang	lin	lin
hui	hui	ling	ling
hun	hun	liu	liu
huo	huo	lo	lo
ji	chi	long	lung

Pinyin	Modified Wade–Giles	Pinyin	Modified Wade–Giles
lou	lou	pa	pha
lu	lu	pai	phai
lü	lü	pan	phan
luan	luan	pang	phang
lüe	lüeh	pao	phao
lun	lun	pei	phei
luo	lo	pen	phên
ma	ma	peng	phêng
mai	mai	pi	phi
man	man	pian	phien
mang	mang	piao	phiao
mao	mao	pie	phieh
mei	mei	pin	phin
men	mên	ping	phing
meng	mêng	po	pho
mi	mi	pou	phou
mian	mien	pu	phu
miao	miao	qi	chhi
mie	mieh	qia	chhia
min	min	qian	chhien
ming	ming	qiang	chhiang
miu	miu	qiao	chhiao
mo	mo	qie	chhieh
mou	mou	qin	chhin
mu	mu	qing	chhing
na	na	qiong	chhiung
nai	nai	qiu	chhiu
nan	nan	qu	chhü
nang	nang	quan	chhüan
nao	nao	que	chhüeh, chhio
nei	nei	qun	chhün
nen	nên	ran	jan
neng	nêng	rang	jang
ng	ng	rao	jao
ni	ni	re	jê
nian	nien	ren	jên
niang	niang	reng	jêng
niao	niao	ri	jih
nie	nieh	rong	jung
nin	nin	rou	jou
ning	ning	ru	ju
niu	niu	rua	jua
nong	nung	ruan	juan
nou	nou	rui	jui
nu	nu	run	jun
nü	nü	ruo	jo
nuan	nuan	sa	sa
nüe	nio	sai	sai
nuo	no	san	san
o	o, ê	sang	sang
ou	ou	sao	sao

Pinyin	Modified Wade–Giles	Pinyin	Modified Wade–Giles
se	sê	wan	wan
sen	sên	wang	wang
seng	sêng	wei	wei
sha	sha	wen	wên
shai	shai	weng	ong
shan	shan	wo	wo
shang	shang	wu	wu
shao	shao	xi	hsi
she	shê	xia	hsia
shei	shei	xian	hsien
shen	shen	xiang	hsiang
sheng	shêng, sêng	xiao	hsiao
shi	shih	xie	hsieh
shou	shou	xin	hsin
shu	shu	xing	hsing
shua	shua	xiong	hsiung
shuai	shuai	xiu	hsiu
shuan	shuan	xu	hsü
shuang	shuang	xuan	hsüan
shui	shui	xue	hsüeh, hsio
shun	shun	xun	hsün
shuo	shuo	ya	ya
si	ssu	yan	yen
song	sung	yang	yang
sou	sou	yao	yao
su	su	ye	yeh
suan	suan	yi	i
sui	sui	yin	yin
sun	sun	ying	ying
suo	so	yo	yo
ta	tha	yong	yung
tai	thai	you	yu
tan	than	yu	yü
tang	thang	yuan	yüan
tao	thao	yue	yüeh, yo
te	thê	yun	yün
teng	thêng	za	tsa
ti	thi	zai	tsai
tian	thien	zan	tsan
tiao	thiao	zang	tsang
tie	thieh	zao	tsao
ting	thing	ze	tsê
tong	thung	zei	tsei
tou	thou	zen	tsên
tu	thu	zeng	tsêng
tuan	thuan	zha	cha
tui	thui	zhai	chai
tun	thun	zhan	chan
tuo	tho	zhang	chang
wa	wa	zhao	chao
wai	wai	zhe	chê

Pinyin	Modified Wade–Giles	Pinyin	Modified Wade–Giles
zhei	chei	zhui	chui
zhen	chên	zhun	chun
zheng	chêng	zhuo	cho
zhi	chih	zi	tzu
zhong	chung	zong	tsung
zhou	chou	zou	tsou
zhu	chu	zu	tsu
zhua	chua	zuan	tsuan
zhuai	chuai	zui	tsui
zhuan	chuan	zun	tsun
zhuang	chuang	zuo	tso

MODIFIED WADE–GILES/PINYIN

Modified Wade–Giles	Pinyin	Modified Wade–Giles	Pinyin
a	a	chhio	que
ai	ai	chhiu	qiu
an	an	chhiung	qiong
ang	ang	chho	chuo
ao	ao	chhou	chou
cha	zha	chhu	chu
chai	chai	chhuai	chuai
chan	zhan	chhuan	chuan
chang	zhang	chhuang	chuang
chao	zhao	chhui	chui
chê	zhe	chhun	chun
chei	zhei	chhung	chong
chên	zhen	chhü	qu
chêng	zheng	chhüan	quan
chha	cha	chhüeh	que
chhai	chai	chhün	qun
chhan	chan	chi	ji
chhang	chang	chia	jia
chhao	chao	chiang	jiang
chhê	che	chiao	jiao
chhên	chen	chieh	jie
chhêng	cheng	chien	jian
chhi	qi	chih	zhi
chhia	qia	chin	jin
chhiang	qiang	ching	jing
chhiao	qiao	chio	jue
chhieh	qie	chiu	jiu
chhien	qian	chiung	jiong
chhih	chi	cho	zhuo
chhin	qin	chou	zhou
chhing	qing	chu	zhu

Modified Wade–Giles	Pinyin	Modified Wade–Giles	Pinyin
chua	zhua	huan	huan
chuai	zhuai	huang	huang
chuan	zhuan	hui	hui
chuang	zhuang	hun	hun
chui	zhui	hung	hong
chun	zhun	huo	huo
chung	zhong	i	yi
chü	ju	jan	ran
chüan	juan	jang	rang
chüeh	jue	jao	rao
chün	jun	jê	re
ê	e, o	jên	ren
ên	en	jêng	reng
êng	eng	jih	ri
êrh	er	jo	ruo
fa	fa	jou	rou
fan	fan	ju	ru
fang	fang	jua	rua
fei	fei	juan	ruan
fên	fen	jui	rui
fêng	feng	jun	run
fo	fo	jung	rong
fou	fou	ka	ga
fu	fu	kai	gai
ha	ha	kan	gan
hai	hai	kang	gang
han	han	kao	gao
hang	hang	kei	gei
hao	hao	kên	gen
hên	hen	kêng	geng
hêng	heng	kha	ka
ho	he	khai	kai
hou	hou	khan	kan
hsi	xi	khang	kang
hsia	xia	khao	kao
hsiang	xiang	khei	kei
hsiao	xiao	khên	ken
hsieh	xie	khêng	keng
hsien	xian	kho	ke
hsin	xin	khou	kou
hsing	xing	khu	ku
hsio	xue	khua	kua
hsiu	xiu	khuai	kuai
hsiung	xiong	khuan	kuan
hsü	xu	khuang	kuang
hsüan	xuan	khuei	kui
hsüeh	xue	khun	kun
hsün	xun	khung	kong
hu	hu	khuo	kuo
hua	hua	ko	ge
huai	huai	kou	gou

Modified Wade–Giles	Pinyin	Modified Wade–Giles	Pinyin
ku	gu	mu	mu
kua	gua	na	na
kuai	guai	nai	nai
kuan	guan	nan	nan
kuang	guang	nang	nang
kuei	gui	nao	nao
kun	gun	nei	nei
kung	gong	nên	nen
kuo	guo	nêng	neng
la	la	ni	ni
lai	lai	niang	niang
lan	lan	niao	niao
lang	lang	nieh	nie
lao	lao	nien	nian
lê	le	nin	nin
lei	lei	ning	ning
lêng	leng	niu	nüe
li	li	niu	niu
lia	lia	no	nuo
liang	liang	nou	nou
liao	liao	nu	nu
lieh	lie	nuan	nuan
lien	lian	nung	nong
lin	lin	nü	nü
ling	ling	o	e, o
liu	liu	ong	weng
lo	luo, lo	ou	ou
lou	lou	pa	ba
lu	lu	pai	bai
luan	luan	pan	ban
lun	lun	pang	bang
lung	long	pao	bao
lü	lü	pei	bei
lüeh	lüe	pên	ben
ma	ma	pêng	beng
mai	mai	pha	pa
man	man	phai	pai
mang	mang	phan	pan
mao	mao	phang	pang
mei	mei	phao	pao
mên	men	phei	pei
mêng	meng	phên	pen
mi	mi	phêng	peng
miao	miao	phi	pi
mieh	mie	phiao	piao
mien	mian	phieh	pie
min	min	phien	pian
ming	ming	phin	pin
miu	miu	phing	ping
mo	mo	pho	po
mou	mou	phou	pou

Modified Wade–Giles	Pinyin	Modified Wade–Giles	Pinyin
phu	pu	tên	den
pi	bi	têng	deng
piao	biao	tha	ta
pieh	bie	thai	tai
pien	bian	than	tan
pin	bin	thang	tang
ping	bing	thao	tao
po	bo	thê	te
pu	bu	thêng	teng
sa	sa	thi	ti
sai	sai	thiao	tiao
san	san	thieh	tie
sang	sang	thien	tian
sao	sao	thing	ting
sê	se	tho	tuo
sên	sen	thou	tou
sêng	seng, sheng	thu	tu
sha	sha	thuan	tuan
shai	shai	thui	tui
shan	shan	thun	tun
shang	shang	thung	tong
shao	shao	ti	di
shê	she	tiao	diao
shei	shei	tieh	die
shên	shen	tien	dian
shêng	sheng	ting	ding
shih	shi	tiu	diu
shou	shou	to	duo
shu	shu	tou	dou
shua	shua	tsa	za
shuai	shuai	tsai	zai
shuan	shuan	tsan	zan
shuang	shuang	tsang	zang
shui	shui	tsao	zao
shun	shun	tsê	ze
shuo	shuo	tsei	zei
so	suo	tsên	zen
sou	sou	tsêng	zeng
ssu	si	tsha	ca
su	su	tshai	cai
suan	suan	tshan	can
sui	sui	tshang	cang
sun	sun	tshao	cao
sung	song	tshê	ce
ta	da	tshên	cen
tai	dai	tshêng	ceng
tan	dan	tsho	cuo
tang	dang	tshou	cou
tao	dao	tshu	cu
tê	de	tshuan	cuan
tei	dei	tshui	cui

Modified Wade–Giles	Pinyin	Modified Wade–Giles	Pinyin
tshun	cun	wang	wang
tshung	cong	wei	wei
tso	zuo	wên	wen
tsou	zou	wo	wo
tsu	zu	wu	wu
tsuan	zuan	ya	ya
tsui	zui	yang	yang
tsun	zun	yao	yao
tsung	zong	yeh	ye
tu	du	yen	yan
tuan	duan	yin	yin
tui	dui	ying	ying
tun	dun	yo	yue, yo
tung	dong	yu	you
tzhu	ci	yung	yong
tzu	zi	yü	yu
wa	wa	yüan	yuan
wai	wai	yüeh	yue
wan	wan	yün	yun